Functional Analysis

THEORY AND APPLICATIONS

R. E. EDWARDS

EMERITUS PROFESSOR OF MATHEMATICS
INSTITUTE OF ADVANCED STUDIES
AUSTRALIAN NATIONAL UNIVERSITY

DOVER PUBLICATIONS, INC.
New York

Copyright

Published in Canada by General Publishing Company, Ltd., 30 Lesmill Road, Don Mills, Toronto, Ontario.
Published in the United Kingdom by Constable and Company, Ltd., 3 The Lanchesters, 162–164 Fulham Palace Road, London W6 9ER.

Bibliographical Note

This Dover edition, first published in 1995, is an unabridged, corrected republication of the work first published by Holt, Rinehart and Winston, New York, 1965. For this edition the author has made numerous corrections in the original text and provided an Appendix with additional corrections.

Library of Congress Cataloging-in-Publication Data

Edwards, R. E. (Robert E.), 1926–
Functional analysis: theory and applications/R. E. Edwards.
p. cm.
Originally published: New York: Holt, Rinehart and Winston, 1965.
Includes bibliographical references and index.
ISBN 0-486-68143-2 (pbk.)
1. Functional analysis. I. Title.
QA320.E34 1994
515′.7–dc20 94-20873
CIP

Manufactured in the United States of America
Dover Publications, Inc., 31 East 2nd Street, Mineola, N.Y. 11501

Preface

At the time of its publication, Banach's classic "Théorie des Opérations Linéaires" presented much of what was then known of both the abstract theory and the applications of functional analysis. Subsequent developments have greatly enlarged the scopes of both aspects of the subject. More recent books on functional analysis have in most cases been biased fairly heavily in favour of the abstract theory, the applications being relegated to the position of exercises for the reader. The alternative has been the sort of book devoted to a particular type of application, and in which the necessary functional analytic background is sketched in rather hastily. I have attempted to give in this book an account of a few of the more recent developments, in which abstract theory and applications share roughly equal roles. If bias there is, it is probably in favour of the applications.

Two suppositions prompt this attempt. The first asserts the possibility of recognizing a tolerably sensible division of functional analysis into "pure" and "applied" components, leaving aside the debatable issue of where the boundary is to be drawn. The second is that a useful purpose can be served by an account in which the two portions of the subject receive roughly equal treatment, with some give-and-take on both sides. I take it for granted that the applications of abstract methods have indeed proved themselves useful, both as unifying agents and as sources of new results and problems; and that something essential is gained by studying a general theorem in close relation to its nontrivial applications.

This avowal of the aim is followed by two disclaimers. First, in respect of the range of topics covered, the book must fail to qualify as a treatise on functional analysis or even on topological vector spaces in particular. Ordered vector spaces, topological algebras, general spectral theory, and topological tensor products (to mention but a few headings) receive little or no attention. Secondly, the aim to cover nontrivial applications of the chosen portions of the abstract theory has taken precedence over giving a fully adequate account of the "interior problems" of the subject. So, for example, many problems pertaining to the abstract structural properties of topological vector spaces, and an exhaustive study of the various categories of such spaces, are left aside. The properties that are, in fact, isolated tend to be those, and only those, that have to date proven to be most useful in applications. In those few cases in which the general theory is developed beyond the point necessary for subsequent applications, personal taste must be the excuse.

Applications of axiomatically based theories seldom follow the same impeccably logical development inherent in the theories themselves, and it has been thought desirable to reverse the logical order on occasions. Thus, were this book about topological vector spaces per se, Chapters 4 and 5 would almost certainly follow Chapters 6, 7, and 8. The existing arrangement, illogical though it may be, is aimed at providing some motivation for the introduction of certain categories of topological vector spaces that might otherwise appear somewhat pointless. Frequently, when no circularity is involved, general and abstract results are freely used prior to their systematic discussion when special situations offer opportunities for their application. This is notably the case with results from duality theory (Chapter 8), which are called upon in Chapters 4, 5, 6, and 7.

In presenting the abstract theory I have relied heavily on several existing texts, notably those of Bourbaki, of Grothendieck, of Dunford and Schwartz, and, to a lesser extent, those of Taylor, of Day, of Köthe, and of Kelley and Namioka. Material concerning applications has come in part from these same sources, but more often from reports of seminars, published research papers, and monographs.

A list of applications which it would have been desirable to include, but that have, in fact, not been included, would certainly contain the linear aspects of analysis on manifolds, including de Rham's theory of currents, the "finiteness theorems" in cohomology theory for complex-analytic manifolds, and more about the linear aspects of potential theory. Their exclusion, as well as that of numerous other topics, is the result of the very considerable preliminary technical equipment they demand, much of which lies outside functional analysis.

The would-be exponent of two viewpoints is always in danger of adequately presenting neither. Nowhere is my fear of this fate more pronounced than in respect to Chapter 5. The functional-analytic approach to linear partial differential equations is currently a field of almost explosive activity. In view of the enormous classical literature, it might be expected that new ideas and methods be constantly compared and combined with old techniques. To do this would, however, entail the inclusion of masses of material with little or no functional analytic content. In ignoring the classical literature I have taken shelter behind some remarks of Gårding quoted at the beginning of Chapter 5. It is hoped that this chapter succeeds, in spite of its obvious shortcomings, in conveying something of the spirit of one of the most recent fields of application.

For the reader's convenience a preliminary Chapter 0 has been devoted to summarizing some results from set theory and general topology. It is assumed that the reader has a rudimentary knowledge of the basic ideas in these fields, and also of a coordinate-free introduction to vector spaces. The assumed knowledge of classical analysis is modest. In particular, although the reader is assumed to be thoroughly familiar with the Riemann integral for continuous functions over a compact interval, detailed reference to the Lebesgue integral is kept to a minimum until Chapter 4, which includes the theory in its entirety and in detail.

Exercises are appended to each chapter and are meant to be treated seriously. Sometimes they appear as devices for shortening the main text by carrying the burden of the proofs of auxiliary results, while at other times they are aimed at extending, developing, or specializing portions of the text. The more difficult exercises are accompanied by hints to their solutions.

The bibliography makes no pretence at completeness. Some items listed receive no specific mention in the text, being included as implied suggestions for further reading in directions that are usually adequately indicated by their titles.

Where it has seemed helpful to mark the end of a proof we have used the symbol "▌."

Besides a general index, also included is a list of frequently used symbols and abbreviations.

Acknowledgments

My sincere thanks are due to several departmental secretaries for their help with the typescript: to Mrs. Jean Chapman of the University of Reading, England, and to Mrs. Margaret Edmonds and Miss Sybil Stuckey of the Institute of Advanced Studies of the Australian National University. Each of these ladies contributed many hours of labour which I now gladly and gratefully acknowledge.

For encouragement and advice my thanks are due to the Technical Editor, Professor Edwin Hewitt. Dr. Martin Dunwoody and Dr. Barron Brainerd contributed welcome and valuable suggestions after reading parts of the typescript. I would like to record my thanks to Professor Lars Gårding and to Dr. J. A. Todd for permission to quote in Section 5.12 some published remarks of the former.

To my wife is due what is perhaps my largest debt of gratitude. She contributed an enormous amount of help with the onerous tasks of proof reading and compiling the index and the bibliography. Perhaps more importantly still, however, I would thank her for encouragement on those distressingly frequent occasions when nothing would go as I wished.

R. E. E.

Canberra,
February, 1965.

Contents

Functional Analysis

THEORY AND APPLICATIONS

CHAPTER 0

Preliminaries on Set Theory and Topology

0.0 Foreword

In compiling within the confines of one chapter a summary of the basic material in set theory and general topology, it is well-nigh impossible to achieve a strict adherence to completeness and difficult even to preserve logical order. A glaring example of divergence from these ideals is the absence of any "official" discussion of basic structures such as the natural numbers and the real and complex number fields. On the contrary, it is necessary to assume that the reader is entirely familiar with the algebraic and topological structures normally imposed on these sets, even though he may not know them under these titles. Illogical though this may be, it corresponds quite closely with the order in which mathematics is normally taught and learnt. Little disappointment or confusion will result if Chapter 0 is accepted for what it is—namely, a summary of concepts and techniques to be called upon, and not a logically coherent exposition of several flourishing fields of study.

0.1 Preliminaries from Set Theory

Our attitude toward set theory is completely naive, and the natural companion text is Halmos [4]. The reader who is not content with this approach may consult Suppes [1] and the bibliography collected there. A bird's-eye view of the axiomatic approach is also available in the Appendix to Kelley [1]. Nowhere do we speak of "classes" in a sense technically distinct from "sets," even though the axiomatics of some versions of set theory show that some such distinction is essential to preserve consistency. The fact is that the underlying axioms of such theories ensure that all the classes we handle are indeed sets in the technical sense.

0.1.1 Basic Set-theoretical Notations. We always use the symbol $\in$ to denote the relation of membership, so that the formula "$x \in A$" signifies that A is a set and that the object x is a point (member, element) of A. (The reader will notice that we are already speaking in a manner that confounds an object with the symbol used to denote it, which is but one form of naivety.) The negation of "$x \in A$" is written "$x \notin A$."

Sets are often specified in terms of a defining property, an object being a member of the set so defined if and only if it possesses the said property. Accordingly we shall write $\{x : P(x)\}$, P denoting a propositional function, for the set whose members are precisely those objects x for which $P(x)$ is true. In order to avoid ambiguities (see Kelley [1], pp. 5–6) we make the convention that this symbolism is to be governed by the rule that the space lying between the first brace and the colon will be occupied by the symbol of a "dummy variable," which symbol may be replaced there and in $P(x)$ by any symbol different from all symbols other than x that appear in the proposition denoted by $P(x)$.

Finite sets are often symbolized by displaying their members individually, enclosed in braces and separated by commas (when there is more than one element). Thus we should write $\{x, y, z\}$ for the set whose members are precisely (the objects denoted by) x, y, and z. In particular, $\{x\}$, called "singleton x," is the set whose only member is x.

The empty, or void, set $\emptyset$ is that set without members. It may be specified as $\{x : x \neq x\}$.

If A and B are sets, we write $A \subset B$ (read as "A is contained in B," or "A is a subset of B," or "B contains A," or "B is a superset of A") if and only if each member of A is a member of B. In this situation we also write $B \supset A$. If $A \subset B$ and $B \subset C$, then $A \subset C$. Also $A \subset B$ and $B \subset A$ together imply that $A = B$.

$A \cup B$, called the union of A and B, is the set

$$\{x : x \in A \text{ or } x \in B \text{ (or both)}\}.$$

$A \cup B$ is the smallest set containing both A and B.

$A \cap B$, called the intersection of A and B, is the set

$$\{x : x \in A \text{ and } x \in B\}.$$

$A \cap B$ is the largest set contained in both A and B.

For all A it is true that $\emptyset \subset A$, $A \cup \emptyset = A$, $A \cap \emptyset = \emptyset$.

A and B are termed disjoint if $A \cap B = \emptyset$. If $\mathscr{A}$ is a set of sets, $\mathscr{A}$ is said to be disjoint if whenever A and B are members of $\mathscr{A}$, one has either $A = B$ or $A \cap B = \emptyset$.

The absolute complement of a set A, written $\sim A$, is $\{x : x \notin A\}$. The relative complement of A in B is $B \setminus A = B \cap (\sim A)$.

It is easily shown that $\cup$ and $\cap$ are commutative and associative. Moreover one has the distributive laws

$$A \cap (B \cup C) = (A \cap B) \cup (A \cap C),$$
$$A \cup (B \cap C) = (A \cup B) \cap (A \cup C),$$

and the de Morgan formulae

$$X \setminus (X \setminus A) = A \cap X,$$
$$X \setminus (A \cup B) = (X \setminus A) \cap (X \setminus B),$$
$$X \setminus (A \cap B) = (X \setminus A) \cup (X \setminus B).$$

One often needs to form unions and intersections of sets of sets. Suppose $\mathscr{A}$ is a set of sets. Then $\bigcup \{A : A \in \mathscr{A}\}$ is by definition the set $\{x : \text{for some } A \in \mathscr{A}, x \in A\}$. Likewise one defines $\bigcap \{A : A \in \mathscr{A}\}$ to be $\{x : \text{for all } A \in \mathscr{A}, x \in A\}$. For families of sets, in which a set A_i is assigned to each member i of some "index set" I, other convenient notations are

$$\bigcup \{A_i : i \in I\} \qquad \text{or} \qquad \bigcup_{i \in I} A_i$$

and

$$\bigcap \{A_i : i \in I\} \qquad \text{or} \qquad \bigcap_{i \in I} A_i.$$

The de Morgan formulae extend to these "infinite" unions and intersections. Naturally, if $\mathscr{A} = \{A, B\}$, then $\bigcup \{A : A \in \mathscr{A}\}$ is simply $A \cup B$. If $\mathscr{A}$ is finite, say $\{A, B, \cdots, X\}$, one often writes $A \cup B \cup \cdots \cup X$ in place of $\bigcup \{A : A \in \mathscr{A}\}$.

0.1.2 **Ordered Pairs and Relations.** The characteristic property of ordered pairs can be expressed unofficially by saying that if (x, y) and (x', y') are ordered pairs, then (x, y) and (x', y') are equal if and only if $x = x'$ and $y = y'$. A definition of ordered pairs in the language of set theory, which fulfils this criterion, is expressed by the defining equation

$$(x, y) = \text{def } \{\{x\}, \{x, y\}\},$$

$\{x, y\}$ being (as explained in Section 0.1.1) the same as $\{x\} \cup \{y\}$. If (x, y) is an ordered pair, x is termed the first coordinate and y the second coordinate of (x, y).

In terms of ordered pairs, a *relation* is a set whose members are ordered pairs. If the relation is denoted by R, we shall often write xRy to mean that $(x, y) \in R$.

The *domain* of the relation R, written Dom R, is

$$\{x : \text{for some } y, (x, y) \in R\};$$

and the *range* of R is

$$\{y : \text{for some } x, (x, y) \in R\}.$$

If R is a relation and A a set, the set

$$\{y : xRy \text{ for some } x \in A\}$$

is denoted by $R[A]$ or $R(A)$. $R[\text{Dom } R]$ is the range of R.

An especially important type of relation is the (Cartesian) *product* $A \times B$ of two sets A and B, which is defined to be

$$\{(x, y) : x \in A, y \in B\}.$$

Relations that are certain subsets of such a product play an especially basic role in handling functions (see Subsection 0.1.3).

If R is a relation, its *inverse* R^{-1} is the relation

$$\{(x, y) : (y, x) \in R\}.$$

If R and S are relations, the *composite relation* $R \circ S$ is the relation

$$\{(x, z) : \text{for some } y, (x, y) \in S \text{ and } (y, z) \in R\}.$$

Composition is not generally commutative; that is, in general $R \circ S$ and $S \circ R$ are different relations.

If X is a set, the *identity* or *diagonal relation on* X is

$$\Delta(X) = \{(x, x) : x \in X\},$$

a subset of $X \times X$. If X is understood we write Δ in place of $\Delta(X)$.

A relation R is *symmetric* if and only if $R = R^{-1}$, and is *antisymmetric* if and only if $R \cap R^{-1} = \emptyset$. R is *reflexive* if and only if $\Delta \subset R$, Δ denoting the diagonal on the union of the domain and range of R. R is *transitive* if and only if $R \circ R \subset R$.

R is an *equivalence relation* if and only if it is reflexive, symmetric, and transitive. In that case, putting $X = \operatorname{Dom} R$, X can be partitioned into the union of disjoint *equivalence classes modulo* R. Such an equivalence class is a set of the type $R(x) = R(\{x\})$, x being a member of X. Two such classes are either identical or disjoint, and X is the union $\bigcup \{R(x) : x \in X\}$. R is expressible in terms of these equivalence classes by the formula

$$R = \bigcup \{R(x) \times R(x) : x \in X\}.$$

Conversely, given a disjoint family $(A_i)_{i \in I}$ of sets, one obtains an equivalence relation R on defining

$$R = \bigcup \{A_i \times A_i : i \in I\};$$

the domain of this relation is $X = \bigcup \{A_i : i \in I\}$.

If R and S are relations, R is said to be a *restriction of* S and S an *extension of* R, if and only if R⊂S. If $X \subset \operatorname{Dom} R$, the *restriction of* R *to* X is the relation

$$R \,|\, X = \{(x, y) : (x, y) \in R \text{ and } X \in X\}.$$

0.1.3 **Functions.** A *function* is a relation f having the property that

$$(x, y) \in f \text{ and } (x, y') \in f \Rightarrow y = y'.$$

The terms "correspondence," "transformation," "map," "mapping," "operator" are each synonymous with "function."

If f is a function, to each x in $\operatorname{Dom} f$ corresponds exactly one object y for which $(x, y) \in f$; this object y is termed the *value* of f at x and is usually denoted by $f(x)$ or f_x. One says that f *maps* or *carries* x into y.

It often happens that one has assigned (by some method or other) an object z_x to each member x of a set X. Then $\{(x, z_x) : x \in X\}$ is a function f with domain X that will sometimes be denoted by

$$f : x \to z_x,$$

and referred to as "the function that maps x into z_x" (it being understood that x varies over the set X). Actually, the traditional view of a function is very close to the "operational" one just explained. If this view be adopted, it is customary to refer to the set

$$\{(x, f(x)) : x \in X\}$$

as the *graph* of the function f, whereas, on the basis of the set-theoretical definition, the function and the graph are the same thing. We shall often indulge in this harmless terminological duplication in deference to accepted usage. In particular we shall often speak later of the graph of a linear operator (see Subsection 1.3.3), this being one context in which the operational connotation of a function or mapping is customarily stressed.

A function f is *one-to-one* if and only if the inverse relation f^{-1} is itself a function.

If f is a function whose range is a set Y, one says that f is "onto Y."

When f is a function, $f(A) = \{y : y = f(x) \text{ for some } x \in A\}$ is termed the *image* of A under (or by) f, while $f^{-1}(A) = \{x : f(x) \in A\}$ is the *inverse image* of A under (or by) f. One has then the formulae

$$\begin{aligned} f(A \cup B) &= f(A) \cup f(B), \\ f(A \cap B) &\subset f(A) \cap f(B), \\ f^{-1}(A \cup B) &= f^{-1}(A) \cup f^{-1}(B), \\ f^{-1}(A \cap B) &= f^{-1}(A) \cap f^{-1}(B), \\ f^{-1}(A \setminus B) &= f^{-1}(A) \setminus f^{-1}(B), \end{aligned}$$

A and B being arbitrary sets. It is important to note that in the second of the above formulae equality does not hold in general.

There is no genuine distinction between a function and a *family* or *indexed set*, save that in the latter the domain of the function is used merely to label the range in a convenient manner. Thus, if the "index set" is I, the family will usually be denoted by $(a_i)_{i \in I}$ or simply (a_i) if I is understood. This concept of family sprang up as an obvious generalization of the older notion of a sequence, in which case the index set is usually the set of natural numbers or a finite subset thereof. If the family (a_i) is such that $a_i \in A$ for each i (that is, has its range contained in A), one speaks of a family of points (or members) of A.

When dealing with real- or complex-valued functions on a topological space X, we define the *support* (sometimes called the *carrier*) of a function f to be the closure in X of the set $\{x \in X : f(x) \neq 0\}$.

We have already (in Subsection 0.1.2) defined the product $A \times B$ of two sets, and the definition is obviously extendible to the product of any finite family of sets. However, need is felt for a product of an arbitrary family $(X_i)_{i \in I}$ of sets. This product, denoted by $\Pi\, X_i$ or $\Pi\, \{X_i : i \in I\}$, is defined to be the set of all families $x = (x_i)$, indexed by I, such that $x_i \in X_i$ for each i. The function π_i on $X = \Pi\, X_i$ into X_i, defined by $\pi_i(x) = x_i$, is termed the ith *projection;* it maps X onto X_i.

If $X_i = Y$ for each i, Y being a fixed set, the product set is none other than the set of functions on I into Y, often denoted by Y^I.

0.1.4 **Orderings.** If X is a set, by a (partial) *order on* X, we mean a relation $\Omega \subset X \times X$ that is reflexive, transitive, and such that $\Omega \cap \Omega^{-1} = \Delta = \Delta(X)$. Kelley's treatment ([1], pp. 13 et seq) assumes a little less of Ω, but the definition above is adequate for all of our needs.

When no confusion is likely, we shall write "$x \leq y$" in place of "$x \, \Omega \, y$," and say that $\leq$ is an order on X or that X is ordered by $\leq$. When this is the case, the following formulae are true

$$x \leq x, x \leq y \quad \text{and} \quad y \leq x \Rightarrow x = y,$$

$$x \leq y \quad \text{and} \quad y \leq z \Rightarrow x \leq z,$$

x, y, and z denoting arbitrary elements of X.

Let X be ordered by $\leq$ and let A be a subset of X. An element x of X *majorizes*, or is a *majorant* of, A if and only if $a \leq x$ for each $a \in A$. If on the other hand $x \leq a$ for each $a \in A$, x is said to *minorize*, or to be a *minorant* of, A. Furthermore an element x_0 of X is the *supremum* or *least upper bound* of A in X if and only if x_0 is a majorant of A and $x_0 \leq x$ is true for any majorant x. The *infimum* or *greatest lower bound* of A is characterized similarly. We denote the supremum of A by Sup A or sometimes Sup $\{x : x \in A\}$ or even $\text{Sup}_{x \in A}\, x$, with an analogous notation for the infimum. If either of Sup A or Inf A exists, it is uniquely determined as a member of X.

An order $\leq$ on a set X is a *total* (or *linear*) *order* on X if and only if $(x, y) \in X \times X$ implies that either $x \leq y$ or that $y \leq x$; that is, if and only if $\Omega \cup \Omega^{-1} = X \times X$. A totally ordered set is often called a *chain*.

A set X is said to be *inductively ordered* by $\leq$ if and only if $\leq$ is an order on X and any totally ordered subset A of X possesses a supremum in X. To say that A is totally ordered means here that the restriction of $\leq$ to A is a total order on A.

An important example of an order is that order on the set X of subsets of a given set S, defined by the convention that $A \leq B$ if and only if $A \subset S$, $B \subset S$, and $A \subset B$. This is spoken of as the order of "set inclusion." If $Y \subset X$, Sup Y will exist and be equal to $\bigcup \{A : A \in Y\}$, whenever this union belongs to X. The opposing order, in which $A \leq B$ signifies that $A \supset B$, is termed the order of "set inclusion reversed" and also arises quite frequently, being the "natural" order to impose on the set of neighbourhoods of any one point in a topological space (see Subsection 0.2.2).

0.1.5 **Zorn's Lemma.** If X is an inductively ordered set, and if x is a given element of X, there exists a maximal element m of X such that $x \leq m$.

To say that m is maximal signifies that there exists in X no element y satisfying $m \leq y$ and $m \neq y$.

In most of the cases we encounter, an appeal to Zorn's lemma provides an alternative to a reasoning that depends on such principles as those of transfinite induction, Zermelo's well-ordering theorem, and the axiom of choice. For a summary of the relations between several such principles, see Kelley [1], pp. 31–36.

Although we have adopted the name that appears to be current, the result was discovered independently in 1923 by R. L. Moore and by Kuratowski, was rediscovered by Zorn in 1935, and then was discovered yet again by Teichmüller a little later. The name was coined by Bourbaki, who was one of the first to make systematic use of the principle. For further remarks, see Rosenbloom [1], p. 21 and p. 150.

Many authors state and prove a variant of Zorn's lemma in which the hypothesis that X be inductively ordered is replaced by the apparently weaker demand that each totally ordered subset of X admits a majorant in X. Proofs of this version appear in Kelley [1], p. 33, and Dunford and Schwartz [1], pp. 4–7. The variation is not important in connection with the applications made in this book.

0.1.6 Countable Sets. While we have little to do with cardinal numbers in general, countability is almost indispensable.

Two sets X and Y are said to be *equipotent*, or to have the same *power* or the same *cardinal number*, if and only if there exists a one-to-one function f with domain X and range Y.

A set X is *infinite* if it is equipotent with some proper subset of itself, otherwise X is *finite*.

Granted the existence and properties of natural numbers, let N denote the set of such numbers and, for any natural number n, let N_n be the set $\{m : m \in N$ and $m \leq n\}$. Then it can be shown that a set is finite if and only if it is equipotent with N_n for some n, in which case X is said to have n elements and can be enumerated as $\{x_1, x_2, \cdots, x_n\}$ or as a family $(x_k)_{1 \leq k \leq n}$.

A set X is *denumerably* or *countably infinite* if it is equipotent with N, in which case X can be arranged (in many ways, of course) as a sequence $(x_n)_{n \in N}$. X is *countable* if it is either finite or countably infinite.

Any subset of a countable set is countable; if the domain of a function f is countable, so too is the range of f; if $\mathscr{A}$ is a countable set of countable sets, the union $\bigcup \{A : A \in \mathscr{A}\}$ is countable; the product of two (or of any finite number of) countable sets is countable; the set of finite subsets of a countable set is countable; the set of *all* subsets of an infinite set is *not* countable; the set of all real numbers is *not* countable.

0.1.7 Directed Sets and Families. By a *directed set* we mean an ordered set (see Subsection 0.1.4) in which any two elements possess a common majorant. A *directed family* is a family whose index set (or domain) is a directed set. Kelley ([1], p. 65) and many writers use the term *net* for a directed family.

Directed families were first introduced as a generalization of (denumerable) sequences called for in handling general topological spaces and convergence notions.

Each denumerable sequence is a directed family. The set of neighbourhoods of a given point in a topological space, ordered by reversed set inclusion (see the end of Subsection 0.1.4) is a directed set that is frequently used.

We shall need a few results and concepts relating to the handling of directed families.

COFINAL SUBFAMILIES. Let $(x_\lambda)_{\lambda \in L}$ be a directed family. A directed family $(y_\mu)_{\mu \in M}$ is a *cofinal subfamily* of $(x_\lambda)_{\lambda \in L}$ if and only if, given $\lambda \in L$, there exists $\mu_\lambda \in M$ such that $\mu' \in M$ and $\mu' \geq \mu_\lambda$ entail that $y_{\mu'} = x_{\lambda'}$ for some $\lambda' \geq \lambda$.

This idea generalizes that of a subsequence of a sequence. However, a sequence may have cofinal subfamilies that are *not* themselves sequences.

PRODUCT OF DIRECTED SETS. We consider for a moment sequences of real numbers. There is no difficulty in defining the sum or product of two sequences (x_n) and (y_n)—namely, the sequences $(x_n + y_n)$ and $(x_n y_n)$. This obvious and natural rule is available because the index sets involved are the same for the two sequences. Failing this, one is forced into adopting a more circuitous path, which calls for use of the product of the two directed index sets.

Suppose that L and M are directed sets. We form the product set $L \times M$ (see Subsection 0.1.2) and order $L \times M$ by the convention that $(\lambda, \mu) \leq (\lambda', \mu')$ if and only if $\lambda \leq \lambda'$ and $\mu \leq \mu'$. It is easy to check that $L \times M$, thus ordered, is a directed set.

0.2 Preliminaries from General Topology

0.2.1 Open-set Definition of a Topology.

Let X be a set. By a *topology* on X is meant a set $\mathfrak{T}$ of subsets of X satisfying the following conditions:

(1) $\emptyset$ and X are members of $\mathfrak{T}$;

(2) if U and U' are members of $\mathfrak{T}$, so too is $U \cap U'$;

(3) the union of any subset of $\mathfrak{T}$ is a member of $\mathfrak{T}$.

Members of $\mathfrak{T}$ are called the *open sets* of (or in, or for) the topology $\mathfrak{T}$, or the $\mathfrak{T}$-*open* sets.

A *topological space* is a pair $(X, \mathfrak{T})$ in which X is a set and $\mathfrak{T}$ a topology on X. When $\mathfrak{T}$ is understood, one often speaks of X as being a topological space.

If $\mathfrak{T}$ is a topology on X, a (open) *base* for $\mathfrak{T}$ is a subset $\mathscr{A}$ of $\mathfrak{T}$ such that each member of $\mathfrak{T}$ is the union of a suitable subset of $\mathscr{A}$. A (open) *subbase* for $\mathfrak{T}$ is a subset $\mathscr{B}$ of $\mathfrak{T}$ such that the set of intersections of finite subsets of $\mathscr{B}$ is a base for $\mathfrak{T}$.

In a topological space $(X, \mathfrak{T})$, by a *neighbourhood* of a point x of X is meant any subset of X containing a member of $\mathfrak{T}$ which in turn contains x. Similarly, by a neighbourhood of a subset A of X is meant any subset of X containing a member of $\mathfrak{T}$ containing A. A subset of X then belongs to $\mathfrak{T}$ if and only if it is a neighbourhood of each of its member points.

Knowledge of the neighbourhoods of points in a topological space fully determines the topology, and opens the way for an alternative method of defining topologies that we shall frequently use. This method is described in Subsection 0.2.2.

Meanwhile it should be remarked that a given set X may be topologized in various ways. For example, there is the *zero* or *null* topology $\{\emptyset, X\}$; and at the other extreme there is the so-called *discrete* topology comprised of *all* subsets of X. These are extremes relative to the scale of comparison explained in Subsection 0.2.6 below.

0.2.2 **Neighbourhood Definition of a Topology.** For the definition of topological vector spaces (abbreviated TVS's, see Section 1.8) the most convenient way of specifying the topology involved is in terms of the neighbourhood system attached to each point. Due to the relationship between the vector space and topological structures, which is an integral part of the concept of a TVS, the neighbourhood system of any given point is determined by appropriate translation of the corresponding system of neighbourhoods of zero. In view of this, it is imperative for us to discuss this way of defining topologies on a set.

Let X be a set. Let us suppose given also a set $\mathfrak{N}$ of ordered pairs (x, N) in which $x \in X$ and N is a subset of X containing x. We shall assume that $\mathfrak{N}$ fulfils the following conditions:

(a) $(x, X) \in \mathfrak{N}$ for each $x \in X$;

(b) if $(x, N') \in \mathfrak{N}$ and $(x, N'') \in \mathfrak{N}$, then $(x, N' \cap N'') \in \mathfrak{N}$;

(c) if $(x, N) \in \mathfrak{N}$ and $X \supset N' \supset N$, then $(x, N') \in \mathfrak{N}$;

(d) if $(x, N) \in \mathfrak{N}$, there exists N' such that $(x, N') \in \mathfrak{N}$ and $(x', N) \in \mathfrak{N}$ for each $x' \in N'$.

A topology $\mathfrak{T}$ on X results if it be agreed that $U \in \mathfrak{T}$ if and only if $(x, U) \in \mathfrak{N}$ for each $x \in U$. Furthermore, the neighbourhoods of x in the topology $\mathfrak{T}$ (as defined in Subsection 0.2.1) are precisely the sets N such that $(x, N) \in \mathfrak{N}$. Thus $\mathfrak{N}$ specifies a topology $\mathfrak{T}$ on X, and constitutes the so-called "neighbourhood definition of $\mathfrak{T}$."

If X is a topological space, x one of its points, and $\mathfrak{N}_x$ the set of neighbourhoods of x, by a *base* of neighbourhoods of, or a *neighbourhood base* at, x (for the given topology) is meant a subset $\mathscr{B}_x$ of $\mathfrak{N}_x$ such that a subset of X belongs to $\mathfrak{N}_x$ if and only if it contains some set belonging to $\mathscr{B}_x$. It is evident that when this is the case, the following conditions are fulfilled:

(a′) $x \in N$ for each $N \in \mathscr{B}_x$;

(b′) if N' and N'' belong to $\mathscr{B}_x$, there exists an $N \in \mathscr{B}_x$ such that $N \subset N' \cap N''$;

(d′) if $N \in \mathscr{B}_x$, there exists $N' \in \mathscr{B}_x$ such that, for each $x' \in N'$, N contains a set belonging to $\mathscr{B}_{x'}$.

Reciprocally, if to each point x of a set X is assigned a set $\mathscr{B}_x$ of subsets of X in such a way that (a′), (b′), and (d′) are fulfilled, then there exists on X a unique topology for which, for each x in X, $\mathscr{B}_x$ is a neighbourhood base at x.

Example. If R is the set of real numbers, the usual topology on R is that for which an open base comprises the so-called open intervals (a, b), where $a < b$. A neighbourhood base at x is formed of all intervals of the type $(x - p, x + p)$, where $p > 0$. Naturally, there exist many systems of neighbourhood bases leading to the same topology. For example, one might restrict p to any sequence of strictly positive numbers that tends to zero; again, one might in each case replace the open interval $(x - p, x + p)$ by the closed interval $[x - p, x + p]$.

0.2.3 Closed Sets. Closure Operation. Suppose that X is a topological space. A subset $A \subset X$ is said to be *closed* (in, or for, the given topology on X) if and only if $X \setminus A$ is open. The set of closed sets accordingly enjoys the following properties:

(a″) $\emptyset$ and X are closed sets;

(b″) any intersection of closed sets is closed;

(c″) the union of any two closed sets is closed.

It is possible to define a topology by specifying its closed sets, which must satisfy (a″)–(c″); see, for example, Kelley [1], p. 40.

If A is a subset of X, its *closure* (or *adherence*) $\bar{A}$ is the smallest closed set containing A—that is (in view of (b″)) the intersection of all closed sets containing A. It is easy to check that

$$\bar{A} = \{x : x \in X \text{ and every neighbourhood of } x \text{ intersects } A\}.$$

The reader will notice that A is closed if and only if $A = \bar{A}$. Furthermore,

$$\bar{\bar{A}} = \bar{A}, \qquad \bar{A} \cup \bar{B} = \overline{A \cup B}, \qquad A \subset B \text{ entails } \bar{A} \subset \bar{B},$$

$$\bar{\emptyset} = \emptyset, \qquad \bar{X} = X, \qquad \overline{A \cap B} \subset \bar{A} \cap \bar{B}.$$

A is said to be *dense* in X if and only if $\bar{A} = X$.

Here again one may define a topology in terms of a closure operation $A \to \bar{A}$, defined for all subsets of X and subject to the above conditions (not all of which are independent of the others); see, for example, Kelley [1], p. 43, or Kuratowski [1], p. 15, to whom the method is due.

0.2.4 Interior of a Set. This is the concept dual to that of closure. If X is a topological space and A a subset of X, the *interior* of A, denoted by $\mathring{A}$ or Int A, may be defined as $X \setminus \overline{(X \setminus A)}$ or, what is equivalent, as the largest (that is, the union of all) open set(s) contained in A. Then A is open if and only if $A = \mathring{A}$.

0.2.5 Frontier or Boundary of a Set. X being again a topological space and A a subset of X, the *frontier* or *boundary* of A is the set of points of X that are adherent to (that is, lie in the closure of) both A and $X \setminus A$. It is denoted by Fr A or sometimes by $\dot{A}$. An equivalent definition is

$$\text{Fr } A = \bar{A} \setminus \mathring{A}.$$

Evidently Fr A is always a closed set; moreover

$$\bar{A} = A \cup \text{Fr } A, \qquad \mathring{A} = A \setminus \text{Fr } A.$$

0.2.6 Comparison of Topologies. Given two topologies, $\mathfrak{T}$ and $\mathfrak{T}'$, on a set X, one says that $\mathfrak{T}$ is *weaker than* $\mathfrak{T}'$, or that $\mathfrak{T}'$ is *stronger than* $\mathfrak{T}$, if and only if $\mathfrak{T} \subset \mathfrak{T}'$, that is, if each $\mathfrak{T}$-open set is also $\mathfrak{T}'$-open. It is equivalent to assert that, for each x in X, any $\mathfrak{T}$-neighbourhood of x is also a $\mathfrak{T}'$-neighbourhood of x; or that each $\mathfrak{T}$-closed subset of X is also $\mathfrak{T}'$-closed.

Of all topologies on X, the weakest is the zero topology and the strongest is the discrete topology (see end of Subsection 0.2.1).

0.2.7 **Induced Topologies and Subspaces.** Let $(X, \mathfrak{T})$ be a topological space, Y a subset of X. The *induced topology* on Y is the one whose open sets are by definition the intersections with Y of the $\mathfrak{T}$-open sets in X. This induced topology is denoted by $\mathfrak{T} \mid Y$.

By a *sub-* (topological) *space* of X is meant a subset Y endowed with its induced topology. The sets that are open (or closed) in this subspace are often said to be *relatively* open (or closed); and the neighbourhoods of a point y of Y, for this induced topology, are often spoken of as *relative neighbourhoods* of y.

0.2.8 **Separation Axioms.** Postulates of the type that assert the existence of "sufficiently many" open sets (or closed sets, or neighbourhoods) in some sense or other are termed *separation axioms*. It is convenient to make here a list of the usual postulates of this nature.

A topological space $(X, \mathfrak{T})$ is said to be:

(1) T_0 if, for any two distinct points x and y of X, there exists *either* a neighbourhood of x not containing y, *or* a neighbourhood of y not containing x; or, what is equivalent: *either* $x \notin \overline{\{y\}}$ *or* $y \notin \overline{\{x\}}$.

(2) T_1 if each one-point set in X is closed; it is equivalent to assert that, for any two distinct points x and y of X, there exists a neighbourhood of x not containing y, and a neighbourhood of y not containing x (cf. T_0).

(3) T_2 (*separated*, or *Hausdorff*) if, for any two distinct points x and y of X, there exist a neighbourhood of x and a neighbourhood of y that are disjoint.

(4) *Regular* if, for each point x of X, the closed neighbourhoods of x form a neighbourhood base at x.

(5) T_3 if it is regular and T_1.

(6) *Normal* if any two disjoint closed sets possess disjoint neighbourhoods.

(7) T_4 if it is normal and T_1.

(8) *Completely regular* if, given a point x of X and a neighbourhood N of x, there exists a continuous function f mapping X into $[0, 1]$ satisfying $f(x) = 0$ and $f = 1$ on $X \setminus N$.

(9) *Tychonoff* if it is completely regular and T_1. The relations

$$T_4 \Rightarrow T_3 \Rightarrow T_2 \Rightarrow T_1 \Rightarrow T_0, \text{ completely regular} \Rightarrow \text{regular}$$

are proved very simply.

Regularity does *not* imply T_1, as is shown by considering a semimetrizable space whose semimetric is not a metric (see Subsection 0.2.21). Also, there exist regular T_2 spaces on which every continuous function is constant; see Kelley [1], p. 117.

For T_1 spaces it is true that

$$\text{normal} \Rightarrow \text{completely regular} \Rightarrow \text{regular} \Rightarrow T_2;$$

the first implication here is not trivial and depends on Urysohn's lemma (Subsection 0.2.12). The other implications are almost immediate from the definitions.

0.2.9 **Continuous Functions.** Let X and Y be topological spaces and f a function on X into Y (that is, the domain of f is X and its range is a subset of Y; see Subsection 0.1.3). f is said to be *continuous* if and only if $f^{-1}(V)$ is open in X whenever V is open in Y. For this it suffices that $f^{-1}(V)$ be open for each V belonging to an open base in Y, or even to an open subbase in Y.

If $x \in X$, f is said to be *continuous at x* if and only if $f^{-1}(V)$ is a neighbourhood of x whenever V is a neighbourhood of $f(x)$ (or, what is equivalent, whenever V belongs to a neighbourhood base at $f(x)$).

In these definitions the domain of f *must* be specified or understood to be X. If f is a function on X into Y, and if A is a subspace of X, then $f \mid A$ may well be continuous (or continuous at a point x) without f sharing this property. On the other hand, continuity of f entails that of $f \mid A$ for any subspace A of X.

If f maps X into Y and g maps Y into Z (Z being a topological space), the continuity of f and of g entails that of $g \circ f$ (which is on X into Z).

f is a *homeomorphism* of X into Y, if and only if f is one-to-one, continuous, and f^{-1} is continuous (qua function on the subspace $f(X)$ of Y into X). If also $f(X) = Y$, f is a homeomorphism of X *onto*, or *with*, Y. X and Y are said to be *homeomorphic* if there exists at least one homeomorphism of X onto Y.

f is *open* on X into Y if and only if $f(U)$ is open in Y for each open set U in X. It is equivalent to demand that, for each $x \in X$ and each member U of a neighbourhood base at x, $f(U)$ is a neighbourhood of $f(x)$.

A one-to-one function f on X onto Y is a homeomorphism of X with Y if and only if f is continuous and open.

If f is not onto, one must take care to distinguish between the assertion that f is open qua function on X into Y and the assertion that f is open qua function f on X onto $f(X)$, the second statement being weaker than the first. For example, if X is a subspace of Y and f is the injection of X into Y, then f, viewed as a function on X into Y, is open if and only if X is an open subset of Y; whereas it is in any case open when viewed as a function on X onto $f(X)$ (here $= X$). In other words, specification of the space in which the range of f lies is an essential factor in determining whether or not f is open. (Compare the analogous precautions necessary in discussing continuity.)

0.2.10 **Semicontinuous Functions.** It is convenient when formulating the definition of semicontinuity to admit functions whose range is the extended real axis. This is a set $\bar{R}$ obtained by adjoining two distinct points, denoted by $-\infty$ and $+\infty$, to R. The usual order on R is extended to $\bar{R}$ by agreeing that

$$-\infty < r < +\infty$$

for each real number r. Notice that any subset of $\bar{R}$ has a supremum and an infimum in $\bar{R}$.

The field operations are only partially extendible from R to $\bar{R}$. It is to be agreed that $-(-\infty) = +\infty$, $-(+\infty) = -\infty$, that addition and multiplication are to be commutative wherever they are defined, that

$$r + (-\infty) = -\infty, \qquad r + (+\infty) = +\infty$$

for any real r, that

$$(-\infty) + (-\infty) = -\infty, \qquad (+\infty) + (+\infty) = +\infty,$$

and that

$$r \cdot (+\infty) = +\infty \text{ or } -\infty \text{ according as } r > 0 \text{ or } r < 0,$$
$$r \cdot (-\infty) = -\infty \text{ or } +\infty \text{ according as } r > 0 \text{ or } r < 0,$$
$$0 \cdot (-\infty) = 0 \cdot (+\infty) = 0.$$

On the other hand $(-\infty) + (+\infty)$, $(-\infty) - (-\infty)$, $(+\infty) - (+\infty)$ are not defined.

The usual topology on R (see the end of Subsection 0.2.2) is "extended" to $\bar{R}$ by assigning to $+\infty$ a base of neighbourhoods formed of intervals $(r, +\infty]$, where $r \in R$, and to $-\infty$ a base of neighbourhoods formed of intervals $[-\infty, r)$, where $r \in R$.

All this being agreed, suppose that X is a topological space. A function f on X into $\bar{R}$ is said to be *lower semicontinuous* (LSC for short) if and only if its range lies in $\bar{R} \setminus \{-\infty\}$ and, for any $r \in \bar{R}$ (or, equivalently, for any $r \in R$) the set

$$\{x : x \in X \text{ and } f(x) > r\}$$

is open in X. *Upper semicontinuous* functions are those whose negatives are LSC.

A function on X into $\bar{R}$ that is both lower and upper semicontinuous has its range in R and is continuous on X into R.

The *upper* [resp. *lower*] *envelope* of any set of LSC [resp. USC] functions on X is itself LSC [resp. USC].

The sum of two LSC [resp. USC] functions is LSC [resp. USC]; the same is true of the product of two *positive* LSC [resp. USC] functions.

In Chapter 4 we shall need a consequence of the following lemma.

Lemma. Let X be a completely regular space, $\mathscr{A}$ a set of subsets of X containing a neighbourhood base at each point of X. Suppose also that f is a positive LSC function on X. Then f is the upper envelope of the set G of all positive continuous functions g on X into R, each of which vanishes outside some set in $\mathscr{A}$, and such that $g \leq f$ (that is, $g(x) \leq f(x)$ for each x in X).

Proof. Let f_0 be the said upper envelope. It is obvious that $f_0 \leq f$. On the other hand, given $x_0 \in X$, choose any real number $r < f(x_0)$. Since f is LSC, there exists a neighbourhood U of x_0 on which $f > r$. By shrinking U if necessary, we may suppose that $U \in \mathscr{A}$. Since X is completely regular, there exists a continuous function g_0 on X into $[0, 1]$ such that $g_0(x_0) = 1$ and $g_0(X \setminus U) = \{0\}$ (see Subsection 0.2.8). If $g = rg_0$, then $g(x_0) = r$, g is continuous on X into R, g vanishes outside the set $U \in \mathscr{A}$, and $g \leq f$. Thus g belongs to G, and so $f_0(x_0) \geq g(x_0) = r$. Since r is arbitrary save for the restriction $r < f(x_0)$, it follows that $f_0(x_0) \geq f(x_0)$, and so (x_0 being arbitrary) $f_0 \geq f$. ▌

We shall see in Subsection 0.2.18 that any separated or regular locally compact space X is completely regular. In this case we may take $\mathscr{A}$ to consist of all compact subsets of X. This is the case we shall need in Chapter 4, G being now the set of positive continuous functions on X into R, each of which has a compact support.

0.2.11 **Directed Families in a Topological Space.** Let X be a topological space. We need to discuss the convergence and related properties that may or may not be possessed by directed families of points of X. In doing this it is helpful to borrow Kelley's terminology ([1], p. 65). If (x_λ) is a directed family in X and A a subset of X, we say that

(1) (x_λ) is *eventually* in A if and only if there exists an index λ_0 such that $x_\lambda \in A$ for each $\lambda \geq \lambda_0$;

(2) (x_λ) is *frequently* in A if and only if, given any index λ, there exists an index $\lambda' \geq \lambda$ for which $x_{\lambda'} \in A$.

Using these concepts we may readily define the convergence and related concepts linked with directed families.

To begin with, (x_λ) is said to *converge* to x, or that x is a *limit* of (x_λ), if and only if, given any neighbourhood U of x, (x_λ) is eventually in U. One then writes

$$\lim_\lambda x_\lambda = x \qquad \text{or} \qquad x_\lambda \to x \qquad \text{as } \lambda \text{ grows.}$$

Secondly, a point $x \in X$ is said to be a *limiting point* (or *cluster point*) of (x_λ) if and only if, given any neighbourhood U of x, (x_λ) is frequently in U.

The ensuing five properties will be frequently used in the sequel.

(a) If (x_λ) converges to x, the same is true of any cofinal subfamily of (x_λ). This is evident from the definitions.

(b) If x is a limiting point of (x_λ), there exists a cofinal subfamily (y_μ) of (x_λ) that converges to x. The converse is also true.

To see this, let $\{U\}$ denote the directed set of neighbourhoods of x (ordered by set-inclusion reversed—see Subsection 0.1.7) and form the product directed set $M = L \times \{U\}$, L being the domain of (x_λ). We know that, given λ and U, there exists $\lambda^* = f(\lambda, U) \geq \lambda$ such that $x_{\lambda^*} \in U$. We define $y_\mu = x_{f(\lambda, U)}$ for $\mu = (\lambda, U) \in M$. Then (y_μ) is a cofinal subfamily of (x_λ), and it is easy to verify that (y_μ) is convergent to x. The converse is trivial. ▌

(c) If A is a subset of X, a point x of X belongs to the closure $\bar{A}$ if and only if *either* there exists a directed family of points of A that converges to x, *or* there exists a directed family of points of A having x as a limiting point.

This also is a direct consequence of the related definitions.

(d) A topological space X is separated ($= T_2$, or Hausdorff) if and only if no directed family of points of X converges to more than one point of X. If X is separated and $(x_\lambda) \to x$, then x is the sole limiting point of (x_λ).

This follows easily on using (a) and (b). (See also Subsection 0.2.17(1).)

(e) If f is a function on X into Y (X and Y being topological spaces), then f is continuous at x if and only if $(f(x_\lambda)) \to f(x)$ for any directed family (x_λ) in X converging to x.

0.2.12 **Urysohn's Lemma.** This asserts that if X is a normal topological space, and if A and B are disjoint closed subsets of X, then there exists a continuous function on X into $[0, 1]$ such that $f(A) \subset \{0\}$ and $f(B) \subset \{1\}$.

For a proof see Kelley [1], pp. 114–115, or Gillman and Jerison [1], pp. 43–44.

0.2.13 **Urysohn-Tietze Extension Theorem.** The Urysohn-Tietze theorem asserts that if X is a normal topological space, A a closed subset of X, and f a continuous function on A into $[-1, 1]$, then there exists a continuous g on X into $[-1, 1]$ such that $g \mid A = f$.

As a corollary it appears that with the same conditions on X and A, any continuous function f on A into R has a continuous extension to X.

For a proof see Kelley [1], p. 242, Problem O, or Gillman and Jerison [1], pp. 18–19. Tietze gave the result for metric spaces, and Urysohn extended it to embrace normal spaces. The property expressed in this theorem is in fact equivalent to normality.

0.2.14 **Topologies Defined by Families of Functions.** We shall frequently have the need to construct a topology on a set X that has the property of rendering continuous one or more functions on X into topological spaces. We suppose therefore that $(Y_i) = (Y_i)_{i \in I}$ is a given family of topological spaces and that f_i is, for each i, a function on X into Y_i. We then wish to consider those topologies $\mathfrak{T}$ on X (if any) for which it is true that f_i is continuous on X into Y_i for each $i \in I$.

First of all, such topologies do exist—for example, the discrete topology on X. Second, there exists a weakest such topology, as we shall now show.

If $\mathfrak{T}$ is any such topology, and if V_i is open in Y_i, then $f_i^{-1}(V_i)$ must be open for $\mathfrak{T}$, and so therefore is any set of the type

$$U = \bigcap \{f_i^{-1}(V_i) : i \in I\},$$

provided V_i is in all cases open in Y_i and $V_i = Y_i$ for all save a finite (possibly void) set of indices i. On the other hand, these sets U are easily verified to form a base of open sets for some topology on X. Evidently, this topology is the weakest among all those that render continuous each f_i. We refer to this topology as that defined by the given family of functions (f_i).

If x is a point of X, a neighbourhood base at x for the topology defined by the family (f_i) is formed of sets $N = \bigcap \{f_i^{-1}(N_i) : i \in I\}$, where N_i is for each i a neighbourhood of $f_i(x)$ and $N_i = Y_i$ for all save a finite (possibly void) set of indices i.

0.2.15 **Product Spaces and Product Topology.** Suppose that $(X_i) = (X_i)_{i \in I}$ is a family of topological spaces. Let us form the product set $X = \Pi X_i$ (see Subsection 0.1.3).

In Subsection 0.2.14 we take $Y_i = X_i$ and $f_i = \pi_i$, the ith projection on X into X_i. The resulting topology on X, defined by the family (π_i), is called *product* of the topologies on the factor spaces X_i; the set X, together with this product topology, is termed the (topological) *product* of the family (X_i).

Each map π_i is open on X onto X_i.

A function f on a topological space S into the product space X is continuous if and only if $\pi_i \circ f$ is continuous on S into X_i for each index i.

If each factor space X_i is separated, so is their product.

In the special case in which each X_i is a fixed topological space Y, the product set $X = \Pi X_i$ is Y^I, the set of all functions on I into Y, and the product topology on X is that sometimes spoken of as the topology of pointwise convergence on I.

The usual topology on R^n (resp. C^n) is the product of the usual topologies on R (resp. C).

0.2.16 **Quotient Spaces.** Consider a situation in which X is a topological space, Y is a set, and f is a function on X into Y. It is then natural to consider those topologies on Y such that f is continuous. Such topologies exist—for example, the zero topology on Y. For any such topology $\mathfrak{T}'$, $f^{-1}(V)$ must be open in X for each $\mathfrak{T}'$-open V. Now the set $\mathfrak{T}$ of all subsets V of Y, such that $f^{-1}(V)$ is open in X, is a topology on Y. This topology is the strongest topology on Y relative to which f is continuous.

In studying this topology it is convenient to write

$$A^* = f^{-1}(f(A))$$

for each subset A of X. Then

$$A^* \supset A, \qquad f(A^*) = f(A), \qquad A^{**} = A^*.$$

We note five statements relating to this topology.

(1) $f(X \setminus A^*) = f(X) \setminus f(A)$, A being any subset of X.

(2) A being a subset of X, $f(A)$ is open if and only if A^* is open.

The first assertion is immediate from the definitions; and (2) follows from the continuity of f.

(3) f is open if and only if U^* is open in X whenever U is open in X, in which case one obtains a neighbourhood base at $y = f(x)$ on taking the sets $f(U)$ when U ranges over some neighbourhood base at x.

(4) If f is onto and open, then $f(A)$ is closed if and only if A^* is closed.

In fact, A^* is in any case closed if $f(A)$ is closed, thanks to continuity of f. On the other hand if f is onto, one has from (1) the formula $f(X \setminus A^*) = Y \setminus f(A)$, so that if A^* is closed and f is open, it appears that $Y \setminus f(A)$ is open and hence $f(A)$ is closed.

(5) A function g on Y into a topological space Z is continuous if and only if $g \circ f$ is continuous on X into Z.

This also is evident from the definition of $\mathfrak{T}$ and of continuity.

Quotient spaces may be treated as a special case of the situation just studied. Let X be any topological space, and let R be an equivalence relation on X. As explained in Subsection 0.1.2, R serves to define a partition of X into equivalence classes modulo R. The set of all such equivalence classes is the *quotient set* of X modulo R, usually denoted by X/R. Let π be the natural projection of X onto X/R, $\pi(x)$ being the equivalence class $R(x)$. The *quotient topology* on X/R is by definition the strongest topology on X/R relative to which π is continuous on X into X/R. In this case, A^* is none other than

$$R(A) = \{y : (x, y) \in R \text{ for some } x \in A\}.$$

If X/R is T_2, then R is closed in the product space $X \times X$; and if π is open and R is closed in $X \times X$, then X/R is T_2.

For us the most important case is that in which X is a topological vector space and R is defined by some vector subspace L of X,

$$R = \{(x, y) : x - y \in L, x \in X, y \in X\}.$$

It is then necessarily the case that U^* is open whenever U is open. So, in accord with (3), π is in this case open; hence (4) also is true in this case; that is, $\pi(A)$ is closed if and only if $A + L$ is closed.

0.2.17 **Compact Spaces.** By a *covering* (or *cover*) of a set X is meant a set Σ (or sometimes a family (A_i)) of subsets of X whose union is X. If X is a topological space, a covering is said to be *open* if each of its members is an open subset of X. Also, if X is a set, a set Σ (or a family (A_i)) of subsets of X is said to have the *finite intersection property* (FIP for short) if each finite subset of Σ (or finite subfamily of (A_i)) has a nonvoid intersection.

X being a topological space, it is easily verified that the following four statements are equivalent:

(a) Every open covering of X contains a finite covering.

(b) If a set Σ (or a family (A_i)) of closed subsets of X has the FIP, then the intersection of Σ (or of (A_i)) is nonvoid.

(c) Every directed family of points in X has a limiting point in X.

(d) Every directed family of points of X has a convergent cofinal subfamily.

The equivalence of (a) and (b) is displayed by taking complements and using the de Morgan formulae (see Subsection 0.1.1), and that of (c) and (d) is seen on reference to Subsection 0.2.11(b). It remains to establish the equivalence of (b) and (c). On the one hand, if (x_λ) is a directed family of points of X, introduce the closed sets $F_\lambda = \overline{\{x_{\lambda'} : \lambda' \geq \lambda\}}$; since λ runs over a directed set, the family (F_λ) has the FIP and so, if (b) is true, there is a point x common to all the F_λ; it is then simple to check that x is a limiting point of (x_λ). Thus (b) implies (c). Reciprocally assume that (c) is true and that (F_λ) is a family of closed subsets of X having the FIP; if here λ runs over an index set L, let I be the set of all finite subsets i of L. I is ordered by set inclusion and is then a directed set. By hypothesis, for each i one may choose x_i from $P_i = \bigcap \{F_\lambda : \lambda \in i\}$, and by

(c) the directed family (x_i) has a limiting point $x \in X$. Since each P_i is closed and $P_{i'} \subset P_i$ if $i' \supset i$, it follows that x lies in P_i for every i and therefore $x \in \bigcap \{F_\lambda : \lambda \in L\}$. Thus (c) implies (b), as we had to show. ▌

Definition. A topological space X is said to be *compact* if and only if it possesses any one (and therefore all) of the equivalent properties specified in (a)–(d) above.

Bourbaki [2] reserves the name "compact" for those spaces that satisfy the above definition and that are also separated.

We proceed to enumerate some basic properties of compact spaces.

(1) If X is a compact space, and if a directed family (x_λ) of points of X possesses at most one limiting point x, then $\lim x_\lambda = x$.

Since X is compact, (x_λ) must have exactly one limiting point x, and it remains to show that (x_λ) converges to x. If this were not the case, however, there would exist an open set U containing x and a cofinal subfamily (y_μ) of (x_λ), each y_μ lying in $X \setminus U$. Since X is compact, (y_μ) itself has a cofinal subfamily (z_ν) converging to a limit z. Since $X \setminus U$ is closed, $z \in X \setminus U$. In particular, $z \neq x$. Yet, since (z_ν) is a cofinal subfamily of (x_λ), z must be a limiting point of (x_λ). This contradiction of the uniqueness of x establishes (1). ▌

A subset A of a topological space X is said to be *compact* if and only if it is a compact space when endowed with the induced topology; A is said to be *relatively compact* (in X) if and only if its closure $\bar{A}$ is compact.

A topological space is said to be *sigma-compact* (often written σ-compact) if it is the union of countably many compact subsets.

It is evident that any closed subset of a compact space is compact, and that a compact subset of a separated space is necessarily closed.

The Weierstrass-Bolzano theorem of real analysis shows that the compact subsets of R^n or C^n are precisely those that are bounded and closed.

(2) If X is a compact space, Y a topological space, and f a continuous function on X into Y, then $f(X)$ is a compact subset of Y. If also f is one-to-one and Y is separated, then f is a homeomorphism of X onto $f(X)$.

If (V_i) is an open covering of $Y_0 = f(X)$, then $(f^{-1}(V_i))$ is an open covering of X, hence contains a finite covering, say $\{f^{-1}(V_i) : i \in J\}$ where J is a finite set of indices. But then $(V_i)_{i \in J}$ is a finite open covering of Y_0. So Y_0 is compact. Suppose now that f is one-to-one and that Y is separated. Since each closed subset of X is compact, its image is compact in Y, hence is closed in Y, hence is closed in Y_0. This shows that f^{-1} is continuous on Y_0 into X, so that f is indeed a homeomorphism of X onto $Y_0 = f(X)$. ▌

An immediate corollary of (2) is the following useful remark.

(3) Let X be a set, and let $\mathfrak{T}$ and $\mathfrak{T}'$ be two topologies on X such that $(X, \mathfrak{T}')$ is compact, $(X, \mathfrak{T})$ is separated, and $\mathfrak{T} \leq \mathfrak{T}'$. Then $\mathfrak{T} = \mathfrak{T}'$.

For the identity map f on $(X, \mathfrak{T}')$ onto $(X, \mathfrak{T})$ is continuous (since $\mathfrak{T}$ is weaker than $\mathfrak{T}'$) and is one-to-one. By (2) and the assumed compactness of $(X, \mathfrak{T}')$ and separatedness of $(X, \mathfrak{T})$, f is a homeomorphism of $(X, \mathfrak{T}')$ onto $(X, \mathfrak{T})$; that is, $\mathfrak{T}$ and $\mathfrak{T}'$ are identical topologies. ▌

(4) Any separated compact space X is normal.

For suppose that A and B are disjoint closed subsets of X. If $a \in A$ and $b \in B$, there exist open neighbourhoods N_a of a and N_b of b that are disjoint. Since B is compact, a finite number of the N_b cover B and the union of these, say V_a, is an open set containing B and such that $V_a \cap N_a = \emptyset$. Since A is compact, a finite number of the N_a—say, $N_{a_1}, \cdots, N_{a_n}$—cover A. The union U of these N_{a_i} is open and contains A. Also, $V = \bigcap \{V_{a_i} : 1 \leq i \leq n\}$ is an open set containing B. Plainly, $U \cap V = \emptyset$. ▮

(5) If X is a regular topological space, A a compact subset of X, and U a neighbourhood of A, then U contains a closed neighbourhood V of A. In particular, any compact and regular space is normal.

Indeed, X being regular, for each $a \in A$ there exists an open neighbourhood W_a of a such that $\bar{W}_a \subset U$. By compactness, a finite number of these W_a cover A, and for V one may take the union of the closures of these selected neighbourhoods W_a. This establishes the first assertion. For the second, suppose that A and B are disjoint closed subsets of a compact and regular space X. Then $X \setminus B$ is a neighbourhood of A and so, by the first part, contains a closed neighbourhood U of A. Since the closed sets B and U are disjoint, the same argument shows that there exists a closed neighbourhood V of B disjoint from U. Thus A and B have disjoint neighbourhoods U and V, respectively, showing that X is normal. ▮

(6) If X is a completely regular topological space, A a compact subset of X, U a neighbourhood of A, then there exists a continuous function f on X into $[0, 1]$ such that $f(A) \subset \{1\}$ and $f(X \setminus U) \subset \{0\}$.

For each $a \in A$ there is a continuous g_a on X into $[0, 1]$ such that $g_a(a) = 1$ and $g_a(X \setminus U) = \{0\}$. The set $\{x : g_a(x) > \frac{1}{2}\}$ is open and so, if we define h_a by $h_a(x) = \text{Inf}\,(2g_a(x), 1)$, then h_a is continuous on X into $[0, 1]$, is zero on $X \setminus U$, and is one on a neighbourhood N_a of a. By compactness, there is a finite family (a_i) of points of A such that the corresponding N_{a_i} cover A. It suffices to define $f = \text{Sup}\,\{h_{a_i} : 1 \leq i \leq n\}$. ▮

Perhaps the most significant single result about compact spaces is the following.

(7) *Tychonoff's theorem.* The topological product of any family of compact spaces is again compact.

Proof. Let $(X_i)_{i \in I}$ be a family of compact spaces and X the product space. It will suffice to show that if Σ is a set of subsets of X having the FIP, then $\bigcap \{\bar{A} : A \in \Sigma\}$ is nonvoid. A simple application of Zorn's lemma 0.1.5 shows that in doing this one may assume that Σ is maximal with respect to the FIP. Assuming maximality leads to the following two statements:

(a) Σ is stable under finite intersections.

(b) Any subset of X that intersects every member of Σ itself belongs to Σ.

Now for each i, $\pi_i(\Sigma)$ is a set of subsets of X_i having the FIP. The assumed compactness of X_i entails, therefore, the existence of $x_i \in X_i$ which is adherent to every set in $\pi_i(\Sigma)$. So if U_i is any neighbourhood in X_i of x_i, U_i intersects $\pi_i(A)$ for every $A \in \Sigma$. In other words, $\pi_i^{-1}(U_i)$ intersects every member of Σ. According to (b), $\pi_i^{-1}(U_i)$ belongs to Σ. Thanks to (a), the same is true of any finite intersection of these sets. But these finite intersections form a neighbourhood base at $x = (x_i)$ for the product topology on X. Thus every neighbourhood of x belongs to Σ and must therefore intersect every member of Σ. Hence x is adherent to every member of Σ. ▌

Remark. Apart from a studied avoidance of the term "ultrafilter," the above proof is that given by Bourbaki ([2], Chapitre 1, p. 63, Théorème 2).

(8) It is sometimes important to be able to express compactness in terms of ordinary (denumerable) sequences.

In Chapter 8 we shall spend some time on dealing with this topic for the special case of topological vector spaces. Meanwhile we note a few simple and general results of this type.

Consider the following three statements concerning a topological space X:

(a) Each sequence of points of X has a limiting point.

(b) Each sequence of points of X has a convergent subsequence.

(c) X is compact.

It is quite evident that (b) implies (a), and that (b) and (a) are equivalent whenever X satisfies the *first countability axiom* (that is, each point of X has a countable neighbourhood base). The definition of compactness shows that (c) implies (a) in all cases, so that (c) implies both (a) and (b) if X satisfies the first countability axiom. Finally, if X satisfies the *second countability axiom* (that is, if there is a countable open base for the topology on X), then all three assertions are equivalent. We leave the proof of this to the reader, with the remark that one may begin by showing that in the presence of the second countability axiom, each open covering has a countable open refinement (a covering (B_j) being a *refinement* of a covering (A_i) if and only if each B_j is a subset of some A_i).

0.2.18 **Locally Compact Spaces.** A topological space X is said to be *locally compact* if and only if each point of X has at least one compact neighbourhood in X. Thus, any compact space is locally compact; any discrete space is locally compact; any closed subspace of a locally compact space is locally compact. The number spaces R^n and C^n are each locally compact, and neither is compact.

We note a few properties of locally compact spaces that are used (explicitly or otherwise) in Chapter 4.

(1) If X is locally compact and also either separated or regular, then the closed, compact neighbourhoods of any point of X form a neighbourhood base at that point. In particular:

(a) Any separated locally compact space is regular.

(b) Any open subspace of a locally compact and regular space is locally compact and regular.

For if $x \in X$, there exists a compact neighbourhood U_0 of x. Let U be any neighbourhood of x. If X is regular, there exists a closed neighbourhood of x contained in $U \cap U_0$, and this neighbourhood will be compact (being a closed subset of U_0). Whence the assertion in case X is regular. Next suppose that X

is separated. Put V for the interior of $U \cap U_0$. V is a neighbourhood of x and $\bar{V}$ is a compact separated space. Also, $\{x\}$ and $\bar{V} \setminus V$ are disjoint closed subsets of the space $\bar{V}$. So, by Subsection 0.2.17(4), there exists a closed, compact neighbourhood W of x in $\bar{V}$ that is contained in V. W is then a neighbourhood of x in V and so, V being open in X, W is a neighbourhood of x in X. Finally, $W \subset V \subset U$ and the assertion is proved. ▌

(2) Suppose that X is locally compact and regular, that A is a closed and compact subset of X, and that U is a neighbourhood of A. Then there exists a closed neighbourhood V of A such that $V \subset U$. Moreover, there exists a continuous function f on X into $[0, 1]$ such that $f(A) \subset \{0\}$ and $f(X \setminus V) \subset \{1\}$. In particular, any locally compact and regular space is completely regular, and any separated locally compact space is a Tychonoff space (that is, is T_1 and completely regular).

In fact, for each $x \in A$ there exists a neighbourhood W_x of x that is contained in U and that is closed and compact. Since A is compact, a finite number of the W_x cover A, and their union is a closed and compact neighbourhood V of A. The subspace V is regular and compact, hence is normal (see Subsection 0.2.17(5)). Thus there exists a continuous function g on V into $[0, 1]$ such that $g(A) \subset \{0\}$ and $g(V \setminus \text{Int } V) \subset \{1\}$. Define f to be equal to g on V and to one on $X \setminus V$. Then f is continuous on X and satisfies all our demands. ▌

There is an analogue for locally compact spaces of Tychonoff's theorem (0.2.17(7)).

(3) If (X_i) is a family of locally compact spaces, the product $X = \Pi X_i$ is locally compact if and only if all but a finite number of the factor spaces is compact.

For a proof see Kelley [1], p. 149, or Bourbaki [2], Chapitre 1, p. 65, or both.

(4) *Compactification processes.* There are numerous constructions that aim at imbedding a given space X into a compact space. We require only the crudest and simplest such process.

Given a topological space X, we consider the space X^* obtained by adjoining to the set X one more point, denoted by ∞, and endowing X^* with the topology whose open sets are the open subsets of X together with all sets of the type $X^* \setminus K$, where K is any closed and compact subset of X. The resulting space X^* is the so-called *Alexandroff* (one-point) *compactification* of X.

The compactness of X^* is almost immediate—if (U_i) is an open covering of X^*, some U_{i_0} must contain ∞ and $X^* \setminus U_{i_0}$ is then a compact subset of X. The latter set is therefore covered by a finite number of the U_i, and so the same is true of X^*.

Also evident is the assertion that the induced topology on X is the initial topology on X (that is, X is topologically imbedded in X^*), and that X is an open subspace of X^*.

X^* is separated if and only if X is separated and locally compact.

For more details see Kelley [1], p. 150 or Bourbaki [2], Chapitre 1, p. 65, or both.

0.2.19 It is convenient to summarize here certain properties of locally compact spaces that are fundamental to the integration theory developed in Chapter 4 and used subsequently.

(1) Any separated locally compact space is regular. From Subsection 0.2.8 we recall also that any regular T_1 space is separated.

(2) Any regular locally compact space is completely regular.

Both of these assertions have been discussed in the preceding section.

It is easily verified that in a regular locally compact space, the set of all compact sets contains a neighbourhood base at each point. So, applying (2) and the lemma in Subsection 0.2.10, one is led to the following statement:

(3) Let X be a regular locally compact space. Each positive LSC function f on X is the upper envelope of those positive continuous functions that vanish outside compact subsets of X and that minorize f.

0.2.20 **Paracompact Spaces.** Compactness has been characterized in Subsection 0.2.17 in terms of open coverings. At one or two places (notably in Chapter 5) we shall need to appeal to another property of certain topological spaces that is related to the behaviour of open coverings.

As a matter of definition, a covering (A_i) of a topological space X is said to be *locally finite* if for each $x \in X$ there exists a neighbourhood U of x that intersects A_i for at most a finite set of indices i (the finite set depending in general on x and on U).

A topological space X is said to be *paracompact* if and only if it is regular and each open covering of X has an open, locally finite refinement.

There are various alternative ways of characterising paracompactness in terms of coverings; see Kelley [1], pp. 156–160, where the following two results are established:

(1) Each paracompact space is normal.

(2) Each semimetrizable space (see Subsection 0.2.21) is paracompact.

It is important for us to establish in addition a property of certain open coverings of normal spaces. Let us agree to say that a covering (A_i) of a topological space X is *point finite* if each point of x belongs to A_i for at most a finite set of indices i (the finite set depending on x).

(3) Let X be a normal topological space and $(U_i)_{i \in I}$ a point finite open covering of X. Then there exists an open covering $(V_i)_{i \in I}$ such that $\bar{V}_i \subset U_i$ for each $i \in I$.

The alleged property of (V_i) in relation to (U_i) is often expressed by saying that (V_i) is *subordinate* to (U_i); it is evidently a stronger condition than refinement.

In order to prove (3), let us consider first the finite case, in which we may suppose that $I = \{1, 2, \cdots, n\}$. The closed sets $F = X \setminus U_1$ and $F' = X \setminus \bigcup \{U_i : 1 < i \leq n\}$ are disjoint and therefore have disjoint open neighbourhoods U and V_1, respectively. Since $V_1 \subset X \setminus U$, so $\bar{V}_1 \subset X \setminus U$. And since $X \setminus \bigcup \{U_i : 1 < i \leq n\} \subset V_1$, the family $(V_1, U_2, \cdots, U_n)$ is an open covering. It suffices now to repeat this process a further $n - 1$ times.

Coming to the general case, let Σ be the set of all functions u on I such that, for each i, $u(i)$ is either U_i or is an open set V_i satisfying $\bar{V}_i \subset U_i$, and such that $(u(i))_{i \in I}$ is a covering of X. We order Σ by writing $u \leq u'$ if and only if $u'(i) = u(i)$ for all indices $i \in I$ for which $u(i)$ is a V_i. We will show first that Σ is inductively ordered.

Suppose that Σ' is a linearly ordered subset of Σ and define u'' by

$$u''(i) = \bigcap \{u'(i) : u' \in \Sigma'\}.$$

This u'' is a covering of X by virtue of the fact that (U_i) is point finite. Moreover, u'' is the supremum of Σ' in Σ.

According to Zorn's Lemma 0.1.5, Σ contains a maximal element, say u^*. We shall show that $u^*(i)$ is a V_i for each i, and then (3) will be established.

Suppose that for some index $j \in I$ one had $u^*(j) = U_j$. Let

$$F = X \setminus \bigcup \{u^*(i) : i \neq j\}.$$

As for the finite case already established, there exists an open V_j such that $F \subset V_j$ and $\bar{V}_j \subset U_j$. But then if u^{**} is defined by $u^{**}(i) = u^*(i)$ for $i \neq j$ and $u^{**}(j) = V_j$, one would have $u^* \leq u^{**}$ and $u^* \neq u^{**}$, contrary to the maximality of u^*. ∎

The preceding argument is suggested by Kelley ([1], p. 171, Problem V).

Combining (1) and (3) leads to the following statement:

(4) If X is a paracompact space and (U_i) an open covering of X, then there exists a locally finite open covering (V_j) such that for each j there exists i for which $\bar{V}_j \subset U_i$.

For, thanks to paracompactness, (U_i) has a locally finite open refinement $(U_j^\dagger)$; and by (3) there exists an open covering (V_j) such that $\bar{V}_j \subset U_j^\dagger$ for each j. Then (V_j) is necessarily locally finite, and for each j one has $\bar{V}_j \subset U_j^\dagger \subset U_i$ for some i (since $(U_j^\dagger)$ refines (U_i)). ∎

The covering properties discussed above have a bearing upon the existence of certain families of continuous functions on X. By a (continuous) *partition of unity* on a topological space X is meant a family (f_i) of continuous functions on X such that $0 \leq f_i \leq 1$ and $\sum f_i = 1$, the sum being locally finite in the sense that each point x of X has a neighbourhood U such that $f_i(U) = \{0\}$ for all but a finite set of indices i.

By using (1), (4), and Urysohn's Lemma 0.2.12, it is possible to establish the existence of an abundance of such partitions of unity.

(5) Let X be paracompact, (U_i) any open covering of X. Then there exists a partition of unity (f_j) such that for each j the support of f_j is contained in some U_i.

In fact, we choose (V_j) as a locally finite open covering such that each $\bar{V}_j$ is contained in some U_i. By (3) there exists an open covering (W_j) that is locally finite and subordinate to (V_j), so that $\bar{W}_j \subset V_j$ for each j. For each j, $\bar{W}_j$ and $X \setminus V_j$ are disjoint closed sets and so Urysohn's lemma affirms the existence of a continuous function g_j on X into $[0, 1]$ such that $g_j(\bar{W}_j) = \{1\}$ and $g_j(X \setminus V_j) \subset \{0\}$. The family (g_j) is then locally finite, so that the sum $\sum g_j$ is continuous. Obviously, $\sum g_j \geq 1$ everywhere. It suffices to define $f_j = g_j/(\sum g_j)$. ▌

For use in Chapter 5 it is important to observe that if $X = R^m$ (or an open subspace of R^m, or a general paracompact real differentiable manifold), one can construct the functions f_j so that each is indefinitely differentiable and has a compact support. This formulation of (5) can be deduced from the original by a simple process of regularizing explained in Chapter 5.

0.2.21 **Semimetrics and Metrics and Associated Topologies.** If X is a set, by a *semimetric* on X is meant a function d on $X \times X$ into R satisfying the following conditions:

(1) $$d(x, y) = d(y, x) \geq 0;$$

(2) $$d(x, z) \leq d(x, y) + d(y, z) \text{ (triangle inequality)};$$

and

(3) $$d(x, x) = 0,$$

x, y, and z denoting arbitrary points of X. If also

(4) $$d(x, y) = 0 \text{ entails } x = y,$$

d is termed a *metric* on X. Some writers (for example, Kelley [1], pp. 118–119) use the term "pseudometric" in place of "semimetric."

By a *semimetric* (or *metric*) *space* is meant a pair (X, d) comprising a set X and a semimetric (or metric) d on X. Usually, when d is understood, we refer to X as the semimetric (or metric) space.

If (X, d) is a semimetric space, $x \in X$, and r is a real number > 0, the set

$$B(x, r) = \{y : y \in X \quad \text{and} \quad d(x, y) < r\}$$

is termed the *open ball with centre x and radius r*, while the set

$$\overline{B(x, r)} = \{y : y \in X \quad \text{and} \quad d(x, y) \leq r\}$$

is the *closed ball with centre x and radius r*. Justification of use of the adjectives "open" and "closed" will appear shortly.

(X, d) being a semimetric space, the properties (1)–(3) arrange that the set of pairs $(x, B(x, r))$ with $x \in X$ and $r > 0$ satisfy the requirements (a′), (b′), (d′) in Subsection 0.2.2. Accordingly there is a unique topology $\mathfrak{T}_d$ on X for which, for each $x \in X$, the open balls $B(x, r)$ form a neighbourhood base at x. Since furthermore $B(x, r) \subset \overline{B(x, r)} \subset B(x, r')$ for $0 < r < r'$, the closed balls with centre x and radii >0 also form a neighbourhood base at x.

Each open (resp. closed) ball is easily verified to be open (resp. closed) for the topology $\mathfrak{T}_d$—hence the reason for the nomenclature. It follows in particular that the topology $\mathfrak{T}_d$ is regular. We leave it as an exercise for the reader to show that $\mathfrak{T}_d$ is, in fact, completely regular. On the other hand, $\mathfrak{T}_d$ satisfies the T_0 separation axiom (see Subsection 0.2.8) if and only if d is a metric, in which case $\mathfrak{T}_d$ is even separated.

The topology $\mathfrak{T}_d$ is also definable in terms of the general process described in Subsection 0.2.15, being the weakest topology on X for which each of the functions $d_x : y \to d(x, y)$ is continuous on X into R (the index x ranging over X).

A topological space $(X, \mathfrak{T})$ is said to be *semimetrizable* (or *metrizable*) if and only if there exists a semimetric (or metric) d on X such that $\mathfrak{T} = \mathfrak{T}_d$. Much study has gone into the characterization of semimetrizable and metrizable spaces (see, for example, Kelley [1], pp. 124–130). We shall be more concerned with a slightly different problem, however, in which topological spaces are replaced by so-called uniform spaces (see Subsections 0.3.5 and 0.3.6).

0.3 Uniform Spaces

0.3.1 Uniformities, or Uniform Structures, on a Set. Let X be a set and $\Delta = \Delta(X) \subset X \times X$ the diagonal on X (see Subsection 0.1.2). By a *uniformity*, or a *uniform structure*, on X is meant a set $\mathfrak{X}$ of subsets of $X \times X$ satisfying the following conditions:

(a) $U \supset \Delta$ for each $U \in \mathfrak{X}$.

(b) If $U \in \mathfrak{X}$, then $U^{-1} \in \mathfrak{X}$.

(c) If $U \in \mathfrak{X}$, there exists $U' \in \mathfrak{X}$ such that $U' \circ U' \subset U$.

(d) The intersection of two members of $\mathfrak{X}$ also belongs to $\mathfrak{X}$.

(e) Any subset of $X \times X$, which contains a member of $\mathfrak{X}$, itself belongs to $\mathfrak{X}$.

The members of $\mathfrak{X}$ are called the *vicinities* (of Δ) of, or in, the uniformity $\mathfrak{X}$.

By a *base* for a uniformity $\mathfrak{X}$ is meant a subset $\mathscr{B}$ of $\mathfrak{X}$ such that a subset of $X \times X$ belongs to $\mathfrak{X}$ if and only if it contains a set belonging to $\mathscr{B}$. $\mathscr{B}$ must then satisfy the following conditions:

(a′) $B \supset \Delta$ for each $B \in \mathscr{B}$.

(b′) If B is a member of $\mathscr{B}$, then B^{-1} contains a member of $\mathscr{B}$.

(c′) If B is a member of $\mathscr{B}$, there exists $B' \in \mathscr{B}$ such that $B' \circ B' \subset B$.

(d′) The intersection of two members of $\mathscr{B}$ contains a member of $\mathscr{B}$.

Conversely, given a set $\mathscr{B}$ of subsets of $X \times X$ that satisfies (a′)–(d′), the set $\mathfrak{X}$ of all subsets of $X \times X$ that contain a (variable) member of $\mathscr{B}$ is a uniformity on X, of which $\mathscr{B}$ is a base.

By a *uniform space* is meant a pair $(X, \mathfrak{X})$ comprising a set X and a uniformity $\mathfrak{X}$ on X. Often, if no confusion is to be feared, one speaks of "the uniform space X"; compare the analogous liberty taken with topological spaces (Subsection 0.2.1).

For a given set X one may (as in the case of topologies, see Subsection 0.2.6) compare certain pairs of uniformities on X. If $\mathfrak{X}$ and $\mathfrak{X}'$ are uniformities on X, we say that $\mathfrak{X}$ is *weaker* than $\mathfrak{X}'$, and that $\mathfrak{X}'$ is *stronger* than $\mathfrak{X}$, if and only if $\mathfrak{X} \subset \mathfrak{X}'$ (qua subsets of the set of all subsets of $X \times X$). There is a weakest uniformity on X, namely, $\{X \times X\}$, and strongest, namely, $\{U : \Delta \subset U \subset X \times X\}$.

Historically the first type of uniformity to be considered was that derived from a semimetric d on X (see Section 0.2.21), a base for the associated uniformity being by definition comprised of the sets

$$B_r = \{(x, x') \in X \times X : d(x, x') < r\},$$

r ranging over all strictly positive numbers (or over any set of strictly positive numbers having zero as its infimum). We shall later need to examine this case in more detail (see Subsections 0.3.5–0.3.7).

0.3.2 **The Topology Defined by a Uniformity.** If $(X, \mathfrak{X})$ is a uniform space, one may define a topology $\mathfrak{T}$ on X by assigning to each point x of X the neighbourhood base comprised of the sets $U(x)$, U ranging over the uniformity; the notation is as in Subsection 0.1.2, so that

$$U(A) = \{y : y \in X \text{ and } (x, y) \in U \text{ for some } x \in A\},$$

A being any subset of X.

At the same time, one may consider the product topology on $X \times X$ (see Subsection 0.2.15). It is important to relate these topologies with the given uniformity. Unless the contrary is specified, X and $X \times X$ are regarded as topological spaces with the above-named topologies.

(1) If $U \in \mathfrak{X}$, then the interior of U also belongs to $\mathfrak{X}$. Hence the open symmetric vicinities form a base for $\mathfrak{X}$.

(2) If A is a subset of X, its closure is identical with $\bigcap \{U(A) : U \in \mathfrak{X}\}$. If S is a subset of $X \times X$, its closure is identical with $\bigcap \{U \circ S \circ U : U \in \mathfrak{X}\}$. Hence the symmetric closed vicinities form a base.

(3) From the last statement we see that $\mathfrak{T}$ is always regular. Moreover, $\mathfrak{T}$ is separated (T_2) if and only if $\bigcap \{U : U \in \mathfrak{X}\} = \Delta$; in which case we say that the uniformity is *separated*.

With a nonseparated space $(X, \mathfrak{X})$ a separated space may be associated in a natural way. Thus, if

$$R = \bigcap \{U : U \in \mathfrak{X}\},$$

then R is an equivalence relation on X. Form the quotient set $\dot{X} = X/R$. For each $U \in \mathfrak{X}$ let

$$\dot{U} = \{(\dot{x}, \dot{y}) : \exists\, x \in \dot{x}, y \in \dot{y} \text{ such that } (x, y) \in U\}.$$

Then the $\dot{U}$ form a base for a uniformity $\dot{\mathfrak{X}}$ on $\dot{X}$, and $(\dot{X}, \dot{\mathfrak{X}})$ is separated: we term it the separated uniform space *associated* with $(X, \mathfrak{X})$.

0.3.3 **Uniform Continuity.** If $(X, \mathfrak{X})$ and $(Y, \mathfrak{Y})$ are uniform spaces, f a function on X into Y, then f is said to be uniformly continuous if for each $V \in \mathfrak{Y}$ the set

$$\{(x, x') \in X \times X : (f(x), f(x')) \in V\}$$

belongs to $\mathfrak{X}$. f is then continuous on X into Y for the associated topologies. The converse of this is generally false.

It is very simple to check that a composite of uniformly continuous functions is again uniformly continuous.

0.3.4 Uniformity Defined by a Family of Functions. Suppose that X is a set and that $((Y_i, \mathfrak{Y}_i))_{i \in I}$ is a family of uniform spaces. Then there exists a weakest uniformity $\mathfrak{X}$ on X such that, for each i, f_i is uniformly continuous on $(X, \mathfrak{X})$ into $(Y_i, \mathfrak{Y}_i)$. Indeed, the finite intersections of sets of the type

$$\{(x, x') \in X \times X : (f_i(x), f_i(x')) \in V_i\},$$

where $V_i \in \mathfrak{Y}_i$, form a base for the said uniformity.

As special cases one may in this way introduce the *induced uniformity* on a subset X of a given uniform space Y, and also the *product uniformity* on the set ΠX_i, where the X_i are uniform spaces. The reader should compare the analogous situation for topological spaces (Subsections 0.2.7 and 0.2.15). The topology associated with a product uniformity is the product of the associated topologies; and that associated with the induced uniformity is the induced topology.

0.3.5 Uniformities and Semimetrics. If X is a set and $(X, \mathfrak{X})$ is a uniform space, a semimetric d on X (see Subsection 0.2.21) is uniformly continuous for the product uniformity if and only if for each $\varepsilon > 0$ the set

$$U_{d,\varepsilon} = \{(x, x') \in X \times X : d(x, x') < \varepsilon\}$$

belongs to $\mathfrak{X}$.

On the other hand if X is a set and d a semimetric on X, it is natural to consider the uniformity $\mathfrak{X}_d$ with a base of vicinities comprised of the sets $U_{d,\varepsilon}$ defined above. This is the weakest uniformity on X such that d is uniformly continuous for the product uniformity on $X \times X$. We notice that $\mathfrak{X}_d$ is separated if and only if d is a metric on X. Uniformities of the type $\mathfrak{X}_d$ were the prototypes of uniformities in general.

A natural question is to demand conditions on a uniform space $(X, \mathfrak{X})$ in order that $\mathfrak{X} = \mathfrak{X}_d$ for some semimetric d; in which case the uniform space $(X, \mathfrak{X})$, or the uniformity $\mathfrak{X}$, are said to be *semimetrizable* (or *metrizable*, if d is a metric).

0.3.6 Semimetrization Lemma. It may be shown (Kelley [1], p. 185; Bourbaki [2], Chapitre IX, n° 4, page 7, Proposition 2) that a uniform space is semimetrizable if and only if it has a countable base of vicinities.

For us by far the most important case is that of a topological vector space, and we shall give the proof for this case (which is similar to the general situation) at the beginning of Chapter 6.

0.3.7 Suppose more generally that one has a family $(d_i)_{i \in I}$ of semimetrics on a set X. The sets U_{d_i,ε_i} form a subbase for a uniformity on X that is characterized as the weakest uniformity $\mathfrak{X}$ for which the identity map of $(X, \mathfrak{X})$ into $(X, \mathfrak{X}_{d_i})$ is uniformly continuous.

0.3.8 It is demonstrable that any given uniformity on a set X is definable as in Subsection 0.3.7, (d_i) being the family of all semimetrics that are uniformly continuous for the product uniformity on $X \times X$. It follows that any uniform space is completely regular as a topological space.

0.3.9 As a corollary to Subsection 0.3.8 one deduces a result due to Weil: Any uniform space is isomorphic (as a uniform space) with a subspace of a product of semimetrizable spaces (metrizable spaces, if the uniform space is separated).

See Kelley [1], p. 188 or Bourbaki, loc. cit. in Subsection 0.3.6, Théorème 1, or both.

0.3.10 **Uniformizable Topological Spaces.** From the preceding results one may deduce that, given a topological space $(X, \mathfrak{T})$, there exists a uniformity $\mathfrak{X}$ on X, whose associated topology is identical with $\mathfrak{T}$, if and only if $(X, \mathfrak{T})$ is completely regular. One then says that the topology $\mathfrak{T}$, and the topological space $(X, \mathfrak{T})$, are *uniformizable*. By Subsection 0.2.19(2) it follows that any locally compact and regular topological space is uniformizable.

The uniformity is uniquely determined by $(X, \mathfrak{T})$ if the latter is a compact space; in this case any continuous function on $(X, \mathfrak{T})$ into a uniform space is uniformly continuous on $(X, \mathfrak{X})$, any semimetric that is continuous on $X \times X$ is uniformly continuous, and each open subset of $X \times X$ that contains Δ is a member of $\mathfrak{X}$.

For these matters see Kelley [1], pp. 188–199 or Bourbaki [2], Chapitre II, pp. 107–110, or both.

0.3.11 **Cauchy Directed Families and Completeness.** Since a uniformity leads to a way of indexing the neighbourhoods of variable points, the index set being the same for all points, it also opens the way for defining "smallness" for sets, irrespective of their position. Thus, if $(X, \mathfrak{X})$ is a uniform space, and if $U \in \mathfrak{X}$, one says that a set $A \subset X$ is *U-small* if and only if $A \times A \subset U$. Comparison with the semimetrizable case will clarify and justify this concept.

In terms of this idea we can easily frame the definition of a *Cauchy directed family* (sometimes "fundamental" is used in place of "Cauchy"): A directed family $(x_\lambda)_{\lambda \in L}$ has this property if and only if for each $U \in \mathfrak{X}$ there exists an index $\lambda_0 = \lambda_0(U)$ such that $(x_\lambda, x_{\lambda'}) \in U$ for all λ and λ' that exceed λ_0. This amounts to saying that the tail end $\{x_\lambda : \lambda > \lambda_0\}$ is U-small, provided λ_0 is large enough.

If the uniformity is derived from a semimetric d, (x_λ) is Cauchy if and only if $\lim d(x_\lambda, x_{\lambda'}) = 0$ as λ and λ' grow independently. In general, the same is true if one demands that this relation is to hold for all semimetrics that are uniformly continuous.

If $\lim_\lambda x_\lambda = x$ for the associated topology, then (x_λ) is Cauchy; and if (x_λ) is Cauchy, it converges to each of its limiting points (if any exist). A Cauchy directed family is not necessarily convergent. (Consider a nonclosed subspace of a separated uniform space.)

A uniform space $(X, \mathfrak{X})$ is termed *complete* if and only if each Cauchy directed family in X is convergent in X.

Any closed subspace of a complete space is itself complete; any complete subspace of a *separated* uniform space is closed. A product of uniform spaces is complete if and only if each factor is complete; and a directed family in a product space is Cauchy if and only if each projection yields a Cauchy family in the corresponding factor space.

A semimetrizable uniform space is complete if and only if each Cauchy (countable) sequence is convergent. Any uniform space satisfying this condition is said to be *sequentially complete*; there exist noncomplete uniform spaces that are yet sequentially complete.

If (x_λ) is Cauchy in X and f is a uniformly continuous map of $(X, \mathfrak{X})$ into $(Y, \mathfrak{Y})$, then $(f(x_\lambda))$ is Cauchy in Y.

0.3.12 **Completion of a Uniform Space.** A process for imbedding any metric space as a dense subspace in a complete metric space was used by Cantor to derive real numbers from rational numbers, and the procedure is by now quite well known. The extension to semimetric spaces is trivial; see, for example, Kelley [1], p. 196.

On occasions we shall need to refer to a similar process applicable to any uniform space. The possibility of doing this can be inferred from the Cantor procedure for semimetric spaces, combined with the result stated in Subsection 0.3.9. However, we shall give a description of a procedure that begins *ab initio* and that embraces the Cantor process as a special case.

Let $(X, \mathfrak{X})$ be a uniform space. By a *completion* of $(X, \mathfrak{X})$ is meant a pair $[(\hat{X}, \hat{\mathfrak{X}}), f]$ consisting of a complete uniform space $(\hat{X}, \hat{\mathfrak{X}})$ and a uniform isomorphism f of X onto a dense subspace of $\hat{X}$.

To construct such an entity in the general case, we may follow the steps described by Bourbaki ([2], pp. 102–105), phrasing the arguments in terms of Cauchy directed families in place of his filters. The first step is to introduce the set $\tilde{X}$ of all Cauchy directed families in X. A uniformity $\tilde{\mathfrak{X}}$ is defined on $\tilde{X}$, having by definition a base of vicinities formed of the sets $\tilde{U}$ where for any symmetric vicinity U of $\mathfrak{X}$, $\tilde{U}$ denotes the set of pairs $((x_\lambda), (y_\mu))$ of Cauchy directed families in X with the property that for some λ_0 and some μ_0, one has $(x_\lambda, y_\mu) \in U$ for $\lambda \geq \lambda_0$ and $\mu \geq \mu_0$. One may verify that the resulting uniform space $(\tilde{X}, \tilde{\mathfrak{X}})$ is complete. For f one takes the map that assigns to $x \in X$ the one-element directed family $x^\cdot = (x)$. This f is then a uniform isomorphism of X onto a dense subspace $X^\cdot$ of $\tilde{X}$, and one has in $[(\tilde{X}, \tilde{\mathfrak{X}}), f]$ a completion of $(X, \mathfrak{X})$. The uniform space $(\tilde{X}, \tilde{\mathfrak{X}})$ is in general not separated.

If $(X, \mathfrak{X})$ is separated, it is natural to take one more step in order to end up with a separated completion. This step consists of forming the separated space $(\hat{X}, \hat{\mathfrak{X}})$ associated with $(\tilde{X}, \tilde{\mathfrak{X}})$ (see Subsection 0.3.2). Then it appears that the natural map g of $\tilde{X}$ onto $\hat{X}$ carries $X^\cdot$ in a one-to-one fashion into a dense subspace of X' of $\hat{X}$, and $g \circ f$ is a uniform isomorphism of X onto X'. The pair

$[(\hat{X}, \hat{\mathfrak{X}}), g \circ f]$ is then a separated completion of $(X, \mathfrak{X})$. It may furthermore be shown (Bourbaki [2], p. 105, Proposition 8) that such a separated completion is unique up to uniform isomorphism, so that it is natural to speak of *the* (separated) completion.

It is worth noting that if $(X, \mathfrak{X})$ is not separated, and if Y is the associated separated space, then $\hat{Y}$ is uniformly isomorphic with $\hat{X}$ (the separated space associated with $\tilde{X}$); in other words, the passage to the associated separated space may be performed either before or after the $\sim$-process. (See Bourbaki [2], p. 107, Exercice 3).)

In case $\mathfrak{X}$ is specified in terms of a given family (d_i) of semimetrics (see Subsections 0.3.7 and 0.3.8), we may describe the completion process rather more concretely. To begin with it is easily verified that if (x_λ) and (y_μ) are Cauchy directed families in X, then the directed family $(d_i(x_\lambda, y_\mu))$ (indexed by pairs (λ, μ)) is Cauchy, so that

$$\tilde{d}_i((x_\lambda), (y_\mu)) = \lim_{\lambda,\mu} d_i(x_\lambda, y_\mu)$$

exists finitely. Thus defined, $\tilde{d}_i$ is a semimetric on $\tilde{X}$. Moreover the uniformity $\tilde{\mathfrak{X}}$ is none other than that defined by the family $(\tilde{d}_i)$. Notice that for $x, y \in X$ one has

$$\tilde{d}_i((x), (y)) = d_i(x, y).$$

One may confirm without difficulty that if $x \in X$, then a base of neighbourhoods in $\tilde{X}$ of $x^\cdot$ is obtained by taking the closures in $\tilde{X}$ of f-images ($f : x \to x^\cdot$, as before) of neighbourhoods in X of x.

The separated space $\hat{X}$ is obtained by identifying (x_λ) and (y_μ) if and only if for each i it is the case that

$$\tilde{d}_i((x_\lambda), (y_\mu)) = \lim_{\lambda,\mu} d_i(x_\lambda, y_\mu) = 0,$$

that is, if and only if (x_λ) and (y_μ) are equivalent Cauchy families. If $\hat{x}$ and $\hat{y}$ are any two such equivalence classes of Cauchy directed families (that is, two points of $\hat{X}$), then we can define

$$\hat{d}_i(\hat{x}, \hat{y}) = \tilde{d}_i((x_\lambda), (y_\mu))$$

for any $(x_\lambda) \in \hat{x}$ and $(y_\mu) \in \hat{y}$. The $\hat{d}_i$ then define the uniformity $\hat{\mathfrak{X}}$ on $\hat{X}$. If X is separated, then $x^\cdot$ and $y^\cdot$ are equivalent (that is, belong to the same element of $\hat{X}$) if and only if $d_i(x, y) = 0$ for all i; that is, if and only if $x = y$. This provides verification of statements made before semimetrics were introduced into the discussion.

0.3.13 Precompactness. A uniform space $(X, \mathfrak{X})$ is said to be *precompact* or *totally bounded* if and only if, for each $U \in \mathfrak{X}$, it is the case that X is the union of a finite number of sets, each of which is U-small. In other words, for each $U \in \mathfrak{X}$, there must exist a finite set $F \subset X$ such that $X = U(F)$.

Plainly, any subspace of a precompact space is precompact.

If $(X, \mathfrak{X})$ is not precompact, one can construct a sequence (x_n) which, for some $U \in \mathfrak{X}$, is such that $x_n \notin U(x_p)$ for $p < n$; the sequence (x_n) therefore has no Cauchy cofinal subfamily.

0.3.14 **Theorem.** A uniform space $(X, \mathfrak{X})$ is precompact if and only if each directed family in X has a Cauchy cofinal subfamily. Therefore in order that $(X, \mathfrak{X})$ be precompact it is sufficient that some completion of $(X, \mathfrak{X})$ be compact, and necessary that every completion of $(X, \mathfrak{X})$ be compact.

0.3.15 **Category and Baire's Theorem.** Let X be a topological space, A a subset of X. A is said to be:

(1) *nowhere dense* in X, if and only if its closure has an empty interior;

(2) *meagre*, or *first category*, in X if and only if it is the union of a countable family sets, each nowhere dense in X;

(3) *nonmeagre*, or *second category*, in X if and only if it is not meagre in X.

As will be seen subsequently in this book, and as is illustrated by Chapter 3 of Kelley and Namioka [1], category enters into numerous proofs in functional analysis and owes much of its importance to the following theorem.

0.3.16 **Theorem** (Baire). If X is either locally compact and regular, or semimetrizable as a complete space, then the complement of any set meagre in X is dense. (Equivalently, the intersection of a countable family of dense open sets is itself dense.) In particular, X is nonmeagre in itself.

A proof will be found in Kelley [1], p. 200.

Very important so-called boundedness principles in functional analysis (discussed in Chapter 7) hinge upon the following result.

0.3.17 **Theorem.** Let X be a nonmeagre topological space and f a finite-valued lower semicontinuous function on X. Then f is bounded above on some nonvoid open subset of X.

Proof. The sets $F_n = \{x : x \in X \text{ and } f(x) \leq n\}$ are closed and cover X. Hence there exists an n such that F_n has a nonvoid interior, Q.E.D. ▌

0.3.18 The above result is often applied when f is the upper envelope, assumed finite valued, of an arbitrary family $(f_i)_{i\in I}$ of lower semicontinuous functions on X; the conclusion is that there exists a nonvoid open set $A \subset X$ and a number $c(< +\infty)$ such that

$$\operatorname{Sup}\{f_i(x) : i \in I \quad \text{and} \quad x \in A\} \leq c.$$

0.3.19 **Topological Groups.** In Chapters 4 and 10 reference will be made to topological groups. It is convenient to collect here the few elementary topological properties we shall require.

By a topological group we shall mean a set X endowed with (1) the structure of a group and (2) a topology in such a way that the group operations are continuous; that is, so that the function $(x, y) \to x^{-1}y$ is continuous on the product space $X \times X$ into X.

It follows from this that, for fixed $a \in X$, the functions $x \to ax$ and $x \to a^{-1}x$ (and likewise the functions $x \to xa$ and $x \to xa^{-1}$) are mutually inverse homeomorphisms of X onto itself. Moreover, if U ranges over a neighbourhood base at e (the neutral element of the group X), then the sets xU, and likewise the sets Ux, range over a neighbourhood base at x.

If to each neighbourhood U of e one orders the subset

$$U_L = \{(x, y) : x^{-1}y \in U\}$$

of $X \times X$, then the U_L form a base of vicinities for a uniformity on X. Obviously, $U_L[x] = xU$, so that the topology derived from this uniformity is the initial one. Thus X is uniformizable and therefore completely regular (see Subsection 0.3.8). Consequently X is T_1 if and only if it is separated.

It should be remarked that the uniformity just defined is termed the *left* uniformity on X. By replacing the sets U_L by the sets

$$U_R = \{(x, y) : yx^{-1} \in U\},$$

one obtains the so-called *right* uniformity on X. Here $U_R[x] = Ux$, so that again the derived topology is the initial one. Nonetheless in general these two uniformities are distinct, even though the function $x \to x^{-1}$ is an isomorphism of one onto the other. The two uniformities are identical if the group X is commutative, and also if the space X is compact (see Subsection 0.3.10).

We shall have no need of a detailed study of these uniformities and refer the interested reader to Bourbaki [2a].

0.4 Ascoli's Theorem

Ascoli's theorem is a most important criterion of compactness for sets of functions, of which we shall make repeated use throughout the book.

0.4.1 If T and X are sets we denote by X^T the set of all X-valued functions on T. If also T and X are topological spaces, we denote by $C(T, X)$ the subset of X^T formed of continuous X-valued functions on T.

0.4.2 If T is a set and X a topological space, we introduce the topology of simple (pointwise) convergence on X^T by assigning to each $f_0 \in X^T$ a base of neighbourhoods comprised of sets of the type

$$W(t_1, \cdots, t_n, U_1, \cdots, U_n) = \{f \in X^T : f(t_i) \in U_i \quad \text{for } i = 1, \cdots, n\},$$

n ranging over all natural numbers, the t_i ranging individually over T, and U_i being for each i a neighbourhood in X of $f_0(t_i)$. We shall denote this topology by τ_s.

0.4.3 If T is a set and X a uniform space, we define sets $V(f_0, S, U)$ thus,

$$V(f_0, S, U) = \{f \in X^T : (f(t), f_0(t)) \in U \qquad \text{for every } t \in S\};$$

here f_0 is any element of X^T, S a subset of T, and U a vicinity of the uniformity on X. These sets $V(f_0, S, U)$—where f_0 is fixed, U ranges over all vicinities (or over a base of vicinities), and S ranges over a given set $\mathfrak{S}$ of subsets of T—form a subbase for a topology $\tau_{\mathfrak{S}}$ on X^T called the topology of convergence uniform on the sets belonging to $\mathfrak{S}$, or simply the $\mathfrak{S}$-topology. If $\mathfrak{S}$ consists of the one-point (or the finite) subsets of T, $\tau_{\mathfrak{S}}$ becomes τ_s. In fact τ_s is the weakest of all $\mathfrak{S}$-topologies for which $\mathfrak{S}$ is a covering of X (that is, which, when X is separated, is itself separated).

0.4.4 If T is a topological space, an especially important $\mathfrak{S}$-topology is that for which $\mathfrak{S}$ comprises all the compact subsets of T; this is τ_c, often termed the topology of compact convergence. Another important case is that in which $\mathfrak{S} = \{T\}$, giving rise to the topology of uniform convergence (on T) denoted by τ_u. τ_u and τ_c are identical if T is compact.

0.4.5 Each $\mathfrak{S}$-topology is derivable from a corresponding uniformity $u_{\mathfrak{S}}$ which has a subbase of vicinities formed of sets of pairs $(f, f_0) \in X^T \times X^T$ satisfying $(f(t), f_0(t)) \in U$ for $t \in S$. The symbol $C_k(T, X)$, where k is c or u, will on occasions be used to denote the topological space $(C(T, X), \tau_k)$ or the uniform space $(C(T, X), u_k)$.

0.4.6 It is evident that in all cases in which T is a topological space and X a uniform space, u_c is stronger than u_s and so that τ_c is stronger than τ_s. Ascoli's theorem aims at picking out certain subsets of $C(T, X)$ on which u_c and u_s induce the same uniformity, and τ_c and τ_s induce the same topology. The essential component here is the idea of equicontinuity.

0.4.7 If T is a topological space, X a uniform space, and F a subset of X^T, F is said to be *equicontinuous* at a point t_0 of T if and only if, given any vicinity U in $X \times X$, there exists a neighbourhood N of t_0 in T such that $t \in N$ and $f \in F$ entails

$$(f(t), f(t_0)) \in U.$$

F is said to be *equicontinuous* (on T) if it is equicontinuous at each point of T, in which case it is necessarily a subset of $C(T, X)$.

We preface our presentation of Ascoli's theorem with some remarks concerning equicontinuity. In these remarks it is assumed (at least) that T is a topological space and X a uniform space.

0.4.8 **Lemma.** If F is equicontinuous, then the τ_s-closure of F in X^T is also equicontinuous.

Proof. This is trivial from the definition, provided one recalls that the closed vicinities in $X \times X$ form a base for the given uniformity. See Subsection 0.3.2(2). ▮

0.4.9 **Proposition.** If F is equicontinuous, then u_s and u_c (resp. τ_s and τ_c) induce on F the same uniformity (resp. topology).

Proof. The assertion in parentheses follows at once from that concerning the uniformities. As for this, let $S \subset T$ be compact and let U be a vicinity in $X \times X$. Take a symmetric vicinity U_1 such that $U_1 \circ U_1 \circ U_1 \subset U$. For each $t \in T$ there exists a neighbourhood N_t of t such that

$$(f(t'), f(t)) \in U_1$$

for $t' \in N_t$ and $f \in F$. Since S is compact one may choose $t_1, \cdots, t_n$ such that the N_{t_i} cover S. It is then easy to check that

$$(F \times F) \cap W(t_1, \cdots, t_n, U_1) \subset (F \times F) \cap W(S, U).$$

This shows that $u_c \mid F$ is weaker than, hence is identical with, $u_s \mid F$. ▮

0.4.10 **Corollary.** If F is equicontinuous, its τ_s-closure and its τ_c-closure (in X^T in each case) are identical.

Proof. Let F_s and F_c be the respective closures, so that $F_c \subset F_s$ is a triviality. On the other hand we know from Lemma 0.4.8 that F_s is equicontinuous. By Proposition 0.4.9, τ_s and τ_c coincide when restricted to F_s:

$$\tau_s \mid F_s = \tau_c \mid F_s.$$

Now the closure of F in the topology $\tau_s \mid F_s$ is just F_s. This must coincide with the closure of F in the topology $\tau_c \mid F_s$. Thus one has

$$(\tau_s \mid F_s)\text{-closure of } F = (\tau_c \mid F_s)\text{-closure of } F$$

that is,

$$F_s \cap F_s = F_c \cap F_s,$$

showing that $F_c \supset F_s$, as alleged. ▮

0.4.11 **Theorem** (Ascoli). Suppose that T is a topological space and X a uniform space. If $F \subset X^T$ is equicontinuous on T and $F(t) = \{f(t) : f \in F\}$ is relatively compact in X for each $t \in T$, then F is relatively compact in X^T for the topology τ_c.

Proof. The simple closure F_s of F satisfies the same conditions as does F. According to Tychonoff's theorem, F_s is relatively compact in X^T for τ_s, and, by Proposition 0.4.9, is therefore relatively compact for τ_c. Finally Corollary 0.4.10 leads from this result to the conclusion that F is relatively compact in X^T for τ_c. ▮

A partial converse of Ascoli's theorem is sometimes useful.

Proposition. Let T be a locally compact space, X a uniform space, and F a subset of $C(T, X)$. If F is relatively compact in $C_c(T, X)$, then $F(t)$ is relatively compact in X for each $t \in T$ and F is equicontinuous.

Proof. The mapping $f \to f(t)$ of $C_c(T, X)$ into X being continuous for each $t \in T$, the image of F, namely, $F(t)$, must be relatively compact for each t.

On the other hand, if $\bar{F}$ is the compact closure of F in $C_c(T, X)$, then the mapping ξ of $\bar{F} \times T$ into X, defined by $\xi(f, t) = f(t)$, is readily verified to be continuous owing to the fact that T is locally compact. We associate with each $t \in T$ the mapping t^* of $\bar{F}$ into X defined by $t^*(f) = f(t)$. To assert that $\bar{F}$ is equicontinuous is equivalent to saying that the mapping $t \to t^*$ is continuous from T into $C_u(\bar{F}, X)$, and so it suffices to prove the latter statement.

Were this not the case, however, there would exist a symmetric vicinity V^2 in $X \times X$, a point $t \in T$, and to every neighbourhood N of t a point $t_N \in N$ and a function $f_N \in \bar{F}$ such that

$$(f_N(t_N), f_N(t)) \notin V^2.$$

By hypothesis, the directed family (f_N) has a limiting point $f \in \bar{F}$. Compactness of $\bar{F}$ shows that by passing to a cofinal subfamily, we may assume that (f_N) converges to f in $\bar{F}$, in which case $(f(t), f_N(t)) \in V$ for all sufficiently small neighbourhoods N. Consequently one would have

$$(f_N(t_N), f(t)) \notin V$$

for all sufficiently small N. Since the directed family $((f_N, t_N))$ converges to $((f, t))$ in $\bar{F} \times T$, this contradicts the continuity of ξ. ▮

0.4.12 **Corollary.** If T is locally compact, X and F being as in Theorem 0.4.11, then F is relatively compact in $C(T, X)$ for τ_c.

Proof. The assertion holds in fact if T is metrizable, or is such that $f \mid K$ continuous for each compact $K \subset T$ entails that f is continuous. For whenever this is the case, $C(T, X)$ is closed in X^T for τ_c. ▮

Remark. Most applications of Ascoli's theorem encountered in this book refer to cases in which F is a set of linear forms on a topological vector space, or a convergent sequence of linear maps of one such space into another. In such cases Proposition 0.4.9 and Theorem 0.4.11 are equivalent.

One other consequence is to be noted. If T and X are uniform spaces, a subset F of X^T is said to be *uniformly equicontinuous* if for each vicinity V in $X \times X$ there exists a vicinity U in $T \times T$ such that $f \in F$ and $(t, t') \in U$ imply that

$$(f(t), f(t')) \in V.$$

Each member of F is then uniformly continuous on T.

0.4.13 **Theorem.** Let T and X be uniform spaces, T being precompact and X separated, and suppose that $F \subset X^T$ is uniformly equicontinuous and such that $F(t)$ is relatively compact in X for each $t \in T$. Then F is relatively compact in $C(T, X)$ for τ_u.

Proof. Since T is precompact, it is possible to construct a compact completion $\hat{T}$ of T. Let $\hat{X}$ be the separated completion of X. Thanks to uniform continuity, each member f of F has a unique continuous extension $\hat{f}$ that maps

$\hat{T}$ into $\hat{X}$. It is readily verifiable that the set $\hat{F}$ of these extensions is equicontinuous on $\hat{T}$, and that $\hat{F}(\hat{t})$ is relatively compact in $\hat{X}$ for each $\hat{t} \in \hat{T}$. It then suffices to apply Theorem 0.4.11 and finally restrict the members of $\hat{F}$ to T. ▌

We note a useful application of the preceding results.

0.4.14 **Proposition.** Suppose that T is a topological space, X a uniform space, and F a subset of X^T that is equicontinuous and such that $F(t)$ is relatively compact in X for each $t \in T$. Let F_c denote the τ_c-closure of F in X^T. Suppose finally that $(\xi_i)_{i \in I}$ is a family of complex-valued functions on F_c, each ξ_i being continuous for $\tau_c \mid F_c$, and suppose that $(\xi_i)_{i \in I}$ separates the points of F_c. If (f_λ) is a directed family of elements of F such that $\lim_\lambda \xi_i(f_\lambda)$ exists for each $i \in I$, then (f_λ) is τ_c-convergent to some element of F_c.

Proof. By Lemma 0.4.8 and Theorem 0.4.11, F_c is compact for τ_c. Hence (f_λ) admits at least one τ_c-limiting point $g \in F_c$. On the other hand, any such limiting point g satisfies the relations $\xi_i(g) = \lim_\lambda \xi_i(f_\lambda)$ $(i \in I)$ and is therefore uniquely determined. The assertion now follows from Subsection 0.2.17(1). ▌

0.4.15 **Remark.** If T is locally compact (or metrizable), the limit g of (f_λ) must belong to $C(T, X)$. Compare Corollary 0.4.12.

0.5 Brouwer's Fixed Point Theorem

In Section 3.6 we shall need to assume as a starting point a famous theorem of Brouwer which may be stated as follows.

0.5.1 **Theorem.** Let B denote the closed unit ball with centre the origin in R^n, and let f be a continuous mapping of B into itself. Then f admits at least one fixed point—that is, a point $x \in B$ satisfying $f(x) = x$.

0.5.2 It is evident that this theorem is purely topological. For a topological proof see, for example, Lefschetz [1], pp. 117–119. A similar proof of a sharper result, conveniently placed for analysts, appears in Graves [1], pp. 145–150, where all the preliminaries from combinatorial topology are included. A purely analytical proof is possible, although perhaps less natural; see Dunford and Schwartz [1], pp. 467–470.

0.5.3 Plainly the Brouwer theorem remains true if therein we replace B by any topological space X homeomorphic with B. In particular one might take for X the closed unit ball in C^n, which is homeomorphic with the closed unit ball in R^{2n}. In Section 3.6 we shall want to call upon both cases.

CHAPTER 1

Vector Spaces and Topological Vector Spaces

1.0 Foreword

This, the first chapter in which we approach closely our main objects of study, falls into two parts.

The first part, comprising Sections 1.1 through 1.7, deals with vector spaces solely. The first four sections cover material that will be fairly familiar to any reader who has followed a course of linear algebra at all. Sections 1.5 through 1.7 cope with slightly less familiar material and are therefore somewhat more detailed. Despite the familiarity of the ground covered in Sections 1.1–1.4, the principal definitions and results are collected here for the reader's convenience. For an exposition at greater length and in more detail we may refer to the admirable book by Halmos entitled *Finite Dimensional Vector Spaces* (Halmos [3]). We shall refer so often to this text that throughout this chapter we shall denote it by the symbol [H].

The second part of this chapter begins with Section 1.8, where topological vector spaces are introduced. Here the reader may use Bourbaki [4] and Taylor [1] to supplement the account we offer. In Section 1.12 we again make contact with material partially and admirably covered by Halmos.

1.1 Vector Spaces ([H], §§1, 2)

The reader will probably be familiar with the concept of a vector space (or, what is synonymous, a linear space) of finite dimension. Provided this concept is encountered and defined in a manner free of initial reference to any selected basis, or to the assumed existence of a finite basis or generating set, no change is required in the formation of the concept of an infinite dimensional space. We shall be concerned solely with vector spaces over either the real field R or the complex field C; often our results and arguments are indifferent to which of these two scalar fields is involved, and so we shall generally denote the scalar field by K. But it is sometimes essential to bear in mind that K is always either R or C. Since K is in particular a commutative ring, we do not have to distinguish between left and right vector spaces over K (see Bourbaki [3], pp. 1–2).

Putting the matter formally, then, a *vector space over* K is a nonvoid set E

equipped with two laws of composition, one internal and the other external and involving K, namely:

(1) E forms an Abelian, additively written group with respect to the internal law of composition which is termed "addition" and written $+$. This law of composition therefore assigns to each pair (x, y) of elements of E the "sum" $x + y$, which is again an element of E. The following axioms are assumed to hold: addition is associative; $x + y = y + x$ (commutativity of addition); there exists a zero element 0 characterized by the demand that $x + 0 = x$ for all x in E; for each x in E there exists at least one additive inverse $-x$ for which $x + (-x) = 0$. It is an easy exercise to verify that 0 is unique, and that $-x$ is uniquely determined by x. The element $x + (-y)$ is denoted by $x - y$.

(2) The external law of composition is a mapping

$$(\alpha, x) \rightarrow \alpha x$$

of $K \times E$ into E satisfying the following axioms:

$$\alpha(x + y) = \alpha x + \alpha y, \qquad (\alpha + \beta)x = \alpha x + \beta x,$$
$$\alpha(\beta x) = (\alpha\beta)x, \qquad 1x = x,$$

these relations being postulated for arbitrary elements x, y of E and arbitrary scalars α, β (that is, elements of K); 1 denotes the unit element of the field K. It then follows that $0x = 0$ and $(-1)x = -x$ identically for x in E.

For further discussion, see Bourbaki ([3], pp. 3–4) and Taylor ([1], pp. 8 et seq.); Taylor uses the term "linear space" in place of "vector space." The reader may find it profitable to consult these and other references as we progress.

We give a few instances of vector spaces of the type that appear frequently in functional analysis. All finite dimensional examples are essentially indistinguishable from K^n (regarded as a vector space over K in the usual way) and we pay no special attention to these; they are contained as special cases (T finite) of Example 1.1.1. In functional analysis most important vector spaces are function spaces or spaces derived from these by various general processes (formation of quotient, or difference, spaces, formation of dual spaces, and so forth) yet to be explained.

1.1.1 **Example.** A function space may be described as follows. We start with an arbitrary set T and form the "product" K^T, whose elements are K-valued functions defined on T. K^T is made into a vector space over K by adopting the rules that $x + y$ and αx are to denote the functions $t \rightarrow x(t) + y(t)$ and $t \rightarrow \alpha \cdot x(t)$, respectively (the $+$ and $\cdot$ in the right-hand members being the field operations in K). K^T, thus equipped, is a function space over T. The same description will also be applied to vector subspaces (see Section 1.2) of K^T and even, by an abuse of language, sometimes to quotient spaces thereof. As we shall see, if T is a topological space, those vector subspaces of K^T obtained by taking only continuous K-valued functions play a fundamental role in functional analysis. So too do vector subspaces singled out by other criteria; see, for example, Chapter 4.

If T is finite and consists of n elements, K^T is, up to isomorphism (see Section 1.3), simply K^n, the space of n-termed sequences of elements of K with the usual "coordinate" definitions of addition and multiplication by scalars. Again, if T is countably infinite, K^T is isomorphic with K^N (N the set of natural numbers), whose elements are infinite sequences of elements of K with the vector-space operations again interpreted "coordinate-wise." Thus K^N and its subspaces are often described as "sequence spaces" which, with or without topological dressage, have played an important role in the development of functional analysis; see F. Riesz [1], Banach [1], Orlicz [1], Mazur and Orlicz [1], Cooke [1], [2] for some indications of the importance of this role.

1.2 Vector Subspaces, Quotient Spaces, Product Spaces ([H]§§10–12, 18, 21)

1.2.1 Vector Subspaces. If E is a vector space and L a subset of E, the operations defined in E will generally lead from elements of L to elements of $E \setminus L$. If, however, this is not the case—that is, if L is stable under addition and multiplication by scalars—and if we restrict these operations from E to L, we thereby endow L with the structure of a vector space. So equipped, L is termed a *vector subspace* of E. Taylor [1], in common with many authors, uses the term "linear manifold in E" in place of "vector subspace of E."

Let us introduce a piece of notation: if A and B are subsets of E and M a subset of K, we define

$$A + B = \{a + b : a \in A, b \in B\},$$

the vectorial sum of A and B, and

$$M \cdot A = \{\mu a : \mu \in M, a \in A\}.$$

With these notations we may express the necessary and sufficient condition that L be a vector subspace of E in the form

$$L + L \subset L, \qquad K \cdot L \subset L. \tag{1.2.1}$$

If 0 is the zero element of E, $\{0\}$ is the smallest of all vector subspaces of E. The intersection of any set of vector subspaces of E is again a vector subspace of E. Consequently, any given subset S of E is contained in a smallest vector subspace of E, termed that *generated* by S. If S is nonvoid, the elements of the vector subspace generated by S are those and only those of the form $\sum_{i=1}^n \alpha_i s_i$ with $\alpha_i \in K$, $s_i \in S$ $(1 \leq i \leq n)$, n being an arbitrary natural number (which may vary with the element in question). Some writers attach the name *linear envelope* (or *hull*) *of S* to the vector subspace generated by S. S is said to *generate* E, or to *form a system of generators* of E, if its linear envelope is the whole of E.

1.2.2 Examples. (1) If from K^T (see Example 1.1.1) we select only those functions that are bounded, we obtain a vector subspace $B(T)$ of K^T. If T is a topological space, and if we select only those functions that are continuous, we obtain a vector subspace $C(T)$ of K^T. Likewise, if we take a vector subspace M of K, M^T is a vector subspace of K^T; thus, if $K = C$ and $M = R$, M^T will consist precisely of those functions in K^T that are real valued.

(2) Again, if E is a given vector subspace of K^T, and if T_0 is a subset of T, we obtain a vector subspace of E (hence also one of K^T) on choosing from E only those functions that vanish on T_0.

(3) More generally, if E is any vector space, and if (f_i) is an arbitrary family of linear maps (see Section 1.3) of E into vector spaces F_i, then the set of elements x of E such that, for each i, $f_i(x)$ belongs to a preassigned vector subspace of F_i, is a vector subspace of E. It will appear from the results in Section 1.4 that *any* vector subspace of E may be prescribed in this way.

1.2.3 **Quotient Spaces.** We turn next to the concept of quotient or difference spaces. Let E be a vector space, L a vector subspace of E. The relation $x \equiv y \pmod{L}$, defined to signify that $x - y \in L$, is an equivalence relation in E. We form the quotient set E/L. This quotient set may be endowed with a vector space structure: denoting by $x_c = x + L$ and $y_c = y + L$ the classes of x and y, respectively, $(x + y)_c$ depends only on x_c and y_c and we may, therefore, define $x_c + y_c = (x + y)_c$; similarly we may justifiably define $\alpha \cdot x_c = (\alpha x)_c$. The vector space thus obtained is the *quotient (vector) space of* E *modulo* L, denoted still by E/L.

1.2.4 **Examples.** (1) Let L denote the set of functions in K^T which vanish on T_0. Then K^T / L is identifiable as a vector space with K^{T_0}. If T is a topological space, it will generally be false to infer by analogy that $C(T)/(L \cap C(T))$ is isomorphic with $C(T_0)$; the absence of a topological isomorphism is more readily apparent (see Subsections 1.8.5 and 1.9.2).

(2) In Chapter 4 we shall encounter various standard spaces that are quotients of function spaces and that play a basic role in integration theory. These are obtainable by following a certain canonical procedure, namely: one begins with a vector space E and a seminorm p on E (see Section 1.6); the kernel $N = p^{-1}(\{0\})$ is a vector subspace of E, and it is entirely natural for many purposes to work with the quotient space E/N.

1.2.5 **Product Spaces.** If $(E_i)_{i \in I}$ is a family of vector spaces over K, the product set $E = \prod_{i \in I} E_i$ can be formed into a vector space over K by defining $(x_i) + (y_i) = (x_i + y_i)$ and $\alpha \cdot (x_i) = (\alpha x_i)$. We call this the *product vector space*.

If in particular all the $E_i = F$, a given vector space over K, we obtain the power space F^I. If I and J have the same cardinal, F^I and F^J are canonically isomorphic. If I has the cardinal number n, a natural number, we usually write F^n in place of F^I: this is really making use of the canonical isomorphism just mentioned and regarding I as the prototype set $\{1, 2, \cdots, n\}$.

1.2.6 **Hyperplanes.** The concept of linear form (or linear functional) is ubiquitous in functional analysis, as the reader will in due course have ample opportunity to observe. The geometrical counterpart of a linear form is a hyperplane, a concept we now define.

We consider all vector subspaces of E, applying the adjective "proper" to those that are distinct from E itself. Proper vector subspaces exist unless E is reduced to its zero element. By a (homogeneous) *hyperplane* in E we mean a maximal proper vector subspace of E; that is, a proper vector subspace H of E such that the only vector subspaces V of E satisfying $H \subset V$ are $V = H$ and $V = E$.

If H is a hyperplane in E, and if x is any element of $E \setminus H$, then the vectorial sum $H + Kx$ must, by virtue of the maximality of H, coincide with E. This entails that the quotient space E/H has dimension precisely 1, that is, has a single generator distinct from the zero element of E/H. Reciprocally, if H is a vector subspace of E such that E/H is of dimension 1, then H is a hyperplane in E.

Given that E/H is of dimension 1, if we choose $x_0 \in E \setminus H$, it follows directly that to each x in E corresponds a unique scalar α such that $x \equiv \alpha x_0 \pmod H$. Consider then the scalar-valued function $f : x \to \alpha$ on E. Plainly, $f(x + y) = f(x) + f(y)$ and $f(\lambda x) = \lambda \cdot f(x)$: these properties characterize f as a linear form on E (see Section 1.3). Since $f(x_0) = 1, f \neq 0$. Moreover,

$$H = \{x \in E : f(x) = 0\} = f^{-1}(\{0\}) = \operatorname{Ker} f.$$

The linear form f thus obtained depends on x_0 as well as on H, but a change of x_0 merely multiplies f by a nonzero scalar.

Conversely, if f is a nonzero linear form on E, then $H = \operatorname{Ker} f$ is easily seen to be a hyperplane in E.

To sum up, the hyperplanes in E are precisely the sets of the form $\operatorname{Ker} f$, where f is a nonzero linear form on E; thus two such linear forms determine the same hyperplane if and only if they are nonzero scalar multiples each of the other.

1.2.7 Linear Varieties. By a *linear variety* in E is meant a translate of some vector subspace of E—that is, a set V of the form $x_0 + L$, where L is some vector subspace of E. V determines L uniquely, but x_0 is determined only modulo L.

Sometimes the term "hyperplane" is applied also to those linear varieties $x_0 + H$ where H is a homogeneous hyperplane in E. These are precisely the sets of the type

$$f^{-1}(\{\alpha\}) = \{x \in E : f(x) = \alpha\},$$

where f is a nonzero linear form on E and α is a scalar.*

1.3 Linear Maps and Linear Forms. Isomorphism. Graphs ([H], §32, 13)

1.3.1 Linear Maps and Linear Forms. Let E and F be vector spaces over K. A *linear map of* or *linear operator from* E into F is simply a function

*See Appendix, p. 771.

(or map) u of E into F such that

$$u(\alpha x + \beta y) = \alpha u(x) + \beta u(y) \tag{1.3.1}$$

for all x, y in E and all α, β in K. In the special case in which $F = K$, we speak of a *linear form* or *linear functional* on E.

A situation somewhat more general in appearance is that in which the domain of definition of u is merely a vector subspace L of E; linearity of u then means that x and y are restricted to L in (1.3.1). Since L is itself a vector space over K, nothing new in principle is involved here.

Returning to the case in which L is taken to be E itself, two linear maps u and v may be added in the usual manner to give a third, $u + v$; similarly αu is defined for any scalar α. It is virtually obvious that in this way we may form the set $L(E, F)$, composed of all linear maps of E into F, into a vector space over K. In particular we may take $F = K$; we write E^* in place of $L(E, K)$ and call E^* the *algebraic dual* of E.†

The reader should note that if $u \in L(E, F)$, then u transforms each vector subspace L (resp. linear variety) in E into a vector subspace (resp. linear variety) $u(L)$ in F; if M is a vector subspace of F, then $u^{-1}(M)$ is a vector subspace of E. In particular, $\text{Ker } u = u^{-1}(\{0\})$ is a vector subspace of E. u is one-to-one if and only if $\text{Ker } u = \{0\}$, in which case it is sometimes called a *monomorphism* of E into F.

If E is a vector space, an element u of $L(E, E)$—that is, a linear map of E into itself—is often termed more briefly an *endomorphism of* E.

1.3.2 **Isomorphism.** ([H], §9). An *isomorphism of E into F* is a map $u \in L(E, F)$ that is one-to-one; the map is said to be *onto* (or to be an *epimorphism*) if $u(E) = F$.

E and F are said to be *isomorphic* (as vector spaces over K) if and only if there exists a $u \in L(E, F)$ that is both a monomorphism and an epimorphism, that is, that is both one-to-one and onto.

Let us suppose that $u \in L(E, F)$ and that L is a vector subspace of E such that $u(L) = \{0\}$. It is then possible to derive from u a member u_c of $L(E/L, F)$ by defining $u_c(x_c) = u(x)$, x_c denoting the class modulo L containing x. Here we may take in particular $L = \text{Ker } u$; then u_c proves to be one-to-one. So we see that any linear map u of E into F defines an isomorphism u_c of $E/\text{Ker } u$ onto $u(E)$. This process of "passage to the quotient" is a useful device on many occasions.

1.3.3 **The Graph of a Linear Map.** We return to consider the case in which u is a linear map of some vector subspace L of E into F. As usual, we refer to L as the *domain* (of definition) of u and denote it briefly by $\text{Dom } u$.

---

† Halmos' notation is E' in place of our E^*; in view of Subsection 1.9.4, confusion must be avoided at all costs.

The set $u(E)$ is variously termed the *counterdomain, range,* or *image* of u; we shall denote it also by Im u.

The third set to be placed alongside Dom u and Im u is the *graph* of u: this is defined to be the subset of $E \times F$ composed of elements of the form $(x, u(x))$ with x in Dom u. We shall denote the graph of u by Gr u. u being linear, Gr u is a vector subspace of $E \times F$.

Given a subset G of $E \times F$, one may ask under what conditions is G the graph of some mapping u with domain in E and range in F. A moment's thought will show that a necessary and sufficient condition is that for each x in E there exists at most one y in F such that $(x, y) \in G$; the corresponding mapping u is then that which has for domain the set of x in E for which at least one, and hence precisely one, element y of F exists for which $(x, y) \in G$, while for each such x, $u(x)$ is this element y.

If we demand further that u be linear, the extra condition to be imposed upon G is simply that it be a vector subspace of $E \times F$. Consequently, a subset G of $E \times F$ is the graph of some linear mapping with domain in E and range in F, if and only if (1) G is a vector subspace of $E \times F$, and (2) $(0, y) \in G$ implies $y = 0$. This characterization will be useful when, later on, we come to consider closed linear maps from one topological vector space into another.

1.4 Linear Dependence. Algebraic Bases. Algebraic Complements ([H], §§5, 6, 7)

1.4.1 Linear Dependence. Let E be a vector space over K. A family $(x_i)_{i \in I}$ of elements of E is said to be *linearly independent* if any relation of the form

$$\sum_{i \in F} \alpha_i x_i = 0,$$

where F is a finite subset of I and $\alpha_i \in K$ for $i \in F$, implies that $\alpha_i = 0$ for $i \in F$. In the contrary case, the family is said to be *linearly dependent.*

Obviously, the linear dependence or otherwise of a family is preserved if the family is permuted in any way; further, any subfamily of a linearly independent family is also linearly independent. No element of a linearly independent family can be zero; and $x_i = x_j$ only if $i = j$.

If $E = \{0\}$, the only linearly independent family in E is empty; otherwise, the family composed of any nonzero element of E is linearly independent.

Any one-to-one linear map transforms linearly independent families into linearly independent families.

We give now two simple but useful results concerning finite linearly independent families in E and in E^*.

1.4.2 Proposition. Let E be a vector space over K, $(f_i)_{1 \le i \le n}$ a finite family in E^*, and f an element of E^*. Then f is a linear combination of the f_i $(1 \le i \le n)$ if and only if

$$\operatorname{Ker} f \supset \bigcap \{\operatorname{Ker} f_i : 1 \le i \le n\}.$$

In particular, if $\bigcap\{\operatorname{Ker} f_i : 1 \leq i \leq n\} = \{0\}$, then the $f_i (1 \leq i \leq n)$ generate E^*, hence $\dim E^* \leq n$ and so also $\dim E \leq n$.

Proof. The last sentence depends on parts of the substance of subsections 1.4.4–1.4.6, but we make this assertion here so as to avoid displaying separately a rather trivial consequence of the rest of the proposition.

For the rest of the proposition, we argue by induction on n. (The "only if" part is evident.)

Suppose first that $n = 1$. Our hypothesis is that $f(x) = 0$ whenever $f_1(x) = 0$. If $f_1 = 0$, this shows already that $f = 0$ and hence $f = 0 \cdot f_1$. Otherwise, there exists $e \in E$ such that $f_1(e) = 1$. Then $f_1(x - f_1(x)e) = 0$ for all x, hence $f(x - f_1(x)e) = 0$ for all x, and so $f = f(e)f_1$. Thus the assertion is true for $n = 1$.

Let us assume that for some $n > 1$, the assertion is proven when $n - 1$ replaces n. We then deduce the truth of the given assertion as follows. If $f_n = 0$, there is nothing to be proved. Otherwise we take e_n in E satisfying $f_n(e_n) = 1$. We apply the case "$n - 1$" to the new linear forms $f' = f - f(e_n)f_n$ and $f_i' = f_i - f_i(e_n)f_n$ $(1 \leq i \leq n - 1)$. Since we are assuming that

$$\operatorname{Ker} f \supset \bigcap\{\operatorname{Ker} f_i : 1 \leq i \leq n\},$$

it follows readily that

$$\operatorname{Ker} f' \supset \bigcap\{\operatorname{Ker} f_i' : 1 \leq i \leq n - 1\} \tag{1.4.1}$$

Indeed, if x belongs to the intersection of the kernels of the f_i' $(1 \leq i \leq n - 1)$, then $y = x - f_n(x)e_n$ belongs to the intersection of the $\operatorname{Ker} f_i$ for $1 \leq i \leq n - 1$; it also belongs to the kernel of f_n (since $f_n(e_n) = 1$). Hence y belongs to $\bigcap\{\operatorname{Ker} f_i : 1 \leq i \leq n\}$. Therefore y belongs to $\operatorname{Ker} f$, and so x belongs to $\operatorname{Ker} f'$, which proves (1.4.1). By inductive hypothesis, therefore, we conclude that f' is a linear combination of the f_i' $(1 \leq i \leq n - 1)$. This implies that f is a linear combination of the f_i $(1 \leq i \leq n)$.

Appeal to the principle of finite induction completes the proof. ▌

Remark. A neater but less obvious proof is based upon the observation that any linear form on a vector subspace of K^n can be extended into a linear form on K^n. This is applied by considering the linear map u of E into K^n defined by $u(x) = (f_1(x), \cdots, f_n(x))$ and the linear form g defined on $u(E)$ by $g(u(x)) = f(x)$. Any linear extension of g to K^n is of the type

$$(\xi_1, \cdots, \xi_n) \in K^n \to \sum_{i=1}^{n} \lambda_i \xi_i.$$

1.4.3 **Proposition.** Let E be a vector space over K.

(1) Let $(f_i)_{1 \leq i \leq n}$ be a finite linearly independent family in E^*. Then there exists a family $(x_i)_{1 \leq i \leq n}$ in E such that

$$f_i(x_j) = \delta_{ij} \qquad (1 \leq i, j \leq n).$$

(2) The analogous statement with E and E^* interchanged is also valid.

Proof. (2) is deduced from (1) via the device, explained in Section 1.4.7, of injecting E into E^{**}: once again we do not wish to defer the statement (2) on these grounds.

To prove (1), we again argue by induction on n. The assertion is trivially true if $n = 1$. We assume it true for n and pass to the case "$n + 1$" as follows. Given that $f_1, \cdots, f_{n+1}$ are linearly independent, the same is *a fortiori* true of $f_1, \cdots, f_n$. So, by inductive hypothesis, we can choose $x_1, \cdots, x_n$ from E so that $f_i(x_j) = \delta_{ij}$ for $1 \leq i, j \leq n$. It only remains to show that there exists x in E such that $f_i(x) = 0$ for $1 \leq i \leq n$ and $f_{n+1}(x) \neq 0$: for then one need only take $x_n = x/f_{n+1}(x)$. Now if no such x existed, we should have

$$\operatorname{Ker} f_{n+1} \supset \bigcap \{\operatorname{Ker} f_i : 1 \leq i \leq n\}.$$

But then Proposition 1.4.2 would entail that f_{n+1} is a linear combination of $f_1, \cdots, f_n$, in direct conflict with the hypothesis that $f_1, \cdots, f_{n+1}$ are linearly independent. The proof is complete. ▌

1.4.4 Algebraic Bases. By an *algebraic basis of* E is meant a family $(e_i)_{i \in I}$ of elements of E that is linearly independent and such that the corresponding set $\{e_i : i \in I\}$ generates E (see Section 1.2). This amounts precisely to saying that each element x of E admits a unique expression $\sum_{i \in I} \alpha_i e_i$, the α_i being scalars that are nonzero for at most a finite subset of I (which subset will in general vary with x).

It can be shown (Day [1], p. 3, Theorem 2) that all bases for E have the same cardinal number, which may be termed the *dimension* of E. The proof is relatively simple if E is finitely generated, in which case all bases have the same finite cardinal number: we assume this case to be known. We shall have little use for the concept of dimensionality in the remaining cases and therefore omit the appropriate proof. A corollary of this result is that two vector spaces over K are isomorphic if and only if they have the same dimension.

We shall need the following result.

1.4.5 Theorem. Every vector space $E \neq \{0\}$ admits at least one algebraic basis. More precisely, given any linearly independent family $(x_i)_{i \in I}$ in E, there exists an algebraic basis of E of the form $(x_j)_{j \in J}$, where J is a superset of I.

Proof. Zorn's lemma (see Subsection 0.1.5) will be called into action. If $L = (x_i)_{i \in I}$ and $L' = (y_j)_{j \in J}$ are linearly independent families in E, we write $L \leq L'$ if and only if $I \subset J$ and $y_i = x_i$ for $i \in I$. In this way we obtain a partial order on the set of all linearly independent families in E. It is also easy to verify that this ordering is inductive. For suppose $L^\alpha = (x_i^\alpha)_{i \in I_\alpha}$ is a linearly ordered family of linearly independent sets, the index α running over an arbitrary set A. Put $I = \bigcup \{I_\alpha : \alpha \in A\}$. If $i \in I$, then $i \in I_\alpha$ for at least one α; the elements x_i^α are coincident for all such indices α, because of the linear ordering. Indeed, if $i \in I_\alpha$ and $i \in I_\beta$, then either $L_\alpha \leq L_\beta$ or $L_\beta \leq L_\alpha$; in the former case $I_\alpha \subset I_\beta$ and $x_i^\alpha = x_i^\beta$ for all $i \in I_\alpha$; and similarly if $L_\beta \leq L_\alpha$. If we now define $x_i = x_i^\alpha$ for $i \in I$, α being arbitrary save for the demand that $i \in I_\alpha$, then $L = (x_i)_{i \in I}$ is easily verified to be a linearly independent family (again thanks to linear ordering of the L^α) and to be the least upper bound of the L^α. Inductive ordering thus obtains.

By Zorn's lemma, therefore, given any linearly independent family $L = (x_i)_{i \in I}$, there exists a maximal linearly independent family $M = (y_j)_{j \in J}$ satisfying $M \geq L$. It remains only to show that M generates E. Putting E_0 for the vector subspace of E generated by M, we have to show that $E_0 = E$. Now, were this not the case, by choosing any x_0 from $E \setminus E_0$, we could adjoin x_0 to the family M and so obtain a linearly independent family M' satisfying $M' \geq M$, $M' \neq M$. This contradicts the maximality of M and leads to the conclusion that $E_0 = E$. ▌

Theorem 1.4.5 has numerous important consequences, two of which we shall now deal with explicitly.

1.4.6 **Proposition.** Let E be a vector space over K. Then $\dim E^* \geq \dim E$, equality holding if either side is finite.

Proof. Let $(e_i)_{i \in I}$ be any basis for E: the existence of at least one such basis is assured by Theorem 1.4.5. For each i we can then define $e_i^* \in E^*$ by defining first $e_i^*(e_j) = \delta_{ij}$ for j in I and then extending e_i^* linearly. It is evident that (e_i^*) is a linearly independent family in E^*. By Theorem 1.4.5 applied to E^*, (e_i^*) can be imbedded into a basis for E^*. It follows that $\dim E^*$ is not less than the cardinal of I, which is $\dim E$.

For the rest, we assume first that $\dim E = n$ is finite. We let $(e_i)_{1 \leq i \leq n}$ be a basis for E and construct the $e_i^* \in E^*$ as above. We know already that the e_i^* are linearly independent in E^*. But also, since it is evident that

$$\bigcap \{\operatorname{Ker} e_i^* : 1 \leq i \leq n\}$$

is $\{0\}$, Proposition 1.4.2 shows that the e_i^* $(1 \leq i \leq n)$ generate, and hence form a basis for, E^*. Thus $\dim E^* = n$ also. On the other hand, if we are given that $\dim E^*$ is finite, the first part of this proposition shows that $\dim E$ is finite and we are back to the instance already dealt with. ▌

Remarks. (1) The family (e_i^*) in E^* defined in terms of the given linearly independent family (e_i) in E via the relations $e_i^*(e_j) = \delta_{ij}$ is often said to be the family *dual* to the given one. If (e_i) is a basis for E and (e_i^*) happens to be a basis for E^* (as is necessarily the case when $\dim E$ is finite), we speak accordingly of *dual bases* for E and E^*. The concept is symmetrical with respect to E and E^*, as is explained in Subsection 1.4.7. See [H], §15.

(2) If $\dim E = n$ is finite, it follows from Proposition 1.4.6 that E and E^* are isomorphic. However, the isomorphism is not "canonical;" that is, it is not determined solely by the vector space structures involved. Rather, there is an isomorphism u of E onto E^* corresponding to each choice of basis (e_i) for E, u being defined by $u(e_i) = e_i^*$ and linear extension. Different bases (e_i) lead thus to different isomorphisms u. To single out and distinguish an isomorphism

between E and E^* is, in fact, tantamount to introducing a suitable bilinear form on $E \times E$. This device lies at the bottom of tensor algebra and the classification of vectors and tensors into contra- and co-variant types.

1.4.7 Injecting E into E.** ([H], §§ 14, 16). Let E be a vector space over K, E^* its algebraic dual. E^* is again a vector space over K, and we may visualize its algebraic dual E^{**}. One obtains a canonical linear map $x \to x^{**}$ of E into E^{**} by assigning to $x \in E$ that linear form x^{**} on E^* defined by

$$x^{**}(f) = f(x) \qquad (f \in E^*).$$

This mapping is one-to-one because, as appears readily from Theorem 1.4.5, given $x \neq 0$ in E we can always find f in E^* such that $f(x) \neq 0$. It is therefore possible and convenient to identify x with x^{**}, thereby causing E to appear as a vector subspace of E^{**}. Hereafter we shall usually assume this done without explicit warning. In this sense, E is usually much smaller than E^{**}; they are the same, however, if $\dim E$ is finite.

In view of this symmetry one often adopts a more symmetric notation, denoting elements of E^* by $x^*, y^*, \cdots$ and writing $\langle x, x^* \rangle$ in place of $x^*(x)$. (Halmos uses $[x, x^*]$ in place of $\langle x, x^* \rangle$.)

A similar device will be used later in connection with topological vector spaces when we work with the "topological dual" E' in place of the algebraic dual E^*.

1.4.8 Direct Sums. Algebraic Complements. ([H], §§19, 20). Let E be a vector space over K, L and M vector subspaces thereof. We shall say that E is the (internal) *direct sum* of L and M if and only if each x in E is uniquely expressible in the form

$$x = a + b, \qquad a \in L, \qquad b \in M. \tag{1.4.2}$$

An equivalent formulation is that

$$L + M = E, \qquad L \cap M = \{0\}. \tag{1.4.3}$$

One then says that each of L and M is an *algebraic complement* (relative to, or in) E of the other.

An immediate corollary of Theorem 1.4.5 is that any given vector subspace of E admits at least one (and usually many) algebraic complements.

Assuming that (1.4.3) holds, the maps

$$p_L : x \to a \qquad \text{and} \qquad p_M : x \to b$$

belong to $L(E, L)$ and $L(E, M)$ respectively,

$$p_L + p_M = 1, \tag{1.4.4}$$

where 1 denotes here the identity map of E onto itself; further,

$$p_L(x) = x \text{ if and only if } x \in L,$$
$$p_M(x) = x \text{ if and only if } x \in M.$$

It follows that p_L and p_M are idempotent:

$$p_L^2 = p_L, \qquad p_M^2 = p_M, \tag{1.4.4}$$

so that each of these maps are projectors (linear and idempotent). p_L is the *projection of E onto L parallel* to M.

Conversely suppose p is a projector on E. If we put L for the set of x in E satisfying $p(x) = x$, then L is a vector subspace of E; $q = 1 - p$ is also a projector, and $M = \{x \in E : q(x) = x\}$ is also a vector subspace of E. Given x in E we have

$$q(x - p(x)) = q^2(x) - q(p(x)) = q^2(x) - 0 = q(x) = x - p(x),$$

showing that $x - p(x) \in M$. Hence $x = p(x) + q(x)$, and so $L + M = E$. Furthermore, if $x = p(x) = q(x)$, then also

$$x = p(x) = p(q(x)) = p(x - p(x)) = p(x) - p^2(x) = 0,$$

which shows that $L \cap M = \{0\}$. Thus (1.4.3) holds and E is the direct sum of L and M. It is easily verified that $p_L = p$ and $p_M = q$.

1.4.9 **Remark.** We have shown that any vector subspace L of E admits a complement and that this complement is generally not unique. Therefore the projector p_L is not uniquely determined by L, but only when one has chosen some definite complement M. Later (Section 1.12) we shall see that in an inner product space E we can single out from all the complements of L that unique one which is orthogonal to L; the related projector is then uniquely determined by L, being the associated orthogonal projector. No such distinguished choice is possible in general.

1.4.10 Despite what was said in Remark 1.4.9, it is nevertheless true that

▶ all algebraic complements relative to E of a given vector subspace L of E are isomorphic, being indeed isomorphic copies of E/L.

For suppose that $E = L + M$, a direct sum. Then $p_M(x)$ depends only on the class x_c of x modulo L, as appears from the uniqueness of the decomposition (1.4.2). So we obtain a linear map u of E/L onto M by setting $u(x_c) = p_M(x)$. Moreover $u(x_c) = 0$ means that $p_M(x) = 0$, that is, that $x \in L$, and that x_c is the zero element of E/L. Thus u is one-to-one. u is therefore an isomorphism of E/L onto M.

1.4.11 **Proposition.** Let the vector space E over K be the direct sum of vector subspaces L and M. Then

(1) $\dim E = \dim L + \dim M$,

(2) $\dim E/L = \dim E - \dim L$ whenever either of $\dim L$ or $\dim E/L$ is finite.

Proof. (1) Let $(e_i)_{i \in I}$ be a basis for L. By Theorem 1.4.5 we can extend it into a basis $(e_i)_{i \in J}$ for E, J being some superset of I. Let M' be the vector subspace of E generated by those e_i with index $i \in J \setminus I$. Since it is obvious that E is the direct sum of L and M', Section 1.4.10 shows that M' and M are isomorphic and thus have the same dimension. On the other hand, $\dim E = \operatorname{card} J$; and $\operatorname{card} J = \operatorname{card} I + \operatorname{card} (J \setminus I) = \dim L + \dim M'$ by definition of dimension. (1) is established.

(2) By Subsection 1.4.10, $\dim E/L = \dim M$. So (1) can be rewritten $\dim E = \dim L + \dim E/L$. From this (2) follows whenever at least one of $\dim L$ or $\dim E/L$ is finite. (The reader will recall that $c - n = c$ if c is any infinite cardinal, n any finite cardinal.) ▌

1.4.12 **Proposition.** Let E and F be vector spaces over K, $u \in L(E, F)$. Then

$$\dim E = \dim \operatorname{Ker} u + \dim \operatorname{Im} u.$$

Proof. Let L denote $\operatorname{Ker} u$. We know (see the end of Subsection 1.3.2) that E/L is isomorphic with $\operatorname{Im} u$. We know also that L admits an algebraic complement M, and that then M is isomorphic with E/L. Hence Proposition 1.4.11 gives

$$\begin{aligned} \dim E &= \dim L + \dim M = \dim L + \dim \operatorname{Im} u \\ &= \dim \operatorname{Ker} u + \dim \operatorname{Im} u, \end{aligned}$$

as alleged. ▌

[*Note:* $\dim \operatorname{Im} u$ is sometimes called the *rank* of u. In view of the remarks in Subsection 1.4.13 immediately below, Proposition 1.4.12 contains the essence of various statements about systems of linear equations and the rank of the matrix concerned.]

1.4.13 **Matrices Used to Represent Linear Maps.** Let E and F be vector spaces over K, $u \in L(E, F)$. We choose bases $(e_i)_{i \in I}$ and $(f_j)_{j \in J}$ for E and F, respectively. Using the defining property of bases and the linearity of u, the relation $y = u(x)$ is seen to be equivalent to the following system of equations:

$$\eta_j = \sum_{i \in I} \alpha_{ij} \xi_i \qquad (j \in J),$$

$$y = \sum_{j \in J} \eta_j f_j, \qquad x = \sum_{i \in I} \xi_i e_i.$$

Here the scalars α_{ij} are determined by u and the chosen bases by the equation

$$u(e_i) = \sum_{j \in J} \alpha_{ij} f_j \qquad (i \in I).$$

It is customary to refer to the family $(\alpha_{ij})_{i \in I, j \in J}$ (an element of $K^{I \times J}$) as a *matrix of type* $I \times J$, or as one having $\operatorname{card} I$ rows and $\operatorname{card} J$ columns. When $\operatorname{card} I$ and $\operatorname{card} J$ are finite, say the natural numbers m and n, respectively, this matrix is often written out as an $m \times n$ rectangular array, thus:

$$\begin{bmatrix} \alpha_{11} & \alpha_{12} & \cdots & \alpha_{1n} \\ \alpha_{21} & \alpha_{22} & \cdots & \alpha_{2n} \\ \cdot & & & \\ \cdot & & & \\ \cdot & & & \\ \alpha_{m1} & \alpha_{m2} & \cdots & \alpha_{mn} \end{bmatrix}.$$

At all events, we have in this fashion prescribed a one-to-one linear correspondence $u \leftrightarrow (\alpha_{ij})$ between $L(E, F)$ and all $I \times J$ matrices. This isomorphism depends, of course, on the selected bases (e_i) and (f_j).

When I and J are finite, this correspondence is often useful in certain circumstances, whence the study of "matrix algebra." The utility of this approach diminishes markedly if either of I or J is infinite, though when I and J are countable the technique is often used in studying linear transformations of sequence spaces (see Cooke [1]). We would, however, insist that the device is no more than a sometimes-convenient fashion of labelling and manipulating linear maps, and that one should in no circumstances identify linear maps with matrices in such a manner as to deprive the former concept of the logical priority proper to it. If the correspondence between linear maps and matrices were canonical, which it is not, there would be some grounds for this misconception: in fact, however, there are none.

1.5 Absorbent Sets. Balanced Sets. Convex Sets

Throughout this section E denotes a vector space over K.

1.5.1 Absorbent Sets. A subset S of E will be termed *absorbent* if to each x in E there corresponds at least one real scalar $\alpha > 0$ such that $\alpha x \in S$. (This definition is not quite the same as that used by Bourbaki ([4], p. 6) nor yet that used by Taylor ([1], p. 124), but agreement is reached in that most important case in which S is convex and balanced (see below).)

1.5.2 Balanced Sets. A subset S of E will be termed *balanced* (Bourbaki's "équilibré": [4], p. 5) if $\alpha x \in S$ whenever $x \in S$ and $\alpha \in K$, $|\alpha| \leq 1$; see also Taylor [1], p. 123.

Since the intersection of balanced sets is balanced, each subset of E is contained in a smallest balanced subset of E; the latter is termed the *balanced envelope* (or *balanced hull*) of the given set. The balanced envelope of the set $A \subset E$ is none other than the set $\{\alpha x : \alpha \in K, |\alpha| \leq 1, x \in A\}$.

Note that any nonvoid balanced set contains 0.

1.5.3 Convex Sets. A subset S of E is said to be *convex* if $(1 - \alpha)x + \alpha y \in S$ whenever $0 \leq \alpha \leq 1$ and $x, y \in S$. The intersection of convex sets is convex. Hence any given subset A of E is contained in a smallest convex subset of E, the latter being termed the *convex envelope* or *convex hull* of A. This envelope is void if A is void; otherwise it consists of all finite sums $\sum_i \alpha_i x_i$ where the x_i are freely chosen from A, the α_i are arbitrary scalars chosen from the real interval $[0, 1]$, and $\sum_i \alpha_i = 1$.

The direct and the inverse images by a linear map of a convex set are convex sets; any translate of a convex set is convex.

Note also that the convex envelope of the balanced envelope of a set A is the smallest set containing A that is at once convex and balanced: it is therefore named the *convex-balanced envelope* (or *hull*) of A. This envelope consists of all finite sums $\sum_i \alpha_i x_i$ with $x_i \in A$, $\alpha_i \in K$, and $\sum_i |\alpha_i| \leq 1$.

The discussion of sets of the types just described is facilitated by the introduction of corresponding types of real-valued functions on E that serve to represent the "equations" of the sets involved.

1.6 Gauges and Seminorms

Throughout this section E denotes a vector space over K.

1.6.1 Gauge Functions. By a *gauge* or *gauge function* on E is meant a real-valued function g on E such that

$$g(x + y) \leq g(x) + g(y), \tag{1.6.1}$$

$$g(\alpha x) = \alpha \cdot g(x) \tag{1.6.2}$$

for arbitrary x, y in E and arbitrary real scalars $\alpha \geq 0$. Then necessarily $g(0) = 0$.

1.6.2 Seminorms (or Prenorms). By a *seminorm* on E we mean a gauge p on E for which (1.6.2) is satisfied in the stronger form

$$p(\alpha x) = |\alpha| \cdot p(x) \tag{1.6.2*}$$

for x in E and arbitrary scalars α.

The relations

$$0 = p(0) = p(x + (-x)) \leq p(x) + p(-x) = 2p(x)$$

show that a seminorm is positive (that is, nonnegative) valued.

Note also that

$$|p(x) - p(y)| \leq p(x - y). \tag{1.6.3}$$

Also, starting from (1.6.1) and (1.6.2*) and using finite induction, one finds that

$$p\left(\sum_{i=1}^{n} \alpha_i x_i\right) \leq \sum_{i=1}^{n} |\alpha_i| \cdot p(x_i) \tag{1.6.4}$$

for arbitrary $x_i \in E$ and $\alpha_i \in K$.

1.6.3 Norms. By a *norm* on E we mean a seminorm p that has the property that $p(x) > 0$ whenever $x \in E$ and $x \neq 0$. Usually, one writes a norm in the form $\|x\|$, using subscripts ($\|x\|_1$, and so forth) to distinguish between different norms.

We shall now show how gauges and seminorms serve in the description of various types of convex subsets of E.

1.6.4 Theorem. Let A be a convex and absorbent subset of E which contains 0. Define the real-valued function g on E by

$$g(x) = \operatorname{Inf}\{\alpha : \alpha \text{ real}, \alpha > 0, x/\alpha \in A\}. \tag{1.6.5}$$

Then g is a positive gauge on E such that

(1) A contains the set $\{x \in E : g(x) < 1\}$;

(2) A is contained in the set $\{x \in E : g(x) \leq 1\}$.

We say that g is the *Minkowski gauge* of A.

Conversely, if g is any positive gauge on E, each of the sets defined by $g(x) < 1$ and by $g(x) \leq 1$ is convex and absorbent and contains 0.

Proof. The final sentence is trival, thanks to the defining properties of a gauge (see Section 1.6.1).

For the rest, observe first that since A is absorbent, the set whose infimum appears in (1.6.5) is nonvoid, so that $g(x)$ exists as a real number, obviously satisfying $g(x) \geq 0$. Property (1.6.2) follows at once from (1.6.5), as also does property (1.6.1) if convexity of A is brought into play.

If $g(x) < 1$, then there exists α satisfying $0 < \alpha < 1$ and $x/\alpha \in A$. Since $0 \in A$ and A is convex, so $x = \alpha(x/\alpha) + (1 - \alpha)0 \in A$. On the other hand, if $x \in A$, then $x/\alpha \in A$ for $\alpha = 1$, and so $g(x) \leq 1$. This completes the proof. ▌

1.6.5 This theorem may be supplemented in the following manner. Let us agree to say that a set $A \subset E$ is *open* (resp. *closed*) *in rays* if for each x in E, the set $I_x = \{\alpha : \alpha \text{ real}, \alpha > 0, x/\alpha \in A\}$ is open (resp. closed) relative to the interval $I = (0, +\infty)$. We may then state a corollary.

1.6.6 Corollary. Let A be a convex and absorbent subset of E containing 0, and let g be its Minkowski gauge. If A is open (resp. closed) in rays, then A is the subset of E defined by the equation $g(x) < 1$ (resp. $g(x) \leq 1$). In either case, A is balanced if and only if g is a seminorm on E.

Proof. Suppose that A is open in rays. We know already that A contains the set defined by $g(x) < 1$. Suppose conversely that $x \in A$. Then $1 \in I_x$ and so, since I_x is open relative to I, I_x contains a number α such that $0 < \alpha < 1$. But then $g(x) \leq \alpha < 1$. Thus A coincides with the set defined by $g(x) < 1$.

Suppose next that A is closed in rays. We know already that A is contained in the set defined by $g(x) \leq 1$. Let us show that conversely any x satisfying $\alpha = g(x) \leq 1$ must belong to A. We consider two cases. The first is that in which $\alpha = 0$. Then $g(x) < 1$ and so $x \in A$ by what we know already from Theorem 1.6.4. Otherwise we have $0 < \alpha \leq 1$, and there exists a sequence

$(\alpha_n) \subset I_x$ converging to α. Since $\alpha \in I$ and I_x is closed relative to I, therefore $\alpha \in I_x$. Then the equality $x = \alpha(x/\alpha) + (1 - \alpha)0$ shows, via convexity of A, that $x \in A$, as we had to establish.

Now it is obvious that if g is a seminorm, then each of the sets defined by $g(x) < 1$ and $g(x) \leq 1$ is balanced. Suppose conversely that A is defined via a positive gauge g as the set for which $g(x) < 1$; we have to show that if A is balanced, then g is a seminorm. Now if $x \in E$ and $\varepsilon > 0$, then $(g(x) + \varepsilon)^{-1}x \in A$ and so, if $\alpha \in K$ and $\alpha \neq 0$, then $\alpha x/|\alpha|(g(x) + \varepsilon) \in A$. This shows that $g(\alpha x) \leq |\alpha|\,(g(x) + \varepsilon)$. Letting $\varepsilon \to 0$ we conclude that

$$g(\alpha x) \leq |\alpha| \cdot g(x)$$

for any scalar $\alpha \neq 0$. If we replace α by α^{-1}, α being any nonzero scalar, and simultaneously replace x by αx, it appears that

$$g(\alpha x) \geq |\alpha| \cdot g(x).$$

It follows that in fact $g(\alpha x) = |\alpha| \cdot g(x)$ for $\alpha \neq 0$. If $\alpha = 0$, both sides are zero; so the equality is true for all scalars without exception. Consequently g is a seminorm.

The argument is similar when A is defined by $g(x) \leq 1$. ▌

1.7 The Hahn-Banach Theorem in Analytic Form

This theorem and its multitudinous corollaries is one of the central results in functional analysis. It is given here in one of the most general of its so-called "analytic," as opposed to "geometric," forms; for the latter, see Chapter 2. The setting for the theorem in this guise is a general vector space endowed with a selected gauge function.

1.7.1 **Theorem.** (1) Let E be a real vector space, p a gauge on E, L a vector subspace of E, and f_0 a linear form on L satisfying

$$f_0(x) \leq p(x) \qquad (x \in L). \tag{1.7.1}$$

Then there exists a linear form $f \in E^*$ satisfying

$$f(x) = f_0(x) \qquad (x \in L), \tag{1.7.2}$$

$$f(x) \leq p(x) \qquad (x \in E). \tag{1.7.3}$$

If p is a seminorm on E, then (1.7.1) and (1.7.3) are, respectively, equivalent to

$$|f_0(x)| \leq p(x) \qquad (x \in L), \tag{1.7.1*}$$

$$|f(x)| \leq p(x) \qquad (x \in E). \tag{1.7.3*}$$

(2) If E is a complex vector space and p a seminorm on E, L and f_0 being otherwise as in (1), a similar theorem holds if we replace (1.7.1) and (1.7.3) by (1.7.1*) and (1.7.3*).

Proof. The equivalence of (1.7.n) with (1.7.n*) for $n = 1$ and 3 is obvious. We shall establish (1) first and then deduce (2) from it.

(1) We shall use Zorn's lemma (Subsection 0.1.5.) applied to the set F of all linear forms f for which $\text{Dom } f$ is a vector subspace of E satisfying $\text{Dom } f \supset L$ and for which $f \mid L = f_0$ and $f \leq p$ on $\text{Dom } f$. By hypothesis, $f_0 \in F$. We partially order F by writing $f \leq g$ to signify that g extends f; that is, that $\text{Dom } g \supset \text{Dom } f$ and $g \mid (\text{Dom } f) = f$. There is no difficulty in verifying that F is thus inductively ordered. Once this is done, Zorn's lemma assures us that maximal elements of F exist. Let f be any such maximal element of F. An argument due to Banach ([1], pp. 28–29), which we reproduce forthwith, then shows that f has the desired properties. Indeed, all we need to establish is that $\text{Dom } f = E$.

Now suppose, if possible, that $\text{Dom } f \neq E$. We choose any $x_0 \in E \setminus \text{Dom } f$. If x' and x'' belong to $\text{Dom } f$, we have

$$\begin{aligned} f(x'') - f(x') = f(x'' - x') \leq p(x'' - x') &= p[(x'' + x_0) + (-x' - x_0)] \\ &\leq p(x'' + x_0) + p(-x' - x_0) \end{aligned}$$

and so

$$-p(-x' - x_0) - f(x') \leq p(x'' + x_0) - f(x'').$$

It follows that

$$\lambda = \text{Sup}\, \{-p(-x - x_0) - f(x) : x \in \text{Dom } f\}$$

and

$$\Lambda = \text{Inf}\, \{p(x + x_0) - f(x) : x \in \text{Dom } f\}$$

are both finite and that $\lambda \leq \Lambda$. We choose freely any real number γ satisfying $\lambda \leq \gamma \leq \Lambda$. Then

$$-p(-x - x_0) - f(x) \leq \gamma \leq p(x + x_0) - f(x) \tag{1.7.4}$$

holds for x in $\text{Dom } f$.

We let $M = Rx_0 + \text{Dom } f$ and define f' on M by

$$f'(x + \xi x_0) = f(x) + \gamma\xi \qquad (x \in \text{Dom } f,\ \xi \in R).$$

It is plain that f' is a linear form with domain $M \supset \text{Dom } f \supset L$. If we show that f' belongs to F, maximality of f would contradict the fact that $M \supset \text{Dom } f$ properly. Since $f' \mid \text{Dom } f = f$, it remains for this purpose only to verify that $f' \leq p$ on M. However, if $\xi \neq 0$, and if in (1.7.4) we replace x by x/ξ, there results

$$-p\left(-\frac{x}{\xi} - x_0\right) - \frac{f(x)}{\xi} \leq \gamma \leq p\left(\frac{x}{\xi} + x_0\right) - \frac{f(x)}{\xi}\cdot$$

If $\xi > 0$, the right-hand inequality gives

$$\gamma\xi \leq p(x + \xi x_0) - f(x),$$

and if $\xi < 0$, the left-hand inequality gives

$$p(x + \xi x_0) - f(x) \geq \gamma\xi.$$

In either case, therefore, we have

$$f(x) + \gamma\xi \le p(x + \xi x_0).$$

This is trivially true if $\xi = 0$. Thus

$$f'(x + \xi x_0) \le p(x + \xi x_0) \qquad (x \in \operatorname{Dom} f,\ \xi \in R).$$

showing that $f' \le p$ on M. The proof of (1) is thereby completed.

(2) The derivation of (2) from (1) depends on a device due to Bohnenblust and Scobczyk ([1]). We write $f_0 = u_0 + iv_0$ where u_0, v_0 are real-valued and real-linear (that is, linear with respect to real scalars) forms on L. Since $f_0(ix) = if_0(x)$, it follows that $v_0(x) = -u_0(ix)$ and so

$$f_0(x) = u_0(x) - iu_0(ix).$$

Then (1.7.1*) gives $|u_0(x)| \le p(x)$ for x in L.

According to (1) we may extend u_0 into a real-valued, real-linear form u on E satisfying $|u| \le p$. If we define f by

$$f(x) = u(x) - iu(ix),$$

then f is a (complex-) linear form on E and $f \mid L = f_0$. It remains only to show that (1.7.3*) is obeyed. But if $x \in E$ and $f(x) = re^{it}$ ($r \ge 0$, t real), then

$$\begin{aligned} |f(x)| &= e^{-it}f(x) = f(e^{-it}x) = \operatorname{Re} f(e^{-it}x) = u(e^{-it}x) \\ &\le |u(e^{-it}x)| \le p(e^{-it}x) = p(x), \end{aligned}$$

as required.

This completes the proof of the theorem. ▌

1.7.2 **Corollary.** (Hahn [2]). Let E be a vector space over K, p a seminorm on E, L a vector subspace of E, $x_0 \in E$, and

$$\delta = \operatorname{Inf}\{p(x_0 + y) : y \in L\}.$$

Then there exists a linear form f on E satisfying

$$\begin{aligned} |f(x)| &\le p(x) \qquad (x \in E), \\ f(L) &= 0, \\ f(x_0) &= \delta. \end{aligned}$$

1.7.3. A systematic study of further corollaries of Theorem 1.7.1, especially in the context of topological vector spaces, is deferred until Chapter 2. There, and also in parts of Chapter 3, we shall deal with some applications and extensions of Theorem 1.7.1.

1.8 Topological Vector Spaces

We shall now introduce a topological vector space as a concept which results from welding together that of a vector space and that of a topological space (see Section 0.2) with the proviso that the two structures shall be interrelated in such a fashion that the algebraic operations are continuous. The phrase "topological vector space" will appear so frequently that we shall abbreviate it to the symbol "TVS." Here is the formal definition of a TVS.

1.8.1 **Definition of a TVS.** By a TVS over K we shall mean a system comprised of (1) a vector space E over K, and (2) a topology $\mathfrak{T}$ on E, the vector space and topological structures carried by E being so related that the maps

$$(1) \quad (x, y) \to x + y \text{ from } E \times E \text{ into } E$$

$$(2) \quad (\alpha, x) \to \alpha \cdot x \quad \text{from } K \times E \text{ into } E$$

are continuous.

Properly one should denote by some symbol such as $(E, \mathfrak{T})$ the TVS thus defined. Indeed, when one wishes to consider simultaneously several TVSs obtained by endowing a given vector space E with several topologies (each assumed to be such that the result is indeed a TVS), we shall resort to this notation. Otherwise, and whenever it is clear from the context which topology is involved, we shall be content to denote simply by E the TVS in question.

Before listing any examples we shall enumerate a few almost obvious consequences of the above definition.

1.8.2 **Remarks.** (1) Taking a fixed $a \in E$, the map $x \to x + a$ (translation by amount a) has as its inverse the map $x \to x - a$. It therefore appears that in a TVS, each translation is a homeomorphism of the space onto itself. Consequently the topology is determined completely by the system of neighbourhoods of 0.

(2) Condition (1) of Subsection 1.8.1 is equivalent to the demand that translation be continuous, and that $(x, y) \to x + y$ be continuous at (0, 0).

(3) If we choose a fixed scalar $\alpha \neq 0$, the map $x \to \alpha \cdot x$ (homothety of ratio α) has as its inverse the map $x \to \alpha^{-1} \cdot x$. Consequently, in a TVS, each homothetic map with nonzero ratio is a homeomorphism of the space onto itself. Since

$$\alpha x - \alpha_0 x_0 = (\alpha - \alpha_0)x_0 + \alpha_0(x - x_0) + (\alpha - \alpha_0)(x - x_0),$$

condition (2) of Subsection 1.8.1 is equivalent to the requirement that $\alpha \cdot x$ be separately continuous in either variable and be continuous in the pair at (0, 0).

(4) The topology of a TVS may or may not be separated. However, if it is T_1 (or even merely T_0), then it is separated. For if it is T_0 and $x \neq y$, there exists a neighbourhood of one of these points, say x, that does not contain the other—that is, does not contain y. By what has been said about the homeomorphic nature of translation, this means that there exists a neighbourhood U of 0 such that y does not belong to $x + U$. Since $x - y$ is continuous, there exists a second neighbourhood V of 0 such that $V - V \subset U$. Then $x + V$ and $y + V$ are disjoint neighbourhoods of x and of y, respectively, showing that the topology is separated. We shall see later how to associate naturally a separated TVS with one that is not separated.

(5) In a TVS there always exists a neighbourhood base at 0 comprised of sets that are all *either* (a) balanced and open, *or* (b) balanced and closed.

To see this, we note first that (as in any topological space) there exists a neighbourhood base at 0 comprised of open sets. Let U be any neighbourhood of 0. By (2) of Subsection 1.8.1 there exists then a number $\varepsilon > 0$ and an open neighbourhood V of 0 such that the conditions $|\alpha| \leq \varepsilon$ and $x \in V$ together entail $\alpha x \in U$. Let W be the balanced envelope of εV (see Subsection 1.5.2). Since V is open, it is easy to check on the basis of (1) and (2) of Subsection 1.8.1 that W is open. Moreover, it is evident that $W \subset U$. (a) is thus established.

To prove (b), we begin again with any neighbourhood U of 0. Let us first show that there exists a closed neighbourhood V of 0 contained in U. Since, by combining (1) and (2) of Subsection 1.8.1, we infer that $x - y$ is continuous in the pair, there exists a neighbourhood U' of 0 such that $U' - U' \subset U$. But then $V = \overline{U'}$ satisfies our demands. This shows that the closed neighbourhoods of 0 form a base at 0.

This being so we may (as in the proof of part (a)) choose a number $\varepsilon > 0$ and a neighbourhood V' of 0 such that the balanced envelope W of $\varepsilon V'$ is contained in V. W is a balanced neighbourhood of 0. Since V is closed, so $\overline{W} \subset V$. Here $\overline{W}$ is a closed neighbourhood of 0. It suffices now to show that $\overline{W}$ is balanced whenever W is balanced. But this is simple: let N be any neighbourhood of 0; choose, as in the proof of part (a), a neighbourhood N' of 0 whose balanced envelope is contained in N. If $x \in \overline{W}$, $x - N'$ must meet W; that is, $x \in W + N'$. Then, if $|\alpha| \leq 1, \alpha x \in \alpha W + \alpha N' \subset W + N$ (W being balanced). Since this is true for each N, so $\alpha x \in \overline{W}$, showing that $\overline{W}$ is balanced.

(6) In a TVS each neighbourhood of 0 is absorbent. This is indeed a trivial consequence of Subsection 1.8.1 (2).

A consequence of the preceding remarks is embodied in the "direct" part of the following proposition.

1.8.3 Proposition. If E is any TVS over K, there exists a neighbourhood base $\mathcal{N}$ at 0 such that:

(1) Each set $U \in \mathcal{N}$ is balanced and absorbent.

(2) If $U \in \mathcal{N}$, $\alpha \in K$, and $\alpha \neq 0$, then $\alpha U \in \mathcal{N}$.

(3) If $U \in \mathcal{N}$, there exists $V \in \mathcal{N}$ such that $V + V \subset U$.

(4) If U_1 and U_2 belong to $\mathcal{N}$, there exists $U \in \mathcal{N}$ satisfying $U \subset U_1 \cap U_2$.

Conversely, given a set $\mathcal{N}$ of subsets of a vector space E over K satisfying (1)–(4), there exists just one topology $\mathfrak{T}$ on E having $\mathcal{N}$ as a neighbourhood base at 0 and such that $(E, \mathfrak{T})$ is a TVS over K.

Proof. Only the "converse" portion has to be dealt with. The uniqueness of $\mathfrak{T}$, if it exists, follows because in a TVS one obtains a neighbourhood base at any point x by taking the translates by amount x of the sets belonging to a neighbourhood base at 0 (see Subsection 1.8.2(1)). On the other hand, if the topology $\mathfrak{T}$ is defined in this manner, the properties (1)–(4) are easily verified to imply that conditions (1) and (2) of Subsection 1.8.1 are fulfilled. ∎

1.8.4 **Examples.** At this stage only a few simple examples will be set before the reader. As we progress, several general processes will come to light that lead from given TVSs to new ones.

(1) Let us consider the function space K^T introduced in Example 1.1.1. Initially we assume merely that T is a set.

If we assign to each finite subset F of T and each number $\varepsilon > 0$ the subset $U(F, \varepsilon)$ of K^T consisting precisely of all functions x in K^T for which

$$p_F(x) = \operatorname{Sup} \{|x(t)| : t \in F\}$$

satisfies $p_F(x) < \varepsilon$, then it is a moment's work to verify that the sets $U(F, \varepsilon)$ (F and ε varying separately) fulfill the conditions of Proposition 1.8.3. The corresponding vector-space topology on K^T is often described as "the topology of pointwise (or simple) convergence on T"; it is none other than the product topology. The reader will notice that each p_F is a seminorm on K^T.

Further TVSs may be obtained by passage to vector subspaces of K^T and endowing these with their induced topologies.

(2) One may seek to generalize (1) by replacing the finite subsets of T by some other system of subsets of T. Provided we use only finite subsets, we shall always obtain a TVS in this fashion—although in general the topology will not be separated. If, however, one includes infinite subsets of T, then the derived topology on K^T will *not* in general make K^T into a TVS.† Thus if there exists at least one x in K^T that is unbounded on some set F of the chosen system, then it will not be the case that $\alpha x \to 0$ as $\alpha \to 0$. This can be remedied by taking the vector subspace of K^T, say E, comprised of those x in K^T that are bounded on each set F of the chosen system.

For instance, suppose our system contains just the set T itself. Then we must take for E the subspace of K^T comprising precisely all bounded K-valued functions on T; we denote this space by $B(T)$ or $B_K(T)$. The corresponding topology is just that normally described as "the topology of uniform convergence on T."

Further instances appear in (4) below.

(3) Suppose T is finite. Then $K^T = B(T)$ (as vector spaces). Pointwise and uniform convergence coincide, of course. If T has cardinal n (a natural number), this topology is the usual metric topology of K^n.

(4) Let us consider now the case in which T is a topological space. We denote by $C(T)$ or $C_K(T)$ the vector subspace of K^T formed of all continuous K-valued functions on T.

If T is compact, $C(T)$ is a vector subspace of $B(T)$ and may be topologized as a subspace thereof (see (2)).

If T is not compact, we can still topologize $C(T)$ by taking the system of all *compact* subsets F of T. For this topology, described as "the topology of compact convergence (or locally uniform convergence, if T is locally compact) on T"

---

† It is still the case, however, that K^T is thus made into a topological group under addition.

or the "compact-open topology," a neighbourhood base at 0 is comprised of the sets $U(F, \varepsilon)$, F ranging over the compact subsets of T (or a base for such sets) and ε over all strictly positive numbers.

Whether or not T is compact, we may topologize $C_{bd}(T)$, the space of all bounded, continuous K-valued functions on T, as a subspace of $B(T)$.

Still other examples will be subjected to study in Chapters 4 and 5.

(5) All the preceding examples can be generalized by considering, in place of the K-valued functions on T, those that take their values in a given TVS X over K. We leave the reader to formulate the details.

(6) Let T be the real interval $(0, 1)$ of the real axis. Let p be a number satisfying $p > 0$. We denote by $\mathscr{L}^p$ the set of all real-valued functions x on T that are measurable and such that $|x|^p$ is Lebesgue integrable over $(0, 1)$. (The reader not familiar with the Lebesgue integral for functions of a real variable may just as well restrict his attention to continuous functions for the purposes of this example. A detailed account of integration theory is given in Chapter 4 below.) Let D_p be the semimetric on $\mathscr{L}^p$ defined by

$$D_p(x, y) = \left\{\int_0^1 |x(t) - y(t)|^p \, dt\right\}^{1/p}.$$

Then $\mathscr{L}^p$, endowed with the resulting topology, is a TVS. We may write $D_p(x, y) = N_p(x - y)$, where

$$N_p(z) = \left\{\int_0^1 |z(t)|^p \, dt\right\}^{1/p};$$

if $p \geq 1$, N_p is a seminorm on $\mathscr{L}^p$, but this is not the case if $0 < p < 1$.

(7) We consider again real-valued functions on $T = (0, 1)$. Let μ denote Lebesgue measure on the real axis. Given numbers $\alpha > 0$, $\beta > 0$, we denote by $U_{\alpha,\beta}$ the set of x in R^T for which the μ-measure of the set $\{t \in T;\ |x(t)| > \alpha\}$ is at most β. Then the $U_{\alpha,\beta}$ may be taken as a neighbourhood base at 0 for a topology $\mathfrak{T}$ on R^T, neighbourhoods at any other point x of R^T being obtained by translation. In this way R^T is made into a TVS. The topology $\mathfrak{T}$ is sometimes described as "the topology of convergence in measure on T."

We now turn to some general processes leading from given TVSs to others.

1.8.5 **Subspaces, Product Spaces, Quotient Spaces.** If E is a TVS and L a vector subspace thereof, then L, endowed with the topology induced upon it by that of E, is itself a TVS. No special comments are required at this stage.

Let $(E_i)_{i \in I}$ be a family of TVSs and let $E = \prod_{i \in I} E_i$ be the product vector space (see Subsection 1.2.5). If E be endowed with the product topology (see 0.2.15), the result is a TVS. For instance, taking $E_i = K$ (regarded as a TVS over K, its topology being the usual one) and $I = T$, an arbitrary set, the product TVS is none other than K^T with the topology of convergence pointwise on T (Example 1.8.4(1)).

Turning to quotient spaces, let E be a TVS and L a vector subspace thereof. We form the quotient vector space E/L (see Subsection 1.2.3). If E/L be endowed with the quotient topology (see Subsection 0.2.16), there results a TVS, which we refer to as the quotient TVS of E modulo L. We denote by x_c the class modulo L containing x, and by f the canonical map $x \to x_c$ of E onto E/L. f is linear. As we know, the quotient topology on E/L is the strongest for which f is continuous. Consequently a neighbourhood base at 0 in E/L is obtained by taking those subsets W of E/L for which $f^{-1}(W)$ is a neighbourhood of 0 in E.

It is frequently useful to remember that

▶ if (U_i) is a neighbourhood base at 0 in E, then the $W_i = f(U_i)$ form a neighbourhood base at 0 in E/L.

Indeed, since $f^{-1}(W_i) \supset U_i$, W_i *is* a neighbourhood of 0 in the quotient space. On the other hand, let W be any neighbourhood of 0 in E/L, so that $U = f^{-1}(W)$ is a neighbourhood of 0 in E. Hence, $U \supset U_i$ for some i. Since f maps E onto E/L, $W = f(f^{-1}(W)) = f(U) \supset f(U_i) = W_i$. This shows that the W_i form a neighbourhood base at 0, as alleged.

This last observation may be expressed by saying that f, the natural map of E onto E/L, is open—or, in the terminology introduced in Section 1.9, that this natural map is a homomorphism of the TVS E onto the TVS E/L.

Whether or not E itself is separated, *the quotient E/L is separated if and only if L is closed in E*. For the natural map f of E onto E/L is continuous and L is its kernel; on the other hand (as for any TVS), E/L is separated if and only if its zero element forms a closed set. So if E/L is separated, $L = \operatorname{Ker} f$ is the inverse image by a continuous map of a closed set, hence itself is closed. On the other hand $f(E \setminus L) = (E/L) \setminus f(L)$, whence it follows that if L is closed in E, then $f(L)$ is closed in E/L; and $f(L)$ is just the one-element subset of E/L formed of its zero element.

If E is not separated, we may obtain separated quotient spaces E/L by taking for L any closed vector subspace of E. Naturally, one would like to do this in such a fashion as to involve as little identification of distinct elements of E as is possible; that is, we would naturally take L as small as possible. The smallest choice is obviously $L = \overline{\{0\}}$. For this reason $E/\overline{\{0\}}$ is called the *separated TVS associated with E*.

1.8.6 **Balanced Sets and Convex Sets in TVSs.** It is time to supplement Section 1.5 by reconsidering some of the concepts introduced there for vector spaces in the context of TVSs.

The reader should verify that the closure (in a TVS) of a vector subspace (resp. convex subset, balanced subset) is a set of the same type. Hence, given a subset A of a TVS E, the closure of the linear (resp. convex, balanced) envelope of A is the smallest closed vector subspace (resp. closed convex set, closed balanced set) in E containing A: we refer to these as the respective *envelopes* of A in E.

We note too that a hyperplane in E is either closed or everywhere dense (as a consequence of its maximal property—see Subsection 1.2.6).

We shall defer until Section 1.9 the supplementation of the substance of Sections 1.3 and 1.4 called for as a consequence of topologization. Meanwhile there is a new concept to be introduced.

1.8.7 **Bounded Sets in a TVS.** Let E be a TVS. A subset B of E is said to be *bounded* if, given any neighbourhood U of 0 in E, there exists a scalar $\alpha > 0$ such that $B \subset \alpha U$.

The reader should observe that this concept of boundedness, when applied to subsets of K^n, agrees with that of metric boundedness.

Two properties of bounded sets should be noted.

(1) If B is bounded, so also is the closed and balanced envelope of B.

(2) The union, and the vectorial sum, of any finite number of bounded sets is again bounded.

In fact, (1) follows at once from Remark 1.8.2(5). The reader is warned that, unless E is locally convex (see Section 1.10), the convex envelope of a bounded set need *not* be bounded. As for (2), consider for example the vectorial sum $B = B' + B''$ of two bounded sets. Given any neighbourhood U of 0, choose (see Proposition 1.8.3) a neighbourhood V of 0 such that V is balanced and $V + V \subset U$. By hypothesis $B' \subset \alpha' U$ and $B'' \subset \alpha'' V$. Then, if $\alpha = \max(\alpha', \alpha'')$, $B \subset \alpha V + \alpha V \subset \alpha(V + V) \subset \alpha U$, showing that B is bounded.

It is otherwise trivial that any finite subset of a TVS is bounded.

1.8.8 **Examples.** In the product space K^T a subset B is bounded if and only if the functions belonging to B are bounded at each point of T; that is, if and only if

$$\text{Sup}\,\{|x(t)| : x \in B\} < +\infty$$

for each t in T. In the space $B(T)$ (Example 1.8.4(2)) a subset B is bounded if and only if the functions belonging to B are uniformly bounded; that is, if and only if

$$\text{Sup}\,\{|x(t)| : x \in B, t \in T\} < +\infty.$$

1.8.9 **TVSs as Uniform Spaces.** Each TVS E carries an associated uniform structure (Section 0.3.1)—namely, that for which a base of vicinities of the diagonal Δ in $E \times E$ is formed of the sets $\Delta(U) = \{(x, y) : x - y \in U\}$ when U ranges over any neighbourhood base at 0 in E. Any two vector-space topologies on a vector space E, which are identical, thus give rise to identical uniform structures on E. This is why there is no harm in confounding the topology of a TVS and its uniform structure, and why, therefore, we may speak of a TVS being complete or not. In general, any notions pertaining to a uniform structure (completeness, uniform continuity, and so forth) are available in connection with a TVS.

1.8.10 **Completion of a TVS.** We shall let E be a TVS, supposing first that E is separated. As for any separated uniform space we may (Subsection

0.3.12) construct a complete uniform space $\hat{E}$ with the property that there exists a one-to-one map ι of E into $\hat{E}$, which is an isomorphism of the uniform space E onto a dense uniform subspace $\iota(E)$ of $\hat{E}$; moreover, $\hat{E}$ is determined up to isomorphism (as a uniform space) by these requirements.

We recall that one more-or-less concrete realization of $\hat{E}$ is the following: The points of $\hat{E}$ are equivalence classes of Cauchy directed families of points of E, two such directed families, say $X = (x_i)$ and $Y = (y_j)$, being regarded as equivalent if for any neighbourhood U of 0 in E and any indices i_0 and j_0 there exist indices $i \geq i_0$ and $j \geq j_0$ such that $x_i - y_j \in U$. Thus the set $\hat{E}$ is constructed as the quotient, modulo this equivalence relation R, of the set $\mathscr{E}$ of all Cauchy directed families $X = (x_i)$ of points of E: $\hat{E} = \mathscr{E}/R$. In $\mathscr{E}$ we introduce a uniform structure by taking as a base of vicinities of the diagonal the sets $W(U)$, where U ranges over a neighbourhood base at 0 in E, and where $W(U)$ is the set of pairs (X, Y) for which $X = (x_i)$, $Y = (y_j)$ are such that $x_i - y_j \in U$ for arbitrarily large (and hence all sufficiently large) indices i and j. The uniform structure of $\hat{E}$ is that obtained by taking as a base for the vicinities of the diagonal the images, by the canonical map of $\mathscr{E}$ onto $\hat{E} = \mathscr{E}/R$, of the sets $W(U)$.

Now if E is a vector space, we may endow $\hat{E}$ with a similar structure in an obvious way, defining, for example, $X + Y$ to be $(x_i + y_j)$ (a Cauchy directed family indexed by the set of pairs (i, j)) and then $x + y$ as the class modulo R of $X + Y$.

Moreover, if E is a TVS, the uniform structure defined on $\hat{E}$ yields a topology that, as one may verify rapidly, is compatible with the vector space structure of $\hat{E}$. Thus $\hat{E}$ is made into a TVS. $\hat{E}$ is separated since E has that property. One may inject E into $\hat{E}$ via the isomorphism ι. A neighbourhood base at 0 in $\hat{E}$ is obtained by taking the closures in $\hat{E}$ of the members of any neighbourhood base at 0 in E. Any continuous linear map of E into a separated TVS F has a unique continuous extension into an element of $L_c(\hat{E}, \hat{F})$; and any continuous seminorm on E has a unique continuous extension into a continuous seminorm on $\hat{E}$.

If E is not separated, the above process may be applied to the associated separated space $E/\overline{\{0\}}$. This preliminary identification leads to no trouble when we are dealing with continuous linear maps or continuous seminorms on E.

1.9 Continuity of Linear Maps. Isomorphisms and Homomorphisms of TVSs. Closed Linear Maps

Throughout this section, unless the contrary is explicitly mentioned, E and F denote TVSs over K and u is a linear map of E into F : $u \in L(E, F)$.

1.9.1. It is evident that u is continuous at some point of E if and only if it is continuous at 0, in which case it is continuous everywhere and indeed uniformly continuous on E.

There is no need to list here those properties of continuous linear maps of one TVS into another that are special cases of simple results true of continuous maps of one topological space into another.

A property that is not of this category, and is important for us, is expressed by the statement that

▶ any continuous linear map u of a TVS E into a TVS F transforms each bounded subset A of E into a bounded subset $B = u(A)$ of F.

By continuity, $U = u^{-1}(V)$ is a neighbourhood of 0 in E if V is a neighbourhood of 0 in F, and therefore $A \subset \lambda \cdot U$ for some $\lambda > 0$; accordingly, $B \subset \lambda \cdot V$. V being any neighbourhood of 0 in F, this shows that B is bounded.

A linear map of E into F that has the property of transforming bounded subsets of E into bounded subsets of F is said to be *bounded*. It is natural to ask whether, as a converse to the result established, any bounded linear map is continuous. This is trivially the case when E and F are seminormable (see Subsection 1.10.2); less obvious is the fact that this proposed converse is true for other categories of TVSs, termed *bornological spaces*. We return to this matter later (see Section 7.3).

1.9.2. By an *isomorphism* of E into F (qua TVSs) we mean an element u of $L(E, F)$ that is an isomorphism of E into F in the algebraic sense (see Subsection 1.3.2) and that is at the same time a homeomorphism of E into F. If u is known to be linear, continuous, and one-to-one, it is an isomorphism of E into F if and only if in addition it is open onto $u(E)$. The adjectives "into" and "onto" are used here in the same sense as they are described in Subsection 1.3.2.

1.9.3. A *homomorphism* of E into F (qua TVSs) is an element u of $L(E, F)$ that is continuous and open. This is equivalent to saying that u is an isomorphism of $E/\text{Ker } u$ into F (or onto $u(E)$), $E/\text{Ker } u$ being the quotient TVS (see Subsection 1.8.5) of E modulo Ker u. Again, a continuous linear map u of E into F is a homomorphism of E into F if and only if it is open onto $u(E)$.

The preceding concepts may be applied in the case in which u has for its domain a proper vector subspace of E: continuity, for example, then means that u is continuous when considered as an element of $L(\text{Dom } u, F)$.

We introduce now two pieces of notation that will remain standard throughout the remainder of this book.

1.9.4 **Definition.** Let E and F be TVSs. We denote by $L_c(E, F)$ the vector subspace of $L(E, F)$ consisting of all those linear maps of E into F that are continuous. If, in particular, $F = K$ (the field of scalars), $L_c(E, F)$ is denoted by E'; E' is thus the vector subspace of E^* comprising exactly the continuous linear forms on E. E' is named the *topological dual* of E. In all general discussions we shall usually denote elements of E' by x', y', $\cdots$ and denote by $x'(x)$, or the more symmetrical $\langle x, x' \rangle$, the value at $x \in E$ of the linear form $x' \in E'$. (Compare the remarks in Subsection 1.4.7.)

1.9.5 **Remarks.** (1) In general E' is very much smaller than E^*; and, although E^* always separates the points of E (which is the justification for injecting E into E^{**} as described in Section 1.4.7), E' may fail to have this property. Failure will occur whenever E is nonseparated as a topological space, of course; but this is not the only, nor even the most interesting, cause of the breakdown. The one large category of spaces, for which we shall learn that E' separates the points of E, is the class of separated spaces that are also "locally convex" (see Section 1.10). Meanwhile we are content with the observation that there are many ways of making E' into a TVS in such a fashion that each x in E generates an element $x' \to \langle x, x' \rangle$ of E'', the associated map $x \to x''$ of E into E'' being in general not one-to-one and therefore not justifying any attempt to inject E into E''.

(2) Agreement on the terminology and notation relating to the topological dual is by no means universal. The terms "dual space," "adjoint space," "conjugate space" have all been used with some frequency for what we choose to term the topological dual. In place of our E', the symbols E^* and $\bar{E}$ are used by various writers. When the reader consults other texts, he must be on his guard against possible misunderstandings. Our notation and terminology agree with those of Bourbaki.

(3) On occasions we shall speak about the vector subspace of $L(E, F)$ formed of the bounded linear maps of E into F. We denote this subspace by $L_b(E, F)$. It is always the case that $L_b(E, F) \supset L_c(E, F)$; in general the inclusion is strict, but we shall see in Chapter 7 that identity obtains for an important category of TVSs E.

1.9.6 **Finite Dimensional TVSs.** It is important that we should settle once and for all the essential uniqueness of the possible ways of making a finite dimensional vector space into a separated TVS. This we do by establishing the following proposition.

Proposition. Let E be a separated TVS over K having finite dimension n. Let $(e_i)_{1 \le i \le n}$ be any algebraic basis for E. The map $u \in L(K^n, E)$ defined by

$$u(\xi) = \sum_{i=1}^{n} \xi_i e_i, \qquad \xi = (\xi_i)_{1 \le i \le n} \in K^n,$$

is an isomorphism of the TVS K^n onto the TVS E.

Proof. It is evident that u is linear, continuous, one-to-one, and onto, so that we have only to show that u is open. We define

$$\|x\| = \operatorname*{Sup}_{1 \le i \le n} |\xi_i|$$

for $x = u(\xi)$, and let B and S be the subsets of E defined, respectively, by the relations $\|x\| \le 1$ and $\|x\| = 1$. Since u is continuous, B and S are each compact. E being separated, S is closed. Since 0 does not belong to S, there exists a neighbourhood U of 0 in E such that $U \cap S$ is void. We may assume that U is closed and balanced (Remark 1.8.2 (5)). It will suffice to show that $U \subset k \cdot B$ for some natural number k.

Now if this were not the case, we could for each k choose $x_k \in k^{-1}U$, $x_k \in E \backslash B$. We put $y_k = x_k/\|x_k\|$. Since $\|x_k\| > 1$ and $k^{-1}U$ is balanced, therefore $y_k \in S \cap U$. Since S is compact, the sequence (y_k) has a limiting point $y \in S$. But then, U being closed, $y \in U$ too. This contradicts the fact that $U \cap S$ is void. ▮

Three corollaries are worth enumerating.

▶ (1) Any finite dimensional vector subspace of a separated TVS is closed.

For the subspace, being isomorphic to K^n for some n, is complete.

▶ (2) Any linear map on a finite-dimensional separated TVS is continuous.

▶ (3) Let E be a TVS, H a hyperplane in E, $f \in E^*$, $f \neq 0$, $\alpha \in K$, and $H = f^{-1}(\{\alpha\})$ (cf. 1.2.6 and 1.2.7). Then H is closed if and only if f is continuous, in which case f is a homomorphism of E onto K.

For it is evident that H is closed if f is continuous. Suppose conversely that H is closed. By translation we may assume that H is a homogeneous hyperplane, so that necessarily $\alpha = 0$. Since H is closed, E/H is separated. It is in any case of dimension 1. f defines a one-to-one linear map of E/H onto K (see Subsection 1.3.2). By the above proposition, this map is an isomorphism (of the TVSs involved) and so f is a homomorphism of E onto K. ▮

There is a characterization of those separated TVSs that are finite dimensional,* first stated for normed spaces by F. Riesz, which is of importance.

F. Riesz' theorem. A separated TVS E is finite dimensional if and only if there exists in E a precompact neighbourhood of zero.

Proof. If E is finite dimensional, it is isomorphic as a TVS with K^n for some n, and is therefore locally compact. Whence the "only if" part of the assertion.

To prove the converse we follow an argument due to J. -P. Serre. Let U be a precompact neighbourhood of 0 in E. One may then choose finitely many points $x_1, \cdots, x_n$ of U such that the sets $x_i + \frac{1}{2}U$ cover U. Let M be the vector subspace of E generated by the x_i, and consider the quotient space E/M. If f denotes the natural mapping of E onto E/M, then (see Subsection 1.8.5) $W = f(U)$ is a neighbourhood of 0 in E/M. Since also f is linear and continuous, and therefore uniformly continuous, W is precompact. The relation $U \subset M + \frac{1}{2}U$ yields, on passage to the quotient space, $W \subset \frac{1}{2}W$. By induction it would appear that $2^k W \subset W$ for all natural numbers k. But, since W is a neighbourhood of zero in E/M, it is then the case that $E/M = W$. Consequently E/M is precompact. M being finite dimensional and E separated, M is closed in E, so that E/M is separated. If E/M were not reduced to its zero element, it would contain

* See Appendix, page 771.

a vector subspace isomorphic with K and would therefore not be precompact. Thus E/M is reduced to its zero element and $E = M$ is finite dimensional. ∎

1.9.7 **Topological Direct Sums and Topological Complements.** The substance of Subsection 1.4.8 must now be supplemented by some remarks bearing upon the case in which TVSs are involved.

Suppose that E is a TVS, L a vector subspace of E, M an algebraic complement of L relative to E. We know that then the map $(a, b) \to a + b$ is an algebraic isomorphism of $L \times M$ onto E. This map is also evidently continuous. If it should happen that this map has a continuous inverse—that is, if it is an isomorphism of the TVSs involved—we say that E is the *topological direct sum* of L and M, and that each of L and M is a *topological complement* of the other relative to E. Introducing the related projectors p_L and p_M, we see that a topological direct sum obtains if and only if either one (and hence, in view of the relation $p_L + p_M = 1$, both) is continuous. For this it is *necessary* (but in general not sufficient) that both L and M be closed in E.

In view of the results of Subsection 1.9.6 we have the following proposition.

Proposition. Let E be a separated TVS, L a vector subspace of E. Suppose that *either* (1) dim $L < +\infty$, or (2) L is closed and codim $L < +\infty$. Then any algebraic complement M of L relative to E is also a topological complement.

Proof. If (1) holds, L is separated and finite dimensional and therefore (Subsection 1.9.6 (2)) p_L is continuous. If (2) holds, E/L and M are both separated and finite dimensional and so (by the same token) p_M is continuous. ∎

1.9.8 **Remarks.** (1) Since the map $(a, b) \to a + b$ is in any case continuous one-to-one, and onto, the inversion theorem of Chapter 6 can sometimes be used to show that any algebraic direct sum $E = L + M$, in which L and M are each closed, is necessarily topological. We shall later see that this is notably the case when E is a complete semimetrizable TVS.

An alternative approach is to note that $p_L : E \to L$ has a closed graph whenever M is closed in E. Then the closed graph theorem, if applicable to the pair E and L, will aver that p_L is continuous and that therefore the direct sum is topological.

The details must wait until the relevant concepts have been defined and examined (see Chapter 6 and Exercise 6.2).

(2) Despite Remark (1), even if E is a Banach space (one of the most favourable situations), there will in general exist in E closed vector subspaces L that admit, relative to E, no topological complement whatsoever; see Newman [1] and also Exercise 7.8.

(3) We notice that if a TVS E is an algebraic direct sum $L + M$, then the associated projectors, p_L and p_M, are open maps of E onto L and M, respectively. Indeed, since $p_L x = x$ for $x \in L$, so $p_L(U) \supset p_L(U \cap L) = U \cap L$ for any neighbourhood U of 0 in E. If also the direct sum decomposition is topological, p_L (resp. p_M) is a continuous, open map of E onto L (resp. M).

1.9.9 Closed and Preclosed Linear Maps. In this section we turn our attention to the case of a linear map u defined on a vector subspace $\mathrm{Dom}\, u$ of one TVS E into a TVS F. Let $\mathrm{Gr}\, u \subset E \times F$ be the graph of u (defined in Subsection 1.3.3).

We shall say that u is *closed* (from E into F) if $\mathrm{Gr}\, u$ is a closed vector subspace of $E \times F$. This depends *a priori* on both E and F and their topologies (and not merely on the topologies they induce on $\mathrm{Dom}\, u$ and $\mathrm{Im}\, u$, respectively). Any admissible narrowing of E or of F, as also any refinement of their topologies, can only make it easier for u to be closed.

Alongside this idea it is useful to introduce a slightly less exacting condition. u is said to be *preclosed* (from E into F) if there exists at least one closed extension of u—that is, at least one map v with domain in E and range in F that extends u in the sense that $\mathrm{Dom}\, u \subset \mathrm{Dom}\, v$ and $u = v \mid (\mathrm{Dom}\, u)$, and that is additionally closed from E into F. Clearly, if one such v exists, there is a smallest such closed extension of u—namely, that map whose graph is $\overline{\mathrm{Gr}\, u}$ (closure in the product space $E \times F$). This map is necessarily linear and is termed the *closure* of u and is denoted by $\bar{u}$. The existence of $\bar{u}$ and its domain, when it exists, depends upon E and F and their respective topologies. Any restriction of a preclosed map is preclosed.

Obviously, if u is closed it is *a fortiori* preclosed and its closure $\bar{u}$ is just u itself.

Notice also that if F is not separated, no linear map u is preclosed from E into F. For if $y \neq 0$, $y \in \overline{\{0\}}$, then $(0, y) \in \overline{\{(0, 0)\}} = \overline{\{(0, u(0))\}} \subset \overline{\mathrm{Gr}\, u}$; which shows that $\overline{\mathrm{Gr}\, u}$ is not a graph.

Since we shall have a good deal to say in the sequel about closed linear maps, we shall set forth here a number of important properties of them.

1.9.10 Proposition. Let E and F be TVSs, u a linear map with domain in E and range in F. Then u is closed from E into F if and only if the following condition is fulfilled: If (x_i) is a directed family extracted from $\mathrm{Dom}\, u$ that converges to x in E, and if the directed family $(u(x_i))$ converges to y in F, then necessarily $x \in \mathrm{Dom}\, u$ and $y = u(x)$. In the special case in which $\mathrm{Dom}\, u = E$, it is equivalent to demand that the above condition be satisfied under the additional hypothesis that $x = 0$ (the conclusion being simply that y is necessarily 0). If E and F are metrizable, it is enough that the preceding condition be satisfied with denumerable sequences in place of directed families.

Proof. This is obvious, the stated condition being necessary and sufficient in order that $\mathrm{Gr}\, u$ be closed in $E \times F$. ▌

1.9.11 Proposition. Let E and F be TVSs, u a linear map with domain in E and range in F. Either of the following conditions is necessary and sufficient in order that u be preclosed from E into F:

(1) If (x_i) is a directed family extracted from $\mathrm{Dom}\, u$ that converges in E to 0, and if $(u(x_i))$ is convergent in F, then the limit of $(u(x_i))$ in F is 0. As before,

if E and F are metrizable, denumerable sequences may replace the directed families.

(2) If (U_j) is any neighbourhood base at 0 in E, then

$$\bigcap_j \overline{u(U_j \cap \operatorname{Dom} u)} = \{0\}.$$

Proof. The necessity of (1) is clear; for if (x_i) has the stated properties, and if $y = \lim u(x_i)$, then $(0, y)$ belongs to the closure of Gr u. Since this closure is assumed to be a graph (u being assumed to be preclosed), it follows that y can only be 0.

Suppose conversely that (1) is satisfied. We have to show that if $(x, y_n) \in \overline{\operatorname{Gr} u}$ for $n = 1, 2$, then $y_1 = y_2$. Now to say that $(x, y_n) \in \overline{\operatorname{Gr} u}$ for a given n signifies that there exists a directed family $(x_{i,n})$ (indexed by i) of points of Dom u such that $\lim_i x_{i,n} = x$ and $\lim_i u(x_{i,n}) = y_n$. By subtraction of these relations pertaining to $n = 1$ and to $n = 2$, respectively, we obtain a directed family (x_j) of points of Dom u (in general with a different index set) such that $\lim x_j = 0$ and $\lim u(x_j) = y_1 - y_2$. But then (1) shows that $y_1 - y_2 = 0$, as we wished to show.

Thus (1) is necessary and sufficient.

Let us next show that (2) is necessary. Suppose then that $\overline{\operatorname{Gr} u}$ is a graph. If y belongs to the intersection mentioned in (2), and if V is any neighbourhood of 0 in F, we shall have that $y + V$ meets $u(U_j \cap \operatorname{Dom} u)$ for each j; that is, there exists $x_j \in U_j \cap \operatorname{Dom} u$ such that $-y + u(x_j) \in V$. Hence $(x_j, u(x_j)) \in (0, y) + (U_j \times V)$. From this it appears that $(0, y)$ is adherent to Gr u. Since $\overline{\operatorname{Gr} u}$ is a graph, y must be 0. This establishes the necessity of (2).

Finally let us show that (2) is sufficient. Since $\overline{\operatorname{Gr} u}$ is a vector subspace of $E \times F$, it will be a graph if and only if $(0, y) \in \overline{\operatorname{Gr} u}$ entails $y = 0$. Now $(0, y) \in \overline{\operatorname{Gr} u}$ signifies that, whenever U and V are neighbourhoods of 0 in E and in F respectively, there exists x_1 in $U \cap \operatorname{Dom} u$ such that $y - u(x_1) \in V$. This in turn entails that $y \in u(U \cap \operatorname{Dom} u) + V$. This being so for each V, therefore $y \in \overline{u(U \cap \operatorname{Dom} u)}$. Hence finally y is seen to belong to the intersection appearing in (2), and must therefore be 0. ▌

1.9.12 **Proposition.** Suppose that E and F are TVSs and that u is a linear map with domain in E and range in F. The following statements are true:

(1) If Dom u is closed in E, and if u is preclosed, then u is closed.

(2) If F is separated and u is continuous, then u is preclosed.

(3) If F is separated, if Dom u is closed in E, and if u is continuous, then u is closed.

(4) If E is separated and F is separated and complete, and if u is closed and continuous, then Dom u is closed in E.

(5) If u is one-to-one and preclosed (resp. closed), then u^{-1} is preclosed (resp. closed)—from F into E, that is.

(6) If E is separated and complete and F is separated, if u is closed, and if u^{-1} exists and is continuous, then Im u is closed in F.

Proofs. (1). This follows by combining Propositions 1.9.10 and 1.9.11.

(2) and (3). These statements follow from Proposition 1.9.11 and 1.9.10, respectively.

(4). Suppose that x belongs to $\overline{\text{Dom } u}$. Then there exists a directed family (x_i) of points of Dom u converging in E to x. Since u is continuous, the family $(u(x_i))$ is Cauchy; and so, F being complete, $y = \lim_i u(x_i)$ exists in F. Then, however, since u is closed, x must belong to Dom u.

(5). This statement is evident since Gr u^{-1} is simply the image, by the map $(x, y) \to (y, x)$ of $E \times F$ into $F \times E$, of Gr u; and the said map is a homeomorphism.

(6). This follows on combining (4) and (5), taking account of the obvious remark that Im u = Dom u^{-1}. ▌

1.9.13 **Remark.** From statement (3) it follows in particular that if F is separated, any $u \in L_c(E, F)$ is closed. There is a very important theorem, known as the "closed graph theorem," which states that for certain types of TVSs E and F, the converse is also true. We shall later devote a great deal of attention to this converse (see Sections 6.4–6.7, and 8.9).

Before turning to some examples we shall record one more proposition that frequently obviates the need for repetitive proofs and that, when applied to one example, often leads to others without further effort.

1.9.14 **Proposition.** Let (E, s) and (F, t) be TVSs and u a linear map with domain $D \subset E$ and range in F. Suppose that (E_0, s_0) and (F_0, t_0) are TVSs such that $E_0 \subset E$, $s_0 \geq s \mid E_0$ and $F_0 \subset F$, $t_0 \geq t \mid F_0$. Finally let u_0 be the restriction of u to $E_0 \cap D \cap u^{-1}(F_0)$, regarded as a linear map with domain in E_0 and range in F_0. Then the following assertions are valid.

(1) If u is preclosed, so too is u_0.

(2) If u is closed, and if further $u(D) \subset F_0$ (so that the domain of u_0 is just $E_0 \cap D$), then u_0 is closed.

Proof. If $(X, \mathfrak{T})$ is a topological space and A a subset of X, we shall temporarily denote by $Cl_{\mathfrak{T}} A$ the closure of A with respect to the topology $\mathfrak{T}$.

(1) Let G_0 = Gr u_0 and G = Gr u, so that

$$G_0 \subset G \cap (E_0 \times F_0).$$

Then

$$\begin{aligned} Cl_{s_0 \times t_0} G_0 &\subset Cl_{(s \times t) \mid E_0 \times F_0} G_0 \\ &= (E_0 \times F_0) \cap Cl_{s \times t} G_0 \\ &\subset (E_0 \times F_0) \cap Cl_{s \times t} G. \end{aligned}$$

By hypothesis, $Cl_{s \times t} G$ is a graph. It follows at once that $Cl_{s_0 \times t_0} G_0$ is also a graph, so that u_0 is preclosed.

(2) In this case one has $G_0 = (E_0 \times F_0) \cap G$ and also $Cl_{s \times t} \; G = G$. So the relations established in (1) lead to

$$Cl_{s_0 \times t_0} G_0 \subset (E_0 \times F_0) \cap G = G_0,$$

showing that u_0 has a closed graph. ▌

We turn now to the consideration of some examples. To begin with, the circumstances described in Proposition 1.9.12 (2) [resp. (3)] frequently arise in analysis and each such case provides an example of a preclosed [resp. closed] linear map. Again, in Chapter 6, in connection with the closed graph theorem, diverse examples of closed linear maps will be recorded (see also Exercise 8.10). In the majority of the examples obtained this way the linear maps are either visibly continuous, or turn out to be so on account of the closed graph theorem. There is ample room in analysis for linear maps that are preclosed or closed and not continuous, however, and it is instructive to examine a few such cases. An especially important source of such instances is the study of linear differential operators of various kinds, and we shall confine ourselves to this source.

The case of ordinary linear differential operators (one independent variable only) is decidedly simpler than that of several independent variables, and we begin with that case.

1.9.15 **Example.** Let Ω be an open interval on the real axis. In Example 1.8.4 (4) we have defined the vector space $C(\Omega)$ formed of all continuous functions on Ω; in addition we described the topology of compact (or locally uniform) convergence on $C(\Omega)$. Here we shall take for our space F the vector space $C(\Omega)$ endowed with a weaker topology—namely, that of convergence in mean over each compact subinterval of Ω. This topology may be defined as the weakest that renders continuous each seminorm of the type

$$g \to \int_a^b |g(x)| \, dx,$$

$[a, b]$ being a compact subinterval of Ω. A base at 0 for this topology is formed of the sets of the type

$$\left\{g \in C(\Omega) : \int_a^b |g(x)| \, dx < \varepsilon\right\},$$

where $\varepsilon > 0$.

Assuming that m is a natural number we shall choose for E the vector space $C^{m-1}(\Omega)$ consisting of all functions f on Ω having $m - 1$ continuous derivatives. (The zeroth derivative of f is to mean f itself, so that $C^0(\Omega) = C(\Omega)$.) On $C^{m-1}(\Omega)$ we take its "natural" topology, the weakest that renders continuous each seminorm of the type

$$\operatorname{Sup} \{|f^{(p)}(x)| : x \in K, 0 \leq p \leq m - 1\},$$

where $f^{(p)}$ denotes the pth derivative of f and K a compact subset of Ω.

We note that both E and F are metrizable TVSs. Both are locally convex (see Section 1.10) and neither is normable, but we shall not use these facts.

Suppose now that $a_0, a_1, \cdots, a_m$ are continuous functions on Ω, a_m being nonvanishing on Ω. The linear map u to be considered is obtained by assigning a specific domain to the formal linear differential operator

$$\sum_{0 \le p \le m} a_p(x) \cdot \left(\frac{d}{dx}\right)^p.$$

Thus, we define the domain D of u to consist precisely of those functions $f \in E$ such that $f^{(m)}$ exists in Ω and is such that

$$\sum_{0 \le p \le m} a_p f^{(p)} \in F;$$

for such functions $f \in D$ we define

$$u(f) = \sum_{0 \le p \le m} a_p f^{(p)}.$$

Our aim is to show that u is closed from E into F.

To do this we must show that if (f_n) is a sequence extracted from D such that $\lim f_n = f$ exists in E and $\lim u(f_n) = g$ exists in F, then $f \in D$ and $u(f) = g$.

We arbitrarily fix a point x_0 of Ω and introduce the integral operator J on $C(\Omega)$ through the definition

$$Jf(x) = \int_{x_0}^{x} f(x')\, dx';$$

J is an endomorphism of $C(\Omega)$ and one may form the successive powers of J. Since a_m is continuous and nonvanishing on Ω, it is evident that if $f \in D$, then $f^{(m)}$ is continuous on Ω and that

$$f(x) = f(x_0) + (x - x_0)f'(x_0) + \cdots + (x - x_0)^{m-1} \frac{f^{(m-1)}(x_0)}{(m-1)!} + J^m f^{(m)}(x). \tag{1.9.1}$$

Consider now the formula (1.9.1) when f is replaced by f_n. Since

$$f_n^{(m)} = a_m^{-1}\left[u(f_n) - \sum_{p<m} a_p f_n^{(p)}\right],$$

and since $u(f_n) \to g$ in F and $f_n \to f$ in E, it follows that

$$\lim_n f_n^{(m)} = a_m^{-1}\left[g - \sum_{p<m} a_p f^{(p)}\right]$$

in the sense of the topology on F. But then

$$\lim_n J^m f_n^{(m)} = J^m \left\{a_m^{-1}\left[g - \sum_{p<m} a_p f^{(p)}\right]\right\}$$

pointwise (and even locally uniformly) on Ω. Thus the formula (1.9.1), with f_n replacing f, leads in the limit as $n \to \infty$ to the relation

$$f(x) = \sum_{p<m} (x - x_0)^p f^{(p)}(x_0)/p! + J^m \left\{a_m^{-1}\left[g - \sum_{p<m} a_p f^{(p)}\right]\right\}(x).$$

Now the expression $[\cdots]$ is a continuous function on Ω, and so the above formula shows that $f^{(m)}$ exists on Ω and that moreover

$$f^{(m)} = a_m^{-1}\left[g - \sum_{p<m} a_p f^{(p)}\right]$$

on Ω. In other words, $f \in D$ and $u(f) = g$, which is what we had to show. ▌

Some comments on a different approach are worth recording, especially in view of Example 1.9.17 to follow. Granted further conditions on the coefficients $a_0, \cdots, a_m$, together with a more refined study of the differential operator u, better results may be obtained. We suppose for this that $\Omega = (a, b)$ is bounded and consider the problem that results when homogeneous linear boundary conditions are to be applied. The precise form of these boundary conditions, which we shall denote temporarily by B, need not be closely specified at the moment. Without loss of generality it may be assumed that the system

$$u(f) = 0, \qquad f \text{ satisfies condition } B$$

has no solutions other than 0; that is, that zero is not an eigenvalue of the homogeneous system. (If this is not already the case, it can be arranged by replacing $u(f)$ by $u(f) + cf$, where c is a suitably chosen number.) Smoothness conditions on the coefficients $a_0, \cdots, a_m$ arrange that the solutions we consider are continuously extendible to the closure $[a, b]$ of Ω.

One may then introduce the Green's function for the problem, $G(x, y)$. This has the property that, for $y \in \Omega$, $x \to G(x, y)$ satisfies conditions B and also satisfies the equation $u(f) = 0$ throughout each of the intervals (a, y) and (y, b); at $x = y$ certain one-sided derivatives of $G(x, y)$ exhibit a suitable singular behaviour of such a type that if g is a given continuous function on $[a, b]$, the solution f of the system

$$u(f) = g, \qquad f \text{ satisfies } B$$

is given by the formula

$$f(x) = \int_\Omega G(x, y)g(y)\,dy, \tag{1.9.2}$$

the integral representing a function that, together with its first m derivatives, is continuously extendible to $[a, b]$. For details see, for example, Coddington and Levinson [1], Chapter 7.

Suppose now we take E and F to be $C([a, b])$ with the uniform norm, and D to be the set of $f \in E$ whose first m derivatives are continuously extendible to $[a, b]$ and that satisfy condition B. It is then a simple matter to show that u, regarded as having domain D, is closed from E into F. For if $f_n \in D$, $f_n \to f$ in E, and $u(f_n) = g_n \to g$ in F, and if we write down formula (1.9.2) with f_n and g_n replacing f and g, a limiting process $n \to \infty$ leads at once to the formula (1.9.2) itself with f and g reinstated. From this it appears that $f \in D$ and that $u(f) = g$.

Should it be desired to consider solutions unrestricted by the boundary conditions B, one might proceed as follows. Assuming that condition B is expressed in the form

$$L_1(f) = 0, \cdots, L_r(f) = 0,$$

where each L_s is a linear form involving the first $(m - 1)$ (one-sided) derivatives of f at a and b, one seeks solutions $e_1, \cdots, e_r$ of the homogeneous equation $u(f) = 0$ for which $L_s(e_t) = \delta_{st}$ for $s, t = 1, \cdots, r$. Then any solution f of $u(f) = g$ can be written in the form

$$f(x) = \sum_{s=1}^{r} L_s(f)e_s(x) + \int_\Omega G(x, y)g(y)\, dy. \tag{1.9.2*}$$

This being the case it would be necessary to refine the topology of E in such a way that each linear form L_s becomes continuous on E. Then the preceding argument can be adapted by use of (1.9.2*) in place of (1.9.2) to show that u, regarded as having for its domain the set D of $f \in E$ whose first m derivatives are continuously extendible to $[a, b]$, is again closed from E into F.

A treatment of first- and second-order operators u of the above type, framed in the context of Hilbert space (L^2-) theory, is developed in great detail in Stone [1], pp. 424–530.

Although it would appear at first sight that the method can be taken over for certain linear partial differential operators, it turns out that the details become somewhat more complex. This is due partly to the generally more singular nature of the Green's functions involved. (See Example 1.9.17.)

1.9.16 **Example.** Turning to consider linear partial differential operators, we begin with the simplest case—that of first-order operators of the type

$$u(f) = \sum_{k=1}^{n} a_k(x) \frac{\partial f}{\partial x_k} + b(x)f,$$

where $x = (x_1, \cdots, x_n) \in R^n$ and it is assumed that the given functions $a_1, \cdots, a_n$ and b are continuous on the domain Ω in R^n. A natural choice for E and F would be $C(\Omega)$, while for the domain D of u one might take $C^1(\Omega)$. We will show first that, quite unlike the situation prevailing when $n = 1$, u is in general not closed when these choices of E, F, and D are made and n exceeds 1.

One begins by choosing a function H in $C^1(\Omega)$ that satisfies throughout Ω the equation

$$\sum_{k=1}^{n} a_k(x) \frac{\partial H}{\partial x_k} = 0.$$

We consider the functions on Ω defined by

$$f_r = \varphi_r \circ H \qquad (r = 1, 2, \cdots),$$

where each φ_r is a continuously differentiable function of a real variable and $\lim_{r\to\infty} \varphi_r = \varphi$ exists locally uniformly. It is evident that each f_r belongs to

$C^1(\Omega)$. Moreover

$$\begin{aligned} u(f_r) &= \sum_{k=1}^{n} a_k(\varphi_r' \circ H) \frac{\partial H}{\partial x_k} + bf_r \\ &= (\varphi_r' \circ H) \sum_{k=1}^{n} a_k \frac{\partial H}{\partial x_k} + bf_r \\ &= bf_r, \end{aligned}$$

thanks to our choice of H. Thus it appears that $f_r \to f = \varphi \circ H$ in E and $u(f_r) \to bf$ in F. Despite this, f will in general not be continuously differentiable on Ω—that is, will not belong to D; in other words u in general is not closed.

For example, if we assume that there exists at least one point x_0 of Ω at which not all the first-order partial derivatives of H vanish, we may assume (by subtraction from H of a constant) that $H(x_0) = 0$. We choose then the φ_r so that their limit φ is such that $\varphi(\xi) = |\xi|$ throughout some neighbourhood of $\xi = 0$. It is then easily seen that $f = \varphi \circ H$ has no differential at x_0, and therefore cannot be of class C^1 throughout any neighbourhood of x_0. Indeed if the differential of H at x_0, which is known to exist since $H \in C^1(\Omega)$, is denoted by L, and if $\varphi \circ H$ had at x_0 a differential L^*, then one would have

$$H(x_0 + x) = L(x) + o(|x|)$$

and also

$$\varphi(H(x_0 + x)) = L^*(x) + o(|x|)$$

as $|x| \to 0$. Since $\varphi(\xi) = |\xi|$ for small ξ, it would follow that

$$|L(x) + o(|x|)| = L^*(x) + o(|x|)$$

as $x \to 0$. Since L and L^* are both linear forms on R^n, it follows first that $|L(x)| = L^*(x)$ for all x and then that $L = 0$. This would contradict the assumption that not all the first-order partial derivatives of H at x_0 are zero.

Despite this the operator u is, under the stated conditions, preclosed. To verify this we assume that $f_r \in D$ $(r = 1, 2, \cdots)$, that $\lim_r f_r = 0$ in E, and that $g_r = u(f_r)$ is such that $\lim_r g_r = g$ in F. It is necessary to show that $g = 0$.

To this end we choose and fix a point x_0 of Ω. We consider a continuous curve with the equation

$$x = \xi(t), \qquad 0 \leq t < \varepsilon,$$

satisfying $\xi(0) = x_0$ and $d\xi_k/dt = a_k(\xi(t))$ for $0 < t < \varepsilon$ and $k = 1, 2, \cdots, n$. For $0 < t < \varepsilon$ one has then

$$\begin{aligned} \left(\frac{d}{dt}\right) f_r(\xi(t)) &= \sum_{k=1}^{n} a_k(\xi(t)) \cdot \frac{\partial f_r}{\partial x_k} \\ &= g_r(\xi(t)) - b(\xi(t)) f_r(\xi(t)), \end{aligned}$$

whence it follows that

$$f_r(\xi(t)) - f_r(x_0) = \int_0^t [g_r(\xi(t')) - b(\xi(t')) f_r(\xi(t'))] \, dt'$$

for $0 < t < \varepsilon$. Letting $r \to \infty$ one is led to the conclusion

$$0 = \int_0^t g(\xi(t'))\, dt'.$$

The continuity of g now entails that $g(\xi(t')) = 0$ for $0 < t' < \varepsilon$ and thence, letting $t' \to 0$, that $g(x_0) = 0$. Since x_0 is an arbitrary point of Ω, this shows that $g = 0$.

1.9.17 **Example.** We shall consider here the especially important Laplace operator Δ in three dimensions acting on functions defined on a domain Ω in R^3. As is suggested by Example 1.9.15, Green's functions play an important role in the discussion. Since we shall not wish to involve any questions of the boundary behaviour of the functions concerned, we shall arrange the arguments so that the Green's functions are employed only for subdomains ω of Ω such that ω is bounded, $\bar{\omega} \subset \Omega$ and the boundary $\partial\omega$ of ω is smooth. We shall denote by $G(x, y)$ the Green's function of ω; see, for example, Kellogg [1], p. 236. It is necessary to consider first two results involving this Green's function.

(1) If $f \in C^2(\Omega)$, a standard application of Green's identities leads to the formula

$$f(x) = \iiint_\omega G(x, y)\, \Delta f(y)\, dy + \iint_{\partial\omega} f(y) \frac{\partial G(x, y)}{\partial \nu_y}\, d\sigma(y), \tag{1.9.3}$$

holding for $x \in \omega$. In this formula, dy denotes the volume (Lebesgue) measure on R^3, ν the outward normal to $\partial\omega$, and σ the surface measure on $\partial\omega$. Compare Kellogg [1], p. 219.

It is necessary for our purposes to consider (1.9.3) when it is known merely that the partials $\partial^2 f/\partial x_k^2$ $(k = 1, 2, 3)$ exist throughout Ω and $\Delta f = \sum_{k=1} \partial^2 f/\partial x_k^2$ is continuous on Ω. An approximation argument will show that (1.9.3) remains valid under these weaker assumptions. To see this we choose a function f^* that coincides with f on the ε-neighbourhood of $\bar{\omega}$ ($\varepsilon > 0$ being so small that this neighbourhood is contained in Ω), which is continuous on R^n, and which vanishes outside some bounded set. The approximation process is one of regularization. One takes a sequence (r_n) of positive functions in $C^\infty(R^n)$ such that $r_n(x) = 0$ if $|x| > \varepsilon/n$ and $\iiint r_n(x)\, dx = 1$. We consider the functions

$$f_n^*(x) = f^* * r_n(x) = \iiint f^*(y) r_n(x - y)\, dy = \iiint f^*(x - y) r_n(y)\, dy.$$

Each of these functions belongs to $C^\infty(R^n)$ and $f_n^*(x) \to f^*(x)$ locally uniformly, thanks to continuity of f^*; in particular, $f_n^*(x) \to f(x)$ uniformly on $\bar{\omega}$. Besides this, $\Delta f^* = \Delta f$ on a neighbourhood of $\bar{\omega}$ and Δf is continuous on ω. Since it can be shown that

$$\Delta f_n^*(x) = \iiint \Delta f^*(x - y) r_n(y)\, dy,$$

it follows that $\Delta f_n^*(x) \to \Delta f(x)$ uniformly on $\bar{\omega}$. Now (1.9.3) is true when f is replaced by f_n^* $(n = 1, 2, \cdots)$. If we let $n \to \infty$ and make use of the convergence properties just recorded, (1.9.3) itself follows as a consequence.

(2) Suppose that $g \in C(\Omega)$ satisfies a Lipschitz (or Hölder) condition at each point of Ω, by which we mean that to each point $x_0 \in \Omega$ there correspond numbers $\varepsilon > 0$, $\alpha > 0$ and $M > 0$ such that

$$|g(x) - g(x_0)| \leq M|x - x_0|^\alpha$$

for $|x - x_0| < \varepsilon$. We consider the function

$$f(x) = \iiint_\omega G(x, y) g(y)\, dy.$$

We claim that the second-order partial derivatives of f exist at points of ω and that the relation $\Delta f = g$ holds throughout ω. In fact, since $G(x, y)$ differs from $-[4\pi\,|x - y|]^{-1}$ by a function that is harmonic on Ω, this statement follows from Kellogg [1], p. 156, Theorem III.

We are now ready to consider the main problem. Take $E = C(\Omega)$ with its natural topology of locally uniform convergence on Ω; and for F take the vector subspace of $C(\Omega)$ formed of those functions in $C(\Omega)$ that satisfy a Lipschitz condition at each point of Ω, F being endowed with the topology induced on it by that of $C(\Omega)$. We consider the operator $u = \Delta$, with domain prescribed as the set D of $f \in E$ such that $\partial^2 f/\partial x_k^2$ exists for $k = 1, 2, 3$ and Δf is continuous and satisfies a Lipschitz condition at each point of Ω. Plainly, u maps D into F. We will show that u is closed from E into F.

Suppose in fact that $f_n \in D$ for $n = 1, 2, \cdots$, that $f_n \to f$ in E, and that $g_n = u(f_n) \to g$ in F. According to (1) one has for each n and each x,

$$f_n(x) = \iiint_\omega G(x, y)\, \Delta f_n(y)\, dy + \iint_{\partial\omega} f_n(y) \frac{\partial G(x, y)}{\partial \nu_y}\, d\sigma(y).$$

On letting $n \to \infty$ one obtains then for $x \in \omega$ the relation

$$f(x) = \iiint_\omega G(x, y) g(y)\, dy + \iint_{\partial\omega} f(y) \frac{\partial G(x, y)}{\partial \nu_y}\, d\sigma(y).$$

The surface integral on the right-hand side is, as a function of x, harmonic on ω. So, by (2), we find that the second-order partial derivatives of f exist at points of ω and that $\Delta f = g$ on ω. This itself shows that Δf is continuous on ω and satisfies a Lipschitz condition at each point of ω. This being true for any subdomain ω of Ω of the specified type, it follows that $f \in D$ and that $u(f) = g$, and thus that u is closed.

1.9.18 **Remarks.** Examples 1.9.16 and 1.9.17 suggest strongly that interesting problems arise in connection with the study of linear partial differential operators if one poses such questions as: Which linear partial differential operators with carefully prescribed domain are closed? Which are preclosed? Can one describe in concrete terms the closure of a given, preclosed linear partial differential operator? In most of the studies made so far the starting point is a given, formal linear partial differential operator of the type

$$P(x, \partial) = \sum_{|p| \leq m} a_p(x)\, \partial^p,$$

where $p = (p_1, \cdots, p_n)$ is an n-uple of nonnegative integers, $|p| = p_1 + \cdots + p_n$,

$$\partial^p = \frac{\partial^{|p|}}{\partial x_1^{p_1} \cdots \partial x_n^{p_n}},$$

and the a_p are given functions defined on a domain Ω in R^n. There arises then the question of assigning an initial domain to $P(x, \partial)$, the choice depending to some extent on the behaviour of the coefficients a_p, regarding $P(x, \partial)$ as a linear mapping from one function space E into another F, and then examining the closure in various senses. The majority of the current interest lies in cases in which for E and F one chooses various Lebesgue spaces $L^q(\Omega)$. The interested reader may consult Halperin [1], Fuglede [1] and the references there cited, especially Friedrichs [2] and [3], and Schwarz [1].

1.10 Locally Convex Topological Vector Spaces

By a *locally convex topological vector space* (known sometimes more briefly as "locally convex space" or even simply "convex space") we mean a TVS E in which there exists a neighbourhood base at 0 comprised of convex subsets of E. We shall consistently use the symbol "LCTVS" as an abbreviation for the title of this section.

The reason why locally convex spaces are so important will appear undisguised in Chapter 2. Suffice it to say here that the reason is basically that in a locally convex space, one is assured of the existence of "sufficiently many" open convex sets. Why this is important becomes clearer when the Hahn-Banach theorem is invoked, for it then follows that consequently there exist "sufficiently many" continuous linear functionals on E—enough in fact to separate the points of E, and indeed to delineate all the closed convex subsets of E. In practice this means that the structure of a LCTVS is pretty faithfully imaged, and largely discoverable and expressible, in terms of continuous linear forms. This situation is not realized for certain TVSs that are not locally convex—for example, the space in Subsection 1.8.4(7), and the space $L^p(0 < p < 1)$ mentioned in Subsection 1.10.2 below.

Reference to Subsection 1.8.4 will provide us with some immediate examples of LCTVSs: the examples under (1)–(4) are all locally convex; those under (5) are likewise locally convex, provided the space X is locally convex; in (6) we have a locally convex space provided $p \geq 1$. On the contrary, however, example (7) is *not* locally convex: it is easily seen in fact that there any nonvoid open convex set coincides with the whole space. (The reader should verify this as an exercise.)

These examples suggest that the topology of a LCTVS can be defined in terms of seminorms, and our first task will be to verify this.

Before embarking upon this, however, let us note these almost obvious facts: A subspace, a quotient space, and a product space of LCTVSs are in each case themselves locally convex.

1.10.1 Locally Convex Topologies and Seminorms. Thanks to Remark (5) of 1.8.2 it is easily seen that, in a LCTVS E, there exists always a neighbourhood base at 0 formed of sets that are closed (or open), convex, and balanced. (For example: given a neighbourhood U, we choose a closed and balanced neighbourhood $U_1 \subset U$, then a convex $U_2 \subset U_1$, and finally a closed and balanced $U_3 \subset U_2$; then the closure of the convex envelope of U_3 is contained in U and is, of course, a neighbourhood of 0.)

This being so, let E be any LCTVS, and let (U_i) be a neighbourhood base at 0 comprised of closed, convex, and balanced subsets of E. For each i, let p_i be the Minkowski gauge of U_i (see Section 1.6). Since U_i is closed, it is closed in rays and so Corollary 1.6.6 informs us that U_i is defined by the equation $p_i(x) \leq 1$; moreover, p_i is a seminorm on E. Obviously, each p_i is continuous for the given topology on E. By construction of the p_i, E's topology is that which has for a neighbourhood base at 0 the sets $p_i(x) \leq \varepsilon$; or again the sets

$$U(F, \varepsilon) = \{x \in E : p_i(x) \leq \varepsilon \text{ for } i \in F\},$$

where $\varepsilon > 0$ and where F is any finite set of indices i.

Conversely, if E is a vector space and (p_i) is a family of seminorms on E, then the sets $U(F, \varepsilon)$ are easily seen to constitute a neighbourhood base at 0 for a vector space topology on E making E into a LCTVS. We shall then say that (p_i) is a *defining family* of seminorms for the corresponding topology (or LCTVS). The reader will note that many different defining families may define the same topology.

Should it be desirable to work with open neighbourhoods of 0, one has merely to replace each wide inequality $p_i(x) \leq \varepsilon$ by the strict one, $p_i(x) < \varepsilon$.

The topology defined by a family (p_i) of seminorms is separated if and only if $\operatorname{Sup}_i p_i(x) > 0$ for each $x \neq 0$ in E.

If the LCTVS E has a topology definable by the family (p_i) of seminorms, and if L is a vector subspace of E, then the subspace L has a topology definable by the seminorms $p_i \mid L$; and, if the family (p_i) contains a majorant of each finite subfamily, the quotient TVS E/L has a topology definable by the seminorms

$$\dot{p}_i(x_c) = \operatorname{Inf} \{p_i(x) : x \in x_c\}.$$

THE STRONGEST LOCALLY CONVEX TOPOLOGY CARRIED BY A GIVEN VECTOR SPACE. Let E be a given vector space. Among all locally convex topologies carried by E, there is a strongest. This topology, t_∞ say, has a neighbourhood base at 0 formed of all convex, balanced, and absorbent subsets W of E. Although t_∞ seldom if ever occurs in the study of problems having their roots in analysis, save when E is of finite dimension, it has certain properties that make it useful on occasions in the general theory. The reader may in addition consult Kelley and Namioka [1], especially Problems 6I, 6D, 14D, and 20G.

Perhaps the foremost of the "extremal properties" of t_∞ is expressed by the assertion that every seminorm, and every linear form, on E is t_∞-continuous. Thus, for instance, if p is any seminorm on E, then the set defined by $p(x) < 1$

is convex, balanced, and absorbent, hence a t_∞-neighbourhood of zero, so that p is indeed continuous for t_∞.

The family of all seminorms on E accordingly forms a defining family for t_∞. However, it is instructive and useful to exhibit smaller defining families. Instances of these may be constructed in terms of algebraic bases for E, much as is the case for finite-dimensional spaces.

Let $(e_\lambda)_{\lambda \in L}$ be a chosen algebraic base for E. To each strictly positive function w on L we order a seminorm N_w on E, defined by the formula

$$N_w(x) = \sum_{\lambda \in L} w(\lambda)^{-1} \, |\hat{x}(\lambda)|,$$

where $\hat{x}$ denotes that scalar-valued function on L having a finite support and such that $x = \sum_{\lambda \in L} \hat{x}(\lambda) e_\lambda$. We aim to show that these seminorms N_w form a defining family for t_∞.

In view of what has already been said, all that remains to be established is this: Given any convex, balanced, and absorbent subset W of E, there exists w such that W contains the set

$$U_w = \{x \in E : N_w(x) \leq 1\}.$$

To construct such a function w, we note that the absorbent character of W entails that for each λ in L one may choose a number $w(\lambda) > 0$ such that $w(\lambda) e_\lambda \in W$. If $N_w(x) \leq 1$ for the w thus defined,

$$x = \sum \hat{x}(\lambda) e_\lambda = \sum w(\lambda)^{-1} \hat{x}(\lambda) \cdot w(\lambda) e_\lambda.$$

Since $w(\lambda) e_\lambda$ belongs to W for each λ, and since W is convex and balanced, it appears at once that x belongs to W. This is what we had to show.

Using the seminorms N_w, let us verify next that E is t_∞-complete. Assuming that (x_i) is a Cauchy directed family in E (for the uniformity defined by t_∞), let us observe first that, since the linear form $x \to \hat{x}(\lambda)$ is t_∞-continuous for each λ in L, therefore the limit $\xi(\lambda) = \lim_i \hat{x}_i(\lambda)$ exists finitely for each λ. The remaining task is to show that the function ξ on L thus defined has a finite support; and that if we then set $x = \sum_{\lambda \in L} \xi(\lambda) e_\lambda$, then $\lim_i x_i = x$ for t_∞. To do this we start from the Cauchy nature of (x_i) expressed in terms of the seminorms N_w, namely: given any w, there exists an index $i_0 = i_0(w)$ such that $i, j \geq i_0$ entails

$$N_w(x_i - x_j) \leq 1,$$

or explicitly

$$\sum_{\lambda \in L} w(\lambda)^{-1} \, |\hat{x}_i(\lambda) - \hat{x}_j(\lambda)| \leq 1 \qquad \text{for } i, j \geq i_0.$$

This inequality holds *a fortiori* if we replace L by any finite subset F thereof. Making this replacement, letting $j \to \infty$, and then allowing F to expand, we arrive at the conclusion that

$$\sum_{\lambda \in L} w(\lambda)^{-1} \, |\hat{x}_i(\lambda) - \xi(\lambda)| \leq 1 \qquad \text{for } i \geq i_0. \tag{1.10.1}$$

This implies *a fortiori* that

$$|\hat{x}_{i_0}(\lambda) - \xi(\lambda)| \leq w(\lambda) \qquad \text{for } \lambda \in L.$$

Putting F for the finite support of $\hat{x}_{i_0}$, we conclude that

$$|\xi(\lambda)| \leq w(\lambda) \qquad \text{for } \lambda \in L \backslash F. \tag{1.10.2}$$

This being the case for any w and a corresponding finite F, one may conclude that ξ has a finite support; for otherwise there would exist distinct indices λ_n $(n = 1, 2, \cdots)$ for which $\xi(\lambda_n) \neq 0$, in which case we could define $w(\lambda)$ to be $\frac{1}{2}|\xi(\lambda_n)|$ for $\lambda = \lambda_n$, $n = 1, 2, \cdots$, and to be 1 for all other indices λ, for which choice (1.10.2) would be false whatever the selection of the finite subset F of L. Thus ξ has a finite support and $x = \sum \xi(\lambda) e_\lambda$ is an element of E for which $\hat{x} = \xi$. Then (1.10.1) is equivalent to

$$N_w(x_i - x) \leq 1 \qquad \text{for } i \geq i_0.$$

Since this is the case for each w, $x = t_\infty\text{–}\lim_i x_i$ and the completeness of E is established.

Remark. In general the topology t_∞ should not be confused with either the strongest vector space topology on E, or with the so-called "finite topology" on E. In the latter the open sets U are precisely those such that $U \cap F$ is relatively open in F (for the unique, separated vector space topology on F) for each finite-dimensional vector subspace F of E. For a discussion of these topologies, see Kakutani and Klee [1].

1.10.2 **Seminormed and Normed Spaces.** Consider the case in which the family (p_i) is composed of a single member, say p. We may refer to the pair (E, p) as a *seminormed vector space*, or a *normed vector space* if p happens to be a norm (that is, $p(x) > 0$ whenever $x \neq 0$). In either case the one-element family (p) is a defining family for a locally convex topology on E, and we may say that the resulting LCTVS is *seminormable* or *normable.* This LCTVS is separated if and only if it is normable.

Subspaces and quotient spaces of seminormable spaces are seminormable; subspaces and quotients modulo closed subspaces of normable spaces are normable.

The earliest examples of function spaces were normable.

Referring again to Examples 1.8.4, $B(T)$ and $C_{bd}(T)$ are normable; $\mathscr{L}^p$ is seminormable if $p \geq 1$.

The separated space associated with a seminormable space is normable. In particular, that associated with $\mathscr{L}^p$ is the Lebesgue space usually denoted by L^p (see Chapter 4). If $0 < p < 1$, $\mathscr{L}^p$ is not even locally convex (see Subsection 4.16.3).

We add here a few more examples of normed spaces.

1.10.3 **Example: The Spaces $\ell^p(T)$, $c(T)$ and $c_0(T)$.** Let T be any set and let $p \geq 1$. $\ell^p(T)$ denotes the vector subspace of K^T formed of those functions

x for which

$$\sum_{t \in T} |x(t)|^p < +\infty.$$

$\ell^p(T)$ is endowed with the norm (see Section 4.2)

$$\|x\|_p = \left\{ \sum_{t \in T} |x(t)|^p \right\}^{1/p}.$$

Sometimes the symbol $\ell^\infty(T)$ is used to denote what we have hitherto (see Examples 1.8.4(3)) written as $B(T)$, the corresponding norm being written then as

$$\|x\|_\infty = \text{Sup}\,\{|x(t)| : t \in T\}.$$

No special interest attaches to the case in which T is finite; if card $T = n$, a natural number, the various LCTVSs $\ell^p(T)$ are all isomorphic with K^n (though naturally their respective norms are distinct).

If T is countably infinite we may identify it with N (the set of natural numbers) and $\ell^p(T)$ with the familiar sequence space $\ell^p = \ell^p(N)$ of pth power summable sequences.

The spaces $\ell^p(T)$ may be defined similarly for any $p > 0$, but if $0 < p < 1$ they are not locally convex (see Section 4.2).

Two subspaces of $\ell^\infty(T)$ are of importance. The first is the subspace $c(T)$ composed of those x in $\ell^\infty(T)$ that tend to finite limits at infinity—that is, for which there exists a number ξ, the limit of $x(t)$ as t tends to infinity, with the property that for each $\varepsilon > 0$ the set of points of T for which $|x(t) - \xi| > \varepsilon$ is finite. The second is the subspace $c_0(T)$ composed of those x in $c(T)$ for which the said limit at infinity takes the value 0. Unless the contrary is explicitly stated, these two spaces are endowed with the norm induced by that on $\ell^\infty(T)$. Each is a closed subspace of $\ell^\infty(T)$ (Exercise 1.4) and therefore each is a Banach space (see 1.10.5).

Some basic properties of the spaces $\ell^p(T)$, $c(T)$, and $c_0(T)$ are established in Exercises 1.1–1.4.

1.10.4 Example: The Hardy Spaces H^p and the Spaces HL^p. Let D denote the open unit disk in the complex plane, and let $0 < p < \infty$. We denote by H^p the set of functions f, holomorphic on D, for which

$$N_p(f) = \underset{0 \le r < 1}{\text{Sup}} \left[\frac{1}{2\pi} \int_{-\pi}^{\pi} |f(re^{i\theta})|^p \, d\theta \right]^{1/p}$$

is finite. For $p = \infty$ the corresponding condition is that

$$N_\infty(f) = \underset{|z|<1}{\text{Sup}} |f(z)|$$

be finite.

If $p \ge 1$, N_p is a norm on H^p. This is no longer the case when $0 < p < 1$, but then the inequality [see Hardy, Littlewood & Polya [1], (6.13–6); compare the inequality (4.2.4)]

$$\int |u + v|^p \le \int |u|^p + \int |v|^p$$

shows that $N_p^p(f-g)$ is a metric on H^p. The same inequality shows that H^p is a vector space when $0 < p < 1$; if $p \geq 1$, this conclusion derives from Minkowski's inequality (see Subsection 4.11.2)—namely,

$$N_p(f+g) \leq N_p(f) + N_p(g).$$

Thus H^p is a normed vector space when $p \geq 1$, and a metrizable TVS when $0 < p < 1$.

It has been shown by Livingstone [1] that H^p is not locally convex when $0 < p < 1$. Further study of the spaces H^p as TVSs has been carried out by Taylor [3] and Walters [1], [2]. Many of the deeper properties of these spaces H^p are related to the study of Fourier series; see Zygmund [1], 7.56 and [2], pp. 271–284. See also Hoffman [1].

Analogous spaces $H^p(D)$ may be defined for more general domains D, but the procedure is more complicated inasmuch as the circumferences $z = r$ have to be replaced by the level lines of the Green's function for D.

A closely related space associated with a domain D is that formed of the functions f, holomorphic on D, for which

$$n_p(f) = \left[\iint_D |f(z)|^p \, dx \, dy\right]^{1/p}$$

is finite. The integral is taken to be the supremum of $\iint_K |f|^p \, dx \, dy$ with K varying over any base for the compact subsets of D (cf. Chapter 4). The resulting space of functions is denoted by $HL^p(D)$. As before, $HL^p(D)$ is a vector space; n_p is a norm on $HL^p(D)$ if $p \geq 1$, and $n_p^p(f-g)$ is a metric if $0 < p < 1$.

If D is the open unit disk centre the origin,

$$\iint_D |f(z)|^p \, dx \, dy = \int_0^1 r \, dr \int_0^{2\pi} |f(re^{i\theta})|^p \, d\theta,$$

whence it follows that

$$n_p(f) \leq \pi^{1/p} N_p(f),$$

showing that $H^p \subset HL^p$.

The proof that the spaces H^p and HL^p are complete is based upon some inequalities. Suppose that $z_0 \in D$ and that $r > 0$ is so small that the closed disk $\Delta(z_0, r)$ with centre z_0, and radius r lies within D. By Cauchy's formula one has

$$|f(z_0)| \leq \frac{1}{2\pi} \int_0^{2\pi} |f(z_0 + re^{i\theta})| \, d\theta.$$

If $p \geq 1$, this yields via Hölder's inequality the relation

$$|f(z_0)|^p \leq \frac{1}{2\pi} \int_0^{2\pi} |f(z_0 + re^{i\theta})|^p \, d\theta. \tag{1.10.3}$$

The same inequality holds for any $p > 0$ by virtue of the fact that $|f|^p$ is subharmonic on D. If we multiply (1.10.3) by $r \, dr$ and then integrate with respect to r over the interval $[0, R]$ we get

$$\tfrac{1}{2}R^2 |f(z_0)|^p \leq \frac{1}{2\pi}\int_0^R r\,dr \int_0^{2\pi} |f(z_0 + re^{i\theta})|^p\,d\theta$$

$$= \frac{1}{2\pi}\iint_{\Delta(z_0,R)} |f|^p\,dx\,dy \leq (2\pi)^{-1}[n_p(f)]^p.$$

Hence

$$|f(z_0)| \leq (\pi R^2)^{-1/p} n_p(f).$$

Here we may take any $R < d(z_0, D^*)$, the distance of z_0 from the frontier D^* of D. So we have for $z \in D$

$$|f(z)| \leq [\pi d^2(z, D^*)]^{-1/p} n_p(f) \leq [d^2(z, D^*)]^{-1/p} N_p(f). \qquad (1.10.4)$$

Armed with (1.10.4), it is quite simple to verify that H^p and HL^p are complete; the details are given for the case $p = 2$ in Subsection 1.12.1(8) and the reader is invited in Exercise 1.5 to make sure that this case is typical.

Another consequence of (1.10.4) is the existence of many nontrivial continuous linear forms on H^p and on HL^p, even when $0 < p < 1$. This is in strong contrast with the situation for a general L^p-space with $0 < p < 1$, for which see Subsection 4.16.3.

1.10.5 Banach Spaces. By a *Banach space* we shall understand a normable TVS E that is complete. For our purposes it generally will not be important which one of the possible equivalent norms is chosen; in other words, we often regard a Banach space as a normable TVS rather than a normed vector space. However, examples are usually specified together with a norm—this being a matter of convenience.

These Banach spaces are the "espaces du type (B)" of Banach's book [1], and they are so known by many writers.

The spaces $B(T)$, $C(T)$ (T compact), and $C_{bd}(T)$ of Section 1.8.4(2) and (4) are Banach spaces; so also is the separated space L^p associated with $\mathscr{L}^p$ (Subsection 1.8.4(6)) and the space $\ell^p(T)$ of Subsection 1.10.3, provided in each case that $p \geq 1$.

A vector subspace of a Banach space is a normable TVS; and a closed vector subspace of a Banach space is a Banach space.

The product TVS of a *finite* family $(E_i)_{1\leq i\leq n}$ of Banach spaces is a Banach space when endowed with the topology defined by the norm

$$\|x\| = \sum_{i=1}^{n} \|x_i\|$$

for $x = (x_i)_{1\leq i\leq n} \in \prod_{1\leq i\leq n} E_i$. (There are, of course, many other equivalent norms that one might utilize to the same end.)

The quotient of a Banach space by a closed vector subspace thereof is a Banach space. Thus, suppose that E is a Banach space and L a closed vector subspace of E. We know already that E/L is a normed vector space, endowed with the norm

$$\|x_c\| = \text{Inf}\,\{\|x\| : x \in x_c\}.$$

In view of the lemma below, to show that E/L is complete, it is necessary and sufficient to prove that any series $\sum_n x_c(n)$ of elements of E/L which is such that $\sum_n \|x_c(n)\| < +\infty$ is convergent in E/L. Now, if $\sum_n \|x_c(n)\| < +\infty$, we can choose elements $x(n) \in x_c(n)$ such that $\sum_n \|x(n)\| < +\infty$. Then, since E is complete, the same lemma (or a familiar and direct argument) shows that $\sum x(n)$ is convergent in E. Let its sum be x, so that $\|x - \sum_{n \le r} x(n)\| \to 0$ as $r \to \infty$. Then

$$\left\| x_c - \sum_{r \le n} x_c(n) \right\| = \left\| \left(x - \sum_{r \le n} x(n) \right)_c \right\|$$
$$\le \left\| x - \sum_{r \le n} x(n) \right\| \to 0$$

as $r \to \infty$, showing that x_c is the sum of $\sum_n x_c(n)$ in E/L. It thus remains only to establish the following lemma.

Lemma. Let (F, p) be a seminormed vector space. In order that F be complete for the structure defined by p, it is necessary and sufficient that the following condition be fulfilled: If (y_n) is a sequence of elements of F such that $\sum_n p(y_n) < +\infty$, then the series $\sum y_n$ is convergent in F.

Proof. Suppose completeness obtains. The sequence (s_r) of partial sums $s_r = \sum_{n \le r} y_n$ is Cauchy by virtue of the inequality

$$p(s_{r+k} - s_r) = p\left(\sum_{r<n\le r+k} y_n \right) \le \sum_{r<n\le r+k} p(y_n)$$

and the assumed convergence of $\sum_n p(y_n)$. Hence (s_r) is convergent in F.

Conversely suppose that our condition is fulfilled and that (s_n) is any Cauchy sequence of elements of F. We can then define inductively natural numbers $n_k < n_{k+1}$ such that $y_k = s_{n_{k+1}} - s_{n_k}$ satisfies $p(y_k) \le 1/k^2$. Then $\sum_k p(y_k) < +\infty$. By hypothesis, therefore, the series $\sum_k y_k$ is convergent in F. This means that (s_{n_k}) is convergent in F. But then the Cauchy character of (s_n) shows that it too is convergent in F. ▌

1.10.6 **The Topological Dual of a Seminormed Space.** If (E, p) is a seminormed space, and if E' is the topological dual of E equipped with the topology defined by p, then E' consists precisely of those linear forms $x^* \in E^*$ that are bounded with respect to p—that is, for which

$$p^*(x^*) = \text{Sup}\,\{|\langle x, x^* \rangle| : x \in E, p(x) \le 1\}$$

is finite.

It is very easy to check that p^* is a norm on E', which we refer to as the *natural norm* on E'. It is not uniquely determined by the topology on E, but if we begin with different seminorms on E, say p and q, defining the same topology (so that p and q are equivalent), then p^* and q^* are equivalent; that is, there exists a number $c > 0$ such that

$$c \cdot p^* \leq q^* \leq c^{-1} \cdot p^*.$$

Accordingly p^* and q^* define the same topology on E'. This topology on E' is normable and is canonically associated with the seminormable TVS E.

It is important to observe that E' is complete for the structure defined by p^*, hence is a Banach space. To see this, we suppose that (x'_n) is a sequence of points of E' that is Cauchy for p^*. Then obviously the sequence $(\langle x, x'_n \rangle)$ of scalars is Cauchy for each x in E, so that the sequence (x'_n) is convergent to some x^* in E^* for the weak topology $\sigma(E^*, E)$ (see 1.11.1). On the other hand we know that given $\varepsilon > 0$, there exists $n_0 = n_0(\varepsilon)$ such that $p^*(x'_m - x'_n) \leq \varepsilon$ whenever $m, n \geq n_0$. Thus, if $B = \{x \in E : p(x) \leq 1\}$ is the closed unit ball in E, we have

$$|\langle x, x'_m \rangle - \langle x, x'_n \rangle| \leq \varepsilon \text{ for } x \in B,\ m, n \geq n_0.$$

If we hold x and $n \geq n_0$ fixed and let $m \to 0$, there follows the relation

$$|\langle x, x^* \rangle - \langle x, x'_n \rangle| \leq \varepsilon \text{ for } x \in B,\ n \geq n_0. \tag{1.10.5}$$

Taking $n = n_0$, and using the obvious remark that $\text{Sup}_n\, p^*(x'_n) = k < +\infty$, this yields $|\langle x, x^* \rangle| \leq k + \varepsilon$ for x in B, showing that $p^*(x^*) < +\infty$ and that therefore x^* belongs to E'. Denoting it by x' and returning to (1.10.5) we infer that $|\langle x, x' - x'_n \rangle| \leq \varepsilon$ for x in B, $n \geq n_0$. That is, $p^*(x' - x'_n) \leq \varepsilon$ for $n \geq n_0$. Thus $\lim_n x'_n = x'$ in E'. Completeness of E' is thereby established.

Remark. This result is a special case of Proposition 1.10.9 below.

1.10.7 **Kolmogorov's Normability Criterion.** Kolmogorov's criterion states (see [1]) that *a separated TVS E is normable if and only if there exists in E at least one nonvoid, bounded, open, convex set.* The necessity of this condition is quite evident. Conversely, suppose the condition to be fulfilled. By translation, it follows that there exists a bounded, open, convex neighbourhood of 0 in E. The closure of this set, say W, is a bounded, convex, closed neighbourhood of 0. There exists then a balanced neighbourhood V of 0 contained in W. Let U be the closed, convex envelope of V. Now the convex envelope of V is balanced and convex and contained in W, so that U is closed, balanced, convex, and contained in W, hence is bounded. So U is a bounded, closed, balanced, and convex neighbourhood of 0 in E.

Let p be the Minkowski gauge of U (Theorem 1.6.4). By Corollary 1.6.6, U has equation $p(x) \leq 1$, and p is a seminorm on E. We shall show that p defines the given topology on E. To do this, we have only to show that the sets λU, with $\lambda > 0$, form a neighbourhood base at 0 in E; and this in turn demands merely a proof that each neighbourhood U' of 0 in E contains some λU. But, U being bounded, $U \subset \alpha U'$ is true for some $\alpha > 0$ and so $U' \supset \alpha^{-1} U$. Normability of E is thus established. ∎

We notice that $\overline{\{0\}}$ is the set defined by $p(x) = 0$—that is, is the set $\bigcap \{\lambda U : \lambda > 0\}$. This in turn is equal to $\bigcap \{\lambda U' : \lambda > 0\}$ for any chosen bounded neighbourhood U' of 0 in E.

It thus follows that Kolmogorov's criterion may be supplemented to this extent:

▶ A TVS E is normable if and only if there exists in E at least one bounded convex neighbourhood of 0, say U, satisfying $\bigcap \{\lambda U : \lambda > 0\} = \{0\}$.

Other criteria for normability have been given recently by Williamson [1], [2].

1.10.8 Spaces of Continuous Linear Maps. We turn now to the consideration of various ways of topologizing vector subspaces of $L(E, F)$ when F is a given TVS. In most cases of interest, E too is a TVS, but a general procedure may be described without assuming this.

Given a subset A of E and a neighbourhood V of 0 in F, we write

$$W(A, V) = \{u \in L(E, F) : u(A) \subset V\}.$$

Supposing specified a set $\mathfrak{S}$ of subsets A of E, assumed to be such that the union of any two sets in $\mathfrak{S}$ is contained in a third, our aim is to consider vector subspaces H of $L(E, F)$ with a view to making H a TVS having a base at 0 composed of the sets $H \cap W(A, V)$, A ranging over $\mathfrak{S}$ and V over a base at 0 in F. The resulting structure may be referred to as "the topology (on H) of convergence uniform on the sets belonging to $\mathfrak{S}$," or briefly "the $\mathfrak{S}$-topology," denoted by $T_{\mathfrak{S}}$.

To ensure that one thus obtains a genuine vector-space topology on H, one must demand that each set $H \cap W(A, V)$ be absorbent in H. Now this will be the case if and only if $u(A)$ is a bounded subset of F for each A in $\mathfrak{S}$. Consequently, if $\mathfrak{S}$ be given, the largest choice of H is the vector subspace $L_{\mathfrak{S}}(E, F)$ of $L(E, F)$ comprising precisely those linear maps u in $L(E, F)$ that transform each member of $\mathfrak{S}$ into a bounded subset of F. Henceforth we shall denote by $T_{\mathfrak{S}}$ the associated topology on $L_{\mathfrak{S}}(E, F)$.

$T_{\mathfrak{S}}$ is separated provided F is separated and $\bigcup \{\lambda A : \lambda > 0, A \in \mathfrak{S}\}$ generates E. It is furthermore obvious that if $\mathfrak{S} \subset \mathfrak{S}'$, then $L_{\mathfrak{S}}(E, F) \supset L_{\mathfrak{S}'}(E, F)$ and $T_{\mathfrak{S}} \mid L_{\mathfrak{S}'}(E, F)$ is weaker than $T_{\mathfrak{S}'}$. On the other hand, $L_{\mathfrak{S}}(E, F)$ and $T_{\mathfrak{S}}$ are unaltered if we replace each A in $\mathfrak{S}$ by its balanced envelope; if F is locally convex, convexifying the sets A in $\mathfrak{S}$ also leaves $L_{\mathfrak{S}}(E, F)$ and $T_{\mathfrak{S}}$ unchanged.

An especially important case is that in which E (as well as F) is a TVS and $\mathfrak{S}$ comprises all the bounded subsets of E. We then write $L_b(E, F)$ in place of $L_{\mathfrak{S}}(E, F)$, and $T_{\mathfrak{S}}$ is spoken of as the *strong topology*. A linear map belonging to $L_b(E, F)$ is said to be *bounded*. $L_b(E, F)$ always contains $L_c(E, F)$ (the space of continuous linear maps of E into F).

Another extreme but important case is that in which $\mathfrak{S}$ comprises all one-point or all finite subsets of E, in which case $L_{\mathfrak{S}}(E, F) = L(E, F)$ and $T_{\mathfrak{S}}$ is simply the topology of pointwise (or simple) convergence.

The reader should note for future reference that a subset $\mathscr{F}$ of $L_{\mathfrak{S}}(E, F)$ is bounded (relative to the topology $T_{\mathfrak{S}}$) if and only if $\bigcup \{u(A) : u \in \mathscr{F}\}$ is a bounded subset of F for each A in $\mathfrak{S}$. Regarding this, see also Theorem 7.3.1 (3).

When F is a LCTVS, the topology $T_{\mathfrak{S}}$ is likewise locally convex.

1.10.9 **Proposition.** Suppose that $\bigcup \{\lambda A : \lambda > 0,\ A \in \mathfrak{S}\}$ generates E and that F is complete. Then $L_{\mathfrak{S}}(E, F)$ is complete (for the uniform structure defined by $T_{\mathfrak{S}}$).

Proof. Let (u_i) be a directed family in $L_{\mathfrak{S}}(E, F)$ which is $T_{\mathfrak{S}}$-Cauchy. Our hypothesis concerning $\mathfrak{S}$ then ensures that $(u_i(x))$ is a Cauchy family in F for each x in E, and the assumed completeness of F arranges that $u(x) = \lim_i u_i(x)$ exists in F for each x in E. It is evident that u, thus defined, is an element of $L(E, F)$.

The Cauchy character of (u_i) signifies that, given A in $\mathfrak{S}$ and any neighbourhood V of 0 in F, there is an index i_0 such that

$$u_j(x) - u_i(x) \in V$$

for x in A, $i \geq i_0$ and $j \geq i_0$. Assuming that V is closed in F, it follows that

$$u(x) - u_i(x) \in V \qquad \text{for } x \in A \text{ and } i \geq i_0.$$

In particular, therefore, $u(A) \subset u_{i_0}(A) + V$. Since $u_{i_0}(A)$ is bounded, it appears that $u(A)$ is bounded. Thus u belongs to $L_{\mathfrak{S}}(E, F)$. Furthermore we see now that $u - u_i$ belongs to $W(A, V)$ whenever $i > i_0$, so that $T_{\mathfrak{S}}\text{-}\lim_i u_i = u$. This establishes the alleged completeness of $L_{\mathfrak{S}}(E, F)$. ∎

1.10.10 **The Case of Seminormed Spaces.** We terminate this section with some remarks pertaining to the important special case in which both E and F are topologized by single seminorms p and q, respectively.

In this case, $L_b(E, F)$ and $L_c(E, F)$ are identical. Each in fact comprises precisely those u in $L(E, F)$ for which

$$n(u) \equiv \text{Sup}\,\{q(u(x)) : x \in E,\ p(x) \leq 1\}$$

is finite.

n is a seminorm on $L_c(E, F)$; and n is a norm if q is a norm. We notice also that if p and q are replaced by equivalent seminorms p' and q', then n is replaced accordingly by a seminorm n' equivalent to n. Thus we may say that $L_c(E, F)$ is a seminormable TVS whenever E and F are each of that type.

It is important to remark that $L_c(E, F)$ is complete whenever F is complete: this is a special case of Proposition 1.10.9.

It follows therefore that $L_c(E, F)$ is a Banach space if E is seminormable and F is a Banach space.

Convergence in $L_c(E, F)$, relative to the topology we have in mind, is often spoken of as "convergence in the uniform operator topology." This is not to be confused with convergence uniform on E, of course, but rather convergence uniform on each *bounded* subset of E.

1.11 Weak Topologies on Dual Spaces

The systematic treatment of this and related topics will be undertaken in Chapter 7, but it is convenient to have available a few definitions and properties. Our remarks specialize and amplify those of Subsection 1.10.8 to the case in which $F = K$.

1.11.1 The Topologies $\sigma(E^*, L)$. Suppose that E is a vector space, no topology on E being involved. Let L be a vector subspace of E. Denoting by E^* the algebraic dual of E, amongst those topologies on E^* relative to which each function $x^* \to \langle x, x^* \rangle$ is continuous, x being fixed but arbitrary in L, there is a weakest. This topology is denoted by $\sigma(E^*, L)$ and is termed the *weak topology on E^* generated by L*. $\sigma(E^*, L)$ is compatible with the vector space structure of E^* and in fact makes E^* into a LCTVS. A neighbourhood base at 0 in E^* is formed of the sets

$$U^*(F, \varepsilon) = \{x^* \in E^* : |\langle x, x^* \rangle| \leq \varepsilon \qquad \text{for } x \in F\},$$

where $\varepsilon > 0$ and F is a finite subset of L. A defining family of seminorms for $\sigma(E^*, L)$ is formed of the seminorms

$$p_F^*(x^*) = \text{Sup}\,\{|\langle x, x^* \rangle| : x \in F\},$$

F ranging over the finite subsets of L.

If we choose $L = E$, we speak simply of the *weak topology on E^**. This topology, $\sigma(E^*, E)$, is always separated. We notice that $E^* = L(E, K)$ and $\sigma(E^*, E)$ is none other than the $T_{\mathfrak{S}}$-topology in which $\mathfrak{S}$ comprises all finite subsets of E (see Subsection 1.10.8).

1.11.2 The Weak Topologies $\sigma(E, M)$. Suppose again that E is a vector space with algebraic dual E^*. We let M be a vector subspace of E^*. The method of Subsection 1.11.1 leads then to the topology $\sigma(E^{**}, M)$. Using the device of injecting E into E^{**} (Subsection 1.4.7), we obtain the induced topology $\sigma(E, M)$, spoken of as the *weak topology on E generated by M*. This has for a defining family of seminorms the functions

$$p_G(x) = \text{Sup}\,\{|\langle x, x^* \rangle| : x^* \in G\},$$

G ranging over the finite subsets of M.

Naturally, $\sigma(E, M)$ can equally well be defined without intermediate reference to E^{**} and the injection of E into E^{**}. It is simply the weakest topology on E relative to which each of the linear forms $x \to \langle x, x^* \rangle$, x^* being fixed but arbitrary in M, is continuous.

Unlike $\sigma(E^*, E)$, $\sigma(E, M)$ is not generally separated. $\sigma(E, E^*)$ is separated, however.

1.11.3 The Weakened Topology of a TVS. Let E be a TVS and let E' be its topological dual (see Subsection 1.9.4); E' is a vector subspace of E^* and so we have a corresponding topology $\sigma(E, E')$ on E; we refer to this as the *weakened topology* of E. The name is justified because $\sigma(E, E')$ is invariably weaker than the given (initial) topology on E.

In extreme cases, $\sigma(E, E')$ can be the zero topology; this happens whenever $E' = \{0\}$ (for example, if $E = \mathscr{L}^p$, $0 < p < 1$; or if E is the TVS defined in Subsection 1.8.4 (7)). It will follow from the Hahn-Banach theorem, however, that $\sigma(E, E')$ is separated whenever E is separated and locally convex (see Corollary 2.2.6).

Remark. Dual to $\sigma(E, E')$ we have $\sigma(E', E)$, the topology induced on $E' \subset E^*$ by $\sigma(E^*, E)$. $\sigma(E', E)$ is always separated.

At this point in the development we shall state and prove just one useful result concerning weak topologies.

1.11.4 Theorem. (1) If E is a vector space, a subset S of E^* is relatively compact for $\sigma(E^*, E)$ if and only if it is weakly bounded; that is,

$$\text{Sup}\,\{|\langle x, x^*\rangle| : x^* \in S\} < +\infty$$

for each x in E.

(2) If E is a TVS, a subset S of E' is relatively compact for $\sigma(E', E)$ provided it is equicontinuous.

Proof. (1) The necessity of the stated condition follows from the general principle according to which a continuous numerically valued function on a compact space is bounded.

To prove the condition sufficient, we have only to appeal to Tychonoff's theorem (see Section 0.2.17) and to the evident fact that E^*, endowed with the weak topology, is a closed subspace of K^E (taken with the product topology).

(2) Equicontinuity of S signifies the existence of a neighbourhood U of 0 in E such that $|\langle x, x^*\rangle| \leq 1$ for x in U and x^* in S. This property, as also that of being weakly bounded, is shared by the weak closure $\bar{S}$ of S in E^*. Moreover, thanks to equicontinuity, $\bar{S}$ is contained in E'. It remains only to appeal to (1). ∎

1.11.5 Remark. If E is seminormable, the equicontinuous subsets of E' are precisely those that are bounded with respect to the natural norm on E' (see Subsection 1.10.6). Combining this remark with part (2) of Theorem 1.11.4 we obtain several important statements about the weak compactness of weakly bounded subsets of various function spaces and spaces of measures. The details will appear in Chapter 4 (see Subsections 4.10.3, 4.12, and 4.16).

1.12 Inner Product Spaces, pre-Hilbert Spaces, Hilbert Spaces

In this section we shall collect together those parts of the elementary aspects of Hilbert spaces that we shall need in the remainder of this book, and that have at the same time proved themselves useful in applications of functional analysis. It is in fact surprising how useful such elementary results turn out to be. Nowhere in this book do we enter into the really detailed study of Hilbert spaces and linear operators therein. There are many excellent texts on these topics, of which we mention here a few: Stone [1]; Riesz and Nagy [1]; Cooke [2]; Halmos [2]; Schmeidler [1]; von Neumann [1]; Dunford and Schwartz [2]. For an excellent account of the algebraic bases, see [H], Chapter III: although this is concerned mainly with the finite-dimensional case, most of the concepts extend (though the results, or the proofs, or both, may not).

1.12.1 **First Definitions and Examples.** ([H]; §§ 23, 26–30). Let E be a vector space over $K(= R$ or $C)$. A K-valued function f on $E \times E$ is termed a *bilinear form on E* if it is linear in each argument; it is *symmetric* if $f(x, y) = f(y, x)$ identically for x, y in E. By a *sesquilinear form* (Grothendieck's terminology) *on E* is meant a K-valued function f on $E \times E$ such that $x \to f(x, y)$ is linear for fixed y and $y \to \overline{f(x, y)}$ is linear for fixed x. (The bar denotes the complex conjugate; the latter condition is sometimes expressed by saying that $y \to f(x, y)$ is *conjugate linear* for fixed x.) If $K = R$, bilinear and sesquilinear are equivalent qualifiers. A K-valued function f on $E \times E$ is termed *Hermitian* if $f(x, y) = \overline{f(y, x)}$ identically for x and y in E; if $K = R$, "Hermitian" and "symmetric" are synonyms.

An *inner product* (often otherwise termed a *scalar product*) on a vector space E over K is defined to be a Hermitian sesquilinear form f on E such that $f(x, x)$ (which is real because of the Hermitian property) is nonnegative for each x in E. The inner product is said to be *strictly positive* if $f(x, x) > 0$ for each $x \neq 0$ in E.

An *inner-product space* (IPS for short) will mean a pair consisting of a vector space E over K, together with a distinguished inner product on E. Usually we shall employ the symbol E to denote the IPS, and write the inner product in the form $(x \mid y)$. In [H], §§ 59, 60 the terms "unitary space" and "Euclidean space" are suggested for finite dimensional complex and real IPSs, respectively.

THE CAUCHY-SCHWARZ INEQUALITY. Suppose that $(x \mid y)$ is an inner product on E. By considering the Hermitian quadratic form $(\alpha x + \beta y \mid \alpha x + \beta y)$ in the scalar variables α and β, which form is always nonnegative, an elementary argument ([H], § 64) leads to the important inequality

$$|(x \mid y)| \leq (x \mid x)^{1/2}(y \mid y)^{1/2}. \tag{1.12.1}$$

A consequence of this is that $N(x) = (x \mid x)^{1/2}$ is a seminorm on E. Consequently any IPS can be viewed as a seminormable TVS.

A *pre-Hilbert space* is defined to be an IPS with a strictly positive inner product; that is, one for which the associated seminorm $N(x) = (x \mid x)^{1/2}$ is in fact a norm. In this case we shall usually write $\|x\|$ in place of $N(x)$.

Finally, a *Hilbert space* is a pre-Hilbert space that is complete for the associated norm. A Hilbert space is, therefore, a Banach space.

Some Examples and Remarks. (1) Let T be any set, p any positive function defined on T. Consider the subset $\ell^2(T, p)$ composed of those x in K^T for which $\sum_{t \in T} p(t) \, |x(t)|^2 < +\infty$. The Cauchy-Schwarz inequality for finite sums of scalars shows that $\ell^2(T, p)$ is a vector subspace of K^T. We form it into an IPS by defining

$$(x \mid y) = \sum_{t \in T} p(t) x(t) \overline{y(t)},$$

the series being automatically absolutely convergent (Cauchy-Schwarz inequality once again). The associated seminorm is thus

$$N(x) = \left\{ \sum_{t \in T} p(t) \, |x(t)|^2 \right\}^{1/2}.$$

N is a norm if and only if $p(t) > 0$ for each t in T.

We shall leave it to the reader to supply the proof of the fact that $\ell^2(T, p)$ is always complete. Consequently, $\ell^2(T, p)$ is a Hilbert space whenever $p(t) > 0$ for each t in T. Failing this, if we put T_0 for the set of t satisfying $p(t) > 0$, there is a natural linear map of $\ell^2(T, p)$ onto $\ell^2(T_0, p_0)$, where $p_0 = p \mid T_0$, which preserves the IPS structures involved and the kernel of which consists of those x in $\ell^2(T, p)$ that vanish on T_0.

If $p = 1$, $\ell^2(T, p)$ is simply $\ell^2(T)$ (see Subsection 1.10.3).

(2) It is a remarkable fact, the proof of which will appear later in this section (see Remark (e) following Corollary 1.12.5), that any Hilbert space is isomorphic as an IPS with $\ell^2(T)$ for a suitable set T. Moreover $\ell^2(T)$ and $\ell^2(T')$ are thus isomorphic if and only if T and T' have the same cardinal number. In this way we may associate a *Hilbert dimension* with any Hilbert space, the specification of which determines the Hilbert space up to isomorphism. Originally, Hilbert spaces were required to be separable (as topological spaces), and these are characterized by having Hilbert dimension equal to aleph-zero (the first infinite cardinal number).

(3) Any finite-dimensional vector space E can be made into an IPS in many ways. The results include the finite-dimensional Hilbert spaces (for which the Hilbert and algebraic dimensions agree).

(4) If E is any IPS with associated seminorm N, we may form the quotient space $E/\text{Ker } N$. This is a pre-Hilbert space. If we regard E and $E/\text{Ker } N$ as TVSs, the latter is simply the separated space associated with E (see Subsection 1.8.5); this is so because $\text{Ker } N$ is none other than the closure (in the TVS E) of $\{0\}$.

(5) From any pre-Hilbert space E one may obtain by completion a Hilbert space.

We turn now to a few more concrete examples.

(6) Apart from the Hilbert sequence space $\ell^2 = \ell^2(N)$, the best-known type of Hilbert space is undoubtedly that formed of square-integrable functions. The details are set forth in Chapter 4 (see especially Sections 4.11 and 4.12).

Meanwhile we may describe here some closely associated IPSs and pre-Hilbert spaces that do not involve any special knowledge of the Lebesgue integral.

Let us suppose T is a real interval, supposed bounded for simplicity. We take any bounded continuous function $p \geq 0$ on T and consider the set E of bounded continuous K-valued functions x on T for which $\int_T p(t) \, |x(t)|^2 \, dt < +\infty$. (Our assumptions on p and x, largely superfluous in reality, arrange that the integrals exist in the elementary Riemann sense.) The Cauchy-Schwarz inequality for integrals show that E is a vector subspace of K^T. We make it into an IPS by defining

$$(x \mid y) = \int_T p(t)x(t)\overline{y(t)} \, dt.$$

The whole process is clearly a "continuous" analogue of that described in (1) above for $\ell^2(T, p)$.

The IPS E constructed in this way is a pre-Hilbert space if and only if the set if zeros of p is nowhere dense in T. Unless $p = 0$, E is never complete; this is indeed one reason why the Riemann theory of integration is unsatisfactory.

It is clear that the construction can be repeated in several variables, replacing T by simple regions in a plane or in space. Much more wide-reaching extensions are covered by the material presented in Chapter 4.

(7) *The Hardy space* H^2 has been defined in Example 1.10.4 and has been shown to be a Banach space with the norm $\|f\| = N_2(f)$ given by the formula

$$\|f\| = \operatorname*{Sup}_{0 \leq r < 1} \frac{1}{2\pi} \int_{-\pi}^{\pi} |f(re^{i\theta})|^2 \, d\theta.$$

Now if f is expanded in a Taylor series about the origin,

$$f(z) = \sum_{n=0}^{\infty} c_n z^n \qquad (|z| < 1),$$

it is easily verified that

$$\frac{1}{2\pi} \int_{-\pi}^{\pi} |f(re^{i\theta})|^2 \, d\theta = \sum_{n=0}^{\infty} |c_n|^2 r^{2n}$$

for $0 \leq r < 1$. It follows that

$$\|f\|^2 = \sum_{n=0}^{\infty} |c_n|^2.$$

The convergence of the series on the right-hand side is indeed a necessary and sufficient condition that a function f, having the c_n as its Taylor coefficients, shall belong to H^2. The mapping $f \to (c_n)$ thus establishes an isometric isomorphism of H^2 onto ℓ^2. This in itself indicates that H^2 becomes a Hilbert space if we define the inner product

$$(f \mid g) = \sum_{n=0}^{\infty} c_n \bar{d}_n$$

whenever f and g have the Taylor coefficients c_n and d_n, respectively. Since the series appearing here is absolutely convergent, it appears that

$$(f \mid g) = \lim_{r \to 1-0} \frac{1}{2\pi} \int_{-\pi}^{\pi} f(re^{i\theta})\overline{g(re^{i\theta})}\, d\theta.$$

(8) *The space* $HL^2(D)$. Let D be a domain in the complex plane of the variable $z = x + iy$. As in Section 1.10.4, $HL^2(D)$ denotes the set of functions f that are holomorphic on D and such that

$$\iint_D |f(z)|^2 \, dx\, dy < +\infty.$$

We can form $HL^2(D)$ into an inner-product space by defining

$$(f \mid g) = \iint_D f(z)\overline{g(z)}\, dx\, dy.$$

Since the functions involved are continuous, the result is indeed a pre-Hilbert space. We shall now carry out the details of the proof that $HL^2(D)$ is complete.

To do this we call on inequality (1.10.4) of Section 1.10.4, rewriting it as

$$|f(z)| \leq \pi^{-1/2}\, d^{-1}\, \|f\|, \tag{1.12.2}$$

where $\|f\| = (f \mid f)^{1/2}$ and $d = d(z, D^*)$ is the distance of z from the frontier of D.

From (1.12.2) we infer at once that if (f_n) is a Cauchy sequence in $HL^2(D)$, then this sequence is also Cauchy for the structure of compact convergence (that is, locally uniform convergence) on D (Subsection 1.8.4 (4)). Consequently $f = \lim f_n$ exists locally uniformly on D. The limit function f is thus holomorphic on D. It remains to show that $f \in HL^2(D)$ and is the limit in this IPS of (f_n).

Since (f_n) is Cauchy, one has $\text{Sup}_n \|f_n\| = M < +\infty$. Let K be any compact subset of D. Then we have

$$\iint_K |f_n|^2 \, dx\, dy \leq \|f_n\|^2 \leq M^2.$$

Keeping K fixed and letting $n \to \infty$, locally uniform convergence shows that

$$\iint_K |f|^2 \, dx\, dy \leq M^2.$$

Since M is independent of K, this now establishes that f belongs to $HL^2(D)$:

$$\iint_D |f|^2 \, dx\, dy = \text{Sup}_K \iint_K |f|^2 \, dx\, dy \leq M^2 < +\infty.$$

Furthermore, if we repeat this type of argument with differences $f_m - f_n$ in place of f_n, and remember that $\lim_{m,n\to\infty} \|f_m - f_n\| = 0$, we arrive easily at the conclusion that $\lim_{n\to\infty} \|f - f_n\| = 0$, which is what we had to prove.

It is interesting to note that in $HL^2(D)$ we have a Hilbert space whose elements really are functions, and not equivalence classes of functions (as is the case of L^2). (In the latter case we obtain individual functions only when the measure involved is discrete—a case that "cheats" in that integrals are really sums anyway.)

Inequality (1.12.2), and its analogues for $HL^p(D)$ with $p \geq 1$, show that there exist many continuous linear forms on these spaces.

For further study of the spaces $HL^2(D)$, see Exercises 1.7–1.11.

1.12.2 **Orthogonality and Orthogonal Projection.** ([*H*], §§ 62–67). In an IPS E two elements x and y are said to be *orthogonal* if and only if $(x \mid y) = 0$; and two subsets A and B of E are said to be orthogonal if $(a \mid b) = 0$ whenever $a \in A$ and $b \in B$. If A is a subset of E, there is a largest subset of E orthogonal to A (to wit, the set of all elements x of E satisfying $(x \mid a) = 0$ for each a in A); we call this set the *orthogonal complement of A in E* and denote it by $A^\perp$. We notice that $A^\perp$ is always a closed vector subspace of E. A family $(x_i)_{i \in I}$ of elements of E is said to be orthogonal if $(x_i \mid x_j) = 0$ whenever $i, j \in I$ and $i \neq j$; sometimes we describe a subset A of E as orthogonal if the corresponding family is orthogonal—that is, if $(x \mid y) = 0$ whenever $x, y \in A$ and $x \neq y$. An *orthonormal family* (or *set*) is one that is orthogonal and such that, in addition, each member of the family (or set) satisfies $\|x\| = (x \mid x)^{1/2} = 1$. Thus for an orthonormal family (x_i) one has $(x_i \mid x_j) = \delta_{ij}$.

In dealing with the question of orthogonal projection we shall follow an account due originally to F. Riesz that takes for its starting point an extremal problem.

1.12.3 **Theorem.** Let E be an IPS, M a nonvoid convex subset of E that is complete (for the uniform structure induced by that of E). Given x in E, there exists at least one element m of M such that

$$\|x - m\| \leq \|x - y\| \qquad (y \in M);$$

when x and M are given, m is uniquely determined modulo $\overline{\{0\}}$. If, therefore, E is a pre-Hilbert space, m is uniquely determined when x and M are given.

Proof. This is based upon Apollonius' identity

$$\|c - a\|^2 + \|c - b\|^2 = \tfrac{1}{2} \|a - b\|^2 + 2 \|c - \tfrac{1}{2}(a + b)\|^2,$$

which may be established by straight-forward juggling with inner products and which is valid for arbitrary a, b, c in E.

Let $\delta = \operatorname{Inf} \{\|x - y\| : y \in M\}$. We choose a sequence (y_n) of points of M such that $\delta_n = \|x - y_n\| \downarrow \delta$. The above identity gives then

$$\|x - y_m\|^2 + \|x - y_n\|^2 = \tfrac{1}{2} \|y_m - y_n\|^2 + 2 \|x - \tfrac{1}{2}(y_m + y_n)\|^2.$$

Now $\frac{1}{2}(y_m + y_n)$ belongs to M, since M is convex, so that

$$\|x - \tfrac{1}{2}(y_m + y_n)\|^2 \geq \delta^2.$$

Hence

$$\|y_m - y_n\|^2 \leq 2(\delta_m^2 + \delta_n^2 - 2\delta^2),$$

which shows that (y_n) is Cauchy. By completeness of M, (y_n) is convergent to some element m of M. Then $\|x - m\| = \lim_{n \to \infty} \|x - y_n\| = \delta$. This establishes the existence of m.

If m' is another minimizing element of M, Apollonius' identity gives

$$\|x - m\|^2 + \|x - m'\|^2 = \tfrac{1}{2}\,\|m - m'\|^2 + 2\,\|x - \tfrac{1}{2}(m + m')\|^2$$

and so

$$\delta^2 + \delta^2 \geq \tfrac{1}{2}\,\|m - m'\|^2 + 2\delta^2.$$

Thus $\|m - m'\| = 0$ and so $m - m' \in \overline{\{0\}}$. If E is a pre-Hilbert space, then in fact $m = m'$. The proof is complete. ∎

This theorem justifies a definition: Given x and M in E, any element m of M with the property specified in Theorem 1.12.3 is called an *orthogonal projection of x onto M*; if E is a pre-Hilbert space, we may speak of *the* orthogonal projection of x onto M. Use of the term "orthogonal" in this connection is justified by the next result.

1.12.4 **Proposition.** Let E be an IPS, M a convex subset of E, and x an element of E. Then:

(1) If $0 \in M$, an element m of M is an orthogonal projection of x onto M if and only if

$$\operatorname{Re}\,(x - m \mid y - m) \leq 0 \qquad \text{for } y \in M. \tag{1.12.3}$$

(2) If $0 \in M$, and if $M + M \subset M$, an element m of M is an orthogonal projection of x onto M if and only if

$$\text{(a)}\ \operatorname{Re}\,(x - m \mid y) \leq 0 \qquad \text{for } y \in M$$

and

$$\text{(b)}\ \operatorname{Re}\,(x - m \mid m) = 0.$$

(3) If M is a vector subspace of E, an element m of M is an orthogonal projection of x onto M if and only if $x - m \in M^{\perp}$.

(4) If M is a linear variety in E, an element m of M is an orthogonal projection of x onto M if and only if

$$(x - m \mid y - m) = 0 \qquad \text{for } y \in M.$$

Proof. (1) We have as an identity

$$\begin{aligned}\|x - y\|^2 &= \|(x - m) - (y - m)\|^2 \\ &= \|x - m\|^2 - 2\operatorname{Re}\,(x - m \mid y - m) + \|m - y\|^2.\end{aligned} \tag{1.12.4}$$

Hence m is minimizing, that is,

$$\|x - y\|^2 \geq \|x - m\|^2 \qquad \text{for } y \in M \tag{1.12.5}$$

holds, if and only if

$$2\operatorname{Re}\,(x - m \mid y - m) \leq \|y - m\|^2 \qquad \text{for } y \in M.$$

Now if $y_1 \in M$ and $0 < \lambda \leq 1$, then $y = m + \lambda(y_1 - m) \in M$, since M is convex and contains 0. So, if m is minimizing, we must have

$$2\lambda\operatorname{Re}\,(x - m \mid y_1 - m) \leq \lambda^2\,\|y_1 - m\|^2.$$

Dividing by λ and then letting $\lambda \to 0$, we conclude that

$$\operatorname{Re}\,(x - m \mid y - m) \leq 0$$

for any y in M, which is (1.12.3). Conversely, if (1.12.3) holds, then (1.12.4) shows that (1.12.5) is valid—that is, that m is minimizing.

(2) If m is minimizing then, by (1), we must have

$$\operatorname{Re}(x - m \mid y - m) \leq 0$$

for y in M. If $y_1 \in M$, our hypotheses on M ensure that $y = y_1 + m \in M$, thus (a) follows. On the other hand, taking $y = 0$ in (1.12.3), we derive

$$\operatorname{Re}(x - m \mid m) \geq 0;$$

so in fact (b) is satisfied. Conversely, if (a) and (b) are true, then obviously (1.12.3) is fulfilled.

(3) This follows directly from (2), since now $-M = M$.

(4) We must start again from the condition

$$2\operatorname{Re}(x - m \mid z - m) \leq \|z - m\|^2 \qquad \text{for } z \in M,$$

characterizing $m \in M$. In this case, since M is a linear variety, $z = \alpha m + (1 - \alpha)y \in M$ whenever $y \in M$ and α is an arbitrary scalar. Then $z - m = (1 - \alpha)(y - m)$ and our condition reads

$$2\operatorname{Re}(x - m \mid (1 - \alpha)(y - m)) \leq |1 - \alpha|^2 \, \|y - m\|^2,$$

that is,

$$2\operatorname{Re} \lambda(x - m \mid y - m) \leq |\lambda|^2 \, \|y - m\|^2$$

for y in M and λ any scalar.

Taking λ real and positive, the argument used in the proof of (1) shows that consequently

$$\operatorname{Re}(x - m \mid y - m) \leq 0 \qquad \text{for } y \in M.$$

But in addition we may take λ to be real and negative, and then a similar procedure (division by λ followed by passage to the limit as $\lambda \to 0$) leads to $\operatorname{Re}(x - m \mid y - m) \geq 0$ for y in M. Hence, in fact

$$\operatorname{Re}(x - m \mid y - m) = 0 \qquad \text{for } y \in M.$$

If the scalar field is real, we have already reached the end. Otherwise, we can go through the preceding arguments taking λ to be pure imaginary and so conclude that

$$\operatorname{Im}(x - m \mid y - m) = 0 \qquad \text{for } y \in M.$$

Combining this with the preceding conclusion yields

$$(x - m \mid y - m) = 0 \qquad \text{for } y \in M.$$

This establishes the necessity of the stated condition. Its sufficiency is clear, by virtue of the identity (1.12.4). ▌

1.12.5 **Corollary.** (1) If E is an IPS and M a complete vector subspace of E distinct from E, then there exists $a \neq 0$ in E orthogonal to M. If further M is closed, then we may choose a so that $\|a\| > 0$.

(2) Let E be a complete IPS, and let $(e_i)_{i \in J}$ be any orthonormal family in E. Then there exists an orthonormal family $(e_i)_{i \in I}$ in E that extends the given one (so that I is a superset of J) and that is *complete* in the following sense: given x in E, there is precisely one family of scalars $(\xi_i)_{i \in I}$ for which the series $\sum_{i \in I} \xi_i e_i$ is convergent (or even merely weakly convergent) to x in E; this family is given by $\xi_i = (x \mid e_i)(i \in I)$, the series $\sum_{i \in I} \xi_i e_i$ is then unconditionally convergent (that is, convergent following the directed set of finite subsets of I), any sum x_∞ satisfies $x - x_\infty \in \overline{\{0\}}$, and one has the Parseval formula

$$\|x\|^2 = \sum_{i \in I} |\xi_i|^2 = \sum_{i \in I} |(x \mid e_i)|^2. \tag{1.12.6}$$

Proof. (1) Take any x in $E \setminus M$, let m be an orthogonal projection of x onto M, and take $a = x - m$. If $\|a\|$ were 0, then the relation $x - a \in M$ would entail that $x \in \bar{M}$, which is a contradiction if M is closed.

(2) The set of all orthonormal sets extending the given one may be partially ordered by inclusion (or extension) and is thereby inductively ordered. (The simple verification is left to the reader.) By Zorn's lemma (Section 0.1.5) there is a maximal element of this set, say $(e_i)_{i \in I}$. It is easily shown that the closed linear envelope M of the set $\{e_i : i \in I\}$ coincides with E; otherwise, in fact, part (1) shows that we could extend the said family, contradicting thereby its maximality.

Now we let $x \in E$. If any series $\sum_{i \in I} \xi_i e_i$ converges weakly in E to x, we should have

$$(x \mid e_j) = \sum_{i \in I} \xi_i (e_i \mid e_j)$$

for each $j \in I$, which, by orthonormality, yields $\xi_j = (x \mid e_j)$ for each j.

Suppose then that the ξ_i are so defined. For any finite $S \subset I$ we have (again by orthonormality), putting $x_S = \sum_{i \in S} \xi_i e_i$,

$$\|x - x_S\|^2 = \|x\|^2 - \sum_{i \in S} |\xi_i|^2. \tag{1.12.7}$$

Hence

$$\|x_S\|^2 = \sum_{i \in S} |\xi_i|^2 \le \|x\|^2,$$

showing that

$$\sum_{i \in I} |\xi_i|^2 \le \|x\|^2.$$

Thus $\sum_{i \in I} |\xi_i|^2$ is convergent. Moreover, if $S \subset S'$, then

$$\|x_{S'} - x_S\|^2 = \sum_{i \in S' \setminus S} |\xi_i|^2$$

and so, by convergence of the series $\sum |\xi_i|^2$, we are led to conclude that the family (x_S) is Cauchy. Completeness of E shows that the series $\sum_{i \in I} \xi_i e_i$ is convergent in E; let x_∞ be any sum of this series. By convergence of the series defining x_∞, and by orthonormality, we find that $(x - x_\infty \mid e_i) = 0$ for all i. Since the linear envelope of the e_i is, as we have seen, dense in E, it follows that $x - x_\infty$ is orthogonal to all elements of E, that is, that $\|x - x_\infty\| = 0$, or $x - x_\infty \in \overline{\{0\}}$. This being so, (1.12.7) yields in the limit the relation (1.12.6). ∎

Remarks. (a) If E is a pre-Hilbert space, a complete subspace of E is necessarily closed.

(b) Because of the results established in (2), an orthonormal family in E which is such that its closed linear envelope coincides with E may sensibly be termed an *orthonormal base* for E. We shall adopt this term rather than "complete orthonormal set" in view of the confusion possible between this use of the word "complete" and other uses.

(c) If E is separated, we may say that $x = \sum_i (x \mid e_i)e_i$ rather than merely $x \equiv \sum_i (x \mid e_i)e_i$ modulo $\overline{\{0\}}$.

(d) Given the relations $x \equiv \sum_i (x \mid e_i)e_i$ modulo $\overline{\{0\}}$ and $y \equiv \sum_i (y \mid e_i)e_i$ modulo $\overline{\{0\}}$, the Parseval formula can be generalized into the shape

$$(x \mid y) = \sum_{i \in I} (x \mid e_i) \cdot \overline{(y \mid e_i)}. \tag{1.12.8}$$

(e) If E is a Hilbert space, the map $x \to ((x \mid e_i))_{i \in I}$ gives an isomorphism of E onto the Hilbert space $\ell^2(I)$.

1.12.6 **Corollary.** Let E be a complete IPS, f a continuous linear form on E. Then there exists $a \in E$ such that

$$f(x) = (x \mid a) \qquad \text{for } x \in E.$$

For any such element a one has

$$\|a\| = \|f\| \equiv \text{Sup}\,\{|f(x)| : \|x\| \leq 1\}.$$

Consequently, if E is also separated (that is, if E is a Hilbert space), a is uniquely determined by f.

Proof. The existence of a is trivial if $f = 0$. Otherwise let $M = \text{Ker}\, f$; this is a closed and complete vector subspace of E, $M \neq E$. By Corollary 1.12.5, there exists $a_1 \in E$ with $\|a_1\| > 0$ which is orthogonal to M. Thus $f(x) = 0$ entails $(x \mid a_1) = 0$. We take a_2 satisfying $f(a_2) = 1$. Then, for any x we have $f(x - f(x)a_2) = 0$ and so $f(x) \cdot (a_2 \mid a_1) = (x \mid a_1)$. We cannot have here $(a_2 \mid a_1) = 0$ (since otherwise, taking $x = a_1$, the contradiction $\|a_1\| = 0$ would appear). Hence $f(x) = (x \mid a)$ with $a = a_1/(a_1 \mid a_2)$.

If a is any element of E such that $f(x) = (x \mid a)$ for all x, we may take $x = a$ to get

$$\|a\|^2 = (a \mid a) = f(a) \leq \|f\| \cdot \|a\|$$

and so $\|a\| \leq \|f\|$. The reverse inequality follows from the Cauchy-Schwarz inequality (see (1.12.1)). Thus indeed $\|a\| = \|f\|$.

The final statement of the corollary follows from this, for if both a and a' correspond to f, then $a - a'$ corresponds to the zero functional, hence $\|a - a'\| = 0$ and so, E being separated, $a = a'$. ▌

Remarks. As a consequence of Corollary 1.12.6 we may say that if E is any complete IPS, there exists a conjugate linear map J of E onto E' (the topological dual of E) such that

$$\langle x, J(y) \rangle = (x \mid y)$$

identically for x, y in E; J preserves norms. Very often when one is concerned solely with Hilbert spaces, one uses this map J in order to dispense entirely with any explicit mention of E'.

There are numerous applications of Corollary 1.12.6, many of which are covered by the so-called "projection principle," which we consider next.

1.12.7 **The Projection Principle.** The principles embodied, respectively, in Corollary 1.12.6 and in Theorem 1.12.3 as applied to closed vector subspaces are, as we shall verify in a moment, equivalent. Either of them may be referred to as the "projection principle;" and the method involving their use may be described as the "method of orthogonal projection." Which formulation is used in any particular instance will vary from case to case.

To verify the equivalence, we consider any IPS E and the following two assertions:

(1) If M is a closed vector subspace of E, any element x of E may be projected orthogonally onto M, that is, there exists an element m of M such that $x - m \in M^{\perp}$.

(2) If M is a closed vector subspace of E, any continuous linear form f on M can be represented as $f(x) = (x \mid a)$ for some a in M.

Since any closed vector subspace of E is a complete IPS (with the induced structure), the argument used to prove Corollary 1.12.6 shows that (1) entails (2). Conversely, we assume that (2) is satisfied. Let M be any closed vector subspace of E and x any element of E; then $y \to (y \mid x)$ is a continuous linear form on M. (2) being satisfied by hypothesis, there exists $a \in M$ such that $(y \mid x) = (y \mid a)$ for y in M, which says exactly that $x - a \in M^{\perp}$. Thus (2) entails (1). Therefore (1) and (2) are equivalent, as we alleged.

Remarks. Seemingly Hermann Weyl was among the first to isolate the projection principle and to show how useful it could be in potential theory and in the proof of existence theorems for linear partial differential equations (see Weyl [1]). Weyl concerned himself originally with the Dirichlet problem, but the method has since been modified in various ways and applied to more general problems. This sort of application is examined at considerable length in Chapter 5 (see especially Subsection 5.13.2). The difficulty which prevents us from dealing with this application at this point is that completeness of the spaces involved is achieved only if one uses the Lebesgue integral, a topic we leave until Chapter 4. Despite this, the broad outlines (neglecting questions of completeness and certain other subtleties) can be sketched in so rapidly that we yield to temptation.

Suppose Ω is a domain in R^2 (the case of n dimensions requires only formal changes) and f is a given function defined in and on the boundary of Ω. The original Dirichlet problem is concerned with the existence of a function u, harmonic on Ω, taking the same boundary values as f. For a long time it has been known that, formally, u is obtainable by minimizing the so-called Dirichlet integral $D(g) = \iint_\Omega |\text{grad } g|^2 \, dx \, dy$ when g ranges over all suitably restricted functions g taking the same boundary values as f. The difficulties in the path of making this approach rigorous are too well known to call for repetition. Let us see in broad outline how the method of orthogonal projection is likely to be useful.

We must first choose suitably an IPS E of functions on Ω, including f as one of its members, and such that $D(g)$ will coincide with the square of the norm of $g : D(g) = (g \mid g)$. (We shall consider only real-valued functions and treat E as a real vector space.) Such a choice of inner product would be

$$(g_1 \mid g_2) = \iint_\Omega \text{grad } g_1 \cdot \text{grad } g_2 \cdot dx \, dy.$$

In minimizing $D(g) = \|g\|^2$, that is, in minimizing $\|g\|$, the contending functions g are those that belong to the linear variety $M \subset E$ formed of those functions taking the same boundary values as f. But this problem is simply that of projecting 0 orthogonally onto M. If this task can be discharged, Proposition 1.12.4 (4) shows that the minimizing function, call it u, satisfies $(u \mid g - u) = 0$ for g in M; that is, $(u \mid h) = 0$ for any h in E having zero boundary values. Now we have formally by Green's identity

$$(u \mid h) = \iint_\Omega \text{grad } u \cdot \text{grad } h \cdot dx \, dy$$
$$= \iint_\Omega h \cdot \Delta u \cdot dx \, dy - \int_\Gamma h \cdot \frac{\partial u}{\partial n} \cdot ds$$

(Γ being the boundary of Ω and n the normal). If h has zero boundary values, then one has

$$(u \mid h) = \iint_\Omega h \cdot \Delta u \cdot dx \, dy.$$

If this is to vanish for all h with zero boundary values, then necessarily $\Delta u = 0$ throughout Ω; that is, u is harmonic in Ω. Thus we shall have in u a solution of the Dirichlet problem for the given boundary values defined by f.

Of the obvious serious "gaps" in this procedure we mention only three.

(a) In order to work with a *complete* IPS (which is essential) one is forced into completing the space of "smooth" functions g on Ω for the structure defined by the inner product defined above. But then we are confronted with the second difficulty.

(b) What sense is to be given to the term "boundary values" as applied to elements of the completion? Specifically, which elements of the completion shall be described as having zero boundary values? It turns out that these will have to be characterized as those belonging to the closure (in the said completion) of the set of functions that are smooth on Ω and that vanish on boundary strips (that is, on sets of the type $\Omega \backslash C$, where C is a variable compact subset of Ω). There is, of course, no assurance whatsoever that such a function has a pointwise boundary limit that is 0.

(c) In the preceding argument we have made use of Green's identity in the form

$$\iint_{\Omega} \operatorname{grad} u \cdot \operatorname{grad} h \cdot dx\, dy = \iint_{\Omega} h \cdot \Delta u \cdot dx\, dy.$$

We should need this to hold for the minimizing function u and for any h in E having zero boundary values in the generalized sense. Clearly, this demands careful investigation. For, to begin with, u is known only to be an element of the completion, so that even the existence of the Laplacian Δu lies in grave doubt. One of Weyl's major contributions is to show that the above integral relation does indeed entail that if we are prepared to correct u on a set of Lebesgue measure 0, then u has derivatives of all orders and Green's identity can be applied without further serious difficulty.

When we come to deal with linear partial differential equations in Chapter 5, we shall use a formulation in terms of Schwartz's distributions. This will allow us to separate clearly those parts of the problems that are readily accessible to abstract methods from those the solutions of which involve something of the nature of Weyl's lemma. At the same time we shall then contemplate a considerably more elaborate application of the projection principle.

1.12.8 **The Hilbert Adjoint of a Continuous Endomorphism.** In subsection 1.12.9 we shall give another application of the projection principle. This application will involve the concept of the Hilbert adjoint u^* of a continuous endomorphism u of a Hilbert space E, which concept may in turn illustrate the significance of Corollary 1.12.6. As we shall point out later, the Hilbert adjoint is a notion applicable to endomorphisms that are not continuous, and it is closely related to a slightly different idea of adjoint linked with endomorphisms of general LCTVSs. For the moment we are considering only the most straightforward situation.

Suppose then that E is a Hilbert space and u a continuous endomorphism of E. If y is a given element of E, the linear form

$$x \to (u(x) \mid y)$$

is then continuous on E. According to Corollary 1.12.6, therefore, there exists a unique element y^* of E such that

$$(u(x) \mid y) = (x \mid y^*)$$

for all x in E. It is evident that the mapping $u^* : y \to y^*$ is linear—that is, is itself an endomorphism of E. This u^* is termed the *Hilbert adjoint* of u. Throughout the remainder of this chapter, where no confusion can arise, we shall speak simply of the "adjoint of u" in place of the "Hilbert adjoint of u."

The defining equation of u^* is thus the identity

$$(u(x) \mid y) = (x \mid u^*(y))$$

for x and y in E. This shows at once that

$$\|u^*\| = \|u\|$$

and that

$$u^{**} = u,$$

where u^{**} means, of course, $(u^*)^*$. From the same source it appears that $u \to u^*$ is a conjugate-linear map of $L_c(E, E)$ onto itself.

In terms of this adjoint process one may single out two very important categories of continuous endomorphisms of E.

First of all there are the so-called *self-adjoint* (continuous) endomorphisms u, characterized by the relation $u^* = u$. Second there are the so-called *unitary* endomorphisms u, characterized by the relation $u^* = u^{-1}$. Thus, a continuous endomorphism u of E is self-adjoint if and only if

$$(u(x) \mid y) = (x \mid u(y))$$

for all x and y in E. (Actually, as we shall show much later in Section 8.11, continuity may be dropped since it is in fact a consequence of the remaining identity.) Analogously, a continuous endomorphism u of E is unitary if and only if it maps E onto itself and

$$(u(x) \mid u(y)) = (x \mid y)$$

for all x and y in E. A little juggling shows that it is equivalent to replace this last identity by the demand that $\|u(x)\| = \|x\|$ for all x in E—that is, that u be isometric (or normpreserving).

A highly detailed study of self-adjoint and unitary endomorphisms of a Hilbert space has been made and is included in many well-known books, the motivation consisting in part of their importance in many fields of applications. A bare knowledge of the meanings of these terms suffices for the program immediately ahead of us.

1.12.9 **The Projection Principle and the Abstract Ergodic Theorem.** Suppose again that E is a Hilbert space and u a continuous endomorphism of E. The abstract ergodic theorem we have in mind asserts that under suitable conditions the arithmetic means of the powers of u converge pointwise to a limit, necessarily a continuous endomorphism of E. Later, in Section 9.13, we shall consider a few of the many extensions of this type of result, but for a fuller and more detailed account the reader is referred to Dunford and Schwartz [1], pp. 657–730. We here confine ourselves to a proof for the Hilbert space set-up that depends solely on the projection principle. In this abstract form the result was first proved (at least for unitary endomorphisms u) by von Neumann [6]. The method of proof we give is due to F. Riesz [7].

What we are about to prove will be termed an "abstract" ergodic theorem on the grounds that it contains as a concrete instance a theorem of the same name that arose out of the considerations of statistical mechanics. In Subsection 1.12.10 we shall try to indicate very briefly the passage from the abstract form of the theorem to this realization of it. For both formulations the reader may be referred to the accounts given by Halmos [5] and by Dunford and Schwartz [1].

Upon the continuous endomorphism u we shall impose one or the other of the following conditions:

(a) u is self-adjoint and

$$c \equiv \operatorname{Sup}\{\|u^n\| : n = 1, 2, \cdots\} < +\infty.$$

(b) u is unitary.

We notice that if (b) holds then so too does the second clause of (a), with $c = 1$, in fact.

THEOREM. Suppose that E is a Hilbert space and u an endomorphism of E satisfying either (a) or (b). We define the endomorphism $A_{m,n}$ for any two natural numbers $m < n$ by

$$A_{m,n} = (n - m)^{-1} \sum_{m \le k < n} u^k.$$

The conclusion is that $P = \lim_{n-m\to\infty} A_{m,n}$ exists pointwise on E—that is, that the sliding arithmetic means of the powers of u converge pointwise on E. The limit endomorphism P is an orthogonal projector that satisfies the relation

$$uP = Pu = P.$$

Proof. Let us first show that $\lim_{n-m\to\infty} A_{m,n}(x)$ exists in E for each x in E—that is, that the pointwise limit P exists. This is done in three stages and in such a manner that certain properties of P become evident in the course of the proof.

We notice first that the $A_{m,n}$ are equicontinuous:

$$\|A_{m,n}\| \le c \qquad (m < n).$$

(1) Let us suppose that $x = u(y) - y$ for some y in E. Then

$$A_{m,n}(x) = (n - m)^{-1}[u^n(y) - u^m(y)]$$

and so $\|A_{m,n}(x)\| \le 2c\,\|y\|/(n - m)$. It thus appears that $P(x)$ exists and equals 0 for such x. The same is therefore true whenever x belongs to the vector subspace M of E generated by such elements $u(y) - y$.

(2) Equicontinuity of the $A_{m,n}$ now ensures that $P(x)$ exists and equals 0 for each x in $\bar{M}$, the closure in E of M.

(3) By the projection principle, if N is the orthocomplement in E of $\bar{M}$, any x in E may be written $x = a + b$ with a in $\bar{M}$ and b in N. To establish the existence of the limit $P(x)$, it now suffices to prove that of $P(b)$. Now for any y in E we have (N being the orthocomplement of M and $u(y) - y$ belonging to M)

$$\begin{aligned} 0 &= (b \mid u(y) - y) = (b \mid u(y)) - (b \mid y) \\ &= (u^*(b) \mid y) - (b \mid y) = (u^*(b) - b \mid y). \end{aligned}$$

This shows that $u^*(b) = b$. Consequently under either hypothesis (a) or (b) it appears that $u(b) = b$. But then $A_{m,n}(b) = b$ for all pairs (m, n) with $m < n$, and so obviously $P(b)$ exists and equals b.

This establishes the existence of the pointwise limit P, and also the property that $P(x) = b$ for any x admitting the decomposition $x = a + b$ with a in $\bar{M}$ and b in N. P is thus the orthogonal projector of E onto N.

The representation of P as the limit of the sliding average $A_{m,n}$, each $A_{m,n}$ commuting with u, shows that P commutes with u. Since also $A_{m,n}u = A_{m+1,n+1}$, it follows that $Pu = P$. Hence $Pu = uP = P$, and the proof is complete. ▌

Remark. Since $uP = P$ it follows that $y = P(x)$ is, for each x, an element of E that is invariant under u; that is,

$$y = u(y) = u^2(y) = \cdots.$$

It is frequently the case (see Subsection 1.12.10) that this invariance property is strong enough to severely restrict the nature of y.

THE CASE OF CONTINUOUS AVERAGES. As will be seen in Subsection 1.12.10 it is often the case that the powers u^n are naturally imbedded in a continuous family u_t of endomorphisms depending on a real parameter t (often interpreted as a time coordinate) and having the semigroup property

$$u_{t+t'} = u_t u_{t'}, \tag{1.12.9}$$

the imbedding being such that $u = u_1$ and therefore $u_n = u^n$. It is enough for our purposes to assume that u_t is defined for $t > 0$. A natural step is to attempt a replacement of the discrete sliding averages $A_{m,n}$ hitherto employed by the continuous sliding averages

$$A_{s,t} = (t - s)^{-1} \int_s^t u_\xi \, d\xi \qquad (0 < s < t) \tag{1.12.10}$$

and to ask about the existence of the pointwise limit of $A_{s,t}$ as $t - s \to \infty$. We shall show that, subject to mild precautions, this problem can be reduced to the preceding discrete case and thereby answered affirmatively.

To begin with we must consider the definition of such integrals as appear above in the definition of $A_{s,t}$. In doing this we shall make two assumptions about the variation of u_t with t. The first is that this family be weakly continuous, by which we mean that for each pair (x, y) of elements of E the function $t \to (u_t(x) \mid y)$ be continuous for $t > 0$. (We could make do with the weaker assumption that it be Lebesgue measurable for $t > 0$.) The second requirement is that this same function be bounded for $t > 0$ (again for each pair (x, y)). From this it may be deduced via the boundedness principles dealt with in Chapter 7 that there exists a number c such that

$$|(u_t(x) \mid y)| \leq c \, \|x\| \, \|y\|$$

for $t > 0$ and x, y in E—or, what is equivalent, that

$$\|u_t\| \leq c \qquad (t > 0).$$

This second demand is, naturally, expressed by saying that the family is bounded. Taken together, therefore, our assumptions are that the family u_t $(t > 0)$ is bounded and weakly continuous.

These restrictions are enough to ensure the existence of the integral $\int_s^t u_\xi \, d\xi = v$ whenever $t > s > 0$. (The definition of u_t for $t = 0$ is, of course, irrelevant as regards this integral.) The meaning of v is as follows:† For each pair x, y of elements of E we may form the integral

$$B(x, y) = \int_s^t (u_\xi(x) \mid y) \, d\xi,$$

which exists in the Riemann sense. For fixed x, $y \to \overline{B(x, y)}$ is a linear form on E that is continuous—

$$\overline{|B(x, y)|} \leq \int_s^t |(u_\xi(x) \mid y)| \, d\xi \leq c(t - s)\|x\| \, \|y\|.$$

By Corollary 1.12.6, therefore, there exists a unique element z of E such that $B(x, y) = (z \mid y)$ for all y in E and $\|z\| \leq c(t - s) \, \|x\|$. This z is naturally written symbolically as $z = \int_s^t u_\xi(x) \, d\xi$. It is evident that $x \to \int_s^t u_\xi(x) \, d\xi$ is an endomorphism of E, say v, and that

$$\|v\| \leq c(t - s).$$

Equally naturally one denotes this v by $\int_s^t u_\xi \, d\xi$, and in this way we have the desired interpretation of the integral in question as a continuous endomorphism of E.

It is important to observe that our definition of the integral is such that

$$T \cdot \int_s^t u_\xi \, d\xi = \int_s^t (T u_\xi) \, d\xi \tag{1.12.11}$$

and

$$\int_s^t u_\xi \, d\xi \cdot T = \int_s^t (u_\xi T) \, d\xi \tag{1.12.12}$$

for any continuous endomorphism T of E. Thus, taking the first identity for example, the preceding definition of adjoints and of integrals give for arbitrary x and y in E

$$\begin{aligned}(Tv(x) \mid y) = (v(x) \mid T^*(y)) &= \int_s^t (u_\xi(x) \mid T^*(y)) \, d\xi \\ &= \int_s^t (T u_\xi(x) \mid y) \, d\xi,\end{aligned}$$

which is $(v'(x) \mid y)$, where $v' = \int_s^t (T u_\xi) \, d\xi$. Consequently $Tv = v'$, which is precisely what is asserted by (1.12.11).

We can now handle the problem of the pointwise limit of the continuous sliding averages $A_{s,t}$, assuming that the family u_t is bounded and weakly continuous, and that $u = u_1$ satisfies either of (a) or (b) (as before). We put $A^0_{m,n}$ for the discrete sliding averages of u. We know that $P^0 = \lim A^0_{m,n}$ exists pointwise on E. If we now define

$$S = \int_0^1 u_\xi \, d\xi,$$

† See Sections 8.14 and 8.15.

then by (1.12.10), (1.12.11), and (1.12.12) we derive the relation

$$A_{m,n} = (n - m)^{-1}\int_m^n u_\xi \, d\xi = A^0_{m,n}S = SA^0_{m,n}$$

for any two natural numbers $m < n$. Thus $\lim A_{m,n}$ exists and equals $P^0S = SP^0$. On the other hand, if $0 < s < t$, and if $m = [s]$ and $n = [t]$ (square brackets denoting integer parts, as usual), then

$$A_{s,t} = -(t - s)^{-1}\int_{[s]}^s u_\xi \, d\xi + (t - s)^{-1}\int_{[s]}^{[t]} u_\xi \, d\xi + (t - s)^{-1}\int_{[t]}^t u_\xi \, d\xi.$$

Now the first and third terms on the right each have norms at most equal to $c(t - s)^{-1}$ and so tend to zero as $t - s \to \infty$. The middle term is equal to

$$(n - m)(t - s)^{-1} A_{m,n},$$

and, since $(n - m)(t - s)^{-1} \to 1$ as $t - s \to \infty$, it therefore converges pointwise to P^0S. So

$$\lim_{t-s\to\infty} A_{s,t} = P^0S = SP^0 = P, \tag{1.12.13}$$

say, in the pointwise sense.

We notice that, on account of (1.12.10), (1.12.11), and (1.12.12), each of S, P^0, P, $A_{s,t}$ $(0 < s < t)$ commutes with every u_ξ $(\xi > 0)$. Moreover it is easily verified, again because of (1.12.10) and the simplest properties of the integrals, that

$$u_tP = Pu_t = P \qquad (t > 0). \tag{1.12.14}$$

Thus, for each x in E, the limit $y = P(x) = \lim_{t-s\to\infty} A_{s,t}(x)$ is invariant under u_t for every $t > 0$.

If in (1.12.14) we replace t by ξ and then take the mean value with respect to ξ over the interval $s < \xi < t$, we conclude that

$$P^2 = P. \tag{1.12.15}$$

If each u_t is self-adjoint, the same is true of each $A_{s,t}$ and hence also of P. Then (1.12.15) shows that P is an orthogonal projector. If, however u_t is unitary, this argument does not work and—although it is still true that P is an orthogonal projector—a different device is required.

One way out of the difficulty, which has its own interest, is to observe that the argument used to deal with the discrete case can be adapted right from the beginning so as to apply to the continuous case. Indeed the semigroup property (1.12.9) permits one to show that $P(x) = \lim A_{s,t}(x)$ exists and is 0 for each x of the form $u_\alpha(y) - y$, where $\alpha > 0$ and y is in E. The same is therefore true if x belongs to the closure $\bar{M}$ of the vector subspace M of E generated by such elements. This depends as in the discrete case on the equicontinuity of the $A_{s,t}$. We then introduce N, the orthocomplement in E of $\bar{M}$, and decompose any x as a sum $a + b$ with a in $\bar{M}$ and b in N. Then b is orthogonal to $u_\alpha(y) - y$ for every $\alpha > 0$ and every y in E, which shows that $u_\alpha(b) = b$ for every $\alpha > 0$

and therefore $A_{s,t}(b) = b$ for $0 < s < t$. Hence $P(b)$ exists and equals b, and therefore $P(x)$ exists and equals b. In this way it appears that P is the orthogonal projector of E onto N.

The results for the discrete case are thus extended in toto to the continuous case. We now turn to the promised illustration in terms of statistical mechanics.

1.12.10 **An Illustration from Statistical Mechanics.** The origins of what is termed ergodic theory lie in the statistical behaviour of Newtonian mechanical systems of many degrees of freedom. The motion of such a system is visualized in terms of the phase space, in which the coordinates are generalized coordinates and their corresponding generalized momenta. The phase space, which we denote by $Q = \{q\}$, may thus be viewed as a subset of R^n, where n is usually very large indeed. The Newtonian equations of motion then specify the development of the system through time, the state of the system that initially (that is, at time $t = 0$) is represented by the point q_0 of Q being at time t represented by a point q_t of Q. In other words, the development of the system is represented by an "orbit" in Q—that is, the family q_t of points of Q. With systems of an enormous number of degrees of freedom (for example, a macroscopically significant lump of gas), the initial state q_0 is practically never known and one is forced to seek a statistical picture of the processes involving averages over the unknown elements of the initial state. This involves the computation of the expected values of various functions (for example, the total energy) defined on Q.

The mapping $T_t : q_0 \to q_t$ of Q into itself may be expected on physical grounds to be continuous, and we assume it to be so. The semigroup property: $T_{t+t'} = T_t T_{t'}$ is physically evident. A famous theorem due to Liouville (Khinchin [1], pp. 15–19) asserts furthermore that, in all cases in which the Hamiltonian of the system does not depend explicitly on the time, each T_t preserves volumes. More precisely, if μ is the Lebesgue measure on Q, then $\mu[T_t^{-1}(S)] = \mu(S)$ for any measurable subset S of Q. The total measure $\mu(Q)$ may or may not be finite. If it is finite, we shall normalize it to be unity, so that μ may be thought of as a genuine probability measure on Q. If $\mu(Q)$ is infinite, one has to think of μ as a relative probability measure. In either case, the expectation value of a function x on Q will be

$$\langle x \rangle = \int_Q x \, d\mu. \tag{1.12.16}$$

(Here and subsequently we must assume that the reader has at least a nodding acquaintance with integration relative to a positive, countably additive measure; the theory expounded in Chapter 4 is more than adequate to cover all the points we shall need.)

Now the direct evaluation of such expectation values is not feasible in practice and a different approach was suggested long ago in the development of statistical mechanics. The line of reasoning may be expressed very briefly and crudely as follows.

If it should be the case that the "flow"—that is, the family of transformations T_t parametrized by the time variable t—should sufficiently "stir up" the points of Q, then it is plausible to assume that the space average $\langle x \rangle$ should be computable otherwise as a limiting time average

$$\lim_{t \to \infty} t^{-1} \int_0^t x(T_\xi q)\, d\xi, \tag{1.12.17}$$

which, again on the grounds of the assumed "stirring property" of the flow, should prove to be independent of q. That a suitable "stirring property" did in fact obtain is analogous to the original so-called ergodic hypothesis; although the precisely desirable form of the latter remained in some doubt. It is perhaps fair to say that most early attempts to formulate a suitable hypothesis led to postulates that were demonstrably not fulfilled in most instances, or could not be verified in any but the most trivial circumstances. Most took the form of statements about the nature of individual orbits, and their verification thus inevitably met with much the same difficulties that were from the start to be circumnavigated.

The limiting time average (1.12.17) itself is physically significant. If the system is initially in the state (represented by the point) q, its state at time t is $T_t q$, and the value of x at time t is $x(T_t q)$. Now for most systems $x(T_t q)$, qua function of t, is subject to extremely rapid and violent fluctuations. These fluctuations will individually occupy an interval that is minute compared with any experimental procedure designed to measure the value of $x(T_t q)$. Consequently, what the procedure will really determine lies much closer to an average

$$t^{-1} \int_0^t x(T_\xi q)\, d\xi$$

than to an instantaneous value $x(T_t q)$. In fact, t will be so large compared with the microscopic time scale of the system, that little error is incurred by actually taking the limiting expression (1.12.17). Whether or not this limiting time average (1.12.17) bears any relation to the space average (1.12.16), is a matter for the physicist rather than the pure mathematician. As we shall see, however, the mathematician's primary object—the study of the existence of (1.12.17)—leads to some information concerning the equality of (1.12.16) and (1.12.17).

From the mathematical point of view a major advance came when Koopman directed attention away from the T_t themselves to the induced functional operators u_t defined by

$$u_t(x)(q) = x(T_t q), \tag{1.12.18}$$

pointing out that because of the measure-preserving nature of T_t, each u_t can be regarded as an isometry of the Lebesgue space $\mathscr{L}^p = \mathscr{L}^p(Q, \mu)$. The notation here is that introduced in Chapter 4: $\mathscr{L}^p(Q, \mu)$ means the vector space of functions x on Q that are measurable and such that $|x|^p$ is integrable with respect to μ, endowed with the seminorm

$$N_p(x) = \left[\int_Q |x|^p\, d\mu\right]^{1/p}.$$

The case $p = 2$ is especially interesting, $\mathscr{L}^2$ being an IPS with the inner product

$$(x \mid y) = \int_Q xy \, d\mu.$$

(Real-valued functions only are considered here.) We shall establish in Chapter 4 the fact that $\mathscr{L}^2$ is complete. The associated quotient space L^2, whose elements are classes of functions equal except on sets of μ-measure zero, is thus a Hilbert space.

Recognition of these points places one in the position of being able to apply the results of Subsection 1.12.9, taking $E = L^2(Q, \mu)$ and u_t as defined in (1.12.18) above. The conclusion is that for each x in $\mathscr{L}^2$ the sliding time average

$$(t - s)^{-1} \int_s^t x(T_\xi q) \, d\xi \tag{1.12.19}$$

converges in L^2 to (the class determined by) a certain function y in $\mathscr{L}^2$. Moreover, this y is invariant in the sense that for each t one has

$$y(T_t q) = y(q) \tag{1.12.20}$$

for almost all points q of Q [that is, for all q save perhaps those belonging to a subset of Q having μ-measure 0]. Besides this, if $\mu(Q)$ is finite, then

$$\langle x \rangle = \langle y \rangle.$$

Under suitable conditions, the invariance property (1.12.20) will entail that y is (equal almost everywhere to) a constant. This is notably the case if any subset S of Q, invariant in the sense that $S = T_t^{-1}(S)$, satisfies either $\mu(S) = 0$ or $\mu(Q \backslash S) = 0$. The flow is then said to be *ergodic* or *metrically transitive*. Thus, if $\mu(Q)$ is finite and if the flow is ergodic, then for each x in $\mathscr{L}^2$ the sliding time averages, (1.12.19), converge as $t - s \to \infty$ in the Hilbert space L^2 to the constant $\langle x \rangle$. Relative to the stated mode of convergence, this is precisely what was to be hoped for. This assertion is customarily referred to as the "mean ergodic theorem," the adjective "mean" referring to the type of convergence postulated.

An illustration. We return to the discrete case, and assume $\mu(Q) = 1$. We take a measurable subset S of Q and $x(q) = \chi_S(q)$, the characteristic function of S. Then $\langle x \rangle = \mu(S)$. Also

$$A_{1,n+1}(x)(q) = n^{-1} \sum_{1 \le k \le n} \chi_S(T^k q) \;=\; N_n(q),$$

say, represents the fraction of time intervals of unit length lying in the interval from 0 to n and for which the system, initially at q, lies within S. The theorem asserts that

$$\lim_{n \to \infty} N_n(q) = \mu(S)$$

in the mean-square sense. This is entirely in accord with what one would expect. It follows easily (via Lebesgue's convergence theorem for integrals) that

$$\limsup_{n \to \infty} N_n(q) \ge \mu(S)$$

for almost all q.

In view of this last relation, it is natural to ask whether one cannot show that $\lim N_n(q) = \mu(S)$ almost everywhere. More generally one is led to ask whether, in our main theorem, mean-square convergence cannot be replaced by other modes of convergence. Here we can do no more than remark that G. D. Birkhoff [1] established what is now known as the "individual (or pointwise) ergodic theorem" which asserts that if the T_t are measure preserving, and if x belongs to $\mathscr{L}^1$, then the aforesaid sliding time-averages converge pointwise almost everywhere on Q; the limit y is invariant and belongs to $\mathscr{L}^1$, and $\langle x \rangle = \langle y \rangle$ provided $\mu(Q)$ is finite, in which case convergence takes place also in the sense of the seminormed space $\mathscr{L}^1$. The consequences of ergodicity of the flow are as before.

Both forms of the ergodic theorem have received improved and simplified proofs. For a brief review see Halmos [5] and the references cited there; more details are to be found in Dunford and Schwartz ([1], Chapter VIII).

Readers who are especially interested in the applications of ergodic theorems in statistical mechanics, and in the physical aspects generally, will find it profitable to consult Khinchin [1] (where will be found an elegant proof of Birkhoff's result) and Ter Haar [1].

EXERCISES

1.1 Let T be any set and p a number satisfying $1 \leq p < \infty$. Show that $\ell^p(T)$ (Example 1.10.3) is complete for the norm

$$\|x\|_p = \left[\sum_{t \in T} |x(t)|^p\right]^{1/p}.$$

Show also that if e_t is the characteristic function of the one-point subset $\{t\}$ of T, then each x in $\ell^p(T)$ is the limit in $\ell^p(T)$ of the sums

$$x_F = \sum_{t \in F} x(t) e_t,$$

F denoting a variable finite subset of T, these sets being partially ordered by set inclusion.

1.2 The notation is as in the preceding exercise. Let U be any continuous linear form on $\ell^p = \ell^p(T)$. Show that the function u, defined by

$$u(t) = U(e_t) \qquad (t \in T),$$

belongs to $\ell^{p'}$, where $p' = \infty$ if $p = 1$ and $1/p + 1/p' = 1$ for $1 < p < \infty$. Deduce that

$$U(x) = \sum_{t \in T} u(t) x(t)$$

for each $x \in \ell^p$, the series being absolutely convergent.

Prove that conversely each element u of $\ell^{p'}$ defines in this way a continuous linear form U on ℓ^p, and that the norm of the linear form U is none other than $\|u\|_{p'}$.

See also Theorem 4.2.1.

1.3 Show by example that the result of the preceding exercise becomes false when $p = \infty$. (See also Exercises 1.19 and 1.21.)

1.4 T being any set, let $c_0(T)$ and $c(T)$ be defined as in Example 1.10.3. Prove that both these spaces are Banach spaces. By reasoning as in the last two exercises, show that the dual of c_0 is (isomorphic and isometric with) $\ell^1 = \ell^1(T)$, the duality being defined by the expression

$$\langle x, y \rangle = \sum_{t \in T} x(t)y(t)$$

for $x \in c_0$ and $y \in \ell^1$.

1.5 Complete the details in the proof of the completeness of H^p and of HL^p for $1 \leq p \leq \infty$ (see Subsections 1.10.4 and 1.12.1(8)).

The techniques used in this situation suggest the following more general result, which the reader is invited to prove. Let X be a sequentially complete TVS, (N_α) a family of lower semicontinuous seminorms on X. Suppose that to each neighbourhood U of 0 in x corresponds a number $\varepsilon > 0$ such that

$$x \in X, \text{ Sup } N_\alpha(x) \leq \varepsilon \Rightarrow x \in U.$$

Let X_0 be the vector subspace of X formed of those $x \in X$ for which

$$N(x) \equiv \sup_\alpha N_\alpha(x) < +\infty.$$

Prove that X_0 is complete for the uniformity defined by the seminorm N on X_0.

1.6 Let E be a normed vector space in which there exists a sphere (that is, a set of the form

$$\{x \in E: \|x - x_0\| = r\},$$

where $r > 0$) that is precompact. Show that E is finite dimensional. (Compare F. Riesz' Theorem in 1.9.6.)

1.7 The notations are as Example 1.12.1(8). Show that $HL^2(D)$ is a Hilbert space when endowed with the inner product

$$(f \mid g) = \iint_D f\bar{g}\, dx\, dy.$$

1.8 The notation is as in the preceding exercise. Establish the following statements:

(1) If $D = C$, the complex plane, then $HL^2(D)$ is reduced to the zero function.

(2) Starting from a domain D, let A be a subset of D admitting no limit points in D, and let $D' = D \backslash A$ (assumed to be a domain). Then each element of $HL^2(D')$ is extendable into an element of $HL^2(D)$.

[Hints: For (1) modify the usual proof of Liouville's theorem, using integral formulae for the Taylor coefficients of any f in $HL^2(C)$. For (2) represent f near any point z_0 of A by a contour integral around the boundary of an annulus with centre z_0, the smaller radius tending to 0, and show that the singularity at z_0 is at most removable.]

1.9 The notation is as in the two preceding exercises. Show that to each point ξ of D there corresponds just one function $z \to K(z, \xi)$ in $HL^2(D)$ such that

$$f(\xi) = \iint_D f(z)\overline{K(z, \xi)}\, dx\, dy$$

for each f in $HL^2(D)$. (This is Bergmann's kernel for D.) Show also that if (f_n) is any orthonormal base in $HL^2(D)$, then

$$K(z, \xi) = \sum_n f_n(z)\overline{f_n(\xi)},$$

and deduce that

$$\iint_D |K(z, \xi)|^2 \, dx \, dy = \sum |f_n(\xi)|^2 = K(\xi, \xi).$$

1.10 The notation is as in the preceding exercise. Show that the function

$$f_\xi(z) = \frac{K(z, \xi)}{K(\xi, \xi)}$$

has the minimum norm, namely, $K(\xi, \xi)^{-1/2}$, among all functions f in $HL^2(D)$ satisfying $f(\xi) = 1$. (We assume here that $HL^2(D)$ is not reduced to the zero function.) Thus f_ξ is the orthogonal projection of 0 onto the closed convex set of $f \in HL^2(D)$ satisfying $f(\xi) = 1$.

1.11 Notations are as in the preceding four exercises. Let D be the open unit disk with centre at the origin. Determine the numbers α_n $(n = 0, 1, 2, \cdots)$ so that the functions $f_n(z) = \alpha_n z^n$ form an orthonormal base in $HL^2(D)$. Hence (or otherwise) show that the Bergmann kernel is

$$K(z, \xi) = \pi^{-1}(1 - z\bar{\xi})^{-2}.$$

1.12 Let E be a Banach space, F a normed vector space, u a linear topological isomorphism of E onto F, and v a linear map of E into F such that

$$\mu \equiv \|(v - u)u^{-1}\| < 1.$$

Show that v is a topological isomorphism of E onto F.

1.13 Let H be a Hilbert space and (x_n) an orthonormal base in H. Show that if a new sequence (y_n) is defined by $y_n = x_n + z_n$, and if there exists a $\mu < 1$ for which

$$\left\|\sum \xi_n z_n\right\| \leq \mu\left(\sum |\xi_n|^2\right)^{1/2}$$

for all scalar sequences (ξ_n) having finite supports, then each x in H is representable uniquely as a convergent series $\sum \xi_n y_n$ in which $\sum |\xi_n|^2 < +\infty$.

1.14 Let E be a TVS, A an open subset of E, B the convex envelope in E of A. Verify that B is open.

1.15 Let E be a TVS, A a convex subset of E having an interior point a, b any point of $\bar{A}$. Show that $(1 - \alpha)a + \alpha b$ is an interior point of A for $0 \leq \alpha < 1$.

1.16 Show that a TVS E is semimetrizable if there exists in E a bounded neighbourhood of zero. Give an example of a metrizable TVS in which there exists no bounded neighbourhood of zero.

1.17 Let E be an inner product space, (e_n) $(n = 1, 2, \cdots)$ an orthonormal family in E, A the subset of E formed of elements $e_m + me_n$, where m and n are natural numbers and $n > m$. Show that 0 is weakly adherent to A, but that 0 is *not* weakly adherent to any bounded subset of A. (The second assertion implies that there exists no sequence extracted from A which converges weakly to zero.)

This example is due to von Neumann.

1.18 Let E be a TVS, (x_α) a bounded directed family in E. Let H be a subset of E' that generates a strongly dense vector subspace of E'. Show that if $\lim_\alpha \langle x_\alpha, x' \rangle = 0$ for each $x' \in H$, then $\lim_\alpha x_\alpha = 0$ weakly in E.

State and prove an analogue of this for weak convergence of a strongly bounded directed family of elements of the dual of a normed vector space.

1.19 Let T be any set, and let $m(T)$ denote the set of all (finitely) additive scalar-valued functions μ defined for all subsets of T for which the total variation

$$\|\mu\| = \operatorname{Sup} \sum_{k=1}^{n} |\mu(T_k)|$$

is finite, the supremum being taken relative to all finite disjoint families $(T_k)_{1 \leq k \leq n}$ of subsets of T.

Prove the following statements:

(1) $m(T)$ is a Banach space when endowed with the norm defined above.

(2) If to each $f \in \ell^1(T)$ one assigns the set function μ_f defined by

$$\mu_f(X) = \sum_{t \in X} f(t) \qquad (X \subset T),$$

then the mapping $u : f \to \mu_f$ is a linear isometry of $\ell^1(T)$ into $m(T)$.

(3) If $\mu \in m(T)$, then the function $f_\mu : t \to \mu(\{t\})$ $(t \in T)$ belongs to $\ell^1(T)$ and $\|f_\mu\|_{\ell^1(T)} \leq \|\mu\|$.

(4) The image of the map u defined in (2) is the set $m_c(T)$ of $\mu \in m(T)$ with the following two properties:

(a) μ is countably additive.

(b) There exists a countable subset C of T such that $\mu(X) = 0$ for all subsets X of T satisfying $X \subset T \backslash C$.

(5) Defining $m_0(T)$ to consist of those $\mu \in m(T)$ such that $\mu(F) = 0$ for all finite subsets F of T, $m(T)$ is the topological direct sum of $m_0(T)$ and $m_c(T)$. In particular, each of $m_0(T)$ and $m_c(T)$ is a closed vector subspace of $m(T)$.

Remarks. It is evident that $m_0(T) = \{0\}$ if T is finite. On the contrary, if T is infinite, $m_c(T)$ is a proper subset of $m(T)$. This is a corollary of results established in Exercises 1.21, 1.24, and 1.25 to follow; it can also be deduced from the Hahn-Banach theorem by establishing the existence of so-called "generalized limits" in the manner described in Section 3.4.

1.20 The notations are as in the preceding exercise. Let μ be a member of $m(T)$. Show that if $x \in \ell^\infty(T)$ has a range $x(T)$ that is finite, then x can be written as a finite sum $\sum_{k=1}^n \xi_k \cdot \chi_{T_k}$, where the ξ_k are scalars, $(T_k)_{1 \leq k \leq n}$ is a finite partition of T, and χ_{T_k} is the characteristic function of T_k relative to T. Show further that the sum $\sum_{k=1}^n \xi_k \mu(T_k)$ has the same value for all such representations of a given x with a finite range. We define $\int_T x \, d\mu$ for such x as the common value of these sums. Verify that $x \to \int_T x \, d\mu$ is linear for such functions x, and that

$$\|\mu\| = \operatorname{Sup} \left| \int_T x \, d\mu \right|,$$

the supremum being taken with respect to all x having a finite range and satisfying $\|x\|_\infty = \operatorname{Sup}_{t \in T} |x(t)| \leq 1$.

Deduce that the integral has a unique continuous extension to $\ell^\infty(T)$, and that this extension is linear and satisfies

$$\left| \int_T x \, d\mu \right| \leq \|\mu\| \cdot \|x\|_\infty$$

for $x \in \ell^\infty(T)$.

1.21 The notations are as in the preceding exercise. Prove that if to each $\mu \in m(T)$ one assigns the continuous linear form on $\ell^\infty(T)$ defined by $x \to \int_T x \, d\mu$, one obtains thereby a linear isometry of $m(T)$ onto the topological dual of $\ell^\infty(T)$. [Hint: In order to represent a given continuous linear form F on $\ell^\infty(T)$ by an integral, consider the set function μ defined by $\mu(X) = F(\chi_X)$ for subsets X of T.]

1.22 The notations are as in the preceding three exercises. Prove the following statements:

(1) Any norm-bounded subset of $m(T)$ is relatively compact for the topology $\sigma(m(T), \ell^\infty(T))$.

(2) If (μ_i) is a norm-bounded directed family in $m(T)$, then $\lim_i \mu_i = 0$ for $\sigma(m(T), \ell^\infty(T))$ if an only if $\lim_i \mu_i(X) = 0$ for each subset X of T.

[Hints: Use Exercises 1.18 and 1.21.]

Foreword. The following six exercises are concerned with the relationship between certain elements of $m(T)$ and the so-called "ultrafilters on T." The concept of filter is due to H. Cartan. It is adopted and exploited systematically by Bourbaki. In relation to convergence in topological spaces, Bourbaki uses filters as an alternative to directed families. Although the idea of a filter is much less intuitive than that of directed family (at least for those readers already at home with ordinary sequences), filters have several technical advantages. See Bourbaki [2] for his introduction to filters.

1.23 Let T be any set and denote by 2^T the set of subsets of T. By a *filter on* T is meant a nonvoid subset $\mathscr{F}$ of 2^T with the following properties:

(1) Any subset of T that contains a member of $\mathscr{F}$ is itself a member of $\mathscr{F}$;
(2) The intersection of two members of $\mathscr{F}$ is a member of $\mathscr{F}$;
(3) The void subset of T is *not* a member of $\mathscr{F}$.

The set of all filters on T is partially ordered by inclusion; that is, a filter $\mathscr{F}$ precedes a filter $\mathscr{F}'$ if and only if each member of $\mathscr{F}$ is a member of $\mathscr{F}'$. Verify that this partial order is inductive and conclude that any given filter on T is contained in some maximal filter on T. Such a maximal filter on T is termed an *ultrafilter on* T.

Prove the following statements:

(a) If T is any set and $A \in 2^T$, then $\{X : X \in 2^T \text{ and } X \supset A\}$ is a filter on T, provided that A is nonvoid. If T is any infinite set, then $\{X : X \in 2^T$ and $T \backslash X$ is finite$\}$ is a filter on T.
(b) If T is any directed set, then the set of subsets of T, each of which contains some set of the form $S_t = \{t' \in T : t' \geq t\}$ (t being a variable element of T), is a filter on T.
(c) If T is a topological space and $t \in T$, the set of all neighbourhoods in T of t is a filter on T.
(d) If T is any set and $t \in T$, the set of all subsets of T that contain t as a member is an ultrafilter on T. An ultrafilter on T of this type is said to be *trivial* or *fixed*.
(e) Let T be a set and $\mathscr{U}$ an ultrafilter on T. If A and B are subsets of T and $A \cup B \in \mathscr{U}$, then at least one of A and B is a member of $\mathscr{U}$.
(f) Let $\mathscr{F}$ be a filter on a set T with the property that, for any subset A of T, either $A \in \mathscr{F}$ or $T \backslash A \in \mathscr{F}$. Then $\mathscr{F}$ is an ultrafilter on T.
(g) If T is a finite set, the only ultrafilters on T are the trivial ones.
(h) If T is an infinite set, there exists at least one nontrivial ultrafilter on T.

1.24 Let T be any set, X a separated topological space, f a function mapping T into X whose range is relatively compact in X, and $\mathscr{U}$ an ultrafilter on T. Prove that there exists precisely one point $x \in X$ with the property that $f^{-1}(U) \in \mathscr{U}$ for each neighbourhood U of x in X, and that this x belongs to $\overline{f(T)}$. This x will be denoted by $f^*(\mathscr{U})$.

Verify also that if f and g are functions mapping T into X such that $f(T)$ and $g(T)$ are relatively compact in X, and if f and g coincide on a set that belongs to $\mathscr{U}$, then $f^*(\mathscr{U}) = g^*(\mathscr{U})$.

[Hint: Show that the sets $f(A)$, $A \in \mathscr{U}$, have the finite intersection property, and use the fact that $f(T)$ is contained in a compact subset of X.]

1.25 In Exercise 1.24 take X to be the scalar field (real or complex) with its usual topology. Let $\mathscr{U}$ be any ultrafilter on T. Show that $f \to f^*(\mathscr{U})$ is a continuous linear form on $\ell^\infty(T)$ and that moreover

$$(fg)^*(\mathscr{U}) = f^*(\mathscr{U}) \cdot g^*(\mathscr{U})$$

for $f, g \in \ell^\infty(T)$.

[Hint: Consider first functions in $\ell^\infty(T)$ having finite ranges.]

Remarks. If T is infinite and we take for $\mathscr{U}$ an ultrafilter on F containing the filter of complements of finite subsets of T, $f^*(\mathscr{U})$ is a sort of "generalized limit" of $f(t)$ as $t \to \infty$. For a different approach to generalized limits via the Hahn-Banach theorem, see Section 3.4.

1.26 Let T be a set, $m(T)$ the space of additive scalar-valued set functions on T of finite total variation (see Exercise 1.19). Denote by M the set of $\mu \in m(T)$ that take only the values 0 and 1 and satisfy $\mu(T) = 1$. Verify that M is compact for the topology induced by $\sigma(m(T), \ell^\infty(T))$ (see Exercise 1.22).

Show further that M is in one-to-one correspondence with the set of all ultrafilters on T by assigning to an ultrafilter $\mathscr{U}$ on T the set function μ defined for subsets X of T by putting $\mu(X) = 1$ or 0 according as X does or does not belong to $\mathscr{U}$.

Remarks. The above correspondence can be used to transport the topology on M onto the set—call it T^*—of all ultrafilters on T. T^* then becomes a separated compact space. By identifying $t \in T$ with the corresponding trivial ultrafilter $\mathscr{U}_t$ (see Exercise 1.23), we regard T as a subset of T^*. At the same time each $f \in \ell^\infty(T)$ has an extension f^* to T^*.

1.27 With the notations of the preceding exercise, show that if $\mathscr{U}$ is an ultrafilter on T and μ the corresponding member of M, then $\int_T f\, d\mu = f^*(\mathscr{U})$ for each $f \in \ell^\infty(T)$, the integral being defined as in Exercise 1.20. Deduce that $f^* \in C(T^*)$.
[Hint: It suffices to verify equality for those f of the type $f = \sum_{k=1}^n \xi_k \chi_{T_k}$, the T_k forming a partition of T. Then just one $T_k \in \mathscr{U}$, in which case $\mu(T_k) = 1$ and $\mu(T_{k'}) = 0$ for $k' \neq k$, while $f^*(\mathscr{U}) = \xi_k$.]

Remark. $f \to f^*$ is thus seen to be a linear isometry of $\ell^\infty(T)$ into $C(T^*)$ that also preserves products.

1.28 With the notations of Exercise 1.26, show that T is dense in T^*. Deduce that $f \to f^*$ maps $\ell^\infty(T)$ *onto* $C(T^*)$.

Remarks. We have now reached the conclusion that T^* is a separated compact space, containing T as a dense subset, and such that each $f \in \ell^\infty(T)$ has a (necessarily unique) continuous extension $f^* \in C(T^*)$; $f \to f^*$ is a linear isometry of $\ell^\infty(T)$ onto $C(T^*)$, and indeed an isomorphism in almost all conceivable senses of the algebra $\ell^\infty(T)$ onto the algebra $C(T^*)$. Moreover, $f^*(T^*) \subset \overline{f(T)}$. It can be shown that T^* is uniquely determined up to homeomorphism by these properties; it is the so-called *Stone-Čech compactification* of T, sometimes denoted by βT. T^* induces on T the discrete topology.

A similarly named compactification T^* may be constructed for any completely regular topological space T, starting from $C_{bd}(T)$ in place of $\ell^\infty(T)$. Various constructions are possible, one of the neatest being that via the theory of Banach algebras. The essence of this approach lies in showing that any continuous linear form F on $C_{bd}(T)$ with the properties $F(1) = 1$ and $F(fg) = F(f)F(g)$ has the form $F(f) = f^*(\mathscr{U})$ for some ultrafilter $\mathscr{U}$ on T (compare Exercise 1.25). For details see Gillman and Jerison [1], especially Chapters 6 and 7; Loomis [1], p. 88; Rickart [1], p. 190.

CHAPTER 2

The Hahn-Banach Theorem

2.1 Foreword

In the preceding chapter we dealt with the Hahn-Banach theorem in its so-called analytic form and applicable to any (not necessarily topological) vector space. In this chapter we shall work with a topological vector space and we shall see that in this situation the theorem has much to say about the existence of continuous linear forms satisfying certain conditions. All the basic theorems to follow stem from the theorem as stated in Section 1.7, when suitable specialization is introduced. These basic developments occupy Section 2.2. We note in passing that it is possible to formulate geometric versions of the Hahn-Banach theorem that in turn imply the analytic form given in Theorem 1.7.1 (see Exercise 2.19).

In Section 2.3 we are concerned with a so-called *extension principle* dealing in some detail with the possibility of extending a continuous linear form from a vector subspace to the entire TVS in question.

One of the most fruitful fields of application of the Hahn-Banach theorem employs that theorem in the guise of *approximation principles*. This aspect is developed in Section 2.4. Many applications will appear in later chapters, but we shall give here some instances that show the technique in action.

In Section 2.5 we shall revert temporarily to real vector spaces and study certain subsets of these known as cones. The specification of a convex cone of a certain type is equivalent to endowing the space with a partial order compatible with its vector-space structure. This in turn leads to an interest in positive linear forms and extension theorems for these.

Section 2.6 develops the ideas of Section 2.5 in the case of a real, locally convex TVS, giving some results that have proved to be useful in a functional analytic approach to some aspects of the theory of games. Applications in this field are given, in particular functional analytic proofs of a generalized form of Ville's theorem and of the fundamental theorem in the theory of games.

Certain extensions of the Hahn-Banach Theorem and its corollaries are conveniently postponed until Sections 3.3–3.5.

2.2 Geometric Forms of the Hahn-Banach Theorem

While one can transform the results of Section 1.7 into a geometric form valid for arbitrary vector spaces, here we shall combine this transformation with the assumption that we are concerned with a topological vector space. The presence of this added ingredient focuses attention on the *continuous* linear forms and the *closed* vector subspaces wherever possible. It turns out that these topological restrictions correspond closely with the situations encountered in applications to concrete analysis.

The principal outcome of this examination is, broadly speaking, that in a locally convex TVS there exist sufficiently many continuous linear forms to effect a faithful image of the structure of such a space, as expressed in terms of linear and topological properties of closed subspaces and closed convex sets in the space.

2.2.1 **Theorem.** (1) Let E be a TVS (real or complex), A an open convex set in E, M a linear variety in E not meeting A. Then there exists a closed hyperplane H in E that contains M and that does not meet A.

(2) Let E be a real TVS, A an open convex set in E, L a vector subspace of E not meeting A. Then there exists a continuous linear form f on E such that

$$f(L) = 0 \qquad \text{and} \qquad f(A) > 0,$$

that is, $f(x) = 0$ for x in L and $f(a) > 0$ for a in A.

Proof. (1) We deal first with the real case. By translation, we may assume that M is a vector subspace of E. According to the results of Section 1.6, A may be represented by an equation $p(x - x_0) < 1$, where x_0 is a point of E and p is a positive, subadditive, and positive-homogeneous function on E. p is continuous since A is open. Since $M \cap A = \emptyset$, $p(y - x_0) \geq 1$ for y in M. Define the linear form f on $M + Rx_0$ by

$$f(y - \alpha x_0) = \alpha \qquad \text{for } y \in M,\ \alpha \in R.$$

If $\alpha > 0$, we then have

$$f(y - \alpha x_0) = \alpha \leq \alpha \cdot p\left(\frac{y}{\alpha} - x_0\right) = p(y - \alpha x_0);$$

and if $\alpha \leq 0$,

$$f(y - \alpha x_0) = \alpha \leq 0 \leq p(y - \alpha x_0).$$

Thus $f \leq p$ at all points of $M + Rx_0$. Since p is a positive gauge function on E, the analytic form of the Hahn-Banach theorem (Theorem 1.7.1(1)) shows that f may be extended into a linear form f^* on E such that $f^* \leq p$ at all points of E. This entails that f^* is continuous. Moreover, $H = f^{*-1}(\{0\})$ is a closed hyperplane in E which contains $M = (M + Rx_0) \cap f^{-1}(\{0\})$. Finally, $f^*(z) = 0$ for z in H and so

$$\begin{aligned} 0 = f^*(z) &= f^*(z - x_0) + f^*(x_0) = f^*(z - x_0) + f(x_0) \\ &= f^*(z - x_0) - 1 \leq p(z - x_0) - 1, \end{aligned}$$

showing that $p(z - x_0) \geq 1$ for z in H—that is, that A does not meet H. This completes the proof for the real case.

To deal with the complex case we first restrict the scalars to real values and appeal to the result just established. This shows that there exists a closed real hyperplane $H_0 \supset M$ with $H_0 \cap A = \emptyset$. Then iH_0 is a closed real hyperplane containing M (since $M = iM$). It thus suffices to take $H = H_0 \cap iH_0$.

(2) We apply (1) with L in place of M. In this case H is a homogeneous hyperplane and, being closed, has an equation $f = 0$, where $f \neq 0$ is a continuous linear form on E. Since $H \supset L$, so $f(L) = 0$. Since H does not meet A, and since A is convex, $f(A)$ must be a real interval not containing 0. By changing the sign of f, if necessary, we can therefore arrange that $f(A) > 0$. ▌

There are no less than six corollaries of Theorem 2.2.1 that are each of considerable importance.

2.2.2 **Corollary.** Let E be a real TVS, A an open convex set in E, B a convex set in E, and suppose that $A \cap B = \emptyset$. Then there exists a continuous linear form $f \neq 0$ on E and a real number α such that

$$f(A) < \alpha, \qquad f(B) \geq \alpha.$$

(These inequalities may be reversed by changing f into $-f$.)

Proof. The set $K = A - B$ is open and convex and does not contain 0. Hence (Theorem 2.2.1) there exists a continuous linear form $f \neq 0$ on E such that $f(K) < 0$. If $\alpha = \sup \{f(a) : a \in A\}$, then α is finite and

$$f(A) \leq \alpha, \qquad f(B) \geq \alpha.$$

However, since $f \neq 0$, f is an open map of E onto R. Hence from $f(A) \leq \alpha$ follows indeed $f(A) < \alpha$. ▌

2.2.3 **Corollary.** Let E be a real LCTVS, A a closed convex set in E, K a compact convex set in E, and suppose that $A \cap K = \emptyset$. Then there exists a continuous linear form $f \neq 0$ on E and a real number α such that

$$f(A) < \alpha, \qquad f(K) > \alpha.$$

(Again we may reverse the inequalities if we wish.)

Proof. There exists an open convex neighbourhood U of 0 in E such that $(A + U) \cap (K + U) = \emptyset$. Applying Corollary 2.2.2 to the sets $A + U$ and $K + U$, we deduce the existence of a continuous linear form $f \neq 0$ on E and a real number α such that

$$f(A + U) < \alpha, \qquad f(K + U) \geq \alpha.$$

Since $K + U$ is open and f is an open map of E onto R, the second relation implies in fact that $f(K + U) > \alpha$, whence the assertion. ▌

Remark. The closed hyperplane $H = f^{-1}(\{\alpha\})$ strictly separates A and K.

2.2.4 **Corollary.** (1) Let E be a real LCTVS, A a closed convex subset of E, x_0 a point of E not in A. Then there exists a continuous linear form f on E satisfying

$$f(x_0) > \operatorname{Sup} f(A);$$

in particular, A is the intersection of all the closed half spaces in E that contain it.

(2) Let E be a (real or complex) LCTVS, A a closed, convex and balanced subset of E, x_0 a point of E not in A. Then there exists a continuous linear form f on E such that

$$|f(x_0)| > \mathrm{Sup}\,|f(A)|.$$

Proof. (1) It suffices to apply Corollary 2.2.3, taking $K = \{x_0\}$.

(2) According to (1), there exists a continuous real-linear form u on E satisfying

$$u(A) \leq \alpha, \qquad u(x_0) > \alpha$$

for some real number α. Since A is convex and balanced, 0 belongs to A and so necessarily $\alpha \geq 0$. Moreover, since A is balanced, from $u(A) \leq \alpha$ it follows that $|u(A)| \leq \alpha$. If E is real, the proof is complete. Otherwise we define

$$f(x) = u(x) - iu(ix) \qquad \text{for} \quad x \in E,$$

so that $u = \mathrm{Re}\, f$. Then f is complex linear. Further, if x is in A and $f(x) = re^{it}$ ($r \geq 0$, t real), then $e^{-it}x$ belongs to A and so $f(e^{-it}x) = r$ is real and therefore equal to $u(e^{-it}x)$, hence has modulus at most α. Thus $|f(A)| \leq \alpha$. On the other hand,

$$|f(x_0)| \geq |\mathrm{Re}\, f(x_0)| = |u(x_0)| > \alpha. \ \blacksquare$$

2.2.5 **Corollary.** Let E be a (real or complex) LCTVS, L a closed vector subspace of E, x_0 a point of E not in L. Then there exists a continuous linear form f on E satisfying

$$f(L) = 0, \qquad f(x_0) \neq 0.$$

In particular, every closed vector subspace of E is the intersection of all the closed hyperplanes in E containing it.

Proof. In Corollary 2.2.4, we take $A = L$. Alternatively we might appeal to Corollary 1.7.2. $\blacksquare$

2.2.6 **Corollary.** If E is a separated LCTVS, the points of E are separated by the continuous linear forms on E.

Proof. By the preceding corollary if $x_0 \neq 0$, there exists a continuous linear form f on E for which $f(x_0) \neq 0$: we are here using the fact that $L = \{0\}$ is closed in E, the latter being separated. If $x_1 \neq x_2$, then $x_0 = x_1 - x_2 \neq 0$ and $f(x_0) \neq 0$ gives $f(x_1) \neq f(x_2)$. $\blacksquare$

2.2.7 **Corollary.** If E is a LCTVS, each closed convex subset of E is weakly closed.

Proof. If E is real, we have only to apply Corollary 2.2.4(1). If E is complex, it suffices to restrict the scalars to real values and notice that the weak topology on E is identical with that defined by the *real*-linear continuous forms. $\blacksquare$

2.2.8 **An Application of Corollary 2.2.4 to Linear Potential Theory.** Curious though it may seem, the linear aspects of potential theory have been systematically studied only quite recently. Hitherto the nonlinear aspects (involving, for example, quadratic functionals like the mutual energy) have appeared more fundamental. Within the last decade, however, linear aspects have been treated in great generality by Choquet and Deny [1] and [2] and by Hunt [1], [2], and [3]. As an illustrative example we shall apply Corollary 2.2.4 in such a way as to link two basic principles of potential theory, the *domination principle* and the *balayage principle*, which have long been known to be cornerstones in the Newtonian and certain other special theories. The following account is based on that of Deny [1].

This illustration involves the use of Radon measures, which are dealt with systematically in Chapter 4. Relatively few nonobvious facts concerning these objects are called for, so that their introduction here causes little difficulty.

In what follows, T denotes a separated, locally compact space. $C = C(T)$ is the space of continuous real-valued functions on T, $\mathscr{K} = \mathscr{K}(T)$ the subspace formed of functions $f \in C$ whose support S_f is compact. We recall that S_f is the closure of the set of points t for which $f(t) \neq 0$. C is viewed as a LCTVS, being endowed with the topology of locally uniform convergence. As for $\mathscr{K}$, it suffices here to introduce a notion of sequential convergence in $\mathscr{K}$, according to which the symbolism

$$f_n \twoheadrightarrow f$$

signifies that

(1) $\bigcup \{S_{f_n} : n = 1,2, \cdots\}$ is relatively compact in T,

and

(2) $f_n \to f$ uniformly on T.

It is easily verified that $\mathscr{K}$ is dense in C.

By a (real Radon) measure on T is meant a linear form μ on $\mathscr{K}$ that is continuous in the sense that $\lim \mu(f_n) = 0$ whenever $f_n \twoheadrightarrow 0$. The set of such measures is denoted by M. It is in no way essential to know that M can be identified with the topological dual of $\mathscr{K}$ relative to a suitable, locally convex topology on the latter (see the discussion in Section 6.3). What is important is that M is a subspace of the algebraic dual of $\mathscr{K}$, so that on M one has a corresponding weak topology $\sigma(M, \mathscr{K})$.

It is important for us to identify the topological dual of C with a subspace of M. The details appear in Section 4.10, but the main features of the argument may be summarized in the following terms. If μ is a measure and U an open subset of T, μ is said to vanish on U if and only if $\mu(f) = 0$ for each $f \in \mathscr{K}$ satisfying $S_f \subset U$. It can be shown that for a given μ, there exists a largest open subset of T on which μ vanishes; the complement of this set is termed the support of μ and denoted here by S_μ. Let

$$M_c = \{\mu \in M : S_\mu \text{ compact}\}.$$

Then M_c is a vector subspace of M. We will verify that the topological dual C' can be identified with M_c.

To this end, we let $L \in C'$. It is evident that the restriction $\mu = L \mid \mathscr{K}$ is a measure. Moreover the continuity of L on C signifies that there exists a compact subset K of T and a number c such that

$$|\mu(f)| \leq c \cdot \operatorname*{Sup}_K |f|.$$

This shows at once that μ vanishes on $T \backslash K$, and hence that $S_\mu \subset K$ and $\mu \in M_c$. On the other hand, if $\mu \in M_c$, then $\mu(f) = \mu(g)$ whenever f and g belong to $\mathscr{K}$ and agree on a neighbourhood of the compact set S_μ. Let us define L on C by putting $L(f) = \mu(f_0)$ for any $f_0 \in \mathscr{K}$ that agrees with f on a neighbourhood of the support of μ. Then L is a linear form on C and $L \,\big|\, \mathscr{K} = \mu$. Furthermore, it is easy to verify that L is continuous on C. It now remains only to add that, since $\mathscr{K}$ is dense in C, each element of C' is uniquely determined by its restriction to $\mathscr{K}$. Thus each element of M_c is uniquely extendible from $\mathscr{K}$ to C so as to achieve membership of C', and each element of C' is obtainable in this way from a unique element of M_c. The identification of C' with M_c is thus established.

On M_c we shall accordingly use the corresponding weak topology $\sigma(M_c, C)$, which is generally stronger than that induced on M_c by $\sigma(M, \mathscr{K})$.

A measure μ is said to be positive, written $\mu \geq 0$, if and only if $\mu(f) \geq 0$ whenever $f \in \mathscr{K}$ and $f \geq 0$. M^+ will denote the set of positive measures, and we further define $M_c^+ = M_c \cap M^+$. If $\mu \in M_c^+$, and if μ is extended to C as described above, then $\mu(f) \geq 0$ whenever $f \in C$ and $f \geq 0$.

As a convenient notation, we shall often write $\langle f, \mu \rangle$ in place of $\mu(f)$, when $f \in \mathscr{K}$ and $\mu \in M$, or when $f \in C$ and $\mu \in M_c$.

It will be helpful to list at this point a few necessary and less-evident properties of measures.

(a) If $\lambda \in M$, $\mu \in M$ and $\lambda = \mu$ on an open set $U \subset T$ (by which it is meant, of course, that $\lambda - \mu$ vanishes on U), then $\mu(f) = \lambda(f)$ for any $f \in \mathscr{K}$ that vanishes outside U.

(b) If $\mu \in M_c^+$, and if $f, g \in C$ satisfy $f \leq g$ on S_μ, then $\mu(f) \leq \mu(g)$.

(c) If F is a closed subset of T, and if

$$M^+(F) = \{\mu \in M^+ : S_\mu \subset F\},$$

then $M^+(F)$ is closed in M.

(d) If K is a compact subset of T, and if

$$\Sigma = \{\mu \in M_c^+ : S_\mu \subset K, \mu(1) = 1\},$$

then Σ is compact in M_c.

Of these, (a) follows from Propositions 4.5.1 and 4.9.1, while (b) is covered by the opening remarks in Section 4.6. (c) follows directly from the definitions. To prove (d) it suffices, in view of (c) and Theorem 1.11.4 (2), to verify that Σ is an equicontinuous subset of C'. However, we take any $u \in \mathscr{K}$ satisfying $0 \leq u \leq 1$ and $u = 1$ on a compact neighbourhood K' of K. For any $f \in C$ and $\mu \in \Sigma$, $\mu(f) = \mu(fu)$. Also, if $K'' = S_u$,

$$-(\operatorname*{Sup}_{K''} |f|) \cdot u \leq fu \leq (\operatorname*{Sup}_{K''} |f|) \cdot u,$$

so that positivity of μ yields for $f \in C$ and $\mu \in \Sigma$

$$|\mu(f)| \leq \mu(u) \cdot \operatorname*{Sup}_{K''} |f| = \operatorname*{Sup}_{K''} |f|.$$

Since K'' is compact, this formula shows that Σ is equicontinuous.

The link with potential theory is based upon the remark that, in the Newtonian, Riesz, and Greenian theories alike, one associates with each $f \in \mathscr{K}$, representing a mass or a charge distribution, a function $Gf \in C$, representing the corresponding potential. The mapping G is derived in these cases from a positive "kernel" $k(t, s)$ via an integration process,

$$Gf(t) = \int k(t, s) f(s)\, ds,$$

but this particularity is not important. The important feature is that G is linear and positive, the latter qualification meaning that $Gf \geq 0$ if $f \geq 0$. The positive nature of G ensures its continuity, in the sense that $f_n \twoheadrightarrow 0$ entails $Gf_n \to 0$ in C. In fact, if K is a compact set containing S_{f_n} for all n, and if u satisfies $u \geq 0$ and $u = 1$ on K, then there exists a sequence (α_n) of numbers such that $\alpha_n \to 0$ and $|f_n| \leq \alpha_n u$. So, by positivity of G, $|Gf_n| \leq \alpha_n \cdot Gu$, which shows that $Gf_n \to 0$ in C.

Thanks to this continuity property of G, one may introduce the adjoint map G' of M_c into M, defined by

$$\langle Gf, \mu \rangle = \langle f, G'\mu \rangle \tag{2.2.1}$$

for $f \in \mathscr{K}$ and $\mu \in M_c$. The reasoning is virtually the same as that produced in a general setting in Section 8.6. Thus, $f \to \langle Gf, \mu \rangle$ is, for a given $\mu \in M_c$, a measure; this measure is just $G'\mu$. The formula (2.2.1) shows that G' is continuous for the topologies $\sigma(M_c, C)$ and $\sigma(M, \mathscr{K})$, and also that G' is positive—that is, maps M_c^+ into M^+.

The two principles in potential theory that we aim to study can now be formulated entirely in terms of G and G'. The first, sometimes known also as Frostman's maximum principle, may be enunciated as follows.

The domination principle. If f and g are positive functions in $\mathscr{K}$, and if $Gf \leq Gg$ at all points of S_f, then the same is true at all points of T.

The second principle concerns the possibility of applying Poincaré's balayage process.

The balayage principle. Given a measure $\mu \in M_c^+$ and a relatively compact open set $U \subset T$, there exists a measure $\mu^* \in M^+$ such that $S_\mu^* \subset \bar{U}$, $G'\mu^* = G'\mu$ at points of U, and $G'\mu^* \leq G'\mu$ at all points of T.

It is not possible here to explain the general significance of either of the two principles above. There is a vast literature, as a sample of which we mention de la Vallée Poussin [1], Frostman [1] and [2], Cartan [5], [6], and [7], and Brelot

[1], [2], and [3]. The reader may also consult the expository paper Ohtsuka [1], which offers an extensive bibliography.

In discussing the relationship between the two principles, one further definition is required. G' is said to be strictly positive, if and only if $G'\mu \neq 0$ whenever $\mu \in M_c^+$ and $\mu \neq 0$. It can be shown that this is so if and only if $G'\varepsilon_t \neq 0$ for each $t \in T$, ε_t being the Dirac measure at t, defined by the formula $\varepsilon_t(f) = f(t)$ for $f \in \mathscr{K}$. No use will be made of this fact, however.

A clear-cut relationship between the two principles is evidenced in the following theorem.

Theorem. If G' satisfies the balayage principle, then G satisfies the domination principle. The converse is also true, provided G' is strictly positive.

Proof. (1) We assume that G' satisfies the balayage principle, and that f and g are as in the hypothesis of the domination principle. For each $t \in T$, let ε_t^* be a measure of the type postulated by the balayage principle when therein one takes $\mu = \varepsilon_t$ and U to be the set of points $t \in T$ at which $f(t) > 0$. According to the postulated properties of ε_t^*, one then has

$$Gf(t) = \langle Gf, \varepsilon_t \rangle = \langle f, G'\varepsilon_t \rangle = \langle f, G'\varepsilon_t^* \rangle,$$

the last step by (2.2.1) and the defining properties of ε_t^*. So, by (b),

$$Gf(t) = \langle Gf, \varepsilon_t^* \rangle \leq \langle Gg, \varepsilon_t^* \rangle.$$

The final expression here is $\langle g, G' \varepsilon_t^* \rangle$. By the defining properties of $G'\varepsilon_t^*$, this is majorized by

$$\langle g, G'\varepsilon_t \rangle = \langle Gg, \varepsilon_t \rangle = Gg(t).$$

Thus $Gf(t) \leq Gg(t)$ for each $t \in T$, and the domination principle is shown to be valid.

(2) We suppose now that G satisfies the domination principle and that G' is strictly positive. In order to deduce the balayage principle, we first recast the latter into a different form.

We write $A = G'(M_c^+)$ and $A(K) = G'(M^+(K))$ for each compact subset K of T. With this notation one may verify easily that G' satisfies the balayage principle if and only if

$$A \subset A(\bar{U}) + M^+(T \backslash U) \tag{2.2.2}$$

whenever U is a relatively compact open subset of T.

Now $M^+(T \backslash U)$ is convex and, by (c), closed in M. Also, $A(\bar{U}) = \bigcup \{cB : c \geq 0\}$, where $B = G'(\Sigma)$ and Σ is defined as in (d) with K therein taken to be $\bar{U}$. According to (d), Σ is compact in M_c. Since G' is continuous, B is compact in M. Moreover, since G' is strictly positive, $0 \notin B$. A simple lemma, which will be established in a moment, shows that the right-hand member of (2.2.2), which is plainly convex, is closed in M. Therefore, by Corollary 2.2.4, this set is precisely the intersection of all closed half spaces that contain it.

To establish (2.2.2) and with it the balayage principle, it now suffices to show that if $h \in \mathscr{K}$ satisfies

$$\langle h, \mu \rangle \geq 0 \qquad \text{for } \mu \in M^+(T \backslash U)$$

and

$$\langle h, G'\lambda \rangle \geq 0 \qquad \text{for } \lambda \in M^+(\bar{U}),$$

then

$$\langle h, G'\beta \rangle \geq 0 \qquad \text{for all } \beta \in M_c^+. \tag{2.2.3}$$

In view of (b), the hypotheses here signify that

$$h \geq 0 \qquad \text{on } T \backslash U$$

and

$$Gh \geq 0 \qquad \text{on } \bar{U},$$

$$f = h^- = \text{Sup}\,(-h, 0), \qquad g = h^+ = \text{Sup}\,(h, 0).$$

Then $S_f \subset \bar{U}$ and so, since $h = g - f$,

$$Gg - Gf = Gh \geq 0 \qquad \text{on } S_f.$$

According to the domination principle, therefore, $Gh \geq 0$ everywhere. But then $\langle Gh, \beta \rangle \geq 0$ for all $\beta \in M_c^+$, which is equivalent to (2.2.3). Thus the balayage principle is valid. ▌

It remains only to establish the following lemma.

Lemma. Suppose that B is a compact, and C a closed, subset of M^+, and that $0 \notin B$. Then

$$A = \bigcup \{cB: c \geq 0\} \cup C$$

is closed in M.

Proof. It has to be shown that if (α_i) is a directed family of points of A that converges in M to ξ, say, then $\xi \in A$. We choose numbers $c_i \geq 0$, elements β_i of B, and elements γ_i of C, such that $\alpha_i = c_i\beta_i + \gamma_i$. It is necessarily the case that $\limsup_i c_i < +\infty$, for otherwise, by passage to a cofinal subfamily, one could assume that $\lim_i c_i = +\infty$. Then

$$0 \leq \beta_i = c_i^{-1}(c_i\beta_i) \leq c_i^{-1}\alpha_i,$$

which would entail that $\lim_i \beta_i = 0$ and so, since B is closed, that $0 \in B$. This would contradict our hypothesis. So $\limsup_i c_i < +\infty$. In that case, by passage to a suitable cofinal subfamily, we may assume that (c_i) converges to c, say, and at the same time that (β_i) converges to $\beta \in B$. Then $\gamma_i = \alpha_i - c_i\beta_i$ is convergent, and its limit, γ, belongs to C. Consequently $\xi = c\beta + \gamma \in A$, as we wished to show. ▌

2.3 The Extension Principle

2.3.1 **Theorem.** Let E be a TVS, S a subset of E, g a scalar-valued function defined on S. In order that there shall exist a continuous linear form f on E such that $f \mid S = g$, it is necessary and sufficient that there exist a continuous seminorm p on E such that

$$\left| \sum_{i=1}^{n} \alpha_i g(x_i) \right| \leq p\left(\sum_{i=1}^{n} \alpha_i x_i \right) \tag{2.3.1}$$

holds for all finite sequences $(\alpha_i)_{1\le i\le n}$ of scalars and $(x_i)_{1\le i\le n}$ of points of S.

Proof. The necessity is trivial since if f exists with the stated conditions satisfied, then $p = |f|$ is a continuous seminorm on E for which (2.3.1) is true.

To prove the sufficiency, let L be the vector subspace of E generated by S. If x belongs to L, it has at least one representation $x = \sum_{i=1}^n \alpha_i x_i$, the α_i being scalars and the x_i elements of S. Moreover, (2.3.1) shows that $f_0(x) = \sum_{i=1}^n \alpha_i g(x_i)$ depends only upon x (and not on the representation chosen). f_0, so defined, is obviously a linear form on L satisfying $|f_0| \le p$ there. It remains only to apply the analytic form of the Hahn-Banach theorem, ensuring the possibility of extending f_0 into a continuous linear form f on E. ∎

Remarks. The most usual form of Theorem 2.3.1 is that in which S is itself a vector subspace of E (and so $S = L$) and g is a linear form on S. In this case, (2.3.1) amounts simply to the demand that g be continuous on S. Furthermore the extension can in all cases be made in such a way that any continuous seminorm on E that majorizes $|g|$ on S continues to majorize $|f|$ on E.

The reader will observe that the theorem becomes applicable to any vector space E on endowing it with the topology $\sigma(E, E^*)$.

We shall give first an application of Theorem 2.3.1 to the proof of a theorem due to Toeplitz that is important in the study of summability methods.

2.3.2 **Theorem** (Toeplitz). Let S and T be arbitrary sets and k a scalar-valued function on $S \times T$ such that for each s in S, the set $\{t \in T: k(s, t) \neq 0\}$ is finite. Given a scalar-valued function y on S, in order that there shall exist a scalar-valued function x on T such that

$$y(s) = \sum_{t\in T} k(s, t)x(t) \qquad (s \in S), \tag{2.3.2}$$

it is necessary and sufficient that for each finite sequence $(\alpha_i)_{1\le i\le n}$ of scalars and each finite sequence $(s_i)_{1\le i\le n}$ of elements of S, the relations

$$\sum_{i=1}^n \alpha_i \cdot k(s_i, t) = 0 \qquad (t \in T) \tag{2.3.3}$$

imply the relation

$$\sum_{i=1}^n \alpha_i \cdot y(s_i) = 0. \tag{2.3.3'}$$

Proof. The necessity is obvious. As for sufficiency, we denote by E the vector space of scalar-valued functions on T that vanish outside finite subsets of T (the subset varying with the function in question). By hypothesis if s is in S, the function $u_s: t \to k(s, t)$ is a member of E. Now the algebraic dual of E may be identified with the space K^T (K denoting the scalar field) by associating with each x in K^T the linear form $u \to \sum_{t\in T} u(t)x(t)$ on E. We therefore seek a linear form f on E such that $f(u_s) = y(s)$ for each s in S. Applying Theorem 2.3.1 with E endowed with the topology $\sigma(E, E^*)$, we see that such an f exists if and only if the relation $\sum_{i=1}^n \alpha_i \cdot u_{s_i} = 0$ implies $\sum_{i=1}^n \alpha_i y(s_i) = 0$,

(α_i) and (s_i) being arbitrary finite sequences of scalars and of elements of S, respectively. Since the first relation here signifies exactly that

$$\sum_{i=1}^{n} \alpha_i u_{s_i}(t) = \sum_{i=1}^{n} \alpha_i \cdot k(s_i, t)$$

be 0 for each t in T, the sufficiency is established. ▮

2.3.3 **Corollary.** Let S, T, and k be as in Theorem 2.3.2. The set of scalar-valued functions y on S having the form (2.3.2), for some scalar-valued function x on T, is closed in K^S (K the scalar field) for the topology of pointwise convergence on S.

Proof. This is immediate since if y is the pointwise limit of a directed family (y_α), each y_α being such that (2.3.3) implies (2.3.3′) (with y_α in place of y), it is evident that y has the same property. ▮

Remarks. Both Theorem 2.3.2 and Corollary 2.3.3 remain true if we substitute for scalar-valued functions ones that take their values in a given vector space. We leave to the reader the simple task of reformulating the results.

2.3.4 **Application of the Extension Principle to the Green's Function.** Throughout this subsection Ω will denote a fixed bounded domain in R^n ($n \geq 2$) and F will denote the (compact) frontier of Ω relative to R^n. The classical Dirichlet problem (first boundary value problem of potential theory) consists of finding a function that is harmonic on Ω and that assumes prescribed continuous boundary values on F. In other words, given a continuous real-valued function f on F, one seeks a function u_f that is harmonic on Ω and such that

$$\lim_{x \in \Omega, x \to t} u_f(x) = f(t)$$

for each point t of F. The maximum principle for harmonic functions shows that the solution, if it exists, is uniquely determined by f. It is also known that if F is sufficiently smooth, it is possible to compute u_f for any given f in terms of the solution corresponding to certain particular boundary values. Specifically, suppose that for each $x_0 \in \Omega$, the solution $U(x, x_0)$ is known for the boundary values $h(t, x_0)$ of the function $h(x, x_0)$ defined for $x \neq x_0$ by

$$h(x, x_0) = \begin{cases} \log |x - x_0|^{-1} & \text{if } n = 2, \\ |x - x_0|^{2-n} & \text{if } n > 2, \end{cases}$$

where in each case $|x - x_0|$ denotes the Euclidean distance in R^n between x and x_0. Suppose then we define the Green's function $G(x, x_0)$ for Ω with pole at x_0 by

$$G(x, x_0) = h(x, x_0) - U(x, x_0).$$

As a function of x, G is harmonic on $\Omega \backslash \{x_0\}$ and has zero boundary values; near x_0, G has the same type of singularity as has $h(x, x_0)$. An application of Green's formula for Ω then shows that

$$u_f(x_0) = -\omega_n^{-1}\int_F f(t) \cdot \partial G(t, x_0)\, d\sigma(t),$$

where ω_n is the hypersurface area of the unit sphere in R^n, σ is the surface measure on F, and ∂ denotes partial differentiation along the normal to F. Naturally, there are points of detail to be examined here, all of which are fully dealt with in Kellogg's well known book (Kellogg [1]; see especially pp. 236–237).

It is our aim to show how Theorem 2.3.1 may serve to establish the existence of the Green's function $G(x, x_0)$. It may be used to show, in fact, that $G(x, x_0)$ exists for each bounded domain, harmonic on $\Omega\backslash\{x_0\}$ and having the prescribed singular behaviour at $x = x_0$, though perhaps not fulfilling the desired boundary condition

$$\lim_{x \in \Omega, x \to t} G(x, x_0) = 0$$

for t in F. This last condition is satisfied provided F is sufficiently smooth but, as has been shown by later researches, is inevitably violated in the contrary case. Even then, however, it is true that in a sense the boundary condition is fulfilled at "most" points of F, failure taking place at the points of a subset of F that has zero capacity, being therefore negligible in potential theory. These later developments are due to Wiener, Perron, Frostman, de la Vallée Poussin, and many others.

To apply Theorem 2.3.1 we introduce the Banach space $C(F)$ of real-valued continuous functions on F, the algebraic operations being pointwise and the norm being equal to the maximum modulus of the function in question. We denote by E the vector subspace of $C(F)$ formed of the restrictions to F of functions continuous on $\Omega \cup F$ and harmonic on Ω. If f belongs to E, it is the restriction of a unique function of the latter type, which is none other than u_f (the solution of the Dirichlet problem for the boundary values f). Given x_0 in Ω we define a linear form L on E by setting $L(f) = u_f(x_0)$ for f in E. The maximum principle shows that L is continuous on E (its norm being in fact just unity). According to Theorem 2.3.1, L has an extension L^* to the whole of $C(F)$, L^* being linear and continuous. In terms of L^* we define functions U and G on Ω by

$$U(x) = L^*(h_x),$$
$$G(x) = h_{x_0}(x) - U(x),$$

where $h_p(y) = h(y, p)$ for $y \neq p$. We have to show that $U(x)$ and $G(x)$ just defined play the roles of $U(x, x_0)$ and $G(x, x_0)$ defined hitherto.

Apart from consideration of the boundary behaviour of G, it suffices to show that U is harmonic on Ω. It is plain that the mapping $x \to h_x$ of Ω into $C(F)$ is continuous. Since L^* is continuous on $C(F)$, it follows that U is continuous on Ω. To show that U is harmonic on Ω it therefore suffices to show that for each point x of Ω, $U(x)$ is equal to the mean value of U taken over balls with centre x and radius r sufficiently small. Now if B is the ball with centre x and radius r so small that $B \subset \Omega$, the mean value m of U over B is

$$m = \int_B U(y)\, d\mu(y),$$

where μ represents Lebesgue measure (that is, $d\mu$ is the element of volume in R^n). Since U is continuous on Ω, if we partition B into a finite number of subsets B_k of sufficiently small diameter and choose points $y_k \in B_k$, this integral differs by less than any preassigned $\varepsilon > 0$ from the sum

$$\begin{aligned} s &= \sum_k U(y_k) \cdot \mu(B_k) \\ &= \sum_k L^*(h_{y_k}) \cdot \mu(B_k) \\ &= L^*\left[\sum_k h_{y_k} \cdot \mu(B_k)\right]. \end{aligned}$$

On the other hand, it is easily seen that the difference

$$\sum_k h_{y_k}(t) \cdot \mu(B_k) - \int_B h_y(t)\, d\mu(y)$$

tends to zero, uniformly for $t \in F$, as the maximum diameter of the B_k tends to zero. Therefore, by continuity of L^* on $C(F)$, we can arrange that $s - L^*(z)$ is in modulus at most ε, where z is the element of $C(F)$ defined by

$$z(t) = \int_B h_y(t)\, d\mu(y).$$

Now, for each t in F, $y \to h_y(t) = h(y, t)$ is harmonic for $y \in \Omega$, and therefore $z(t) = h(x, t) = h_x(t)$; that is, $z = h_x$. Thus, if the partition is fine enough, one has $|s - L^*(h_x)| \leq \varepsilon$; that is, $|s - U(x)| \leq \varepsilon$. Consequently $|m - U(x)| \leq 2\varepsilon$. Letting $\varepsilon \to 0$, we see that $m = U(x)$, showing that U is harmonic on Ω.

To show that $G \to 0$ at all boundary points, one has to assume smoothness on the part of F. We shall not enter into the details here.

It may be noted that the developments of Chapter 4 to follow show that L^* is represented by integration with respect to a Radon measure λ_{x_0} on F; λ_{x_0} is a positive measure of total mass 1. Correspondingly

$$G(x) = h(x, x_0) - \int_F h(x, t) d\lambda_{x_0}(t),$$

and this gives a definition of $G(x) = G(x, x_0)$ whether or not F is smooth. This formula may also be written

$$G(x, x_0) = U^{\varepsilon_{x_0}}(x) - U^{\lambda_{x_0}}(x),$$

the difference between the potentials (logarithmic if $n = 2$, Newtonian if $n = 3$) of the Dirac measure ε_{x_0} at x_0 and of the measure λ_{x_0} on F. It turns out that λ_{x_0} is the measure obtained by applying to ε_{x_0} Poincare's *balayage* (sweeping) process onto the frontier F.

2.3.5 The extension principle as formulated in Theorem 2.3.1 is the simplest possible, in the sense that the extension is required to satisfy broadly only the conditions of linearity and continuity. We shall later (in Section 2.6 and again in Section 3.3) consider analogous results pertaining to cases in which further requirements are imposed upon the extension sought. These extra conditions are related either to a partial ordering on the vector space, or to invariance of the desired extension relative to a given group of automorphisms of the vector space.

2.4 The Approximation Principle

In this section we discuss the role of the Hahn-Banach theorem as a basis of a general technique for the study of approximation problems. This is one of the most fruitful fields of practical application of the theorem.

The situation to be handled is that in which one has a LCTVS E and a subset S of E, and one seeks criteria that are necessary and sufficient in order that a given element x_0 of E be approximable, arbitrarily closely in E, by certain combinations of elements of S. The permitted combinations may be of three types, namely:

(1) Arbitrary finite linear combinations $\sum_{i=1}^{n} \lambda_i x_i$, the λ_i being arbitrary scalars and the x_i arbitrary members of S;

(2) Arbitrary convex combinations $\sum_{i=1}^{n} \alpha_i x_i$, the scalars α_i being positive and such that $\sum_{i=1}^{n} \alpha_i = 1$ and the x_i arbitrary members of S;

(3) Arbitrary convex and balanced combinations $\sum_{i=1}^{n} \beta_i x_i$, the β_i being scalars satisfying $\sum_{i=1}^{n} |\beta_i| \leq 1$ and the x_i arbitrary members of S.

To say that x_0 is approximable, arbitrarily closely in E, by combinations of type (1) (resp. (2), (3)) amounts to saying precisely that x_0 belongs to the closed vector subspace (resp. closed convex envelope, closed convex balanced envelope) in E of S. The appropriate criteria may be read off from Corollaries 2.2.4(1), (2), and 2.2.5 and the result may be formulated as follows.

2.4.1 **Theorem.** Let E be a LCTVS, S a subset of E, x_0 a point of E. Then:

(a) x_0 is the limit in E of finite linear combinations (1) of elements of S if and only if each continuous linear form f on E satisfying $f(S) = 0$ satisfies also $f(x_0) = 0$.

(b) x_0 is the limit in E of convex combinations (2) of elements of S if and only if for each real-linear continuous functional f on E one has

$$f(x_0) \leq \operatorname{Sup} f(S).$$

(c) x_0 is the limit in E of convex and balanced combinations (3) of elements of S if and only if for each continuous linear form f on E one has

$$|f(x_0)| \leq \operatorname{Sup} |f(S)|.$$

2.4.2 **Remarks.** (1) The theorem shows that in each case, if the approximation is possible in the sense of the weak topology of E, it is then possible in terms of the initial topology of E. In particular, if a directed family (x_n) of elements of E is such that $\lim x_n = 0$ weakly in E (that is, $\lim f(x_n) = 0$ for each continuous linear form f on E), then one may construct a directed family (y_i), each y_i being a convex combination of a finite number of the x_n, such that $\lim y_i = 0$ in the sense of the initial topology of E.

(2) Banach introduced the terms "total" and "fundamental" for subsets S of a TVS E: S is said to be "total in E" if each continuous linear form f on E satisfying $f(S) = 0$ necessarily satisfies $f(E) = 0$; and S is "fundamental in E" if the smallest closed vector subspace of E containing S is E itself. Using these terms, Theorem 2.4.1(a) implies exactly this: a subset S of a LCTVS E is total in E if and only if it is fundamental in E.

2.4.3 The useful applications of Theorem 2.4.1 are multitudinous, so much so that it is quite impossible to assemble here even a representative collection. A systematic exploitation of Theorem 2.4.1(a) for a detailed study of families of functions $e^{-\alpha x}$ (for varying values of the parameter α) in several familiar spaces of functions on the half line $(0, \infty)$ has been given by L. Schwartz [15]. The interested reader will find there an example of the use of functional analytic methods in the study of a problem arising in classical analysis. We shall give some examples of applications of a more restricted type, although nevertheless each is typical of one category of problem.

2.4.4 **Example.** Schwartz has given [3] a very neat application of Theorem 2.4.1(a)—the statement of the identity of total and of fundamental subsets of a LCTVS—to the study of particular families of continuous functions on R^n.

Let us denote by $C(R^n)$ the vector space of all continuous real-valued functions on R^n. (Complex-valued functions could be treated in the same fashion.) $C(R^n)$ is made into a metrizable LCTVS by using the seminorms

$$N_k(f) = \text{Sup}\,\{|f(x)|: |x| \leq k\};$$

these define the so-called "compact-open" topology, or the topology of convergence uniform on compact subsets of R^n. Given f in $C(R^n)$, a number $\alpha > 0$, and a point b of R^n, we denote by $f_{\alpha,b}$ the function $x \to f(\alpha x + b)$. Each $f_{\alpha,b}$ belongs to $C(R^n)$. We consider first under what conditions upon f can it happen that the $f_{\alpha,b}$ $(\alpha > 0,\ b \in R^n)$ be nonfundamental in $C(R^n)$.

Theorem A. Suppose that $f \in C(R^n)$ is such that the functions $f_{\alpha,b}$ $(\alpha \in R,$ $b \in R^n)$ are nonfundamental in $C(R^n)$. Then there exists a linear partial differential operator of the form

$$L = \sum_{p_1 + \cdots + p_n = N} a_{p_1 \cdots p_n} \frac{\partial^N}{\partial x_1^{p_1} \cdots \partial x_n^{p_n}}, \tag{2.4.1}$$

the p_k being integers ≥ 0 and the $a_{p_1, \cdots, p_n}$ real numbers, not all zero, such that f is a generalized solution of the equation $Lf = 0 \cdots$ by which we mean that f is the limit in $C(R^n)$ of indefinitely differentiable functions that satisfy the said equation.

Proof. We must use the fact (a consequence of Weierstrass' famous polynomial approximation theorem, itself a special case of what is established in Subsection 4.10.5) that the functions $x^p = x_1^{p_1} \cdots x_n^{p_n}$, where the x_k denote the natural coordinate functions on R^n and where the p_k range separately over all integers ≥ 0, are fundamental in $C(R^n)$. If the $f_{\alpha,b}$ are not fundamental in $C(R^n)$, Theorem 2.4.1(a) informs us of the existence of a continuous linear form F on $C(R^n)$ satisfying $F \neq 0$ and $F(f_{\alpha,b}) = 0$ for all $\alpha > 0$ and all b in R^n. Since $F \neq 0$, there must exist a monomial x^q, q being an n-uple $(q_1, \cdots, q_n)$ of integers $q_k \geq 0$, such that $F(x^q) \neq 0$. Let $N = q_1 + \cdots + q_n$ and define

$$a_{p_1, \cdots, p_n} = \frac{N!}{p_1! \cdots p_n!} F(x^p) = \frac{N!\, F(x_1^p \cdots x_n^p)}{p_1! \cdots p_n!}$$

for all n-uples $p = (p_1, \cdots, p_n)$ of integers $p_k \geq 0$. L is then defined by (2.4.1) and is clearly of the stated form.

We have now to circumvent the difficulty that f may not be differentiable. To do this, we take any indefinitely differentiable function r on R^n that vanishes outside a bounded set and consider

$$g(x) = \int f(x + y) r(y)\, d\mu(y), \tag{2.4.2}$$

μ denoting Lebesgue measure on R^n. It is clear that g is indefinitely differentiable on R^n. Further, it is easily seen that since $F(f_{\alpha,b}) = 0$ for $\alpha > 0$ and b in R^n, the same is true when f is replaced by g:

$$F(g_{\alpha,b}) = 0 \qquad (\alpha > 0,\ b \in R^n). \tag{2.4.3}$$

This is perhaps most easily seen by approximating g arbitrarily closely in $C(R^n)$ by sums $\sum_k f(x + y_k) r(y_k) \cdot \mu(S_k)$, the S_k forming a finite partition of the bounded set on which r is nonzero, and the y_k arbitrarily selected points of the S_k. Continuity of F on $C(R^n)$ is called into play here.

In (2.4.3) we take the Nth derivative with respect to α and then let $\alpha \to 0$. Using the indefinite differentiability of g and the continuity of F, the result is seen to be

$$Lg(b) = 0,$$

and this for all b in R^n. Thus g is a solution of the equation $Lg = 0$.

Finally, we have only to observe that we can approximate f arbitrarily closely in $C(R^n)$ by functions of the type (2.4.2), taking a sequence (r_i) of functions r satisfying $r_i \geq 0$, $\int r_i(y)\, d\mu(y) = 1$, and arranging that r_i is zero outside a neighbourhood U_i of 0 in R^n such that the diameter of U_i tends to zero as $i \to \infty$. This proves Theorem A. ▌

It is worthwhile pursuing the question a little further on the supposition that the set of all similitudes of f is not fundamental. By a similitude (or similarity transformation) of R^n we mean a transformation belonging to the group generated by the translations, rotations about 0, and homothetic transformations with centre 0 and positive ratios. We denote by S this group of transformations. If s is a member of S and f a function of R^n, we denote by f_s the function $f \circ s$; the functions f_s ($s \in S$) are termed the similitudes of f.

Theorem B. Let f belong to $C(R^n)$ and be such that the similitudes of f are nonfundamental in $C(R^n)$. Then there exists an integer $m \geq 0$ such that f is a generalized solution of the polyharmonic equation

$$\Delta^m f = 0, \tag{2.4.4}$$

Δ denoting the Laplacian on R^n.

Proof. Since S contains all translations and homothetic transformations with centre 0 and positive ratio, Theorem A informs us that f is a generalized solution of $Lf = 0$, where L is the differential operator (2.4.1) and

$$a_{p_1, \cdots, p_n} = \frac{N!\, F(x_1^p \cdots x_n^p)}{p_1! \cdots p_n!}. \tag{2.4.5}$$

Since S contains all rotations, we can arrange that F is invariant under rotations in the sense that $F(g_r) = F(g)$ for all g in $C(R^n)$ and all rotations r. To see this we regard the set G of all rotations as a compact group. (Its topology may be defined, for instance, by representing each rotation by a matrix relative to a fixed base for R^n, and taking the natural topology in which convergence means convergence of each entry in the matrix.) One then takes the bi-invariant Haar measure λ on G (see Section 4.18) and replaces F by the linear form F^* defined by

$$F^*(g) = \int_G F(g_r)\, d\lambda(r),$$

which is easily seen to be invariant under rotations in the sense described; a suitable choice of the centre of rotation will arrange that $F^* \neq 0$.

Having arranged rotational invariance of F in this way, we consider the algebraic form

$$\sum_{p_1 + \cdots + p_1 = N} a_{p_1, \cdots, p_n} X_1^{p_1} \cdots X_n^{p_n}$$

which, by (2.4.5) is equal to

$$F[(X_1 x_1 + \cdots + X_n x_n)^N].$$

Since F is invariant under rotations, this expression is a function of

$$(X_1^2 + \cdots + X_n^2)^{1/2}$$

only, which can only be of the type

$$(X_1^2 + \cdots + X_n^2)^{\frac{1}{2}N},$$

where $\frac{1}{2}N$ is necessarily an integer, say m. But then clearly $Lf = 0$ is equivalent to $\Delta^m f = 0$, as alleged. ∎

Remarks. (1) If $n = 2$, we can parametrize G by the usual angular variable θ ranging over $(0, 2\pi)$, and then

$$\int_G \cdots d\lambda(r) = \frac{1}{2\pi} \int_0^{2\pi} \cdots d\theta.$$

If $n = 1$, G is composed of just two elements—multiplication by 1 and by -1 —and integration with respect to λ means simply taking the arithmetic mean taken over these two elements.

(2) If $n = 1$, we may omit any reference to the group G and rotational invariance. The result here would be that if the $f_{\alpha,b}$ are nonfundamental, α ranging through any set in R having 0 as a limit point and b over any nonvoid open interval (b', b'') in R, then $f \mid (b', b'')$ is a polynomial of degree at most m, where m depends only upon f (and not on the interval (b', b'')). The argument is a slight modification of the proof of Theorem A, using Rolle's theorem to infer that if an indefinitely differentiable function of one variable vanishes at points of a set having 0 as a limit point, then that function and all its derivatives vanish at the origin.

2.5 Supplements to, and Further Applications of, Theorem 2.4.1

So far we have concentrated on part (a) of Theorem 2.4.1. In this section we wish to supplement parts (b) and (c) and provide some applications thereof.

To begin with, parts (b) and (c) are often enriched by an independent assertion about the possibility of giving series and integral representations of elements of the closed, convex envelope (or the closed, convex, balanced envelope) in E of a given family of elements of E. We shall first consider the simpler case involving series.

Suppose given in E a family $(x_t)_{t\in T}$, T being an arbitrary set. We denote by $\ell^1(T)$ the Banach space of scalar functions σ on T for which

$$\|\sigma\|_1 = \sum_{t\in T} |\sigma(t)| < +\infty.$$

We shall make the following assumption:

Assumption. There exists a total subset F of E' such that for each σ in $\ell^1(T)$ the series $\sum_{t\in T} \sigma(t)x_t$ is convergent for the weak topology $\sigma(E, F)$, and such that $\lim_{t\to\infty} f(x_t) = 0$ for each f in F (the limit $\lim_{t\to\infty}$ being taken according to the increasing directed set of finite subsets of T).

2.5.1 **Theorem.** Suppose that the family $(x_t)_{t\in T}$ in E and the total subset F of E' are such that the Assumption above is fulfilled. Then each element x of the closed, convex envelope A (resp. the closed, convex, balanced envelope B) in E of $\{x_t : t \in T\}$ can be represented in the form

$$x = \sum_{t\in T} \sigma(t)x_t, \tag{2.5.1}$$

where $\sigma \in \ell^1(T)$ satisfies $\sigma \geq 0$ and $\|\sigma\|_1 \leq 1$ (resp. $\|\sigma\|_1 \leq 1$). Moreover, each x of the form (2.5.1), with $\|\sigma\|_1 \leq 1$, belongs to B.

Proof. Let A_0 (resp. B_0) be the set of σ in $\ell^1(T)$ such that $\sigma \geq 0$, $\|\sigma\|_1 \leq 1$ (resp. $\|\sigma\|_1 \leq 1$). A_0 is convex and B_0 is convex and balanced. We shall see later (Theorem 4.2.2) that $\ell^1(T)$ may be identified with the dual of the Banach space $c_0(T)$ of scalar functions on T that tend to zero at infinity, endowed with the norm equal to the supremum of the modulus of the function in question. Then Theorem 1.11.4(2) entails that each of A_0 and B_0 is relatively compact for the topology $\sigma(\ell^1(T), c_0(T)) = \mathcal{T}$, say. However, it is evident that both A_0 and B_0 are closed for $\mathcal{T}$. Hence both are compact for $\mathcal{T}$.

Consider now the linear map u of $\ell^1(T)$ into E defined by $u(\sigma) = \sum_{t \in T} \sigma(t)x_t$. Thanks to our Assumption, u is continuous for the topology $\mathcal{T}$ on $\ell^1(T)$ and the topology $\sigma(E, F)$ on E. It follows that each of $u(A_0)$ and $u(B_0)$ is compact for $\sigma(E, F)$, hence closed for this topology (which is separated since F is a total subset of E').

It is evident on the other hand that the $\sigma(E, F)$-closure of $u(A_0)$ (resp. $u(B_0)$) contains A (resp. B). Hence in fact $u(A_0) \supset A$ and $u(B_0) \supset B$. Finally, it is evident that $u(B_0) \subset B$ and hence $u(B_0) = B$. This completes the proof. ▌

2.5.2 **Remarks.** (1) There is no reason for expecting that $u(A_0) \subset A$. What is evident is that if A_1 denotes the set of σ in $\ell^1(T)$ such that $\sigma \geq 0$ and $\|\sigma\|_1 = 1$, then $u(A_1) \subset A$ and $u(A_1)$ is dense in A (for the topology $\sigma(E, F)$). But A_1 is not generally compact for $\mathcal{T}$, so that we cannot infer that $u(A_1)$ is closed for $\sigma(E, F)$ and therefore coincides with A. It may, however, be possible to show that $u(A_1) = A$ by showing that if $\sigma \in A_0$ and $x = u(\sigma) \in A$, then necessarily $\|\sigma\|_1 = 1$ (that is, that $\sigma \in A_1$). Whether or not this deduction is feasible, depends primarily on how large is the maximal subset G of E' for which it is true that the series defining $u(\sigma)$ is convergent for the topology $\sigma(E, G)$ $\cdots$ or, somewhat more directly, on how extensive is the set $\hat{G}$ of real-valued functions $\hat{g}$ on T that are of the form $\hat{g}(t) = g(x_t)$ with g in G. For if $\sigma \in A_0$ and $x = u(\sigma) \in A$, it is true that $g(x) \geq \operatorname{Inf}\{\hat{g}(t) : t \in T\}$ for each g in E', and therefore

$$\sum_{t \in T} \sigma(t)\hat{g}(t) \geq \operatorname{Inf}\{\hat{g}(t) : t \in T\} \tag{2.5.2}$$

for each g in G. It is sometimes the case that (2.5.2) entails that

$$\sum_{t \in T} \sigma(t) \geq 1,$$

hence that σ belongs to A_1 and x belongs to $u(A_1)$.

As a simple instance, consider the case in which T is the set of integers ≥ 0, E is the space of real-valued continuous functions of a real variable having period 2π, endowed with the topology defined by the uniform norm

$$\|x\| = \operatorname{Sup}\{|x(s)| : s \text{ real}\},$$

x_t being the function $s \to \cos ts$. In this case, it is plain that we may take $G = E'$ and F equal to the set of linear forms of the type $x \to \int_0^{2\pi} x(s)f(s)\,ds$ where f is (say) continuous and with period 2π. In (2.5.2) we may take for g the linear form $x \to x(0)$, which shows at once that $\sum_{t=0}^{\infty} \sigma(t) \geq 1$. Thus in this case we may assert that the closed convex envelope in E of the x_t consists of precisely those functions of the form $x(s) = \sum_{t=0}^{\infty} \sigma(t) \cos ts$ with $\sigma(t) \geq 0$ and $\sum_{t=0}^{\infty} \sigma(t) = 1$.

(2) There are analogous results concerned with suitable "continuous" families $(x_t)_{t \in T}$. In such cases T is usually a locally compact space and the mapping $t \to x_t$ is usually continuous and such that $\lim_{t \to \infty} x_t = 0$, in each instance relative to a suitable separated topology on E. The space $\ell^1(T)$ would then be replaced by the space of bounded Radon measures on T (see Chapter 4). We do not propose to enter into details now. A similar line of argument appears explicitly, and in a similar connection, in Theorems 8.13.1 and 10.1.7.

2.6 Extension of Positive Linear Forms

2.6.1 Convex Cones and Partial Ordering. Let E be a real vector space. A subset K of E is termed a *cone* if $\lambda K \subset K$ for each scalar $\lambda > 0$; a cone K is said to be *pointed* if $0 \in K$. A pointed cone K is termed *salient* if K contains no 1-dimensional vector subspace of E. A cone K that is also a convex subset of E is termed a *convex cone*: thus a subset K of E is a convex cone if and only if $\lambda K \subset K$ for each $\lambda > 0$ and $K + K \subset K$. In this case, the vector subspace of E generated by K is simply $K - K$; and if K is pointed, then $K \cap (-K)$ is the largest vector subspace of E contained in K, so that a pointed convex cone K is salient if and only if $K \cap (-K) = \{0\}$.

If the vector space E is endowed with a partial order, written "$x \leq y$" (or equivalently, "$y \geq x$"), and if the vector space structure and the partial order are so interrelated that

(1) $x \leq y$ implies $x + z \leq y + z$ (x, y, z in E),

and

(2) $\lambda \geq 0$, $x \geq 0$ imply $\lambda x \geq 0$ (λ in R, x in E),

then we say that E is an *ordered vector space* (OVS for short). The partial order is, of course, assumed to satisfy the usual axioms, viz. $x \leq x$ (reflexivity), $x \leq y$ and $y \leq z$ together imply $x \leq z$ (transitivity), and $x \leq y$ and $y \leq x$ together imply $x = y$.

If E is an OVS, it is clear that the so-called "*positive cone*"

$$P = \{x : x \in E, x \geq 0\}$$

is a salient, pointed, convex cone in E. Conversely, given such a subset P of E, we get a partial order by defining $x \geq y$ to signify that $x - y \in P$. This partial order turns E into an OVS, and this is the only structure of an OVS on E for which P is precisely the set of positive elements.

It is natural to ask under what conditions it is the case that given a positive linear form f on a vector subspace of E, one can find a linear form on E extending f and positive on E. We shall first give a result (Theorem 2.6.2) of this nature that is applicable when E is any OVS and M is a "sufficiently large" vector subspace of E. From this we shall pass easily to a similar result (Theorem 2.6.3) due to M. Krein and applying to the case in which the OVS E is also a TVS, provided $P \cap M$ has interior points.

2.6.2 Theorem. Let E be an OVS, M a vector subspace of E with the property that to each x in E corresponds at least one element m of M for which

$x \leq m$. If f is a positive linear form on M (for the induced structure), then f can be extended into a positive linear form f^* on E.

Proof. We define p on E by

$$p(x) = \text{Inf}\,\{f(y): y \in M,\, y \geq x\}.$$

Our hypothesis on M entails that to each x in E correspond m and m' in M such that $m' \leq x \leq m$. It follows that p is finite-valued—in fact,

$$f(m') \leq p(x) \leq f(m).$$

Since P is a convex cone, p is a gauge function on E (see Section 1.6). If x is in M, it is evident that $p(x) \leq f(x)$. On the other hand, if x and y are in M and $y \geq x$, then $f(y) \geq f(x)$ (f being positive on M); so $p(x) \geq f(x)$ for x in M. Thus in fact $f(x) = p(x)$ for x in M.

Applying Theorem 1.7.1(1), we deduce that there exists a linear form f^* on E that extends f and that satisfies $f^*(x) \leq p(x)$ for x in E. Then, if $x \geq 0$, $f^*(-x) \leq p(-x) = \text{Inf}\,\{f(y): y \in M, y + x \geq 0\} \leq f(0) = 0$ and so $f^*(x) \geq 0$. Thus f^* is a positive linear form on E. ▌

2.6.3 **Theorem** (M. Krein). Suppose that E is an OVS with positive cone P, and that E is also a TVS. Let M be a vector subspace of E such that at least one point x_0 of $P \cap M$ is interior to P. Each positive linear form f on M (for the induced structure) can be extended into a continuous positive linear form f^* on E.

Proof. Let U be any balanced neighbourhood of 0 in E such that $x_0 + U \subset P$, If x belongs to E and $\lambda > 0$ is such that $x \in \lambda U$, then $-x/\lambda \in U$ and so $x_0 - x/\lambda \in P$; that is, $x \leq \lambda x_0 = m$. The hypothesis of Theorem 2.6.2 is fulfilled and the construction used in its proof shows that $p(x) \leq \lambda f(x_0)$ if $x \in \lambda U$ and $\lambda > 0$. According to Theorem 2.6.2 we can extend f into a positive linear form f^* on E such that $f^*(x) \leq p(x)$ for all x in E. Hence $f^*(x) \leq \lambda f(x_0)$ if $x \in \lambda U$, $\lambda > 0$. Since U is balanced, this entails that $|f^*(x)| \leq \lambda\, |f(x_0)|$ if $x \in \lambda U$, $\lambda > 0$. Thus f^* is continuous on E. ▌

2.6.4 As has been stated in Subsection 2.3.5, one may seek analogous extension theorems in which further restrictions are imposed upon the given linear form, and the same demands made upon the extension. A return is made to this point in Section 3.3.

2.7 Some Results with Applications to the Theory of Games

As we shall show later in this section, it has been possible to frame certain basic principles in the theory of games in terms of compact convex sets and convex cones in real LCTVSs. For the time being we shall take on trust the existence of such applications and attend to the abstract results behind them, in doing which we follow quite closely an exposition given by Bohnenblust and Karlin (On a theorem of Ville—The Theory of games, Annals of Math.

Study No. 24, pp. 155–160). For the theory of finite discrete games the underlying principle corresponds to the case of finite-dimensional vector spaces and can be formulated thus: Suppose we are given a finite matrix (a_{ij}) with real entries, and suppose also that to each family (q_j) with $q_j \geq 0$ and $\sum_j q_j = 1$ there corresponds a family (p_i) with $p_i \geq 0$ and $\sum_i p_i = 1$ for which

$$\sum_i \sum_j a_{ij} p_i q_j \geq 0;$$

then there exists a family (p_i) of the same type for which

$$\sum_i \sum_j a_{ij} p_i q_j \geq 0$$

for all families (q_j) of the stated type. The extension we have in mind amounts to replacing the expression

$$\sum_i \sum_j a_{ij} p_i q_j$$

by a fairly general type of bilinear form on the product of two LCTVSs. Initially, however, it is convenient to work with one such space and its topological dual.

Throughout this section, unless the contrary is explicitly stated, E and F denote separated real LCTVSs whose topological duals we denote by E' and F', respectively; it is convenient in the general work to use the symmetric notation $\langle x, x' \rangle$ to denote the value at $x \in E$ of the linear form $x' \in E'$, with an analogous notation for other LCTVSs.

Our first goal is concerned with the following situation. Suppose we are given a subset K of E and a convex cone Γ in E'; in all cases we shall assume that Γ is weakly closed (that is, closed for the topology $\sigma(E', E)$), although one may equally well replace this topology on E' by any locally convex topology that is compatible with the duality between E and E' (that is, relative to which the topological dual of E' is precisely E; see Section 8.3). The main hypothesis will read as follows:

Hypothesis. To each x' in Γ corresponds an x in K (which may *a priori* depend upon x') such that $\langle x, x' \rangle \geq 0$.

The conclusion we wish to draw is this:

Conclusion. There exists an x_0 in K such that $\langle x_0, x' \rangle \geq 0$ for all x' in Γ.

Our first result will show that this conclusion is valid whenever K is compact and convex, and we shall then show that this condition on K may be modified. The former result is relatively simple to establish and itself is adequate to cover some applications in the theory of games.

2.7.1 **Theorem.** Let K be a compact convex set in E, Γ a weakly closed convex cone in E', and suppose that the Hypothesis above is true. Then the Conclusion above is also true.

Proof. Consider the set

$$A = \{x \in E: \langle x, x' \rangle \geq 0 \text{ for all } x' \in \Gamma\};$$

our task is to show that $A \cap K$ is not void. It is clear that A is a pointed, closed, convex cone in E. Suppose if possible that $A \cap K$ were void. By Corollary 2.2.3, there would exist x'_0 in E' and a real number α such that $\langle x, x'_0 \rangle > \alpha$ for x in A and $\langle x, x'_0 \rangle < \alpha$ for x in K. Since A contains 0, the first relation shows that $\alpha < 0$. Since A is a cone, the first relation shows that necessarily $\langle x, x'_0 \rangle \geq 0$ for x in A. We deduce that x'_0 belongs to Γ: indeed, if this were not so, Corollary 2.2.4 (applied to E' in place of E) would entail the existence of an element c of E such that $\langle c, x'_0 \rangle < \text{Inf } \{\langle c, x' \rangle : x' \in \Gamma\}$; since Γ is a cone, this would entail $\langle c, x' \rangle \geq 0$ for x' in Γ, which shows that c belongs to A and yet also $\langle c, x'_0 \rangle < 0$, in conflict with the definition of A. But then the relation $\langle x, x'_0 \rangle < \alpha < 0$ for x in K contradicts our Hypothesis, and so completes the proof. ▌

We now turn at once to a modified form of Theorem 2.7.1.

2.7.2 **Theorem.** Let K be a subset of E satisfying the following conditions:
(1) K is compact.
(2) $Q = \bigcup_{\lambda \geq 0} \lambda K$ is convex.
(3) No segment joining two points of K contains 0.
Let Γ be a weakly closed convex cone in E'. If the original Hypothesis is true, then so too is the Conclusion.

The proof of Theorem 2.7.2 will be based on three lemmas.

2.7.3 **Lemma.** Let K be a compact subset of E that does not contain 0, and let Q be as in (2) above. Then Q is closed. (If E were not separated, the same conclusion follows if the condition $0 \notin K$ is replaced by $\overline{\{0\}} \cap K = \emptyset$.)

Proof. Suppose x is a point of E adherent to Q. Then there exist directed families (λ_i) and (x_i), with the same index set, such that $\lambda_i \geq 0$ and $x_i \in K$ and $\lim_i \lambda_i x_i = x$. Since K is compact, (x_i) has a cofinal subfamily (y_j) such that $y = \lim_j y_j$ exists and belongs to K. Let (μ_j) be the corresponding subfamily of (λ_i). Then $\lim_j \mu_j y_j = x$. Now $y \neq 0$, and so there exists a continuous seminorm p on E such that $p(y) \neq 0$ (if E were not separated, this would still follow when we know that $y \notin \overline{\{0\}}$). Since then $p(y_j) \to p(y)$ and $\mu_j p(y_j) = p(\mu_j y_j) \to p(x)$, it follows that $\mu = \lim \mu_j$ exists finitely. Necessarily $\mu \geq 0$. Since now $\mu_j \to \mu$ and $y_j \to y$, so $x = \lim (\mu_j y_j) = \mu y$, showing that x belongs to Q and that Q is closed. ▌

Remark. The restriction that 0 be not a point of K cannot be dropped in Lemma 2.7.3. For example, let $E = R^2$ and let K consist of $0 = (0, 0)$ and the points $x_n = (n^{-1}, (n + 2)^{-1/2})$ $(n = 1, 2, \cdots)$. Then $(0, 1)$ does not belong to Q and yet the points $n^{1/2} x_n$ belong to Q and converge to $(0, 1)$, showing that Q is not closed.

2.7.4 **Lemma.** Suppose that K satisfies conditions (1), (2), and (3) of Theorem 2.7.2. Then there exists a number $\delta > 0$ and an element x'_0 of E such that $\langle x, x'_0 \rangle \geq \delta$ for x in K.

Proof. By the last lemma, Q is closed; it is convex by condition (2). According to condition (3) if x belongs to K, then $-x$ does not belong to Q; for otherwise we should have $x = -\lambda y$ ($y \in K$, $\lambda \geq 0$), hence $x + \lambda y = 0$, and so 0 would be the point $x/(1 + \lambda) + (\lambda/1 + \lambda)y$ of the segment joining the points x and y of K, contrary to condition (1). By Corollary 2.2.3 again, for each x_0 in K there exists x' in E' and a real number α such that

$$\langle -x_0, x' \rangle < \alpha \quad \text{and} \quad \langle y, x' \rangle > \alpha \qquad \text{for } y \in Q.$$

Taking $y = 0$ we conclude that $\alpha < 0$, and so $\langle x_0, x' \rangle > 0$. At the same time, Q being a cone, the second relation entails that $\langle y, x' \rangle \geq 0$ for y in K.

Using the compactness of K, it is easily seen now that we may choose $x'_1, \cdots, x'_n$ from E' such that $\sum_{i=1}^{n} \langle x, x'_i \rangle > 0$ for all x in K. Putting $x'_0 = \sum_{i=1}^{n} x'_i$ and using compactness of K once more, we see that $\langle x, x'_0 \rangle$ has a strictly positive infimum when x ranges over K. ▌

2.7.5 **Lemma.** Suppose that K satisfies conditions (1), (2), and (3) of Theorem 2.7.2. Then there exists a compact convex set K_0 in E such that $x \neq 0$ belongs to Q if and only if λx belongs to K_0 for some $\lambda > 0$.

Proof. We choose x'_0 and $\delta > 0$ as in Lemma 2.7.4. Let K_0 be the intersection of Q with the hyperplane $\langle x, x'_0 \rangle = 1$. This K_0 is obviously closed and convex. If x belongs to K then $x/\langle x, x'_0 \rangle$ belongs to K_0; and if y belongs to K_0, then $y \neq 0$ and y belongs to Q, hence $y = \lambda x$ for some $\lambda > 0$ and some x in K. It thus remains only to show that K_0 is compact. But if x belongs to K_0, then $x = \lambda y$ with $\lambda > 0$ and y in K; since $\langle y, x'_0 \rangle \geq \delta$ and $\langle x, x'_0 \rangle = 1$, it appears that $\lambda \leq \delta^{-1}$. Thus K_0 is contained in $\bigcup_{0 \leq \lambda \leq \mu} \lambda K$, where $\mu = \delta^{-1}$. The map $f: R \times K \to E$ defined by $(\lambda, y) \to \lambda y$ is continuous and so maps compact sets into compact sets. K_0, being contained in the f-image of the compact set $[0, \mu] \times K$, is therefore relatively compact; being closed in E, it is compact. ▌

Proof of Theorem 2.7.2. It is now easy to prove Theorem 2.7.2. We take K_0 as in Lemma 2.7.5. The hypotheses of Theorem 2.7.1 are fulfilled when K is replaced by K_0. Hence, by that theorem, there exists y_0 in K_0 such that $\langle y_0, x' \rangle \geq 0$ for x' in Γ. But there exists $\lambda > 0$ such that $x_0 = \lambda y_0$ belongs to K, and then $\langle x_0, x' \rangle \geq 0$ for x' in Γ, Q.E.D. ▌

On the basis of Theorems 2.7.1 and 2.7.2 we may now deduce a direct extension of Ville's theorem in the theory of games.

2.7.6 **Theorem.** Let E and F be real separated LCTVSs, and let L and M be subsets of E' and F', respectively. Let u be a map of L into F with the property that to each y' in M corresponds an x' in L such that

$$\langle u(x'), y' \rangle \geq 0.$$

Suppose also that the set $K = u(L)$ is *either* compact and convex *or* fulfills the conditions (1), (2), and (3) of Theorem 2.7.2 (E being therein everywhere replaced by F), and that the set $\bigcup_{\lambda \geq 0} \lambda M$ is convex and weakly closed in F'. The conclusion is that there exists an element x_0' of L such that

$$\langle u(x_0'), y' \rangle \geq 0$$

for all y' in M.

Proof. We apply Theorem 2.7.1 or Theorem 2.7.2 with F in place of E, $K = u(L)$, and $\Gamma = \bigcup_{\lambda \geq 0} \lambda M$. ▮

2.7.7 **Applications to the Theory of Games.** The aim now is to provide some *a posteriori* motivation for the preceding results of this section by showing their connection with the theory of games. In order to do this we first make some remarks about the latter subject. For the details the reader is recommended to consult von Neumann and Morgenstern [1] and McKinsey [1].

Many of the problems in the theory of games can be formulated, directly or otherwise, in the following terms. Two players, A and B, say, are involved; a "game" (or a single "play" of the said game) amounts to a choice by A and B, respectively, of elements s and t from preassigned sets S and T. According to the "rules of the game," at the end of a game resulting in the choice (s, t), B is to pay A an amount $P(s, t)$, P being a preassigned function on $S \times T$. By allowing P to take negative values, account is taken of the case in which the payment may be in the reverse direction. We are here contemplating only "zero-sum" games, in which there is no extraneous source of wealth: at each stage each player's gain is the other's loss.

The natural question is whether or not there exists always a "best" or "most advantageous" choice for each player to make; in examining this we may adopt the viewpoint of A. Let us assume for the moment that S and T are finite sets, and introduce the two numbers

$$W_* = \operatorname*{Max}_s \operatorname*{Min}_t P(s, t), \qquad W^* = \operatorname*{Min}_t \operatorname*{Max}_s P(s, t),$$

the "maximin" and the "minimax," respectively, of P (defined on $S \times T$). Suppose that $s_0 \in S$ and $t_0 \in T$ have the respective properties that

$$W_* = \operatorname*{Min}_t P(s_0, t), \qquad W^* = \operatorname*{Max}_s P(s, t_0).$$

A moment's thought shows that (1) if A chooses s_0, he will win W_* independently of what B chooses; and in no case can he win more than W_*; (2) if B chooses t_0, then A can in no case win more than W^*; and by appropriate choice A can win W^*. Supposing that in fact A chooses s_0 and B chooses t_0, we see that $W_* \leq W^*$. (This may of course be established quite independently of the present interpretation, thus: $\operatorname{Min}_{t'} P(s', t') \leq P(s', t')$, hence *a fortiori* $\operatorname{Min}_{t'} P(s', t') \leq \operatorname{Max}_s P(s, t)$; the first member here is independent of t and so $\operatorname{Min}_{t'} P(s', t') \leq \operatorname{Min}_t \operatorname{Max}_s P(s, t) = W^*$; the second member here is independent of s' and so

$$W_* = \operatorname*{Max,}_{s'} \operatorname*{Min}_{t'} P(s', t') \leq W^*.)$$

Now if it should happen that $W_* = W^* = W$, say, then A, by choosing s_0, can win at least W, while if B plays suitably (by choosing t_0) he can prevent A from winning any more than W. In an obvious sense, therefore, s_0 represents the "optimal strategy" for A. Similarly, t_0 represents B's optimal strategy. One may refer to (s_0, t_0) as a "strategic saddle point" for the game in question. We see that (s_0, t_0) has the property that

$$W = W_* = \operatorname*{Min}_t P(s_0, t) \leq P(s_0, t)$$

and

$$W = W^* = \operatorname*{Max}_s P(s, t_0) \geq P(s, t_0);$$

hence $W \leq P(s_0, t_0)$ and also $W \geq P(s_0, t_0)$. Thus

$$W = P(s_0, t_0)$$

and

$$P(s, t_0) \leq P(s_0, t_0) \leq P(s_0, t). \tag{2.7.1}$$

Reciprocally if (2.7.1) holds, we find that

$$\operatorname*{Max}_s P(s, t_0) \leq P(s_0, t_0) \leq \operatorname*{Min}_t P(s_0, t)$$

and therefore

$$W_* = \operatorname*{Max}_s \operatorname*{Min}_t P(s, t) \geq \operatorname*{Min}_t P(s_0, t) \geq P(s_0, t_0)$$

and

$$W^* = \operatorname*{Min}_t \operatorname*{Max}_s P(s, t) \leq \operatorname*{Max}_s P(s, t_0) \leq P(s_0, t_0),$$

that is, $W^* \leq P(s_0, t_0) \leq W_*$. Since we know that always $W^* \geq W_*$, it follows that $W_* = W^* = P(s_0, t_0)$. In other words, (2.7.1) characterizes (s_0, t_0) as a strategic saddle point, hence characterizes s_0 (resp. t_0) as optimal strategies for A (resp. B).

If now we contemplate the case in which S and T are not both finite, there arises the separate and preliminary question as to whether the expressions

$$\operatorname*{Sup}_s \operatorname*{Inf}_t P(s, t),$$

generalizing W_*, and

$$\operatorname*{Inf}_t \operatorname*{Sup}_s P(s, t),$$

generalizing W^*, exist finitely, and if so, whether each is actually assumed.

A point of greater significance, however, is that even in the simplest case in which S and T are finite, the numbers W_* and W^* are not generally equal. For instance, if $S = T = \{1, 2\}$ and $P(1, 1) = P(2, 2) = 1$, $P(1, 2) = P(2, 1) = -1$, then clearly $W_* = -1$ and $W^* = +1$. In such a case optimal strategies will not exist, and this is indeed the exception rather than the rule.

In view of this, one is driven to ask whether, even so, there may not be strategies relevant to a long sequence of games and the outcome in "the long run." One can interpret this mathematically by replacing individual choices (s, t) by pairs (σ, τ) in which σ (resp. τ) is a probability measure on S (resp. T), the "pay-off function" $P(s, t)$ being now replaced by the quantity

$$B(\sigma, \tau) = \int_S \int_T P(s, t)\, d\sigma(s)\, d\tau(t).$$

Most of the interesting cases are covered by that in which S and T are locally compact spaces and σ (resp. τ) is taken to mean a positive Radon measure on S (resp. T) with total mass exactly unity. Such objects are treated in great detail in Chapter 4. The search is now for "optimal mixed strategies" σ_0 and τ_0 having the property (entirely analogous to (2.7.1)) that

$$B(\sigma, \tau_0) \leq B(\sigma_0, \tau_0) \leq B(\sigma_0, \tau). \tag{2.7.2}$$

The fundamental theorem of the theory of games states that under pretty wide conditions on S, T and $P(s, t)$, such optimal mixed strategies σ_0 and τ_0 do indeed exist. We proceed to show how this may be deduced rapidly from Theorem 2.7.6, provided one makes use of a few simple properties of Radon measures.

We shall suppose in doing this that S and T are compact spaces (in practice usually compact real intervals) and that $P(s, t)$ is continuous on $S \times T$. From results established in Chapter 4, it follows that the integral defining $B(s, t)$ exists finitely and may be interpreted indifferently as an iterated itegral with either order of integrations adopted. In Theorem 2.7.6 we shall take $E = C(S)$ and $F = C(T)$, which are, respectively, the spaces of real-valued continuous functions on S (resp. T), made into Banach spaces with the appropriate uniform norm. As will appear in Section 4.10 E' (resp. F') may then be identified with the space of real Radon measures σ (resp. τ) on S (resp. T); in this identification the measure σ (resp. τ) corresponds in a one-to-one fashion with the linear form $x \to \int_S x(s)\, d\sigma(s)$ [resp. $y \to \int_T y(t)\, d\tau(t)$] on $C(S)$ [resp. $C(T)$]. For L (resp. M) in Theorem 2.7.6 we take the set of positive Radon measures σ (resp. τ) on S (resp. T) that have total mass $\sigma(S)$ (resp. $\tau(T)$) equal to unity. According to Theorem 4.10.3, each of L and M is weakly compact; and it is almost obvious that $\Gamma = \bigcup_{\lambda \geq 0} \lambda M$, the set of all positive Radon measures on T, is a weakly closed convex cone in F'. Finally, the map u involved in Theorem 2.7.6 will be that which carries the measure σ into the function.

$$\hat{\sigma}(t) = \int_S P(s, t)\, d\sigma(s),$$

which is easily seen to be an element of F (that is, a continuous real-valued function on T).

Appeal to Theorem 2.7.6 demands that we verify the compactness of $K = u(L)$ as a subspace of $F = C(T)$. Now if $|P(s, t)| \leq m$ on $S \times T$, then it is plain that $|\hat{\sigma}(t)| \leq m$ for t in T and $\hat{\sigma}$ in L; the functions belonging to K are therefore uniformly bounded. The next step is to show that they are equicontinuous. However, using the continuity of P on $S \times T$ and the compactness of S and T, it is easily seen that, given $\varepsilon > 0$ and a point t_0 of T, there exists a neighbourhood V of t_0 such that $|P(s, t) - P(s, t_0)| \leq \varepsilon$ for s in S and t in V. Then $|\hat{\sigma}(t) - \hat{\sigma}(t_0)| \leq \varepsilon$ for t in V and any σ in L, establishing the equicontinuity at t_0 of the members of K. From Ascoli's theorem it now follows that K is relatively compact in $C(T)$, and so it suffices to show that K is closed in $C(T)$. But, for a fixed t in T, it is obvious that $\sigma \to \hat{\sigma}(t)$ is continuous on L (relative to the weak topology); since L is weakly compact, $K = u(L)$ is compact for the topology of simple convergence on T, hence closed for this topology (the latter being separated), and *a fortiori* closed for the stronger topology of convergence uniform on T.

As a final preliminary to the application of Theorem 2.7.6, we wish to establish that the maximin

$$W_* = \operatorname*{Max}_{\sigma} \operatorname*{Min}_{\tau} B(\sigma, \tau)$$

and the minimax

$$W^* = \operatorname*{Min}_{\tau} \operatorname*{Max}_{\sigma} B(\sigma, \tau)$$

both exist finitely. Since L and M are weakly compact, so is $L \times M$ for the product topology. It is therefore enough to show that $B(\sigma, \tau)$ is continuous on $L \times M$ for the product topology—for then $\operatorname{Min}_\tau B(\sigma, \tau)$ will be upper semicontinuous (and finite) on L and will therefore admit a finite attained maximum W_* on L; likewise $\operatorname{Max}_\sigma B(\sigma, \tau)$ will be finite and lower semicontinuous on M and will there admit a finite minimum W^*. Now, as regards the continuity of $B(\sigma, \tau)$ on $L \times M$, this follows from the fact that the continuous function P on $S \times T$ is the uniform limit of finite sums of the type

$$P_0(s, t) = \sum_{i=1}^{n} u_i(s) v_i(t),$$

where u_i and v_i are chosen from $C(S)$ and $C(T)$, respectively. (If S and T are compact real intervals, this follows from Weierstrass' polynomial approximation theorem for functions of two real variables; in the general case it is an instance of Stone's extension of Weierstrass' theorem given in Subsection 4.10.5.) If P_0 is thus chosen so that $|P - P_0| \leq \varepsilon$ uniformly on $S \times T$, then $|B(\sigma, \tau) - \int_S \int_T P_0 \, d\sigma \, d\tau| \leq \varepsilon$ uniformly for (σ, τ) in $L \times M$. On the other hand,

$$\int_S \int_T P_0 \, d\sigma \, d\tau = \sum_{i=1}^{n} \int_S u_i \, d\sigma \cdot \int_T v_i \, d\tau$$

is obviously continuous in (σ, τ) for the product topology. Consequently, $B(\sigma, \tau)$, being the uniform limit of continuous functions, is continuous.

Now we apply Theorem 2.7.6. Since

$$B(\sigma, \tau) = \langle u(\sigma), \tau \rangle,$$

the duality being that between $C(T) = F$ and F', the main Hypothesis in that theorem says exactly that $W^* \geq 0$, while the Conclusion reads precisely $W_* \geq 0$. Thus for any given continuous P, $W^* \geq 0$ entails $W_* \geq 0$. Since we may here replace P by $P - k$, k being any real constant, which change replaces $B(\sigma, \tau)$ by $B(\sigma, \tau) - k$, we see that by the same token $W^* \geq k$ entails $W_* \geq k$. It follows directly that $W_* \geq W^*$, and hence $W_* = W^*$.

In this way we have established the existence of optimal mixed strategies for any game for which S and T are compact and the payoff function P is continuous. This is the fundamental theorem in the theory of games. Further details and developments will be found in the books of von Neumann and Morgenstern and of McKinsey already cited.

2.7.9 **Remarks.** The "minimax theorem" discussed in the preceding subsection has been formulated in even more general terms by Fan [1]. Fan considers the situation in which one has separated LCTVSs E and F, compact convex subsets H and K of E and F, respectively, and a real-valued continuous function f on $H \times K$; it is assumed that for each $(x_0, y_0) \in H \times K$ the sets

$$\left\{x \in H : f(x, y_0) = \operatorname*{Max}_{x' \in H} f(x', y_0)\right\}$$

and

$$\left\{y \in K : f(x_0, y) = \operatorname*{Min}_{y' \in K} f(x_0, y')\right\}$$

are each convex. The conclusion is that then

$$\operatorname*{Max}_{x \in H} \operatorname*{Min}_{y \in K} f(x, y) = \operatorname*{Min}_{y \in K} \operatorname*{Max}_{x \in H} f(x, y).$$

Fan's proof depends on a generalized version of Kakutani's fixed point theorem mentioned in Subsection 3.6.6.

EXERCISES

2.1 Let $T = N$, the set of natural numbers, and consider the space $c_0 = c_0(N)$. For q complex and subject to Re $q > 0$, let x_q be that element of c_0 defined by $x_q(t) = e^{-qt}$; and for any $n \in N$, let y_n be that element of c_0 defined by $y_n(t) = t^{n-1}e^{-t}$. Show that the x_q and the y_n each form families that are fundamental in c_0 (that is, the finite linear combinations of the x_q are dense in c_0, and likewise with the finite linear combinations of the y_n).

The x_q and the y_n belong also to the spaces $\ell^p(N)$ when $1 \leq p < \infty$. Discuss whether they are fundamental in these spaces.

2.2 Let E be a vector space, F a LCTVS, and $L = L(E, F)$ the vector space of all linear maps of E into F. On L consider two topologies, namely, (1) the topology t_1 having as a base at 0 the sets of the type

$$W(A, V) = \{u \in L : u(A) \subset V\},$$

where A is a finite subset of E and V a neighbourhood of 0 in F; and (2) the topology t_2 having a base a 0 formed of the sets $W(A, U)$, where again A is a finite subset of E and where U is a weak neighbourhood of 0 in F. Show that any linear form f on L that is continuous for t_1 is also continuous for t_2.

2.3 Let E be a vector space and F a LCTVS, f a linear form on $L(E, F)$ continuous for the topology of simple convergence. Show that there exist finite families $(x_i)_{1 \leq i \leq n}$ in E and $(y'_i)_{1 \leq i \leq n}$ in F' such that

$$f(u) = \sum_{1 \leq i \leq n} \langle u(x_i), y'_i \rangle$$

for all u in $L(E, F)$. Deduce that f is continuous for the topology of simple convergence when F is taken with its weak topology $\sigma(F, F')$.

2.4 Let E be the space of entire functions of one complex variable, endowed with the topology of compact convergence (for which a defining family of seminorms is given by

$$p_n(f) = \operatorname{Sup}_{|t| \leq n} |f(t)| \qquad (n = 1, 2, \cdots)).$$

For $f \in E$ and ξ a complex number, let $f_\xi \in E$ be fined by

$$f_\xi(t) = f(\xi t).$$

Show that if F is a continuous linear form on E, and if $f \in E$, then $F_0(\xi) = F(f_\xi)$ is an entire function of order not exceeding that of f, and that

$$F_0^{(n)}(0) = f^{(n)}(0) \cdot F(t^n).$$

Deduce that the f_ξ are fundamental in E, provided $f^{(n)}(0) \neq 0$ for $n = 0, 1, \cdots$.

2.5 Let E be the space of all continuous, complex-valued functions of a real variable having period 2π, taken with the norm $\|f\| = \operatorname{Sup} |f(t)|$. Verify that E is a Banach space.

If $f \in E$ and a is real we define the a-translate of f by $f_a(t) = f(t + a)$. Let F be a continuous linear form on E and put $F_0(a) = F(f_a)$, where $f \in E$ is given. Show that the nth Fourier coefficient of F_0, namely,

$$\hat{F}_0(n) = \left(\frac{1}{2\pi}\right) \int_{-\pi}^{\pi} F_0(a) e^{-ina} \, da,$$

has the value $\hat{f}(n) \cdot F(e^{int})$.

Let T_f denote the closed vector subspace of E generated by all translates f_a of f. Prove that $e^{int} \in T_f$ if and only if $\hat{f}(n) \neq 0$. Assuming that the e^{int} are fundamental in E, deduce that $T_f = E$ if and only if $\hat{f}(n) \neq 0$ for $n = 0, \pm 1, \pm 2, \cdots$.

[Notes: This is an analogue for Fourier series of a theorem of Wiener for Fourier integrals (see Wiener [1] p. 100; Loomis [1] p. 148; Rudin [9], Chapter 7.) See also Exercises 4.16, 10.13, and 10.28. There are many approximation problems related to translates; see for example Edwards [1], [2], [12], Agnew [1], [2] and Schwartz [7], [9] and [11].]

2.6 Let T be an arbitrary set, and let $L \subset M$ be vector subspaces of R^T such that for each $x \in M$ there exist elements x' and x'' of L such that $x' \leq x \leq x''$. Show that any positive linear form f on L admits a positive linear extension to M.

[Hint: Define p on M by

$$p(x) = \operatorname{Inf} \{f(y) : y \in L, y \geq x\}$$

and then use the Hahn-Banach theorem.]

2.7 Let E be a vector space endowed with a topology for which $x \to \lambda x$ is continuous at $x = 0$ for any scalar λ. Show that in order that there shall exist a linear form $f \neq 0$ on E that is continuous at 0, it is necessary and sufficient that there exist in E a convex and absorbent neighbourhood U of 0 distinct from E.

[Hint: If f exists, the set $U = \{x \in E : |f(x)| < 1\}$ has the said properties. Conversely, suppose U exists. Let p be its Minkowski functional; p is subadditive and positive homogeneous on E, $p(U) \leq 1$, and $p \neq 0$. Choose x_0 with $p(x_0) = 1$. Define the linear form f_0 on the vector subspace generated by x_0 by $f(\lambda x_0) = \lambda$. Apply the Hahn-Banach theorem to extend f_0 into a linear form f on E such that $f(x_0) = f_0(x_0) = 1$ and $f(x) \leq p(x)$. f is continuous at 0 since $f(x) \leq \alpha$ if $x \in \alpha U$, and αU is a neighbourhood of 0 for any $\alpha > 0$. Since $x \to -x$ is continuous at 0, $-\alpha U$ is also a neighbourhood of 0. One has $|f(x)| \leq \alpha$ if $x \in (\alpha U) \cap (-\alpha U)$.]

2.8 Let E be a real pre-Hilbert space, K a complete and convex subset of E, x_0 a point of $E \backslash K$. Choose $a \in K$ at minimum distance from x_0 and put

$$y = (x_0 - a)/\|x_0 - a\|.$$

Show that

$$(x \mid y) < (x_0 \mid y) \qquad \text{for } x \in K.$$

Compare with Corollary 2.2.4.

2.9 Let E be a real separated TVS of finite dimension, K a convex subset of E, x_0 a frontier point of K. Show that there exists a hyperplane of support of K passing through x_0. *

Remark. If E is of infinite dimension, the result is in general false, even if K is assumed to be compact and x_0 to be an extreme point of K; see Bourbaki [4], p. 85.

[Hint: One may argue with $\overline{K}$ in place of K and so assume that K is closed. Take a sequence $x_n \in E \backslash K$ with $x_n \to x_0$. Apply Exercise 2.8, using the compactness of bounded, closed subsets of the dual space.]

2.10 Let K be a convex subset of a real vector space E. A point x of E (necessarily in K) is said to be *internal to* K (relative to E) if and only if to each $y \in E$ corresponds a number $\varepsilon > 0$ such that $x + \alpha y \in K$ whenever $|\alpha| \leq \varepsilon$. This signifies that $K - x$ is absorbent.

Show that if K possesses internal points, all of which lie on one side of a hyperplane H in E, then K itself lies on one side of H.

2.11 Let K be a convex set in a real vector space E. Denote by K_i the set of points of E that are internal to K. Show that all points of K_i are internal to K_i.

2.12 Let E be a real vector space, K a nonvoid convex set all of whose points are internal to K. Let V be a vector subspace of E disjoint from K. Show that there exists a hyperplane H in E containing V and disjoint from K. Compare with Theorem 2.2.1.

2.13 Let E be a real vector space, E' a vector subspace of E, H' a hyperplane in E', and K a convex set in E such that E' contains at least one point internal to K relative to E. Given that $K \cap E'$ lies on one side of H', show that there exists a hyperplane H in E such that $H' = H \cap E'$ and K lies on one side of H.

[Hint: Let K_i be the set of points of E that are internal to K relative to E. Show that K_i does not meet H', using Exercise 2.11. Apply Exercise 2.12 with K there replaced by K_i and V by H'.]

2.14 Let E be a real vector space, E' a vector subspace of E, and H' a hyperplane in E'. Suppose that K is a convex set in E such that $K \cap E'$ lies on one side of H'. Suppose further that for each $y \in E$ there exists $x \in E'$ such that $x + \alpha y \in K$ for $|\alpha|$ sufficiently small. Show that there exists a hyperplane H in E such that $H' = H \cap E'$ and such that K lies on one side of H. (Dixmier [3]. This generalizes the result of Exercise 2.13.)

[Hints: Consider the set of all pairs (E^*, H^*), where E^* is a vector subspace of E, H^* is a hyperplane in E^*, $K \cap E^*$ lies on one side of H^*, and $E^* \supset E'$ and $H^* \cap E' = H'$. Partially order this set so that $(E_1, H_1) \leq (E_2, H_2)$ signifies that $E_1 \subset E_2$ and $H_1 \subset H_2$. Verify that this partially ordered set is inductively ordered. Apply Zorn's lemma to establish the existence of a maximal element (E_0, H_0). It suffices to show that $E_0 = E$.

* See Appendix, page 771.

Assuming the contrary choose E_0' so that $\dim E_0'/E_0 = 1$ and put $K_0 = E_0' \cap K$. Passing to the quotient modulo H_0, reduce the problem to a two-dimensional one and use Exercise 2.9 to deduce that there exists a hyperplane H_0' in E_0' such that the pair (E_0', H_0') is "bigger than" (E_0, H_0), contradicting maximality of the latter.]

2.15 Let T be an arbitrary set and G a semigroup of maps $t \to u \cdot t$ of T into itself. Suppose that there exists a left-invariant mean μ on G—that is, a positive linear form on the space of all bounded real-valued functions on G that is invariant under left translations and such that $\mu(1) = 1$.

Write E for the vector space of all bounded real-valued functions on T, and for $f \in E$ and $u \in G$ define $f_u = f \circ u$. Take any $g \in E$ with $g \geq 0$. Write E' for the vector subspace of E generated by the functions g_u as u ranges over G, and write E'' for the vector subspace of E generated by positive functions that are majorized by members of E' (so that $E' \subset E'' \subset E$). Suppose there exists on E' a positive linear form $\varphi \neq 0$ that is invariant under G.

Show that there exists on E'' a positive linear form Φ that is invariant under G and such that $\Phi(g) = 1$. (Dixmier [3], Théorème 8.)

Note that μ exists provided G is an Abelian semigroup or a solvable group (see Section 3.5 and Exercise 3.12).

[Hint: Apply the result of Exercise 2.14 (with E there replaced by E'') to show that there exists a PLF ω on E'' such that $\omega \mid E'$ differs from φ by a nonzero constant factor. So arrange that $\omega(g_u) = 1$ for all $u \in G$. Consider then $\Phi(f) = \mu[\omega(f_u)]$, where $u \to \omega(f_u)$ is viewed as a bounded real-valued function on G.]

2.16 Show that there exists a positive, finitely additive set function m defined for all subsets of R^2 such that $m(E) = +\infty$ if $E \subset R^2$ is unbounded, $m(Q) = 1$ for any square Q of side 1, and m is invariant under all isometries of R^2. (Dixmier [3], p. 226.)

[Hint: In the preceding exercise take $T = R^2$, G the group of all isometries of R^2. G is solvable since, if G_1 is the normal subgroup of translations, both G_1 and G/G_1 are Abelian. Take $g = \chi_Q$, the characteristic function of Q.]

2.17 Let E be any normed vector space. Show that there exists a separated compact space T and a linear isometry of E onto a vector subspace of $C(T)$. (Here $C(T)$ is taken to mean the space of scalar-valued continuous functions on T, the norm being the maximum modulus of the function in question.) Show further that if E is separable, one may suppose that T is metrizable. (For a more general result, see Exercise 6.17.)

[Hint: Take for T the closed unit ball in the dual space E', endowed with the topology induced by $\sigma(E', E)$.]

2.18 Let E and F be vector subspaces of a vector space G, each endowed with a locally convex, vector-space topology. Consider the vector-space topology on $E \cap F$ having a base at 0 formed of the sets $U \cap V$, where U (resp. V) is a neighbourhood of 0 in E (resp. F). Show that each continuous linear form on $E \cap F$ is of the type

$$z \to \langle z, x' \rangle + \langle z, y' \rangle$$

for $z \in E \cap F$, where $x' \in E'$ and $y' \in F'$.

[Hint: Consider the linear map $J: z \to (z, z)$ of $E \cap F$ into $E \times F$ and use the Hahn-Banach theorem and Exercise 1.29.]

2.19 Use Exercise 2.13 to derive part (1) of Theorem 1.7.1, that is, the following assertion: If E is a real vector space, p a gauge function on E, E_0 a vector subspace of E, and f_0 a linear form on E_0 satisfying $f_0(x) \leq p(x)$ for $x \in E_0$, then there exists on E a linear form f extending f_0 and such that $f(x) \leq p(x)$ for $x \in E$.

[Hints (adapted from an argument due to Klee and indicated to me by D. Fearnley-Sander): Form the vector space $E \times R$ and consider in it the convex set

$$K = \{(x, \xi): p(x) \leq \xi\}$$

and the homogeneous hyperplane H_0 in the vector subspace $E_0 \times R$ defined by

$$H_0 = \{(x, \xi): x \in E_0, f_0(x) = \xi\}.$$

Verify that the point (0, 1) is internal to K and that $K \cap (E_0 \times R)$ lies on one side of H_0. Apply Exercise 2.13 to obtain a homogeneous hyperplane H in $E \times R$, on one side of which K lies. Show that H has an equation of the form $F(x, \xi) = 0$, where F is a linear form on $E \times R$ satisfying $F(0, 1) = -1$, and deduce that $f(x) = F(x, 0)$ satisfies all the requirements.]

CHAPTER 3

Fixed-point Theorems

3.0 Foreword

By a fixed-point theorem (FPT) we shall understand a statement which asserts that under certain conditions a mapping u of a set E into itself admits one or more fixed points—that is, points x of E for which $u(x) = x$. The first theorems of this type applied to cases in which E is a topologically simple subset of R^n and u is a continuous map of E into itself. Thus, for example, Brouwer's FPT (see Section 0.5) asserts the existence of a fixed point whenever E is the closed unit ball in R^n and u is continuous. Clearly, in any such FPT, one may replace E by any homeomorph thereof. Theorems of this sort, restricted to spaces E that are subspaces of R^n, are of little immediate use in functional analysis, where usually one is concerned with the case in which E is an infinite dimensional subset of some function space. The first infinite dimensional FPTs appeared some forty years ago in Birkhoff and Kellogg [1]. Later Schauder [3], [6] extended Brouwer's theorem to the case in which E is a compact convex set in a normed vector space. Shortly afterwards Tychonoff [1] extended Schauder's result from normed spaces to arbitrary locally convex TVSs. In both cases, Brouwer's theorem was used as a starting point.

We shall in this chapter begin with two FPTs that can be approached without any combinatorial topology as background. The first result (Theorem 3.1.1) applies to contraction (that is, distance-diminishing) maps of a complete metric space into itself. The second result (Theorem 3.2.1), due to Markov and Kakutani, refers to simultaneous fixed points of suitable families of continuous maps of a compact convex subset of a TVS into itself. This last will be used to yield a further theorem of the Hahn-Banach type (see Subsection 2.3.5).

The maps in the Markov-Kakutani theorem must satisfy a condition close to linearity. Markov's proof [2] used the Schauder-Tychonoff theorem, but we use an approach avoiding any such appeal and which is close to that given by Kakutani [1].

Our third main result (Theorem 3.6.1) is the Schauder-Tychonoff theorem already mentioned. This relaxes all conditions of linearity on the mapping involved, hence its great generality. Even before Schauder's formulations (for separable Banach spaces in [3] and Banach spaces in general in [6]), Birkhoff and Kellogg [1] gave special cases of the theorem for the spaces $C^n[0, 1]$ and $L^2(0, 1)$. Applications of the Schauder-Tychonoff theorem to differential equations have been discussed by many writers, amongst the first being Graves [2] and Leray [3]. Further developments appear in Bonsall [1].

3.1 FPTs for Contraction Maps

Let E be a semimetric space—that is, a set equipped with a semimetric d; d is thus a real-valued function on $E \times E$ such that

$$d(x, y) = d(y, x) \geq 0, \; d(x, x) = 0, \; d(x, y) \leq d(x, z) + d(z, y),$$

for arbitrary elements x, y, z of E. (d is a metric if in addition $d(x, y) \neq 0$ for $x \neq y$.) A map u of E into itself is termed a *contraction of* E if there exists a number k satisfying $0 \leq k < 1$ such that $d(u(x), u(y)) \leq k \cdot d(x, y)$ for arbitrary x and y in E. Any such contraction of E is obviously a continuous map of E into itself.

3.1.1 Theorem. (1) Let (E, d) be a complete semimetric space, u a continuous map of E into itself such that for some natural number p, u^p is a contraction. Then there exists at least one point x of E satisfying

$$d(u(x), x) = 0. \tag{3.1.1}$$

Moreover, if y is any point of E satisfying $d(u(y), y) = 0$, then $d(x, y) = 0$.

(2) If (E, d) is a complete metric space and u is as in (1), then u admits exactly one fixed point in E.

Proof. (2) is an immediate corollary of (1).

Let us next establish the second assertion of (1). To do this, we notice that (3.1.1) signifies that $u(x) \in \overline{\{x\}}$. Then, by continuity of u,

$$u^2(x) \in u(\overline{\{x\}}) \subset \overline{\{u(x)\}} \subset \overline{\overline{\{x\}}} = \overline{\{x\}}.$$

By repetition of this calculation we find that $u^p(x) \in \overline{\{x\}}$. Similarly, $u^p(y) \in \overline{\{y\}}$. It follows thence that

$$d(x, y) = d(u^p(x), u^p(y)) \leq k \cdot d(x, y),$$

where $k < 1$ since u^p is a contraction. Hence $d(x, y) = 0$, as alleged.

To prove the existence of a solution of (3.1.1), we consider first the case in which $p = 1$, so that u itself is a contraction. We choose freely any point x_0 of E and define the sequence (x_n) by the recurrence relation $x_n = u(x_{n-1})$ for $n = 1, 2, \cdots$. If n and m are natural numbers,

$$d(x_{n+m}, x_n) = d(u^{n+m}(x_0), u^n(x_0)) \leq k^n \cdot d(x_m, x_0)$$

and

$$\begin{aligned} d(x_m, x_0) &\leq d(x_m, x_{m-1}) + \cdots + d(x_1, x_0) \\ &\leq k^{m-1} d(x_1, x_0) + \cdots + d(x_1, x_0) \leq (1 - k)^{-1} d(x_1, x_0). \end{aligned}$$

Hence

$$d(x_{n+m}, x_n) \leq k^n(1 - k)^{-1} d(x_1, x_0),$$

which, since $k < 1$, shows that the sequence (x_n) is Cauchy in E. E being complete, the sequence (x_n) converges to some x in E. The sequence $(u(x_n)) = (x_{n+1})$, being a subsequence of (x_n), then also converges to x. But, since u is continuous, $(u(x_n))$ converges to $u(x)$, and (3.1.1) follows.

Passing finally to the case in which we know merely that $v = u^p$ is contraction, we repeat the preceding construction with v in place of u. This leads to a point x of E satisfying $d(v(x), x) = 0$. Moreover, for the sequence $x_n = v^n(x_0)$ thus formed, we have $u(x_n) = u(v(x_{n-1})) = v(u(x_{n-1}))$. Since (x_n) converges to x and both u and v are continuous, it follows that $u(x)$ and $v(u(x))$ are at zero distance. But then we have

$$\begin{aligned} d(u(x), x) &\leq d(u(x), v(u(x)) + d(v(u(x)), v(x)) + d(v(x), x)) \\ &= d(v(u(x)), v(x)) \leq k \cdot d(u(x), x), \end{aligned}$$

which, since $k < 1$, entails that $d(u(x), x) = 0$. ∎

Remarks. The construction of the sequence (x_n) and the study of its convergence are often spoken of as the method of successive approximations. It is usually known by Picard's name (see Picard [1]) but was used earlier by Peano [1]. The abstract setting for the method is due to Banach [2] and Cacciop-poli [1]. See also the more recent work of Kantorovich [1].

The fixed point theorem we have established represents only one of the applications of this sort of technique. An excellent exposition of other developments is to be found in Chapter X of Dieudonné [13]. Extensions of the contraction principle itself have been given recently by Edelstein [1], [2], one of which is discussed in Exercise 3.7.

As we shall now verify by examples, the above theorem, simple though it is, has useful applications in analysis.

3.1.2 **Applications to Integral equations.** Let T be an interval of the real axis, and let E denote the vector space of bounded, continuous, K-valued functions on T (K being the real or the complex field). E is formed into a complete metric space by introducing the metric $d(x, y) = \|x - y\|$, where

$$\|x\| = \text{Sup}\,\{|x(t)| : t \in T\}.$$

Suppose given a K-valued function f on $T \times T \times K$ such that for each x in E and each t in T, the function

$$s \to f(t, s, x(s))$$

is integrable over T, while the function

$$t \to \int_T f(t, s, x(s))\, ds$$

is bounded and continuous on T. Choosing any element b of E, we consider the map u of E into itself defined by

$$u(x)(t) = -\int_T f(t, s, x(s))\, ds + b(t).$$

A fixed point of u is thus a solution x of the nonhomogeneous and generally nonlinear integral equation

$$x(t) + \int_T f(t, s, x(s))\, ds = b(t), \tag{3.1.2}$$

subject to the condition $x \in E$.

Under suitable conditions, u will be a contraction map of E, and the existence of a unique solution in E of (3.1.2) will then follow from Theorem 3.1.1.

Suppose, for instance, that f satisfies an inequality of the form

$$|f(t, s, \xi) - f(t, s, \xi')| \leq F(t, s) \cdot |\xi - \xi'|. \tag{3.1.3}$$

Then, if x_1 and x_2 belong to E and $y_1 = u(x_1)$ and $y_2 = u(x_2)$, we find that

$$d(y_1, y_2) \leq k \cdot d(x_1, x_2),$$

where

$$k = \text{Sup}\left\{\int_T F(t, s)\, ds:\ t \in T\right\}.$$

If $k < 1$, u is a contraction map and Theorem 3.1.1 applies.

Were we to introduce an "eigenvalue parameter" λ and replace f by λf, we should be able to conclude the existence and uniqueness of a solution x in E of the equation

$$x(t) + \lambda \int_T f(t, s, x(s))\, ds = b(t) \tag{3.1.4}$$

for any given b in E, provided λ is small enough to make $k\lambda < 1$.

If f takes the form

$$f(t, s, \xi) = K(t, s) \cdot \xi,$$

then (3.1.4) assumes the guise of a nonhomogeneous (linear) Fredholm equation with kernel K. Solubility obtains at least when λ is sufficiently small. In general, for certain values of λ, the equation will be insoluble for certain choices of b. This aspect forms part of the general and highly developed theory of such equations, for which see Smithies [2], Zaanen [1], Dieudonné [13].

An interesting variant arises when T is a bounded interval with extremities $\alpha < \beta$ and the kernel K is bounded on $T \times T$ and $K(t, s) = 0$ for $s > t$. Then (3.1.3) takes the form

$$x(t) + \lambda \int_\alpha^t K(t, s)x(s)\, ds = b(t), \tag{3.1.5}$$

a nonhomogeneous (linear) Volterra equation. The corresponding form taken by u is

$$u(x)(t) = -\lambda \int_\alpha^t K(t, s)x(s)\, ds + b(t).$$

If we suppose that $|K(t, s)| \leq M$ and we construct the iterated transforms $x_n = u^n(x)$ and $y_n = u^n(y)$ of any two chosen functions x and y in E, we find step by step that

$$d(x_n, y_n) \leq |\lambda|^n M^n (\beta - \alpha)^n \frac{d(x, y)}{n!}.$$

Now the cofactor of $d(x, y)$ on the right-hand side tends to zero as $n \to \infty$. Thus, no matter what the value of λ, there will exist a natural number p such that u^p is a contraction. The second half of Theorem 3.1.1 accordingly affirms the existence and uniqueness of a solution in E of (3.1.5) for any given λ and any given b in E. The absence of any restriction on the magnitude of λ is peculiar to the Volterra type of equation.

It will soon become clear that the setup described above may be modified in numerous ways. We have so far imposed conditions of continuity on the functions involved in order to avoid any questions about the existence of the integrals. However, granted further developments in integration theory (exemplified in Chapter 4), one may take for E the Lebesgue space $\mathscr{L}^p(T, \mu)$ and carry out a similar analysis in that setting.

The arguments may further be extended to include the case in which the functions x on T take their values in a given semimetric space, while f takes its values in a complete TVS whose topology is defined by a semimetric. Such extensions will, of course, depend upon the use of techniques for handling the integration of vector-valued functions. The reader may care to consider the matter after studying the relevant portions of Chapter 8.

3.1.3 Applications to Differential Equations. Fixed point theorems for integral operators can usually be turned around so as to yield at least local existence and uniqueness theorems for ordinary differential equations of the type

$$\frac{dx}{dt} = f(t, x).$$

The general idea is that if f is continuous, the solution of this differential equation and an initial condition of the type $x(t_0) = y_0$, t_0 and y_0 being given, is none other than a solution of the integral equation

$$x(t) = y_0 + \int_{t_0}^{t} f(s, x(s))\, ds.$$

If f is assumed to be Lipschitzian in its second argument, the results of Subsection 3.1.2 will apply, a contraction mapping being obtained provided the Lipschitzian constant and the length of the fundamental interval of t-values are small enough. This leads to local existence and uniqueness theorems. A separate argument is required to piece these local solutions together into a global one.

Quite recently it has been shown that a more effective method results if one takes a new definition of distance, depending on the Lipschitzian constant. This method, due to Bielecki [1], gives directly a global theorem. A description of the method follows.

Suppose that F is a Banach space and T a real interval. Since we shall wish to integrate F-valued functions, it is necessary at this stage to assume that F is finite-dimensional, the theory for vector-valued functions in infinite-dimensional spaces appearing only in Chapter 8. We consider in any case the differential equation

$$\frac{dx}{dt} = f(t, x) \tag{3.1.6}$$

and the initial condition

$$x(t_0) = y_0, \tag{3.1.7}$$

$t_0 \in T$ and $y_0 \in F$ being preassigned. By a solution of (3.1.6) we shall mean a continuous function x on T into F that has at each interior point t of T a derivative equal to $f(t, x(t))$. (If F is of infinite dimension it would be necessary to choose between weak and strong differentiability of F-valued functions. The present method is adequate to show that strong differentiability obtains.) It is assumed that f is continuous on $T \times F$ into F, that it satisfies a Lipschitz condition

$$\|f(t, y_1) - f(t, y_2)\| \leq L\|y_1 - y_2\|$$

for $t \in T$ and $y_1, y_2 \in F$, the number L being independent of these variables, and that f is bounded on each set of the type $T \times B$, where B is a bounded subset of F.

In the first instance we shall assume that T is bounded and introduce the Banach space E of bounded continuous functions on T into F, the norm being

$$N(x) = \text{Sup}\,\{e^{-\lambda L|t-t_0|}\,\|x(t)\| : t \in T\}$$

and λ being a fixed number satisfying $\lambda > 1$.

We note that the bounded solutions of (3.1.6) and (3.1.7) are precisely those elements x of E satisfying

$$u(x) = x,$$

u being defined by

$$u(x)(t) = y_0 + \int_{t_0}^{t} f(s, x(s))\,ds \qquad (t \in T).$$

It is evident that u maps E into itself. If we can show that u is a contraction of E, the existence and uniqueness of a bounded solution of (3.1.6) and (3.1.7) will be established by appeal to Theorem 3.1.1. The verification that u is a contraction is quite straightforward and we merely sketch in an outline.

Suppose that x_1 and x_2 belong to E, so that

$$[u(x_1) - u(x_2)](t) = \int_{t_0}^{t} [f(s, x_1(s)) - f(s, x_2(s))]\,ds.$$

So

$$\|[u(x_1) - u(x_2)](t)\| \leq L \int_{I_t} \|x_1(s) - x_2(s)\|\,ds,$$

where I_t is the interval (t_0, t) or (t, t_0) according as $t > t_0$ or $t \leq t_0$. The term on the right-hand side is majorized by

$$L \cdot N(x_1 - x_2) \cdot \int_{I_t} e^{\lambda L|s-t_0|}\,ds,$$

and the integral remaining is majorized by

$$(\lambda L)^{-1}[e^{\lambda L|t-t_0|} - 1] < (\lambda L)^{-1} e^{\lambda L|t-t_0|}.$$

It follows thus that

$$N[u(x_1) - u(x_2)] \leq N(x_1 - x_2) \cdot \lambda^{-1}.$$

Since $\lambda > 1$, u is a contraction.

It remains to examine the case in which T is unbounded. To discuss this we introduce an increasing sequence (T_k) of compact intervals containing t_0 as an interior point (relative to the subspace T) and having T as their union. By taking the bounded solutions for the various intervals T_k, the uniqueness portion of the result already established shows that these various solutions "piece themselves together" to form a solution throughout T which is locally bounded on T. Conversely if x is a locally bounded solution on T, then uniqueness shows that $x \mid T_k$ is the only bounded solution throughout T_k. This being the case for each k, uniqueness of the solution throughout T is established.

For further details concerning abstract-valued solutions of differential equations, see Bourbaki's treatise "Eléments de mathématique, Livre IV, Fonctions d'une variable réelle. Chapter IV" (Act. Sci. et Ind. no. 1132, Paris 1951).

It should be noted that, as much for differential as for integral equations, the contraction mapping method is practically restricted in its applications to those cases in which $f(t, y)$ is Lipschitzian in y. To handle cases in which this Lipschitzian condition may not be satisfied, an improved fixed-point theorem is necessary. This matter receives further attention in Section 3.6.

3.2 The Markov-Kakutani Theorem

We are here concerned with the existence of simultaneous fixed points of suitable families of continuous maps of a compact convex set into itself.

3.2.1 Theorem. Let E be a separated TVS, K a nonvoid compact convex set in E. Suppose that Γ is a set of continuous maps of K into itself satisfying the following two conditions:

(a) If $u \in \Gamma$, x and y belong to K, and α and β are two positive real numbers satisfying $\alpha + \beta = 1$, then

$$u(\alpha x + \beta y) = \alpha u(x) + \beta u(y).$$

(b) There exists a natural number n and subsets Γ_i $(0 \leq i \leq n - 1)$ of Γ for which

$$\{1\} = \Gamma_n \subset \Gamma_{n-1} \subset \cdots \subset \Gamma_0 = \Gamma,$$

where 1 denotes the identity map of K, while to each pair u', u'' of elements of Γ_{i-1} corresponds an element s of Γ_i such that

$$u'u'' = u''u's. \tag{3.2.1}$$

Then there exists a point x_0 of K such that $u(x_0) = x_0$ for all u in Γ.

Proof. This proceeds by induction on n.

(1) Consider the case $n = 1$. In this case (b) asserts that the elements of Γ commute in pairs. We follow Bourbaki's reasoning ([4], pp. 114–115). For each u in Γ and each natural number r, define

$$u_r = r^{-1}(1 + u + \cdots + u^{r-1}).$$

Since K is convex, $u_r(K) \subset K$. Let Γ^* stand for the set of maps v that are products of a finite number of maps u_r with u in Γ and r a natural number. Each $v \in \Gamma^*$ has property (a), the elements of Γ^* commute in pairs, and each v maps K into itself.

Let us examine the family of sets $v(K)$ as v ranges over Γ^*. Each such set is compact and contained in K. Moreover if v' and v'' belong to Γ^*, then $v = v'v'' = v''v'$ belongs to Γ^* and $v(K) \subset v'(K)$, $v(K) \subset v''(K)$. Since no $v(K)$ is void, and since E is separated, we see that the family of sets $v(K)$ is a family of closed subsets of K having the finite intersection property. By compactness of K, therefore, the intersection of all the $v(K)$ contains at least one point x_0. We show that any such point x_0 satisfies $u(x_0) = x_0$ for all u in Γ.

For any natural number r, $x_0 \in u_r(K)$ and so there exists y in K such that

$$x_0 = r^{-1}(y + u(y) + \cdots + u^{r-1}(y)).$$

Thus

$$u(x_0) - x_0 = r^{-1}(u^r(y) - y) \in r^{-1}(K - K).$$

Let V be any balanced neighbourhood of 0 in E, W a similar such neighbourhood for which $W + W \subset V$. Since $H = K - K$ is compact, there exists a finite number of points a_k $(1 \leq k \leq m)$ of H such that H is contained in the union of the sets $a_k + W$. Also, there exists a real number α satisfying $0 < \alpha \leq 1$ such that $\alpha a_k \in W$ for each k. Hence $\alpha H \subset W + \alpha W \subset W + W \subset V$. Accordingly, if we choose $r > \alpha^{-1}$, we see that $u(x_0) - x_0 \in V$. Since E is separated and V is any balanced neighbourhood of 0 in E, this forces upon us the conclusion that $u(x_0) - x_0 = 0$, as we wished to show.

(2) Let us assume now that the theorem is true whenever Γ satisfies (b) for a given value of n, and show that the theorem remains valid whenever Γ satisfies (b) with $n + 1$ in place of n. Thus, in the sequence

$$\{1\} = \Gamma_{n+1} \subset \Gamma_n \subset \cdots \subset \Gamma_1 \subset \Gamma_0 = \Gamma.$$

we shall be sure that the theorem applies to Γ_1; that is, the set K_1 of points of K left fixed by the elements of Γ_1 is not void. K_1 is clearly closed and convex, hence compact.

By hypothesis, if u' and u'' belong to Γ $(= \Gamma_0)$, there exists s in Γ_1 such that $u'u'' = u''u's$. Then $u'u''(x) = u''u'(x)$ whenever x belongs to K_1. Hence the restrictions $u \mid K_1$ $(u \in \Gamma)$ commute in pairs. Moreover, if $u \in \Gamma$ and $u_1 \in \Gamma_1$ and $x \in K_1$, we have $u_1 u = u u_1 s$ for some s in Γ_1, and so

$$u_1 u(x) = u u_1 s(x) = u u_1(x) = u(x).$$

Thus $u_1(u(x)) = u(x)$ for x in K_1, showing that $u \mid K_1$ maps K_1 into itself.

This being so, it remains only to apply case (1) to the maps $u \mid K_1$, mapping K_1 into itself. The conclusion is that there exists x_0 in K_1 such that $u(x_0) = x_0$ for all u in Γ.

The principle of finite induction now completes the proof of Theorem 3.2.1.

3.2.2 **Remarks.** There are two notable cases in which hypothesis (b) is satisfied:

(1) The case in which Γ is an Abelian semigroup of continuous maps of K into itself, Γ containing the identity map. Actually, if the identity does not already belong to Γ, it may always be adjoined without disturbing the remaining hypotheses of the theorem.

(2) The case in which Γ is a solvable group of one-to-one continuous maps of K onto itself. We recall that a group is solvable if it has a composition series $(\Gamma_i)_{0 \leq i \leq n}$ formed of subgroups such that $\Gamma_0 = \Gamma$, $\Gamma_n = \{1\}$ (the trivial subgroup comprising only the neutral element of Γ), Γ_{i+1} is a normal (= invariant) subgroup of Γ_i, and Γ_i/Γ_{i+1} is Abelian. Thus every Abelian group is solvable; the group of similitudes of R^2 is solvable but not Abelian; if $n \geq 3$ the orthogonal group in R^n (and a fortiori the group of similitudes in R^n) is not solvable. For results of this type the reader may consult Dieudonné [16] and Artin [1].

We proceed to a discussion of some applications of Theorem 3.2.1, beginning with an extension of the Hahn-Banach theorem.

3.3 An Extension of the Hahn-Banach Theorem

The result in question concerns the possibility of extending linear forms subject to a condition of invariance with respect to a suitable collection of automorphisms of the given vector space. A result of this type was given by Agnew and Morse [1].

3.3.1 Theorem. Let E be a vector space. Suppose that Γ is *either* a solvable group of automorphisms of E *or* an Abelian semigroup of endomorphisms of E. Let L be a vector subspace of E that is stable under Γ (that is, $u(L) \subset L$ for each u in Γ), and p a seminorm on E such that $p \circ u \leq p$ for each u in Γ. Given any linear form f_0 on L satisfying

$$|f_0(x)| \leq p(x) \qquad (x \in L) \tag{3.3.1}$$

and

$$f_0(u(x)) = f_0(x) \quad (x \in L,\ u \in \Gamma), \tag{3.3.2}$$

there exists a linear form f on E that extends f_0 and for which

$$|f(x)| \leq p(x) \qquad (x \in E) \tag{3.3.3}$$

and

$$f(u(x)) = f(x) \qquad (x \in E,\ u \in \Gamma). \tag{3.3.4}$$

Remark. If E is a real vector space, there is an analogous result in which the seminorm p is replaced by any gauge function on E (see Section 1.6) and the moduli are removed from the inequalities (3.3.1) and (3.3.3); cf. part (1) of Theorem 1.7.1.

Proof. In the algebraic dual E^* of E, equipped with the weak topology $\sigma(E^*, E)$, let K be the set of linear forms f on E that extend f_0 and that satisfy (3.3.3). By Theorem 1.7.1, K is not void. Moreover, it is clear that K is compact and convex. Equally obvious is the fact that K is stable under the maps $f \to f_u$, where

$$f_u(x) = f(u(x))$$

for u in Γ and x in E. It now suffices to appeal to Theorem 3.2.1, Γ being the set of maps $f \to f_u$ with u in Γ: this leads to the conclusion that there exists in K at least one linear form f such that $f_u = f$ for all u in Γ. Any such f answers our requirements.

3.3.2 **Remark.** It is plain that a similar result is valid whenever Γ is a set of linear maps of E into itself satisfying the condition (b) of Theorem 3.2.1. For most applications, however, the two cases cited are sufficiently general.

In the remaining sections of this chapter we shall discuss some applications of Theorem 3.3.1.

3.3.3. For further reading concerning extended forms of the Hahn-Banach Theorem and means on semigroups (see Section 3.5), see Dunford [3], Woodbury [1], Yood [1], Klee [3], Silverman [1], [2], Ti Yen [1], Hewitt and Ross [1], §17. In Silverman [1], [2] extensions of the theorem to ordered vector spaces are also considered.

3.4 Banach's Generalized Limits

Banach ([1], pp. 33–34) used the Hahn-Banach theorem in order to deduce the existence of a translation-invariant "generalized limit" applicable to arbitrary bounded functions on the semi-axis $(0, \infty)$ (or to arbitrary sequences). We shall derive this result from Theorem 3.3.1, the use of which streamlines the arguments used in Banach's original derivation.

To do this, we take for E the vector space of all real-valued bounded functions on $(0, \infty)$, regarded as a real vector space; for L we take the vector subspace of E comprised of those functions x on $(0, \infty)$ for which $\lim_{t\to\infty} x(t)$ exists finitely, and f_0 is defined on L to be this limit. For Γ we take the Abelian semigroup of translations $x \to x_a$, where $a \geq 0$ and $x_a(t) = x(t + a)$ for t in $(0, \infty)$. Finally, p is the gauge function on E defined by

$$p(x) = \limsup_{t\to\infty} x(t).$$

It is trivial to verify that all the hypotheses of Theorem 3.3.1 are fulfilled. The conclusion postulates the existence of an extension f of f_0 to E such that (3.3.3) and (3.3.4) hold. Following Banach, we use the suggestive notation $\mathrm{Lim}_{t\to\infty} x(t)$ for $f(x)$. Thus we have

$$\operatorname*{Lim}_{t\to\infty} x(t) = \lim_{t\to\infty} x(t) \tag{3.4.1}$$

for each bounded, real-valued function x on $(0, \infty)$ for which the right-hand side exists;

$$\operatorname*{Lim}_{t\to\infty} x(t + a) = \operatorname*{Lim}_{t\to\infty} x(t) \tag{3.4.2}$$

for each bounded, real-valued function x on $(0, \infty)$ and each $a \geq 0$; and

$$\operatorname*{Lim}_{t\to\infty} x(t) \leq \limsup_{t\to\infty} x(t)$$

for each x of the stated nature. Using linearity of Lim, it follows that, in fact,

$$\liminf_{t\to\infty} x(t) \leq \operatorname*{Lim}_{t\to\infty} x(t) \leq \limsup_{t\to\infty} x(t) \tag{3.4.3}$$

for each bounded, real-valued function x on $(0, \infty)$. (Clearly, (3.4.3) implies (3.4.1).)

An exactly similar construction pertains to the case of bounded, real-valued sequences. Also, it is obvious how to extend Lim to complex-valued bounded functions on $(0, \infty)$ (or sequences).

The reader will note that the existence of such generalized limits shows that not every continuous linear form on the Banach space ℓ^∞ of bounded sequences (with norm equal to the supremum of the moduli of the terms) can be represented in the form

$$x \to \sum_{n=1}^{\infty} c_n x(n) \qquad (x \in \ell^\infty)$$

with (c_n) a sequence of numbers. For, on choosing x to be that member of ℓ^∞ that takes the value unity at k and 0 elsewhere, (3.4.1) would show that $c_k = 0$ for all k, in which case the linear form would be everywhere 0 on ℓ^∞. Yet (3.4.1), with $x = 1$, shows that the generalized limit does not have this property.

3.5 Invariant Means on Semigroups

Let T be initially an arbitrary set and let E be a vector space of bounded, real-valued functions on T. By a *mean-value* on E we mean a linear form M on E such that

$$\operatorname*{Inf}_T x \leq M(x) \leq \operatorname*{Sup}_T x \tag{3.5.1}$$

for each x in E. If E contains the constant function 1, a linear form M on E is a mean value in the above sense if and only if M is positive (for the order structure induced on E by that on the vector space of all real-valued functions on T, that is,

$$M(x) \geq 0 \qquad \text{for } x \in E,\ x \geq 0)$$

and

$$M(1) = 1.$$

If we are given in addition a collection of maps u of T into itself, we have the associated endomorphisms $x \to x_u = x \circ u$ of E, and it makes sense to discuss those mean values M on E which are invariant in the sense that $M(x_u) = M(x)$ for x in E and all u of the specified collection. When the collection of maps satisfies suitable conditions, Theorem 3.3.1 is often useful in the discussion of the existence of such invariant means. A case of special interest is that in which T is a semigroup, its law of composition being written multiplicatively, and the maps u considered are either left translations $t \to at$ or right translations $t \to ta$, or perhaps both. In this case we shall write ${}_a x$ for the left translate $t \to x(at)$ of x, x_a for its right translate $t \to x(ta)$. The mean value M is said to be left (right, bi-) invariant if it is invariant with respect to left (right, two-sided) translations.

By taking E to be the vector space of all bounded, real-valued functions on T and $p(x) = \operatorname{Sup}_T x$, Theorem 3.3.1 shows that there exist bi-invariant mean values defined for all bounded, real-valued functions on any Abelian semigroup T.

Let us now specialize still further to the case in which T is a group (not merely a semigroup). Translation invariance is defined exactly as for semigroups. In addition, we may now introduce the endomorphism $x \to \check{x}$ of E defined by $\check{x}(t) = x(t^{-1})$. If M is any mean value, $\check{M}$ is defined by

$$\check{M}(x) = M(\check{x})$$

for x in E; $\check{M}$ is also a mean value. If M is left invariant, and if we define $x'(t) = M(x_t)$ and $M_1(x) = \check{M}(x')$, then it is easily verified that M_1 is bi-invariant. Left invariance of M_1 is a consequence of the relation $({}_a x)' = x'$, thanks to left invariance of M; right invariance of M_1 follows from the identities $(x_s)' = (x')_s$, $((x')_s)^\vee = {}_{s^{-1}}((x')^\vee)$, combined again with left invariance of M.

With these remarks in mind, Theorem 3.3.1 is easily applied to show the existence of at least one bi-invariant mean value defined for all bounded, real functions on the group T, provided T is solvable. Furthermore, one might apply the same reasoning to the space of all real-valued, continuous, and bounded functions on T, if T is a topological group that is solvable.

Warning. Suppose T happens to be a locally compact topological group, M a left-invariant mean defined (say) for all bounded, continuous, real-valued functions on T. It is necessary to avoid confusing such an M with the left-invariant Haar measure on T, introduced and described at some length later in Section 4.18. Indeed, unless T is compact, the restriction of M to the space $\mathscr{K}(T)$ of real-valued continuous functions on T having compact supports turns out to be 0. For suppose that x is any such function: we may assume it to be positive. We choose a symmetric compact set K in T outside which x vanishes. If T is noncompact, we can then choose elements $a_n (n = 1, 2, \cdots)$ of T such that the sets $a_n^{-1}K$ are disjoint. Then we have

$$\sum_{n=1}^{r} x(a_n t) \leq m = \operatorname*{Sup}_{T} x$$

for all t in T and all natural numbers r. Applying M to both sides and using left invariance, we conclude that

$$r \cdot M(x) \leq M(m) = m \cdot M(1)$$

for all natural numbers r. On the other hand, x being positive, we have $M(x) \geq 0$. It follows, on letting $r \to \infty$, that $M(x) = 0$, showing that $M \mid \mathscr{K}(T) = 0$.

If, however, T is compact, any such left-invariant M, defined for all continuous, real-valued functions on T, may be shown to be identical with the left-Haar measure on T.

For further discussion of invariant means on groups see Hewitt and Ross [1], §17. The case of invariant means on (not necessarily Abelian) semigroups is dealt with in some detail in Dixmier [3]. See, too, the references listed in Subsection 3.3.3. A basic result concerning the existence of invariant means on semigroups, due to Dixmier, is given in Exercise 3.12. Results of this nature can be utilized in establishing results like Theorem 3.3.1; see Exercise 2.15 and Exercise 3.12.

3.6 The Schauder-Tychonoff Theorem

The Schauder-Tychonoff theorem differs from the Markov-Kakutani theorem in that only one map u is involved, but repayment for this concession comes in the form of a complete absence of hypotheses on u save continuity.

3.6.1 Theorem (Schauder-Tychonoff). Let E be a separated LCTVS, K a nonvoid compact convex subset of E, u any continuous map of K into itself. Then u admits at least one fixed point.

The proof of the theorem, which is taken from Dunford and Schwartz ([1], pp. 453–456), is contained in three lemmas, the third of which is the most difficult and also the critical step. For brevity we shall say that a topological space has the *fixed point property* if any continuous map of that space into itself admits at least one fixed point.

The first two lemmas establish that certain subsets of the Hilbert sequence space ℓ^2 have the fixed point property. By the Hilbert cube Q we shall mean the subset of ℓ^2 formed of sequences x in ℓ^2 for which $|x(n)| \leq 1/n$ for all n. It is very simple to verify that Q is compact and convex.

Lemma 1. Q has the fixed point property.

Proof. Let u be any continuous map of Q into itself.
Let π_n be the projection defined by

$$\pi_n(x)(k) = \begin{cases} x(k) & \text{for } 1 \leq k \leq n, \\ 0 & \text{for } k > n. \end{cases}$$

The set $Q_n = \pi_n(Q)$ is then homeomorphic with the closed unit ball in K^n (K being the scalar field, real or complex). Since $\pi_n \circ u$ is a continuous map of Q_n into itself, the Brouwer fixed point theorem (see Section 0.5) implies the existence of $x_n \in Q_n \subset Q$ such that $\pi_n(u(x_n)) = x_n$. Since Q is compact, some subsequence (x_{n_r}) is convergent to a point x of Q. Since

$$\| \pi_n(u(x_n)) - u(x_n) \| \leq \left[\sum_{k>n} k^{-2} \right]^{1/2},$$

it appears at once that $u(x) = x$. Thus x is a fixed point of u. ∎

Lemma 2. Any nonvoid closed convex subset K of Q has the fixed point property.

Proof. According to Theorem 1.12.3, each x in Q determines a unique nearest point $f(x)$ of K. The map f is continuous. For suppose that $x_n \to x$ in Q and yet $f(x_n) \not\to f(x)$. Since K is compact, some subsequence $f(x_{n_k})$ would converge to a point y of K distinct from $f(x)$. Then

$$\|x_{n_k} - f(x_{n_k})\| \leq \|x_{n_k} - f(x)\| \leq \|x_{n_k} - x\| + \|x - f(x)\|$$

and so, in the limit as $k \to \infty$, $\|x - y\| \leq \|x - f(x)\|$. This would, however, contradict the definition of $f(x)$. Thus f is continuous from Q into K, and is the identity on K.

Suppose now that u is any continuous map of K into itself. Then $u \circ f$ is continuous from Q into K and so, by Lemma 1, admits a fixed point x in Q. Thus $u(f(x)) = x$, which shows that x belongs to $u(K) \subset K$. So in fact x is a fixed point of u. ▌

Lemma 3. Let E be a separated LCTVS, K a compact convex subset of E possessing at least two distinct points, u a continuous map of K into itself. Then there exists a nonvoid closed convex set K_1, a proper subset of K, such that $u(K_1) \subset K_1$.

Proof. Since K is compact and E is separated and locally convex, the initial and weakened topologies on E induce on K one and the same topology. Thus u is continuous from K into itself when the latter is endowed with the weakened topology. K being compact, and a fortiori weakly compact, u will be uniformly continuous on K for the weakened topology.

We now introduce a new concept. If F and G are subsets of E', G is said to determine F if the following is true. Given f in F and $\varepsilon > 0$, there exists a finite subset G_0 of G and a number $\delta > 0$ such that if x and y lie in K and

$$x - y \in W(G_0, \delta) \equiv \{z \in E: |\langle z, g\rangle| < \delta \qquad \text{for } g \in G_0\},$$

then

$$|\langle u(x), f\rangle - \langle u(y), f\rangle| < \varepsilon.$$

This says in effect that u is continuous for the weak topologies $\sigma(E, G) \mid K$ and $\sigma(E, F) \mid K$. Three comments are needed.

(1) Each f in E' is determined by a suitable countable set G. For, since we have seen that $x \to \langle u(x), f\rangle$ is uniformly continuous on K for the weakened topology, to each natural number n corresponds a finite set $G_n \subset E'$ and a number $\delta_n > 0$ such that $(x, y) \in K \times K$ and $x - y \in W(G_n, \delta_n)$ together imply that $|\langle u(x), f\rangle - \langle u(y), f\rangle| < 1/n$. The countable set $G = \bigcup G_n$ then determines f.

(2) From (1) we conclude at once that each countable subset F of E' is determined by some countable subset G of E'. More than this is true, however.

(3) Each f in E' is contained in some self-determining countable subset G of E'. Indeed, if f is determined by a countable set G_1 (possible by (1)), and G_1 is determined by a countable set G_2 (possible by (2)), G_2 by G_3, and so on, then it suffices to take $G = \{f\} \cup \bigcup \{G_n: n = 1, 2, \cdots\}$.

With these remarks made, we can proceed with the proof.

Suppose that a and b are distinct points of K and that we choose f in E' such that $\langle a, f\rangle \neq \langle b, f\rangle$, as is possible since E is separated and locally convex. We choose, as is possible by (3) above, a countable self-determining set G containing f; and we enumerate G as (g_k). Since K is bounded we may assume that $|g_k(K)| \leq 1/k$ for all k. We map K into ℓ^2 by setting $t(x)(k) = \langle x, g_k\rangle$. t is a continuous map of K onto a compact convex subset K_0 of the Hilbert cube Q. Since $t(a) \neq t(b)$, K_0 contains at least two distinct points.

Since G is self-determining, it is seen that (even though t may not be one to one) the map $v = t \circ u \circ t^{-1}$ is single valued from K_0 into itself. Moreover, v is continuous. For suppose that $p_0 \in K_0$ and $0 < \varepsilon < 1$. We choose N so large that $\sum_{k>N} k^{-2} < \varepsilon$. Thanks to the self-determining property of G, there exists $\delta > 0$ and a natural number m such that $(x, y) \in K \times K$ and $|\langle x - y, g_j\rangle| < \delta$ for $1 \leq j \leq m$ together entail that

$$|\langle u(x), g_k\rangle - \langle u(y), g_k\rangle| < \frac{\varepsilon^{1/2}}{N^{1/2}} \quad \text{for} \quad 1 \leq k \leq N.$$

Then if $p \in K_0$ and $\|p - p_0\| < \delta$, and if x and y are chosen from K to satisfy $t(x) = p$ and $t(y) = p_0$, the preceding inequality gives

$$\begin{aligned} \|v(p) - v(p_0)\|^2 &= \sum_k |\langle u(x), g_k\rangle - \langle u(y), g_k\rangle|^2 \\ &= \sum_{k \leq N} + \sum_{k>N} \\ &\leq \sum_{k \leq N} \frac{\varepsilon}{N} + \sum_{k>N} \left(\frac{2}{k}\right)^2 \\ &< \varepsilon + 4\varepsilon = 5\varepsilon, \end{aligned}$$

hence the alleged continuity of v.

Lemma 2 can now be applied to conclude that v has a fixed point p_0 in K_0. Then $u(t^{-1}(p_0)) \subset t^{-1}(p_0)$. The set $K_1 = t^{-1}(p_0)$ is a proper closed convex subset of K and $u(K_1) \subset K_1$, which completes the proof. ▌

Proof of Theorem 3.6.1. Consider the set of all compact convex subsets K' of K such that $u(K') \subset K'$. If this set be partially ordered by set-inclusion reversed, it is evidently inductively ordered. By Zorn's lemma, there is thus a minimal compact convex subset K^* of K satisfying $u(K^*) \subset K^*$. Lemma 3 shows that K^* possesses just one point, which is thus a fixed point of u. ▌

3.6.2 **Corollary.** Let E be a separated LCTVS space satisfying the condition:

CONDITION. In E, the closed convex envelope of any compact set is compact. Suppose that A is a nonvoid, closed, convex subset of E, and u a continuous map of A into itself such that $u(A)$ is relatively compact in E. Then u admits at least one fixed point.

Proof. Let B be the convex envelope in E of $u(A)$, and let K be the closure in E of B. Then $K \subset A$. Since $u(A)$ is relatively compact, it follows from the above condition that K is compact. K is also nonvoid and convex. Since $K \subset A$, $u(K) \subset u(A)$; and $u(A) \subset K$ by definition of K. Thus $u(K) \subset K$.

It only remains to apply Theorem 3.6.1 to $u \mid K$. ▌

Remarks. The condition is fulfilled whenever E is complete, or even quasi-complete—in the sense that each bounded, closed subset of E is complete. For if P is a compact subset of E and Q its closed convex envelope, then Q is precompact (Proposition 9.1.3). Further, since P is bounded, so too is Q. If E is quasi-complete, Q is complete and therefore compact.

3.6.3 Application to Differential Equations. Suppose that T is an interval on the real axis, F a finite-dimensional normed vector space, B the ball $\|y - y_0\| \le r$ in F. Given a point t_0 of T and a function f, mapping $T \times B$ into F, we are concerned with the differential equation

$$\frac{dx}{dt} = f(t, x), \qquad x(t_0) = y_0. \tag{3.6.1}$$

A solution of (3.6.1) throughout T will signify a function x mapping T into B, taking at t_0 the value y_0, and possessing at each point t of T a derivative equal to $f(t, x(t))$. We have indicated in Subsection 3.1.3 how the contraction mapping method will ensure the existence and uniqueness of such a solution, provided f is continuous and satisfies some sort of Lipschitz condition with respect to its second variable. It remained for Peano [2] to succeed first in showing that a local solution exists, albeit possibly nonunique, when f is assumed merely to be continuous and bounded, say $\|f(t, y)\| \le M$ $(t \in T, y \in B)$. It is now known how to derive this result on the basis of Ascoli's theorem; see, for example, Kolmogorov and Fomin [1], Vol. 1, pp. 56–57. A proof having the same roots but based upon Corollary 3.6.2 will now be given.

Put $c = r/M$, and let $T_1 = T \cap [t_0 - c, t_0 + c]$.

We introduce the Banach space $E = C(T_1, F)$ of continuous F-valued functions on T_1, the norm being

$$\|x\| = \text{Sup}\,\{\|x(t)\| : t \in T_1\}.$$

For x in E we define $u(x)$, likewise an element of E, by

$$u(x)(t) = y_0 + \int_{t_0}^{t} f(s, x(s))\, ds. \tag{3.6.2}$$

If we can show that u has a fixed point x, then this x will be a solution of (3.6.1) throughout the subinterval T_1 of T.

To establish the existence of a fixed point, we apply Corollary 3.6.2 to the set A composed of those x in E such that $\|x - y_0\| \le r$; here y_0 is identified with the constant function taking the value y_0 at all points of T_1. Let us verify that the hypotheses of Corollary 3.6.2 are satisfied.

It is obvious that A is closed and convex.

If x belongs to A, we have $\|x(s) - y_0\| \le r$ whenever $t \in T_1$ and s belongs to the interval with extremities at t_0 and t. So $\|f(s, x(s))\| \le M$ throughout the range of integration in (3.6.2), and therefore

$$\|u(x)(t) - y_0\| \le M\,|t - t_0| \le Mc = r.$$

Thus u maps A into itself.

As for continuity of u on A, it is clear that if a sequence $(x_n) \subset A$ converges in E to $x \in A$, then by uniform continuity of f on $T_1 \times B$, $u(x_n)(t) \to u(x)(t)$ in F, uniformly for t in T_1. So u is continuous on A.

It remains to verify that $u(A)$ is relatively compact in E. Since $u(A) \subset A$ is bounded, and since F is finite dimensional, it suffices (Ascoli's theorem) to show that $u(A)$ is equicontinuous. But

$$\|u(x)(t) - u(x)(t')\| = \left\|\int_{t'}^{t} f(s, x(s))\, ds\right\| \leq M\,|t - t'|$$

whenever x belongs to A and t and t' to T_1.

An application of Corollary 3.6.2 is thus justified, and the Peano theorem is established.

Remarks. (1) It is known that the hypothesis that F be finite dimensional cannot be dispensed with.

(2) We have assumed that t_0 is interior to T. If on the contrary t_0 is an extremity of T, a few minor modifications suffice to rehabilitate the argument.

3.6.4 A Second Application to Differential Equations. We consider this time a second-order equation of the type

$$\frac{d^2x}{dt^2} = f\left(t, x, \frac{dx}{dt}\right) \tag{3.6.3}$$

over an interval $0 \leq t \leq T$, together with boundary conditions

$$x(0) = x_1, \qquad x(T) = x_2. \tag{3.6.4}$$

By reproducing an argument given by R. Bass [1], we will show that if f is continuous and bounded on $[0, T] \times R \times R$, then (3.6.3) and (3.6.4) admit at least one solution.

The proof is based upon the observation that if we define

$$G(t, s) = \begin{cases} \dfrac{s(T - t)}{T} & \text{for } 0 \leq s \leq t \leq T, \\[2ex] \dfrac{t(T - s)}{T} & \text{for } 0 \leq t \leq s \leq T, \end{cases}$$

then any solution x of the integral equation

$$x(t) = x_1 + \frac{(x_2 - x_1)t}{T} - \int_0^T G(t, s) f[s, x(s), x'(s)]\, ds,$$

the prime denoting derivation, is a solution of the given problem.

The solubility of the integral equation may be established by an appeal to Corollary 3.6.2. In doing this we take for E the Banach space of continuously differentiable functions on $[0, T]$ (one-sided derivatives being involved at each end point), the norm being

$$\|x\| = \operatorname*{Sup}_{0 \leq t \leq T} \{|x(t)| + |x'(t)|\}.$$

For A we shall take E itself, and for u we take the map defined by

$$u(x)(t) = x_1 + \frac{(x_2 - x_1)t}{T} - \int_0^T G(t, s) f[s, x(s), x'(s)]\, ds.$$

It is easy to verify that u maps E continuously into itself. In view of Corollary 3.6.2, it remains only to show that $u(E)$ is relatively compact in E. Using Ascoli's theorem one sees that for this it suffices to show that the members of $u(E)$ are bounded at each point, that their derivatives are bounded at each point, and that these derivatives are equicontinuous. Now, if we suppose that $|f| \leq M$ on $[0, T] \times R \times R$; and if we put $y = u(x)$, then it is evident that

$$|y(t)| \leq |x_1| + |x_2 - x_1| + \frac{MT^2}{4},$$

since $0 \leq G \leq T/4$. Moreover direct computation leads to the formula

$$y'(t) = \frac{(x_2 - x_1)}{T} + \int_0^t \left(\frac{s}{T}\right) f[\cdot \cdot \cdot]\, ds - \int_t^T \left(1 - \frac{s}{T}\right) f[\cdot \cdot \cdot]\, ds,$$

whence it follows at once that

$$|y'(t)| \leq \frac{|x_2 - x_1|}{T} + 2MT$$

and

$$|y'(t_1) - y'(t_2)| \leq 2M\, |t_1 - t_2|.$$

The functions $y = u(x)$, x ranging over E, thus have the desired properties, and the proof is complete.

Remarks. For further applications of fixed-point theorems to differential equations, see Corduneanu [1], [2], [3], and [4]. The existence of applications to differential equations of other functional analytic principles may also be noted here. See, for example, Bellman [1], Massera and Schäffer [1], [2], [3] and [4] and the references there cited, and also Example 7.2.6 below.

3.6.5 **An Application to Reflexive Banach Spaces.** Let E be a Banach space and f a map of E into itself. Denoting by 1 the identity map of E onto itself, one may ask for conditions under which $1 + f$ maps E onto itself. If f is linear, a well-known sufficient condition is that $\|f\| < 1$: in this case, in fact, if y is a given element of E, the series $\sum_{n=0}^{\infty} (-1)^n f^n(y)$ is convergent in E and has a sum x that evidently satisfies $x + f(x) = y$. However, we are here concerned with applying Corollary 3.6.2 to cases in which f is not necessarily linear. We will in fact establish the following result.

(1) Assume that f is weakly continuous, transforms bounded sets into weakly relatively compact sets, and satisfies

$$\limsup_{\|x\| \to \infty} \frac{\|f(x)\|}{\|x\|} < 1. \tag{3.6.5}$$

Then $1 + f$ maps E onto E.

To see this, we choose any y in E. It suffices to show that the map u defined by $u(x) = y - f(x)$ has a fixed point in E. This we shall do by applying Corollary 3.6.2 to E with its weakened topology. The set A will be a ball $\{x: \|x\| \leq r\}$ with r suitably chosen. Evidently, whatever the value of r, $u \mid A$ is continuous and $u(A)$ is relatively compact (for the weakened topology in each case), owing to our hypotheses on f. It remains to take care of the demand that $u(A)$ be contained in A.

Now, if $M(r)$ denotes the supremum of $\|f(x)\|$ for $\|x\| \leq r$, then (3.6.5) and the fact that f is locally bounded together entail that $\limsup_{r\to\infty} M(r)/r < 1$. On the other hand, if $\|x\| \leq r$, then $\|u(x)\| \leq \|y\| + M(r)$. So, if r be chosen large enough, $u(x)$ will belong to A whenever x has that property.

Theorems 8.1.1, 8.3.5, 8.13.1 and Corollary 3.6.2 now lead to (1).

(2) A simplification results if it be supposed that E is reflexive, for then the weak continuity of f itself ensures that f transforms bounded sets into weakly relatively compact sets (each bounded set being itself weakly relatively compact as a consequence of the assumed reflexivity; see Section 8.4.)

3.6.6 The Fixed-Point Theorems of Kakutani and Fan. Kakutani [7] established a fixed-point theorem for mappings carrying points into sets. Subsequently Fan [1] gave a theorem that simultaneously generalizes Kakutani's theorem (from finite to infinite dimensional spaces) and the Schauder-Tychonoff theorem. We state Fan's theorem without proof; it will not be used in this book.

Theorem. Let E be a separated LCTVS, K a nonvoid compact convex subset of E, and f a mapping that assigns to each $x \in K$ a nonvoid closed convex subset $f(x)$ of K. Assume that f is upper semicontinuous, in the sense that if $x_0 \in K$ and U is a neighbourhood of $f(x_0)$, then there exists a neighbourhood V of x_0 such that $f(x) \subset U$ for each $x \in K \cap V$. Then there exists an $x_0 \in K$ that is a fixed point of f, in the sense that $x_0 \in f(x_0)$.

3.7 The Work of Leray and Schauder

As long ago as 1934 a novel approach towards fixed-point theorems was developed and published by Leray and Schauder [1]. We do not intend to reproduce their work here, referring the interested reader to the original paper. Yet it is perhaps of interest to indicate the principles underlying the Schauder-Leray theory for comparison with the methods discussed earlier in this chapter.

The first stage of the Schauder-Leray technique was an extension of Brouwer's concept of topological degree, initially defined for maps from one finite dimensional space E into itself, to the case in which E is a (possibly infinite dimensional) normed vector space. Included in this investigation is (a) the relationship between the solubility of an equation of the type

$$x - f(x) = 0$$

and the topological degree at $y = 0$ of the mapping $x \to y = x - f(x)$; and (b) the invariance of the topological degree under continuous deformation of the map involved. This program was carried out under the assumption that f is defined and continuous on the closure of a bounded open set Ω in E and that $f(\bar{\Omega})$ is relatively compact in E.

The second stage amounts to the expression of a given problem in the form of the solubility of an equation of the type

$$x - f_\beta(x) = 0,$$

f_β being a member of a continuous family (f_λ) of maps of the preceding type. Success depends on the family being such that, for some member f_α of the family, the solubility of the corresponding equation

$$x - f_\alpha(x) = 0$$

can be discussed with a minimum of difficulty.

Schauder and Leray succeeded in carrying out this program in such a way as to derive existence theorems for general types of partial differential equations, the second stage of the operation demanding a great deal of ingenuity and considerable prior knowledge of more specialized (though still quite general) types of partial differential equations. A summary is given by Miranda ([1], p. 140 et seq.). A somewhat simplified and restricted application, in which the only fixed-point theorem involved is Theorem 3.6.1, is given in Courant-Hilbert [2], pp. 357–362.

It is plain that the principles of the Schauder-Leray method go beyond those used earlier in this chapter. In operation it is more complex and more difficult to apply, but its range of applicability covers much new ground.

The original paper of Schauder and Leray has inspired numerous extensions and simplifications; see Leray [3], Nagumo [1], Altman [1], [2], Granas [1], and Klee [4].

3.8 Fixed-Point Theorems in Ordered Sets

Picard, in his fundamental paper [1], indicated the use of the method of successive approximations in forming theorems about fixed points of suitable self-maps of ordered sets. The method was later developed in more general terms by Kantorovich [1]. One result of this nature will be given here.

Suppose that E is a (partially) ordered set (see Subsection 0.1.4). We assume that any majorized, increasing sequence (x_n) of elements of E has a supremum in E; and that any minorized, decreasing sequence (y_n) of elements of E has an infimum in E. If, under these conditions, x (resp. y) is the supremum (resp. infimum) of (x_n) (resp. (y_n)), we shall write $x_n \uparrow x$ (resp. $y_n \downarrow y$). The reader will notice that then $x_{n+1} \uparrow x$ and $y_{n+1} \downarrow y$.

We consider a self-map u of E that is increasing—that is, $u(x) \leq u(y)$ whenever $x \leq y$—and that is such that $u(x_n) \uparrow u(x)$ and $u(y_n) \downarrow u(y)$ whenever $x_n \uparrow x$ and $y_n \downarrow y$.

3.8.1 **Theorem.** Suppose that E and u are as explained immediately above. Suppose further that there exist elements x_0 and y_0 of E such that

$$x_0 \leq y_0, \qquad x_0 \leq u(x_0), \qquad u(y_0) \leq y_0. \tag{3.8.1}$$

Define (x_n) and (y_n) by

$$x_{n+1} = u(x_n), \qquad y_{n+1} = u(y_n). \tag{3.8.2}$$

Then there exist elements x and y of E such that

$$x_n \uparrow x, \qquad y_n \downarrow y, \qquad x \leq y, \tag{3.8.3}$$

and x and y are each fixed points of u. Moreover, if x^* (resp. y^*) is a fixed point of u satisfying $x^* \geq x_0$ (resp. $y^* \leq y_0$), then $x^* \geq x$ (resp. $y^* \leq y$).

Proof. From the first clause of (3.8.1) it follows that $u(x_0) \leq u(y_0)$, and from this and the remainder of (3.8.1) it follows by induction that

$$x_0 \leq x_1 \leq \cdots \leq x_n \leq y_n \leq y_{n-1} \leq \cdots \leq y_0. \tag{3.8.4}$$

Thus (x_n) is increasing and majorized by y_0, and therefore $x_n \uparrow x$ for some $x \in E$. Similarly $y_n \downarrow y$ for some $y \in E$. Moreover, one sees from (3.8.4) that $x_n \leq y_m$ if $n \geq m$, which entails that $x \leq y_m$ for all m, and so that $x \leq y$. Equation (3.8.3) is thus established. Next, since $x_n \uparrow x$, so $u(x_n) \uparrow u(x)$. On the other hand, $x_{n+1} \uparrow x$ and $x_{n+1} = u(x_n)$. It follows that $x = u(x)$. Similarly, $y = u(y)$. Each of x and y is thus a fixed point of u.

Finally, suppose that x^* is a fixed point of u satisfying $x^* \geq x_0$. Then $x^* = u(x^*) \geq u(x_0) = x_1$, which argument can now be repeated to show that $x^* \geq x_n$ for all n. Hence $x^* \geq x$. Similarly, if y^* is a fixed point of u satisfying $y^* \leq y_0$, then $y^* \leq y$.

3.8.2 **An Example.** Let us consider maps u of the type introduced in 3.1.3, namely,

$$u(x)(t) = \eta_0 + \int_{t_0}^{t} f(s, x(s))\, ds, \tag{3.8.5}$$

where η_0 is a real number and f a function on $T \times R^1$, T being an interval containing t_0. In (3.8.5) x denotes a variable real-valued function on T, and t a variable point of T. We assume f to be real valued and measurable on $T \times R^1$, such that $x \to f(t, x)$ is increasing for each $t \in T$, and such that

$$\int_T |f(t, c)|\, dt < +\infty$$

for each real number c. For E one may take (for example) the set of all bounded, real-valued, measurable functions on T, endowed with the natural partial order. It is then fairly simple to verify that E satisfies the hypotheses needed in Theorem 3.8.1, and that u is an increasing self-map of E. Moreover, if x_n converges boundedly and monotonely to x, then $u(x_n)$ converges likewise to $u(x)$ (see Theorem 4.6.6). In order to apply Theorem 3.8.1, it remains only to make sure that there exist elements x_0 and y_0 of E satisfying (3.8.1). We leave the reader to investigate this matter.

We notice that, subject to quite mild restrictions on f, u will actually transform each $x \in E$ into a continuous function. In that case, if f is continuous, the relation $u(x) = x$ will entail that x satisfies the differential equation

$$\frac{dx}{dt} = f(t, x(t))$$

at all interior points of T. Even without assuming the continuity of f, it will still be true that this differential equation is satisfied at almost all points of T.

EXERCISES

3.1 Let E be a Banach space, A a convex open subset of E, f a continuous map of $\bar{A}$ into itself. Suppose further that f is continuously differentiable on A in the sense that to each $x \in A$ corresponds a continuous endomorphism A_x of E, varying continuously with $x \in A$, such that

$$f(x + h) - f(x) = A_x(h) + o(\|h\|) \qquad \text{as } h \to 0;$$

the constant implied in the o-notation may depend upon x. Suppose finally that $\|A_x\| \leq \lambda$ for $x \in A$, where $\lambda < 1$.

Prove that f admits a unique fixed point in $\bar{A}$.
[Hint: Show that f is a contraction map. To do this show that if $t \to x(t)$ is any differentiable map of a real interval into A, then $(d/dt)f(x(t)) = A_{x(t)}(dx/dt)$, and so that

$$f(x_1) - f(x_0) = \int_0^1 A_{x(t)}(dx/dt)\, dt$$

for any two points x_0 and x_1 of A and any differentiable "curve" $t \to x(t)$ lying in A and such that $x(0) = x_0$ and $x(1) = x_1$.]

Remark. The hypothesis of convexity on A can be weakened given further information about λ; such a weakening may also be effected sometimes by appeal to the results of Edelstein [1] and [2].

3.2 Let E be a Banach space and T a continuous endomorphism of E satisfying $\|T\| \leq 1$. By applying the contraction principle to the mapping $u(x) = Tx + y$ restricted to balls with centre 0 and varying radii, show that $1 - T$ has an inverse that is a continuous endomorphism of E having norm at most $(1 - \|T\|)^{-1}$.

Reach the same conclusion by considering the so-called Neumann series $\sum_{n=0}^{\infty} T^n$.
[Hint: u is a contraction map of the ball centre 0 and radius r into itself, provided $\|y\| \leq (1 - \|T\|)r$.]

3.3 Let X, Y, and Z be normed vector spaces, X being complete, f and g linear maps of X into Y and of X into Z, respectively. Suppose that the pair of equations

$$f(x) = y, \qquad g(x) = z$$

has a unique solution x in X for given y in Y and z in Z, and that this solution satisfies

$$\|x\|_X \leq p\, \|y\|_Y + q\, \|z\|_Z$$

for suitable constants p and q. Let F be a map of the ball

$$S = \{x \in X : \|x\|_X \leq r\}$$

into Y such that for some number λ one has

$$\|F(x) - F(x')\|_Y \leq \lambda\, \|x - x'\|_X \qquad (x, x' \in S).$$

Show that if

$$q\lambda < 1 \quad \text{and} \quad p\,\|x\|_Z + q\,\|F(0)\|_Y \leq (1 - q\lambda)r,$$

then the pair of equations

$$f(x) = F(x), \qquad g(x) = z$$

admits a unique solution $x \in S$.
[Hint: Consider the map T of S into X defined in a such a way that, for each $u \in S$, $x = Tu$ is the unique solution of the system

$$f(x) = F(u), \qquad g(x) = z.$$

Prove that under the stated conditions T is a contraction map of S into itself, and apply the contraction principle.]

NOTES. This process is applied to problems relating to ordinary differential equations by Corduneanu ([5], Théorème 1).

3.4 A subspace Y of a topological space X is said to be a *retract* of X if there exists a continuous map f of X into Y that leaves fixed each point of Y.

Show that if X has the fixed-point property, and if Y is a retract of X, then Y has the fixed-point property.

Remark. Note that the crux of the proof of Lemma 2 in Section 3.6 amounts to showing that in a pre-Hilbert space, any closed convex subset is a retract of any compact set containing it.

3.5 Let E be a separated LCTVS that is quasi-complete, in the sense that each bounded, closed subset of E is complete. Let A be a compact subset of E, and let K be the closed, convex envelope in E of A. Show that if A is a retract of K, then A has the fixed-point property.
[Hint: Show that K is compact and apply the preceding exercise and the Schauder-Tychonoff theorem.]

3.6 Let F denote the real or complex scalar field and K a continuous F-valued function on $[0, 1] \times [0, 1] \times F$. Put for $r \geq 0$

$$M(r) = \text{Sup}\,\{|K(t, s, \xi)| : 0 \leq t \leq 1,\ 0 \leq s \leq 1,\ \xi \in F,\ |\xi| \leq r\}.$$

Let

$$a \in C([0, 1], F)$$

and

$$\|a\| = \text{Sup}\,\{|a(t)| : 0 \leq t \leq 1\}.$$

Show that if $M(\|a\| + r) \leq r$ holds for some $r \geq 0$, then the equation

$$x(t) + \int_0^1 K(t, s, x(s))\,ds = a(t) \qquad (0 \leq t \leq 1)$$

has a solution $x \in C([0, 1], F)$.

Deduce that the said equation is soluble for any $a \in C([0, 1], F)$, provided

$$\liminf_{r \to \infty} r^{-1} M(r) < 1.$$

[Hint: Adapt the method used in Subsection 3.6.3.]

3.7 Let E be a metric space, f a self-map of E such that

$$d(f(x), f(y)) < d(x, y)$$

for $x, y \in E$, $x \neq y$. Suppose that $x_0 \in E$ is such that the sequence (x_n) of iterates $x_n = f^n(x_0)$ possesses a subsequence convergent to a point x_∞ of E. Prove that x_∞ is the unique fixed point of f. (Edelstein [2].)
[Hint: Uniqueness is evident. For the rest argue by contradiction. If $f(x_\infty) \neq x_\infty$, show, by considering the function $d(f(x), f(y))/d(x, y)$ on $(E \times E)\backslash\Delta$ (Δ the diagonal

in $E \times E$), that there exists a number $r < 1$ and a neighbourhood U of $(x_\infty, f(x_\infty))$ such that

$$d(f(x), f(y)) < r \cdot d(x, y)$$

for $(x, y) \in U$. Choose also $r' > 0$ so small that $B(x_\infty, r') \times B(f(x_\infty), r') \subset U$, and $r' < d(f(x_\infty), x_\infty)/3$. Taking a subsequence (x_{n_k}) converging to x_∞, show that $d(x_{n_k}, x_{n_k+1}) > r'$ for $k > k_0$, and that for $h > k > k_0$

$$d(x_{n_h}, x_{n_h+1}) < r^{h-k} \cdot d(x_{n_k}, x_{n_k+1}),$$

which tends to 0 as $h \to \infty$. This is the required contradiction.]

3.8 Let E be a TVS, M a subset of E, (x'_m) a strongly bounded sequence in E' generating a vector subspace that is strongly dense in E'. Suppose that f is a self-map of M such that for some $x_0 \in M$, the sequence (x_n) of iterates $x_n = f^n(x_0)$ has a bounded subsequence admitting a weak limiting point x_∞ in M. Suppose finally that

$$|\langle f(x) - f(y), x'_m \rangle| \leq |\langle x - y, x'_m \rangle|$$

for all $x, y \in E$ and all m, while if $x \neq y$ strict inequality obtains for at least one value of m. Prove that x_∞ is the unique fixed point of f in M.
[Hint: Introduce on E the metric

$$d(x, y) = \sum m^{-2} \, |\langle x - y, x'_m \rangle|$$

and apply the preceding exercise.]

3.9 Let E be a TVS, A a continuous endomorphism of E whose adjoint A' has a sequence of eigenvectors, each belonging to an eigenvalue strictly inferior to 1 in modulus, the sequence being strongly bounded, total over E, and generating a strongly dense vector subspace of E'. Let $y \in E$ be given. Suppose there exists an $x_0 \in E$ such that the sequence (x_n) defined by

$$x_n = A^n x_0 + y + Ay + \cdots + A^{n-1} y$$

admits a bounded subsequence possessing a weak limiting point x_∞ in E. Prove that x_∞ is the unique solution in E of the equation $x - Ax = y$.
[Hint: Suppose the eigenvectors of A' are x'_m. Put $f(x) = Ax + y$ and apply Exercise 3.8.]

3.10 Suppose that $1 < p < \infty$, that $y \in L^p(0, \infty)$ and $K \in L^1(0, \infty)$ are given. Define for $x \in L^p(0, \infty)$

$$Kx(t) = \int_0^t x(t - s) K(s) \, ds.$$

Show that K is a continuous endomorphism of $L^p(0, \infty)$ and that at least one of two following assertions, (a) or (b), is true:

(a) For each $x \in L^p(0, \infty)$

$$\lim_{n \to \infty} \|K^n x + y + Ky + \cdots + K^{n-1} y\|_{L^p} = +\infty.$$

(b) The equation $x - Kx = y$ has unique solution $x \in L^p(0, \infty)$.

NOTE. Assume the weak relative compactness of bounded subsets of L^p, together with the fact that the dual of L^p is identifiable with $L^{p'}$, whenever $1 < p < \infty$ and $1/p + 1/p' = 1$. These facts are established in Chapter 4.
[Hint: Apply Exercise 3.9, taking for x'_m the functions

$$t \to \exp(-\xi_m t),$$

where (ξ_m) is a suitably chosen sequence of real numbers $\xi_m > 0$.]

3.11 Let X be a topological semigroup, $C_{bd}(X)$ the space of bounded real-valued continuous functions on X. For $f \in C_{bd}(X)$ and $a \in X$ define ${}_af(x) = f(ax)$ for $x \in X$. By a left-invariant mean on X we understand a linear form $f \to \mu(f)$ on $C_{bd}(X)$ that is positive, satisfies $\mu(1) = 1$, and is left invariant in the sense that $\mu({}_af) = \mu(f)$ for all $f \in C_{bd}(X)$ and all $a \in X$. [Right-invariant means are defined analogously, replacing the left translate ${}_af$ by the right translate $f_a(x) = f(xa)$.]

Show that for the existence of a left-invariant mean on X it is necessary and sufficient that X fulfill the following condition:

If $f_1, \cdots, f_n \in C_{bd}(X)$ and $a_1, \cdots, a_n \in X$, then

$$\text{Sup}\,(f_1 - {}_{a_1}f_1 + \cdots + f_n - {}_{a_n}f_n) \geq 0.$$

[Hint: For sufficiency consider the vector subspace L of X generated by elements of the form $f - {}_af$, and the open convex set K formed of elements g such that $\text{Sup}\, g < 0$. Apply Theorem 2.2.1.]

Notes. The result is due to Dixmier ([3], Théorème 1) and is used by him to establish a number of statements concerning the existence or otherwise of invariant means. Compare also with Exercise 2.15.

3.12 This exercise is concerned with a variant of Theorem 3.3.1.

Let S be a semigroup with a right-invariant mean which we here denote by $\int \cdots ds$. (The definition is as in the preceding exercise with S in place of X and with discrete topology.) Let E be a real vector space, and suppose that to each $s \in S$ is assigned an endomorphism T_s of E in such a way that $T_{ss'} = T_sT_{s'}$ for $s, s' \in S$. Suppose L is a vector subspace of E stable under each T_s, and that p is a gauge function on E such that $p(T_sx) \leq p(x)$ for $s \in S$ and $x \in X$. Suppose finally that f_0 is a linear form on L such that

$$f_0(x) \leq p(x), \qquad f_0(T_sx) = f_0(x) \qquad (s \in S,\ x \in L).$$

The conclusion is that there exists a linear form f on E satisfying

$$f(x) \leq p(x), \qquad f(T_sx) = f(x) \qquad (x \in S,\ x \in E)$$

and $f \mid L = f_0$.

[Hint: First extend f_0 into a linear form f_1 on E such that $f_1 \leq p$. Then consider

$$f(x) = \int f_1(T_sx)\, ds.]$$

Notes. Compare with Exercise 2.15. The present result is less refined and applies only when the element g there specified satisfies $\text{Inf}\, g > 0$. On the other hand the present result has a wider range of application. That Exercise 2.15 calls for the existence of a *left*-invariant mean is explained by the fact that there T_s acts on functions x on a set Q according to the rule $T_sx = x \circ s$, each s being a self-map of Q; accordingly $T_sT_{s'} = x \circ (s' \circ s)$, so that $T_sT_{s'} = T_{ss'}$ holds only if the semigroup product ss' signifies $s' \circ s$ (rather than the more usual $s \circ s'$). This reversal of order of the factors accounts for the interchange of "left" and "right."

CHAPTER 4

Topological Duals of Certain Spaces: Radon Measures

4.1 Plan of the Chapter

Suppose that $T = \{t\}$ is a topological space. Our aim is to study in considerable detail some very important TVSs that are subspaces of K^T or quotient spaces of these, together with their duals. For applications to functional analysis it is no great handicap to assume that T is locally compact. While Radon measures and Schwartz distributions are closely akin, it is convenient to review them separately in this chapter and the next.

In respect of Radon measures, the TVSs we shall consider in detail will be as follows. One begins with a "minimal" space $\mathscr{K}(T)$, consisting of the continuous scalar-valued functions on T having compact supports. If T is compact this is identical with the more familiar Banach space $C(T)$ of continuous scalar-valued functions, normally furnished with the norm

$$\|x\|_\infty = \text{Sup}\,\{|x(t)| : t \in T\}.$$

An appropriate locally convex topology on $\mathscr{K}(T)$ is of the inductive limit variety introduced later in Section 6.3, but meanwhile it is possible to work with a suitable definition of sequential convergence in $\mathscr{K}(T)$. The linear forms on $\mathscr{K}(T)$ that are continuous relative to this concept of sequential convergence, and that subsequently show up as precisely the members of the topological dual of $\mathscr{K}(T)$, are the so-called Radon measures on T. It is part of our task to show that each Radon measure thus defined can be split into a linear combination of positive Radon measures (that is, ones that assign to each positive function in $\mathscr{K}(T)$ a positive number) and that the action of such a positive Radon measure is that of integration with respect to an associated measure function defined at least for all Borel subsets of T. This will provide some justification for attaching the name "measure" to what is initially introduced as a linear form on $\mathscr{K}(T)$.

In doing this we shall be involved in developing the integration theory linked with a given positive Radon measure μ on T, so following a selection from Bourbaki's very extensive development of ideas of Daniell and F. Riesz (Bourbaki [6]), together with minor variations presented in Edwards [3]; see also (Hewitt and Ross [1]).

At first sight the problem, which is one of extending the linear form μ from $\mathscr{K}(T)$ to a bigger subspace of K^T, appears to be solved by appeal to the Hahn-Banach theorem. Indeed, such a solution is possible. It results, however, in a nonunique extension that is analytically most unsatisfactory—an integration process based upon a finitely additive measure, lacking the powerful convergence theorems characteristic of a Lebesgue-like integral. Because of this a constructive extension is made at considerable cost in labour. As an outcome one is led to the Lebesgue spaces $L^p(T, \mu)$ associated with the given positive Radon measure μ, and a second phase amounts to a study of these spaces and their duals.

Our study will commence with the especially simple case in which T is discrete, so that all topological restrictions become void (all functions are continuous, all sets open, and so forth). Only measures remain to be considered here. The most important instance is that in which $T = N$, the set of natural numbers, in which case the Lebesgue spaces reduce essentially to the sequence spaces ℓ^p. This case merits separate study owing to its relatively great simplicity. It is dealt with in Section 4.2, the main problems being commenced in Section 4.3.

4.2 The Banach Spaces $\ell^p(T)$ and $c_0(T)$

Let T be an arbitrary set. T may be considered as a locally compact space with discrete topology, in which case it fits into the general scheme of this chapter.

If $p > 0$ is given, we denote by $\ell^p(T)$ the subspace of K^T comprising those x in K^T for which

$$\|x\|_p \equiv \left[\sum_{t \in T} |x(t)|^p\right]^{1/p} < +\infty;$$

if $p = \infty$, $\ell^\infty(T)$ consists precisely of those x in K^T that are bounded, that is, for which

$$\|x\|_\infty \equiv \operatorname{Sup}_{t \in T} |x(t)| < +\infty.$$

Suppose first that $1 \leq p \leq \infty$, and define the conjugate exponent p' by the relation $1/p + 1/p' = 1$, setting $p' = 1$ if $p = \infty$ and $p' = \infty$ if $p = 1$. The elementary forms of the inequalities of Hölder and of Minkowski state then that

$$\left|\sum_{t \in T} x(t)y(t)\right| \leq \sum_{t \in T} |x(t)y(t)| \leq \|x\|_p \cdot \|y\|_{p'} \tag{4.2.1}$$

and

$$\|x + y\|_p \leq \|x\|_p + \|y\|_p \tag{4.2.2}$$

for x and y in K^T having finite supports. (See also Subsection 4.11.2.) An obvious limiting process shows that (4.2.1) continues to hold whenever x and y belong, respectively, to $\ell^p(T)$ and $\ell^{p'}(T)$, while (4.2.2) continues to hold whenever x and y belong to $\ell^p(T)$.

Thanks to (4.2.2) we see that $x \to \|x\|_p$ is a norm on $\ell^p(T)$, and it is very simple to verify that accordingly $\ell^p(T)$ is a Banach space whenever $1 \le p \le \infty$.

$c_0(T)$ may be defined as the closure in $\ell^\infty(T)$ of the set of functions x having finite supports; it too is a Banach space when equipped with the norm induced upon it by $\|x\|_\infty$. The space $c_0(T)$ comprises precisely those x in K^T that tend to zero at infinity—that is, having the property that to each number $\varepsilon > 0$ corresponds a finite (=compact) subset F of T such that $|x(t)| < \varepsilon$ whenever $t \in T \backslash F$.

Let us suppose now that $0 < p < 1$. Then p' is negative and there is no way of salvaging (4.2.1). Equation (4.2.2) is generally false and must in fact be replaced by

$$\|x + y\|_p \ge \|x\|_p + \|y\|_p \tag{4.2.2*}$$

with equality only if $x = \alpha y$ for some scalar α; see Hardy, Littlewood and Pólya [1], p. 30 and p. 124.

There is, however, a partial substitute for (4.2.2) obtainable by starting from the elementary numerical inequalities

$$2^{p-1}(a^p + b^p) \le (a + b)^p \le a^p + b^p, \tag{4.2.3}$$

valid for $0 < p < 1$, $a \ge 0$, and $b \ge 0$. From these it is easily deduced that

$$\|x + y\|_p^p \le \|x\|_p^p + \|y\|_p^p \tag{4.2.4}$$

and

$$\|x + y\|_p \le 2^{(1/p)-1}(\|x\|_p + \|y\|_p). \tag{4.2.5}$$

These inequalities are the best-possible ones. From them it appears that $x \to \|x\|_p$ is no longer a norm on $\ell^p(T)$. Nevertheless, $\ell^p(T)$ is a complete metrizable TVS when endowed with the topology defined by the metric $d(x, y) = \|x - y\|_p^p$.

If $0 < p < 1$ and T is infinite, the TVS $\ell^p(T)$ thus obtained is not locally convex. To see this it suffices to exhibit a subset X of $\ell^p(T)$ that is bounded but whose convex envelope Y is not bounded. Now, T being infinite, we may choose a sequence (t_n) of elements of T such that $t_m \ne t_n$ whenever $m \ne n$. Let x_n be that element of $l^p(T)$ taking at t_n the value 1 and at each $t \ne t_n$ the value 0. Then $\|x_n\|_p = 1$, so that the set $X = \{x_n : n = 1, 2, \cdots\}$ is bounded in $\ell^p(T)$. The convex envelope Y of X contains each of the elements

$$y_n = n^{-1}(x_1 + \cdots + x_n).$$

Since, as one may compute directly, $\|y_n\|_p = n^{1/p-1}$ and thus tends to infinity with n, Y is not bounded in $\ell^p(T)$.

The case $1 \le p \le \infty$ is by far the most important and one has here the important result:

4.2.1 **Theorem.** If $1 \le p < \infty$, the dual of ℓ^p may be identified as a Banach space with $\ell^{p'}$ in such a way that the element x' of $\ell^{p'}$ generates the linear form

$$x'(x) = \sum_{t \in T} x(t)x'(t). \tag{4.2.6}$$

Proof. For each t let e_t be defined by $e_t(s) = 1$ if $s = t$, and 0 otherwise. e_t belongs to every ℓ^p. Since $p \neq \infty$, each x in ℓ^p can be written as $\sum_{t \in T} x(t)e_t$, the series converging in ℓ^p. Consequently, if f is a continuous linear form on ℓ^p, and if we define $x'(t) = f(e_t)$, then

$$f(x) = \sum_{t \in T} x(t)x'(t),$$

the series on the right being automatically convergent for each x in ℓ^p. Moreover, since f is continuous,

$$\left| \sum_{t \in T} x(t)x'(t) \right| \leq \text{const } \|x\|_p.$$

Choosing x suitably, it is easily seen that consequently x' belongs to $\ell^{p'}$. Reciprocally, if x' belongs to $\ell^{p'}$, Hölder's inequality for sums shows that the sum on the right of (4.2.6) is a continuous linear form on ℓ^p, the norm of which is precisely the $\ell^{p'}$-norm of x'. ▮

Remark. Theorem 4.2.1 is false when $p = \infty$: the dual of ℓ^∞ "contains" ℓ^1 as a proper closed vector subspace.

4.2.2 **Theorem.** The dual of c_0 may be identified as a Banach space with ℓ^1 via (4.2.6).

Proof. The argument is very similar to that used in the proof of the preceding theorem; the details are left to the reader. See Exercise 1.4. ▮

4.3 The Dual of $\mathscr{K}(T)$: Radon Measures

Throughout the rest of this chapter T will denote a separated locally compact space; any further specialization will be stated explicitly. $\mathscr{K}(T)$ or, more precisely, $\mathscr{K}_K(T)$, is the vector subspace of K^T consisting of the continuous functions with compact supports. K may be R or C. In studying a linear form on $\mathscr{K}_K(T)$ it is enough to examine its behaviour on $\mathscr{K}_R(T)$; moreover, it is enough to examine real-valued (real-) linear forms on $\mathscr{K}_R(T)$. In future we shall use $\mathscr{K}(T)$ to denote $\mathscr{K}_C(T)$.

The type of continuity demanded of such linear forms characterizing them as Radon measures will now be described. (Compare Subsection 2.2.8.)

Definition. If (x_n) is a sequence (or a directed family) in $\mathscr{K}_K(T)$, we shall write $x_n \twoheadrightarrow 0$ if (a) there exists a compact set E in T containing the support of x_n for every n, and (b) $x_n \to 0$ uniformly on T. We shall then say that (x_n) tends to 0 in (or in the sense of) $\mathscr{K}_K(T)$.

Obviously, if the x_n are real valued, this sense of convergence is the same whether the x_n are regarded as elements of $\mathscr{K}(T)$ or of $\mathscr{K}_R(T)$.

4.3.1 **Definition.** A complex Radon measure ϕ on T is a linear form on $\mathscr{K}_C(T)$ which is continuous in the sense that $(x_n) \subset \mathscr{K}_C(T)$ and $x_n \twoheadrightarrow 0$ together imply $\lim \phi(x_n) = 0$. If $\mathscr{K}_C(T)$ is replaced by $\mathscr{K}_R(T)$, regarded as a real vector space, the corresponding concept is that of a real Radon measure on T.

The reader will verify easily that a linear form on $\mathscr{K}_C(T)$ (resp. $\mathscr{K}_R(T)$) satisfies the conditions of Definition 4.3.1 if and only if to each compact set K in T corresponds a number m_K such that

$$|\phi(x)| \leq m_K \cdot \|x\|_\infty \tag{4.3.1}$$

for each x in $\mathscr{K}_C(T)$ (resp. $\mathscr{K}_R(T)$) with support contained in K. Anticipating the results of Section 6.3, it may be stated that there is a locally convex topology on $\mathscr{K}_C(T)$ (or $\mathscr{K}_R(T)$) relative to which the continuity of linear forms amounts precisely to the condition (4.3.1). Accordingly we shall speak of Definition 4.3.1, or the equivalent (4.3.1), as expressing the characteristic continuity of Radon measures.

Aside from any topological questions, it is plain that each real Radon measure may be extended from $\mathscr{K}_R(T)$ to $\mathscr{K}_C(T)$ so as to appear as a complex Radon measure; and that each complex Radon measure can be expressed uniquely as $\alpha + i\beta$, where α and β are real Radon measures. It is therefore enough to discuss in detail real Radon measures.

A less obvious fact is the possibility of decomposing any real Radon measure into the difference of two positive Radon measures.

4.3.2 **Theorem.** Let λ be a real Radon measure on T. There exist unique positive Radon measures λ^+ and λ^- on T with the following properties:

(a) $\lambda = \lambda^+ - \lambda^-$;

(b) if α and β are positive Radon measures on T for which $\lambda = \alpha - \beta$, then $\alpha - \lambda^+$ and $\beta - \lambda^-$ are positive Radon measures on T.

Accordingly, (a) is said to express the "minimal decomposition" of λ, and λ^+ and λ^- are termed the positive and negative parts of λ.

Proof. It is clear that uniqueness obtains, granted the existence of such a decomposition.

Turning to the proof of existence, we define

$$\lambda^+(x) = \text{Sup}\,\{\lambda(y) : y \in \mathscr{K}_+(T),\, y \leq x\}$$

for each x in $\mathscr{K}_+(T)$, the set of positive functions in $\mathscr{K}(T)$. For such x one has clearly $\lambda^+(x) \geq 0$ and $\lambda(x) \leq \lambda^+(x)$. We show next that λ^+ is additive on $\mathscr{K}_+(T)$, that is, that

$$\lambda^+(x_1 + x_2) = \lambda^+(x_1) + \lambda^+(x_2)$$

for x_1 and x_2 in $\mathscr{K}_+(T)$. If y_i is in $\mathscr{K}_+(T)$ and $y_i \leq x_i$ $(i = 1, 2)$, then $y_1 + y_2$ is in $\mathscr{K}_+(T)$ and $y_1 + y_2 \leq x_1 + x_2$, whence it follows easily that

$$\lambda^+(x_1 + x_2) \geq \lambda^+(x_1) + \lambda^+(x_2).$$

On the other hand, suppose that y is in $\mathscr{K}_+(T)$ and $y \leq x_1 + x_2$. Put $y_1 = \text{Sup}\,(y - x_2, 0)$ and $y_2 = \text{Inf}\,(y, x_2)$. These belong to $\mathscr{K}_+(T)$ and

$$y_1 \leq \text{Sup}\,(x_1, 0) = x_1,\ y_2 \leq x_2 \qquad \text{and} \qquad y = y_1 + y_2.$$

Thus

$$\lambda^+(x_1) + \lambda^+(x_2) \geq \lambda(y_1) + \lambda(y_2) = \lambda(y_1 + y_2) = \lambda(y).$$

Taking the supremum on the right, additivity is proved.

A similar argument shows that $\lambda^+(c \cdot x) = c \cdot \lambda^+(x)$ if $c \geq 0$ and x is in $\mathscr{K}_+(T)$.

These properties make it possible to extend λ^+ to $\mathcal{K}_R(T)$ so as to be linear: any x in $\mathscr{K}(T)$ can be decomposed in many ways into the difference $x_1 - x_2$ of two functions in $\mathscr{K}_+(T)$; by additivity, the value of $\lambda^+(x_1) - \lambda^+(x_2)$ is independent of which such decomposition is used and may be taken as the definition of $\lambda^+(x)$. This λ^+ is a positive linear form on $\mathcal{K}_R(T)$, and its positivity ensures that (4.3.1) is fulfilled. Thus λ^+ is a positive Radon measure. Since also $\lambda^+(x) \geq \lambda(x)$ for x in $\mathscr{K}_+(T)$, $\lambda^- = \lambda^+ - \lambda$ is also a positive Radon measure. (a) holds for this pair (λ^+, λ^-).

It remains to establish (b). If x and y are in $\mathscr{K}_+(T)$ and $y \leq x$, however, one has

$$\lambda(y) = \alpha(y) - \beta(y) \leq \alpha(y) \leq \alpha(x),$$

and so

$$\lambda^+(x) \leq \alpha(x).$$

Thus $\alpha - \lambda^+$ is positive. Similarly, $\beta - \lambda^-$ is positive. The proof is thus complete. ▌

In view of this last theorem our investigation is narrowed down to a detailed study of positive Radon measures on T—that is, to linear forms on $\mathcal{K}_R(T)$ that are positive. Throughout the remainder of this chapter the phrase "positive measure" will abbreviate "positive Radon measure."

4.4 Some Examples

Before beginning the rather lengthy theory of integration associated with a given positive measure, we give a few examples of Radon measures.

4.4.1 Atomic Measures. T being any separated locally compact space, we denote by ε_t the linear form on $\mathcal{K}_R(T)$ (or its natural extension to $\mathscr{K}_C(T)$) defined by

$$\varepsilon_t(x) = x(t); \tag{4.4.1}$$

ε_t is a positive measure on T, often called the Dirac measure placed at the point t. Physicists often deal formally with ε_t as if it were a function, T being in this case a number space R^n. In this situation, ε_t is not a function; it is, however, the limit (weakly in the dual of $\mathscr{K}(T)$) of functions, and it is this principle that dictates the physicist's use of the Dirac measures. See also Subsection 4.4.3 below.

More generally, one may choose a function f in C^T and seek to define a Radon measure μ on T via the formula

$$\mu(x) = \sum_{t \in T} f(t)x(t), \tag{4.4.2}$$

it being assumed that the series on the right is unconditionally convergent for each x in $\mathscr{K}(T)$. It is then natural to write

$$\mu = \sum_{t \in T} f(t)\varepsilon_t; \tag{4.4.3}$$

this series is unconditionally convergent in the sense of the weak topology of the dual of $\mathscr{K}(T)$. Measures of this type are termed atomic. The desired convergence of (4.4.2) signifies exactly that

$$\sum_{t \in K} |f(t)| < +\infty \tag{4.4.4}$$

for each compact set K in T.

For further details, see Bourbaki [6], Chapitre V, § 5.10; and Edwards [16], pp. 377–84.

4.4.2 Lebesgue Measures. Suppose T is R^n. If x is in $\mathscr{K}(T)$ and we choose any hypercube Q outside which x vanishes, the Riemann integral

$$\int \cdots \int_Q x(t_1, \cdots, t_n)\, dt_1 \cdots dt_n = \int_Q x(t)\, dt$$

exists; its value is independent of Q and may be denoted by $\int x(t)\, dt$ simply. As a function of x, it is plainly a positive measure. We call it the Lebesgue measure on R^n, even though Lebesgue's name is normally associated with the extended integration theory yet to follow.

The peculiar significance of this measure on R^n is due to its translation invariance: the Lebesgue measure on R^n is, apart from a multiplicative constant, the only Radon measure on R^n for which it is true that the integrals of all translates of an arbitrary given function x in $\mathscr{K}(R^n)$ have the same value.

Quite generally if T is a locally compact group, there is a similarly unique Radon measure on T that is invariant with respect to left translations. This is the so-called left Haar measure on T. Likewise for right Haar measures (see Section 4.18). These Haar measures dominate the functional analytic study of various functional spaces over groups.

4.4.3 Densities. T being an arbitrary separated locally compact space, suppose μ is a given Radon measure on T and f a given continuous function on T. The formula

$$\nu(x) = \mu(fx) \tag{4.4.5}$$

is easily seen to define a new Radon measure ν on T that is denoted by $f \cdot \mu$. In fact, if we anticipate the results of Section 4.13, $f \cdot \mu$ is well defined for any f that is locally integrable for μ. One may say in any case that $f \cdot \mu$ is the measure with density f relative to μ, or that $\nu = f \cdot \mu$ has f as its "derivative" with respect to $\mu : f = d\nu/d\mu$. The complex of ideas associated with this situation is discussed at greater length in Sections 4.13–4.15.

Taking $T = R^n$ and μ the Lebesgue measure, when one asserts (as in Subsection 4.4.1 above) that the Dirac measure ε_t is not a function, one means precisely that there exists no function f (locally integrable for Lebesgue measure) such that $\varepsilon_t = f \cdot \mu$. On the other hand, as has been said in Subsection 4.4.1, one can find sequences (f_n) of continuous functions (even functions in $\mathscr{K}_+(R^n)$) such that $\varepsilon_t = \lim f_n \cdot \mu$ weakly in the dual of $\mathscr{K}(R^n)$: it is in fact sufficient to take continuous functions $f_n \geq 0$ such that

$$\mu(f_n) = 1$$

and such that f_n vanishes outside an interval with centre t whose length tends to zero as $n \to \infty$. Another popular choice of f_n is

$$f_n(t') = c_n \cdot \exp(-n\,|t' - t|^2),$$

where $|t' - t|$ is the usual distance between t and t', and where the numbers $c_n > 0$ are chosen so that $\mu(f_n) = 1$. The functions f_n converge pointwise to 0 at points distinct from t and to $+\infty$ at t itself, and this in precisely such a way that $\mu(f_n x)$ converges to $x(t) = \varepsilon_t(x)$ for each x in $\mathscr{K}(R^n)$.

4.4.4 Lebesgue-Stieltjes Measures. There are important extensions of Subsection 4.4.2. Although these can be carried through in several dimensions, the multidimensional case is notationally rather cumbersome and we shall deal with the one-dimensional situation only. Throughout this subsection, T denotes an interval of R. Our aim is to associate with each function f that is locally of bounded variation on T (a concept defined below) a Radon measure on T; Radon measures obtained in this way are termed Stieltjes or Lebesgue-Stieltjes measures. The Lebesgue measure introduced in Subsection 4.4.2 arises in this way when the original function f is defined by $f(t) = t$ for t in T. It will suffice to handle the case in which f is real valued, the corresponding measure being then a real Radon measure.

Suppose then that f is a real-valued function on T. If I is a bounded interval contained in T with extremities $t_1 \leq t_2$, we write $\Delta_f(I) = f(t_2) - f(t_1)$. If K is a compact subinterval of T, the total variation of f on K is

$$V_f(K) = \operatorname{Sup} \sum_r |\Delta_f(I_r)|,$$

the supremum being taken with respect to all finite families of disjoint intervals $I_r \subset K$. $V_f(K)$ may be $+\infty$. If $V_f(K) < +\infty$, we say that f is of bounded variation (BV) on K. If $V_f(K) < +\infty$ for each compact interval $K \subset T$, we say that f is of locally bounded variation (LBV) on T. Any monotone function on T is LBV on T, and the same is true of the difference of two such functions. Conversely, if f is BV on K, it can be represented on K as the difference of two monotone functions on K (see, for example, Natanson [1], p. 218). From this we deduce the existence of one-sided limits of f at points of T and its extremities.

The first step towards defining the Radon measure associated with f is to define the Riemann-Stieltjes integral with respect to f, reducing to the Riemann integral when $f(t) = t$. Since we need the Riemann-Stieltjes integral only for integrands that are continuous (in $\mathscr{K}(T)$ in fact), the briefest summary suffices; for details see, for example, Natanson [1], Ch. VIII. Supposing x to belong to

$\mathscr{K}(T)$, we choose a compact interval $K \subset T$ outside which x vanishes. Then one can show, using the fact that $V_f(K) < +\infty$, that there exists a unique number i with the following property: Given $\varepsilon > 0$, there exists $\delta > 0$ such that

$$\left| i - \sum_r x(t_r) \Delta_f(I_r) \right| \leq \varepsilon$$

for all finite partitions of K into intervals I_r and all choices of t_r from I_r, provided only that the maximum length of the various I_r does not exceed δ. This number i does not depend upon the choice of K, provided only that x vanishes outside K. Thus i may be denoted by $RS\text{-}\int_T x\, df$, the Riemann-Stieltjes integral of x with respect to the integrator f. It is easily seen that the mapping

$$x \to RS\text{-}\int_T x\, df$$

is linear, and that

$$\left| RS\text{-}\int_T x\, df \right| \leq V_f(T) \cdot \|x\|_\infty,$$

where $V_f(T) = \operatorname{Sup}_K V_f(K)$ is the total variation of f on T. This mapping, call it μ_f, is the Radon measure associated with f; it is also termed the Lebesgue-Stieltjes (or simply Stieltjes) measure on T generated by f. Clearly $\mu_f \geq 0$ if f is $\uparrow$.

At this point we should introduce a more restricted class of integrators f which will assume more significance in connection with Sections 4.13–4.15. A function f on T is said to be locally absolutely continuous (LAC) on T if, for each compact interval $K \subset T$ and each number $\varepsilon > 0$, there exists a number $\delta > 0$ such that $\sum_r |\Delta_f(I_r)| \leq \varepsilon$ for any finite family (I_r) of disjoint subintervals of K, the sum of whose lengths does not exceed δ. If this condition is fulfilled for some particular K, we say that f is absolutely continuous (AC) on K. When we come to Sections 4.13–4.15, we shall see that this concept is the exact analogue of absolute continuity of the measure μ_f with respect to the Lebesgue measure. Meanwhile, the following fact should be noted: if f is LAC on T, then it is locally uniformly continuous on T and LBV on T. However, there exist functions that are continuous and LBV but not LAC. In real function theory the importance of LAC functions is that they possess derivatives "almost everywhere" that are integrable and that, upon integration (in the Lebesgue sense), lead back to the primitive function. Sections 4.13–4.15 examine this question in the language of measures and for arbitrary locally compact spaces T.

Further discussion of the Lebesgue-Stieltjes measure μ_f will appear as examples as we proceed; see Examples 4.5.6, 4.7.7, 4.7.8, 4.9.2, 4.15.10.

4.5 Integration Theory Associated with a Positive Measure

From this point onward until the end of Section 4.10, μ will denote a fixed positive Radon measure on T. We are concerned with a constructive extension of the functional μ from $\mathscr{K}(T)$ to several wider sets of real- or complex-valued functions on T.

The first stage of the extension amounts to defining a so-called "upper integral" μ^* for arbitrary positive functions on T. In this connection it is convenient to admit infinite-valued functions and we commence by describing in more detail how this is to be done.

The set R of real numbers is enlarged by the adjunction of two further elements denoted by $-\infty$ and $+\infty$, respectively, and we label the enlarged set $\bar{R}$. We order $\bar{R}$ by leaving undisturbed the natural order of R and defining further $-\infty < a < +\infty$ for all a in R. Then every nonvoid subset of $\bar{R}$ has both a supremum and an infimum. As regards topology, that of R is left undisturbed, while we take as a base for the neighbourhoods of $+\infty$ the intervals of the form $(a, +\infty]$ with a in R, and for $-\infty$ the intervals $[-\infty, a)$ with a in R. It is easily seen that $\bar{R}$ is a compact space of which R is a dense open subspace. Some, but not all, operations defined for real numbers can be extended by continuity for elements of $\bar{R}$. Thus Sup (x, y) and Inf (x, y) are continuous on $\bar{R} \times \bar{R}$; $-x$ can be extended by continuity to $\bar{R}$; $x + y$ can be extended by continuity to $(-\infty, +\infty] \times (-\infty, +\infty]$ and to $[-\infty, +\infty) \times [-\infty, +\infty)$ in such a way that

$$x + (+\infty) = (+\infty) + x = +\infty \qquad \text{if } x \neq -\infty,$$
$$x + (-\infty) = (-\infty) + x = -\infty \qquad \text{if } x \neq +\infty.$$

$x^+ = \text{Sup}\,(x, 0)$, $x^- = \text{Sup}\,(-x, 0)$, and $|x| = \text{Sup}\,(x, -x)$ are defined on $\bar{R}$, as also are $x^+ + x^-$ and $x^+ - x^-$, which are seen to be none other than $|x|$ and x, respectively. Also, $|x + y| \leq |x| + |y|$ whenever $x + y$ is defined. Similarly, xy can be extended by continuity to $(\bar{R} \setminus \{0\}) \times (\bar{R} \setminus \{0\})$ in such a way that

$$x \cdot (+\infty) = (+\infty) \cdot x = \begin{cases} +\infty & \text{if } x > 0 \\ -\infty & \text{if } x < 0, \end{cases}$$

$$x \cdot (-\infty) = (-\infty) \cdot x = \begin{cases} -\infty & \text{if } x > 0 \\ +\infty & \text{if } x < 0. \end{cases}$$

However, $0 \cdot (+\infty)$, $(+\infty) \cdot 0$, $0 \cdot (-\infty)$, $(-\infty) \cdot 0$ are not so definable by continuity: we set them all equal to 0 by convention. Then $|xy| = |x|\,|y|$ always. Finally, $x(y + z) = xy + xz$ whenever all the operations involved are separately defined. The manipulation of $\bar{R}$-valued functions is, of course, subject to the conventions just made. In addition, we shall assume that a lower (resp. upper) semicontinuous $\bar{R}$-valued function never takes the value $-\infty$ (resp. $+\infty$); a continuous $\bar{R}$-valued function, being both lower and upper semicontinuous, is therefore in fact R-valued (that is, finite-valued).

A major role will be played by the set $\Phi = \Phi(T)$ of positive lower semicontinuous functions on T. According to our definitions, Φ is stable under addition and multiplication by constants $c \geq 0$. See also Subsection 0.2.19.

For ϕ in Φ we define

$$\mu^*(\phi) = \text{Sup}\,\{\mu(x) : x \in \mathscr{K}_R(T),\quad x \leq \phi\}, \tag{4.5.1}$$

which is possibly $+\infty$ and is always positive (that is, ≥ 0). It is clear that

$$\mu^*(c \cdot \phi) = c \cdot \mu^*(\phi) \tag{4.5.2}$$

if $c \geq 0$ is a real number, and that

$$\mu^*(\phi_1) \leq \mu^*(\phi_2) \tag{4.5.3}$$

if $\phi_1 \leq \phi_2$ both belong to Φ. It is evident also that $\mu^*(x) = \mu(x)$ for any positive x in $\mathscr{K}(T)$.

The remaining results of this section are less obvious.

4.5.1 **Proposition.** Let Γ be an increasing directed set of functions in Φ, and let $\phi_0 = \text{Sup}\,\{\phi : \phi \in \Gamma\}$. Then

$$\mu^*(\phi_0) = \text{Sup}\,\{\mu^*(\phi) : \phi \in \Gamma\}.$$

Proof. It is evident that, in the stated equality, the left-hand side is at least as large as the right-hand side, so it remains to prove the reverse inequality. Now, given any $c < \mu^*(\phi_0)$, one can choose x in $\mathscr{K}_+(T)$ satisfying $x \leq \phi_0$ and $\mu(x) = c' > c$. We put $p = c' - c$. We choose any compact set K in T outside which x vanishes, and then x_0 in $\mathscr{K}_+(T)$ taking the value unity at all points of K. Using compactness of K and the fact that Γ is an increasing directed set, it is easily seen that, no matter how small we choose $\varepsilon > 0$ in advance, there exists ϕ in Γ such that $\phi \geq x - \varepsilon x_0$ at all points of K and hence everywhere. Then $\mu^*(\phi) \geq \mu(x - \varepsilon x_0) = c' - \varepsilon \cdot \mu(x_0)$. If we choose ε so small that $\varepsilon \cdot \mu(x_0) < p$, then $\mu^*(\phi) > c$. Since c is arbitrary save for the restriction $c < \mu^*(\phi_0)$, the desired inequality is established. ▌

4.5.2 **Corollary.** If ϕ_1 and ϕ_2 are in Φ, then

$$\mu^*(\phi_1 + \phi_2) = \mu^*(\phi_1) + \mu^*(\phi_2).$$

Proof. For $i = 1, 2$, ϕ_i is the upper envelope of the increasing directed set Γ_i of functions in $\mathscr{K}_+(T)$ minorizing ϕ_i. Consequently, $\phi_1 + \phi_2$ is the upper envelope of the set $\Gamma = \{x_1 + x_2 : x_1 \in \Gamma_1,\ x_2 \in \Gamma_2\}$. The corollary therefore follows from Proposition 4.5.1, together with the additivity of μ. ▌

At this stage we extend the definition of μ^* to an arbitrary function f on T taking values in the interval $[0, +\infty]$ of $\bar{R}$ by setting

$$\mu^*(f) = \text{Inf}\,\{\mu^*(\phi) : \phi \in \Phi,\ \phi \geq f\}.$$

This definition and the above results make it clear that $0 \leq \mu^*(f) \leq +\infty$; $\mu^*(f_1 + f_2) \leq \mu^*(f_1) + \mu^*(f_2)$; $\mu^*(c \cdot f) = c \cdot \mu^*(f)$ ($c = \text{const.} \geq 0$); and $\mu^*(f_1) \leq \mu^*(f_2)$ whenever $f_1 \leq f_2$.

We come now to two of the most fundamental convergence theorems.

4.5.3 Proposition. If the positive functions f_n $(n = 1, 2, \cdots)$ increase monotonely to f, then the upper integrals $\mu^*(f_n)$ increase monotonely to $\mu^*(f)$.

Proof. Since it is evident that $\mu^*(f) \geq \mu^*(f_{n+1}) \geq \mu^*(f_n)$, it appears at once that the $\mu^*(f_n)$ increase monotonely to a limit not exceeding $\mu^*(f)$. It therefore remains only to prove that $\mu^*(f) \leq \lim \mu^*(f_n)$, in doing which we may assume without loss of generality that $\mu^*(f_n)$ is finite for each n. We will in fact show that: For any $\varepsilon > 0$ one may choose an increasing sequence of functions ϕ_n in Φ such that $f_n \leq \phi_n$ and $\mu^*(\phi_n) \leq \mu^*(f_n) + \varepsilon$. If this is done, and if we put $\phi = \lim \phi_n$, then ϕ will belong to Φ, will majorize f, and Proposition 4.5.1 will show that $\mu^*(\phi) = \lim \mu^*(\phi_n) \leq \lim \mu^*(f_n) + \varepsilon$. Hence we shall have $\mu^*(f) \leq \lim \mu^*(f_n) + \varepsilon$. The proof will be completed by letting $\varepsilon \to 0$.

To construct the ϕ_n, we begin by choosing ϕ'_n in Φ majorizing f_n and such that $\mu^*(f_n) \leq \mu^*(\phi'_n) \leq \mu^*(f_n) + \varepsilon/2^n$, and then show that the

$$\phi_n = \mathrm{Sup}(\phi'_1, \cdots, \phi'_n)$$

satisfy our demands. To begin with, ϕ_n evidently belongs to Φ and majorizes f_n. As a final step we prove by induction that

$$\mu^*(\phi_n) \leq \mu^*(f_n) + \varepsilon(1 - 2^{-n}). \tag{4.5.4}$$

This is true for $n = 1$. On the other hand

$$\phi_{n+1} = \mathrm{Sup}(\phi_n, \phi'_{n+1}).$$

Since $\phi_n \geq f_n$ and $\phi'_{n+1} \geq f_{n+1} \geq f_n$, therefore

$$\mathrm{Inf}\,(\phi_n, \phi'_{n+1}) \geq f_n. \tag{4.5.5}$$

Now

$$\mathrm{Inf}(\phi_n, \phi'_{n+1}) + \mathrm{Sup}(\phi_n, \phi'_{n+1}) = \phi_n + \phi'_{n+1}.$$

Using (4.5.5), (4.5.3), and Corollary 4.5.2 there results

$$\mu^*(\phi_n) + \mu^*(\phi'_{n+1}) \geq \mu^*(\phi_{n+1}) + \mu^*(f_n).$$

So assuming (4.5.4) we obtain

$$\begin{aligned} \mu^*(\phi_{n+1}) &\leq \mu^*(\phi_n) + \mu^*(\phi'_{n+1}) - \mu^*(f_n) \\ &\leq \mu^*(f_{n+1}) + \varepsilon(1 - 2^{-n} + 2^{-n-1}), \end{aligned}$$

by choice of ϕ'_{n+1}. Thus (4.5.4) is established with n replaced by $n + 1$, and the inductive proof is complete. ∎

4.5.4 Proposition (Fatou's Lemma). If the functions f_n $(n = 1, 2, \cdots)$ satisfy $0 \leq f_n \leq +\infty$, then

$$\mu^*(\liminf f_n) \leq \liminf \mu^*(f_n).$$

Proof. Put $F_n = \mathrm{Inf}\,\{f_m\colon m \geq n\}$; these are positive functions that increase monotonely to $\liminf f_n$. The preceding proposition gives therefore

$$\mu^*(\liminf f_n) = \lim \mu^*(F_n).$$

On the other hand, $F_n \le f_m$ and so $\mu^*(F_n) \le \mu^*(f_m)$, whenever $m \ge n$. Therefore

$$\lim \mu^*(F_n) \le \liminf \mu^*(f_n),$$

as required. ▌

4.5.5 Remarks. The definition we have used for $\mu^*(f)$ works well for any f, positive or not, provided we enlarge Φ so as to include all LSC functions on T that majorize functions in $\mathscr{K}(T)$. It is easily seen that Propositions 4.5.3 and 4.5.4 then remain valid, provided in the first case $\mu^*(f_n) > -\infty$ for some n, and in the second case that all but a finite number of the f_n majorize a function g for which $\mu^*(g) > -\infty$. Without these additional restrictions the propositions become false.

4.5.6 Example. We return to the situation described in Subsection 4.4.4. Let T be a real interval with extremities a and b $(a < b)$, and let f be an increasing function on T. We aim to determine $\mu_f^*(\chi_I)$ for relatively open intervals $I \subset T$. Anticipating the notation and terminology of Sections 4.6 and 4.7, we shall write $\mu_f^*(\chi_I)$ in the form $\mu_f^*(I)$ and call it the *exterior μ_f-measure* of I.

We assume first that $I = (\alpha, \beta)$, where $a < \alpha \le \beta < b$. We shall show that

$$\begin{aligned} \mu_f^*(I) = 0 \qquad &\text{if} \qquad \alpha = \beta, \\ f(\beta - 0) - f(\alpha + 0) \qquad &\text{if} \qquad \alpha < \beta. \end{aligned} \tag{4.5.6}$$

Of these, the first is obvious since in that case $I = \emptyset$ and $\chi_I = 0$. Otherwise—that is, if $\alpha < \beta$—we suppose that x belongs to $\mathscr{K}_+(T)$ and $x \le \chi_I$. Thus $x \le 1$ on T and $x(\alpha) = x(\beta) = 0$. Since x is continuous, given $\varepsilon > 0$, there exists $\delta > 0$ such that $x(t) \le \varepsilon$ for $\alpha \le t \le \alpha + \delta$ and for $\beta - \delta \le t \le \beta$. Thus

$$\begin{aligned} \mu_f(x) = RS\text{-}\int_T x\, df &= RS\text{-}\int_\alpha^{\alpha+\delta} x\, df + RS\text{-}\int_{\beta-\delta}^{\beta} x\, df + RS\text{-}\int_{\alpha+\delta}^{\beta-\delta} x\, df \\ &\le \varepsilon[f(\alpha + \delta) - f(\alpha)] + \varepsilon[f(\beta) - f(\beta - \delta)] + 1[f(\beta - \delta) - f(\alpha + \delta)]. \end{aligned}$$

Letting $\delta \to 0$ and then $\varepsilon \to 0$, we conclude that

$$\mu_f^*(I) = \operatorname{Sup} \mu_f(x) \le f(\beta - 0) - f(\alpha + 0).$$

To prove the opposing inequality, we use the function x shown in the accompanying graph. For this function we have

$$\begin{aligned} \mu_f(x) &= RS\text{-}\int_T x\, df \ge RS\text{-}\int_{\alpha+\delta}^{\beta-\delta} x\, df \\ &= RS\text{-}\int_{\alpha+\delta}^{\beta-\delta} 1\, df = f(\beta - \delta) - f(\alpha + \delta). \end{aligned}$$

Hence

$$\mu_f^*(I) \ge f(\beta - \delta) - f(\alpha + \delta)$$

and so, letting $\delta \to 0$,

$$\mu_f^*(I) \ge f(\beta - 0) - f(\alpha + 0).$$

This establishes our assertion (4.5.6).

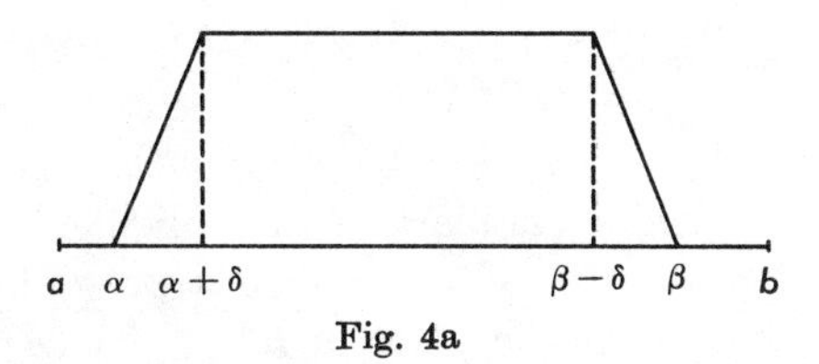

Fig. 4a

Thus far (4.5.6) is shown to be true when $a < \alpha \leq \beta < b$, and indeed the preceding proof breaks down if either $\alpha = a$ or $\beta = b$ (or both). Nevertheless, the formula (4.5.6) remains valid in these cases too: this appears by using the formula in the restricted case in combination with Proposition 4.5.3. Thus (4.5.6) is valid for $a \leq \alpha \leq \beta \leq b$, and this even if $a = -\infty$ or $b = +\infty$ (or both). Furthermore, the result extends to the case in which f is LBV on T, at least whenever I is bounded: this depends upon writing f as the difference of two increasing functions.

If $a \in T$, the interval $I = [a, \beta)$ is also open relative to T, when $a \leq \beta < b$. If $\beta = a$, $I = \emptyset$ and $\mu_f^*(I) = 0$. Otherwise an argument similar to that used to establish (4.5.6) shows that

$$\mu_f^*([a, \beta)) = f(\beta - 0) - f(a) \qquad \text{if} \qquad a \in T, \qquad a < \beta < b. \tag{4.5.7}$$

Similarly

$$\mu_f^*((\alpha, b]) = f(b) - f(\alpha + 0) \qquad \text{if} \qquad b \in T, \qquad a < \alpha < b. \tag{4.5.8}$$

By combination of these results with the additivity of μ_f^* (a consequence of Corollary 4.5.2) we deduce that

$$\mu_f^*((a, b)) = \mu_f^*(T) = f(b) - f(a) \qquad \text{if} \qquad T = [a, b]. \tag{4.5.9}$$

The formulae (4.5.6)–(4.5.9) between them determine $\mu_f^*(I)$ for all intervals $I \subset T$ open relative to T. These relations extend to the case in which f is LBV on T, at least whenever the interval I involved is bounded.

The case which is simplest to remember is that in which $T = R$, in which case

$$\mu_f^*((\alpha, \beta)) = f(\beta - 0) - f(\alpha + 0)$$

for all open intervals (α, β) with $\alpha < \beta$.

Another case which is important in practice is that in which $T = [a, b]$ is a compact real interval. It then turns out that μ_f^* is simply the restriction to subintervals of T of μ_g^*, where g is the increasing function on R obtained by extending f by setting it equal to $f(a)$ on $(-\infty, a)$ and to $f(b)$ on $(b, +\infty)$.

4.6 The Space $\mathscr{L}^1 = \mathscr{L}^1(T, \mu)$ of μ-integrable Functions

Our plan is to use μ^*, defined now for positive functions, to introduce a seminorm N_1 on a vector subspace of K^T and then to introduce $\mathscr{L}^1$ by what amounts to completing $\mathscr{K}_K(T)$ with respect to the structure defined by N_1.

For x in K^T ($K = R$ or C, as usual), we define

$$N_1(x) = \mu^*(|x|),$$

and write $\mathscr{F}^1_K = \mathscr{F}^1_K(T, \mu)$ for the set of x in K^T for which $N_1(x) < +\infty$. The elementary properties of μ^* show that $\mathscr{F}^1_K$ is a vector subspace of K^T and that N_1 is a seminorm on $\mathscr{F}^1_K$. x is said to be *μ-negligible* if $N_1(x) = 0$; when μ is understood we say "negligible" in place of "μ-negligible."

In particular, if A is a subset of T, we shall write χ_A for the characteristic function of A and write $\mu^*(A)$ in place of $\mu^*(\chi_A)$. $\mu^*(A)$ is the *exterior* (μ-) *measure* of A. The elementary properties of μ^* show that $0 \leq \mu^*(A) \leq +\infty$, $\mu^*(A) \leq \mu^*(B)$ if $A \subset B$, and

$$\mu^*\left(\bigcup_n A_n\right) \leq \sum_n \mu^*(A_n), \tag{4.6.1}$$

first of all for finite unions. Then, however, Proposition 4.5.3 shows that

$$\mu^*\left(\bigcup_n A_n\right) = \lim \mu^*(A_n) \tag{4.6.2}$$

for any increasing sequence of sets A_n; and (4.6.2) in turn shows that (4.6.1) extends to arbitrary countable unions of sets. The sets A for which χ_A is negligible are termed negligible sets. We see then that any subset of a negligible set is negligible, and any countable union of negligible sets is negligible.

A property of points of T is said to hold μ-almost everywhere (μ-a.e.) if the set of points of T for which the property is false is μ-negligible. Once more, when μ is understood, we use the phrase "almost everywhere" and the symbol "a.e."

If f and g are positive functions, and if $f \leq g$ a.e., then $\mu^*(f) \leq \mu^*(g)$—which explains the use of the term "negligible." This statement follows from the fact that if A is the negligible set of points at which $f \leq g$ is false, then one has $f \leq \lim\,(g + n \cdot \chi_A)$ at all points; so

$$\begin{aligned}\mu^*(f) &\leq \mu^*(\lim\,(g + n\,\chi_A)) \\ &= \lim \mu^*(g + n\,\chi_A) \leq \lim\,(\mu^*(g) + n \cdot \mu^*(A)) = \mu^*(g).\end{aligned}$$

Consequently, if f and g are positive and $f = g$ a.e., then $\mu^*(f) = \mu^*(g)$.

Again, a function f is negligible if and only if $f = 0$ a.e. Indeed, if $f = 0$ a.e., then $|f| = 0$ a.e. and so, by what we have just proved, $\mu^*(|f|) = \mu^*(0) = 0$. On the other hand, if $\mu^*(|f|) = 0$, it is evident that, for any natural number n, the set where $|f| \geq n^{-1}$ is negligible; and then the union of these sets for $n = 1, 2, \cdots$ is also negligible—that is, the set where $f \neq 0$ is negligible.

We come now to the principal definition of this section.

4.6.1 Definition. $\mathscr{L}^1 = \mathscr{L}^1(T, \mu)$ will denote the closure, relative to the seminorm N_1 on $\mathscr{F}^1_C = \mathscr{F}^1_C(T, \mu)$, of $\mathscr{K}_C(T)$; a function is said to be (μ-) integrable if it belongs to $\mathscr{L}^1(T, \mu)$. It is obvious that one might introduce $\mathscr{L}^1_R$ as the closure in $\mathscr{F}^1_R$ of $\mathscr{K}_R(T)$; the reader will verify easily that the functions thus obtained are just the real-valued functions belonging to $\mathscr{L}^1$.

We know that the inequality

$$|\mu(x)| \leq N_1(x) \tag{4.6.3}$$

holds for x in $\mathscr{K}_R(T)$. If x is in $\mathscr{K}_C(T)$, it is still true but not quite so obvious:

we write $x = u + iv$ with u and v in $\mathscr{K}_R(T)$ and observe that if a and b are real numbers, then

$$|a + ib| = \text{Max}\,\{a \cdot \cos r + b \cdot \sin r : r \text{ real}\}.$$

Thus

$$\begin{aligned} |\mu(x)| = |\mu(u) + i\mu(v)| &= \text{Max}\,\{\mu(u) \cos r + \mu(v) \sin r\} \\ &= \text{Max}\,\mu\,(u \cdot \cos r + v \cdot \sin r) \\ &\leq \mu(|u + iv|) = \mu(|x|) = N_1(x). \end{aligned}$$

In view of (4.6.3), μ admits a unique continuous extension from $\mathscr{K}_C(T)$ to $\mathscr{L}^1$ which we continue to denote by μ. This extended μ is a linear form on $\mathscr{L}^1$ and (4.6.3) continues to hold for x in $\mathscr{L}^1$.

If f and g belong to C^T and are equal almost everywhere, then they are together integrable or not; and if they are integrable, then $\mu(f) = \mu(g)$.

If f and g belong to C^T, one has $|f| \leq |g| + |f - g|$ and so

$$\mu^*(|f|) \leq \mu^*(|g|) + N_1(f - g).$$

Combining this inequality with that which results from an interchange of f and g, we get

$$|\mu^*(|f|) - \mu^*(|g|)| \leq N_1(f - g) \tag{4.6.4}$$

This shows that the function $f \to \mu^*(|f|)$ is uniformly continuous on $\mathscr{F}^1_C$. Since it coincides on $\mathscr{K}_C(T)$ with the function $f \to \mu(|f|)$, if follows that

$$\mu(|f|) = \mu^*(|f|) \quad (f \in \mathscr{L}^1). \tag{4.6.5}$$

In particular, $\mu(f) = \mu^*(f)$ for any positive integrable function.

The inequality (4.6.4) shows also that $|f|$ is integrable whenever f is integrable.

4.6.2 **Lemma.** A function ϕ in Φ is integrable if and only if $\mu^*(\phi) < +\infty$, in which case $\mu^*(\phi) = \mu(\phi)$.†

Proof. The last clause is a special case of what has just been established. On the other hand if ϕ is in Φ and $\mu^*(\phi) < +\infty$, one may choose a sequence of functions x_n in $\mathscr{K}(T)$ satisfying $x_n \leq \phi$ and $\mu^*(\phi) - \mu(x_n) \leq 1/n$. Since x_n and $\phi - x_n$ are in Φ, Corollary 4.5.2 shows that $\mu^*(\phi - x_n) = N_1(\phi - x_n) \leq 1/n$. This shows that ϕ is integrable. The converse is true by definition of $\mathscr{F}^1_C$. ∎

We show next that the real functions in $\mathscr{L}^1$ form a lattice.

4.6.3 **Proposition.** If f_1 and f_2 are real integrable functions, then $f = \text{Sup}\,(f_1, f_2)$ and $g = \text{Inf}\,(f_1, f_2)$ are also integrable.

Proof. Since $g = -\text{Sup}\,(-f_1, -f_2)$, it is enough to prove that f is integrable. Since $|h|$ is integrable whenever h has this property, we can choose an integrable function p (for example, $-|f_1| - |f_2|$) such that $f_i \geq p$ for $i = 1, 2$. Then $f - p = \text{Sup}\,(f_1 - p, f_2 - p)$. By virtue of this, we may assume that f_1 and f_2 are positive. But then, given $\varepsilon > 0$, we can choose functions ϕ_i in Φ such that $\phi_i \geq f_i$ and

† This statement requires a minor qualification; see Subsection 4.6.11.

$\mu^*(\phi_i) \leq \mu^*(f_i) + \frac{1}{2}\varepsilon$. The preceding lemma shows that ϕ_i is integrable; hence we shall have $N_1(\phi_i - f_i) \leq \frac{1}{2}\varepsilon$. By Lemma 4.6.2 again, $\phi = \text{Sup}\,(\phi_1, \phi_2)$ is integrable; it clearly majorizes f; and the inequality

$$\phi - f \leq (\phi_1 - f_1) + (\phi_2 - f_2)$$

leads to $N_1(\phi - f) \leq \frac{1}{2}\varepsilon + \frac{1}{2}\varepsilon = \varepsilon$. Thus f is approximable, arbitrarily closely in the sense of N_1, by integrable functions, hence is itself integrable. ▌

4.6.4 Corollary. A real function f is integrable if and only if both $f^+ = \text{Sup}\,(f, 0)$ and $f^- = \text{Sup}\,(-f, 0)$ are integrable, in which case $\mu(f) = \mu(f^+) - \mu(f^-)$.

We shall in the sequel need a similar result not implied by Proposition 4.6.3.

4.6.5 Lemma. If f is a positive integrable function and c a number ≥ 0, then $f' = \text{Inf}\,(f, c)$ is integrable.

Proof. Given $\varepsilon > 0$, choose ϕ in Φ majorizing f and such that $\mu^*(\phi) \leq \mu^*(f) + \varepsilon$. Since f is integrable, Lemma 4.6.2 shows that this may be written $N_1(\phi - f) \leq \varepsilon$. Then $\phi' = \text{Inf}\,(\phi, c)$ belongs to Φ and is integrable. Further, $\phi' \geq f'$ and

$$\phi' - f' = \text{Inf}\,(\phi, c) - \text{Inf}\,(f, c).$$

If $f \geq c$, this last expression is $c - c = 0 \leq \phi - f$; if $f < c$, it is $\text{Inf}\,(\phi, c) - f \leq \phi - f$. So $\phi' - f' \leq \phi - f$ and therefore $N_1(\phi' - f') \leq \varepsilon$. Since ϕ' is integrable and ε is arbitrarily small, we conclude that f' is integrable. ▌

This brings us to one of the central convergence theorems for integrable functions.

4.6.6 Theorem. If the real integrable functions f_n $(n = 1, 2, \cdots)$ are monotone increasing, then $f = \lim f_n$ is integrable if and only if $\lim \mu(f_n) < +\infty$, in which case $\mu(f) = \lim \mu(f_n)$.†

Proof. By considering the functions $f_n - f_1$ in place of f_n, one may reduce oneself to the case in which $f_n \geq 0$ for all n. Then $\mu^*(f_n) = \mu(f_n)$ for all n. Proposition 4.5.3 gives $\mu^*(f) = \lim \mu(f_n)$. It suffices to show that f is integrable, given that $\mu^*(f) < +\infty$. Granted this and given any $\varepsilon > 0$, we can choose ϕ in Φ majorizing f such that $\mu^*(\phi) \leq \mu^*(f) + \varepsilon$. By Lemma 4.6.2, ϕ is integrable; so $\phi - f_n$ is integrable and positive. Hence

$$\mu^*(\phi - f_n) = \mu(\phi - f_n) = \mu(\phi) - \mu(f_n) \leq \varepsilon + o(1) \text{ as } n \to \infty.$$

Since $f - f_n \geq 0$ we have

$$N_1(f - f_n) = \mu^*(f - f_n) \leq \mu^*(\phi - f_n) \leq \varepsilon + o(1).$$

Thus $N_1(f - f_n) \leq 2\varepsilon$ if n is large enough, showing that f is integrable. ▌

The next theorem, again a major one, shows in effect that $\mathscr{L}^1$ is obtained from $\mathscr{K}_C(T)$ by completing the latter for the structure defined by N_1.

---

† See Subsection 4.6.11 for a minor qualification needed in this statement.

4.6.7 Theorem. If a sequence (f_n) of functions in C^T is N_1-Cauchy, a subsequence (f_{n_k}) may be extracted which converges almost everywhere to a function f for which $N_1(f - f_n) \to 0$ as $n \to \infty$. In particular, $\mathscr{L}^1$ is complete.

Proof. We may choose natural numbers $n_1 < n_2 < \cdots$ so that, putting $g_k = f_{n_k}$, one has $N_1(g_{k+1} - g_k) \leq 4^{-k}$. Let A_k be the set defined by the inequality $|g_{k+1} - g_k| \geq 2^{-k}$. Then it follows that $N_1(g_{k+1} - g_k) \geq 2^{-k}\, \mu^*(A_k)$ and therefore $\mu^*(A_k) \leq 2^{-k}$. If $B_k = \bigcup \{A_i : i \geq k\}$, then $\mu^*(B_k) \leq 2^{-k+1}$ by (4.6.1), and so the intersection N of the B_k is negligible. If t belongs to $T \backslash N$ we shall have $|g_{k+1}(t) - g_k(t)| < 2^{-k}$ for all k exceeding some $k_0 = k_0(t)$. It follows that the numerical sequel $(g_k(t))$ is Cauchy for almost all t. Define $f(t)$ to be the limit of $(g_k(t))$ for such t, and to be 0 elsewhere. By Fatou's lemma (Proposition 4.5.4):

$$N_1(f - g_k) = \mu^*(|f - g_k|) = \mu^*(\lim_{i\to\infty} |g_i - g_k|)$$
$$\leq \liminf_{i\to\infty} \mu^*(|g_i - g_k|) = \liminf_{i\to\infty} N_1(g_i - g_k),$$

which tends to 0 as $k \to \infty$ by virtue of the Cauchy character of the sequence (g_k). The proof is complete. ▌

It is worth noting two corollaries.

4.6.8 Corollary. If f, f', and $f_n (n = 1, 2, \cdots)$ belong to C^T, and if $f_n \to f$ a.e. and $N_1(f' - f_n) \to 0$ as $n \to \infty$, then $f = f'$ a.e.

Proof. The proof is immediate since, by the theorem, a subsequence of (f_n) exists that converges almost everywhere to f'. ▌

4.6.9 Corollary. In order that a function f in C^T be integrable it is

(a) **necessary** that there exist an N_1-Cauchy sequence (x_n) extracted from $\mathscr{K}_C(T)$ such that $\lim x_n = f$ a.e.

(b) **sufficient** that there exist an N_1-Cauchy sequence (f_n) extracted from $\mathscr{L}^1$ such that $\lim f_n = f$ a.e.

Proof. (a) follows from the definition of $\mathscr{L}^1$ as the closure in $\mathscr{F}^1_C$ of $\mathscr{K}_C(T)$, together with Theorem 4.6.7. On the other hand, if (b) holds, the same theorem ensures that a function f' in C^T and a subsequence (f_{n_k}) exist such that $f_{n_k} \to f'$ a.e. and $N_1(f' - f_{n_k}) \to 0$ as $k \to \infty$. The second relation here implies that f' is integrable, and the first implies that $f' = f$ a.e., so that f itself is integrable. ▌

We end this section with a provisional result, easily deducible from Corollary 4.6.9 and later to be entirely superseded (Corollary 4.11.8 to follow).

4.6.10 Lemma. If f and g are integrable and bounded, then $h = fg$ is integrable.

Proof. By (a) of Corollary 4.6.9, we may write $f = \lim x_n$ a.e., (x_n) being an N_1-Cauchy sequence extracted from $\mathscr{K}_C(T)$. Since f is bounded, it is evident that we may assume that for some number M one has $|x_n| \leq M$. We choose also a similar sequence (y_n) approximating g. Then it is clear that the $z_n = x_n y_n$ form an N_1-Cauchy sequence in $\mathscr{K}_C(T)$ converging almost everywhere to fg. It remains but to appeal to (b) of Corollary 4.6.9. ▌

4.6.11 Functions Defined Almost Everywhere. Thus far we have reserved the name "integrable functions" for those functions that are defined everywhere, are everywhere finite valued, and that satisfy the criterion laid down at the beginning of this section. As we have observed in connection with Lemma 4.6.2 and Theorem 4.6.6, this attitude results in annoying restrictions because if one starts with functions satisfying these conditions, one is led naturally to functions that may well take infinite values, and even so may reasonably be described as being integrable.

It is therefore convenient at times to apply the term "integrable" to a function f under the following conditions: f is defined almost everywhere, possibly infinite at some points, and there exists a function g, defined and finite everywhere, and integrable in the preceding sense, such that $f = g$ a.e. If one such function g exists, there are evidently many satisfying the same conditions. However their integrals have a common value, which we take to be that of $\mu(f)$.

The convenience obtained in this way has to be paid for. The functions that are integrable in this wider sense do *not* form a vector space, nor even a group. In practice, this is not a serious impediment since one can replace such a function by one that is everywhere defined and finite without disturbing any relations involving integrals. The formal escape from the difficulty lies in using the quotient space L^1 introduced in Subsection 4.11.10. The space obtained in this way, starting from functions that are integrable in the wider sense, is isomorphic in every conceivable sense with that obtained by starting from the functions that are integrable in the strict sense.

4.7 Integrable Sets, Measurable Sets, Locally Negligible Sets

4.7.1 Integrable Sets. A set $A \subset T$ is said to be *integrable* (for, or relative to, the positive Radon measure μ) if and only if the function χ_A is integrable (for μ). From the preceding results about integrable functions one obtains at once the following basic properties.

► (1) If A_n $(n = 1, 2, \cdots)$ are integrable and $A_n \uparrow A$, then A is integrable if and only if $\lim \mu^*(A_n) < +\infty$, in which case $\mu^*(A) = \lim \mu^*(A_n)$. Cf. (4.6.2).

► (2) If A_n $(n = 1, 2, \cdots)$ are integrable and $A_n \downarrow A$, then A is integrable and $\mu^*(A) = \lim \mu^*(A_n)$.

► (3) If A and B are integrable, so are $A \cap B$, $A \cup B$, and $A \backslash B$.

► (4) Any relatively compact open set, and any compact set, is integrable.

Of these, (1) and (2) follow from the principle of monotone convergence. For (3) we observe that $\chi_{A\cap B} = \chi_A \cdot \chi_B$ and $\chi_{A\backslash B} = \chi_A - \chi_A \cdot \chi_B$, while $A \cup B = (A\backslash B) \cup (B\backslash A) \cup (A \cap B) = C \cup C' \cup C''$, the sets C, C', and C'' being disjoint and integrable, so that $\chi_{A\cup B} = \chi_C + \chi_{C'} + \chi_{C''}$ is the sum of three integrable functions and is therefore itself integrable. As for (4), the characteristic function of any relatively compact open set is in Φ and has a finite upper integral, hence is integrable. If K is compact, it has a relatively compact open neighbourhood U; then $U \backslash K$ is relatively compact and open, hence integrable; and $K = U \backslash (U \backslash K)$ is therefore integrable by the last part of (3). ∎

Unless T is itself integrable (that is, μ is "bounded" or "finite"), the complement of an integrable set is never integrable. Accordingly we introduce a second and weaker concept that is purely local in character and that is preserved under complementation.

4.7.2 **Measurable Sets.** A set $A \subset T$ is said to be *measurable* (for μ) if and only if $A \cap K$ is integrable (for μ) for each compact set $K \subset T$. One might therefore describe the measurable sets as those that are "locally integrable." We proceed to list some properties of measurable sets.

▶ (5) A is integrable if and only if A is measurable and $\mu^*(A) < +\infty$.

If A is integrable, so is $A \cap K$ by (3) and (4); and certainly $\mu^*(A) < +\infty$. Conversely suppose A is measurable and $\mu^*(A) < +\infty$. Then there exists a negligible set N and an increasing sequence of compact sets $K_n (n = 1, 2, \cdots)$ such that $A \subset (\bigcup K_n) \cup N$ and therefore $A = \bigcup (A \cap K_n) \cup (A \cap N)$. Now $A \cap N$ is negligible and hence integrable. Each $A \cap K_n$ is integrable, they increase, and $\mu^*(A \cap K_n) \leq \mu^*(A) < +\infty$. So $\bigcup (A \cap K_n) = \lim (A \cap K_n)$ is integrable by (1). Therefore, by (3) again, A is integrable.

▶ (6) *If* A and A_n $(n = 1, 2, \cdots)$ are measurable, so too are $T \backslash A$, $\bigcup A_n$, $\bigcap A_n$, lim sup A_n, and lim inf A_n.

For $\chi_{(T\backslash A) \cap K} = \chi_K - \chi_K \chi_A$ is the difference of two integrable functions by hypothesis and (4), so that $(T \backslash A) \cap K$ is integrable for every compact K. From (3) and (4) it follows that any finite union of measurable sets is again measurable; and from (1) it follows that the limit of any increasing sequence of measurable sets is measurable. Combining these two assertions, we see that the union of any sequence of measurable sets is measurable. The measurability of $\bigcap A_n$ follows similarly on using (3) and (2). Finally

$$\limsup A_n = \bigcap_m \bigcup_{n \geq m} A \qquad \liminf A_n = \bigcup_m \bigcap_{n \geq m} A_n$$

are both measurable by what we have just established concerning unions and intersections.

▶ (7) Any open set, and any closed set, is measurable.

This is apparent from (4): if F is closed, $F \cap K$ is compact and therefore integrable (K being any compact set in T); if U is open, $U = T \backslash F$, where $F = T \backslash U$ is closed.

▶ (8) If A_n $(n = 1, 2, \cdots)$ are disjoint and measurable, and if $A = \bigcup A_n$, then

$$\mu^*(A) = \sum_n \mu^*(A_n).$$

For we know already that $\mu^*(A) \leq \sum_n \mu^*(A_n)$ whether or not the A_n are disjoint and measurable. So we have to show that $\mu^*(A) \geq \sum_n \mu^*(A_n)$ under the stated hypotheses. This is trivial if $\mu^*(A) = +\infty$. Otherwise, (5) and (6) show that A and each A_n are integrable. So $\chi_A = \sum_n \chi_{A_n}$ and the assertion follows on integration.

4.7.3 Locally Negligible Sets. A set $L \subset T$ is said to be locally negligible (for μ) if $L \cap K$ is negligible (for μ) for each compact set $K \subset T$. L is then necessarily measurable. Any negligible set is locally negligible, but the converse is generally false. This difference is the root of many minor complications arising when T is not "σ-finite" in the sense to be described.

A subset of T is said to be μ-σ-finite if it is contained in the union of countably many integrable open sets—or, equivalently, in the union of a negligible set and of countably many compact sets. It is easy to see that any μ-σ-finite set that is locally negligible is also negligible. Hence if T is μ-σ-finite (a fortiori if T is σ-compact), then "negligible" and "locally negligible" (for μ) are equivalent concepts. Otherwise there will in general exist locally negligible sets L for which $\mu^*(L) > 0$; for such a set one must have $\mu^*(L) = +\infty$ (for if $\mu^*(L) < +\infty$, L would be μ-σ-finite).

By analogy with the terminology introduced in Section 4.6, we shall say that a property of points of T holds locally almost everywhere for μ (or μ-locally almost everywhere) if the set of points of T at which the said property does not hold is locally negligible for μ. In formulae, the phrase "locally almost everywhere" will be abbreviated to "l.a.e."

Clearly, if the sets L_n $(n = 1, 2, \cdots)$ are each locally negligible, the same is true of any subset of $\bigcup L_n$.

4.7.4 Essential Outer Measure. The concept of locally negligible set suggests the introduction of the set function

$$\bar{\mu}^*(A) = \text{Sup}\,\{\mu^*(A \cap K) : K \text{ compact}, K \subset T\}.$$

In the notation of Section 4.14, this is just the essential upper integral $\bar{\mu}^*(\chi_A)$; we therefore call $\bar{\mu}^*(A)$ the *essential outer* (μ-) *measure* of A. L is locally negligible (for μ) if and only if $\bar{\mu}^*(L) = 0$. It is clear that $0 \leq \bar{\mu}^*(A) \leq \mu^*(A) \leq +\infty$ for any set $A \subset T$, and that

$$\bar{\mu}^*\left(\bigcup_n A_n\right) \leq \sum_n \bar{\mu}^*(A_n) \tag{4.7.1}$$

for any sequence (A_n) of subsets of T. As will follow from the results established in Section 4.14, $\bar{\mu}^*(A) = \mu^*(A)$ whenever A is open. From the same source it appears that A is measurable and $\bar{\mu}^*(A) < +\infty$ if and only if there exists an integrable set $B \subset A$ such that $A \backslash B$ is locally negligible, in which case $\bar{\mu}^*(A) = \mu^*(B)$. From this we deduce that

$$\bar{\mu}^*\left(\bigcup_n A_n\right) = \sum_n \bar{\mu}^*(A_n) \tag{4.7.2}$$

whenever the sets A_n $(n = 1, 2, \cdots)$ are disjoint and measurable.

4.7.5 Numerical Measure of Measurable Sets. As we have seen, both μ^* and $\bar{\mu}^*$ are countably additive set functions on the σ-field of μ-measurable sets; in general, these are distinct set functions. When restricted to measurable sets, we denote these set functions by μ and $\bar{\mu}$, respectively. The sets A that are measurable and such that $\mu(A) < +\infty$ (resp. $\bar{\mu}(A) < +\infty$) are the integrable (resp. essentially integrable) sets.

▶ (9) We observe that for any set A one has

$$\mu^*(A) = \text{Inf}\,\{\mu(U) : U \text{ open}, U \supset A\}. \tag{4.7.3}$$

Since each U is measurable and $\mu(U) = \mu^*(U)$, while μ^* is an increasing set function, it is clear that $\mu^*(A) \leq \text{Inf}\,\mu(U)$. In proving the opposing inequality, we may evidently assume that $\mu^*(A) < +\infty$. Then, given $\varepsilon > 0$, there exists a positive lower semicontinuous function $\phi \geq \chi_A$ such that $\mu^*(\phi) \leq \mu^*(A) + \varepsilon$. The set U of points t for which $\phi(t) > 1 - \varepsilon$ is then open, $U \supset A$, and $\chi_U \leq (1 - \varepsilon)^{-1}\phi$; so

$$\mu(U) = \mu^*(\chi_U) \leq (1 - \varepsilon)^{-1}\,\mu^*(\phi) \leq (1 - \varepsilon)^{-1}\,\{\mu^*(A) + \varepsilon\}.$$

Since $\varepsilon > 0$ is arbitrarily small, the proof is complete.

4.7.6 Interior or Inner Measure. By analogy with (4.7.3) we introduce

$$\mu_*(A) = \text{Sup}\,\{\mu(K) : K \text{ compact}, K \subset A\} \tag{4.7.4}$$

and call $\mu_*(A)$ the *interior* or *inner* μ-measure of A. Plainly, μ_* is a positive increasing set function and

$$0 \leq \mu_*(A) \leq \bar{\mu}^*(A) \leq \mu^*(A) \leq +\infty. \tag{4.7.5}$$

From the countable additivity of μ on measurable sets, it is easily deduced that

▶ (10) If the sets A_n $(n = 1, 2, \cdots)$ are disjoint, then

$$\mu_*(\bigcup A_n) \geq \sum_n \mu_*(A_n).$$

▶ (11) If A is measurable, then $\mu_*(A) = \bar{\mu}(A)$.

To prove this it suffices, in view of (4.7.5), to show that $\mu_*(A) \geq \bar{\mu}^*(A)$ when A is measurable. Supposing this established when A is in addition relatively compact, we shall have $\bar{\mu}(A \cap K) \leq \mu_*(A \cap K) \leq \mu_*(A)$ for each compact set K. Since $A \cap K$ is integrable, it follows that $\mu^*(A \cap K) \leq \mu_*(A)$. Taking the supremum with respect to K, this gives $\bar{\mu}^*(A) \leq \mu_*(A)$, as required. It thus suffices to deal with the case in which A is relatively compact and measurable. We take any relatively compact open set $U \supset \bar{A}$. Then $U \backslash A$ is integrable and so, by (9), there exists an open set $V \supset U \backslash A$ satisfying $\mu(V) \leq \mu(U \backslash A) + \varepsilon$. We may clearly suppose that $V \subset U$. Consider $K = U \backslash V$. Then it is evident that $K \subset A$. Therefore any adherence point of K belongs to $\bar{A} \subset U$ and to the closed complement of V, so that K is compact. Now $V = (U \backslash A) \cup (A \cap V)$, the two sets on the right being disjoint. Hence

$$\mu(V) = \mu(U \backslash A) + \mu(A \cap V)$$

and so $\mu(A \cap V) \leq \varepsilon$. But $A \backslash K = A \cap V$. Thus $K \subset A$, K is compact, and

$$\mu(K) = \mu(A) - \mu(A \backslash K) \geq \mu(A) - \varepsilon = \bar{\mu}^*(A) - \varepsilon.$$

Letting $\varepsilon \to 0$, the result follows. ▌

▶ (12) A is integrable if and only if $\mu_*(A) = \mu^*(A) < +\infty$, in which case the common value is equal to $\mu(A) = \bar{\mu}(A)$.

For, if A is integrable, $\mu_*(A) = \bar{\mu}(A)$ by (11), which is also equal to $\bar{\mu}^*(A) = \mu^*(A)$ because A is measurable and μ-σ-finite. Conversely, suppose $\mu_*(A) = \mu^*(A) < +\infty$. Then there exists a set $X \subset A$, a countable union of compact sets, and a set $Y \supset A$, a countable intersection of open sets, such that

$$\mu^*(X) = \mu(X) = \mu_*(A) = \mu^*(A) = \mu(Y) = \mu^*(Y) < +\infty.$$

It follows that $\mu^*(Y \backslash X) = 0$; that is, $Y \backslash X$ is negligible. So $A = X \cup (A \backslash X)$, where $A \backslash X \subset Y \backslash X$ is negligible. Since X is measurable, so too is A. Since $\mu^*(A) < +\infty$, integrability of A follows from (5). ▌

We end this section with two examples that illustrate the preceding results in terms of Lebesgue-Stieltjes measures on a real interval.

4.7.7 **Example.** We take up the notations and results of Subsections 4.4.4 and 4.5.6, supposing that T is a real interval with extremities $a = \operatorname{Inf} T$ and $b = \operatorname{Sup} T$, $a < b$, and f an LBV function on T. It is enough to consider the case in which f is increasing, so that the associated Lebesgue-Stieltjes measure μ_f is positive. In Subsection 4.5.6 we determined $\mu_f(I)$ for each relatively open interval in T. From this and countable additivity of μ_f, we find easily $\mu_f(I)$ for *any* interval $I \subset T$. The reader will notice first that T is σ-compact, so that μ_f-negligible and μ_f-locally negligible are equivalent qualifications, and there is no distinction between μ_f^* and $\bar{\mu}_f^*$.

To begin with, if $a < t < b$, $\{t\}$ is the limit of a decreasing sequence of relatively compact open intervals whose closures lie in T. So the results of Subsection 4.5.6 combine with (2) to show that

$$\mu_f(\{t\}) = f(t+0) - f(t-0) \qquad \text{if } a < t < b. \tag{4.7.6}$$

Similarly,

$$\mu_f(\{a\}) = f(a+0) - f(a) \qquad \text{if } a \in T, \tag{4.7.6$'$}$$

$$\mu_f(\{b\}) = f(b) - f(b-0) \qquad \text{if } b \in T. \tag{4.7.6$''$}$$

Next we consider any compact interval $[\alpha, \beta]$ contained in T. If $a < \alpha \leq \beta < b$, we represent $[\alpha, \beta]$ as the intersection of a decreasing sequence of relatively compact open intervals (α', β') with $a < \alpha' < \beta' < b$. By the results of Subsection 4.5.6 and (2), it results that

$$\mu_f([\alpha, \beta]) = f(\beta+0) - f(\alpha-0) \qquad \text{if } a < \alpha \leq \beta < b. \tag{4.7.7}$$

If $a \in T$ and $a < \beta < b$, we write $[a, \beta] = [a, \beta') - (\beta, \beta')$ with $\beta < \beta' < b$ and so deduce via additivity that

$$\mu_f([a, \beta]) = f(\beta+0) - f(a) \qquad \text{if } a \in T, \quad a < \beta < b. \tag{4.7.7$'$}$$

Similarly,

$$\mu_f([\alpha, b]) = f(b) - f(\alpha+0) \qquad \text{if } b \in T, \quad a < \alpha < b, \tag{4.7.7$''$}$$

$$\mu_f([a, b]) = f(b) - f(a) \qquad \text{if } T = [a, b]. \tag{4.7.7$'''$}$$

These results bring to light a fact that needs to be remembered. If $T \subset T^*$ and f^* is given on T^*, and if we denote by f the restriction $f^* \mid T$, then μ_f is *not* necessarily the restriction of μ_{f^*} to subsets of T. For example, if $T = [a, b]$ while T^* is open, then

$$\mu_f([a, b]) = f(b) - f(a) = f^*(b) - f^*(a)$$

while

$$\mu_{f^*}([a, b]) = f^*(b+0) - f^*(a-0),$$

and agreement obtains only if f^* is right continuous at b and left continuous at a.

4.7.8 **Example.** We are now in a position to show that, given any real interval T and any Radon measure μ on T, then $\mu = \mu_f$ for some LBV function f on T. In doing this we may assume that μ is positive, and show that f may be chosen to be increasing on T.

Suppose first that $a = \operatorname{Inf} T \in T$; let $b = \operatorname{Sup} T$. Define

$$f(t) = \begin{cases} \mu([a, t)) & \text{if } t \in T, t < b \\ \mu([a, b]) & \text{if } t = b \in T. \end{cases}$$

Then f is increasing on T, finite valued there, and left continuous for $a < t < b$. Now if $t \in T$ and $t < b$, we have from the results of Subsection 4.7.7:

$$\mu_f([a, t)) = \begin{cases} f(t-0) - f(a) & \text{if } t > a \\ 0 & \text{if } t = a \end{cases}$$

and this is equal to $\mu([a, t))$ since f is left continuous. If $t = b \in T$, then

$$\mu_f([a, b]) = f(b) - f(a) = f(b) = \mu([a, b]).$$

Since $\{b\} = [a, b] \setminus [a, b)$, we see that $\mu_f(\{b\}) = \mu(\{b\})$ if $b \in T$. It is easily seen now that $\mu_f(x) = \mu(x)$ for any x in $\mathscr{K}(T)$. To do this we partition T into subintervals $[t, t')$ (and perhaps $\{b\}$ if $b \in T$) the maximum length of which tends to 0; x is then approximated uniformly by finite linear combinations of characteristic functions of such intervals. The μ- and μ_f-integrals of such functions are equal since we know that $\mu_f([t, t')) = \mu([t, t'))$ (and $\mu_f(\{b\}) = \mu(\{b\})$ if $b \in T$); equality of $\mu(x)$ and $\mu_f(x)$ then follows in the limit.

Suppose finally that a does not belong to T. We take a sequence $a_n \downarrow a$, $a_n \in T$. We define f on T in the following way: any t in T satisfies $t \geq a_n$ for some (actually many) n, and for all such n the numbers

$$\mu([a_n, t)) - \mu([a_n, a_1))$$

are independent of n (by additivity of μ); this common value is denoted by $f(t)$ when $t < b$; if b belongs to T, we put $f(b) = \mu([a_1, b])$. Then for $a < \alpha < \beta < b$, we have

$$\mu_f([\alpha, \beta)) = f(\beta - 0) - f(\alpha - 0) = \mu([a_n, \beta)) - \mu([a_n, \alpha))$$

if n is large enough, and this is just $\mu([\alpha, \beta))$. The same equality holds with $\beta = b$ by countable additivity. If b belongs to T, we have

$$\begin{aligned}\mu_f(\{b\}) = f(b) - f(b - 0) &= \mu([a_1, b]) - \mu([a_1, b)) \\ &= \mu(\{b\}).\end{aligned}$$

The same argument as before now shows that $\mu(x) = \mu_f(x)$ for each x in $\mathscr{K}(T)$; that is, $\mu = \mu_f$.

4.8 Measurable Functions

Having introduced the concept of a measurable set and established the properties given in Section 4.7 of such sets, the notion of measurability for functions may be defined and studied in a manner that is readily accessible in any one of several works on real analysis or integration theory (for example: McShane [1], Ch. III; L. M. Graves [1], Ch. X, §4; J. C. Burkhill [1], §2.12; C. Goffman [1], Ch. 15; I. P. Natanson [1], Ch. IV; S. Saks [1], §§7, 8; P. R. Halmos [1], Ch. IV). It suffices for us to recall the definitions and principal properties, leaving the proofs as easy exercises for the reader.

Beginning with real-valued functions f, we say that such a function is measurable (that is, μ-measurable) if for every real number α, the set of points t of T for which $f(t) > \alpha$ is measurable. (One might equally well use the sets where $f \geq \alpha$, or $f < \alpha$, or $f \leq \alpha$ in this definition.) A similar definition applies to functions defined on a given measurable set A. The sum and product of two measurable functions are measurable; if f is measurable and nonvanishing on a measurable set A, then $1/f$ is measurable on A. The upper and lower envelopes, and the upper and lower limits, of any countable sequence of measurable functions are measurable. Any semicontinuous function is measurable. Any positive measurable function is the limit of an increasing sequence of measurable functions, each a finite linear combination of characteristic functions of measurable sets.

As for complex-valued functions, we say that such a function is measurable if and only if its real and imaginary parts are measurable.

In either case, if two functions are equal locally almost everywhere (l.a.e.), then they are together measurable or not.

So much for the basic definition and properties of measurable functions. We pass now to showing how measurability fits into the integration theory, providing an essential component of the criterion for integrability.

4.8.1 **Theorem.** A function f is integrable if and only if (a) $N_1(f) \equiv \mu^*(|f|) < +\infty$ and (b) f is measurable.

Proof. NECESSITY. The necessity of (a) is part of the very definition of integrability (Section 4.6). As for (b), *necessity* is shown by Corollary 4.6.9(a), together with the fact that continuous functions are measurable.

SUFFICIENCY. Suppose f satisfies (a) and (b); we have to show that f is integrable. In doing this there is no loss of generality in assuming that f is real and positive. The functions $f_n = \text{Inf}(f, n)$ are then measurable, $N_1(f_n) \leq N_1(f) < +\infty$, and they increase to f. So (Theorem 4.6.6) it suffices to show that each f_n is integrable. In other words we may assume that f is bounded. In this case, we may indeed assume that $0 \leq f < 1$. Since furthermore (a) implies that f vanishes almost everywhere outside a σ-finite set, we may (on appeal to Theorem 4.6.6 once more) reduce ourselves to the case in which f vanishes outside an integrable set A. This being so, we take numbers $c_0(=0) < c_1 < \cdots < c_n = 1$ and let A_k be the set of points of A defined by the inequalities $c_{k-1} \leq f < c_k$. Each A_k is measurable and contained in A, hence is integrable. So the function

$$g = \sum_{k=1}^{n} c_{k-1}\chi_{A_k}$$

is integrable. Evidently, $g \leq f \leq g + \varepsilon\chi_A$, where ε is the maximum of the numbers $(c_k - c_{k-1})$. Since ε may be made as small as we wish, integrability of f follows from the relation $N_1(f - g) \leq N_1(\varepsilon\chi_A) = \varepsilon \cdot \mu(A)$. ∎

Remark. The above construction suggests how we should approximate the integral $\mu(f)$ whenever f is integrable, vanishes outside an integrable set A, and is such that one can partition A by measurable subsets A_k on each of which the oscillation of f is very small: if one chooses t_k in A_k, the sums $\sum_k f(t_k)\mu(A_k)$ approximate $\mu(f)$. This remark in turn suggests the notation $\int f\, d\mu$ and $\int_A f\, d\mu$. In future we shall sometimes use $\int f\, d\mu$ and $\int^* f\, d\mu$ in place of $\mu(f)$ and $\mu^*(f)$, respectively.

We close this section with Lebesgue's famous convergence theorem.

4.8.2 **Theorem** (Lebesgue). Let us suppose the real-valued functions f_n $(n = 1, 2, \cdots)$ to be measurable, g to be integrable, and the inequality $|f_n| \leq g$ to hold almost everywhere for each n. Then the functions $\liminf f_n$ and $\limsup f_n$ are integrable and

$$\int (\liminf f_n)\, d\mu \leq \liminf \int f_n\, d\mu,$$
$$\int (\limsup f_n)\, d\mu \geq \limsup \int f_n\, d\mu.$$

If therefore $\lim f_n$ exists almost everywhere, then

$$\int (\lim f_n)\, d\mu = \lim \int f_n\, d\mu.$$

Proof. The second half follows from the first by change of sign. So we consider $f = \liminf f_n$. f is measurable and $|f| \leq g$ a.e. Hence (Theorem 4.8.1), f is integrable. If $F_n = \operatorname{Inf}\{f_m : m \geq n\}$, then F_n is measurable and $|F_n| \leq g$ a.e. By the same token, F_n is integrable. Also $F_n \uparrow f$. The functions $F_n + g$ are thus positive, integrable, and increase to the limit $f + g$. Hence (Proposition 4.5.3), $\mu(F_n + g) \uparrow \mu(f + g)$ and so, since μ is additive and $\mu(g)$ finite, $\mu(F_n) \uparrow \mu(f)$. Finally, since $F_n \leq f_m$ $(m \geq n)$, we have $\mu(F_n) \leq \mu(f_m)$ $(m \geq n)$ and therefore $\lim \mu(F_n) \leq \liminf \mu(f_n)$. ∎

We turn next to some properties of measurable functions in general, beginning with an important convergence theorem.

4.8.3 **Theorem** (Egoroff). Let A be an integrable set and let f_n $(n = 1, 2, \cdots)$ and f be functions defined and measurable on A such that $f_n \to f$ a.e. on A. Then convergence takes place "almost uniformly on A"; that is, given $\varepsilon > 0$ there exists a measurable set $B \subset A$ (which may be assumed to be compact) such that $\mu(A \backslash B) \leq \varepsilon$ and $f_n \to f$ uniformly on B.

Proof. We may assume that $f_n \to f$ everywhere on A. For any two natural numbers n and k we define $Q_{n,k}$ to be the set of points t of A for which

$$\operatorname*{Sup}_{m \geq n} |f_m(t) - f(t)| > \frac{1}{k}.$$

The sets $Q_{n,k}$ are measurable, $Q_{n+1,k} \subset Q_{n,k}$, and $\bigcap_n Q_{n,k} = \emptyset$. It follows that $\mu(Q_{n,k}) \to 0$ as $n \to \infty$. One can therefore choose n_k so that, if $P_k = Q_{n_k,k}$, then $\mu(P_k) \leq \varepsilon/2^k$. If t belongs to $A \backslash P_k$, we have

$$|f_m(t) - f(t)| \leq \frac{1}{k}$$

for all $m \geq n_k$. It suffices to take $B = A \backslash \bigcup_k P_k$. (If it is desired that B be compact, one has simply to proceed one stage further and select a compact subset K of B such that $\mu(B \backslash K)$ is arbitrarily small.) ∎

We proceed to show next how measurable functions may be approximated by simple functions.

4.8.4 **Theorem.** Let A be an integrable set, f a function defined and measurable on A. Then f is the limit, almost uniformly on A (and hence pointwise almost everywhere on A), of a sequence (f_n), each f_n being a finite linear combination of characteristic functions of integrable subsets of A.

Proof. Let (α_n) be a dense sequence of scalars (real or complex). Let $A_{n,p}$ be the set of points t of A for which $|f(t) - \alpha_n| \leq 1/p$; these sets are integrable. We define $B_{1,p} = A_{1,p}$ and

$$B_{n+1,p} = A_{n+1,p} \setminus \bigcup_{1 \leq k \leq n} A_{k,p};$$

these sets are integrable and form a partition of A when p is fixed and n varies. For each p we may choose n_p so that $A \setminus \bigcup_{n \leq n_p} B_{n,p}$ has measure at most 2^{-p}. We define f_p to be equal to α_n on $B_{n,p}$ $(1 \leq n \leq n_p)$ and to be 0 elsewhere. Then f_p is a function of the type specified in the theorem, and we have $|f - f_p| \leq 1/p$ at all points of A save perhaps those of a set N_p satisfying $\mu(N_p) \leq 2^{-p}$. The set $N_q^* = \bigcup_{p \geq q} N_p$ satisfies $\mu(N_q^*) \leq 2^{1-q}$ and we have

$$|f_p(t) - f(t)| \leq \frac{1}{p} \qquad (t \in A \setminus N_q^*, \qquad p \geq q).$$

Thus $f_p \to f$ uniformly on $A \setminus N_q^*$, hence $f_p \to f$ almost uniformly on A. ∎

4.8.5 **Corollary** (Lusin). Let A be an integrable set, f a function defined and measurable on A. Given $\varepsilon > 0$, there exists a compact set $K \subset A$ such that $\mu(A \setminus K) \leq \varepsilon$ and $f \mid K$ is continuous.

Proof. By virtue of Theorem 4.8.4, it suffices to prove the result when f is a finite linear combination of characteristic functions of integrable subsets A_i of A $(1 \leq i \leq n)$. We may suppose that the A_i are disjoint and form a partition of A. However, we can choose compact sets $K_i \subset A_i$ such that their union K satisfies $\mu(A \setminus K) \leq \varepsilon$. Obviously, $f \mid K$ is continuous. ∎

4.8.6 **Remarks.** It is evident that conversely any function f satisfying Lusin's condition is measurable. Bourbaki ([6], Ch. IV) chooses Lusin's condition as the definition of measurability when considering functions whose values lie in a topological space T^*. It is only when T^* satisfies suitable countability conditions that the two concepts of measurability coincide, and there is little doubt that Lusin's criterion is the appropriate one for abstract-valued functions.

Having reached this stage, we might proceed forthwith to the topics that in fact appear in Section 4.11 and thereafter. Instead, we break the development of the general integration theory in order to show that certain sets of measures just suffice to represent the duals of spaces of continuous functions. This topic is dealt with in Section 4.10. As a preliminary to this, we introduce in Section 4.9 the idea of the support or carrier of a measure.

4.9 The Support of a Measure

Let T be a locally compact space and μ a positive Radon measure on T. Referring to Proposition 4.5.1 we see that the union of the set (countable or not) of negligible open sets is the *maximal negligible open set* U; a point of T belongs to U if and only if there is a negligible neighbourhood of that point. The complement F of U is termed the *support* (or *carrier*) of μ; F is necessarily closed in T. A point t of T belongs to F if and only if every neighbourhood of t is non-negligible.

If now μ is a complex Radon measure on T, its support is by definition that of the positive measure

$$|\mu| = \alpha^+ + \alpha^- + \beta^+ + \beta^-,$$

where α and β are the real and imaginary parts of μ and $\alpha = \alpha^+ - \alpha^-$, $\beta = \beta^+ - \beta^-$ are the minimal decompositions of α and β (see Theorem 4.3.2). Evidently, the support of μ is the union of the supports of α^+, α^-, β^+, and β^-. For convenience, we write μ_1, μ_2, μ_3, μ_4 for α^+, α^-, β^+, β^- and refer to $\mu = \mu_1 - \mu_2 + i\mu_3 - i\mu_4$ as the minimal decomposition of μ.

4.9.1 Proposition. Let μ be a complex Radon measure, F its support. An open subset U of T is negligible (that is, is contained in $T \backslash F$) if and only if $\mu(f) = 0$ for every $f \in \mathscr{K}_+(T)$ whose support is contained in U.

Proof. The condition is evidently necessary. For if U is negligible, then $\mu_k^*(U) = 0$ for $k = 1, 2, 3, 4$; and so, since $0 \leq f \leq \|f\|_\infty \cdot \chi_U$ whenever f is in $\mathscr{K}_+(T)$ and has support contained in U, therefore $\mu_k(f) = 0$ ($k = 1, 2, 3, 4$) and so $\mu(f) = 0$.

Conversely, suppose $\mu(f) = 0$ for all f of the stated category. The same is then true when μ is replaced by its real and imaginary parts separately. But then (cf. the proof of Theorem 4.3.2) $\mu_k(f) = 0$ for $k = 1, 2, 3, 4$. Now χ_U is the upper envelope of certain functions f in $\mathscr{K}_+(T)$ with supports in U and therefore (Proposition 4.5.1) $\mu_k^*(U) = 0$ for $k = 1, 2, 3, 4$. Thus U is negligible. ▌

Remark. If F is the support of the positive measure μ, there may well be μ-negligible nonempty sets contained in F. For instance, if μ is the Lebesgue measure on R, its support is R itself (since obviously every nonempty open set has strictly positive measure); yet there exist nonempty negligible sets.

4.9.2 Example. Let T be a real interval, f an LBV function of T, μ_f the associated Lebesgue-Stieltjes measure (see Subsections 4.4.4, 4.7.7, 4.7.8). We determine the support S of μ_f. Let $U = T \backslash S$, which is open relative to T. Suppose first that f is $\uparrow$, so that $\mu_f \geq 0$. If I is any bounded interval contained in U, $\mu_f(I) = 0$. By Subsection 4.7.7, taking I a single point, we see that f must be continuous at all points of U; and then that f must be constant on each connected component of U; that is, f is locally constant on U. Conversely, if f is locally constant on U, then $\mu_f(U) = 0$. Thus the support S of μ_f is the complement relative to T of the largest relatively open set $U \subset T$ on which f is locally constant. The same applies if f is LBV—as is seen by decomposing f into the difference of two $\uparrow$ functions or by appealing to Proposition 4.9.1.

As an example, suppose T is $[a, b]$ and that f is constant on each of the intervals (c_{i-1}, c_i) $(1 \leq i \leq m)$ where $c_0 = a < c_1 < \cdots < c_m = b$. It is then easily shown from first principles (see Natanson [1], p. 231, Th. 3) that

$$RS\text{-}\int_a^b x\,df = [f(a+0) - f(a)]x(a) + \sum_{i=1}^{m} [f(c_i+0) - f(c_i-0)]x(c_i) + [f(b) - f(b-0)]x(b)$$

for x in $\mathscr{K}(T)$. This can also be inferred from the results of Subsection 4.7.7. Thus

$$\mu_f = [f(a+0) - f(a)]\varepsilon_a + \sum_{i=1}^{m} [f(c_i+0) - f(c_i-0)]\varepsilon_{c_i} + [f(b) - f(b-0)]\varepsilon_b.$$

It is clear from this that the support S of μ_f consists of those c_i $(1 \leq i \leq m)$ for which $f(c_i+0) - f(c_i-0) \neq 0$, together with a if $f(a+0) \neq f(a)$ and with b if $f(b) - f(b-0) \neq 0$.

4.10 The Space $C(T)$ and Its Dual

Let T be a locally compact space. We here denote by $C(T)$ the vector space of all complex-valued continuous functions on T, and furnish it with the topology of convergence uniform on compact subsets of T. This topology may be defined by the seminorms

$$p_K(x) = \operatorname{Sup}\{|x(t)| : t \in K\}, \tag{4.10.1}$$

K ranging over any preassigned base for the compact subsets of T. When T is compact, $C(T)$ is normable as a Banach space, the norm being $\|x\|_\infty$.

Our aim is to show how the dual of $C(T)$ may be represented in terms of certain complex Radon measures on T. The same may be done if we consider in place of $C(T)$ the real vector space of real-valued continuous functions on T: one will then be concerned only with real Radon measures. The details attendant upon this change may be left to the reader. We shall feel free to make the change without further special comment.

4.10.1 Theorem. Let L be a continuous linear form on $C(T)$. There exists then a uniquely determined complex Radon measure μ on T having a compact support and such that for all $x \in C(T)$

$$L(x) = \mu(x) = \int_T x(t)\,d\mu(t). \tag{4.10.2}$$

Conversely, if μ is a complex Radon measure on T with a compact support, then (4.10.2) defines L as a continuous linear form on $C(T)$.

Proof. The converse is evident: If K is the compact support of μ, then any x in $C(T)$ is integrable for $|\mu|$, hence also for μ_k $(k = 1, 2, 3, 4)$, so that

$$\mu(x) = \mu_1(x) - \mu_2(x) + i\mu_3(x) - i\mu_4(x)$$

is a linear form on $C(T)$. Moreover,

$$|\mu(x)| = |\mu(x\chi_K)| \leq |\mu|\,(K)\cdot p_K(x),$$

showing continuity of the linear form L defined by (4.10.2).

Reciprocally, let L be a continuous linear form on $C(T)$. Then there exists a compact set K and a number c such that $|L(x)| \leq c \cdot p_K(x)$ for all x in $C(T)$. Let μ be the restriction of L to $\mathscr{K}(T)$. It is evident that μ is a complex Radon measure on T, and Proposition 4.9.1 shows that the support of μ is contained in K, hence is compact. This being so, the relation $L(x) = \mu(x)$, true by definition for x in $\mathscr{K}(T)$, now extends at once to all x in $C(T)$ (both sides being unaltered when x is replaced by fx, where f is any chosen element of $\mathscr{K}_+(T)$ which is identically 1 on K). ▌

The result just established may be formulated in the following manner. If we denote by $M(T)$ the complex vector space of complex Radon measures on T, and by $M_c(T)$ the subspace formed of measures with compact supports, then *the dual of the* LCTVS $C(T)$ *may be identified with* $M_c(T)$, the bilinear form defining the duality being

$$\langle x, \mu \rangle = \mu(x) = \int_T x\, d\mu \tag{4.10.3}$$

for x in $C(T)$ and μ in $M_c(T)$. When T is compact, $M_c(T) = M(T)$ and we recover the classical theorem of F. Riesz in a generalized form. The history of various generalizations of Riesz's result, which referred to the case in which T is the compact interval $[0, 1]$ of the real axis, is rich.† The reader will find some historical notes in Subsection 4.10.7. More recently, the possibility of removing the restriction that T be locally compact has also been considered by various authors (see Subsection 4.10.7).

It is worth recording here two consequences of the duality just established. The first is a criterion for weak convergence in $C(T)$, and the second a criterion for weak compactness in $M_c(T)$.

4.10.2 **Theorem.** In order that a sequence (x_n) in $C(T)$ be weakly convergent to x in $C(T)$, it is necessary and sufficient that

(a) $\lim x_n = x$ pointwise on T;

(b) $\operatorname{Sup}\{|x_n(t)| : t \in K,\ n = 1, 2, \cdots\} < +\infty$ for each compact set K in T.

Proof. The sufficiency follows at once from Theorem 4.8.2 and the preceding identification of the dual of $C(T)$.

The necessity of (a) is evident, $x \to x(t)$ being a continuous linear form on $C(T)$ for each point t of T (the corresponding representative measure being, of course, ε_t, the Dirac measure at t). The necessity of (b) follows as a special case of Theorem 8.2.2, since any weakly convergent sequence is weakly bounded. ▌

† For the theorem in its original form, see, for example, Natanson [1], p. 236.

4.10.3 **Theorem.** In order that a subset L of $M_c(T)$ be relatively weakly compact, it is sufficient that there exist a compact subset K of T such that each μ in L has its support contained in K and further

$$\text{Sup}\,\{|\mu|\,(T) : \mu \in L\} < +\infty.$$

If T is σ-compact, this condition is also necessary.

Proof. The stated condition signifies precisely that L is equicontinuous (considered as a subset of the dual of $C(T)$), hence certainly relatively weakly compact (Theorem 1.11.4(2)).

On the other hand, if T is σ-compact, $C(T)$ is a Fréchet space (see Section 6.1 and Example 6.1.3); it suffices then to apply Corollary 7.1.2 and the observation that any relatively weakly compact set is a fortiori weakly bounded. ▌

Remarks. If $C(T)$ is separable (for example, if T is metrizable and σ-compact) and L satisfies the stated condition, Theorem 4.10.3 amounts to the assertion that from any sequence (μ_n) extracted from L one may in turn extract a subsequence that converges weakly to a complex Radon measure whose support is necessarily contained in K.

If T is a compact interval of real numbers, this statement is closely related to some theorems of Helly stated in terms of functions of bounded variation, the connection involving Riemann-Stieltjes integrals; for details, see Natanson [1], pp. 222 and 233. If T is a number space R^n, the result was announced as a "choice principle" by de la Vallée Poussin and put to good use by him and by Frostman in potential theory.

We observe that when T is compact, $M_c(T) = M(T)$ is a Banach space for the norm $\|\mu\| = |\mu|\,(T)$, or for the equivalent norm

$$\|\mu\|' = \text{Sup}\left\{\left|\int_T x\,d\mu\right| : x \in C(T),\ \|x\|_\infty \leq 1\right\}.$$

The theorem states in this case that a subset L of $M(T)$ is weakly relatively compact if and only if it is norm bounded. There are many useful applications of this principle. It is used in Section 8.13 and again in Chapter 10 (see especially the accounts of the representation theorems of Bernstein and of Bochner in Sections 10.2 and 10.3, respectively). A similar and more immediate application follows as an example.

4.10.4 **Example. Poisson Integral for Harmonic Functions.** We consider the Poisson integral representation of harmonic functions in the unit disk D of the complex plane. Let T be the frontier of D, that is, the unit circumference. We start from the fact that if u is a function continuous on $\bar{D} = D \cup T$ and harmonic on D, then

$$u(z) = \int_T P(z, t)u(t)\,d\alpha(t) \qquad (z \in D) \tag{4.10.4}$$

where α is the positive Radon measure on T defined by

$$\alpha(x) = (2\pi)^{-1} \int_0^{2\pi} x(e^{i\phi})\,d\phi$$

(an ordinary Riemann integral) for x in $C(T)$, and where P is the Poisson kernel

$$P(z, t) = \operatorname{Re} \frac{t + z}{t - z}.$$

A short sketch proof of (4.10.4) runs somewhat as follows. The right-hand side, call it U, is harmonic on D; a standard piece of analysis shows that $U(z) \to u(t)$ as the point z of D tends to the frontier point t; thus $u - U$ is harmonic on D and has a frontier limit that is everywhere zero; the maximum modulus principle then asserts that $u = U$ throughout D, which is (4.10.4).

Now we observe that $P(0, t) = 1$, so that (4.10.4) gives

$$u(0) = \int_T u(t)\, d\alpha(t), \tag{4.10.5}$$

which is none other than the mean value property of harmonic functions.

If we introduce the measure $\mu_u = u \cdot \alpha$, then we may write

$$u(z) = \int_T P(z, t)\, d\mu_u(t) \qquad (z \in D) \tag{4.10.6}$$

and

$$u(0) = \int_T d\mu_u(t). \tag{4.10.7}$$

The measure μ_u is positive if u is positive.

Suppose now that u is harmonic and positive on D (but not necessarily continuous $\cdots$ or even defined $\cdots$ on T). For $n = 1, 2, \cdots$, let $u_n(z) = u(nz/n + 1)$. Then u_n is harmonic and positive on D and continuous on $\bar{D}$. We let $\mu_n = \mu_{u_n}$ be the associated measure. Since (4.10.7) gives

$$\int_T d\mu_n(t) = u_n(0) = u(0),$$

and since the measures μ_n are all positive, the choice principle affirms that a subsequence may be extracted from (μ_n) which converges weakly to some measure μ on T, necessarily positive. For fixed z in D, $P(z, t)$ is a member of $C(T)$. Hence for the chosen subsequence one has

$$\begin{aligned} u_n(z) &= \int_T P(z, t)\, d\mu_n(t) \\ &\to \int_T P(z, t)\, d\mu(t). \end{aligned}$$

Evidently, the whole sequence (u_n) converges to u pointwise on D. Thus we conclude that

$$u(z) = \int_T P(z, t)\, d\mu(t), \tag{4.10.8}$$

a Poisson integral formula for u. This is now established for every *positive* harmonic function u on D.

If u is not positive, the preceding argument breaks down, the measures μ_n being not necessarily positive and (4.10.7) then yielding directly no information about the norms $\|\mu_n\| = |\mu_n|\,(T)$. However, μ_n is in any case the measure $u_n \cdot \alpha$ and so

$$|\mu_n|(T) = \int_T |u_n| \, d\alpha = \int_T \left| u \quad (nt/n + 1) \right| d\alpha(t).$$

Thus the preceding argument can be reinstated whenever we know that

$$\liminf_{r \uparrow 1} \int_T |u(rt)| \, d\alpha(t) = \liminf_{r \uparrow 1} (2\pi)^{-1} \int_0^{2\pi} |u(re^{i\phi})| \, d\phi < +\infty \tag{4.10.9}$$

Naturally, the measure μ in the resulting formula (4.10.8) is no longer generally positive.

It is easily seen that, conversely, if (4.10.8) holds for some measure μ on T, then (4.10.9) holds: indeed since $P \geq 0$, we have from (4.10.8)

$$|u(z)| \leq \int_T P(z, t) \, d\,|\mu|(t);$$

if $0 \leq r < 1$ and t^* lies in T, we get thence

$$\int_T |u(rt^*)| \, d\alpha(t^*) \leq \int_T d\alpha(t^*) \int_T P(rt^*, t) \, d\,|\mu|\,(t).$$

Inversion of the order of integrations is easily justified and shows that the right-hand side here is equal to

$$\int_T d\,|\mu|\,(t) \int_T P(rt^*, t) \, d\alpha(t^*).$$

Now for fixed r and t, the function $z \to P(rz, t)$ is harmonic on D and continuous on $\bar{D}$; so the inner integral, being the mean value of this function, is none other than $P(0, t) = 1$. Thus $\int_T |u(rt^*)| \, d\alpha(t^*) \leq |\mu|\,(T)$ for all r satisfying $0 \leq r < 1$, so that (4.10.9) holds. We have shown incidentally that (4.10.9) is equivalent to

$$\operatorname*{Sup}_{0 \leq r < 1} \int_T |u(rt)| \, d\alpha(t) < +\infty. \tag{4.10.9'}$$

To sum up:

▶ In order that a function u on D be representable in the form (4.10.8), μ being a suitable Radon measure on T, it is necessary and sufficient that u be harmonic on D and satisfy (4.10.9′). The latter is automatically fulfilled whenever u is positive and harmonic on D.

We observe finally that such representation theorems for harmonic functions on other domains are obtainable by similar arguments (or in some cases by conformal transformation applied to (4.10.8) itself), and that they yield analogous formulae for holomorphic functions associated with the names Herglotz, Riesz, Nevanlinna, et alia. The representation (4.10.8) itself is due to Evans, Bray, and Plessner.

4.10.5 $C(T)$ as an Algebra; the Stone-Weierstrass Theorem. In this and the following subsection we shall indicate how the notion of the support of a measure, together with Theorem 4.10.1, may be used to elucidate further the structure of $C(T)$ when account is taken of the possibility of multiplying functions in $C(T)$; that is, we regard $C(T)$ as an algebra and not merely a vector space. In this subsection we give a proof of the famous Stone-Weierstrass theorem regarding subalgebras of $C(T)$, and in Subsection 4.10.6 we are concerned with closed ideals in the algebra $C(T)$.

If T is a compact interval of the real axis, which we may without any loss of generality take to be $[0, 1]$, Weierstrass' classical approximation theorem states that every continuous real-valued function on T is a uniform limit of polynomials. From this it is evident that similar approximation is possible for continuous complex-valued functions. In a now-famous pioneering paper, M. H. Stone [2] (especially pp. 466–469) began the study of wide-sweeping extensions of Weierstrass' theorem in which T is taken to be an arbitrary separated compact space.

Throughout this subsection we shall use T to denote a separated compact space; $C_R(T)$ (resp. $C(T)$) is the vector space over R (resp. over C) of real- (resp. complex-) valued continuous functions on T; $M_R(T)$ (resp. $M(T)$) is the vector space over R (resp. over C) of real (resp. complex) bounded Radon measures on T. Each of $C_R(T)$ and $C(T)$ is topologized by use of the "uniform norm"

$$\|x\| = \text{Sup}\,\{|x(t)| : t \in T\}.$$

We know (Theorem 4.10.1) that $M_R(T)$ (resp. $M(T)$) can be identified with the topological dual of $C_R(T)$ (resp. $C(T)$).

By a *separating subset* of $C(T)$ we mean a subset thereof with the property that, given distinct points t, t' of T, there exists a member of the set taking distinct values at t and t'.

For any function x on T (real or complex valued) we denote by $Z(x)$ the zero set, $\{t \in T : x(t) = 0\}$, of x; and if F is a set of such functions, $Z(F)$ is defined to be the intersection $\bigcap \{Z(x) : x \in F\}$ of the zero sets of members of F.

A *subring* of $C_R(T)$ (resp. $C(T)$), say A, is an additive subgroup of $C_R(T)$ (resp. $C(T)$) which is such that $xy \in A$ whenever $x \in A$ and $y \in A$. If, besides this, A is a vector subspace of $C_R(T)$ (resp. $C(T)$), then A is a *subalgebra* of $C_R(T)$ (resp. $C(T)$). In either case A is said to be unitary if it contains the constant function 1.

We consider first the case of $C_R(T)$. The theorem we aim to establish states that any separating subalgebra A of $C_R(T)$ for which $Z(A) = \emptyset$ is dense in $C_R(T)$. Analogous results for $C(T)$ (Gelfand and Šilov [1]) will be derived therefrom. Now some of the very elegant proofs of this and related results make use of the partial ordering of $C_R(T)$ (see, for example, Bourbaki [2], Chapitre X, p. 55; Gillman and Jerison [1], p. 242; Kakutani [1]). The arguments we shall give, due to de Branges [1], make very little explicit use of this partial order (though this structure is implicit inasmuch as it plays some role in integration theory).

On the other hand de Branges' proof makes use of the Krein-Milman theorem concerning compact convex sets in a TVS. Although the proof of this is delayed until Chapter 10, appeal to it at this point involves no circularity. However, it is worthwhile explaining here and now the content of this theorem.

Suppose that X is a separated real LCTVS and Q a convex subset of X. By an *extreme* (or *extremal*) *point* of Q is meant a point $x \in Q$ with the property that if $x = \alpha y + (1 - \alpha)z$ for some number α satisfying $0 \leq \alpha \leq 1$ and for some y, z in Q, then x is either y or z; in other words, x is an end point of any segment containing it and lying entirely in Q. The Krein-Milman theorem (Theorem 10.1.2 of this book) states now that if Q is compact and convex, then it is the closed convex envelope of the set of its own extreme points; in particular, if $Q \neq \{0\}$, then there must exist nonzero extreme points of Q.

The situation to which this theorem will be applied is as follows. A is a subset of $C_R(T)$, A^0 the set of measures in $M_R(T)$ vanishing on A, B the closed unit ball in $M_R(T)$, and $Q = A^0 \cap B$. We endow $M_R(T)$ with the weak topology, considering it as the topological dual of $C_R(T)$ (cf. Theorem 4.10.1). By Theorem 4.10.3, B is compact; and since A^0 is closed, Q is compact. Q is obviously convex. So we can apply the Krein-Milman theorem to Q and deduce in particular that Q, if not reduced to $\{0\}$, has extreme points different from 0. We note that, in fact, any extreme point $\lambda \neq 0$ of Q must be such that $\|\lambda\| = 1$; for if $\lambda \neq 0$, then $\lambda/\|\lambda\| \in Q$ and we have $\lambda = \|\lambda\|(\lambda/\|\lambda\|) + (1 - \|\lambda\|)0$ which, since λ is extremal, entails that $\lambda = \lambda/\|\lambda\| \neq 0$, and so $\|\lambda\| = 1$.

With these preliminary remarks in mind, we begin by establishing a lemma.

Lemma. Suppose that λ is an extreme point of Q and that g is a bounded real-valued universally measurable function on T such that $\int xg\, d\lambda = 0$ for all x in A. Then there exists a real number k such that $g \cdot \lambda = k\lambda$.

Proof. If $\lambda = 0$, there is nothing to prove. Also, if $\int xg\, d\lambda = 0$ for x in A, then the same is true when g is replaced by $a(g + b)$, no matter what the real numbers a and b. We may therefore suppose throughout that $0 \leq g \leq 1$.

We define $\lambda^* = g \cdot \lambda$ and put $m = \|\lambda^*\|$. If $m = 0$, then $\lambda^* = 0 = 0 \cdot \lambda$ and nothing remains to be proved. If $m = 1$, then we have $1 = m = \int g\, d\,|\lambda|$ which, since $0 \leq g \leq 1$, shows that $g = 1$ a.e. (λ) and therefore $\lambda^* = \lambda = 1 \cdot \lambda$; again there is nothing left to prove.

We assume then that $0 < m < 1$ and put

$$\lambda_1 = \frac{\lambda^*}{m}, \qquad \lambda_2 = \frac{\lambda - \lambda^*}{1 - m}.$$

It is simple to verify that λ_1 and λ_2 belong to Q. Thus $\|\lambda_1\| = 1$ and

$$\int x\, d\lambda_1 = m^{-1} \int x\, d\lambda^* = m^{-1} \int xg\, d\lambda = 0$$

for x in A. Thus λ_1 belongs to Q. This shows that $\lambda^* = m\lambda_1 \in A^0$; on the other hand

$$\lambda_2 = (1 - m)^{-1}(1 - g) \cdot \lambda$$

and so

$$\|\lambda_2\| = (1 - m)^{-1} \int (1 - g)\, d\,|\lambda| = (1 - m)^{-1}(\|\lambda\| - \|\lambda^*\|)$$
$$= (1 - m)^{-1}(1 - m) = 1;$$

moreover, since λ and λ^* belong to A^0, so also does λ_2. Thus λ_2 belongs to Q.

The equation

$$\lambda = (1 - m)\lambda_2 + m\lambda_1$$

shows now, via the extremal nature of λ, that λ is either λ_1 or λ_2. In either case, we derive at once that $\lambda_1 = \lambda$. Accordingly, $g \cdot \lambda = \lambda^* = m\lambda_1 = m\lambda$, as was alleged (with m in place of k). ▌

Now we can establish the main result for $C_R(T)$.

Theorem A. (Stone-Weierstrass). Let A be a separating subalgebra of $C_R(T)$ for which $Z(A) = \emptyset$ (that is, the functions in A have no common zeros in T). Then A is dense in $C_R(T)$.

Proof. A being a subalgebra of $C_R(T)$, it is a vector subspace thereof. So (Hahn-Banach theorem) A is dense if and only if $A^0 = \{0\}$; that is, if and only if $Q = \{0\}$.

Suppose, however, that Q were not $\{0\}$. Then the Krein-Milman theorem implies (as we have seen) that Q has at least one extreme point $\lambda \neq 0$. We will show, by using the preceding lemma, that any such λ is necessarily a nonzero scalar multiple of ε_t for some t in T. But then, since $\lambda \in Q \subset A^0$, we should be forced to the conclusion that $x(t) = \int x\, d\lambda$ vanishes for all x in A; that is, that $t \in Z(A)$, a contradiction.

To show that λ must have the stated form we note that, A being a subring of $C_R(T)$, $g \cdot \lambda \in A^0$ whenever $g \in A$. But then the lemma asserts that $g \cdot \lambda =$ const. λ and so that g is constant on the support of λ. A being separating, it follows that the support of λ is reduced to one point, say t, of T. Accordingly, $\lambda = c \cdot \varepsilon_t$, c being a number that, since $\lambda \neq 0$, must be nonzero. This completes the proof. ▌

From this theorem it is very simple to deduce an analogous result for $C(T)$.

Theorem B. Let A be a separating subalgebra of $C(T)$ for which $Z(A) = \emptyset$, and suppose that $x \in A$ entails $\bar{x} \in A$. Then A is dense in $C(T)$.

Proof. Our hypotheses ensure that $\operatorname{Re} x$ and $\operatorname{Im} x$ belong to A whenever $x \in A$. We let A_R be the set of real functions in A. A_R fulfills the conditions of Theorem A. Hence A_R is dense in $C_R(T)$, and so A is dense in $C(T)$. ▌

Remarks. If T is locally compact and noncompact, it is easy to deduce analogous results for the spaces of real- (or complex-) valued continuous functions on T that vanish at ∞. One has only to extend all the functions to the Alexandroff compactification $\tilde{T}$ of T and consider the subring of $C_R(\tilde{T})$ (or $C(\tilde{T})$) generated by $A \cup \{1\}$. The conclusion is that the direct analogous of Theorems A and B are valid when we assume merely that T is locally compact and we consider only those continuous functions (real or complex valued as the case may be) that tend to 0 at ∞.

There exists a very rich literature dealing with various extensions of the Stone-Weierstrass theorem; we mention only a few items: Bourbaki [2], Chapitre X; Arens [2]; Kaplansky [1] and [2]; Stone [3]; Hewitt [3]; and the long expository article by M. H. Stone appearing in Buck [1].

4.10.6 **Closed Ideals in** $C(T)$. We shall indicate here two further uses of the concept of support of a measure and of Theorem 4.10.1. Both are concerned with $C(T)$ regarded as an *algebra*, the product xy meaning the function $t \to x(t)y(t)$. Since, as is easily verified, the mapping $(x, y) \to xy$ is continuous from $C(T) \times C(T)$ into $C(T)$, the latter is a topological algebra as well as a TVS. We consider two problems relating to this additional structure (Edwards [7]).

By a *character* of $C(T)$ we mean a homomorphism X of $C(T)$ onto the complex field C (actually the latter is better regarded here as an algebra over itself)—that is, a mapping of $C(T)$ into C that is linear and multiplicative ($X(xy) = X(x)X(y)$) and not identically 0. Then, if e denotes the constant function with unit value everywhere, we have $X(e) = 1$. The first of the two results we aim to prove reads thus:

▶ (a) If X is a continuous character of $C(T)$, there exists a uniquely determined point t of T such that $X(x) = x(t)$ for all x in $C(T)$. If T is compact, any character of $C(T)$ is continuous and therefore of the above type.

The converse is trivial.

The second result concerns the notion of an *ideal* I in the algebra $C(T)$. The characteristic properties of an ideal I are that I is a vector subspace of $C(T)$, and that $xI \subset I$ for each x in $C(T)$ (that is, x in $C(T)$ and y in I together imply that xy is in I). For example, if F is a subset of T, the set $I(F)$ of x in $C(T)$ that vanish on F is obviously an ideal in $C(T)$. $I(F)$ is unaltered if F is replaced by its closure in T. Besides this, $I(F)$ is always a closed subset of $C(T)$. We aim to prove that every closed ideal in $C(T)$ is obtainable in this way starting from a suitably chosen closed subset F of T.

This allegation may be stated in a different way. Given I (which may be any subset of $C(T)$ for the moment), we let $Z(I)$ be the set of all common zeros of members of I. $Z(I)$ is obviously a closed subset of T. Equally obviously, $I \subset I(Z(I))$. The second result we have to prove can be formulated thus:

▶ (b) If I is any closed ideal in $C(T)$, then $I = I(Z(I))$.

It is easily seen a priori that if $I = I(F)$ is true for any closed $F \subset T$, then F must be $Z(I)$; however our proof will not make use of this remark. Since $I(\emptyset) = C(T)$, we see that an ideal in $C(T)$ is dense in $C(T)$ if (and, obviously, only if) the members of I have no common zeros in T.

Proof of (a). Let X be a continuous character of $C(T)$. By Theorem 4.10.1 there exists a Radon measure μ on T, whose support K is compact, such that

$X(x) = \mu(x)$ for all x in $C(T)$. It will suffice to show that K is reduced to a single point of T. Since $X \neq 0$, K is not empty. Let t belong to K and let t' be any other point of T. Then there exist disjoint open neighbourhoods U and U' of t and t', respectively. Since t belongs to K, we can choose x in $C(T)$ with support contained in U and such that $X(x) \neq 0$. If y is in $C(T)$ and has support contained in U', then $xy = 0$ and hence $X(xy) = X(x)X(y) = 0$, showing that $X(y) = 0$. So Proposition 4.9.1 shows that t' does *not* belong to K. K is therefore the one-point set $\{t\}$, μ is a multiple of ε_t, the multiple being necessarily unity since X is a homomorphism, and thus $X(x) = \varepsilon_t(x) = x(t)$.

For the second part we have to show that the hyperplane H defined by $X(x) = 0$ is closed in $C(T)$, or equivalently, that H is not dense in $C(T)$. But if H were dense in $C(T)$, it would contain an x_0 such that $\|x_0 - e\|_\infty \leq \frac{1}{2}$; x_0 would therefore be nonvanishing and so, T being compact, $1/x_0$ would belong to $C(T)$. The relation $e = x_0 \cdot 1/x_0$ would then give $1 = X(e) = X(x_0) \cdot X(1/x_0) = 0 \cdot X(1/x_0)$, which is absurd. ▌

Proof of (b). Let $J = I(Z(I))$. It is plain that $I \subset J$, and so it remains only to prove that $J \subset I$. This is done by appeal to the Hahn-Banach theorem combined with Theorem 4.10.1.

Let μ belong to $M_c(T)$ and vanish on J; that is, $\mu(j) = 0$ for each j in J; we must show that μ vanishes on I. Since J is an ideal, our hypothesis entails that $\mu(jx) = 0$ (j in J, x in $C(T)$). This entails that the open set $U = T \backslash Z(I)$ is negligible for $|\mu|$. For if t belongs to U, there exists j in J that is nonvanishing at t, hence nonvanishing throughout a neighbourhood V of t. Then, however, any y in $C(T)$ with support contained in V can be written in the form $y = jx$ for some x in $C(T)$, so that $\mu(y) = 0$. That U is indeed negligible for $|\mu|$ follows from Proposition 4.9.1. The support K of μ is thus contained in $T \backslash U = Z(I)$. But then evidently $\mu(x) = 0$ for any x in I, since x vanishes on $Z(I)$. ▌

Remarks. It is natural to consider whether assertions (a) and (b) remain true for various other algebras of functions, in particular for algebras of differentiable and of holomorphic functions. In each of these cases the problems require much more subtle arguments; in particular, the direct analogue of (b) is generally false. Regarding differentiable functions, see Whitney [1], Schwartz [11]; for holomorphic functions see Kakutani [3] and the references there cited, and Edwards [14].

4.10.7 Alternatives to, and Extensions of, the Preceding Theory.

The chosen technique for dealing with integration theory has now reached a stage where one may conveniently and profitably "place" it in relation to alternatives and review possible extensions. The growth of the theories is a complex story; the reader will find much of interest in the "Note Historique" at the end of Bourbaki [6], Chapitre 5.

The more orthodox treatment of integration theory, typified by Halmos [1] and Saks [1], founds the concept of integral on that of a given measure function —that is, a function defined on a suitable set of subsets of the "base space" T

that is at least finitely additive. The Lebesgue-like theories assume, in fact, that the measure is countably additive. In any case one can usually reduce the study to that case in which the measure is positive valued. We, on the contrary, have taken the integral as the basic concept, defining it as a linear form on a certain vector space of functions. In doing this we have limited our study to the case in which T is a locally compact space and the function space is $\mathscr{K}(T)$. It is, however, possible to remove these hypotheses to some extent (see Loomis [1], Chapter III; Stone [4]; Bourbaki [6], Chapitre IV, pp. 114 et seq.)

That case of Theorem 4.10.1 in which T is compact (technically the easiest to handle) may be termed for convenience the "Riesz representation theorem," or RRT for short, though this name neglects the many writers who have contributed essentially to the development of this result into its present form. Our approach to integration theory is so arranged that the RRT is almost a tautology. On the other hand, we have had to extract from our functional definition of an integral the corresponding measure.

The alternative is to adopt the more orthodox, measure-based integration theory, in which case the RRT is no longer a tautology. This approach is typified by Halmos [1], §56. See also Hewitt [2] and Hewitt and Zuckerman [1].

So much for the alternative approaches.

Regarding supplements and extensions, one has first to look at the RRT and various ways of proving it. One recent and refreshingly different method of attack is that due to Varadarajan [1], in which the idea is to represent any compact T as a product of compact spaces, each so simple that the RRT is almost obvious for each such factor, and then to show how to handle the theorem for the product space in terms of the theorem for each factor.

Apart from this we may seek extensions in which local compactness of T is relaxed. The case of completely regular spaces T and analogues of the RRT has been discussed by various writers; we mention here only Hewitt [1], [6], Glicksberg [1], and Gould and Mahowald [2]. The latter seek to reduce the problem for completely regular T to the compact case by imbedding T into its Stone-Čech compactification, βT. Similar extensions have been considered by Stone and Bourbaki (loc cit above), Arens [3], Loomis [4], and others.

What of the functionals corresponding to finitely additive measures? These have been discussed by Fichtenholz and Kantorovitch [1] and more recently by Hewitt and Yosida [1]. See also Hewitt [5]. The utility of analogues of the RRT in which integrals with respect to finitely additive measures is contemplated is rather limited. Finitely additive measures can exhibit behaviour that is almost barbaric. Compare also our reasons, given in Section 4.1, for rejecting a simple appeal to the Hahn-Banach theorem in favour of a laborious but constructive extension process leading to countably additive measures.

Another type of extension is that in which $\mathscr{K}(T)$ is replaced by some other vector space of functions on T. In certain cases such a change can be handled by reducing it in some fashion to the basic one. This is not always the case, however. An interesting example where the reduction is either impossible—or, better, successful only in a superficial sense—is that in which T is a group and we replace $\mathscr{K}(T)$ by the vector space of almost periodic functions on T (see Hewitt [9]).

An entirely different sort of development is that in which T is undisturbed, but we consider either vector-valued (rather than scalar-valued) functions as integrands, or vector-valued measures, or perhaps both (if the range space is an algebra). The case of vector-valued functions and a scalar-valued measure represents a less radical extension than vice versa. This problem enjoyed great popularity some thirty years ago and typical of the contributions of numerous writers is the "Bochner integral" (see Bochner [3]), for an exposition of which see Hille [1], Chapter III or Dunford and Schwartz [1], Chapter III (or both). However, time has shown that the simpler and more manageable "weak" theories serve equally well in most circumstances. We shall in Section 8.14 begin the rudiments of the theory, treating it as an offshoot of duality theory, upon which it is heavily dependent. Use will be made of weak integrals in Sections 8.13 and 10.1.

As regards integration with respect to vector-valued measures, one has again the choice of starting point. If we are concerned with measures taking their values in E, we may adopt either of the following alternative approaches: (1) The measure means from the outset a countably additive function defined on suitably restricted subsets of T. (2) The measure means a suitably restricted linear map of $\mathscr{K}(T)$ into E. An exposition of the former approach will be found in Dunford and Schwartz [1], Chapters III and IV. The objects described in (2) are naturally described as E-valued Radon measures on T. The question is, however, in precisely what way one should restrict the maps referred to in (2) in order that they shall correspond (as in the scalar-valued case) to measures as described in (1). The answer to this question is due largely to Grothendieck [4], who showed that, at least when T is compact, the desired requirement is that the map be compact (= completely continuous) from $C(T)$ into E. Some aspects of this problem are discussed in Section 8.19 and Subsection 9.4.14.

Nearly all integration theories for numerically valued functions make heavy use of the natural partial order amongst real-valued functions. The essential contributions of this structure are brought to the fore in McShane [2] and Christian [1].

The reader who wishes to delve further into matters relating to vector-valued integrands, or measures, or both, may find it a helpful start to consult Hildebrandt's expository article [1].

Historical notes on the RRT. As early as 1903 and 1904 Hadamard [1] and Fréchet [1], [2], [3] had given representation theorems for continuous linear forms on the Banach space $C = C[0, 1]$, but in both cases there were unsatisfactory features. In 1909 Riesz gave the theorem in a form that is now recognized as the "right" one, differing only from that obtained by specialization of current versions in the use of Riemann-Stieltjes integrals rather than Lebesgue-Stieltjes integrals. Other proofs followed by Riesz himself (Riesz [4], [5], [6] in 1911, 1914, and 1952, respectively) and by Helly [1] in 1912.

The next major advance amounted to an extension of the theorem for $C(T)$ when T is a compact subset of R^n, given by Radon [1] in 1913. This paper contained a great deal besides the RRT, of course.

Over twenty years later Banach, in Note II at the end of Saks [1] (published in 1937) showed how to extend the theorem to the case of compact metric spaces T, using Daniell's ideas on abstract integration. Saks [2] himself in 1938 dealt with related topics.

Finally, in 1941, Kakutani [6] gave the theorem for compact Hausdorff spaces in general, thus completing one chapter of the story.

Already in 1938, however, Markov [1] made the first attempt to deal with noncompact spaces, considering in place of $C(T)$ the space of bounded, continuous functions on a space T that is normal but not necessarily T_1. Finitely additive measures appeared in his account. Similar work was undertaken by Alexandroff [1], [2], [3], covering the period 1940–1943. We have already commented on some of the more recent work along these lines.

4.11 The Seminorms N_p; Inequalities of Hölder and Minkowski; the Spaces $\mathscr{L}^p$ and L^p

At this point we return to the development of integration theory (for scalar-valued functions) at the stage where it was left in Section 4.8. We have to deal with the "continuous" analogue of the spaces ℓ^p introduced in Section 4.2. As in that section we use p to denote a number > 0 or ∞ (that is, $+\infty$); p' is defined to be 1 if $p = \infty$ and in all other cases is defined by the equation $1/p + 1/p' = 1$. T is a locally compact space and μ a positive Radon measure on T, fixed throughout this section.

4.11.1 If f is any (real- or complex-) valued function on T, and if $p \neq \infty$, we define

$$N_p(f) = \mu^*(|f|^p)^{1/p}, \tag{4.11.1}$$

with the convention that $(+\infty)^a = +\infty$ if $a > 0$. If $p = 1$, N_p is identical with N_1 introduced in Section 4.6. We define also $\mathscr{F}^p = \mathscr{F}^p(T, \mu)$ to be the set of functions f for which $N_p(f) < +\infty$; for $p = 1$ we recover $\mathscr{F}^1$ defined in Section 4.6. While it was evident that N_1 is a seminorm on $\mathscr{F}^1$, this is not necessarily the case of N_p on $\mathscr{F}^p$ when $p < 1$; it is so when $p > 1$, but then the statement amounts to Minkowski's inequality and calls for proof. Before doing this, we complete our definitions by turning to the case $p = \infty$.

There are at least two possible definitions when $p = \infty$ that are divergent when (T, μ) is not σ-finite but accordant otherwise. The first definition makes $N_\infty(f)$ the essential supremum of $|f|$—that is, the least number m (possibly $+\infty$) such that $|f| \leq m$ holds except on a negligible set (a priori dependent upon m). The second definition, for which we use the notation $\bar{N}_\infty$, makes $\bar{N}_\infty(f)$ the *local* essential supremum of $|f|$—that is, the least m (again possibly $+\infty$) such that $|f| \leq m$ holds except on a locally negligible set (again a priori dependent upon m). It is evident that $\bar{N}_\infty(f) \leq N_\infty(f)$ always; strict inequality will hold for suitable f whenever there exist subsets of T that are locally negligible and not negligible (and in this case only). The two agree, therefore, whenever (T, μ) is σ-finite. The corresponding function spaces, $\mathscr{F}^\infty = \mathscr{F}^\infty(T, \mu)$ and $\overline{\mathscr{F}}^\infty = \overline{\mathscr{F}}^\infty(T, \mu)$, satisfy $\mathscr{F}^\infty \subset \overline{\mathscr{F}}^\infty$, identity being the case at least whenever (T, μ) is σ-finite.

Remark. The distinction between $\mathscr{F}^\infty$ and $\overline{\mathscr{F}}^\infty$ is the forerunner of one to to be made between the existing $\mathscr{F}^1$ and an $\overline{\mathscr{F}}^1$ not yet introduced (and, more generally, between $\mathscr{F}^p$ and a new $\overline{\mathscr{F}}^p$ for $1 \leq p < \infty$). It seems that the need for such a distinction, which is in any case confined entirely to cases in which (T, μ) is not σ-finite, is first felt when $p = \infty$. Rather than deal here and now with finite values of p, we prefer to wait until the need comes to be felt in the discussion of the product of a measure by a function (Section 4.13 below). The distinction seems to have been introduced first by Bourbaki ([6], Ch. 5).

With either choice in the case $p = \infty$, it is clear that N_∞ and $\bar{N}_\infty$ are seminorms on $\mathscr{F}^\infty$ and $\overline{\mathscr{F}}^\infty$, respectively.

When T is discrete and μ assigns unit mass to each point of T, the spaces $\mathscr{F}^p(T, \mu)$ are identical with $\ell^p(T)$ and the seminorms N_p identical with the norms $\| \cdot \|_p$ used in Section 4.2. Furthermore, there is in this case no distinction between N_p and $\bar{N}_p$ or between $\mathscr{F}^p$ and $\overline{\mathscr{F}}^p$. This state of affairs is, in fact, attained whenever the measure μ is "atomic" (see Example 4.4.1). For the details, see Edwards [16].

From Section 4.2 and the remarks just made, it is evident that N_p is not in general a seminorm on $\mathscr{F}^p$ when $0 < p < 1$. Our next step is to examine the case $p \geq 1$; we shall discover that the seminorm property obtains in that case.

4.11.2 **Inequalities of Hölder and Minkowski.** These read as follows, f and g denoting arbitrary functions on T:

$$N_1(fg) \leq N_p(f)N_{p'}(g) \qquad (1 \leq p \leq \infty); \tag{H}$$

$$N_1(fg) \leq N_1(f)\bar{N}_\infty(g) \qquad \text{if } N_1(f) < +\infty; \tag{H_1}$$

$$N_p(f + g) \leq N_p(f) + N_p(g) \qquad (1 \leq p \leq \infty); \tag{M}$$

$$\bar{N}_\infty(f + g) \leq \bar{N}_\infty(f) + \bar{N}_\infty(g). \tag{M_∞}$$

Of these we know already the truth of (H) when $p = 1$ or $p = \infty$ (see (4.6.3)), of (M) when $p = 1$, and of (M_∞). It therefore remains to discuss (H) when $1 < p < \infty$, (H_1), and (M) when $1 < p < \infty$.

4.11.3 **Proof of** (H_1). If $N_1(f) < +\infty$, f vanishes outside a set P that is σ-finite for μ. On the other hand, $|g| \leq \bar{N}_\infty(g)$ holds except on a locally negligible set $S \subset T$. But then $|fg| \leq \bar{N}_\infty(g) \cdot |f|$ holds outside $P \cap S$, which is negligible. So

$$N_1(fg) = \mu^*(|fg|) \leq \bar{N}_\infty(g) \cdot \mu^*(|f|) = \bar{N}_\infty(g) \cdot N_1(f),$$

which is (H_1). ▌

4.11.4 Proof of (*H*) for $1 < p < \infty$. Although we might make use of Hölder's inequality for finite sums, we give here a proof from "first principles." This is based upon the elementary numerical inequality

$$xy \leq \frac{x^p}{p} + \frac{y^{p'}}{p'} \qquad (x \geq 0, y \geq 0, 1 < p < \infty), \tag{4.11.2}$$

the proof of which is left to the reader.

In our proof of (*H*) we may assume that neither factor on the right-hand side is 0, since otherwise the corresponding function (f or g), and in any case the product fg, would be negligible. This being so, we may assume further that each factor on the right-hand side is finite; the reader will recall the convention that $c(+\infty) = +\infty$ for $c > 0$ (Section 4.5).

Assume then that $0 < N_p(f) < +\infty$ and $0 < N_{p'}(g) < +\infty$. We apply (4.11.2) for each point t of T, taking

$$x = \frac{|f(t)|}{N_p(f)}, \qquad y = \frac{|g(t)|}{N_{p'}(g)},$$

and then apply μ^*. Since $1/p + 1/p' = 1$, we conclude that

$$\frac{\mu^*(|fg|)}{N_p(f)N_{p'}(g)} \leq 1,$$

which yields (*H*). ▮

4.11.5 Proof of (*M*) for $1 < p < \infty$. We may assume that f and g are positive functions, that both summands on the right are finite, and that $N_p(f + g) > 0$ (since otherwise f and g are negligible and (*M*) is trivially true). The inequality $(x + y)^p \leq 2^p(x^p + y^p)$, valid for $x \geq 0$ and $y \geq 0$, shows then that $N_p(f + g)$ is finite.

We let $h = f + g$, write $h^p = h^{p-1}f + h^{p-1}g$ and $\mu^*(h^p) \leq \mu^*(h^{p-1}f) + \mu^*(h^{p-1}g)$, and apply (*H*) to each summand on the right separately. This leads to

$$\mu^*(h^p) \leq \mu^*(h^p)^{1/p'} \cdot [N_p(f) + N_p(g)].$$

Since $0 < \mu^*(h^p) < +\infty$ and $1/p + 1/p' = 1$, this is equivalent to (*M*). ▮

4.11.6 Corollary. N_p is a seminorm on $\mathscr{F}^p$ $(1 \leq p \leq \infty)$ and $\bar{N}_\infty$ is a seminorm on $\overline{\mathscr{F}}^\infty$.

4.11.7 The Spaces $\mathscr{L}^p(T, \mu)$ and $\overline{\mathscr{L}}^\infty(T, \mu)$. We can now mimic the definition of $\mathscr{L}^1$ given in Section 4.6, introducing $\mathscr{L}^p = \mathscr{L}^p(T, \mu)$ $(1 \leq p < \infty)$ as the closure in $\mathscr{F}^p$, relative to the seminorm N_p, of $\mathscr{K}(T)$. For $p = \infty$, $\mathscr{L}^\infty = \mathscr{L}^\infty(T, \mu)$ and $\overline{\mathscr{L}}^\infty = \overline{\mathscr{L}}^\infty(T, \mu)$ are defined directly as the set of measurable functions in $\mathscr{F}^\infty$ and in $\overline{\mathscr{F}}^\infty$, respectively. These are quite different from the N_∞- and $\bar{N}_\infty$-closures of $\mathscr{K}(T)$.

We leave to the reader the easy task of verifying that Theorem 4.6.7 and Corollaries 4.6.8 and 4.6.9 extend to the case in which $\mathscr{L}^1$ and N_1 are replaced by $\mathscr{L}^p$ and N_p $(1 \leq p < \infty)$; and that Theorem 4.8.1 likewise extends for any p satisfying $1 \leq p \leq \infty$ (when $p = \infty$ there are two analogues and both are valid). Thus, if $\mathscr{M}$ denotes the set of measurable functions, then $\mathscr{L}^p = \mathscr{F}^p \cap \mathscr{M}$ and $\overline{\mathscr{L}}^\infty = \overline{\mathscr{F}}^\infty \cap \mathscr{M}$. As a consequence, a complex-valued function belongs to $\mathscr{L}^p$ (or to $\overline{\mathscr{L}}^\infty$) if and only if its real and imaginary parts enjoy this property. (In this connection the reader will observe that if $f = u + iv$ with u and v real, then $\text{Max}\,(|u|^p, |v|^p) \leq |f|^p \leq 2^p(|u|^p + |v|^p)$.)

4.11.8 Corollary. If f belongs to $\mathscr{L}^p$ and g to $\mathscr{L}^{p'}$ $(1 \leq p \leq \infty)$, then fg belongs to $\mathscr{L}^1$ and

$$|\mu(fg)| \leq N_1(fg) \leq N_p(f)N_{p'}(g).$$

If $p = 1$, we can here replace $\mathscr{L}^\infty$ by $\overline{\mathscr{L}}^\infty$ and $N_\infty(g)$ by $\bar{N}_\infty(g)$.

Proof. The product fg is measurable whenever f and g have this property. It remains only to apply (H) or (H_1) as the case may be. ▌

4.11.9 Integrability for a Complex Radon Measure. If λ is a complex Radon measure on T,

$$\lambda = \lambda_1 - \lambda_2 + i\lambda_3 - i\lambda_4,$$

its minimal decomposition (see Theorem 4.3.2), measurability, integrability, negligibility, and local negligibility relative to λ is by definition equivalent to the conjunction of the same property relative to λ_k $(k = 1, 2, 3, 4)$. In particular, $\mathscr{L}^1(\lambda) = \bigcap_{1 \leq k \leq 4} \mathscr{L}^1(\lambda_k)$. For any f in $\mathscr{L}^1(\lambda)$ we define

$$\int f\,d\lambda = \int f\,d\lambda_1 - \int f\,d\lambda_2 + i\int f\,d\lambda_3 - i\int f\,d\lambda_4.$$

The spaces $\mathscr{L}^p(\lambda)$ are defined similarly. If we put $|\lambda| = \lambda_1 + \lambda_2 + \lambda_3 + \lambda_4$, then $\mathscr{L}^p(\lambda) = \mathscr{L}^p(|\lambda|)$, and λ-measurability is the same as $|\lambda|$-measurability. Beginning from (4.6.3) it is easy to see that

$$\left|\int f\,d\lambda\right| \leq \int |f|\,d\,|\lambda|$$

for any f in $\mathscr{L}^1(\lambda)$, and hence that

$$\left|\int fg\,d\lambda\right| \leq \left[\int |f|^p\,d\,|\lambda|\right]^{1/p} \cdot \left[\int |g|^{p'}\,d\,|\lambda|\right]^{1/p'}$$

whenever f is in $\mathscr{L}^p(\lambda)$ and g is in $\mathscr{L}^{p'}(\lambda)$. The function

$$f \to \left[\int |f|^p\,d\,|\lambda|\right]^{1/p}$$

is an appropriate seminorm for the space $\mathscr{L}^p(\lambda)$, and $\mathscr{L}^p(\lambda)$ is complete for the structure defined by this seminorm.

In Subsections 4.15.6 and 4.15.13 we shall see that an alternative definition of $|\lambda|$ has certain formal advantages: this amounts to taking $|\lambda|$ equal to $(\alpha^2 + \beta^2)^{1/2}$, where α and β are the real Radon measures satisfying $\lambda = \alpha + i\beta$. The resulting seminorm is equivalent to the preceding one, and (H), (H_1), and (M) continue to hold with this new meaning for $|\lambda|$.

4.11.10 $\mathscr{L}^{p'}$ as a Subset of the Dual of $\mathscr{L}^p$; the Spaces L^p. Suppose that $1 \leq p \leq \infty$ and that $q = p'$. Let g belong to $\mathscr{L}^q$ (or to $\overline{\mathscr{L}}^\infty$ if $p = 1$). According to Corollary 4.11.8, g generates a continuous linear form L_g on $\mathscr{L}^p$ (or on $\overline{\mathscr{L}}^\infty$ if $p = \infty$) defined by

$$L_g(f) = \mu(fg).$$

Moreover, the norm of L_g, that is, the number

$$L_g = \operatorname{Sup}\{|L_g(f)| : f \in \mathscr{L}^p, N_p(f) \leq 1\}$$
$$(\text{or } \operatorname{Sup}\{|L_g(f)| : f \in \overline{\mathscr{L}}^\infty, \bar{N}_\infty(f) \leq 1\}, \text{ if } p = \infty),$$

satisfies $\|L_g\| \leq N_q(g)$ (or $\|L_g\| \leq \bar{N}_\infty(g)$ if $p = 1$). Actually, if $p = \infty$, it is immaterial for the evaluation of $\|L_g\|$ whether we regard it as a linear form on $\mathscr{L}^\infty$ or on $\overline{\mathscr{L}}^\infty$, the corresponding norms being equal in this case. An easy argument shows in fact that

$$\|L_g\| = \begin{cases} N_q(g) & \text{if } 1 < p \leq \infty, \\ \bar{N}_\infty(g) & \text{if } p = 1; \end{cases}$$

$\bar{N}_\infty(g)$ cannot here be replaced by $N_\infty(g)$ in general. On the other hand, to each g in $\overline{\mathscr{L}}^\infty$ corresponds at least one g' in $\mathscr{L}^\infty$ such that $g' = g$ l.a.e. and $N_\infty(g') = \bar{N}_\infty(g)$; for any such g' we have $L_g(f) = L_{g'}(f)$ for each f in $\mathscr{L}^1$. This means that when $p = 1$, we get the same set of linear forms on $\mathscr{L}^1$ by taking the L_g with g in $\mathscr{L}^\infty$, or the L_g with g in $\overline{\mathscr{L}}^\infty$. Even so, the fact that $\|L_g\|$ is equal to $\bar{N}_\infty(g)$, rather than $N_\infty(g)$, is one of the reasons why $\overline{\mathscr{L}}^\infty$ and its seminorm $\bar{N}_\infty$ are important.

At all events we have a natural linear mapping $g \to L_g$ of $\mathscr{L}^q$ (or of $\overline{\mathscr{L}}^\infty$ if $p = 1$) into the dual of $\mathscr{L}^p$ (or of $\overline{\mathscr{L}}^\infty$ if $p = \infty$). This mapping is not one to one in general, its kernel being the set of g in $\mathscr{L}^q$ (or $\overline{\mathscr{L}}^\infty$ if $p = 1$) satisfying $N_q(g) = 0$ (or $\bar{N}_\infty(g) = 0$ if $p = 1$). This suggests introducing the associated quotient spaces, thus rendering the mapping one to one.

The desired quotient space of $\mathscr{L}^q$ (or of $\overline{\mathscr{L}}^\infty$) necessary to render the mapping one to one is precisely the associated separated quotient space of the LCTVS $\mathscr{L}^q$ (or $\overline{\mathscr{L}}^\infty$) with the topology defined by the seminorm N_q (or $\bar{N}_\infty$). This quotient space is customarily denoted by L^q (or $\bar{L}^\infty$) and the associated norm by $\|\cdot\|_q$ (or $\overline{\|\cdot\|}_\infty$). All of these quotient spaces are Banach spaces. The reader should remember that in general the elements of L^q (or of $\bar{L}^\infty$) are *not* functions, but classes of functions modulo equality almost everywhere (or locally almost everywhere). If g is an element of $\mathscr{L}^q$ (or $\overline{\mathscr{L}}^\infty$), we denote here by $\dot{g}$ its class, an element of L^q (or $\bar{L}^\infty$). These definitions and notations apply for any q satisfying $1 \leq q \leq \infty$.

Returning to the case in which $q = p'$, we see that L_g depends only on the class determined by g in L^q (or in $\bar{L}^\infty$ if $p = 1$), and so we get a corresponding linear map $\dot{g} \to L_g = L_{\dot{g}*}$. This map is one to one by arrangement.

It should be remarked that if $\dot{g}$ belongs to L^q (or $\bar{L}^\infty$ if $p = 1$), then $L_{\dot{g}}(f) = \mu(fg)$ depends only on the class $\dot{f}$ in L^p (or $\bar{L}^\infty$ if $p = \infty$) determined by f. Thus we might also map L^q (or $\bar{L}^\infty$ if $p = 1$) isometrically and linearly into the dual of L^p (or $\bar{L}^\infty$ if $p = \infty$).

It is of the greatest importance to consider under what conditions the above maps of an $\mathscr{L}$-space (or an L-space) into the dual of the space of similar type with conjugate exponent are in fact *onto*. Nothing essential is lost in formulating the problem in terms of the $\mathscr{L}$-spaces, the corresponding results in terms of L-spaces being obvious deductions.

If $1 \leq p < \infty$, it turns out that the mapping is indeed onto. On the contrary, if $p = \infty$, the mapping is onto only in the trivial case in which T is discrete and finite. The second assertion is much the easier to prove, and is in any case of little general significance. The first statement is, however, fundamental. When T is a bounded interval on the real axis and μ the Lebesgue measure, the result has been known from the earliest days of functional analysis. Since then, the result has been extended step by step. For the original case (T a real interval), the proof (see, for example, Banach [1], p. 64) has been based upon properties of absolutely continuous functions. Extensions have depended upon the study of absolute continuity for countably additive set functions in general. The possibility that (T, μ) may not be σ-finite introduces some complications.

Our discussion will proceed in several steps. First we shall deal with the special case $p = 2$. $\mathscr{L}^2$ is a complete inner-product space (see Section 1.12) and it is indeed easy to prove the desired result in this setting. This is done in Section 4.12. The next step involves a continuation of integration theory in the direction of absolute continuity and the central theorem related to this concept, the Lebesgue-Radon-Nikodym theorem. This occupies Sections 4.13, 4.14, and 4.15. Finally, in Section 4.16, we shall see how the Lebesgue-Radon-Nikodym theorem leads easily to the desired result. We shall see en route that the results of Section 4.12 and the Lebesgue-Radon-Nikodym theorem are important for other purposes too.

4.12 The Space $\mathscr{L}^2$ and Its Dual

We are now in a position to appreciate in full detail the statements that $\mathscr{L}^2$ is a complete inner-product space (nonseparated in general), and that consequently the associated separated space L^2 is a Hilbert space. These spaces may or may not be separable.

The same is true of the spaces $\mathscr{L}^2(A, \mu)$ and $L^2(A, \mu)$, A being any μ-measurable subset of T; the case already contemplated is that in which A is T itself. In general, $\mathscr{L}^2(A, \mu)$ consists of those μ-measurable functions f defined on A and such that $\int_A^* |f(t)|^2 \, d\mu(t) < +\infty$. The functions considered may be real or complex valued; according to this choice it is customary to regard $\mathscr{L}^2(A, \mu)$ as a real or as a complex vector space. To cover both cases we may define the inner product as

$$(f \mid g) = \int_A f\bar{g} \, d\mu,$$

the bar denoting complex conjugation. That $f\bar{g}$ is integrable over A follows at once from Corollary 4.11.8 with $p = p' = 2$. The corresponding seminorm is

$$N_{2,A}(f) = \left[\int_A |f|^2 \, d\mu\right]^{1/2}.$$

By the remarks preceding Corollary 4.11.8, $\mathscr{L}^2(A, \mu)$ is complete for this seminorm. The associated separated quotient space, $L^2(A, \mu)$, has for elements equivalence classes of functions, two functions in $\mathscr{L}^2(A, \mu)$ falling into the same class if and only if they are equal almost everywhere on A.

Having said thus much, the reader may be left to interpret the results of Section 1.12 in this particular setting. The principal consequence is summarized in

4.12.1 **Theorem.** If L is any continuous linear form on $\mathscr{L}^2(A, \mu)$, there exists g in $\mathscr{L}^2(A, \mu)$ such that

$$L(f) = \int_A f\bar{g} \, d\mu \qquad (f \in \mathscr{L}^2(A, \mu)),$$

and conversely. Moreover, for any such g one has

$$\|L\| = N_{2,A}(g) = \left[\int_A |g|^2 \, d\mu\right]^{1/2}$$

4.13 Product of a Measure by a Function

Throughout this section, μ again denotes a positive Radon measure on T.

4.13.1 Let g be a function defined on T. In order that gk be μ-integrable for each k in $\mathscr{K}_+(T)$, it is obviously necessary and sufficient that g be μ-integrable over each compact subset of T. This situation is described by saying that g is *locally μ-integrable.* Then gk is μ-integrable for each k in $\mathscr{K}(T)$, and a new Radon measure λ is obtained by defining

$$\lambda(x) = \mu(gx) \qquad (x \in \mathscr{K}(T)). \tag{4.13.1}$$

This measure is denoted by $g \cdot \mu$. It is evident that $g \cdot \mu$ is real (resp. positive) if and only if g is real (resp. positive) locally almost everywhere for μ. Consequently, each decomposition $g = g_1 - g_2 + ig_3 - ig_4$, in which g_k is positive locally almost everywhere, yields a decomposition $\lambda = \lambda_1 - \lambda_2 + i\lambda_3 - i\lambda_4$ in which the λ_k are positive measures. We do not yet know that by choosing $g_1 = (\operatorname{Re} g)^+$, $g_2 = (\operatorname{Re} g)^-$, $g_3 = (\operatorname{Im} g)^+$, $g_4 = (\operatorname{Im} g)^-$ we are led to the minimal decomposition of λ, though this is in fact the case; what is clear and useful to recognize is that the λ_k derived in this way at any rate majorize the corresponding positive measures appearing in the minimal decomposition.

The problem before us is that of characterizing in some way those measures λ on T that are of the form $g \cdot \mu$. It is precisely this problem that is solved by the Lebesgue-Radon-Nikodym theorem.

We shall begin by recording certain properties shared by all the measures $g \cdot \mu$. The Lebesgue-Radon-Nikodym theorem will show that just one, apparently much weaker, property suffices to ensure that a given measure is of the stated type.

4.13.2 **Proposition.** Let $\lambda = g \cdot \mu$, where g is μ-locally integrable. Then:

(1) If $g \geq 0$ μ-l.a.e., for each positive lower semicontinuous function ϕ one has $\lambda^*(\phi) \leq \mu^*(g\phi)$; equality holds whenever ϕ is μ-σ-finite.

(2) Each relatively compact μ-negligible set is λ-negligible; in particular, each μ-locally negligible set is λ-locally negligible; and if T is σ-compact, each μ-negligible set is λ-negligible.

(3) If f is bounded, μ-measurable, and vanishes outside a compact set, then f is λ-integrable and $\lambda(f) = \mu(gf)$. In particular, if P is the set where $g \neq 0$, then $T \backslash P$ is λ-locally negligible.

(4) If gf is μ-integrable, then there exists a function f^* such that $f = f^*$ l.a.e. (λ), f^* is λ-integrable, and $\lambda(f^*) = \mu(gf)$.

(5) If f is equal l.a.e. (λ) to a λ-integrable function then there exists a λ-integrable function $f^* = f$ l.a.e. (λ), such that gf^* is μ-integrable, equal l.a.e. (μ) to gf, and

$$\lambda(f^*) = \mu(gf^*).$$

Proof. By the preceding remarks, we may assume throughout that $g \geq 0$, hence that λ is a positive measure.

(1) Since λ is positive,

$$\lambda^*(\phi) = \text{Sup}\, \lambda(x) = \text{Sup}\, \mu\,(gx) \leq \mu^*(g\phi),$$

x ranging over all functions in $\mathscr{K}_+(T)$ minorizing ϕ, gx being by hypothesis μ-integrable for each such x. To prove equality under the stated conditions, it is enough to prove it when $\mu^*(\phi) < +\infty$. For in any case we can write $\phi = \lim \phi_n$, where the ϕ_n increase and each is μ-integrable; monotone convergence then leads to the desired result. Assuming, then, that ϕ is μ-integrable, we can choose x_n in $\mathscr{K}_+(T)$ such that $x_n \uparrow x$, x_n minorizes ϕ, and $\mu(x_n) \uparrow \mu^*(\phi)$. By dropping terms we may suppose (Theorem 4.6.7) that $x_n \uparrow \phi$ a.e.(μ). So $gx_n \uparrow g\phi$ a.e. (μ). Then

$$\lambda^*(\phi) \geq \lambda(x_n) = \mu(gx_n) \uparrow \mu^*(g\phi)$$

by monotone convergence, and the result is established.

(2) Suppose that N is relatively compact and μ-negligible. Then we can find a decreasing sequence (ϕ_n) of μ-integrable lower semicontinuous functions such that ϕ_n majorizes χ_N and $\mu^*(\phi_n) \downarrow 0$. By dropping terms, we may assume that $\phi_n \downarrow 0$ a.e. (μ). We may also assume that each ϕ_n vanishes outside some fixed compact neighbourhood of N. It then follows that $g\phi_1$ is integrable and (for example from Theorem 4.8.2) that $\mu^*(g\phi_n) \downarrow 0$. At this point we apply (1) to derive $\lambda^*(\phi_n) \leq \mu^*(g\phi_n) \downarrow 0$. But $\lambda^*(N) \leq \lambda^*(\phi_n)$ and so N is λ-negligible. The rest of (2) follows directly.

(3) Suppose f vanishes outside the compact set K. f is then μ-integrable, as also is gf (the latter as a consequence of local integrability of g together with Theorem 4.8.1). We may then take a sequence (x_n) in $\mathscr{K}(T)$ such that $x_n \to f$ a.e. (μ). It can evidently be supposed that the x_n are uniformly bounded and vanish outside a fixed compact neighbourhood of K, thanks to which it will be the case that $gx_n \to gf$ in $\mathscr{L}^1(\mu)$. At the same time, it appears that (x_n) is Cauchy in $\mathscr{L}^1(\lambda)$ so that, by dropping terms if need be, we may assume that $x_n \to f'$ a.e. (λ). (2) now shows that necessarily $f' = f$ a.e. (λ). Thus f is λ-integrable and

$$\lambda(f) = \lambda(f') = \lim \lambda(x_n) = \lim \mu(gx_n) = \mu(gf),$$

as claimed. The remaining statement is a corollary.

(4) Let P be the set where $g \neq 0$; by (3), $T \setminus P$ is λ-locally negligible. Since fg is μ-integrable, the set S where $fg \neq 0$ is μ-σ-finite. Further, $S \subset P$ and $f = 0$ on $P \setminus S$. Let f' be set equal to f on S and to 0 elsewhere; equivalently, $f' = f$ on P and $f' = 0$ elsewhere. Then $f' = f$ l.a.e. (λ) and $gf' = gf$. Since f is μ-measurable on P, f' is μ-measurable. f' being zero outside the μ-σ-finite set S, we may write $f' = \lim f_n$ a.e. (μ), where each f_n is a finite linear combination of characteristic functions of relatively compact μ-measurable sets such that $|f_n| \leq |f'|$. Then $|gf_n| \leq |gf'| \leq |gf|$ and $gf_n \to gf$ a.e. (μ). Theorem 4.8.2 shows that $gf_n \to gf$ in $\mathscr{L}^1(\mu)$. On the other hand, (3) shows that each f_n is λ-integrable and

$$\lambda(|f_m - f_n|) = \mu(g\,|f_m - f_n|).$$

The sequence (f_n) is therefore Cauchy in $\mathscr{L}^1(\lambda)$. There exists therefore in $\mathscr{L}^1(\lambda)$ a function f^* such that $f_n \to f^*$ in $\mathscr{L}^1(\lambda)$; by dropping terms we may also arrange that $f_n \to f^*$ a.e. (λ). Since $f_n \to f'$ a.e. (μ), (2) forces upon us the conclusion that $f' = f^*$ l.a.e. (λ), hence $f = f^*$ l.a.e. (λ). Finally, since $gf' = gf$,

$$\lambda(f^*) = \lim \lambda(f_n) = \lim \mu(gf_n) = \mu(gf)$$

because of convergence $f_n \to f^*$ in $\mathscr{L}^1(\lambda)$ and $gf_n \to gf$ in $\mathscr{L}^1(\mu)$. This completes the proof of (4).

(5) There is evidently no loss of generality in assuming that $f \geq 0$. Then, since $T \setminus P$ is λ-locally negligible, amongst the λ-integrable functions equal l.a.e. (λ) to f there exists at least one, say f', that vanishes outside P. The relationship $f = f'$ l.a.e. (λ) shows, via (3), that $f = f'$ l.a.e. (μ) on P, and so that $gf = gf'$ l.a.e. (μ). We may choose an increasing sequence (f_n) converging a.e. (λ) to f', each f_n being bounded, λ-measurable, and 0 outside a relatively compact subset of P. By (2), the λ-measurability of f_n implies its μ-measurability (first on P, then on T itself). So (3) entails that $\lambda(f_n) = \mu(gf_n)$ for each n. By monotone convergence, $\lambda(f') = \lim \lambda(f_n) = \lim \mu(gf_n)$. Now we know that $\lim f_n = f'$ outside a λ-negligible set N, so that $\lim gf_n = gf'$ outside $P \cap N$. Since $g > 0$ on P, (3) shows that $P \cap N$ is μ-locally negligible. Thus $\lim gf_n = gf' = gf$ l.a.e. (μ). Monotone convergence now implies that $\lim gf_n$ is μ-integrable, the gf_n being μ-integrable, increasing, and $\mu(gf_n) = \lambda(f_n)$ being majorized by $\lambda(f') < +\infty$; moreover

$$\mu(\lim gf_n) = \lim \mu(gf_n) = \lambda(f').$$

Thus $h = \lim gf_n$ is μ-integrable and 0 outside a σ-compact set. Put f^* equal to $g^{-1}h$ on P and to 0 elsewhere, so that $gf^* = h$ everywhere and $= gf' = gf$ l.a.e. (μ). This implies that $f^* = f' = f$ l.a.e. (λ). Since f^* vanishes outside a σ-compact set and f' is λ-integrable, it follows that $f^* = f'$ a.e. (λ). Hence f^* is λ-integrable and

$$\lambda(f^*) = \lambda(f') = \mu(gf^*),$$

$gf^* = h$ being μ-integrable and equal l.a.e. (μ) to gf. The proof is thus complete. ▮

4.13.3 **Remarks.** If (T, μ) is not σ-finite, equality does not generally hold in Proposition 4.13.2(1). Thus, if g is μ-locally negligible and not μ-negligible, then $\mu^*(g) > 0$ and yet $\lambda^*(1) = \text{Sup } \lambda(k)$ (k varying over $\mathscr{K}_+(T)$ subject to $k \leq 1$), which is equal to $\text{Sup } \mu(gk) = \text{Sup } 0 = 0 < \mu^*(g)$. Again, although the statements (4) and (5) are annoyingly cumbersome, this feature is in general unavoidable. Nevertheless, they *do* suggest the introduction of a modified concept of integral that would permit simplified statements. It seems clear already, and will be verified in the next section, that in place of integrability for μ (or λ) one needs the concept of essential integrability, amounting to equality locally almost everywhere with an integrable function. Taking this cue, we proceed to devote a section to developing this idea in a fashion entirely parallel to that used in connection with integrability itself. Having done this, it will be easy to reformulate the contents of Proposition 4.13.2 in a more satisfying manner.

4.14 Essentially Integrable Functions

The modifications we have in mind relate to any given positive Radon measure μ on T. The procedure mimics almost exactly those used in Sections 4.6 and 4.11.

We begin by defining the essential upper integral $\bar{\mu}^*(f)$ for any function $f \geq 0$:

$$\bar{\mu}^*(f) = \text{Sup } \{\mu^*(f\chi_K) : K \subset T, K \text{ compact}\}. \tag{4.14.1}$$

Evidently, $0 \leq \bar{\mu}^*(f) \leq \mu^*(f) \leq +\infty$. A set N is locally negligible if and only if $\bar{\mu}^*(\chi_N) = 0$. Equally obvious are the following properties: if f and g are positive and $f \leq g$ l.a.e., then $\bar{\mu}^*(f) \leq \bar{\mu}^*(g)$; $\bar{\mu}^*(\alpha f) = \alpha \cdot \bar{\mu}^*(f)$ if $f \geq 0$ and α is a positive real number; $\bar{\mu}^*(f + g) \leq \bar{\mu}^*(f) + \bar{\mu}^*(g)$ for any two positive functions.

4.14.1 **Proposition.** If $f \geq 0$, the equality

$$\bar{\mu}^*(f) = \mu^*(f) \tag{4.14.2}$$

holds if *either* (a) f is lower semicontinuous, *or* (b) f vanishes outside a μ-σ-finite set.

Proof. (a) If f is lower semicontinuous, the definition of $\mu^*(f)$ assigns to it the value $\text{Sup } \mu(k)$ for k in $\mathscr{K}_+(T)$ minorizing f; if k vanishes outside the compact set K, then $k \leq f\chi_K$, hence $\mu(k) \leq \mu^*(f\chi_K) \leq \bar{\mu}^*(f)$. It follows that $\mu^*(f) \leq \bar{\mu}^*(f)$, and so (4.14.2) holds.

(b) Under the stated conditions, there exists an $\uparrow$ sequence (K_n) of compact sets such that $f = \lim f\chi_{K_n}$ a.e. (μ) and so $\mu^*(f) = \lim \mu^*(f\chi_{K_n})$, which shows at once that $\mu^*(f) \leq \bar{\mu}^*(f)$. Thus (4.14.2) is again true. ▌

The next step is define $\bar{N}_1(f)$, for any complex-valued function f on T, as $\bar{\mu}^*(|f|)$. Thus $\bar{N}_1(f) \leq N_1(f) \leq +\infty$, $\bar{N}_1$ is subadditive and $\bar{N}_1(\alpha f) = |\alpha| \cdot \bar{N}_1(f)$ for any complex number α.

$\overline{\mathscr{F}}^1 = \overline{\mathscr{F}}^1(T, \mu)$ is now defined to be the set of complex-valued functions f on T such that $\bar{N}_1(f) < +\infty$. $\overline{\mathscr{F}}^1$ is a vector subspace of C^T containing $\mathscr{F}^1 = \mathscr{F}^1(T, \mu)$, and $\bar{N}_1$ is a seminorm on $\overline{\mathscr{F}}^1$. The resulting topology is generally non-separated since $\bar{N}_1(f) = 0$ if and only if $f = 0$ l.a.e.

4.14.2 Proposition. A complex-valued function f on T belongs to $\overline{\mathscr{F}}^1$ if and only if there exists an f' in $\mathscr{F}^1$ that is equal locally almost everywhere to f, in which case $\bar{N}_1(f) = N_1(f')$.

Proof. If f' exists with the stated properties, $+\infty > N_1(f') = \bar{N}_1(f')$ (Proposition 4.14.1(b)) $= \bar{N}_1(f)$ (since $f = f'$ l.a.e.), showing that f belongs to $\overline{\mathscr{F}}^1$.

Conversely suppose that f' belongs to $\overline{\mathscr{F}}^1$. There exists then an $\uparrow$ sequence (K_n) of compact subsets of T such that $\bar{N}_1(f) = \lim \mu^*(|f|\,\chi_{K_n})$. Let A be the union of the K_n. By monotone convergence one has

$$\mu^*(|f|\,\chi_A) = \bar{N}_1(f),$$

so that $f' = f\chi_A$ belongs to $\mathscr{F}^1$. It remains only to prove that $f' = f$ l.a.e. Now if this were not the case, there would exist a compact set $K \subset T\backslash A$ such that $\alpha = \mu^*(|f|\,\chi_K) > 0$. Then $L_n = K \cup K_n$ is compact and

$$\mu^*(|f|\,\chi_{L_n}) = \mu^*(|f|\,\chi_K + |f|\,\chi_{K_n}).$$

The sets K and K_n being compact and disjoint, it is easily shown that the right-hand side here is equal to

$$\mu^*(|f|\,\chi_K) + \mu^*(|f|\,\chi_{K_n});$$

for the details, see Lemma 4.14.3 below. It appears then that

$$\mu^*(|f|\,\chi_{L_n}) \to \alpha + \bar{N}_1(f) > \bar{N}_1(f).$$

Then, however, one would be forced to conclude that

$$\mu^*(|f|\,\chi_{L_n}) > \bar{N}_1(f)$$

for all sufficiently large n, which is absurd. ▌

4.14.3 Lemma. If $f \geq 0$ and if K is the union of two disjoint compact sets K_1 and K_2, then

$$\mu^*(f\chi_K) = \mu^*(f\chi_{K_1}) + \mu^*(f\chi_{K_2}).$$

Proof. Since μ^* is subadditive, we have only to prove that the right-hand side does not exceed the left-hand side. In doing this we may assume that

$\mu^*(f\chi_K) < +\infty$. Then, given any $\varepsilon > 0$, we can choose a positive lower semicontinuous function ϕ majorizing $f\chi_K$ and such that $\mu^*(\phi) \leq \mu^*(f\chi_K) + \varepsilon$. We take disjoint open neighbourhoods U_1 and U_2 of K_1 and K_2, respectively, and put $\phi_1 = \phi\chi_{U_1}$ and $\phi_2 = \phi\chi_{U_2}$. These are positive lower semicontinuous, majorize $f\chi_{K_1}$ and $f\chi_{K_2}$, respectively, and $\phi_1 + \phi_2 \leq \phi$. Hence

$$\mu^*(f\chi_{K_1}) + \mu^*(f\chi_{K_2}) \leq \mu^*(\phi_1) + \mu^*(\phi_2) = \mu^*(\phi_1 + \phi_2) \leq \mu^*(\phi)$$
$$\leq \mu^*(f\chi_K) + \varepsilon.$$

Letting $\varepsilon \to 0$, the assertion follows. ▌

Now is the point at which to introduce $\overline{\mathscr{L}^1} = \overline{\mathscr{L}^1}(T, \mu)$ as the closure in $\overline{\mathscr{F}^1}$, relative to the topology defined by the seminorm $\bar{N}_1$, of $\mathscr{K}_C(T)$. For x in $\mathscr{K}_C(T)$ one has

$$|\mu(x)| \leq \mu(|x|) = \mu^*(|x|) = N_1(x) = \bar{N}_1(x),$$

so that μ has a unique continuous extension to $\overline{\mathscr{L}^1}$ that we denote by $\bar{\mu}$. Then

$$|\bar{\mu}(f)| \leq \bar{N}_1(f) \qquad (f \in \overline{\mathscr{L}^1}). \tag{4.14.3}$$

It is clear that $\mathscr{L}^1 = \mathscr{L}^1(T, \mu) \subset \overline{\mathscr{L}^1}$; moreover, since $\bar{\mu} = \mu$ on $\mathscr{K}_C(T)$ and both are uniquely extendible to $\mathscr{L}^1$ by continuity with respect to the seminorm N_1, it follows that

$$\bar{\mu}(f) = \mu(f) \qquad (f \in \mathscr{L}^1). \tag{4.14.4}$$

We notice too that

$$\lim_K f\chi_K = f \text{ in } \overline{\mathscr{F}^1} \qquad (f \in \overline{\mathscr{F}^1}), \tag{4.14.5}$$

whence it follows that the same is true for f in $\overline{\mathscr{L}^1}$ and hence that

$$\bar{\mu}(f) = \bar{\mu}^*(f) \qquad (f \in \overline{\mathscr{L}^1}). \tag{4.14.6}$$

The limit in (4.14.5) is to be taken following the increasing directed set of compact subsets K of T.

If a set S is μ-measurable, then $\bar{\mu}(S)$ (as defined in Subsection 4.7.5) is equal to $\bar{\mu}^*(\chi_S)$, so that S is measurable and $\bar{\mu}(S) < +\infty$ if and only if χ_S belongs to $\overline{\mathscr{L}^1}(\mu)$, in which case $\bar{\mu}(S) = \bar{\mu}(\chi_S)$.

We give now the analogue of Theorem 4.8.1.

4.14.4 **Proposition.** A complex-valued function f on T belongs to $\overline{\mathscr{L}^1}$ if and only if it is μ-measurable and $\bar{N}_1(f) < +\infty$.

Proof. If $\bar{N}_1(f) < +\infty$, then (Proposition 4.14.2) there exists $f' = f$ l.a.e. with $N_1(f') = \bar{N}_1(f) < +\infty$. Since f is measurable, so too is f'. Hence (Theorem 4.8.1), f' belongs to $\mathscr{L}^1$; but then f' belongs to $\overline{\mathscr{L}^1}$ and f, equal locally almost everywhere to f', has the same property.

Conversely, if f belongs to $\overline{\mathscr{L}^1}$, then $\bar{N}_1(f) < +\infty$ by definition. Further, there exists k in $\mathscr{K}_C(T)$ such that $\bar{N}_1(f - k) \leq \varepsilon$, $\varepsilon > 0$ being arbitrarily preassigned.

Proposition 4.14.2 again applies; let f' be as in the first part of the proof. Then $N_1(f' - k) = \bar{N}_1(f - k) \leq \varepsilon$. This shows that f' belongs to $\mathscr{L}^1$ and so f, being equal locally almost everywhere to f', is measurable. ▮

4.14.5 Corollary. In order that a complex valued function f on T belong to $\overline{\mathscr{L}}^1$, it is necessary and sufficient that there exists in $\mathscr{L}^1$ a function f' equal locally almost everywhere to f; then $\bar{\mu}(f) = \mu(f')$.

Proof. The sufficiency is evident. On the other hand, if f belongs to $\overline{\mathscr{L}}^1$, it is measurable (Proposition 4.14.4) and (Proposition 4.14.2) there exists an f' in $\mathscr{F}^1$ equal locally almost everywhere to f. f' is then measurable and so (Theorem 4.8.1) belongs to $\mathscr{L}^1$. ▮

The functions in $\overline{\mathscr{L}}^1(T, \mu)$ are said to be *essentially integrable* for μ, and $\bar{\mu}$ is the *essential integral.* The last result shows that they are characterized in precisely the manner that seemed desirable in Remarks 4.13.3. It will be evident how to introduce the spaces $\overline{\mathscr{L}}^p(T, \mu)$ for $1 < p < \infty$, and we leave to the reader the simple task of verifying for these modified notions the analogues of Fatou's lemma, Theorem 4.8.2, and the results of Section 4.11. Finally, the substance of Subsection 4.6.11 regarding the use of the term "integrable" has its analogue in respect of "essentially integrable," "a.e." being replaced by "l.a.e." throughout.

Let us now return to the matter discussed in Section 4.13 and in particular to a more satisfactory formulation of Proposition 4.13.2 in terms of essential integrability.

4.14.6 Proposition. Let μ be a positive Radon measure on T, g a positive function locally integrable for μ, P the set where $g \neq 0$, λ the positive measure $g \cdot \mu$. Then:

(1) Each μ-locally negligible set is λ-locally negligible.

(1′) A subset S of T is λ-locally negligible if and only if $P \cap S$ is μ-locally negligible.

(2) A complex-valued function f on T belongs to $\overline{\mathscr{L}}^1(\lambda)$ if and only if gf belongs to $\overline{\mathscr{L}}^1(\mu)$, in which case

$$\bar{\lambda}(f) = \bar{\mu}(gf). \tag{4.14.7}$$

(3) For any positive function f on T one has

$$\bar{\lambda}^*(f) = \bar{\mu}^*(gf). \tag{4.14.8}$$

(4) If f is a complex-valued function on T, then f is locally integrable for $\lambda = g \cdot \mu$ if and only if fg is locally integrable for μ, in which case $f \cdot \lambda = f \cdot (g \cdot \mu) = (fg) \cdot \mu$.

Proof. It is clear that (3) implies (1′), which in turn implies (1). We shall first prove (2) and then show that it implies (3). Our proof of (2) makes explicit use only of parts (4), (5), and the last part of (3) (that is, $\bar{\lambda}^*(T \backslash P) = 0$) of Proposition 4.13.2.

To prove (2) let us begin by assuming that gf belongs to $\overline{\mathscr{L}^1}(\mu)$. According to Corollary 4.14.5 there exists in $\mathscr{L}^1(\mu)$ a function $h = gf$ l.a.e. (μ). We define f' to be $g^{-1}h$ on P and to be 0 elsewhere. Then $f' = f$ l.a.e. (μ) on P and so, since $\bar{\lambda}^*(T \setminus P) = 0$, $f' = f$ l.a.e.(λ). Also $gf' = h$ belongs to $\mathscr{L}^1(\mu)$, so that (Proposition 4.13.2(4)) there exists in $\mathscr{L}^1(\lambda)$ a function $f^* = f'$ l.a.e. (λ) for which $\lambda(f^*) = \mu(gf')$. Since $gf' = h$ is equal l.a.e. (μ) to gf and $f^* = f$ l.a.e. (λ), we see that f belongs to $\overline{\mathscr{L}^1}(\lambda)$ and that (4.14.7) holds. The converse portion of (2) follows likewise from Proposition 4.13.2(5).

The deduction of (3) from (2) rests upon two simple observations refering to an arbitrary positive Radon measure μ and an arbitrary positive function f. The first is that

$$\mu^*(f) = \operatorname{Inf} \{\mu(h) : h \in \mathscr{L}^1(\mu), h \geq f\},$$

it being understood that $\operatorname{Inf} \emptyset = +\infty$ and $\operatorname{Inf} \{+\infty\} = +\infty$; moreover, one may alter the relation $h \geq f$ into $h \geq f$ a.e. (μ). The second observation, which follows at once from this combined with Proposition 4.14.2, is that

$$\bar{\mu}^*(f) = \operatorname{Inf} \{\bar{\mu}(h) : h \in \overline{\mathscr{L}^1}(\mu), h \geq f \text{ l.a.e. } (\mu)\},$$

with the same conventions as before. In view of this it remains to remark merely that if v is any positive function and we define u to be $g^{-1}v$ on P and to be 0 elsewhere, then $u \geq f$ l.a.e. (λ) if and only if $v \geq gf$ l.a.e. (μ) and (by (2)) u belongs to $\overline{\mathscr{L}^1}(\lambda)$ if and only if v belongs to $\overline{\mathscr{L}^1}(\mu)$, in which case $\bar{\lambda}(u) = \bar{\mu}(v)$. Accordingly

$$\bar{\lambda}^*(f) = \operatorname{Inf} \bar{\lambda}(u) = \operatorname{Inf} \bar{\mu}(v) = \bar{\mu}^*(gf),$$

which is (3).

It remains to establish (4). Now f is locally integrable for $\lambda = g \cdot \mu$ if and only if, for each x in $\mathscr{K}(T)$, fx is integrable for $g \cdot \mu$; since x has a compact support, and so therefore has fx, this is so if and only if fx is essentially integrable for $g \cdot \mu$; by (2) this is so if and only if gfx is essentially integrable for μ; that is (gfx having a compact support) if and only if gfx is integrable for μ; finally, this signifies precisely that gf is locally integrable for μ. Moreover we have by definition for any x in $\mathscr{K}(T)$:

$$f \cdot \lambda(x) = \lambda(fx) = \mu(gfx),$$

thanks to (4.14.7) and the fact that fx and gfx have compact supports, and the last term is $(gf) \cdot \mu(x)$. Thus $f \cdot \lambda = f \cdot (g \cdot \mu) = (fg) \cdot \mu$, as asserted. ▌

The Lebesgue-Radon-Nikodym theorem, the proof of which will be tackled in the next section, asserts that condition (1) alone is enough to ensure that the positive measure λ is of the form $g \cdot \mu$ for some $g \geq 0$ locally integrable for μ, so that (1) in fact already implies all the remaining assertions of Proposition 4.14.6 and Proposition 4.13.2. There are some consequences of Proposition 4.14.6 that will be useful in the next section and elsewhere and that we propose to discuss forthwith.

4.14.7 Restriction of a Measure. Let μ be a positive measure on T, M a subset of T. The function $g = \chi_M$ is locally integrable for μ if and only if M is μ-measurable. In this case we may apply Proposition 4.14.6 to the measure

$$\mu_M = \chi_M \cdot \mu, \tag{4.14.9}$$

which is termed the *restriction* of μ to M. According to the said proposition,

$$\bar{\mu}_M^*(S) = \bar{\mu}^*(M \cap S) \tag{4.14.10}$$

for each subset S of T. If S is open we may (Proposition 4.14.1(a)) write $\mu_M^*(S)$ in place of $\bar{\mu}_M^*(S)$, whence it follows that the support of μ_M is the smallest closed set S such that $M \cap (T \backslash S)$ is μ-locally negligible. In particular, if M is closed, the support of μ_M is contained in M. Moreover, a set S is μ_M-measurable if and only if $M \cap S$ is μ-measurable; and a function f is μ_M-measurable if and only if f is μ-measurable on M.

4.14.8 It is fundamental in the next section to know that one can always, at the expense of neglecting locally negligible sets, decompose T in a certain way into the union of disjoint compact subsets. That this is possible when (T, μ) is σ-finite is rather easily established, but we have to reckon with the contrary case.

A set of subsets of T is said to be locally countable if each compact subset of T meets at most countably many members of the given set.

4.14.9 Proposition. Let μ be a positive Radon measure on T. There exists a locally countable set $\mathfrak{X}$ of compact subsets of T with the following properties:

(1) Any two distinct sets belonging to $\mathfrak{X}$ are disjoint.

(2) $\mu(X \cap U) > 0$ for each X in $\mathfrak{X}$ and each open set U meeting X.

(3) $T \backslash \bigcup_{X \in \mathfrak{X}} X$ is μ-locally negligible.

Proof. We consider all sets of compact subsets of T having properties (1) and (2). Such sets do exist: for example, the set consisting solely of any one compact subset of the support of μ. These sets may be partially ordered by inclusion, and it is evident that they are thus inductively ordered. Choose any maximal set of the stated type, say $\mathfrak{X}$. We have to show that $\mathfrak{X}$ is locally countable and that (3) is true.

Let U be any relatively compact open set. Then $\sum_i \mu(X_i \cap U) \le \mu(U) < +\infty$ for any finite number of X_i in $\mathfrak{X}$, that is, that $\sum_{X \in \mathfrak{X}} \mu(X \cap U) \le \mu(U) < +\infty$. But then $\mu(X \cap U) > 0$ for at most countably many X which, in view of (2), implies that $\mathfrak{X}$ is locally countable.

Finally, if $N = T \backslash \bigcup_{X \in \mathfrak{X}} X$ were not locally negligible, there would exist a compact set $K \subset N$ for which $\mu(K) > 0$. The restriction $\mu_K = \chi_K \cdot \mu$ would be nonzero, and its support S would be a nonvoid compact set contained in K. For any open set U meeting S,

$$\mu_K(U) = \mu_K(S \cap U) > 0$$

and so

$$\mu(S \cap U) \ge \mu_K(S \cap U) > 0.$$

Thus $\mathfrak{X}' = \mathfrak{X} \cup \{S\}$ would enjoy properties (1) and (2), so that maximality of $\mathfrak{X}$ would demand that S should already belong to $\mathfrak{X}$. Since this conclusion is obviously absurd, the proof is completed by contradiction. ▌

4.15 The Lebesgue-Radon-Nikodym Theorem

This central result amounts essentially to Proposition 4.14.6 together with its (deeper) converse.

4.15.1 **Theorem.** Let λ and μ be positive Radon measures on a locally compact space T. In order that there should exist a function $g \geq 0$ locally integrable for μ such that $\lambda = g \cdot \mu$, it is necessary and sufficient that each μ-locally negligible set be λ-locally negligible (that is, that condition (1) of Proposition 4.14.6 be fulfilled).

Proof. The necessity has already been established in Proposition 4.14.6. To prove sufficiency, let it be observed first that the stated condition arranges that μ-measurability implies λ-measurability. Indeed, if S is a μ-measurable set and K any compact set, then $S \cap K$ is μ-integrable. Hence, as is easily seen, there exists a relatively compact set $P \supset S \cap K$ that is the intersection of a sequence of relatively compact open sets and such that $\mu(P\backslash(S \cap K)) = 0$. Consequently, $\lambda(P\backslash(S \cap K)) = 0$ also. P is integrable for any positive Radon measure, and it therefore follows that $S \cap K$ is λ-integrable, so that S is λ-measurable.

If we introduce the positive measure $\rho = \mu + \lambda$, then ρ-measurability is equivalent to μ-measurability, and $\lambda \leq \rho$. Let A be any compact set (or, more generally, any ρ-integrable set). The mapping $f \to \lambda(f)$ is a continuous linear form on $\mathscr{L}^2(A, \rho)$ since

$$\begin{aligned}|\lambda(f)| &\leq \lambda(|f|) \leq \lambda(|f|^2)^{1/2} \cdot \lambda(\chi_A)^{1/2} \\ &\leq \rho(A)^{1/2} \cdot \rho(|f|^2)^{1/2}.\end{aligned}$$

By Theorem 4.12.1, there exists a function ϕ_A in $\mathscr{L}^2(A, \rho)$ such that

$$\lambda(f) = \rho(\phi_A f) \qquad (f \in \mathscr{L}^2(A, \rho)).$$

Since $\lambda \leq \rho$, if we put $\phi_A = 0$ outside A, we shall have $0 \leq \phi_A \leq 1$ a.e. (ρ).

We show next that $\phi_A < 1$ a.e. (ρ). Let E, necessarily a subset of A, be the set where $\phi_A = 1$. E is ρ-integrable. Taking $f = \chi_E$ we shall have

$$\lambda(E) = \lambda(\chi_E) = \rho(\phi_A \chi_E) = \rho(\chi_E) = \rho(E) = \mu(E) + \lambda(E).$$

If $\rho(E)$ were strictly positive, this would imply that $\mu(E) = 0$ and yet $\lambda(E) > 0$, violating our hypothesis. Thus $\rho(E) = 0$ and we may arrange that $0 \leq \phi_A < 1$ everywhere.

Now the equality $\lambda(f) = \rho(\phi_A f)$ may be written as

$$\lambda((1 - \phi_A)f) = \mu(\phi_A f) \qquad (f \in \mathscr{L}^2(A, \rho))$$

or, putting $g_A = \phi_A/(1 - \phi_A)$, g_A is positive and μ-measurable and

$$\lambda(h) = \mu(g_A h) \tag{4.15.1}$$

for all h such that $(1 - \phi_A)^{-1}h$ belongs to $\mathscr{L}^2(A, \rho)$. The reader will notice that g_A vanishes outside A and is μ-integrable.

The set A can be represented as the union of an increasing sequence (A_n) of ρ-integrable subsets on each of which $1 - \phi_A$ is bounded away from zero. Accordingly, if h is bounded and μ-measurable and vanishes outside some A_n, then $(1 - \phi_A)^{-1}\, h$ belongs to $\mathscr{L}^2(A, \rho)$ and so (4.15.1) applies. If f is positive, μ-measurable and 0 outside A, $h_n = \text{Inf}\,(f\chi_{A_n}, n)$ is bounded, ρ-measurable and 0 outside A_n; moreover, $h_n \uparrow f$. Monotone convergence leads now from (4.15.1), valid for $h = h_n$ $(n = 1, 2, \cdots)$, to the relation

$$\lambda^*(f) = \mu^*(g_A f). \tag{4.15.2}$$

Formula (4.15.2) holds for any ρ-integrable A and any positive f that is μ-measurable and 0 outside A.

At this stage we appeal to Proposition 4.14.9. We take $\mathfrak{X}$ as there stipulated. The preceding portion of this proof allows us to assign a function g_X to each X in $\mathfrak{X}$, g_X being positive, μ-integrable, and 0 outside X. We define g to be equal to g_X on X, for each X in $\mathfrak{X}$, and to 0 on $N = T\backslash \bigcup_{X \in \mathfrak{X}} X$. It is plain that g is positive and μ-measurable. Let f be any positive, bounded, μ-measurable function vanishing outside some compact set K. The set of X's meeting K is at most countable, say $X_1, X_2, \cdots$. Then $f = \sum_n f\chi_{X_n}$, $gf = \sum_n g_{X_n} f$, $f\chi_{X_n}$ is λ-integrable, and $g_{X_n} f$ is μ-integrable. Hence (4.15.2) and the principle of monotone convergence give

$$\lambda(f) = \sum_n \lambda(f\chi_{X_n}) = \sum_n \mu(g_{X_n} f) = \mu(gf), \tag{4.15.3}$$

gf being necessarily μ-integrable. In particular, g is locally integrable for μ, as follows on taking $f = \chi_H$ (H any compact set) in (4.15.3). This last formula, being valid for f in $\mathscr{K}_+(T)$, shows too that $\lambda = g \cdot \mu$, as had to be established. ∎

4.15.2 Local Absolute Continuity; Mutually Singular Measures. Whenever it is the case that two complex Radon measures λ and μ are so related that each μ-locally negligible set is λ-locally negligible, it is usual to say that λ is *locally absolutely continuous relative to* μ, in symbols: λ is LAC(μ). When λ has the minimal decomposition $\lambda_1 - \lambda_2 + i\lambda_3 - i\lambda_4$, λ is LAC(μ) if and only if each positive measure λ_k is LAC($|\mu|$), or, equivalently, $|\lambda|$ is LAC($|\mu|$). In this case it follows easily from Theorem 4.15.1 that $\lambda = g \cdot \mu$, where g is a complex-valued function locally integrable for μ (that is, what is the same thing by definition, for $|\mu|$); it is moreover easily deduced from Proposition 4.15.3 below that if $\mu \geq 0$ the minimal decomposition of λ is given by $\lambda_k = g_k \cdot \mu$, where

$$g_1 = (\text{Re}\, g)^+, \qquad g_2 = (\text{Re}\, g)^-, \qquad g_3 = (\text{Im}\, g)^+, \qquad g_4 = (\text{Im}\, g)^-.$$

The case in which λ is real and μ positive is especially important and forms the subject of the said Proposition 4.15.3. Before dealing with the details, however, we wish to add a comment regarding the above definition of local absolute continuity and also introduce a relationship between two measures that is, so to speak, the complete antithesis of local absolute continuity. With the aid of this latter concept we can supplement the Lebesgue-Radon-Nikodym theorem in a manner due to Lebesgue (see Subsections 4.15.7, 4.15.8, and 4.15.9).

To begin with it may strike the reader as rather odd that the implication "$\bar{\mu}^*(E) = 0 \Rightarrow \bar{\lambda}^*(E) = 0$" should be described as a variety of continuity—namely, the local absolute continuity of the positive measure λ relative to the positive measure μ. It would seem more natural to attach such a name to an implication of the type $\bar{\lambda}^*(E) \to 0$ as $\bar{\mu}^*(E) \to 0$. While this is not precisely the formulation required (unless λ is bounded), it is very easy to prove the following fact:

► Given two Radon measures λ and μ, λ is LAC(μ) if and only if to each compact set K and each $\varepsilon > 0$ corresponds $\delta > 0$ such that $E \subset K$ and $|\mu|^*(E) < \delta$ together entail $|\lambda|^*(E) < \varepsilon$.

(It is this latter condition, referring to K, that qualifies for the title "absolute continuity on K"; local absolute continuity amounts to absolute continuity on each compact set K.) Indeed, it is evident that this condition ensures that any μ-locally negligible set is λ-locally negligible, that is, that λ is LAC(μ). In proving the converse we may assume that λ and μ are positive. In this case, if λ is LAC(μ), we may write $\lambda = g \cdot \mu$ where $g \geq 0$ is locally integrable for μ. In particular, $g\chi_K$ is μ-integrable. We can therefore choose f in $\mathscr{K}_+(T)$ such that $\mu^*(|g - f|\,\chi_K) < \frac{1}{2}\varepsilon$. For any $E \subset K$ we have $\lambda^*(E) = \mu^*(g\chi_E)$ and hence $|\lambda^*(E) - \mu^*(f\chi_E)| < \frac{1}{2}\varepsilon$. On the other hand, $\mu^*(f\chi_E) \leq M \cdot \mu^*(E)$, where M is the maximum of f. Thus $\lambda^*(E) < \frac{1}{2}\varepsilon + M \cdot \mu^*(E)$ for any $E \subset K$, whence the assertion follows. In proving this converse, it suffices to assume merely that K is essentially λ-integrable (not necessarily compact), replacing λ^* and μ^* by $\bar{\lambda}^*$ and $\bar{\mu}^*$ throughout.

We turn next to the introduction of the concept of singularity of measures. Let λ and μ be two complex Radon measures on T. We say that λ and μ are *mutually singular* (*estranged* in Bourbaki's terminology) if there exist disjoint sets L and M carrying λ and μ, respectively—that is, such that $T \backslash L$ and $T \backslash M$ are $|\lambda|$- and $|\mu|$-locally negligible, respectively. (The reader should beware the error involved in confusing the statement that L carries λ in the above sense—that is, $T \backslash L$ is $|\lambda|$-locally negligible—with the assertion that L is, or contains, the support of λ; the latter set is precisely the smallest *closed* subset of T that carries λ.) It is evident that if λ and μ are mutually singular, while also λ is LAC(μ), then $\lambda = 0$. Also, if λ and μ and also λ' and μ are mutually singular, then $\lambda + \lambda'$ and μ are mutually singular. Illustrations of this new concept will appear throughout the remainder of this section.

4.15.3 **Proposition.** Let μ be a positive and λ a real Radon measure on T, and suppose that λ is LAC(μ) and has the minimal decomposition $\lambda = \lambda^+ - \lambda^-$.

Then $\lambda = g \cdot \mu$, $\lambda^+ = g^+ \cdot \mu$, and $\lambda^- = g^- \cdot \mu$ where g is a real-valued function locally integrable for μ. The measures λ^+ and λ^- are mutually singular.

Proof. Since λ is LAC(μ), the same is true of λ^+ and λ^- separately. So (Theorem 4.15.1) we have $\lambda^+ = p \cdot \mu$ and $\lambda^- = q \cdot \mu$, where p and q are positive and locally integrable for μ. Correspondingly, $\lambda = g \cdot \mu$ where $g = p - q$ is locally integrable for μ. Let P and Q be the sets where $p > 0$ and $q > 0$, respectively: we claim that $P \cap Q$ is μ-locally negligible. Indeed if this were not so, there would exist a compact set K and a number $c > 0$ such that $\mu(K) > 0$ and $p \geq c$, $q \geq c$ on K. Then the measures

$$\alpha = (p - c\chi_K) \cdot \mu, \qquad \beta = (q - c\chi_K) \cdot \mu$$

are positive, $\lambda = \alpha - \beta$, and yet $\alpha - \lambda^+$ and $\beta - \lambda^-$ are not positive. This would contradict the minimality of the decomposition $\lambda = \lambda^+ - \lambda^-$. Thus $P \cap Q$ is μ-locally negligible. By altering p and q on μ-locally negligible sets, thereby leaving unaltered the measures $p \cdot \mu$ and $q \cdot \mu$, we may arrange that $P \cap Q = \emptyset$. Then, however, $g = p - q$ is equal to p on P, to $-q$ on Q, and to 0 elsewhere. Thus $p = g^+$ and $q = g^-$. Finally, $\lambda^+ = g^+ \cdot \mu$ and $\lambda^- = g^- \cdot \mu$ are carried by the sets P and Q, respectively; these sets being disjoint, λ^+ and λ^- are mutually singular. ▌

4.15.4 **Corollary.** If λ and μ are as in Proposition 4.15.3, then $|\lambda|^*(1) = \bar{\mu}^*(|g|)$. In particular, if λ is bounded (that is, $|\lambda|^*(1) < +\infty$), then g is equal l.a.e. (μ) to a function g_0 in $\mathscr{L}^1(\mu)$ and $\mu^*(|g_0|) = |\lambda|(1)$.

4.15.5 **Corollary.** Let λ be any real Radon measure on T. Then T is the union of disjoint λ-measurable sets P_+ and P_- such that $\lambda^+ = |\lambda|_{P_+}$ and $\lambda^- = |\lambda|_{P_-}$, so that

$$\lambda_{P_+} = |\lambda|_{P_+} \geq 0, \quad \lambda_{P_-} = -|\lambda|_{P_-} \leq 0.$$

Proof. We apply Proposition 4.15.3, taking therein $\mu = |\lambda|$. The sets P and Q appearing in the proof of that proposition are none other than the sets $P_+ = \{g > 0\} = \{g^+ > 0\}$ and $P_- = \{g < 0\} = \{g^- > 0\}$, respectively. Since $|\lambda| = \lambda^+ + \lambda^- = (g^+ + g^-) \cdot |\lambda|$, it follows that now $g^+ + g^- = 1$ l.a.e. (λ). Supposing that $g^+ + g^- = 1$ except on N, where N is λ-locally negligible, we leave g unaltered on $T \backslash N$ and place it equal to 1 on N. It is then still true that $\lambda^+ = g^+ \cdot |\lambda|$ and $\lambda^- = g^- \cdot |\lambda|$, and further $g^+ + g^- = 1$ everywhere. Consequently at points of P_+, where $g^- = 0$, g^+ must take the value 1. Thus $g^+ = \chi_{P_+}$. Similarly $g^- = \chi_{P_-}$. Hence $\lambda^+ = \chi_{P_+} \cdot |\lambda| = |\lambda|_{P_+}$, and so on. Since P_+ and P_- are disjoint and have union equal to T, the proof is complete. ▌

4.15.6 **Remark.** Suppose λ is a real Radon measure LAC(μ). From Corollary 4.15.4 we have $\lambda = g \cdot \mu$, where g is real valued and locally integrable for μ. In this case $|\lambda|^*(1) = \bar{\mu}^*(|g|)$ is equal to

$$n(\lambda) = \text{Sup}\, |\lambda(x)|$$

x ranging over the functions in $\mathscr{K}_C(T)$ (or $\mathscr{K}_R(T)$) satisfying $\|x\|_\infty = \operatorname{Sup}_{t \in T} |x(t)| \leq 1$. If, however, λ is a complex measure, g is complex valued, say $g = u + iv$. Then $|\lambda|^* (1) = \bar{\mu}^*(|u|) + \bar{\mu}^*(|v|)$ is in general bigger than $n(\lambda)$. Nevertheless, we still have $n(\lambda) \geq 2^{-1/2} |\lambda|^*(1)$, so that $n(\lambda)$ and $|\lambda|^*(1)$ define equivalent norms. With these remarks in mind one often identifies the space of bounded measures that are LAC(μ) with the Lebesgue space $L^1(\mu)$. A way of avoiding the discrepancy is to redefine $|\lambda|$, where $\lambda = \alpha + i\beta$, as $\sqrt{\alpha^2 + \beta^2}$ (see Subsection 4.15.13).

4.15.7 The Lebesgue Decomposition. Suppose we are given two complex Radon measures λ and μ on T. The Lebesgue-Radon-Nikodym theorem investigates the case in which λ is LAC(μ). We shall now show that in any case λ can be decomposed uniquely into the sum of two measures λ_1 and λ_2 such that λ_1 and μ are mutually singular while λ_2 is LAC(μ).

4.15.8 Theorem. Let λ and μ be complex Radon measures on T. There exists a unique decomposition

$$\lambda = \lambda_1 + \lambda_2, \tag{4.15.4}$$

where λ_1 and μ are mutually singular and λ_2 is LAC(μ).

Proof. It is clear from the definitions in Subsection 4.15.2 that there can exist at most one such decomposition, and that in proving the existence of such a decomposition it suffices to consider the case in which both λ and μ are positive. In this case we can modify the proof of Theorem 4.15.1 in the following way.

Let us introduce as before the measure $\rho = \mu + \lambda$ and let A be any compact set. We conclude as before that there exists a function ϕ_A in $\mathscr{L}^2(A, \rho)$ such that

$$\lambda(f) = \rho(\phi_A f) \tag{4.15.5}$$

for f in $\mathscr{L}^2(A, \rho)$. It may be supposed, as before, that $0 \leq \phi_A \leq 1$ and $\phi_A = 0$ outside A. Let A_1 (denoted by E in the proof of Theorem 4.15.1) be the set where $\phi_A = 1$, A_2 the remainder of A. We can no longer use local absolute continuity of λ relative to μ to show that $\rho(A_1) = 0$. Nevertheless (4.15.5) gives

$$\lambda(A_1) = \rho(\phi_A \chi_{A_1}) = \rho(\chi_{A_1}) = \rho(A_1) = \lambda(A_1) + \mu(A_1)$$

and hence $\mu(A_1) = 0$. Thus A_1 is μ-negligible.

On the other hand, suppose $S \subset A_2$ satisfies $\mu(S) = 0$. Then

$$\lambda(S) = \rho(\phi_A \chi_S) = \mu(\phi_A \chi_S) + \lambda(\phi_A \chi_S) = \lambda(\phi_A \chi_S).$$

Since $1 - \phi_A > 0$ on S, it follows that $\lambda(S) = 0$.

At this point we introduce the partition $\mathfrak{X}$ of T into compact subsets X as before. Each X we partition into X_1 and X_2 according to the manner just described for A. We know that $N = T \setminus \bigcup X$ is μ-locally negligible. We put $T_1 = \bigcup X_1 \cup N$, $T_2 = \bigcup X_2$. Since $\mathfrak{X}$ is locally countable, T_1 and T_2 are μ-measurable. They are disjoint and have union equal to T. Let $\lambda_1 = \lambda_{T_1}$ and $\lambda_2 = \lambda_{T_2}$. Obviously, $\lambda = \lambda_1 + \lambda_2$.

Now T_1 is μ-locally negligible since, as we have shown above, $\mu(X_1) = 0$ and since $\mathfrak{X}$ is locally countable. Hence λ_1 and μ are mutually singular, λ_1 being carried by T_1.

Finally, λ_2 is LAC(μ). For suppose that we let S be μ-locally negligible and let K be any compact set. Then we have $\mu(S \cap K \cap T_2) \leq \mu(S \cap K) = 0$ and $\lambda_2^*(S \cap K) = \lambda^*(S \cap K \cap T_2)$. Now the compact set K meets only a countable number of sets X, say, X^n $(n = 1, 2, \cdots)$, and so $S \cap K \cap T_2$ is contained in the union of the $S \cap K \cap X_2^n$. Since $\mu(S \cap K \cap X_2^n) = 0$ for each n, so $\lambda^*(S \cap K \cap X_2^n) = 0$ for each n (as was proved earlier with X_2^n replaced by A), and hence $\lambda_2^*(S \cap K) = 0$. Thus S is λ_2-locally negligible and local absolute continuity is proved. ▌

4.15.9 **Remark.** With the above notations, the Lebesgue-Radon-Nikodym theorem shows that $\lambda_2 = g \cdot \mu$, where g is locally integrable for μ and may be assumed to vanish outside T_2—that is, on T_1; and we have

$$\lambda = \lambda_1 + g \cdot \mu = \lambda_{T_1} + g \cdot \mu_{T_2},$$

showing explicitly how λ is the sum of two measures carried by T_1 and by T_2, respectively.

4.15.10 **LBV and LAC Functions of a Real Variable.** The Lebesgue-Radon-Nikodym theorem and the Lebesgue decomposition may be applied to Lebesgue-Stieltjes measures in such a way as to relate these theorems with several well-known results in real function theory. We adopt the notations used in Subsection 4.4.4.

Suppose first that f is LAC on T. It is then evident from Example 4.7.7 that the associated measure μ_f is LAC relative to Lebesgue measure on T. Consequently the Lebesgue-Radon-Nikodym theorem gives $\mu_f = g \cdot \mu$, where μ is Lebesgue measure on T and g is locally integrable for μ. Since f is continuous, the results of Example 4.7.7 show that

$$f(t) = f(a) + \int_a^t g(s)\, ds \tag{4.15.6}$$

whenever t and a are points of T. (In the integral we write ds in place of $d\mu(s)$, following the usual notation.)

Now it is shown in books on real function theory (for example, Natanson [1], p. 253, Th. 2) that the function $\int_a^t g(s)\, ds$ has almost everywhere a finite derivative which is equal almost everywhere to g. We conclude therefore that if f is LAC on T, then the derivative f' exists almost everywhere, is equal almost everywhere to a function that is locally integrable on T and that we may denote still by f', and that

$$f(t) = f(a) + \int_a^t f'(s)\, ds \tag{4.15.6'}$$

for any two points of T. Conversely—and this is evident—the formula (4.15.6) holds only if f is LAC on T.

It is easily verified that the product fg is LAC on T whenever both f and g have this property, and that $(fg)' = f'g + fg'$ a.e. (in fact at all points where both f' and g' exist). By integration one obtains the usual formula for integration by parts:

$$\int_a^b f(t)g'(t)\,dt = f(b)g(b) - f(a)g(a) - \int_a^b f'(t)g(t)\,dt \tag{4.15.7}$$

for any compact interval $[a, b]$ lying in T.

Let us consider next the more general case in which we know merely that f is LBV on T. In this case Remark 4.15.9 leads to a decomposition $\mu_f = \sigma + g \cdot \mu$, where g is locally integrable (for μ) and σ is singular (relative to μ). The passage from this relationship between measures to the corresponding relationship between functions, analogous to (4.15.6′), is somewhat more complex. Since the result will never be needed in this book, we merely indicate the steps; for more details see Natanson [1], Ch. IX. To begin with, and constituting the major step, is the fact that the singularity of σ relative to μ signifies that for each point t one has

$$\lim \frac{\sigma(E_n)}{\mu(E_n)} = 0$$

for any sequence (E_n) of closed sets such that, for some sequence (I_n) of open intervals containing t, it is true that

$$E_n \subset I_n, \qquad \mu(I_n) \to 0, \qquad \operatorname{Inf} \frac{\mu(E_n)}{\mu(I_n)} > 0.$$

This says that a sort of generalized derivative of σ with respect to μ vanishes at t. (For a proof see, for example, McShane [1], p. 381.) Using this result one finds that, corresponding to (4.15.6′), one has a formula

$$f(t) = S(t) + \int_a^t f'(s)\,ds, \tag{4.15.6″}$$

wherein the function f' is locally integrable and is equal almost everywhere to the ordinary derivative of f, and the LBV function S is singular in the sense that $S' = 0$ a.e. This result can be, and usually is, treated from the point of real variable theory without any reference to Lebesgue-Stieltjes measures; such a treatment is found, for example, in Natanson loc. cit.

4.15.11 **Application of the Lebesgue-Radon-Nikodym Theorem: Functions of Measures.** The aim is to use the Lebesgue-Radon-Nikodym theorem in order to define $H(\mu_1, \cdots, \mu_n)$ when $(\mu_i)_{1 \le i \le n}$ is a given family of real measures on T and $H = H(x_1, \cdots, x_n)$ is a given positive-homogeneous function on R^n. (The case of complex measures may be dealt with likewise, provided H is defined on C^n.) Positive homogeneity of H means that

$$H(\alpha x_1, \cdots, \alpha x_n) = \alpha \cdot H(x_1, \cdots, x_n)$$

for $x = (x_1, \cdots, x_n)$ in R^n and positive numbers α.

Our definition is based upon the ensuing proposition.

4.15.12 **Proposition.** Let H and the μ_i $(1 \leq i \leq n)$ be as above. Choose any two positive Radon measures λ_1 and λ_2 on T such that $|\mu_i| \leq \lambda_k$ $(1 \leq i \leq n,$ $1 \leq k \leq 2)$ and write correspondingly $\mu_i = f_{ik} \cdot \lambda_k$, where f_{ik} is λ_k-measurable and $|f_{ik}| \leq 1$. Then $H(f_{11}, \cdots, f_{n1})$ is locally integrable for λ_1 if and only if $H(f_{12}, \cdots, f_{n2})$ is locally integrable for λ_2, in which case

$$H(f_{11}, \cdots, f_{n1}) \cdot \lambda_1 = H(f_{12}, \cdots, f_{n2}) \cdot \lambda_2. \tag{4.15.8}$$

Proof. Put $\lambda = \lambda_1 + \lambda_2$. We may compare the situation involving λ_k ($k = 1$ or 2) with that involving λ, and thus indirectly compare the situations involving λ_1 and λ_2. In this way one is reduced to the case in which $\lambda_1 \leq \lambda_2$. This being so, we may write $\lambda_1 = g \cdot \lambda_2$ where g is λ_2-measurable and $0 \leq g \leq 1$. So $\mu_i = f_{i1} \cdot \lambda_1 = f_{i1} \cdot (g \cdot \lambda_2) = (f_{i1}g) \cdot \lambda_2$ by Proposition 4.14.6(4). Therefore $f_{i2} = f_{i1}g$ l.a.e. (λ_2) and hence also

$$H(f_{12}, \cdots, f_{n2}) = g \cdot H(f_{11}, \cdots, f_{n1}) \quad \text{l.a.e. } (\lambda_2).$$

Using Proposition 4.14.6(4) again we see then that $H(f_{12}, \cdots, f_{n2})$ is locally integrable for λ_2 if and only if $H(f_{11}, \cdots, f_{n1})$ is locally integrable for $g \cdot \lambda_2 = \lambda_1$, in which case (4.15.8) holds. ▌

As a result of this proposition we see that the $H(f_1, \cdots, f_n) \cdot \lambda$ depends only on H and the family (μ_i), being independent of the choice of the positive measure λ satisfying $|\mu_i| \leq \lambda$ for each i and the subsidiary choice of functions f_i such that $\mu_i = f_i \cdot \lambda$ (the f_i existing as a consequence of the Lebesgue-Radon-Nikodym theorem). This uniquely determined measure is denoted by $H(\mu_1, \cdots, \mu_n)$.

The reader will verify easily that one obtains by this process the measures μ^+, μ^-, $|\mu|$ (when $n = 1$) and $\mu_1 + \mu_2$ (when $n = 2$), as well as various other combinations such as Sup (μ_1, μ_2) and Inf (μ_1, μ_2); and that this mode of definition is consistent with the more direct ones given earlier.

4.15.13 **Example.** We consider the case in which

$$H(x_1, \cdots, x_n) = (x_1^2 + \cdots + x_n^2)^{1/2}.$$

Since $H \geq 0$, the process leads to a positive measure

$$\mu = (\mu_1^2 + \cdots + \mu_n^2)^{1/2}$$

which, since $|x_i| \leq H(x)$, satisfies $|\mu_i| \leq \mu$. In this case therefore we can take $\lambda = \mu$ in the construction, writing $\mu_i = f_i \cdot \mu$. Then

$$\mu = (f_1^2 + \cdots + f_n^2) \cdot \mu$$

and so $(f_1^2 + \cdots + f_n^2)^{1/2} = 1$ l.a.e. (μ).

Conversely, suppose that α is a positive measure such that $\mu_i = g_i \cdot \alpha$ and $g_1^2 + \cdots + g_n^2 = 1$ l.a.e. (α). Then $|\mu_i| = |g_i| \cdot \alpha \leq \alpha$ and so

$$\mu = (\mu_1^2 + \cdots + \mu_n^2)^{1/2} = (g_1^2 + \cdots + g_n^2) \cdot \alpha = \alpha,$$

that is, $\mu = \alpha$. Furthermore $\mu_i = f_i \cdot \mu = g_i \cdot \alpha = g_i \cdot \mu$, so that $f_i = g_i$ l.a.e. (μ).

Thus, given $\mu_1, \cdots, \mu_n$, there exists just one positive measure μ such that $\mu_i = f_i \cdot \mu$ $(1 \leq i \leq n)$ with $f_1^2 + \cdots + f_n^2 = 1$ l.a.e. (μ); this μ is $(\mu_1^2 + \cdots + \mu_n^2)^{1/2}$.

4.15.14 Application to Surface Measures. Let G be a bounded open set in R^n with frontier F, and let E be the vector space of real-valued indefinitely differentiable functions on R^n having compact supports in R^n. We denote by λ the Lebesgue measure on R^n and by x_i $(1 \leq i \leq n)$ the natural coordinate functions on R^n. We put D_i for the partial differential operator $\partial/\partial x_i$. We suppose, finally, that there exists a number $M \geq 0$ such that

$$\left| \int_G D_i u \cdot d\lambda \right| \leq M \cdot \operatorname*{Sup}_{t \in F} |u(t)| \qquad (u \in E, 1 \leq i \leq n). \tag{4.15.9}$$

This condition is fulfilled for all sufficiently simple bounded open sets G, being true if, for example, there exists a natural number r such that every (or even merely "almost every") straight line parallel to a coordinate axis meets F in at most r points. In this case one of the standard proofs of Green's formula applies. In any case, (4.15.9) is clearly a quantitatively crude form of Green's formula in which the boundary values of u enter only to the extent of a majorization. We shall see how to apply the result of Subsection 4.15.13 so as to introduce a surface measure on F, a unit normal defined almost everywhere on F, and a more precise form of Green's formula.

To this end we introduce the Banach space $C(F)$ formed of real-valued continuous functions on F, with the norm equal to the maximum modulus on F. For u in E, let u_0 be the restriction of u to F. Then $u \to u_0$ is a linear map of E into $C(F)$, the range of which we denote by E_0.

We allege that E_0 is dense in $C(F)$. This can be established in several ways. To begin with we might argue that any v in $C(F)$ can, according to a theorem of Lebesgue, be extended continuously to R^n; since F is compact, we may assume that this extension has a compact support. It suffices to apply to this extension $\bar{v}$ the process of regularization. We choose a sequence k_n $(n = 1, 2, \cdots)$ of positive functions in E such that $\int k_n \, d\lambda = 1$ and k_n vanishes outside a neighbourhood of 0 in R^n whose diameter tends to 0 as $n \to \infty$. A very simple argument shows that the functions $\bar{v}_n = \bar{v} * k_n$, defined by

$$\bar{v}_n(t) = \int \bar{v}(t - s) k_n(s) \, d\lambda(s),$$

belong to E and converge uniformly to $\bar{v}$; in particular, $(\bar{v}_n)_0 \to v$ in $C(F)$. Alternatively we can appeal directly to the Weierstrass-Stone theorem (Theorem A in Subsection 4.10.5), which is plainly applicable to the subalgebra E_0 of $C(F) = C_R(F)$.

The second phase of the argument rests upon the observation that, thanks to (4.15.9), $\int_G D_i u \cdot d\lambda$ depends only on u_0 and that the linear form

$$u_0 \to \int_G D_i u \cdot d\lambda$$

is continuous on E_0 (qua subspace of $C(F)$). According to the Hahn-Banach theorem and the results of Section 4.10, there exists a measure σ_i on F such that

$$\int_G D_i u \cdot d\lambda = \int_F u \, d\sigma_i \tag{4.15.10}$$

for u in E. Moreover, having shown that E_0 is dense in $C(F)$, we see that σ_i is uniquely determined by G.

Finally, by Example 4.15.13, we know that there exists a unique positive measure $\sigma = (\sigma_1^2 + \cdots + \sigma_n^2)^{1/2}$ on F such that $\sigma_i = N_i \cdot \sigma$ $(1 \le i \le n)$, the N_i being real-valued functions on F such that $N_1^2 + \cdots + N_n^2 = 1$ a.e.(σ). This measure σ is uniquely determined by G; it is the surface measure on F. In terms of it we can write (4.15.10) in the form

$$\int_G D_i u \cdot d\lambda = \int_F u N_i \, d\sigma, \tag{4.15.11}$$

which has exactly the shape of Green's formula. This is so far proved only for u in E. However it is easily extended (by regularization, for example) to all functions u that are continuously differentiable on some neighbourhood in R^n of $\bar{G}$.

4.16 The Dual of $\mathscr{L}^p$ and of L^p $(1 \le p < \infty)$

We can now give a complete justification of the assertions, made without proof in Section 4.11, concerning the dual of $\mathscr{L}^p$ and of L^p when $1 \le p < \infty$. The representation theorem about to be established can be worded so as to apply to either space. We choose to word it in terms of $\mathscr{L}^p$, adding as remarks some of the possible variants.

4.16.1 Theorem. Let T be a locally compact space, μ a positive Radon measure on T, L a continuous linear form on $\mathscr{L}^p = \mathscr{L}^p(T, \mu)$, where $1 \le p < \infty$. Then there exists a function g in $\mathscr{L}^{p'} = \mathscr{L}^{p'}(T, \mu)$ such that

$$L(f) = \mu(gf) \tag{4.16.1}$$

for all f in $\mathscr{L}^p$. For any such g one has

$$N_{p'}(g) = \text{Sup}\,\{|L(f)| : f \in \mathscr{L}^p, N_p(f) \le 1\}; \tag{4.16.2}$$

in particular, the element of $L^{p'}$ corresponding to g is uniquely determined by L.

Proof. Continuity of L signifies the existence of a number $c \ge 0$ such that

$$|L(f)| \le c \cdot N_p(f) \tag{4.16.3}$$

for all f in $\mathscr{L}^p$. Let λ be the restriction of L to $\mathscr{K}(T)$. Then (4.16.3) shows that λ is a complex Radon measure and furthermore that if λ has the minimal decomposition $\lambda_1 - \lambda_2 + i\lambda_3 - i\lambda_3$, then each λ_k satisfies an inequality of the form (4.16.3). It follows then that $\lambda_k^*(U) \le c \cdot \mu^*(U)^{1/p}$ for each relatively compact open set U. This shows that λ_k is LAC(μ) for each k and so that λ is LAC(μ). According to the Lebesgue-Radon-Nikodym theorem, $\lambda = g \cdot \mu$, where g is locally integrable for μ. Thus

$$L(f) = \lambda(f) = \mu(gf) \qquad (f \in \mathscr{K}(T)). \tag{4.16.4}$$

We have therefore $|\mu(gf)| \leq c \cdot N_p(f)$ for f in $\mathscr{K}(T)$, hence it follows easily that $\bar{N}_{p'}(g) < +\infty$. By altering g on a μ-locally negligible set, we may arrange that g belongs to $\mathscr{L}^{p'}$; such a change will not affect (4.16.4). Knowing that g belongs to $\mathscr{L}^{p'}$ we may assert that $f \to \mu(gf)$ is a continuous linear form on $\mathscr{L}^p$ with norm equal to $N_{p'}(g)$. Since $\mathscr{K}(T)$ is dense in $\mathscr{L}^p$ (p being $\neq \infty$), (4.16.4) extends by continuity to give (4.16.1). ▌

4.16.2 **Remarks.** It may be said that the dual of $\mathscr{L}^p$ may be identified, together with its natural norm, with $L^{p'}$. Alternatively, since we may identify in an obvious way continuous linear forms on $\mathscr{L}^p$ with those on L^p, we may say that the dual of L^p "is" $L^{p'}$. It is even possible to reformulate Theorem 4.16.1 in terms of the spaces $\overline{\mathscr{L}}^p$ and $\overline{L}^p$ (see the remarks in Section 4.11). Only in the most trivial cases does Theorem 4.16.1 and its variants remain true when $p = \infty$.

4.16.3 **Failure when** $0 < p < 1$. If $0 < p < 1$, $\mathscr{L}^p$ is a complete semi-metrizable TVS; N_p is no longer a seminorm, but (cf. (4.2.4)) N_p^p is a translation-invariant semimetric defining the topology on $\mathscr{L}^p$. In general Theorem 4.16.1 fails spectacularly when $0 < p < 1$ as we shall now show.

Concerning μ we assume throughout this subsection the following hypothesis:

▶ There exists a number c with the property that arbitrarily large compact sets $K \subset T$ exist that, for arbitrarily large natural numbers n, admit partitions $(A_r)_{1 \leq r \leq n}$ into n μ-integrable sets satisfying $\mu(A_r) \leq cn^{-1}$ $(1 \leq r \leq n)$.

It may be shown that this hypothesis is fulfilled if and only if μ is continuous in the sense that $\mu(\{t\}) = 0$ for each $t \in T$.

Our aim is to establish the following four statements, in each of which it is assumed that μ satisfies the hypothesis and that $0 < p < 1$.

▶ (1) Any convex subset of $\mathscr{L}^p$ containing interior points is identical with $\mathscr{L}^p$.

▶ (2) Any closed, convex and absorbent subset of $\mathscr{L}^p$ is identical with $\mathscr{L}^p$.

▶ (3) There exists no nonzero continuous linear map of $\mathscr{L}^p$ into any LCTVS. (See Exercises 4.29 and 4.33; see also Day [2].)

▶ (4) There exists no nonzero continuous linear form on $\mathscr{L}^p$.

It is evident that (2) implies (3), which in turn implies (4). Also (1) implies (2) by virtue of the Baire category theorem and completeness of $\mathscr{L}^p$. It thus remains only to establish (1).

Let then $A \subset \mathscr{L}^p$ be convex and have interior points. To show that $A = \mathscr{L}^p$ it suffices (Exercise 1.15) to show that A is dense in $\mathscr{L}^p$. By translation, we may assume that A is a neighbourhood of 0. We will show that A contains any $f \in \mathscr{L}^p$ that is bounded and has a compact support. It is plain that such functions are dense in $\mathscr{L}^p$. There exists a number M such that $|f| \leq M$, and a compact set K of the type specified in the hypothesis above outside which f vanishes. Since A is a neighbourhood of 0, there exists a number $d > 0$ such that A contains all $g \in \mathscr{L}^p$ satisfying $N_p^p(g) < d$. We write now

$$f = n^{-1} \sum_{r=1}^{n} (nf\chi_{A_r}) = n^{-1} \sum_{r=1}^{n} f_r,$$

say. One has

$$N_p^p(f_r) = \mu(n^p \chi_{A_r} |f|^p) \leq n^p M^p \mu(A_r) < \frac{cM^p}{n^{1-p}},$$

which, since $p < 1$, can be made as small as we wish for all r by taking n sufficiently large. In particular we can ensure that $f_r \in A$ for $1 \leq r \leq n$ by taking n large enough, in which case the convexity of A entails that $f \in A$ too. Thus (1) is verified. ▌

The situation described by (1)–(4) should be compared with that which obtains (see Exercise 6.14) for the superficially similar spaces H^p with $0 < p < 1$.

See now Exercises 4.29 and 4.33.

4.16.4 Perhaps the most direct practical consequences of Theorem 4.16.1 arise from combining it with the Hahn-Banach theorem as applied to linear approximation problems by the methods set forth in general terms in Section 2.4.

Suppose that $1 \leq p < \infty$ and that V is a vector subspace of $\mathscr{L}^p = \mathscr{L}^p(T, \mu)$ or of the associated separated space L^p. We may then assert that the closure of V consists precisely of those elements f_0 of $\mathscr{L}^p$ (or of L^p) satisfying

$$\mu(f_0 g) = 0 \tag{4.16.5}$$

whenever g belongs to $\mathscr{L}^{p'}$ and satisfies

$$\mu(fg) = 0 \qquad \text{for all } f \in V. \tag{4.16.6}$$

If V is not a vector subspace, the same criterion is necessary and sufficient in order that f_0 be the limit in $\mathscr{L}^p$ (or L^p) of finite linear combinations of elements f of V.

4.17 Product Measures and the Fubini Theorem

Let T and T' be separated locally compact spaces, $P = T \times T'$ their product, which is a space of the same type.† Further, let μ and μ' be positive Radon

† Throughout this section, the prime in T' has no significance other than that of a distinguishing mark; we temporarily waive the convention introduced in Definition 1.9.4 and generally adopted elsewhere in this book.

measures on T and on T', respectively. Starting from a function f on P one may endeavour to apply to f the iterated integrations with respect to μ and to μ', recognizing that a priori the order in which these integrations are performed must be taken into account. As we shall see, this leads to just one product measure on the space P—in other words, that the order of integration is immaterial when f belongs to $\mathscr{K}(P)$ and the result defines a positive Radon measure on P. Once this is established, it is inevitable that one should ask to what extent does the a priori ambiguity in order of integrations prove to be ultimately irrelevant for more general functions f. Alternatively, one may ask how the integral of a "general" function f on P with respect to the product measure is to be expressed in terms of the "partial' integrals with respect to μ and μ'. The answer to this is provided by the Lebesgue-Fubini theorem, together with a complementary result due to Tonelli. (Fubini's association with the result was limited to the case of Riemann integrals in real analysis; the primitive form of the theorem, dealing essentially with the case in which f belongs to $\mathscr{K}(P)$ was indeed known to Cauchy.)

4.17.1 **Definition of the Product Measure.** Here we need concern ourselves only with the case in which f belongs to $\mathscr{K}(P)$. In this case the partial function $f_t : t' \to f(t, t')$ belongs to $\mathscr{K}(T')$, so that

$$\mu'(f_t) = \int f(t, t')\, d\mu'(t')$$

is certainly defined. Moreover, $t \to \mu'(f_t)$ is easily seen to belong to $\mathscr{K}(T)$, so that the integral

$$\int \mu'(f_t) \cdot d\mu(t) = \int d\mu(t) \int f(t, t')\, d\mu'(t')$$

is defined. If f is positive, this iterated integral is positive. Accordingly the map

$$\pi_1 : f \to \int d\mu(t) \int f(t, t')\, d\mu'(t') \tag{4.17.1}$$

is a positive Radon measure on P.

Reversal of the order of integrations with respect to μ and μ' and interchange of the roles of T and T' leads likewise to the positive Radon measure π_2 on P defined by

$$\pi_2 : f \to \int d\mu'(t') \int f(t, t')\, d\mu(t). \tag{4.17.2}$$

Our first task is to show that the measures π_1 and π_2 are in fact identical: this measure will then be by definition the (tensor) product $\mu \otimes \mu'$ of μ and μ'.

It is possible and logically most satisfactory to prove this identity by using only integrals of functions in the spaces $\mathscr{K}(T)$ and $\mathscr{K}(T')$. This is done in Edwards [3] in the following manner. If x (resp. x') belongs to $\mathscr{K}(T)$ (resp. $\mathscr{K}(T')$), we denote by $x \otimes x'$ the element of $\mathscr{K}(P)$ defined by $(t, t') \to x(t) \cdot x'(t')$, and write temporarily $\mathscr{K}(T) \otimes \mathscr{K}(T')$ for the vector subspaces of $\mathscr{K}(P)$ generated by such functions. It is evident that π_1 and π_2 give the same result when applied to $x \otimes x'$ and hence also when applied to any member of $\mathscr{K}(T) \otimes \mathscr{K}(T')$. This being so, it suffices to show that $\mathscr{K}(T) \otimes \mathscr{K}(T')$ is in a suitable sense dense in $\mathscr{K}(P)$. That this is the case may be deduced from the Weierstrass-Stone theorem.

An alternative method, which avoids explicit dependence on the Weierstrass-Stone theorem and is yet formally very similar to the above, proceeds as follows. (It is logically less satisfactory since it introduces integrals with respect to μ and μ' of discontinuous functions.) Choose compact subsets K and K' of T and T' such that f vanishes outside $K \times K'$. By using compactness one can show that for any preassigned $\varepsilon > 0$, K (resp. K') can be partitioned into finitely many universally integrable sets A_i (resp. A'_j) such that

$$|f(t_1, t') - f(t_2, t')| \le \varepsilon \qquad (t_1, t_2 \in A_i, t' \in K'),$$
$$|f(t, t'_1) - f(t, t'_2)| \le \varepsilon \qquad (t'_1, t'_2 \in A'_j, t \in K).$$

(Both assertions, in fact, follow from uniform continuity of f on $K \times K'$, provided one invokes the existence of the unique uniform structure on the compact space $K \times K'$ yielding its induced topology.) The sets $P_{ij} = A_i \times A'_j$ then partition $K \times K'$. If (t_1, t'_1) and (t_2, t'_2) belong to any one P_{ij}, the triangle inequality shows that

$$|f(t_1, t'_1) - f(t_2, t'_2)| \le 2\varepsilon.$$

Choose arbitrarily points p_{ij} from P_{ij}. Then we find that

$$\left| \int_{A_i} f(t, t')\, d\mu(t) - f(p_{ij})\mu(A_i) \right| \le 2\varepsilon \cdot \mu(A_i)$$

for t' in A'_j and so

$$\left| \int_{A_j'} d\mu'(t') \int_{A_i} f(t, t')\, d\mu(t) - f(p_{ij})\mu(A_i)\mu'(A'_j) \right| \le 2\varepsilon \cdot \mu(A_i)\mu'(A'_j).$$

Summing with respect to i and j, we conclude that

$$\left| \int d\mu'(t') \int f(t, t')\, d\mu(t) - \sum f(p_{ij})\mu(A_i)\mu'(A'_j) \right| \le 2\varepsilon \cdot \mu(K)\mu'(K').$$

By inverting the order of integrations we find that a similar inequality is satisfied by $\int d\mu(t) \int f(t, t')\, d\mu'(t')$. So the two iterated integrals of f differ by at most $4\varepsilon \cdot \mu(K)\mu'(K')$. Since ε is arbitrarily small, these two integrals are equal, as we had to show.

4.17.2 Semicontinuous Functions. Let ϕ be any positive lower semicontinuous function on P. Each partial function $\phi_t : t' \to \phi(t, t')$ is then lower semicontinuous on T'. Moreover, on representing ϕ as the upper envelope of functions in $\mathscr{K}(P)$ and using the definition in (4.5.1), we see that the function $t \to \mu'^*(\phi_t)$ is lower semicontinuous on T. Finally, appealing to Proposition 4.5.1, we see that

$$\pi^*(\phi) = \int^* d\mu(t) \int^* \phi(t, t')\, d\mu'(t'). \tag{4.17.3}$$

A similar argument shows that the same equality obtains when, on the right, the order of integrations is reversed.

4.17.3 Upper Integrals of Arbitrary Functions. Let f be any positive function on P. The reader will recall that the upper integral of f (relative to any positive measure) is by definition the infimum of upper integrals of positive lower semicontinuous functions majorizing f. Combining this with the remarks in Subsection 4.17.2 above, we conclude that in all cases

$$\pi^*(f) \geq \int^* d\mu(t) \int^* f(t, t')\, d\mu'(t'). \tag{4.17.4}$$

Once again, the order of integration on the right-hand side may be reversed.

This puts us in a position to prove the principal theorem.

4.17.4 Theorem (Lebesgue-Fubini). Let f be a function on $P = T \times T'$ integrable for the product measure $\pi = \mu \otimes \mu'$. Then, for μ-almost all t in T, the function $f_t : t' \to f(t, t')$ is integrable for μ'; the function $t \to \int f(t, t')\, d\mu'(t')$, defined for μ-almost all t in T, is integrable for μ, and

$$\pi(f) = \int d\mu(t) \int f(t, t')\, d\mu'(t'). \tag{4.17.5}$$

The same is true with the roles of T and T' interchanged.

Proof. Without loss of generality we may assume that f is positive. Since f is integrable for π we may choose a decreasing sequence (ϕ_n) of positive lower semicontinuous functions on P, each majorizing f, such that $\lim \phi_n = f$ in $\mathscr{L}^1(P, \pi)$. Then $g = \lim \phi_n$ pointwise is integrable for π and $f = g$ a.e.(π). By monotone convergence applied twice over, together with (4.17.3), we have

$$\begin{aligned}\pi(f) = \lim \pi(\phi_n) &= \lim \int^* d\mu(t) \int^* \phi_n(t, t')\, d\mu'(t') \\ &= \int^* d\mu(t) \int^* g(t, t')\, d\mu'(t'). \end{aligned} \tag{4.17.6}$$

Also, g_t is measurable for μ' for each t in T, and $t \to \int g(t, t')\, d\mu'(t')$ is measurable for μ. Because of this, (4.17.6) shows that g_t is integrable for μ' for almost all t, and that $t \to \int g(t, t')\, d\mu'(t')$, defined for almost all t, is integrable for μ. Finally, (4.17.4) shows in particular that if a set $N \subset P$ is π-negligible, then $N_t = \{t' : t' \in T',\ (t, t') \in N\}$ is μ'-negligible for almost all t in T. It follows that for almost all t, $f_t = g_t$ except on a subset of T' negligible for μ'. The theorem now follows from (4.17.6) and the preceding remarks. ▌

4.17.5 Remark. The theorem collapses almost entirely if one tries to replace integrability by essential integrability. It can happen that f is essentially integrable for π, and yet f_t be essentially integrable for μ' for *no* value of t whatsoever. However, see Corollary 4.17.12.

4.17.6 Notation. It is customary and convenient to write

$$\iint f(t, t')\, d\mu(t)\, d\mu'(t') \qquad \text{and} \qquad \iint^* f(t, t')\, d\mu(t)\, d\mu'(t')$$

in place of $\pi(f)$ and $\pi^*(f)$, respectively. The reader should take care to distinguish conceptually these "double integrals" from the iterated ones, since otherwise the Lebesgue-Fubini theorem seems quite pointless.

4.17.7 Corollary. (1) If f is positive and measurable for $\pi = \mu \otimes \mu'$ and if f vanishes outside a π-σ-finite set, then the functions $t \to \int^* f(t, t')\, d\mu'(t')$ and $t' \to \int^* f(t, t')\, d\mu(t)$ are measurable for μ and μ', respectively, and

$$\iint^* f(t, t')\, d\mu(t)\, d\mu'(t') = \int^* d\mu(t) \int^* f(t, t')\, d\mu'(t')$$
$$= \int^* d\mu'(t') \int^* f(t, t')\, d\mu(t).$$

(2) If f is positive and measurable for $\pi = \mu \otimes \mu'$, then the function $t \to \overline{\int}^* f(t, t')\, d\mu'(t')$ and $t' \to \overline{\int}^* f(t, t')\, d\mu(t)$ are measurable for μ and μ', respectively, and

$$\overline{\iint}^* f(t, t')\, d\mu(t)\, d\mu'(t') \leq \overline{\int}^* d\mu(t) \overline{\int}^* f(t, t')\, d\mu'(t').$$

Proof. One may write $f = \lim f_n$, where (f_n) is an increasing sequence of π-integrable functions. The conclusion (1) follows from Theorem 4.17.4 combined with monotone convergence applied several times. Conclusion (2) follows directly from (1) applied to $f(t, t')\chi_K(t)\chi_{K'}(t')$, where K and K' are arbitrary compact subsets of T and T' respectively. ∎

4.17.8 Theorem (Tonelli). Let f be measurable for $\pi = \mu \otimes \mu'$ and zero outside a π-σ-finite set. If either of the iterated upper integrals

$$\int^* d\mu(t) \int^* |f(t, t')|\, d\mu'(t'), \qquad \int^* d\mu'(t') \int^* |f(t, t')|\, d\mu(t)$$

is finite, then f is π-integrable and the conclusions of Theorem 4.17.4 apply.

Proof. Apply Corollary 4.17.7 to $|f|$. ∎

4.17.9 Measurability for the Product Measure. In many instances Tonelli's theorem is used to justify the application of the Lebesgue-Fubini theorem. It is therefore important to have some criteria for measurability with respect to $\mu \otimes \mu'$.

4.17.10 Proposition. If g and g' are positive functions on T and T', respectively, then

$$\iint^* g(t)g'(t')\, d\mu(t)\, d\mu'(t') = \mu^*(g) \cdot \mu'^*(g') \tag{4.17.8}$$

save perhaps when one factor on the right is 0 and the other is $+\infty$.

Proof. Notice that, according to the convention (made in Section 4.5) that $0 \cdot +\infty = 0$, the formula $\lambda^*(\alpha f) = \alpha \cdot \lambda^*(f)$ holds for any positive measure λ, any function f satisfying $0 \leq f \leq +\infty$, and any number $\alpha \geq 0$ or $+\infty$. [When $\alpha = +\infty$, $\lambda^*(\alpha f)$ is 0 or $+\infty$ according to whether f is or is not negligible for λ; and the same applies to $\alpha \cdot \lambda^*(f)$.]

We denote by I the upper double integral in (4.17.8). By Subsection 4.17.3 we have $I \geq \int^* d\mu(t) \int^* g(t)g'(t')\, d\mu'(t')$. Also, by the opening remarks,

$$\int^* g(t)g'(t')\, d\mu'(t') = g(t) \cdot \int^* g'(t')\, d\mu'(t'),$$

and so likewise

$$\int^* d\mu(t) \int^* g(t)g'(t')\, d\mu'(t') = \mu'^*(g') \cdot \mu^*(g).$$

It thus suffices to show that

$$I \leq \mu^*(g) \cdot \mu'^*(g'). \tag{4.17.8'}$$

In proving this we may assume that the right-hand side is finite, and so it remains to consider only the case in which both factors are finite (all others being excluded by hypothesis). In this case, the right-hand side of (4.17.8′) is the infimum of $\mu^*(\phi) \cdot \mu'^*(\phi')$, where ϕ and ϕ' are lower semicontinuous on T and T' and majorize g and g', respectively. The function $(t, t') \to \phi(t) \cdot \phi'(t')$ is lower semicontinuous on $T \times T'$ and majorizes $g(t) \cdot g'(t')$. It now suffices to apply Subsection 4.17.2 and the obvious inequality $I \leq \pi^*(\phi \otimes \phi')$. ▌

4.17.11 Corollary. If $A \subset T$ and $A' \subset T'$, then $A \times A'$ is negligible for $\mu \otimes \mu'$ provided *either* $\mu^*(A) = 0$ and $\mu'^*(A') < +\infty$ *or* $\mu^*(A) < +\infty$ and $\mu'^*(A') = 0$.

4.17.12 Proposition. If g and g' are positive functions on T and T', respectively, then

$$\overline{\iint}^* g(t)g'(t')\, d\mu(t)\, d\mu'(t') = \bar{\mu}^*(g) \cdot \bar{\mu}'^*(g'). \tag{4.17.9}$$

Proof. Let K and K' be arbitrary compact subsets of T and T', respectively. By Proposition 4.17.10 we have

$$\iint^* g(t)\chi_K(t)g(t')\chi_{K'}(t')\, d\mu(t)\, d\mu'(t') = \mu^*(g\chi_K) \cdot \mu'^*(g'\chi_{K'}) \tag{4.17.9'}$$

provided one term on the right-hand side is not 0, the other being $+\infty$. However if, for example, $\mu^*(g\chi_K) = 0$, then the integrand in the upper double integral vanishes outside a set that is negligible for $\mu \otimes \mu'$ (Corollary 4.17.11), and so the equality (4.17.9′) still holds. Thus (4.17.9′) holds in all cases. The result now follows from the definition of the essential upper integrals involved (4.14.1), the sets $K \times K'$ forming a base for the compact subsets of $T \times T'$. ▌

4.17.13 Proposition. If g (resp. g') is measurable for μ (resp. μ'), and if u is a continuous function on R^2, then the function

$$u(g, g') : (t, t') \to u(g(t), g'(t)) \text{ is measurable for } \mu \otimes \mu'.$$

Proof. According to Lusin's criterion (Corollary 4.8.5), if K and K' are compact subsets of T and T', respectively, we can partition K (resp. K') by compact sets K_n (resp. K'_n) ($n = 1, 2, \cdots$) and a set N (resp. N') negligible for μ (resp. μ') in such a way that $g \mid K_n$ (resp. $g' \mid K'_n$) is continuous. Let $H_{mn} = K_m \times K'_n$ and let M be the union of the sets $N \times K'_n$, $K_n \times N'$ and $N \times N'$ ($n = 1, 2, \cdots$).

Then (Corollary 4.17.11) M is negligible for $\mu \otimes \mu'$. The H_{mn} $(m, n = 1, 2, \cdots)$ and M partition $K \times K'$. Since $u(g, g') \mid H_{mn}$ is evidently continuous, it follows that $u(g, g')$ is measurable for $\mu \otimes \mu'$, as asserted. ▌

4.17.14 **Corollary.** If $A \subset T$ and $A' \subset T'$ are measurable (resp. integrable) for μ and μ', then $A \times A'$ is measurable (resp. integrable) for $\mu \otimes \mu'$ and

$$\overline{(\mu \otimes \mu')}(A \times A') = \bar{\mu}(A) \cdot \bar{\mu}'(A'). \tag{4.17.10}$$

Proof. Proposition 4.17.13 shows that in either case $A \times A'$ is measurable for $\mu \otimes \mu'$. In the case of integrable sets, (4.17.10) follows from Proposition 4.17.10, which shows incidentally that $A \times A'$ is integrable for $\mu \otimes \mu'$. In the remaining case we appeal to Proposition 4.17.12 and recall that the measure of a measurable set is equal to the essential upper integral of its characteristic function (see Sections 4.7 and 4.14). ▌

4.17.15 **Corollary.** If $A \subset T$ and $A' \subset T'$, and if A (resp. A') is locally negligible for μ (resp. μ'), then $A \times T'$ (resp. $T \times A'$) is locally negligible for $\mu \otimes \mu'$.

Proof. The proof is immediate from Corollary 4.17.14 in view of our adopted convention that $0 \cdot +\infty = 0$. ▌

4.18 Locally Compact Groups and Haar Measures

In this section T will denote a topological group (Subsection 0.3.19). This means that T is a group in the algebraic sense and also a topological space, the two structures being so related that the map $(t, t') \to t^{-1}t'$ of $T \times T$ into T is continuous. We shall assume also that T is separated. This concept includes all finite groups; in fact any group may be endowed with the discrete topology, in which guise it appears as a separated and locally compact topological group. Any separated TVS is, of course, a topological group when multiplication by scalars is ignored; such a topological group is commutative (Abelian); it is locally compact if and only if it is of finite dimension. This is the case with R^n and C^n. Quite large sections of classical analysis depend on the fact that R^n is a locally compact group; indeed, some portions (such as the broad aspects of harmonic analysis) are now known to depend on little besides this property. The set of complex numbers of unit modulus is a compact commutative topological group (the group operation being multiplication and the topology the usual one) that is isomorphic to the quotient group $R/2\pi$ (the group operation being addition modulo 2π and the topology the quotient of the usual topology on R). Those portions of classical analysis dealing with functions of one or more real variables that are periodic or multiply periodic are equivalent to the study of functions on $R/2\pi$ or on powers $(R/2\pi)^n$; in particular, many of the basic results about Fourier series (simple or multiple) depend solely on the nature of $R/2\pi$ and its powers as compact commutative topological groups. A detailed account of those aspects of topological groups that first received attention is given by Pontryagin [1], while a more recent account of the basic concepts appears in Bourbaki's treatise ([2], Ch. III). The appropriate integration theory and related matters is treated systematically and in detail in Hewitt and Ross [1], §§15–20.

Our concern with topological groups is a strictly limited one, being confined to that facet of the subject owing its existence of an invariant integration process for locally compact groups. It is through the study of various spaces of functions and measures more or less closely related to this invariant measure that functional analysis contributes to the theory of locally compact groups.

In the latter half of the nineteenth century finite groups were studied algebraically. The invariant measure for such a finite group T is simply the finite sum $\sum_{t \in T} \varepsilon_t$ (ε_t = the Dirac measure placed at t) or some positive multiple thereof. Naturally, the analysis involved in such cases is rather trivial. The next step historically was the relatively explicit construction of the invariant integral for compact Lie groups given in 1927 by Peter and Weyl. That this case merited special attention is due to the fact that Lie transformation groups were the first topological groups, other than R^n and its quotients, to attract interest; and the construction of the invariant integral in such cases is made easier by the existence of local coordinate systems. In 1934 von Neumann gave a construction for arbitrary compact groups (in fact for the mean value of almost periodic functions on arbitrary groups), while in 1933 Haar gave a construction for any locally compact group satisfying the second countability axiom. Included in this category is, of course, the group R^n, for which the invariant measure had been known for long as the Lebesgue measure. Since that time Haar's method has been modified in such a way as to dispense entirely with countability restrictions, so that one now knows of the existence and essential uniqueness of left- (and of right-) invariant measures on any separated locally compact topological group. A condensed account of many of the related topics is given by Weil [1], while a more leisurely account of some more recent developments is to be found in Loomis [1], Naĭmark [1], and Hewitt and Ross [1]; the last three books illustrate abundantly the use of functional analytic techniques.

4.18.1 **Definition of Haar Measures.** Henceforth and until the end of Section 4.19 we use T to denote a separated locally compact topological group. Amongst the nonzero positive Radon measures on T one may seek hopefully for those that are left (translation) invariant in the following sense. With each function f on T and each group element a we associate a new function, called the left translate of f by amount a and denoted by $L_a f$ or ${}_a f$, defined by $L_a f(t) = {}_a f(t) = f(at)$. Right translates $R_a f$ or f_a are defined analogously, ta replacing at. There is, of course, no distinction if T is commutative. By a left (translation) invariant, or left Haar, measure on T is meant a nonzero positive Radon measure μ on T such that

$$\mu(L_a x) = \mu(x) \tag{4.18.1}$$

for x in $\mathscr{K}(T)$ and a in T; right Haar measures are defined analogously, $L_a x$ being replaced by $R_a x$.

Let us first dispose of any necessity to consider separately both left and right Haar measures. If we define the reflection $\check{f}$ of a function f on T by $\check{f}(t) = f(t^{-1})$, then it is easily seen that, whenever μ is a left Haar measure, then $\nu = \check{\mu} : x \to \mu(\check{x})$ is a right Haar measure, and vice versa. Thus it suffices to consider only left Haar measures.

In terms of the set function defined by μ, (4.18.1) implies that

$$\mu^*(aA) = \mu^*(A) \quad \text{and} \quad \bar{\mu}^*(aA) = \bar{\mu}^*(A)$$

for arbitrary sets $A \subset T$ and arbitrary group elements a; and conversely, (4.18.1) is implied by the condition that $\mu(aA) = \mu(A)$ for all compact (or all open) sets $A \subset T$. In particular, if μ is left invariant and N is negligible for μ, then aN is negligible for μ; if the function f is measurable for μ, so is L_af; for any positive function f, $\mu^*(L_af) = \mu^*(f)$ and $\bar{\mu}^*(L_af) = \bar{\mu}^*(f)$; f is integrable (resp. essentially integrable) for μ if and only if the same is true of L_af, in which case the integrals of these functions are equivalued. Thus invariance in the form expressed by (4.18.1) implies invariance in the widest reasonable sense.

4.18.2 Existence and Essential Uniqueness of Haar Measures. The basic facts about Haar measures are these:

▶ (1) There exists a left Haar measure on T.

▶ (2) If μ and μ' are two left Haar measures on T, then $\mu' = c \cdot \mu$ for some number $c > 0$.

We shall not give the proof of (1) and (2), referring the reader to Weil (loc. cit.) and the references cited there, to Halmos [1], to Naĭmark [1], and to Hewitt and Ross [1], §§15, 16. There are analogous assertions referring to right Haar measures. We assume (1) and (2) henceforth.

There is in general no especially significant way of avoiding the ambiguity due to the constant of proportionality involved in (2). However, if T is discrete, one may normalize the left Haar measure to that each point of T has unit mass; if T is compact, we may adopt the normalization that makes $\mu(T) = 1$. (If T is finite and discrete, these two choices conflict unless T is reduced to its neutral element.)

If $T = R/2\pi$, the normalizing condition $\mu(T) = 1$ leads to

$$\mu(x) = (2\pi)^{-1} \int_0^{2\pi} x(t)\, dt,$$

the Lebesgue (or Riemann) integral being applied to x regarded as a function of a real variable with period 2π; the interval $(0, 2\pi)$ of integration can, of course, be replaced by any other interval of length 2π.

We shall give two less trivial examples of Haar measure in Subsections 4.18.6 and 4.18.7.

4.18.3 **First Properties of Haar Measure.** We shall use properties (1) and (2) of Subsection 4.18.2 in order to deduce the important properties (3)–(6) listed immediately below.

▶ (3) There exists a positive continuous function Δ on T, termed the modular function for T, such that

$$\Delta(t) > 0, \qquad \Delta(e) = 1, \qquad \Delta(tt') = \Delta(t)\,\Delta(t') \tag{4.18.2}$$

and such that if μ is any left Haar measure on T, then

$$\mu(R_a x) = \Delta(a) \cdot \mu(x),$$

that is,

$$\int x(ta)\,d\mu(t) = \Delta(a) \cdot \int x(t)\,d\mu(t) \tag{4.18.3}$$

for all x in $\mathscr{K}(T)$.

▶ (4) If Δ is the modular function for T, and if μ is any left Haar measure on T, then $\check{\mu} = \Delta \cdot \mu$; that is,

$$\int x(t^{-1})\,d\mu(t) = \int x(t)\,\Delta(t)\,d\mu(t) \tag{4.18.4}$$

for all x in $\mathscr{K}(T)$.

▶ (5) If Δ is the modular function for T, and if μ is any left Haar measure and ρ any right Haar measure on T, then there exists a number $c > 0$ such that $\rho = c\Delta \cdot \mu$; that is,

$$\int x(t)\,d\rho(t) = c \cdot \int x(t)\,\Delta(t)\,d\mu(t) \tag{4.18.5}$$

for all x in $\mathscr{K}(T)$; furthermore $\rho(L_a x) = \Delta(a)^{-1} \cdot \rho(x)$; that is,

$$\int x(at)\,d\rho(t) = \Delta(a)^{-1} \cdot \int x(t)\,d\rho(t) \tag{4.18.6}$$

for all x in $\mathscr{K}(T)$.

▶ (6) Let μ be any left or right Haar measure on T; then $\mu(U) > 0$ for any nonvoid open set U of T, and $\mu(T) < +\infty$ if and only if T is compact.

Proofs. Ad (3), if μ is any left Haar measure on T, then $\lambda : x \to \mu(R_a x)$ is a nonzero positive Radon measure on T. Since $R_a L_b x = L_b R_a x$, left invariance of μ entails that of λ. According to (2), therefore, there exists a number $\Delta(a) > 0$ such that $\lambda = \Delta(a) \cdot \mu$; and (2) shows further that the same function Δ attaches to any two left Haar measures μ and μ' from which one starts. The remaining properties of Δ are evident.

Ad (4), if μ is any left Haar measure on T, then $\nu : x \to \mu(\Delta\check{x})$ is a nonzero positive Radon measure on T. Since $(L_a x)^{\vee} = R_{a^{-1}}\check{x}$, (4.18.2) leads to

$$\begin{aligned}\nu(L_a x) &= \mu((L_a x)^{\vee}\Delta) = \mu((R_{a^{-1}}\check{x})\Delta)\\ &= \Delta(a) \cdot \mu(R_{a^{-1}}(\Delta\check{x})) = \Delta(a) \cdot \Delta(a^{-1}) \cdot \mu(\Delta\check{x}),\end{aligned}$$

the last step coming from (3). Thus ν is left invariant and so, by (2), must be $c \cdot \mu$ for some number $c > 0$. Changing x into $\check{x}$, we have then $\mu(\Delta x) = c \cdot \mu(\check{x})$. Taking x to have a support contained in an arbitrarily small neighbourhood of e, we see that c must equal $\Delta(e) = 1$. This proves (4).

Ad (5), we note that $\check{\rho}$ is left invariant and therefore by (2) is $c \cdot \mu$ for some number $c > 0$. By (4), this means that $\rho = c \cdot \check{\mu} = c\Delta \cdot \mu$. But then

$$\begin{aligned}\rho(L_a x) &= c(\Delta \cdot \mu)(L_a x) = c \cdot \mu(\Delta \cdot L_a x) = c \cdot \Delta(a^{-1}) \cdot \mu(L_a(\Delta x)) \\ &= \Delta(a^{-1}) \cdot c\mu(\Delta x) = \Delta(a^{-1}) \cdot \rho(x),\end{aligned}$$

as alleged.

Ad (6), it is enough to deal with the case in which μ is a left Haar measure, the remaining case being entirely analogous or deducible from the preceding results. If U is any nonvoid open set such that $\mu(U) = 0$, then each compact set K in T can be convered by finitely many left translates aU. But then, by left invariance of μ, it would follow that $\mu(K) = 0$ for all K; that is, $\mu = 0$, contrary to hypothesis.

If T is compact, $\mu(T) < +\infty$ for *any* positive Radon measure μ on T. Conversely, if T is noncompact and if V is any compact neighbourhood of e, one can define inductively elements a_n $(n = 1, 2, \cdots)$ of T such that a_n does not belong to the union of the sets $a_i V$ $(1 \leq i < n)$. If U is a neighbourhood of e such that $U^2 \subset V$, then the sets $a_n U$ are disjoint. Their common measure is strictly positive. Therefore $\mu(T) = +\infty$. ∎

Remarks. The preceding results show that the concepts of local negligibility, negligibility, and σ-finiteness are, respectively, the same for all Haar measures (left and right simultaneously). As regards local negligibility, this follows at once from (2) and (5). Further, since Δ is strictly positive and continuous, (4.18.5) shows that

$$\rho^*(\phi) = c \cdot \mu^*(\Delta\phi) \tag{4.18.7}$$

for any positive lower semicontinuous function ϕ on T. If N is ρ-negligible, then $N \subset M = \bigcap_{n=1}^{\infty} U_n$, where the U_n are open, decrease and $\rho^*(U_n) \to 0$. It follows that the functions $\chi_{U_n} \cdot \Delta$ are μ-integrable and dominated convergence shows that

$$\mu^*(\chi_M \Delta) = \lim \mu^*(\chi_{U_n} \Delta) = \lim c^{-1}\rho^*(U_n) = 0.$$

Thus Δ must vanish μ-almost everywhere on M. Since Δ is nonvanishing so $\mu^*(M)$ must be zero, and so $\mu^*(N) = 0$. Similarly, since (4.18.7) is equivalent to

$$\mu^*(\phi) = c^{-1} \cdot \rho^*(\Delta^{-1}\phi), \tag{4.18.7'}$$

we see that $\mu^*(N) = 0$ implies $\rho^*(N) = 0$. Thus negligibility is the same for all Haar measures. Again, if U is open and $\rho^*(U) < +\infty$, (4.18.7) shows that $\chi_U \Delta$ is μ-integrable and therefore vanishes outside a μ-σ-finite set. Since Δ is nonvanishing, U must be μ-σ-finite. Thus any ρ-σ-finite set, being contained in the union of countably many open sets of finite ρ-measure, is also μ-σ-finite. The converse is established analogously.

Since local negligibility means the same for all Haar measures, the same is true of measurability.

The integral relationships (4.18.3)–(4.18.6) can, of course, be extended to cases in which x is replaced by much more general functions and upper or essential upper integrals appear. In particular, they can be formulated in terms of the associated measures of sets. The details may be left to the reader.

4.18.4 **Unimodular Groups.** T is said to be unimodular if there exists a bi-invariant nonzero positive Radon measure on T; that is, a measure that is at once a left Haar measure and a right Haar measure. This is so if and only if the modular function $\Delta \equiv 1$. Thus, any locally compact commutative group is unimodular. Also, taking $x = 1$ in (4.18.3), we see that any compact group is unimodular.

In any case the equation $\Delta(a) = 1$ defines a closed invariant subgroup G of T and Δ defines an algebraic isomorphism of T/G onto a subgroup of the multiplicative group of strictly positive real numbers. So T/G is commutative. If T is a semisimple Lie group, this is so if and only if $G = T$, so that any such group is unimodular.

4.18.5 **Relatively Invariant Measures.** A complex Radon measure λ on T is said to be relatively left invariant if there exists a complex-valued function D on T such that

$$\lambda(L_a x) = D(a) \cdot \lambda(x) \tag{4.18.8}$$

for all x in $\mathscr{K}(T)$. Relatively right-invariant measures ρ are defined analogously:

$$\rho(R_a x) = D'(a) \cdot \rho(x),$$

D' being a complex-valued function on T. The study of such relatively right-invariant measures can be reduced to that of relatively left-invariant measures by observing that if ρ is relatively right invariant, then $\lambda = \check{\rho}$ is relatively left invariant, (4.18.8) being true with $D(a) = D'(a^{-1})$. This depends on the identity $(L_a x)^{\vee} = R_{a^{-1}}\check{x}$.

Suppose then that λ is relatively left invariant. Neglecting the trivial case $\lambda = 0$, (4.18.8) shows that D is continuous, $D(e) = 1$, and $D(ab) = D(a)D(b)$. The measure $\alpha = D \cdot \lambda$ is then nonzero and

$$\begin{aligned}\alpha(L_a x) &= \lambda(D \cdot L_a x) = \lambda[D(a)^{-1} \cdot L_a(Dx)] \\ &= D(a)^{-1} \cdot \lambda(L_a(Dx)) = D(a)^{-1} \cdot D(a)\lambda(Dx) \\ &= \alpha(x).\end{aligned}$$

Thus α is left invariant. The same is therefore true of the real and imaginary parts of α, and also of the positive and negative parts of each of these. By (2), it follows that $\alpha = c \cdot \mu$, μ being any chosen left Haar measure and c a suitable complex number. Thus $\lambda = (cD^{-1}) \cdot \mu$. Conversely, any λ of this form is obviously relatively left invariant.

4.18.6 **Example.** We consider the group T whose elements are ordered pairs $t = (a, b)$ of real numbers with $a \neq 0$, and whose law of composition is defined as follows: if $t = (a, b)$ and $t' = (a', b')$, then

$$tt' = (aa', ab' + b).$$

T is topologized as a subspace of R^2. It is easily seen that T is a locally compact group that is neither compact nor commutative.

The elements of T may be thought of as one-to-one mappings of R onto itself, $t = (a, b)$ being visualized as the mapping $u \to au + b$, and the law of composition of T is that dictated by the composition of the corresponding maps.

In order to determine the left Haar measures on T we make the assumption that any such measure, μ, is LAC relative to the restriction to T, considered as a subset of R^2, of the Lebesgue measure on R^2. This means that we assume the existence of a function F on R^2 such that

$$\mu(x) = \iint x(a, b)F(a, b)\, da\, db$$

for all x in $\mathscr{K}(T)$, and then use left invariance of μ to determine F.

The assumed left invariance of μ is expressed by the equation

$$\iint x(a_0 a, a_0 b + b_0)F(a, b)\, da\, db = \iint x(a, b)F(a, b)\, da\, db$$

holding for all x in $\mathscr{K}(T)$ and all (a_0, b_0) in T. The simple change of variables $b' = a_0 b + b_0$, $a' = a_0 a$ may be applied to the left-hand member of this equation, leading to the expression

$$\iint x(a', b') \cdot F(a_0^{-1}a', a_0^{-1}(b' - b_0))a_0^{-2}\, da'\, db'.$$

It follows that

$$F(a_0^{-1}a, a_0^{-1}(b - b_0)) = a^2 F(a, b)$$

for almost all (a, b). Assuming that this holds everywhere and setting $a_0 = a$, $b_0 = b$, we get

$$F(a, b) = \frac{F(1, 0)}{a^2} = \text{const.}\ a^{-2}.$$

It can now be checked that any such F arranges left invariance of μ. Thus (2) of Subsection 4.18.2 shows that necessarily

$$\mu(x) = \text{const.} \iint x(a, b) \cdot a^{-2}\, da\, db,$$

the constant being strictly positive.

Having determined the left Haar measures on T, we can use (4.18.3) to determine the modular function Δ. The result is $\Delta(a, b) = |a|$. Then (4.18.5) determines the form of any right Haar measure ρ, viz.

$$\rho(x) = \text{const.} \iint x(a, b)\, |a|^{-1}\, da\, db.$$

4.18.7 **Example.** Let T be the group of real triangular matrices of the form

$$t = \begin{bmatrix} a & b \\ 0 & a^{-1} \end{bmatrix},$$

where $a \neq 0$ and b are real, the law of composition being matrix multiplication. Once again T may be thought of as a subset of R^2 and topologized accordingly. The result is a locally compact group. The method of the preceding example may be employed to determine the left Haar measures, the modular function, and the right Haar measures. The left Haar measures prove to be of the form

$$\mu(x) = \text{const.} \iint x(a, b)a^{-2}\, da\, db,$$

the modular function to be $\Delta(a, b) = a^2$, and the right Haar measures therefore of the form

$$\rho(x) = \text{const.} \iint x(a, b)\, da\, db,$$

all functions on T being here expressed as functions of the corresponding point (a, b) of R^2.

For further examples, see Hewitt and Ross [1], (15.17).

4.19 Group Algebras and Convolution

In the study of finite groups T, it was found useful to introduce the so-called "group algebra" or "group ring." By this we mean an algebra A over the complex (or real) field and a map θ of T into A such that (1) $\theta(tt') = \theta(t)\theta(t')$ for arbitrary group elements t and t', (2) $\theta(T)$ generates A, and (3) the $\theta(t)$ are linearly independent in A. These requirements ensure that θ is one to one and that A is uniquely determined up to isomorphism. Each element of A has just one representation of the form $\sum_{t\in T} f(t)\theta(t)$, f being a complex-(or real-)valued function on T. Because of (1) the product, in A, of two elements corresponding, respectively, to the functions f and g is that element of A corresponding to the function $f * g$ defined by

$$f * g(t) = \sum_{s\in T} f(s)g(s^{-1}t) = \sum_{s\in T} f(ts^{-1})g(s);$$

$f * g$ is termed the convolution of f and g in that order. Thus one realization of A is the algebra of complex-(or real-)valued functions on T with pointwise addition and multiplication by scalars and with convolution as the multiplication, the map θ being that which associates with t the function e_t defined to be 1 at t and 0 at all other group elements. This algebra is commutative if and only if T is commutative.

A second realization of the group algebra is obtained by pairing off the function f with the measure $\sum_{t\in T} f(t)\varepsilon_t$, ε_t being the Dirac measure placed at t. Apart from a positive factor, the (bi-invariant) Haar measure on T is $\mu = \sum_{t\in T} \varepsilon_t$, corresponding to the constant function 1. The measure associated with the function f is just $f \cdot \mu$. If α and β are the measures associated with f and g, respectively, the measure associated with $f * g$ will be denoted by $\alpha * \beta$. Expressed directly in terms of α and β, one finds easily the relation

$$\alpha * \beta(x) = \int d\alpha(t) \int x(tt')\, d\beta(t'),$$

that is,

$$\alpha * \beta(x) = \alpha \otimes \beta(\tilde{x}), \tag{4.19.1}$$

where x is an arbitrary element of $\mathscr{K}(T)$ (that is, an arbitrary complex-valued function on T) and $\tilde{x}$ denotes the function $(t, t') \to x(tt')$ on $T \times T$. In this realization, θ maps t onto ε_t.

We may now attempt to generalize these ideas to the case in which T is any separated locally compact topological group. Experience proves that, with this degree of generality, it is unprofitable to adopt the structural characterization of the group algebra afforded by (1), (2), and (3) when T is finite. In practice one extends (4.19.1) so as to define the convolution of suitably restricted measures, and the group algebras hitherto studied are suitable sets of measures that form algebras with convolution as the multiplication. In recovering the concept of convolution of functions from that for measures there is generally some ambiguity owing to the fact that one has the choice between associating with the function f the measure $f \cdot \lambda$, where λ is a left Haar measure on T, or the measure $f \cdot \rho$, where ρ is a right Haar measure on T. This ambiguity disappears if and only if T is unimodular (as is the case when T is finite), when it is natural to choose any bi-invariant Haar measure. In the contrary case we shall, in fact, use the left Haar measure for this purpose.

When one takes over the definition afforded by (4.19.1) to the case where T is noncompact, the measures α and β must be suitably restricted because the function $\tilde{x}$, although continuous, has not in general a compact support. Consequently we have to frame our definition in the following manner.

4.19.1 **Definition.** (1) If α and β are positive Radon measures on T, the convolution $\alpha * \beta$ is said to exist if and only if, for each x in $\mathscr{K}(T)$, the function $\tilde{x}$ is integrable for the product measure $\alpha \otimes \beta$; in that case $\alpha * \beta$ is the positive Radon measure on T defined by (4.19.1).

(2) If α and β are complex Radon measures on T with minimal decompositions $\alpha = \alpha_1 - \alpha_2 + i\alpha_3 - i\alpha_4$ and $\beta = \beta_1 - \beta_2 + i\beta_3 - i\beta_4$, $\alpha * \beta$ is said to exist if and only if each of $\alpha_r * \beta_s$ exist ($r, s = 1, 2, 3, 4$); in that case $\alpha * \beta$ is defined to be the measure $\sum_{r,s=1}^{4} c_r c_s(\alpha_r * \beta_s)$, where $c_r = 1, -1, i, -i$ according as $r = 1, 2, 3, 4$. (This formula defining $\alpha * \beta$ is plainly the only one possible if we wish $\alpha * \beta$ to be bilinear.)

In respect of (1) it is evident that $\alpha * \beta$ exists (α and β positive) if and only if $\tilde{x}$ is integrable for $\alpha \otimes \beta$ whenever x belongs to $\mathscr{K}_+(T)$. In that case $\tilde{x}$ is positive and lower semicontinuous, and so Subsection 4.17.2 allows us to write the condition in the form

$$\left.\begin{aligned} \int^* d\alpha(t) \int^* x(tt')\, d\beta(t') = \int^* d\beta(t') \int^* x(tt')\, d\alpha(t) < +\infty \\ \text{for } x \in \mathscr{K}_+(T). \end{aligned}\right\} \tag{4.19.2}$$

This condition may in turn be rephrased in terms of the α- and β-measures of subsets of T. To do this we introduce some notation.

If E is any subset of T, $\tilde{E}$ denotes the subset of $T \times T$ consisting of pairs (t, t') such that tt' belongs to E. Besides this, supposing still that α and β are positive Radon measures T, we introduce three set functions on T defined as follows:

$$\gamma_1(E) = (\alpha \otimes \beta)^*(\tilde{E}),$$

$$\gamma_2(E) = \int^* \beta^*(t^{-1}E)\, d\alpha(t), \qquad \gamma_3(E) = \int^* \alpha^*(Et'^{-1})\, d\beta(t').$$

It is evident that each of these is increasing and countably subadditive.

4.19.2 **Proposition.** Let α and β be positive Radon measures on T, and let γ_k $(k = 1, 2, 3)$ be defined as above. Then:

(1) $\gamma_1(U) = \gamma_2(U) = \gamma_3(U)$ for each open set $U \subset T$;

(2) $\alpha * \beta$ exists if and only if $\gamma_1(U) = \gamma_2(U) = \gamma_3(U) < +\infty$ for each relatively compact open set $U \subset T$;

(3) $\alpha * \beta$ exists if and only if

$$\operatorname{Inf}\, (\gamma_1(K), \gamma_2(K), \gamma_3(K)) < +\infty$$

for each compact set $K \subset T$, in which case

$$\gamma_1(K) = \gamma_2(K) = \gamma_3(K)$$

for all such K;

(4) $\alpha * \beta$ exists if and only if there exist "arbitrarily large" relatively compact sets $A \subset T$ such that

$$\operatorname{Inf}\, (\gamma_1(A), \gamma_2(A), \gamma_3(A)) < +\infty.$$

(In (4), it is meant that any preassigned compact set $K \subset T$ is contained in some A satisfying the stated condition.)

Proof. (1) follows at once from Subsection 4.17.2.

As regards (2), each $x \in \mathscr{K}_+(T)$ satisfies an inequality $x \leq \text{const.}\ \chi_U$ where U is relatively compact and open; conversely, if U is relatively compact and open, $\chi_U \leq x$ for some x in $\mathscr{K}_+(T)$. Now we refer to the criterion (4.19.2).

Re (3): Assuming (1), let K be compact. Then $K \subset U$ for some relatively compact open U, and so $\chi_{\tilde{K}} \leq \chi_{\tilde{U}}$. Now $\chi_{\tilde{K}}$ is upper semicontinuous, hence measurable for $\alpha \otimes \beta$. But then (1) shows that $\chi_{\tilde{K}}$ is integrable for $\alpha \otimes \beta$. Fubini's theorem gives, therefore,

$$\gamma_1(K) = (\alpha \otimes \beta)^*(\tilde{K}) = \gamma_2(K) = \gamma_3(K),$$

the common value being at most $\gamma_1(U) = (\alpha \otimes \beta)^*(\tilde{U}) < +\infty$.

Reciprocally, if U is relatively compact and open, then $U \subset K$ where $K = \overline{U}$ is compact. So, if (3) is satisfied, (2) ensures that $\alpha * \beta$ exists.

Finally, a similar argument shows that (4) is equivalent to (3). The proof is complete. ▌

4.19.3 **Corollary.** Suppose that α, α', β, β' are positive Radon measures on T. Then:

(1) if $\alpha \leq \alpha'$ and $\beta \leq \beta'$, and if $\alpha' * \beta'$ exists, then $\alpha * \beta$ exists and $\alpha * \beta \leq \alpha' * \beta'$;

(2) if $\alpha * \beta$ and $\alpha' * \beta$ exist, then $(\alpha + \alpha') * \beta$ exists and equals $\alpha * \beta + \alpha' * \beta$; likewise $\alpha * (\beta + \beta') = \alpha * \beta + \alpha * \beta'$ if both terms on the right exist;

(3) $\alpha * \beta$ exists if *either* one of α or β has a compact support, *or* each of α and β is bounded; in the latter case $\alpha * \beta$ is bounded and $(\alpha * \beta)(T) = \alpha(T)\beta(T)$.

Proof. All of these properties follow directly from Definition 4.19.1 (1) and Proposition 4.19.2 when one takes account of the bilinear dependence of each of γ_k $(k = 1, 2, 3)$ on α and β. ▌

4.19.4 **Corollary.** If α and β are complex Radon measures on T, then

(1) $\alpha * \beta$ exists if and only if $|\alpha| * |\beta|$ exists, in which case

$$|\alpha * \beta| \leq |\alpha| * |\beta|;$$

(2) $\alpha * \beta$ exists if *either* one of α or β has a compact support, *or* each of α and β is bounded.

Proof. The proof is immediate from Definition 4.19.1 (2) and Corollary 4.19.3. ▌

We wish now to consider whether the defining Formula (4.19.1) remains true when x is replaced by more general functions.

4.19.5 **Theorem.** Let α and β be positive Radon measures on T, and let $\pi = \alpha \otimes \beta$. Suppose that $\gamma = \alpha * \beta$ exists. Then:

(1) The formulae

$$\gamma^*(f) = \pi^*(\tilde{f}) = \int^* d\alpha(t) \int^* f(tt')\, d\beta(t')$$
$$= \int^* d\beta(t') \int^* f(tt')\, d\alpha(t) \qquad (4.19.3)$$

hold for any f that is lower semicontinuous and positive, and for any $f \geq 0$ that is γ-measurable and γ-σ-finite.

(2) If a function f is γ-negligible, then $\tilde{f}$ is π-negligible; if f is γ-measurable, then $\tilde{f}$ is π-measurable;

(3) The inequalities

$$\gamma^*(f) \geq \pi^*(\tilde{f}) \geq \text{Sup}\left[\int^* d\alpha(t) \int^* f(tt')\, d\beta(t'), \int^* d\beta(t') \int^* f(tt')\, d\alpha(t)\right] \qquad (4.19.4)$$

are valid for any $f \geq 0$.

(4) The formula

$$\gamma^*(f) = \pi^*(\tilde{f}) \qquad (4.19.5)$$

holds for any $f \geq 0$ that is γ-σ-finite.

Proof. If $f \geq 0$ is lower semicontinuous, then f is the upper envelope of those $x \geq 0$ in $\mathscr{K}$ that are majorized by f; and $\tilde{f}$ is the upper envelope of the corresponding $\tilde{x}$. The first part of (1) follows on applying Subsections 4.5.1 and 4.17.2. (3) follows from this on using Subsection 4.17.3. A special case of (3) entails the first part of (2). As regards the second part of (2), it suffices to show that if $A \subset T$ is γ-measurable, then $\tilde{A}$ is π-measurable. Suppose then that $H \subset T \times T$ is compact. Then there exists a compact set $K \subset T$ such that $H \subset \tilde{K}$. Accordingly $\tilde{A} \cap H = \tilde{A} \cap \tilde{K} \cap H = (A \cap K)^{\smile} \cap H$. Now $A \cap K$ is γ-integrable, and so there exists a γ-integrable G_δ-set S containing $A \cap K$ and such that $S \backslash (A \cap K)$ is γ-negligible. Then $\tilde{S} \backslash (A \cap K)^{\smile}$ is π-negligible. $\tilde{S}$ is a G_δ-set in $T \times T$, hence π-measurable. Thus $(A \cap K)^{\smile}$ is π-measurable, and the same is therefore true of $\tilde{A} \cap H = (A \cap K)^{\smile} \cap H$. (2) is thereby established.

We now turn to (4). By (2) and the principle of monotone convergence, it is enough to consider the case in which f vanishes outside a relatively compact open neighbourhood V of e. In view of (3), it remains only to show that $\gamma^*(f) \leq \pi^*(\tilde{f})$. In doing this we may assume that $I = \pi^*(\tilde{f}) < +\infty$. Given $\varepsilon > 0$ there exists a lower semicontinuous $\Phi \geq \tilde{f}$ such that $\pi^*(\Phi) \leq I + \varepsilon$. Let g be defined on T to be 0 outside $V \cdot V$ and to be

$$\operatorname{Inf} \{\Phi(t, t') : t \in \bar{V}, t' \in \bar{V}, tt' = s\}$$

at any point s of $V \cdot V$. Then $\tilde{g}(t, t') = g(tt') \leq \Phi(t, t')$ on $V \times V$, and so $\tilde{g} \leq \Phi$ everywhere on $T \times T$. Further, $g(s) \geq f(s)$ on $V \cdot V \supset V$ and so $g \geq f$ everywhere on T. Thus $f \leq g$ and $\tilde{f} \leq \tilde{g} \leq \Phi$.

We claim that $g \mid (V \cdot V)$ is lower semicontinuous; since $g = 0$ outside $V \cdot V$, it follows that g itself is lower semicontinuous. To prove our assertion, suppose that s_0 is a point of $V \cdot V$ and that $g(s_0) > c$. We take $c' > c$ such that $g(s_0) > c'$. Then $\Phi(t, t') > c'$ on $L_{s_0} \cap (\bar{V} \times \bar{V})$, where $L_s = \{(t, t') : tt' = s\}$. Now $L_{s_0} \cap (\bar{V} \times \bar{V})$ is compact and Φ is lower semicontinuous. Hence $\Phi(t, t') > c'$ on some open set W containing $L_{s_0} \cap (\bar{V} \times \bar{V})$. If s is close enough to s_0, W will contain $L_s \cap (\bar{V} \times \bar{V})$ and so $g(s)$ will be at least $c' > c$ for all s in $V \cdot V$ sufficiently close to s_0. This shows that $g \mid (V \cdot V)$ is lower semicontinuous.

It now follows that $\gamma^*(f) \leq \gamma^*(g)$ which, by the first part of (1), is equal to $\pi^*(\tilde{g}) \leq \pi^*(\Phi) \leq I + \varepsilon$. Letting $\varepsilon \to 0$, (4) follows. ▮

4.19.6 **Corollary.** (a) If α and β are bounded, then $\gamma = \alpha * \beta$ exists, is bounded, and (4.19.5) holds for all $f \geq 0$, while (4.19.3) holds for all $f \geq 0$ that are γ-measurable.

(b) If $\gamma = \alpha * \beta$ exists, if T is γ-σ-finite, then (4.19.5) holds for all $f \geq 0$, and (4.19.3) holds for all γ-measurable $f \geq 0$.

Proof. These statements follow at once from (4) and (1) of Theorem 4.19.5, since in either case any f is γ-σ-finite. ▮

Concerning associativity of convolution. Let α, β, and γ be three Radon measures on T. Given that $\alpha * \beta$ and $(\alpha * \beta) * \gamma$ exist, we cannot infer that $\beta * \gamma$ and $\alpha * (\beta * \gamma)$ exist. Indeed, the first two exist whenever $\alpha = 0$, no matter what β and γ are. As the next result shows, however, associativity does obtain for α, β, and γ suitably restricted.

4.19.7 Proposition. Let α, β, and γ be complex Radon measures on T. Suppose that $\alpha * \beta$, $(\alpha * \beta) * \gamma$, $\beta * \gamma$, and $\alpha * (\beta * \gamma)$ all exist. Then

$$(\alpha * \beta) * \gamma = \alpha * (\beta * \gamma). \tag{4.19.6}$$

Proof. It is enough to consider the case in which all three measures are positive. If x belongs to $\mathscr{K}_+(T)$, the values of $(\alpha * \beta) * \gamma(x)$ and $\alpha * (\beta * \gamma)(x)$ are, respectively,

$$\int d\gamma(t'') \int d\alpha(t) \int x(tt't'') \, d\beta(t')$$

and

$$\int d\alpha(t) \int d\beta(t') \int x(tt't'') \, d\gamma(t'').$$

The equality of these follows from the fact that $x(tt't'')$ is lower semicontinuous on $T \times T \times T$, together with repeated application of Subsection 4.17.2. The identity of the two measures in question is thus demonstrated. ▌

4.19.8 Corollary. Convolution is associative on the set of bounded Radon measures.

Proof. The proof follows from Corollary 4.19.6 (a) and Proposition 4.19.7. ▌

This is perhaps the most important case in practice.

4.19.9 Corollary. If at least two of α, β, and γ have compact supports, then (4.19.6) holds. In particular, convolution is associative on the set of Radon measures with compact supports.

4.19.10 Remarks. Let $M(T)$ (resp. $M_c(T)$) represent the set of Radon measures on T that are bounded (resp. that have compact supports). The last two corollaries show that both M and M_c are algebras when the ring product is taken to be convolution. In either case, $t \to \varepsilon_t$ is a one-to-one map of T into the algebra in question such that $\varepsilon_t * \varepsilon_{t'} = \varepsilon_{tt'}$. Either algebra therefore seems to qualify for the title of group algebra of T. If T is compact, the two algebras coincide. Otherwise, however, they are structurally quite different (even as topological vector spaces). There is at this stage little upon which to base a choice of either one in preference to the other. As we shall see, there are still more algebras that also qualify for the title of group algebra.

4.19.11 Convolutions Involving Functions. Hitherto convolution has been regarded as a process applying exclusively to measures. By identifying functions with those measures that are LAC(μ), where μ is a chosen left Haar measure, the concept can be extended to include functions. (One might equally well employ the right Haar measure to obtain a parallel development, but the differences would be rather trivial.)

Let μ denote any fixed left Haar measure on T. We consider first the convolution $\alpha * \beta$, where $\alpha = f \cdot \mu$, the function f being assumed to be locally integrable for μ. This convolution exists if and only if $|\alpha| * |\beta| = (|f| \cdot \mu) * |\beta|$

exists. By Proposition 4.19.2(2), this is the case if and only if

$$\int^* |\alpha|^*(Ut'^{-1})\, d\, |\beta|\, (t) < +\infty$$

for each relatively compact open set U. Moreover, by Proposition 4.13.2(1),

$$|\alpha|^*(Ut'^{-1}) = \int \chi_{Ut'^{-1}}(t)\, |f(t)|\, d\mu(t)$$
$$= \int \chi_U(tt')\, |f(t)|\, d\mu(t).$$

Thus $\alpha * \beta$ exists if and only if

$$\int^* d\, |\beta|\, (t') \int \chi_U(tt')\, |f(t)|\, d\mu(t) < +\infty,$$

that is, if and only if

$$\int^* d\, |\beta|\, (t') \int \chi_U(s)\, |f(st'^{-1})|\, \Delta(t')\, d\mu(s) < +\infty$$

for each relatively compact open set $U \subset T$. Assuming this to be so, the theorems of Tonelli and Fubini show that if x belongs to $\mathscr{K}(T)$, then

$$(\alpha * \beta)(x) = \int d\beta(t') \int x(tt') f(t)\, d\mu(t)$$
$$= \int d\beta(t') \int x(s) f(st'^{-1})\, \Delta(t')\, d\mu(s)$$
$$= \int x(s)\, d\mu(s) \int f(st'^{-1})\, \Delta(t')\, d\beta(t'),$$

the function $t' \to f(st'^{-1})\, \Delta(t')$ being β-integrable for almost all (μ) values of s. Thus the function g, defined a.e. (μ) by the formula

$$g(t) = \int f(ts^{-1})\, \Delta(s)\, d\beta(s), \tag{4.19.7}$$

is locally integrable for μ, and $(f \cdot \mu) * \beta = g \cdot \beta$. In view of this we regard the convolution $f * \beta$ as being the function g, defined almost everywhere by (4.19.7).

The above argument may be repeated under the assumption that β, rather than α, has the form $f \cdot \mu$. This would lead to defining $\alpha * f$ as the function h defined a.e. (μ) by the formula

$$h(t) = \int f(s^{-1}t)\, d\alpha(s), \tag{4.19.8}$$

this h being locally integrable for μ as a consequence of the assumed existence of $\alpha * \beta = \alpha * (f \cdot \mu)$.

One may also consider the case in which $\alpha = f \cdot \mu$ and $\beta = g \cdot \mu$, f and g being locally integrable for μ. The convolution $\alpha * \beta$ exists if and only if

$$\int^* |f(t)|\, d\mu(t) \int^* \chi_U(s)\, |g(t^{-1}s)|\, d\mu(s) < +\infty \tag{4.19.9}$$

for each relatively compact open set $U \subset T$. In this case the function

$$h(t) = \int f(ts) g(s^{-1})\, d\mu(s) = \int f(s) g(s^{-1}t)\, d\mu(s)$$
$$= \int f(ts^{-1}) g(s)\, \Delta(s)\, d\mu(s) \tag{4.19.10}$$

is defined a.e. (μ), is locally integrable for μ, and $\alpha * \beta = (f \cdot \mu) * (g \cdot \mu) = h \cdot \mu$. Accordingly, it is natural to regard h as the convolution $f * g$ of the functions f and g (in that order). Perhaps the most important case here is that in which f and g are μ-integrable, in which case $\alpha = f \cdot \mu$ and $\beta = g \cdot \mu$ are bounded measures; then $\alpha * \beta$ exists and is bounded, so that h is μ-integrable; in fact, one finds that

$$N_1(h) = N_1(f * g) \leq N_1(f)N_1(g). \tag{4.19.11}$$

Thus in $\mathscr{L}^1 = \mathscr{L}^1(T, \mu)$ (or its quotient space $L^1 = L^1(T, \mu)$) one has another group algebra that is, as it turns out, usually the most manageable of all.

Two other cases of $f * g$ are of frequent use, and will be discussed in the next section.

Meanwhile we may note that with the definition (4.19.10), $\mathscr{K}(T)$ itself forms an algebra under convolution. Here is yet another contender for the title of group algebra.

4.19.12. Suppose f belongs to $\mathscr{L}^1$ and g to $\mathscr{L}^p$ $(1 \leq p \leq \infty)$ (or $g \in \overline{\mathscr{L}}^\infty$ if $p = \infty$). Then Hölder's inequality shows that the inner integral in (4.19.9) is a bounded function of t, so that (4.19.9) is satisfied. Consequently, $h = f * g$ exists. Moreover, if x belongs to $\mathscr{K}(T)$, one has

$$\left| \int hx \, d\mu \right| \leq \int^* |x(t)| \, d\mu(t) \int^* |f(s)g(s^{-1}t)| \, d\mu(s)$$
$$= \int^* |f(s)| \, d\mu(s) \int^* |g(s^{-1}t)| \, |x(t)| \, d\mu(t);$$

Hölder's inequality shows that the inner integral here is majorized by $N_p(g) \cdot N_{p'}(x)$. It follows that h belongs to $\mathscr{L}^p$ and

$$N_p(f * g) \leq N_1(f) \cdot N_p(g). \tag{4.19.12}$$

If $p = \infty$, one may replace $N_\infty(g)$ by $\bar{N}_\infty(g)$.

4.19.13. Suppose that f belongs to $\mathscr{L}^{p'}$ and $\check{g}$ to $\mathscr{L}^p$ (that is, $\Delta^{1/p}g$ belongs to $\mathscr{L}^p$). The left-hand side of (4.19.9) is equal to

$$\int^* \chi_U(s) \, d\mu(s) \int^* |f(t)| \, |g(t^{-1}s)| \, d\mu(t),$$

and Hölder's inequality shows that the inner integral here is majorized by

$$N_{p'}(f) \cdot \int \chi_U(s)^{1/p} \, d\mu(s) \cdot N_p(g) < +\infty.$$

Thus $h = f * g$ exists. Furthermore, Hölder's inequality applied to the right-hand side of (4.19.10) shows that the integral defining h exists everywhere, and that $h(t)$ is majorized by $N_{p'}(f) \cdot N_p(g)$. In particular, $f * g$ belongs to $\mathscr{L}^\infty$ and

$$N_\infty(f * g) \leq N_{p'}(f)N_p(\check{g}). \tag{4.19.13}$$

Actually, (4.19.10) can be used to show that $f * g$ is continuous; and if $1 < p < \infty$, $f * g$ tends to 0 at infinity.

4.19.14. From (4.19.12) it follows that $\mathscr{L}^1 \cap \mathscr{L}^p$ forms an algebra under convolution. When T is compact, this reduces to $\mathscr{L}^p$ itself. All these algebras might qualify for the title of group algebra.

4.19.15 Convolution and Translation. The left and right translates by amount a of a complex Radon measure α on T are defined by the formulae

$$(L'_a\alpha)(x) = \alpha(L_{a^{-1}}x), \qquad (R'_a\alpha)(x) = \alpha(R_{a^{-1}}x)$$

for x in $\mathscr{K}(T)$; or, what is equivalent, by the formulae

$$(L'_a\alpha)^*(E) = \alpha^*(aE), \qquad (R'_a\alpha)^*(E) = \alpha^*(Ea)$$

for arbitrary subsets E of T. For example: $L'_a\varepsilon_b = \varepsilon_{a^{-1}b}$. With this definition we verify easily that $L'_a(f \cdot \mu) = (L_af) \cdot \mu$ for any function f that is locally integrable for μ. Nevertheless, a different notation is called for because, while $L_{ab} = L_aL_b$, one has $L'_{ab} = L'_bL'_a$. In fact, L'_a is the adjoint of $L_{a^{-1}}$ in the duality between $\mathscr{K}(T)$ and the space of measures on T.

Reference to Proposition 4.19.2 shows then that $(L'_a\alpha) * \beta$ exists whenever $\alpha * \beta$ exists, in which case $(L'_a\alpha) * \beta = L'_a(\alpha * \beta)$. Similarly, $\alpha * (R'_a\beta)$ exists whenever $\alpha * \beta$ exists, in which case $\alpha * (R'_a\beta) = R'_a(\alpha * \beta)$. Thus left translation commutes with convolution on the left, and right translation with convolution on the right. These properties come close to being characteristic of convolution (on the left or on the right) considered as a linear mapping of various classes of measures (or functions) into themselves. For more details, see Subsection 5.11.3.

4.19.16 Convolution as a Smoothing Process. We return to the formulae (4.19.7) and (4.19.8) defining the functions $f * \beta$ and $\alpha * f$, respectively. Let us take the former for definiteness. If f is continuous, and if for each compact $K \subset T$,

$$|f(ts^{-1})|\,\Delta(s) \le \phi(s) \qquad (t \in K,\ s \in T \backslash A_t)$$

where ϕ (which may be taken to be lower semicontinuous) is β-integrable and A_t is β-negligible, then $f * \beta$ exists; further, the integral in (4.19.7) exists for all t and represents a continuous function of t. We may say, therefore, that $f * \beta$ is a continuous function whenever the stated conditions are fulfilled. Thus convolution on the left with f "smooths" the measure β into the continuous function $f * \beta$.

If T is R or $R/2\pi$, one may go further. For if it is the case that f has n continuous derivatives, each of which fulfills the same type of majorization applying to f itself, then $f * \beta$ will have n continuous derivatives, and the process of differentiation commutes with convolution:

$$(f * \beta)' = f' * \beta, \qquad (f * \beta)'' = f'' * \beta, \qquad \text{and so forth.}$$

When T is a product of finitely many groups R or $R/2\pi$, similar remarks apply to partial differentiation. Analogous assertions are valid for $\alpha * f$.

This type of smoothing process owes much of its interest and usefulness to the fact that, since $\varepsilon * \beta = \beta$, one may reasonably expect that $f * \beta$ will converge to β in some sense when f converges suitably to ε. This expectation is justified, the sense of convergence depending to some extent on further restrictions upon f and β. We consider a few typical cases.

(1) If the measures α_i have supports contained in a fixed compact set, and if $\lim \alpha_i = \varepsilon$ vaguely, then $\lim (\alpha_i * \beta) = \beta$ vaguely. For the hypotheses entail that $\lim \alpha_i(T) = 1$. Moreover if x belongs to $\mathscr{K}(T)$, one has

$$\alpha_i * \beta(x) - \alpha_i(T) \cdot \beta(x) = \int d\alpha_i(t) \int \{x(tt') - x(t')\}\, d\beta(t')$$
$$= \alpha_i(z),$$

where

$$z(t) = \int \{x(tt') - x(t')\}\, d\beta(t').$$

The function z is continuous, hence coincides on the compact set containing the supports of all the α_i with some function in $\mathscr{K}(T)$. Hence $\lim \alpha_i(z) = \varepsilon(z) = z(e) = 0$, whence our assertion.

(2) By applying (1) to the case in which $\alpha_i = f_i \cdot \mu$, the f_i being continuous functions vanishing outside a fixed compact set (and such that the α_i satisfy the other hypotheses appearing in (1)), the continuous functions $f_i * \beta = b_i$ converge vaguely to β (that is, the measures $b_i \cdot \mu$ converge vaguely to β).

(3) Let us consider the case in which $\alpha_i = f_i \cdot \mu$ and $\beta = g \cdot \mu$, where g belongs to $\mathscr{L}^p$. We suppose (cf. Subsection 4.19.12) that each f_i belongs to $\mathscr{L}^1$, so that $g_i = f_i * g$ belongs to $\mathscr{L}^p$. Putting $c_i = \int f_i\, d\mu$, we have for x in $\mathscr{K}(T)$

$$\int x(t)[g_i(t) - c_i g(t)]\, d\mu(t) = \int x(t)\, d\mu(t) \int f_i(s)[g(s^{-1}t) - g(t)]\, d\mu(s)$$

the modulus of which is majorized by

$$\int |x(t)|\, d\mu(t) \int |f_i(s)|\, |g(s^{-1}t) - g(t)|\, d\mu(s)$$
$$= \int |f_i(s)|\, d\mu(s) \int |g(s^{-1}t) - g(t)|\, |x(t)|\, d\mu(t).$$

Applying Hölder's inequality to the inner integral, we find that

$$\left| \int (g_i - c_i g)x\, d\mu \right| \leq \int |f_i(s)| \cdot N_p(L_{s^{-1}}g - g)\, d\mu(s) \cdot N_{p'}(x).$$

It follows from this that

$$N_p(g_i - c_i g) \leq \int |f_i(s)|\, N_p(L_{s^{-1}}g - g) \cdot d\mu(s)$$
$$= \int |f_i(s)|\, n(s)\, d\mu(s).$$

The function n is bounded, $|n(s)| \leq 2N_p(g)$; if $p \neq \infty$, n is continuous and $n(e) = 0$. If we assume that

$$\left.\begin{aligned} &\lim_i \int f_i(s)\, d(s) = 1, \\ &\limsup_i \int |f_i(s)|\, d\mu(s) < +\infty, \\ \text{and} \qquad &\lim_i \int_{T \setminus V} |f_i(s)|\, d\mu(s) = 0, \end{aligned}\right\} \tag{4.19.14}$$

for any neighbourhood V of e in T, it follows that $N_p(g_i - g) \to 0$. In other words, under these conditions on the family (f_i), and provided $1 \leq p < \infty$, $\lim (f_i * g) = g$ in the sense of $\mathscr{L}^p$ whenever g belongs to $\mathscr{L}^p$. The assertion is false when $p = \infty$, though it remains true that $\lim (f_i * g) = g$ weakly in $\mathscr{L}^\infty$; see also (4) immediately below.

(4) Lastly we examine the case in which $\alpha_i = f_i \cdot \mu$, the f_i being as in (3), and $\beta = g \cdot \mu$ with g continuous. Putting $g_i = f_i * g$ we find

$$|g_i(t) - c_i g(t)| \leq \int |f_i(s)| \cdot |g(s^{-1}t) - g(t)| \, d\mu(s).$$

We conclude that if the f_i satisfy (4.19.14), and if the f_i have supports contained in a fixed compact set, then $f_i * g \to g$ uniformly on compact sets. If g is bounded, and if the f_i satisfy (4.19.14) alone, the same is true. Finally, if g is bounded and uniformly continuous, and if the f_i satisfy (4.19.14), then $f_i * g \to g$ uniformly on T.

4.19.17. **Further Group Algebras.** We have so far met no less that five examples of group algebras associated with a given T, namely: $M(T)$ and $M_c(T)$ (4.19.10), $\mathscr{L}^1 = \mathscr{L}^1(T, \mu)$ and $\mathscr{K}(T)$ (4.19.11), and $\mathscr{L}^1 \cap \mathscr{L}^p$ (4.19.14). We now describe briefly a means of constructing infinitely many more group algebras that have proved to be of some interest.

Each of these algebras is characterized by a "weight function" w, which is to be a strictly positive function on T that is bounded away from 0 on each compact subset of T; w may take the value $+\infty$ at some or all points of T. We assume further that there exists a number $k > 0$ such that $w(tt') \leq k \cdot w(t)w(t')$ for arbitrary points t and t' of T. It turns out that we shall not alter the algebra to be associated with w if we replace w by any strictly positive multiple thereof. Since this multiple may be chosen so that the new function satisfies

$$w(tt') \leq w(t)w(t'), \tag{4.19.15}$$

we may as well assume that w satisfies (4.19.15) to begin with. We then say that w is a positive subcharacter of T.

Given $w > 0$ subject to (4.19.15), we denote by $M\{w\}$ the set of complex Radon measures α on T for which

$$N_w(\alpha) \equiv \int^* w \, d|\alpha| < +\infty. \tag{4.19.16}$$

$M\{w\}$ is obviously a complex vector space of Radon measures on which N_w is a seminorm.

To verify that $M\{w\}$ is an algebra under convolution, let U be any relatively compact open set in T. Then, by (4.19.15), one has

$$\begin{aligned} \chi_U(tt') &= w(t)w(t') \cdot \frac{\chi_U(tt')}{w(t)w(t')} \\ &\leq w(t)w(t') \cdot \frac{\chi_U(tt')}{w(tt')}; \end{aligned}$$

since w is bounded away from 0 on $U \cdot U$, we see that

$$\chi_U(tt') \leq \text{const. } w(t)w(t'),$$

the constant depending perhaps upon U. Applying Theorem 4.19.5(1) and Proposition 4.19.2(3), we see that $\alpha * \beta$ exists whenever α and β belong to $M\{w\}$. Moreover, if $\gamma = \alpha * \beta$, then $|\gamma| = |\alpha| * |\beta|$ and so (Theorem 4.19.5(1))

$$\begin{aligned} N_w(\gamma) = \int^* w \, d\,|\gamma| &= \int^* d\,|\alpha|\,(t) \int^* w(tt')\, d\,|\beta|\,(t') \\ &\leq \int^* d\,|\alpha|\,(t) \int^* w(t)w(t')\, d\,|\beta|\,(t') \\ &= N_w(\alpha) \cdot N_w(\beta), \end{aligned}$$

at any rate, if w is lower semicontinuous; the conclusion remains valid if T is σ-compact (Theorem 4.19.5(4)) or if w is bounded away from 0 (since then α and β are necessarily bounded, the same is true of γ, and Theorem 4.19.5(4) is again applicable). In all these cases, therefore, $\alpha * \beta$ again belongs to $M\{w\}$ and

$$N_w(\alpha * \beta) \leq N_w(\alpha) \cdot N_w(\beta). \tag{4.19.17}$$

An interesting subalgebra of $M\{w\}$ is obtained if we consider only those α in $M\{w\}$ that are LAC(μ). Any such α can be written $\alpha = f \cdot \mu$, where f is locally integrable for μ. Then, if w is lower semicontinuous or if T is α-σ-finite (which is necessarily the case if w is bounded away from 0), we have

$$N_w(\alpha) = \int^* w \, d\,|\alpha| = \overline{\int}^* w\,|f|\,d\mu < +\infty.$$

It follows that f is equal locally almost everywhere to a function f' such that wf' is in $\mathscr{L}^1(\mu)$ and for which

$$N_w(\alpha) = N_1(wf').$$

The subalgebra in question may therefore be identified with the set $\mathscr{L}^1\{w\}$ of functions f, measurable for μ and such that

$$N_w(f) = N_1(wf) < +\infty. \tag{4.19.18}$$

Those f for which $N_w(f) = 0$ are precisely the μ-negligible functions, and the associated quotient space $L^1\{w\}$ is a normed algebra. When $w = 1$, $\mathscr{L}^1\{w\}$ and $L^1\{w\}$ reduce to $\mathscr{L}^1(\mu)$ and $L^1(\mu)$, respectively.

It should be noticed that if E is the set of points t of T at which $w(t) = +\infty$, then E is α-negligible for each α in $M\{w\}$; and each f in $\mathscr{L}^1\{w\}$ vanishes a.e. (μ) on E.

4.19.18 **Example.** If we take $T = R$ and $w(t) = e^{k|t|}$, where k is real and $k \geq 0$, then all our requirements are satisfied.

4.19.19 Example. We take again $T = R$, k a real number. Define

$$w(t) = +\infty \ (t \leq 0), \ e^{kt} \ (t > 0).$$

or

$$w(t) = +\infty \ (t \leq 0), \ e^{kt^\alpha} \ (t > 0), \text{ where } k \geq 0 \text{ and } 0 \leq \alpha < 1.$$

This w satisfies all requirements. Each measure α in $M\{w\}$ is carried by $(0, +\infty)$, and each f in $\mathscr{L}^1\{w\}$ vanishes almost everywhere on $(-\infty, 0)$. One may thus think of $\mathscr{L}^1\{w\}$ as consisting of functions f on $(0, +\infty)$, μ-measurable there, and such that

$$N_w(f) = \int^* e^{kt} \, |f(t)| \, d\mu(t) < +\infty,$$

μ being now Lebesgue measure on R_+. The convolution of two such functions, f and g, is then given by

$$f * g(t) = \int_0^t f(t - s) g(s) \, d\mu(s) \qquad (t > 0);$$

in relation to convolution for functions on $(-\infty, +\infty)$, this is sometimes referred to as "truncated convolution" and is a natural operation when one is concerned with one-sided Laplace transforms $\int_0^\infty e^{-pt} f(t) \, d\mu(t)$ defined for p real and $p \geq k$ (or for complex p with $\operatorname{Re} p \geq k$).

4.19.20 Homomorphisms of Group Algebras. The well-developed study of special types of algebras (especially those that are complete, normed, and commutative), among which one finds many group algebras of commutative groups, indicates that the homomorphisms of such algebras onto the complex field play a crucial role. These homomorphisms play a role analogous to that played by the continuous linear forms on a LCTVS. Justification of this remark would take us far afield, and the reader must be referred to Naĭmark [1], Loomis [1], Rickart [1]. However, we wish to see how this remark, if accepted, leads one in the case of at least some group algebras to a natural extension of Fourier integrals and series and of Laplace transforms.

T denoting again a separated locally compact group, by a *character* of T is meant a complex-valued function χ on T such that

$$\chi(e) = 1, \qquad \chi(tt') = \chi(t)\chi(t'); \tag{4.19.19}$$

in other words, χ is a homomorphism of the group T onto the multiplicative group C^* of nonzero complex numbers. If χ is bounded, then (4.19.19) shows that necessarily $|\chi(t)| = 1$ for all t, and so χ is a homomorphism of T into the multiplicative group of complex numbers having unit modulus. We shall be concerned mainly with those characters that are continuous; but see Exercise 4.22.

It is easily seen that the continuous characters of R are precisely the functions of the form $\chi(t) = e^{ict}$ where c is a complex number; if χ is bounded, c is real, and conversely. The continuous characters of $R/2\pi$ are derived from the above functions by restricting c to be an integer.

If χ is any character of T, we may in Subsection 4.19.17 take $w = |\chi|$ and construct the associated algebras $M\{w\}$ and $\mathscr{L}^1\{w\}$, which we now denote by $M\{\chi\}$ and $\mathscr{L}^1\{\chi\}$. If χ is bounded, these become simply M and $\mathscr{L}^1$. In any case, if χ is continuous, the mappings

$$\alpha \to \hat{\alpha}(\chi) = \int \bar{\chi}\, d\alpha$$

or (4.19.20)

$$f \to \hat{f}(\chi) = \int \overline{\chi(t)} \cdot f(t)\, d\mu(t)$$

of $M\{\chi\}$ or $\mathscr{L}^1\{\chi\}$ are homomorphisms of these algebras into C. If the algebra involved is not $\{0\}$, the homomorphism is actually *onto* C. (To see this it suffices to show that the algebra is not mapped into $\{0\}$. Suppose it were. If the measure or function α belongs to the algebra, so does $f \cdot \alpha$ for any f in $\mathscr{K}(T)$. Our supposition therefore involves that $\int \bar{\chi} f\, d\alpha = 0$ for all f, hence $\bar{\chi} \cdot \alpha = 0$; since χ is continuous and nonvanishing, this entails that $\alpha = 0$, contrary to the assumption that the algebra is not $\{0\}$.)

More generally one may take a set S of continuous characters and form the algebras

$$M\{S\} = \bigcap_{\chi \in S} M\{\chi\}, \qquad \mathscr{L}^1\{S\} = \bigcap_{\chi \in S} \mathscr{L}^1\{\chi\}.$$

Then each character belonging to S generates via (4.19.20) a homomorphism of $M\{S\}$ or of $\mathscr{L}^1\{S\}$ onto C. Especially important is the case in which S consists of all bounded continuous characters, when $M\{S\} = M$ and $\mathscr{L}^1\{S\} = \mathscr{L}^1$.

It is entirely natural to ask whether, conversely, every homomorphism of $M\{S\}$ or $\mathscr{L}^1\{S\}$ onto C is obtainable in the manner just described. In *general* the answer is "No" in the case of $M\{S\}$ and "Yes" in the case of $\mathscr{L}^1\{S\}$. In particular, taking $T = R$, it is known (Rudin [9], §5.3; Hewitt and Ross [1]) that not all homomorphisms of M are obtained by this process, and the situation remains obscure. On the other hand, it is known that every homomorphism of $\mathscr{L}^1$ onto C is obtainable from some bounded, continuous character of T, and this whenever T is commutative. Taking $T = R$, the function $\hat{f}$, defined by (4.19.20) for all bounded continuous characters χ, is just the classical Fourier transform of f. In support of our positive assertion, we shall consider briefly the case of $\mathscr{L}^1\{w\}$.

For simplicity we shall restrict the discussion to the case in which w is finite everywhere and locally integrable for μ. (Certain other cases, including that described in Example 4.19.19, can be dealt with by suitable modification of the reasoning to follow.) In this case, $\mathscr{K}(T) \subset \mathscr{L}^1\{w\}$ and an easy application of the Hahn-Banach theorem coupled with Theorem 4.16.1 will show that $\mathscr{K}(T)$ is dense in $\mathscr{L}^1\{w\}$ (with respect to the seminorm N_w). $\mathscr{L}^1\{w\}$ is also complete. When T is commutative, $\mathscr{L}^1\{w\}$—and likewise all the group algebras listed so far—are commutative. For algebras of this type it is known (Rickart [1], Corollary (3.1.7)) that every homomorphism is continuous. Thus, if X denotes any homomorphism of $\mathscr{L}^1\{w\}$ onto C, then

$$|X(f)| \leq \text{const. } N_w(f).$$

Under the stated conditions upon w, $\mathscr{L}^1\{w\}$ is stable under translation: if f belongs to $\mathscr{L}^1\{w\}$, so too does $L_a f$ for any group element a. Since X is not identically vanishing, we can choose an f in $\mathscr{K}(T)$ with $X(f) \neq 0$. We consider the function on T defined by

$$\chi(a) = \overline{X(L_{a^{-1}}f)/X(f)}.$$

Our conditions upon w ensure that $L_{a^{-1}}f$ depends continuously upon a, so that χ is continuous. Further, whenever T is commutative,

$$L_{a^{-1}}f * L_{b^{-1}}f = L_{(ab)^{-1}}f * f,$$

as is easily checked. It follows that

$$\chi(ab) = \chi(a)\chi(b),$$

so that χ is a continuous character of T. Moreover,

$$|\chi(a)| \leq \text{const. } N_w(L_{a^{-1}}f) \leq \text{const. } w(a). \tag{4.19.21}$$

Suppose now that g belongs to $\mathscr{K}(T)$. The convolution

$$f * g(t) = \int f(ts^{-1})g(s)\, d\mu(s)$$

can then be approximated arbitrarily closely in $\mathscr{L}^1\{w\}$ by finite sums

$$\sum_i L_{s_i^{-1}}f \cdot g(s_i)\mu(A_i)$$

of such a form that the sums $\sum_i \overline{\chi(s_i)}g(s_i)\mu(A_i)$ tend to $\int \overline{\chi(s)}g(s)\, d\mu(s)$. The continuity of X shows that then

$$X(f * g) = X(f)X(g)$$

is the limit of the sums $\sum_i X(L_{s_i^{-1}}f) \cdot g(s_i)\mu(A_i)$, so that division by $X(f) \neq 0$ leads to the conclusion that

$$X(g) = \int \overline{\chi(s)}g(s)\, d\mu(s).$$

This is true for each g in $\mathscr{K}(T)$. Both sides are continuous on $\mathscr{L}^1\{w\}$ (since $|\chi(s)| \leq \text{const} \cdot w(s)$), and so the equality must be true by continuity for each g in $\mathscr{L}^1\{w\}$. Thus

$$X(f) = \hat{f}(\chi)$$

for f in $\mathscr{L}^1\{w\}$, showing that the homomorphism X is derived via (4.19.20) from the continuous character χ satisfying (4.19.21). Conversely (what is trivial), each continuous character χ satisfying (4.19.21) generates in this fashion a continuous homomorphism of $\mathscr{L}^1\{w\}$ onto C.

4.19.21 **Example.** Taking T and W as in Example 4.19.18 we see that $\chi(t) = e^{ict}$ and $|e^{ict}| \leq \text{const} \cdot e^{k|t|}$. Thus $|\text{Im } c| \leq k$ and

$$\hat{f}(\chi) = \int e^{-ict}f(t)\, d\mu(t).$$

When $k = 0$, c is real and $\hat{f}$ is just the classical Fourier transform of $f \in \mathscr{L}^1(\mu)$.

We shall return in Sections 10.3 and 10.4 to the problem of developing the theory of such Fourier transforms on locally compact commutative groups.

4.19.22 **Notes.** The various convolution algebras over a locally compact group T present a number of problems, solved and unsolved, concerning their structure. See Segal [1] and [3]; Reiter [1]–[6]; Rudin [1], [2], [5], [6], [7], and [9]; Hewitt [11]; Hewitt and Ross [1]; and Malliavin [1], [2], [3].

For several function spaces A over T it is the case that each bounded measure μ defines an endomorphism of A when it acts through convolution. For example, each μ defines an endomorphism $C_{\mu,p}$ of L^p, $C_{\mu,p}f$ being the class of the function $\mu * f$ (which is equal almost everywhere to a function in $\mathscr{L}^p$). Each $C_{\mu,p}$ is continuous and has a norm at most equal to $\|\mu\|$ (the total mass of $|\mu|$). There are some unsolved problems relating to these endomorphisms $C_{\mu,p}$, for which see Dieudonné [12] and [12a].

In Chapter 5 we shall study briefly the convolution for entities (distributions) more general than functions and measures, restricting ourselves to the case $T = R^m$.

A somewhat different approach to convolutions and their properties is given in Hewitt and Ross [1], §§19–20.

4.20 Weil's Criterion for Compactness in $\mathscr{L}^p$ Spaces over a Group

In this section T denotes a separated locally compact group with left Haar measure μ. The reader is reminded that whenever f is a function on T and a is a group element, ${}_af$ denotes the left translate $t \to f(at)$ of f. Weil's useful criterion runs as follows.

4.20.1 **Theorem.** Suppose $1 \le p < \infty$. A subset F of $\mathscr{L}^p = \mathscr{L}^p(T, \mu)$ is relatively compact in $\mathscr{L}^p$ if and only if it satisfies the following three conditions:

(1) Sup $\{N_p(f) : f \in F\} < +\infty$;

(2) Given $\varepsilon > 0$, there exists a compact set $K \subset T$ such that

$$\operatorname{Sup}\left\{\int_{T\setminus K} |f|^p \, d\mu : f \in F\right\} \le \varepsilon;$$

(3) Given any $\varepsilon > 0$, there exists a neighbourhood U of the neutral element of T such that $a \in U$ implies

$$\operatorname{Sup}\{N_p({}_af - f) : f \in F\} \le \varepsilon$$

Remarks. (1) says that F is a bounded subset of $\mathscr{L}^p$; (2) says that the functions in F are "equivanishing at ∞" in the $\mathscr{L}^p$ sense; and (3) says that the functions in F are equicontinuous in the $\mathscr{L}^p$ sense. There are, therefore, obvious similarities with Ascoli's theorem.

Proof. The necessity of all the conditions is an almost obvious consequence of precompactness (that is, total boundedness) of F: this means that, given any $\varepsilon > 0$, there exist $f_1, \cdots, f_n$ in F such that each f in F is subject to $N_p(f - f_i) \leq \varepsilon$ for some i, $1 \leq i \leq n$, depending on f. Each of (1), (2), and (3) is satisfied for a one-element subset of $\mathscr{L}^p$, hence by any finite subset $\{f_1, \cdots, f_n\}$, and hence by approximation by any precompact subset F of $\mathscr{L}^p$.

Let us pass to the proof of sufficiency. Given any $\varepsilon > 0$ choose K as in (2) and let F^* be the set of functions $f^* = f\chi_K$ obtained as f ranges over F. According to (3), if we choose a function r in $\mathscr{K}(T)$, $r \geq 0$, $r = 0$ outside a sufficiently small symmetric compact neighbourhood U of the neutral element of T, and $\int r\, d\mu = 1$, then we shall have

$$N_p(f^* - f^{**}) \leq \varepsilon \tag{4.20.1}$$

uniformly for f in F, where we have put $f^{**} = f^* * r$. Let F^{**} be the set of all such f^{**} when f ranges over F. These f^{**} vanish outside the compact set $K \cdot U$. Moreover Hölder's inequality yields

$$\begin{aligned} N_\infty(f^{**}) &\leq N_p(f^*) \cdot N_{p'}(r) \leq \text{const. } N_p(f^*) \\ &\leq \text{const. } (N_p(f) + \varepsilon) \leq \text{const.} \end{aligned}$$

by (1), the constant possibly depending on ε. By (3), the functions f^{**} are equicontinuous. By Ascoli's theorem, F^{**} is relatively compact in $C(\tilde{T})$, where $\tilde{T}$ is the Alexandroff compactification of T. Hence F^{**} is totally bounded in $C(\tilde{T})$. This, together with the fact that the functions in F^{**} all vanish outside of $K \cdot U$, shows that a fortiori F^{**} is totally bounded in $\mathscr{L}^p$. But then (4.20.1) shows that F^* is totally bounded, and then (2) shows likewise that F itself is totally bounded. $\mathscr{L}^p$ being complete, F is therefore relatively compact. ∎

Remark. Weil ([1], p. 53) states the criterion for $p = \infty$ as well, but Weil's L^∞ is what we have denoted by $C_0(T)$; in this case the criterion is essentially Ascoli's theorem itself.

The case in which T is the real axis is due to M. Riesz [1]. The result is extended to $0 < p < 1$ by Tsuji [1].

4.21 Weak Compactness in $\mathscr{L}^1$; the Dunford-Pettis Criterion

We know already that if $1 < p \leq \infty$, the weakly relatively compact subsets of $\mathscr{L}^p$ are simply the bounded subsets thereof: this is a consequence of Theorem 4.16.1 in conjunction with Theorem 1.11.4(2). The analogous statement for $\mathscr{L}^1$ is false, save in the trivial case in which $\mathscr{L}^1 = \mathscr{L}^1(\mu)$ and the measure μ has a finite support. On the other hand, Weil's Theorem 4.20.1 gives a criterion for compactness in each of the spaces $\mathscr{L}^p$ with $1 < p < \infty$. It is natural to seek an analogous criterion for weak compactness in $\mathscr{L}^1$. Such a criterion is due to Dunford and Pettis, given as Theorem 3.2.1 of their joint paper [1]. In this section we shall deal with this criterion, following in the main a later treatment due to Dieudonné [3]. The reader is recommended to examine also the treatment given by Grothendieck ([7], Ch. V, pp. 27 et seq.). Both of these accounts influence the approach sketched by Bourbaki ([6], Chapitre 5, p. 67, Exercice 17).

Throughout this section T denotes a separated locally compact space and μ a fixed positive Radon measure on T; we shall frequently write $\mathscr{L}^1$ as an abbreviation for $\mathscr{L}^1(T, \mu)$.

Our labour is lightened by making appeal to results belonging to general duality theory—specifically Theorem 1.11.4(2), Theorem 8.2.2, and the theorems of Šmulian and of Eberlein (see Section 8.12). It is hoped that the reader will not be disconcerted by this long glance ahead.

4.21.1 **Proposition.** Let (f_n) be a sequence extracted from $\mathscr{L}^1$.

(1) If $\lim_{n\to\infty} \int_A f_n \, d\mu$ exists finitely for each integrable set $A \subset T$, then, given $\varepsilon > 0$, there exists $\delta > 0$ such that

$$\operatorname*{Sup}_n \int_A |f_n| \, d\mu \leq \varepsilon$$

holds for each integrable set $A \subset T$ satisfying $\mu(A) \leq \delta$.

(2) If $\lim_{n\to\infty} \int_A f_n \, d\mu$ exists finitely for each σ-finite measurable set $A \subset T$, then, given any $\varepsilon > 0$, there exists a compact set $K \subset T$ such that

$$\operatorname*{Sup}_n \int_{T\backslash K} |f_n| \, d\mu \leq \varepsilon.$$

Proof. (a) We shall deal first with the case in which the limits are zero.

(1) Let L be the subset of $\mathscr{L}^1$ formed of all functions equal almost everywhere to characteristic functions of integrable subsets of T. L is closed in $\mathscr{L}^1$, hence is a complete semimetrizable space. Thus (Baire's theorem) L is nonmeagre.

Let us show first that for each f in $\mathscr{L}^1$, the function $g \to \int fg \, d\mu$ is continuous on L. To see this, we notice that f is the limit in $\mathscr{L}^1$ of bounded functions f^*, so that $\int fg \, d\mu$ is the limit, uniformly on L, of the functions $\int f^*g \, d\mu$. It is obvious that $\int f^*g \, d\mu$ is continuous on L. Hence continuity of $\int fg \, d\mu$.

We now put

$$L_n = \left\{g \in L : \left| \int g f_m \, d\mu \right| \leq \frac{\varepsilon}{16} \text{ for } m \geq n\right\}.$$

L_n is closed in L and, by hypothesis, $L = \bigcup\{L_n : n \in N\}$. Since L is nonmeagre, there exists a natural number k such that L_k contains a nonvoid open subset of L. This signifies that there exists an integrable set $A_0 \subset T$ and a number $\delta > 0$ such that

$$\operatorname{Sup} \left\{\left| \int_A f_n \, d\mu \right| : n \geq k\right\} \leq \frac{\varepsilon}{16}$$

holds for all integrable sets $A \subset T$ such that $\int |\chi_A - \chi_{A_0}| \, d\mu \leq \delta$. On the other hand, by enlarging k if necessary, we may assume that

$$\left| \int_{A_0} f_n \, d\mu \right| \leq \frac{\varepsilon}{16} \qquad \text{for } n \geq k.$$

Now if $B \subset T$ is integrable, $B \subset A_0$, and $\mu(B) \leq \delta$, then

$$\int |\chi_{A_0} - \chi_{A_0 \backslash B}| \, d\mu \leq \delta$$

and so, by the above inequalities,

$$\left| \int_B f_n \, d\mu \right| \leq \frac{\varepsilon}{8} \qquad \text{for } n \geq k.$$

This holds a fortiori if f_n is replaced by $\operatorname{Re} f_n$ and $\operatorname{Im} f_n$; and, by replacing B by the sets $B \cap \{\operatorname{Re} f_n \geq 0\}$, $B \cap \{\operatorname{Re} f_n < 0\}$, and so forth, we find that

$$\int_B |f_n| \, d\mu \leq \frac{\varepsilon}{2}$$

whenever $\mu(B) \leq \delta$ and $B \subset A_0$.

A similar result obtains for those integrable sets $B \subset T \backslash A_0$ with $\mu(B) \leq \delta$.

If A is integrable and $\mu(A) \leq \delta$, we may apply the preceding inequalities to the case of $A \cap A_0$ and $A \backslash A_0$ in turn and so obtain $\int_A |f_n| \, d\mu \leq \varepsilon$ for $n \geq k$. Moreover, by decreasing δ if necessary, we can arrange that the same holds for $n = 1, 2, \cdots, k - 1$. This will complete the proof of (1).

(2) We use techniques similar to that used to prove (1). We note first that there exists an increasing sequence (K_r) of compact sets and a negligible set E such that all the f_n vanish outside the union of $S = \bigcup K_r$ and E.

We now replace L by the set M of functions equal almost everywhere to characteristic functions of measurable subsets of S. We define

$$d_r(g, g') = \int_{K_r} |g - g'| \, d\mu.$$

These semimetrics d_r $(r = 1, 2, \cdots)$ make M into a complete semimetrizable space. (To prove completeness, we use Theorem 4.6.7 and the diagonal process.)

If f belongs to $\mathscr{L}^1$ and vanishes almost everywhere on $T \backslash S$, then $g \to \int fg \, d\mu$ is continuous on M. For, given $\varepsilon > 0$, there exists r such that $\int_{T \backslash K_r} |f| \, d\mu \leq \varepsilon$; and then we may approximate f in $\mathscr{L}^1(K_r)$ by a bounded function f^*, and so on.

Applying Baire's theorem as in the proof of (1) we infer the existence of a measurable set $A_0 \subset S$ and natural numbers s and k and a number $\delta > 0$ such that

$$\left| \int_A f_n \, d\mu \right| \leq \frac{\varepsilon}{8} \qquad \text{for } n \geq k$$

whenever A is measurable, $A \subset S$, and $\int_{K_s} |\chi_A - \chi_{A_0}| \, d\mu \leq \delta$. By a little juggling like that used in the proof of (1), this leads to the desired conclusion. Indeed, we first enlarge k if necessary so as to arrange that $|\int_{A_0} f_n \, d\mu| \leq \varepsilon/8$ for

$n \geq k$. Then if B is measurable and $B \subset T \backslash K_s$ the preceding inequalities may be applied to $A = B \cup (A_0 \cap K_s)$, which satisfies

$$\int_{K_s} |\chi_A - \chi_{A_0}| \, d\mu = 0,$$

to get

$$\left| \int_A f_n \, d\mu \right| \leq \frac{\varepsilon}{8} \qquad \text{for } n \geq k.$$

The same is true with $A_0 \cap K_s$ in place of A. Thence it follows, since A is the disjoint union of B and $A_0 \cap K_s$, that

$$\left| \int_B f_n \, d\mu \right| \leq \frac{\varepsilon}{4} \qquad \text{for } n \geq k.$$

Hence

$$\int_B |f_n| \, d\mu \leq \varepsilon \qquad \text{for } n \geq k.$$

This being true for each measurable $B \subset T \backslash K_s$, it follows that

$$\int_{T \backslash K_s} |f_n| \, d\mu \leq \varepsilon \qquad \text{for } n \geq k.$$

On the other hand, for each $n < k$ there exists r_n such that $\int_{T \backslash K_{r_n}} |f_n| \, d\mu \leq \varepsilon$ —this because $f_n \in \mathscr{L}^1$ is 0 outside S. So our assertion is established with $K = K_r$ where $r = \text{Sup}\,(r_1, \cdots, r_{k-1}, s)$.

(b) We have yet to deal with the case in which the limits are not assumed to be zero. In describing the passage to this more general case, it suffices to consider (1), the technique for handling (2) being entirely similar.

Assuming the hypothesis of (1), let us show first that, given any $\varepsilon > 0$, there exists a number $\delta' > 0$ and a natural number k such that

$$\int_A |f_m - f_n| \, d\mu \leq \tfrac{1}{2}\varepsilon \tag{4.21.1}$$

whenever $m \geq k$, $n \geq k$ and $\mu(A) \leq \delta'$. In fact, if this were not so, there would exist a number $\varepsilon_0 > 0$ and sequences (m_i), (n_i), and (A_i) such that $m_i \geq i$, $n_i \geq i$ and $\mu(A_i) \leq i^{-1}$, and for which

$$\int_{A_i} |f_{m_i} - f_{n_i}| \, d\mu \geq \varepsilon_0.$$

However, the sequence $(g_i) = (f_{m_i} - f_{n_i})$ is such that $\lim_i \int_A g_i \, d\mu = 0$ for each integrable set A. And thus the last-written inequality leads to a contradiction of (1) in the form already established. Thus (4.21.1) is established.

Finally, having chosen $\delta' > 0$ and k so that (4.21.1) holds, we choose $\delta'' > 0$ so that $\int_A |f_n| \, d\mu \leq \frac{1}{2}\varepsilon$ for $\mu(A) \leq \delta''$ and $n \leq k$. These inequalities together with (4.21.1) show that then

$$\int_A |f_n| \, d\mu \leq \tfrac{1}{2}\varepsilon + \tfrac{1}{2}\varepsilon = \varepsilon$$

for all n, provided $\mu(A) \leq \delta = \text{Inf}\,(\delta', \delta'')$, and the argument is complete. ▮

4.21.2 **Theorem** (Dunford-Pettis). In order that a subset P of $\mathcal{L}^1$ be weakly relatively compact, it is necessary and sufficient that the following three conditions be fulfilled:

(1) $\operatorname{Sup}\left\{\int |f|\, d\mu \colon f \in P\right\} < +\infty$.

(2) Given $\varepsilon > 0$, there exists a number $\delta > 0$ such that

$$\operatorname{Sup}\left\{\int_A |f|\, d\mu : f \in P\right\} \leq \varepsilon$$

provided $A \subset T$ is integrable and $\mu(A) \leq \delta$.

(3) Given any $\varepsilon > 0$, there exists a compact set $K \subset T$ such that

$$\operatorname{Sup}\left\{\int_{T \setminus K} |f|\, d\mu : f \in P\right\} \leq \varepsilon.$$

Proof. Let us first establish the necessity of the conditions. That of (1) follows from Theorem 8.2.2. Suppose next that P were weakly relatively compact in $\mathcal{L}^1$ and yet that (2) fails to hold. Then we could extract from P a sequence (f_n) and find a sequence (A_n) of integrable sets such that

$$\mu(A_n) \leq n^{-1}, \qquad \int_{A_n} |f_n|\, d\mu \geq \alpha > 0,$$

α being independent of n. By Šmulian's theorem (Theorem 8.12.1), there exists a subsequence $(g_k) = (f_{n_k})$ that is weakly convergent to some f in $\mathcal{L}^1$. Then $f - g_k \to 0$ weakly. In that case, however, the above inequalities contradict Proposition 4.21.1(1).

In a similar way we find that (3) is necessary, relying this time on Proposition 4.21.1(2).

Turning to sufficiency, (1) combines with Theorem 1.11.4(2) to show that $\bar{P}$, the closure of P in the dual of $\mathcal{L}^\infty$, is compact for the weak topology in that dual space. Thus everything hinges on showing that $\bar{P}$ is a subset of $\mathcal{L}^1$. (To speak more precisely, one should really pass from $\mathcal{L}^1$ to the quotient space L^1, which is injected into the dual of $\mathcal{L}^\infty$.) We suppose then that L is a continuous linear form on $\mathcal{L}^\infty$, which is weakly adherent to P. From (2) it follows that to each $\varepsilon > 0$ corresponds a $\delta > 0$ such that if $\mu(A) \leq \delta$ and if $g \in \mathcal{L}^\infty$ satisfies $|g| \leq \chi_A$, then $|\int fg\, d\mu| \leq \varepsilon$ for all f in P, and hence that $|L(g)| \leq \varepsilon$ in the same circumstances. It is clear on the other hand that $\lambda = L \mid \mathcal{K}(T)$ is a Radon measure on T. We therefore have

$$\left|\int g\, d\lambda\right| = |L(g)| \leq \varepsilon$$

whenever $g \in \mathcal{K}(T)$, $|g| \leq \chi_A$, and $\mu(A) \leq \delta$. An easy argument now shows that λ is LAC relative to μ. By the Lebesgue-Radon-Nikodym Theorem (Theorem 4.15.1), there exists a locally integrable function f_0 such that $\lambda = f_0 \cdot \mu$. Condition (1) may be used to show that this f_0 is necessarily integrable; that is, $f_0 \in \mathcal{L}^1$. We now have

$$L(g) = \int f_0 g\, d\mu$$

for g in $\mathscr{K}(T)$. Using again the property that $|L(g)| \leq \varepsilon$ whenever $g \in \mathscr{L}^\infty$, $|g| \leq \chi_A$, and $\mu(A) \leq \delta$, it is an easy step to show that $L(g) = \int f_0 g \, d\mu$ holds good for each g in $\mathscr{L}^\infty$, showing that L is (that is, is generated by) the element f_0 of $\mathscr{L}^1$, which is what we wished to establish. ▌

Remark. We have used Theorem 8.2.2 to deduce the necessity of (1). A direct proof is also possible, using Proposition 4.21.1; see Proposition 4.21.8 below.

4.21.3 **Corollary.** If a subset P of $\mathscr{L}^1$ is weakly relatively compact in $\mathscr{L}^1$, so too is the closed, convex, and balanced envelope of P in $\mathscr{L}^1$.

Proof. It is evident that if P fulfills the conditions of Theorem 4.21.2, so too does the said envelope. ▌

Apart from the Dunford-Pettis theorem itself, there are other important corollaries of Proposition 4.21.1.

4.21.4 **Theorem.** $\mathscr{L}^1$ is weakly sequentially complete; that is, each weak Cauchy sequence in $\mathscr{L}^1$ is weakly convergent in $\mathscr{L}^1$.

Proof. If (f_n) is a weak Cauchy sequence in $\mathscr{L}^1$, so that $\lim_{n\to\infty} \int f_n g \, d\mu$ exists finitely for each g in $\mathscr{L}^\infty$, then (f_n) is bounded (Theorem 8.2.2 again). Moreover, conclusion (1) of Proposition 4.21.1 is available. It shows that if we write

$$L(g) = \lim_{n\to\infty} \int f_n g \, d\mu$$

for g in $\mathscr{L}^\infty$, so that L is a continuous linear form on $\mathscr{L}^\infty$, then the Radon measure $\lambda = L \mid \mathscr{K}(T)$ is LAC relative to μ. (Compare the proof of the Dunford-Pettis theorem.) Thus there exists an f in $\mathscr{L}^1$ such that

$$L(g) = \int fg \, d\mu,$$

first for g in $\mathscr{K}(T)$. But then, as in the proof of the Dunford-Pettis theorem, this equality continues to hold for g in $\mathscr{L}^\infty$. Thus $\lim f_n = f$ weakly in $\mathscr{L}^1$. ▌

Remark. It is only in the trivial case in which the support of μ is finite that $\mathscr{L}^1(\mu)$ is weakly complete (that is, that each Cauchy directed family in $\mathscr{L}^1$ is weakly convergent in $\mathscr{L}^1$).

A further consequence is one expressing a relationship between convergence and weak convergence in $\mathscr{L}^1$.

4.21.5 **Theorem.** A sequence (f_n) converges in $\mathscr{L}^1$ to f if and only if

(1) $\lim_{n\to\infty} f_n = f$ weakly in $\mathscr{L}^1$,

and

(2) $\lim_{n\to\infty} f_n = f$ locally in measure, that is, for each compact set $K \subset T$ and each number $\varepsilon > 0$ one has

$$\lim_{n\to\infty} \mu(K \cap \{|f_n - f| \geq \varepsilon\}) = 0.$$

Proof. We consider first the necessity of the conditions. That of (1) is trivial. On the other hand, let $K \subset T$ be compact and let $\varepsilon > 0$ be given. We put $A_n = K \cap \{|f_n - f| \geq \varepsilon\}$. Then $\chi_{A_n} \leq \varepsilon^{-1} |f - f_n|$, whence it follows immediately that $\mu(A_n) \leq \varepsilon^{-1} \int |f - f_n| \, d\mu \to 0$, so that (2) follows.

Turning to sufficiency, we note that (1) and Proposition 4.21.1(2) combine to show that for any given $\varepsilon > 0$ we can choose a compact set $K \subset T$ such that $\int_{T \setminus K} |f_n - f| \, d\mu \leq \varepsilon$. This being so, to show that $f_n \to f$, it remains merely to show that $\int_K |f_n - f| \, d\mu \to 0$ as $n \to \infty$. But for any $\alpha > 0$ one has

$$\int_K |f - f_n| \, d\mu \leq \alpha\mu(K) + \int_{K \cap \{|f - f_n| > \alpha\}} |f - f_n| \, d\mu$$

and so, by (2) and Proposition 4.21.1(1),

$$\limsup_{n \to \infty} \int_K |f - f_n| \, d\mu \leq \alpha\mu(K)$$

Letting $\alpha \to 0$ we conclude that $f_n \to f$ in $\mathscr{L}^1$. ▮

4.21.6 **Corollary.** If $f_n \to f$ weakly in $\mathscr{L}^1$, and if also $f_n \to f$ a.e., then $f_n \to f$ in $\mathscr{L}^1$.

Proof. One has only to observe that $f_n \to f$ a.e. entails that $f_n \to f$ locally in measure (see Theorem 4.8.3). ▮

4.21.7 **Corollary.** Suppose there exists a subset T_0 of T that is locally negligible, while $\mu(\{t\}) > 0$ for each $t \in T \setminus T_0$. Then $f_n \to f$ in $\mathscr{L}^1$ if and only if $f_n \to f$ weakly in $\mathscr{L}^1$; and a weak Cauchy sequence in $\mathscr{L}^1$ is a Cauchy sequence.

Proof. For now weak convergence entails pointwise convergence on $T \setminus T_0$, while

$$\int |g| \, d\mu = \int_{T \setminus T_0} |g| \, d\mu$$

for each g in $\mathscr{L}^1$. Moreover, as regards Cauchy sequences we have Theorem 4.21.4. ▮

Remark. Corollary 4.21.7 applies in particular when T is discrete. For the case in which $T = N$ (the discrete space of natural numbers) and $\mathscr{L}^1 = \ell^1$, the result is due to Banach (see Banach [1], p. 137).

We proceed to derive two more criteria for weak convergence of sequences in $\mathscr{L}^1$. As a starting point we shall need yet another deduction from Proposition 4.21.1.

4.21.8 **Proposition.** If (f_n) is a sequence in $\mathscr{L}^1$ such that $\lim_{n \to \infty} \int_A f_n \, d\mu$ exists finitely for each σ-finite measurable set $A \subset T$, then $\operatorname{Sup}_n \int |f_n| \, d\mu < +\infty$ and (f_n) is weakly convergent in $\mathscr{L}^1$.

Proof. Both conclusions of Proposition 4.21.1 apply to the sequence (f_n). Suppose we can show that $\int |f_n| \, d\mu$ is bounded with respect to n. Then, using conclusions (1) and (2) of Proposition 4.21.1, together with the fact that each g in $\mathscr{L}^\infty$ can be approximated arbitrarily closely by finite linear combinations of characteristic functions χ_A, it will follow readily that $\lim_{n\to\infty} \int f_n g \, d\mu$ exists finitely for each such g—that is, that (f_n) is a weak Cauchy sequence in $\mathscr{L}^1$. But then (Theorem 4.21.4) this sequence is weakly convergent in $\mathscr{L}^1$, and we shall have reached our goal. Thus everything hinges on showing that (f_n) is bounded in $\mathscr{L}^1$.

In view of Proposition 4.21.1(2), we are reduced to showing that $\int_K |f_n| \, d\mu$ is bounded with respect to n for each compact set $K \subset T$. By virtue of conclusion (1) of that same proposition, this will in turn follow if we can show that there exists a measurable subset D of K such that $\int_D |f_n| \, d\mu$ is bounded with respect to n, while $K \backslash D$ can be covered by a finite number of measurable sets each of arbitrarily small measure.

We are going to show that such a set D may be obtained as the set of points t of K for which $\mu(\{t\}) > 0$. It is evident that D is countable. Given $\delta > 0$, there exists a finite set $F \subset D$ such that $\mu(D \backslash F) \leq \delta$. By (1) of Proposition 4.21.1, we can be sure that $\int_{D\backslash F} |f_n| \, d\mu$ is bounded with respect to n. On the other hand, we can be sure that $\int_F |f_n| \, d\mu$ is bounded. For, by treating the real and imaginary parts of f_n separately, we may assume that f_n is real valued; and then for each point t of F we have

$$\int_{\{t\}} |f_n| \, d\mu = \pm \int_{\{t\}} f_n \, d\mu,$$

which is bounded since it has (by hypothesis) a finite limit as $n \to \infty$. Thus boundedness of $\int_D |f_n| \, d\mu$ is established.

The last stage is to show that $K \backslash D$ may be partitioned into a finite union of measurable sets, each of measure at most δ ($\delta > 0$ being arbitrarily preassigned). This will follow from a lemma. ▌

Lemma. Let S be any measurable set, all of whose points have zero measure. Let $0 \leq \beta < \mu(S)$. Then there exists an integrable set $M \subset S$ such that $\mu(M) = \beta$.

Proof. By (11) of Subsection 4.7.3 we can choose a compact set $K \subset S$ so that $\beta < \mu(K)$. We may therefore suppose from the outset that S is compact.

We will indicate how to construct integrable sets M_n such that $S \supset M_1 \supset M_2 \supset \cdots$ and $\beta \leq \mu(M_n) < \beta + \alpha n^{-1}$, where $\alpha = \mu(S)$. Once this is done, we need only take $M = \bigcap M_n$. For M_1 we take S itself. Suppose $M_1, \cdots, M_n$ are satisfactorily defined. For each point t of S we may choose a neighbourhood U_t of t for which $\mu(U_t) \leq \alpha/(n+1)$. From these we may select a finite number, say $U_1, \cdots, U_m$, which cover S. We put $V_1 = U_1$, $V_2 = U_2 \backslash U_1$, $V_3 = U_3 \backslash (U_1 \cup U_2)$, $\cdots$, and so forth. The V_i are disjoint and cover S. Hence the $W_i = V_i \cap M_n$ partition M_n and each W_i has measure at most $\alpha/(n+1)$. On the other hand

$$\sum_{i=1}^{m} \mu(W_i) = \mu(M_n) \geq \beta.$$

From this it follows readily that a certain subfamily of (W_i) can be chosen, say $X_1 = W_{i_1}, \cdots, X_r = W_{i_r}$, such that

$$\beta \leq \sum_{j=1}^{r} \mu(X_j) \leq \beta + \frac{\alpha}{n+1}.\dagger$$

It suffices to take $M_{n+1} = \bigcup\{X_j : 1 \leq j \leq r\}$. ▌

4.21.9 **Theorem.** Let $\mathscr{M}$ be a set of measurable subsets of T such that for each σ-finite measurable set A and each $\delta > 0$, there exists $M \in \mathscr{M}$ such that $\mu(A \backslash M) + \mu(M \backslash A) \leq \delta$. In order that a sequence $(f_n) \subset \mathscr{L}^1$ be weakly convergent in $\mathscr{L}^1$, it is necessary and sufficient that the following three conditions be fulfilled:

$$\text{(1) } \operatorname*{Sup}_n \int |f_n| \, d\mu < +\infty.$$

(2) Given $\varepsilon > 0$, there exists a number $\delta > 0$

such that

$$\operatorname*{Sup}_n \int_A |f_n| \, d\mu \leq \varepsilon$$

whenever A is σ-finite and measurable and $\mu(A) \leq \delta$.

(3) $\lim\limits_{n \to \infty} \int_M f_n \, d\mu$ exists finitely for each $M \in \mathscr{M}$.

Proof. The necessity of (1) follows from the preceding proposition; that of (2) follows from Proposition 4.21.1(1); and that of (3) from the definition of weak convergence, since

$$\int_M f_n \, d\mu = \int f_n \chi_M \, d\mu \qquad \text{and} \qquad \chi_M \in \mathscr{L}^\infty.$$

Consider the sufficiency. It is plain that (2) and (3) together show that $\lim_{n \to \infty} \int_A f_n \, d\mu$ exists finitely for each σ-finite measurable set A. But then (Proposition 4.21.1(2)) to each $\varepsilon > 0$ corresponds a compact set $K \subset T$ such that $\int_{T \backslash K} |f_n| \, d\mu \leq \varepsilon$ for all n.

Let $g \in \mathscr{L}^\infty$. To show that $\lim_{n \to \infty} \int f_n g \, d\mu$ exists finitely, it is enough (in view of the last remark) to show that $\lim_n \int_K f_n g \, d\mu$ exists finitely for each compact set $K \subset T$. But we can approximate g in $\mathscr{L}^\infty(K)$ by finite linear combinations of characteristic functions of integrable sets $A \subset K$. Then, using (1) and our hypothesis concerning $\mathscr{M}$, we see that (3) entails the convergence (to a finite limit) of $\int_K f_n g \, d\mu$.

Weak convergence of (f_n) now follows from Theorem 4.21.4. ▌

---

† Let $P = \{1, 2, \cdots, m\}$; if $Q \subset P$, put $s(Q) = \sum_{i \in Q} \mu(W_i)$ so that $s(P) \geq \beta$. We consider all Q for which $s(Q) \geq \beta$. Only a finite number of such Q exist (in fact P has only a finite number of distinct subsets). So among such Q there is a minimal one, say Q_0. If $s(Q_0)$ were greater than $\beta + \alpha/(n+1)$, we could remove from Q_0 one element i to obtain $Q_1 \subset Q_0$ with

$$s(Q_1) = s(Q_0) - \mu(W_i) \geq \left(\beta + \frac{\alpha}{n+1}\right) - \frac{\alpha}{n+1} = \beta,$$

contradicting minimality of Q_0. Thus $\beta \leq s(Q_0) \leq \beta + \alpha/(n+1)$ and so, if $Q_0 = \{i_j : 1 \leq j \leq r\}$, then $(X_j) = (W_{i_j})$ is a subfamily of the desired type.

Remark. Suppose, for example, that T is a real interval, and μ is the restriction of Lebesgue measure to T. One might then take for $\mathscr{M}$ the set of all finite disjoint unions of subintervals of T (or all open, or all closed, such subintervals). In this way we should recover from Theorem 4.21.9 that special case of it established by Banach (see Banach [1], p. 136).

We terminate this section with a result needed in Section 9.4. It was first recorded by Grothendieck ([4], p. 139, Proposition 3).

4.21.10 **Theorem.** If a sequence (f_n) in $\overline{\mathscr{L}}^\infty$ is uniformly bounded and converges locally in measure to f (necessarily in $\overline{\mathscr{L}}^\infty$), then

$$\lim_{n\to\infty} \int hf_n\, d\mu = \int hf\, d\mu$$

uniformly when h ranges over any weakly relatively compact subset P of $\mathscr{L}^1$.

Remark. In the language of duality theory (see Chapter 8, especially Subsection 8.3.3), the conclusion expresses the fact that the classes $\dot{f}_n$ (elements of L^∞) converge to the class $\dot{f}$ in the sense of the Mackey topology $\tau(L^\infty, L^1)$.

Proof. This is a direct application of Theorem 4.21.2. Given $\varepsilon > 0$ we first choose a compact set K such that

$$\int_{T\setminus K} |h|\, d\mu < \frac{\varepsilon}{6c}$$

for all h in P, c being chosen so that $|f_n| < c$ l.a.e. for all n. Next we choose $\delta > 0$ so that for any integrable set A satisfying $\mu(A) < \delta$, one has

$$\int_A |h|\, d\mu < \frac{\varepsilon}{6c}$$

for all h in P.

By the assumed local convergence in measure, we may choose our integrable set $A \subset K$ such that $\mu(A) < \delta$ and $f_n \to f$ uniformly on $K\setminus A$. If we then put

$$c' = \operatorname{Sup}\left\{\int |h|\, d\mu : h \in P\right\} < +\infty$$

and $g_n = f - f_n$, we shall have $|g_n| < 2c$ l.a.e. for all n. It follows that for all h in P one has

$$\left|\int g_n h\, d\mu\right| \le \int_{T\setminus K} |g_n h|\, d\mu + \int_A |g_n h|\, d\mu + \int_{K\setminus A} |g_n h|\, d\mu$$

$$< 2c\cdot\frac{\varepsilon}{6c} + 2c\cdot\frac{\varepsilon}{6c} + c'\cdot \operatorname*{Sup}_{K\setminus A} |g_n|.$$

The last term on the right-hand side tends to 0 as $n \to \infty$, thanks to uniform convergence. It follows that we can determine n_0 independent of h in P and such that $n > n_0$ entails

$$\left|\int g_n h\, d\mu\right| < \varepsilon$$

for all h in P. This concludes the proof. ▌

4.22 Weakly Compact Sets of Bounded Measures

From this point onward we denote by $M(T)$ the Banach space of bounded Radon measures on T, the norm being

$$\|\mu\| = \int d\,|\mu|.$$

$M(T)$ is isomorphic to the dual of $C_0(T)$ (see Exercise 4.45). Necessary and sufficient conditions in order that a subset A of $M(T)$ be relatively compact for the weak topology $\sigma(M, C_0)$ flow from the general theory: from Theorem 1.11.4(2) we see that it is sufficient that A be bounded; and, since $C_0(T)$ is a Banach space, Corollary 6.2.3 and Theorem 7.1.1(1)(b) combine to show that this condition is also necessary.

In this section, however, we are concerned with criteria for relative compactness for the weakened topology $\sigma(M, M')$, which is (save in trivial cases) strictly stronger than $\sigma(M, C_0)$. Henceforth in this section when we speak of weak convergence or weak relative compactness in connection with elements or subsets of M, it is the weakened topology $\sigma(M, M')$ that is involved.

The results we present are analogues for $M(T)$ of Theorem 4.21.2 and are due to Dieudonné [14] and Grothendieck ([4], p. 146, Théorème 2 and p. 151, Lemme 5). For similar work on spaces of finitely additive and countably additive set functions in general, see Dunford and Schwartz [1], pp. 305 et seq. The assertions involving Radon measures will find their use in Section 9.4.

4.22.1 **Theorem.** Let A be a bounded subset of $M(T)$. In order that A be weakly relatively compact in $M(T)$, any one of the following five (equivalent) conditions is necessary and sufficient:

(1) If (f_n) is a uniformly bounded sequence of functions, each measurable for each μ in A, and if $f_n \to f$ pointwise on T, then $\int f_n\, d\mu \to \int f\, d\mu$ uniformly for μ in A.

(1′) As (1), except that it is assumed that $f_n \to f$ locally in measure for each μ in A.

(2) If a sequence (f_n) converges weakly to 0 in $C_0(T)$, then $\int f_n\, d\mu \to 0$ uniformly for μ in A.

(3) If (U_n) is a sequence of disjoint open subsets of T, then $\mu(U_n) \to 0$ uniformly for μ in A.

(4) One has both:

(a) For each compact set $K \subset T$ and each $\varepsilon > 0$, there exists an open set $U \supset K$ such that $|\mu|\,(U \backslash K) < \varepsilon$ for all μ in A.

(b) For each $\varepsilon > 0$, there exists a compact set $K \subset T$ such that $|\mu|(T \backslash K) < \varepsilon$ for all μ in A.

Remark. Criterion (4) is strikingly similar to the criteria for weak relative compactness in $\mathscr{L}^1$ (or in L^1) afforded by Theorem 4.21.2. Indeed, there is one case that is directly reducible to the latter situation. Since the device involved in this reduction is useful in the proof of the present theorem, and again in Chapter 9 (especially Theorem 9.4.4), we shall discuss it here.

Suppose that $A \subset M(T)$ is separable (for the normed topology). We claim that there exists a bounded positive Radon measure μ on T and an isometric isomorphism u of $L^1(\mu)$ onto a closed vector subspace L of $M(T)$ that contains A. To see this, we choose a countable set $\{\mu_n\}$ dense in A and then numbers $c_n > 0$ such that $\sum c_n \|\mu_n\| < +\infty$. Then $\mu = \sum c_n |\mu_n|$ is a bounded positive Radon measure on T. We define u by $u(\dot{h}) = h \cdot \mu$, $\dot{h}$ being any element of $L^1(\mu)$ and h any representative of $\dot{h}$. Since it is clear that each μ_n is absolutely continuous relative to μ, Theorem 4.15.1 ensures that $u(L^1)$ contains each μ_n. Moreover (see Remark 4.15.6), u is an isometric isomorphism of $L^1(\mu)$ into $M(T)$. Since $L^1(\mu)$ is complete, $L = u(L^1)$ is a closed vector subspace of $M(T)$. But then, since L contains each μ_n, L must contain A.

This being so, if we assume furthermore that A is weakly relatively compact in $M(T)$, it will follow (using the Hahn-Banach theorem to infer that $\sigma(L, L')$ is identical with the topology induced on L by $\sigma(M, M')$) that A is weakly relatively compact in L, and conversely. Thanks to the fact that u is an isomorphism of $L^1(\mu)$ onto L, we may infer that A is weakly relatively compact in $M(T)$ if and only if $u^{-1}(A)$ is weakly relatively compact in $L^1(\mu)$. In other words, A is weakly relatively compact in $M(T)$ if and only if it can be written as $\{h \cdot \mu : h \in P\}$, where P is weakly relatively compact in $\mathscr{L}^1(\mu)$.

If the hypothesis of separability of A is relaxed, there remains a substitute result, the foundations of which lie a good deal deeper and which will be described briefly in Section 4.23.

Proof of Theorem 4.22.1. It is evident that (1′) implies (1), pointwise convergence entailing local uniform convergence for any measure; and that (1) implies (2). If we denote temporarily by "(0)" the assertion "A is weakly relatively compact in $M(T)$," it will suffice to establish that (0) implies (1′), that (2) implies (3), that (3) implies (4), and that (4) implies (0), since we shall then have the cyclic scheme.

$$(0) \Rightarrow (1') \Rightarrow (1) \Rightarrow (2) \Rightarrow (3) \Rightarrow (4) \Rightarrow (0).$$

(a) (0) implies (1′). If (1′) were false, there would exist a number $\varepsilon > 0$, a subsequence (g_n) of (f_n), and a sequence (μ_n) extracted from A, such that $|\int (g_n - f)\, d\mu_n| > \varepsilon$ for all n. We let $\mu = \sum_n 2^{-n} |\mu_n|$. Then $g_n \to f$ locally in measure for μ, as is easily verified. Yet, by the above remark, one may write $\mu_n = h_n \cdot \mu$, the set $\{h_n\}$ being weakly relatively compact in $\mathscr{L}^1(\mu)$. The preceding inequalities then constitute a contradiction of Theorem 4.21.10.

(b) (2) implies (3). If (3) were false one would have $|\mu_n(U_{k_n})| > \varepsilon$, for some $\varepsilon > 0$, some subsequence (U_{k_n}) of (U_n), and some sequence (μ_n) extracted from A. Then, for each n, there would exist a function f_n in $\mathscr{K}(T) \subset C_0(T)$ satisfying $|f_n| \le 1$, having its support contained in U_{k_n}, and for which $|\int f_n\, d\mu_n| > \varepsilon$. Since the sequence (f_n) then converges weakly to zero in $C_0(T)$, a contradiction of (2) would be apparent.

(c) (3) implies (4).* It is evidently enough to consider the case in which only real Radon measures are involved. Suppose, for example, that (4)(a) were violated. One could then construct inductively a decreasing sequence (V_n) of open

* See Appendix, p. 771.

neighbourhoods of K, a sequence (μ_n) extracted from A, and a sequence (U_n) of relatively compact open sets, such that

$$\bar{U}_n \subset V_{n-1} \backslash V_n, \qquad |\mu_n(U_n)| > \frac{\varepsilon}{2};$$

and such a construction would conflict with (3). For suppose the construction has been successfully carried to the nth stage. Then there exists by hypothesis an element μ_{n+1} of A such that $|\mu_{n+1}|(V_n \backslash K) > \varepsilon/2$, hence a compact set $K_{n+1} \subset V_n \backslash K$ such that $|\mu_{n+1}|\,(K_{n+1}) > \varepsilon/2$, and therefore a relatively compact open set W such that $K_{n+1} \subset W \subset \overline{W} \subset V_n \backslash K$ and $|\mu_{n+1}|(W) > \varepsilon/2$. Then since (as follows easily from Corollary 4.15.5; see Exercise 4.42) $|\mu_{n+1}|\,(W) = \text{Sup}\, |\mu_{n+1}(S)|$ for S measurable and $S \subset W$, we shall have $|\mu_{n+1}(S)| > \varepsilon/2$ for some such S. We can then determine an open set U_{n+1} such that $S \subset U_{n+1} \subset W$ and such that $\mu_{n+1}(U_{n+1})$ differs by as little as we please from $\mu_{n+1}(S)$; in particular, we can arrange that $|\mu_{n+1}(U_{n+1})| > \varepsilon/2$. It now suffices to take for V_{n+1} an open neighbourhood of K that is contained in V_n and that does not meet $\bar{U}_{n+1}$.

A similar argument shows that any violation of (4)(b) is incompatible with (3).

(d) (4) implies (0). According to Eberlein's Theorem 8.12.7, to show that A is weakly relatively compact, it suffices to show that each sequence (μ_n) extracted from A admits a weak limiting point. Using the device explained in the remark preceding the present proof, we may choose a bounded positive Radon measure μ such that $\mu_n = h_n \cdot \mu$ and show that $\{h_n\}$ is weakly relatively compact in $\mathscr{L}^1(\mu)$—that is, that $\{h_n\}$ satisfies the conditions of Theorem 4.21.2.

Of these conditions, (1) holds since A is bounded, and (3) is equivalent to (4)(b). It will therefore suffice to show that $\int_U |h_n|\, d\mu \to 0$ when U is open and $\mu(U) \to 0$, and this uniformly with respect to n.

Now if this were false, we should have —on replacing (h_n) by a suitable subsequence if necessary—

$$\int_{U_n} |h_n|\, d\mu > \varepsilon$$

for some $\varepsilon > 0$ and all n, (U_n) being a suitable sequence of open sets satisfying $\mu(U_n) < 2^{-n}$. Put $V_n = \bigcup\{U_k : k \geq n\}$, so that V_n is open, $V_n \supset V_{n+1}$, $\mu(V_n) \to 0$, and

$$\int_{V_n} |h_n|\, d\mu > \varepsilon. \tag{4.22.1}$$

Observing that (4)(a) entails (on passing to complementary sets) that for each open set V and $\alpha > 0$ there exists a compact set $K \subset V$ such that $\int_{V\backslash K} |h_m|\, d\mu < \alpha$ for all m, we infer that a compact set $K_n \subset V_n$ may be chosen for which

$$\int_{V_n \backslash K_n} |h_m|\, d\mu < 2^{-n-1}\varepsilon \qquad \text{for all } m. \tag{4.22.2}$$

We put $K_n^* = \bigcap\{K_k : 1 \leq k \leq n\}$. From (4.22.1) we obtain

$$\int_{K_n^*} |h_n|\, d\mu = \int_{V_n} - \int_{V_n \backslash K_n^*} > \varepsilon - \int_{V_n \backslash K_n^*} |h_n|\, d\mu. \tag{4.22.3}$$

Since

$$V_n \backslash K_n^* = \bigcup \{V_n \backslash K_k : 1 \leq k \leq n\} \subset \bigcup \{V_k \backslash K_k : 1 \leq k \leq n\},$$

(4.22.2) shows that

$$\int_{V_n \backslash K_n^*} |h_n| \, d\mu < \sum_{k=1}^{n} 2^{-k-1}\varepsilon < \tfrac{1}{2}\varepsilon,$$

hence via (4.22.3) one obtains

$$\int_{K_n^*} |h_n| \, d\mu > \varepsilon - \tfrac{1}{2}\varepsilon = \tfrac{1}{2}\varepsilon. \tag{4.22.4}$$

Now the K_n^* decrease and $\mu(K_n^*) \to 0$, and so their intersection K is negligible. According to (4)(a), there exists an open set $U \supset K$ such that

$$\int_{U \backslash K} |h_n| \, d\mu < \tfrac{1}{2}\varepsilon \qquad \text{for all } n,$$

that is, K being negligible,

$$\int_U |h_n| \, d\mu < \tfrac{1}{2}\varepsilon \qquad \text{for all } n. \tag{4.22.5}$$

On the other hand, since U contains K, it must contain K_r^* for some r, and then (4.22.4) yields

$$\int_U |h_r| \, d\mu \geq \int_{K_r^*} |h_r| \, d\mu > \tfrac{1}{2}\varepsilon,$$

which contradicts (4.22.5). This conflict ends the proof. ▌

Remark. A result to be established much later (Corollary 9.2.2) combines with criterion (2) of the present theorem to show that a set $A \subset M(T)$ is weakly relatively compact if and only if it is compact for the Mackey topology $\tau(M, C_0)$ (see Subsection 8.3.3).

4.22.2 **Corollary.** A bounded sequence (μ_n) in $M(T)$ is weakly convergent if and only if $(\mu_n(U))$ is convergent (to a finite limit) for each open subset U of T.

Proof. It is evidently enough to show that $\{\mu_n\}$ is weakly relatively compact in $M(T)$. To do this, we verify that criterion (3) of the preceding theorem is fulfilled.

For each μ in $M(T)$ denote by μ^* the element of ℓ^1 defined by $\mu^*(n) = \mu(U_n)$. If S is any set of natural numbers,

$$\sum_{k \in S} \mu_n^*(k) = \mu_n\left(\bigcup_{k \in S} U_k\right),$$

which converges (as $n \to \infty$) to a finite limit, by hypothesis. Proposition 4.21.8 shows that (μ_n^*) is weakly convergent in ℓ^1. Hence $\{\mu_n^*\}$ is weakly relatively compact in ℓ^1 and so, by Theorem 4.21.2, given $\varepsilon > 0$, there exists r such that

$$\sum_{k > r} |\mu_n^*(k)| < \varepsilon \qquad \text{for all } n.$$

A fortiori $k > r$ entails $|\mu_n^*(k)| = |\mu_n(U_k)| < \varepsilon$ for all n, whence the verification of (3). ▌

Remark. When T is metrizable, the result is due to Dieudonné ([14], Proposition 8), who shows also that the hypothesis of boundedness can, in fact, be dropped. This is of little consequence for our purposes, and we omit the proof.

4.22.3 **Theorem.** Let Q denote the vector space of functions on T generated by characteristic functions of closed sets with countable neighbourhood bases, A a bounded subset of $M(T)$. Then A is weakly relatively compact in $M(T)$ if and only if

(5) A is relatively compact for $\sigma(M, Q)$.

Remark. In interpreting $\sigma(M, Q)$ it is to be understood that each function f belonging to Q (and in general each universally measurable and bounded function f on T) is identified with the linear form $\mu \to \int f \, d\mu$ on $M(T)$.

Proof. It is manifestly adequate to show that (5) implies the weak relative compactness of A.

We begin by showing that one may reduce the problem to the case in which T is compact. For consider the Alexandroff compactification $T^* = T \cup \{\infty\}$ of T. As will be seen in Exercise 4.40, there is an isometric isomorphism $\mu \to \mu^*$ of $M(T)$ onto the closed vector subspace N of $M(T^*)$ comprising those λ in $M(T^*)$ for which $\{\infty\}$ is negligible (that is, for which $\lambda(\{\infty\}) = 0$). In this isomorphism the topologies $\sigma(M, Q)$ and $\sigma(N, Q^*)$ correspond, Q^* being defined relative to T^* as is Q relative to T. Moreover, $\sigma(N, Q^*)$ is the topology on N induced by $\sigma(M^*, Q^*)$, where $M^* = M(T^*)$. Hence follows the said reduction.

We assume hereafter, then, that T is compact.

The next step is to consider the case in which T is metrizable, in which case each closed subset of T has a countable neighbourhood base. To show that A is weakly relatively compact it suffices (Eberlein's Theorem 8.12.7) to establish that each sequence (μ_n) extracted from A admits a weakly convergent subsequence. Consider the $\sigma(M, C)$-closure A_1 of A. Since A_1 is bounded, $\sigma(M, Q)$ and $\sigma(M, \bar{Q})$ induce on A_1 the same topology, $\bar{Q}$ denoting the uniform closure of Q. But (Exercise 4.41) $\bar{Q} \supset C = C(T)$, so that $\sigma(M, \bar{Q})$ is stronger than $\sigma(M, C)$. Again since A_1 is bounded, it is compact for $\sigma(M, C)$. Since $C(T)$ is separable (T being compact and metrizable), $\sigma(M, C)$ induces on A_1 a metrizable topology. Consequently there exists a subsequence (λ_n) of (μ_n) that is $\sigma(M, C)$-convergent to some limit μ in A_1. On the other hand, A being relatively compact for $\sigma(M, \bar{Q})$, the sequence (λ_n) admits at least one limiting point for this topology. Moreover, if λ is any such limiting point, then λ must coincide with μ—for λ necessarily belongs to A_1 and, if V is any $\sigma(M, C)$-closed neighbourhood of λ for the topology $\sigma(M, C)$, then $V \cap A_1$ is also a neighbourhood of λ for the induced topology $\sigma(M, \bar{Q}) \mid A_1$, so that for some subsequence (λ_{n_k}) one has $\lambda_{n_k} \in V$ for all k, and therefore in the limit $\mu \in V$, which, since $\sigma(M, C)$ is separated, forces upon us the conclusion that $\lambda = \mu$. Thus μ is the unique limiting point of the sequence (λ_n) for the topology $\sigma(M, \bar{Q})$, a fortiori the unique limiting point of (λ_n) in the compact space formed of the $\sigma(M, \bar{Q})$-closure of A endowed with the topology induced by $\sigma(M, \bar{Q})$. It follows that (λ_n) is $\sigma(M, \bar{Q})$-convergent to μ, a fortiori $\sigma(M, Q)$-convergent to μ. But then (Corollary 4.22.2) $\lambda_n \to \mu$ weakly, which is what we wished to establish.

We come now to the case in which T, although still assumed to be compact, is perhaps not metrizable. This will be handled through the medium of Proposition 4.22.4 immediately below, coupled with the following observations involving metrizable quotient spaces of T.

Let T^* denote any separated quotient space of T, π the natural map of T onto T^*. π is both continuous and open, and T^* is necessarily itself compact. There is an associated linear map P of $M(T)$ into $M(T^*)$, $\mu^* = P\mu$ being the measure on T^* defined by

$$\int g \, d\mu^* = \int (g \circ \pi) \, d\mu.$$

The reader will notice that if S^* is a closed subset of T^*, then $S = \pi^{-1}(S^*)$ is a closed subset of T; moreover, if U^* ranges over a neighbourhood base of S^*, the sets $U = \pi^{-1}(U^*)$ range over a neighbourhood base of S. In particular, if T^* is metrizable, $S = \pi^{-1}(S^*)$ is closed and has a countable neighbourhood base in T whenever S^* is closed in T^*. Since the defining formula for μ^* in terms of μ extends to show that

$$\int \chi_{S^*} \, d\mu^* = \int (\chi_{S^*} \circ \pi) \, d\mu = \int \chi_S \, d\mu,$$

it becomes plain that $P : \mu \to \mu^*$ is continuous for $\sigma(M, Q)$ and $\sigma(M^*, Q^*)$, where $M^* = M(T^*)$ and Q^* is defined relative to T^* as is Q to T. Thus, if A satisfies condition (5), so too does its image $A^* = P(A)$: if T^* is metrizable, this image A^* will then (by what we have already proved) be weakly relatively compact in $M(T^*)$.

In view of all this, the proof of Theorem 4.22.3 will be completed by establishing the next proposition. ▌

4.22.4 **Proposition.** Let T be a separated compact space, A a bounded subset of $M(T)$. In order that A be weakly relatively compact in $M(T)$, it is necessary and sufficient that for each metrizable quotient space T^* of T, the image $A^* = P(A)$ be weakly relatively compact in $M(T^*)$.

Proof. Since P is evidently continuous from the Banach space $M(T)$ into the Banach space $M(T^*)$, the condition is necessary.

In proving sufficiency we may assume without loss of generality that A is convex, balanced, and closed for $\sigma(M, C)$.

We introduce as an auxiliary a Banach space F, defined in the following manner. On $E = C(T)$ we consider the seminorm p, equal to the gauge of A^0, form the quotient space $E/\text{Ker}\, p$, and obtain F as the completion of the latter normed space. If u is natural map of E into F, then u is continuous from E into F, and its adjoint u' maps the closed unit ball of F' onto A.

This being so, to show that A is weakly relatively compact in $M(T)$ (that is, is relatively compact for $\sigma(E', E'')$), it suffices (Corollary 9.3.2) to show that u is weakly compact. By virtue of Eberlein's Theorem 8.12.7, it is enough for this to show that $(u(f_n))$ has a weak limiting point in F whenever (f_n) is a sequence extracted from the unit ball of $E = C(T)$.

Let T^* be the quotient of T modulo the equivalence relation expressed by "$f_n(t_1) = f_n(t_2)$ for all n." T^* is metrizable, its topology being the weakest that renders continuous each of the functions on T^* obtained by passage to the quotient of the f_n. Moreover, $C(T^*)$ can be identified with the closed vector subspace of $C(T)$ comprising those f in $C(T)$ for which $f(t_1) = f(t_2)$ whenever $f_n(t_1) = f_n(t_2)$ for all n. Our task is thus reduced to showing that $v = u \mid C(T^*)$ is weakly compact, that is, (Corollary 9.3.2 again) that v' transforms the closed unit ball B of F' into a subset of the dual $G' = M(T^*)$ of $G = C(T^*)$ which is relatively compact for $\sigma(G', G'')$. However, it is easily verified that the said image $v'(B)$ is precisely A^*, and the proof is complete. ▌

4.23 A Theorem of Kakutani and Its Consequences

Nowhere in this book do we exploit fully and systematically the partial order structure on the set of real Radon measures on a given separated locally compact space T, nor that on the set of real-valued continuous functions on T. As a matter of fact, a great deal of work has been done on these aspects from the point of view of a general theory of (partially) ordered vector spaces, especially those that are at the same time Banach spaces. See, for example, Bourbaki [6], Chapitre II; Nakano [2]; Birkhoff [2]; Dunford and Schwartz [1], pp. 394–398 and the references cited there; and the Appendix in Kelley and Namioka [1]. Much of the pioneering work is due to F. Riesz [8] and to Kakutani [5], [6]. The major result of Kakutani [5] is the object of our present interest: it furnishes an abstract characterization of the real spaces L^1. The second of Kakutani's papers cited does likewise for the real spaces $C(T)$; with this we are not at the moment concerned, but see Remark (2) of Subsection 9.4.15.

4.23.1. We must first describe axiomatically the type of structure in question.

The reader will recall first that an *ordered vector space* is a real vector space E which, as a set, is partially ordered in such a way that the following conditions are fulfilled:

$$x \geq 0 \text{ and } -x \geq 0 \text{ imply } x = 0;$$
$$x \geq y \text{ and } \quad y \geq z \text{ imply } x \geq z;$$
$$x \geq 0 \text{ and } \quad \lambda \geq 0 \text{ imply } \lambda x \geq 0;$$
$$x \geq y \text{ implies } x + z \geq y + z.$$

If, besides this, the partial ordering is that of a lattice—that is, if each pair of elements x and y of E admits a supremum $x \vee y$ and an infimum $x \wedge y$—E is spoken of as a *vector lattice* or (Bourbaki) a *Riesz space*.

If a Riesz space E is endowed with a norm such that

(N) E is complete;

(N') the set of positive elements of E is closed;

(N'') $x \wedge y = 0$ implies $\|x + y\| = \|x - y\|$;

and

(L) $x > 0$, $y > 0$ imply $\|x + y\| = \|x\| + \|y\|$;

then E is termed an *abstract* (L)-*space*.

An element e of abstract (L)-space E is termed an F-(= Freudenthal) *unit* if

$$x > 0 \text{ implies } x \wedge e > 0.$$

4.23.2 **Theorem** (Kakutani). Any abstract (L)-space E is isometric and isomorphic (both as a normed vector space and as a partially ordered set) with a space $L^1_R(S, \sigma)$, where S is a separated locally compact space and σ a positive Radon measure on S. The space S may be chosen to be totally disconnected (that is, the only connected subsets of S are one-point subsets). If E possesses an F-unit, S may be chosen to be in addition compact.

Remarks. Here $L^1_R(S, \sigma)$ is used to denote the real vector space of (equivalence classes of) real-valued σ-integrable functions on S, normed in the usual way and partially ordered by the convention that $\dot{f} \geq \dot{g}$ if and only if $f \geq g$ a.e. for some one (and hence any) function f (resp. g) chosen from the class $\dot{f}$ (resp. $\dot{g}$). The case in which E possesses an F-unit is proved in Kakutani [5], the more general theorem being given in Kakutani [6] as a relatively easy extension of the more restricted case. The complete proof is lengthy and will not be attempted here.

The result is attainable as a culmination of several of Bourbaki's exercises (Bourbaki [6], Chapitres I–IV, p. 30, Exercice 12); p. 31, Exercice 13); p. 175, Exercice 10)). In this formulation the space S turns out to be separated, locally compact and *Stonian*. The last condition means that in S the closure of any open set is again open and, in the presence of the other conditions on S, implies that S is totally disconnected.

4.23.3 **Application to Spaces of Measures.** Consider the set $M_R(T)$ of real, bounded Radon measures on a separated locally compact space T. It is easy to verify (using the results established in the present chapter) that $M_R(T)$ is an abstract (L)-space. According to Kakutani's theorem, therefore, $M_R(T)$ is isomorphic with $L^1_R(S, \sigma)$ for a suitably chosen S and σ. One of the many corollaries of this has been heralded in the remarks following the statement of Theorem 4.22.1.

4.23.4. **Corollary.** If A is weakly relatively compact set in $M(T)$, there exists a positive measure μ_0 in $M(T)$ such that each μ is A is of the form $\mu = f \cdot \mu_0$ for some f in $\mathscr{L}^1(T, \mu_0)$.

Proof. It is evident from Theorem 4.22.1 that one may assume A to comprise positive measures only. Moreover, according to the Lebesgue-Radon-Nikodym Theorem 4.15.1, it suffices to show that each member of A is absolutely continuous relative to μ_0.

We choose S and σ so that $M_R(T)$ is isomorphic with $L^1_R(S, \sigma)$. In this isomorphism A will correspond to a weakly relatively compact subset P of $L^1(S, \sigma)$. Applying Theorem 4.21.2 to P, we infer that there exists an increasing sequence (K_n) of compact subsets of S such that each element of P has a representative function f vanishing outside the union of the K_n. We may assume that $\sigma(K_n) > 0$ for each n and then define the function f_0 on S by

$$f_0 = \sum_{n=1}^{\infty} n^{-2}\sigma(K_n)^{-1}\chi_{K_n}.$$

f_0 is positive and σ-integrable. Each element of P has a representative f that is the limit in $\mathscr{L}^1(S, \sigma)$ of the sequence (f_n) defined by

$$f_n = \text{Inf}\,(f, nf_0).$$

If we suppose that μ_0 is the element of $M_R(T)$ corresponding to f_0, it follows that each μ in A is the limit in $M_R(T)$ of a sequence (μ_n) for which $\mu_n \leq n\mu_0$. From this it is immediate that μ is absolutely continuous relative to μ_0. ▌

4.24 Remarks on Convexity Theorems

The detailed exposition of those parts of integration theory used in this book is now finished, but we intend to insert here a few very brief remarks about two results peculiar to, and holding a central position in, the study of linear operators between Lebesgue spaces. The results referred to are the so-called convexity theorems of M. Riesz and of Marcinkiewicz, respectively, of which a brief description follows.

Let S and T be separated locally compact spaces with associated positive Radon measures σ and τ. The linear operators J involved are assumed to be defined initially on some vector subspace D of each of the Lebesgue spaces $L^p(S, \sigma)$; this exponent p, and others to follow, are asumed to lie in the interval $[1, \infty]$. A possible choice of D would be $\mathscr{K}(S)$ or, what is often a more convenient choice, the set of finite linear combinations of characteristic functions of σ-integrable subsets of S. Either of these choices is such that D is dense in $L^p(S, \sigma)$ for $p < \infty$. The range of J is initially assumed to consist of functions (or equivalence classes of functions) which are τ-measurable and finite τ-a.e. The operator J is said to be of the (strong) type $(p, q,)$ if there exists a number C such that

$$\|Jf\|_q \leq C\,\|f\|_p \qquad (f \in D),$$

where

$$\|f\|_p = \left(\int^* |f|^p\, d\sigma\right)^{1/p}, \qquad \|g\|_q = \left(\int^* |g|^q\, d\tau\right)^{1/q}.$$

The smallest admissible C is denoted by $\|J\|_{p,q}$ and is termed the (strong) (p, q)-norm of J. If no such C exists, we write $\|J\|_{p,q} = \infty$. When D is dense in $L^p(S, \sigma)$, it is plain that J is of type (p, q) if and only if it admits an extension that is continuous from $L^p(S, \sigma)$ into $L^q(T, \tau)$.

The first (Riesz) convexity theorem, proved originally in M. Riesz [2], may be summarized by the inequality

$$\|J\|_{p,q} \leq \|J\|_{p_1,q_1}^{t} \cdot \|J\|_{p_2,q_2}^{1-t}, \tag{4.24.1}$$

holding whenever

$$\frac{1}{p} = \frac{t}{p_1} + \frac{1-t}{p_2}, \qquad \frac{1}{q} = \frac{t}{q_1} + \frac{1-t}{q_2}, \qquad 0 \le t \le 1;$$

the right-hand side of (4.24.1) is to be interpreted as ∞ if either of $\|J\|_{p_1,q_1}$ or $\|J\|_{p_2,q_2}$ is ∞.

For proofs, variants, extensions and applications of this result, the reader may consult Dunford and Schwartz [1]; Zygmund [2], Chapter XII and the Notes attached thereto; Hardy, Littlewood, and Pólya [1], Chapter VIII; Cotlar [1], [2]; Weil [1], Chapitres II and VI; Thorin [1], [2]; and Stein [1]. We note in particular the extension to sublinear operators given by Calderón and Zygmund [6], the operator J being termed sublinear if

$$|J(f_1 + f_2)(t)| \le |Jf_1(t)| + |Jf_2(t)| \qquad \tau\text{-a.e.}$$

A well-known and useful opportunity for applying the theorem is noted in Subsection 10.4.11 *infra*.

The second convexity theorem, due to Marcinkiewicz [1], refers to a weaker type condition on J. The operator J is said to be of weak type (p, q), where $q < \infty$, if there exists a number C such that

$$\tau(\{t \in T : |Jf(t)| > \alpha\}) \le (C\alpha^{-1} \|f\|_p)^q \qquad (f \in D, \alpha > 0);$$

and weak type (p, ∞) is defined to be synonymous with (strong) type (p, ∞). The least admissible value of C is here denoted by $w\,\|J\|_{p,q}$ and is termed the weak (p, q)-norm of J. We write $w\,\|J\|_{p,q} = \infty$ if J is not of weak type (p, q). It is very simple to show that an operator of type (p, q) is also of weak type (p, q). The unrestricted converse of this statement is false, but Marcinkiewicz's convexity theorem is a remarkable partial converse inasmuch as it asserts the inequality

$$\|J\|_{p,q} \le K \cdot w\,\|J\|_{p_1,q_1}^{t} \cdot w\,\|J\|_{p_2,q}^{1-t} \tag{4.24.2}$$

whenever $p_1 \ge q_1$, $p_2 \ge q_2$ and

$$\frac{1}{p} = \frac{t}{p_1} + \frac{1-t}{p_2}, \quad \frac{1}{q} = \frac{t}{q_1} + \frac{1-t}{q_2}, \quad 0 < t < 1.$$

In (4.24.2), K depends on p_1, q_1, p_2, q_2, and t, but remains bounded for fixed p_1, q_1, p_2, q_2 when t is bounded away from both 0 and 1.

Proofs, variants, extensions, and applications will be found in Zygmund [2], Chapter XII and the attached Notes; Cotlar [1], [2]; Calderón and Zygmund [3], [4]; Stein [2]; and J. T. Schwartz [1].

In general, properties of operators J which hinge upon their weak type lie deeper than those involving only their strong type. The concept of weak type seems to have entered real-variable analysis first in a now-famous paper of

Hardy and Littlewood [1]. In its simplest guise their work asserts that the sublinear operator J, defined for integrable functions f of a real variable by

$$Jf(x) = \operatorname*{Sup}_{h>0} \left| h^{-1} \int_x^{x+h} f(\xi)\, d\xi \right|$$

is of type (p, p) if $1 < p \leq \infty$ and of weak type (1, 1) (but not of strong type (1, 1)). The same is true of functions of several real variables (see Zygmund [2], Chapters I and XII) and indeed in even more general situations (see Smith [1] and Rauch [1]). Such results are of paramount importance in the deeper parts of harmonic analysis, complex variable theory, and ergodic theory.

EXERCISES

Note. Failing any indication to the contrary, in the following exercises T denotes a separated locally compact space, μ a positive Radon measure on T, and measurability and integrability are interpreted relative to μ.

4.1 Let ϕ be a positive lower semicontinuous function on R^n, μ a positive Radon measure on R^n. Show that the function

$$U(x) = \int^* \phi(x - y)\, d\mu(y)$$

is lower semicontinuous. [Hint: Use Proposition 4.5.1.]

4.2 Let μ be a positive Radon measure on T. Show that one can write $\mu = \alpha + \beta$, where α and β are positive Radon measures, α is diffuse (that is, $\alpha(\{t\}) = 0$ for each $t \in T$), and β is atomic (see Subsection 4.4.1).

[Hint: One has $\sum_{t \in K} \mu(\{t\}) < +\infty$ for each compact set $K \subset T$. Consider $\beta = \sum_{t \in T} \mu(\{t\}) \varepsilon_t$.]

4.3 Let $\mu = \sum_{t \in T} m(t) \varepsilon_t$ be a positive atomic Radon measure on T. Show that

$$\bar{\mu}^*(A) = \sum_{t \in A} m(t)$$

for any subset A of T.

4.4 Show that all subsets of T, and therefore all functions on T, are measurable for any positive atomic Radon measure on T.

4.5 (Lebesgue "Ladder Process.") Let A be an integrable set, f a real-valued, bounded, and measurable function on A, say, $|f| < M$. By a ladder is meant a finite sequence $(l_k)_{0 \leq k \leq n}$ of numbers such that

$$-M = l_0 < l_1 < \cdots < l_n = M$$

Define

$$A_k = \{t \in A : l_{k-1} \leq f(t) < l_k\} \qquad (1 \leq k \leq n)$$

and suppose that

$$\varepsilon = \operatorname*{Sup}_{1 \leq k \leq n} (l_k - l_{k-1}).$$

Show that

$$\left| \int_A f\, d\mu - \sum_{k=1}^{n} l_k \mu(A_k) \right| \leq \varepsilon \cdot \mu(A).$$

4.6 Show that

$$\int^* (f+g)\,d\mu = \int^* f\,d\mu + \int^* g\,d\mu$$

for any two positive measurable functions f and g.

4.7 (Jensen's inequality). A real-valued function F on $[0, \infty)$ is said to be *convex* if

$$F(a_1x_1 + \cdots + a_nx_n) \le a_1F(x_1) + \cdots + a_nF(x_n)$$

for any two finite sequences (a_k) and (x_k) of positive numbers such that $a_1 + \cdots + a_n = 1$.

Suppose that F is positive, increasing, and convex on $[0, \infty)$. Show that if A is an integrable set satisfying $\mu(A) > 0$, and if f is a positive, measurable function on A, then

$$F\left[\mu(A)^{-1}\int_A^* f\,d\mu\right] \le \mu(A)^{-1}\int_A^* (F \circ f)\,d\mu,$$

$F(+\infty)$ being understood to mean

$$\lim_{x\to+\infty} F(x) = \operatorname{Sup}_{x\ge 0} F(x) \le +\infty.$$

[Hint: Let M denote the set of functions f, positive and measurable on A, for which the inequality holds. Show that M is stable under passage to the limit of any increasing sequence of its members. Apply this, starting from functions that are finite linear combinations of characteristic functions of integrable subsets of A.]

4.8 Suppose that f and A are as in Exercise 4.7. Show that if $p \ge 1$, then

$$\int_A^* f\,d\mu \le \mu(A)^{1-1/p}\left[\int_A^* f^p\,d\mu\right]^{1/p}.$$

Show also that if $\mu(A) = 1$, then

$$\int_A^* f\,d\mu \le \log\int_A^* e^f\,d\mu.$$

Note. The first result may also be derived from Hölder's inequality (see Section 4.11).

4.9 Suppose that μ is bounded, that $0 < p < q$, and that f is a positive function on T. Show that

$$\left[\int^* f^p\,d\mu\right]^{1/p} \le \mu(T)^{1/p-1/q}\left[\int^* f^q\,d\mu\right]^{1/q}.$$

4.10 Suppose that $0 < p < 1$. Define on $\mathscr{L}^p = \mathscr{L}^p(T, \mu)$ the semimetric

$$d_p(f, g) = \int^* |f-g|^p\,d\mu.$$

Verify that $\mathscr{L}^p$ is a vector space and that the above semimetric defines on $\mathscr{L}^p$ a vector space topology. Show that $\mathscr{L}^p$ is complete for this semimetric.

4.11 Prove that if μ is bounded and $0 < p < q$, then $\mathscr{L}^p \supset \mathscr{L}^q$ and that the injection map of $\mathscr{L}^q$ into $\mathscr{L}^p$ is continuous.

4.12 Consider the Hilbert space $L^2(0, 1)$ (the basic measure being Lebesgue measure on R restricted to $(0, 1)$) and the linear map S of $L^2(0, 1)$ into itself that carries the function class $\dot{x}$ into the function class $(fx)^{\cdot}$; where f is a given bounded and measurable function on $(0, 1)$. Prove the following statements:

(1) S is continuous and $\|S\| = N_\infty(f)$.

(2) The adjoint map S^* corresponds to the function $\bar{f}$.

(3) S is unitary if and only if $|f(t)| = 1$ a.e. on $(0, 1)$.

(4) S^{-1} exists and is continuous if and only if f^{-1} is essentially bounded.

(5) The spectrum of S is nonvoid; that is, there exist scalars λ such that $S - \lambda I$ has no continuous inverse (I the identity endomorphism of $L^2(0, 1)$).

(6) If $f^{-1}(\{\lambda\})$ is negligible for each scalar λ, then S has no eigenvectors.

4.13 Prove that any positive function integrable for a given positive Radon measure μ is the limit a.e. (μ) of an increasing sequence (f_n), where f_n is positive, bounded, universally measurable, and zero outside a compact set.

[Hint: Notice that the set F of positive functions that are integrable for μ and that have the stated property is stable under the formation of limits of increasing sequences. Use the fact that any μ-integrable set differs by a μ-negligible set from some F_σ-set contained in it.]

4.14 Let $\mu = \sum_{t \in T} m(t)\varepsilon_t$ be a positive atomic Radon measure. Show that

$$\bar{\mu}^*(f) = \sum_{t \in T} m(t)f(t)$$

for any positive function f (finite or not).

4.15 Let λ be a real Radon measure on T, S its support, and t a point of S. Show that at least one of the following three assertions is true:

(1) $\lambda^+(U) > 0$ and $\lambda^-(U) > 0$ for each neighbourhood U of t (that is, t belongs to the support of λ^+ and the support of λ^-).

(2) There exists a neighbourhood U_0 of t such that

$$\lambda(X) = \lambda^+(X) \geq 0$$

for any measurable set $X \subset U_0$, while $\lambda(U) = \lambda^+(U) > 0$ for any neighbourhood U of t satisfying $U \subset U_0$ (in particular, t belongs to the support of λ^+).

(3) There exists a neighbourhood U_0 of t such that

$$\lambda(X) = -\lambda^-(X) \leq 0$$

for any measurable set $X \subset U_0$, while

$$\lambda(U) = -\lambda^-(U) < 0$$

for any neighbourhood U of t satisfying $U \subset U_0$ (in particular, t belongs to the support of λ^-).

4.16 Investigate the analogues of the results of Exercise 2.5 for the case in which the space of continuous periodic functions is replaced by the spaces $L^p(-\pi, \pi)$ (formed relative to Lebesgue measure) with $1 \leq p < \infty$.

4.17 Suppose that $c_1, \cdots, c_n$ are real or complex numbers, that μ denotes Lebesgue measure on R^n, and that $g \in \mathscr{K}(R^n)$ satisfies

$$\int e^{-(c_1t_1 + \cdots + c_nt_n)}g(t)\, d\mu(t) = 0.$$

Show that there exist continuously differentiable functions $f_1, \cdots, f_n$ in $\mathscr{K}(R^n)$ such that

$$g = \sum_{k=1}^{n} \left(\frac{\partial f_k}{\partial t_k} - c_k f_k \right),$$

the t_k being the natural coordinate functions on R^n.

[Hint: First reduce the problem to the case in which all the c_k are zero. Then proceed by induction on n.]

4.18 Let μ denote Lebesgue measure on R^n and λ a Radon measure on R^n that is relatively invariant under translations. Show that there exist real or complex numbers $c_1, \cdots, c_n$ such that

$$\lambda = \text{const.}\, e^{-(c_1t_1 + \cdots + c_nt_n)}\mu.$$

[Hint: If $\lambda = 0$ there is nothing to prove. Otherwise, if $\lambda(f_a) = D(a)\lambda(f)$, show that $D(0) = 1$, $D(a + b) = D(a)D(b)$, D is continuously differentiable, and $\partial D/\partial t_k = c_k D$ for suitable numbers c_k. Hence $\lambda(\partial f/\partial t_k) = c_k\lambda(f)$ if $f \in \mathscr{K}(R^n)$ is continuously differentiable. Apply the preceding exercise.]

4.19 Determine all real Radon measures on the real axis that are relatively invariant under the homothetic transformations $t \to at$, a ranging over all nonzero real numbers.

4.20 Let T be a locally compact group, μ its left Haar measure and $1 \leq p \leq \infty$. Show that if $f \in \mathscr{L}^{p'}$ and $g \in \mathscr{L}^p$, then $f * g$ is (equal locally almost everywhere to) a continuous function. Show also that if furthermore $1 < p < \infty$, then this continuous function tends to zero at infinity.

4.21 Let T be a locally compact group, E a subset of T such that $\mu_*(E) > 0$, μ being a left Haar measure on T. Prove that $E \cdot E^{-1}$ is a neighbourhood of the neutral element of T.

[Hint: One may assume that E is compact. Consider the function

$$f(t) = \int \chi_E(ts)\chi_E(s)\, d\mu(s).$$

Show that f is continuous and is nonzero at the neutral element.]

4.22 Let T be a locally compact group, χ a character of T. Show that if χ is measurable (for left Haar measure), then it is continuous.

[Hint: First use Exercise 4.21 to show that χ is locally bounded. Then choose a function $g \in K(T)$ so that $\int \chi^{-1} g \Delta\, d\mu \neq 0$ and consider the function $g * \chi$.]

4.23 Let R denote the real axis and μ its Lebesgue measure. Take $k \in \mathscr{L}^1(R)$ and consider the endomorphism S of $\mathscr{L}^2(R)$ defined by $Sf = k * f$. Show that

(1) S is continuous on $\mathscr{L}^2(R)$ and $\| S \| \leq N_1(k)$.

(2) The spectrum of S is the set $\overline{\hat{k}(R)}$, where

$$\hat{k}(s) = \int e^{-2\pi i s t} k(t)\, d\mu(t),$$

the Fourier transform of k.

(3) A complex number λ is an eigenvalue of S if and only if $\hat{k}(s) = \lambda$ holds for s is some set of positive measure.

Note. Using the Parseval equation one can see that $\|S\| = N_\infty(\hat{k})$.

4.24 Let P be a weakly relatively compact subset of $\mathscr{L}^1 = \mathscr{L}^1(T, \mu)$. If (f_n) is a bounded sequence in $\mathscr{L}^\infty$ that converges locally in measure to a limit f (necessarily in $\mathscr{L}^\infty$), show that

$$\lim_n \int g f_n\, d\mu = \int g f\, d\mu$$

uniformly for $g \in P$.

Remark. In the next four exercises assume the result (a corollary of Theorem 8.2.2) that in any LCTVS a weakly convergent sequence is bounded for the initial topology.

4.25 Suppose that $1 < p < \infty$. Show that a sequence (f_n) in $\mathscr{L}^p$ is weakly convergent in that space to f if and only if the following two conditions are fulfilled:

(1) $\mathrm{Sup}_n\, N_p(f_n) < +\infty$.

(2) $\lim_n \int_A f_n\, d\mu = \int_A f\, d\mu$ for each integrable set A.

Show that $\lim f_n = f$ weakly in $\mathscr{L}^1$ if and only if the following two conditions are fulfilled:

(1′) As (1) above, with $p = 1$.

(2′) $\lim_n \int_A f_n\, d\mu = \int_A f\, d\mu$ for each measurable set A.

4.26 Show that $\lim f_n = f$ weakly in $\mathscr{L}^1$ if and only if the following three conditions are fulfilled:

(1′) As in Exercise 4.25.

(2) As in Exercise 4.25.

(3) Given $\varepsilon > 0$, there exists an integrable set K such that $\mathrm{Sup}_n \int_{T \setminus K} |f_n|\, d\mu \leq \varepsilon$.

4.27 Suppose that T is a real interval and that μ is the restriction of Lebesgue measure to T. Show that $\lim f_n = f$ weakly in $\mathscr{L}^1$ if and only if the following four conditions are fulfilled:

(1′) As in Exercise 4.25.

(2″) $\lim_n \int_a^b f_n \, d\mu = \int_a^b f \, d\mu$ for each bounded interval $(a, b) \subset T$.

(3) As in Exercise 4.26.

(4) Given $\varepsilon > 0$ there exists $\delta > 0$ such that $\text{Sup}_n \int_A |f_n| \, d\mu < \varepsilon$ for any integrable set $A \subset T$ satisfying $\mu(A) < \delta$.

[Hints: The necessity of (1′), (2), (2′), and (2″) follows from the remark preceding these exercises and the definition of weak convergence. The necessity of (3) and (4) comes from Theorem 4.21.2. The sufficiency of the stated conditions comes in each case from an application of Exercise 1.18, the subset H of $\mathscr{L}^\infty$ being suitably chosen.]

4.28 Show that the unit ball in the space $\mathscr{L}^1 = \mathscr{L}^1(T, \mu)$ is weakly relatively compact if and only if the support of μ is finite (in which case $\mathscr{L}^1$ is finite dimensional).

[Hints: Let S be the support of μ. Show first that $\mu(\{t\}) > 0$ for each $t \in S$. Then use Theorem 4.21.2 to show in succession that there exists a number $\delta > 0$ such that $\mu(\{t\}) > \delta$ for each $t \in S$, so that S is locally finite, and then that there exists a compact set $K \subset T$ such that $T \backslash K$ is locally negligible. Consequently $S = S \cap K$ is finite.]

Remark. As we shall see in Chapter 8, $\mathscr{L}^1$ is reflexive (that is, each continuous linear form on $(\mathscr{L}^1)' = \mathscr{L}^\infty$ is expressible in the manner $g \to \int fg \, d\mu$ for some $f \in \mathscr{L}^1$) if and only if the unit ball in $\mathscr{L}^1$ is weakly relatively compact. The present exercise shows that this is the case only if $\mathscr{L}^1$ is finite dimensional. On the contrary, the results of Section 4.16 show that $\mathscr{L}^p$ is reflexive whenever $1 < p < \infty$.

4.29 Let μ be a continuous positive Radon measure on a separated locally compact space T, and let p be a number satisfying $0 < p < 1$. Show that any positive linear form L on $\mathscr{L}^p = \mathscr{L}^p(T, \mu)$ is identically 0.

[Hint: Consider $\lambda = L \mid \mathscr{K}(T)$; show that the positive Radon measure λ is LAC(μ), hence of the form $\lambda = g \cdot \mu$, where g is locally integrable for μ. Then show that g is 0 l.a.e. (μ).]

4.30 Let E be a Riesz space (see Subsection 4.23.1) and let f be a linear form on E that is relatively bounded in the sense that for each $x \geq 0$ in E the set $\{f(y) : y \in E, 0 \leq y \leq x\}$ is bounded. Show that f is the difference of two positive linear forms on E.

[Hint: Adapt the argument used to prove Theorem 4.3.2.]

4.31 Let E be a Riesz space endowed with a vector-space topology having a base at 0 formed of sets W such that $x \in W$ and $0 \leq y \leq x$ imply that $y \in W$. Show that any bounded linear form on E is relatively bounded.

4.32 Let E be a Riesz space endowed with a vector-space topology. Suppose we are given also a subset N of E formed of so-called "negligible" elements on E. Suppose also that the following condition is fulfilled: To each bounded set B in E corresponds a sequence of numbers $c_n > 0$ such that if (x_n) is any sequence extracted from B, one may find $y \in E$ and a sequence (z_n) of negligible elements of E such that

$$\sum_{n=1}^{m} c_n(|x_n| + z_n) \leq y$$

for all natural numbers m.

Show that any relatively bounded linear form on E, vanishing on all negligible elements, is bounded.

[Hint: By Exercise 4.30 it suffices to consider a positive linear form f on E. If f were not bounded there would exist a bounded set $B \in E$ such that $f(B)$ is not bounded. Choose $x_n \in B$ so that $|f(x_n)| > c_n^{-1}$ and therefore $f(|x_n|) > c_n^{-1}$. Show that this leads to a contradiction.]

4.33 Using Exercises 4.31 and 4.32, compare the result of Exercise 4.29 with that stated in Subsection 4.16.3.

[Hint: Notice that if (f_n) is a bounded sequence in $\mathscr{L}^p$, and if $\sum c_n^p < +\infty$, then $\sum c_n |f_n|$ is convergent almost everywhere to a finite limit, since

$$\left(\sum_{n=1}^{m} c_n |f_n|\right)^p \leq \sum_{n=1}^{m} c_n^p |f_n|^p$$

and so

$$\int^* \left(\sum c_n |f_n|\right)^p d\mu \leq \sum c_n^p \int |f_n|^p \, d\mu < +\infty.]$$

4.34 Show that any integrable function is equal almost everywhere to the difference of two positive, integrable, lower semicontinuous functions. (Compare Riesz-Nagy [1], pp. 29–33.)

4.35 Let T be the separated locally compact space $[0, +\infty)$, and let $\alpha > 0$. Show that the functions $e^{-\alpha nt}$ $(n = 1, 2, \cdots)$ are total in $C_0(T)$, and deduce that the functions $e^{-\alpha t}P(t)$, where P is a polynomial, are dense in $C_0(T)$.

4.36 Let T and α be as in the preceding exercise. Show that the functions $P(t) \cdot \exp(-\alpha t^2)$, where P is a polynomial, are dense in $C_0(T)$.

4.37 Show that if $\alpha > 0$ and $1 \leq p < \infty$, then the functions $e^{-\alpha t}P(t)$, where P is a polynomial, are dense in $L^p(0, \infty)$. Show that the same is true of the functions $e^{-\alpha t^2}P(t)$.

4.38 Let T be an interval in R^n, μ a positive Radon measure on T. By a step function we here mean a finite linear combination of characteristic functions of bounded subintervals of T.

Prove that a function f is μ-integrable if and only if it satisfies the following condition: For each $\varepsilon > 0$ there exist sequences $(s_n)_{n \geq 1}$ and $(s'_n)_{n \geq 1}$ of step functions such that $s_n \geq 0$, $s'_n \geq 0$ for $n > 1$,

$$\sum \int s_n \, d\mu < +\infty, \qquad \sum \int s'_n \, d\mu < +\infty,$$

$$\sum s_n \leq f \leq \sum s'_n,$$

and

$$\sum \int (s'_n - s_n) \, d\mu \leq \varepsilon.$$

4.39 Let μ be a positive Radon measure on a separated locally compact space T. Suppose that s is a finitely additive set function defined (finitely) for all μ-integrable subsets of T and such that $\mu(A_n) \to 0$ entails $s(A_n) \to 0$. Prove that there exists a positive locally μ-integrable function f on T such that

$$s(A) = \int_A f \, d\mu.$$

[Hint: Define a positive Radon measure σ on T in the following way. If g is a continuous real-valued function on T vanishing outside a compact set $K \subset T$, put $\sigma(g)$ for the limit of sums $\sum_k c_k s(A_k)$ as $\max_k (c_k - c_{k-1}) \to 0$, where

$$c = \operatorname{Max} |g|, \qquad -c = c_0 < c_1 < \cdots < c_n = c + 1$$

and

$$A_k = \{t : t \in K, c_{k-1} \leq g(t) < c_k\}.$$

Show that $\sigma(U) \leq s(U)$ if U is a relatively compact open set, hence that σ is LAC(μ). Use the Lebesgue-Radon-Nikodym theorem.]

4.40 Let T be a separated locally compact space and $T^* = T \cup \{\infty\}$ its Alexandroff compactification (see Subsection 0.2.18(4)). Show that there exists an isometric isomorphism $\mu \to \mu^*$ of the space of bounded Radon measures on T onto the subspace N of the space of all Radon measures on T^* comprising all Radon measures λ on T^* such that $|\lambda|(\infty) = 0$.

4.41 Let T be a locally compact space. Show that any continuous function on T that tends to zero at infinity is the uniform limit on T of finite linear combinations of characteristic functions χ_A of compact sets $A \subset T$ having countable neighbourhood bases.

4.42 Let λ be a real Radon measure on a separated locally compact space T. Prove that for any measurable set $S \subset T$ one has

$$\lambda^+(S) = \text{Sup}\, \lambda(M), \qquad \lambda^-(S) = -\text{Inf}\, \lambda(M),$$

$$\tfrac{1}{2}|\lambda|(S) \leq \text{Sup}\, |\lambda(M)| \leq |\lambda|(S)$$

where in each case M ranges over all measurable subsets of S.
[Hint: Use Corollary 4.15.5.]

4.43 Let T be a separated locally compact space, $M_b(T)$ the space of bounded Radon measures on T. Without using Kakutani's Theorem 4.23.2, show that M is sequentially weakly complete.
[Hint: Use the device explained in the Remark following Theorem 4.22.1.]

4.44 The notations being as in the preceding exercise, show that $M_b(T)$ is reflexive if and only if T is finite.
[Hint: Use Exercise 4.28.]

4.45 Let T be a separated locally compact space, and let $C_0(T)$ be the Banach space of continuous scalar-valued functions x on T that tend to zero at infinity (in the sense that the set

$$\{t \in T: |x(t)| > \varepsilon\}$$

is relatively compact in T for each $\varepsilon > 0$), the norm on $C_0(T)$ being defined by

$$\|x\| = \text{Sup}\, \{|x(t)|: t \in T\}.$$

Show that the topological dual of $C_0(T)$ may be identified with $M_b(T)$ by associating with each $\mu \in M_b(T)$ the linear form $x \to \int_T x\, d\mu$ on $C_0(T)$.

Remarks. This is the continuous analogue of Theorem 4.2.2. Regarding the nature of the associated dual norm on $M_b(T)$, see Remark 4.15.6. From Theorem 1.11.4(2) it follows that the norm-bounded subsets of $M_b(T)$ are relatively compact for $\sigma(M_b, C_0)$.

CHAPTER 5

Distributions and Linear Partial Differential Equations

5.0 Foreword

In the last chapter we studied in considerable detail those objects (Radon measures) arising out of the representation of continuous linear forms on certain TVSs of continuous functions. Continuity of the linear form has been interpreted relative to the uniform convergence of functions restricted to the extent that their supports lie in a fixed but arbitrary compact set. In this chapter we seek and study those linear forms which are initially defined only for functions that are indefinitely differentiable, and which are continuous in a weaker sense. The new mode of continuity of the linear form is to be interpreted relative to functions having supports in a fixed but arbitrary compact set and which, together with each of their derivatives, converge uniformly. Such linear forms demand for their representation the distributions introduced by Schwartz.

Distributions introduced in this (or some equivalent) fashion generalize both functions and measures, and to them one may extend some of the operations applied to functions and measures. Above all, distributions may be differentiated as many times as one may wish and differentiation becomes a continuous process. It follows that among the most natural equations involving distributions will appear all linear partial differential equations with constant coefficients (or even with coefficients that are functions, provided these are restricted by suitable smoothness conditions depending on the type of distribution involved). The study of such equations within the enlarged domain of distributions has gathered great impetus and represents a flourishing field of research. We have endeavoured to discuss at some length a few of the aspects of these developments. The result of this discussion will, it is hoped, simultaneously motivate the study of distributions and encourage the reader to follow up recent research in applied functional analysis.

The introduction of distributions into differential equations naturally means that one will have to admit as solutions things which, even if they are still functions, will generally lack the smoothness implied by point-wise differentiability in the elementary sense. To do this may at first seem entirely wrong, but a little reflection indicates that the change proves to be not without some foundation. To regard a function as a distribution means that it is characterized no longer by its values at individual points, but instead by its spatial averages represented by the result of multiplying it by an arbitrary "test function" and then integrating. These test functions are to be indefinitely differentiable and to have compact supports; they form, in fact, the function space on which distributions are defined to act as linear forms. Now when one attempts to apply mathematics to the natural sciences, the observable quantities are indeed measured by such averages rather than by their values at isolated points. In classical field theories it is well known that the partial differential equations can always be replaced by relations involving integrals. In the theory of distributions, a partial differential equation expresses precisely such a relation and thus represents more closely the real situation than does a differential equation interpreted in the old-fashioned way.

Any selection from the following list of texts will constitute a profitable accompaniment to the whole of this chapter: Schwartz [1], [2]; Gelfand and Šilov [2], [3]; Hörmander [7]; and Friedman [1]. Special mention should be made of Courant-Hilbert [2] which offers, in addition to large portions of the classical theory of partial differential equations, an excellent bridge from there to the modern theory. See also Dunford and Schwartz [2].

5.1 Distributions

The definition given in Subsection 4.3.1 of Radon measures on R^n, which introduces them as linear forms on $\mathscr{K}(R^n)$ satisfying certain continuity requirements, sets the pattern for the introduction of distributions on R^n or on open subsets Ω of R^n in the manner described by Schwartz [1], [2]. To do this we pass from the space $\mathscr{K}(\Omega)$ to a vector subspace $\mathscr{D}(\Omega)$ thereof, at the same time endowing the latter with a topology strictly stronger than that induced by the topology of $\mathscr{K}(\Omega)$. The Hahn-Banach theorem warns us that some refinement of the topology is essential if we are to encompass amongst the continuous linear forms on $\mathscr{D}(\Omega)$ objects more general than Radon measures on Ω.

This process may also be described roughly in physical terms, as follows. Real Radon measures on R^3 can be thought of as the mathematical objects appropriate for the description of those distributions of electric charges, magnetic poles, or gravitating matter that involve only "simple poles"; this view is borne out by the developments in potential theory during the last 30 years or so. In mathematical physics, however, it is customary to introduce, alongside those distributions described by Radon measures, certain idealized limits of these—notably those distributions involving multipoles of various types. One can regard Schwartz distributions as the mathematical entities corresponding to extremely general distributions of charge or matter involving multipoles of arbitrarily high, or even infinite, order.

Just as in Section 4.3 for measures, so here for distributions, it is convenient to use provisionally a notion of convergence of sequences in $\mathscr{D}(\Omega)$ in order to describe the continuity requirement to be characteristic of distributions. The problem of making $\mathscr{D}(\Omega)$ into a LCTVS whose dual is realized by the distributions so defined will be solved later in terms of inductive limit spaces (see Section 6.3, Example 2).

We shall find that distributions can be differentiated as often as we please, all partial differentiations commuting in pairs, and that differentiation is a continuous process. As a consequence, the theory of partial differential equations takes on a new lease of life: the use of distributions raises new problems and offers new methods of attack. During the past decade, progress has been rapid. Historically, many of the basic ideas concerning distributions were first used in studying the so-called "weak solutions" of partial differential equations, and the growth of the systematic theory is the result of contributions of many workers.

The following account seeks to do little more than to present enough of the fundamental ideas, themselves of great intrinsic interest in functional analysis, to make it possible to see how the new ideas combine with general theorems in functional analysis so as to supply new approaches and new viewpoints toward problems in analysis. In Subsection 5.11.5 we indicate some further reading that may interest the reader.

5.1.1 **Notations.** In all that follows, Ω is an open subset of R^n and $x_1, \cdots, x_n$, denote the natural coordinate functions on R^n.

We denote by $\mathscr{D}(\Omega)$ the vector subspace of $\mathscr{K}(\Omega)$ formed of functions that have partial derivatives of all orders throughout Ω and that vanish outside compact subsets of Ω (depending on the function in question). We shall write ∂_k $(k = 1, \cdots, n)$ for the partial differential operator $\partial/\partial x_k$ acting on $\mathscr{D}(\Omega)$, and the higher-order operator $\partial_1^{p_1} \cdots \partial_n^{p_n}$ is denoted by ∂^p, p representing the n-uple $(p_1, \cdots, p_n)$ of integers $p_k \geq 0$; $|p|$ will stand for $p_1 + \cdots + p_n$, the order of ∂^p. Since each function in $\mathscr{D}(\Omega)$ is of class $C^\infty(\Omega)$, the ∂_k commute as endomorphisms of $\mathscr{D}(\Omega)$.

If S is a subset of Ω, $\mathscr{D}(\Omega, S)$ is the subspace of $\mathscr{D}(\Omega)$ formed of those functions ϕ in $\mathscr{D}(\Omega)$ with support contained in S. We define positive functions N_m^S on $C^\infty(\Omega)$ as follows:

$$N_m^S(\phi) = \text{Sup}\ \{|\partial^p \phi(x)| : x \in S, |p| \leq m\},$$

$m \geq 0$ being an integer and S a subset of Ω. It is clear that $N_m^S(\phi) < +\infty$ if S is relatively compact in Ω (that is, $\bar{S} \cap \Omega$ is compact) and ϕ belongs to $C^\infty(\Omega)$, or for arbitrary S if ϕ is in $\mathscr{D}(\Omega)$. N_m^S and $N_m \equiv N_m^\Omega$ coincide on $\mathscr{D}(\Omega, S)$.

If S is relatively compact in Ω, the N_m or the N_m^S are norms on $\mathscr{D}(\Omega, S)$ and serve to define a complete, metrizable, locally convex topology on $\mathscr{D}(\Omega, S)$: this makes $\mathscr{D}(\Omega, S)$ into a Fréchet space.

If S is allowed to range over a base for the relatively compact subsets of Ω, the N_m^S serve to define on $C^\infty(\Omega)$ a topology making it also a Fréchet space. The topology induced on $\mathscr{D}(\Omega)$ is *not* one relative to which the dual of $\mathscr{D}(\Omega)$ is composed of distributions on Ω. As we shall see a little later, we obtain thus only distributions having compact supports in Ω. The desired topology on $\mathscr{D}(\Omega)$ will in fact be obtained by taking the inductive limit of the Fréchet spaces $\mathscr{D}(\Omega, S)$

with S ranging over a base for the relatively compact (or compact) subsets of Ω (see Section 6.3). Meanwhile, the norms N_m help to specify notions of sequential convergence and boundedness in $\mathscr{D}(\Omega)$ which prove later to be identical with those derived from the appropriate inductive limit topology (see Subsections 6.6.1 and 7.4.3).

Here are the definitions of these concepts. A sequence (ϕ_n) in $\mathscr{D}(\Omega)$ is said to converge to 0 in $\mathscr{D}(\Omega)$ if and only if (a) $\lim_n N_m(\phi_n) = 0$ for each $m \geq 0$, and (b) there exists a relatively compact (or compact) set $S \subset \Omega$ such that $(\phi_n) \subset \mathscr{D}(\Omega, S)$. Similarly, a subset B of $\mathscr{D}(\Omega)$ is said to be bounded if and only if (a) $\operatorname{Sup}_B N_m(\phi) < +\infty$ for each $m \geq 0$, and (b) $B \subset \mathscr{D}(\Omega, S)$ for some relatively compact (or compact) $S \subset \Omega$. In other words (ϕ_n) converges to 0 in $\mathscr{D}(\Omega)$ (resp. B is bounded in $\mathscr{D}(\Omega)$) if and only if there exists a relatively compact (or compact) subset S of Ω such that all the ϕ_n (resp. B) fall into $\mathscr{D}(\Omega, S)$, and ϕ_n converges to 0 (resp. B is bounded) in the sense of the Fréchet space $\mathscr{D}(\Omega, S)$.

5.1.2 **Definition of Distributions.** We can now frame the definition of a distribution on Ω: By this we mean a linear form X on $\mathscr{D}(\Omega)$ such that the restriction $X \mid \mathscr{D}(\Omega, S)$ is continuous on the Fréchet space $\mathscr{D}(\Omega, S)$ for each relatively compact (or compact) subset S of Ω. Clearly, the content of this condition is unaltered if we suppose S to range over any assigned base for the relatively compact (or compact) subsets of Ω. Equivalent conditions are that $\lim X(\phi_n) = 0$ for each sequence $\phi_n \to 0$ in $\mathscr{D}(\Omega)$; or again that X transforms each bounded subset of $\mathscr{D}(\Omega, S)$ (S any relatively compact subset of Ω) into a bounded set of scalars.

As for measures, each distribution X on Ω is fully determined by its behaviour on the real-valued functions in $\mathscr{D}(\Omega)$. Moreover, X can always be written uniquely as $A + iB$, where A and B are distributions on Ω which are real in the sense that they assign real values to real-valued functions in $\mathscr{D}(\Omega)$. Unlike measures, however, a real distribution is not generally the difference of two positive distributions. In fact, it is easily shown (Schwartz [1], pp. 28–29) that a distribution X on Ω is positive (that is, $X(\phi) \geq 0$ for each $\phi \geq 0$ in $\mathscr{D}(\Omega)$) if and only if it is a positive Radon measure on Ω.

We shall denote by $\mathscr{D}'(\Omega)$ the set of all distributions on Ω; it is a vector subspace of the algebraic dual of $\mathscr{D}(\Omega)$. Anticipating the duality between $\mathscr{D}(\Omega)$ and $\mathscr{D}'(\Omega)$, we shall often write $\langle \phi, X \rangle$ in place of $X(\phi)$.

5.2 Measures and Functions as Distributions

Let λ be any Radon measure on Ω. It is obvious that $L = \lambda \mid \mathscr{D}(\Omega)$ is a distribution on Ω. In addition L determines λ uniquely. This is so because each ϕ in $\mathscr{K}(\Omega)$ is the limit in $\mathscr{K}(\Omega)$ of a sequence $(\phi_n) \subset \mathscr{D}(\Omega)$. To see this one has only to use the process of "regularization"—that is, take a sequence (r_n) of positive functions in $\mathscr{D}(\Omega)$, having supports that are smaller and smaller neighbourhoods of 0, and such that $\int r_n \, d\mu = 1$ (μ = Lebesgue measure on R^n);

if we define ϕ to be 0 outside Ω, the functions

$$\phi_n(x) = r_n * \phi(x) = \int r_n(y)\phi(x - y)\, d\mu(y)$$

belong to $\mathscr{D}(\Omega)$ and $\phi - \phi_n \to 0$ in $\mathscr{K}(\Omega)$.

Since the correspondence $\lambda \leftrightarrow L$ is one to one, there is no harm in identifying the measure λ with the distribution L and dropping the notational distinction.

If f is a function that is locally integrable for μ on Ω, we can apply the preceding remarks to the measure $f \cdot \mu$ on Ω. In this case, knowledge of the measure implies only knowledge of the class of f (modulo functions negligible on Ω). Thus it is possible to identify each locally integrable function f on Ω (or, better, each class of such functions) with the distribution on Ω defined by

$$\phi \to \int_\Omega f\phi\, d\mu.$$

In the case of functions that are continuous on Ω, the correspondence between functions and distributions is genuinely one to one.

Examples of distributions on Ω that are neither functions nor measures (after the identifications just described) arise from the differentiation and division of these same special distributions, processes to be defined in Sections 5.4 and 5.5.

5.3 Convergence of Distributions

Regarding $\mathscr{D}'(\Omega)$ as the topological dual of $\mathscr{D}(\Omega)$, there are at least two natural topologies on $\mathscr{D}'(\Omega)$. These are (1) the weak topology $\sigma(\mathscr{D}', \mathscr{D})$, and (2) the strong topology $\beta(\mathscr{D}', \mathscr{D})$. We have already met the weak topology σ in its general setting in Section 1.11; a parallel treatment of β in the language of general duality theory will be undertaken in Section 8.4. Meanwhile we shall merely say enough about it in the present special case to permit sensible reference to it, and then summarize certain of its properties. Proofs of these properties flow in the main from general theorems to be established in Section 8.4.

By definition, a base of neighbourhoods at 0 for $\beta = \beta(\mathscr{D}', \mathscr{D})$ is formed of the polar sets B^0 when B ranges over the bounded subsets of $\mathscr{D}(\Omega)$. In other words, if (X_i) is a directed family of distributions on Ω and X a distribution on Ω, the relation $\lim_i X_i = X$ (relative to, or for) β signifies that

$$\lim_i \text{Sup}\, \{|X(\varphi) - X_i(\varphi)| : \varphi \in B\} = 0$$

holds for each bounded subset B of $\mathscr{D}(\Omega)$. A better grasp of the significance of this condition is obtained when one has in mind some characterization of the bounded subsets of $\mathscr{D}(\Omega)$ such as is recalled immediately below. For either topology, σ or β, $\mathscr{D}'(\Omega)$ is a separated LCTVS.

Let us now comment upon the nature of the bounded subsets of $\mathscr{D}(\Omega)$. The topology of $\mathscr{D}(\Omega)$ will receive special mention in Example (2) of Subsection 6.3.3, and it will be shown in Section 6.6 that a subset B of $\mathscr{D}(\Omega)$ is bounded if and only if there exists a compact subset K of Ω such that B is contained and bounded in the Fréchet space $\mathscr{D}(\Omega, K)$. This in turn can be expressed as the conjunction of two conditions, viz.

(a) $\varphi = 0$ on $\Omega \backslash K$ for each $\varphi \in B$;

(b) $\text{Sup}\ \{|\partial^p \varphi(x)| : x \in \Omega,\ \varphi \in B\} < +\infty$

for each exponent $p = (p_1, \cdots, p_n)$ (the p_k being integers ≥ 0, as usual).

Example. Suppose $n = 1$, and that (f_i) is a directed family in $L^1_{\text{loc}}(\Omega)$. We let k denote a nonnegative integer and suppose that for each i, F_i is a k-fold indefinite integral of f_i. The preceding criterion for bounded subsets of $\mathscr{D}(\Omega)$ then allows us to conclude that (f_i) is weakly [resp. strongly] convergent in $\mathscr{D}'(\Omega)$ provided that for each compact subset K of Ω, the sequence (F_i) is weakly [resp. strongly] convergent in $L^1(K)$. It is, in fact, the case that all distributions are obtainable locally as limits of sequences (f_i) that are convergent in this sense, this circumstance providing the basis for some alternative approaches to the theory. The idea may be developed so as to apply in several dimensions.

Bearing this characterization in mind we can now proceed to record a number of important properties of $\mathscr{D}(\Omega)$ and $\mathscr{D}'(\Omega)$.

(1) $\mathscr{D}(\Omega)$ is barrelled (see Section 6.2 and Subsection 6.3.3). This means that each subset of $\mathscr{D}(\Omega)$ that is closed, convex, balanced, and absorbent is a neighbourhood of 0 in $\mathscr{D}(\Omega)$. Equivalently (Theorem 7.1.1) each weakly bounded subset of $\mathscr{D}'(\Omega)$ is equicontinuous, a fortiori strongly bounded.

(2) $\mathscr{D}(\Omega)$ is a Montel space (see Subsection 8.4.7); that is, each bounded, closed subset of $\mathscr{D}(\Omega)$ is compact. The proof of this depends on the preceding characterization of the bounded subsets of $\mathscr{D}(\Omega)$, coupled with Ascoli's theorem. It follows (Theorem 8.4.8 and (1)) that on each bounded subset B of $\mathscr{D}(\Omega)$, the initial topology of $\mathscr{D}(\Omega)$ and the weak topology $\sigma(\mathscr{D}, \mathscr{D}')$ induce the same topology; and that on each bounded (weakly or strongly) subset of $\mathscr{D}'(\Omega)$, the weak and the strong topologies coincide. In particular, a weakly convergent sequence in $\mathscr{D}(\Omega)$ [resp. in $\mathscr{D}'(\Omega)$] is convergent [resp. strongly convergent]. Furthermore, $\mathscr{D}'(\Omega)$ is also a Montel space when equipped with its strong topology β (see Theorem 8.4.11).

(3) $\mathscr{D}(\Omega)$ is reflexive (see Section 8.4, especially Theorems 8.4.2 and 8.4.5), and therefore $\mathscr{D}'(\Omega)$ is barrelled when endowed with β (see Theorem 8.4.3). Reflexivity of $\mathscr{D}(\Omega)$ means that each linear form L on $\mathscr{D}'(\Omega)$ that is continuous for β can be written (necessarily uniquely) as $L(X) = X(\varphi)$ for some φ in $\mathscr{D}(\Omega)$; and that further the convergence of φ in $\mathscr{D}(\Omega)$ is equivalent to the convergence, uniformly on each bounded subset of $\mathscr{D}'(\Omega)$, of the linear form L.

(4) $\mathscr{D}(\Omega)$ is bornological (see Section 7.3, Theorems 7.3.2 and 7.3.3(1)). This signifies that each convex, balanced subset of $\mathscr{D}(\Omega)$ that absorbs each bounded subset of $\mathscr{D}(\Omega)$ is a neighbourhood of 0 in $\mathscr{D}(\Omega)$. Consequently (see Theorem 7.3.1, Remark (a)) a linear form X on $\mathscr{D}(\Omega)$ belongs to $\mathscr{D}'(\Omega)$ if (and only if) $\text{Sup}\ \{|X(\varphi)| : \varphi \in B\} < +\infty$ for each bounded subset B of $\mathscr{D}(\Omega)$.

(5) $\mathscr{D}'(\Omega)$ is complete for β. This follows from (4) according to Subsection 8.4.13.

5.4 Differentiation of Distributions

Suppose that $n = 1$, that Ω is a real open interval, and that the function f is LAC on Ω. Then a partial integration tells us that for ϕ in $\mathscr{D}(\Omega)$ one has

$$\langle \phi, f' \rangle = \int_\Omega f' \phi \, d\mu = -\int_\Omega f \phi' \, d\mu = -\langle \phi', f \rangle,$$

f' denoting the derivative of f (defined almost everywhere and locally integrable on Ω); the integrated part vanishes because ϕ vanishes outside a compact subinterval of Ω. Thus f', as a distribution on Ω, satisfies

$$\langle \phi, f' \rangle = -\langle \phi', f \rangle$$

for ϕ in $\mathscr{D}(\Omega)$.

We adopt this method of defining the derivative of *any* distribution X on Ω; that is, in the general case, we define $\partial^p X$ to be that distribution satisfying

$$\langle \phi, \partial^p X \rangle = (-1)^{|p|} \langle \partial^p \phi, X \rangle$$

for all ϕ in $\mathscr{D}(\Omega)$. It is plain that this does indeed result in $\partial^p X$ being a distribution on Ω, since $\phi \to \partial^p \phi$ is a continuous endomorphism of $\mathscr{D}(\Omega)$. Also, since $\phi \to \partial^p \phi$ transforms bounded subsets of $\mathscr{D}(\Omega)$ into bounded subsets thereof, it appears at once that $X \to \partial^p X$ is continuous for both the weak and the strong topologies on $\mathscr{D}'(\Omega)$.

In general terms, ∂^p acting on distributions is just $(-1)^{|p|}$ times the adjoint of ∂^p qua map of $\mathscr{D}(\Omega)$ into itself.

As an example let us take X to be ε_a, the Dirac measure placed at the point a of Ω. In the one-dimensional case we shall have then

$$\langle \phi, \partial \varepsilon_a \rangle = -\langle \phi', \varepsilon_a \rangle = -\phi'(a)$$

for all ϕ in $\mathscr{D}(\Omega)$. Since we can plainly choose sequences (ϕ_n) converging to 0 in $\mathscr{K}(\Omega)$ for which $\phi_n'(a)$ does not tend to zero, the distribution $\partial \varepsilon_a$ is not a measure (a fortiori, not a function). In physical terms, the distributions of the type $\partial^p \varepsilon_a$ correspond exactly to multipoles placed at a.

Many further examples are to be found in Gelfand and Šilov [2], pp. 31–53.

WARNING. Taking again the case in which Ω is a real, open interval, if f is a locally integrable function on Ω that is LBV but not LAC on Ω, the ordinary derivative f' (defined almost everywhere and locally integrable on Ω) yields a distribution on Ω that is generally quite different from the distribution derivative ∂f. For example, suppose $\Omega = R$ and f is the Heaviside unit function ($f(x) = 0$ for $x < 0$, and $= 1$ for $x \geq 0$). Then the function f' is 0 a.e. and so defines the zero distribution. But ∂f is determined by

$$\langle \phi, \partial f \rangle = -\langle \phi', f \rangle = -\int_{-\infty}^{\infty} \phi' f \, d\mu = -\int_0^\infty \phi' \, d\mu = \phi(0),$$

so that $\partial f = \varepsilon_0$, the Dirac measure placed at 0.

5.5 Multiplication and Division of a Distribution by a Function. Pseudofunctions

5.5.1. If f belongs to $C^\infty(\Omega)$, then $\varphi \to f\varphi$ is a continuous endomorphism of $\mathscr{D}(\Omega)$, and we may define the product of f with a distribution X on Ω as the result of applying the adjoint of this endomorphism to X. In other words, fX is the distribution on Ω defined by

$$\langle \varphi, fX \rangle = \langle f\varphi, X \rangle$$

for φ in $\mathscr{D}(\Omega)$. It is evident that the mapping $(f, X) \to fX$ is bilinear. It is also continuous in the pair for the product of the Fréchet space topology on $C^\infty(\Omega)$ (see Subsection 5.1.1) and the strong topology on $\mathscr{D}'(\Omega)$. Moreover, by using the ordinary rule for differentiating $f\varphi$, one sees that the same rule applies to fX—namely,

$$\partial_k(fX) = (\partial_k f)X + f \cdot \partial_k X.$$

Leibnitz' formula follows at once.

If X is suitably restricted, fX may be defined for more general functions f. This is notably the case when X is a measure, in which case fX is defined for any f that is locally integrable for the measure X. In Section 5.8 we shall introduce the concept of a distribution of finite order m (m an integer ≥ 0), and it will then become apparent that fX is defined for any f in $C^m(\Omega)$, fX being then a distribution of order at most m.

The problem of defining fX under the widest possible conditions is of interest to the quantum physicist as well as to the pure mathematician; see König [2] and Schwartz [19].

5.5.2. We come now to the inverse problem of division, that of solving an equation

$$fX = A \tag{5.5.1}$$

for the unknown $X \in \mathscr{D}'(\Omega)$ when $A \in \mathscr{D}'(\Omega)$ and the function f on Ω are given. One might, of course, seek solutions in $\mathscr{D}'^m(\Omega)$, but for definiteness and simplicity we confine our remarks to the case in which the distribution X is left unrestricted and where f is assumed to be in $C^\infty(\Omega)$.

The substance of Section 5.6 shows that the problem is in any case a local one, at least in so far as existence theorems are concerned.

On the other hand the dimension n affects critically the facility of the discussion; see Schwartz [1], pp. 121–126. Let us therefore begin with the one-dimensional case.

5.5.3 **Division: the Case $n = 1$.** Irrespective of the dimension n, if f is nonvanishing on Ω, no problem remains. In this case f^{-1} belongs to $C^\infty(\Omega)$ and the unique solution is $f^{-1}A$.

Admitting the existence of zeros of f, one will consider first the case in which these are isolated and each of finite order. If this condition is violated, (5.5.1) will in general be insoluble. If $n = 1$, this violation does not often arise in practice and we shall have little to say about it; unfortunately such exceptions are much more prevalent if $n > 1$.

Assuming that the zeros of f are isolated and of finite order, the local character of the problem allows one to make a reduction in principle to the case in which Ω contains but one zero of finite order of f. A simple translation brings one back to the situation in which Ω is a neighbourhood of 0 and $f = x^m g$, m being the order of the sole zero $x = 0$ of f, and g a nonvanishing function in $C^\infty(\Omega)$. Since division by g is always possible, we may concentrate hereafter on division by the monomial x^m. This in turn is reducible to repeated division by x itself. Thus finally we have to consider the simpler equation

$$xX = A. \tag{5.5.2}$$

If A itself is a function in $C^\infty(\Omega)$, the solubility of (5.5.2) would follow already from that of $xX = 1$.

Now (5.5.2), if soluble at all, has infinitely many solutions. The difference S between any two solutions satisfies the homogeneous equation

$$xS = 0. \tag{5.5.3}$$

If S satisfies (5.5.3), then $S(x\varphi) = 0$ for each φ in $\mathscr{D}(\Omega)$. On the other hand, if ψ belongs to $\mathscr{D}(\Omega)$ and $\psi(0) = 0$, then ψ can be written as $x\varphi$ for some φ in $\mathscr{D}(\Omega)$; φ is, in fact, uniquely determined by ψ via the relations $\varphi(0) = \psi'(0)$ and $\varphi(x) = x^{-1}\psi(x)$ for $x \neq 0$. Accordingly $S(\psi) = 0$ whenever $\psi(0) = 0$, and therefore S must be a multiple of ε (the Dirac measure at 0). Conversely any such multiple, $S = c\varepsilon$, gives a solution of (5.5.3).

One may note in passing that a similar argument shows that the solutions of $x^m S = 0$ are precisely the distributions of the form $S = \sum_{p<m} c_p \, \partial^p \varepsilon$.

With this information available, it remains only to consider the existence and nature of any one solution of (5.5.2). Now if X is a solution, and if φ and ψ are related as above, then necessarily $X(\psi) = A(\varphi)$. This determines X on the vector subspace $H = \{\psi \in \mathscr{D}(\Omega) : \psi(0) = 0\}$ of $\mathscr{D}(\Omega)$. H is a hyperplane in $\mathscr{D}(\Omega)$. The explicit dependence of ψ on φ shows that $\lim \psi = 0$ whenever $\lim \varphi = 0$ (in $\mathscr{D}(\Omega)$ in each case). If, therefore, we *define* X on H by $X(\psi) = A(\varphi)$, X is continuous for the topology induced on H by that of $\mathscr{D}(\Omega)$. According to the Hahn-Banach theorem, therefore, this linear form has at least one continuous extension to $\mathscr{D}(\Omega)$; and any such extension is an element of $\mathscr{D}'(\Omega)$ satisfying (5.5.2). The existence of a solution is thus settled.

By repeated division by x one reaches the conclusion that the equation

$$x^m X = A \tag{5.5.4}$$

has, for given A in $\mathscr{D}'(\Omega)$, infinitely many solutions any two of which differ by a distribution of the type $\sum_{p<m} c_p \, \partial^p \varepsilon$ (an empty sum being interpreted, as usual, to be 0).

We shall see below how to express explicitly certain solutions of (5.5.4) in terms of a new type of distribution.

5.5.4 Division: the Case $n > 1$. To assume here that the zeros of f are isolated is entirely unrealistic, being violated even in such simple cases as those in which f is a polynomial. Nevertheless the solubility of (5.5.1) can be established under hypotheses on f wide enough to cover many instances of practical importance.

To begin with, division is possible when $f = x_k$. Here one may adapt the argument used above for the case $n = 1$ and $f = x$.

If division is possible by each of a finite number of functions, say $f_1, \cdots, f_r$, then it is possible by their product $f = f_1 \cdots f_r$.

Division is possible by any function f in $C^\infty(\Omega)$ which, if it has zeros, is such that $f, \partial_1 f, \cdots, \partial_n f$ have no common zeros. For then in the neighbourhood of any zero of f one may treat the problem locally and make a local change of coordinates in such a way as to reduce the problem to that of division by x_n. Thus, for example, division is possible by any polynomial of the type $f(x) = (c_1 + |x|^2)^{m_1} \cdots (c_r + |x|^2)^{m_r}$, the c_j being nonzero numbers.

The functions $|x|^m$ do not generally satisfy the preceding conditions and yet are nevertheless perhaps the most important divisors of all. For this reason alone it is worthwhile making a rather lengthy digression on techniques permitting one to study division by these functions. In any case, the techniques involved lead to new and important examples of distributions and would merit attention for their own sake.

Regarding the general theory of division problems, see the solutions presented by Łojasiewicz [1], [2], and Hörmander [4].

5.5.5 Finite Parts of Divergent Integrals. For simplicity let us first look at the one-dimensional case ($n = 1$).

Suppose $m \geq 1$. Pointwise division of 1 by the function x^m leads to the function x^{-m}. This last is not locally integrable over any open set containing the origin and thus fails to define a distribution on any such open set. Despite this we know that there exist infinitely many distribution solutions X of $x^m X = 1$. [If $0 < m < 1$, then the solutions would be $x^{-m} + c\varepsilon$ (c an arbitrary constant).] We shall see that when $m \geq 1$, we can still define a distribution related to the function x^{-m}—called the "finite part of x^{-m}" and typical of a class of so-called "pseudofunction distributions"—that will play the role of the function distribution x^{-m} when $m < 1$.

It is not surprising to learn that the definition of this pseudofunction x^{-m} is intimately related to the study of the (generally divergent) integral $\int x^{-m}\varphi \, d\mu$ as a functional of $\varphi \in \mathscr{D}$. The appropriate technique for the study of such divergent integrals was seemingly initiated by Hadamard [2] in connection with partial differential equations.

Our plan is to set forth the general ideas for arbitrary dimension n. In dealing with examples, however, it is often convenient to separate the cases $n = 1$ and $n > 1$, the former being often considerably simpler from a computational point of view.

As regards notation, β will denote a strictly positive number that will ultimately be allowed to tend to 0. It will sometimes be convenient to make a change of variable from β to $t = \log (1/\beta)$, so that t is real and will ultimately tend to $+\infty$. $B(\beta)$ denotes the ball (open or closed, it makes no difference) in R^n with centre 0 and radius β, and $A(\beta) = R^n \backslash B(\beta)$.

We shall begin by dealing with divergent integrals of the type $\int f\, d\mu$, where it is assumed that f is integrable over $A(\beta)$ for each $\beta > 0$. (This means that the divergence of $\int f\, d\mu$ is due to misbehaviour of the integrand f at the origin only. If one wished, one could obviously deal with misbehaviour at other points of R^n.)

The underlying idea is that under certain further conditions upon f one may write

$$\int_{A(\beta)} f\, d\mu = I(\beta) + F(\beta) \qquad (\beta > 0), \tag{5.5.5}$$

where $I(\beta)$ is a restricted type of function exhibiting singular behaviour as $\beta \to 0$, while $F(0) = \lim_{\beta\to 0} F(\beta)$ exists finitely. The component $I(\beta)$—the "infinite part"—will be so narrowly restricted in advance that there is at most one such decomposition (5.5.5), and the so-called "finite part" of the integral $\int f\, d\mu$ is then defined as $F(0)$ and denoted by F.P. $\int f\, d\mu$.

The pseudofunction distribution F.P.f will subsequently be defined by the relation $\langle \varphi, \text{F.P.}f \rangle = \text{F.P.} \int f\varphi\, d\mu$ whenever the right-hand side exists for each φ in $\mathscr{D}$ and represents a continuous function of φ.

The first detail demanding attention is the a priori specification of the allowed infinite parts $I(\beta)$ and the verification of uniqueness in (5.5.5). As to the first point we insist that an allowed infinite part $I(\beta)$ is to be a finite linear combination of powers $\beta^{-\lambda}$, where λ is complex, $\lambda \neq 0$ and $\operatorname{Re} \lambda \geq 0$, together possibly with a logarithmic term $\log (1/\beta)$:

$$I(\beta) = \sum_k A_k \left(\frac{1}{\beta}\right)^{\lambda_k} + B \log \frac{1}{\beta}, \tag{5.5.6}$$

the sum being finite and the coefficients A_k and B possibly complex. In terms of the variable $t = \log(1/\beta)$, one will have

$$I(e^{-t}) = \sum_k A_k e^{\lambda_k t} + Bt. \tag{5.5.6$'$}$$

Correspondingly (5.5.5) reads

$$\int_{A(\beta)} f\, d\mu = \sum_k A_k e^{\lambda_k t} + Bt + F(e^{-t}), \qquad \beta = e^{-t}. \tag{5.5.5$'$}$$

The second point is now seen to be a thinly disguised form of the following problem. Given a function $g(t)$, defined for real t and known to be expressible in at least one way in the form

$$g(t) = \sum_{r=1}^{N} e^{\alpha_r t} P_r(t) + P_0(t) + Bt + h(t), \tag{5.5.7}$$

where

(1) $\alpha_1 > \alpha_2 > \cdots > \alpha_N > 0$;

(2) each $P_r(t)$ $(r = 0, 1, \cdots, N)$ is $\not\equiv 0$ and is a finite linear combination of functions $e^{i\xi t}$ with ξ real; and in P_0 no term with $\xi = 0$ appears;

(3) B is a complex number;

and

(4) $\lim_{t\to+\infty} h(t)$ exists finitely;

it is necessary to show that the α_r, the P_r, B and h are uniquely determined by g. (Actually, all we need is the uniqueness of $\lim_{t\to+\infty} h(t)$. The argument to follow, however, showing how the various unknowns are to be determined, produces h only as the final step.)

The process of determination depends on the observation that if $P(t)$ is a finite sum $\sum_j K_j e^{i\xi_j t}$, the ξ_j being real, and if $P(t) \not\equiv 0$, then $\limsup_{t\to+\infty} |P(t)| > 0$. To see this one makes use of the formula

$$\lim_{T\to+\infty} T^{-1} \int_T^{2T} e^{i\xi t}\, dt = \begin{cases} 1 & \text{if } \xi = 0, \\ 0 & \text{if } \xi \neq 0 \text{ is real.} \end{cases}$$

(Here and elsewhere in the sequel we shall often write dt in place of $d\mu(t)$ for the element of Lebesgue measure on R^1.) This formula shows that

$$\lim_{T\to+\infty} \int_T^{2T} P(t)e^{-i\xi t}\, dt = \begin{cases} K_j & \text{if } \xi = \xi_j \text{ for some } j, \\ 0 & \text{for all other real } \xi. \end{cases} \tag{5.5.8}$$

Thus, if $\limsup_{t\to+\infty} |P(t)|$ were 0, we should conclude that all the K_j are 0, leading to the contradiction $P(t) \equiv 0$.

To proceed with the determination of the unknown elements in the expression for $g(t)$, notice first that α_1 is the sole real number α for which

$$0 < \limsup e^{-\alpha t} |g(t)| < +\infty,$$

the limiting process being as $t \to +\infty$. This follows immediately from conditions (1)–(4) and the preliminary remark immediately above.

Having thus determined α_1 we can determine P_1. For we have

$$e^{-\alpha_1 t} g(t) = P_1(t) + o(1).$$

The formula (5.5.8) can be applied with P_1 in place of P, noticing as one goes that

$$\lim_{T\to+\infty} T^{-1} \int_T^{2T} o(1)\, dt = 0.$$

This leads to a determination of the ξ_j and K_j in the finite expansion $\sum_j K_j e^{i\xi_j t}$ of $P_1(t)$, and thereby to the determination of P_1 itself.

With α_1 and P_1 known, one then begins afresh and determines likewise α_2 and P_2 from the known function $g(t) - e^{-\alpha_1 t} P_1(t)$.

Proceeding thus one determines $\alpha_1, \cdots, \alpha_N, P_1, \cdots, P_N$. At this stage, since $P_0 + h$ is bounded as $t \to +\infty$, B is uniquely determined by the equation

$$B = \lim_{t\to+\infty} t^{-1}\left[g - \sum_{r=1}^{N} e^{\alpha_r t} P_r \right].$$

As the final step we use (5.5.8) once again, the integrand being taken as $[g - \Sigma - Bt]$, where Σ stands for the (already determined) sum $\sum_r e^{\alpha_r t} P_r$. Since (by (2)) $P(t)$ is a finite sum $\Sigma\, C_\xi e^{i\xi t}$ with $C_0 = 0$ and (by (4)) $h(t) = L + o(1)$, we obtain

$$\lim_{T\to+\infty} T^{-1} \int_T^{2T} [g - \sum - Bt]\, dt = L.$$

And this L is precisely the data required. We may note, however, that P_0 (and therefore h too) is fully determined [by (5.5.8) once again]:

$$\lim_{T\to+\infty} T^{-1} \int_T^{2T} [g - \sum - Bt] e^{-i\xi t}\, dt = C_\xi,$$

ξ being real and nonzero.

At this point we may justifiably frame the formal definition of the finite part of an integral.

Definition. Suppose that the function f on R^n is integrable over $A(\beta)$ for each $\beta > 0$. The finite part F.P. $\int f\, d\mu$ exists if and only if there exists a decomposition (5.5.5), wherein the infinite part $I(\beta)$ is of the type (5.5.6), while $\lim_{\beta\to 0} F(\beta)$ exists finitely. When this is the case, we define

$$\text{F.P.}\int f\, d\mu = \lim_{\beta\to 0} F(\beta).$$

By way of a supplement we add that if $n = 1$ the finite parts F.P. $\int_0^\infty f\, d\mu$ and F.P.$\int_{-\infty}^0 f\, d\mu$ are often of interest. These are defined to be F.P. $\int g\, d\mu$ where $g = f$ on $(0, +\infty)$ and 0 on $(-\infty, 0)$ and $g = 0$ on $(0, +\infty)$ and f on $(-\infty, 0)$, respectively.

Remarks. It is obvious that if f is integrable, then F.P. $\int f\, d\mu$ exists and equals $\int f\, d\mu$. More generally, if the Cauchy principal value

$$\text{P.V.} \int f\, d\mu = \lim_{\beta\to 0} \int_{A(\beta)} f\, d\mu$$

exists finitely, then the finite part exists and has the same value. If $n = 1$, then

$$\text{F.P.} \int f\, d\mu = \text{F.P.} \int_0^\infty f\, d\mu + \text{F.P.} \int_{-\infty}^0 f\, d\mu$$

whenever both terms on the right exist.

5.5.6 **Simple Criteria for the Existence of F.P. $\int f\, d\mu$.** We suppose throughout that f is integrable over $A(\beta)$ for each $\beta > 0$. We assume in addition that for some (and therefore any) number $\beta_0 > 0$ one may write

$$f = \sum_{p,\lambda} A_{p,\lambda} x^p r^{-\lambda} + g, \tag{5.5.9}$$

where $r = |x|$ and g is integrable over $B(\beta_0)$. In this expression one may in principle always take $\beta_0 = 1$ and suppose that the summation is restricted to pairs (p, λ) satisfying $|p| \leq \text{Re}\,\lambda - n$. For, in respect of the last point, since $|x^p r^{-\lambda}| \leq r^{|p| - \text{Re}\lambda}$, this term is integrable over $B = B(1)$ whenever $\text{Re}\,\lambda - |p| < n$ and may then be absorbed into g.

Granted (5.5.9) we shall have for $0 < \beta < \beta_0$

$$\int_{A(\beta)} f\,d\mu = \int_{A(\beta_0)} f\,d\mu + \int_{\beta<r<\beta_0} f\,d\mu$$
$$= \int_{A(\beta_0)} f\,d\mu + \int_{\beta<r<\beta_0} \left[\sum \cdots\right] d\mu + \int_{\beta<r<\beta_0} g\,d\mu.$$

In relation to the limiting process $\beta \to 0$, $\int_{A(\beta_0)} f\,d\mu$ is fixed and $\int_{\beta<r<\beta_0} f\,d\mu$ tends to the finite limit $\int_{r<\beta_0} g\,d\mu$. It thus remains to consider the integral $\int_{\beta<r<\beta_0} [\Sigma \cdots]\,d\mu$, which is a finite linear combination of integrals

$$\int_{\beta<r<\beta_0} x^p r^{-\lambda}\,d\mu(x). \tag{5.5.10}$$

As we shall see, these contribute the right sort of infinite part.

We denote by S the unit sphere $|x| = 1$ in R^n and by σ its surface measure. The formula

$$\int_{\beta<r<\beta_0} F(x)\,d\mu(x) = \int_\beta^{\beta_0} r^{n-1}\,dr \int_S F(ry)\,d\sigma(y)$$

is well known and easy to establish for (say) continuous functions F. It shows at once that the integral (5.5.10) has the value

$$W_p \int_\beta^{\beta_0} \frac{dr}{r^{\lambda-|p|-n+1}},$$

where

$$W_p = W_{p,n} = \int_S y^p\,d\sigma(y).$$

Neglecting the factor W_p we therefore have an integral which, as $\beta \to 0$, tends to a finite limit, has a logarithmic pole, or behaves like $(1/\beta)^{\lambda-|p|-n}$ according as $\operatorname{Re} \lambda - |p| - n$ is <0, $= 0$ or >0.

This shows that a decomposition (5.5.5) is possible and establishes the existence of F.P.$\int f\,d\mu$ under the stated hypotheses on f.

For $n = 1$ there is an analogous and slightly more general result in which the sum on the right-hand side of (5.5.9) is augmented by powers x^{-m}, where m is a natural number. The discussion runs almost exactly as before.

For most simple functions f it is an easy matter to find a decomposition (5.5.9) and thereby evaluate F.P. $\int f\,d\mu$ in terms of convergent integrals (which may themselves be difficult to evaluate, of course).

5.5.7 **Pseudofunction Distributions.** If we suppose that f is as in Subsection 5.5.6, it is then easy to see that, together with f itself, each function $f\varphi$ (φ in $\mathscr{D}$) admits a decomposition of the type (5.5.9). Moreover, if we absorb into the "g-part" of (5.5.9) all locally integrable terms of the summation, a little thought will show that the summation will extend over at most a finite set of pairs (p, λ) independent of φ. The corresponding coefficients $A_{p,\lambda}$ will further be seen to be finite linear combinations of the partial derivatives $\partial^p\varphi(0)$ of orders $|p| \le N$, where N depends only on f.

All of these allegations are easily verifiable if we use the Taylor expansion

$$\varphi(x) = \sum_{|p| \leq N} \partial^p \varphi(0) \frac{x^p}{p!} + O(|x|^{N+1}),$$

valid for any natural number N, $p!$ denoting $p_1! \cdots p_n!$

This being so, the finite part F.P. $\int f\varphi \, d\mu$ exists for each φ in $\mathscr{D}$ and will admit an expression of the type

$$\text{F.P.} \int f\varphi \, d\mu = \lim_{\beta \to 0} \left[\int_{A(\beta)} f\varphi \, d\mu - \sum_{|p| \leq N} A_p(\beta) \partial^p \varphi(0) \right].$$

Now, for each $\beta > 0$ the term $[\cdots]$ is a linear functional of φ that is plainly defined and continuous on $\mathscr{D}^N = \mathscr{D}^N(R^n)$. Since $\mathscr{D}^N$ is barrelled, the pointwise limit of these linear forms, that is, the functional $\varphi \to$ F.P. $\int f\varphi \, d\mu$, is continuous and linear on $\mathscr{D}^N$. This signifies precisely that the mapping

$$\varphi \to \text{F.P.} \int f\varphi \, d\mu \tag{5.5.11}$$

is a distribution on R^n of order at most N (see Section 5.8). This distribution is denoted by F.P.f.

The same notation is employed in all cases in which the linear form (5.5.11) is defined and continuous on $\mathscr{D}$, F.P.f denoting then the distribution on R^n so defined. Distributions obtained in this way are termed collectively *pseudofunctions.*

In the one-dimensional case ($n = 1$) one may define likewise (for suitably restricted functions f) the pseudofunction distributions F.P.$(f)_{x>0}$ and F.P. $(f)_{x<0}$ by the formulae

$$\text{F.P.} \int_0^\infty f\varphi \, d\mu \quad \text{and} \quad \text{F.P.} \int_{-\infty}^0 f\varphi \, d\mu,$$

respectively.

5.5.8 **Example.** We shall consider the computation of F.P.x^{-m} in one dimension, m being a natural number. If φ belongs to $\mathscr{D}$, we have for small x the expression

$$x^{-m}\varphi(x) = x^{-m} \sum_{p \leq m-1} \partial^p \varphi(0) \frac{x^p}{p!} + g(x)$$

where g is integrable. If $\beta > 0$ one has therefore

$$\int_{A(\beta)} x^{-m} \varphi \, dx = \sum_{p \leq m-2} \partial^p \varphi(0) \frac{(1/\beta)^{m-p-1}}{p!} (m - p - 1) + \partial^{m-1}\varphi(0) \frac{\log (1/\beta)}{(m-1)!} + F(\beta),$$

where $F(\beta)$ has a finite limit as $\beta \to 0$. So

$$\langle \varphi, \text{F.P.}x^{-m} \rangle = \lim_{\beta \to 0} \left\{ \int_{A(\beta)} x^{-m} \varphi \, dx - \sum_{p \leq m-2} [\cdots] - \partial^{m-1}\varphi(0) \frac{\log (1/\beta)}{(m-1)!} \right\}, \tag{5.5.12}$$

which constitutes an explicit formula defining F.P.x^{-m}.

If in (5.5.12) we replace φ by $\psi = x^m\varphi$, the summation and the term involving $\log(1/\beta)$ vanish identically for $\beta > 0$. Moreover, $\int_{A(\beta)} x^{-m}\psi\, dx = \int_{A(\beta)} \varphi\, dx$, which tends to $\int \varphi\, dx = \langle \varphi, 1\rangle$ as $\beta \to 0$. Thus

$$\langle x^m\varphi, \text{F.P.}x^{-m}\rangle = \langle \varphi, 1\rangle$$

for each φ in $\mathscr{D}$, showing that

$$x^m\text{F.P.}x^{-m} = 1.$$

In other words, $X = \text{F.P.}x^{-m}$ is, as one would hope, a solution of $x^mX = 1$.

A similar argument may be used to show that (whatever the dimension n) $X = \text{F.P.}r^{-m}$ is a solution of $r^mX = 1$.

5.5.9 **Example.** When $n = 1$, it is often possible to compute finite parts by repeated partial integrations. By way of illustration let us verify the formula

$$\text{F.P.}x^{-1} = \frac{d(\log|x|)}{dx}. \tag{5.5.13}$$

If φ belongs to $\mathscr{D}$ one has

$$\int_{A(\beta)} \varphi \frac{dx}{x} = \int_\beta^\infty \varphi \frac{dx}{x} - \int_\beta^\infty \varphi(-x)\frac{dx}{x}.$$

In each integral on the right one integrates partially just once. After a little manipulation and juggling there results the formula

$$\begin{aligned}\int_{A(\beta)} \varphi \frac{dx}{x} &= \varphi(\beta)\log\frac{1}{\beta} - \varphi(-\beta)\log\frac{1}{\beta} - \int_\beta^\infty \varphi'(x)\log x\, dx \\ &\qquad - \int_\beta^\infty \varphi'(-x)\log x\, dx \\ &= -\int \varphi'(x)\log|x|\, dx + o(1)\end{aligned}$$

as $\beta \to 0$. Thus

$$\text{F.P.}\int \varphi\frac{dx}{x} = -\int \varphi'(x)\log|x|\, dx$$

or

$$\langle \varphi, \text{F.P.}x^{-1}\rangle = -\left\langle \frac{d\varphi}{dx}, \log|x|\right\rangle.$$

This is equivalent to (5.5.13).

In an exactly similar way it may be shown that

$$\frac{d}{dx}\text{F.P.}x^{-m} = -m\text{F.P.}x^{-m-1}$$

for any natural number m. This is a consistent extension of the formula obtained by removing the symbol "F.P." throughout and assuming that the powers of x appearing are positive.

5.5.10 **Example.** If $n > 1$, one may often replace partial integration with respect to a single variable by some form of Green's formula applied to the domain $A(\beta)$. (Actually, since the integrands involve a member of $\mathscr{D}$ as a factor, one is really using the Green's formula only for an annulus $B(\beta') \cap A(\beta)$.) This device is frequently useful for the computation of finite parts, especially for functions of $r = |x|$ only.

If we assume that φ belongs to $\mathscr{D}$ and that the function F is (say) of class C^2 on $R^n \backslash \{0\}$, then the Green's formula (derived from Subsection 4.15.14) may be written

$$\int_{A(\beta)} (\varphi\, \Delta F - F\, \Delta\varphi)\, d\mu = \int_{S(\beta)} \left[\varphi \frac{-\partial F}{\partial r} - F \frac{-\partial \varphi}{\partial r}\right] d\sigma_\beta,$$

$S(\beta)$ denoting the sphere with centre 0 and radius β and σ_β its surface measure. If we denote by primes partial derivatives with respect to r, this result may be rewritten as

$$\int_{A(\beta)} (\varphi\, \Delta F - F\, \Delta\varphi)\, d\mu = -\beta^{n-1} \int_S [\varphi(\beta y) F'(\beta y) - F(\beta y)\varphi'(\beta y)]\, d\sigma(y),$$

where $S = S(1)$ and $\sigma = \sigma_1$. In particular, if F is a function of r alone, we shall have

$$\int_{A(\beta)} (\varphi\, \Delta F - F\, \Delta\varphi)\, d\mu = -\beta^{n-1} F'(\beta) \int_S \varphi(\beta y)\, d\sigma(y) + \beta^{n-1} F(\beta) \int_S \varphi'(\beta y)\, d\sigma(y). \tag{5.5.14}$$

By choosing $F = F(r)$ suitably we can utilize (5.5.14) to compute the finite parts of various simple functions of r that are not locally integrable.

Suppose, for example, that we are interested in computing F.P.r^{-2}. If $n > 2$, r^{-2} is locally integrable and F.P.$r^{-2} = r^{-2}$. Despite this we will follow a procedure valid for general values of n. The first step is to seek a function F satisfying $\Delta F = r^{-2}$ for $r \neq 0$ (the Laplacian being here understood in the classical sense). Since $\Delta F = r^{-n+1} d/dr(r^{n-1} F)$ if F is a function of r only, we may take for F the function defined by

$$F = \begin{cases} (n-2)^{-1} \log r & \text{if } n \neq 2, \\ \frac{1}{2} \log^2 r & \text{if } n = 2. \end{cases}$$

This F is locally integrable.

Applying (5.5.14) with this choice of F, we obtain

$$\int_{A(\beta)} \varphi \frac{d\mu}{r^2} = \int_{A(\beta)} F\, \Delta\varphi\, d\mu - J(\beta)$$

where

$$J(\beta) = \beta^{n-1} F'(\beta) \int_S \varphi(\beta y)\, d\sigma(y) - \beta^{n-1} F(\beta) \int \varphi'(\beta y)\, d\sigma(y).$$

Now if $n > 2$, both $\beta^{n-1} F'(\beta)$ and $\beta^{n-1} F(\beta)$ tend to 0 with β and so we find that

$$\text{F.P.} r^{-2} = r^{-2} = (n-2)^{-1}\, \Delta \log r \qquad (n > 2). \tag{5.5.15}$$

The Laplacian here is interpreted in the sense of distributions theory.

If $n = 2$, $\beta^{n-1}F'(\beta) = -\log(1/\beta)$ and $\beta^{n-1}F(\beta) \to 0$ with β, so that

$$J(\beta) = -\log\frac{1}{\beta}\cdot\int_S \varphi(\beta y)\,d\sigma(y) + o(1).$$

Thus

$$\int_{A(\beta)} \varphi\,\frac{d\mu}{r^2} = \log\frac{1}{\beta}\int_S \varphi(\beta y)\,d\sigma(y) + o(1) + \int F\,\Delta\varphi\,d\mu + o(1)$$
$$= \log\frac{1}{\beta}\,\varphi(0)\int_S d\sigma(y) + \int F\,\Delta\varphi\,d\mu + o(1).$$

Since $\int_S d\sigma = 2\pi$ we infer that

$$\langle \varphi, \text{F.P.}r^{-2}\rangle = \lim_{\beta\to 0}\left[\int_{A(\beta)} \varphi\,\frac{d\mu}{r^2} - 2\pi\varphi(0)\log\frac{1}{\beta}\right] = \langle \Delta\varphi, F\rangle. \tag{5.5.16}$$

and therefore

$$\text{F.P.}r^{-2} = \tfrac{1}{2}\Delta(\log^2 r) \qquad (n = 2). \tag{5.5.17}$$

Here again the Laplacian is interpreted according to the theory of distributions. The formula (5.5.16) allows one to check that r^2 F.P. $r^{-2} = 1$.

For many more examples the reader may consult Gelfand and Šilov [2], pp. 53–88.

5.6 Restriction of Distributions; Localization; Support of a Distribution

Suppose that Ω and Ω' are open subsets of R^n and that $\Omega \subset \Omega'$. Suppose too that X belongs to $\mathscr{D}'(\Omega')$: we aim to define the result of restricting X to Ω. To this end, we observe that each ϕ in $\mathscr{D}(\Omega)$ has a natural extension as an element of $\mathscr{D}(\Omega')$, namely, the function ϕ' on Ω' defined to equal ϕ in Ω and to be 0 elsewhere in Ω'. Then $\phi \to \phi'$ is a continuous linear map of $\mathscr{D}(\Omega)$ into $\mathscr{D}(\Omega')$, the image being formed precisely of those functions in $\mathscr{D}(\Omega')$ the supports of which are contained in Ω. The operation of restriction of distributions from Ω' to Ω is just the adjoint of $\phi \to \phi'$: thus $X \mid \Omega$ is the distribution on Ω defined by

$$\phi \to \langle \phi', X\rangle$$

for ϕ in $\mathscr{D}(\Omega)$. The mapping $X \to X \mid \Omega$ is linear and continuous (for either the weak or the strong topologies on $\mathscr{D}(\Omega')$ and $\mathscr{D}(\Omega)$). If $X \in \mathscr{D}'(\Omega')$, we shall say that X is 0 on Ω, and write $X = 0$ on Ω, if and only if $X \mid \Omega = 0$—that is, if and only if $\langle \phi, X\rangle = 0$ for each ϕ in $\mathscr{D}(\Omega')$ whose support is contained in Ω. This agrees with the convention employed in the case of measures (see Section 4.9), as is seen by recalling from Section 5.2 that $\mathscr{D}(\Omega)$ is dense in $\mathscr{K}(\Omega)$.

The converse problem—that of piecing together a distribution from its local constituents—is effected in terms of open coverings and partitions of unity. Each open subset Ω of R^n (like any subspace of R^n) is paracompact and hence normal (Subsection 0.2.20). Being locally compact, Ω has also the property that each open covering of Ω has a refinement formed of relatively compact open subsets of Ω. Thanks to these two properties, we know that given any open covering (Ω_i) of Ω there exists a locally finite open covering (Ω'_j) of Ω such that each Ω'_j is relatively compact in Ω and is contained in some Ω_i.

5.6.1 **Proposition.** Let Ω be an open subset of R^n and (Ω_i) any open covering of Ω. Then there exists a family (ϕ_j) of functions in $\mathscr{D}(\Omega)$ such that

(a) $\phi_j \geq 0$, support $\phi_j \subset \Omega_i$ for some i;

(b) the sum $\sum_j \phi_j$ is locally finite and everywhere equal to 1 on Ω.

If (Ω_i) is locally finite and each Ω_i is relatively compact in Ω, the family of functions ϕ_j may be assumed to have the same index set as (Ω_i).

Proof. By our preliminary remarks we can in any case find a locally finite covering (Ω'_j), each Ω'_j being relatively compact in Ω and Ω'_j being contained in some Ω_i. Then, again since Ω is paracompact, we can find an open covering (Ω''_j), with the same index set as (Ω'_j), such that $\overline{\Omega''_j} \subset \Omega'_j$. By normality of Ω and Urysohn's lemma (see Subsection 0.2.12) there exists for each j a continuous function α_j on Ω such that $0 \leq \alpha_j \leq 1$, $\alpha_j = 1$ on Ω''_j and $\alpha_j = 0$ outside Ω'_j. Since $\overline{\Omega''_j}$ and the frontier of Ω'_j are compact and disjoint, they are at positive distance ε_j apart and so we can regularize α_j to get $\beta_j = \alpha_j * r_j$, where r_j belongs to $\mathscr{D}(R^n)$, is positive, has its support in the ball centre 0 and radius $\frac{1}{2}\varepsilon_j$, and $\int_{R^n} r_j \, d\mu = 1$. Then β_j belongs to $\mathscr{D}(R^n)$, $\beta_j > 0$ on Ω''_j, and has its support contained in $\Omega'_j \subset \Omega_i$. The sum $\sum_j \beta_j$ is thus locally finite on Ω; it is everywhere strictly positive; and (by local finiteness) $\sum_j \beta_j \in \mathscr{D}(R^n)$. The functions $\phi_j = \beta_j/(\sum_j \beta_j)$ answer all requirements. ∎

We speak of a family (ϕ_j) of the type described in Proposition 5.6.1 as a *partition of unity in* $\mathscr{D}(\Omega)$ *subordinate to the covering* (Ω_i). By using such partitions of unity we can tackle the problem of piecing together a distribution from its local constituents.

5.6.2 **Proposition** (Principle of Localization). Let Ω be an open subset of R^n, (Ω_i) an open covering of Ω. Suppose (X_i) is a family of distributions, $X_i \in \mathscr{D}'(\Omega_i)$, such that $X_i \mid \Omega_i \cap \Omega_j = X_j \mid \Omega_i \cap \Omega_j$ whenever $\Omega_i \cap \Omega_j$ is nonvoid. Then there exists a unique distribution X on Ω such that $X \mid \Omega_i = X_i$ for each i.

Proof. Let (Ω'_j) be an open covering of Ω that is locally finite, each Ω'_j being relatively compact in Ω and such that $\Omega'_j \subset \Omega_i$ for some i. We can define X'_j as a distribution on Ω'_j by setting $X'_j = X_i \mid \Omega'_j$ for any i such that $\Omega_i \supset \Omega'_j$: our compatability hypothesis on (X_i) ensures that this definition is effective. We take a partition of unity (ϕ_j) in $\mathscr{D}(\Omega)$ with the support of ϕ_j contained in Ω'_j for each j. If ϕ belongs to $\mathscr{D}(\Omega)$, we have

$$\phi = \sum_j (\phi_j \phi),$$

a finite sum only. Hence if X exists, then necessarily

$$X(\phi) = \sum_j X(\phi_j\phi) = \sum_j X'_j(\phi_j\phi),$$

showing that X is uniquely determined when the family (X_i) is given. Reciprocally, this formula defines a distribution X on Ω. If the support of ϕ lies in Ω_i, that of $\phi_j\phi$ lies in $\Omega_i \cap \Omega'_j$ and so, since $X'_j = X_i \mid \Omega'_j$, therefore $X'_j(\phi_j\phi) = X_i(\phi_j\phi)$. Thus

$$X_i(\phi) = \sum_j X_i(\phi_j\phi) = \sum_j X'_j(\phi_j\phi) = X(\phi)$$

is true for any ϕ in $\mathscr{D}(\Omega)$ with support in Ω_i. In other words, $X_i = X \mid \Omega_i$, as required. ▌

5.6.3 **Corollary.** If Ω is an open subset of R_n and (Ω_i) is any family of open subsets of Ω, and if the distribution X on Ω is 0 on each Ω_i, then it is 0 on $\bigcup_i \Omega_i$. In particular, for each X in $\mathscr{D}'(\Omega)$, there is a maximal open subset U of Ω on which X is 0; the relatively closed set $F = \Omega \backslash U$ is the *support* of X.

The reader should observe that if X happens to be a Radon measure on Ω, then this definition of support is an agreement with that given in Section 4.9.

The following relations should be noted:

support $\partial^p \varepsilon_x = \{x\}$, support $\partial^p X \subset$ support X,

support $fX \subset$ (support f) $\cap$ (support X).

Also $X(\phi)$ depends only on the restriction of ϕ to any neighbourhood of the support of X: that is, $X(\phi_1) = X(\phi_2)$ whenever ϕ_1 and ϕ_2 agree on some neighbourhood of support X. If the directed family (X_i) of distributions converges weakly to X, and if support $X_i \subset F$ for all i, F being a relatively closed subset of Ω, then support $X \subset F$.

5.7 Distributions with Compact Supports

The distributions in $\mathscr{D}'(\Omega)$ having compact supports can be seen to represent precisely the elements of the dual of $C^\infty(\Omega)$ with its Fréchet space topology defined in Subsection 5.1.1.

Indeed, regularization shows that $\mathscr{D}(\Omega)$ is dense in $C^\infty(\Omega)$. Thus any continuous linear form L on $C^\infty(\Omega)$ is fully determined by its restriction X to $\mathscr{D}(\Omega)$. X is an element of $\mathscr{D}'(\Omega)$, and the continuity of L signifies that there exists a compact set $K \subset \Omega$ and an integer $m \geq 0$ such that

$$|L(f)| \leq \text{const. Sup}\,\{|\partial^p f(x)| : x \in K, |p| \leq m\}$$

for f in $C^\infty(\Omega)$. Hence

$$\begin{aligned}|X(\phi)| &\leq \text{const. Sup}\,\{|\partial^p \phi(x)| : x \in K, |p| \leq m\}\\ &= \text{const. } N_m^K(\phi)\end{aligned}$$

for ϕ in $\mathscr{D}(\Omega)$. This shows that the support of X lies in K, and also that X is of finite order $\leq m$ (see Section 5.8). Conversely, if X in $\mathscr{D}'(\Omega)$ has a compact support K, then X has a natural extension to $C^\infty(\Omega)$ that is continuous on the latter space—for example, one may define

$$X(f) = X(\alpha f)$$

for all f in $C^\infty(\Omega)$, α being any element of $\mathscr{D}(\Omega)$ that is equal to unity on a neighbourhood of K. (Within this class, the choice of α is immaterial owing to the assumption that support $X \subset K$.)

If $X \in \mathscr{D}'(\Omega)$ has a compact support $K \subset \Omega$, then X is the restriction to Ω of a distribution $Y \in \mathscr{D}'(R^n)$ with support K: indeed, we have only to define

$$Y(\phi) = X(\phi^*)$$

for each ϕ in $\mathscr{D}(R^n)$, ϕ^* being any element of $\mathscr{D}(\Omega)$ that coincides with ϕ on some neighbourhood of K. For example, we might choose once and for all a function α in $\mathscr{D}(\Omega)$ that is equal to unity on some neighbourhood of K, and then put $\phi^* = \alpha(\phi \mid \Omega)$. The fact that distributions on Ω with a compact support can be thus extended into distributions on R^n with the same compact support often simplifies the handling of such distributions.

5.8 Distributions of Finite Order

Let $m \geq 0$ be an integer, and let $\mathscr{D}^m(\Omega)$ be the set of functions ϕ on Ω having compact supports and such that $\partial^p \phi$ is continuous on Ω for each p with $|p| \leq m$. Thus $\mathscr{D}^0(\Omega) = \mathscr{K}(\Omega)$ and $\mathscr{D}(\Omega) = \bigcap_{m \geq 0} \mathscr{D}^m(\Omega)$. A sequence (ϕ_n) in $\mathscr{D}^m(\Omega)$ is said to converge to 0 in $\mathscr{D}^m(\Omega)$ if and only if the supports of the ϕ_n lie in a compact set K independent of n, and $\lim_n \partial^p \phi_n = 0$ uniformly on Ω for each p satisfying $|p| \leq m$.

By the process of regularization, it is easily seen that $\mathscr{D}(\Omega)$ is dense in $\mathscr{D}^m(\Omega)$. Hence the dual of $\mathscr{D}^m(\Omega)$ can be identified with a subset of $\mathscr{D}'(\Omega)$:

$$\mathscr{D}'^m(\Omega) \subset \mathscr{D}'(\Omega).$$

An $X \in \mathscr{D}'(\Omega)$ belongs to $\mathscr{D}'^m(\Omega)$ if and only if it is continuous on $\mathscr{D}(\Omega)$ for the above-defined concept of convergence in $\mathscr{D}^m(\Omega)$. In other words, X belongs to $\mathscr{D}'^m(\Omega)$ if and only if for each compact set $K \subset \Omega$ there exists a number $c_K \geq 0$ such that

$$|X(\phi)| \leq c_K \cdot N_m^\Omega(\phi) = c_K \cdot N_m^K(\phi)$$

for all ϕ in $\mathscr{D}(\Omega)$ with support in K. Such distributions X on Ω are said to be *of order at most m*. A distribution X on Ω is *of order m* if it is of order at most m but not of order at most m' for $m' < m$.

Thus, for example, the distributions on Ω of order 0 (equivalently: of order at most 0) are precisely the Radon measures on Ω. The distribution $\partial^p \varepsilon_x$ (x any point of Ω) is of order $m = |p|$. If X is of order at most m, then $\partial^p X$ is of order at most $m + |p|$. If X is of order at most m, and if f belongs to $C^m(\Omega)$, then fX is defined and of order at most m.

We write $\mathscr{D}'_F(\Omega) = \bigcup_{m\geq 0}\mathscr{D}'^m(\Omega)$ for the set of distributions on Ω of finite order. There always exist distributions on Ω that are not of finite order: for example, we may take a sequence (x_n) of points of Ω having no limit point in Ω, and let (c_n) be an arbitrary sequence of nonzero numbers. Then the series $\sum_n c_n \cdot \partial^{p_n}\varepsilon_{x_n}$ is convergent (weakly or strongly, it comes to the same thing) in $\mathscr{D}'(\Omega)$ and, if $\sup_n |p_n| = +\infty$, the sum distribution X is not of finite order.

As we have observed in Section 5.7, any distribution with a compact support is necessarily of finite order.

We shall now give an interesting and important structure theorm for distributions of finite order. This will apply in particular to distributions with compact supports. Furthermore, it will apply to the restrictions, to relatively compact open subsets of Ω, of any $X \in \mathscr{D}'(\Omega)$, and shows that any $X \in \mathscr{D}'(\Omega)$ is equal *locally* (though perhaps not globally) to a finite sum of derivatives of Radon measures.

5.8.1 **Theorem.** Let $X \in \mathscr{D}'^m(\Omega)$. Then X is equal to a finite sum

$$\sum_{|p|\leq m} \partial^p\alpha_p,$$

where the α_p are Radon measures on Ω; if further X has a compact support K, the α_p may be assumed to have their supports contained in any preassigned neighbourhood of K. Conversely, any such finite sum of derivatives of orders $|p| \leq m$ of Radon measures on Ω is an element of $\mathscr{D}'^m(\Omega)$.

Proof. The converse statement is evident.

Suppose then that $X \in \mathscr{D}'^m(\Omega)$. Since X is continuous for the topology of $\mathscr{D}^m(\Omega)$,† it follows that if a sequence (ϕ_n) in $\mathscr{D}(\Omega)$ is such that $(\partial^p\phi_n)$ converges to 0 in $\mathscr{K}(\Omega)$ for each p satisfying $|p| \leq m$, then $X(\phi_n) \to 0$. We associate with each ϕ in $\mathscr{D}(\Omega)$ the finite sequence $\Phi = (\partial^p \phi)_{|p|\leq m}$, which is an element of the product of N copies of $\mathscr{K}(\Omega)$ (N being the number of n-uples p satisfying $|p| \leq m$). We equip this product space with the product topology. By what we have said, $\Phi \to X(\phi)$ is a linear form defined on the vector subspace of $\mathscr{K}^N$ formed of the Φ's when ϕ ranges over $\mathscr{D}(\Omega)$, and this linear form is continuous for the topology induced by the product topology of $\mathscr{K}^N$. By the Hahn-Banach theorem, therefore, this linear form can be extended continuously to the whole of $\mathscr{K}^N$. This extension is thus an element of the dual of $\mathscr{K}^N$. Now each element of the dual $\mathscr{K}^N$ is representable by a sequence $(\alpha_p)_{|p|\leq m}$ of Radon measures on Ω in such a way that a general element $(f_p)_{|p|\leq m}$ of $\mathscr{K}^N$ is mapped onto the number $\sum_{|p|\leq m}(-1)^{|p|} \int f_p\, d\alpha_p$. So in particular we shall have

$$X(\phi) = \sum_{|p|\leq m} (-1)^{|p|} \int \partial^p\phi \cdot d\alpha_p;$$

in other words, $X = \sum_{|p|\leq m} \partial^p\alpha_p$, as asserted.

† In order that we may appeal to the Hahn-Banach theorem, it is essential that we assume here the existence of appropriate locally convex topologies on $\mathscr{D}(\Omega)$, $\mathscr{D}^m(\Omega)$, and $\mathscr{K}(\Omega)$ Their existence is established in Section 6.3.

Suppose lastly that in addition X has a compact support $K \subset \Omega$. Let U be any relatively compact open neighbourhood of K. We choose and fix any function g in $\mathscr{D}(\Omega)$ that is unity on a neighbourhood of K and has a compact support $F \subset U$. If ϕ belongs to $\mathscr{D}(\Omega)$, ϕ and $g\phi$ agree on a neighbourhood of K, and so

$$X(\phi) = X(g\phi) = \sum_{|p| \leq m} \langle g\phi, \partial^p \alpha_p \rangle.$$

Now

$$\langle g\phi, \partial^p \alpha_p \rangle = (-1)^{|p|} \langle \partial^p (g\phi), \alpha_p \rangle$$
$$= (-1)^{|p|} \left\langle \sum_{q \leq p} C_p^q \partial^{p-q} g \cdot \partial^q \phi, \alpha_p \right\rangle,$$

where the C_p^q are the multinomial coefficients; thus

$$\langle g\phi, \partial^p \alpha_p \rangle = \left\langle \phi, \sum_{q \leq p} (-1)^{|p|+|q|} C_p^q \partial^q \alpha_{p,q} \right\rangle,$$

where $\alpha_{p,q} = (\partial^{p-q} g)\alpha$ is a Radon measure on Ω. Now the support of $\alpha_{p,q}$ is contained in that of $\partial^{p-q} g$, which is contained in that of g, hence is contained in $F \subset U$. And we now have

$$X = \sum_{|p| \leq m} \sum_{q \leq p} (-1)^{|p|+|q|} C_p^q \partial^q \alpha_{p,q},$$

a sum of derivatives of orders at most m of Radon measures on Ω with supports lying in U. The proof is complete. ▮

Remark. At the expense of involving higher-order derivatives, we may in Theorem 5.8.1 replace the Radon measures by smooth functions. Consider for definiteness the case $\Omega = R^n$, and let α be any Radon measure on R^n. Without any loss of generality we may assume that α is positive. If K is compact, we have then

$$\left| \int \phi \, d\alpha \right| \leq \alpha(K) \cdot \operatorname{Sup} |\phi|$$

for ϕ in $\mathscr{D} = \mathscr{D}(R^n)$ with support in K. Also if $\partial = \partial_1 \cdots \partial_n$, we have

$$\phi(x) = \int_{-\infty}^{x_1} \cdots \int_{-\infty}^{x_n} \partial\phi \cdot d\mu$$

and therefore

$$\left| \int \phi \, d\alpha \right| \leq c_K \cdot \|\partial\phi\|_{L^1}.$$

From this we may infer (via the Hahn-Banach theorem) that

$$\int \phi \, d\alpha = (-1)^n \int \partial\phi \cdot F \, d\mu,$$

where the function F is locally bounded and measurable. Hence $\alpha = \partial F$ in the sense of distributions. If we now define functions G_r on

$$\Omega_r = \{x : x_k > -r, 1 \leq k \leq n\} \qquad \text{for } r = 1, 2, \cdots$$

by setting

$$G_r(x) = \int_{-r}^{x_1} \cdots \int_{-r}^{x_n} F \, d\mu \qquad \text{for } x \in \Omega_r,$$

then we see that each G_r is continuous on Ω_r and $\partial G_r = F$ on Ω_r. So it follows that $G_{r+1} - G_r = c_{r+1} =$ constant on Ω_r. If we define G on R^n to be G_1 on Ω_1, $G_2 - c_2$ on Ω_2, $G_3 - c_2 - c_3$ on Ω_3, and so on, we see that G is continuous and $\partial G = F$ everywhere. Consequently $\alpha = \partial^2 G$. Finally if we iterate κ times the process leading from F to G, starting now from F, we shall end up with a function $f \in C^\kappa$ such that $\alpha = \partial^{\kappa+2} f$. So Theorem 5.8.1 will show that each $X \in \mathscr{D}'^m$ is a finite sum of derivatives of order at most $m + (\kappa + 2)n$ of functions belonging to C^κ.

5.9 Distributions Whose First Derivatives Are Known

Suppose X is a distribution on Ω and that each partial derivative $\partial_k X = \partial X / \partial x_k$ is known and equal to a distribution A_k on Ω. What can be said about X? Under what conditions can the simultaneous equations $\partial_k X = A_k$ $(1 \leq k \leq n)$ be solved for X when the A_k are preassigned?

Turning to the first question, the most interesting and important case in practice is that in which the A_k are known to be functions f_k (locally integrable on Ω). Apart from any explicit method of solution, it is important to know whether each solution X of $\partial_k X = f_k$ $(1 \leq k \leq n)$ is necessarily a function. We consider first the case in which the f_k are continuous.

5.9.1 **Proposition.** (1) If a distribution X on a domain Ω satisfies the equations

$$\partial_k X = 0 \qquad (1 \leq k \leq n),$$

then X is a constant function on Ω.

(2) If a distribution X on Ω satisfies the equations

$$\partial_k X = f_k \qquad (1 \leq k \leq n),$$

where the f_k are continuous functions on Ω, then X is a function f in $C^1(\Omega)$ such that $\partial_k f = f_k$ $(1 \leq k \leq n)$ in the ordinary pointwise sense. Consequently, if Ω is a domain,

$$f(x) - f(x_0) = \int_{x_0}^{x} [f_1\, dt_1 + \cdots + f_n\, dt_n],$$

the path of integration being any rectifiable arc in Ω joining x_0 and x.

(3) If a distribution X on Ω is such that its partial derivatives $\partial^p X$ of all orders are continuous functions on Ω, then X is a function of class $C^\infty(\Omega)$.

Proof. We begin with (2). We cover Ω with open balls B. If a is the centre of B, we define g_B on B by

$$g_B(x) = \int_{a_1}^{x_1} f_1(t_1, x_2, \cdots, x_n)\, dt_1 + \int_{a_2}^{x_2} f_2(a_1, t_2, x_3, \cdots, x_n)\, dt_2 + \cdots + \int_{a_n}^{x_n} f_n(a_1, \cdots, a_{n-1}, t_n)\, dt_n.$$

The paths of integration here all lie in B, hence in Ω. g_B belongs to $C^1(B)$ and $\partial_k g_B = f_k$ on B.* If (1) is assumed, it follows that $X - g_B$ is a constant, say c_B, on B. Then $f_B = g_B + c_B$ belongs to $C^1(B)$ and $f_B = f_{B'} = X$ on

* See Schwartz [1], pp. 62–63.

$B \cap B'$ whenever B and B' intersect. It follows that the f_B are simply the restrictions to B of a function f in $C^1(\Omega)$ such that $X = f$ on Ω, as alleged. Since $\partial_k f = f_k$ throughout (in the pointwise sense), f is recovered by integrating $[f_1\, dt_1 + \cdots + f_n\, dt_n]$ along arbitrary rectifiable arcs in Ω.

It remains to establish (1). Suppose then that $X \in \mathscr{D}'(\Omega)$ satisfies $\partial_k X = 0$ on Ω for $k = 1, 2, \cdots, n$. Let ω denote any subset of Ω that is a product of n open intervals in R^1 and let $X_\omega = X \mid \omega$. Then it is clear that $\partial_k X_\omega = 0$ on ω for $k = 1, 2, \cdots, n$. It suffices to show that X_ω is a constant function on ω, for it then follows (by covering the domain Ω with various sets ω) that X is a constant function on Ω. We are thus reduced to the case in which Ω is itself a product of open intervals in R^1.

To say that $\partial_k X = 0$ on Ω for $k = 1, 2, \cdots, n$ signifies exactly that $X(\phi) = 0$ whenever $\phi \in \mathscr{D}(\Omega)$ is expressible in the form

$$\phi = \sum_{k=1}^{n} \partial_k \phi_k \tag{5.9.1}$$

for suitable ϕ_k in $\mathscr{D}(\Omega)$. It is evident that any ϕ of this form satisfies

$$\int_\Omega \phi\, d\mu = 0, \tag{5.9.2}$$

μ being Lebesgue measure on R^n. Supposing the converse to be true, we should conclude that $X(\phi) = 0$ whenever $\phi \in \mathscr{D}(\Omega)$ satisfies $1_\Omega(\phi) = 0$, where 1_Ω denotes the constant function on Ω with value unity. And from this it follows that $X = c \cdot 1_\Omega$ for some number c, the desired result.

We leave to the reader the task of showing that (5.9.2) implies (5.9.1) for functions in $\mathscr{D}(\Omega)$, Ω being a product of n open intervals in R^1. The result is very simple to prove when $n = 1$; and the general result may then be established by induction on n.

(3) follows by repeated application of (2). ▌

5.9.2 **Remarks.** The case in which, in (2), the f_k are assumed to be merely locally integrable (or to be pth power locally integrable) over Ω cannot be treated in so direct a fashion. Instead, one must use the idea of convolution of two distributions and appeal to some results about convolutions of functions due to Soboleff and similar to those given in Section 9.5. The line of attack will be sketched in Section 5.11 after we have introduced the idea of convolution of distributions.

Let us now turn to a brief comment on the second problem, that of the existence of a solution of the simultaneous equations $\partial_k X = A_k$ $(1 \le k \le n)$ when the A_k are given. It is evident that a necessary condition for solubility is the fulfilment of the system of equations

$$\partial_i A_k - \partial_k A_i = 0 \qquad (1 \le i, k \le n). \tag{5.9.3}$$

The proof of the sufficiency of these conditions is far from trivial, especially for domains Ω different from R^n itself. We state the result.

5.9.3 **Theorem.** Given distributions A_k $(1 \leq k \leq n)$ on Ω, the system (5.9.3) is necessary and sufficient for the existence of a solution $X \in \mathscr{D}'(\Omega)$ of the simultaneous equations

$$\partial_k X = A_k \qquad (1 \leq k \leq n). \tag{5.9.4}$$

When this condition is fulfilled, the solution is unique up to additive constants.

Proof. We refer the reader to Schwartz [1], p. 60, Théorème VI for the case in which $\Omega = R^n$. For general Ω (which may be any real differentiable manifold of class C^∞), the result is a consequence of a theorem of de Rham ([1], p. 114, Théorème 17′) which is given also in de Rham-Kodaira ([1], p. 40, Theorem C). ▌

5.9.4 **Remark.** Concerning the nature of the solution X, assumed to exist, see Section 5.11.

5.10 Convolution of Distributions on R^n

The concept of convolution of distributions on R^n may, as in the case of measures (see Section 4.19), be formulated in terms of the direct or tensor product of distributions. Since we shall be concerned here only with the case of a convolution $X * Y$ in which at least one of X and Y has a compact support, a more direct approach is feasible and satisfactory. For a more general exposition, see Schwartz [1], Chapitre IV and [2] Chapitre VI. Alternative accounts are given by Chevalley [1], Hirata [1], Shiraishi [1] (see also Exercise 7.19).

By direct analogy with the case of measures, we should expect to define the convolution $X * Y = Z$ by the convention that $Z(\phi)$ shall be the common value of $X_{(x)}\{Y_{(y)}[\phi(x + y)]$ and $Y_{(y)}\{X_{(x)}[\phi(x + y)]\}$ for ϕ in $\mathscr{D} = \mathscr{D}(R^n)$. We are here indicating by $Y_{(y)}[\phi(x + y)]$ the result (a function of x) obtained by applying the distribution Y to the function $y \to \phi(x + y) \in \mathscr{D}$; and analogously for the other terms. Now here, as in the case of measures, neither of the above expressions makes sense if X and Y are arbitrary distributions. If, however, one of X and Y—say X—has a compact support, both expressions make sense: the first since $Y_{(y)}[\phi(x + y)]$ belongs to C^∞ as a function of x, so that X may be applied to it; the second because $X_{(x)}[\phi(x + y)]$ belongs to $\mathscr{D}$ qua function of y, so that Y may be applied to it. We aim to show further that, in this case, the two expressions are equal. To this end, let us denote by $Z'(\phi)$ and $Z''(\phi)$, respectively, the first and second expressions. It is evident that for given ϕ, both $Z'(\phi)$ and $Z''(\phi)$ are bilinear in the pair (X, Y). By Theorem 5.8.1, X can be represented as a finite sum of derivatives $\partial^p\alpha$, where α is a Radon measure with a compact support. So it suffices to deal with the case in which $X = \partial^p\alpha$ itself. In this case, to establish that $Z'(\phi)$ and $Z''(\phi)$ are equal amounts to establishing the equality of $Z'(\partial^p\phi)$ and $Z''(\partial^p\phi)$ when we take $X = \alpha$. Since $\partial^p\phi$ is in $\mathscr{D}$ whenever ϕ has this property, we are thereby reduced to the case in which $X = \alpha$, a measure with a compact support. In other words we wish to show that

$$\int Y_{(y)}[\phi(x + y)]\, d\alpha(x) = Y_{(y)}\Big[\int \phi(x + y)\, d\alpha(x)\Big]$$

whenever $Y \in \mathscr{D}' = \mathscr{D}'(R^n)$, $\phi \in \mathscr{D}$, and α is a measure with a compact support. Now, if Y also has a compact support, we can again use Theorem 5.8.1 to reduce ourselves to the case in which Y is a measure with a compact support. Equality in this case is contained in the theory of product measures (Section 4.17).

Finally, if Y has not a compact support, we can represent it as the strong limit of a sequence (Y_n) of distributions with compact supports—for example, $Y_n = \alpha_n Y$, where $\alpha_n \in \mathscr{D}$ and $\alpha_n = 1$ on the ball in R^n with centre 0 and radius n. Then

$$Y_{n(y)}[\phi(x + y)] \to Y_{(y)}[\phi(x + y)]$$

uniformly when x ranges over any compact set (the corresponding functions $y \to \phi(x + y)$ then falling into a bounded subset of $\mathscr{D}$), so that

$$\int Y_{n(y)}[\phi(x + y)]\, d\alpha(x) \to \int Y_{(y)}[\phi(x + y)]\, d\alpha(x).$$

At the same time, since $\int \phi(x + y)\, d\alpha(x)$ belongs to $\mathscr{D}$, so

$$Y_{n(y)}\Big[\int \phi(x + y)\, d\alpha(x)\Big] \to Y_{(y)}\Big[\int \phi(x + y)\, d\alpha(x)\Big].$$

Thus the desired equality follows in the limit, and the proof is complete.

It is now easy to see how to define a convolution $X * Y * Z * \cdots$ (any finite number of factors), provided all but at most one of $X, Y, Z, \cdots$ have compact supports; the commutative and associative laws are obeyed by such expressions. (Associativity may *not* obtain if the conditions on the supports are relaxed.) Moreover, $X * Y$ is bilinear in the pair (X, Y), one at least of X and Y having a compact support.

The following formulae may be verified at once:

$$\varepsilon * X = X \quad \text{for any } X \in \mathscr{D}'; \quad \partial^p(X * Y) = (\partial^p X) * Y = X * (\partial^p Y)$$

whenever at least one of X and Y has a compact support;

$$\partial^p X = (\partial^p \varepsilon) * X \quad \text{for any } X \in \mathscr{D}',$$

showing that differentiation is itself a special case of convolution with a distribution $\partial^p \varepsilon$ having its support equal to $\{0\}$. Thus the convolution type of equation $A * X = B$, where A is given and has a compact support, B is given, and X is the unknown, includes all linear partial differential equations with constant coefficients; also included are certain types of integral equation.

The reader will also verify that

$$\text{support } (X * Y) \subset (\text{support } X) + (\text{support } Y)$$

whenever at least one of the sets on the right is compact; in particular, $X * Y$ has a compact support whenever both X and Y share this property.

Regarding the continuity of $X * Y$ in each factor, it is very simple to verify that

$$X_i * Y \to X * Y \qquad \text{strongly in } \mathscr{D}'$$

whenever the distributions X_i have their supports contained in a fixed compact set (independent of i) and $X_i \to X$ strongly in $\mathscr{D}'$. Similarly $X * Y_i \to X * Y$ strongly in $\mathscr{D}'$ whenever X has a compact support and $Y_i \to Y$ strongly in $\mathscr{D}'$. To prove the first relation, for example, we examine the functions

$$\psi_i(y) = X_{i(x)}\,[\phi(x + y)].$$

If the supports of the X_i lie in a fixed compact set, while ϕ describes a bounded subset B of $\mathscr{D}$, then the functions ψ_i have uniformly bounded supports and the ψ_i converge in $\mathscr{D}(R^n, S)$ (S a suitable bounded set) to $X_{(x)}\,[\phi(x + y)]$ as $X_i \to X$ strongly; whence it appears that $X_i * Y(\phi) \to X * Y(\phi)$ uniformly with respect to ϕ in B, so that $X_i * Y \to X * Y$ strongly in $\mathscr{D}'$. The second relation is established by a similar reasoning.

If X is a function ρ in $\mathscr{D}$, then $\rho * Y$ is (the distribution generated by) the function

$$x \to Y_{(y)}[\rho(x - y)],$$

which function belongs to C^∞. Similarly, if $\rho \in \mathscr{D}^m$ and $Y \in \mathscr{D}'^m$, then $\rho * Y$ is the continuous function

$$x \to Y_{(y)}[\rho(x - y)].$$

If Y has a compact support, then $\rho * Y$ also has a compact support. By appealing to the results in the preceding paragraph, we can show that the functions $\rho_n (n = 1, 2, \cdots)$ may be chosen from $\mathscr{D}$, having their supports contained in any preassigned neighbourhood of 0, and such that the distributions ρ_n converge strongly to ε. (We need only arrange that $\rho_n \geq 0$, $\int \rho_n \, d\mu = 1$, and $\rho_n = 0$ outside the ball with centre 0 and radius $1/n$.) Then we see that $\rho_n * Y \to Y$ strongly in $\mathscr{D}'$ for each Y in $\mathscr{D}'$. If Y has a compact support K, then the $\rho_n * Y$ are functions in $\mathscr{D}$ whose supports lie in any preassigned neighbourhood of K. In particular, $\mathscr{D}$ is strongly dense in $\mathscr{D}'$ (functions in $\mathscr{D}$ being identified with the distributions they generate). We have here the idea of "regularization" or "smoothing," introduced already in Subsection 4.19.16 for the case of measures. In this connection we may note the formula

$$f * Y(\phi) = Y(\check{f} * \phi)$$

for ϕ in $\mathscr{D}$, Y in $\mathscr{D}'$, and f any integrable function with a compact support.

Finally, it may be observed that our procedure for defining $X * Y$ when at least one of X and Y has a compact support is effective whenever, more generally, the supports A and B of X and Y are so related that, for each compact set K in R^n, the subset $(A \times B) \cap \tilde{K}$ in $R^n \times R^n$ is likewise compact; here $\tilde{K}$ is the set of points (x, y) in $R^n \times R^n$ for which $x + y \in K$ (cf. the discussion in Section 4.19 applying to the case of measures). This is the case if, for example, the supports of X and of Y are both contained in half spaces of the type $x_1 \geq a_1, \cdots, x_n \geq a_n$, a case which is of importance in connection with Laplace transforms. An introductory account of the latter topic for the one-dimensional case is to be found in Garnir [1], pp. 154–168.

Various more refined properties of distributions may be developed by combining the use of convolutions with the known existence and simple properties of elementary solutions of Laplace's equation. In the next section we shall illustrate the use of the combined techniques, beginning with a further study of the problems encountered in Section 5.9 above.

For a small but important extension of the concept of convolution, see Exercise 5.19.

5.11 Further Properties of Distributions

5.11.1 Return to Distributions Whose First Derivatives Are Known. Our interest lies in those distributions X on Ω about which we know that the first partials $\partial_k X$ are functions f_k $(k = 1, \cdots, n)$. The problem is a local one, inasmuch as our main concern is to determine whether or not X is itself necessarily a function. Unfortunately, however, this problem does not permit an immediate reduction to the case in which X has a compact support and has therefore an extension belonging to $\mathscr{D}' = \mathscr{D}'(R^n)$. Still, whether or not X has a compact support, the first step amounts to an explicit solution for X of the simultaneous equations $\partial_k X = f_k$ $(k = 1, \cdots, n)$.

The principal tool depends on the existence of a distribution E on R^n that satisfies

$$\Delta E = \varepsilon, \tag{5.11.1}$$

where $\Delta = \sum_{k=1}^n \partial_k^2$ is the Laplacian. Such a distribution E is termed an "elementary solution"† for the partial differential operator Δ. If E is known, and if X has a compact support and satisfies $\partial_k X = f_k$ $(k = 1, \cdots, n)$, then we have

$$\sum_k (\partial_k E) * f_k = \sum_k (\partial_k E) * (\partial_k X) = \sum_k \partial_k(\partial_k E) * X$$
$$= \Delta E * X = \varepsilon * X = X,$$

that is,

$$X = \sum_{k=1}^n (\partial_k E) * f_k. \tag{5.11.2}$$

Clearly, (5.11.2) holds when $f_k = \partial_k X$ are arbitrary distributions (not necessarily functions), provided only that X (and therefore f_k, too) has a compact support. Equation (5.11.2) gives an explicit solution for X. If X has not a compact support, the above calculations are not justified. For it is impossible that E, satisfying (5.11.1), should have a compact support, so that we should not be sure in advance that the convolutions involved even make sense. (The impossibility of finding a solution E of $\Delta E = \varepsilon$ having a compact support is assured by reference to the following observation: If A and B are distributions

† We shall later have much to say about elementary solutions of linear partial differential operators; see Sections 5.12, 5.14, and 5.18.

with compact supports, and if $B(1) = 0$, then $A * B(1) = 0$; to apply this we take $A = E$ and $B = \Delta\varepsilon$. The proof is simple: We choose relatively compact open neighbourhoods U and V of the supports of A and B, respectively. Then

$$A * B(1) = A * B(\phi) = A_{(x)}\{B_{(y)}[\phi(x + y)]\}$$

for any ϕ in $\mathscr{D}$ that is equal to unity on $U + V$. For any such ϕ and any x in U, $y \to \phi(x + y)$ is equal to unity on V, so that $B_{(y)}[\phi(x + y)] = B(1)$ for x in U. But then

$$A_{(x)}\{B_{(y)}[\phi(x + y)]\} = A_{(x)}\{B(1)\} = A(1) \cdot B(1) = A(1) \cdot 0 = 0,$$

as asserted.)

To take care of this situation, we take any function h in $\mathscr{D}$ that is unity on some neighbourhood of 0. Consider $F = hE$: this has a compact support and

$$\Delta F = (\Delta h) \cdot E + 2\sum_{k=1}^{n} \partial_k h \cdot \partial_k E + h \cdot \Delta E.$$

Now E can be chosen so that it is a function of class C^∞ on $R^n\backslash\{0\}$; then, since $\Delta h = \partial_k h = 0$ on a neighbourhood of 0, the first two terms on the right are functions in C^∞ and have supports contained in that of h—that is, they belong to $\mathscr{D}$. Also $h \cdot \Delta E = h \cdot \varepsilon = h(0) \cdot \varepsilon = \varepsilon$. Thus

$$\Delta F = \varepsilon + \zeta, \tag{5.11.3}$$

where ζ belongs to $\mathscr{D}$, has its support contained in that of h, and vanishes on a neighbourhood of 0. F is a "parametrix" for Δ—unlike E, the elementary solution, it has a compact support.†

Now, supposing first that $\Omega = R^n$, replacing E by F in the preceding calculations leads to

$$X = -\zeta * X + \sum_{k=1}^{n} \partial_k F * f_k; \tag{5.11.4}$$

here $-\zeta * X$ is a function of class C^∞ (since ζ belongs to $\mathscr{D}$).

If Ω is arbitrary, we can still use the same type of equation as (5.11.4) to study X locally in Ω as a function of the f_k. To do this, suppose ω_1 and ω are relatively compact open subsets of Ω with $\bar{\omega}_1 \subset \omega$. We take any α in $\mathscr{D}$ that is unity on ω and has its support contained in Ω. Then αX can be extended into a distribution Y on R^n having a compact support. At the same time we extend αf_k into a function g_k on R^n having a compact support. Then X and Y agree on ω, as also do f_k and g_k. So $\partial_k Y = g_k$ on ω. If we now choose the support of h to be so small a neighbourhood V of 0 that $\omega_1 + V \subset \omega$, then we shall have

$$\partial_k F * g_k = \partial_k F * \partial_k Y = \partial_k^2 F * Y \qquad \text{on } \omega_1,$$

and hence (5.11.4) will hold on ω_1 when Y and g_k replace X and f_k, respectively. From this one may infer the nature of Y on ω_1, hence that of X on ω_1. Thus the local nature of X is determined.

---

† We shall have occasion to use the parametrix device in the discussion of linear partial differential operators; See Subsection 5.18.4.

Since the first term on the right-hand side of (5.11.4) belongs to $\mathscr{D}$, we see that in all cases the local nature of X is that of

$$\partial_k F * f = \partial_k(F * f)$$

or of

$$\partial_k E * f = \partial_k(E * f),$$

where f is a function with a compact support. And, as far as local behaviour is concerned, the difference between E and $F = hE$ is insignificant. Thus one has only to examine $\partial_k E * f$ with f a function with a compact support (necessarily integrable).

Further progress depends on the nature of E, the selected elementary solution. Naturally, E is not uniquely determined: one may add to E any distribution H on R^n that is harmonic in the sense that $\Delta H = 0$. Now it can be shown ("Weyl's lemma"; see Subsection 5.18.4 and also Schwartz [1], p. 136) that any harmonic distribution H is in fact a harmonic function in the ordinary sense, hence is a function of class C^∞. The contribution of this component of E to $\partial_k E * f$ is thus itself a function of class C^∞, and is insignificant for our purposes; any distribution E satisfying $\Delta E = \varepsilon$ is as good as any other.

Perhaps the simplest choice of E is the following:

$$E = \begin{cases} -[(n-2)\omega_n]^{-1} r^{2-n} & \text{if } n > 2, \\ -(2\pi)^{-1} \cdot \log(1/r) & \text{if } n = 2, \\ x \cdot Y(x) & \text{if } n = 1, \end{cases} \tag{5.11.5}$$

where Y is the unit Heaviside function on R^1 ($Y(x) = 0$ or 1 according as $x < 0$ or $x > 0$), and where in any case $r = |x| = (x_1^2 + \cdots + x_n^2)^{1/2}$ and ω_n is the surface area of the unit sphere in R^n. (It is quite unnecessary to use the present technique for the case $n = 1$, where much more direct methods suffice. It is included here merely to make it clear that a uniform approach is possible.) In all cases, E is a function that is of class C^∞ on $R^n \backslash \{0\}$; when $n = 1$, E is moreover a function that is everywhere continuous. The verification that E, defined in (5.11.5), satisfies $\Delta E = \varepsilon$, is trivial when $n = 1$. For $n \geq 2$, the verification depends on the use of Green's formula for simple domains and very smooth functions. We sketch the proof for the case $n > 2$.

In Subsection 4.15.14 we have obtained Green's formula

$$\int_\Omega \partial_k u \cdot d\mu = \int_F u \, d\sigma_k = \int_F u N_k \, d\sigma$$

for any sufficiently simple bounded domain in R^n with frontier F and any function u that is of class C^1 on some neighbourhood of $\overline{\Omega}$; $N = (N_1, \cdots, N_n)$ is the unit normal to F and σ is the area measure on F. In this formula we take $u = \phi \cdot \partial_k \psi$ to get

$$\int_\Omega \partial_k \phi \cdot \partial_k \psi \cdot d\mu + \int_\Omega \phi \partial_k^2 \psi \, d\mu = \int_F \phi \cdot \partial_k \psi \cdot N_k \, d\sigma$$

whenever ϕ is of class C^1 and ψ of class C^2 on a neighbourhood of $\bar{\Omega}$. Summing with respect to k, interchanging ϕ and ψ, and then subtracting, we obtain (with the usual notation)

$$\int_\Omega (\phi\,\Delta\psi - \psi\,\Delta\phi)\,d\mu = \int_F (\phi \cdot \operatorname{grad}\psi - \psi \cdot \operatorname{grad}\phi) \cdot N \cdot d\sigma,$$

whenever ϕ and ψ are of class C^2 on a neighbourhood of $\bar{\Omega}$. In this formula we choose for Ω an annulus $\alpha < r \equiv |x| < \beta$, where $\alpha > 0$ and the set $\{x: r < \beta\}$ to contain the support of $\phi \in \mathscr{D}$, while $\psi = r^{2-n}$. Then ϕ and grad ϕ are 0 on the outer boundary $|x| = \beta$, and $\Delta\phi = 0$ on Ω. We thus obtain

$$-\int_\Omega r^{2-n}\,\Delta\phi\,d\mu = \int_{S_\alpha} (\phi \cdot \operatorname{grad}(r^{2-n}) - r^{2-n} \cdot \operatorname{grad}\phi) \cdot N \cdot d\sigma$$

where S_α is the sphere centre 0 and radius α. If we allow α to tend to 0, the left-hand side tends to $-\int r^{2-n}\,\Delta\phi\,d\mu = -\langle\phi, \Delta(r^{2-n})\rangle$; by a familiar argument, the right-hand side tends to $(n-2)\omega_n\phi(0)$. It therefore follows that $\Delta(r^{2-n}) = -(n-2)\omega_n\varepsilon$ in the sense of distributions, so that the first formula in (5.11.5) follows. The case $n = 2$ is discussed similarly.

The next step is to discuss the nature of the distribution $\partial_k E$. When $n = 1$, it is trivial to verify that ∂E is the function Y. Consider the case $n \geq 2$. Here, the pointwise partial derivative of E exists for $x \neq 0$ and is equal to $-\omega_n^{-1}x_k r^{-n}$ if $n > 2$ and to $(2\omega_1)^{-1}x_k r^{-2}$ if $n = 2$. It is natural to expect these locally integrable functions to generate the distribution $\partial_k E$—that is, that

$$-\langle\partial_k\phi, E\rangle = -\omega_n^{-1}\langle\phi, x_k r^{-n}\rangle$$

for all ϕ in $\mathscr{D}$ if $n > 2$, with a similar formula when $n = 2$. The verification that this is indeed the case proceeds in the same manner for $n = 2$ as for $n > 2$: we outline the argument for the latter case. It has to be shown that

$$\int \partial_k\phi \cdot r^{2-n}\,d\mu = (n-2)\int \phi x_k r^{-n}\,d\mu$$

for each ϕ in $\mathscr{D}$. To do this, we take any β so large that the support of ϕ is contained in the ball with centre 0 and radius β. Since both integrands are integrable, each integral is the same as that obtained by integrating over the annulus $\Omega: \alpha < r < \beta$, where $\alpha > 0$, and letting $\alpha \downarrow 0$. To the integrals extended over this annulus we can apply Green's formula, noticing that $\partial_k(\phi \cdot r^{2-n}) = r^{2-n}\,\partial_k\phi - (n-2)x_k\phi r^{-n}$ in the pointwise sense throughout the annulus. From this we deduce that

$$\int_\Omega \partial_k\phi \cdot r^{2-n}\,d\mu = \int_\Omega \partial_k(r^{2-n}\phi)\,d\mu + (n-2)\int_\Omega x_k\phi r^{-n}\,d\mu,$$

the first term on the right of which is shown by Green's formula to be

$$\int_{S_\alpha} N_k\phi r^{2-n}\,d\sigma,$$

where S_α is the sphere with centre 0 and radius α. When $\alpha \downarrow 0$, this surface integral tends to 0, and the desired formula follows.

The final step is to show that each term $\partial_k E * f$ (f being integrable and having a compact support) is a locally integrable function. When $n = 1$, this is obvious because ∂E is the bounded function Y and so $E * f$ is itself a bounded and continuous function. If $n \geq 2$, $\partial_k E$ is, as we have just shown, a constant multiple of the function $x_k r^{-n} = g$, say. We have, therefore, only to show that the function $g * f$ is locally integrable. Since $|g| \leq r^{1-n}$, it is enough to show that $r^{1-n} * f$ is locally integrable, in doing which we may assume that $f \geq 0$ is integrable and has a compact support contained in the ball B_β, say. However, for any $\alpha > 0$ one has

$$\int_{B_\alpha} (r^{1-n} * f)\, d\mu = \int_{B_\alpha} d\mu(x) \int_{B_\beta} f(y)\, |x - y|^{1-n}\, d\mu(y)$$
$$= \int_{B_\beta} f(y)\, d\mu(y) \int_{B_\alpha} |x - y|^{1-n}\, d\mu(x);$$

the inner integral is

$$\int_{B_{\alpha - y}} |z|^{1-n}\, d\mu(z) \leq \int_{B_{\alpha + |y|}} |z|^{1-n}\, d\mu(z)$$
$$= \omega_n(\alpha + |y|).$$

Hence,

$$\int_{B_\alpha} (r^{1-n} * f)\, d\mu \leq \int_{B_\beta} f(y)\omega_n(\alpha + |y|)\, d\mu(y) < +\infty,$$

showing that indeed $r^{1-n} * f$ is locally integrable.

In this way we have shown that if the distribution X on Ω is such that $\partial_k X$ is a function ($k = 1, \cdots, n$), then X is also a function on Ω.

If we are given about the f_k further information, we can say more about the local nature of X. One source of such results is Theorem 9.5.10, which is concerned specifically with statements about $r^{1-n} * f$. In this theorem we have to take $\lambda = n - 1$. There results then the following conclusions:

(a) If each f_k is locally pth power integrable and $1 \leq p \leq +\infty$, $0 < q < +\infty$, $1/p - 1/q < 1/n$, then X is locally qth power integrable;

(b) If $p > n$ and each f_k is locally pth power integrable, then X is a continuous function.

With suitable refinements of Theorem 9.5.10 due to Soboleff, these results may be further strengthened; for details see Schwartz [2], pp. 37–40, and Deny-Lions [1], Théorème 2.1. It is thus possible to prove that

(a′) If $1 < p \leq +\infty$, then in (a) one may take q such that $0 < q < +\infty$ and $1/p - 1/q = 1/n$.

(c) If $p = n > 1$, then $X \in L^r_{\text{loc}}(\Omega)$ for every finite r.

Similar problems arise if one is given information about all the partial derivatives $\partial^q X$ of order $|q| \leq h$. This topic is the subject of several so-called "imbedding theorems" of S. L. Soboleff [1], [1a]. Soboleff denotes by $W_p^{(h)}(\Omega)$ the space of functions $f \in L^1_{\text{loc}}(\Omega)$ for which $\partial^q f \in L^p(\Omega)$ for $|q| \leq h$, and the simplest of his imbedding theorems asserts that if Ω is bounded and $n < ph$, then $W_p^{(h)}(\Omega) \subset C_b(\Omega)$ (the space of bounded continuous functions on Ω). We shall find no explicit use for these interesting theorems and content ourselves with the remark that if h is even, this theorem can be deduced from Theorem 9.5.10(c) by use of the parametrix P introduced in the proof of Theorem A of 5.11.3 *infra*.

All these results assert something about the local nature of X and it is natural to ask whether it is possible to establish results concerning the global nature of X on the basis of the assumption that each $\partial_k X$ belongs to $L^p(\Omega)$ for certain values of p. Two special problems of this sort have proved themselves to be of particular importance, namely

(1) For which domains Ω in R^n is it true that the relations

$$X \in \mathscr{D}'(\Omega),\ \partial_k X \in L^2(\Omega) \quad (k = 1, 2, \cdots, n) \tag{5.11.6}$$

imply the existence of a constant $c = c(X)$ such that $X + c \in L^q(\Omega)$, where it is assumed that $n > 2$ and $q = 2n/(n - 2)$?

(2) For which domains $\Omega \subset R^n$ having finite measure is it true that the relations (5.11.6) entail that $X \in L^2(\Omega)$?

The two types of domains characterized by (1) and (2) are termed *Soboleff domains* and *Nikodym domains*, respectively. Clearly, any Soboleff domain of finite measure is a Nikodym domain; the converse is false, however.

Some connections between Nikodym domains and the study of boundary value problems for linear partial differential equations is already evident in Courant-Hilbert [1], Ch. VII. See also Deny-Lions [1], Ch. I, §10, and Subsection 9.12.2 below.

The distributions X on Ω that satisfy (5.11.6) are said to belong to the (generalized) Beppo Levi space $BL(\Omega)$, while the intersection $L^2(\Omega) \cap BL(\Omega)$ is denoted by $\mathscr{E}^1_{L^2}(\Omega)$. This latter space will play an important role in the study of linear partial differential equations undertaken later in this chapter (see especially Section 5.13).

The reader will notice that if Ω has finite measure, then Ω is a Soboleff domain if and only if $BL(\Omega) \subset L^q(\Omega)$; and Ω is a Nikodym domain if and only if $BL(\Omega) \subset \mathscr{E}^1_{L^2}(\Omega)$. It can be shown (Exercises 6.12 and 6.13) that the Soboleff and Nikodym domains of finite measure can each be characterized by means of inequalities, which summarize quite effectively the special functional analytic properties of these domains.

For a detailed study of these questions, see Deny-Lions [1], Ch. I.

5.11.2 **Concerning Bounded Sets of Distributions.** We shall be concerned here with criteria for the boundedness of subsets Σ of $\mathscr{D}'(\Omega)$, Ω being a given open subset of R^n. As has been pointed out in Section 5.3(1), weak boundedness, strong boundedness, and equicontinuity are all equivalent for subsets of $\mathscr{D}'(\Omega)$. We may therefore speak simply of boundedness.

Let us begin by recording necessary conditions for boundedness of Σ. Suppose ω is any open set that is relatively compact in Ω (which signifies that ω is bounded and $\bar{\omega}$, its closure in R^n, is contained in Ω). Then ω is at distance $r > 0$ from the frontier of Ω. Now there is an obvious and natural way of attaching a meaning to $X * Y$, making it an element of $\mathscr{D}'(\omega)$, for each Y in $\mathscr{D}'$ whose support is contained in the ball B_r with centre 0 and radius r. In fact, if φ belongs to $\mathscr{D}(\omega)$ and y belongs to B_r, then $x \to \varphi(x + y)$ has its support in Ω, and so $X_{(x)}[\varphi(x + y)]$ is defined. As a function of y, the result is of class C^∞ on

B_r. So, the support of Y lying within B_r, $Y_{(y)}\{X_{(x)}[\varphi(x+y)]\}$ makes sense. $X * Y$ is defined by $X * Y(\varphi) = Y_{(y)}\{X_{(x)}[\varphi(x+y)]\}$. The reader may check that the same result is obtained by taking $X_{(x)}\{Y_{(y)}[\varphi(x+y)]\}$, the expression inside the braces $\{\cdots\}$ being an element of $\mathscr{D}(\Omega)$. See Exercise 5.19.

In this way we have a meaning assigned to $X * f * g$ for all X in Σ and all distributions f and g with supports lying in a sufficiently small (depending on Ω and ω) neighbourhood U of 0, $X * f * g$ being in each case an element of $\mathscr{D}'(\omega)$. The definition makes it clear that, f and g being thus fixed, if X ranges over a bounded subset Σ of $\mathscr{D}'(\Omega)$, then $X * f * g$ ranges accordingly over a bounded subset of $\mathscr{D}'(\omega)$. We aim to establish a strong converse of this, which will form the first of two boundedness criteria.

Theorem A. Let Ω be an open subset of R^n, Σ a subset of $\mathscr{D}'(\Omega)$. In order that Σ be bounded in $\mathscr{D}'(\Omega)$, it is necessary and sufficient that it fulfil the following condition: for each relatively compact open subset ω of Ω, there exist an integer $m \geq 0$ and a neighbourhood U of 0 in R^n, such that the set $\Sigma * f * g = \{X * f * g : X \in \Sigma\}$ is bounded in $\mathscr{D}'(\omega)$ for each f and g in $\mathscr{D}^m(U)$.

Proof. The necessity is a trivial consequence of the preceding remarks.

The proof of sufficiency depends on the existence and nature of an elementary solution for the iterated Laplacian Δ^k. Specifically, we must recognize the fact that there exists a distribution E on R^n satisfying $\Delta^k E = \varepsilon$, and that moreover E may be chosen to be of the form $|x|^{2k-n}(A \log |x| + B)$. Explicit examples (which are of no importance for us) will be found in Schwartz [1], p. 48. Given any m, we may arrange that E belongs to C^m by choosing k sufficiently large.

From E we form a parametrix $P = fE$, where f belongs to $\mathscr{D}$, $f = 1$ on a neighbourhood of 0, and f has its support in U. Then P belongs to $\mathscr{D}^m(U)$ and

$$\Delta^k P = \varepsilon - \xi,$$

where ξ belongs to $\mathscr{D}(U) \subset \mathscr{D}^m(U)$.

If we now shrink U, if necessary, in order that the convolutions appearing below are all defined on ω, we shall have on ω the relations

$$X = X * \xi + \Delta^k(X * P)$$

and, by repeating the process,

$$X = X * \xi * \xi + 2\Delta^k(X * \xi * P) + \Delta^{2k}(X * P * P).$$

According to hypothesis, since ξ and P belong to $\mathscr{D}^m(U)$, each of the triple convolutions featuring in the right-hand member here remains bounded when X ranges over Σ. Both Δ^k and Δ^{2k} transform bounded sets into bounded sets. Thus we see that Σ is bounded in $\mathscr{D}'(\omega)$ and the proof is complete. ▌

Remarks. It is clear that an analogous result holds when we replace the sets $\Sigma * f * g$ by sets of the type

$$\Sigma * f_1 * \cdots * f_s,$$

where s is fixed. To reach this conclusion one may either reshape the preceding proof so as to apply to such multiple convolutions, or proceed by induction on s from Theorem A itself.

To Theorem A we add a further and similar criterion.

Theorem B. Let Ω be an open subset of R^n, Σ a subset of $\mathscr{D}'(\Omega)$. In order that Σ be bounded in $\mathscr{D}'(\Omega)$ it is necessary and sufficient that the following condition be fulfilled: to each open set ω, relatively compact in Ω, corresponds a neighbourhood U of 0 such that $\Sigma * f$ is bounded in $\mathscr{D}'(\omega)$ for each f in $\mathscr{D}(U)$.

Proof. The necessity is obvious. To establish the sufficiency of the condition, we show that it entails the condition in Theorem A. The crux of the matter lies in showing that $\mathscr{D}(U)$ may be enlarged to $\mathscr{D}^m(U)$ for some finite m, without disturbing boundedness. The problem is a local one and the proof for the case $\Omega = R^n$ is typical. This will be deferred until Subsection 7.8.2, at which stage the proof serves as an illustration of some general functional analytic principles. ▌

Remarks. Theorems A and B—and, apart from minor details, their proofs—are due to Schwartz [2], Chapitre VI, pp. 47–52.

5.11.3 **Linear Maps Commuting with Derivations.** Using Theorem B of Subsection 5.11.2 we proceed to derive a central theorem concerning continuous linear maps u of $\mathscr{D}$ into $\mathscr{D}'$ which satisfy a certain simple condition. This theorem, due to Schwartz ([2], Chapitre VI, pp. 53–54), is suggested by and includes as special cases a number of similar assertions.

From the earliest days of functional analysis interest has been shown in what Volterra termed "operators of the closed cycle"—that is, operators from functions to functions that commute with translations. Thus, such an operator u is assumed to transform suitably restricted functions f of one or more real variables into another such function $g = u(f)$, and to have the property that a translation of f leads via u to the same translation of g. It is evident that if L is a function, suitably restricted in relation to the given class of functions f, then $u(f) = L * f$ is a continuous operator of the specified type.

In Example 6.4.9 below we establish a result which, for the special categories of functions there involved, shows that conversely each operator u of the said type is representable as a convolution. This and many other similar results [see the references cited in Example 6.4.9] are all derivable from Schwartz' theorem, much of the interest and utility of which is due to the weak assumptions imposed upon u.

Theorem. Let u be a continuous linear map of $\mathscr{D}$ into $\mathscr{D}'$. Then the following four conditions are equivalent:

(1) u commutes with derivations.

(2) u commutes with translations.

(3) u commutes with convolutions.

(4) There exists a distribution L on R^n such that

$$u(f) = L * f \qquad (f \in \mathscr{D}).$$

Proof. If f is a function on R^n and x a point of R^n, we denote by $t_x f$ the x-translate of f, defined as the function $y \to f(y + x)$. Acting on distributions t_x is defined by the adjoint process: in other words, $\langle f, t_x X\rangle = \langle t_{-x} f, X\rangle$ for f in $\mathscr{D}$ and X in $\mathscr{D}'$.

It is important to notice that

$$\partial_k f = \lim_{\varepsilon \to 0} \varepsilon^{-1}(t_{\varepsilon e} f - f)$$

in the sense of $\mathscr{D}$ whenever f is in $\mathscr{D}$, e denoting the kth unit vector in R^n. Similarly

$$\partial_k X = \lim_{\varepsilon \to 0} \varepsilon^{-1}(t_{\varepsilon e} X - X)$$

in the sense of $\mathscr{D}'$ whenever X is in $\mathscr{D}'$.

These remarks show already that (2) implies (1).

To show that (1) implies, hence is equivalent to, (2) it suffices to establish that

$$F(x) = \langle t_x g, u(t_x f) \rangle$$

is independent of x for each pair f and g in $\mathscr{D}$. We show, in fact, that $\partial_k F = 0$ for each $k = 1, 2, \cdots, n$. Now

$$\partial_k F(x) = \left\langle \frac{\partial}{\partial x_k}(t_x g), u(t_x f) \right\rangle + \left\langle t_x g, \frac{\partial}{\partial x_k}[u(t_x f)] \right\rangle$$

where, in order not to miss the point of the proof, it must be understood that (for example) $\partial/\partial x_k(t_x g)$ denotes that element of $\mathscr{D}$ defined by

$$\lim_{\varepsilon \to 0} \varepsilon^{-1}(t_{x+\varepsilon e} g - t_x g),$$

a similar remark applying to $\partial/\partial x_k[u(t_x f)]$. However, it is a simple matter to verify that $\partial/\partial x_k(t_x g)$ so defined is none other than $\partial_k t_x g$, and that

$$\partial/\partial x_k[u(t_x f)] = u(\partial_k t_x f).$$

By (1) this last term is $\partial_k[u(t_x f)]$. Thus the second summand in the above expression for $\partial_k F(x)$ is equal to

$$-\langle \partial_k t_x g, u(t_x f) \rangle,$$

which is the negative of the first. So $\partial_k F = 0$, as alleged.

Thus (1) and (2) are equivalent.

Next let us check that (2) implies (3). This is seen to be the case, if we approximate in $\mathscr{D}$ the convolution $f * g$ by Riemann sums $\sum_r t_{-x_r} f \cdot g(x_r) \cdot \Delta x_r$. Then continuity of u and (T) combine to show that

$$u(f * g) = \lim \sum_r t_{-x_r} u(f) \cdot g(x_r) \cdot \Delta x_r \qquad \text{in } \mathscr{D}',$$

this limit being none other than $u(f) * g$. Thus $u(f * g) = u(f) * g$, which is the content of (3).

Since it is obvious that (4) implies each of (1), (2), and (3), we may close the proof by showing that (3) implies (4). To this end we take any sequence (f_r) in $\mathscr{D}$ that converges in $\mathscr{D}'$ to ε. Then $f_r * g$ tends to g in $\mathscr{D}$ whenever g belongs to $\mathscr{D}$. Accordingly, thanks to (3),

$$\lim u(f_r) * g = \lim u(f_r * g) = u(g)$$

in $\mathscr{D}'$ for each g in $\mathscr{D}$. Appeal to Theorem B of Subsection 5.11.2 shows that the sequence $L_r = u(f_r)$ is bounded in $\mathscr{D}'$. This sequence therefore falls into an equicontinuous subset of $\mathscr{D}'$ and so admits a weak limiting point L in $\mathscr{D}'$. But then $L * g$ can be none other than the limit of $L_r * g$, the latter limit being $u(g)$. This establishes (4). ▮

Remarks. It appears from (4) that u has a natural extension from $\mathscr{D}$ to the space $(C^\infty)'$ of all distributions with compact supports, and that this extension satisfies (3) still:

$$u(X * Y) = u(X) * Y.$$

On the other hand, if (3) is given to hold in this stronger form, then (1) and (2) are trivial consequences, since $\partial_k X = \partial_k \varepsilon * X$ and $t_x X = \varepsilon_x * X$ for arbitrary distributions X.

By imposing further conditions on u one obtains various corollaries of the theorem. Here are some examples.

Corollary 1. If u satisfies the hypotheses of the theorem, and if further the support of $u(f)$ is contained in that of f whenever f belongs to $\mathscr{D}$, then u is a linear partial differential operator with constant coefficients:

$$u(f) = \sum_{|p| \leq m} c_p \partial^p f.$$

Conversely, any such operator satisfies the stated conditions on u.

Proof. The converse is evident. As for the direct assertion we refer to the construction of L as the limit of $L_r = u(f_r)$. Since we may assume all the f_r to have their supports inside an arbitrarily small preassigned neighbourhood U of 0 in R^n, the same must be true of L. Hence in fact the support of L is contained in $\{0\}$, in which case L is a finite linear combination of derivatives $\partial^p \varepsilon$ and therefore u is of the stated type. ▮

The next corollary is utilized in Example 6.4.9, where a more direct proof (albeit following much the same lines of thought) is presented.

Corollary 2. Let u be a continuous linear map of L^1 into L^p $[1 \leq p \leq \infty]$ that commutes with translations. Then

$$u(f) = L * f,$$

where $L \in L^p$ if $p > 1$ and L is a bounded Radon measure if $p = 1$. The converse is true and almost trivial.

Proof. Consider the restriction of u to $\mathscr{D}$. (We identify a function in $\mathscr{D}$ with the class—element of L^1—it defines.) The distribution L figuring in (4) must now have the property that $L * f$ is (the distribution generated by) a function class in L^p and, using $L * f$ to denote this function class,

$$\|L * f\|_p \leq \text{const. } \|f\|_1.$$

Allowing f to vary along a sequence (f_r) in $\mathscr{D}$ satisfying $f_r \geq 0$, $\int f_r \, d\mu = 1$, $f_r(x) = 0$ for $|x| > 1/r$, we see that the sequence $L * f_r$, which converges in $\mathscr{D}'$ to L, is bounded in L^p. If $p > 1$, the bounded subsets of L^p are weakly relatively compact, and so the sequence $L * f_r$ admits in L^p a weak limiting point. This limiting point cannot be other than L, so that L belongs to L^p. The argument fails if $p = 1$, but instead of L^1 we may work with the enlarged space of bounded Radon measures. In that space we find that the sequence $L * f_r$ has a weak limiting point, and so L must be a bounded Radon measure.

In either case, therefore, $u(f) = L * f$ with L in L^p if $p > 1$ and a bounded measure if $p = 1$, in each case f being assumed to be in $\mathscr{D}$. But in either case $f \to L * f$ is continuous on L^1 and $\mathfrak{D}$ is dense in L^1. Continuity of u therefore ensures that the equality continues to hold for any f in L^1. ∎

Some rather similar results are given in Edwards [9].

Remarks. We should not leave the circle of ideas centred around the last theorem without mention of Schwartz' "kernel theorem" concerning continuous linear maps u of $\mathscr{D}$ into $\mathscr{D}'$ which are not assumed to fulfil (1), (2), or (3). The theorem asserts that to each u corresponds a "kernel," that is, a distribution K on the product space, such that

$$\langle f, u(g)\rangle = \langle f \otimes g, K\rangle,$$

$f \otimes g$ denoting the function on the product space defined by

$$(x, y) \to f(x) \cdot g(y).$$

This theorem is, of course, more general than that in the text and gives a basic representation theorem for a vast class of functional operators u. For a discussion of this result, see Schwartz [18] and Ehrenpreis [16].

5.11.4. There are other applications of the techniques illustrated in Subsections 5.11.1–5.11.3. One is a characterization in terms of the convolutions $X * f$ of those distributions X that are analytic functions (Schwartz [2], p. 54). Another is an application to F. Riesz' representation theorem for subharmonic functions (Schwartz [2], pp. 74–78). See also Exercises 7.17–7.27.

5.11.5 **A Pause for Review and Reorientation.** Enough general and basic information about distributions has now been assembled to make it possible (a) to profitably indicate in broad outline various extensions and modifications of the theory that the reader may wish to pursue elsewhere, and (b) to embark on our account of the use of distributions in the theory of linear partial differential equations. A start on the new topic is made in Section 5.12, while item (a) is dealt with forthwith.

As we have remarked in Section 5.1, there are numerous approaches to distributions theory and its development. See, amongst other references, Temple [1], [2], Lighthill [1], e Silva [2], König [1], Mikusiński [1], [2], Gårding-Lions [1], Halperin [3], and Gelfand and Šilov [2]. The last-named treatise is noteworthy for its very detailed treatment of the computational aspects of distributions theory. The second volume of this work (Gelfand and Šilov [3]) handles the theoretical and functional analytic aspects of the subject.

One type of extension is that in which the "base space" R^n is replaced by something more general. Little apart from detail changes are needed if we envisage replacing R^n by a real differentiable manifold of class C^∞. The requisite ideas are indicated at various points throughout Schwartz [1] and [2] and full details are presented in de Rham [1]. In fact, de Rham considers a case incorporating this type of extension with that in which the distributions act on differential forms of arbitrary degree rather than on functions: these are his so-called "currents," the theory of which is due to de Rham and which was developed by him as an independent effort. Even more sweeping extensions are due to Riss [1], the base space being in his case a locally compact Abelian group, and Ehrenpreis [1], who took for his base space a general separated locally compact space and showed how much of Schwartz's theory may be paralleled. In Ehrenpreis' theory there are, of course, no partial differential operators parallel to the ∂_k available in advance; instead one assumes given certain operators of "local type" that shall take over the role of the ∂_k and that are unspecified save for a few general postulates.

One powerful motive for extensions such as those of Riss and Ehrenpreis is to gain ever-increasing scope for suitably generalized concepts of differentiation and of the Fourier transformation; see Ehrenpreis [2], [3], [4], [5], [6], [7], [8].

While it is in no sense coextensive with the theory of distributions, mention should be made of Mikusiński's development ([3], [4], [5]) of his operational calculus. The two techniques have a good deal in common and present similar problems of a functional analytic character.

Another type of extension of the theory is that in which one studies, in place of scalar-valued distributions (that is, continuous linear forms on $\mathscr{D}$), continuous linear maps of $\mathscr{D}$ into a given TVS E. The latter objects are termed "E-valued distributions." The development in the case of Radon measures, commented upon in Subsection 4.10.7, is dealt with in Section 8.14 onward. For details of the case of distributions, see Schwartz [4], [5], and [6].

It is interesting to note that motivation for this last type of extension is to be found in the current development of the quantum theory of fields. In that domain it has been somewhat reluctantly accepted that the field variables may be operator-valued distributions. As a result, here and in particle quantum mechanics, the use of distributions aids in the legitimization of some hitherto rather suspect procedures. In the field theories, however, it is only fair to add that the use of distributions raises almost as many thorny problems as it solves. For some details accessible to mathematicians, see Schwartz [14], Wightman [1], and Friedrichs [1].

We have already referred in passing to the enlarged scope offered to harmonic analysis by the use of distributions; in this respect the theory develops ideas heralded by Bochner [1]. We turn to this topic in Section 5.15.

5.12 Linear Partial Differential Equations

The remaining sections of this chapter are devoted entirely to some of the modern developments in the theory of linear partial differential equations. The reader acquainted with the classical study of partial differential equations will find a lack of continuity on making the step from the classical to the modern approach. It is true to say that perhaps the earliest external motivation for the development of a theory of "generalized functions" or distributions is to be found amid certain writings on linear partial differential equations belonging to a rather ill-defined transition period typified, for instance, by Soboleff [1], and some of the relevant portions of Schwartz [1] and [2]. Regarding this discontinuity, let us quote Gårding [4]:

"The guide to fruitful problems in this connection (that of linear partial differential operators with constant coefficients) has been the theory of distributions rather than classical physics, and this shows that a new theory does not have to solve old problems to motivate its existence. It is quite enough to state and solve new ones."

The reader is recommended to read Gårding's exposition from which this quotation is taken.

One of the rather surprising characteristics of the new theory is the fundamental role played by inequalities (see Sections 5.14 and 5.19). Quoting Gårding again: "When a problem about linear partial differential operators has been fitted into the abstract theory, all that remains is usually to prove a certain inequality and much of our new knowledge is, in fact, essentially contained in such inequalities." In establishing such inequalities it is often useful (though not by any means always essential) to use Fourier transforms, and here one finds additional motivation for extending the Fourier transformation to distributions, a sketch of which we give in Section 5.15.

One of the earliest methods based ultimately on inequalities is the "method of orthogonal projection" due to Hermann Weyl, designed originally to cope with the Dirichlet (first boundary value) problem for Laplace's equation and since greatly extended by various authors. This is dealt with in Section 5.13. The abstract basis here amounts to simple properties of Hilbert spaces set forth in Chapter 1. A recent exposition of this and similar techniques is given by Garnir [1].

In the course of the present section we shall formulate a number of problems concerning linear partial differential equations to which we shall return in subsequent sections of this chapter.

Concerning our treatment of LPDEs (linear partial differential equations) one general point should be stressed. We are primarily concerned with the functional analytic approach to and content of the modern theory. No attempt can be made within the space at our disposal to follow up the aspects involving detailed analysis of a classical nature. For example, when dealing with the orthogonal projection method in Section 5.13, we do not proceed to seek integral representations of the Green's operators involved. Such cavalier treatment is, of course, not intended to underrate the interest and importance of these problems. The interested reader will no doubt seek to rectify the omissions. In the cited instance, redress may be found in Garnir [1] and the references cited there. That the problems we have in mind can indeed be treated in distinct stages, the first of which is predominantly functional analytic in nature and the second principally composed of classical analysis, is well illustrated in the case of the Dirichlet problem by the treatment presented on pp. 183–199 of Epstein [1].

A second limitation demands mention. Save in those cases where the methods utilized for handling equations with constant coefficients are effective also for equations with variable coefficients, the latter case receives no adequate treatment. For the treatment of the more general case the interested reader is referred to research papers mentioned at the appropriate places, and to Hörmander [7]. The latter is indeed a reference appropriate to almost all the material covered in the present chapter, and to much else besides.

5.12.1 **Definitions, Notations, and Preliminary Remarks.** In what follows we suppose that $P(x, Z) = P(x_1, \cdots, x_n, Z_1, \cdots, Z_n)$ is a polynomial in the indeterminates $Z_1, \cdots, Z_n$ with coefficients that are complex-valued functions defined and locally integrable on a domain Ω in R^n, $x = (x_1, \cdots, x_n)$ denoting a variable point of Ω. Thus

$$P(x, Z) = \sum_{|p| \leq m} a_p(x) Z^p,$$

$$Z^p = Z_1^{p_1} \cdots Z_n^{p_n}.$$

As a polynomial in Z, $P(x, Z)$ is of degree at most m; the degree will be said to be precisely m if there exists an index p with $|p| = m$ for which the function a_p is not negligible on Ω.

If for $Z = (Z_1, \cdots, Z_n)$ we formally substitute $\partial = (\partial_1, \cdots, \partial_n)$, we obtain a LPDO on Ω, viz.,

$$P(x, \partial) = \sum_{|p| \leq m} a_p(x)\, \partial^p, \tag{5.12.0}$$

the order of which is the same as the degree of $P(x, Z)$. There is no interest attached to the case $m = 0$, and we assume $m \geq 1$ throughout.

When, from Section 5.14 onward, we make use of the Fourier transformation, it is notationally advantageous in the general theory to express our LPDOs as polynomials in the operators $D_k = (2\pi i)^{-1} \partial_k$ in place of the ∂_k themselves. Corresponding to $P(x, Z)$ one will then have the LPDO

$$P(x, D) = \sum_{|p| \leq m} a_p(x) D^p. \tag{5.12.0'}$$

In any instance we shall feel free to adopt whichever notation seems temporarily most convenient.

If the a_p are constant functions on Ω, the LPDO $P(x, D)$ is said to have "constant coefficients" and is usually written simply $P(D)$; in this case it is natural to take $\Omega = R^n$, though this is by no means obligatory. Indeed, unless $n = 1$, solutions of $P(D)X = A$ in an open set Ω are in general not (even when $A = 0$) extendible into solutions of the same equation throughout R^n. Consider, for example, the harmonic functions in a domain, which are simply the solutions of $P(D)X = 0$ there when $P(D) = \Delta = -4\pi^2 \sum_{k=1}^n D_k^2$ is the Laplacian on R^n.

Depending on the nature of the coefficients a_p, $P(x, D)$ may be viewed as a continuous linear map of various subspaces of $\mathscr{D}'(\Omega)$ into itself. Accordingly there will be variations in the interpretation of the so-called "formal" or "Lagrange" adjoint of $P(x, D)$. This formal adjoint is itself a LPDO $\check{P}(x, D)$ characterized by the property that

$$\int_\Omega P(x, D)f \cdot g \, d\mu = \int_\Omega f \cdot \check{P}(x, D)g \cdot d\mu \tag{5.12.1}$$

for arbitrary f and g in $\mathscr{D}(\Omega)$. Using the notation $<, >$ as in the theory of distributions, this is equivalent to

$$\langle P(x, D)f, g\rangle = \langle f, \check{P}(x, D)g\rangle. \tag{5.12.1'}$$

Expressed explicitly one finds easily that

$$\check{P}(x, D)g = \sum_{|p| \leq m} (-1)^{|p|} D^p(a_p g) \tag{5.12.1''}$$

for g in $\mathscr{D}(\Omega)$. On the right-hand side one is forced in general to interpret D^p in the sense of distributions, though the classical sense is legitimate if, for example, a_p belongs to $C^{|p|}(\Omega)$.

The above formulae succeed in defining $\check{P}(x, D)$ in its action on functions in $\mathscr{D}(\Omega)$. If we use Leibnitz' formula to express the right-hand side of (5.12.1″) in the form

$$\sum_{|p| \leq m} b_p D^p g,$$

it appears that the b_p involve each a_p together with its derivatives $D^q a_p$ with $|q| \leq |p|$. Accordingly, without further restrictions of smoothness on the a_p, $\check{P}(x, D)$ will not be a LPDO of the same type as $P(x, D)$, its coefficients b_p failing to be locally integrable functions. The situation is retrieved by imposing further restrictions on the a_p.

For example, if each a_p belongs to $C^\infty(\Omega)$, then $P(x, D)$ maps $\mathscr{D}(\Omega)$ continuously into itself, so that $\check{P}(x, D)$ is extendible into a continuous endomorphism of $\mathscr{D}'(\Omega)$. The extension is given by (5.12.1″) after replacing g by a general X in $\mathscr{D}'(\Omega)$. Similarly, if a_p belongs to $C^{|p|+k}(\Omega)$, where k is a given integer, then $P(x, D)$ is continuous from $\mathscr{D}(\Omega)$ into $\mathscr{D}^k(\Omega)$, and so $\check{P}(x, D)$ is extendible into a continuous map of $\mathscr{D}'^k(\Omega)$ into $\mathscr{D}'(\Omega)$, again expressed by extending (5.12.1″) in the obvious fashion.

In each of these cases formula (5.12.1′) extends by continuity into

$$\langle P(x, D)f, X\rangle = \langle f, \check{P}(x, D)X\rangle, \tag{5.12.1'''}$$

valid for f in $\mathscr{D}(\Omega)$ and X in $\mathscr{D}'(\Omega)$ or $\mathscr{D}'^k(\Omega)$, respectively.

The preceding remarks concerning extensions of $\check{P}(x, D)$ may profitably be viewed as special instances of general results concerning adjoint maps set out in Section 8.6, taking into account special properties of the space $\mathscr{D}(\Omega)$ noted in Section 5.3 (see also Subsection 8.4.7).

It is well to point out here that the adjoint $\check{P}$ defined above differs from that used by Hörmander [1], who works with the Hilbert adjoint $\tilde{P}$ introduced temporarily in Subsection 5.14.4. The difference is not very profound, however, being merely the result of using the Hilbert inner product $(\mid)_{L^2}$ in place of the bilinear form $<, >$ expressing the duality between $\mathscr{D}(\Omega)$ and $\mathscr{D}'(\Omega)$.

The study of the linear partial differential equation

$$\check{P}(x, D)X = A, \tag{5.12.2}$$

A being given, and in particular of the corresponding homogeneous equation

$$\check{P}(x, D)X = 0, \tag{5.12.3}$$

has been intimately connected with the development of mathematical physics and applied mathematics ever since the days of the Bernoulli family, Euler and d'Alembert. In what may be termed the classical era, it was assumed that A was a function on Ω (usually continuous and at all events locally integrable) and the solutions were sought among the functions belonging to $C(\Omega)$ that satisfy (5.12.2) in the pointwise sense throughout Ω. It was in the field of ordinary differential equations (one independent variable, $n = 1$) that the break with these smoothness assumptions first appeared, abstract methods based on Hilbert space theory proving themselves useful. In this connection the reader may consult Stone [1], Chapter IX, where the abstract basis is a relatively detailed slice of the spectral theory of endomorphisms of a Hilbert space. Another approach, closer to ones we intend to study, is set out in Chapters 7 and 9 of Coddington and Levinson [1]. This depends on recasting the combined differential equation plus boundary conditions into the form of an equivalent integral equation, to which one may apply the general Riesz theory for compact endomorphisms; see Chapter 9, especially Section 9.12. Both of these techniques differ in important respects from the orthogonal projection method set forth in Section 5.13. For one thing, the latter applies even to the case of singular problems (though of course the conclusions are then not as strong), whereas [cf. Coddington and Levinson, loc. cit.] the approach based on the Riesz theory applies directly only to the simpler regular case and independent arguments are necessary to pass then to the singular case.

The import of abstract methods into the theory of partial differential equations (the case $n > 1$) came rather later, the majority of the techniques we aim to discuss having appeared within the past 20 years and much within the past decade. This importation of new methods itself initiated many ideas subsequently swallowed up in the theory of distributions.

To illustrate this point, so-called "weak solutions" of (5.12.2) with A a locally integrable function were introduced as locally integrable functions X on Ω for which

$$\int_\Omega P(x, D)f \cdot X \, d\mu = \int_\Omega Af \, d\mu$$

for each f in $\mathscr{D}(\Omega)$. Reference to (5.12.1''') shows that this is tantamount to considering those distribution solutions X of (5.12.2) which happen to be (locally integrable) functions. At the same time, of course, the use of such generalized solutions called for reformulation of the boundary conditions. Garnir [1] uses weak solutions systematically.

Weyl's fundamental paper [1] illustrates this half-way stage of generalization. Reluctance to losing all connections with the restricted, classical theory accounts for an extremely fruitful and significant portion of Weyl's work, namely, the demonstration that any weak solution of Laplace's equation is (after adjustment on a negligible set) already a classical C^∞ solution. This circumstance has led to the definition and study of the class of so-called "hypoelliptic" LPDOs $P(x, D)$, namely, those for which the local C^∞ nature of $P(x, D)X$ entails that of X itself. We shall dwell at some length on this point in Section 5.18 below.

It will become apparent as we proceed that the one-dimensional case is genuinely exceptional, and not merely relatively easy to handle: if the coefficients a_p and A are smooth functions, the only distribution solutions of (5.12.2) prove to be the classical ones. This is far from being the case when $n > 1$, as we shall see by example in Subsection 5.12.2.

It is plain that all LPDOs $P(x, D)$ are of "local type," in the sense that the support of $P(x, D)X$ is contained in that of X. It turns out that this property is characteristic, a result established by Peetre [1] after examination of the homomorphisms of sheaves (=stacks, faisceaux) of germs of C^∞ functions on differentiable manifolds. This intrinsic characterization of the maps to be studied is satisfying.

Throughout the remaining subsections of Section 5.12 we shall speak only of LPDOs with constant coefficients. We shall examine briefly some of the ways in which the one-dimensional case, although in most respects too simple to be typical, nevertheless suggests nontrivial problems for the multidimensional case. Some such problems will be formulated at length pending detailed discussion in subsequent sections.

5.12.2 **Preview of Examples and Suggested Problems.** As has been said we shall confine our attention throughout this preview to equations

$$P(D)X = A \tag{5.12.4}$$

with constant coefficients, and with the associated homogeneous equation

$$P(D)X = 0. \tag{5.12.5}$$

In the present instance, $P(D) = \sum_{|p| \leq m} a_p D^p$, the a_p being constants. We assume that $m \geq 1$ and that $a_p \neq 0$ for at least one index p satisfying $|p| = m$.

Even when the coefficients a_p are quite general smooth functions, the one-dimensional case is extraordinarily simple in structure. Assuming that a_m is nonvanishing throughout the open set Ω, it may be shown (Schwartz [1], pp. 126 et seq) that all distribution solutions of (5.12.5) are necessarily smooth functions and therefore solutions in the classical sense. The same is true of

(5.12.4), provided A itself is a smooth function. To this one must add, however, that if a_m vanishes at any point of the open set in question, (5.12.4) may fail to have any distribution solutions at all. Consider, for example, the equation

$$x^3 \, \partial X + 2X = 0.$$

Throughout the open set $\Omega = R^1 \backslash \{0\}$, its solutions are the functions

$$X = a \cdot \exp(x^{-2}) \quad \text{for } x < 0, \qquad = b \cdot \exp(x^{-2}) \quad \text{for } x > 0,$$

where a and b are constants. No other solutions are available. It is quite easily seen that there exists no distribution solution at all throughout any open neighbourhood of the origin.

For the case of constant coefficients it costs very little effort to be more explicit. Suppose $n = 1$ and that A is a continuous function on the interval Ω in R^1. $P(D)$ may be factorized into $(D - \lambda_1)^{m_1} \cdots (D - \lambda_r)^{m_r}$, where the λ's are distinct complex numbers, the m's natural numbers, and $m = m_1 + \cdots + m_r$. Since one has identically

$$D(e^{-2\pi i\lambda x}X) = e^{-2\pi i\lambda x}(D - \lambda)X,$$

it is easily seen that the only distribution solutions X of $(D - \lambda)X = A$ are the functions

$$ae^{2\pi i\lambda x} + \int_{x_0}^{x} e^{2\pi i\lambda(x-y)}A(y)\, dy,$$

where a is a constant and x_0 any fixed point of Ω. By repetition of this argument we are led to the following conclusion.

(1) Suppose $n = 1$, Ω being an interval in R^1, $P(D) = \sum_{p \le m} a_p D^p$ a LPDO with constant coefficients of order m (so that $a_m \neq 0$). Then $P(D)$ is hypoelliptic in the sense that if ω is any subinterval of Ω, and if A is a C^∞ function on ω, then any solution X of (5.12.4) has a restriction to ω that is likewise a C^∞ function. If A belongs to $C^\infty(\Omega)$, all solutions of (5.12.4) are members of $C^\infty(\Omega)$ and therefore classical solutions thereof.

In particular we have also the following conclusion.

(2) The hypotheses being as in (1), every solution of (5.12.5) is a finite linear combination of exponential-monomial solutions $x^p e^{2\pi i\lambda x}$, λ being a zero of P of order q, and p satisfying $|p| < q$.

We now ask ourselves: what happens to the assertions (1) and (2) when the dimension n exceeds unity? Let us consider these questions separately.

(a) The answer is that (1) is entirely false when $n > 1$ and $P(D)$ is unrestricted—that is, that there is now room for a special category of LPDOs $P(D)$ which shall be termed the hypoelliptic ones and which will deserve special attention. The study of this category is reserved until later. Meanwhile we exhibit an example that is not hypoelliptic.

We begin by observing that if $\Omega = R^n$, and if X satisfies (5.12.5), then so also does $K * X$ for any distribution K with a compact support. (This property comes very close to characterizing the LPDOs with constant coefficients (see Subsection 5.11.3).)

For our example we take $n = 2$ and the operator

$$P(D) = \partial_1 + \partial_2 = 2\pi i(D_1 + D_2).$$

If f belongs to $C^1(R^1)$, and if we define g on R^2 by

$$g(x_1, x_2) = f(x_1 - x_2),$$

then g belongs to $C^1(R^2)$ and satisfies $P(D)g = 0$ in the pointwise sense. Therefore g satisfies the same equation in the sense of distributions.

Suppose now that f is merely locally integrable. Then g has the same property and therefore generates a distribution on R^2. We aim to show that $P(D)g = 0$ in the sense of distributions. To this end, we take a regularizing sequence (r_i) in $\mathscr{D} = \mathscr{D}(R^2)$ and define $G_i = g * r_i$. Direct calculation shows that

$$G_i(x_1, x_2) = F_i(x_1 - x_2),$$

where

$$F_i(x) = -\iint f(y) r_i(y' - y, y' - x)\, dy\, dy'.$$

It is evident that F_i belongs to $C^\infty(R^1)$, so that G_i belongs to $C^\infty(R^2)$. By what has been said, $P(D)G_i = 0$ for each i. As i tends to infinity, G_i tends to g in $\mathscr{D}'(R^2)$. So, since any LPDO with constant coefficients is a continuous endomorphism of $\mathscr{D}'(R^n)$, it follows that $P(D)g = 0$.

On the other hand it is plain that, f being merely locally integrable, the function g need not be in $C^1(R^2)$. Moreover, since $D^pX = (D^p\varepsilon) * X$ is a solution whenever X is a solution, one may derive from g solutions of $P(D)X = 0$ that are distributions of arbitrarily high order. $P(D)$ is therefore certainly not hypoelliptic. For another example, see Exercise 5.3.

(b) The answer is again negative: the direct analogue of (2) is false when $n > 1$, the exponential-monomial $x^p e^{2\pi i \lambda \cdot x}$ being now the function

$$x_1^{p_1} \cdots x_n^{p_n} \cdot \exp\left(2\pi i \sum_{k=1}^{n} \lambda_k x_k\right)$$

wherein $\lambda = (\lambda_1, \cdots, \lambda_n)$ is a point of C^n. The example used in (a) shows at once that "most" solutions of the equation considered there fail to be finite linear combinations of such exponential monomials no matter how the various points λ of C^n are selected.

It turns out, however, that although the finite linear combinations of exponential monomials generally fail to exhaust all the solutions, they are nevertheless dense amongst all solutions. We shall take this topic as the subject of the first of several problems to be stated and studied (Problem 1 of Subsection 5.12.3).

Granted the density of the finite linear combinations of exponential-monomial solutions, it is natural to ask whether and under what conditions the polynomial solutions alone share this density property. We shall in Subsection 5.12.3 formulate this as Problem 1′. Much later (see Subsection 5.16.4) we shall see that this problem also is a fruitful one. For the moment let us see how a special case of it is related to complex function theory.

We take again $n = 2$ and make the usual identification between points of R^2 and complex numbers. Let Ω be a domain in R^2 and denote by $H(\Omega)$ the set of functions on Ω that are holomorphic there. Each f in $H(\Omega)$ generates a distribution (a measure, in fact), namely, $f \cdot \mu$, where μ is the restriction to Ω of Lebesgue measure on R^2. Since f is continuous, the correspondence $f \leftrightarrow f \cdot \mu$ is one to one, and so we regard $H(\Omega)$ as a vector subspace of $\mathscr{D}'(\Omega)$. This time the chosen LPDO is

$$P(D) = \frac{\partial}{\partial \bar{z}} = \partial_1 + i\partial_2,$$

z denoting the complex coordinate function $x_1 + ix_2$ and $\bar{z}$ its complex conjugate.

It is an elementary result that any function f in $C^1(\Omega)$ satisfying $P(D)f = 0$ throughout Ω is a member of $H(\Omega)$. We wish first to show that the same is true of any distribution solution of $P(D)X = 0$—by which we mean, of course, that any such X is generated by a locally integrable function f which, after correction on a negligible subset of Ω, belongs to $H(\Omega)$. This we shall achieve by modifying slightly the regularization process used in (a) above.

If Ω were the whole of R^2, we should seek to approximate X by the functions $X * r_i$, (r_i) being as before a regularizing sequence chosen from $\mathscr{D}$. As it is, however, X is defined only in Ω, and a little care is necessary.

We take advantage of the remark that it suffices to show that the restriction of X to each relatively compact open subset ω of Ω is a member of $H(\omega)$. Then, fixing any such ω, we suppose the r_i chosen to have their supports within the ball with centre 0 and radius c, where $2c$ is the distance between ω and the frontier of Ω. For fixed x in ω, the function $y \to r_i(x - y)$ will consequently have its support in Ω and so $f_i(x) = X_{(y)}[r_k(x - y)]$ will be defined. It is evident that the functions f_i thus defined belong to $C^\infty(\omega)$, and it is not difficult to verify that $f_i \to X$ in $\mathscr{D}'(\omega)$ as $i \to \infty$. As we should hope and expect, f_i satisfies $P(D)f_i = 0$ on ω. In fact one has the formula

$$P(D)f_i(x) = \langle R_x, P(D)X \rangle$$

for any fixed x in ω, R_x denoting the function $y \to r_i(x - y)$. (This formula will follow once it is established for the special LPDO $P(D) = \partial_k$, and for this special case one need merely check that $f_i(x) = \langle R_x, X \rangle$ and that for a small increment h in x_k leading from $x = (x_1, \cdots, x_k, \cdots, x_n)$ to $x' = (x_1, \cdots, x_k + h, \cdots, x_n)$ the difference quotient $h^{-1}(R_{x'} - R_x)$ converges in $\mathscr{D}(\omega)$ to the function $y \to \partial_k r_i(x - y) = -(\partial/\partial y_k)r_i(x - y)$.)

We have thus shown that X is the limit in $\mathscr{D}'(\omega)$ of a sequence (f_i) of functions in $C^\infty(\omega)$ that satisfy $P(D)f_i = 0$ on ω. Each f_i is therefore a member of $H(\omega)$. If we can now show that the uniform structure induced on $H(\omega)$ by that of $\mathscr{D}'(\omega)$ is not weaker than (and therefore is identical with) that of locally uniform convergence, it would follow that the sequence (f_i) is Cauchy for the latter structure. But it is an elementary fact that $H(\omega)$ is complete for the latter structure, hence it would appear that X itself must belong to $H(\omega)$.

The comparison of the two structures in question is easily effected by making a small modification in the Cauchy integral formula. We let f belong to $H(\omega)$, and consider any open disk with centre $x_0 \in \omega$ and radius $c > 0$ so small that the corresponding closed disk is contained in ω. By the integral formula,

$$f(z) = (2\pi i)^{-1} \int_{|x-x_0|=r} f(x) \frac{dx}{x - z}$$

for $|z - x_0| < r < c$. We take now any function q in $\mathscr{D}(R^1)$ having its support in $(3c/4, c)$ and such that $\int_{-\infty}^{\infty} q(r)\, dr = 1$. If $|z - x_0| < c/2$, we shall have

$$\begin{aligned} f(z) &= (2\pi)^{-1} \int_{-\infty}^{\infty} q(r)\, dr \int_0^{2\pi} f(x_0 + re^{i\alpha}) re^{i\alpha} \frac{d\alpha}{x_0 + re^{i\alpha} - z} \\ &= (2\pi)^{-1} \int_\omega f(x)(x - x_0) q(|x - x_0|) \frac{d\mu(x)}{x - z} \\ &= \int_\omega f(x) g_z(x)\, d\mu(x), \end{aligned}$$

where μ is Lebesgue measure on R^2 and

$$g_z(x) = (x - x_0) \frac{q(|x - x_0|)}{x - z}.$$

Since $q(|x - x_0|) = 0$ for $|x - x_0| < 3c/4$ and for $|x - x_0| > c$, it is easy to verify that the set $\{g_z : |z - x_0| < \frac{1}{2}c\}$ is bounded in $\mathscr{D}(\omega)$. Consequently, if the sequence (f_i) is Cauchy in $\mathscr{D}'(\omega)$, this same sequence is Cauchy for the structure of convergence uniform on the disk $|z - x_0| < \frac{1}{2}c$. Each compact subset of ω being coverable by finitely many such disks, the desired comparison is effected.

We now know, then, that any solution in $\mathscr{D}'(\Omega)$ of the equation $P(D)X = (\partial/\partial\bar{z})X = 0$ is a member of $H(\Omega)$—that is, is holomorphic on Ω. This being so, if Ω is simply connected, Runge's theorem entails that any solution X is the limit (in $H(\Omega)$ and a fortiori in $\mathscr{D}'(\Omega)$) of polynomials in the complex coordinate $z = x_1 + ix_2$—that is, of polynomial solutions of the same equation. In other words we may in this instance replace, in the approximation theorem (stated earlier without proof), exponential monomials by monomials in the ordinary sense. Further study of such replacement in more general situations (see 5.16.4) leads to rather sweeping extensions of Runge's theorem.

We have now reached a stage where we may begin to formulate more precisely the several natural problems suggested by the preceding cursory glance at special examples. The study of these problems will occupy our attention throughout the remainder of this chapter.

5.12.3 Statement of Problems 1 and 1′

Problem 1. Given an open set Ω in R^n and a LPDO $P(D)$ with constant coefficients, under what conditions is it true that each X in $\mathscr{D}'(\Omega)$ satisfying $P(D)X = 0$ is the limit in $\mathscr{D}'(\Omega)$ of finite linear combinations of exponential monomial solutions of this equation?

Problem 1'. Ω and $P(D)$ being as in Problem 1, under what conditions is it true that each X in $\mathscr{D}'(\Omega)$ satisfying $P(D)X = 0$ is the limit in $\mathscr{D}'(\Omega)$ of polynomial solutions of this equation?

It is natural also to pose the analogous problems after replacement of $\mathscr{D}'(\Omega)$ by certain other spaces—$C^\infty(\Omega)$, for example.

In each case, conditions are known that are sufficient to ensure the possibility of the desired approximations, but it is not yet entirely clear to what extent such conditions are also necessary.

Partial solutions will be discussed at some length in Section 5.16.

5.12.4 **Existence of Elementary Solutions: Statement of Problem 2.** Let Ω be an open subset of R^n, t a point of Ω, and $P(D)$ a LPDO. By an elementary (or fundamental) solution, in Ω and relative to t, for $P(D)$, is meant a distribution E_t in $\mathscr{D}'(\Omega)$ such that

$$P(D)E_t = \varepsilon_t, \tag{5.12.6}$$

where (as always) ε_t denotes the Dirac measure placed at the point t.

If, as we shall assume henceforth, $\Omega = R^n$ and $P(D)$ has constant coefficients, and if $E_0 = E$ is an elementary solution relative to the origin, that is, if

$$P(D)E = \varepsilon, \tag{5.12.7}$$

then $E_t = \varepsilon_t * E$ is an elementary solution relative to the point t. So it suffices in this case to deal only with $E = E_0$, which we shall speak of simply as an elementary solution for $P(D)$.

Now if $P(D)$ is hypoelliptic, any elementary solution E is necessarily a function of class C^∞ on $R^n \backslash \{0\}$. Outside the origin, $P(D)E = 0$ in the pointwise sense. In any neighbourhood of the origin, E will usually be singular to just such an extent that in place of $P(D)E = 0$ one will have $P(D)E = \varepsilon$. For this reason one has in the past spoken of E as an elementary (or fundamental) *solution* of the LPDE $P(D)E = 0$. Of course, considered as a distribution on R^n, E is *not* a solution of this homogeneous equation.

The significance of elementary solutions is already pretty clear from the substance of Section 5.11, where we have seen how the existence and properties of an elementary solution for the Laplacian $\Delta = \sum_{k=1}^n \partial_k^2$ had a good deal of influence on general topics in the theory of distributions. An elementary solution for Δ is given by (5.11.5). The reader is advised to return for a moment to Section 5.11 and to see how it turns out that E, thus defined, fills the role of an elementary solution for Δ. Replacing the formula $P(D)E = 0$—true in the classical sense on $R^n \backslash \{0\}$—by the formula (5.12.7), which is valid globally and in the sense of distributions, summarizes the manipulations involving Green's formula and subsequent passage to the limit which appear in Section 5.11, and which would appear frequently in any arguments based upon the classical interpretation of $P(D)E$.

There is another sense in which a knowledge of an elementary solution and its properties governs the study of the LPDO $P(D)$. Granted (5.12.7), since

$P(D)$ commutes with convolution, a solution of $P(D)X = A$ is given by $X = E * A$, at least whenever A has a compact support.

An existence theorem for elementary solutions is clearly desirable, and so we formulate:

Problem 2. To prove the existence of an elementary solution for as many LPDOs $P(D)$ as possible.

This problem will be examined and solved in Section 5.14 by the use of the Fourier transformation combined with complex variable theory, thus following techniques exploited by Malgrange ([1], Chapitre II, §1) and expounded there and also in Exposé 2 of Schwartz [8].

A different method of proof and a closer look at the properties of elementary solutions (especially in the hypoelliptic case) forms the subject matter of Section 5.18.

5.12.5 **The Solubility of $P(D)X = A$: Statement of Problem 3.** We shall discover that quite wide-sweeping results regarding the solubility of the equation $P(D)X = A$ may be obtained on the basis of even partial solutions of Problems 1 and 2. We formulate the task in the form of:

Problem 3. To examine the truth or otherwise of such relations as $P(D)C^\infty = C^\infty$, $P(D)\mathscr{D}' = \mathscr{D}'$, and so forth.

Numerous results of a positive nature are given in Section 5.17, together with references to further developments. Here the results are due principally to Malgrange ([1], Chapitre II, §3) and, independently, Ehrenpreis.

5.13 The Method of Orthogonal Projection and Its Developments

In this section our aim is to indicate in some detail how the technique of orthogonal projection in inner-product spaces, laid out in Section 1.12, can be used to yield results about boundary value problems of the Dirichlet and Neumann types, linked with linear partial differential equations. The method and its applications in this direction seem first to have been conceived and exploited by Weyl [1], and a bird's-eye view of Weyl's approach has been noted in Subsection 1.12.7. Subsequently the method underwent modifications and extensions at the hands of many workers. The exposition we follow is suggested by those of Garnir [1], Schwartz [8] (Exposés 11 onwards) and [10], and Lions [1].

Throughout this section, Ω will denote an open subset of R^n; with no loss of generality we may assume that Ω is connected (hence a domain in R^n). All the function spaces we need to consider ($\mathscr{D}(\Omega)$, $L^p(\Omega)$, $L^p_{\text{loc}}(\Omega)$, and so on) may and will be regarded as subspaces of $\mathscr{D}'(\Omega)$.

Let us begin by attempting to describe the method in general terms. A central role will be played by a space V satisfying

$$\mathscr{D}(\Omega) \subset V \subset \mathscr{D}'(\Omega) \tag{5.13.1}$$

and formed into an inner-product space with a suitably chosen inner product $(\ |\ )_V$. The choice of V is to be made (whenever this is possible) in such a way that a preassigned formal LPDO $P = P(x, \partial)$ may be interpreted as a linear map of V into $\mathscr{D}'(\Omega)$ such that

$$(v \mid \varphi)_V = \langle \bar{\varphi}, Pv \rangle \qquad \text{for } \varphi \in \mathscr{D}(\Omega) \text{ and } v \in V. \tag{5.13.2}$$

The method will seek to express boundary conditions in terms of suitable vector subspaces N of V. It is not intended that boundary conditions thus expressed shall be manifestly equivalent to ones expressed in classical terms and applied to smooth functions. This is an issue that may be broken off from the rest of the problem and examined separately. It will receive totally inadequate attention in this book owing to lack of space. Our remarks pertaining to this and related problems will be deferred until Subsection 5.13.9, where we shall collect a number of statements; in many cases the proofs would take us too far afield and we do no more than quote suitable references.

The principal aim of this section is to establish the existence and uniqueness of the solution of the system

$$u \in N, \qquad Pu = f,$$

together with a brief study of the functional analytic nature of the dependence of this solution upon the data f. The system is solved by an application of the projection principle stated in Subsection 1.12.7, hence the title of this section. Special instances of the problem reduce to abstract formulations of the Dirichlet problem attacked by development of the methods sketched in Subsection 1.12.7.

We proceed to describe in precise terms a space V that will remain the standard type throughout this section.

5.13.1 The Choice of V. Let us suppose we are given in advance functions $p_0, \cdots, p_n$ which are assumed always to be positive and locally integrable over Ω.

We define V to be the set of functions (or, rather, function classes) f on Ω such that

$$f \in L^1_{\text{loc}}(\Omega), \quad \partial_k f \in L^1_{\text{loc}}(\Omega) \quad (k = 1, 2, \cdots, n) \tag{5.13.3}$$

and

$$p_0^{1/2} f \in L^2(\Omega), \quad p_k^{1/2} \partial_k f \in L^2(\Omega) \quad (k = 1, 2, \cdots, n). \tag{5.13.3'}$$

On V we introduce the inner product

$$\begin{aligned}(f \mid g)_V &= (p_0^{1/2} f \mid p_0^{1/2} g)_{L^2} + \sum_{k=1}^{n} (p_k^{1/2}\, \partial_k f \mid p_k^{1/2}\, \partial_k g)_{L^2} \\ &= \int_\Omega p_0 f \bar{g}\, d\mu + \sum_{k=1}^{n} \int_\Omega p_k\, \partial_k f\, \partial_k \bar{g}\, d\mu.\end{aligned} \tag{5.13.4}$$

Since the functions $p_0, \cdots, p_n$ are locally integrable over Ω, it follows that $\mathscr{D}(\Omega) \subset V$. Moreover, if $f \in V$, one has $p_0^{1/2} \in L^2_{\text{loc}}(\Omega)$ and $p_0^{1/2} f \in L^2(\Omega)$, so that $p_0 f = p_0^{1/2} \cdot p_0^{1/2} f$ is locally integrable. Similarly, $p_k\, \partial_k f$ is locally integrable. It follows that if $\varphi \in \mathscr{D}(\Omega)$, then

$$\int_\Omega p_k\, \partial_k f \cdot \partial_k \bar{\varphi}\, d\mu = -\langle \bar{\varphi}, \partial_k (p_k\, \partial_k f) \rangle$$

in the distributions sense. Consequently (5.13.2) holds with

$$Pf = p_0 f - \sum_{k=1}^{n} \partial_k (p_k \, \partial_k f). \tag{5.13.5}$$

Some remarks on the nature of V are to be found in Subsection 5.13.9.

5.13.2 Statement and Solution of the Dirichlet-Neumann Problem. We consider the situation described in Subsection 5.13.1, the aim being to study the equation $Pu = 0$ when P is given by (5.13.5). To this equation are to be added boundary conditions of mixed Dirichlet and Neumann types. These boundary conditions will be formulated in an abstract manner to be described forthwith.

Let $\dot{\Omega}$ denote the frontier of Ω and let Σ be a subset of $\dot{\Omega}$. We define V_Σ to be the closure in V of the set of functions $v \in V$ that vanish (almost everywhere) on the intersection with Ω of some neighbourhood in R^n of Σ, this neighbourhood possibly depending upon the function v. The reader will notice that $\mathscr{D}(\Omega) \subset V_\Sigma$. Membership to V_Σ is intended to convey in a generalized sense the property of having vanishing boundary values on the part Σ of $\dot{\Omega}$. This point will be discussed later.

To formulate the Dirichlet-Neumann boundary conditions involves partitioning $\dot{\Omega}$ into two sets $\dot{\Omega}_d$ and $\dot{\Omega}_n$. We then write for brevity V_0 in place of $V_{\dot{\Omega}_d}$. We introduce two conditions—namely,

$$(B_d) \qquad u \in V_0,$$

and

$$(B_n) \qquad u \in V, \quad Pu \in L^2(\Omega), \quad \text{and}$$

$$(Pu \mid v)_{L^2(\Omega)} = (u \mid v)_V \qquad \text{for } v \in V_0.$$

A function u is said to satisfy the Dirichlet condition on $\dot{\Omega}_d$, if and only if it satisfies (B_d). Similarly, a function u is said to satisfy the Neumann condition on $\dot{\Omega}_n$, if and only if it satisfies (B_n).

We introduce finally two subspaces of V, namely,

$$H = \{u \in V: Pu \in L^2(\Omega)\}, \tag{5.13.6}$$

and

$$N = \{u \in V: Pu \in L^2(\Omega) \text{ and } u \text{ satisfies } (B_d) \text{ and } (B_n)\}. \tag{5.13.6$'$}$$

Thus N is the set of $u \in V_0 \cap H$ such that

$$(Pu \mid v)_{L^2(\Omega)} = (u \mid v)_V \qquad \text{for } v \in V_0. \tag{5.13.6$''$}$$

Membership of N involves a global condition (membership of $V_0 \cap H$) and the demand that (5.13.6″) be fulfilled. As we shall see later, (5.13.6″) has the nature of a boundary condition.

The Dirichlet-Neumann problem can now be stated in the following terms: Given $f \in L^2(\Omega)$, it is required to solve the system

$$u \in N, \ Pu = f. \tag{5.13.7}$$

EXISTENCE AND UNIQUENESS OF THE SOLUTION. Suppose that $f \in L^2(\Omega)$ is given. We assume that

(1) V_0 is complete for the structure induced on it by that of V, and

(2) $V_0 \subset L^2(\Omega)$ and there exists a number c such that

$$\|v\|_{L^2(\Omega)} \leq c \cdot \|v\|_V \qquad \text{for } v \in V_0. \tag{5.13.8}$$

Then the system (5.13.7) admits precisely one solution.

Proof. To establish the existence of a solution we make use of the projection principle. The inequality (5.13.8) shows that for $v \in V_0$ one has

$$|(f \mid v)_{L^2(\Omega)}|^2 \leq \|f\|^2_{L^2(\Omega)} \cdot \|v\|^2_{L^2(\Omega)} \leq \text{const. } \|v\|^2_V,$$

the constant possibly depending upon f and not exceeding $c^2 \|f\|^2_{L^2(\Omega)}$. Since V_0 is complete, by hypothesis (1), Subsection 1.12.7(2) shows that there exists an element u of V_0 such that

$$(f \mid v)_{L^2(\Omega)} = (u \mid v)_V \qquad \text{for } v \in V_0, \tag{5.13.9}$$

and

$$\|u\|_V \leq c \, \|f\|_{L^2(\Omega)}.$$

If in the first equation we take $v = \varphi \in \mathscr{D}(\Omega)$ and use (5.13.2), one obtains

$$(f \mid \varphi)_{L^2(\Omega)} = (u \mid \varphi)_V = \langle \bar{\varphi}, Pu \rangle,$$

or

$$\langle \bar{\varphi}, f \rangle = \langle \bar{\varphi}, Pu \rangle,$$

hence it follows that $Pu = f$. It remains only to verify that (5.13.6″) holds. But in (5.13.9) we may now write $f = Pu$, in which case (5.13.6″) appears at once.

As for the uniqueness, the difference, z, of two solutions of the system (5.13.7) satisfies $z \in N \subset V_0$ and $Pz = 0$. But then by (5.13.6″) one has

$$0 = (Pz \mid v)_{L^2(\Omega)} = (z \mid v)_V \qquad \text{for all } v \in V_0.$$

Taking $v = z$ there follows $\|z\|_V = 0$ and therefore, by, (5.13.8), $\|z\|_{L^2(\Omega)} = 0$, and so finally $z = 0$. ▌

5.13.3 The Pure Dirichlet Problem. The reader will notice that if (5.13.8) holds, then a function $u \in V$ such that $Pu \in L^2(\Omega)$ satisfies (B_n) provided

$$(Pu \mid v)_{L^2(\Omega)} = (u \mid v)_V$$

holds for all v in a dense subset of V_0. Moreover, if we take $\dot{\Omega}_d = \dot{\Omega}$, and if the $p_0, \cdots, p_n$ satisfy suitable conditions, it is possible to show that $\mathscr{D}(\Omega)$ is already dense in V_0. (See Garnir [1], pp. 94–97 for the discussion of a particular case.) By (5.13.2), N becomes simply the set of $u \in V_0$ such that $Pu \in L^2(\Omega)$, and the system (5.13.7) reduces to

$$u \in V_0, \; Pu = f.$$

This is a "pure" Dirichlet problem with vanishing boundary values. Here the boundary conditions are homogeneous and linear, while the LPDE is non-homogeneous.

Let us look now at the system

$$u \in V, u - g \in V_0, \qquad Pu = 0, \tag{5.13.10}$$

which is again a "pure" Dirichlet problem, this time the boundary conditions being nonhomogeneous and the LPDE being homogeneous. When $P = -\Delta$, this is the exact analogue of the classical Dirichlet problem for harmonic functions with assigned boundary values. As is shown in Exercise 5.8, this type of problem can be reduced to one of type (5.13.7). But, as we aim to show now, it may also be approached directly and more simply in a way that involves fewer hypotheses. In particular, we need not assume explicitly that (5.13.8) holds, nor need V_0 be specified in the preceding manner.

We assume merely that V is defined as in Subsection 5.13.1. We then have the following existence theorem.

(a) Let V_0 be any vector subspace of V which is such that $\mathscr{D}(\Omega) \subset V_0$ and which is complete for the structure induced by that of V. Let $g \in V$ be given. Then the system (5.13.10) possesses at least one solution.

Proof. By Theorem 1.12.3 and Proposition 1.12.4 one may project orthogonally g onto V_0 to get g', say. We put $u = g - g'$. Then $u \in V$ and $u - g = -g' \in V_0$. Also $(u \mid w)_V = 0$ for all $w \in V_0$. Taking $w = \varphi \in \mathscr{D}(\Omega)$ and using (5.13.2), it appears that $Pu = 0$. Thus u is a solution of (5.13.10). ▌

In addition one has the following uniqueness theorem.

(b) Let V_0 be a vector subspace of V that contains $\mathscr{D}(\Omega)$ as a dense subset, and suppose that V_0 is separated for the induced structure. We let $g \in V$ be given. Then the system (5.13.10) has at most one solution.

Proof. The difference, z, of two solutions satisfies $z \in V_0$ and $Pz = 0$. Then (5.13.2) shows that $(z \mid \varphi)_V = 0$ for all φ in $\mathscr{D}(\Omega)$. Since $\mathscr{D}(\Omega)$ is dense in V_0, it follows that $\|z\|_V^2 = 0$ and so, V_0 being separated, that $z = 0$. Hence the uniqueness of the solution. ▌

To consider an especially important special case, we take $p_0 = 0$ and $p_1 = \cdots = p_n = 1$, so that $P = -\Delta$. V is now the space of function classes $f \in L^1_{\text{loc}}(\Omega)$ for which $\partial_k f \in L^2(\Omega)$ for $k = 1, \cdots, n$. The inner product on V is

$$(f \mid g)_V = \int_\Omega \left(\sum_{k=1}^n \partial_k f \cdot \partial_k \bar{g} \right) d\mu = D(f, g)$$

and the square of the norm is

$$\|f\|_V^2 = \int_\Omega \left[\sum_{k=1}^n |\partial_k f|^2 \right] d\mu = D(f),$$

the usual Dirichlet integral expressions.

By virtue of the minimal distance property of orthogonal projection, the solution of (5.13.10) postulated in (a) above can be characterized by the relations

$$u \in V, u - g \in V_0, D(u) \leq D(w) \qquad \text{for } w \in V, w - g \in V_0.$$

In other words, if membership to V_0 is taken to signify the possession of vanishing boundary values in some generalized sense, then u minimizes $D(w)$ among all members w of V that assume the same boundary values as g.

This minimal property was used by Riemann in his proposed proof of the existence theorem, the difference being that he had in mind the variation of $D(w)$ as w ranges over all functions of class $C^1(\Omega)$ that are continuous on $\bar{\Omega}$ and that coincide on $\dot{\Omega}$ with a given continuous function on $\dot{\Omega}$. The difficulties attached to carrying out Riemann's proposal, pointed out by Weierstrass and crystallized by Hadamard, are too well known to call for comment here. The present technique circumvents these difficulties at the expense of modifying the meaning of both terms "solution" and "boundary values." It involves also the penalty that the new way of specifying boundary values excludes cases in which the classical solution exists. To see this one has only to consider an example produced by Hadamard, in which Ω is the unit disk with centre the origin in R^2 and g is the function

$$g = \sum_{r=1}^{\infty} 2^{-r} \rho^{2^{2r}} \cos(2^r \theta),$$

ρ and θ being polar coordinates. This function g is continuous on $\bar{\Omega}$ and harmonic in Ω, but $D(g) = +\infty$. Thus g does not belong to V and our existence theorem is not applicable. Equally, Riemann's proposed method fails since it can be shown that there exist no functions w that are continuous on $\bar{\Omega}$, belong to $C^1(\Omega)$, assume the same boundary values as g, and for which $D(w) < +\infty$.

It is thus fair to conclude that neither method based on consideration of the Dirichlet integral proves to be as satisfactory as alternative methods that have been developed for this particular problem by Wiener, Perron, and others.

There is an excellent connected account of the basic facts concerning each of the various approaches to the classical Dirichlet problem (the methods of balayage, of Perron, of integral equations, and of Riemann) in Chapter 7 of Epstein [1]. This account contains a detailed discussion of the rehabilitation of Riemann's proposed existence proof, together with an account of the relationship between abstractly formulated boundary conditions and their classical counterparts in the case of domains with sufficiently smooth frontiers (loc. cit. pp. 196–198).

5.13.4 **Continuation of Subsection 5.13.1.** Our aim now is to apply the results of Subsections 5.13.2 and 5.13.3 to the situation described in Subsection 5.13.1. With this end in view we must investigate the separatedness and completeness of the space V_0. Actually we shall give sufficient conditions in order that the larger space V be complete. Since V_0 is closed in V, the completeness of V entails that of V_0.

It is assumed throughout this subsection that the functions $p_0, \cdots, p_n$ satisfy these conditions:

$$p_0, \cdots, p_n, p_1^{-1}, \cdots, p_n^{-1} \text{ are locally integrable over } \Omega. \qquad (5.13.11)$$

Regarding separatedness, we observe that (5.13.11) ensures that $p_k > 0$ a.e. on Ω for $k = 1, \cdots, n$. Thus if f belongs to V and satisfies $\|f\|_V = 0$, then $\partial_k f = 0$ for $k = 1, \cdots, n$. Consequently f is a constant function c on Ω. The relation $\|f\|_V = 0$ then imposes the further condition that $|c|^2 \int p_0 \, d\mu = 0$. If p_0 is nonnegligible, this entails that $c = 0$ and hence $f = 0$. Thus V, and therefore V_0 too, is separated if p_0 is nonnegligible. Even if p_0 is negligible, it may still be true that V_0 is separated (without V being so, of course) by virtue of generalized boundary conditions satisfied by members of V_0.

The relation $V_0 \subset L^2(\Omega)$ and the inequality (5.13.8) are valid whenever ess $\inf_\Omega p_0 > 0$.

Turning to the question of completeness, we shall approach our conclusions via a number of lemmas, some of which have their own interest. The space $\mathscr{L}^1_{\text{loc}}(\Omega)$ of functions on Ω that are there locally integrable is a semimetrizable LCTVS when furnished with the seminorms

$$N_K(f) = \int_K |f| \, d\mu$$

obtained as K ranges over any base (K_j) for the compact subsets of Ω. That $\mathscr{L}^1_{\text{loc}}(\Omega)$ is complete is easily deduced from Theorem 4.6.7. The associated separated space (whose elements are classes of functions equal almost everywhere on Ω), denoted by $L^1_{\text{loc}}(\Omega)$, is a Fréchet space. We may, and will, regard $L^1_{\text{loc}}(\Omega)$ as a subspace of $\mathscr{D}'(\Omega)$.

Our main conclusion is as follows.

Proposition. If (5.13.11) holds, then V is complete.

The auxiliary lemmas we need are each special cases of results given by Deny and Lions ([1], p. 308, Théorème 1.1; p. 311, Théorème 2.2).

Lemma A. If (X_α) is a directed family in $\mathscr{D}'(\Omega)$ such that $(\partial_k X_\alpha)$ is strongly Cauchy in $\mathscr{D}'(\Omega)$ for $k = 1, \cdots, n$, then there exists an X in $\mathscr{D}'(\Omega)$ such that $\lim_\alpha \partial_k X_\alpha = \partial_k X$ for $k = 1, \cdots, n$.

Proof. $\mathscr{D}'(\Omega)$ is complete: this is a property that it shares with the strong dual of any bornological space (here $\mathscr{D}(\Omega)$). Regarding these points, see Section 7.3 and Theorem 8.4.13. It follows that $Y_k = \lim_\alpha \partial_k X_\alpha$ exists in $\mathscr{D}'(\Omega)$ for each k. Since differentiation is continuous,

$$\partial_j Y_k = \lim_\alpha \partial_j \, \partial_k X_\alpha = \lim_\alpha \partial_k \, \partial_j X_\alpha = \partial_k Y_j$$

for $k, j = 1, \cdots, n$. So (Theorem 5.9.3) there exists an X in $\mathscr{D}'(\Omega)$ such that $Y_k = \partial_k X$, as asserted. ▌

Lemma B. Let M denote the vector subspace of $\mathscr{D}'(\Omega)$ formed of distributions X on Ω such that $\partial_k X \in L^1_{\text{loc}}(\Omega)$ for $k = 1, 2, \cdots, n$. Then M is complete for the structure defined by the seminorms

$$p_K(X) = \sum_{k=1}^n N_K(\partial_k X) = \sum_{k=1}^n \int_K |\partial_k X| \, d\mu,$$

K ranging over any base (K_j) for the compact subsets of Ω. The associated separated space $M^\cdot$ is the quotient M/C, where C denotes the space of constant functions on Ω, and $M^\cdot$ is a Fréchet space.

Proof. It is clear that the aforesaid structure on M is that of a semi-metrizable LCTVS. In view of Proposition 5.9.1, it suffices to establish the completeness of M. Suppose then that (X_n) is a Cauchy sequence in M. By Lemma A above, there exists a distribution X on Ω such that $\partial_k X_n \to \partial_k X$ in $\mathscr{D}'(\Omega)$ as $n \to \infty$, and this for each $k = 1, \cdots, n$. For each k, however, the sequence $(\partial_k X_n)$ is Cauchy in $L^1_{\text{loc}}(\Omega)$, hence convergent in that space. Its limit in that space can only be $\partial_k X$. Thus $\partial_k X$ belongs to $L^1_{\text{loc}}(\Omega)$, and is the limit in that space of $(\partial_k X_n)$. Thus (X_n) is convergent in M to X and completeness is established. ▌

Our final lemma demands a few preliminary remarks. In view of the conclusions reached in Section 5.11, we may say that each element of M is necessarily a member of $L^1_{\text{loc}}(\Omega)$. This being so, we may consider on M the structure defined by the seminorms

$$q_K(f) = N_K(f) + p_K(f),$$

K again ranging over any base for the compact subsets of Ω. On the quotient space $M^\cdot = M/C$ we shall have correspondingly two structures: that derived from the structure derived from the p_K, which we may term the p-structure, and that derived from the q_K, which we term the q-structure. Completeness for the p-structure has already been established in Lemma B. Completeness for the q-structure, which is obviously the stronger of the two, is easier to establish; we leave this task to the reader.

Lemma C. On $M^\cdot$, the p- and the q-structures are identical.

Proof. We consider the identity map of $(M^\cdot, q)$ onto $(M^\cdot, p)$. This is one to one and continuous. Since both spaces are Fréchet spaces, the open mapping theorem (in the form of Theorem 6.4.4, for example) asserts just what is required. ▌

Remark. We shall need the implication of Lemma C in the following guise. If (f_n) is a sequence in M that is Cauchy for the p-structure, then there exists a sequence (c_n) of constants such that $(f_n + c_n)$ is Cauchy, and therefore convergent, in $L^1_{\text{loc}}(\Omega)$.

Proof of the Proposition. Suppose that (f_n) is a Cauchy sequence in V. Let us observe first that the Cauchy-Schwartz inequality yields

$$\int_K |\partial_k f| \, d\mu \leq \int_K p_k^{-1} \, d\mu \cdot \int_K p_k \, |\partial_k f|^2 \, d\mu$$

for any compact subset $K \subset \Omega$ and for $k = 1, 2, \cdots, n$. This, by virtue of our condition (5.13.11), shows that $V \subset M$ and that V's structure is finer than that induced by the p-structure on M. Therefore, by the remark following Lemma C, there is a sequence (c_n) of constants such that $(f_n + c_n)$ is convergent in $L^1_{\text{loc}}(\Omega)$. At the same time, of course, we infer from the completeness of M the existence of an element f^a of M such that $\lim_{n\to\infty} \partial_k f_n = \partial_k f^a$ in $L^1_{\text{loc}}(\Omega)$ for each k; f^a belongs to $L^1_{\text{loc}}(\Omega)$.

Since $(p_k^{1/2} \, \partial_k f_n)$ is Cauchy in $L^2(\Omega)$, we may conclude (using Fatou's Lemma

4.5.4) that indeed $p_k^{1/2}\,\partial_k f^a$ belongs to $L^2(\Omega)$ and is the limit in that space of the sequence $(p_k^{1/2}\,\partial_k f_n)$: this for each k.

Now if p_0 is negligible, there is nothing more to prove: (f_n) is then convergent in V to f^a. So we shall suppose hereafter that p_0 is nonnegligible.

The V-Cauchy character of (f_n) involves, in addition to what we have thus far used, the Cauchy character of $(p_0^{1/2}f_n)$ in $L^2(\Omega)$. This being so we may call upon Theorem 4.6.7, modified as indicated in Subsection 4.11.7 so as to apply to $L^2(\Omega)$, and upon Fatou's lemma once again, to deduce the existence of a measurable function f^b on Ω such that $p_0^{1/2}f_n \to p_0^{1/2}f^b$ in $L^2(\Omega)$.

Since $(f_n + c_n)$ is convergent in $L^1_{\text{loc}}(\Omega)$, Theorem 4.6.7 can be used to establish the possibility of extracting a subsequence $(f'_r + c'_r) = (f_{n_r} + c_{n_r})$ which converges almost everywhere on Ω. (We are here sinning to the extent of using the symbol f_{n_r} to denote a representative function chosen from the class f_{n_r}: the reader is asked for his indulgence here and elsewhere, the alternative being an extremely cumbersome notation.) Then the sequence of terms $p_0^{1/2}f'_r + p_0^{1/2}c'_r$ is convergent almost everywhere on Ω. But $p_0^{1/2}f'_r$ converges to $p_0^{1/2}f^b$ in $L^2(\Omega)$, and so a sub-subsequence $(f''_s) = (f'_{n_{r_s}})$ exists such that $p_0^{1/2}f''_s$ converges almost everywhere on Ω to $p_0^{1/2}f^b$. It appears then that $(p_0^{1/2}c''_s)$ is convergent almost everywhere on Ω and so, p_0 being nonnegligible, (c''_s) is convergent. Consequently (f''_s) is convergent in $L^1_{\text{loc}}(\Omega)$ and almost everywhere on Ω; let f be its limit.

We have $p_0^{1/2}f''_s \to p_0^{1/2}f$ a.e. on Ω and, at the same time, this same sequence converges almost everywhere on Ω to $p_0^{1/2}f^b$. The conclusion is that $p_0^{1/2}f$ and $p_0^{1/2}f^b$ are the same (classes), so that $p_0^{1/2}f''_s \to p_0^{1/2}f$ in $L^2(\Omega)$.

Since f''_s converges to f in $L^1_{\text{loc}}(\Omega)$, a fortiori in $\mathscr{D}'(\Omega)$, so $\partial_k f''_s \to \partial_k f$ in $\mathscr{D}'(\Omega)$. But $\partial_k f''_s$ (as a subsequence of $\partial_k f_n$) converges in $L^1_{\text{loc}}(\Omega)$ to $\partial_k f^a$. Therefore $\partial_k f$ and $\partial_k f^a$ are the same (classes), and this for each k.

Finally, therefore, we have that $p_0^{1/2}f''_s \to p_0^{1/2}f$ and $p_k^{1/2}\,\partial_k f''_s \to p_k^{1/2}\,\partial_k f$ in $L^2(\Omega)$, for $k = 1, 2, \cdots, n$. Since we know that f belongs to $L^1_{\text{loc}}(\Omega)$, it appears that f belongs to V and is the limit in that space of the subsequence (f''_s) of (f_n). (f_n) being Cauchy in V, it follows on general grounds that it too is convergent in V to f.

Having thus shown that each Cauchy sequence in V is convergent in V, the proposition is proved. ▌

Remarks. A little thought will make it clear that further hypotheses on p_0 and the p_k would render the proof a good deal simpler. Most of the difficulty attaches to the case in which p_0, although nonnegligible, is not "sufficiently distinct from being negligible" for one to be able to infer at once from the Cauchy character of (f_n) in V the convergence of (f_n) in $L^1_{\text{loc}}(\Omega)$. If we assume that p_0^{-1} is locally integrable, this inference is valid and the proof proceeds without complication.

5.13.5 **The Green's Operator.** The hypotheses are as in Subsections 5.13.1 and 5.13.2. We introduce the map G of $L^2(\Omega)$ into N which assigns to

$f \in L^2(\Omega)$ the solution $u = Gf$ of the system (5.13.7). It is evident that G is linear. Also, as appears from the arguments given in Subsection 5.13.2,

$$\|Gf\|_V \leq c \, \|f\|_{L^2(\Omega)},$$

and

$$(Gf \mid v)_V = (f \mid v)_{L^2(\Omega)} \qquad \text{for } f \in L^2(\Omega),\ v \in V_0. \tag{5.13.12}$$

This last relation suffices to characterize Gf in V_0. By definition of G one has

$$PGf = f \qquad \text{for } f \in L^2(\Omega). \tag{5.13.13}$$

Uniqueness of the solution of the Dirichlet-Neumann problem implies furthermore that

$$GPu = u \qquad \text{for } u \in N. \tag{5.13.14}$$

It thus appears that P is an isomorphism (algebraic and topological) of N, endowed with the structure defined by the inner product

$$(u \mid u')_N = (u \mid u')_V + (Pu \mid Pu')_{L^2(\Omega)},$$

onto $L^2(\Omega)$, its inverse being G.

If in (5.13.12) we take $v = Gf$ and use (5.13.8), it becomes apparent that $(Gf \mid f)_{L^2(\Omega)}$ is real and that

$$(Gf \mid f)_{L^2(\Omega)} \geq c^{-2} \, \|Gf\|^2_{L^2(\Omega)} \qquad \text{for } f \in L^2(\Omega). \tag{5.13.15}$$

Again, if one takes in (5.13.12) $v = Gg$ with $g \in L^2(\Omega)$, it appears that

$$(f \mid Gg)_{L^2(\Omega)} = \overline{(g \mid Gf)}_{L^2(\Omega)} = (Gf \mid g)_{L^2(\Omega)},$$

showing that G, considered as an endomorphism of $L^2(\Omega)$, is self-adjoint. Then (5.13.15) shows that G is a continuous endomorphism of $L^2(\Omega)$,

$$\|Gf\|_{L^2(\Omega)} \leq c^2 \, \|f\|_{L^2(\Omega)} \qquad \text{for } f \in L^2(\Omega), \tag{5.13.16}$$

and that moreover G is a strictly positive self-adjoint endomorphism of $L^2(\Omega)$.

As we have already noted,

$$\|Gf\|_V \leq c \, \|f\|_{L^2(\Omega)} \qquad \text{for } f \in L^2(\Omega), \tag{5.13.17}$$

and therefore, by (5.13.8),

$$\|Gv\|_V \leq c^2 \, \|v\|_V \qquad \text{for } v \in V_0. \tag{5.13.18}$$

We call G the Green's operator for the system (5.13.7).

Since, under the hypotheses given in Subsection 5.13.2, V_0 is contained in $L^2(\Omega)$, we are at liberty to consider the restriction $\mathscr{G}$ of G to V_0. Since G maps L^2 onto $N \subset V_0$, we may regard $\mathscr{G}$ as an endomorphism of V_0. From (5.13.12) it then appears that

$$(\mathscr{G}v \mid v')_V = (v \mid v')_{L^2} \qquad \text{for } v, v' \in V_0. \tag{5.13.19}$$

By (5.13.18), $\mathscr{G}$ is continuous:

$$\|\mathscr{G}v\|_V \leq c^2 \, \|v\|_V \qquad \text{for } v \in V_0. \tag{5.13.20}$$

At the same time, (5.13.19) shows that $\mathscr{G}$ is positive and self-adjoint. It is an algebraic isomorphism of V_0 into $N \subset V_0$ and one has

$$P\mathscr{G}v = v \qquad \text{for } v \in V_0, \tag{5.13.21}$$

$$\mathscr{G}Pu = u \qquad \text{for } u \in N,\ Pu \in V_0. \tag{5.13.22}$$

5.13.6 It is sometimes the case that not only is the injection of V_0 into L^2 continuous (as expressed by (5.13.8)), but that this injection is even compact (=completely continuous). Whenever this is so, $\mathscr{G}$ proves to be a compact endomorphism of V_0 since $\mathscr{G} = G \circ i$, where G is continuous from L^2 into V_0 and i is the compact injection of V_0 into L^2. When additionally V_0 is a Hilbert space, the general theory of compact endomorphisms of such spaces furnishes a good deal more information about $\mathscr{G}$ and its eigenvectors (or eigenfunctions in the case of linear partial differential operators). The general theory of compact linear maps is dealt with in Chapter 9, and the aspect just mentioned taken up in Section 9.12.

5.13.7 For the sake of easy reference from the next subsection, we collect here simple conditions sufficient to ensure that the results of Subsections 5.13.2 and 5.13.5 are applicable.

First, to ensure that V_0 is complete, it suffices to assume that

$$p_0, \cdots, p_n, p_1^{-1}, \cdots, p_n^{-1} \in \mathscr{L}^1_{\text{loc}}(\Omega). \tag{5.13.11}$$

Second, to ensure that $V_0 \subset L^2(\Omega)$ and that (5.13.8) is valid, it suffices to assume that

$$m_0 = \text{ess inf}_\Omega\, p_0 > 0; \tag{5.13.23}$$

then (5.13.8) holds with $c = m_0^{-1/2}$.

5.13.8 **The Equation** $Pu + \lambda u = f$. It sometimes happens that interest is pointed not only towards the equation $Pu = f$, but also towards the equation $Pu + \lambda u = f$, where λ is a scalar (real or complex).

For example, if the primary concern is the wave equation

$$\Delta U = \frac{\partial^2 U}{\partial t^2},$$

Δ being the Laplacian in the space variables, the first step is often the search for simple harmonic solutions $U(x, t) = u(x)e^{ikt}$. Then u is required to satisfy the so-called metaharmonic equation

$$(-\Delta + \lambda)u = 0,$$

where $\lambda = -k^2$, which contains the harmonic (Laplace's) equation as a special case ($\lambda = 0$).

The abstract treatment of the equation $Pu + \lambda u = f$ will break quite naturally into two stages. In the first stage it is assumed that λ is real and positive. Then the method of orthogonal projection applies without modification: indeed, the method often applies to the case when $\lambda > 0$ more directly than to the case $\lambda = 0$ (the conditions enumerated in Subsection 5.13.7 being easier to fulfill when $\lambda > 0$ than when $\lambda = 0$). The next stage uses simple properties of endomorphisms of a Hilbert space to extend the family of associated Green's operators G_λ, obtained for $\lambda > 0$ in the first stage, to certain complex values of λ as well. The family G_λ, thus extended, continues to resolve the associated system (5.13.7) ($P + \lambda$ replacing P). The reader is referred to Garnir [1] for a more detailed study of the family G_λ than we have time to undertake.

In this subsection it will suffice to assume that (5.13.11) holds. Then V is defined as at the outset of Subsection 5.13.1, so that (5.13.5) holds. V may or may not be either complete or separated.

This being so, let $\lambda > 0$ and let V^λ be constructed as was V, with the sole difference that p_0 is replaced by $p_0 + \lambda$. Then $V^\lambda \subset V$, $V^\lambda \subset L^2(\Omega)$, and we define the inner product on V^λ by the equation

$$(f \mid g)_\lambda = \lambda(f \mid g)_{L^2(\Omega)} + (f \mid g)_V.$$

Then (5.13.8) is satisfied when V^λ replaces V and $v \in V^\lambda$, the number c being at most $\lambda^{-1/2}$. Furthermore (5.13.2) holds now with $P + \lambda$ in place of P.

In view of the inequality

$$\frac{\lambda}{\lambda'} \leq \frac{t + \lambda}{t + \lambda'} \leq 1,$$

valid for $0 < \lambda < \lambda'$ and $t \geq 0$, one sees that the V^λ are identical as sets and that their uniform structures are identical. We therefore write V^+ in place of V^λ. The structure of V^+ is stronger than that induced by the structure of V.

V_0^+, the closure in V^+ of the set of functions in V^+ that vanish on the intersection of Ω with variable neighbourhoods of $\dot{\Omega}_d$, is no larger than V_0. The appropriate boundary conditions can now be written

$$(B_d^+) \qquad u \in V_0^+,$$

$$(B_n^+) \qquad u \in V^+,\ Pu \in L^2(\Omega), \text{ and}$$

$$(Pu \mid v)_{L^2(\Omega)} = (u \mid v)_V \qquad \text{for } v \in V_0^+.$$

Correspondingly we write

$$H^+ = \{u \in V^+ \colon Pu \in L^2(\Omega)\},$$

$$N^+ = \{u \in H^+ \colon u \text{ satisfies } (B_d^+) \text{ and } (B_n^+)\};$$

The reader will notice that $H^+ \subset H$ and $N^+ \subset N$.

The results of Subsections 5.13.2 and 5.13.5 apply to this new situation and we denote by G_λ and $\mathscr{G}_\lambda$ the corresponding Green's operators. These have the following properties, the first of which serves to characterize $G_\lambda f$:

$$\lambda(G_\lambda f \mid v)_{L^2(\Omega)} + (G_\lambda f \mid v)_V = (f \mid v)_{L^2(\Omega)} \qquad (f \in L^2(\Omega),\ v \in V_0^+);$$

$$(P + \lambda)G_\lambda f = f \qquad (f \in L^2(\Omega));$$

$$G_\lambda(P + \lambda)u = u \qquad (u \in N^+);$$

$$(P + \lambda)\mathscr{G}_\lambda v = v \qquad (v \in V_0^+);$$

$$\mathscr{G}_\lambda(P + \lambda)u = u \qquad (u \in N^+,\ Pu \in V_0^+);$$

$$\|G_\lambda\|_{L^2(\Omega),V^+} \leq \lambda^{1/2}, \qquad \|G_\lambda\|_{L^2(\Omega),L^2(\Omega)} \leq \lambda^{-1}, \qquad \|\mathscr{G}_\lambda\|_{V_0^+,V_0^+} \leq \lambda^{-1};$$

in the last three inequalities we are using the notation $\|T\|_{A,B}$ to denote the norm of the linear map T from one normed vector space, A, into another such space, B. Moreover, G_λ is a continuous, positive self-adjoint endomorphism of $L^2(\Omega)$, since, in fact,

$$(G_\lambda f \mid f)_{L^2(\Omega)} \geq \lambda \, \|G_\lambda f\|^2_{L^2(\Omega)}.$$

Actually G_λ is strictly positive since $G_\lambda f = 0$ gives $f = (P + \lambda)G_\lambda f = 0$, so that

$$(G_\lambda f \mid f)_{L^2(\Omega)} > 0 \qquad \text{if } f \in L^2(\Omega), f \neq 0.$$

From these identities there follows a relationship between the various G_λ with $\lambda > 0$ that is crucial for the second stage extension of the family G_λ. If f belongs to L^2 and $\lambda > 0$, $\lambda_0 > 0$, one has

$$\begin{aligned} G_\lambda f &= G_\lambda(P + \lambda_0)G_{\lambda_0} f = G_\lambda[(P + \lambda) - (\lambda - \lambda_0)]G_{\lambda_0} f \\ &= G_\lambda(P + \lambda)G_{\lambda_0} f - (\lambda - \lambda_0)G_\lambda G_{\lambda_0} f \end{aligned}$$

and so, since $u = G_{\lambda_0} f$ belongs to N^+,

$$G_\lambda - G_{\lambda_0} = -(\lambda - \lambda_0)G_\lambda G_{\lambda_0} \qquad \text{for } \lambda > 0,\ \lambda_0 > 0.$$

In particular, the various $G_\lambda (\lambda > 0)$ commute. Moreover

$$G_\lambda + (\lambda - \lambda_0)G_\lambda G_{\lambda_0} = G_{\lambda_0} = G_\lambda + (\lambda - \lambda_0)G_{\lambda_0} G_\lambda \tag{5.13.24}$$

whenever $\lambda > 0$ and $\lambda_0 > 0$.

As we shall now see, this relationship allows us to extend the definition of G_λ from $\lambda > 0$ to more general complex values of λ.

Lemma. Let E be a complete inner-product space and T a positive self-adjoint endomorphism of E. If γ is a complex number that is not real and strictly negative, then $(1 + \gamma T)^{-1}$ exists as a continuous endomorphism of E.

Proof. γ may be written as $\gamma = \alpha + i\beta$ where α and β are real and *either* (1) $\beta \neq 0$, *or* (2) $\beta = 0$ and $\alpha \geq 0$. For any f in E,

$$\|(1 + \gamma T)f\|^2 = \|f\|^2 + 2\alpha(Tf \mid f) + |\gamma|^2 \, \|Tf\|^2,$$

since $(Tf \mid f) = (f \mid Tf)$ is real. Thus,

$$\|(1 + \gamma T)f\|^2 \geq \|f\|^2 - 2\,|\alpha|\, \|Tf\| \cdot \|f\| + (\alpha^2 + \beta^2)\, \|Tf\|^2.$$

If $\|f\| \neq 0$, this gives

$$\|(1 + \gamma T)f\|^2 \geq \|f\|^2\{1 - 2\,|\alpha|\, t + (\alpha^2 + \beta^2)t^2\},$$

where $t = \|Tf\|/\|f\|$. The expression on the right-hand side is never less than $\beta^2(\alpha^2 + \beta^2)^{-1} \|f\|^2$. If $\beta = 0$, so that $\alpha \geq 0$, the same expression is never less than $\|f\|^2$, thanks to the fact that $(Tf \mid f) \geq 0$. Thus in either case, (1) or (2), we have

$$\|(1 + \gamma T)f\|^2 \geq C\, \|f\|^2, \tag{5.13.25}$$

where $C = C(\gamma) > 0$. So in either case the range M of $1 + \gamma T$ is closed in E. On the other hand M is dense in E because if g is an element of E orthogonal to M, the relation $((1 + \gamma T)f \mid g) = 0$ gives $(f \mid (1 + \bar{\gamma} T)g) = 0$, and this for all f—that is, $\|(1 + \bar{\gamma}T)g\| = 0$. But then (5.13.25), which is valid equally well for $\bar{\gamma}$ as for γ, would yield the conclusion $\|g\| = 0$. That M is dense in E follows then from the Hahn-Banach theorem.

M must therefore coincide with E, and then (5.13.25) affirms that $(1 + \gamma T)^{-1}$ is a continuous endomorphism of E. ▌

EXTENSION OF THE FAMILY G_λ. Let R_- denote the set of real numbers r satisfying $r \leq 0$, C the set of complex numbers. Our object is to define G_λ for λ in $C \backslash R_-$.

For any such λ one may choose $\lambda_0 > 0$ in many ways to arrange that $\lambda - \lambda_0$ is not real and strictly negative. If this is done, since G_{λ_0} is a positive self-adjoint endomorphism of L^2, the lemma shows that

$$[1 + (\lambda - \lambda_0)G_{\lambda_0}]^{-1}$$

exists as a continuous endomorphism of L^2. So, taking our cue from (5.13.24), we agree to *define* G_λ by the relation

$$G_\lambda = G_{\lambda_0}[1 + (\lambda - \lambda_0)G_{\lambda_0}]^{-1} = [1 + (\lambda - \lambda_0)G_{\lambda_0}]^{-1}G_{\lambda_0}, \qquad (5.13.26)$$

which will be consistent with the existing definition of G_λ whenever λ is real and strictly positive (as follows from the relation (5.13.24), established for $\lambda > 0$ and $\lambda_0 > 0$). It is easy to verify that this method of defining G_λ does not depend on the particular choice of $\lambda_0 > 0$ making $\lambda - \lambda_0$ distinct from all real and strictly negative numbers. (Here again the verification depends on (5.13.24).)

Now (5.13.26) will show that G_λ is holomorphic on $C \backslash R_-$, in the sense that for each pair (f, g) in $L^2 \times L^2$ the complex-valued function $\lambda \to (G_\lambda f \mid g)_{L^2}$ is holomorphic on $C \backslash R_-$. To see this, we suppose λ is a point of $C \backslash R_-$ and choose λ_0 as has been said. Then, if $\lambda' = \lambda + \varepsilon$ with $|\varepsilon|$ sufficiently small, the same λ_0 can be used in defining $G_{\lambda'}$. We write temporarily $T = 1 + (\lambda - \lambda_0)G_{\lambda_0}$. If f and g belong to L^2, (5.13.26) gives

$$(G_{\lambda'}f \mid g)_{L^2} - (G_\lambda f \mid g)_{L^2} = (G_{\lambda_0}(T + \varepsilon G_{\lambda_0})^{-1}f \mid g)_{L^2} - (G_{\lambda_0}T^{-1}f \mid g)_{L^2}.$$

If $|\varepsilon|$ is so small that $|\varepsilon| \cdot \|G_{\lambda_0}\|_{L^2,L^2} < \|T^{-1}\|_{L^2,L^2}^{-1}$, the series

$$T^{-1}\sum_{n \geq 0}(-1)^n(\varepsilon T^{-1}G_{\lambda_0})^n$$

is convergent in the sense of the norm $\| \cdot \|_{L^2,L^2}$, and its sum is none other than $(T + \varepsilon G_{\lambda_0})^{-1}$. As a consequence we see that

$$(G_{\lambda'}f \mid g)_{L^2} - (G_\lambda f \mid g)_{L^2} = \sum_{n \geq 0}(-1)^n\varepsilon^n(G_{\lambda_0}T^{-1}(T^{-1}G_{\lambda_0})^n f \mid g)_{L^2},$$

the series on the right being necessarily convergent for sufficiently small $|\varepsilon|$. Holomorphy in a neighbourhood of λ is thereby established.

Once this point is made, we can apply the principle of analytic continuation to deduce that (5.13.24) continues to hold whenever λ and λ_0 both belong to $C \backslash R_-$ (extension of the admissible values of λ and of λ_0 being made in sequence).

Finally, we infer that $u = G_\lambda f$ is a solution of the system

$$u \in N^+, \ (P + \lambda)u = f \tag{5.13.7'}$$

whenever λ belongs to $C\backslash R_-$ and f to L^2. To verify this we need only show that

$$((P + \lambda)G_\lambda f \mid g)_{L^2} = (f \mid g)_{L^2} \tag{5.13.27}$$

for all such λ and all pairs (f, g) in $L^2 \times L^2$. Now the left-hand side here is, for fixed f and g, holomorphic on $C\backslash R_-$. The relation is known to hold when λ is real and strictly positive. The principle of analytic continuation therefore ensures its validity for λ in $C\backslash R_-$. On the other hand, that $G_\lambda f$ belongs to N^+ whenever λ belongs to $C\backslash R_-$ and f to L^2 follows from (5.13.26) if we take therein λ_0 to be real and strictly positive, for G_{λ_0} is then known already to map L^2 into N^+.

5.13.9 **Remarks on the Spaces V and on the Boundary Conditions.** We begin with some comments on the nature of the elements of V. Let us recall from Subsection 5.13.1 that V is the set of function classes f in $L^1_{\text{loc}}(\Omega)$ such that

(1) $\partial_k f \in L^1_{\text{loc}}(\Omega)$ for $k = 1, 2, \cdots, n$,

and

(2) $p_0^{1/2} f \in L^2(\Omega)$, $p_k^{1/2}\, \partial_k f \in L^2(\Omega)$ for $k = 1, 2, \cdots, n$.

(a) NATURE OF THE ELEMENTS OF V WHEN $n = 1$. In this case, as we shall see directly, the whole procedure can be discussed quite satisfactorily without explicit mention of distributions or of any related concept of "weak" derivatives.

The domain Ω is now simply a linear interval (α, β), bounded or not. In place of p_0 and p_1 we shall adopt a more usual notation—namely, q and p.

The definition of V makes it the set of function classes f belonging to $L^1_{\text{loc}}(\Omega)$ such that ∂f belongs to $L^1_{\text{loc}}(\Omega)$, while $q^{1/2} f$ and $p^{1/2}\, \partial f$ belong to $L^2(\Omega)$. Here ∂f is initially interpreted in the sense of distributions, but we shall see that we can legitimately regard elements of V as single functions and interpret the derivatives in the almost everywhere pointwise sense.

Indeed, since the class ∂f belongs to $L^1_{\text{loc}}(\Omega)$, the class f contains precisely one continuous function, which we use to represent this class and which we shall continue to denote by f. Moreover, this representative function f is even locally absolutely continuous† on Ω, so that there exists a locally integrable function f' on Ω such that

$$f(x) - f(x') = \int_{x'}^{x} f'\, d\mu \qquad (x, x' \in \Omega); \tag{5.13.28}$$

the continuous function f has almost everywhere on Ω a pointwise derivative that is equal almost everywhere on Ω to f'. Only the class of f' is uniquely determined, of course; but whichever representative function f' we choose, it

† By this we mean that f is absolutely continuous in the usual sense on each compact subinterval of Ω.

must be the case that $p^{1/2}f'$ and $q^{1/2}f$ are members of $\mathscr{L}^2(\Omega)$. All these facts flow from the results of Subsection 4.15.10, which themselves entail that ∂f (the distribution derivative of the class of the locally absolutely continuous function f) is (the class of) f'.

If we choose continuous (hence locally absolutely continuous) functions f and g representing two elements of V, the corresponding inner product is

$$(f \mid g)_V = \int_\Omega (pf'\bar{g}' + qf\bar{g})\, d\mu.$$

In view of these remarks there is, as we have predicted, no harm in regarding the elements of V as individual functions on Ω. We adopt this convention henceforth.

As we have seen in Subsections 5.13.2 and 5.13.3, the solubility and uniqueness of the solution of the various boundary value problems call for the completeness and separatedness of various subspaces of V. Despite the general results of Subsection 5.13.4, the case $n = 1$ is worth special consideration.

To begin with, the problems become almost entirely trivial (when $n = 1$) if q is negligible, the differential equation reducing then to the form $(pf')' = 0$. We shall suppose from now on that q is nonnegligible on Ω.

If the element f of V has zero norm, then

$$\int_\Omega p\, |f'|^2\, d\mu = \int_\Omega q\, |f|^2\, d\mu = 0.$$

If $p > 0$ a.e. on Ω, the first integral vanishes only if $f' = 0$ a.e.; then (5.13.28) shows that f is constant on Ω. Vanishing of the second integral then shows that $f = 0$. Thus V is separated if $p > 0$ a.e. on Ω; and this property is inherited by any vector subspace of V.

Again, let I be any subinterval of Ω and suppose that $\int_I p^{-1}\, d\mu < +\infty$. Then (5.13.28) and the Cauchy-Schwarz inequality gives

$$|f(x) - f(x_0)| \leq \left[\int_I p^{-1}\, du\right]^{1/2} \|f\|_V \qquad (x, x_0 \in I) \tag{5.13.29}$$

and therefore also

$$|f(x)| \leq |f(x_0)| + \left[\int_I p^{-1}\, d\mu\right]^{1/2} \|f\|_V \qquad (x, x_0 \in I). \tag{5.13.30}$$

If at the same time q is nonnegligible on I it may be deduced (Exercise 5.4) that

$$|f(x)| \leq \left[\left(\int_I q\, d\mu\right)^{-1} + \left(\int_I p^{-1}\, d\mu\right)^{1/2}\right] \|f\|_V \qquad (x \in I). \tag{5.13.31}$$

In particular, if q is nonnegligible on Ω and p^{-1} is locally integrable on Ω, then to each sufficiently large (and therefore any) compact interval $K \subset \Omega$ corresponds a number m_K such that

$$|f(x)| \leq m_K \|f\|_V \qquad (x \in K). \tag{5.13.31$'$}$$

Moreover, if q is nonnegligible on Ω and p^{-1} is integrable on Ω, then there exists a number m such that

$$|f(x)| \leq m \|f\|_V \qquad (x \in \Omega). \tag{5.13.31$''$}$$

Using these inequalities it is relatively simple to show that V is complete whenever q is nonnegligible and p^{-1} locally integrable. The same result has been proved for general values of n and under weaker assumptions in Subsection 5.13.4.

(b) NATURE OF THE ELEMENTS OF V WHEN $n > 1$. The results established in Subsection 5.11.1 show that any distribution $f \in \mathscr{D}'(\Omega)$ satisfying (1) above necessarily belongs to $L^1_{\text{loc}}(\Omega)$. Moreover for suitable domains Ω (for example, all Nikodym domains) it is the case that the relations

$$\text{ess inf}_\Omega \, p_k > 0 \qquad (k = 1, 2, \cdots, n)$$

entail that $V \subset L^2(\Omega)$. On the other hand this last inclusion is trivially true for any Ω, if

$$\text{ess inf}_\Omega \, p_0 > 0.$$

An especially important special case arises when

$$p_0 = p_1 = \cdots = p_n = 1;$$

the resulting space V is then $\mathscr{E}^1_{L^2}(\Omega)$, a Hilbert space with the inner product

$$(f \mid g)_{\mathscr{E}^1_{L^2}(\Omega)} = \int_\Omega f\bar{g} \, d\mu + \sum_{k=1}^n \int_\Omega \partial_k f \cdot \partial_k \bar{g} \, d\mu.$$

If

$$\operatorname*{ess\,inf}_\Omega p_r > 0 \qquad (r = 0, 1, \cdots, n),$$

then $V \subset \mathscr{E}^1_{L^2}(\Omega)$ with a stronger uniform structure. If also

$$\operatorname*{ess\,sup}_\Omega p_r < +\infty \qquad (r = 0, 1, \cdots, n),$$

then V and $\mathscr{E}^1_{L^2}(\Omega)$, together with their uniform structures, are identical (though their inner products may be numerically distinct).

Various structural properties of $V = \mathscr{E}^1_{L^2}(\Omega)$ and of its subspaces V_Σ are given in Garnir [1] (see especially pp. 87–101). Some of these results bear upon functions $f \in L^1_{\text{loc}}(\Omega)$ which are known merely to satisfy (1), and so apply to V whatever the choice of the p_r. A typical such result is the relation

$$\partial_k f^+ = (\partial_k f)\chi_{\{f>0\}},$$

and the conclusion that if f is a real function in V, then f^+, f^- and $|f|$ likewise belong to V.

(c) NATURE OF THE BOUNDARY CONDITIONS WHEN $n = 1$. As has been seen under (a), each element u of V may be regarded as an individual LAC function on $\Omega = (\alpha, \beta)$. The definition (5.13.6) of H shows that if $u \in H$, then $-\partial(p \, \partial u) + qu \in L^2(\Omega)$, gence it follows much as in (a) that the class ∂u contains just one function u' such that pu' is LAC; then pu' has almost everywhere a derivative g, and $-(pu)' + qu$ is the class of g. Thus the solutions of (5.13.7) may now be thought of as LAC functions u such that for some determination of u' the function pu' is LAC and such that $-(pu')' + qu$ is equal almost everywhere to f.

Let us next consider the nature of the subspace $V_0 = V_{\dot{\Omega}_d}$. In the present case $\dot{\Omega}_d$ is one of $\emptyset$, $\{\alpha\}$, $\{\beta\}$, $\{\alpha, \beta\}$. Nothing remains to be said if $\dot{\Omega}_d = \emptyset$. For the rest it suffices to suppose that $\alpha \in \dot{\Omega}_d$ and examine the significance of this assumption in terms of the boundary behaviour of functions u in V_0. The assumption that $\alpha \in \dot{\Omega}_d$ is meant to imply that $\alpha > -\infty$.

If we suppose that $\int_\alpha^{x_0} p^{-1}\, d\mu < +\infty$ for some (and hence all) $x_0 \in \Omega$, then inequality (5.13.29) in (a) shows that for each $u \in V_0$ the boundary limit

$$u(\alpha) = \lim_{x \downarrow \alpha} u(x)$$

exists finitely. If furthermore q is nonnegligible on Ω, then (5.13.31) of (a) can be used to show that necessarily $u(\alpha) = 0$ whenever $u \in V_0$. Whether or not any $u \in V$ satisfying $u(\alpha) = 0$ belongs to V_0 depends on the behaviour of p and q in the neighbourhood of α.

Let us next proceed to consider the meaning of the relationship $u \in N$, taking again $\dot{\Omega}_d = \{\alpha\}$. For $u \in H$ and $v \in V$ one has

$$(u \mid v)_V = \int_\Omega (pu'\bar{v}' + qu\bar{v})\, d\mu,$$

$$(pu \mid v)_{L^2(\Omega)} = \int_\Omega [-(pu')' + qu]\bar{v}\, d\mu.$$

By definition, $u \in N$ if and only if $u \in V_0$, $Pu \in L^2(\Omega)$ and $(u \mid v)_V = (Pu \mid v)_{L^2(\Omega)}$ for all $v \in V_0$. This last equality means that

$$\int_\Omega pu'\bar{v}'\, d\mu = -\int_\Omega (pu')'\bar{v}\, d\mu.$$

Now if $[a, b]$ is a compact subinterval of Ω, partial integration gives

$$\int_a^b (pu')'\bar{v}\, d\mu = [pu'v]_a^b - \int_a^b pu'\bar{v}'\, d\mu.$$

It follows that $u \in N$ if and only if $u \in V_0$, $Pu \in L^2(\Omega)$, and

$$\lim_{a \downarrow \alpha, b \uparrow \beta} [pu'v]_a^b = 0 \qquad \text{for } v \in V_0.$$

It thus appears that, subject to mild restrictions on p and q, the part of the relation $u \in N$ bearing upon the boundary behaviour of u amounts to the demand that

$$\lim_{x \downarrow \alpha} u(x) = 0, \qquad \lim_{x \uparrow \beta} p(x)u'(x) = 0,$$

two genuine boundary conditions on u. (We recall that u' is to be chosen so that pu' is LAC on Ω.)

Similar remarks apply if α and β are interchanged.

If $\dot{\Omega}_d = \{\alpha, \beta\}$, one arrives at the boundary conditions

$$\lim_{x \downarrow \alpha} u(x) = \lim_{x \uparrow \beta} u(x) = 0,$$

while if $\dot{\Omega}_d = \emptyset$, the boundary conditions are

$$\lim_{x \downarrow \alpha} p(x)u'(x) = \lim_{x \uparrow \beta} p(x)u'(x) = 0.$$

(d) Nature of the boundary conditions when $n > 1$. Here again the first question arising concerns the boundary behaviour of the elements of $V_0 = V_{\dot{\Omega}_d}$ for a given subset $\dot{\Omega}_d$ of $\dot{\Omega}$.

(1) First let us observe that if $f \in V$, then the membership of f to V_0 is genuinely a restriction on the behaviour of f only in the immediate neighbourhood of $\dot{\Omega}_d$. More precisely, let U be any neighbourhood in R^n of $\dot{\Omega}_d$, and let α be a function belonging to $C^1(\Omega)$ such that α and $\partial_k\alpha$ ($k = 1, 2, \cdots, n$) is bounded on Ω and such that $\alpha = 1$ on $\Omega \cap U$. It is claimed that if $f \in V$, then $f \in V_0$ if and only if $\alpha f \in V_0$. In fact if $f \in V$, then αf and $(1 - \alpha)f$ both belong to V; moreover $(1 - \alpha)f$ vanishes on $\Omega \cap U$ and therefore certainly belongs to V_0. The relationship $\alpha f = f - (1 - \alpha)f$ then shows that $\alpha f \in V_0$ if $f \in V_0$. Reciprocally, if $f \in V$ and $\alpha f \in V_0$, the equation $f = \alpha f + (1 - \alpha)f$ shows that $f \in V_0$. This establishes the assertion claimed.

It follows easily that if $f, g \in V$ are such that $f = g$ on $\Omega \backslash K$, where K is any compact subset of Ω, then f and g together belong to V_0 or not. This is seen on taking α in $C^1(R^n)$ with α and $\partial_k\alpha$ ($k = 1, 2, \cdots, n$) bounded, $\alpha = 0$ on K and $\alpha = 1$ on a neighbourhood U of $\dot{\Omega}$. Then $\alpha f = \alpha g$ on Ω, while $f \in V_0$ if and only if $\alpha f \in V_0$ and $g \in V_0$ if and only if $\alpha g \in V_0$.

(2) The boundary behaviour of elements of V_0 has been studied closely only in the case in which $p_r = 1$ ($r = 0, 1, \cdots, n$), so that $V = \mathscr{E}^1_{L^2}(\Omega)$, and $\dot{\Omega}_d = \dot{\Omega}$ (pure Dirichlet condition). The results we quote for this case apply equally whenever

$$0 < \text{ess inf}_{\Omega}\, p_r \leq \text{ess sup}_{\Omega}\, p_r < +\infty$$

for each r. In these cases $\mathscr{D}(\Omega)$ is dense in V_0, so that $V_0 = \mathscr{D}^1_{L^2}(\Omega)$, the closure in $\mathscr{E}^1_{L^2}(\Omega)$ of $\mathscr{D}(\Omega)$. For a proof of this see Garnir [1], p. 97.

For the detailed study of the elements of $\mathscr{E}^1_{L^2}(\Omega)$, see Schwartz [8], Exposés 12 and 13, Part 1, and also Deny-Lions [1], Ch. I, §7. The principal relevant result is that if the frontier $\dot{\Omega}$ is sufficiently smooth, there exists a continuous linear map b of $\mathscr{E}^1_{L^2}(\Omega)$ into $L^1_{\text{loc}}(\dot{\Omega}, \sigma)$ which, for functions of class C^∞ on some neighbourhood in R^n of Ω, coincides with the operation of restriction to $\dot{\Omega}$. (We use σ to denote the surface measure on $\dot{\Omega}$.) Besides this, $\mathscr{D}^1_{L^2}(\Omega)$ coincides exactly with the set of $f \in \mathscr{E}^1_{L^2}(\Omega)$ for which $bf = 0$. For some more general results of this type, see Calderón [1].

It is worth noting that $\mathscr{D}^1_{L^2}(\Omega)$ may coincide with $\mathscr{E}^1_{L^2}(\Omega)$; this is so if and only if $R^n \backslash \Omega$ has zero capacity (see Schwartz [8], Exposé 15). When this happens, there is in reality no freedom in the choice of V_0 in Subsection 5.13.3.

(3) Turning to the consideration of the condition (B_n), let us first attend to cases in which $\dot{\Omega}_d = \dot{\Omega}$ and $\mathscr{D}(\Omega)$ is dense in V_0 (see (2) above). One would then expect that the condition

$$(Pu \mid v)_{L^2(\Omega)} = (u \mid v)_V \qquad \text{for } v \in V_0$$

would be satisfied by any $u \in V_0$ such that $Pu \in L^2(\Omega)$. Now (5.13.2) asserts that the said relation holds for v in $\mathscr{D}(\Omega)$. Therefore if (5.13.8) holds, so that both sides are continuous in $v \in L^2(\Omega)$, the relation will indeed continue to hold for all v in V_0. In this case therefore

$$N = \{u \in V_0 \colon Pu \in L^2(\Omega)\}$$

and the system (5.13.7) is equivalent to

$$u \in V_0, \ Pu = f,$$

a pure Dirichlet problem.

(4) The purely formal approach to the meaning of (B_n) proceeds by use of Green's formula, which gives

$$\begin{aligned}(u \mid v)_V &= \int_\Omega p_0 u\bar{v}\, d\mu + \sum_{k=1}^{n} \int_\Omega p_k\, \partial_k u \cdot \partial_k \bar{v}\, d\mu \\ &= \int_\Omega p_0 u\bar{v}\, d\mu - \int_\Omega \sum_k \partial_k(p_k\, \partial_k u) \cdot \bar{v}\, d\mu + \int_{\dot{\Omega}} \sum_k N_k p_k\, \partial_k u \cdot \bar{v}\, d\sigma.\end{aligned}$$

where σ is the surface measure on $\dot{\Omega}$ and $(N_1, \cdots, N_n)$ the unit normal to $\dot{\Omega}$. Accordingly

$$(u \mid v)_V = (Pu \mid v)_{L^2(\Omega)} + \int_{\dot{\Omega}} \left(\sum_k N_k p_k\, \partial_k u\right) \bar{v}\, d\sigma.$$

Thus $u \in N$ signifies formally that $u \in V_0$, $Pu \in L^2(\Omega)$ and

$$\int_{\dot{\Omega}} \left(\sum_k N_k p_k\, \partial_k u\right) \bar{v}\, d\sigma = 0 \qquad \text{for } v \in V.$$

Since $v \in V_0$ has zero boundary values on $\dot{\Omega}_d$, one would anticipate that this last formula amounts to demanding that

$$\sum_k N_k p_k\, \partial_k u = 0 \qquad \text{on } \dot{\Omega}_n.$$

When p_k is a nonzero constant independent of k, this means the vanishing on $\dot{\Omega}_n$ of the normal derivative of u, the original Neumann-type boundary condition.

5.13.10 Remarks. The methods used in this section to deal with certain second-order linear partial differential equations can be extended to some types of equations of higher order. See Vischik [1], Lions [1]. A simple example is dealt with in Exercise 5.24.

5.14 Existence of an Elementary Solution; Discussion of Problem 2

Suppose the domain Ω in R^n is given, together with a linear partial differential operator

$$P(x, D) = \sum_{|p| \leq m} a_p(x) D^p, \qquad (5.14.1)$$

the a_p being given (sufficiently smooth) functions on Ω. Then $P(x, D)$ may be applied to any distribution on Ω. We denote by P the endomorphism of $\mathscr{D}'(\Omega)$ thus obtained. The solubility of the equation

$$PX = P(x, D)X = A, \tag{5.14.2}$$

for given A in $\mathscr{D}'(\Omega)$, will be examined further in Section 5.17. Meanwhile we are primarily interested in the case in which $A = \varepsilon_a$, a being a specified point of Ω. In this case a solution X of (5.14.2) is termed an elementary (or fundamental) solution for $P(x, D)$ relative to a. The techniques that have been developed a propos the search for elementary solutions of linear partial differential operators P with constant coefficients, and that will be set forth in part in this section, do not extend directly either to the case in which A is more general or to that in which $P(x, D)$ has variable coefficients. This explains the separate treatment of the restricted problem. A slightly different and more constructive approach is discussed in Subsection 5.18.3 below. See also Hörmander [7].

The methods described in this section, however, do extend satisfactorily to the case in which A has a compact support in Ω. This is evident when $\Omega = R^n$ and the a_p are constants, for then from $PX = \varepsilon$ follows at once $P(X * A) = A$.

The work of the present section is founded largely on the efforts of Malgrange [1], an alternative account of which appears in Schwartz [8]. Much of this material is also covered by Hörmander [1], whose approach will receive closer attention in Subsection 5.18.3. The important contributions of Ehrenpreis ([2], [3], [10], [11]) bear more directly on the general equation (5.14.2) and are thus to be consulted in connection with Section 5.17 below.

5.14.1 **Description of the Method.** In opposition to the techniques described in Subsection 5.18.3, those of the present section are "existential" in nature. They illustrate clearly the fundamental role played by inequalities in the modern (functional analytic) theory of linear differential equations, already mentioned in Section 5.12.

It is perhaps worth giving in advance an outline of the general principles involved in establishing the existence of solutions of (5.14.2) when P is a linear partial differential operator.

The general situation may be described in the following terms:

(a) $\mathscr{E}$ will be a LCTVS containing $\mathscr{D}(\Omega)$ as a dense subspace and such that the injection of $\mathscr{D}(\Omega)$ into $\mathscr{E}$ is continuous. This will arrange that each member of $\mathscr{E}'$, the topological dual of $\mathscr{E}$, is representable by precisely one distribution $X \in \mathscr{D}'(\Omega)$ and we may identify $\mathscr{E}'$ (vectorially) with a subspace of $\mathscr{D}'(\Omega)$.

(b) Q (resp. P) is a linear map of $\mathscr{D}(\Omega)$ (resp. $\mathscr{E}'$) into $\mathscr{E}$ (resp. $\mathscr{D}'(\Omega)$) such that

$$\langle \varphi, PX \rangle = \langle Q\varphi, X \rangle \qquad \text{for } \varphi \in \mathscr{D}(\Omega),\ X \in \mathscr{E}'. \tag{5.14.3}$$

(c) The distribution $A \in \mathscr{D}'(\Omega)$ is such that

$$|\langle \varphi, A \rangle| \leq p(Q\varphi) \qquad \text{for } \varphi \in \mathscr{D}(\Omega), \tag{5.14.4}$$

where p is a continuous seminorm on $\mathscr{E}$.

(d) The conclusion is that there exists an X in $\mathscr{E}'$ satisfying (5.14.2).

In our applications, P will be a linear partial differential operator with constant coefficients:

$$PX = P(D)X = \sum_{|p| \leq m} a_p D^p X, \tag{5.14.5}$$

and Q will be its formal adjoint, likewise a linear partial differential operator with constant coefficients:

$$Q\varphi = Q(D)\varphi = \sum_{|p| \leq m} a_p (-1)^{|p|} D^p \varphi. \tag{5.14.6}$$

Thus $Q(D) = \check{P}(D)$, as defined in Subsection 5.12.1.

It turns out that for this case it is possible to establish inequalities (5.14.4) of the desired form. In seeking to apply these it is convenient to state as a lemma the conclusions to which one is led on applying the general principles described above to appropriate special cases.

Lemma. Suppose we are given a finite number λ_k of continuous linear maps of $\mathscr{D}(\Omega)$ into L^{r_k}, where each r_k satisfies $1 \leq r_k < \infty$. Suppose further that P is a LPDO with constant coefficients such that the following inequality is known to be valid for φ in $\mathscr{D}(\Omega)$

$$\|\varphi\|_{L^\infty} \leq \text{const.} \sum_k \|\lambda_k \check{P} \varphi\|_{L^{r_k}}.$$

The conclusion is that if A is a bounded Radon measure on Ω, then the equation $PX = A$ has a distribution solution X of the form

$$X = \sum_k \lambda_k' f_k,$$

where $f_k \in L^{r_k'}$ for each k, and where λ_k' is the linear map of $L^{r_k'}$ into $\mathscr{D}'(\Omega)$ defined by

$$\langle \lambda_k \varphi, g \rangle = \langle \varphi, \lambda_k' g \rangle$$

for φ in $\mathscr{D}(\Omega)$ and g in $L^{r_k'}$.

Proof. We shall take $\mathscr{E} = \mathscr{D}(\Omega)$ as a vector space, endowing $\mathscr{E}$ with the topology defined by the single seminorm

$$N(\varphi) = \sum_k \|\lambda_k \varphi\|_{L^{r_k}}.$$

Since A is a bounded measure one will have

$$|\langle \varphi, A \rangle| \leq \text{const.} \ \|\varphi\|_{L^\infty} \leq \text{const.} \ N(\check{P}\varphi).$$

Thus $\check{P}\varphi \to \langle \varphi, A \rangle$ is a continuous linear form on the vector subspace $\check{P}(\mathscr{D}(\Omega))$ of $\mathscr{E}$. According to the Hahn-Banach theorem this linear form therefore has a continuous extension to $\mathscr{E}$. Moreover, by considering the linear map $\varphi \to (\lambda_k \varphi)$ of $\mathscr{E}$ into the product space $\prod_k L^{r_k}$ and applying the Hahn-Banach theorem once more, we infer that each continuous linear form on $\mathscr{E}$ is expressible as

$$\varphi \to \sum_k \int \lambda_k \varphi \cdot f_k \, d\mu = \sum_k \langle \lambda_k \varphi, f_k \rangle$$

$$= \left\langle \varphi, \sum_k \lambda_k' f_k \right\rangle,$$

where $f_k \in L^{r'_k}$ for each k. Thus finally it appears that for arbitrary φ in $\mathscr{D}(\Omega)$ one has

$$\langle \varphi, A \rangle = \langle \check{P}\varphi, X \rangle = \langle \varphi, PX \rangle,$$

where $X = \sum_k \lambda'_k f_k$. Whence it follows that $A = PX$, and the proof is complete. ▌

As appears from Exercises 8.34–8.39, the methods described above are easily generalized.

5.14.2 Statement of the Auxiliary Inequalities and of the Existence Theorem. We shall indicate two methods of attack, the second and less general of which applies effectively only to bounded domains Ω, but which is of interest in other connections too (see Subsection 5.19.3). The two methods involve inequalities, and lead to conclusions, of the same general type but differing in detail. In either case, one terminates with the solubility of (5.14.2) when A is a given bounded Radon measure on Ω, the solution X being a distribution of order at most $n + 1$ on Ω. Supplementary arguments permit passage from this case to that in which A is more general, but for the moment we leave the reader to consider this point since Section 5.17 deals with the matter at some length.

(1) The first method depends upon but one auxiliary inequality, the proof of which uses the Fourier transformation as applied to functions in $\mathscr{D} = \mathscr{D}(R^n)$. Here is the inequality in question.

Inequality A. Let $Q \neq 0$ be a linear partial differential operator with constant coefficients. There exist similar operators $L_1, \cdots, L_{n+1}$ of orders at most $n + 1$, and a linear form λ on R^n, with the property that to any $\varepsilon > 0$ corresponds a number $c(Q, n, \varepsilon)$ such that

$$\|\varphi\|_{L^\infty} \leq c(Q, n, \varepsilon) \cdot \sum_{k=1}^{n+1} \|e^{\varepsilon|\lambda|} L_k Q\varphi\|_{L^1}$$

for each φ in $\mathscr{D}$. (All Lebesgue spaces L^p are constructed relative to Lebesgue measure μ on R^n.)

This inequality, and the proof we offer in Subsection 5.14.3, are due to Malgrange (Schwartz [8], Exposé 2; compare Malgrange [1], Ch. I, Proposition 1).

At this stage one may appeal to the lemma in Subsection 5.14.1, taking $Q = \check{P}$, $\lambda_k\varphi = e^{\varepsilon|\lambda|}L_k\varphi$, and $r_k = 1$, $r'_k = \infty$. The conclusion is the following existence theorem.

Theorem A. Let $P \neq 0$ be a linear partial differential operator with constant coefficients; let $\varepsilon > 0$ be given; and let $L_1, \cdots, L_{n+1}$ and λ be related to $Q = \check{P}$ as in Inequality A. If Ω is any domain in R^n and A a bounded measure on Ω, then (5.14.2) has a solution X of the form

$$X = \sum_{k=1}^{n+1} \check{L}_k f_k,$$

the functions f_k (or, better, their classes) being such that $e^{-\varepsilon|\lambda|}f_k \in L^\infty(\Omega)$. In particular, X is of order at most $n + 1$ on Ω (and is indeed the restriction to Ω of a distribution of order at most $n + 1$ on R^n).

Remark. In any case A can be extended into a bounded Radon measure on R^n, so there is really no loss of generality in taking Ω to be R^n from the start.

(2) For the second approach two auxiliary inequalities are needed. Taken together these yield a result essentially equivalent to Inequality A when Ω is bounded and φ belongs to $\mathscr{D}(\Omega)$. The first of the two inequalities, which is due to Hörmander ([1], Theorem 2.1), runs as follows.

Inequality B. Let Q be as in Inequality A. There exists a number $c(Q, d)$ such that

$$\|\varphi\|_{L^2} \leq c(Q, d)\, \|Q\varphi\|_{L^2}$$

for each φ in $\mathscr{D}$ whose support has diameter at most d.

The proof of this may be effected without any use of the Fourier transformation (Hörmander [1], § 2.6), or by use of that technique (Hörmander [1], Theorem 2.1). We shall give the former proof.

Use of the Fourier transformation will be made in proving the second inequality, which bounds $\|\varphi\|_{L^\infty}$ in terms of $\|\varphi\|_{L^2}$.

Inequality B′. To each natural number $r > \frac{1}{2}n$ corresponds a number $C(n, r)$ such that

$$\|\varphi\|_{L^\infty} \leq C(n, r) \left\| \left(1 - \frac{\Delta}{4\pi^2}\right)^r \varphi \right\|_{L^2}$$

for each φ in $\mathscr{D}$.

If Ω is taken to be bounded with diameter d, the two preceding inequalities combine to show that

$$\|\varphi\|_{L^\infty} \leq c(Q, n, r, d) \cdot \left\| \left(1 - \frac{\Delta}{4\pi^2}\right)^r Q\varphi \right\|_{L^2}$$

for φ in $\mathscr{D}$. From this the lemma in 5.14.1 leads to the following theorem.

Theorem B. Let Ω be a bounded domain in R^n, $P \neq 0$ a linear partial differential operator with constant coefficients, and A a bounded Radon measure on Ω. Then (5.14.2) has a solution X of the form

$$X = \left(1 - \frac{\Delta}{4\pi^2}\right)^r g,$$

where r is the least natural number exceeding $\frac{1}{2}n$, and where $g \in L^2(\Omega)$; in particular, X is of order at most $n + 2$ if n is even and at most $n + 1$ if n is odd.

Remarks. Before embarking on the proofs of the inequalities leading to Theorems A and B, we should indicate that the results can be improved. It is known, for example, that there exists always an elementary solution X for $P(D)$ that is a finite linear combination of derivatives of orders at most $[\frac{1}{2}n] + 1$ of functions in $L^2_{\text{loc}}(\Omega)$, and indeed (what says more) such that $X * g \in L^2_{\text{loc}}$ whenever $g \in L^2$ has a compact support. See Malgrange [1], pp. 21–22; Schwartz [8], Exposé 3, Théorème 2 and Exposé 3 bis, pp. 3–17; Hörmander [5]. Further properties of the elementary solution will be discussed in Section 5.18.

5.14.3 Proof of Inequality ***A.*** We must commence by summarizing a few properties of the Fourier transformation as applied to functions in $\mathscr{D}$. All the results we need are special cases of the material presented in Sections 10.3 and 10.4. However, on account of the special properties of functions in $\mathscr{D}$, simpler and more direct proofs are possible and familiar.

Suppose φ belongs to $\mathscr{D}$. Its Fourier transform Φ is the function, defined initially on R^n, by the formula

$$\Phi(\xi) = \int \varphi(x) e^{-2\pi i \xi \cdot x} \, d\mu(x), \tag{5.14.7}$$

μ denoting Lebesgue measure on R^n and

$$\xi \cdot x = \sum_{n=1}^{n} \xi_k x_k.$$

Often it will be more convenient to use the systematic notation $\hat{\varphi}$ for the Fourier transform of φ, but there is little advantage at the moment.

Since φ has a compact support, the integral appearing in (5.14.7) is absolutely convergent even if we replace the point ξ of R^n by the point $\zeta = \xi + i\eta$ of C^n. It follows that we may regard Φ as an entire analytic function on C^n; as such it is, of course, uniquely determined by its restriction to R^n. For future reference we note the inequality

$$|\Phi(\zeta_1, \xi_2, \cdots, \xi_n)| \le \int |\varphi(x)| \, e^{2\pi\alpha|x_1|} \, d\mu(x), \tag{5.14.8}$$

holding for $\zeta_1 = \xi_1 + i\eta_1$, where the ξ_k are real and η_1 is real and subject to $|\eta_1| \le \alpha$.

Partial integrations applied to (5.14.7) show that Φ is "rapidly decreasing on R^n" in the sense that

$$\lim_{\xi \in R^n, |\xi| \to \infty} p(\xi)\Phi(\xi) = 0$$

for each polynomial p on R^n; in particular, Φ is integrable (for μ). Knowing this, there is little difficulty in establishing the Fourier inversion formula:

$$\varphi(x) = \int \Phi(\xi) e^{2\pi i x \cdot \xi} \, d\mu(\xi); \tag{5.14.9}$$

this is a special case of Lemma 10.4.5.

One of the most important of all properties for our purposes is the Parseval formula:

$$\|\varphi\|_{L^2} = \|\Phi\|_{L^2}, \tag{5.14.10}$$

which is proved in sufficient generality in Subsection 10.4.4.

For the rest, we shall need to know that the Fourier transform of $Q(D)\varphi$ is $Q(\xi) \cdot \Phi(\xi)$. This flows from (5.14.7) by iterated partial integrations.

To prove Inequality A we shall have to combine the preceding properties of the Fourier transformation with a lemma belonging to the theory of functions of one complex variable, due to Malgrange [1], Ch. I, Lemme 1; compare with Hörmander [1], Lemma 2.1.

Lemma. Suppose that f and g are entire functions of a complex variable; that $q(z) = z^m + c_1 z^{m-1} + \cdots + c_m$ is a unitary polynomial with complex coefficients; and that $f = qg$. Then, given $r > 0$, one has for all complex z_0 the inequality

$$|g(z_0)| \leq r^{-m} \cdot \sup_{|z-z_0|\leq 2mr} |f(z)|. \tag{5.14.11}$$

Proof. By factorizing q and then proceeding by induction on m, it suffices to deal with the case in which $q(z) = z - a$. (The case $m = 0$, $q = 1$ is trivial.) If $|z_0 - a| \geq r$ one has

$$|g(z_0)| = \frac{|f(z_0)|}{|z_0 - a|} \leq \frac{|f(z_0)|}{r},$$

and the inequality is obviously fulfilled. If $|z_0 - a| = r$ we have

$$|g(z_0)| = \frac{|f(z_0)|}{r} \leq r^{-1} \cdot \sup_{|z-a|\leq r} |f(z)|.$$

By the maximum modulus principle, this same inequality persists for all z_0 satisfying $|z_0 - a| < r$. For any such z_0 one has $|z - z_0| \leq 2r$ whenever $|z - a| \leq r$, and so the alleged inequality holds in this case too. It is therefore valid for all z_0, Q.E.D. ▌

The proof of Inequality A can now begin. If φ belongs to $\mathscr{D}$, (5.14.9) shows at once that

$$\|\varphi\|_{L^\infty} \leq \int |\Phi(\xi)| \, d\mu(\xi),$$

and so

$$\|\varphi\|_{L^\infty} \leq c_n \cdot \sup_{\xi \in R^n} |(1 + |\xi_1|^{n+1} + \cdots + |\xi_n|^{n+1})\Phi(\xi)|, \tag{5.14.12}$$

where

$$c_n = \int (1 + |\xi_1|^{n+1} + \cdots + |\xi_n|^{n+1})^{-1} \, d\mu(\xi) < +\infty.$$

At this point we observe that since $Q \neq 0$, a change of linear coordinates will arrange that

$$Q(\xi) = b\xi_1^m + Q_0(\xi),$$

where $Q_0(\xi)$ is a polynomial of degree at most $m - 1$ in ξ_1, the coefficients being functions of $\xi_2, \cdots, \xi_n$, and where $b = b(Q)$ is a nonzero number. The lemma now comes into play and shows first that

$$|\Phi(\xi)| \leq r^{-m} \, |b|^{-1} \sup_{|\zeta_1 - \xi_1| \leq 2mr} |Q(\zeta_1, \xi')\Phi(\zeta_1, \xi')|,$$

where, on the right, ζ_1 is complex and where $\xi' = (\xi_2, \cdots, \xi_n)$. Similarly we have

$$|\xi_1^{n+1}\Phi(\xi)| \leq r^{-m} \, |b|^{-1} \sup_{|\zeta_1 - \xi_1| \leq 2mr} |\zeta_1^{n+1} Q(\zeta_1, \xi')\Phi(\zeta_1, \xi')|$$

and

$$|\xi_k^{n+1}\Phi(\xi)| \leq r^{-m} \, |b|^{-1} \sup_{|\zeta_1 - \xi_1| \leq 2mr} |\xi_k^{n+1} Q(\zeta_1, \xi')\Phi(\zeta_1, \xi')|$$

for $k = 2, \cdots, n$.

Now $Q(\zeta)\Phi(\zeta)$ and $\zeta_k^{n+1}Q(\zeta)\Phi(\zeta)$ are, respectively, the Fourier transforms of $Q\varphi$ and $D_k^{n+1}Q\varphi$. Thus an appeal to (5.14.8) with φ replaced in turn by φ, $Q\varphi$, and $D_k^{n+1}Q\varphi$, leads from (5.14.12) to the inequality

$$\|\varphi\|_{L^\infty} \le c_n r^{-m} |b|^{-1} \left\{ \|e^{\varepsilon|x_1|}Q\varphi\|_{L^1} + \sum_{k=1}^{n} \|e^{\varepsilon|x_1|}D_k^{n+1}Q\varphi\|_{L^1} \right\},$$

where $\varepsilon = 4\pi mr$ is as small as we please with r. This is Inequality A with $\lambda(x) = x_1$, $L_k = D_k^{n+1}$ for $k = 1, \cdots, n$ and $L_{n+1} = 1$ (the identity operator). Bearing in mind the linear change of coordinates necessary to reach this conclusion, we see that the general form of Inequality A is justified. ∎

5.14.4 **Proof of Inequality *B*.** The procedure here is based on a lemma (Hörmander [1], § 2.6), whence Inequality B will result by induction on the degree m of Q.

Starting from the polynomial $Q(\xi) = \sum_{|p|\le m} b_p \xi^p$, we introduce the polynomial $Q_1(\xi) = \partial Q/\partial \xi_1$, and note the identity

$$Q(D)(x_1\varphi) = x_1 Q(D)\varphi + Q_1(D)\varphi, \tag{5.14.13}$$

which follows from Leibnitz' formula.

In what follows we use Q and Q_1 for the operators $Q(D)$ and $Q_1(D)$.

Lemma. Let d_1 be the supremum of $|x_1|$ for $x = (x_1, \cdots, x_n)$ ranging over the support of $Q\varphi$. Then, for φ in $\mathscr{D}$, one has

$$\|Q_1\varphi\|_{L^2} \le 2md_1 \|Q\varphi\|_{L^2}, \tag{5.14.14}$$

m being the degree of Q.

Proof. We use $\tilde{Q}$ to denote the L^2-adjoint of Q : $\tilde{Q}$, which is Hörmander's $\bar{Q}$, is the linear partial differential operator with constant coefficients such that

$$(Q\varphi \mid \psi)_{L^2} = (\varphi \mid \tilde{Q}\psi)_{L^2} \tag{5.14.15}$$

for φ and ψ in $\mathscr{D}$. Specifically,

$$\tilde{Q}\psi = \sum_{|p|\le m} (-1)^{|p|}\bar{b}_p D^p \psi.$$

The reader will notice that Q, Q_1, $\tilde{Q}$, and $\tilde{Q}_1$ commute since they have constant coefficients; he will notice too that $\|Q\varphi\|_{L^2} = \|\tilde{Q}\varphi\|_{L^2}$.

Using (5.14.13), we have

$$(x_1 Q\varphi \mid Q_1\varphi) = (Q(x_1\varphi) - Q_1\varphi \mid Q_1\varphi) = (Q(x_1\varphi) \mid Q_1\varphi) - \|Q_1\varphi\|^2,$$

where we have temporarily dropped the subscript L^2. Thus

$$\begin{aligned}\|Q_1\varphi\|^2 &= (Q(x_1\varphi) \mid Q_1\varphi) - (x_1 Q\varphi \mid Q_1\varphi) \\ &= (Q(x_1\varphi) \mid Q_1\varphi) - (x_1 Q\varphi \mid Q_1\varphi).\end{aligned}$$

Using (5.14.13) with Q replaced by Q_1, this leads to

$$\begin{aligned}\|Q_1\varphi\|^2 &= (x_1\tilde{Q}_1\varphi + \tilde{Q}_{11}\varphi \mid \tilde{Q}\varphi) - (x_1 Q\varphi \mid Q_1\varphi) \\ &= (x_1\tilde{Q}_1\varphi \mid \tilde{Q}\varphi) + (\tilde{Q}_{11}\varphi \mid \tilde{Q}\varphi) - (x_1 Q\varphi \mid Q_1\varphi).\end{aligned}$$

Hence, using the simplest estimates in conjunction with the Cauchy-Schwarz inequality, we deduce that

$$\begin{aligned}\|Q_1\varphi\|^2 &\leq d_1 \|\tilde{Q}_1\varphi\| \cdot \|\tilde{Q}\varphi\| + \|\tilde{Q}_{11}\varphi\| \cdot \|\tilde{Q}\varphi\| + d_1 \|Q_2\varphi\| \cdot \|Q_1\varphi\| \\ &= 2d_1 \|Q_1\varphi\| \cdot \|Q\varphi\| + \|Q_{11}\varphi\| \cdot \|Q\varphi\|. \end{aligned} \tag{5.14.16}$$

Now the assertion of the lemma is visibly true when $m = 0$, for then $Q_1 = 0$. We assume the lemma established for polynomials Q of degree at most $m - 1$, where $m \geq 1$. Then, if Q is of degree at most m, Q_1 is of degree at most $m - 1$, and so (by inductive hypothesis)

$$\|Q_{11}\varphi\| \leq 2d_1(m - 1) \|Q_1\varphi\|.$$

Insertion of this estimate into (5.14.16) yields

$$\|Q_1\varphi\| \leq 2d_1 \|Q\varphi\| + 2d_1(m - 1) \|Q\varphi\| = 2d_1 m \|Q\varphi\|.$$

In other words, the inequality is derived for polynomials Q of degree at most m. Induction now closes the proof. ▮

To deduce Inequality B, we first make a change of linear coordinates in order to arrange that $Q(\xi) = b\xi_1^m + \sum_{|p|\leq m, p\neq(m,0,\ldots,0)} b_p\xi^p$, where $b \neq 0$. By repeated use of the lemma, substituting in turn for Q the operators Q_1, Q_{11}, $\cdots$, and so terminating with the scalar operator $b \cdot m!$, we are led to Inequality B.

5.14.5 Proof of Inequality B'. From (5.14.9) we have for φ in $\mathscr{D}$ the inequality

$$\|\varphi\|_{L^\infty} \leq \|\Phi\|_{L^1};$$

by the Cauchy-Schwarz inequality this in turn leads to

$$\|\varphi\|_{L^\infty} \leq c(n, r) \, \|(1 + |\xi|^2)^r\Phi\|_{L^2},$$

where

$$c(n, r) = \left\{\int (1 + |\xi|^2)^{-r} \, d\mu(\xi)\right\}^{1/2}.$$

$c(n, r)$ is finite if $r > \frac{1}{2}n$. Now $(1 + |\xi|^2)^r\Phi(\xi)$ is (when r is a natural number) the Fourier transform of $(1 - \Delta/4\pi^2)^r\varphi$, Δ denoting as usual the Laplacian $-4\pi^2(D_1^2 + \cdots + D_n^2)$. So, using the Parseval formula (5.14.10), we obtain

$$\|\varphi\|_{L^\infty} \leq c(n, r) \left\|\left(1 - \frac{\Delta}{4\pi^2}\right)^r \varphi \right\|_{L_2},$$

as is alleged by Inequality B'. ▮

5.15 Fourier Transforms of Distributions

It is our aim to pursue the study of LPDOs with constant coefficients by methods using the Fourier transform. Classically this transformation is defined for functions that are severely restricted in their rate of growth at infinity. Such restrictions make it extremely difficult, if not impossible, to use the Fourier transform technique in a satisfactory way. The remedy lies in defining the

Fourier transform for at least some distributions. One such extension is due to Schwartz and is applicable only to distributions that are roughly speaking of polynomial order of growth at infinity. With this condition the process retains great flexibility and ease of handling. A different approach, due to Ehrenpreis and applying to all distributions at the expense of some flexibility, will be described briefly at the end of this section.

5.15.1 **The Space $\mathscr{S} = \mathscr{S}(R^n)$ and its Fourier Image.** We define $\mathscr{S} = \mathscr{S}(R^n)$ to be the vector subspace of $C^\infty = C^\infty(R^n)$ formed of those functions u in C^∞ for which

$$S_m(u) = \text{Sup}\,\{(1 + |x|)^m\,|D^p u(x)| : |p| \leq m,\, x \in R^n\}$$

is finite for $m = 1, 2, \cdots$, (or, equivalently, for arbitrarily large m); cf. the examples in Section 6.1. The norms S_m on $\mathscr{S}$ defined a vector-space topology on $\mathscr{S}$, and it is an easy exercise to show that $\mathscr{S}$ is thus made into a Fréchet space. What is more, $\mathscr{S}$ is a Montel space (see Section 8.4); that is, each bounded closed subset of $\mathscr{S}$ is compact. This is a direct consequence of Ascoli's theorem.

It is obvious that $\mathscr{D} = \mathscr{D}(R^n)$ is a vector subspace of $\mathscr{S}$, and that the topology of $\mathscr{S}$ induces on $\mathscr{D}$ a topology that is strictly weaker than the inductive limit topology of $\mathscr{D}$. It is very simple to see that one may choose sequences (β_n) in $\mathscr{D}$ such that $\lim_{n\to\infty} \beta_n u = u$ in $\mathscr{S}$ for each u in $\mathscr{S}$. Thus $\mathscr{D}$ is dense in $\mathscr{S}$.

It is fundamental to Schwartz's theory of generalized Fourier transforms that this same transformation behaves suitably in its action on $\mathscr{S}$. We turn to this matter at once.

Since each u in $\mathscr{S}$ is integrable, the Fourier transform

$$\hat{u}(\xi) = \mathscr{F}u(\xi) = \int e^{-2\pi i\xi\cdot x} u(x)\, d\mu(x) \tag{5.15.1}$$

is a continuous function of ξ in R^n. (As usual, μ denotes Lebesgue measure on R^n.) More than this is true, however. Since u decreases more rapidly than any negative power of $|x|$, as $|x| \to \infty$, it follows that $\hat{u}$ belongs to C^∞ and that its partial derivatives may be computed by differentiations under the integral sign.

On the other hand, since each partial derivative of u is integrable, repeated partial integrations show that $\hat{u}(\xi)$ decreases more rapidly than any negative power of $|\xi|$, as $|\xi| \to \infty$.

By combining these two remarks it is seen that $\mathscr{F}$ maps $\mathscr{S}$ into itself. The simplest estimates made after partial integrations and differentiations under the integral sign show, moreover, that $\mathscr{F}$ is a continuous endomorphism of $\mathscr{S}$.

It is important for us to verify that $\mathscr{F}$ is in fact a (topological as well as algebraic) automorphism of $\mathscr{S}$. This may be done by introducing the transformation $\mathscr{F}^{-1}$ defined on $\mathscr{S}$ by

$$\mathscr{F}^{-1}v(x) = \int e^{+2\pi i x\cdot\xi} v(\xi)\, d\mu(\xi). \tag{5.15.2}$$

The Fourier inversion formula asserts that

$$u = \mathscr{F}^{-1}(\mathscr{F}u) \qquad \text{and} \qquad v = \mathscr{F}(\mathscr{F}^{-1}v)$$

for u and v in $\mathscr{S}$, results that are established under much weaker assumptions on u and v in Section 10.4 for the case in which R^n is replaced by more general groups. For the special case envisaged here the formula follows from standard analysis of singular integrals. Granted these relations, we see that $\mathscr{F}$ is an algebraic automorphism of $\mathscr{S}$ whose inverse is $\mathscr{F}^{-1}$ (hence the notation.) Moreover the similarity between $\mathscr{F}^{-1}$ and $\mathscr{F}$—the difference amounting to no more than a change of sign in the exponential factor—makes it clear that the same arguments as those used to establish the continuity of $\mathscr{F}$ will establish that of $\mathscr{F}^{-1}$. Thus $\mathscr{F}$ has the continuous inverse $\mathscr{F}^{-1}$ and is an automorphism, as announced.

It is worth noting that once we have established that $\mathscr{F}$ is continuous, onto, and one to one, the open mapping theorem for Fréchet spaces (included in Theorem 6.4.4) asserts the continuity of the inverse $\mathscr{F}^{-1}$.

There are two further formulae to be recorded. In the first place, if u belongs to $\mathscr{S}$ and if P is any polynomial on R^n, then Pu belongs to $\mathscr{S}$ and

$$\mathscr{F}(Pu) = \check{P}(D)\mathscr{F}u. \tag{5.15.3}$$

Second, if P and u are as just specified, then $P(D)u$ belongs to $\mathscr{S}$ and

$$\mathscr{F}P(D)u = P \cdot \mathscr{F}u. \tag{5.15.4}$$

By linearity, it suffices to verify these formulae for the case in which P is monomial, $P(x) = x^p$, in which case they follow from differentiations under the integral sign and by partial integrations, respectively.

The last property of $\mathscr{F}$ that we must list is the Parseval formula, viz.,

$$\int u \cdot \bar{v}\, d\mu = \int (\mathscr{F}u) \cdot \overline{(\mathscr{F}v)}\, d\mu, \tag{5.15.5}$$

valid for u and v in $\mathscr{S}$. This, like the Fourier inversion formula, is established under much weaker conditions in Section 10.4 (see especially equation (10.4.7)). For the special functions used here one may again establish the result along classical lines.

5.15.2 **The Space $\mathscr{S}' = \mathscr{S}'(R^n)$ of Temperate Distributions.** Just as $\mathscr{D}'$ was defined as the topological dual of $\mathscr{D}$, we now introduce $\mathscr{S}' = \mathscr{S}'(R^n)$ as the topological dual of $\mathscr{S}(R^n)$. Since the topology of $\mathscr{S}$ induces on $\mathscr{D}$ a topology weaker than the inductive limit topology on $\mathscr{D}$, each element X of $\mathscr{S}'$ defines by restriction to $\mathscr{D}$ a distribution on R^n. Since furthermore $\mathscr{D}$ is dense in $\mathscr{S}$, X is fully determined by this restriction. Accordingly we may regard $\mathscr{S}'$ as a vector subspace of $\mathscr{D}'$. The members of $\mathscr{S}'$ are termed the *temperate distributions on R^n*.

As a TVS $\mathscr{S}$ is simpler to handle than $\mathscr{D}$, being a Fréchet space rather than an inductive limit of such spaces. Since, as follows easily from Ascoli's theorem, $\mathscr{S}$ is a Montel space (each bounded, closed subset of $\mathscr{S}$ being compact), all the results (1) through (5) set forth in Section 5.3 for $\mathscr{D}$ and $\mathscr{D}'$ are true equally for $\mathscr{S}$ and $\mathscr{S}'$. It is, of course, understood that $\mathscr{S}'$ is to be endowed with its weak or strong topologies determined by the duality between $\mathscr{S}$ and $\mathscr{S}'$. In particular, therefore, each of $\mathscr{S}$ and $\mathscr{S}'$ (the latter with its strong topology) is complete, barrelled, and bornological. Also, $\mathscr{S}$ is reflexive; that is, each strongly continuous linear form on $\mathscr{S}'$ can be written in just one way as $X \to \langle u, X \rangle$ for some suitably chosen u in $\mathscr{S}$, and u converges in the sense of $\mathscr{S}$ if and only if the corresponding linear form converges uniformly on the strongly bounded subsets of $\mathscr{S}'$.

It is immediately obvious that $\mathscr{S}'$, considered as a subspace of $\mathscr{D}'$, is stable under partial differentiation and under multiplication by functions f in C^∞ which are of polynomial order at infinity.

5.15.3 Fourier Transforms of Temperate Distributions. We propose to use the Parseval formula (5.15.5) as a guide to extending $\mathscr{F}$ from $\mathscr{S}$ to $\mathscr{S}'$. The general underlying principle is that of taking the adjoint of a linear map of one LCTVS into another, a process that is discussed in general terms in Section 8.6.

Since $\mathscr{F}$ is a continuous endomorphism of $\mathscr{S}$, if we fix X in $\mathscr{S}'$, then $v \to \langle \mathscr{F}v, X \rangle$ is a continuous linear form on $\mathscr{S}$. By definition of $\mathscr{S}'$, therefore, there exists a uniquely determined element $\mathscr{F}'X$ of $\mathscr{S}'$ such that

$$\langle \mathscr{F}v, X \rangle = \langle v, \mathscr{F}'X \rangle$$

for all v in $\mathscr{S}$. In this way we obtain an algebraic endomorphism $\mathscr{F}'$ of $\mathscr{S}'$, termed the adjoint of $\mathscr{F}$. According to Corollary 8.6.6, $\mathscr{F}'$ is continuous for the weak and for the strong topologies on $\mathscr{S}'$. By (d) of Section 8.6, the fact that $\mathscr{F}(\mathscr{S}) = \mathscr{S}$ ensures that $\mathscr{F}'$ is one to one. At the same time, since $\mathscr{S}$ is reflexive and $\mathscr{F}$ is one to one, Theorem 8.6.13 shows that $\mathscr{F}'(\mathscr{S}') = \mathscr{S}'$. Thus $\mathscr{F}'$ is one-to-one linear map of $\mathscr{S}'$ onto itself that is continuous (strongly and weakly). Now $\mathscr{F}^{-1}$ has the same properties as $\mathscr{F}$, and its adjoint is $(\mathscr{F}')^{-1}$. It follows that $(\mathscr{F}')^{-1}$ is also continuous (weakly and strongly). Thus finally $\mathscr{F}'$ is an automorphism (algebraic and topological) of $\mathscr{S}'$ (the latter taken with either its weak or its strong topology).

For $\mathscr{F}^{-1}$ and its adjoint $(\mathscr{F}^{-1})' = \mathscr{F}'^{-1}$ we have the defining formula

$$\langle \mathscr{F}^{-1}v, X \rangle = \langle v, \mathscr{F}'^{-1} X \rangle.$$

If we here replace v by $\mathscr{F}v$, and then change v into $\bar{v}$ and therefore $\mathscr{F}v$ into $\overline{(\mathscr{F}v)}^{\vee}$, the result is the formula

$$\langle \bar{v}, X \rangle = \langle \overline{\mathscr{F}v}, \mathscr{F}_1 X \rangle, \tag{5.15.6}$$

where we have written temporarily

$$\mathscr{F}_1 X = (\mathscr{F}'^{-1} X)^{\vee}.$$

Now each element u of $\mathscr{S}$ generates, and may be identified with, a temperate distribution (the process being that by which any locally integrable function generates a distribution). If we replace X by u in this way in (5.15.6) and compare the resulting equation with the Parseval formula (5.15.5), we see that $\mathscr{F}_1 u = \mathscr{F} u$ for u in $\mathscr{S}$. Because of this we shall henceforth write $\mathscr{F} X$ in place of $\mathscr{F}_1 X$, even for a general temperate distribution X, calling $\mathscr{F} X$ the *Fourier transform of* X. (A better name might be the *Fourier-Schwartz* transform.) As before we shall sometimes write $\hat{X}$ in place of $\mathscr{F} X$.

Thus $\mathscr{F} X$ is determined for each X in $\mathscr{S}'$ as the unique temperate distribution for which the Parseval formula

$$\langle \bar{v}, X \rangle = \langle \overline{\mathscr{F} v}, \mathscr{F} X \rangle \tag{5.15.7}$$

holds for each v in $\mathscr{S}$. Actually, $\mathscr{F} X$ is the unique distribution (temperate or otherwise) such that

$$\langle \overline{\mathscr{F}^{-1} \varphi}, X \rangle = \langle \bar{\varphi}, \mathscr{F} X \rangle \tag{5.15.8}$$

holds for each φ in $\mathscr{D}$. (For if (5.15.8) holds for φ in $\mathscr{D}$, then, since $\mathscr{F}$ is an automorphism of $\mathscr{S}$, the distribution X is continuous for the topology induced on $\mathscr{D}$ by that of $\mathscr{S}$ and therefore has a unique continuous extension from $\mathscr{D}$ to $\mathscr{S}$ as a temperate distribution. By continuity, then, (5.15.8) will continue to hold if we therein replace φ by any element u of $\mathscr{S}$. The resulting formula is equivalent to (5.15.7) on writing $u = \mathscr{F} v$.)

Our mode of construction shows that $\mathscr{F}$ is an automorphism of $\mathscr{S}'$ (for either the weak or the strong topology).

We have already noted that one may regard $\mathscr{S}$ as a vector subspace of $\mathscr{S}'$. Reflexivity of $\mathscr{S}$ combines with the Hahn-Banach theorem to show that $\mathscr{S}$ is dense in $\mathscr{S}'$. Then, since we have adopted the Parseval formula as the defining property of $\mathscr{F}$ as applied to $\mathscr{S}'$, it follows that this map is that obtained by continuously extending $\mathscr{F}$ from $\mathscr{S}$ to $\mathscr{S}'$.

A consequence of this observation is the continued validity of the (5.15.3) and (5.15.4) when u is replaced by an arbitrary $X \in \mathscr{S}'$. (It is necessary merely to observe that partial differentiation and multiplication by a polynomial are each continuous on $\mathscr{S}'$.) We shall indicate below how further extensions may be made to cover the case in which $P(D)X = P(D)\varepsilon * X$ is replaced by a more general convolution $S * X$.

5.15.4 The Effect of Mappings of R^n on the Fourier Transform.

Suppose that H is a one-to-one map of R^n onto itself which, together with its inverse H^{-1}, is of class C^∞. Then H induces mappings H_* and H^* of functions and of distributions, respectively, defined by

$$H_* f = f \circ H$$

and by

$$\langle u, H^* X \rangle = \langle H_* u, X \rangle,$$

u varying over $\mathscr{D}$ (or over $\mathscr{S}$, if one wishes, if X is temperate). If f is a locally integrable function on R^n, it may be thought of as a distribution; in general the distribution $H^* f$ is itself generated by and identified with a function, but this function is in most cases distinct from $H_* f$—which explains the different positioning of the asterisk.

As simple but important instances of maps H one has the translation maps T_a, where a is a point of R^n and $T_a x = x - a$, and the homothetic maps H_λ, where $\lambda \neq 0$ is a real number and $H_\lambda x = \lambda x$. These, together with rotations, are in fact the only specific examples needed in the sequel.

Our interest lies in relating the Fourier transform of H^*X to that of X. We shall do this first for the translations and then for the case in which H is a vector space automorphism of R^n (for example, a homothety H_λ).

If f is a locally integrable function and u belongs to $\mathscr{D}$ one has

$$\begin{aligned}\langle u, T_a^* f\rangle = \langle T_{a*} u, f\rangle &= \int u(x-a) f(x)\, d\mu(x)\\ &= \int u(x) f(x+a)\, d\mu(x),\end{aligned}$$

by translation invariance of the measure μ, which is none other than $\langle u, T_{-a*} f\rangle$. Thus

$$T_a^* f = T_{-a*} f. \tag{5.15.9}$$

A similar argument shows that

$$\mathscr{F} T_{a*} u = e^{-2\pi i a \cdot \xi} \mathscr{F} u \tag{5.15.10}$$

for u in $\mathscr{S}$. To calculate $\mathscr{F} T_a^* X$ for X in $\mathscr{S}'$ one now uses the Parseval formula twice over:

$$\begin{aligned}\langle \overline{\mathscr{F} u}, \mathscr{F} T_a^* X\rangle = \langle \bar{u}, T_a^* X\rangle &= \langle T_{a*} \bar{u}, X\rangle\\ &= \langle \overline{\mathscr{F} T_{a*} u}, \mathscr{F} X\rangle\\ &= \langle e^{2\pi i a \cdot \xi} \overline{\mathscr{F} u}, \mathscr{F} X\rangle \qquad \text{(using (5.15.10))}\\ &= \langle \overline{\mathscr{F} u}, e^{2\pi i a \cdot \xi} \mathscr{F} X\rangle,\end{aligned}$$

showing that

$$\mathscr{F} T_a^* X = e^{2\pi i a \cdot \xi} \mathscr{F} X. \tag{5.15.11}$$

Formulae (5.15.9)–(5.15.11) together cover the case of translation maps.

Let us next consider the case in which H is a vector-space automorphism of R^n. The analogue of (5.15.9) requires in this case the transformation rule for change of variables in a multiple integral. For an automorphism H this takes the simple form

$$\int (f \circ H)\, d\mu = |\det H|^{-1} \int f\, d\mu,$$

where $\det H$ stands for the determinant of the matrix representing H in terms of the canonical basis for R^n. This result is proved in any one of many text books on analysis. If one specializes still further by taking $H = H_\lambda$, so that $|\det H| = |\lambda|^n$, the result may be reached quite painlessly by repeated application of the formula for the 1-dimensional case, or by use of the characteristic translation invariance of μ. For the formula $f \circ T_{a*} \circ H = f \circ H \circ T_{b*}$, where $b = H^{-1}(a)$, combined with translation invariance of μ, shows that $\mu'(f) = \mu(f \circ H)$ defines μ' as a translation-invariant positive Radon measure on R^n; consequently $\mu' = D(H)\mu$, where $D(H)$ is a strictly positive function of H; it is clear that $D(HH') = D(H)D(H')$ for any two automorphisms H and H' and that $D(I) = 1$, I being the identity automorphism of R^n; thus $d(\lambda) = D(H_\lambda)$ satisfies $d(1) = 1$ and $d(\lambda\lambda') = d(\lambda)d(\lambda')$ for any two nonzero real numbers λ and λ'; hence an easy argument shows, that $d(\lambda) = |\lambda|^k$ for some real number k; thus $\int f(\lambda x)\, d\mu(x) = |\lambda|^k \int f(x)\, d\mu(x)$ holds for all integrable functions f and all real $\lambda \neq 0$, and there is no difficulty in fixing the value of k as $-n$ by considering special functions f.

Granted this transformation formula, one finds that

$$H^*f = |\det H|^{-1} H_*^{-1} f; \tag{5.15.12}$$

compare Schwartz [2], formula (VII, 6; 14).

The analogue of (5.15.10) depends on this same transformation rule, together with the introduction of the adjoint H', the latter being the automorphism of R^n (identified with the dual of R^n via the bilinear form $x \cdot \xi = \sum_{k=1}^n x_k \xi_k$) determined by the identity $Hx \cdot \xi = x \cdot H'\xi$. We notice in passing that $(H^{-1})' = (H')^{-1}$. We find in this way that

$$\mathscr{F} H_* u = |\det H|^{-1} \cdot H_*^{-1\prime} \mathscr{F} u \tag{5.15.13}$$

for u in $\mathscr{S}$.

Finally, (5.15.13) leads to the formula

$$\mathscr{F} H^* X = |\det H|^{-1} \cdot H_*^{-1\prime} \mathscr{F} X \tag{5.15.14}$$

for X in $\mathscr{S}'$, in the same manner as (5.15.10) led to (5.15.11).

The reader will notice that if X belongs to $\mathscr{S}$, and if $\mathscr{F}X$ is a function, then (5.15.14) and (5.15.12) combine to yield

$$\mathscr{F} H^* X = H'_* \mathscr{F} X. \tag{5.15.14$'$}$$

If furthermore X itself is a function, one obtains likewise the result

$$\mathscr{F} H_* X = |\det H|^{-1} \cdot H_*'^{-1} \mathscr{F} X, \tag{5.15.14$''$}$$

extending (5.15.13). Compare Schwartz [2], formulae (VII, 6; 15) and (VII, 6; 16) for the case $H = H_\lambda$, for which $H' = H$.

5.15.5 Homogeneous Functions and Distributions. It is usual to say that a function f on R^n is homogeneous of degree β if $H_{\lambda *} f = |\lambda|^\beta f$ for nonzero real values of λ. In terms of the distribution X generated by f (the latter being assumed to be locally integrable), this is equivalent by (5.15.12) and the fact that $\det H_\lambda = \lambda^n$ to the demand that

$$H_\lambda^* X = |\lambda|^{-n-\beta} X.$$

We accordingly adopt this last relation, verified for all nonzero real λ, as the meaning of the statement that the distribution X is homogeneous of degree β.

When this is the case, (5.15.14) shows that $\mathscr{F}X$ is a distribution which is homogeneous of degree $-n - \beta$ whenever X is homogeneous of degree β.

5.15.6 Functions and Distributions Invariant under Rotations. From the automorphisms of R^n we single out the rotations R by the demand that $Rx \cdot R\xi = x \cdot \xi$ identically in x and ξ. This is equivalent to the condition $R' = R^{-1}$, hence it follows that $\det R = \pm 1$. Sometimes one further subdivides the rotations into those that are "proper" and those which are "improper," the former being precisely those with determinant $+1$, but we shall have no need to make the distinction.

It is clear that the rotationally invariant functions f—that is, those for which $R_* f = f$ for all rotations R—are precisely those expressible as functions of the distance $|x|$ only. A distribution X is rotationally invariant, in the sense that

$$\langle R_* u, X \rangle = \langle u, X \rangle$$

for all rotations R and all u in $\mathscr{D}$, if and only if $R^* X = X$ for all rotations R.

Since $R' = R^{-1}$ and $|\det R| = 1$, (5.15.14) shows that a temperate distribution X is rotationally invariant if and only if its Fourier transform $\mathscr{F}X$ has that same property.

As we shall see in Subsection 5.15.8, the preceding simple remarks concerning homogeneity and rotational invariance are useful aids towards the computation of the Fourier transforms of certain functions and distributions.

5.15.7 **Questions of Consistency.** In what may conveniently be termed the "classical" theory one finds several definitions of the Fourier transform of a function or of a measure, the definitions varying with the assumptions made on the function or measure involved. The two most venerable are those applying to integrable functions and bounded measures on the one hand, and to square-integrable functions on the other. Let us consider these first.

If f is an integrable function on R^n, it is customary to define the Fourier transform $\mathscr{F}f$ as the continuous function on R^n defined by the absolutely convergent integral

$$\mathscr{F}f(\xi) = \hat{f}(\xi) = \int f(x) e^{-2\pi i \xi \cdot x} \, d\mu(x).$$

Our definition of $\mathscr{F}$ on $\mathscr{S}$ is, of course, just a restriction of this one. We have already noted that the Fourier-Schwartz transformation is consistent with this elementary definition inasmuch as concerns functions belonging to $\mathscr{S}$. Precisely similar arguments show that consistency obtains for integrable functions in general. In fact, each integrable f is the limit in $\mathscr{L}^1 = \mathscr{L}^1(R^n, \mu)$ of a sequence (u_n) extracted from $\mathscr{S}$. A fortiori, $\lim u_n = f$ in $\mathscr{S}'$. From convergence in $\mathscr{L}^1$ follows $\lim \mathscr{F}u_n = \mathscr{F}f$ uniformly on R^n. From convergence in $\mathscr{S}'$ follows $\lim \mathscr{F}u_n = \mathscr{F}_1 f$ in $\mathscr{S}'$, where we have temporarily reinstated the suffix in $\mathscr{F}_1$ to distinguish the Fourier-Schwartz transformation. From these two relations we infer that $\mathscr{F}_1 f$ is (the distribution generated by) the function $\mathscr{F}f$, which is what is asserted by the allegation that consistency obtains.

An exactly similar argument covers the case in which the integrable function f gives place to a bounded Radon measure λ, the Fourier transform being now directly definable as the absolutely convergent integral

$$\mathscr{F}\lambda(\xi) = \hat{\lambda}(\xi) = \int e^{-2\pi i \xi \cdot x} \, d\lambda(x).$$

Suppose next that f belongs to $\mathscr{L}^2(R^n)$. The Fourier integral now no longer exists in the pointwise sense. The Plancherel theory, however, discussed in some detail in Section 10.4, provides one solution of this difficulty. If we take any sequence (f_n) in $\mathscr{L}^1 \cap \mathscr{L}^2$ which converges in $\mathscr{L}^2$ to f, then the functions $\mathscr{F} f_n$ converge in $\mathscr{L}^2$; the class (element of L^2) determined by this limit depends only on the class of f (and in no way on the chosen sequence) and is adopted as the meaning to be assigned to $\mathscr{F} f$. This limiting process itself makes certain that consistency with $\mathscr{F}_1 f$ still obtains, a conclusion that might otherwise be reached on the grounds of the Parseval formula being available for either definition.

The classical scene embraces yet other definitions, notably that one which applies to functions belonging to $\mathscr{L}^p$ for $1 < p < 2$ and results from interpolation (by means of the Riesz convexity theorem or some other device) between the preceding constructions for $p = 1$ and $p = 2$. This too is consistent with Schwartz's definition, the reasons being as before. The transform in this case is a function class belonging to $L^{p'}$, where $1/p + 1/p' = 1$. See, for example, Titchmarsh [1], Chapter IV, for the one-dimensional case.

The least satisfactory aspect of any of these theories is the relatively very heavy restrictions they impose on the rate of growth at infinity of the function f to be transformed, and no amount of juggling with the permitted value of the exponent p in $\mathscr{L}^p$ will result in very much improvement. Bochner [1] was seemingly the first to make a break with tradition and showed how one might define the Fourier transform for functions f that are of polynomial order at infinity. As we shall see in a moment, Bochner's ideas are implicitly dependent on some sort of generalized differentiation and the practical value of his scheme was limited by the lack of such techniques at that time. More recently definitions of the Fourier transform have been evolved that allow the function f to have an exponential rate of growth at infinity; see, for example, Carleman [1], Titchmarsh [1]. These depend on the use of complex-variable methods and, speaking generally, share a number of drawbacks inasmuch as the transform so defined lacks some of the basic properties expected of the Fourier transform. We shall leave aside these complex-variable definitions and confine our remarks to Bochner's method.

Bochner's ideas were developed for the case $n = 1$ only and we shall retain this restriction. Suppose to begin with that f is an integrable function. Its Fourier transform may be taken to be the continuous function

$$F(\xi) = \int f(x) e^{-2\pi i \xi x} \, d\mu(x).$$

Let us write $F_0(\xi) = F(\xi)$ and then define F_k inductively by

$$F_k(\xi) = \int_0^{\xi} F_{k-1}(\xi') \, d\mu(\xi').$$

It is then easily verified that

$$F_k(\xi) = \int f(x) E_k(\xi, x) \, d\mu(x), \tag{5.15.15}$$

where

$$E_k(\xi, x) = (-2\pi i x)^{-k} \left[e^{-2\pi i \xi x} - \sum_{0 \le r < k} \frac{-2\pi i \xi x)^r}{r!} \right]. \tag{5.15.16}$$

Now we can write (5.15.15) in the form

$$F_k(\xi) = \int_{-1}^{+1} f(x)E_k(\xi, x)\, d\mu(x) + \int_{|x|\geq 1} f(x)(-2\pi i x)^{-k} e^{-2\pi i \xi x}\, d\mu(x) + P_{k-1}(\xi), \tag{5.15.17}$$

where P_r stands for a polynomial of degree at most r. The recovery of $F = F_0$ from F_k is a simple matter. For, since F is continuous (f being integrable), F_k is of class C^k and

$$F = \frac{d^k F_k}{d\xi^k}. \tag{5.15.18}$$

Bochner's observation amounted to this: first, the sum of the first two terms on the right of (5.15.17) is defined as a continuous function of ξ whenever $(1 + |x|)^{-k} f$ is integrable; second, if we denote this sum by $F_k^*(\xi)$ then, since P_{k-1} is a polynomial of degree at most $k - 1$, (5.15.18) may be replaced by

$$F = \frac{d^k F_k^*}{d\xi^k} \tag{5.15.19}$$

when f is integrable.

He therefore reasoned that (5.15.19) might justifiably be used to *define* F when f is no longer itself integrable but is such that $(1 + |x|)^{-k} f$ has that property.

To do this clearly brings one face to face with the problem of generalized differentiation, for F_k^* will now not generally possess k derivatives (nor even one) in the classical sense. On the other hand one perceives how a satisfactory technique of generalized differentiation contributes towards a generalized concept of the Fourier transformation.

One may also remark at this stage that Bochner's proposed definition appears to share with some of the other more recent ones certain drawbacks that derive from the passage from F_k to F_k^*. For instance, the effect on F_k^* of translating f, or of taking the convolution of f with other functions or measures, is neither as clearly visible nor as simple as one desires (see Subsection 5.15.4).

If one takes the trouble to compare Bochner's proposed definition with Schwartz's, however, interpreting the derivatives appearing in (5.15.19) in the sense of distributions, one finds that F, defined by (5.15.19), is indeed the Fourier-Schwartz transform $\mathscr{F} f = \hat{f}$ of f. Thus the apparently dislikable features of Bochner's definition are illusory. But one must remember that this satisfying conclusion is reached only with the help of the Fourier-Schwartz transformation.

This question of consistency is of sufficient interest to call for closer examination. Suppose then that $(1 + |x|)^{-k} f$ is integrable, that F_k^* is the function defined as the sum of the first two integrals on the right-hand side of (5.15.17), and that F is the distribution defined by (5.15.19). Since the function F_k^* is bounded and continuous, it is evident that F is temperate. Our hypothesis on f ensures that it too is a temperate distribution. Thus the Fourier-Schwartz transform $\hat{f} = \mathscr{F} f$ is defined; it also is a temperate distribution. We aim to show that F and $\hat{f}$ are identical distributions.

To establish this identity it suffices to show that

$$\left\langle \bar{\hat{u}}, \frac{d^k F_k^*}{d\xi^k} \right\rangle = \langle \bar{\hat{u}}, \hat{f} \rangle$$

for each u in a dense subset of $\mathscr{S}$. Applying to the left-hand side the definition of differentiation in the sense of distributions, and to the right-hand side the Parseval formula, we see that our aim is the identity

$$(-1)^k \left\langle \left(\frac{d}{d\xi}\right)^k \bar{\hat{u}}, F_k^* \right\rangle = \langle \bar{u}, f \rangle.$$

This is required to hold for a set of u dense in $\mathscr{D}$. Such a set is composed of the inverse Fourier transforms of functions v in $\mathscr{D}$. Thus we have to show that

$$(-1)^k \int \frac{d^k \bar{v}}{d\xi^k} \cdot F_k^*(\xi)\, d\mu(\xi) = \int f(x)\, d\mu(x) \int \bar{v}(\xi) e^{-2\pi i x \xi}\, d\mu(\xi) \qquad (5.15.20)$$

is valid for each v in $\mathscr{D}$.

To do this we must first observe the relations

$$\int \xi^r \cdot \frac{d^k \bar{v}}{d\xi^k} \cdot d\mu(\xi) = 0 \qquad \text{for } 0 \le r < k, \qquad (5.15.21)$$

it being understood that r and k are integers. Formula (5.15.21) results from repeated partial integrations, all integrated parts vanishing since v belongs to $\mathscr{D}$.

Proceeding with the proof of (5.15.20), we recall that

$$e^{-2\pi i x \xi} = \left(\frac{d}{d\xi}\right)^k E_k(\xi, x) = \left(\frac{d}{d\xi}\right)^k [(-2\pi i x)^{-k} e^{-2\pi i x \xi} + P_{k-1}(\xi, x)],$$

where $P_{k-1}(\xi, x)$ denotes a polynomial in ξ of degree at most $k - 1$ with coefficients that are continuous functions of $x \neq 0$. Thus, bearing in mind relations (5.15.21), it is seen that the inner integral on the right-hand side of (5.15.20) may be transformed by iterated partial integrations into

$$(-1)^k \int \frac{d^k \bar{v}}{d\xi^k} \cdot E_k(\xi, x)\, d\mu(\xi),$$

or, when $x \neq 0$, into

$$(-1)^k \int \frac{d^k \bar{v}}{d\xi^k} \cdot (-2\pi i x)^{-k} e^{-2\pi i x \xi}\, d\mu(\xi).$$

We now split the range of integration in the outer integral into the interval $|x| < 1$ and its complement and insert the first of the above two expressions for the inner integral when $|x| < 1$, and insert the second expression when x belongs to the complementary set. To each repeated integral thus obtained one may apply the Fubini-Tonelli theorem to justify an exchange in the order of the integrations. As a result one obtains for the right-hand side of (5.15.20) the expression

$$(-1)^k \int \frac{d^k \bar{v}}{d\xi^k} \cdot d\mu(\xi) \left[\int_{-1}^{1} f(x) E_k(\xi, x)\, d\mu(x) + \int_{|x|>1} f(x)(-2\pi i x)^{-k} e^{-2\pi i x \xi}\, d\mu(x) \right].$$

The term $[\cdots]$ is visibly none other than $F_k^*(\xi)$, so that (5.15.20) is established.

5.15.8 **Some Examples.** The reader will appreciate that Schwartz's theory does not aim to provide any quick methods for computing the Fourier transforms of specific functions or distributions. The position is in this respect no better and no worse than it is when classical interpretations only are envisaged. We list here merely a few simple examples of Fourier transforms of functions or distributions that are not transformable in the classical sense and some of which will prove to be useful.

In this category of useful results we must certainly include (5.15.3) and (5.15.4), generalized so that one may therein replace u by any temperate distribution (see the end of Subsection 5.15.3).

Since it is obvious that

$$\mathscr{F}\varepsilon = 1, \tag{5.15.22}$$

(5.15.4) leads at once to the following particular cases:

$$\mathscr{F}(D^p\varepsilon) = \xi^p, \tag{5.15.23}$$

$$\mathscr{F}(\Delta\varepsilon) = -4\pi^2 \, |\xi|^2, \tag{5.15.24}$$

where Δ is the Laplacian operator on R^n.

Examples in one dimension. The following examples are presented, sometimes in a slightly different manner, in Chapter 3 of Lighthill [1]; see also Schwartz [2], pp. 113–114, and Gelfand and Šilov [2], Kapitel II, especially pp. 164–193.

We denote by H the Heaviside function—that is, the characteristic function of the interval $(0, \infty)$. The first formula to be established gives the Fourier transform of the function $x^\alpha H$, where $-1 < \alpha < 0$. Thus

$$\mathscr{F}(x^\alpha H) = e^{-(1/2)\pi i(\alpha+1) sgn\, \xi} \alpha! \, (2\pi \, |\xi|)^{-\alpha-1} \tag{5.15.25}$$

Here and elsewhere $\alpha! = \Gamma(\alpha + 1)$ is defined for Re $\alpha > -1$ by the customary integral

$$\int_0^\infty y^\alpha e^{-y} \, d\mu(y)$$

and for other values of α by analytic continuation.

Armed with (5.15.25) and the identity

$$|x|^\alpha = (x^\alpha H) + (x^\alpha H)^\vee,$$

one may infer without difficulty that

$$\mathscr{F}(|x|^\alpha) = -2\alpha! \sin \tfrac{1}{2}\pi\alpha \cdot (2\pi \, |\xi|)^{-\alpha-1} \tag{5.15.26}$$

for $-1 < \alpha < 0$.

As a third example we shall establish the formula

$$\mathscr{F}(\log |x|) = -\tfrac{1}{2}F.P. \, |\xi|^{-1} - (\gamma + \log 2\pi)\varepsilon, \tag{5.15.27}$$

where γ denotes Euler's constant.

Before indicating the proofs of (5.15.25), (5.15.26), and (5.15.27) a word about the range of validity of the formulae (5.15.25) and (5.15.26) is called for. (5.15.25) will be established under the assumption that α is real and

$-1 < \alpha < 0$. Failing this, at least one of $x^\alpha Y$ and $|\xi|^{-\alpha-1}$ ceases to be defined as a distribution. The natural way around this difficulty, however, is to replace each by the corresponding pseudofunction distribution, $F.P.(x^\alpha H)$ and $F.P.\,|\xi|^{-\alpha-1}$, respectively. When this is done, one may extend the formula (5.15.25) by analytic continuation to complex values of α. Similar remarks apply to (5.15.26).

Proof of (5.15.25). Since

$$x^\alpha H = \lim_{t\to 0} x^\alpha e^{-tx} H$$

in $\mathscr{S}'$, so

$$\mathscr{F}(x^\alpha H) = \lim_{t\to 0} \mathscr{F}(x^\alpha e^{-tx} H).$$

On the other hand, if $t > 0$, the function $x^\alpha e^{-tx} H$ is integrable and so by consistency its Fourier transform is the function

$$\int_0^\infty x^\alpha e^{-tx} e^{-2\pi i\xi x}\, d\mu(x),$$

the value of which is

$$\alpha!\,(t + 2\pi i\xi)^{-\alpha-1},$$

principal values being taken for fractional powers of a complex number. On allowing t to approach zero through positive values, one obtains for the limit the right-hand side of (5.15.25). ▮

Proof of (5.15.26). The method of proof of (5.15.26) has already been indicated. It is necessary merely to observe the general rule that the transform of $\check{X}$ is $(\mathscr{F}X)^\vee$. ▮

Proof of (5.15.27). Since

$$\log|x| = \lim_{\alpha\to 0} \alpha^{-1}(1 - |x|^{-\alpha})$$

in $\mathscr{S}'$, so formula (5.15.26) leads to

$$\mathscr{F}(\log|x|) = \lim_{\alpha\to 0} \alpha^{-1}[\varepsilon - 2(-\alpha)!\,\sin\tfrac{1}{2}\pi\alpha\cdot(2\pi\,|\xi|)^{\alpha-1}].$$

To handle this we make use of the expansions

$$(-\alpha)! = 1 - \gamma\alpha + O(\alpha^2),$$
$$\alpha^{-1}\sin\tfrac{1}{2}\pi\alpha = \tfrac{1}{2}\pi[1 + O(\alpha^2)],$$
$$(2\pi)^{\alpha-1} = (2\pi)^{-1}[1 + \alpha\log 2\pi + O(\alpha^2)].$$

Thus

$$\alpha^{-1}\varepsilon - 2(-\alpha)!\,\alpha^{-1}\sin\tfrac{1}{2}\pi\alpha(2\pi\,|\xi|)^{\alpha-1}$$
$$= \alpha^{-1}\varepsilon - \tfrac{1}{2}[1 + (\gamma + \log 2\pi)\alpha + O(\alpha^2)]\,|\xi|^{\alpha-1}.$$

A little juggling with partial integration shows that

$$\lim_{\alpha\to 0} \alpha\,|\xi|^{\alpha-1} = 2\varepsilon$$

in $\mathscr{S}'$, while

$$O(\alpha^2)\,|\xi|^{\alpha-1} \to 0$$

in the same sense. Thus

$$\begin{aligned}\mathscr{F}(\log|x|) &= -(\gamma + \log 2\pi)\varepsilon - \tfrac{1}{2}\lim_{\alpha\to 0}\alpha^{-1}(\alpha\,|\xi|^{\alpha-1} - 2\varepsilon)\\ &= -(\gamma + \log 2\pi)\varepsilon - \tfrac{1}{2}\lim_{\alpha\to 0} X_\alpha,\end{aligned}$$

say, in the sense of $\mathscr{S}'$. To establish (5.15.27), it therefore suffices to show that $\lim_{\alpha\to 0} X_\alpha$ (known to exist) is $F.P.\ |\xi|^{-1}$.

Now it is easy to verify that

$$F.P.\ |\xi|^{-1} = \frac{d(sgn\ \xi \cdot \log|\xi|)}{d\xi}.$$

On the other hand if u belongs to $\mathscr{D}$ one has

$$\langle u, X_\alpha\rangle = \int u(\xi)\,|\xi|^{\alpha-1}\,d\mu(\xi) - 2\alpha^{-1}u(0).$$

The integral is transformed by partial integration, treating separately the intervals $(-\infty, 0)$ and $(0, \infty)$. As a result one finds that

$$\langle u, X_\alpha\rangle = -\int_0^\infty u'(\xi)\cdot\alpha^{-1}(\xi^\alpha - 1)\,d\mu(\xi) + \int_0^\infty u'(-\xi)\cdot\alpha^{-1}(\xi^\alpha - 1)\,d\mu(\xi).$$

When α tends to zero, this clearly tends to

$$\begin{aligned}-\int_0^\infty u'(\xi)\log\xi\,d\mu(\xi) + \int_0^\infty u'(-\xi)\log\xi\,d\mu(\xi)\qquad&\\ = -\int u'(\xi)\,sgn\ \xi\cdot\log|\xi|\,d\mu(\xi)&\\ = \left\langle u, \frac{d}{d\xi}sgn\ \xi\cdot\log|\xi|\right\rangle = \langle u, F.P.\ |\xi|^{-1}\rangle.&\end{aligned}$$

Accordingly $\lim_{\alpha\to 0} X_\alpha = F.P.\ |\xi|^{-1}$, at any rate weakly in $\mathscr{S}'$, which is all that is required. This proves (5.15.27). ▮

THE FUNCTION $|x|^{-\alpha}$ IN SEVERAL DIMENSIONS. We consider the function $|x|^{-\alpha}$ on R^n. This is locally integrable provided $\text{Re}\ \alpha < n$. Since it is rotationally invariant, its Fourier transform has that same property (see Subsection 5.15.6). Moreover, since $|x|^{-\alpha}$ is homogeneous of degree $-\alpha$, the results derived in Subsection 5.15.5 show that the Fourier transform is homogeneous of degree $\alpha - n$.

Now for certain values of α, including at any rate those for which $\text{Re}\ \alpha > \frac{1}{2}n$, $\mathscr{F}(|x|^{-\alpha})$ can be shown to be a function. When this is so, homogeneity and rotational invariance fix this transform up to a constant factor: it must in fact be $C_{n,\alpha}\,|\xi|^{\alpha-n}$, where $C_{n,\alpha}$ is a number depending only on n and α. The value of $C_{n,\alpha}$ may be determined by using the Parseval formula (5.15.7) with v taken to be $\exp(-\pi|x|^2)$ and therefore $\mathscr{F}v = \exp(-\pi|\xi|^2)$. The result reads

$$C_{n,\alpha} = \pi^{-(1/2)n+\alpha}\Gamma[\tfrac{1}{2}(n-\alpha)]/\Gamma(\tfrac{1}{2}\alpha).$$

For other values of α one must proceed by considering analytic continuation and the introduction of pseudofunctions. See Schwartz [2], pp. 113 et seq.

5.15.9 **The Exchange Formula.** Schwartz ([2], p. 124) attaches this description to a pair of formulae, only one of which need concern us in the sequel. The formula is one expressing the exchange between convolutions and ordinary products that is effected by the Fourier transformation:

$$\mathscr{F}(S * X) = \mathscr{F}S \cdot \mathscr{F}X. \tag{5.15.28}$$

The formula is classical if S and X are integrable functions or bounded measures: In these cases the Fourier transform may be defined in a pointwise sense by absolutely convergent integrals and (5.15.28) is a consequence of the Fubini-Tonelli theorems. It is, however, clear that (5.15.28) does not even make sense if S and X are unrestricted temperate distributions. Schwartz (loc. cit.) gives a set of sufficient conditions for its validity; others are given in Edwards [4]; and still more are given in Hirata & Ogata [1].

For our purposes it is enough to deal with the case in which S has a compact support and X is temperate. A proof of the formula for this case is our immediate concern.

(1) According to Theorem 5.8.1, S is a finite sum of terms $D^p\lambda$, where λ is a Radon measure with a compact support. By linearity it is therefore plainly enough to verify the formula when $S = D^p\lambda$.

On the other hand, if S takes this form, (5.15.4) shows that

$$\mathscr{F}[(D^p\lambda) * X] = \mathscr{F}[D^p(\lambda * X)] = \xi^p \cdot \mathscr{F}(\lambda * X)$$

and that

$$\mathscr{F}(D^p\lambda) = \xi^p \cdot \mathscr{F}\lambda.$$

Consequently the problem is reduced to proving the formula in the case where $S = \lambda$.

(2) Choose any directed family (u_i) in $\mathscr{S}$ that converges in $\mathscr{S}'$ to X. (Even weak convergence is enough here.) Then one has for each i

$$\mathscr{F}(\lambda * u_i) = \mathscr{F}\lambda \cdot \mathscr{F}u_i,$$

which is a consequence of the pointwise integral representations of the Fourier transforms involved, together with the Fubini-Tonelli theorem.

Now, since λ has a compact support, the formula

$$\langle v, \lambda * u_i \rangle = \langle \check{v} * \lambda, u_i \rangle$$

shows at once that $\lambda * u_i \to \lambda * X$ weakly in $\mathscr{S}'$. Hence

$$\mathscr{F}(\lambda * X) = \lim \mathscr{F}(\lambda * u_i)$$

weakly in $\mathscr{S}'$. At the same time, since $X = \lim u_i$, therefore $\mathscr{F}X = \lim \mathscr{F}u_i$, in each case weakly in $\mathscr{S}'$. That

$$\mathscr{F}(\lambda * X) = \mathscr{F}\lambda \cdot \mathscr{F}X$$

now follows on passage to the limit.

This establishes (5.15.28) under the stated hypotheses.

5.15.10 **The Paley-Wiener Theorem and Lions' "Supports Theorem."** These two theorems are essential in the study of linear partial differential equations. Although Lions' theorem makes no reference to the Fourier transformation, both its proof and its applications have a close affinity with Fourier techniques. The Paley-Wiener theorem is, however, a central theorem in harmonic analysis. We now have enough material concerning the Fourier transformation to make the result fully intelligible. The theorems will be stated without proofs here.

LIONS' "SUPPORTS THEOREM." Let X and Y be distributions on R^n having compact supports. The closed convex envelope of the support of $X * Y$ is precisely the vectorial sum of the closed convex envelopes of the supports of X and of Y.

This theorem and its proof are to be found in Lions [4].

The Paley-Wiener theorem in its original form (Paley and Wiener [1], p. 12; Zygmund [2], Vol. II, p. 272) gave a characterization of those functions of one complex variable obtainable as Fourier transforms of functions in $L^2(R^1)$ having compact supports. It remained for Schwartz ([2], p. 128) to remove the L^2 hypothesis and to extend the theorem to the case of several variables; extended thus, the theorem is obviously more natural and useful.

Let us first make a few observations about the Fourier transform $\hat{X} = \mathscr{F}X$ of a distribution X on R^n having a compact support contained in the hypercube with centre 0 and side $2c$.

To begin with, we know that X is a finite sum of derivatives $D^p\lambda$, where λ is a Radon measure on R^n with a compact support. Also, as we have remarked in Subsection 5.15.7, the transform of λ is (the distribution generated by) the function

$$\hat{\lambda}(\xi) = \int e^{-2\pi i\xi \cdot x}\, d\lambda(x).$$

According to (5.15.4) in generalized form, the transform of X is therefore a finite sum of functions $\xi^p \cdot \hat{\lambda}(\xi)$. Thus $\hat{X}$ is seen to be extendible into an entire function on C^n that is of polynomial order at infinity on R^n. In addition to this, since each λ may be chosen to have its support contained in the hypercube with centre 0 and side $2c'$, for any $c' > c$, one sees directly that for ζ in C^n

$$|\hat{\lambda}(\zeta)| \leq \int d\,|\lambda|\,(x) \cdot \exp\left(2\pi c' \sum_{k=1}^{n} |\zeta_k|\right).$$

Hence

$$\limsup_{|\zeta|\to\infty} \frac{\log^+ |\hat{X}(\zeta)|}{|\zeta|_1 + \cdots + |\zeta_n|} \leq 2\pi c.$$

In other words, the entire function $\hat{X}$ is of exponential type at most $2\pi c$.

Schwartz's form of the Paley-Wiener theorem amounts to the assertion that these conditions on X serve to characterize it as the transform of some distribution with a compact support.

PALEY-WIENER-SCHWARTZ THEOREM. Let X belong to $\mathscr{S}'$. In order that X shall have a compact support contained in the hypercube with centre 0 and side $2c$, it is necessary and sufficient that its Fourier transform $\hat{X}$ be an entire function of exponential type at most $2\pi c$.

Alternatively: In order that a function F on R^n shall be the Fourier transform of a distribution on R^n with a compact support contained in the hypercube with centre 0 and side $2c$, it is necessary and sufficient that (1) F be of polynomial order at infinity on R^n and (2) F be extendible into a function on C^n that is entire and of exponential type at most $2\pi c$.

5.15.11. Before leaving the theory of the generalized Fourier transform mention should be made of Ehrenpreis' theory [7]. Here one introduces the space $\underline{D} = \mathscr{F}(\mathscr{D})$, the space of all functions that are Fourier transforms of functions belonging to $\mathscr{D} = \mathscr{D}(R^n)$. Functions in $\underline{D}$ are naturally regarded as functions of n complex variables. As such they are characterized as comprising exactly those entire analytic functions on C^n of exponential type and rapidly decreasing on R^n. This characterization is based on an appeal to the Paley-Wiener-Schwartz theorem. $\underline{D}$ is topologized by requiring that $\mathscr{F}$ shall be a homeomorphism of $\mathscr{D}$ onto $\underline{D}$. It is not difficult to express this topology directly in terms of the functional values of elements of $\underline{D}$, but this is largely irrelevant to the definition to be made. By using the closed graph theorem one can verify that $\mathscr{F}^{-1}$ is continuous on $\underline{D}$ into $\mathscr{D}$, so that $\mathscr{F}$ is a topological isomorphism of $\mathscr{D}$ onto $\underline{D}$. This being so one can define the Fourier transform acting on arbitrary distributions to be $(\mathscr{F}^{-1})'$, the transform being a member of $\underline{D}'$. In other words, Ehrenpreis' transform $\mathscr{F}T$ is defined for T in $\mathscr{D}'$ by

$$\langle f, \mathscr{F}T\rangle = \langle \mathscr{F}^{-1}f, T\rangle$$

holding for each f in $\underline{D}$. We compare this with the definition in Subsection 5.15.3 of Schwartz's transform when T is temperate; the two definitions are consistent. For details see Ehrenpreis [7] and Gelfand and Šilov [2], Kapitel II.

As remarked in Subsection 5.17.3, Ehrenpreis uses his concept of the Fourier transform in the study of convolution equations, including LPDEs as special cases.

The additional generality in Ehrenpreis' approach has to be paid for in the sense that one cannot compare locally two elements of $\underline{D}'$ as one can do with Schwartz distributions. This is because the new test functions (members of $\underline{D}$) cannot vanish on a nonvoid open set without vanishing everywhere.

5.16 Discussion of Problem 1

We shall present here a few results bearing upon the solution of Problem 1 of Subsection 5.12.3. Although we deal in detail with only part of the information currently available, and that from only one of several possible viewpoints, we shall subsequently comment on further results and alternative approaches.

For our purposes it seems best to follow quite closely the exposition given by Malgrange [1], or Schwartz [8], or by both. A systematic treatment is contained in Hörmander [7].

5.16.1 Remarks on Power Series. Malgrange's approach to Problem 1 and to other more general similar questions makes use of several nontrivial properties of power series in several indeterminates, convergent or formal. For Problem 1 itself some of these results are avoidable. We comment briefly on, and give references to proofs of, those that remain indispensable.

We consider first formal power series in n indeterminates. For the basic definitions and properties, the reader may turn to Bourbaki [3a], pp. 52–69. We shall be concerned solely with the case of series with coefficients in the field C of complex numbers. The algebra of formal power series in n indeterminates (with coefficients in C), hereafter denoted by F_n, may be constructed as follows. We denote as usual by p an n-uple of integers $p_k \geq 0$, $1 \leq k \leq n$; thus, if I denotes the set of integers ≥ 0, p is an element of I^n. $|p|$ stands for $p_1 + \cdots + p_n$. Elements of F_n are families $f = (a_p)$, the index p ranging over I^n, each a_p being a complex number; F_n is, as a set, simply $C^{(I^n)}$. As a power set of C it has a natural vector-space structure over C. We turn it into an algebra in the ensuing manner. For each p, we denote by X^p that element of F_n that is the family taking the value 1 at p and the value 0 at all other indices. Then $f = (a_p)$ can be represented formally as a sum $\sum a_p X^p$. The product of X^p and X^q is *defined* to be X^{p+q}, and this operation is extended to an arbitrary pair of elements of F_n by postulating that it shall be bilinear with respect to formal sums. In other words, if $f = (a_p)$ and $g = (b_p)$, the product fg is (c_p), where $c_p = \sum_{r+s=p} a_r b_s$.

It is evident that this algebraic structure is dictated by thinking of (a_p) as a "power series"

$$\sum a_p X^p = \sum a_{p_1, \cdots, p_n} X_1^{p_1} \cdots X_n^{p_n}$$

in n "indeterminates" and multiplying these series quite formally in accordance with Cauchy's rule. The formal definition given above avoids the tricky problem of defining, in mathematical terms, what is meant by "an indeterminate."

The series $\sum a_p X^p$, so far regarded as purely formal, can be regarded as a series of elements of F_n that converges relative to the topology of convergence of each coefficient—that is, the locally convex vector-space topology on F_n having a base at 0 comprised of the sets $\{(a_p) : |a_p| \leq \varepsilon \text{ for } p \in J\}$, where ε ranges over strictly positive numbers and J over the finite subsets of I^n. In this way F_n is made into a separated LCTVS; it is easy to verify that F_n is in fact a Fréchet space. F_n is indeed a topological algebra. The dual of F_n may be identified with $\mathcal{K}(I^n)$ (the vector space of complex-valued functions on I^n having finite supports, thinking of I^n as a discrete space) in such a way that $u = (u_p)$ in $\mathcal{K}(I^n)$ generates the linear form

$$(a_p) \to \sum u_p a_p.$$

If $f = (a_p) = \sum a_p X^p$ is an element of F_n, and if $\zeta = (\zeta_1, \cdots, \zeta_n)$ is a point of C^n, it makes sense to ask whether or not the series of complex numbers $\sum a_p \zeta^p$ (where $\zeta^p = \zeta_1^{p_1} \cdots \zeta_n^{p_n}$) is absolutely convergent. We shall denote by CF_n the set of f in F_n for which this series of complex numbers is absolutely convergent for each ζ in some neighbourhood of 0 in C^n: we then speak of f as a convergent power series in n indeterminates (or variables), and it is natural to denote by $f(\zeta)$ or $\sum a_p \zeta^p$ the sum of the formal series f, whenever ζ is such that $\sum a_p \zeta^p$ is an absolutely series of complex numbers.

It is obvious that CF_n is a subalgebra of F_n that is everywhere dense in F_n (since it contains all finite series, that is, all polynomials, in n indeterminates).

CF_n may be identified, as an algebra, with the algebra of germs of holomorphic functions at 0, defined in Example 3 of Section 6.3. However, the topology induced on CF_n by that of F_n is strictly weaker than the inductive limit topology defined in Section 6.3.

We may now describe the results concerning F_n and CF_n needed to follow Malgrange's approach to Problem 1 and similar questions.

Theorem a. If $f, g_1, \cdots, g_r$ belong to CF_n, in order that f shall belong to the ideal in CF_n generated by $g_1, \cdots, g_r$, it is sufficient (and trivially necessary) that it belong to the ideal in F_n generated by $g_1, \cdots, g_r$. In other words, if

$$f = \sum_{j=1}^{r} h_j g_j$$

is soluble with h_j in F_n, then it is soluble with h_j in CF_n.

Theorem b (Krull's Lemma). Every ideal in F_n is closed.

For proofs of these results, see Cartan [3], Exposé XI, Théoremè III, Corollaire and Théoremè IV, Corollaire. In addition to these "local" theorems there is a related "global" theorem due to Cartan [4].

Theorem c. Let $f, g_1, \cdots, g_r$ be entire functions on C^n. Suppose that for each α in C^n the element of CF_n defined by $f(\alpha + \zeta)$ belongs to the ideal in F_n generated by the elements of CF_n defined by $g_1(\alpha + \zeta), \cdots, g_r(\alpha + \zeta)$. Then f belongs to the global ideal generated by $g_1, \cdots, g_r$—that is, $f = \sum_{j=1}^{r} h_j g_j$ for suitably chosen entire functions h_r on C^n.

For the general problems dealt with by Malgrange the full force of each of Theorems a, b, and c is necessary. For a study of Problem 1 alone, which is our primary aim, we need only invoke Theorem b for principal ideals (ones with a single generator) and only the case $r = 1$ of Theorems a and c. These special cases are simpler to handle, although they are still not obvious.

We have to add one remark concerning the representation of the dual of F_n. If $f = \sum a_p X^p$ belongs to CF_n, then a_p may be expressed in terms of f by the usual Taylor formula

$$\begin{aligned} a_p &= (p!)^{-1}\, \partial^p f(0) \\ &= (p!)^{-1}(-1)^{|p|}\langle f, \partial^p \varepsilon \rangle \end{aligned}$$

where, on the right-hand side, f denotes the function (having as domain some neighbourhood of 0 in C^n) defined by the convergent power series f. For a general f in F_n, there is no such function, but we may define $\partial^p f(0)$ in terms of a_p by the above formulae. This means that the distribution $\partial^p \varepsilon$ is defined in its action as a linear form on F_n by

$$\langle f, \partial^p \varepsilon \rangle = (-1)^{|p|} p!\, a_p.$$

What has been said already about the topological dual of F_n may accordingly be translated into saying that the dual of F_n may be identified with the vector space of distributions having $\{0\}$ as their support, any such distribution being expressible uniquely as a finite linear combination of derivatives $\partial^p \varepsilon$ of the Dirac measure ε placed at the origin.

5.16.2 **Preliminaries Concerning Convolution Equations.** By using the results about power series described in Subsection 5.16.1 we may now establish assertions concerning convolution equations involving distributions with compact supports that will lead directly to a solution of Problem 1. The link between distributions with compact supports and power series is provided by the Fourier transformation. As we have seen in Subsection 5.15.10, if X is a distribution on R^n with a compact support, its Fourier transform $\hat{X}$ is (extendible into) an entire function on C^n; as such it may be identified with its Taylor series about the origin, which is an element of CF_n, and this correspondence is one to one. On the other hand, if Y is another such distribution, the Fourier transform of $X * Y$ is $\hat{X} \cdot \hat{Y}$. This transformation therefore turns convolutions into ordinary products and is thus a mechanism whereby statements about ideals in F_n and CF_n yield information about convolution equations.

We proceed to make some relevant deductions along these lines. For brevity we write C^∞ for $C^\infty(R^n)$ and $C'^\infty = C'^\infty(R^n)$, the latter being the space of distributions with compact supports.

Lemma 1. Suppose that A and B belong to C'^∞ and that

(α) g a polynomial and $A * g = 0$

implies

(β) $B * g = 0$.

Then $\hat{B}$ belongs to the ideal in CF_n generated by $\hat{A}$.

Conversely, if $\hat{B}$ belongs to this ideal, then $(\alpha) \Rightarrow (\beta)$.

Proof. Let M be the ideal in F_n generated by $\hat{A}$. According to Theorem a, Subsection 5.16.1, it suffices to show that $\hat{B}$ belongs to M. By Theorem b, Subsection 5.16.1, M is closed in F_n. Thus we may appeal to the Hahn-Banach theorem and thus reduce the problem to that of showing that any continuous linear form on F_n that is orthogonal to M is also orthogonal to $\hat{B}$. Now by taking Fourier transforms the implication $(\alpha) \Rightarrow (\beta)$ amounts to this: if X is any distribution of the type $P(D)\varepsilon$, $P(D)$ being any LPDO with constant coefficients, and if $\hat{A} \cdot X = 0$, then $\hat{B} \cdot X = 0$. Thus $\hat{A} \cdot X = 0$ implies in particular $\langle \hat{B}, X \rangle = \langle 1, \hat{B} \cdot X \rangle = 0$. In other words, if $\langle \hat{A}\varphi, X \rangle = 0$ for all φ in $\mathscr{D} = \mathscr{D}(R^n)$, then $\langle \hat{B}, X \rangle = 0$. However, by virtue of the special nature of X, to say that $\langle \hat{A}\varphi, X \rangle = 0$ for all φ in $\mathscr{D}$ is equivalent to saying that

$\langle \hat{A}F, X\rangle = 0$ for all F in F_n—that is, that X is orthogonal to M. Since, as we have observed, these X generate the dual of F_n, we have reached the desired conclusion.

The arguments may be reversed to show that the implication $(\alpha) \Rightarrow (\beta)$ is necessary, as well as sufficient, to ensure that $\hat{B}$ belongs to the ideal in CF_n generated by $\hat{A}$. ▮

Lemma 2. Let $\lambda = (\lambda_1, \cdots, \lambda_n)$ be a point of C^n. If A and B are distributions on R^n with compact supports, in order that $\hat{B} = F\hat{A}$ for some function F holomorphic on a neighbourhood of λ, it is necessary and sufficient that

(α') g a polynomial and $A * (ge^{2\pi i\lambda\cdot x}) = 0$

implies

(β') $B * (ge^{2\pi i\lambda\cdot x}) = 0$.

Proof. One may reduce this to Lemma 1 by considering $A' = e^{-2\pi i\lambda\cdot x} A$ and $B' = e^{-2\pi i\lambda\cdot x} B$, observing that $\hat{A}'(\zeta) = \hat{A}(\lambda + \zeta)$, $\hat{B}'(\zeta) = \hat{B}(\lambda + \zeta)$, and that

$$A * (ge^{2\pi i\lambda\cdot x}) = e^{2\pi i\lambda\cdot x}(A' * g), \tag{5.16.1}$$

and likewise with B and B' replacing A and A' respectively. ▮

By combining these lemmas we obtain the two principal auxiliary results.

Proposition 1. If A and B are distributions with compact supports, in order that $\hat{B} = F\hat{A}$ for some entire function F on C^n, it is necessary and sufficient that, for any λ in C^n, one has $(\alpha') \Rightarrow (\beta')$ (the notation being as in Lemma 2).

Proof. The necessity appears at once from the necessity part of Lemma 2. As for sufficiency, if $(\alpha') \Rightarrow (\beta')$ for each λ, then the sufficiency part of Lemma 2 shows that $\hat{B} = F_\lambda\hat{A}$ for some F_λ that is holomorphic on a neighbourhood of λ. Theorem c of Subsection 5.16.1 is now immediately applicable to show that $\hat{B} = F\hat{A}$ for some entire function F on C^n. ▮

Proposition 2. Let $P(D)$ be a LPDO with constant coefficients, and put $F = \widehat{P(D)\varepsilon}$. Suppose that X is a distribution with a compact support. In order that there shall exist a distribution Y with a compact support such that $P(D)Y = X$, it is necessary and sufficient that $\hat{X}/F$ be an entire function on C^n.

Proof. The necessity is evident since $X = P(D)Y$ is equivalent to $\hat{X} = F\cdot\hat{Y}$, while $\hat{Y}$ is entire.

As for sufficiency, suppose that $\hat{X} = F \cdot G$, where G is an entire function on C^n. We know that here $\hat{X}$ is entire, of exponential type, and of polynomial order at infinity on R^n (see Subsection 5.15.10). According to the Paley-Wiener-Schwartz theorem (Subsection 5.15.10) it suffices to show that the entire function $G = \hat{X}/F$ is of exponential type and of polynomial order on R^n.

Excluding the trivial case in which $P(D) = 0$ we may assume, by a preliminary linear change of coordinates, that the polynomial F is of the type

$$F(\zeta_1, \cdots, \zeta_n) = c\zeta_1^m + \sum_{r=1}^{m} \zeta_1^{m-r}F_r(\zeta_2, \cdots, \zeta_n),$$

c being a nonzero complex number and the F_r polynomials on C^{n-1}. The desired properties of G then flow at once from the Lemma of Subsection 5.14.3. ▮

Remarks. Suppose we assume that X belongs to $\mathscr{D}$, the other hypotheses remaining undisturbed. We may then conclude that $X = P(D)Y$ for some Y in $\mathscr{D}$. For we know already that Y can be chosen from C'^{∞}. Using Theorem A of Section 5.14, we choose an elementary solution E for $P(D)$. Then $X = P(D)Y$ gives $E * X = Y$ which, since X belongs to $\mathscr{D}$, shows that Y belongs to $C^{\infty} \cap C'^{\infty} = \mathscr{D}$.

Other variants become available if we use the substance of the Remarks at the end of Subsection 5.14.2, according to which E may be chosen so that $E * h$ belongs to L^2_{loc} whenever $h \in L^2$ has a compact support. Thus, if we know that X has a compact support and that $D^p X \in L^2_{\text{loc}}$ for $|p| \leq k$, then Y may be chosen $(= E * X)$ so as to satisfy the same conditions. Recall that

$$D^p Y = D^p(E * X) = E * D^p X.$$

5.16.3 **Solution to Problem 1.** We shall give several variants differing in respect to the spaces of functions or distributions involved.

Theorem A. Let Ω be a convex domain in R^n, $P(D)$ a LPDO with constant coefficients on R^n. Each f in $C^{\infty}(\Omega)$ satisfying $P(D)f = 0$ on Ω is the limit in $C^{\infty}(\Omega)$ of finite linear combinations of exponential polynomials satisfying the same equation.

Proof. Suppose $X \in C'^{\infty}(\Omega)$ is orthogonal to all exponential polynomials g satisfying $P(D)g = 0$. It suffices (Hahn-Banach theorem) to show that X is orthogonal to any solution $f \in C^{\infty}(\Omega)$ of the same equation. Now our hypothesis says that $\langle e^{2\pi i\lambda\cdot x}g, X\rangle = 0$ whenever $\lambda \in C^n$, g is a polynomial, and

$$P(D)(e^{2\pi i\lambda\cdot x}g) = 0.$$

In other words, the relation

$$P(D)\,\varepsilon * (e^{2\pi i\lambda\cdot x}g) = 0$$

entails

$$\check{X} * (e^{2\pi i\lambda\cdot x}g) = 0.$$

According to Propositions 1 and 2 of Subsection 5.16.2, therefore, $\check{X} = P(D)Y$, or equivalently $X = \check{P}(D)\check{Y}$, for some distribution Y with a compact support. Also, since Ω is convex, Lions' "supports theorem" (Subsection 5.15.10) shows that the support of $\check{Y}$ is contained in Ω. So if $f \in C^{\infty}(\Omega)$ and $P(D)f = 0$ on Ω, we have $\langle f, X\rangle = \langle f, \check{P}(D)\check{Y}\rangle = \langle P(D)f, \check{Y} = \langle 0, \check{Y}\rangle = 0$, as required. ▌

Theorem A'. Theorem A remains valid if throughout we replace $C^{\infty}(\Omega)$ by $C^k(\Omega)$ $(k = 0, 1, 2, \cdots)$.

Proof. It suffices to regularize f, thus obtaining a solution $f' \in C^{\infty}(\Omega)$, and then to apply Theorem A itself. ▌

Theorem A". Theorem A remains valid if throughout we replace $C^{\infty}(\Omega)$ by $\mathscr{D}'(\Omega)$ or by $\mathscr{D}'^k(\Omega)$ $(k = 0, 1, 2, \cdots)$.

Proof. The proof of Theorem *A* may be repeated, except that now X will belong to $\mathscr{D}(\Omega)$ or to $\mathscr{D}^k(\Omega)$. Therefore, according to the remarks following the proof of Proposition 2 of Subsection 5.16.2, Y may likewise be chosen from $\mathscr{D}$ or $\mathscr{D}^k$ as the case may be. ▌

These basic results are due to Malgrange; see Malgrange [1], pp. 24–26, or Schwartz [8], Exposé 1, or both. For $\Omega = R^n$, a simplified proof of Theorem A has been given recently by Malgrange.

5.16.4 **Supplements.** Malgrange has pursued the study of both Problems 1 and 1′ (and of Problem 3: see Section 5.17) much further, and we shall attempt now to describe briefly some of the advances due to him.

We have seen in Subsection 5.12.2 how holomorphic functions (of one complex variable) may be thought of as the solutions of $P(D)X = 0$ when, with our general notation, $n = 2$ and $P(D) = (\partial_1 + i\partial_2)$. At the same time we commented on the relationship between a solution of Problem 1′ in this case and Runge's theorem. The latter result may indeed be formulated thus:

▶ If $\Omega \subset R_2$ is simply connected, any solution f of

$$P(D)f = \partial f/\partial \bar{z} = (\partial_1 + i\partial_2)f = 0$$

in Ω is the limit of solutions of $P(D)g = 0$ throughout the whole of R^2 (the latter being in this case simply the polynomials g in the complex coordinate $z = x_1 + ix_2$).

It is obvious that this is a refinement of theorems like A, A', and A'' above in at least two respects—viz. (1) the convexity hypothesis on Ω is weakened, and (2) exponential polynomials are replaced by (ordinary) polynomials. This special case proves to be indicative of what Malgrange has discovered to be true for quite large categories of LPDOs $P(D)$ and domains Ω in R^n. We should mention, too, that Malgrange considers even more general situations in which Ω is allowed to be a domain in an indefinitely differentiable manifold M (rather than merely a domain in R^n); lack of space prevents us from describing the formalities that must precede this sort of generalization.

In Proposition 4, p. 66, and Théorème 6, p. 73, of [1] Malgrange gives sufficient conditions on $P(D)$ and on the domain Ω in M in order that each f in $C^\infty(\Omega)$ satisfying $P(D)f = 0$ in Ω shall be the limit in $C^\infty(\Omega)$ of solutions g of $P(D)g = 0$ in M; and in Proposition 8, p. 68, it is shown that in certain stated circumstances these sufficient conditions are also necessary. The conditions on $P(D)$ involved are rather complex and include among other things the assumption that $P(D)$ is elliptic in the sense that it has an elementary solution that is real analytic except at the origin. The condition concerning Ω is one introduced first by Grothendieck [8]. It amounts to the demand that $M \backslash \Omega$ possess no connected components that are relatively compact in M. These results are also discussed in part in Schwartz [8], Exposé 3 bis.

If results of this calibre are applied to the case in which M is a noncompact Riemann surface (Malgrange [1], pp. 75–77), one obtains the form of Runge's theorem for such surfaces due to Behnke and Stein [1], which itself entails that the surface is a Stein manifold. This is a conclusion of major importance and has many implications.

5.17 Solution of Problem 3

By taking together the results of Sections 5.14 and 5.16, we are able now to follow Malgrange ([1], Chapitre III, § 3) in contributing something towards the solution of Problem 3, that is, the problem of the solubility of the equation $P(D)X = A$.

Our detailed account will be confined to the case in which the underlying domain Ω is a convex domain in R^n, but in Subsection 5.17.4 we shall indicate briefly what is known in more general situations. See also Hörmander [7].

Throughout this section $P(D)$ denotes a LPDO with constant coefficients, $P(D) \neq 0$.

5.17.1 The Case Ω Convex. As in Subsection 5.16.3 we shall give several theorems, each concerned with the solubility of $P(D)X = A$, which differ in respect to the spaces of functions or distributions concerned.

Theorem A. Let k be an integer ≥ 0 or ∞, and let Ω be a convex domain in R^n. If f belongs to $C^{k+n+1}(\Omega)$, the equation $P(D)g = f$ is soluble with g in $C^k(\Omega)$. (If $k = \infty$, $k + n + 1$ is understood to be ∞ also.)

Proof. We exhaust Ω by an increasing sequence (ω_r) of convex open sets, each relatively compact in Ω, $\bar{\omega}_{r-1} \subset \omega_r$, $\omega_0 = \emptyset$. By Proposition 5.6.1, we may choose functions α_r in $\mathscr{D}$ forming a partition of unity and such that the support of α_r lies within $\Omega\backslash\bar{\omega}_r$. Accordingly $f = \sum_r f_r$, where $f_r = \alpha_r f$ belongs to C^{k+n+1} and has its support in $\Omega\backslash\bar{\omega}_r$.

According to Theorem A of Subsection 5.14.2, there exists an elementary solution for $P(D)$, say E, which is of order at most $n + 1$. Then, since $P(D)E = \varepsilon$, $E * f$ belongs to C^k and

$$P(D)(E * f_r) = P(D)E * f_r = f_r.$$

A formal solution of the given equation, $P(D)g = f$, would thus appear to be given by $g = \sum_r (E * f_r)$. However, the series $\sum_r (E * f_r)$ is in general not convergent in $C^k(\Omega)$. It is at this point that Theorem A′ of Subsection 5.16.3 comes into play and allows one to mimic the Mittag-Leffler construction of meromorphic functions with preassigned principal parts (as described in any text on complex function theory). Thus, since $P(D)(E * f_r) = 0$ on ω_r, the cited theorem affirms that $E * f_r$ may be approximated arbitrarily closely in $C^k(\omega_r)$ by finite linear combinations of exponential-polynomial solutions of the given LPDE. In particular we may choose such a solution p_r for which $(E * f_r - p_r)$ and its partial derivatives of orders at most k (or r if $k = \infty$) are in absolute value at most 2^{-r}, uniformly on ω_{r-1}. The series $\sum_r (E * f_r - p_r)$ is then convergent in $C^k(\Omega)$. Denoting by g its sum, we shall have

$$P(D)g = \sum_r P(D)(E * f_r - p_r) = \sum_r (f_r - 0) = f;$$

$P(D)$ is, of course, interpreted here in the sense of the theory of distributions; it may be interpreted classically only if k exceeds m (the order of $P(D)$). ▌

Remark. Any improved estimate of the order of E will lead to a corresponding refinement of Theorem A, $n + 1$ being replaced by the order of E. Thus it is known that in fact $n + 1$ may be replaced by $[\frac{1}{2}n] + 1$ on these grounds; see the remarks at the end of 5.14.2 above.

Corollary. If Ω is a convex domain in R^n, then $P(D)C^\infty(\Omega) = C^\infty(\Omega)$ (and here $P(D)$ may be interpreted classically).

Theorem A'. Let $s \geq 0$ be an integer (not ∞) and let Ω be a convex domain in R^n. If A belongs to $\mathscr{D}'^s(\Omega)$, there exists a solution X of $P(D)X = A$ belonging to $\mathscr{D}'^{s+n(n+3)}(\Omega)$. In particular, $P(D)\mathscr{D}'_F(\Omega) = \mathscr{D}'_F(\Omega)$.

Proof. In view of the remark following Theorem 5.8.1, taking therein $\kappa = n + 1$, it suffices to handle the case in which $A = D^pf$, where $|p| \leq s + n(n + 3)$ and f belongs to $C^{n+1}(\Omega)$. By the preceding theorem, with k taken to be 0, there exists a continuous function g on Ω satisfying $P(D)g = f$. Then $P(D)(D^pg) = D^pP(D)g = D^pf = A$, and so we may take $X = D^pg$, which is of order at most $|p|$. ▌

5.17.2 **The Relation** $P(D)\mathscr{D}'(\Omega) = \mathscr{D}'(\Omega)$. A complete discussion of necessary and sufficient conditions for this relation to hold is given in Hörmander [7], Chapter III, and depends on a refined version of the Paley-Wiener-Schwartz theorem. It is shown there, in particular, that the relation holds for any convex Ω. A partial solution only is forthcoming from our preceding results, inasmuch as we need to assume also that $P(D)$ is hypoelliptic. More will be said about this category of operators in Section 15.8 *infra*; meanwhile we take it to signify that there exists an elementary solution E of the equation $P(D)E = \varepsilon$ which belongs to $C^\infty(R^n\backslash\{0\})$.

Theorem A''. If Ω is a convex domain in R^n and if $P(D)$ is hypoelliptic then $P(D)\mathscr{D}'(\Omega) = \mathscr{D}'(\Omega)$.

Proof. Using the partition of unity (α_r) introduced in the proof of Theorem A above, any given A is expressible as the sum $\sum_r A_r$, where $A_r = \alpha_r A$ has a compact support contained in $\Omega\backslash\bar{\omega}_r$. Then $E * A_r$ belongs to $C^\infty(\omega_r)$ thanks to our hypothesis about E and the fact that $A_r = 0$ on $\bar{\omega}_r$. Moreover

$$P(D)(E * A_r) = A_r,$$

which is 0 on ω_r. By Theorem A' of Subsection 5.16.3, $E * A_r$ can be approximated as closely as we wish in $C^\infty(\omega_r)$ by finite linear combinations of exponential polynomials p_r satisfying $P(D)p_r = 0$. In particular, we may choose the p_r so that the series $\sum_r (E * A_r - p_r)$ converges in $\mathscr{D}'(\Omega)$. Let X be its sum. Then

$$P(D)X = \sum_r [P(D)(E * A_r) - P(D)p_r] = \sum_r [A_r - 0] = A,$$

and the proof is complete. ▌

Remarks. When $n = 1$, any $P(D) \neq 0$ is hypoelliptic. When $n > 2$, $P(D) = \Delta$ (the Laplacian) is hypoelliptic. Both of these statements flow from (5.11.5). Again, if $n = 2$, $P(D) = \partial/\partial\bar{z} = \partial_1 + i\partial_2$ is hypoelliptic, as appears from the formula $\partial/\partial\bar{z}(z^{-1}) = \pi\varepsilon$ (Schwartz [1], p. 49).

5.17.3. Results of the same general type as Theorems A, A′, and A″ are established in Chapitre II, § 2 of Malgrange [1] for the case in which $\Omega = R^n$ and $P(D)\varepsilon$ is replaced by a general distribution with a compact support. There are also analogues of the results described in Section 5.16 above.

All the main results of this category have been established independently by Ehrenpreis [2], [3], [10], [11], who formulates the task as a "problem of division." This he can do more incisively than we because of his extended concept of the Fourier transformation applying to arbitrary distributions on R^n; see our remarks at the end of Section 5.15.

As one would expect, the results applying to such general convolution equations are often less precise than those for LPDOs.

Rather surprisingly, the analogue of Theorem A″ for operators with variable analytic coefficients is false. This was first brought to light in 1957 by an example due to H. Lewy: in R^3, if we take

$$P(x, D) = -\frac{\partial}{\partial x_1} - i\frac{\partial}{\partial x_2} + 2i(x_1 + x_2)\frac{\partial}{\partial x_3},$$

and if $\Omega \supset \omega$ are any two nonvoid open subsets of R^3, then there exists a function (and actually a second category subset of functions) f in $\mathscr{S}(R^3)$ such that the equation $P(x, D)g = f$ on ω has no solution $g \in \mathscr{D}'(\omega)$. A full discussion of this rather unexpected phenomenon was given by Hörmander in 1960, gaining for him one of the Fields Medals for 1963. See Hörmander [7], Chapter VI.

5.17.4. There remains still the problem of the solubility of $P(D)X = A$ considered on nonconvex open sets Ω in R^n, A being a given distribution on Ω of some specified type. For a detailed discussion of this we must refer the reader to Schwartz [8], Exposé 3, or Malgrange [1] (or both); in the latter case the portions that are especially relevant are Théorème 4, p. 27 and Proposition 7, p. 68.

It appears that the following conditions are equivalent:

(1) $P(D)C^\infty(\Omega) = C^\infty(\Omega)$;

(2) $P(D)\mathscr{D}'_F(\Omega) = \mathscr{D}'_F(\Omega)$;

(3) To each compact set $K \subset \Omega$ corresponds a compact set $K' \subset \Omega$ such that the relations

$$X \in C'^\infty(\Omega), \qquad \text{support } \check{P}(D)X \subset K$$

together imply that

$$\text{support } X \subset K'.$$

Moreover, if $P(D)$ is hypoelliptic, the preceding three conditions are equivalent also to

(4) $P(D)\,\mathscr{D}'(\Omega) = \mathscr{D}'(\Omega)$.

If $P(D)$ is even elliptic (in the sense that it possesses an elementary solution E which is real analytic on $R^n \backslash \{0\}$), then all four assertions are valid for *any* open set Ω in R^n.

These facts would indicate that, at least for the restricted categories of LPDOs mentioned, the hypothesis of convexity of appearing in Theorems A, A′ and A″ results from defects in the method of proof adopted above, which makes appeal to the results of Section 5.16 in which convexity or some other restriction on Ω is doubtless essential.

5.18 Further Properties of Elementary Solutions; Hypoelliptic and Hyperbolic Operators and Equations

5.18.1 Constructive Approach to Elementary Solutions. The proofs given in Section 5.14 of the existence of an elementary solution for any linear partial differential operator $P = P(D) \neq 0$ with constant coefficients have two characteristic features, namely: (a) they underline the fundamental importance of inequalities, coupled with an appeal to the Hahn-Banach theorem; and (b) they make no indispensable use of the Fourier transformation, even in its classical form applicable to functions in $\mathscr{D} = \mathscr{D}(R^n)$. An alternative approach is possible which is more constructive at the expense of forfeiting the characteristics (a) and (b). One advantage of this approach is that, being relatively explicit, it leads to further important properties of the elementary solution. For more details of this method, a sketch of which follows, see some (as yet unpublished) lecture notes of Gårding [5] prepared for a conference on functional analysis held at London in April 1961, and the references there cited. Much of the basic work was carried out by Hörmander [1], though Hörmander's setting for the problems is guided more closely by the ideas of Hilbert space than by the theory of distributions. The differences are not important for this survey, however. We shall denote by $[G]$ the lecture notes by Gårding just cited; the reader is asked to note that Gårding's symbol D_k differs from ours by a factor 2π. The ensuing technique for the construction of an elementary solution for $P(D)$ amounts precisely to expressing the inversion of the Fourier transform in terms of classical integration processes in the special situation before us. It will be as well to recall from (5.15.7) the defining property of the Fourier transformation, expressed as a Parseval equation, namely,

$$\langle \bar{\varphi}, X \rangle = \langle \bar{\hat{\varphi}}, \hat{X} \rangle \tag{5.18.1}$$

for $\varphi \in \mathscr{D}$ and $X \in \mathscr{S}'$.

Suppose that $P(D)$ admits an elementary solution E that is temperate. The defining property

$$P(D)E = \varepsilon$$

is equivalent, via a Fourier transformation, to

$$P(\xi) \cdot \hat{E}(\xi) = 1; \tag{5.18.2}$$

this gives formally $\hat{E}(\xi) = 1/P(\xi)$, from which we require to construct E. The construction amounts to the use of various devices for "summing" a certain breed of divergent integral.

Let us replace in (5.18.1) φ by $\bar{\varphi}$ and X by E, assumed to be temperate; we obtain then

$$\langle \varphi, E \rangle = \langle (\hat{\varphi})^{\vee}, \hat{E} \rangle.$$

Since we have decided on formal grounds that $\hat{E}(\xi) = 1/P(\xi)$, we are thus led to consider the expression

$$I(\varphi) = \int \hat{\varphi}(-\xi) \frac{d\mu(\xi)}{P(\xi)} \tag{5.18.3}$$

as a function of the variable φ ranging over $\mathscr{D}$. One may take this as the starting point in the construction of an elementary solution for $P(D)$.

5.18.2 Summing the Integral $I(\varphi)$. We begin with the following remark. Suppose we devise a method of interpreting the integral $I(\varphi)$ for each φ in $\mathscr{D}$ that makes it continuous in φ. Then there will exist a unique distribution E (though perhaps not evidently a temperate one) such that $I(\varphi) = \langle \varphi, E \rangle$. If we now replace φ by $P(-D)\varphi$, and if we make the natural assumption that our interpretation of $J(\varphi) = I(P(—D)\varphi)$ reduces merely to the (absolutely convergent) integral $\int \hat{\varphi}(-\xi)\, d\mu(\xi)$, we shall conclude that

$$\begin{aligned}\langle \varphi, P(D)E \rangle = \langle P(-D)\varphi, E \rangle &= I(P(-D)\varphi) \\ &= \int \hat{\varphi}(-\xi)\, d\mu(\xi) = \varphi(0) = \langle \varphi, \varepsilon \rangle,\end{aligned}$$

and so $P(D)E = \varepsilon$: E will be an elementary solution. Thus all depends on devising a suitable and natural interpretation of $I(\varphi)$.

In certain special cases, of course, no subtlety is required. This is the situation, for example, with the meta-polyharmonic operator

$$P(D) = (D^2 + \lambda)^k = \left(-\frac{\Delta^2}{4\pi^2} + \lambda\right)^k$$

for $\lambda > 0$ and k any natural number exceeding $\frac{1}{2}n$. In this case the integral in (5.18.3) exists in the Lebesgue (absolutely convergent) sense and everything is clear; indeed one may even define E quite directly as the continuous function

$$E(x) = \int e^{2\pi i x \cdot \xi} \frac{d\mu(\xi)}{P(\xi)}, \tag{5.18.4}$$

the inverse Fourier transform of the integrable function $1/P(\xi) = 1/(|\xi|^2 + \lambda)^k$.

If caution be thrown to the winds one may adopt (5.18.4) as a starting point in other cases too. It would then be necessary to apply to the integral appearing therein the same sort of summability processes we have in mind for $I(\varphi)$. (Retention of the latter has certain advantages when one wishes to fill in details: one is that, although both $e^{2\pi i x \cdot \xi}$ and $\hat{\varphi}(-\xi)$ are obviously extendible into entire analytic functions of $\xi \in C^n$, the latter vanishes as $\operatorname{Re} \xi \to \infty$ while the former does not. This factor is significant in questions of convergence.) In the quantum theory of fields it is customary to use precisely this technique for constructing the so-called Green's potentials; for the relatively simple case of scalar and pseudo-scalar meson fields, for example, the appropriate operator is that of Klein-Gordon, namely,

$$\begin{aligned}\square - \mu^2 &= \Delta - c^{-2}\,\partial_4^2 - \mu^2 \\ &= (4\pi^2c^2)^{-1}[D_4^2 - c^2(D_1^2 + D_2^2 + D_3^2)] - \mu^2,\end{aligned}$$

the Δ referring only to the space coordinates x_1, x_2, x_3; c being the velocity of light, x_4 the time coordinate and $\mu = 2\pi mc/h$ with m the meson mass and h Planck's constant.

One useful technique for interpreting $I(\varphi)$ depends upon the fact that $\hat{\varphi}$ is extendible into an entire analytic function on C^n, while $1/P$ is likewise extendible into a meromorphic function on C^n. Since all the difficulties arise from the possibly too rapid rate of growth of $1/P$ on the "real axis" $R^n \subset C^n$, and especially from the possible existence of real zeros, it is natural to try to use complex function theory in connection with a change of the "path of integration" in (5.18.3) from the real into the complex domain. With this in mind, suppose we take a point η of R^n and look at

$$I_\eta(\varphi) = \int \hat{\varphi}(-\xi + i\eta)\,\frac{d\mu(\xi)}{P(\xi + i\eta)}. \tag{5.18.5}$$

If we can show that I_η is defined and continuous on $\mathscr{D}$, and if we can verify that

$$J_\eta(\varphi) = I_\eta(P(-D)\varphi)$$

reduces to $\int \hat{\varphi}(-\xi + i\eta)\,d\mu(\xi)$, then we shall have established the existence of a distribution E_η such that

$$\langle \varphi, P(D)E_\eta \rangle = \int \hat{\varphi}(-\xi + i\eta)\,d\mu(\xi).$$

Now it is an easy matter to show that the right-hand side here is independent of η and has therefore the value $\varphi(0) = \langle \varphi, \varepsilon \rangle$. So $P(D)E_\eta = \varepsilon$ and E_η is an elementary solution. At the same time, our construction of E_η will often yield useful information concerning its support.

5.18.3 Existence of the Elementary Solution for General $P(D) \neq 0$. Without further restrictions on P we seek a satisfactory interpretation of $I(\varphi)$. It turns out that this end is attained by setting

$$I(\varphi) = \int d\mu'(\xi') \int_{\Gamma(\xi')} \hat{\varphi}(-\xi)\,\frac{d\xi_1}{P(\xi)}, \tag{5.18.6}$$

where $\xi' = (\xi_2, \cdots, \xi_n) \in R^{n-1}$, μ' is Lebesgue measure on R^{n-1}, and $\Gamma(\xi')$ is the path of integration in the complex ξ_1-plane defined in the following fashion. We choose and fix any $\delta > 0$ and let S be the strip $0 \leq \operatorname{Im} \xi_1 \leq \delta$. Having chosen coordinates so that $P(\xi) = c\xi_1^m + \cdots$, m being the degree of P, we factorize $P(\xi) = c. \prod_{k=1}^{m} (\xi_1 - z_k)$, where the $z_k = z_k(\xi')$ are algebraic functions of ξ'. We let Δ_k be the disk in complex ξ_1-plane with centre z_k and radius δ, $\Delta = \bigcup_{k=1}^{m} \Delta_k$. Finally, we let $\Gamma(\xi')$ be the lower boundary of $S \backslash S \cap \Delta$. Although it is clear that in general $\Gamma(\xi')$ does not depend continuously on ξ', yet it may be shown that the inner integral in (5.18.6) is a measurable function of ξ'.

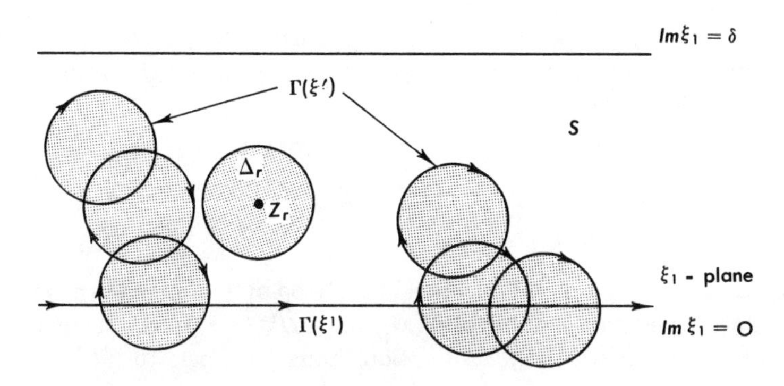

Fig. 5a

By construction of $\Gamma(\xi')$ it is arranged that

$$|P(\xi)| \geq |c| \, \delta^m$$

for $\xi' \in R^{n-1}$ and ξ_1 on $\Gamma(\xi')$. Using this inequality it is easy (even if a little tedious) to verify that (5.18.6) defines a continuous linear form on $\mathscr{D}$, and that $J(\varphi) = I(P(-D)\varphi)$ reduces to $\int \hat{\varphi}(-\xi) \, d\mu(\xi)$ by virtue of Cauchy's theorem applied to $\hat{\varphi}(-\xi_1, -\xi')$ *qua* function of ξ_1 for each fixed ξ' in R^{n-1}. In this way we establish the existence of an elementary solution E.

For a study of the use of the preceding formulae representing the elementary solution E, see Agranovič [1], who applies the method to various problems of division and approximation.

5.18.4 **Hypoelliptic Operators.** Consider the following assertions about $P(D)$:

(1) If X belongs to $\mathscr{D}'(\Omega)$, if $A \in C^\infty(\Omega)$ (Ω an open subset of R^n), and if $P(D)X = A$ on Ω, then $X \in C^\infty(\Omega)$.

(2) There exists an elementary solution E for $P(D)$ whose restriction to $R^n \backslash \{0\}$ belongs to $C^\infty(R^n \backslash \{0\})$.

We assert that these assertions are equivalent.

It is indeed obvious that (1) implies (2), the existence of an elementary solution being established in Section 5.14.

Conversely let us assume the validity of (2), and that the hypotheses of (1) are fulfilled: we have to conclude that $X \in C^\infty(\Omega)$. We begin by choosing any $\varepsilon > 0$ and then any $\alpha \in \mathscr{D}(B_\varepsilon)$ (B_ε the open ball with centre 0 and radius ε) such that

$\alpha = 1$ on a neighbourhood of 0, and consider $E^* = \alpha E$. By Leibnitz' formula we have

$$D^p E^* = \sum_{q \leq p} \binom{p}{q} (D^{p-q}\alpha) D^q E$$
$$= \alpha D^p E + \sum_{\substack{q \leq p \\ q \neq p}} \binom{p}{q} (D^{p-q}\alpha) D^q E;$$

the relations $q \leq p$, $q \neq p$ mean that $q_k \leq p_k$ for $1 \leq k \leq n$ and $q_k < p_k$ for at least one such k. For any such q we have $D^{p-q}\alpha = 0$ on a neighbourhood of 0. It therefore follows that

$$P(D)E^* = \alpha P(D)E + \sum_{|p| \leq m} a_p \sum_{\substack{q \leq p \\ q \neq p}} \binom{p}{q} (D^{p-q}\alpha) D^q E$$
$$= \alpha \cdot \varepsilon + f,$$

say. Here $\alpha \cdot \varepsilon = \varepsilon$ since $\alpha = 1$ on a neighbourhood of 0; moreover the inner sum on the right represents a member of $\mathscr{D}(B_\varepsilon)$ since $D^{p-q}\alpha = 0$ on a neighbourhood of 0 while $E \in C^\infty(R^n \backslash \{0\})$ by assumption; thus f belongs to $\mathscr{D}(B_\varepsilon)$. To sum up, then:

$$P(D)E^* = \varepsilon + f, \quad f \in \mathscr{D}(B_\varepsilon); \tag{5.18.7}$$

here $E^* = \alpha E$ has its support contained in B_ε. In fact, E^* is an example of a "parametrix" for $P(D)$—that is, an approximate elementary solution having a compact support. Also E^* shares with E the property of belonging to $C^\infty(R^n \backslash \{0\})$.

Since E^* and f have their supports in B_ε, we may (see Exercise 5.19) take the convolution with E^* of both sides of the equation $P(D)X = A$ and use associativity and (5.18.7) to derive

$$X = E^* * A - f * X \text{ on } \Omega_\varepsilon \tag{5.18.8}$$

The term $f * X$ belongs to $C^\infty(\Omega_\varepsilon)$, since $f \in \mathscr{D}$, and causes no further trouble. As for the first term, we submit it to further treatment by taking any relatively compact open set $\Omega' \subset (\Omega_\varepsilon)_\delta$ and then $\beta \in \mathscr{D}$ such that support $\beta \subset \Omega$ and $\beta = 1$ on Ω'. Writing $A = \beta A + (1 - \beta)A$, the first term is seen to equal

$$E^* * (\beta A) + E^* * [(1 - \beta)A] = Y + Z,$$

say. Now $\beta A \in \mathscr{D}$ because $A \in C^\infty(\Omega)$ and β has its support contained in Ω. Hence Y belongs to $\mathscr{D}$. On the other hand, the support of $(1 - \beta)A$ is contained in $R^n \backslash \Omega'$; this, together with the fact that E^* belongs to $C^\infty(R^n \backslash \{0\})$, is easily seen to arrange that Z belongs to $C^\infty(\Omega')$. Thus the first term of the right of (5.18.8), and therefore X itself, belongs to $C^\infty(\Omega')$. Since this is the case for any relatively compact open $\Omega' \subset (\Omega_\varepsilon)_\delta$, it follows on letting $\delta \to 0$ and $\varepsilon \to 0$ that $X \in C^\infty(\Omega)$, as we had to establish. The equivalence of (1) and (2) is thus demonstrated. ▌

The reader will note that (1) is equivalent to the apparently stronger condition

(1′) If X, $A \in \mathscr{D}'(\Omega)$, $P(D)X = A$ on Ω, $\omega \subset \Omega$ and $A \mid \omega \in C^\infty(\omega)$, then $X \mid \omega \in C^\infty(\omega)$.

Having established the equivalence of (1), (1′), and (2), we introduce the term *hypoelliptic* to describe those operators $P(D)$ that satisfy any one (and therefore all) of the conditions (1), (1′), and (2).

Examples. If $n = 1$, any $P(D) \neq 0$ is hypoelliptic. (Either of the conditions (1) or (2) is very easily verified, using the simplest results about ordinary linear differential equations with constant coefficients.) Taking $n = 2$: $P(D) = \partial_1 - \partial_2^2$ (the heat operator) is hypoelliptic since it has the elementary solution

$$E = (2\pi x_1)^{-1/2} H(x_1) \exp \frac{-x_2^2}{4x_1}, \tag{5.18.9}$$

H being the Heaviside function on R^1:

$$H(x) = \tfrac{1}{2}(1 + \operatorname{sgn} x);$$

$P(D) = i\partial_1 + \partial_2^2$ is hypoelliptic. For any n the Laplacian Δ is hypoelliptic, having an elementary solution proportional to $|x|^{2-n}$ if $n > 2$ and to $\log |x|$ if $n = 2$. For other examples, see Exercise 5.20.

It is a natural question to ask for simple conditions on the polynomial function $P(\xi)$ that are necessary and sufficient in order that $P(D)$ shall be hypoelliptic. Various such criteria have been given by Hörmander ([1], Theorems 3.3 and 3.4) and we comment on some of these. Other criteria are due to Ehrenpreis [9], who considers convolution operators more general than our $P(D)$.

The following two conditions (a) and (b) are shown to be equivalent (Hörmander [1], Theorem 3.3) and to be sufficient to ensure that $P(D)$ be hypoelliptic:

(a) To each $a > 0$ corresponds $b = b(a) > 0$ such that $P(\xi + i\eta) \neq 0$ whenever $\xi, \eta \in R^n$, $|\eta| < a$ and $|\xi| > b$.

(b) For each $a > 0$ one has

$$\lim_{\xi \in R^n, \xi \to \infty} \inf_{\eta \in R^n, |\eta| > a} |P(\xi + i\eta)| = +\infty.$$

On the other hand Gårding ([G], §11) shows that if $P(D)$ is hypoelliptic, then the following condition is fulfilled:

(a′) There exist numbers $c > 0$, $d > 0$ such that

$$|\eta| > c^{-1} |\xi|^d - c$$

whenever $\xi, \eta \in R^n$ satisfy $P(\xi + i\eta) = 0$.

Since it is clear that (a′) implies (a), we conclude that:

▶ The conditions (a), (a′), and (b) are equivalent, each being necessary and sufficient in order that $P(D)$ be hypoelliptic.

Hörmander's contribution, showing that (b) implies hypoellipticity, uses the techniques of Section 5.18.2 above; we sketch his construction of an appropriate elementary solution E. In this case we adopt the interpretation (5.18.6) of $I(\varphi)$, save that $\Gamma(\xi')$ is replaced by the path $C(\xi')$ chosen in the following manner.

As before, we begin by choosing coordinates so that $P(\xi) = c\xi_1^m + \cdots$, where $c \neq 0$. According to (b) there exists $r > 0$ such that $|P(\xi)| \geq 1$ for ξ in R^n and $|\xi| \geq r$. From this it may be inferred that there exists an $r^* > 0$ such that $|P(\xi_1, \xi')| \geq 1$ for any ξ' in R^{n-1} with $|\xi'| \leq r$ and any complex ξ_1 with $|\xi_1| > r^*$. This being so, $C(\xi')$ will be the real axis in the complex ξ_1-plane whenever $|\xi'| \geq r$; for $|\xi'| < r$, $C(\xi')$ is the real axis modified by replacing the segment from $-r^*$ to r^* by the lower semicircle having this segment as a diameter. Then $|P(\xi)| \geq 1$ for all ξ involved in the integral (5.18.6). The essence of the proof is to show that the resulting elementary solution E satisfies condition (2); for details we refer the reader to Hörmander's account.

We add to this the remark that Petrowsky [1] has given an equivalent purely algebraic condition for hypoellipticity, viz.,

(c) $P_0(\xi) \neq 0$ for $\xi \in R^n$, $\xi \neq 0$;

here $P_0(\xi) = \sum_{|p|=m} C_p \xi^p$ is the principal part of $P(\xi)$.

To conclude our remarks about hypoelliptic operators we should add that John [1] gave a proof of the existence of an elementary solution for operators $P(D)$ satisfying (c). The study of hypoelliptic LPDOs $P(x, D)$ with C^∞ variable coefficients has been pursued by [among others] Malgrange [2], Hörmander [3] and [7], Gårding and Malgrange [1], Trèves [5], and Peetre [2].

5.18.5 **Hyperbolic Operators.** We now return to (5.18.5) as a tool for the study of elementary solutions of a given $P(D)$, granted certain properties of the polynomial $P(\xi)$.

Let us start with the assumption that a point η of R^n and a real number t_0 exist for which the following condition is fulfilled:

(*) $P(\xi + it\eta) \neq 0$ for $\xi \in R^n$ and $t \geq t_0$.

Our allegation is that:

▶ The elementary solution E, defined via (5.18.5) with η there replaced by $t\eta$ for any $t \geq t_0$, has its support contained in the halfspace

$$H_\eta = \{x \in R^n : \eta \cdot x \leq 0\}.$$

The proof of this is based upon a lemma, which we shall admit.

Lemma. Suppose (*) holds. Let P_0 denote the principal part of P. Then $P_0(i\eta) \neq 0$ and

$$|P(\xi + it\eta)| \geq |P_0(i\eta)| \, (t - t_0)^m \tag{5.18.10}$$

for ξ in R^n and real $t \geq t_0$. Furthermore, there exist numbers $a > 0$ and $b > 0$ such that

$$|P(\xi + it\eta)| \geq a[1 + |\xi| + (t - t_0)]^b \tag{5.18.11}$$

for ξ in R^n and $t \geq t_0$.

For a proof of this lemma, see ([G], §11).

Assuming this lemma at hand, the inequalities it contains show that (5.18.5) is effective in defining an elementary solution E for $P(D)$ that is independent of t for $t \geq t_0$. Thus we have

$$\langle \varphi, E \rangle = \int \hat{\varphi}(-\xi - it\eta) \frac{d\mu(\xi)}{P(\xi + it\eta)} \qquad (t \geq t_0)$$

for φ in $\mathscr{D}$. Suppose now that the support K of φ is contained in the open set $R^n \backslash H_\eta$. Then there exists a number $c > 0$ such that $\eta \cdot x \geq c$ for x in K. If we take $t = t_0 + s$, where $s \geq 0$, and notice that

$$\hat{\varphi}(-\xi - it\eta) = \hat{\varphi}_s(-\xi - it_0\eta),$$

where

$$\varphi_s(x) = e^{-2\pi s\eta \cdot x} \cdot \varphi(x),$$

it appears that $\langle \varphi, E \rangle = \langle \varphi_s, E \rangle$. But, because $\eta \cdot x \geq c(>0)$ on the support K of φ, this leads easily to the relation $|\langle \varphi, E \rangle| = O(e^{-2\pi cs})$. Letting $s \to \infty$, we infer that $\langle \varphi, E \rangle = 0$—and this for each φ whose support lies in $R^n \backslash H_\eta$. Thus $E = 0$ on $R^n \backslash H_\eta$; that is, the support of E lies in H_η, as alleged.

While on this matter we note that it has been shown by Gårding [1], that the relation (support E) $\cap H_{-\eta} = \{0\}$ is equivalent to the relations $P_0(i\eta) \neq 0$ and $P(\xi + it\eta) \neq 0$ for ξ in R^n and sufficiently large real t.

We introduce the term *hyperbolic*, as applied to operators $P(D)$, to mean that $P(D)$ has an elementary solution E whose support is contained in some proper cone with vertex at the origin.

The result we have just established shows that a *sufficient* condition in order that $P(D)$ be hyperbolic is that there exist a nonvoid open set Ω in R^n and a real number t_0 such that

$$P(\xi + it\eta) \neq 0 \qquad \text{for } \xi \in R^n, \eta \in \Omega \text{ and real } t \geq t_0. \qquad (5.18.12)$$

Gårding shows ([G], § 11) that if $P(D)$ is hyperbolic, and if we choose coordinates so that the support of E is contained in the half space $x_1 \leq 0$, then

$$\xi \in R^n, \eta \in R^n, \ P(\xi + i\eta) = 0 \Rightarrow \eta_1 \leq \text{const.} \{1 + \text{Max}(|\eta_2|, \cdots, |\eta_n|)\}; \qquad (5.18.13)$$

moreover, if the coordinates be so chosen, then (5.18.13) is equivalent to (5.18.12).

Thus we see that $P(D)$ *is hyperbolic if and only if* (5.18.12) *is satisfied*, giving us a neat and simple criterion for hyperbolicity.

By application of this criterion, or by knowledge of explicit elementary solutions, one can verify the following examples: if $n = 1$, $P(D) = -\partial_1$ is hyperbolic, having the elementary solution $E = H(x_1)$ with support contained in $x_1 \geq 0$; if $n = 2$, $P(D) = \partial_1^2 - \partial_2^2$ (wave operator) is hyperbolic, having the elementary solution

$$E = \tfrac{1}{2} H(x_1)\{H(x_1 + x_2) - H(x_2 - x_1)\}$$

with support contained in the set $x_1 \geq 0$, $|x_2| \leq x_1$; the same is true of the wave operator in any number of variables.

5.19 Characterization of Certain Types of Linear Partial Differential Operators by Inequalities and Order Relations

We have remarked on several occasions that the modern theory of linear partial differential operators is founded largely on inequalities combined with general principles of functional analysis. In this section we aim to indicate how various important types of such operators have been shown to be distinguishable on the basis of inequalities and associated relations of partial order.

In brief one may say that two important classes of linear partial differential operators have been characterized in the fashion just described, namely, the elliptic operators and the hyperbolic operators. (The former category we have not yet defined; hyperbolic operators were defined in Subsection 5.18.5.) Hörmander [1] contributed the first of these two characterizations, and Trèves ([2], [3], [4]) the second. Hörmander's work also provides a characterization of this nature, albeit a good deal more complicated than that for the elliptic case, of hypoelliptic operators (defined in Subsection 5.18.4). We aim to give a short account of the work of Hörmander on elliptic operators and to merely sketch a small part of Trèves' results concerning hyperbolic operators, but we shall leave aside here the hypoelliptic case; the reader will note that the last two categories have been characterized quasi-algebraically in the fashion described in Subsections 5.18.4 and 5.18.5, respectively.

Although in all cases the work has been carried forward so as to apply at least in some measure to operators with variable coefficients (see Hörmander [7]), we intend limiting our remarks to the case of constant coefficients.

For both the elliptic and hyperbolic cases the characterization is phrased in terms of a suitable relation of partial preorder; that of "strength" in Hörmander's case and that of "dominance" in Trèves'. We shall deal with these in turn.

5.19.1 Strength of Operators.

In what follows we shall use the term *operator* to signify a linear partial differential operator with constant coefficients (even though the latter restriction is by no means always necessary).

In following Hörmander's investigations, which are framed in an L^2-context throughout, it is necessary to distinguish the so-called "minimal" and "maximal" operators P_0 and P associated with the differential operator $P(D) = \sum_{|p| \le m} a_p D^p$.

THE MINIMAL OPERATOR P_0. Consider $P(D)$ as a mapping of $\mathscr{D}(\Omega)$ into $L^2(\Omega)$. It is easy to show (Hörmander [1], Lemma 1.4) that $P(D)$ is preclosed from $L^2(\Omega)$ into itself (the closure in $L^2(\Omega) \times L^2(\Omega)$ of its graph is again a graph). P_0 is by definition the closure of this mapping. In other words, the domain of P_0, henceforth denoted by Dom P_0, consists precisely of those $u \in L^2(\Omega)$ for which $P(D)u$ (taken in the sense of distributions) belongs to $L^2(\Omega)$, and for which there exists at least one sequence (φ_n) extracted from $\mathscr{D}(\Omega)$ such that $\varphi_n \to u$ and $P(D)\varphi_n \to P(D)u$ in $L^2(\Omega)$; for any such u, $P_0 u$ is defined to be $P(D)u$.

THE MAXIMAL OPERATOR P. Let $\tilde{P}(D) = \sum_{|p| \le m}(-1)^{|p|}\bar{a}_p D^p$ be the adjoint (relative to the $L^2(\Omega)$ structure) of $P(D)$, and let $\tilde{P}_0$ be the associated minimal operator. Then the maximal operator P associated with $P(D)$ is defined to be the adjoint (again relative to the Hilbert space structure of $L^2(\Omega)$) of $\tilde{P}_0$. It comes to the same thing to say that P is that operator from $L^2(\Omega)$ into itself having domain Dom P formed of those u in $L^2(\Omega)$ for which $P(D)u$ (in the sense of distributions) belongs to $L^2(\Omega)$; for such u, $Pu = P(D)u$.

Hörmander's comments on P_0 and P may be usefully supplemented by the account given by Fuglede [1].

The concept of strength as used by Hörmander is tantamount to that of inclusion as applied to the domains of minimal operators. Specifically, given two operators $P(D)$ and $Q(D)$ and a domain $\Omega \subset R^n$, we say that $P(D)$ is *stronger than* $Q(D)$ *on* Ω—in symbols: $P(D) \ge Q(D)$ *on* Ω—if and only if Dom $P_0 \subset$ Dom Q_0.

Since P_0 and Q_0 are, by definition, closed operators from $L^2(\Omega)$ into itself, the substance of Subsection 6.4.11 furnishes us with necessary and sufficient conditions in order that $P(D)$ be stronger than $Q(D)$, namely, the validity of the inequality

$$\|Q_0 u\|^2 \le C(\|u\|^2 + \|P_0 u\|^2), \tag{5.19.1}$$

where C is a suitable number, u a variable element of Dom P_0, and the norms being those in $L^2(\Omega)$. On the other hand, by definition of the domains of the minimal operators involved, this is equivalent to the inequality

$$\|Q(D)\varphi\|^2 \le C(\|\varphi\|^2 + \|P(D)\varphi\|^2) \tag{5.19.2}$$

for all φ in $\mathscr{D}(\Omega)$. Indeed, suppose that (5.19.2) holds. If u belongs to Dom P_0, there exists a sequence $(\varphi_n) \subset \mathscr{D}(\Omega)$ such that $\varphi_n \to u$ and $P(D)\varphi_n \to P_0 u$ in $L^2(\Omega)$; then (5.19.2) shows that the sequence $(Q(D)\varphi_n)$ is convergent in $L^2(\Omega)$ and, since $Q(D)$ is preclosed, its limit can be none other than $Q_0 u$. Thus, taking $\varphi = \varphi_n$ in (5.19.2) and then taking the limit as $n \to \infty$, we arrive at (5.19.1).

If Ω is *bounded* we know (Inequality B of Subsection 5.14.2; Hörmander [1], Theorem 2.1) that

$$\|\varphi\| \le \text{const. } \|P(D)\varphi\|$$

for φ in $\mathscr{D}(\Omega)$. In this case, therefore, the relation $P(D) \ge Q(D)$ on Ω implies that

$$\|Q(D)\varphi\| \le C\, \|P(D)\varphi\| \tag{5.19.3}$$

for φ in $\mathscr{D}(\Omega)$; the constant C has not in general the same value as in (5.19.2), of course. Conversely, whether or not Ω is bounded, (5.19.3) implies (5.19.2). Thus if Ω is bounded, $P(D) \ge Q(D)$ on Ω holds if and only if (5.19.3) is valid for arbitrary φ in $\mathscr{D}(\Omega)$.

In this way we have described Hörmander's relation of strength in terms of inequalities.

We may note in passing that a slightly different concept of strength is obtained if we replace everywhere the minimal operators P_0 and Q_0 by the maximal operators P and Q. This would be equivalent to the inequality (5.19.1) after P and Q are inserted in place of P_0 and Q_0, respectively, the inequality being now postulated to hold for each u in Dom P. In this case, however, it is not generally possible to pass from this to (5.19.2).

5.19.2 **Definition of Elliptic Operators.** Let us consider an operator $P(D) = \sum_{|p| \leq m} a_p D^p$ of degree exactly m (so that $a_p \not\equiv 0$ for at least one p satisfying $|p| = m$). We denote by $p(\xi)$ the *principal part* of $P(\xi)$—that is, the term in $P(\xi)$ which is homogeneous of degree m:

$$p(\xi) = \sum_{|p|=m} a_p \xi^p.$$

Following the classical terminology we define $P(D)$ to be *elliptic* if and only if

$$p(\xi) \neq 0 \qquad \text{for } \xi \in R^n,\ \xi \neq 0. \tag{5.19.4}$$

Since it is obvious that (5.19.4) arranges that condition (b) of Subsection 5.18.4 is fulfilled, we see that any elliptic operator is hypoelliptic (thus justifying the latter name). The converse is false: The heat operator $P(D) = \partial_1 - \partial_2^2$ is, as we have seen in Subsection 5.18.4, hypoelliptic; but since here $p(\xi) = -2\xi_2^2$, this operator is not elliptic.

Inasmuch as we have defined elliptic operators algebraically, in order to characterize them in terms of strength we must express the latter concept algebraically. This we shall do forthwith, following Hörmander ([1], Theorem 2.2). The desired characterization of ellipticity in terms of strength (loc. cit. Theorem 2.4) is an immediate consequence.

5.19.3 **Strength Expressed Algebraically.** We aim to establish the following assertion.

▶ Let Ω be a bounded domain in R^n; in order that $P(D) \geq Q(D)$ on Ω, it is necessary and sufficient that either one of the following two conditions be fulfilled:

$$\operatorname*{Sup}_{\xi \in R^n} \frac{Q^0(\xi)}{P^0(\xi)} < +\infty, \tag{5.19.5}$$

$$\operatorname*{Sup}_{\xi \in R^n} \frac{|Q(\xi)|}{P^0(\xi)} < +\infty; \tag{5.19.6}$$

in these formulae we have written

$$P^0(\xi) = \left\{ \sum_p |P^{(p)}(\xi)|^2 \right\}^{1/2},$$

$$P^{(p)}(\xi) = (\partial^p P)(\xi),$$

the sum running over all p (only a finite number of terms being nonzero for given P, of course); $Q^0(\xi)$ is defined analogously in terms of Q.

Proof. (1) Suppose that $P(D) \geq Q(D)$ on Ω. We shall show that then (5.19.5) is a consequence of (5.19.3). To see this, we take any f in $\mathscr{D}(\Omega)$ and apply (5.19.3) to $\varphi(x) = e^{i\xi \cdot x}f(x)$. By Leibnitz' formula applied repeatedly one finds that

$$P(D)\varphi = e^{i\xi \cdot x} \sum_p P^{(p)}(\xi) \cdot \frac{\partial^p f(x)}{|p|!},$$

and similarly with Q in place of P. If we write temporarily

$$J_{pq} = (|p|\,!)^{-1}(|q|\,!)^{-1} \int \partial^p f \cdot \overline{\partial^q f}\, d\mu,$$

(5.19.3) yields

$$\sum_{p,q} J_{pq} Q^{(p)}(\xi)\overline{Q^{(q)}(\xi)} \leq C^2 \sum_{p,q} J_{pq} P^{(p)}(\xi)\overline{P^{(q)}(\xi)}. \tag{5.19.7}$$

These sums contain no terms (other than zeros) apart from those for which $|p|$ and $|q|$ are majorized by the maximum of the degrees of P and Q, say N. Let $z = (z_p)_{|p| \leq N}$ be any family of complex numbers such that $z_p = z_q$ whenever q is a permutation of p. The quadratic form

$$\sum_{\substack{|p| \leq N \\ |q| \leq N}} J_{pq} z_p \bar{z}_q = \int \left| \sum_{|p| \leq N} (|p|\,!)^{-1} z_p \cdot \partial^p f \right|^2 d\mu$$

$$= \int \left| \sum_{|p| \leq N} (|p|\,!)^{-1} z_p \xi^p \right|^2 |F(\xi)|^2\, d\mu(\xi),$$

by Parseval's formula, $F(\xi)$ being

$$(2\pi)^{-(1/2)n} \int e^{-i\xi \cdot x} f(x)\, d\mu(x),$$

is thus visibly strictly positive unless $z_p = 0$ for all p. From this it follows that

$$\sum_{|p| \leq N} |z_p|^2 \leq C' \sum_{\substack{|p| \leq N \\ |q| \leq N}} J_{pq} z_p \bar{z}_q.$$

Taking herein $z_p = Q^{(p)}(\xi)$ and using (5.19.7) we obtain

$$Q^0(\xi)^2 \leq C^2 C' \sum_{p,q} J_{pq} P^{(p)}(\xi)\overline{P^{(q)}(\xi)}$$

and thence, by the Cauchy-Schwarz inequality, that

$$Q^0(\xi)^2 \leq C'' P^0(\xi)^2,$$

which is equivalent to (5.19.5).

(2) Suppose now that (5.19.6) holds: we shall show that then (5.19.3) is valid. To this end we recall that since Ω is bounded, the Lemma of Subsection 5.14.4 gives

$$\|P^{(p)}(D)\varphi\| \leq C_p \|P(D)\varphi\| \tag{5.19.8}$$

for each p and each φ in $\mathscr{D}(\Omega)$. If we introduce

$$\Phi(\xi) = (2\pi)^{-(1/2)n}\int e^{-i\xi\cdot x}\,\varphi(x)\,d\mu(x),$$

Parseval's formula gives

$$\|Q(D)\varphi\|^2 = \int |Q(\xi)|^2\,|\Phi(\xi)|^2\,d\mu(\xi),$$

and this, by virtue of (5.19.6), is majorized by

$$C^2\sum_p \int |P^{(p)}(\xi)|^2\,|\Phi(\xi)|^2\,d\mu(\xi).$$

By Parseval's formula again this is equal to

$$C^2\sum_p \|P^{(p)}(D)\varphi\|^2.$$

Finally, by (5.19.8), this is majorized by $C'^2\,\|P(D)\varphi\|^2$. Hence we have (5.19.3). The proof is thus complete. ▌

5.19.4 **Characterization of Elliptic Operators.** It is evident that if $P(D)$ is elliptic in the sense prescribed in Subsection 5.19.2 (5.19.4), then (5.19.6) is fulfilled for any polynomial Q of degree at most m, the degre P. Thus $P(D)$ is stronger than any such $Q(D)$. Conversely, if $P(D)$ is stronger than any such $Q(D)$, (5.19.5) shows that $p(\xi)$ must be nonvanishing for nonzero ξ in R^n, that is, $P(D)$ must be elliptic.

Thus:

► $P(D)$ is elliptic if and only if $P(D) \geq Q(D)$ whenever the degree of Q is majorized by that of P.

This is the desired characterization of elliptic operators in terms of relative strength.

5.19.5 **Remark.** If $P(D)$ is elliptic, it is a fortiori hypoelliptic. Hence there exists an elementary solution E for $P(D)$ that belongs to $C^\infty(R^n\backslash\{0\})$. Actually (see Schwartz [1], pp. 138–143 and the references there cited, and also the results established in Hörmander [1], Theorem 3.5) the fact that $P(D)$ is elliptic can be shown to imply that E may be chosen to be real analytic on $R^n\backslash\{0\}$. From this the argument used in Subsection 5.18.4 shows that the following property is enjoyed by all elliptic operators:

(1*) If X and A belong to $\mathscr{D}'$, if A is real analytic on an open set Ω in R^n, and if $P(D)X = A$, then X too is real analytic on Ω.

Thus the elliptic operators, defined in the classical fashion adopted in Subsection 5.19.2, satisfy a condition analogous to the condition (1) used in 5.18.4 to define the hypoelliptic ones. The author does not know whether, conversely, (1*) implies ellipticity on the part of $P(D)$. Of course, if $n = 1$, the two classes of operators are identical.

5.19.5 **Trèves' relation of domination and characterization of hyperbolic operators.** We here present no more than a snap view of a fragment of Trèves' results that is expressible in terms of a special case of his concept of domination. The portion, although small, is reasonably typical. Our remarks are confined to LPDOs with constant coefficients.

Suppose that Ω is an open subset of R^n, and that $P_i(D)$ $(i \in I)$ and $Q_j(D)$ $(j \in J)$ are two families of LPDOs. The family $(P_i(D))$ is said to (exponentially) *equidominate on* Ω the family $(Q_j(D))$ if, for each bounded open set $\omega \subset \Omega$ and each $\varepsilon > 0$, there exists a point h of C^n (or, equivalently, of R^n) such that

$$\operatorname*{Sup}_{j \in J} \|e^{-x \cdot h} Q_j(D)\varphi\|_{L^2} \leq \varepsilon \cdot \operatorname*{Sup}_{i \in I} \|e^{-x \cdot h} P_i(D)\varphi\|_{L^2}$$

for all φ in $\mathscr{D}(\omega)$.

Herein we have, as with Hörmander's idea of strength, a relation of partial order between LPDOs. Unlike Hörmander's concept, however, domination proves to have little relationship with ellipticity. Instead, as appears from Trèves' investigations, domination has close connections with hyperbolicity [and also with parabolicity—an aspect we here ignore].

Specifically, Trèves shows ([4], Théorème 2.1) that $P(D)$ is hyperbolic if and only if it equidominates the D^p with $|p| < m$ (m being as usual the order of $P(D)$). The same theorem incorporates too a quasi-algebraic characterization of hyperbolic operators similar to that appearing in Subsection 5.18.5.

Results of this type illustrate once more the vital and fundamental role played by a priori functional inequalities in the modern theory of LPDOs.

5.20 The Work of F. E. Browder

The foregoing account of some of the problems concerned with LPDEs has been planned and executed from the viewpoint of distributions theory, and in this it conforms closely with the treatment favoured (broadly speaking) by European authorities (Malgrange, Lions, Schwartz, Hörmander, Gårding, et alia). Despite this it would be rash to conclude that this is the only, or even the best, approach. Indeed F. E. Browder [1], a front-rank representative of the American school, advocates and pursues a somewhat different approach, a full and detailed account of which cannot be included here.

The use of distributions theory arranges that the problems can be formulated in terms of continuous operators acting on rather complicated LCTVSs. Browder's approach, inspiration for which is accredited to J. von Neumann, seeks to handle closed, discontinuous operators on relatively simple spaces. [1] contains a very general and incisive discussion of the analogues of a number of the results of Sections 8.6 and 8.7 for closed, densely defined operators, and this even for very general types of LCTVSs. The Banach space formulation of one such result is given in Theorem 8.7.9 and asserts that a closed, densely defined linear operator from one Banach space, E, into another, F, is onto if and only if its adjoint has a continuous inverse. To this extent both methods have in common the reduction of existence theorems to corresponding "a priori inequalities." The work set out in [1] is intended as a preliminary to its application to LPDEs. The latter task is commenced in Browder [1a] and is to be completed in further articles to appear under the same general title.

EXERCISES

5.1 Give a direct proof of (2) of Subsection 5.12.2 by using the fact that the set S of ordinary solutions of $P(D)X = 0$ is of finite dimension. (Gårding and Lions [1], p. 33.)

[Hint: Recall that the topology and hence also the uniform structure of a separated finite dimensional TVS are unique. Observe that if $X \in \mathscr{D}'$ is a solution of $P(D)X = 0$, and if $\varphi \in \mathscr{D}$, then $\varphi * X$ is an ordinary solution of the same equation.]

5.2 Suppose $n = 1$ and $P(D) = D^m + \cdots$. Show that $P(D)X = \varepsilon$ if and only if X is a function in $C^\infty(R\backslash\{0\})$ and such that

$$\lim_{x \downarrow 0} X^{(j)}(x) - \lim_{x \uparrow 0} X^{(j)}(x) = \begin{cases} 0 & \text{for } 0 \le j < m \\ (-2\pi i)^m & \text{for } j = m. \end{cases}$$

Such functions appear in the theory of ordinary LDEs, where $X(x - y)$ is usually termed the Green's function for the said equation. (Gårding and Lions [1], p. 33.)

5.3 Let X_1 and $X_2 \in \mathscr{D}'(R^1)$ and let X be the element of $\mathscr{D}'(R^2)$ defined by

$$X = X_1 \otimes 1 + 1 \otimes X_2,$$

that is,

$$\langle f, X\rangle = \left\langle \int f(x_1, x_2)\, dx_2, X_1 \right\rangle + \left\langle \int f(x_1, x_2)\, dx_1, X_2 \right\rangle$$

for f in $\mathscr{D}(R^2)$. Show that $D_1D_2X = 0$. This shows that D_1D_2 is not hypoelliptic.

5.4 The notations and assumptions are as in Subsections 5.13.1 and 5.13.9(a). Take the case $n = 1$ and $\Omega = (\alpha, \beta)$ where $\alpha > -\infty$. Assume that q is nonnegligible and that $\int_\alpha^x p^{-1}\, d\mu < +\infty$ for some (and therefore all) x in Ω. Show that to each bounded interval $I \subset \Omega$ there correspond numbers M_I and N_I such that

$$\operatorname*{Sup}_{x', x'' \in I} |f(x') - f(x'')| \le M_I \, \|f\|_V,$$

$$\operatorname*{Sup}_{x \in I} |f(x)| \le N_I \, \|f\|_V.$$

Deduce that

(1) $\lim_{x \downarrow \alpha} f(x)$ exists finitely for each $f \in V$;
(2) If $f_n \to f$ in V, then for any $b \in \Omega$ one has $f_n \to f$ uniformly on $[\alpha, b]$;
(3) $\lim_{x \downarrow \alpha} f(x) = 0$ for each f in the closure $\mathscr{D}_V$ in V of $\mathscr{D}(\Omega)$.

Analagous results are true if $\beta < +\infty$ and $\int_x^\beta p^{-1}\, d\mu < +\infty$ for some (and therefore all) $x \in \Omega$.

5.5 Using the notations of Subsection 5.13.9(a), show that V is complete in the case $n = 1$ provided p, p^{-1} and q lie in $\mathscr{L}^1_{\text{loc}}(\Omega)$ and q is nonnegligible on Ω. (See also the remarks at the end of Subsection 5.13.4.)

5.6 Using the notations of Subsection 5.13.9(a) (case $n = 1$), define

$$V_\alpha = \{f \in V : f(\alpha) = \lim_{x \downarrow \alpha} f(x) = 0\}.$$

Show that

(1) V_α is separated if $p > 0$ a.e. on Ω;

(2) If $\int_\alpha^x p^{-1}\, d\mu < +\infty$ for some (and hence all) $x \in \Omega$, then V_α is closed in V (hence is complete whenever V is complete). Show also that convergence (resp. boundedness) in V_α entails convergence (resp. boundedness) uniformly on (α, x_0) for each $x_0 \in \Omega$.

5.7 The notation is that of Section 5.13. Show that, if $E = L^2(\Omega)$,

$$2 \operatorname{Re} (f \mid v)_E - \|v\|_V^2$$

assumes its maximum for $v \in V$ when $v = Gf$, and that the maximum value is $(Gf \mid f)_E = \|Gf\|_V^2$.

5.8 The hypotheses are those of Section 5.13. Show that given $f \in L^2(\Omega)$ and $g \in H$, the unique solution of the system.

$$u - g \in N, \qquad Pu = f$$

is given by

$$u = g + G(f - Pg);$$

this solution belongs to $H + N = H$.

Remark. For $P = -\Delta + \lambda$ a different approach is given by Gårding [5], pp. 12–13. His conditions on f and g differ from ours, and the boundary values are specified in terms of the closure in V of $\mathscr{D}(\Omega)$.

NOTE. In Exercises 5.9–5.15 the notations and assumptions of Subsection 5.13.8 are adopted.

5.9 Take E to be $L^2(\Omega) = L^2$. By considering the defining property of G_λ, show that $G_{\bar\lambda}\bar f = \overline{G_\lambda f}$ provided $\bar v \in V$ whenever $v \in V$.

5.10 Using (5.13.24), verify the relation

$$\frac{dG_\lambda}{d\lambda} = G_\lambda^2$$

for $\lambda \in C\backslash R_-$, the derivative being formed on the sense of the normed topology on the space of endomorphisms of E.

5.11 Deduce from (5.13.26) the fact that G_λ is self-adjoint when considered as an endomorphism of E or of V_λ, λ being real and strictly positive.

5.12 Prove that the relation

$$\lambda(G_\lambda f \mid v)_E + (G_\lambda f \mid v)_V = (f \mid v)_E \qquad (v \in V)$$

continues to hold for $\lambda \in C\backslash R_-$.

5.13 Consider the situation described in Subsection 5.13.8.

Suppose that $E = L^2 = L^2(\Omega)$ and that $\lambda = \alpha + i\beta$ belongs to $C\backslash R_-$. Establish the following inequalities:

(1) If $\alpha + p_0 \geq 0$ and $(\alpha + p_0)^{-1} \in \mathscr{L}^1(\Omega)$, then

$$\|G_\lambda f\|_{L^2} \leq \|(a + p_0)^{-1}\|_{L^1} \cdot \|f\|_{L^2}.$$

(2) If $\alpha > 0$ then

$$\|G_\lambda f\|_{L^2} \leq |\lambda|^{-1} \cdot \|f\|_{L^2}.$$

(3) If $\alpha > 0$ and $p_k^{-1} \in \mathscr{L}^1(\Omega)$ then

$$\|\partial_k G_\lambda f\|_{L^2} \leq |\lambda|^{-1/2}\, \|p_k^{-1}\|_{L^1} \cdot \|f\|_{L^2}.$$

(4) If $\alpha \leq 0$ and $\beta \neq 0$ then

$$\|G_\lambda f\|_{L^2} \leq |\beta|^{-1} \cdot \|f\|_{L^2}.$$

(5) If $\alpha \leq 0$, $\beta \neq 0$, and $p_k^{-1} \in \mathscr{L}^1(\Omega)$, then

$$\|\partial_k G_\lambda f\|_{L^2} \leq (|\beta|^{-1} + |\alpha\beta^{-2}|) \, \|p_k^{-1}\|_{L^1}^{1/2} \cdot \|f\|_{L^2}.$$

5.14 Show that

$$\lim \lambda G_\lambda f = f$$

in L^2 as $\lambda \to \infty$ along any ray different from the negative real axis.

5.15 Consider the situation described in Subsection 5.13.8. Show that the problem

$$u \in N, \; Pu + \lambda u + \sum_{k+1}^{n} c_k \partial_k u = f$$

is soluble for each given $f \in L^2$, provided

$$c^2 = \operatorname*{ess\,sup}_{\Omega} \sum_{k=1}^{n} p_k^{-1} \, |c_k|^2 < \|G_\lambda\|_{L^2, V}^{2}.$$

[Hint: Rewrite the given equation in the form

$$u + G_\lambda\left(\sum_{k=1}^{n} c_k \partial_k u\right) = G_\lambda f.]$$

5.16 Examine the wave equation

$$\frac{\partial X}{\partial t} - \frac{\partial^2 X}{\partial t^2} = 0, \qquad X(x, 0) = f(x)$$

by seeking a $\mathscr{D}'$-valued function $t \to X_t$ such that

$$\left(\frac{\partial}{\partial t}\right) X_t - \frac{\partial^2 X_t}{x^2} = 0, \qquad X_0 = f.$$

Assuming that X_t and f are temperate, use the Fourier transform to show formally that

$$X_t = f * E_t,$$

E_t being the inverse Fourier transform of $\exp[-4\pi^2\xi^2 t]$, namely, $\exp[-x^2/4t]$.

5.17 Show that the equation

$$(-\Delta + a)X = A,$$

where $a > 0$, has exactly one solution $X \in \mathscr{S}'$ for given $A \in \mathscr{S}'$. In particular, $X = 0$ is the only solution in $\mathscr{S}'$ of $(-\Delta + a)X = 0$.

5.18 Show that the only temperate solutions of $\Delta X = 0$ are polynomials, but that there exist other solutions in $\mathscr{D}'$.

5.19 Let Ω be a nonvoid open subset of R^n and choose $\varepsilon > 0$ so small that $\Omega_\varepsilon = \{x \in \Omega\colon \, d(x, \dot{\Omega}) > \varepsilon\}$ is nonvoid ($\dot{\Omega}$ being the frontier of Ω). Suppose $A \in \mathscr{D}'(\Omega)$ and $B \in \mathscr{D}'(R^n)$, the latter having its support contained in the ball $B(0, \varepsilon)$ with centre 0 and radius ε. Show that if $\varphi \in \mathscr{D}(\Omega_\varepsilon)$ then

$$\chi(x) = B_{(y)}[\varphi(x + y)]$$

belongs to $\mathscr{D}(\Omega)$, and that the formula

$$\langle \varphi, A * B \rangle = \langle \chi, A \rangle$$

defines $A * B$ as an element of $\mathscr{D}'(\Omega_\varepsilon)$.

5.20 Let E be a function on R^n that depends on $|x| = r$ only, which is locally integrable over R^n, belongs to $C^2(R^n \backslash \{0\})$, and satisfies

$$\frac{d^2E}{dr^2} + (n-1)r^{-1}\frac{dE}{dr} - \lambda E = 0$$

for $r > 0$,

$$\lim_{r \downarrow 0} r^{n-1}E = 0,$$

$$\lim_{r \downarrow 0} r^{n-1}\frac{dE}{dr} = -w_n^{-1},$$

where w_n is the area of the unit sphere in R^n, $= 2\pi^{1/2n}/\Gamma(\frac{1}{2}n)$. Verify that E is an elementary solution for $-\Delta + \lambda$.

[Hint: Argue as in Section 5.11.]

Notes: A suitable choice of E for $n \geq 2$ and λ not real and ≤ 0 is

$$[w_n(n-2)!]^{-1}a^{n-2}\int_1^\infty e^{-art}(t^2-1)^{1/2(n-3)}\,dt,$$

where $a^2 = \lambda$ and $\operatorname{Re} a > 0$; for $n = 3$ this reduces to $e^{-ar}/4\pi r$. For $n = 1$ and $\lambda \neq 0$ one may take $E = \exp(-a|x|)/2a$ for any a satisfying $a^2 = \lambda$.

It follows that $-\Delta + \lambda$ is hypoelliptic whenever λ is not real and ≤ 0.

5.21 The notations are as in Subsection 5.13.1. Suppose that $0 < \operatorname{ess\,inf}_\Omega p_r \leq \operatorname{ess\,sup}_\Omega p_r < +\infty$ $(0 \leq r \leq n)$. Show that the following subsets of V are dense in V:

(1) $C^\infty(\Omega) \cap V$;

(2) The functions in $C^\infty(\Omega) \cap V$ having bounded supports.

[Hint: For the first part use the Hahn-Banach theorem and the hypoellipticity of $-\Delta + 1$; see the preceding exercise.]

Notes: Since Δ itself is hypoelliptic, the same arguments show that even if $p_0 = 0$ (the other hypotheses standing) the functions in $C^\infty(\Omega) \cap V$ having bounded supports are dense in V. V is here the Beppo-Levi space $BL(\Omega)$ (Deny-Lions [1]) formed of functions f in $L^1_{\text{loc}}(\Omega)$ for which $\partial_k f \in L^2(\Omega)$ for $k = 1, \cdots n$, formed into an inner-product space with

$$(f \mid g) = \sum_{k=1}^{n} (\partial_k f \mid \partial_k g)_{L^2(\Omega)}.$$

5.22 Let Ω be a domain in R^n and P a polynomial on R^n of the type

$$P(\xi) = \sum_{|p|=|q|=m} a_{pq}\xi^{p+q}$$

such that $P(\xi) > 0$ for $\xi \in R^n$, $\xi \neq 0$. For f and g in $\mathscr{D}(\Omega)$ define

$$(f \mid g)_m = \sum_{|p|=m} \int_\Omega \partial^p f \cdot \partial^p \bar{g}\,d\mu,$$

$$(f \mid g)_P = \sum_{|p|=|q|=m} a_{pq}\int_\Omega \partial^p f \cdot \partial^q \bar{g}\,d\mu.$$

Show (for example by use of the Fourier transform) that there exists a number $c > 0$ such that

$$c^{-1}\|f\|_m \leq \|f\|_P \leq c\,\|f\|_m$$

for $f \in \mathscr{D}(\Omega)$, where $\|f\|_m = +(f \mid f)_m^{1/2}$, and so on.

Assuming now that Ω is bounded, deduce from Inequality B of Subsection 5.14.2 that there exists a number $c_0 > 0$ such that

$$\|f\|_{L^2(\Omega)} \leq c_0 \|f\|_P$$

for $f \in \mathscr{D}(\Omega)$.

5.23 The notations and assumptions are as in Exercise 5.22. Show that the completion E of $\mathscr{D}(\Omega)$ with respect to the norm $\|\cdot\|_P$ may be identified with a subspace of the space $\mathscr{E}^m_{L^2}(\Omega)$ of functions (or function classes) $f \in L^2(\Omega)$ for which $\partial^p f \in L^2(\Omega)$ for each p satisfying $|p| = m$.

5.24 The notations and assumptions are as in the two preceding exercises. Show that, given $g \in \mathscr{E}^m_{L^2}(\Omega)$, there exists $u \in \mathscr{E}^m_{L^2}(\Omega)$ such that

$$u - g \in E \text{ and } P(\partial)u = 0 \text{ on } \Omega.$$

[Hint: Use the orthogonal projection method.]

5.25 Let Ω be a domain in R^n. Show that if f is real valued on Ω, and if f and $\partial_k f$ are locally integrable on Ω ($k = 1, \cdots, n$), then the same is true of f^+ and

$$\partial_k f^+ = (\partial_k f)\alpha,$$

where α is the characteristic function of the set where $f > 0$.

Deduce also that if g satisfies the same conditions as does f, then Sup (f, g) and Inf (f, g) have first-order partials that are locally integrable functions.

Conclude that if f is real valued and belongs to $BL(\Omega)$, and if

$$f_r = \text{Inf}\,[\text{Sup}\,(f, -r), r],$$

then $f_r \in BL(\Omega)$ and $\partial_k f_r = (\partial_k f)\beta_r$, where β_r is the characteristic function of the set where $|f| \leq r$. Show also that

$$\lim_{r \to \infty} f_r = f \text{ in } BL(\Omega).$$

5.26 Suppose that $X \in \mathscr{D}'(\Omega)$ is such that $\Delta^k X \in \mathscr{D}'^m(\Omega)$ for some fixed m and arbitrarily large k. Prove that $X \in C^\infty(\Omega)$. (Schwartz [2], Théorème XIX, p. 47.)

[Hint: Use the techniques introduced in Subsection 5.11.2.]

5.27 Suppose that $X \in \mathscr{D}'(\Omega)$ is such that for some m the following is true—for each bounded open set ω with $\bar{\omega} \subset \Omega$, there exists a neighbourhood U of 0 such that $X * \alpha \in C^\infty(\omega)$ for $\alpha \in \mathscr{D}^m(U)$. Show that $X \in C^\infty(\Omega)$. (Schwartz [2], Théorème XXI, p. 50.)

[Hint: As for the preceding exercise.]

CHAPTER 6

The Open Mapping and Closed Graph Theorems

6.0 Foreword

The theorems entitling this chapter date from the early days of functional analysis (Banach [1]), at which time they were established for complete metrizable TVSs. In this form they assert the following:

▶ Let E and F be complete metrizable TVSs, u and v linear maps of E into F and of F into E, respectively; if u is continuous and onto, it is open; and if v has a closed graph, it is continuous.

Since then, and especially in recent years, the original proofs of these theorems have been subjected to a close analysis. This task has been undertaken partly because of its intrinsic interest, and partly because the desire to apply such theorems in other recently developed branches of mathematics (integration theory, Schwartz's distributions, de Rham's currents, partial differential equations, and so forth) calls for the use of TVSs of increasingly general type. In particular, metrizability has come to be seen to be less dominating than was at first anticipated.

In this chapter we shall learn a little about some of the types of TVS that are currently important: metrizable TVSs, barrelled and ultrabarrelled spaces. The importance of these will continue to be illustrated in Chapters 7 and 8. The process of forming inductive limit spaces will also be introduced.

Regarding the open mapping and closed graph theorems (OMT and CGT, for short), only half the story thus far written can appear in this chapter.† Broadly speaking, the present situation divides into (a) general TVSs, not necessarily locally convex, and (b) LCTVs. Study of the latter case is best made in a manner largely independent of (a) but depending heavily on duality

† A possible exception to this dichotomy is provided by the results of Sections 6.5, 6.6, which can be applied to certain mixed categories of locally convex and nonlocally convex spaces, and which have roots in metrizability.

theory. For this reason it must be left until Chapter 8 to follow. Success here is due in large measure to Pták ([1], [2], [3], [4]).†

As regards case (a)—that of not necessarily locally convex TVSs—the situation is somewhat less satisfactory. Here some recent developments are due to W. Robertson ([1]). Unlike the case of LCTVSs, some traces of metrizability are explicitly retained in respect of the space E, and the proofs are more "traditional" in that they depend explicitly on "category arguments." The present chapter contains all that we intend to say about case (a).

A few words concerning the hypothesis of completeness of E and F in the OMT and CGT are not out of place. In case (a), it makes no explicit appearance in the latest forms as regards F; while it can be removed from E, this is done only at the expense of supplementary conditions that "cheat" in the sense that their role is to ensure that E and F may be replaced by their completions. On the contrary, in case (b) it will appear that a definite and precise form of completeness (stronger than the usual one) is essential for E, while another variety of restriction is needed for F.

It is clear from these general remarks that the original symmetry between E and F is lost. Rather, it will appear that our theorems will run in pairs in which the OMT is asserted for maps $u : E \to F$ and the CGT for maps $v : F \to E$. Our notation will be guided by this cue.

An alternative account of various forms of the OMT and the CGT is to be found in Kelley and Namioka [1], Section 11 and Problems 11D, 12E, 13F, 13G, 18I, 18J, 19B, and 20J.

6.1 Semimetrizable TVSs and Fréchet Spaces

If a TVS E has a countable neighbourhood base at 0, the associated uniform structure has a countable base of vicinities and so, according to Subsection 0.3.6, this uniform structure is semimetrizable. If, furthermore, E is separated, its uniform structure is metrizable. We here give a proof of Lemma 0.3.6 for TVSs.

6.1.1 **Theorem.** If E is a TVS, in order that its uniform structure (and hence also its topology) be semimetrizable, it is necessary and sufficient that there exist in E a countable neighbourhood base at 0. In that case, its uniform structure is definable by a translation-invariant semimetric d, so that $d(x, y) = d(x - y, 0)$.

Proof. The necessity is obvious. To prove sufficiency, let (V_n) be any countable base at 0, and then define by recurrence a second base (U_n) of balanced neighbourhoods of 0 satisfying $U_1 \subset V_1$ and

$$U_{n+1} + U_{n+1} + U_{n+1} \subset U_n \cap \left\{\bigcap_{p=1}^{n} V_p\right\} \qquad \text{for } n \geq 1.$$

† Pták uses the terms "open" and "nearly open" in a sense slightly different from that which we adopt: as applied to a map u: $E \to F$, his meaning is ours as applied with $u(F)$ replacing F.

Next we define the function $g \geq 0$ on E in the following fashion: $g(x) = 0$ if x belongs to $\bigcap_{n=1}^{\infty} U_n$; $g(x) = 2^{-k}$ if x belongs to $U_k \backslash U_{k+1}$; and $g(x) = 1$ if x belongs to $E \backslash U_1$. Obviously, $g(0) = 0$. In terms of g, we define a second function f by

$$f(x) = \text{Inf} \sum_{i=1}^{p} g(x_i),$$

the infimum being taken with respect to all finite sequences $(x_i)_{1 \leq i \leq p}$ (p unrestricted) for which $x = \sum_{i=1}^{p} x_i$. It is clear that f is positive and subadditive, and that $f \leq g$.

We shall show that $f \geq \frac{1}{2} g$, to do which we show by induction on p that $x = \sum_{i=1}^{p} x_i$ entails

$$\sum_{i=1}^{p} g(x_i) \geq \tfrac{1}{2} g(x). \tag{6.1.1}$$

The truth of (6.1.1) is evident if $p = 1$. We will suppose that $p' > 1$ and that (6.1.1) has been established for values of $p < p'$. We put

$$\alpha = \sum_{i=1}^{p'} g(x_i).$$

Then (6.1.1) holds, with p' in place of p, if $\alpha = 0$, or if $\alpha \geq \frac{1}{2}$; this is so since evidently $g \leq 1$. We assume then that $0 < \alpha < \frac{1}{2}$. We let h be the largest natural number r for which

$$\sum_{i<r} g(x_i) \leq \tfrac{1}{2}\alpha,$$

empty sums being interpreted to be 0. Then $1 \leq h \leq p'$,

$$\sum_{i<h} g(x_i) \leq \tfrac{1}{2}\alpha, \qquad \sum_{i<h+1} g(x_i) > \tfrac{1}{2}\alpha.$$

Consequently,

$$\sum_{i>h} g(x_i) \leq \tfrac{1}{2}\alpha.$$

By inductive hypothesis, since $h - 1 < p'$ and $p' - h < p'$, one has

$$g\left(\sum_{i<h} x_i\right) \leq 2 \sum_{i<h} g(x_i) \leq 2 \cdot \tfrac{1}{2}\alpha = \alpha$$

and likewise

$$g\left(\sum_{i>h} x_i\right) \leq \alpha.$$

On the other hand, $g(x_h) \leq \alpha$.

Introducing the smallest natural number k for which $2^{-k} \leq \alpha$, we see that $k > 1$, $\sum_{i<h} x_i \in U_k$, $\sum_{i>h} x_i \in U_k$, and $x_h \in U_k$, as appears from the preceding inequalities and the definition of g. Thus $x = (\sum_{i<h} x_i) + x_h + (\sum_{i<h} x_i)$ belongs to $U_k + U_k + U_k$, which is a subset of U_{k-1}. But then $g(x) \leq 2^{1-k} \leq 2\alpha$, which gives (6.1.1) with p' in place of p. The inductive proof of (6.1.1) is thus complete.

We thus have $f \geq \frac{1}{2}g$. From this it follows that U_k contains all elements x of E satisfying $f(x) \leq 2^{-k-1}$; while if $\beta > 0$, the set $\{x \in E: f(x) \leq \beta\}$ contains U_k whenever $2^{-k} < \beta$. Accordingly, $d(x, y) = f(x - y)$ is a semimetric that defines the uniform structure on E. ▌

Remarks. (1) The above proof is due to Bourbaki (Topologie Générale, Ch. IX, pp. 7–8) and is adaptable to uniform spaces so as to constitute a proof of Lemma 0.3.6.

(2) If E is locally convex, a more direct construction is available. We take any base at 0, say (U_n), formed of open (or closed), convex, and balanced sets. We let p_n be the gauge of U_n, and define

$$d(x, y) = \sum_{n=1}^{\infty} 2^{-n} \frac{p_n(x - y)}{1 + p_n(x - y)}.$$

(3) The most important, and historically the first, metrizable [resp. semimetrizable] TVSs are the normed [resp. seminormed] spaces. These are locally convex, of course.

(4) Following Bourbaki we shall term a *Fréchet space* any metrizable and complete LCTVS. Every Banach space is a Fréchet space. The so-called (F)-spaces of Banach ([1], Chapitre III) are metrizable and complete, but not necessarily locally convex. In Subsections 6.1.3 and 6.1.4 we shall exhibit a few important concrete instances of Fréchet spaces that are not Banach spaces.

Quotients of semimetrizable spaces. Let E be a semimetrizable TVS, M a vector subspace of E, f the natural map of E onto E/M. Since f transforms any neighbourhood base at 0 in E into a neighbourhood base at 0 in E/M, Theorem 6.1.1 implies that E/M is semimetrizable (and therefore metrizable if and only if M is closed in E). Furthermore, there is a positive subadditive function $x \to N(x)$ on E such that $N(x - y)$ is a semimetric defining the uniform structure of E. Suppose now we define N' on E/M by

$$N'(\xi) = \operatorname{Inf}\{N(x): x \in f^{-1}(\{\xi\})\}.$$

Then N' is positive and subadditive on E/M. Also, if we define for each $\varepsilon > 0$ the sets $U_\varepsilon = \{x \in E: N(x) < \varepsilon\}$ and $U'_\varepsilon = \{\xi \in E/M: N'(\xi) < \varepsilon\}$, then it is easily verified that $f(U_\varepsilon) = U'_\varepsilon$. From this it follows that the U'_ε constitute a base at 0 for the quotient topology on E/M, so that the semimetric derived from N' defines the uniform structure of E/M.

From these remarks we may derive an important result.

6.1.2 **Theorem.** If E is a complete semimetrizable TVS, M any vector subspace of E, then E/M is complete semimetrizable. In particular, if E is a Fréchet space and M a closed vector subspace of E, then E/M is a Fréchet space.

Proof. The second assertion follows at once from the first.

To establish the first statement, suppose that (ξ_n) is a Cauchy sequence in E/M. In order to show that (ξ_n) is convergent in E/M, it suffices to show that some subsequence is convergent. Now we are given that $N'(\xi_m - \xi_n) \to 0$ as $m, n \to \infty$. So we may select a subsequence (ξ_{n_k}) such that

$$\sum_k N'(\xi_{n_{k+1}} - \xi_{n_k}) < +\infty.$$

We put $\alpha_k = \xi_{n_{k+1}} - \xi_{n_k}$, and then choose a_k in $f^{-1}(\{\alpha_k\})$ so that

$$N(a_k) < N'(\alpha_k) + 2^{-k}.$$

Then $\sum_k N(a_k) < +\infty$. If $s_k = a_1 + \cdots + a_k$, the subadditivity of N shows that the sequence (s_k) is Cauchy in E. Since E is complete, (s_k) is convergent to some element of E. f being continuous, $(f(s_k))$ is convergent in E/M. But $f(s_k) = \alpha_1 + \cdots + \alpha_k = \xi_{n_{k+1}} - \xi_{n_1}$. Thus the convergence of (ξ_{n_k}) is established. ▌

Remark. It is not true in general that the quotient of a complete TVS by a vector subspace (closed or not) is again complete. A counterexample is given by Köthe [4]; see also Grothendieck [7], p. 145, Exercice 5.

6.1.3 **Examples.** If T is a σ-compact locally compact space, the space $C(T)$ of continuous scalar-valued functions on T is a Fréchet space when equipped with the topology of locally uniform convergence (convergence uniform on compact subsets of T, sometimes called the "compact-open topology"). A defining system of seminorms is obtained by taking a base K_n $(n = 1, 2, \cdots)$ of compact subsets of T and introducing the associated seminorms

$$p_n(x) = \operatorname*{Sup}_{K_n} |x|.$$

When T is further specialized, various important subspaces of $C(T)$ are Fréchet spaces when equipped with the induced topology or with a stronger one. As instances:

(a) If T is an open subset of the complex plane (or more generally a complex analytic manifold), the space of functions holomorphic on T is a closed subspace of $C(T)$, hence a Fréchet space for the induced topology.

(b) If $T = R^n$ or an open subset of R^n, the space $C^\infty(T)$ of all indefinitely differentiable functions on T is a vector subspace of $C(T)$. It is not closed in $C(T)$, hence is not a Fréchet space for the induced topology. It is a Fréchet space, however, if we strengthen the topology into that of locally uniform convergence of x and of each of its partial derivatives. Denoting by $t_1, \cdots, t_n$ the natural coordinate functions on R^n, by k an n-uple $(k_1, \cdots, k_n)$ of integers ≥ 0, and by ∂^k the partial differential operator

$$\frac{\partial^{k_1+\cdots k_n}}{\partial t_1^{k_1} \cdots \partial t_n^{k_n}},$$

a system of defining seminorms for the said topology on $C^\infty(T)$ is composed of the functions

$$p_r(x) = \operatorname{Sup} \{|\partial^k x(t)| : |t| \leq r, k_1 + \cdots + k_n \leq r\}$$

with $r = 1, 2, \cdots$. By using local coordinates, we can topologize in a similar fashion the space $C^\infty(T)$ for any real differentiable manifold T of class C^∞. If T is Riemannian, from $C^\infty(T)$ one may single out the harmonic functions; these form a closed subspace of $C^\infty(T)$, hence a Fréchet space for the induced topology.

(c) A vector subspace of $C^\infty(R^n)$ that plays an important role in the theory of Schwartz distributions is the space $\mathscr{S}(R^n)$ of functions x in $C^\infty(R^n)$ for which each of the functions

$$S_r(x) = \text{Sup}\,\{(1 + |t|)^r\,|\partial^k x(t)| : t \in R^n, k_1 + \cdots + k_n \leq r\}$$

($r = 1, 2, \cdots$) is finite. The S_r are seminorms on $\mathscr{S}(R^n)$ and the latter is a Fréchet space for the topology they define. $\mathscr{S}(R^n)$ contains the set $\mathscr{D}(R^n)$ of functions in $C^\infty(R^n)$ having compact supports, and $\mathscr{D}(R^n)$ is everywhere dense in $\mathscr{S}(R^n)$; see Section 5.15 above and Schwartz [1], [2].

6.1.4 **Examples from Summability Theory.** Let N denote the set of natural numbers, C^N the vector space of all scalar-valued sequences (that is, functions on N). C^N is a Fréchet space with the topology of pointwise convergence, being in fact identical with the space $C(N)$ introduced in Example 6.1.3. Suppose that A is an infinite matrix—that is, a scalar-valued function on $N \times N$. We consider the set $D(A)$ of all x in C^N such that the series $\sum_n A(m, n)x(n)$ converges for each m in N. Obviously, $D(A)$ is a vector subspace of C^N and A defines a linear map u_A of $D(A)$ into C^N that assigns to each x in $D(A)$ the sequence $y = u_A(x)$ defined by

$$y(m) = \sum_n A(m, n)x(n) \qquad (m \in N). \tag{6.1.1}$$

Under suitable conditions (which we shall investigate later in Example 7.2.4) u_A will transform convergent sequences into convergent sequences, and it may transform certain nonconvergent sequences into convergent ones. In this case we shall have a nontrivial process of attaching a generalized limit to certain nonconvergent sequences; that is, we shall have what is (somewhat oddly) called a "summability method" associated with the matrix A. Such summability methods have been the object of great interest and functional analysis proves to be a useful tool in this field of study. (K. Zeller, [1], [2], [3], [4], S. Mazur and W. Orlicz [1]; J. B. Tatchell [1]). We shall show here how certain useful Fréchet spaces are associated with each summability method. Here and subsequently we shall give instances of the use of general theorems from functional analysis. Zeller's book [4], with its extensive bibliography, is a standard reference on these topics. See also Cooke [1], [2].

As usual we denote by ℓ^∞, c, and c_0 the spaces of bounded, convergent, and convergent-to-zero sequences, respectively. ℓ^∞ is a Banach space for the norm

$$\|x\|_\infty = \text{Sup}\,|x(n)|,$$

and c and c_0 are both closed vector subspaces of ℓ^∞.

Given a matrix A, we denote by $B(A)$, $C(A)$, and $C_0(A)$ the subsets of $D(A)$ formed of sequences x such that $u_A(x)$ belongs to ℓ^∞, c, and c_0, respectively. Thus $C_0(A) \subset C(A) \subset B(A) \subset D(A)$ and all are vector subspaces of C^N.

If x is a sequence and k a natural number, we denote by $s_k x$ the kth section of x, defined by

$$s_k x(n) = x(n) \qquad (n \leq k), = 0 \qquad (n > k).$$

Obviously, $s_k x$ belongs to $D(A)$ always; and

$$\lim_{k \to \infty} u_A(s_k x) = u_A(x)$$

in the pointwise sense for each x in $D(A)$.

As seminorms defining the natural topology of C^N we use

$$p_n(x) = |x(n)| \qquad (n = 1, 2, \cdots). \tag{6.1.2}$$

For use in summability theory we introduce further functions to be used as seminorms on $D(A)$. To begin with we define

$$\begin{aligned} q_{A,m}(x) &= \operatorname{Sup}_k |u_A(s_k x)(m)| \\ &= \operatorname{Sup}_k \left| \sum_{n \leq k} A(m, n)x(n) \right| \leq +\infty \end{aligned} \tag{6.1.3}$$

for x in C^N and m any natural number. It is clear that $q_{A,m}$ is a seminorm on $D(A)$. Also put

$$q_A(x) = \|u_A(x)\|_\infty \leq +\infty \qquad (x \in D(A)). \tag{6.1.3'}$$

Then $B(A)$ is precisely the set of x in $D(A)$ for which $q_A(x) < +\infty$ and q_A is a seminorm on $B(A)$.

We shall consider the topology T_A on $B(A)$ defined by the system of seminorms p_n, $q_{A,m}$, and q_A, m and n ranging separately over N. The basic result is

▶ (1) $B(A)$ is a Fréchet space when equipped with T_A. $C(A)$ and $C_0(A)$ are closed vector subspaces of $B(A)$, hence are Fréchet spaces for the induced topology.

Proof. That $C_0(A)$ and $C(A)$ are closed in $B(A)$ follows from the fact that c_0 and c are closed in ℓ^∞. It remains therefore to prove that $B(A)$ is complete for T_A. Let (x_i) be a T_A-Cauchy sequence in $B(A)$ and put $y_i = u_A(x_i)$. Then (y_i) is Cauchy in ℓ^∞ and so there exists y in ℓ^∞ such that $\|y - y_i\|_\infty \to 0$ as $i \to \infty$. At the same time (x_i) is Cauchy for the pointwise topology on C^N, so that $x = \lim_{i \to \infty} x_i$ exists pointwise in C^N. A third consequence of the Cauchy nature of (x_i) is that

$$\operatorname{Sup}_k \left| \sum_{n \leq k} A(m, n)[x_i(n) - x_j(n)] \right| = R_{m,i,j}$$

where $\lim_{i,j \to \infty} R_{m,ij} = 0$ for each m. Letting $j \to \infty$ we deduce that

$$\operatorname{Sup}_k \left| \sum_{n \leq k} A(m, n)[x_i(n) - x(n)] \right| = R_{m,i} \tag{6.1.4}$$

where $\lim_{i\to\infty} R_{m,i} = 0$ for each m. From this it appears first (on fixing i) that x belongs to $D(A)$, and then that $u_A(x) = \lim_{i\to\infty} u_A(x_i) = \lim_{i\to\infty} y_i$ pointwise. But $y_i \to y$ in ℓ^∞, and so necessarily $y = u_A(x)$. Hence x belongs to $B(A)$ and $u_A(x_i) \to u_A(x)$ in ℓ^∞. We now have $x_i \to x$ pointwise, $u_A(x_i) \to u_A(x)$ in ℓ^∞, and (6.1.4) with $\lim_{i\to\infty} R_{m,i} = 0$ for each m. Thus $x_i \to x$ in the sense of T_A and completeness of $B(A)$ is established. ▌

We introduce now the set $K_A = K$ of natural numbers n for which

$$\operatorname{Sup}_m |A(m, n)| > 0.$$

It is evident that $u_A(x)$ depends only on the restriction $x \mid K$. Moreover one has always

$$|A(m, n)|\, p_m(x) \leq 2q_{A,m}(x).$$

Hence if n belongs to K_A, there exists a number $\alpha = \alpha(A, n)$ and a natural number $m = m(n)$ such that

$$p_m(x) \leq \alpha \cdot q_{A,m}(x) \tag{6.1.5}$$

As a corollary we derive

▶ (2) If $B^*(A)$ denotes the set of x in $B(A)$ such that $x \mid N \setminus K_A = 0$, then T_A induces on $B^*(A)$ a topology identical with that defined by the seminorms q_A and $q_{A,m}$ $(m = 1, 2, \cdots)$. In particular, if $K_A = N$, these two topologies coincide on $B(A)$ itself.

There is a similar assertion applicable when A is row finite—that is, when to each m there corresponds n_m such that $A(m, n) = 0$ for $n > n_m$. In this case, $D(A) = C^N$. Moreover we have

$$q_{A,m}(x) \leq \beta(A, m) \cdot \operatorname*{Sup}_{n \leq n_m} p_n(x), \tag{6.1.6}$$

where

$$\beta(A, m) = \sum_{n \leq n_m} |A(m, n)|.$$

Thus

▶ (3) If A is row finite, the topology T_A is identical with that defined by the seminorms q_A and p_n $(n = 1, 2, \cdots)$.

It is important to observe that all the preceding results can be extended without trouble to the case in which one wishes to consider simultaneously a sequence A_r $(r = 1, 2, \cdots)$ of matrices and the methods they define. We then define $D(A_1, A_2, \cdots) = \bigcap_r D(A_r)$, $B(A_1, A_2, \cdots) = \bigcap_r B(A_r)$, and so on, and replace K_A by $K(A_1, A_2, \cdots) = \bigcap_r K(A_r)$. T_A is replaced by the topology $T(A_1, A_2, \cdots)$ defined by the seminorms p_n, q_{A_r}, $q_{A_r,m}$ with n, m, and r separately ranging over all natural numbers. *The analogues of* (1), (2), *and* (3) *remain valid*; the proofs are evident modifications of the arguments set forth already.

We shall return to this and related topics in Examples 6.4.10, 7.2.4, and Remark (3) following Proposition 9.3.9.

6.2 Barrelled, Infrabarrelled, and Ultrabarrelled Spaces

Bourbaki ([5], p. 2) was the first to single out from the category of LCTVSs a subtype whose members share some of the important properties of Banach and Fréchet spaces without being bound by metrizability hypotheses. These spaces are characterizable in terms of the behaviour of certain of their subsets, termed "tonneaux" by Bourbaki and hereafter called "barrels." The spaces themselves are described by Bourbaki as "espaces tonnelés," and we shall call them "barrelled spaces."

Definition. If E is a TVS, by a barrel in E is meant a subset that is closed, convex, balanced, and absorbent. A TVS E is said to be barrelled if it is locally convex, and if in E each barrel is a neighbourhood of 0.

The reader will notice that in every LCTVS E there exists a base at 0 composed of barrels in E.

A weaker property which is often useful is the following one.

Definition. A LCTVS E is said to be infrabarrelled if each barrel in E that absorbs each bounded subset of E is a neighbourhood of 0 in E.

A different characterization of barrelled spaces, avoiding explicit mention of barrels, is contained in Theorem 6.2.1(2). This alternative formulation owes its existence in large part to the nature of the third type of space to be introduced.

For nonlocally convex spaces, W. Robertson ([1], p. 249) introduced and studied the property of being ultrabarrelled. It proves to be inconvenient to attempt a definition of these in terms of certain of their subsets (analogous to barrels). For this reason, and for convenience of future reference, we shall incorporate the precise definition in part (1) of Theorem 6.2.1. Part (2) of the same theorem gives the analogous characterization of barrelled spaces.

By way of an extreme example, if E is any vector space and t_∞ is the strongest locally convex topology carried by E, then (E, t_∞) is barrelled (see Subsection 1.10.1).

6.2.1 **Theorem.** (1) A TVS (E, t) is ultrabarrelled if and only if any vector space topology t' on E, having a base at 0 formed of t-closed sets, is weaker than t.

(2) A LCTVS (E, t) is barrelled if and only if any locally convex vector space topology t' on E, having a base at 0 formed of t-closed sets, is weaker than t.

Proof. (1) This is a definition.

(2) Suppose first that (E, t) is barrelled, and that t' has the stated properties. Let V be any t'-neighbourhood of 0. V contains a t'-neighbourhood of 0, say V', that is t-closed. Since t' is locally convex, V' contains a t'-neighbourhood V'' of 0 that is convex and balanced. The t-closure of V'' is then a barrel in (E, t), hence a t-neighbourhood of 0. Since this t-closure of V'' is contained in V', and therefore in V, it follows that V is a t-neighbourhood of 0. Thus t' is weaker than t, and (2) is satisfied.

Conversely, suppose that (E, t) satisfies (2). Let A be any barrel in E. The sets αA, with $\alpha > 0$, forms a base at 0 for a locally convex topology t' that satisfies the hypothesis in (2). Accordingly, t' is weaker than t, which signifies that A is a t-neighbourhood of 0. Thus (E, t) is barrelled. ▌

Remark. (1) It is a trivial consequence of Theorem 6.2.1 that any ultrabarrelled LCTVS is barrelled. For LCTVSs we therefore have

$$\text{ultrabarrelled} \Rightarrow \text{barrelled} \Rightarrow \text{infrabarrelled},$$

the double arrows denoting implication. An example presented by Robertson ([1], p. 256) and due to J. D. Weston shows that a barrelled space may fail to be ultrabarrelled.

(2) In Subsections 7.1.1 and 8.9.6 we shall encounter yet other characterizations of barrelled spaces, each of interest and importance. Meanwhile we shall proceed to develop some properties and produce important instances of spaces of the categories in question.

6.2.2 **Theorem.** Any nonmeagre TVS is ultrabarrelled; any nonmeagre LCTVS is barrelled.

Proof. The second statement follows directly from the first. Suppose that (E, t) is a nonmeagre TVS, t' any vector-space topology having a base at 0 formed of t-closed sets. Let U be any t'-neighbourhood of 0, and let us show that U is a t-neighbourhood of 0. By hypothesis on t', there exists a t-closed t'-neighbourhood of 0, say V, such that V is balanced and $V + V \subset U$. Since E is the union of the sets nV $(n = 1, 2, \cdots)$, each t-closed, and since (E, t) is nonmeagre, there exists n such that the t-interior of nV is not empty. It follows that V contains $x + W$ for some x in E and some t-neighbourhood W of 0. Then

$$U \supset V + V = V + (-V) \supset W + (-W),$$

showing that U is a t-neighbourhood of 0. Thus t' is weaker than t. ▌

6.2.3 **Corollary** Any complete semimetrizable TVS is ultrabarrelled.

Remark. It is very significant that (see Sections 6.3 and Subsection 8.3.7) the nonmeagre TVSs are not the only ones that are ultrabarrelled. Upon this hinges a great deal of the practical importance of this concept. We note also at this point that since one may show (Bourbaki [5], p. 4, Exercice 7)) that any product of complete metric spaces is a Baire space, one may infer the existence of complete ultrabarrelled TVSs that are not metrizable.

6.2.4 **Theorem.** (1) Let E and F be TVSs, f a continuous, linear, and open map of E onto F. If E is ultrabarrelled (resp. barrelled), then F is ultrabarrelled (resp. barrelled). In particular, the quotient by a vector subspace of an ultrabarrelled (resp. barrelled) space is ultrabarrelled (resp. barrelled).

(2) Let E be a TVS, $\hat{E}$ its completion, F a vector subspace of $\hat{E}$ containing E. If E is ultrabarrelled (resp. barrelled), then F is ultrabarrelled (resp. barrelled).

Proof. In each case it will suffice to deal with the ultrabarrelled instance, the proof for barrelled spaces being very similar.

(1) Let s and t denote the initial topologies on E and F, respectively, and let t' be a vector-space topology on F having a base at 0 formed of t-closed sets V. The corresponding sets $f^{-1}(V)$ are then s-closed and form a base at 0 for a topology s'. Since E is ultrabarrelled, s' is weaker than s. Thus each $f^{-1}(V)$ is an s-neighbourhood of 0. But then, f being open, $f(f^{-1}(V))$ is a t-neighbourhood of 0, and its superset V has the same property. Thus t' is weaker than t, showing that F is ultrabarrelled.

(2) Let t be the initial topology on E, $\hat{t}$ that on $\hat{E}$, and let t' be a vector space topology on F having a base at 0 formed of sets V closed for $\hat{t} \mid F$. We will show that V is a neighbourhood of 0 for $\hat{t} \mid F$. Now the sets $V \cap E$ are t-closed and form a base at 0 for a topology that, since E is ultrabarrelled, is weaker than t. Thus each $V \cap E$ contains a t-neighbourhood of 0. Therefore $\overline{V \cap E}$ (closure in $\hat{E}$) is a $\hat{t}$-neighbourhood of 0, and so $\overline{V \cap E} \cap F$ is a $\hat{t} \mid F$-neighbourhood of 0. A fortiori, $\bar{V} \cap F$ is a $\hat{t} \mid F$-neighbourhood of 0. But $\bar{V} \cap F = V$ since V is closed for $\hat{t} \mid F$. Thus finally V is a $\hat{t} \mid F$-neighbourhood of 0. ∎

6.2.5 **Theorem.** Any product of barrelled [resp. infrabarrelled] spaces is barrelled [resp. infrabarrelled].

We shall nowhere make essential use of this result and the proof is relegated to Exercise 7.23.

6.2.6. It is not generally true that a closed vector subspace of a barrelled space is barrelled (for the induced topology). This follows on combining Exercises 6.17 and 6.18, the second of which shows also that there exist complete (nonmetrizable) separated LCTVSs that are not barrelled. (See also Remark (4) following Theorem 7.3.3.)

If E is barrelled and M a vector subspace of E admitting a topological complement in E, however, then M is barrelled (Exercise 6.19).

There exist noncomplete normed vector spaces that are nonmeagre and therefore barrelled (Exercise 6.23); and also normed vector spaces that are barrelled and meagre (Bourbaki [5], p. 157, Exercice 10)).

6.2.7. To proceed with further examples of barrelled spaces we shall digress to the extent of introducing an extremely important process—that of taking so-called inductive limits—leading from a given family of TVSs to a new TVS. It is by application of this process that one obtains an effective means of handling the TVSs that arise naturally in the theory of distributions.

6.3 Inductive Limit Spaces

Suppose that E_i $(i \in I)$ is a family of TVSs, E a vector space, φ_i a linear map of E_i into E, and $E = \bigcup_i \varphi_i(E_i)$. Among the locally convex topologies on E relative to which each φ_i is continuous there is a strongest, say, $\mathscr{T}$. A base at 0 for $\mathscr{T}$ is formed of the convex and absorbent sets $W \subset E$ such that, for each i, $\varphi_i^{-1}(W)$ is a neighbourhood of 0 in E_i. (That $\mathscr{T}$ *is* a vector-space topology stems from the fact that W is convex and so $\frac{1}{2}W + \frac{1}{2}W \subset W$. It is this property that causes trouble if we drop the condition of local convexity.) $\mathscr{T}$ is said to be the *inductive limit* of the topologies on the E_i, relative to the maps φ_i; and E, equipped with $\mathscr{T}$, is called the inductive limit of the TVSs E_i, relative to the maps φ_i: Bourbaki ([4], p. 61) restricts the E_i to be locally convex, and this is certainly the most important case in practice. See also J. Sebastião e Silva [1].

In the special case in which each E_i is a vector subspace of E and φ_i the injection map of E_i into E, we speak of an *internal inductive limit*.

We shall see in Subsection 7.3.5 that quite wide classes of LCTVSs are obtainable as internal inductive limits of seminormable spaces.

The inductive limits we have just defined are closely related to the direct limits used in algebra; they run parallel to the so-called projective, or inverse, limits. As applied to groups and modules (for use in cohomology theory) they are fully explained in Eilenberg and Steenrod [1], Chapter VIII; see also Kelley and Namioka [1], pp. 9–11 and Problems 8I, 16C, 19A, 19B, and 22C.

We proceed to give some very important examples of inductive limit spaces.

(1) THE SPACE $\mathscr{K}(T)$. The notation is as in Section 4.1. If we suppose S is any relatively compact subset of T, and if we put $\mathscr{K}(T, S)$ for the set of functions f in $\mathscr{K}(T)$ having support contained in S, then $\mathscr{K}(T, S)$ is a vector subspace of $\mathscr{K}(T)$, and $\mathscr{K}(T)$ is the union of the $\mathscr{K}(T, S_i)$ whenever (S_i) is a base for the relatively compact subsets of T (that is, each S_i is relatively compact, and each such subset of T is contained in S_i for some i). Each $\mathscr{K}(T, S)$ is a Banach space when equipped with the "uniform norm."

$$\|f\| = \operatorname{Sup}\{|f(t)| : t \in T\}.$$

Taking a base (S_i) for the relatively compact subsets of T, we may thus endow $\mathscr{K}(T)$ with the inductive limit topology determined by the subspaces $\mathscr{K}(T, S_i)$ and the injections maps ϕ_i of $\mathscr{K}(T, S_i)$ into $\mathscr{K}(T)$.

There is no difficulty in showing that the resulting inductive limit topology is independent of which base (S_i) is chosen: One has only to observe that if $S \subset S'$, then the topology of $\mathscr{K}(T, S')$ induces on $\mathscr{K}(T, S)$ the same topology as the latter carries a priori.

According to the general Theorem 6.3.2, the dual of $\mathscr{K}(T)$, thus topologized as an inductive limit of Banach spaces, comprises exactly the Radon measures as defined in Section 4.3.

(2) SPACES OF TEST FUNCTIONS IN THE THEORY OF DISTRIBUTIONS. We return to the circumstances described in Section 5.1 and use the notations explained there. Our first task to make $\mathscr{D}(\Omega)$ into an inductive limit space. As we have seen in Section 5.1, each of the vector subspaces $\mathscr{D}(\Omega, S)$ of $\mathscr{D}(\Omega)$ (S a relatively compact subset of Ω) is a Fréchet space when endowed with the topology defined by the seminorms

$$N_m(f) = \operatorname{Sup}\{|\partial^p f(x)| : x \in \Omega, |p| \leq m\},$$

when m ranges over all (or over arbitrarily large) natural numbers. If now we take any base (S_i) for the relatively compact subsets of Ω, then $\mathscr{D}(\Omega) = \bigcup_i \mathscr{D}(\Omega, S_i)$. If we denote by φ_i the injection map of $\mathscr{D}(\Omega, S_i)$ into $\mathscr{D}(\Omega)$, we may then endow $\mathscr{D}(\Omega)$ with the inductive limit topology determined by the $\mathscr{D}(\Omega, S_i)$ and the maps φ_i. It is a very simple matter to check that the resulting topology on $\mathscr{D}(\Omega)$ is, in fact, independent of the chosen base (S_i). In particular, one may use suitable countable bases (S_i). Thus $\mathscr{D}(\Omega)$ appears as the inductive limit of a sequence of Fréchet spaces.

In a similar way we may topologize $\mathscr{D}^m(\Omega)$ (m a natural number or 0; see Section 5.8) as an inductive limit of a sequence of Banach spaces $\mathscr{D}^m(\Omega, S_i)$ $(i = 1, 2, \cdots)$.

By Theorem 6.3.2 the dual of $\mathscr{D}^m(\Omega)$ is just $\mathscr{D}'^m(\Omega)$ $(0 \leq m \leq \infty)$.

It is not difficult to specify a defining family of seminorms for the inductive limit topology on $\mathscr{D}^m(\Omega)$ for $0 \leq m \leq \infty$. One way of doing this (Schwartz [1], p. 67) is to select arbitrarily and then hold fixed a sequence $(\Omega_r)_{r \geq 0}$ of sets, each open and relatively compact in ϕ, such that $\Omega_0 = \Omega$, $\bar{\Omega}_r \subset \Omega_{r+1}$, and $\Omega = \bigcup_{r \geq 0} \Omega_r$; then, corresponding to each increasing sequence $s = (s_r)$ of integers ≥ 0 and each decreasing sequence $\varepsilon = (\varepsilon_r)$ of strictly positive numbers tending to zero, assign the seminorm

$$N_{s,\varepsilon}(\varphi) = \operatorname*{Sup}_{r \geq 0} \operatorname{Sup} \{\varepsilon_r^{-1} |\partial^p \varphi(x)| : x \in \Omega \backslash \Omega_r, |p| \leq s_r\}.$$

It is claimed that these $N_{s,\varepsilon}$ form a defining family of seminorms for $\mathscr{D}(\Omega)$; and that for $m < \infty$ a similar role is played by those $N_{s,\varepsilon}$ for which $s_r = m$ for all r.

We set out the proof relevant to the case $m = \infty$. To begin with, it is almost evident that each $N_{s,\varepsilon}$ is continuous on $\mathscr{D}(\Omega)$. To verify this, it suffices to check that $N_{s,\varepsilon} \mid \mathscr{D}(\Omega, K)$ is continuous for each compact set K in Ω, which is quite clear. For the rest we may show that if U is any closed convex neighbourhood of 0 for the inductive limit topology, then s and ε can be chosen so that U contains each φ satisfying $N_{s,\varepsilon}(\varphi) \leq 1$. Now U has the property that to each r corresponds an integer $s_r \geq 0$ and a number $\delta_r > 0$, such that U contains each φ in $\mathscr{D}(\Omega)$ having its support contained in Ω_{r+2} and satisfying $|\partial^p \varphi| \leq \delta_r$ for $|p| \leq s_r$. It may be assumed that the s_r increase, and the ε_r decrease, as r increases. We choose and fix positive functions β_r in $\mathscr{D}(\Omega)$ such that $\sum_r \beta_r = 1$ on Ω and such that the support of β_r lies in $\Omega_{r+2} \backslash \Omega_r$. Then φ may be written as the sum of the series

$$\sum_r 2^{-r-1}(2^{r+1}\beta_r \varphi),$$

this series being finite for each φ in $\mathscr{D}(\Omega)$. Accordingly, U being closed and convex, φ will belong to U provided $2^{r+1} \beta_r \varphi$ belongs to U for all r.

Now $\partial^p(\beta_r \varphi)$ is a finite linear combination of derivatives $\partial^q \beta_r$ and $\partial^{q'} \varphi$ with $|q| \leq |p|$ and $|q'| \leq |p|$. Since only the values of φ in $\Omega_{r+2} \backslash \Omega_r$ are involved in $\beta_r \varphi$, a little thought shows that to each r corresponds a number k_r such that the relations

$$|\partial^p \varphi| \leq \varepsilon_r \qquad \text{for } x \in \Omega \backslash \Omega_r \text{ and } |p| \leq s_r$$

entail the relations

$$|2^{r+1}\,\partial^p(\beta_r\varphi)| \le k_r\varepsilon_r \qquad \text{for } |p| \le s_r.$$

If therefore we choose ε_r so that $k_r\varepsilon_r \le \delta_r$, then $N_{s,\varepsilon}(\varphi) \le 1$ will entail that $2^{r+1}\beta_r\varphi \in U$ for all r and therefore $\varphi \in U$, which is what we wished to establish.

A different choice of defining family of seminorms is described by Gårding and Lions ([1], p. 21).

The knowledge that the inductive limit topology on $\mathscr{D}^m(\Omega)$ is defined by the seminorms $N_{s,\varepsilon}$ paves the way for certain important corollaries of which we shall list three here.

(a) For each relatively compact subset S of Ω, the inductive limit topology of $\mathscr{D}^m(\Omega)$ induces on $\mathscr{D}^m(\Omega, S)$ the initial topology of the latter space.

Proof. There exists an index $r^* = r^*(S)$ such that $S \subset \Omega_{r^*}$. Then for φ in $\mathscr{D}^m(\Omega, S)$ we have

$$N_{s,\varepsilon}(\varphi) = \sup_{0\le r\le r^*} \sup\,\{\varepsilon_r{}^{-1}|\partial^p\varphi(x)| : x \in \Omega\backslash\Omega_r, |p| \le s_r\}.$$

Reference to Section 5.1 now makes it plain that the restrictions

$$N_{s,\varepsilon} \mid \mathscr{D}^m(\Omega, S)$$

define on the space $\mathscr{D}^m(\Omega, S)$ its initial topology. ▮

(b) A subset of $\mathscr{D}^m(\Omega)$ is bounded for the inductive limit topology if and only if it is contained and bounded in $\mathscr{D}^m(\Omega, S)$ for a suitable relatively compact subset S of Ω. (See also Exercise 6.24.)

Proof. By the definition of the inductive limit topology, any bounded subset of any $\mathscr{D}^m(\Omega, S)$ is bounded in the inductive limit space. To prove the converse it suffices, in view of (a), to show that a bounded subset B of $\mathscr{D}^m(\Omega)$ is contained in some subspace $\mathscr{D}^m(\Omega, S)$. Were this not the case, however, one could choose points x_r in $\Omega\backslash\Omega_r$ and functions φ_r in B such that $\varphi_r(x_r) \ne 0$. If we define $\varepsilon_r = r^{-1}\,|\varphi_r(x_r)|$, the seminorm $N_{s,\varepsilon}$ would be unbounded on B, contradicting the assumption that B is bounded in $\mathscr{D}^m(\Omega)$. ▮

(c) $\mathscr{D}^m(\Omega)$ is complete.

Proof. Suppose that (φ_i) is a Cauchy directed family in $\mathscr{D}^m(\Omega)$. This signifies that given any $N_{s,\varepsilon}$ of the appropriate defining family, there exists an index i_0 such that

$$N_{s,\varepsilon}(\varphi_i - \varphi_j) \le 1 \tag{6.3.1}$$

provided neither i nor j is below i_0. (It would be customary to find the "1" on the right-hand side of (6.3.1) replaced by any arbitrary preassigned number $t > 0$, but this would be equivalent to (6.3.1) after ε has been replaced by $t\varepsilon = (t\varepsilon_0, t\varepsilon_1, \cdots)$.)

By suitable choice of s and ε, this shows that $(\partial^p\varphi_i)$ is locally uniformly convergent on Ω for each p (or, if $m < \infty$, for each p satisfying $|p| \le m$), hence it is plain that there exists a function φ in $C^m(\Omega)$ such that $\partial^p\varphi_i \to \partial^p\varphi$ locally uniformly for each p. To confirm that this φ has a compact support in Ω, and hence belongs to $\mathscr{D}^m(\Omega)$, we return to (6.3.1) and apply it with $s = (0, 0, \cdots)$.

It follows that for any ε,

$$|\varphi_i(x) - \varphi_j(x)| \le \varepsilon_r \qquad \text{for } x \in \Omega\backslash\Omega_r,$$

provided i and j are not below i_0. Letting $j \to \infty$ it follows that

$$|\varphi_i(x) - \varphi(x)| \le \varepsilon_r \qquad \text{for } x \in \Omega\backslash\Omega_r$$

provided $i \ge i_0$. Since φ_{i_0} has a compact support, we conclude that

$$|\varphi(x)| \le \varepsilon_r \qquad \text{on } \Omega\backslash\Omega_r$$

for all sufficiently large r. The sequence ε being arbitrary, this entails that φ has a compact support. Indeed, were this not the case, there would exist points x_r in $\Omega\backslash\Omega_r$ such that $\varphi(x_r) \ne 0$, and a contradiction would result on taking $\varepsilon_r = \frac{1}{2}|\varphi(x_r)|$. Thus φ belongs to $\mathscr{D}^m(\Omega)$.

It remains to show that $\lim_i \varphi_i = \varphi$ for the inductive limit topology, to do which it is enough to show that $N_{s,\varepsilon}(\varphi_i - \varphi) \le 1$ for sufficiently large i and any choice of s and ε. But this follows from (6.3.1) coupled with the pointwise convergence on Ω of $\partial^p\varphi_i$ to $\partial^p\varphi$ for any given p, an implication that becomes evident when (6.3.1) is written out explicitly in the form

$$|\partial^p\varphi_i(x) - \partial^p\varphi_j(x)| \le \varepsilon_r \qquad \text{for } x \in \Omega\backslash\Omega_r \qquad \text{and } |p| \le s_r.$$

Completeness is thereby established. ∎

Remarks. (i) If T is σ-compact the same arguments apply to $\mathscr{K}(T)$, which is formally the same thing as $\mathscr{D}^0$.

(ii) Neither assertion (a) nor (b) is true of more general countable internal inductive limits of Fréchet spaces. A counterexample to (a) is exhibited in (3) below. In Section 6.5 we shall see that (b) remains valid for a fairly wide class of such inductive limit spaces. The arguments used there are of a more general nature.

(3) Spaces of holomorphic functions on a closed set. If Ω is an open subset of the complex plane (or, more generally, any complex-analytic manifold), the space $H(\Omega)$ of functions holomorphic on Ω is topologized in a natural and direct fashion with the topology of locally uniform convergence (see Example 6.1.3(a)). Suppose now that T represents a closed subset of the complex plane (or of a complex-analytic manifold). It is customary to say that a function f (taking complex values) is holomorphic on T, if it is defined and holomorphic on some open subset of the plane (or the manifold) containing T, this open set depending upon f. Then (see Grothendieck [5], [6], for example) it is useful to be able to treat this set of functions as a LCTVS.

The first difficulty is that this set is not itself even a vector space, owing to the fact that its members are functions with variable domains of definition. (This leads to trouble when one wishes to specify the zero element of the space and its relation to subtraction.) The difficulty is removable, however, by the simple device of taking quotients modulo an equivalence relation. If Ω denotes any open set containing T, we let $H(\Omega)$ denote the vector space of functions holomorphic on Ω. If f and f' belong to $H(\Omega)$ and to $H(\Omega')$, respectively, we say that they are equivalent if they agree on some open set containing T. A class modulo this equivalence relation will be termed a (holomorphic) *function germ* on T. The set of all such function germs on T is a vector space: algebraic operations are defined in an obvious fashion by taking representatives having a common domain of definition. For example, to form the sum of two function germs, say $\tilde{f}$ and $\tilde{g}$, we choose representatives f and g of $\tilde{f}$ and $\tilde{g}$, respectively, which have a common domain (for example, take any two representatives f_0 and g_0 and let f and g be, respectively, the restrictions of f_0 and g_0 to (Dom f_0) $\cap$ (Dom g_0)); then $\tilde{f} + \tilde{g}$ is by definition the class of the function $f + g$.

We denote by $H(T)$ this vector space of function germs on T. Any base (Ω_i) is taken for the open sets containing T. For each i, we let ϕ_i be the map of $H(\Omega_i)$ into $H(T)$ that carries a function f in $H(\Omega_i)$ into its class $\tilde{f}$. One may then topologize $H(T)$ as the inductive limit of the $H(\Omega_i)$, each topologized as we have said above, relative to the maps ϕ_i.

If T happens to be extensive enough for any function in $H(\Omega)$ ($\Omega \supset T$, Ω open) that vanishes on T to be necessarily zero throughout some (possibly smaller) open set containing T, there is no harm in thinking of the elements of $H(T)$ as functions defined on T. However, an especially important case is that in which T is reduced to a single point—in which case the picture just drawn is entirely misleading.

The reader may also note that one can represent $H(T)$ as an inductive limit of Banach spaces, at least if T is relatively compact. For this it suffices to use the spaces $H_{bd}(\Omega_i)$ of bounded, holomorphic functions on Ω_i, endowed with the topology of convergence uniform on Ω_i, and let Ω_i range over a base of relatively compact open sets containing T. One might even use in place of $H_{bd}(\Omega_i)$ the space of functions holomorphic on Ω_i and continuously extendible to $\bar{\Omega}_i$. If T is relatively compact, either choice leads to the same topology on $H(T)$ as before.

Remarks. Examples (1) and (2) above illustrate the case of what we have termed an *internal* inductive limit—that in which each E_i is a vector subspace of E and ϕ_i is the corresponding injection map. There are various other specializations, notably the strict internal inductive limits of Bourbaki ([4], p. 65) and even more especially the LF-spaces of Dieudonné and Schwartz [1]. At the moment we are concerned with basic properties common to all inductive limits.

6.3.1 **Theorem.** Any inductive limit of barrelled [resp. infrabarrelled] spaces is barrelled [resp. infrabarrelled].

Proof. Let E be the inductive limit of spaces E_i relative to maps φ_i.

If A is a barrel in E, so is $\varphi_i^{-1}(A)$ relative to E_i. So the latter is a neighbourhood of 0 in E_i. Hence A is a neighbourhood of 0 for the inductive limit topology. The infrabarrelled case is treated analogously. ∎

6.3.2 **Theorem.** Let E be the inductive limit of the E_i relative to maps φ_i, F any LCTVS, u a linear map of E into F. u is continuous if (and only if) $u \circ \varphi_i$ is continuous on E_i for each i.

Proof. Let V be a convex neighbourhood of 0 in F. By hypothesis, $U = u^{-1}(V)$ has the property that $\varphi_i^{-1}(U)$ is a neighbourhood of 0 in E_i, for each i. Since U is convex and absorbent, U is a neighbourhood of 0 in E. ▌

6.3.3 **Some Examples.** Consider the spaces $\mathscr{D}(\Omega)$ and $\mathscr{D}^m(\Omega)$ appearing in the theory of distributions and topologized as LCTVSs in the manner described in Example (2) above. Each is an inductive limit of Fréchet spaces (even of Banach spaces in the second instance). Hence (Corollary 6.2.3 and Theorem 6.3.1) these spaces are barrelled. The same is true of the space $\mathscr{K}(T)$, topologized as in Example (1) above. None of these spaces is metrizable save $\mathscr{K}(T)$ in the case in which T is compact (when $\mathscr{K}(T)$ is a Banach space). The theorems of Section 6.5 suffice to disprove metrizability in all other cases.

6.4 Forms of the Open Mapping and Closed Graph Theorems

We begin with a general lemma that incorporates all the traces of metrizability hypotheses involved in the subsequent forms of the theorems in question.

6.4.1 **Lemma.** Suppose that X is a complete semimetrizable space, Y a uniform space, d any semimetric for X, $B_r(x)$ the open ball in X with centre x and radius $r > 0$, π_X and π_Y the projections of $X \times Y$ onto X and Y, respectively. Let S be a closed subset of $X \times Y$ with the following property: for each $r > 0$ there is a vicinity $\mathscr{V}$ (of the uniform structure on Y) such that $(x, y) \in S$ implies

$$\overline{\pi_Y(S \cap \pi_X^{-1}(B_r(x)))} \supset \mathscr{V}(y). \tag{6.4.1}$$

The conclusion is that for each $r > 0$ and each $\varepsilon > 0$, $(x, y) \in S$ implies

$$\pi_Y(S \cap \pi_X^{-1}(B_{r+\varepsilon}(x))) \supset \overline{\pi_Y(S \cap \pi_X^{-1}(B_r(x)))} \supset \mathscr{V}(y) \tag{6.4.2}$$

If further $\pi_Y(S)$ is dense in Y, then $\pi_Y(S) = Y$.

Proof. This is due to Kelley ([1], p. 202) and is presented in two parts, the first of which is an auxiliary. For brevity, if $A \subset X$, we write $S_Y(A)$ for $\pi_Y(S \cap \pi_X^{-1}(A))$.

(1) Given $A \subset X$ and $y \in \overline{S_Y(A)}$, there exist sets $B \subset X$ of arbitrarily small diameter satisfying $A \cap B \neq \emptyset$ and $y \in \overline{S_Y(B)}$.

Indeed, let $r > 0$. We choose a symmetric vicinity $\mathscr{V}$ such that $(x, y) \in S$ implies (6.4.1). We choose then $y' \in S_Y(A) \cap \mathscr{V}(y)$, and subsequently $a \in A$ such that $(a, y') \in S$. We shall then have $y \in \mathscr{V}(y')$, which is contained in $\overline{S_Y(B_r(a))}$, and $B_r(a)$ has diameter $2r$. It remains therefore only to take $B = B_r(a)$, with r sufficiently small.

(2) We shall show that

$$y \in \overline{S_Y(B_r(x))} \quad \text{entails } y \in S_Y(B_{r+\varepsilon}(x)).$$

Let $A_0 = B_r(x)$ and select inductively sets $A_n \subset X$ such that $A_n \cap A_{n+1} \neq \emptyset$, diameter $A_n < \varepsilon/2^{n+2}$, $y \in \overline{S_Y(A_n)}$. This selection is possible by (1). Choosing $x_n \in A_n \cap A_{n+1}$, we see easily that

$$d(x_n, x_{n+p}) < \frac{\varepsilon}{2^{n+1}} + \cdots + \frac{\varepsilon}{2^{n+p}} < \frac{\varepsilon}{2^n}$$

for $p > 0$. So (x_n) is Cauchy. Since X is complete, (x_n) converges to some point $a \in X$; and clearly $d(x, a) < r + \varepsilon$. We take any neighbourhood U of a. Since $y \in \overline{S_Y(A_n)}$, every neighbourhood N of y contains some $y' \in S_Y(A_n)$. Then there exists $x' \in A_n$ such that $(x', y') \in S$. Now

$$d(x', a) \leq d(x_n, x') + d(x_n, a) < \text{diameter } A_n + \varepsilon/2^n.$$

It follows, if n is sufficiently large, that $x' \in U$ and so $y' \in S_Y(U)$. Thus $y \in \overline{S_Y(U)}$.

If N is any neighbourhood of y, $S_Y(U)$ meets N. Hence there exists points $x'' \in U$, $y'' \in N$ such that $(x'', y'') \in S$; that is, $S \cap (U \times N) \neq \emptyset$. S being closed, it follows that $(a, y) \in S$. Thus, since $a \in B_{r+\varepsilon}(x)$, so $y \in S_Y(B_{r+\varepsilon}(x))$, as asserted.

Finally, if $\pi_Y(S)$ is dense in Y, we may for any $y \in Y$ choose x_1 and y_1 so that $(x_1, y_1) \in S$ and $y \in \mathscr{V}(y_1)$. Then (6.4.2) shows that $y \in \pi_Y(S)$. ∎

From this lemma it is possible to derive forms of the open mapping and closed graph theorems applicable to topological groups (Kelley [1], p. 213). We shall, however, pass straight to the case of TVSs.

Notation. From now on to Subsection 6.4.7, E and F will denote TVSs, E_0 a vector subspace of E, F_0 a dense vector subspace of F, u a linear map of E_0 into F, v a linear map of F_0 into E. (From time to time E_0 will be assumed to coincide with E or F_0 with F, or both.) u will be said to be *nearly open* if $\overline{u(E_0 \cap U)}$ (closure in F) is a neighbourhood of 0 in F for each neighbourhood U of 0 in E. v will be said to be *nearly continuous* if $\overline{v^{-1}(U)}$ is a neighbourhood of 0 in F for each neighbourhood U of 0 in E. The reader will observe that u is nearly open whenever $u(E_0)$ is nonmeagre (that is, second category) in F, and that v is nearly continuous whenever F_0 is nonmeagre in F.

6.4.2 **Theorem.** With the above notations and terminology, together with the hypothesis that E is complete and semimetrizable, the following two statements are true:

(a) If u has a graph closed in $E \times F$, and if u is nearly open, then u is onto and open (that is, $u(E_0) = F$ and u transforms relative neighbourhoods of 0 in E_0 into neighbourhoods of 0 in F.)

(b) If v has a graph closed in $F \times E$, and if v is nearly continuous, then $F_0 = F$ and v is continuous (that is, $v^{-1}(U)$ is a neighbourhood of 0 in F for each neighbourhood U of 0 in E).

Proof. (a) In Lemma 6.4.1 we take $X = E$, $Y = F$, and $S = Gr\, u = \{(x, u(x)) : x \in E_0\}$. Then what we have denoted by $S_Y(A)$ or $\pi_Y(S \cap \pi_X^{-1}(A))$ is none other than $u(E_0 \cap A)$, A being any subset of E. The hypothesis expressed by (6.4.1) is easily seen to be equivalent to the assumption that u is nearly open. Also, to say that the graph of u is closed in $E \times F$, signifies exactly that S is a closed subset of $E \times F$. Finally, the conclusion of Lemma 6.4.1 asserts that u transforms relative neighbourhoods of 0 in E_0 into neighbourhoods of 0 in F, which signifies that u is onto and open.

(b) We make a second application of Lemma 6.4.1, S being this time taken as the set $\{(v(y), y) : y \in F_0\}$—that is, the diagonal reflection of the graph of v. In this case $S_Y(A)$ is $v^{-1}(A)$.

The reasoning now proceeds much as in (a). ▮

6.4.3 Corollary. If E is complete and semimetrizable and F is separated, any continuous and nearly open linear map u of E into F is onto and open.

Proof. In this case $E_0 = E$ and, since u is continuous and F is separated, the graph of u is closed in $E \times F$. ▮

We now pass to the next stage of the proceedings, in which we impose upon F conditions that will permit us to by-pass the explicit demand that u be nearly open and v be nearly continuous.

6.4.4 Theorem. Suppose that E is complete and semimetrizable.

(a) If the graph of u is closed in $E \times F$, and if *either* F is ultrabarrelled and $u(E_0) = F$, *or* $u(E_0)$ is nonmeagre in F, then u is onto and open.

(b) If the graph of v is closed in $F \times E$, and if *either* F is ultrabarrelled and $F_0 = F$, *or* F_0 is nonmeagre in F, then $F_0 = F$ and v is continuous.

Proof. In view of Theorem 6.4.2, it will suffice to show that if U is a balanced neighbourhood of 0 in E, then each of the sets $V = \overline{u(E_0 \cap U)}$ and $W = \overline{v^{-1}(U)}$ is a neighbourhood of 0 in F. We choose a balanced neighbourhood U_1 of 0 in E such that $U_1 + U_1 \subset U$, and define V_1 and W_1 in terms of U_1 as V and W were defined in terms of U.

Now $u(E_0) \subset \bigcup \{nV_1 : n = 1, 2, \cdots\}$ and $F_0 \subset \bigcup \{nW_1 : n = 1, 2, \cdots\}$. If $u(E_0)$ is nonmeagre in F, we conclude that for some n the set nV_1, hence also V_1 itself, has interior points. Likewise in respect to W_1 if F_0 is nonmeagre in F. Since each of V_1 and W_1 is balanced, it follows that $V_1 + V_1$ and $W_1 + W_1$ are neighbourhoods of 0 in F. But $V_1 + V_1 \subset V$ and $W_1 + W_1 \subset W$, so that V and W are indeed neighbourhoods of 0 in F.

Suppose on the other hand that $u(E_0) = F$ and that F is ultrabarrelled. In this case we remark that the sets V, obtained when U ranges over the neighbourhoods of 0 in E, may be taken as a base at 0 for a vector-space topology on F. Each V is closed for the initial topology on F and so, since F is ultrabarrelled, the resulting topology is weaker than the initial (Theorem 6.2.1(1)). Thus each V is a neighbourhood of 0 in F. Likewise, if $F_0 = F$, we establish that each W is a neighbourhood of 0 in F whenever the latter space is ultrabarrelled. ▮

Remarks. (1) In Theorem 6.4.4(a) one may seek to drop the assumption that E be complete at the expense of additional demands upon u. W. Robertson

[1] has shown how to do this, imposing upon u an extra condition ensuring that it may be extended in a suitable way to the completion of E. To this extension one then seeks to apply the existing form of (a).

(2) If in either of (a) or (b) we add the hypothesis that E be locally convex, the demand that F be ultrabarrelled may be modified to read that F shall be barrelled.

(3) Part (a) [resp. (b)] will be rendered inapplicable if F [resp. E] is not separated, merely because then no u [resp. v] can have a graph closed in $E \times F$ [resp. in $F \times E$]. One may then seek the following way out of the difficulty.

Taking (a) first, we form the separated quotient space $F_c = F/\overline{\{0\}}$, let f be the natural map of F onto F_c, and consider $u_c = f \circ u$ in place of u. Since f is continuous and open, it will transform nonmeagre sets into nonmeagre sets. F_c will be barrelled or ultrabarrelled whenever F has that same property (Theorem 6.2.4(1)). If the graph of u_c is closed in $E \times F_c$, we may infer that u_c is open, that is, that $u(E_0 \cap U) + \overline{\{0\}}$ is a neighbourhood of 0 in F whenever U is a neighbourhood of 0 in E. If, therefore, $u(E_0 \cap \overline{\{0\}}) \supset \overline{\{0\}}$, the open character of u will follow.

In the case of (b) we form the quotient $E_c = E/\overline{\{0\}}$, let e be the natural map of E onto E_c, and consider $v_c = e \circ v$ in place of v. If v_c has a graph closed in $F \times E_c$, its continuity will result, whence also the continuity of v.

(4) Theorem 6.4.4 contains already those forms of the OMT and the CGT, applicable to complete metrizable TVSs, stated at the beginning of this chapter. The implication is immediate by virtue of Corollary 6.2.3.

An important special case arises when u is assumed to be one to one.

6.4.5 **Theorem (Inversion Theorem).** If E, E_0, F, and u are as in Theorem 6.4.4(a), and if also u is one to one, then $u(E_0) = F$ and u^{-1} is continuous from F into E.

Proof. Theorem 6.4.4(a) asserts that u maps E_0 onto F and is open. Since u is one to one, this is equivalent to saying that u^{-1} is defined and continuous on F. ▌

Remark. As pointed out in Remark (2) following Theorem 6.4.4, if we add the hypothesis that E be locally convex, it is enough that F be barrelled.

6.4.6 **Corollary.** Suppose that a vector space E becomes a complete semimetrizable TVS when endowed with each of two topologies t and t'. Suppose too that whenever x and x' are distinct points of E, there is a t-neighbourhood U of x and a t'-neighbourhood U' of x' that are disjoint. Then t and t' are identical separated topologies.

Proof. Let u be the identity map of (E, t) onto (E, t'). We aim to apply Theorem 6.4.5, E, E_0, and F being the same sets. The hypothesis on t and t' entails at once that u has a graph closed in $E \times E$ (taken with the product of the topologies t and t'). This already shows that t' is separated. Since u is one to one and onto, we conclude that u^{-1} is continuous. In other words, t is weaker than t'. A similar argument shows that t' is weaker than t. Hence t and t' are identical and separated. ▌

Remark. The hypothesis on t and t' is satisfied whenever they are comparable and the weaker is separated; compare Banach [1], p. 41, Théorème 6. It is also satisfied whenever there exists a separating set of linear forms on E, each of which is continuous for both topologies (see Exercise 6.3). This state of affairs is realized notably when E is a commutative semisimple Banach algebra with unit, the separating set of linear forms being constituted by the homomorphisms defined by the maximal ideals in E; the conclusion is that the norm of such an algebra is unique up to equivalence. See Loomis [1], p. 77; also Rickart [1], p. 75.

6.4.7. That form of the inversion theorem applicable to complete metrizable TVSs is easily seen to be equivalent to either of the corresponding restricted formulations of the OMT and the CGT. For such spaces the three theorems form an almost inseparable trio. With more general types of TVSs involved, however, the equivalence is upset and the inversion theorem is rather subordinated to the other two.

As with the OMT and the CGT, the inversion theorem will receive further attention (in its case sometimes merely implicit) in Section 6.7 and again in Chapter 8.

We shall now give some examples applying the preceding general theorems. Other illustrations appear as exercises at the end of this chapter.

6.4.8 **Example.** Let I be a compact interval $[a, b]$ of the real line, $Q = I \times I$ (a plane square), and suppose that K is a function defined on Q. For definiteness and simplicity we suppose that K is continuous, but the reader will find it easy to verify that numerous other less stringent conditions would suffice. Suppose that $C(I)$ denotes the space of continuous functions on I; this is a Banach space when equipped with the "uniform" norm

$$\|x\| = \operatorname{Sup}_{t \in I} |x(t)|.$$

If λ is a number, the formula

$$y(t) = x(t) + \lambda \int_I K(t, s)x(s)\, ds$$

yields a linear map u of $C(I)$ into itself: $y = u(x)$. It is clear that u is continuous; in fact

$$\|u(x)\| \leq (1 + m\,|\lambda|)\,\|x\|,$$

if m is the maximum modulus of K on Q.

If we suppose that $m\,|\lambda| < 1$, there are available several processes of stock analytical nature that show that u is both onto and one to one. (For example, u is a contraction if $m\,|\lambda| < 1$ and we may then apply the general results of Subsection 3.1.2.) Once this is done, the inversion theorem shows that u^{-1} is continuous. Taking account of the results of Chapter 4, this means that to each point t of I corresponds a Radon measure μ_t on I such that the unique solution x of $u(x) = y$ is given by

$$x(t) = \int_I y(s)\, d\mu_t(s) \qquad (t \in I).$$

Exactly similar processes apply to spaces other than $C(I)$. For instance, we may replace I by any locally compact space equipped with a positive Radon measure μ. In place of $C(I)$ we may use $\mathscr{L}^2(I, \mu)$ (or, equally well, the associated separated space $L^2(I, \mu)$). If we assume that $K \in \mathscr{L}^2(Q, \mu \otimes \mu)$, the same formula as before defines a continuous linear map u of $\mathscr{L}^2(I, \mu)$ into itself. If we now take

$$m = \left\{\int\int |K(s, t)|^2\, d\mu(s)\, d\mu(t)\right\}^{1/2}$$

and assume that $m\,|\lambda| < 1$, u is again a contraction mapping. The conclusion would be that u^{-1} is again continuous.

6.4.9 **Example.** We consider so-called Fourier factor sequences. All functions here are defined on the real axis and have period 2π, and Lebesgue spaces are constructed relative to Lebesgue measure restricted to $(-\pi, +\pi)$. If $x \in \mathscr{L}^1$, its nth Fourier coefficient is defined by

$$\hat{x}(n) = (2\pi)^{-1} \int_{-\pi}^{\pi} x(t) e^{-int}\, dt$$

for $n = 0, \pm 1, \pm 2, \cdots$.

Suppose now that we specify two subsets X and Y of $\mathscr{L}^1$. A sequence (f_n) $(n = 0, \pm 1, \pm 2, \cdots)$ is said to be a *Fourier factor sequence of type* (X, Y) if the following is true: For each x in X, the sequence $(f_n \hat{x}(n))$ is of the form $(\hat{y}(n))$ for some y in Y (depending upon x, of course). The problem of characterizing the factor sequences of type (X, Y) has been considered for various choices of X and Y, notably $X = Y = C$ (continuous periodic functions), $X = \mathscr{L}^1$, and $X = Y = \mathscr{L}^p$; see Zygmund [1], [2], Kaczmarz-Steinhaus [1], Edwards [6].

Progress can be made in certain cases by using simple general arguments. We illustrate by taking the case $X = \mathscr{L}^1$, $Y = \mathscr{L}^p$ (Edwards [6]).

The key to the solution is the observation that y depends only on the class (modulo negligible functions) determined by x. Thus, if (f_n) is a factor sequence of type $(\mathscr{L}^1, \mathscr{L}^p)$, it determines a map u of L^1 into L^p by the requirement that u maps the class determined by x into the class determined by y, where $f_n \hat{x}(n) = \hat{y}(n)$ for all n. Now, since convergence in L^1 or in L^p obviously entails the convergence of the corresponding Fourier coefficients of any fixed rank n, it is easily seen that u has a closed graph. Linearity of u is obvious. The CGT theorem assures us that u is continuous.

The remaining essential property of u is that it commutes with convolution (that is, what comes to the same thing in the presence of continuity of u, it commutes with translations). If we now go back to regarding u as a map of $\mathscr{L}^1$ into $\mathscr{L}^p$, this means that

$$u(k * x) = u(k) * x$$

for $k, x \in \mathscr{L}^1$. We allow k to vary along a sequence (k_r) that is bounded in $\mathscr{L}^1$ and such that $k_r * x \to x$ in $\mathscr{L}^1$ as $r \to \infty$. Then the $h_r = u(k_r)$ remain bounded in $\mathscr{L}^p$, u being continuous. We have then

$$u(x) = \lim u(k_r * x) = \lim h_r * x$$

in $\mathscr{L}^p$.

If $p > 1$, the h_r have a weak limiting point $h \in \mathscr{L}^p$, and then the $h_r * x$ have $h * x$ as a limiting point in the sense of point-wise convergence, at least if x is continuous. It follows that $u(x) = h * x$ if x is continuous, hence by continuity of u for all x in $\mathscr{L}^1$. (The reader will recall Theorems 1.11.4 and 4.16.1.)

If $p = 1$, the above argument breaks down since the bounded subsets of $\mathscr{L}^1$ are not weakly relatively compact. However, the space of bounded Radon measures has this property and we may imbed $\mathscr{L}^1$ in it. The same conclusion results, except that h is a bounded Radon measure.

If we now take Fourier transforms of the relation $u(x) = h * x$, we see that (f_n) must be the sequence of Fourier coefficients of a function in $\mathscr{L}^p$ if $p > 1$, or of a bounded Radon measure if $p = 1$. This necessary condition is very easily seen to be sufficient, and so we have a complete solution in this case.

For further examples, see Edwards [6] and also Exercise 9.10 below.

6.4.10 **Example.** We adopt the notations of Example 6.1.4. Let A be a row-finite matrix, u_A the corresponding linear map of C^N into C^N. In Example 6.1.4 we introduced the space $B(A)$, showing that it is a Fréchet space for the topology T_A defined by the seminorms

$$p_n(x) = x(n)$$

$$q_{A,m}(x) = \operatorname*{Sup}_k \left| \sum_{n \le k} A(m, n)x(n) \right|,$$

$$q_A(x) = \|u_A(x)\|_\infty.$$

Our aim now is to show that when A is row finite and u_A is one to one, T_A is identical with the topology defined by q_A alone, thus supplementing (3) of Example 6.1.4. More generally we have the following assertion:

▶ (1) Suppose that A is row finite. Let L be a vector subspace of $B(A)$ that is closed relative to q_A (for example, $B(A)$, $C(A)$, or $C_0(A)$), and let Q be the kernel of $u_A \mid L$. On L/Q the quotient of $T_A \mid L$ is identical with the quotient of the topology defined by $q_A \mid L$; that is, for arbitrary numbers n and m there exist numbers α_n and β_m such that for x in L one has

$$\operatorname*{Inf}_{z \in Q} |x(n) + z(n)| \le \alpha_n \cdot \|u_A(x)\|_\infty, \tag{6.4.3}$$

$$\operatorname*{Inf}_{z \in Q} \operatorname*{Sup}_k \left| \sum_{n \le k} A(m, n)(x(n) + z(n)) \right| \le \beta_m \cdot \|u_A(x)\|_\infty. \tag{6.4.4}$$

L is a complete seminormed space for $q_A \mid L$, and L/Q is a Banach space for $q_A \mid L$. If u_A is one to one, then $T_A \mid L$ is identical with the topology defined by the norm $q_A \mid L$.

Proof. We show first that L is complete for q_A. Let (x_i) be q_A-Cauchy in L. So $y_i = u_A(x_i)$ is convergent in ℓ^∞ to y. Since A is row finite, Toeplitz' theorem (Theorem 2.3.2) implies that $y = u_A(x)$ for some x in C^N. Necessarily x lies in $B(A)$ and $q_A(x_i - x) = \|y_i - y\|_\infty \to 0$. Hence x belongs to L and completeness is proved.

We can now consider L/Q with either of two topologies: the quotient of $T_A \mid L$, or the quotient of the topology defined by q_A. Since L is q_A-closed in $B(A)$, it is a fortiori T_A-closed; so L/Q is a Fréchet space with the former topology. By what we have just proven, it is a Banach space with the second topology. The identity map is continuous and onto. It is therefore open (Theorem 6.4.4(a)), so that the two topologies are identical.

The remaining statements follow immediately. ▮

This result has interesting consequences when combined with the boundedness principles discussed in Chapter 7. A more immediate application is the derivation of an analogue of Toeplitz' theorem used in the preceding proof.

► (2) Suppose that A is row finite, that

$$\lim_m A(m, n) = 0 \qquad (n = 1, 2, \cdots), \tag{6.4.5}$$

and that the relations

$$s \in l^1, \sum_m A(m, n)s(m) = 0 \qquad (n = 1, 2, \cdots) \tag{6.4.6}$$

imply that $s = 0$. Then u_A maps $C_0(A)$ onto c_0; that is, for each y in c_0 there exists x in C^N such that

$$\sum_n A(m, n)x(n) = y(m) \qquad (m = 1, 2, \cdots).$$

Proof. The relations (6.4.5) show that $C_0(A)$ contains every sequence x with a finite support. Let $F \subset c_0$ be the range of u_A. By (1), $C_0(A)$ is complete for the seminorm $q_A(x) = \|u_A(x)\|_\infty$, hence it follows by (6.4.3) that F is a closed vector subspace of c_0. On the other hand, since (6.4.6) entails $s = 0$, the Hahn-Banach theorem shows that F is dense in c_0. Thus $F = c_0$. ▮

6.4.11 An Application to a Theorem of Hörmander. In this subsection we shall use the CGT to prove a result given by Hörmander ([1], Theorem 1.1) and used by him in the theory of linear partial differential equations, a development that has been described already in Section 5.19.

We suppose that E, F_1, and F_2 are complete metrizable TVSs. (All that is needed is that E, F_1, and F_2 be TVSs such that the CGT applies to linear maps of closed vector subspaces of $E \times F_1$ into F_2; in Section 5.19 each of these spaces is a Banach space.) For $i = 1, 2$ we assume that u_i is a linear map from E into F_i (the domain of u_i need not be the whole of E). We make the following assumptions:

(1) u_1 is closed, (2) u_2 is preclosed, (3) Dom $u_1 \subset$ Dom u_2.

The conclusion is that given any neighbourhood V_2 of 0 in F_2, there exist neighbourhoods U and V_1 of 0 in E and in F_1, respectively, such that

$$u_2(\text{Dom } u_1 \cap U \cap u_1^{-1}(V_1)) \subset V_2. \tag{6.4.7}$$

We note that if E, F_1, and F_2 are Banach spaces, the conclusion amounts to saying that a number $c \geq 0$ exists such that the inequality

$$\|u_2(x)\| \leq c(\|x\| + \|u_1(x)\|),$$

or, equivalently (though perhaps with a different value of c),

$$\|u_2(x)\|^2 \leq c(\|x\|^2 + \|u_1(x)\|^2),$$

is valid for x in Dom u_1.

Proof of (6.4.7). Let G denote the graph of u_1, regarded as a subspace of $E \times F_1$; as such it is a complete metrizable TVS, since u_1 has a closed graph by hypothesis (1). Consider the linear map v of G into F_2 defined by

$$v((x, u_1(x)) = u_2(x)$$

for x in Dom u_1: the definition is justified by hypothesis (3). The conclusion (6.4.7) amounts to the allegation that v is continuous. Since all the spaces involved are complete and metrizable, our task is reduced to showing that v is closed (see Theorem 6.4.4).

To this end let (z_k) be a directed family of points of G converging in G to z. So we may write $z_k = (x_k, u_1(x_k))$, $z = (x, u_1(x))$, where x belongs to Dom u_1, x_k is a directed family of points of Dom u_1, and $x_k \to x$ in E and $u_1(x_k) \to u_1(x)$ in F_1. In addition to this, suppose that $v(z_k) \to q$ in F_2. We need to show that $q = v(z)$. Thus we know additionally that $u_2(x_k) = v(x_k) \to q$ in F_2, and we wish to conclude that $q = u_2(x)$. But we are assured that $x - x_k \in \text{Dom } u_1 \subset$ Dom u_2 and $u_2(x - x_k) = u_2(x) - u_2(x_k) \to u_2(x) - q$ in F_2. Referring to Proposition 1.9.11(1), and using hypothesis (2), the desired result, viz., $u_2(x) - q = 0$, is forthcoming. The proof is thus complete. ▌

6.5 Some Results Concerning Sequences of Complete Semimetrizable TVSs

The results to follow are of a somewhat "mixed" type, having something in common with the OMT and in addition some bearing upon the boundedness principles that will occupy our attention in Chapter 7. They are suggested by one of Bourbaki's exercises (Bourbaki [5], p. 36, Exercice 13)); see also Grothendieck [1], p. 16, Théorème A.

6.5.1 **Theorem.** Suppose that E is a TVS, that F and F_n $(n = 1, 2, \cdots)$ are complete semimetrizable TVSs, F being separated, that u and u_n $(n = 1, 2, \cdots)$ are continuous linear maps of F and of F_n, respectively, into E. The main hypothesis is that

$$u(F) \subset \bigcup_{n=1}^{\infty} u_n(F_n). \tag{6.5.1}$$

The conclusion is that there exists an n with the following property: $u(F) \subset u_n(F_n)$ and if U_n is any neighbourhood of 0 in F_n, $u^{-1}(u_n(U_n)) = U$ is a neighbourhood of 0 in F. The same conclusion holds if $E = F$, u is the identity map, and if (6.5.1) is replaced by the assumption that $\bigcup_{n=1}^{\infty} u_n(F_n)$ is nonmeagre in F.

Proof (after Bourbaki, loc. cit.). Let H_n be the vector subspace of $F \times F_n$ formed of pairs (y, y_n) such that $u(y) = u_n(y_n)$. H_n is closed in $F \times F_n$ and is therefore complete semimetrizable. Let P_n be the natural projection of $F \times F_n$ onto F. Then (6.5.1) says that

$$F = \bigcup_{n=1}^{\infty} P_n(H_n).$$

It follows that, for some n, $P_n(H_n)$ is nonmeagre in F. Consequently, P_n is nearly open. By Corollary 6.4.3, P_n is an open map of H_n onto F, which justifies the conclusion of the theorem. ∎

It will next be shown that a similar result is available for certain bounded subsets of E. The deduction involves an interesting device amounting to forming the vector subspace of E generated by the bounded set into a semimetrizable space with its own topology (not necessarily that induced by E's topology). The idea is contained in Bourbaki (*loc. cit.* pp. 21–22, Lemme 1), but we shall make a slight weakening of the hypothesis.

6.5.2 **Lemma.** Let E be a TVS, A a bounded, convex, balanced, sequentially complete, and sequentially closed subset of E. Let E_A be the vector subspace of E generated by A. Define p on E_A by

$$p(x) = \text{Inf}\left\{r > 0 : \frac{x}{r} \in A\right\}.$$

Then p is a seminorm on E_A (a norm if E is separated) for which E_A is complete. The topology on E_A defined by p is stronger than that induced on E_A by the topology of E.

Proof. It is easily verified that A is defined relative to E_A by the equation $p(x) \leq 1$. A base of neighbourhoods of 0 for the topology defined by p is formed of the sets εA ($\varepsilon > 0$). This, together with the fact that A is bounded in E, shows that the p-topology is stronger than that induced by the topology of E. It shows also that p is a norm on E_A if E is separated.

It remains to prove p-completeness of E_A. Let then (x_n) be a p-Cauchy sequence in E_A. It follows that $\text{Sup}_n\, p(x_n) = k < +\infty$, so that $x_n \in kA$. Since kA, like A, is sequentially complete in E, and since E's topology induces a topology on E_A weaker than that defined by p, it follows that there exists $x \in kA$ such that $x_n \to x$ in the sense of E. Now we know that

$$p(x_m - x_n) = \varepsilon_{mn} \to 0 \text{ as } m, n \to \infty.$$

So $(x_m - x_n)/\varepsilon_{mn} \in A$. Letting $m \to \infty$ and using the fact that A is sequentially closed in E, we see that $(x - x_n)/\varepsilon_n \in A$, where $\varepsilon_n = n^{-1} + \limsup_{m\to\infty}\varepsilon_{mn} \to 0$ as $n \to \infty$. So $x \in E_A$ and $p(x - x_n) \to 0$; that is, $x_n \to x$ in E_A. ∎

Remark. If E is separated, sequential completeness of A implies sequential closure of A.

6.5.3 **Theorem.** Let E be a TVS, A a subset of E that is bounded, convex, balanced, sequentially complete, and sequentially closed. Let F_n $(n = 1, 2, \cdots)$ be a complete semimetrizable TVS and u_n a continuous linear map of F_n into E. The main hypothesis is that

$$A \subset \bigcup_{n=1}^{\infty} u_n(F_n). \tag{6.5.2}$$

The conclusion is that there exists n with the property that for each neighbourhood U_n of 0 in F_n,

$$A \subset \mu \cdot u_n(U_n) \tag{6.5.3}$$

for some $\mu = \mu(U_n) > 0$.

Proof. E_A is defined as in Lemma 6.5.2. In Theorem 6.5.1 we take $F = E_A$ and u the injection map of F into E. Relation (6.5.2) entails that (6.5.1) is valid. So the said theorem affirms that $u^{-1}(u_n(U_n))$ is a neighbourhood of 0 in F, hence absorbs the bounded set A. This yields (6.5.3). ▮

It is worth recording the negative form of a special case of Theorem 6.5.3.

6.5.4 **Corollary.** Suppose that E, F_n, and u_n are as in Theorem 6.5.3, and that E is sequentially complete. Assume that $A \subset E$ is bounded, convex, and balanced, and that for each n these exists a neighbourhood U_n of 0 in F_n such that $u_n(U_n)$ does *not* absorb A. Then there exists some $x \in E$ belonging to no $u_n(F_n)$ and that is the limit in E of a sequence extracted from A.

Proof. Apply Theorem 6.5.3 to the sequential closure of A. ▮

We close this section with an analogue of Theorem 6.5.3 in which each F_n is replaced by a suitable bounded subset thereof.

6.5.5 **Theorem.** Suppose that E and E_n $(n = 1, 2, \cdots)$ are TVSs and that A (resp. A_n) is a bounded, convex, balanced, sequentially closed, and sequentially complete subset of E (resp. E_n). Let u_n be a continuous linear map of E_n into E. The main hypothesis is that A is absorbed by $\bigcup_{n=1}^{\infty} u_n(A_n)$. The conclusion is that there exist n and a number $\mu > 0$ such that $A \subset \mu \cdot u_n(A_n)$.

Proof. We apply Theorem 6.5.3 taking for F_n the space $(E_n)_{A_n}$ — cf. Lemma 6.5.2 once more. We put $v_n = u_n \mid F_n$. v_n is continuous from F_n into E. Our main hypothesis entails the inclusion $A \subset \bigcup_{n=1}^{\infty} v_n(F_n)$, so Theorem 6.5.3 entails the existence of n such that $v_n(U_n)$ absorbs A whenever U_n is a neighbourhood of 0 in F_n. In this case, we may take $U_n = A_n$. Hence the theorem. ▮

6.5.6 **Example.** We take a problem concerned with functions of several complex variables. This question, together with several similar ones, receives a similar discussion in Edwards [11].

If M is any complex-analytic manifold, we denote by $H(M)$ the space of all functions holomorphic on M; $H(M)$ is a complex vector space; it becomes a locally convex space when equipped with the topology of convergence uniform on compact subsets of M. If M is σ-compact, $H(M)$ is a Fréchet space. If $A \subset M$, the so-called $H(M)$-*envelope* of A in M is the set of points m_0 in M for which

$$|z(m_0)| \leq \operatorname*{Sup}_{m \in A} |z(m)|$$

for each z in $H(M)$.

Suppose that S and T are complex-analytic manifolds, each assumed to be σ-compact; suppose too that f is a local analytic isomorphism of S into T. The image $S' = f(S)$ is then a domain in T.

Given a function x in $H(S)$, there may or may not exist a domain D in T containing S' properly and a function y in $H(D)$ such that $x = y \circ f$. If y exists, it is uniquely determined by x (f being fixed in advance); it may be termed the *holomorphic continuation into D, relative to f, of x*. Naturally, D may depend upon x—at least a priori. The point of our example, however, is precisely that if continuation is possible for each x in $H(S)$, then in fact there is a D, independent of x, into which continuation is possible. More specifically, we shall prove the following:

▶ Suppose that for each x in $H(S)$ there exists a domain D in T satisfying

$$D \supset S', \qquad D \neq S' \tag{6.5.4}$$

and a function y in $H(D)$ satisfying

$$x = y \circ f. \tag{6.5.5}$$

Then, in fact, D may be chosen independently of x, satisfying (6.5.4) and such that a representation (6.5.4) is possible for each x in $H(S)$ with a suitable y in $H(D)$. Moreover, to each compact set K in D corresponds a compact set K_0 in S and a number $c \geq 0$ such that y may be selected in $H(D)$ satisfying (6.5.5) and also

$$\operatorname*{Sup}_{t \in K} |y(t)| \leq c \cdot \operatorname*{Sup}_{s \in K_0} |x(s)|. \tag{6.5.6}$$

Finally, there exists in S at least one compact set K_0 whose $H(S)$-envelope is not compact.

The proof of this runs as follows. Since any σ-compact manifold is separable, there exists a sequence (D_n) of domains satisfying (6.5.4) such that any D satisfying (6.5.4) contains some D_n. In Theorem 6.5.1 we shall take $E = F = H(S)$, u the identity map, $F_n = H(D_n)$, and u_n the map defined by composition with f:

$$u_n(y) = y \circ f.$$

It is clear that u_n is a continuous linear map of F_n into F. The principal hypothesis signifies that F is the union of the $u_n(F_n)$. Theorem 6.5.1 informs us that for some n, $F = u_n(F_n)$ and u_n is open map. This yields the first assertion, if we take D to be D_n for this particular n, as well as (6.5.6). Moreover, with this choice of D, there exists a point t common to D and the frontier of S' relative to T. We choose a sequence (s_i) in S such that $t = \lim_i f(s_i)$. For the K in (6.5.6) we choose a compact neighbourhood of t containing $f(s_i)$ for all i. Then $|x(s_i)| \leq c \cdot \mathrm{Sup}_{s \in K_0} |x(s)|$ follows for all i and all x in $H(S)$. If in this we replace x by its kth power ($k = 1, 2, \cdots$), the kth root extracted, and then k allowed to tend to infinity, we see that

$$|x(s_i)| \leq \operatorname*{Sup}_{s \in K_0} |x(s)|$$

for all i and all x in $H(S)$. Thus the $H(S)$-envelope of K_0 contains the sequence (s_i). Yet, since f is continuous and $f(s_i) \to t$, which does not belong to $S' = f(S)$, the sequence (s_i) can admit no limiting point in S. The said envelope of K_0 is therefore not compact. The proof is thus complete. ▌

6.6 Bounded Sets in $\mathscr{D}^m(\Omega)$ and in $\mathscr{K}(T)$

(1) Our aim is to utilize Theorem 6.5.3 to recover the results of Section 6.3(2) concerning the nature of the bounded subsets of $\mathscr{D}^m(\Omega)$ (Ω an open subset of R^n, m a natural number or 0 or ∞, understanding $\mathscr{D}^\infty(\Omega)$ to mean $\mathscr{D}(\Omega)$; see Sections 5.1 and 6.3) and of $\mathscr{K}(T)$ when T is σ-compact (see Sections 4.3 and 6.3). That these spaces are sequentially complete follows from the results of Section 6.3, but we give here a somewhat different argument that proceeds without reference to a defining system of seminorms in these spaces.

Now it is obvious that the Cauchy nature (and even the weak Cauchy nature) in any one of these spaces entails convergence pointwise on Ω (or T, as the case may be): this because each Dirac measure belongs to the dual space. Besides this, if x ranges over any compact subset K of Ω (or of T), the Dirac measure ε_x, which is weakly continuous as a function of x, ranges over a weakly compact subset of the dual. Since (see Subsection 6.3.3) the spaces involved are barrelled, this set of Dirac measures ε_x is equicontinuous. By Ascoli's theorem, then, the Cauchy character of a sequence (f_n) in any one of the spaces $\mathscr{D}^m(\Omega)$ or $\mathscr{K}(T)$ entails its convergence uniform on compact subsets of Ω (or of T).

In the case of $\mathscr{D}^m(\Omega)$ the same argument can be used when the derivatives $\partial^p \varepsilon_x$ ($|p| \leq m$) replace the ε_x. The conclusion is then that for each such p the sequence $(\partial^p f_n)$ is locally uniformly convergent on Ω.

Because of this, to establish sequential completeness, it suffices to show that the limit function f has a compact support in Ω (or in T), continuity of f and of those of its derivatives that are involved following from locally uniform convergence.

To establish this point we must assume that T is σ-compact (a condition automatically satisfied by any subset Ω of R^n). Granted this, we assume, if possible, that the f_n fail to vanish outside a common compact subset of Ω (or of T). Then, by σ-compactness, we may choose a sequence (x_k) of points of Ω

(or T) having no limiting point in Ω (or T) and such that $f_{n_k}(x_k) = c_k \neq 0$, where $n_1 < n_2 < \cdots$ are natural numbers. The linear forms $c_k^{-1}\, \varepsilon_{x_k} = \lambda_k$ then converge weakly to 0 in the dual space (each compact subset of Ω or T containing x_k for at most a finite number of values of k). Since our spaces are barrelled, these linear forms are equicontinuous. So, since (f_n) is a Cauchy sequence, it must be that $\lambda_k(f_n) \to 0$ as $k \to \infty$, uniformly with respect to n. (Ascoli's theorem is being applied once again, the Cauchy sequence (f_n) forming a precompact set.) But this is absurd since $\lambda_k(f_{n_k}) = 1$ for all k. This contradiction establishes our point.

Now we may apply Theorem 6.5.3 to derive the following conclusion:

(2) Let B be a bounded subset of $\mathscr{D}^m(\Omega)$ (or of $\mathscr{K}(T)$, T being σ-compact). Then there exists a compact subset K of Ω (or of T) such that B is contained in, and is a bounded subset of, $\mathscr{D}^m(\Omega, K)$ (or $\mathscr{K}(T, K)$).

Remarks. (a) Dieudonné and Schwartz give a different type of proof valid for their LF-spaces; see their paper [1], p. 70. For $\mathscr{K}(T)$, see also Bourbaki [6], Ch. III, p. 64, Exercice 2).

(b) Consider statement (2) in relation to $\mathscr{K}(T)$, when T is not assumed to be σ-compact. There exist pathological non-σ-compact, locally compact spaces T for which (2) is false. Such an example is given by Bourbaki (loc. cit., p. 65, Exercice 3)). For this space T one may show that the inductive limit topology on $\mathscr{K}(T)$ (see Section 6.3) is none other than that of convergence uniform on T, so that the bounded subsets of $\mathscr{K}(T)$ are simply those defined by an inequality

$$\|f\| = \operatorname{Sup}\,\{|f(t)| : t \in T\} \leq c.$$

Despite this peculiarity, it remains true that the dual of $\mathscr{K}(T)$ consists of those linear forms μ on $\mathscr{K}(T)$ for which $\mu \,|\, \mathscr{K}(T, K)$ is continuous for each compact set $K \subset T$,—that is, consists of the Radon measures on T as defined in Section 4.3. Indeed, this is a consequence of the general property of inductive limit spaces expressed in Theorem 6.3.2. For the odd space T in question, it turns out that every Radon measure on T has a compact support (see Bourbaki (loc. cit., p. 78, Exercice 4)).

In view of this state of affairs, the reader is cautioned about Bourbaki's use (loc. cit., p. 65, Exercice 4)) of the term "strong topology" on the space of measures, when T is not σ-compact: this is not generally accordant with the use of this term in duality theory (see Section 8.4). There is no discrepancy when T is σ-compact, however.

6.7 Another Version of the Closed Graph Theorem

We aim to prove a rather specialized theorem having something in common with both the CGT and the results of Section 6.5. See also Grothendieck [1], p. 17, Théorème B.

6.7.1 **Theorem.** We suppose that F is the inductive limit of complete semimetrizable TVSs, and that E is the inductive limit of a sequence $E_n (n = 1,$

$2, \cdots$) of complete semimetrizable TVSs relative to maps $u_n : E_n \to E$. If a linear map v of F into E has a closed (or even merely sequentially closed) graph, then v is continuous.†

Proof. This proof begins with two preliminary reductions. In the first place, supposing that F is the inductive limit of complete semimetrizable TVSs F_i relative to maps f_i, continuity of $v : F \to E$ signifies the continuity of $v \circ f_i : F_i \to E$ for each i. Also, since $f_i : F_i \to F$ is continuous, if v has a (sequentially) closed graph, the same is true of $v \circ f_i$. Thus it is enough to deal with the case in which F is itself complete and semimetrizable.

Next, by taking the quotient spaces $\dot{E}_n = E/\text{Ker } u_n$ and the quotient maps $\dot{u}_n : \dot{E}_n \to E$, and using the $\dot{u}_n$ to transport the topology of E_n (which is complete semimetrizable) to $\dot{u}_n(\dot{E}_n) = u_n(E_n)$, we see (on the basis of Exercise 6.20) that it is enough to deal with the case in which E is the internal inductive limit of vector subspaces E_n relative to the injection maps, each E_n being complete semimetrizable. Naturally

$$E = \bigcup_{n=1}^{\infty} E_n. \tag{6.7.1}$$

These reductions having been made, we shall have for the graph G of v,

$$G = \bigcup_{n=1}^{\infty} \{G \cap (F \times E_n)\} = \bigcup_{n=1}^{\infty} G_n, \text{ say.} \tag{6.7.2}$$

The space $F \times E_n$ is complete semimetrizable for the product topology.

Assuming that G is sequentially closed in $F \times E$, we show first that G_n is sequentially closed (and therefore closed) in $F \times E_n$. To see this we suppose that (z_k) is a sequence extracted from G_n that converges in $F \times E_n$ to z. Thus,

$$\begin{aligned} z_k &= (y_k, v(y_k)) \qquad (y_k \in F,\ v(y_k) \in E_n), \\ z &= (y, x) \qquad (y \in F,\ x \in E_n). \end{aligned}$$

By hypothesis, $y_k \to y$ in F and $v(y_k) \to x$ in E_n. A fortiori, $v(y_k) \to x$ in E. Hence $z_k \to z$ in $F \times E$ which, since G is sequentially closed in $F \times E$, shows that $z \in G$; that is, $x = v(y)$. Thus $z = (y, v(y))$ with $y \in F$ and $v(y) = x \in E_n$; that is, $z \in G \cap (F \times E_n) = G_n$. This shows that G_n is closed in $F \times E_n$.

G_n is therefore complete semimetrizable as a closed subspace of $F \times E_n$.

We consider now the projections $P_n : G_n \to F$. These are continuous, and (6.7.1) implies

$$F = \bigcup_{n=1}^{\infty} P_n(G_n).$$

By Theorem 6.5.1, therefore, there exists n such that $F = P_n(G_n)$, which implies that $v(F) \subset E_n$.

---

† If E is the internal inductive limit of the E_n, it suffices that for any sequence (y_k) converging to 0 in F for which $(v(y_k))$ converges in some E_n to the limit x, one has necessarily $x = v(y)$.

It is easily verified that v, considered as a linear map of F into E_n, has a closed graph: one simply uses the fact that G is sequentially closed in $F \times E$ and remembers that the topology of E_n is stronger than that induced on it by the topology of E. By the closed graph theorem for complete semimetrizable TVSs (Theorem 6.4.4(b) and Corollary 6.2.3), v is continuous from F into E_n, a fortiori continuous from F into E. This completes the proof. ▌

The preceding theorem leads to a companion variant of the open mapping theorem.

6.7.2 **Theorem.** Suppose that F is the inductive limit of complete semimetrizable TVSs, and that E is the inductive limit of a sequence E_n $(n = 1, 2, \cdots)$ of complete semimetrizable TVSs relative to maps $u_n: E_n \to E$. Suppose also that u is a linear map of E onto F whose graph is sequentially closed in $E \times F$. Then u is open.

Proof. We begin by reducing ourselves to the case in which u is one-to-one. To do this we write M for the kernel of u and consider the associated map $\dot{u}$ of the quotient space E/M onto F defined by $\dot{u}(\dot{x}) = u(x)$, x being any element of the class $\dot{x}$ modulo M. Then $\dot{u}$ is one-to-one. According to Exercise 6.20, E/M satisfies the same hypotheses as does E. Therefore, if the theorem be granted for the case in which u is one-to-one, we shall conclude that $\dot{u}$ is open from E/M onto F. Since the natural map π of E onto E/M is open, and since $u = \dot{u} \circ \pi$, it will follow that u is open from E onto F. It thus suffices to handle the case in which u is one-to-one.

Suppose now that u is one-to-one. Introduce the map $v = u^{-1}$ of F into E. Since u has by assumption a graph which is sequentially closed in $E \times F$, the graph of v is easily verified to be sequentially closed in $F \times E$. Hence, by Theorem 6.7.1, v is continuous from F into E. This signifies that u is open from E onto F and so completes the proof. ▌

Remarks. (1) The spaces E and F appearing in Theorems 6.7.1 and 6.7.2 being locally convex (as inductive limit spaces), it is natural to ponder how they are related to those forms of the OMT and the CGT specially associated with such spaces, notably Theorem 8.9.3. The situation requires clarification, partly because an inductive limit of a family of fully complete (even Banach) spaces may fail to be complete (Köthe [5], Kap. V; Grothendieck [7], p. 145, Exercice 4)).

(2) Our hypotheses on F, although appearing rather restrictive, are really quite mild. For we shall see in Subsection 7.3.5 that any sequentially complete and bornological space F is the internal inductive limit of complete seminormable spaces (Banach spaces if F is separated).

(3) There are occasions when one would like to apply Theorems 6.7.1 and 6.7.2 to the case in which E and F are replaced by their strong duals E' and F', respectively. The pressing question is then: under what conditions are E' and F' inductive limits of the appropriate type? We shall discuss this question in Subsection 8.4.14.

6.7.3 **Bourbaki's (GDF) Property.** In connection with the integration of vector-valued functions Bourbaki ([6], Chapitre 6, p. 18; also Section 8.16 below) introduces what he calls the (GDF) property for LCTVSs, GDF standing for *graphe dénombrablement fermée* (which we translate as "sequentially closed graph"). This notion is otherwise strongly suggested by Theorem 6.7.1 above.

A separated LCTVS F is said to possess the (GDF) property if any linear map v of F into a Banach space X, having a graph that is sequentially closed in $F \times X$, is continuous.

By Theorem 6.4.4(b) and Corollary 6.2.3, every Fréchet space has the (GDF) property; by Theorem 6.7.1, so too has any inductive limit of Fréchet spaces. We shall recover this latter result in a different way (due to Bourbaki) in Section 8.16.

6.8 Applications to Topological Bases

In Section 1.4 we discussed the concept of an algebraic base in a vector space E—namely, a family (a_i) of elements of E such that to each element x of E corresponds exactly one family (ξ_i) of scalars, zero for all but at most a finite set of indices i (depending upon x), such that $x = \sum \xi_i a_i$. Such algebraic bases always exist.

If E is an infinite dimensional TVS, such algebraic bases are virtually useless in connection with problems involving the topology of E. (Consider the case of a base for R over the rational field, for example.) Instead, one has sought to modify the concept of base in such a way as to take stock of the topology on E from the outset. For such topological bases one allows infinite sums with the obvious proviso of convergence.

We shall restrict our detailed remarks here to topological bases that are countable and enumerated as a sequence in a definite way. Unfortunately, the mere existence of topological bases (countable or otherwise) presents great difficulties in general. For particular function spaces, however, examples appear in classical analysis with sufficient frequency to have led to an abstract study of the concept.

Only certain aspects concern us here; for other problems references are cited.

6.8.1 **Definition.** Let E be a TVS. By a *topological base* or *topological basis* in (or for) E we mean a sequence (e_n) of elements of E with the property that to each x in E corresponds a unique sequence (ξ_n) of scalars such that the series $\sum_n \xi_n e_n$ converges to sum x. If one demands merely that the series be weakly convergent to x, one speaks of a *weak topological base*.

The qualification "topological" is often absent; its inclusion is intended to stress the fact that the entities in question are not generally bases in the algebraic sense explained in Section 1.4.

To say that (e_n) is a (weak) topological base in E signifies, therefore, that there is a corresponding sequence (e_n^*) in E^* such that $x = \sum_n \langle x, e_n^* \rangle e_n$ for each x in E, the series being (weakly) convergent and the sequences (e_n) and (e_n^*)

being "biorthogonal" in the sense that

$$\langle e_m, e_n^* \rangle = \delta_{mn}.$$

It may or may not happen that the e_n^* lie in E': this is a question to which we shall turn shortly. In any case, we speak of the e_n^* as the "coefficient functionals" defined by, or associated with, the base (e_n).

6.8.2 **Schauder Bases.** A (weak) Schauder base in E is a (weak) topological base in E, the associated coefficient functionals being all continuous on E.

In this case we shall write e_n' in place of e_n^* for the nth coefficient functional. It is evident that (e_n') constitutes a topological base in E', the latter being endowed with the weak topology $\sigma(E', E)$.

Notes. (1) The concept of a topological base in a Banach space E goes back to Schauder [3]. As we shall see, for Banach spaces any weak topological base is necessarily a Schauder base (for the initial, normed, topology). Schauder [3], [5] constructed a topological base for $C[0, 1]$ and showed that a certain orthogonal system of functions (the Haar system) constitutes a topological base in $L^p(0, 1)$ whenever $1 \leq p < \infty$. The idea of a topological base has undergone a good many extensions; see, for example, Markushevich [1], [2], Arsove [1], Arsove and Edwards [1], and Edwards [13]. Further comments and references will be found in Dunford and Schwartz [1], p. 7 and pp. 93–94; Day [1], pp. 67–77; Banach [1], pp. 110–114 and p. 238; and Taylor [1], p. 207 and 220.

(2) It is not difficult to see that there exist biorthogonal systems (e_n), (e_n') in which (e_n) does not constitute a weak topological base. An example is given in Exercise 7.15. Nevertheless, biorthogonal systems in general have their interest. For an incisive study of some of their properties, see Dieudonné [15]. The standard work on orthogonal series is Kaczmarz and Steinhaus [1]. We shall not dwell further upon them.

6.8.3 **Examples.** (1) If E is a Hilbert space and (e_n) an orthonormal base in E (see Section 1.12), then (e_n) is a topological base in E. The separable Hilbert spaces (separability being necessary and sufficient to ensure the existence of a countable orthonormal base) form the only known general class of concrete TVSs for which the existence of a topological base is certain.

It is evident that any TVS admitting a topological base is of necessity separable, but it is an open problem whether every separable Banach space possesses at least one topological base.

(2) In each of the sequence spaces c_0 and $\ell^p(1 \leq p < \infty)$ the e_n, defined by $e_n(k) = \delta_{nk}$, form a topological base. ℓ^∞, being nonseparable, admits no topological base (for its normed topology), though the e_n constitute a base relative to the topology $\sigma(\ell^\infty, \ell^1)$.

(3) We consider the function spaces C and $L^p(1 \leq p \leq \infty)$ built over the circle group (or, equivalently, the corresponding spaces of periodic functions on the line), and denote by e_n the function $t \to e^{int}$ (or its class).

It is true, but lies fairly deep in Fourier series lore, that the e_n ($n = 0, \pm 1, \pm 2, \cdots$) form a topological base in L^p when $1 < p < \infty$; see Zygmund [2], Vol. I, p. 266. On the contrary, they do not form a topological base in any one of C, L^1, or L^∞ (see Exercises 7.15 and 7.16).

In studying the basis properties of the e_n it is understood that $\sum_{-\infty}^{\infty} \xi_n e_n$ is to mean $\lim_{N \to \infty} \sum_{|n| \leq N} \xi_n e_n$.

(4) It is quite easy (see Exercise 6.15) to construct an unlimited number of normed vector spaces E and topological bases for E such that no one of the associated coefficient functionals is continuous.

6.8.4 **Some General Problems.** E being a given TVS, we shall consider the following three statements:

(1) Any topological base in E is a Schauder base.

(2) Any weak topological base in E is a Schauder base (the weak basis theorem).

(3) Any weak Schauder base in E is a topological base (and therefore a Schauder base).

Banach ([1], pp. 110–11) gives a proof of (1) for Banach spaces, and notes (loc. cit. p. 238) that (2) is also true for Banach spaces. Newns ([1], pp. 431–432) extended (1) to Fréchet spaces, and Arsove [2] dealt successfully with the case of complete metrizable TVSs. Quite recently Bessaga and Pelzcyński [1] established (2) for Fréchet spaces, and Arsove and Edwards [1] and Edwards [13] have considered various extensions of (1) to certain inductive limits of Fréchet spaces by using the results of Section 6.5. Dieudonné [15] recorded (3) for all barrelled spaces E.

6.8.5. We propose to give here, in the shape of Theorem 6.8.6, a proof of the Bessaga-Pelczyński result. Our proof will make use of the fact that (3) is true for all barrelled spaces E, which will be established in Example 7.2.2. It suffices here to remark that if (e_n) is a weak Schauder base in E, the corresponding coefficient functionals belong to E' by hypothesis; if they be denoted by (e'_n), then one has

$$x = \sum_{n=1}^{\infty} e'_n(x) e_n$$

for each $x \in E$, the series being weakly convergent. At this point the result of Example 7.2.2 is applicable.

6.8.6 **Theorem.** If E is a Fréchet space, then (2) is true.

In view of Subsection 6.8.5, we have only to show that if (e_n) is a weak topological base in a Fréchet space E, then each of the associated coefficient functionals is continuous on E. This we shall do by adapting and extending the arguments used by Banach ([1], pp. 110–111). The adaptation is developed through some preliminary lemmas.

The lemmas are ultimately concerned with two TVSs Z and E and with a sequence u_n($n = 1, 2, \cdots$) of continuous linear maps of Z into E, but we begin with one referring solely to E and to weakly convergent sequences therein.

Lemma 1. Let E be a complete, separated LCTVS, $(a_{i,n})$ a double sequence, and (b_i) and (a_n) sequences, of elements of E. Suppose that

(a) $\lim_n a_{in} = b_i$ weakly in E for each i;

(b) $\lim_i a_{in} = a_n$, uniformly with respect to n.

Then there exists $a \in E$ such that $\lim a_n = a$ weakly in E, and $\lim b_i = a$ in E.

Proof. Let U be any closed, convex, balanced neighbourhood of 0 in E. According to (b), there is a sequence (ε_i) of positive numbers converging to 0, such that

$$a_{in} - a_n \in \varepsilon_i U \tag{6.8.1}$$

for all n. If x' belongs to U^0 (see Subsection 8.1.4),

$$|\langle a_{in}, x' \rangle - \langle a_n, x' \rangle| \leq \varepsilon_i$$

for all n, and so

$$|\langle a_m - a_n, x' \rangle| \leq |\langle a_{im} - a_{in}, x' \rangle| + 2\varepsilon_i.$$

This combines with (a) to show that the sequence $(\langle a_n, x' \rangle)$ is Cauchy for each x' in U^0. The sets U^0 covering E', (a_n) is weak Cauchy in E and therefore weakly convergent in E'' to some element a, say. We must show that a belongs to E.

To do this we shall make use of the following remark. Let (c_n) be any sequence converging weakly to 0 in E. For any natural number r, we denote by L_r the set of all scalar sequences $\lambda = (\lambda(n))$ with a finite support contained in the set $n > r$, such that $\lambda(n) \geq 0$ for all n, and such that $\sum_n \lambda(n) = 1$. We claim that, given any $\varepsilon > 0$, one can choose λ from L_r such that

$$\sum_n \lambda(n) c_n \in \varepsilon \cdot U,$$

U being any preassigned neighbourhood of 0 in E. To justify this claim, we consider the convex envelope G in E of the c_n with $n > r$. By hypothesis, 0 is weakly adherent to G and therefore also adherent to G for the initial topology. On the other hand, each element of G is expressible as a sum $\sum_n \lambda(n) c_n$ with λ suitably selected from L_r. Hence the assertion.

Now by (6.8.1) we have

$$(a_{in} - b_i) - (a_n - b_i) \in \varepsilon_i U.$$

Since U is convex, we shall therefore have, for any λ in any L_r,

$$\sum_n \lambda(n)(a_{in} - b_i) - \sum_n \lambda(n) a_n + b_i \in \varepsilon_i U.$$

Applying the above remark to $c_n = a_{in} - b_i$, we may select λ_{ri} from L_r so that

$$\sum_n \lambda_{ri}(n)(a_{in} - b_i) \in r^{-1} U.$$

In that case,

$$\sum_n \lambda_{ri}(n) a_n - b_i \in (\varepsilon_i + r^{-1}) U.$$

Thus also

$$\left| \sum_n \lambda_{ri}(n)\langle a_n, x' \rangle - \langle b_i, x' \rangle \right| \leq (\varepsilon_i + r^{-1}) \tag{6.8.2}$$

for x' in U^0:

Putting $s_{ri} = \sum_n \lambda_{ri}(n)a_n$, it is easy to verify that $\lim_r s_{ri} = a$ weakly in E''. Thus for x' in E' we have

$$\begin{aligned} |\langle s_{ri} - a, x' \rangle| &\leq \sum_n \lambda_{ri}(n) \, |\langle a_n - a, x' \rangle|, \\ &= \sum_{n>r} \lambda_{ri}(n) \, |\langle a_n - a, x' \rangle|, \\ &\leq \operatorname{Sup} \{ |\langle a_n - a, x' \rangle| : n > r \}, \end{aligned}$$

which tends to zero as $r \to \infty$ since $a_n \to a$ weakly in E''.

We thus infer from (6.8.2) that

$$|\langle a, x' \rangle - \langle b_i, x' \rangle| \leq \varepsilon_i \tag{6.8.3}$$

for x' in U^0. Therefore

$$|\langle a, x' - y' \rangle| \leq |\langle b_i, x' - y' \rangle| + 2\varepsilon_i$$

whenever x' and y' lie in U^0. This shows at once that the linear form $x' \to \langle a, x' \rangle$ has a restriction to U^0 that is continuous for the topology induced by $\sigma(E', E)$. So (Theorem 8.5.1) a belongs to E, E being complete by hypothesis.

The proof is now completed by reference to the Bipolar Theorem 8.1.5, which combines with (6.8.3), to show that $a - b_i \in \varepsilon_i U$. This shows indeed that $\lim_i b_i = a$ in E. ▮

We now introduce some notation. Z is a TVS, E a LCTVS, and u_n ($n = 1, 2, \cdots$) a sequence of continuous linear maps of Z into E. We write Z_b, Z_{wc}, and Z_c for the vector subspaces of Z comprising those z for which the sequence $(u_n(z))$ is, respectively, bounded, weakly convergent, and convergent in E. Evidently

$$Z_c \subset Z_{wc} \subset Z_b \subset Z.$$

The following hypothesis will be made—

▶ **Hypothesis.** If a directed family (z_i) in Z_b is such that $\lim_i u_n(z_i)$ exists in E, uniformly with respect to n, then (z_i) is convergent in Z.

Lastly, we define a topology on Z_b in the following fashion. Choose any defining system $\{p_\alpha\}$ of seminorms for the topology on E, and to each p_α make correspond the seminorm

$$q_\alpha(z) = \operatorname{Sup} \{ p_\alpha(u_n(z)) : n = 1, 2, \cdots \}$$

on Z_b. The q_α shall, by definition, constitute a defining family of seminorms for the said topology on Z_b. For brevity, we shall denote this topology on Z_b by t.

Lemma 2. The notations being as above, we assume that E is complete and that the Hypothesis is satisfied. Then Z_b is complete for t.

Proof. We suppose that (z_i) is a t-Cauchy directed family in Z_b. This means that

$$p_\alpha(u_n(z_i) - u_n(z_j)) \leq \varepsilon^\alpha_{ij}, \tag{6.8.4}$$

where $\lim_{i,j} \varepsilon^\alpha_{ij} = 0$ for each α. Since E is complete, it follows that $\lim_i u_n(z_i)$ exists in E for each n and so, according to the Hypothesis, that $\lim_i z_i = z$ exists in Z. It remains to show that z belongs to Z_b and that $\lim_i q_\alpha(z - z_i) = 0$ for each α.

Putting $x_n = \lim_i u_n(z_i)$, the relations (6.8.4) show that

$$p_\alpha(x_n - u_n(z_i)) \leq \varepsilon^\alpha_i, \tag{6.8.5}$$

where $\lim_i \varepsilon^\alpha_i = 0$ for each α. If we choose and fix an index i_0 so that $\varepsilon^\alpha_{i_0} \leq 1$, it will follow that

$$p_\alpha(x_n) \leq p_\alpha(u_n(z_{i_0})) + 1,$$

which, since z_{i_0} lies in Z_b, shows that the x_n are bounded in E. But $x_n = \lim_i u_n(z_i)$ is none other than $u_n(z)$, each u_n being continuous. Thus the boundedness of the x_n shows that z belongs to Z_b.

Finally, (6.8.5) can be written as $q_\alpha(z - z_i) \leq \varepsilon^\alpha_i$, showing that $z = t\text{-}\lim z_i$. ▌

Lemma 3. The notation and hypotheses being as in Lemma 2, Z_c is t-closed in Z_b, hence complete for the topology induced by t.

Proof. Suppose that (z_i) is a directed family in Z_c that converges for t to some z in Z_b. We must show that z belongs to Z_c. Now we have by hypothesis

$$p_\alpha(u_n(z_i) - u_n(z)) \leq \varepsilon^\alpha_i,$$

where $\lim_i \varepsilon^\alpha_i = 0$ for each α. Given α and any $\varepsilon > 0$, we choose and fix i so that $\varepsilon^\alpha_i < \varepsilon/4$. We shall then have

$$p_\alpha(u_m(z) - u_n(z)) < \tfrac{1}{2}\varepsilon + p_\alpha(u_m(z_i) - u_n(z_i))$$

uniformly in m and n. Since z_i belongs to Z_c, it is deducible at once that

$$p_\alpha(u_m(z) - u_n(z)) < \varepsilon,$$

provided m and n are sufficiently large. Thus $(u_n(z))$ is Cauchy, and therefore convergent, in E. Consequently, z belongs to Z_c. ▌

Lemma 4. Under the same conditions as in Lemma 2, Z_{wc} is t-closed in Z_b and therefore t-complete.

Proof. Suppose that each z_i belongs to Z_{wc}, that z belongs to Z_b, and that $t\text{-}\lim z_i = z$. We must show that z belongs to Z_{wc}.

To do this, we apply Lemma 1 taking $a_{in} = u_n(z_i)$. Since z_i lies in Z_{wc}, so $\lim_n a_{in} = b_i$ exists weakly in E for each i. Moreover, if we set a_n equal to $u_n(z)$, the fact that $t\text{-}\lim z_i = z$ signifies that $\lim_i a_{in} = a_n$ in E, uniformly with respect to n. The conclusion of Lemma 1 asserts that $(a_n) = (u_n(z))$ is weakly convergent in E—that is, that z lies in Z_{wc}. ▌

Proof of Theorem 6.8.6. Let E be a Fréchet space, (e_n) a weak topological base in E. We wish to show that each coefficient functional is continuous on E, which will be done by use of Lemma 4 and the inversion theorem.

For Z we shall take K^N (K the scalar field, N the set of natural numbers) with its product topology, and for u_n we take

$$u_n(z) = \sum_{k=1}^{n} z(k)e_k.$$

Given n, since $e_n \neq 0$ (being a member of a topological base), there must exist $\alpha = \alpha(n)$ such that $\beta \doteq p_\alpha(e_n) \neq 0$. It then appears that

$$|z(n)| \leq 2\beta^{-1}q_\alpha(z). \tag{6.8.6}$$

Since K^N is complete, it follows that the Hypothesis formulated above is fulfilled. Appeal to Lemma 4 shows that Z_{wc} is t-complete. Moreover, since E is metrizable, so too is each of Z_b, Z_c, and Z_{wc}. So Z_{wc} is a Fréchet space.

We consider now the linear map u of Z_{wc} into E defined by

$$u(z) = \sum_n z(n)e_n.$$

Since trivially $p_\alpha(u(z)) \leq \operatorname{Sup}_n p_\alpha(u_n(z))$, as a consequence of the weak convergence of $(u_n(z))$ to $u(z)$, we have $p_\alpha(u(z)) \leq q_\alpha(z)$. Therefore u is continuous. On the other hand, since (e_n) is a weak topological base in E, u is one to one and onto.

E and Z_{wc} being Fréchet spaces, the Inversion Theorem 6.4.5 shows that u^{-1} is continuous.

This being so, it remains only to remark that in the expansion $x = \sum \xi_n e_n$ one has

$$\xi_n = u^{-1}(x)(n).$$

Since (6.8.6) exhibits the continuity of the coordinate functional $z \to z(n)$ on Z_{wc}, the established continuity of u^{-1} implies that of the nth coefficient functional $x \to \xi_n$. ∎

6.8.7 **Possible Extensions.** Apart from the extensions already mentioned in Subsection 6.8.4, it is clear that the preceding proof can be adapted to the case in which (e_n) is a "weak summability base" in a suitable sense.

Suppose, for example, that there is a family of "summation factors" λ_{nk} such that to each x in E corresponds exactly one sequence (ξ_n) of scalars for which

$$x = \lim_n \sum_k \lambda_{nk}\xi_k e_k \tag{6.8.7}$$

weakly in E, it being supposed that the series $\sum_k \lambda_{nk}\xi_k e_k$ is itself weakly convergent in E for each n.

One would then seek to take

$$u_n(z) = \sum_k \lambda_{nk}z(k)e_k.$$

For many well-known summability methods it is true that $\lambda_{nk} = 0$ for $k > k_0(n)$. The convergence of each series in (6.8.7) is then assured, as also is the continuity of u_n from K^N into E.

It would be necessary to investigate the Hypothesis formulated above (a consequence of (6.8.7) when $\lambda_{nk} = 1$ for $k \leq n$ and to 0 for $k > n$). We leave it to the reader to examine the possibilities.

EXERCISES

6.1 T is a separated locally compact space and μ a positive Radon measure on T. f is a function on T with the property that fg is integrable for each $g \in \mathscr{L}^p(T, \mu)$. Use the closed graph theorem to show that $f \in \mathscr{L}^{p'}(T, \mu)$, where (as usual) p' is defined by $1/p + 1/p' = 1$. (Compare with Exercise 7.1.)

6.2 Prove that if a Fréchet space is the algebraic direct sum of closed vector subspaces L and M, then it is the topological direct sum of L and M.

6.3 E is a vector space. Suppose that E is a Fréchet space when endowed with each of two topologies t and t', and that there exists a separating set of linear forms, each of which is continuous for each of t and t'. Show that t and t' are identicals.

6.4 Let $f \in C^\infty(R^n)$ have the property that $fu \in \mathscr{S}$ whenever $u \in \mathscr{S}$. Show that for each p, $\partial^p f$ is in modulus majorized by a polynomial.

6.5 E and F are complete semimetrizable TVSs and G_n $(n = 1, 2, \cdots)$ is an arbitrary separated TVS. Suppose that for each n, f_n is a continuous linear map of F into G_n, and that the f_n separate points of F. Show that a linear map u of E into F is continuous, provided $f_n \circ u$ is continuous on E into G_n for each n.

6.6 Suppose that E and F are complete metrizable TVSs, that G_n $(n = 1, 2, \cdots)$ are arbitrary TVSs, and that the u_n [resp. v_n] are continuous linear maps of E [resp. F] into G_n. Suppose further that for each $x \in E$ there exists precisely one $y = f(x) \in F$ such that

$$v_n(y) = u_n(x) \qquad (n = 1, 2, \cdots).$$

Show that f is a continuous linear map of E into F.

6.7 Let Ω be a bounded open interval of the real line, $t_0 \in \Omega, p$ and q elements, of $C(\overline{\Omega})$, $\overline{\Omega} = [a, b]$. Consider the differential equation

$$x'' + px' + qx = y.$$

Write B^k for the space of functions f on such that $f, f' \cdots f^{(k)}$ are continuous and bounded on Ω, with the norm

$$\|f\|_k = \text{Sup}\,\{|f^{(h)}(t)| : t \in \Omega,\ 0 \leq h \leq k\}.$$

Assuming the standard existence and uniqueness theorems for differential equations of the above type, show that there exists a unique solution $x \in B^2$ corresponding to any given $y \in B^0$ and satisfying the initial conditions $x(t_0) = x'(t_0) = 0$. Let E be the subspace of B^2 defined by these initial conditions and consider the map u of E into B^0 defined by

$$u(x) = x'' + px' + qx.$$

Verify that u is continuous and maps E onto B^0. Infer that u^{-1} is continuous from B^0 into E, and also that $u^{-1} : B^0 \to B^1$ is compact (that is, transforms the unit ball of B^0 into a relatively compact subset of B^1).

6.8 The notation is as in the last exercise. Examine what happens if we replace the initial conditions $x(t_0) = x'(t_0) = 0$ by the boundary conditions $x(a) = x(b) = 0$. Show in particular that u^{-1} exists provided $q(t) < 0$ for $t \in \Omega$.

6.9 Let E be a semimetrizable LCTVS of infinite dimension, (α_n) and (β_n) two sequences of positive numbers such that $\lim \alpha_n = \lim \beta_n = +\infty$. Show that there exists in E a sequence (x_n) and a linear form f on E such that

$$\lim \alpha_n x_n = 0,$$
$$\lim \frac{f(x_n)}{\beta_n} = +\infty.$$

Note. The next four exercises are based upon Deny-Lions [1].

6.10 The notation is as in Subsection 5.11.1. Prove that $BL(\Omega)$ is complete for the norm defined by

$$\|X\|_1^2 = \sum_{k=1}^{n} \|\partial_k X\|_{L^2(\Omega)}^2.$$

[Hint: Use the method of Lemma A of Subsection 5.13.4.]

6.11 Let Ω be a domain in R^n, and let $\mathscr{E}^1_{q,2}(\Omega) = BL(\Omega) \cap L^q(\Omega)$. Show that $\mathscr{E}^1_{q,2}(\Omega)$ is a Banach space for the norm

$$\|u\|_{\mathscr{E}^1_{q,2}(\Omega)} = \|u\|_{L^q(\Omega)} + \|u\|_1,$$

where $\|u\|_1$ is defined as in the preceding exercise.

6.12 Let $n > 2$, Ω a domain in R^n having finite measure, and let $q = 2n/(n-2)$. Show that Ω is a Soboleff domain if and only if there exists a number $S(\Omega)$ such that

$$\operatorname*{Inf}_{c=\text{const.}} \|u + c\|_{L^q(\Omega)} \le S(\Omega) \cdot \|u\|_1$$

for $u \in \mathscr{E}^1_{q,2}(\Omega)$.

[Hints: Form the quotient spaces $(\mathscr{E}^1_{q,2}(\Omega))^\bullet = E$ and $BL^\bullet(\Omega)$ modulo the constant functions and consider the identity map of E into $BL^\bullet(\Omega)$. This map is onto if and only if Ω is a Soboleff domain. Use the inversion theorem. Show also that the image of E is dense in $BL^\bullet(\Omega)$; see Exercise 5.25.]

6.13 Let Ω be a domain in R^n having finite measure. Show that Ω is a Nikodym domain if and only if there exists a number $P(\Omega)$ such that

$$\|u\|_{L^2(\Omega)}^2 - \mu(\Omega)^{-1} \left| \int_\Omega u \, d\mu \right|^2 \le P(\Omega) \, \|u\|_1^2 \text{ for } u \in \mathscr{E}^1_{L^2}(\Omega).$$

(This is Poincaré's inequality and the smallest admissible value of $P(\Omega)$ is termed the Poincaré constant of Ω.)

[Hint: Argue as in the preceding exercise, replacing $\mathscr{E}^1_{q,2}(\Omega)$ throughout by $\mathscr{E}^1_{L^2}(\Omega)$.]

The results stated in Exercises 6.11–6.13, as also the proofs, are due to Deny and Lions ([1], Théorèmes 5.1, 5.3).

6.14 Assume that $0 < p \le \infty$ and that H^p is defined as in Subsection 1.10.4. Show that if K is a compact subset of the open unit disk in the complex plane with centre at the origin, then the seminorm

$$f \to \operatorname*{Sup}_K |f|$$

is continuous on H^p.

[Hint: Use the completeness of H^p—see Subsection 1.10.4—and the closed graph theorem.]

6.15 Let E be a barrelled space, (e_n) a Schauder base in E with associated coefficient functionals (e'_n). Assume that there exists a number $r > 0$ such that

$$\sum_n |\langle x, e'_n \rangle| \, r^n < +\infty$$

for each $x \in E$. Define $x^*(t) = \sum_n \langle x, e'_n \rangle t^n$ for real t satisfying $|t| \leq r$. Let (t_k) be a 1-1 sequence of real numbers satisfying $0 < |t_k| \leq r$ and $t_k \to t_\infty$ where $0 < |t_\infty| < r$. For x in E define

$$\|x\| = \operatorname*{Sup}_k |x^*(t_k)|.$$

Verify that $\|\cdot\|$ is a norm on E, and denote by E_0 the corresponding normed vector space. Show that (e_n) is a topological base in E_0 such that no one of the associated coefficient functionals is continuous on E_0.

6.16 Let E be a Banach space, T any continuous linear map of E onto $\ell^1 = \ell^1(N)$. Show that $E = M + \text{Ker } T$, a topological direct sum, for some closed vector subspace M of E.

6.17 Extend the result given in Exercise 2.17 in the following fashion:

Let E be any separated LCTVS. Choose a family (U_i) of neighbourhoods of 0 such that the homothetic images of the U_i form a base at 0. For each i let Q_i be the polar in E' of U_i. With the topology induced by $\sigma(E', E)$, Q_i is separated and compact. Consider the mapping u of E into the product $P = \prod_i C(Q_i)$ of the Banach spaces $C(Q_i)$ that assigns to $x \in E$ that element of P whose ith coordinate is the member of $C(Q_i)$ obtained by restricting to Q_i the function $x' \to \langle x, x' \rangle$ on E'.

Verify that u is an isomorphism of E into P.

Deduce that any complete separated LCTVS E is isomorphic with a closed vector subspace of a product of Banach spaces, this product space being therefore barrelled.

6.18 Let T be an arbitrary set, regarded as a discrete topological space, and let $\mathscr{K}(T)$ be the vector space of scalar-valued functions on T with finite supports. Endow $\mathscr{K}(T)$ with the topology defined by the seminorms

$$N_p(x) = \operatorname*{Sup}_{t \in T} p(t) \, |x(t)|,$$

p denoting any positive function on T.

Prove that $\mathscr{K}(T)$ is complete.

By considering the subset

$$A = \left\{ x \in \mathscr{K}(T) : \sum_{t \in T} |x(t)| \leq 1 \right\}$$

show that if T is discrete and *uncountable*, then $\mathscr{K}(T)$ is not barrelled.

(This example is due to Bourbaki [5], p. 3, Exercice 5).)

6.19 Show that any homomorphic image of a barrelled space is barrelled—that is, if E is barrelled, and if there exists a linear continuous and open map u of E onto a LCTVS F, then F is barrelled.

Deduce that if E is barrelled and M a vector subspace of E that admits in E a topological supplement, then M is barrelled.

6.20 Suppose that E is the inductive limit of spaces E_i relative to maps φ_i, that M is a vector subspace of E, and that $F = E/M$ with the quotient topology. Define $M_i = \varphi_i^{-1}(M)$, let $F_i = E_i/M_i$ with the quotient topology, and let f_i and f be the natural maps of E_i onto F_i and of E onto F, respectively. Finally, define $\psi_i : F_i \to F$ by

$$\psi_i(f_i(x)) = f(\varphi_i(x)) \qquad (x \in E_i).$$

Verify that F is the inductive limit of the spaces F_i relative to the maps ψ_i.

Notice that if M is closed in E, then M_i is closed in E_i for each i.

6.21 Let X be a topological space, Y a nonmeagre subset of X. Show that if Y be considered as a subspace of X, it is nonmeagre in itself.

6.22 Show that if E is a Banach space of infinite dimension, then any algebraic base for E is uncountably infinite.

6.23 Let E be a Banach space in which there exists an infinite sequence (a_n) of linearly independent elements that generate a dense vector subspace of E (for example, $E = \ell^1$). Extend (a_n) into an algebraic base $(a_i)_{i \in I}$, I being some superset of the set N of natural numbers. By the preceding exercise, I is uncountable. Choose distinct suffices i_n $(n = 1, 2, \cdots)$ belonging to $I \backslash N$. Let F_k be the vector subspace of E generated by those a_i with i different from all the i_n and the a_{i_n} with $1 \leq n \leq k$. Verify that each F_k is dense in, but distinct from, E. Show also that, for some k, F_k is nonmeagre in E. For any such k, F_k is a noncomplete normed vector space that is barrelled.

This construction is due to Bourbaki [5], p. 3, Exercice 6).

[Hints: Notice that $E = \bigcup_{k=1}^{\infty} F_k$ and use Baire's Theorem 0.3.16 and Exercise 6.21.]

6.24 Establish the result (b) of Section 6.3 (2) without making explicit use of the seminorms $N_{s,\varepsilon}$.

[Hints: Suppose B is a bounded subset of $\mathscr{D}^m(\Omega)$. If the supports of members of B were not contained in a fixed compact subset of Ω, show that there would exist a sequence (x_n) extracted from Ω and a sequence (φ_n) extracted from B such that

$$\operatorname*{Sup}_n \frac{n\,|\varphi(x_n)|}{|\varphi_n(x_n)|}$$

is finite for each $\varphi \in \mathscr{D}^m(\Omega)$ and yet is unbounded on B. Use the fact that $\mathscr{D}^m(\Omega)$ is barrelled to show that this constitutes a contradiction.]

6.25 Let E be a normed vector space and A a subset of E with the following property: there exists a number $\lambda > 1$ such that to each $x \in E$ satisfying $\|x\| \leq 1$ corresponds an $a \in A$ for which $\|x - a\| \leq \lambda^{-1}$. Prove that to each $x \in E$ satisfying $\|x\| \leq 1$ there corresponds a sequence $(a_n)_{n \geq 0}$ of points of A such that

$$x = \sum_{n=0}^{\infty} \frac{a_n}{\lambda^n}.$$

6.26 Use the preceding exercise to prove that a closed linear map u of a Banach space E into a Banach space F is continuous.

[Hint: Use the category theorem to show that for a suitable number k, the set

$$A = \{x \in E : \|u(x)\| \leq k\}$$

is dense in the closed unit ball in E. Apply the preceding exercise, taking (say) $\lambda = 2$.]

CHAPTER 7

Boundedness Principles

7.0 Foreword

The general theme running through this chapter is a type of result that says that if a set of linear or bilinear maps from one or more TVSs into another is bounded in one sense, then it is already bounded in some stronger sense. This stronger sense will often include some sort of uniformity that is not apparent in the hypotheses, which explains the name "uniform boundedness principles" sometimes applied to such results. These principles apply only to certain categories of TVSs. To some extent the barrelled and ultrabarrelled spaces introduced in Chapter 6 were originally delineated precisely as a result of analyzing the proofs of these principles, hitherto enunciated only for Banach spaces or complete metrizable ones.

Broadly speaking, we shall derive four main boundedness principles for linear maps, appearing as Theorems 7.1.1, 7.1.3, 7.3.1, and 7.4.3, respectively. These differ both in respect of the type of space to which they are applicable, and in relation to the type of boundedness assumed at the outset. Thus in Section 7.1 we consider cases where pointwise boundedness implies equicontinuity (barrelled and ultrabarrelled spaces), while in Section 7.3 we are concerned with the passage from strong boundedness to equicontinuity (bornological and infrabarrelled spaces). They imply hosts of minor corollaries, each more or less specially adapted to specific situations. In Section 7.7 we shall consider bilinear maps.

One further category of LCTVS appears in a natural way: the bornological spaces introduced in Section 7.3.

For some developments that are not dealt with in this book see W. L. C. Sargent [2] and R. Henstock [1].

7.1 Boundedness Principles for Barrelled and Ultrabarrelled Spaces

Working independently, Orlicz [1], Gelfand ([1], [2]), and Bosanquet and Kestelman [1] showed that a lower semicontinuous seminorm p on a Banach space E is necessarily continuous. Eberlein [3] replaced lower semicontinuity by the Baire property without prejudicing the conclusion. (A function p on a topological space E with values in a topological space is said to have the Baire property if there exists a meagre subset M of E such that $f \mid (E \backslash M)$ is continuous; see Banach [1], p. 17). Following similar lines, Mehdi [1] has shown that continuity of p results whenever E is a nonmeagre TVS and the set $\{x : x \in E, p(x) \leq 1\}$ is a Baire set (see Banach [1], p. 15; Kuratowski [1], p. 55). As we shall see, any principle of this sort justifies its existence on the grounds of utility alone, leaving aside its obvious intrinsic interest.

The preceding investigations depend solely on the Baire category theorem and similar arguments, so that the results remained in force for complete metrizable TVSs.

Without doubt Bourbaki has this principle in mind when he introduces ([5], p. 1) the barrelled spaces; for it is very simple to show (Theorem 7.1.1) that a LCTVS E is barrelled if and only if each lower semicontinuous seminorm on E is continuous. We shall refer to this principle as the "OGBK principle."

Viewing the situation in relation to the earlier results, Bourbaki's introduction of this new category of LCTVSs is justified by Theorem 6.3.1, which shows that there exist many nonmetrizable LCTVSs for which the OGBK principle is valid. Such spaces lie beyond the range of the original arguments.

We shall also formulate an analogous boundedness principle for ultrabarrelled spaces.

The general boundedness principles are summarized in the results numbered 7.1.1 through 7.1.5, 7.3.1, 7.4.3, 7.4.4, and 7.7.4 through 7.7.9.

Our first result affords an analytical expression of some boundedness principles for the spaces of the type mentioned, and even provides in analytical terms a characterization of both barrelled and infrabarrelled spaces.

7.1.1 **Theorem.** (1) A LCTVS E is barrelled if and only if either of the following two equivalent conditions is fulfilled:

(a) Every lower semicontinuous seminorm on E is continuous (the OGBK principle).

(b) Every weakly bounded subset of E' is equicontinuous.

(2) Suppose that E is an ultrabarrelled TVS, and that f is a positive lower semicontinuous function on E such that for λ in K (the scalar field, real or complex) and x, y in E one has

$$f(\lambda x) \leq A(\lambda, f(x)),$$
$$f(x-y) \leq B(f(x), f(y)),$$

where A and B are functions defined on $K \times R_+$ and $R_+ \times R_+$, respectively, and satisfy the conditions

$$A(\lambda, t) \to 0, \ B(s, t) \to 0$$

as λ, s, and t tend to 0. Then f is continuous at 0.

(3) A LCTVS E is infrabarrelled if and only if *either* every bounded, lower semicontinuous seminorm on E is continuous, *or* every strongly bounded subset of E' is equicontinuous.

Remarks. In the first condition of (3), to say that a seminorm is bounded means that it is bounded on each bounded subset of E. In the second condition of (3), to say that a set $S \subset E'$ is strongly bounded means that it is bounded for the strong topology on E'. This last is defined in Subsection 1.10.8, taking there $F = K$. Thus S is strongly bounded if and only if $\operatorname{Sup} \{|\langle x, x' \rangle| : x \in B, x' \in S\} < +\infty$ for each bounded subset B of E.

Proof. (1) Let us remark first of all that by virtue of the substance of Section 1.6, the barrels in E are precisely the sets of the form $p(x) \leq 1$, where p is a lower semicontinuous seminorm on E.

This remark, together with the definition of barrelled spaces (in Section 6.2), leads at once to (a).

As for (b), the reader will observe that since E' with its weak topology is locally convex, it is enough to deal with those weakly bounded sets B' that are also weakly closed, convex, and balanced. Then (Section 8.1) its polar B'^0 in E is a barrel in E. Moreover, $B' = B'^{00}$. If therefore E is barrelled, B'^0 is a neighbourhood of 0 in E, and so B' is equicontinuous. Conversely, if T is a barrel in E, its polar T^0 in E' is weakly bounded. If (b) holds, T^0 is equicontinuous, and T^{00} is a neighbourhood of 0 in E. But $T = T^{00}$ (Theorem 8.1.5), and so E is barrelled.

(3) This is proved in a similar fashion. The reader will notice that if a barrel in E has the equation $p(x) \leq 1$, then p is bounded if and only if the said barrel absorbs each bounded subset of E. Again, if B' is strongly bounded in E', then B'^0 is a barrel in E absorbing each bounded set; and if T is a barrel in E absorbing the bounded sets, then T^0 is strongly bounded in E'.

(2) For $\varepsilon > 0$ we define

$$V_\varepsilon = \{x \in E : f(x) \leq \varepsilon\}.$$

V_ε is closed since f is lower semicontinuous. The remaining hypotheses on f ensure that the V_ε may be taken as a base at 0 for a vector-space topology on E. If E is ultrabarrelled, this topology is weaker than the initial one. Thus each V_ε is a neighbourhood of 0 in E, which is what we had to prove. ▌

The most frequently used form of the OGBK principle is worth stating separately.

7.1.2 **Corollary.** If E is a Fréchet space, any lower semicontinuous seminorm on E is continuous, and any weakly bounded subset of E' is equicontinuous.

We turn next to one of the generalized versions of the Banach-Steinhaus theorem, the original version for Banach spaces of which is to be found in Banach [1].

7.1.3 **Theorem (Banach-Steinhaus).** Let E and F be TVSs, E being ultrabarrelled. If a set $\mathscr{F}$ of continuous linear maps of E into F is bounded at each point of E, then $\mathscr{F}$ is equicontinuous. The same conclusion follows if E is barrelled and F is locally convex.

Proof. Suppose that V ranges over the closed neighbourhoods of 0 in F. For each such V we define

$$V^* = \bigcap \{u^{-1}(V) : u \in \mathscr{F}\}.$$

Since $\mathscr{F}$ is bounded at each point, V^* is absorbent; and since each u is continuous, V^* is closed in E. The sets V^* may be taken as a base at 0 for a vector-space topology on E. E being ultrabarrelled and the V^* closed, this topology is weaker than the initial topology on E, so that each V^* is a neighbourhood of 0 in E. Thus $\mathscr{F}$ is equicontinuous.

If F is locally convex, we may assume that each V is convex, in which case each V^* has the same property. Thus V^* is a barrel in E, hence a neighbourhood of 0 in E, and the conclusion appears once again. ▌

Remark. If F is locally convex, the boundedness hypothesis on F follows already from the condition that

$$\operatorname*{Sup}_{u \in \mathscr{F}} |x'(u(x))| < +\infty$$

for each $x' \in F'$. The sufficiency of this apparently weaker hypothesis appears from Theorem 8.2.2.

7.1.4 **Corollary.** Suppose *either* that E is ultrabarrelled, *or* that E is barrelled and F is locally convex, and that in either case F is quasi-complete.† Let $(u_i)_{i \in I}$ be a directed family of continuous linear maps of E into F that is bounded at each point of E. Then the set E_0, of $x \in E$ for which $\lim_i u_i(x) = u(x)$ exists in F, is a closed vector subspace of E, u is continuous on E_0, and $\lim_i u_i = u$ holds *uniformly* on each precompact (= totally bounded) subset of E_0. Finally, if the directed family is replaced by a sequence (u_n), it is enough that F be sequentially complete.

Proof. By Theorem 7.1.3, the u_i are equicontinuous. It is clear that E_0 is a vector subspace of E. Equicontinuity shows that, if x is adherent to E_0, then the directed family $(u_i(x))$ is Cauchy in F. Since this family is bounded in F, and since F is quasi-complete, $u(x) = \lim_i u_i(x)$ exists in F. Thus E_0 is closed. Uniformity of convergence on precompact subsets of E_0 is a direct consequence of Ascoli's theorem. ▌

We shall conclude this section by dealing with an application of the Banach-Steinhaus theorem to bilinear maps, but we shall first formulate a theorem of still wider scope.

7.1.5 **Theorem.** Let E be a semimetrizable TVS, G any TVS. Suppose that *either* (a) E is ultrabarrelled, *or* (b) E is barrelled and G is locally convex.

† Quasi-completeness is defined and studied in the Section 7.4; it means that the bounded closed sets are complete and is therefore implied by completeness. The premature use of this term here should not worry the reader for long.

Let T be any topological space satisfying the first countability axiom. A set $\mathscr{F}$ of maps of $E \times T$ into G is equicontinuous if (and only if) the following conditions are fulfilled:

(1) For each u in $\mathscr{F}$ and each t in T, $x \to u(x, t)$ is a continuous linear map of E into G.

(2) For each x in E, the maps $t \to u(x, t)$ $(u \in \mathscr{F})$ are equicontinuous at each point of T.

(3) For each $(x, t) \in E \times T$, the set $\{u(x, t) : u \in \mathscr{F}\}$ is bounded in G.

Proof. The "only if" assertion is trivial, and this without special assumptions on E, G, or T.

For the rest we begin with the following observation: If P is a topological space satisfying the first countability axiom, G a uniform space, $\mathscr{F}$ a set of maps of P into G, then $\mathscr{F}$ is equicontinuous at a given point p of P if (and only if) for each sequence $p_n \to p$, the restrictions of the u in $\mathscr{F}$ to the set $\{p, p_1, p_2, \cdots\}$ are equicontinuous at p. Applying this observation to the case in which $P = E \times T$, we see that it is enough to deal with the case in which T is compact.

This being so, conditions (2) and (3) together entail that the set

$$S_x = \{u(x, t) : t \in T,\ u \in \mathscr{F}\}$$

is bounded in G for each x in E. Indeed, thanks to (2), to each t in T corresponds an open neighbourhood N_t thereof such that $u(x, t) - u(x, t') \in W$ whenever $t' \in N_t$, W being any preassigned balanced neighbourhood of 0 in G. By compactness of T, we may choose t_i $(1 \le i \le n)$ in T such that the N_{t_i} cover T. According to (3), the set

$$\{u(x, t) : t = t_i\ (1 \le i \le n),\ u \in \mathscr{F}\}$$

is bounded in G and hence is contained in kW for some $k > 0$. But then $S_x \subset (k + 1)W$. This shows that S_x is bounded in G, as alleged.

Condition (1) now combines with Theorem 7.1.3 to show that the set of linear maps $x \to u(x, t)$ $(t \in T,\ u \in \mathscr{F})$ is equicontinuous. Thus, given a neighbourhood W of 0 in G and any (x_0, t_0) in $E \times T$, there exists a neighbourhood U of x_0 such that

$$u(x, t) - u(x_0, t) \in W$$

for u in $\mathscr{F}$ and t in T. Using (2) once more, there exists a neighbourhood N of t_0 such that

$$u(x_0, t) - u(x_0, t_0) \in W$$

for u in $\mathscr{F}$ and t in N. But then

$$u(x, t) - u(x_0, t_0) \in W + W$$

for u in $\mathscr{F}$, x in U, and t in N. This completes the proof. ▌

7.2 Some Applications and Examples

We can give some fairly direct applications of the results collected in Section 7.1.

7.2.1 Example. We begin with a minor theorem useful as a stepping stone to its own generalization, which will appear as Theorem 8.2.2.

Let E be any seminormed vector space, A a subset of E that is weakly bounded—that is, for which

$$\operatorname{Sup}_{x \in A} |x'(x)| < +\infty$$

for each $x' \in E'$. We assert that A is *necessarily bounded*; that is, that $p(A)$ is bounded, p being the seminorm defining the topology of E.

To see this, we note that E' is a B-space with the closed unit ball S defined as the set of x' such that

$$|x'(x)| \leq p(x) \qquad (x \in E).$$

According to the Hahn-Banach theorem,

$$p(x) = \operatorname{Sup}_{x' \in S} |x'(x)|. \tag{7.2.1}$$

Now each $x \in E$ generates a continuous linear form $x' \to x'(x)$ on E', and our hypothesis signifies that the elements of A thus generate a subset of E'' that is bounded at each point. By Corollary 7.1.2, with E replaced there by E', this image of A is equicontinuous. In view of (7.2.1), this means that $p(A)$ is bounded.

As we have said, Theorem 8.2.2 will affirm that weak boundedness is equivalent to boundedness in any LCTVS.

7.2.2 Example. We return to Problem (3) raised in Subsection 6.8.4 in order to establish the result spoken of there and already used as an auxiliary in the proof of Theorem 6.8.6.

The situation is that a TVS E and a weak Schauder base (e_n) in E are given, and our aim is to show that:

▶ If E is barrelled, then the series

$$\sum_{n=1}^{\infty} e'_n(x) e_n, \tag{7.2.2}$$

which by hypothesis converges weakly to x for each $x \in E$, in fact converges to x in the sense of the initial topology on E.

To see this we shall apply Theorem 7.1.3, taking therein $F = E$ and

$$u_n(x) = \sum_{m=1}^{n} e'_m(x) e_m - x.$$

At present we may infer the boundedness of $(u_n(x))$ for each $x \in E$ from the weak convergence of (7.2.2) only when E is seminormed, in which case Example 7.2.1 applies. We may, however, anticipate Theorem 8.2.2, which sanctions this inference for any LCTVS E. This being so, Theorem 7.1.3 shows that the u_n are equicontinuous.

Now it is trivially the case that $u_n(x) \to 0$ for the initial topology whenever x lies in the vector subspace E_0 of E generated by the e_n. Also the weak convergence of (7.2.2) to x shows that E_0 is weakly dense in E. So, according to the Hahn-Banach theorem, E_0 is dense in E.

Thanks to equicontinuity, given a convex neighbourhood V of 0 in E, there exists a neighbourhood U of 0 in E such that $u_n(U) \subset \frac{1}{2}V$ for all n. For given $x \in E$ we may then choose $x_0 \in E_0$ such that $x - x_0 \in U$. It follows that $u_n(x) \in u_n(x_0) + \frac{1}{2}V$ for all n. But, since $x_0 \in E_0$, $u_n(x_0) \in \frac{1}{2}V$ for all $n > n_0$. Hence $u_n(x) \in V$ for $n > n_0$. This shows that the series (7.2.2) converges to x for the initial topology, as we wished to show.

7.2.3 **Example.** We shall look at some properties of Fourier coefficients of periodic functions of a real variable belonging to various function spaces. In this example, all functions are assumed to have period 2π; the Lebesgue spaces $\mathscr{L}^p$ are constructed relative to the Lebesgue measure restricted to some fundamental interval of length 2π—say $(-\pi, +\pi)$. If x is an integrable function, we define its Fourier coefficients by

$$\hat{x}(n) = (2\pi)^{-1}\int_{-\pi}^{\pi} x(t)e^{-int}\,dt,$$

n being any integer (zero or negative as well as positive).

The famous Riemann-Lebesgue lemma states that

$$\lim_{|n|\to\infty} \hat{x}(n) = 0$$

whenever $x \in \mathscr{L}^1$. We shall show that this is the best possible result in the following sense:

▶ Given any sequence (a_k) of positive numbers such that $\lim_{k\to\infty} a_k = 0$, and any sequence (n_k) of integers such that $\lim_{k\to\infty} |n_k| = +\infty$, there exists $x \in \mathscr{L}^1$ such that

$$\limsup_{k\to\infty} \frac{|\hat{x}(n_k)|}{a_k} = +\infty; \tag{7.2.3}$$

in fact the functions x satisfying (7.2.3) form a nonmeagre set in $\mathscr{L}^1$.

If (7.2.3) were false for each $x \in \mathscr{L}^1$, then

$$p(x) = \operatorname{Sup}_k \frac{|\hat{x}(n_k)|}{a_k}$$

would be finite, and hence a seminorm on $\mathscr{L}^1$. In any case p is lower semi-continuous. But then Corollary 7.1.2 would require p to be continuous; that is, there exists a number c such that

$$\frac{|\hat{x}(n_k)|}{a_k} \leq c \cdot N_1(x) = c \cdot (2\pi)^{-1} \int_{-\pi}^{\pi} |x(t)| \, dt$$

for all k and all $x \in \mathscr{L}^1$. Choosing a sequence (x_r) such that $N(x_r) \leq 1$ and $\hat{x}_r(n) \to 1$ as $r \to \infty$ for each n, we should derive $1/a_k \leq c$ for all k, contrary to $\lim_{k\to\infty} u_k = 0$. The stronger category assertion follows likewise from refining Corollary 7.1.2 into Theorem 7.5.1.

The argument is easily extended from $\mathscr{L}^1$ to $\mathscr{L}^p$ $(1 \leq p \leq +\infty)$:

▶ If (a_k) is a sequence of positive numbers such that

$$a_k = o(n_k^{-1/p'}) \qquad (k \to \infty),$$

then there exists an $x \in \mathscr{L}^p$ such that (7.2.3) holds; and indeed such functions x form a nonmeagre subset of $\mathscr{L}^p$.

(Here p' is defined by $1/p + 1/p' = 1$, p' being 1 if $p = +\infty$, and $+\infty$ if $p = 1$, while $1/(+\infty) = 0$.) This statement counter balances the Hausdorff-Young-Riesz theorem, asserting that if $x \in \mathscr{L}^p$ with $1 \leq p \leq 2$, then $\hat{x} \in \ell^{p'}$, that is,

$$\sum_{n=-\infty}^{\infty} |\hat{x}(n)|^{p'} < +\infty;$$

this is thus seen to be the best possible theorem in a certain sense.

We terminate this subsection by discussing the following related problem:

▶ For what sequences (c_n) is it true that

$$\sum_{n=-\infty}^{\infty} |c_n \hat{x}(n)| < +\infty \tag{7.2.4}$$

holds for all $x \in \mathscr{L}^p$?

Take first the case $p = 1$. It is obvious that

$$\sum_{n=1}^{\infty} |c_n| < +\infty \tag{7.2.5}$$

is *sufficient*. That it is also *necessary* follows quickly from an application of Corollary 7.1.2 and use of a suitable sequence $(x_r) \subset \mathscr{L}^1$ as before.

The case $p > 1$ is more complicated, and the writer is not aware of any necessary and sufficient conditions that are easily verifiable. An application of Corollary 7.1.2 shows that (7.2.4) holding for each $x \in \mathscr{L}^p$ is equivalent to the existence of a number k such that

$$\sum |c_n \hat{x}(n)| \leq k \cdot N_p(x) \tag{7.2.6}$$

for $x \in \mathscr{L}^p$ (and this whether or not $p > 1$). If we define s_F for each finite set F of integers by

$$s_F(t) = \sum_{n \in F} c_n e^{int},$$

then (7.2.6) implies

$$N_{p'}(s_F) \leq K, \tag{7.2.7}$$

that is, to the boundedness of the s_F in $\mathscr{L}^{p'}$. Also, (7.2.6) can be shown to be equivalent to the following:

► For each sequence (θ_n) satisfying $|\theta_n| \leq 1$ for all n, $(\theta_n c_n)$ is the sequence of Fourier coefficients of a function in $\mathscr{L}^{p'}$ if $p < +\infty$; if $p = +\infty$, (7.2.6) implies that $(\theta_n c_n)$ is the sequence of Fourier coefficients of a measure.

This implies that $(c_n) \in \ell^2$; cf. Zygmund [2], Vol. II, p. 215.

Clearly, by taking x suitably in $\mathscr{L}^p$, (7.2.6) entails that

$$\sum_{n \neq 0} \frac{c_n}{n} < +\infty, \tag{7.2.8}$$

but this is evidently far from being a sufficient condition.

7.2.4 **Example: Applications to Summability Theory.** We return to the topics dealt with in Examples 6.1.4 and 6.4.10 and see how boundedness principles help in the solution of several questions arising in summability theory. In general we adhere to the notations introduced in the examples cited.

It is natural to ask for necessary and sufficient conditions on the matrix A in order that it shall transform convergent sequences into convergent sequences (that is, that $c \subset C(A)$) or convergent-to-zero sequences into convergent-to-zero sequences (that is, $c_0 \subset C_0(A)$). In the former case it is equally natural to ask when limits are preserved by the transformation defined by A. The answers to these (and other similar) questions are known (Cooke [1]).

Boundedness principles may be profitably employed in the solution. We illustrate by dealing with the case $c_0 \subset C_0(A)$ and prove the following results.

► (1) In order that A transform each convergent-to-zero sequence into a bounded sequence—that is, that $c_0 \subset B(A)$—it is necessary and sufficient that

$$\operatorname*{Sup}_m \sum_n |A(m, n)| < +\infty. \tag{7.2.10}$$

In this case A transforms each bounded sequence into a bounded sequence—that is, $\ell^\infty \subset B(A)$—and

$$\|u_A(x)\|_\infty \leq \text{const. } \|x\|_\infty \qquad (x \in \ell^\infty).$$

▶ (2) In order that A transform each convergent-to-zero sequence into one of the same nature—that is, that $c_0 \subset C_0(A)$—it is necessary and sufficient that it satisfy (7.2.10) and in addition

$$\lim_m A(m, n) = 0 \qquad (n = 1, 2, \cdots). \tag{7.2.11}$$

Proofs. In respect to (1) it is evident that (7.2.10) arranges that $\ell^\infty \subset B(A)$ and that $\|u_A(x)\|_\infty \leq \text{const.} \|x\|_\infty$ for x in ℓ^∞. Suppose conversely that $c_0 \subset B(A)$, so that u_A maps c_0 into ℓ^∞. Then

$$q_{A,m}(x) = \operatorname{Sup}_k \left| \sum_{n \leq k} A(m, n)x(n) \right|$$

is finite for each m and each x in c_0. It is obviously a lower semicontinuous seminorm on c_0 and hence (Corollary 7.1.2) is continuous. This shows that $x \to u_A(x)(m)$ is continuous for each m. Since $u_A(x)$ belongs to ℓ^∞ by hypothesis, the same corollary shows that $x \to \operatorname{Sup}_m |u_A(x)(m)|$ is continuous on c_0, so that

$$\operatorname{Sup}_m \left| \sum_n A(m, n)x(n) \right| \leq \text{const.} \|x\|_\infty$$

for x in c_0. From this (7.2.10) follows.

Passing to (2), the necessity of (7.2.10) and (7.2.11) is obvious. Conversely if these conditions are fulfilled, then u_A is continuous from c_0 into ℓ^∞, and $u_A(x)$ belongs to c_0 whenever x has a finite support. These x are dense in c_0, and c_0 is closed in ℓ^∞, and so u_A maps c_0 into c_0. The proof is complete. ▌

It is interesting, and will later be useful, to record a characterization of the linear maps of ℓ^∞ into itself which are derivable from matrices.

▶ (3) A linear map u of ℓ^∞ into itself is of the form $u = u_A$ for some matrix A if and only if it is continuous for the weak topology $\sigma(\ell^\infty, \ell^1)$.

Proof. If $u = u_A$ for some A, then (7.2.10) holds. Then, if $y = u(x)$ and if s is in ℓ^1,

$$\sum_m y(m)s(m) = \sum_m s(m) \sum_n A(m, n)x(n) = \sum_n x(n)t(n)$$

where

$$t(n) = \sum_m A(m, n)s(m).$$

Now

$$\sum_n |t(n)| \leq \sum_n \sum_m |A(m, n)| \cdot |s(m)|$$
$$= \sum_m |s(m)| \cdot \sum_n |A(m, n)| \leq \text{const.} \sum_m |s(m)|$$

by (7.2.10), so that t belongs to ℓ^1. Continuity of u_A relative to $\sigma(\ell^\infty, \ell^1)$ is thus established.

Conversely, suppose u is continuous in this sense. Then since each x in ℓ^∞ can be written $x = \sum_n x(n)e_n$, the series converging for $\sigma(\ell^\infty, \ell^1)$, we have

$$u(x) = \sum_n x(n)u(e_n),$$

the series converging in the same sense. In particular,

$$u(x)(m) = \sum_n x(n) \cdot u(e_n)(m) = \sum_n A(m, n)x(n),$$

where

$$A(m, n) = u(e_n)(m).$$

Thus

$$u = u_A.$$

This completes the proof of (3). ▮

In Example 6.1.4, we have introduced a topology T_A that evidently renders continuous the map u_A of $B(A)$ into ℓ^∞ (or of $C(A)$ into c, or $C_0(A)$ into c_0). On the other hand, by (1) above, if u_A maps c_0 into ℓ^∞, then it is continuous for the normed topologies (and even, by (3) above, for the weak topologies). It is, in fact, the case that if u_A maps a vector subspace E of $D(A)$ into a vector subspace F of C^N, and if E and F carry vector space topologies of a pretty general type, then u_A is automatically continuous from E into F. More precisely:

▶ (4) Suppose that E is a vector subspace of $D(A)$ and F a vector subspace of C^N which are TVSs such that
(a) F is separated and each y in F is the limit of its finite sections $s_k y$;
(b) E is ultrabarrelled and its topology is stronger than that of pointwise convergence.
Then if $u_A(E) \subset F$, u_A is continuous from E into F.

Proof. By (b) and Corollary 7.1.4, $x \to u_A(x)(m)$ is continuous on E for each m. Let u_k $(k = 1, 2, \cdots)$ be the map of E into F defined by $u_k(x) = s_k u_A(x)$, the kth section of $u_A(x)$. By (a), $\lim u_k = u_A$ pointwise on E. If F_k is the subspace of F formed of those y in F such that $y = s_k y$ (that is, $y(n) = 0$ for $n > k$), then F_k is finite dimensional and can carry but one separated vector-space topology, say T. This topology T is that defined by any total set of linear forms on F_k, for example, by the linear forms $y \to y(m)$ $(m = 1, 2, \cdots, k)$. By the opening remark, u_k is continuous for this topology T. On the other hand, since F is separated, T must be identical with the topology induced by that of F. Thus u_k is continuous from E into F. Using again the fact that E is ultrabarrelled it now follows that u_A, the pointwise limit of the u_k, is continuous (and also that the u_k are equicontinuous). The proof is thus complete. ▮

Remark. If we assume that F is in addition locally convex, then it suffices to assume in (a) that $s_k y \to y$ *weakly* in F and in (b) that E is barrelled. This is so because, F being locally convex, the weak convergence of $s_k y$ to y implies

(Theorem 8.2.2) that the $s_k y$ remain bounded in F. The maps u_k are thus bounded at each point, and the preceding argument proceeds satisfactorily. Other hypotheses on E and F will lead to the same conclusion (see Exercise 7.4). Related results are given for Banach spaces E and F by Zeller [3].

As a final application of boundedness principles we shall show that for a wide class of summability methods, including the usual ones, there always exist unbounded sequences that are transformed into convergent ones.

▶ (5) Let A_i $(i = 1, 2, \cdots)$ be matrices such that

$$\lim_m A_i(m, n) \text{ exists finitely} \qquad (i, n = 1, 2, \cdots) \tag{7.2.12}$$

and

$$\lim_n \operatorname{Sup}_m A_i(m, n) = 0 \qquad (i = 1, 2, \cdots). \tag{7.2.13}$$

Then $C(A_1, A_2, \cdots)$ contains unbounded sequences.

Proof. We use the notation of Example 6.1.4. We put $K = K_{A_1} \cap K_{A_2} \cap \cdots$. If K is finite, the assertion is trivial since $C = C(A_1, A_2, \cdots)$ contains any sequence vanishing on K. We assume therefore that K is infinite. Let C^* be the set of x in C vanishing outside K. Referring to the results (1) and (2) and the closing remarks of Example 6.1.4, we see that C is a Fréchet space for the topology induced by $T = T(A_1, A_2, \cdots)$ so that the closed subspace C^* of C is also a Fréchet space for $T \mid C^*$, and that $T \mid C^*$ is defined by the seminorms

$$q_{A_i} \text{ and } q_{A_i,m} \qquad (i, m = 1, 2, \cdots).$$

Moreover, by (7.2.12), e_n belongs to C^* whenever n belongs to K, which is true for a sequence of n tending to infinity.

Now suppose if possible that $C \subset \ell^\infty$. Then the function $\|x\|_\infty$ would be finite for x in C^*. Since $x \to |x(n)|$ is continuous on C^* for each n, Corollary 7.1.2 shows that the seminorm $x \to \|x\|_\infty$ would be continuous on C^*. However, (7.2.13) shows that

$$q_{A_i,m}(e_n) = |A_i(m, n)| \le \alpha_i(n),$$

where $\lim_n \alpha_i(n) = 0$; hence also

$$q_{A_i}(e_n) = \operatorname{Sup}_m |A_i(m, n)| \to 0 \text{ as } n \to \infty.$$

So it would be the case that $e_n \to 0$ in C^* as $n \in K$ converges to infinity, and therefore that $\|e_n\|_\infty \to 0$ in the same circumstances. But this is absurd since $\|e_n\|_\infty = 1$ for all n. This contradiction establishes (5). ▮

Remark. If in place of (7.2.12) one assumed that $\lim_m A_i(m, n) = 0$ $(i, n = 1, 2, \cdots)$, it would follow likewise that $C_0(A_1, A_2, \cdots)$ must contain unbounded sequences. These conditions are fulfilled if, for example, the A_i are the iterated arithmetic means or the iterated Abel means.

7.2.5 Example. The identity, in the dual of a barrelled space, of weakly bounded and equicontinuous sets will now be used to establish a proposition that in turn is useful in connection with the study of the boundedness of solutions of differential equations, as will be seen in the next example.

Let X be a vector space of finite dimension n. In what follows (K_α) is a family of functions of a real variable $t \geq 0$ taking values in $L(X, X)$ (the space of endomorphisms of X). If a basis be chosen in X, each endomorphism $K_\alpha(t)$ will be represented by a matrix whose ijth entry will be denoted by $K_{\alpha,ij}(t)$. We shall assume that each function $K_{\alpha,ij}$ is integrable over $(0, \infty)$ for Lebesgue measure.

We denote by $C_0([0, \infty), X)$ the space of continuous vector-valued functions on $[0, \infty)$ into X that tend to zero at infinity. Let it be supposed that

$$\operatorname*{Sup}_\alpha \left\| \int_0^\infty K_\alpha(t)\varphi(t)\, dt \right\| < +\infty \tag{7.2.14}$$

for each φ in $C_0([0, \infty), X)$, $\int_0^\infty K_\alpha(t)\varphi(t)\, dt$ denoting the element of X whose ith coordinate is

$$\sum_{j=1}^n \int_0^\infty K_{\alpha,ij}(t)\varphi_j(t)\, dt.$$

The norm involved in (7.2.14) may be any one on the finite dimensional vector space X. To be specific we may take the one defined by

$$\|x\| = \operatorname*{Sup}_i |\xi_i|$$

whenever x has the coordinates ξ_i relative to the chosen base. The corresponding norm on $L(X, X)$ will be

$$\|M\| = \operatorname*{Sup}_i \sum_{j=1}^n |M_{ij}|$$

the endomorphism M being represented by the matrix (M_{ij}).

Our aim is to show that from (7.2.14) follows the conclusion that

$$\operatorname*{Sup}_\alpha \int_0^\infty \|K_\alpha(t)\|\, dt < +\infty. \tag{7.2.15}$$

To see this we treat $C_0([0, \infty), X)$ as a Banach space with the norm

$$\|\varphi\| = \operatorname*{Sup}_{i,t\geq 0} \|\varphi_i(t)\|.$$

Now (7.2.14) signifies that

$$\operatorname*{Sup}_\alpha \operatorname*{Sup}_i \sum_j \left| \int_0^\infty K_{\alpha,ij}(t)\varphi_j(t)\, dt \right| < +\infty \tag{7.2.16}$$

for each $\varphi = (\varphi_1, \cdots, \varphi_n)$ in $C_0([0, \infty), X)$. Now for each α and each i the mapping

$$\varphi \to \sum_j \int_0^\infty K_{\alpha,ij}(t)\varphi_j(t)\, dt$$

is a continuous linear form, and (7.2.16) signifies that these linear forms are weakly bounded when α and i vary. It follows that the said linear forms are equicontinuous, which signifies that there exists a number k such that

$$\left|\sum_j \int_0^\infty K_{\alpha,ij}(t)\varphi_j(t)\,dt\right| \le k\|\phi\|,$$

for all α and all i, k being independent of α and of i. If we here allow the φ_j to vary independently, we may conclude that

$$\sum_j \int_0^\infty |K_{\alpha,ij}(t)|\,dt \le k$$

for all α and all i. So

$$\int_0^\infty \|K_\alpha(t)\|\,dt = \int_0^\infty \operatorname{Sup}_i \sum_j |K_{\alpha,ij}(t)|\,dt \le \int_0^\infty \sum_i \sum_j |K_{\alpha,ij}(t)|\,dt \le nk$$

for all α, which implies (7.2.15). ▌

Remarks. There is an analogous result in which we assume that each $K_{\alpha,ij}$ belongs to $\mathscr{L}^{p'}(0, \infty)$ and that (7.2.14) holds for each φ such that $\varphi_i \in \mathscr{L}^p(0, \infty)$ for $1 \le i \le n$. The conclusion is that

$$\operatorname{Sup}_\alpha \int_0^\infty \|K_\alpha(t)\|^{p'}\,dt < +\infty.$$

7.2.6 **Example.** Let X be as in the preceding example. We shall consider X-valued solutions of the differential equations

$$\frac{dx}{dt} = A(t)x + \varphi(t) \tag{7.2.17}$$

and

$$\frac{dx}{dt} = A(t)x + \psi(t, x), \tag{7.2.18}$$

where A is a function on $[0, \infty)$ with values in $L(X, X)$; φ is a function on $[0, \infty)$ with values in X; and ψ a function on $[0, \infty) \times X$ with values in X. For simplicity we shall assume that A, φ, and ψ are each continuous.

Following R. Bellman [1] we aim to show that if for each $\varphi \in C_0([0, \infty), X)$ all solutions of (7.2.17) are bounded, then all solutions of (7.2.18) are likewise bounded, provided ψ is bounded.

The proof is based partly on the fact, well-known and easy to verify, that the solutions of (7.2.17) are of the form

$$x(t) = U(t)\,x(0) + \int_0^t U(t)U(t')^{-1}\varphi(t')\,dt',$$

where U is the function on $[0, \infty)$ with values in $L(X, X)$ defined as the solution of the equations

$$\frac{dU}{dt} = A(t)\,U, \qquad U(0) = I,$$

I denoting the identity endomorphism of X. This being so, our hypothesis signifies that

$$\int_0^t U(t)U(t')^{-1}\varphi(t')\,dt'$$

is bounded with respect to $t \geq 0$ for each φ in $C_0([0, \infty), X)$.

At this point we appeal to the preceding example, permitting the inference that

$$\operatorname*{Sup}_{t\geq 0} \int_0^t \|U(t)U(t')^{-1}\|\,dt' = C < +\infty. \tag{7.2.19}$$

Turning now to (7.2.18) and putting $x(t) = U(t)y(t)$, we find that y satisfies the equation

$$\frac{dy}{dt} = U(t)^{-1}\psi(t, x(t)).$$

Consequently any solution x of (7.2.18) satisfies the relation

$$\begin{aligned} x(t) &= U(t)y(0) + \int_0^t U(t)U(t')^{-1}\psi(t', x(t'))\,dt' \\ &= U(t)x(0) + \int_0^t U(t)U(t')^{-1}\psi(t', x(t'))\,dt'. \end{aligned}$$

Now $U(t)x(0)$, being a solution of (7.2.17) with $\varphi = 0$, is by hypothesis bounded for $t \geq 0$. Furthermore, if

$$\|\psi(t, x(t))\| \leq m \qquad \text{for } t \geq 0,$$

then (7.2.19) shows that

$$\left\| \int_0^t U(t)U(t')^{-1}\psi(t', x(t'))\,dt' \right\| \leq m \cdot \int_0^t \|U(t)U(t')^{-1}\|\,dt' \leq mC,$$

whence it follows that $x(t)$ is bounded for $t \geq 0$. This establishes the result claimed. ▌

7.3 Bornological Spaces

Bornological spaces will be defined in a manner reminiscent of the definition of barrelled and infrabarrelled spaces given in Section 6.2. A LCTVS E is said to be *bornological* if each convex, balanced set $A \subset E$ that absorbs each bounded subset of E is a neighbourhood of 0 in E. The reader will notice that A is not a barrel inasmuch as it may fail to be closed; on the other hand it is assumed to absorb each bounded set, and not merely each one-point set. It is known that there exist barrelled spaces that are not bornological (Remark (3) after Theorem 7.3.3); on the other hand Theorem 7.3.2 leads to examples of bornological spaces that are not barrelled. Corollary 7.4.2 shows, however, that any complete bornological space is barrelled. It is immediate that any bornological space is infrabarrelled.

The next theorem incorporates in part (2) a third boundedness principle.

7.3.1 **Theorem.** (1) A LCTVS E is bornological if and only if each seminorm on E that is bounded (that is, bounded on each bounded subset of E) is continuous.

(2) If E is bornological and F any LCTVS, and if $\mathscr{F}$ is a set of linear maps of E into F that is strongly bounded, then $\mathscr{F}$ is equicontinuous. The same conclusion holds if E is infrabarrelled and each member of $\mathscr{F}$ is continuous [cf. Theorem 7.1.1(3)].

Proof. (1) Suppose that E is bornological and that p is a bounded seminorm on E. Then the set A, defined by $p(x) \leq 1$, is convex, balanced, and absorbs each bounded set; A is thus a neighbourhood of 0 and p therefore continuous. Reciprocally, if A has the stated properties, its gauge function p is a bounded seminorm; and continuity of p signifies that A (which in any case contains the set defined by $p(x) < 1$) is a neighbourhood of 0.

(2) It is necessary merely to apply the definition to

$$A = \bigcap_{u \in \mathscr{F}} u^{-1}(V),$$

V being any given convex and balanced neighbourhood of 0 in F; the statement that $\mathscr{F}$ is strongly bounded signifies that A, which is in any case convex and balanced, absorbs each bounded subset of E. For the second part we observe that A is a barrel if each member of $\mathscr{F}$ is continuous. ∎

Remarks. (a) For $\mathscr{F}$ to be strongly bounded, it suffices that the following condition be fulfilled:

(3) Whatever the sequence $x_n \to 0$ in E and sequence $u_n \in \mathscr{F}$, the sequence $u_n(x_n)$ be bounded in F. Suppose indeed that (3) is fulfilled. If $\mathscr{F}$ were not strongly bounded, there would exist in E a bounded set B such that the set $\{u(x) : x \in B, u \in \mathscr{F}\}$ is unbounded in F. But then there would exist sequences $y_n \in B$ and $u_n \in \mathscr{F}$ such that $u_n(y_n)$ is unbounded in F, and then a sequence of scalars $\lambda_n \to 0$ such that $\lambda_n u_n(y_n)$ is still unbounded in F. Putting $x_n = \lambda_n y_n$ would then lead to a contradiction of (3).

(b) Included in (2) is the assertion that if E is bornological and F is locally convex, any bounded linear map of E into F is continuous. (The converse is true and trivial for any two TVSs E and F.) In view of Proposition 1.10.9 we conclude that $L_c(E, F)$ is strongly complete whenever F is complete and E is bornological. More particularly still we see that if E is bornological, then E' coincides with the space $\bar{E}$ of bounded linear forms on E. The relationship between E' and $\bar{E}$ in general is discussed by Shirota [1], who shows that $E' = \bar{E}$ if and only if $\tau(E, E')$ is bornological.

Our next step is to delineate some general types of spaces that are known to be bornological.

7.3.2 **Theorem.** Any semimetrizable LCTVS is bornological; in fact, if E is a semimetrizable TVS, any balanced subset of E that absorbs each sequence converging to 0 is a neighbourhood of 0.

Proof. Let $A \subset E$ have the stated property. Let $U_1 \supset U_2 \supset \cdots$ be a base at 0. If A were not a neighbourhood of 0, the relation $A \supset rU_n$ would be false for every n and every $r > 0$. Hence there would exist $x_n \in n^{-1}U_n$, $x_n \notin A$. Then $y_n = nx_n$ converges to 0 and yet the sequence (x_n) would not be absorbed by A. ▌

7.3.3 **Theorem.** (1) Any inductive limit of bornological spaces is bornological.

(2) The quotient of a bornological space by any vector subspace thereof is again bornological.

Proof. (1) Let E be the inductive limit of bornological spaces E_i relative to maps φ_i. Let $A \subset E$ be convex and balanced and absorb each bounded set in E. For each i, $A_i = \varphi_i^{-1}(A) \subset E_i$ is convex and balanced. Further, if B_i is bounded in E_i, $\varphi_i(B_i)$ is bounded in E, hence is absorbed by A, and so B_i is absorbed by A_i. Therefore A_i is a neighbourhood of 0 in E_i. But then A is a neighbourhood of 0 in E.

(2) Suppose E is bornological. Let M be a vector subspace of E. Let f be the natural map $E \to E/M$. Suppose A is convex and balanced and absorbs the bounded sets in E/M. If $B \subset E$ is bounded, $f(B)$ is bounded in E/M and so $f(B) \subset \lambda A$ for some λ, hence $B \subset \lambda f^{-1}(A)$. It follows that $f^{-1}(A)$ is a neighbourhood of 0 in E. But f is open, and $f(f^{-1}(A)) \subset A$. So A is a neighbourhood of 0 in E/M. ▌

Remarks. (1) The last two theorems lead easily to examples of bornological spaces that are not barrelled.

(2) It may be shown that a product $\prod_{i \in I} E_i$ of bornological spaces is bornological, provided K^I is bornological (Bourbaki [5], p. 15, Exercice 18b)). So (Theorem 7.3.2) the product of countably many bornological spaces is bornological.

The question of the bornological nature of K^I has been considered (though naturally expressed differently) for a long time. Its relations with measure theory were discovered and explained by Ulam [1] and Mackey [5]. A novel approach has been given more recently by Simons [1] who showed that R^I is bornological if and only if every δ-ultrafilter on I is trivial. (A δ-ultrafilter on a set I is a filter on I that is stable under countable intersections; and a filter on I is trivial if it is generated by some one-point subset of I.)

(3) The bornological nature of $C_R(T)$ (the space of real-valued continuous functions on T, endowed with the topology of locally uniform convergence) has been examined by Nachbin [2] and by Shirota [1]; see also Mahowald and Gould [1]. This same space (and its complex analogue) has been examined exhaustively by Warner [1], who relates properties of T with such properties of $C_R(T)$ as the following: being metrizable, bornological, barrelled, infrabarrelled, complete, quasi-complete, separable, a Fréchet space, a Montel space, reflexive, semireflexive, nuclear, a (DF) space, and so on.

Assuming that T is completely regular, Nachbin and Shirota show (*loc. cit.*) independently that

(a) $C_R(T)$ is barrelled if and only if for each noncompact, closed set $S \subset T$ there exists $f \in C_R(T)$ such that $\operatorname{Sup}_{t \in S} |f(t)| = \infty$;

(b) $C_R(T)$ is bornological if and only if T is a Q-space in Hewitt's sense (that is, a real-compact space in the terminology of Gillman and Jerison [1], Chapter 8). It is possible to produce spaces T having the first property but not the second, in which case $C_R(T)$ is barrelled but not bornological; the space T involved is a certain space of transfinite ordinals.

(4) A closed vector subspace of a bornological space need not be bornological (compare Subsection 6.2.6). Indeed Bourbaki ([5], p. 124, Exercice 21b)) produces an example of a space that is both barrelled and bornological and having a closed vector subspace that is not infrabarrelled; the said subspace in this example is sequentially complete and must therefore (by Corollary 7.4.2(d)) fail to be bornological. See also Köthe [5], Kap. V.

7.3.4 Application. Thanks to Theorem 7.3.3(1) and the topologization described in Section 6.3, we conclude that the spaces $\mathscr{K}(T)$ of integration theory and $\mathscr{D}^m(\Omega)$ $(0 \leq m \leq \infty)$ of distributions theory are bornological.

7.3.5 Bornological Spaces and Inductive Limits. We learn from Theorems 7.3.2 and 7.3.3(1) that any inductive limit of normable spaces is bornological. We turn now to a partial converse of this statement.

Let (E, t) be any LCTVS, and suppose that (B_i) is a family of subsets of E that are each convex, bounded, and balanced and such that any bounded subset of E is absorbed by some B_i. We consider the vector subspace

$$E_i = \bigcup_{\lambda > 0} \lambda B_i \, ;$$

this may be regarded as a seminormed space in which the sets εB_i $(\varepsilon > 0)$ form a base at 0 (cf. Lemma 6.5.2). If E is separated, E_i is thus normed. Furthermore, if B_i is sequentially complete (for t), then E_i is complete. Let f_i denote the injection map of E_i into E. If E_i is taken with the seminormed topology t_i just described, then f_i is continuous (since B_i is t-bounded). One may thus consider (E, t'), the inductive limit of the (E_i, t_i) relative to the injection maps f_i. It is plain that t is in any case weaker than t': for any t-neighbourhood U of 0 in E contains εB_i for some $\varepsilon > 0$ (B_i being t-bounded), so that $U \cap E_i$ is a t_i neighbourhood of 0 for each i.

It is easily seen that if t is bornological, then t and t' are identical. For suppose that V is any t'-neighbourhood of 0. This implies that to each i corresponds a number $\alpha_i > 0$ such that $V \cap E_i \supset \alpha_i B_i$. On the other hand, if B is t-bounded, there exists an index i and a number $\lambda > 0$ such that $B \subset \lambda B_i$. Hence for this index i one has $B \subset \lambda \alpha_i^{-1} V$. V thus absorbs each t-bounded set. Since one may assume that V is convex and balanced, and since (E, t) is bornological by hypothesis, V is a t-neighbourhood of 0. Thus t' is weaker than, and hence identical with, t.

The reader will note that the preceding argument shows that, whether or not (E, t) is bornological, a subset of E is t-bounded if and only if it is t'-bounded. We have established the following facts:

▶ Any bornological space E is the internal inductive limit of seminormable spaces, of normable spaces if E is separated, of complete seminormable spaces if E is sequentially complete, and of Banach spaces if E is separated and sequentially complete.

7.4 Results for Sequentially Complete and Quasi-complete Spaces

A TVS E is said to be sequentially [resp. quasi-] complete if each Cauchy sequence in E is convergent [resp. each bounded, closed subset of E is complete]. Since each Cauchy sequence is bounded, any quasi-complete TVS is sequentially complete. The converse statement is false, as is seen by considering any LCTVS E that is not semireflexive and yet is sequentially weakly complete (a space L^1, for example): endowed with the topology $\sigma(E, E')$, E is sequentially complete and yet fails to be quasi-complete because if x'' is an element of E'' not generated by an element of E, then x'' is the $\sigma(E'', E')$-limit of a bounded directed family (x_i) in E, and then the (weakly) closed convex envelope in E of $\{x_i\}$ is (weakly) bounded and (weakly) closed and not (weakly) complete.

An important property of sequentially complete spaces, which underlies further boundedness principles, stems from the next result.

7.4.1 **Proposition.** Let E be any TVS, Q a subset of E that is sequentially closed, convex, and such that for some $\beta > 0$

$$Q \supset \bigcup \{\alpha Q : |\alpha| < \beta\}.$$

Suppose further that $A \subset E$ is bounded, convex, balanced, and sequentially complete. If Q absorbs each point of A, then Q absorbs A.

Proof. We introduce the space E_A defined in Lemma 6.5.2. This space is complete and seminormed, hence barrelled.

The sets $\varepsilon Q \cap E_A$, with $\varepsilon > 0$, form a base at 0 for a locally convex topology t on E_A. Since the topology of E_A is stronger than that induced by E, and since Q is sequentially closed in E, it follows that t has a base at 0 formed of sets closed for the seminormed topology on E_A. Hence (Theorem 6.2.1(2)) t is weaker than the seminormed topology. In particular $Q \cap E_A$ is a neighbourhood of 0 in E_A, that is, $Q \cap E_A \supset cA$ for some $c > 0$, and so Q absorbs A. ▌

7.4.2 **Corollary.** Let E be any sequentially complete TVS.

(a) If $Q \subset E$ is sequentially closed, convex, balanced, and absorbent, then Q absorbs each bounded, convex, balanced subset of E.

(b) Any barrel in E absorbs each bounded, convex, balanced subset of E.

(c) If E is infrabarrelled, it is barrelled.

(d) If E is bornological, it is barrelled.

Proof. (a) Suppose that $A \subset E$ is bounded, convex, and balanced. Then $\bar{A}$ is bounded, convex, balanced, and sequentially complete. By Proposition 7.4.1, Q absorbs $\bar{A}$, hence absorbs A.

(b) This is a trivial consequence of (a).

(c) If E is infrabarrelled, it is locally convex. It follows then from (b) that each barrel in E absorbs each bounded set and so, E being infrabarrelled, is a neighbourhood of 0. Thus E is in fact barrelled.

(d) This is a consequence of (c) because in any case a bornological space is infrabarrelled. ▌

From this corollary we may derive two more boundedness principles.

7.4.3 **Theorem.** Let E and F be LCTVSs, E being sequentially complete and infrabarrelled, $\mathscr{F}$ a set of continuous linear maps of E into F which is bounded at each point of E. Then $\mathscr{F}$ is equicontinuous.

Proof. In view of Corollary 7.4.2(c), this is a special case of Theorem 7.1.3. ▌

Remark. Compared with Theorem 7.3.1(2), we see that the additional hypothesis of sequential completeness of E allows one to weaken the hypothesis on $\mathscr{F}$ from strong boundedness to pointwise boundedness.

For sequentially complete spaces E we have a weaker principle.

7.4.4 **Theorem.** Let E be a sequentially complete LCTVS, F a LCTVS, $\mathscr{F}$ a set of continuous linear maps of E into F. If $\mathscr{F}$ is bounded at each point, then it is strongly bounded; that is, the set $\mathscr{F}(B) = \{u(x) : u \in \mathscr{F}, x \in B\}$ is bounded in F whenever B is a bounded subset of E. In particular, each weakly bounded subset of E' is strongly bounded.

Proof. Let V be any closed, convex, and balanced neighbourhood of 0 in F and consider $\mathscr{Q} = \bigcap\{u^{-1}(V) : u \in \mathscr{F}\}$. It suffices to show that $\mathscr{Q}$ absorbs any preassigned bounded subset B of E. Now, $\mathscr{F}$ being bounded at each point, $\mathscr{Q}$ is absorbent. Each u in $\mathscr{F}$ being continuous, $\mathscr{Q}$ is closed. It is convex and balanced since V has those properties. $\mathscr{Q}$ is thus a barrel in E. By Corollary 7.4.2(b), $\mathscr{Q}$ absorbs each bounded subset of E, which implies the stated conclusion. ▌

Remark. Strong boundedness of a subset $\mathscr{F}$ of $L_c(E, F)$, as defined above, is tantamount to boundedness of $\mathscr{F}$ relative to the strong topology on $L_c(E, F)$. The reader will recall that this topology has, by definition (see Subsection 1.10.8), a base at 0 formed of the sets

$$W(B, V) = \{u \in L_c(E, F) : u(B) \subset V\},$$

B ranging over the bounded subsets of E and V over any base at 0 in F. When $F = K$ (the scalar field), $L_c(E, F) = E'$, and this strong topology will be encountered frequently in Chapter 8, where it is denoted by $\beta(E', E)$.

7.5 Boundedness Principles for Complete Semimetrizable Spaces

We shall now discuss some refinements of the OGBK principle for Fréchet spaces (Corollary 7.1.2) that stem from the results of Section 6.5.

7.5.1 **Theorem.** Let F be a complete semimetrizable TVS, A a convex bounded, closed, balanced subset of F. For each n let f_n be a function on F with the following properties:

$$0 \leq f_n \leq +\infty;$$

$$f_n(x + y) \leq |\lambda|\, f_n(x) + f_n(y), \quad f_n(\lambda x) \leq |\lambda|\, f_n(x)$$

whenever $\lambda \in K$, $|\lambda| \leq 1$, $x, y \in F$, and $f_n(x)$, $f_n(y)$ are finite; f_n is lower semicontinuous.
Then
either
(a) for some n, f_n is finite and continuous on F,
or
(b) $f_n(x) = +\infty$ for all n and all x belonging to a set $S \subset F$ such that $F\backslash S$ is meagre in F.† Similarly
either
(a′) for some n, $f_n(A)$ is bounded,
or
(b′) for some $x \in A$, $f_n(x) = +\infty$ for all n.

Proof. The first pair will come from Theorem 6.5.1, and the second from Theorem 6.5.3. We illustrate by taking the first pair.

Let F_n consist of those $x \in F$ for which $f_n(x) < +\infty$. F_n is a vector subspace of F, and we topologize it by means of the metric $d(x, y) + f_n(x - y)$, where d is a metric defining the topology of F. The injection map $u_n : F_n \to F$ is thereby rendered continuous. Since f_n is lower semicontinuous, F_n is complete. This puts us in a position to invoke Theorem 6.5.1, taking therein $E = F$ and u the identity map. It then appears that if (b) is false, there exists n such that $F = u_n(F_n) = F_n$ and u_n is a topological isomorphism of F with F_n, which entails (a).

The proof of the disjunction "(a′) or (b′)" proceeds similarly. ▌

It is perhaps worthwhile to state Theorem 7.5.1 in "geometrical" form.

7.5.2 **Theorem.** Let F and A be as in Theorem 7.5.1. Let T_n ($n = 1, 2, \cdots$) be bounded, closed, convex, and balanced subsets of F. Then:
(1) if $\bigcup_{n=1}^{\infty} T_n$ is absorbent, some T_n is a neighbourhood of 0 in F;
(2) if $\bigcup_{n=1}^{\infty} T_n$ absorbs A, some T_n absorbs A.

Proof. Let

$$f_n(x) = \operatorname{Inf}\left\{r > 0 : \frac{x}{r} \in T_n\right\},$$

† In particular, S itself is nonmeagre in F.

interpreted as $+\infty$ if T_n does not absorb x. These f_n satisfy the conditions of Theorem 7.5.1, and T_n is defined by $f_n(x) \leq 1$. If (1) holds, then (b) is false and so (a) is true, and therefore T_n is a neighbourhood of 0. Likewise if (2) holds, then (b′) is false and (a′) is true, and so T_n absorbs A. ▌

7.6 An Application of Theorem 7.5.1 to Summation of Fourier Series

Suppose that for each pair (k, j) of positive integers we have defined a sequence $a_{kj}(n)$ $(n = 0, \pm 1, \cdots)$ of numbers. We consider Fourier series of periodic functions and use the notations introduced in Example 7.2.3. For any $x \in \mathscr{L}^1$ we write

$$f_{kj}(x) = \left| \sum_n a_{kj}(n)\hat{x}(n) \right|, \tag{7.6.1}$$

assuming that

$$\sum_n |a_{kj}(n)| < +\infty \tag{7.6.2}$$

for each pair (k, j); and further

$$f_j(x) = \operatorname*{Sup}_{k \geq 1} |f_{kj}(x)|. \tag{7.6.3}$$

One possibility is that $a_{kj}(n)$ be of the form $s_k(n) \exp(it_j n)$, where the $s_k(n)$ are "summation factors" and the t_j are points of the period interval. In this case, (7.6.1) would be the result of summing the Fourier series of x at the point t_j by the given summation process. Moreover, it will usually be the case that $\lim_k s_k(n) = 1$ for each n, and so that in this case

$$\lim_k |a_{kj}(n)| = 1 \tag{7.6.4}$$

for each j and each n.

Our aim is to prove the following result:

▶ Suppose that (7.6.1), (7.6.2), and (7.6.3) are true. The principal hypothesis is that for each function x belonging simultaneously to each space $\mathscr{L}^p$ $(1 \leq p < +\infty)$, there exists j (possibly depending on x) for which

$$f_j(x) < +\infty. \tag{7.6.5}$$

The conclusion is that there exists a fixed j and a fixed p $(1 \leq p < +\infty)$ and a function $y \in \mathscr{L}^{p'}$ $(1/p + 1/p' = 1)$ with the property that for each n, $\hat{y}(n)$ is a limiting point of the sequence $(a_{kj}(n))_{k=1,2,\cdots}$.

To see this, we observe that the functions x described in the principal hypothesis form a complete semimetrizable TVS with a topology defined by the seminorms N_p (p running through any sequence tending to $+\infty$, say $p = 1$.

$2, \cdots$). It is clear that each f_j satisfies the demands made in Theorem 7.5.1. Since our main hypothesis rules out (b) of that theorem, (a) must be true. This amounts to saying that there exists j and p such that

$$f_j(x) \leq \text{const. } N_p(x),$$

that is, that

$$f_{kj}(x) \leq \text{const. } N_p(x),$$

the constant being independent of k. By the converse of Hölder's inequality, if we put

$$y_k(t) = \sum_n a_{kj}(n)e^{int}$$

(j being fixed), it follows that $N_{p'}(y_k) \leq$ const. Since $p' > 1$, the sequence (y_k) has a weak limiting point $y \in \mathscr{L}^{p'}$, and from this our conclusion follows.

If (7.6.4) holds, we derive in this way the existence of functions x in every $\mathscr{L}^p$ $(1 \leq p < +\infty)$ simultaneously whose Fourier series are nonsummable at all points of any preassigned countable set.

The arguments extend with only verbal changes to the case of Fourier series on any compact Abelian group, and even to some extent to non-Abelian compact groups. Extensions are also possible to expansions in terms of other orthogonal systems of functions.

7.7 Boundedness Principles for Bilinear Maps

We now turn our attention away from linear maps of one TVS into another and towards bilinear maps of a product $E \times F$ of two TVSs into a third TVS G. It is occasionally useful to have boundedness principles for such bilinear maps, or for families of such maps.

If f is a bilinear map of $E \times F$ into G, and if x and y are, respectively, elements of E and of F, we denote by $f_{x\cdot}$ that element of $L(F, G)$ defined by $y \to f(x, y)$ and by $f_{\cdot y}$ that element of $L(E, G)$ defined by $x \to f(x, y)$. It is important to distinguish between continuity of f (understood by regarding f as a function on $E \times F$—that is, as a function of the pair (x, y)) and what we shall term *separate continuity* (continuity in each variable separately, the other being held fixed). This last means, in other words, that $f_{x\cdot}$ belongs to $L_c(F, G)$ for each x in E and $f_{\cdot y}$ belongs to $L_c(E, G)$ for each y in F. Obviously, continuity implies separate continuity.

There is a like distinction between equicontinuity and separate equicontinuity as applied to sets $\mathscr{F}$ of bilinear maps. Our boundedness principles often depend upon relations between these two concepts.

We preface our main results with two illustrative observations, one making a positive and the other a negative assertion about the relationship between equicontinuity and separate equicontinuity.

THE POSITIVE ASSERTION. Although a special case of more general results (specifically Theorems 7.7.3 and 7.7.8), the positive assertion is worth handling separately inasmuch as its special hypotheses allow one to deduce facts about bilinear maps from related ones about linear maps. Besides this, the special assumptions we make—seminormability of the spaces involved—corresponds to historical order; the fact that they are not essential need not concern us here.

Suppose then that E, F, and G are topologized by means of seminorms p, q, and r, respectively; that E is complete (and therefore barrelled); and that $\mathscr{F}$ is a family of bilinear maps of $E \times F$ into G. Our assertion is that *equicontinuity of $\mathscr{F}$ results from its assumed separate equicontinuity.* In fact, we need only assume that (1) for each y in F, $f_{\cdot y}$ is continuous on E, and (2) for each x in E, the set $\{f_{x\cdot} : f \in \mathscr{F}\}$ is equicontinuous on F.

Proof. Hypothesis (2) shows that

$$N(x) = \text{Sup}\,\{r[f(x, y)] : f \in \mathscr{F}, y \in F, q(y) \leq 1\}$$

is finite. N is therefore a seminorm on E. By (1), N is lower semicontinuous on E. Since E is barrelled, N is continuous (Theorem 7.1.1(1)). This signifies precisely that $\mathscr{F}$ is equicontinuous on $E \times F$.

As a corollary, *if a bilinear map f is separately continuous, then it is continuous* (E, F, and G being as above).

THE NEGATIVE ASSERTION. Our negative assertion is intended to lay to rest any hope that the closing sentence of the above discussion may be unrestrictedly true. This is done by exhibiting a case in which E is barrelled, F is a Fréchet space, G is the scalar field, and in which the bilinear form f is separately continuous (and even E-hypocontinuous, in the sense to be defined below) and yet not continuous. We shall learn in Theorem 7.7.9 that the cause of failure is the nonsemimetrizability of E.

For our example we take $E = \mathscr{K}_R(N)$ with its inductive limit topology (described in Example (1) of Section 6.3, T being taken equal to N, the set of natural numbers; the subscript R indicates that we take only real-valued functions on N, though this restriction is without significance at this point). For F we take R^N with its product topology. F is a Fréchet space. G is taken to be R, the field of real scalars. f is defined by

$$f(x, y) = \sum_{t \in N} x(t)y(t).$$

Now f is E-hypocontinuous, that is, $f(x, y) \to 0$ as $y \to 0$, uniformly when x ranges over any bounded subset of E; a fortiori, $f(x, y)$ is continuous in y for each fixed x. For (see Section 6.6(2) the boundedness of A in E entails that there exist a natural number n and a number c such that $x(t) = 0$ for $t > n$ and $|x(t)| < c$ for all t, and this for all x in A. So if $p > 0$ is given, $|f(x, y)| < p$ whenever x belongs to A and $\sum_{t \leq n} |y(t)| < c^{-1}p$, and so certainly whenever x belongs to A and $|y(t)| < p/nc$ for $t \leq n$. The set V of y in F satisfying these last inequalities is a neighbourhod of 0 in F, and we have $|f(x, y)| < p$ for x in A and y in V, thus establishing E-hypocontinuity of f.

It remains to show that f is not continuous on $E \times F$. But if it were, there would exist a neighbourhood U of 0 in E, a finite set $H \subset N$, and a number $p > 0$, such that

$$\left| \sum_{t \in N} x(t) y(t) \right| < 1$$

whenever x belongs to U and $y(t)| < p$ for t in H. But then

$$\left| \sum_{t \in N} x(t) y(t) \right| < p^{-1} \operatorname{Sup}_{t \in H} |y(t)|$$

for all x in U and all y in R^N. This would entail that every x in U vanishes outside H which, since H is finite, is impossible (U being a neighbourhood of 0 in E). Thus f is not continuous.

7.7.1 **Hypocontinuity.** The negative assertion above indicates that in general—and even in the presence of restrictions on E and F—separate continuity does not entail continuity, while indicating at the same time the form of an intermediary property (hypocontinuity) that *is* derivable. We now formulate this property in precise terms.

We consider first a single bilinear map f of $E \times F$ into G. It is easily verified that the following three conditions are equivalent:

(a) For each neighbourhood W of 0 in G and each bounded set A in E, there exists a neighbourhood V of 0 in F such that $f(A \times V) \subset W$.

(b) For each bounded set A in E, the image of A by the map $x \to f_{x\cdot}$ is an equicontinuous subset of $L_c(F, G)$.

(c) The map $y \to f_{\cdot y}$ is continuous from F into $L_c(F, G)$, the latter endowed with its strong topology.

If any one, and hence all, of these conditions is fulfilled, we say that f is *E-hypocontinuous*.

A set $\mathscr{F}$ of bilinear maps of $E \times F$ into G is said to be *E-equihypocontinuous* if condition (a) above is fulfilled for each f in $\mathscr{F}$, and if furthermore the choice of V may be made independently of f varying over $\mathscr{F}$.

There are analogous definitions for F-hypocontinuity and F-equihypocontinuity.

Finally, we say that f is *hypocontinuous* (resp. $\mathscr{F}$ is *equihypocontinuous*) if it is both E- and F- hypocontinuous (resp. both E- and F-equihypocontinuous].

7.7.2 **Preliminary Remarks.** In the remarks that follow it suffices to make statements about equihypocontinuity of various types of a set $\mathscr{F}$, the corresponding statements about hypocontinuity (of the appropriate types) of a bilinear map f arising from the special case in which $\mathscr{F}$ is $\{f\}$.

It is quite obvious that E-equihypocontinuity of $\mathscr{F}$ implies the equicontinuity of the set $\{f_{x\cdot} : f \in \mathscr{F}\}$ for each fixed x in E. Consequently, equihypocontinuity of $\mathscr{F}$ implies its separate equicontinuity. On the other hand, and equally evident, is the fact that equicontinuity of $\mathscr{F}$ implies its equihypocontinuity (hence the justification of the qualifying "hypo"). Besides this, if E is seminormable, E-equihypocontinuity is equivalent to equicontinuity.

A little less immediate are the following conclusions.

▶ (1) If $\mathscr{F}$ is E-equihypocontinuous, and if A and B are bounded subsets of E and of F, respectively, then (a) $\mathscr{F}$ is equicontinuous on $A \times F$, and (b) $\bigcup\{f(A \times B) : f \in \mathscr{F}\}$ is a bounded subset of G.

▶ (2) If $\mathscr{F}$ is equihypocontinuous, then it is equiuniformly continuous on $A \times B$, A and B being arbitrary bounded subsets of E and F, respectively.

Proof. We leave (1) as an exercise for the reader, and sketch the proof of (2). Since $A - A$ is bounded in E, given a neighbourhood W of 0 in G, one may choose neighbourhoods U and V of 0 in E and F, respectively, such that $f[(A - A) \times V] \subset W$ and $f(U \times B) \subset W$. Suppose that x and $x + a$ belong to A, y and $y + b$ to B, a to U, and b to V. Then

$$f(x + a, y + b) - f(x, y) = f(x, b) + f(a, y) + f(a, b).$$

The first term on the right belongs to W since (x, b) belongs to $A \times V$ and we may suppose that A contains 0, which arranges that $A \subset A - A$; the second term belongs to W because (a, y) belongs to $U \times B$; and the third term also belongs to W because (a, b) belongs to $U \times B$. Thus $f(x + a, y + b) - f(x, y)$ belongs to $W + W + W$, which is an arbitrarily small neighbourhood of 0 in G. Equiuniform continuity of $\mathscr{F}$ on $A \times B$ is thus established. ▌

Notation. In the statements of our results it will be a help to use the following notation, additional to that already introduced. $\mathscr{F}$ being a set of bilinear maps of $E \times F$ into G, we write $\mathscr{F}_{x\cdot}$ for $\{f_{x\cdot} : f \in \mathscr{F}\}$ whenever x belongs to E, and $\mathscr{F}_{\cdot y}$ for $\{f_{\cdot y} : f \in \mathscr{F}\}$ whenever y belongs to F. Thus $\mathscr{F}_{x\cdot}$ is a subset of $L(F, G)$ and $\mathscr{F}_{\cdot y}$ a subset of $L(E, G)$.

To say that $\mathscr{F}$ is separately continuous—that is, that each member f of $\mathscr{F}$ is separately continuous—amounts to saying that $\mathscr{F}_{x\cdot} \subset L_c(F, G)$ for each x in E and $\mathscr{F}_{\cdot y} \subset L_c(E, G)$ for each y in F. Likewise, to say that $\mathscr{F}$ is separately equicontinuous signifies that $\mathscr{F}_{x\cdot}$ is an equicontinuous subset of $L_c(F, G)$ for each x in E and $\mathscr{F}_{\cdot y}$ is an equicontinuous subset of $L_c(E, G)$ for each y in F.

Before embarking on the details, it is worth commenting on the tactics to be adopted. The search for boundedness principles for bilinear maps is not as direct as is the case for linear maps. The strongest form of the principle we have in mind is that which says "pointwise boundedness implies equicontinuity" (compare Theorem 7.1.3) for linear maps. For bilinear maps it is best to regard this as compounded of partial results, not all of the latter being boundedness principles at all, but rather statements about the relationship between various types of equicontinuity for suitably restricted types of spaces. Theorem 7.7.3,

Proposition 7.7.4, Theorem 7.7.5, and Corollary 7.7.6 are all of the latter type. It is by combining these with boundedness principles for linear maps that one reaches boundedness principles for bilinear maps, of which Theorems 7.7.7, 7.7.8, and 7.7.9 are examples.

7.7.3 **Theorem.** Suppose that F is ultrabarrelled (or barrelled, if G is locally convex) and that (a) $\mathscr{F}_{x\cdot} \subset L_c(F, G)$ for each x in E, and (b) $\mathscr{F}_{\cdot y}$ is an equicontinuous subset of $L_c(E, G)$ for each y in F. Then $\mathscr{F}$ is E-equihypocontinuous.

Proof. Let W be any neighbourhood of 0 in F, A any bounded subset of E. We may assume that W is closed and balanced in any case, and further convex if G is locally convex.

If y is fixed in F, (b) shows that there exists a neighbourhood U_y of 0 in E such that $f(x, y) \in W$ for x in U_y and f in $\mathscr{F}$. Since A is absorbed by U_y, there exists $\lambda > 0$ such that

$$f(A, \lambda y) \subset W \qquad (f \in \mathscr{F}).$$

Consequently, if we define

$$\begin{aligned} V(A, W) &= \{y \in F : f(A, y) \subset W \text{ for } f \in \mathscr{F}\} \\ &= \bigcap\{f_x^{-1}(W) : x \in A\}, \end{aligned}$$

then each set $V(A, W)$ is absorbent in F.

Apart from this, (a) shows that $V(A, W)$ is closed in F. Also, $V(A, W)$ is balanced in any case, and convex if W is convex.

So if G is locally convex it is thus arranged that each $V(A, W)$ is a barrel in F, hence a neighbourhood of 0 in F, and this signifies that $\mathscr{F}$ is E-equihypocontinuous.

Otherwise we observe that the various $V(A, W)$ (with A and W varying) define a base at 0 for a vector space topology t on F, each of these sets being closed in F. Since F is ultrabarrelled, we face again the conclusion that each $V(A, W)$ is a neighbourhood of 0 in F, and so forth. ▌

Remark. If E is seminormable, the conclusion says that $\mathscr{F}$ is equicontinuous, and one has already an extension of the result given in our opening discussion.

7.7.4 **Proposition.** Suppose that E, F, and G are LCTVSs, E being infrabarrelled. If $\mathscr{F}$ is separately continuous and E-equihypocontinuous, then it is equihypocontinuous.

Proof. It remains to be shown that $\mathscr{F}$ is F-equihypocontinuous; that is, that if B is a bounded subset of F and W a neighbourhood of 0 in G, then

$$U = \{x \in E : f(x, B) \in W \text{ for } f \in \mathscr{F}\}$$

is a neighbourhood of 0 in E. Now we may assume that W is a barrel in G, in which case the separate continuity of f ensures that U is a barrel in E. Furthermore, since $\mathscr{F}$ is E-equihypocontinuous, U absorbs the bounded subsets of E. E being infrabarrelled, the desired conclusion follows. ▌

Remarks. The reader should compare this result with Theorem 7.3.1(2) for linear maps. We have previously remarked that the conclusion is valid whenever E is seminormable, and then F and G may be arbitrary TVSs.

For linear maps it is generally the case that the hypothesis of the barrelled or ultrabarrelled nature of the domain space suffices for the strongest sort of conclusion—in this case, equicontinuity. The two preceding results lead only to equihypocontinuity, however; and our example of the negative assertion shows that we cannot do better than this in general. It seems that semimetrizability plays a role in leading to the strongest results for bilinear maps, as is shown by the next two results.

7.7.5 **Theorem.** Let E be a semimetrizable TVS, G any TVS, T a topological space satisfying the first countability axiom. Suppose too that *either* (a) E is ultrabarrelled, *or* (b) E is barrelled and G is locally convex. Let $\mathscr{F}$ be a set of maps of $E \times T$ into G such that $f_{.t} : x \to f(x, t)$ is linear for each t in T and each f in $\mathscr{F}$. If $\mathscr{F}$ is separately equicontinuous, then it is equicontinuous.

Proof. This is deducible from Theorem 7.1.5. It is clear that conditions (1) and (2) of that theorem are fulfilled, and so it remains only to verify that $\mathscr{F}$ is pointwise bounded.

To this end let (x_0, t_0) be any point of $E \times T$. By equicontinuity in the variable x, given any neighbourhood W of 0 in G, there exists a neighbourhood U of 0 in E such that

$$f(x, t_0) \in W \qquad \text{for } x \in U \text{ and } f \in \mathscr{F}.$$

There exists a number λ such that $x_0 \in \lambda U$, and so

$$f(x_0, t_0) \in \lambda W \qquad \text{for } f \in \mathscr{F}.$$

This shows that the set $\{f(x_0, t_0) : f \in \mathscr{F}\}$ is bounded in G, as required. ▌

7.7.6 **Corollary.** Let E and F be semimetrizable TVSs, G any TVS. Suppose that *either* (a) E is ultrabarrelled, *or* (b) E is barrelled and G is locally convex. If a set $\mathscr{F}$ of bilinear maps of $E \times F$ into G is separately equicontinuous, then it is equicontinuous.

Proof. The proof is immediate from Theorem 7.7.5. ▌

Remark. If we assume that F is seminormable, we can drop the assumption of semimetrizability of E; the conclusion then follows from Theorem 7.7.3 upon interchanging E and F therein.

At this point we can use the above results to derive boundedness principles for sets of bilinear maps.

7.7.7 **Theorem.** Suppose that *either* (a) E and F are ultrabarrelled, *or* (b) E and F are barrelled and G is locally convex. If $\mathscr{F}$ is separately continuous and pointwise bounded, then it is E-equihypocontinuous.

Proof. We appeal to Theorem 7.7.3, to do which we have only to verify that $\mathscr{F}_{.y}$ is an equicontinuous subset of $L_c(E, G)$ for each y in F. But this follows from pointwise boundedness by virtue of Theorem 7.1.3. ▌

7.7.8 Theorem. Suppose that E and F are barrelled, and that G is locally convex. If $\mathscr{F}$ is separately continuous, and pointwise bounded, then it is equihypocontinuous.

Proof. This amounts to combining Theorem 7.7.7 and Proposition 7.7.4. ▌

By adding semimetrizability as a hypothesis we can obtain a "strong" form of the boundedness principle, as follows.

7.7.9 Theorem. Suppose that each of E and F is semimetrizable and ultrabarrelled (or barrelled, if G is locally convex). If $\mathscr{F}$ is separately continuous and pointwise bounded, then it is equicontinuous.

Proof. We combine Theorem 7.1.3., which leads from pointwise boundedness to separate equicontinuity, with Corollary 7.7.6. ▌

Remark. The conclusion remains valid (with the same hypotheses on $\mathscr{F}$) if E is ultrabarrelled (or barrelled, if G is locally convex) and F is complete and seminormable. This is seen by reference to Theorem 7.7.3 once again and interchanging E and F therein.

7.7.10 Limits of Convergent Sequences of Bilinear Maps. Just as for linear maps (see, for example, Corollary 7.1.4) boundedness principles help to establish the continuity in various senses of the limit of a pointwise convergent sequence (or of a pointwise bounded and convergent directed family) of bilinear maps. The sequential case is simpler only insofar as pointwise convergence then entails pointwise boundedness. Nevertheless we confine our remarks to this case.

Suppose then that (f_n) is a pointwise convergent sequence of separately continuous bilinear maps of $E \times F$ into G. The limit f is then plainly bilinear from $E \times F$ into G. Each of the results numbered 7.7.6 through 7.7.9 leads at once to statements about continuity properties of the limit f, granted suitable hypotheses on E and F. We do not propose to list all the corresponding conclusions. Suffice it to remark that, for example, f will be continuous in the pair of variables whenever the hypotheses of Theorem 7.7.9 are fulfilled; while, under the hypotheses of Theorem 7.7.7, f will be E-hypocontinuous.

7.7.11 The Case of Inductive Limit Spaces. There is an analogue for bilinear maps of Theorem 6.3.2.

Suppose that E (resp. F) is the inductive limit of spaces E_i (resp. F_j) relative to maps φ_i (resp. ψ_j), that G is any LCTVS, and that f is a bilinear map of $E \times F$ into G. It is then the case that f is continuous if and only if, for each pair (i, j) of indices, the map

$$f \circ (\varphi_i \times \psi_j) : (x, y) \to f(\varphi_i(x), \psi_j(y))$$

is continuous from $E_i \times F_j$ into G.

The necessity of this condition is obvious, since each φ_i and each ψ_j is continuous. As for sufficiency, we take any convex neighbourhood W of 0 in

G. For each pair (i, j) we may by hypothesis choose neighbourhoods U_i and V_j of 0 in E_i and F_j, respectively, such that

$$f(\varphi_i(x), \psi_j(y)) \in W \qquad \text{if} \quad x \in U_i \quad \text{and} \quad y \in V_j.$$

So $f(x, y) \in W$ if $x \in \bigcup_i \varphi_i(U_i)$ and $y \in \bigcup_j \psi_j(V_j)$, and hence also if x is in U and y in V, where U and V are, respectively, the convex envelopes of $\bigcup_i \varphi_i(U_i)$ and $\bigcup_j \psi_j(V_j)$. Since $\varphi_i^{-1}(U)$ contains U_i, U is a neighbourhood of 0 in E. Likewise V is a neighbourhood of 0 in F. Hence the continuity of f.

There is, of course, an analogous statement about the equicontinuity of a set of bilinear maps on $E \times F$.

7.7.12 **A Reformulation of Theorem 7.7.9.** There is a striking formulation of Theorem 7.7.9 in terms of linear maps. Suppose that E and F are semi-metrizable and barrelled, and that u is a linear map of E into F' that is continuous for the initial topology on E and the weak topology $\sigma(F', F)$. We claim that there exists a neighbourhood U of 0 in E such that $u(U)$ is equicontinuous in F'.

Indeed, consider the bilinear form f on $E \times F$ defined by

$$f(x, y) = \langle y, u(x) \rangle \qquad (x \in E, y \in F).$$

Our hypotheses on u ensure that f is separately continuous. By Theorem 7.7.9, f is continuous. Thus there exist neighbourhoods U and V of 0 in E and F, respectively, such that $|f(U \times V)| \leq 1$, which signifies exactly that $u(U) \subset V^0$, that is, that $u(U)$ is equicontinuous.

The argument is evidently reversible, so that the above assertion concerning linear maps of E into F' is in fact equivalent to Theorem 7.7.9.

7.8 Some Applications

7.8.1 **Example. The Summation of Double Series.** Suppose we are concerned with the summation of (scalar-valued) double series $\sum z(m, n)$, m and n running separately over the set N of natural numbers. The use of summation factors will be admitted; that is, we take a sequence (f_k) of functions on $N \times N$ satisfying the following conditions:

$$f_k \geq 0, \qquad \sum f_k(m, n) < +\infty, \qquad \lim_k f_k(m, n) = 1 \tag{7.8.1}$$

and use as generalized sum of $\sum z(m, n)$ the limit

$$\lim_k \sum f_k(m, n) z(m, n),$$

whenever this exists finitely. There then arises the question of consistency: For suitably restricted double sequences z, this generalized sum should exist and agree with the usual one. We deal here with one variant of this problem (cf. Bourbaki [5], p. 46, Exercice 10)).

Let E denote the space of sequences x on N such that $\sum x(m)$ is convergent to a (finite) sum $s(x)$. E is a Banach space when equipped with the norm

$$\|x\| = \operatorname*{Sup}_{r \in N} \left| \sum_{m \leq r} x(m) \right|.$$

E contains the space ℓ^1 as a dense vector subspace.

The question we consider is this: Under what additional conditions on (f_k) is it true that

$$\lim_k \sum f_k(m, n)x(m)y(n)$$

exists and is equal to $s(x)s(y)$ for each x in E and each y in ℓ^1? If we write

$$u_k(x, y) = \sum f_k(m, n)x(m)y(n),$$
$$u(x, y) = s(x)s(y)$$

for x in E and y in ℓ^1, our question asks under what conditions does u_k converge to u at each point of $E \times \ell^1$. It is clear that each u_k and u is a bilinear form on $E \times \ell^1$. It is also very easy to show, using (7.8.1), that $\lim u_k = u$ on $\ell^1 \times \ell^1$ (in doing which it is sufficient and convenient to assume that $x \geq 0$ and $y \geq 0$).

Our aim is to show that

▶ $\lim_k u_k(x, y)$ exists finitely for each (x, y) in $E \times \ell^1$ if and only if

$$\operatorname*{Sup}_{k,n} \sum_m |f_k(m, n) - f_k(m + 1, n)| < +\infty; \tag{7.8.2}$$

and then

$$\lim_k u_k = u \text{ on } E \times \ell^1.$$

The proof amounts to several applications of Theorem 7.7.9 and the associated remarks in Subsection 7.7.10.

The reader will observe first that convergence in either space E or ℓ^1 implies pointwise convergence on N. Hence it follows that each finite sum

$$u_{k,p}(x, y) = \sum_{m<p, n<p} f_k(m, n)x(m)y(n)$$

is separately continuous (and even continuous) on $E \times \ell^1$. Moreover,

$$u_k(x, y) = \lim_{p \to \infty} u_{k,p}(x, y)$$

for each k and each (x, y) in $E \times \ell^1$.

A first appeal to Theorem 7.7.9 shows that if $u = \lim_k u_k$ exists pointwise on $E \times \ell^1$, then the u_k $(k = 1, 2, \cdots)$ are equicontinuous. On the other hand, since we know that the u_k converge pointwise on $\ell^1 \times \ell^1$ to u, their equicontinuity (together with the obvious continuity of u) arranges that they converge pointwise on $E \times \ell^1$ to u.

Thus $u = \lim_k u_k$ pointwise on $E \times \ell^1$ if and only if the u_k are equicontinuous. It thus remains to show that the u_k are equicontinuous if and only if (7.8.2) is valid.

Now, since E and ℓ^1 are normed spaces, the u_k are equicontinuous if and only if their norms, namely, the numbers

$$c_k = \text{Sup}\,\{|u_k(x, y)| : x \in E,\ \|x\| \leq 1, y \in \ell^1,\ \|y\|_1 \leq 1\},$$

are bounded with respect to k. We need to show, therefore, that

$$c_k = \text{Sup}_n \sum_m |f_k(m, n) - f_k(m + 1, n)|. \tag{7.8.3}$$

To do this we make use of the map $x \to x'$ defined by

$$x'(m) = \sum_{r=1}^{m} x(r).$$

This is a linear isometry of E onto the subspace c of ℓ^∞ formed of convergent sequences, the inverse map being defined by $x(m) = x'(m) - x'(m - 1)$, provided we make the convention that $x'(0) = 0$. It appears therefore that c_k is the supremum of

$$\sum_{m,n} f_k(m, n)[x'(m) - x'(m - 1)]y(n)$$

as x' ranges over the unit ball of c and y over the unit ball of ℓ^1. We express the double sum as an iterated one (justified since the series is absolutely convergent),

$$\sum_n y(n) \sum_m f_k(m, n)[x'(m) - x'(m - 1)]$$
$$= \sum_n y(n) \sum_m [f_k(m, n) - f_k(m + 1, n)]x'(m).$$

Now the supremum with respect to y of the modulus of this expression is

$$\text{Sup}_n \left| \sum_m [\cdots]x'(m) \right|,$$

and the supremum of this with respect to x' is

$$\text{Sup}_n \sum_m |f_k(m, n) - f_k(m + 1, n)|,$$

which is the right-hand member of (7.8.3). With this the proof is complete.

7.8.2 **Application to the Study of Bounded Sets of Distributions.** As a second application of Theorem 7.7.9 we shall fill a gap left in Subsection 5.11.2 by proving the theorem immediately below. Actually, this theorem asserts a good deal more than is required in Subsection 5.11.2—namely, merely the boundedness in $\mathscr{D}'$ of $\Sigma * f * g$ for each pair $(f, g) \in \mathscr{D}^m(\Omega) \times \mathscr{D}^m(\Omega)$. Apart from explicit reference to Theorem 7.7.9, the proof is the same as that given by Schwartz ([2], Ch. VI).

Theorem Let Σ be a subset of $\mathscr{D}'$ having the property that, for each f in $\mathscr{D}$, the set

$$\Sigma * f = \{X * f : X \in \Sigma\}$$

is bounded in $\mathscr{D}'$. Then, given any compact set K in R^n and any relatively compact open set Ω in R^n, there exist an integer $m \geq 0$ and a number $c \geq 0$ such that

$$\text{Sup}\,\{|X * f * g(x)| : X \in \Sigma, x \in K\} \leq c \cdot N_m(f)N_m(g) \tag{7.8.4}$$

for f and g in $\mathscr{D}^m(\Omega)$, where

$$N_m(f) = \text{Sup}\,\{|D^p f(x)| : x \in R^n, |p| \leq m\}.$$

Proof. For each X we denote by F_X the bilinear map of $\mathscr{D}_H \times \mathscr{D}_H$ into $C(K)$ defined by $F_X(f, g) = X * f * g$, where $\mathscr{D}_H$ is used as an abbreviation for $\mathscr{D}(R^n, H)$, H is a compact neighbourhood of $\bar{\Omega}$, and $C(K)$ is the usual Banach space of continuous functions on K with norm equal to the maximum modulus on K. It is important to recall that $\mathscr{D}_H$ is a Fréchet space (see Example (b) in Subsection 6.1.3).

We first verify that our hypothesis—the boundedness in $\mathscr{D}'$ of $\Sigma * f$ for each f in $\mathscr{D}$—arranges that the F_X are separately equicontinuous. Thus we fix f in $\mathscr{D}_H$ and consider the linear maps u_Y of $\mathscr{D}_H$ into $C(K)$ defined by

$$u_Y(g) = (Y * g) \mid K,$$

where $Y = X * f$ and X ranges over Σ. Y therefore remains bounded in $\mathscr{D}'$. Now each u_Y is plainly continuous. Moreover, the u_Y are pointwise bounded because (a) if g is given in $\mathscr{D}_H$, the functions $y \to g(x - y)$ form a bounded subset of $\mathscr{D}$ when x ranges over K, and (b) Y is bounded in $\mathscr{D}'$ by hypothesis. From Theorem 7.1.3 it appears that the u_Y are equicontinuous on $\mathscr{D}_H$. Since $F_X(f, g)$ is symmetric in f and g, their separate equicontinuity follows.

Theorem 7.7.9 is now applicable to the F_X, showing that they are equicontinuous on $\mathscr{D}_H \times \mathscr{D}_H$. This signifies that m and c exist for which (7.8.4) is valid whenever f and g belong to $\mathscr{D}_H$. It remains to see that f and g may in fact be allowed to range separately over $\mathscr{D}^m(\Omega)$.

Suppose that f and g are given in $\mathscr{D}^m(\Omega)$. By regularization we may approximate f and g in the sense of $\mathscr{D}_H^m$ by sequences $f_k = f * r_k$ and $g_k = g * r_k$ lying within $\mathscr{D}_H$. We then have by (7.8.4)

$$|X * f_h * g_k(x)| \leq c \cdot N_m(f_h)N_m(g_k)$$

for X in Σ, x in K, and $h, k = 1, 2, \cdots$. Hold h fixed and let $k \to \infty$. Then $N_m(g_k) \to N_m(g)$ and, since $X * f_h$ belongs to C^∞, $X * f_h * g_k(x) \to X * f_h * g(x)$ for each x. It follows therefore that

$$|X * f_h * g(x)| \leq c \cdot N_m(f_h)N_m(g) \tag{7.8.5}$$

for X in Σ, x in K, and $h = 1, 2, \cdots$. The argument shows indeed that herein we may replace f_h by *any* member of $\mathscr{D}_H$, g being any member of $\mathscr{D}^m(\Omega)$.

Supposing (as we may) that Ω is symmetric about 0, and writing temporarily Z for $X * g$, the replacement in (7.8.5) of f_h by $\check{\varphi}$, where φ belongs to $\mathscr{D}(\Omega)$, leads to

$$|\langle \varphi(y - x), Z_{(y)} \rangle| \leq c' \cdot N_m(\varphi).$$

This reads

$$|\langle \varphi, t_x Z \rangle| \leq c' \cdot N_m(\varphi),$$

t_x denoting translation by amount x. Thus $t_x Z$ is of order at most m on Ω, and this for each x in K. But then it follows that in (7.8.5) we may allow $h \to \infty$, the left-hand member converging to $|X * f * g(f)|$. As a consequence we are led to (7.8.4) once more, and the proof is complete.

EXERCISES

7.1 f is a function on R (the real line) such that fg is locally integrable (for Lebesgue measure) and such that

$$\lim_n I_n(g) = \lim_n \int_{-n}^{n} fg \, d\mu$$

exists finitely for each $g \in C_0(R)$. Prove that f is integrable.

7.2 Let T be a locally compact space, μ a positive Radon measure on T. Suppose that g is a function on T such that fg is integrable for μ for each $f \in \mathscr{L}^p(\mu)$, p being given $1 \leq p \leq \infty$. By using the fact that $\mathscr{L}^p(\mu)$ is barrelled, show that $g \in \overline{\mathscr{L}}^{p'}(\mu)$, where $1/p + 1/p' = 1$. (Compare with Exercise 6.1.)

7.3 Suppose that for each natural number n one has chosen families

$$(t_i^{(n)})_{1 \leq i \leq i_n} \quad \text{and} \quad (c_i^{(n)})_{1 \leq i \leq i_n}$$

of points of $[0, 1]$ and of real numbers, respectively, in such a way that

$$\int_0^1 x(t)\, dt = \sum_{1 \leq i \leq i_n} c_i^{(n)} x(t_i^{(n)}) = S_n(x) \tag{1}$$

for all polynomials x of degree at most n, while $\lim_n S_n(x)$ exists finitely for each continuous x on $[0, 1]$. Show that

$$\operatorname*{Sup}_n \sum_{1 \leq i \leq i_n} |c_i^{(n)}| < +\infty$$

and

$$\lim_n S_n(x) = \int_0^1 x(t)\, dt \tag{2}$$

for each x continuous on $[0, 1]$.

Show also that if (1) holds for all polynomials x of degree at most n, and if all the $c_i^{(n)}$ are positive, then (2) holds.

7.4 Use the notations of Example 7.2.4(4). Suppose that

(a) E is barrelled and the coordinate functionals are continuous on E;
(b) The coordinate functionals are continuous on F;
(c) The closed graph theorem is valid for linear maps of E into F.

Show that if $u_A(E) \subset F$, then u_A is continuous from E into F.

7.5 Suppose that E is barrelled, that F is any TVS such that the closed graph theorem is valid for linear maps of E into F, that L is a total subset of F', and that (u_i) is a directed family of elements of $L_c(E, F)$ such that for each $x \in E$ the directed family $(u_i(x))$ is bounded and convergent for the topology $\sigma(F, L)$. Define

$$u(x) = \sigma(F, L)\text{-}\lim_i u_i(x).$$

Show that the u_i are equicontinuous, and that u is continuous, from E into F.

If (u_i) is furthermore a denumerable sequence, it is enough to assume that $(u_i(x))$ is $\sigma(F, L)$-convergent for each x in E (the other hypotheses remaining as before).

7.6 Suppose that $1 \leq p < \infty$, $1 \leq q \leq \infty$, that Ω is an open subset of R^n, and that (f_n) is a sequence of elements of $L^q(\Omega)$. Assume that for each $x \in \ell^p$ the series

$$\sum x(n) f_n$$

is convergent in the space $\mathscr{D}'(\Omega)$ of distributions on Ω to a limit that belongs to $L^q(\Omega)$. Show that this series is then convergent in $L^q(\Omega)$ for each $x \in \ell^p$.

[Hint: Apply Exercise 7.5 and notice that each $x \in \ell^p$ is the limit in that space of its finite sections.]

7.7 Let $E = L^1(-\pi, \pi)$, formed relative to Lebesgue measure, and e_n the class of the function e^{int} for $n = 0, \pm 1, \pm 2, \cdots$. For $x \in E$ define

$$\hat{x}(n) = (2\pi)^{-1} \int_{-\pi}^{\pi} x(t) e^{-int}\, dt,$$

the nth Fourier coefficient of x. If P is a set of integers, denote by $E[P]$ the set of $x \in E$ such that $\hat{x}(n) = 0$ for $n \notin P$; and let Q denote the complementary set of integers.

Show that E is the direct sum of $E[P]$ and $E[Q]$ if and only if the characteristic function of P is the Fourier transform of some measure in $E[P]$.

7.8 Use the notation of Exercise 7.7. We say that P is a Szidon set (Zygmund [1], p. 139; Section 8.8 below) if every bounded measurable function on $[-\pi, \pi]$ is equal almost everywhere on $[-\pi, \pi]$ to a continuous function.

Show that if P is an infinite Szidon set, then $E[Q]$ admits no topological complement in E.

7.9 With the notation of Exercise 7.7, show that the e_n $(n \in P)$ form a weak basis in $E[P]$ if and only if

$$\sum \hat{x}(n) \hat{g}(n) \text{ converges}$$

for each $x \in E[P]$ and each bounded and measurable function g on $[-\pi, \pi]$.

7.10 Let T be a separated locally compact space and μ a positive Radon measure on T. Suppose given a sequence K_r $(r = 1, 2, \cdots)$ of functions on T, each belonging to $\mathscr{L}^s = \mathscr{L}^s(T, \mu)$ for some $s = s_r > 1$. Assume that

$$\lim_r \int K_r f\, d\mu = \int f\, d\lambda$$

for each f in a total subset of $\mathscr{K}(T)$, where λ is a Radon measure on T *not* of the type $\lambda = h \cdot \mu$ with $h \in \mathscr{L}^q$ and $q > 1$. Show that there exists a function

$$g \in \bigcap \{\mathscr{L}^p(T, \mu) : 1 \leq p < \infty\}$$

such that

$$\operatorname{Sup}_r \left| \int K_r g \, d\mu \right| = +\infty.$$

7.11 Consider the application of Exercise 7.10 to the convergence and summability of ordinary Fourier series.

7.12 Let E be a TVS, A a convex, balanced, sequentially weakly complete, and weakly bounded subset of E'. Prove that A is strongly bounded in E'. [Cf. Kelley [1], pp. 241–242, note at the end of N.]

[Hint: Consider the Banach space E'_A constructed as in Lemma 6.5.2, the space E figuring therein being now taken to be E' with its weak topology.]

7.13 Let Ω be an open set in R^n and (f_i) a family of functions, each locally integrable over Ω, the family being bounded in $\mathscr{D}'(\Omega)$. Show that to each compact set $K \subset \Omega$ corresponds an integer $m \geq 0$ and a constant M such that

$$\operatorname{Sup}_i \left| \int_\Omega \varphi f_i \, d\mu \right| \leq M \cdot \operatorname{Sup} \{|\partial^p \varphi(x)| : x \in \Omega, |p| \leq m\}$$

for each $\varphi \in \mathscr{D}^m(\Omega, K)$ ($=$ the set of $\varphi \in C^m$ with support contained in K).

[Hint: Use the fact that $\mathscr{D}$ is barrelled, together with Theorem 7.1.1 (b).]

7.14 Let X be a distribution on R^n with the property that $X * \varphi \in L^p(R^n)$ for each $\varphi \in \mathscr{D}(R^n)$. Show that there exists an integer $k \geq 0$ and functions $f, g \in L^p$ such that $X = f + \Delta^k g$; if $n = 1$, Δ may here be replaced by ∂. (Schwartz [2], p. 59.)

[Hint: Let B be the intersection of $\mathscr{D}$ with closed unit ball in $L^{p'}$. The $X * \psi$ are bounded in $\mathscr{D}'$ when ψ ranges over B. Use the preceding exercise to deduce that for each compact K, $\langle \psi, X * \varphi \rangle$ is bounded for $\psi \in B$ and each $\varphi \in \mathscr{D}^m$ with support contained in K, so that $X * \varphi \in L^p$ for each such φ. Finally use the formulae in Subsection 5.11.2 to obtain the desired conclusion.]

7.15 Show that the functions $e_n : t \to e^{int}$ do not form a topological base in the space C of continuous functions of a real variable with period 2π, C being topologized (as usual) by the norm $\|x\| = \operatorname{Sup} |x(t)|$.

7.16 Show that the classes of the functions e_n (as in the preceding exercise) do not form a topological base in the space $L^1(-\pi, \pi)$.

7.17 Suppose that $1 \leq p \leq \infty$ and that X is a distribution on R^n. Show that $X * \varphi \in L^p = L^p(R^n)$ for each $\varphi \in \mathscr{D} = \mathscr{D}(R^n)$ if and only if X is a finite linear combination of derivatives of functions in L^p. (Schwartz [2], p. 57, Théorème XXV).

[Hint: For the "only if" part, show that the map $T : \varphi \to X * \varphi$ is continuous from $\mathscr{D}$ into L^p; for this it suffices to consider the restriction of T to $\mathscr{D}(R^n, K)$ (K compact) and use the CGT. Deduce that for each compact K there exists an integer $m \geq 0$ such that $X * \varphi \in L^p$ for $\varphi \in \mathscr{D}^m(R^n, K)$. Then employ the techniques of Subsection 5.11.2.]

7.18 Let X be a distribution on R^n. Show that $X * \varphi$ is a continuous function for each continuous function φ with a compact support if and only if X is a measure (Schwartz [2], p. 48, Théorème XX. The proof suggested below is different from that given by Schwartz.)

[Hints: For the "only if" part, consider the map T: $\varphi \to X * \varphi$ of $\mathscr{D}^0$ into C. Show that T is continuous, and consider the adjoint T', mapping M_c (the space of measures with compact supports) into M (the space of measures). Prove that $T'(\alpha * \beta) = \alpha * T'\beta$ for $\alpha, \beta \in M_c$ and deduce that there exists $\mu \in M$ such that $T'\alpha = \mu * \alpha$ for all $\alpha \in M_c$.]

7.19 Let X and Y be distributions on R^n. The Chevalley convolution (Chevalley [1]) is said to exist if and only if $(X * \varphi)(\check{Y} * \psi) \in L^1 = L^1(R^n)$ for $\varphi, \psi \in \mathscr{D} = \mathscr{D}(R^n)$. Prove that if this condition is fulfilled, there exists a unique distribution $Z = X * Y$ such that

$$\langle \psi, Z * \varphi \rangle = \int (X * \varphi)(\check{Y} * \psi) \, d\mu$$

for $\varphi, \psi \in \mathscr{D}$, μ denoting Lebesgue measure on R^n.

[Hint: Use the theorem in Subsection 5.11.3 about continuous linear maps of $\mathscr{D}$ into $\mathscr{D}'$ that commute with translations.]

7.20 Verify that if X and Y are distributions on R^n, whose supports A and B are such that $A \cap (K - B)$ is compact for each compact $K \subset R^n$, then $X * Y$ exists in Chevalley's sense. (Compare with Section 5.10.)

Verify also that if X and Y are measures, then Chevalley's definition of $X * Y$ is a consistent generalization of that introduced in Section 4.19.

7.21 Suppose that $1 \leq p \leq \infty$ and that X is a distribution on R^n. Show that $X * Y$ exists in Chevalley's sense (see Exercise 7.19) for each $Y \in L^p$ if and only if X is a finite linear combination of derivatives of functions in $L^{p'}$. Show also that $X * Y$ exists in Chevalley's sense for each bounded measure Y if and only if X is a finite linear combination of derivatives of functions in L^∞.

[Hints: For the "only if" part, show first that $X * \varphi * \psi \in L^{p'}$ for $\varphi, \psi \in \mathscr{D}$. Then apply Exercise 7.17 twice in succession.]

7.22 Let T be an arbitrary set, F a set of scalar-valued functions on T each member of which vanishes outside some finite subset of T. Suppose that

$$\sup_{f \in F} \left| \sum_{t \in T} \xi(t) f(t) \right| < +\infty$$

for each scalar-valued function ξ on T. Prove that there exists a finite subset T_0 of T such that all members of F vanish on $T \backslash T_0$.

7.23 Let T be an arbitrary set, $(E_t)_{t \in T}$ a family of barrelled spaces. Prove that the product space $E = \Pi E_t$ is barrelled.

Prove also the analogous assertion about infrabarrelled spaces.

[Hint: Use the criteria (1) (b) or the second condition of (3) of Theorem 7.1.1. In doing this identify the dual of E with the vector subspace of $\Pi E_t'$ formed of those families $x' = (x_t')$ for which $x_t' = 0$ for all save a finite subset of T (depending on x'), the duality being defined by

$$\langle x, x' \rangle = \sum_{t \in T} \langle x_t, x_t' \rangle.$$

If B is weakly (resp. strongly) bounded in E', the same is true of each projection of B. All therefore depends on showing that there is a finite subset T_0 of T such that $x_t' = 0$ for all $x' \in B$ and all $t \in T \backslash T_0$. To prove this, use the preceding exercise.]

7.24 Let (X_r) be a sequence of distributions on R^n with the property that the set $\{c_r X_r\}$ is weakly bounded in $\mathscr{D}'$ for each sequence (c_r) of scalars. Show that there exists no point of R^n common to the supports of all the X_r.

7.25 Let u be a continuous linear map of C^∞ into $\mathscr{D}'$ that commutes with translations. Prove that there exists a distribution L with a compact support such that

$$u(f) = L * f$$

for $f \in C^\infty$.

[Hints: Combine the results of Subsection 5.11.3 with the preceding exercise.]

7.26 Denote by $\mathscr{D}'_c$ the set of distributions on R^n with compact supports, identifiable with the dual of C^∞. Show that any linear map u of $\mathscr{D}$ into $\mathscr{D}'_c$, which commutes with translations and which is continuous for $\sigma(\mathscr{D}'_c, C^\infty)$, is expressible as $u(f) = L * f$ for a suitably chosen $L \in \mathscr{D}'_c$.

[Hint: Consider the adjoint of u and use the preceding exercise.]

7.27 Suppose that $A \in \mathscr{D}'$ and $B \in \mathscr{S}'$ and that the support of $\hat{B}$ has a nonvoid interior. Show that the relation

$$A * \mathscr{D} \subset B * \mathscr{D}'_c$$

is true if and only if $A \in B * \mathscr{D}'_c$. ($A * \mathscr{D}$ denotes the set of all convolutions $A * f$ when f ranges over $\mathscr{D}$, and similarly for the remaining analogous symbols.)

[Hint: The "if" part is trivial—indeed if $A \in B * \mathscr{D}'_c$, then $A * \mathscr{D} \subset B * \mathscr{D}$. For the converse construct a linear map u of $\mathscr{D}$ into $\mathscr{D}'_c$ such that $A * f = B * u(f)$ and show that the preceding exercise is applicable to this u.]

CHAPTER 8

Duality Theory

8.0 Foreword

The present chapter is devoted to the theory and applications of duality, especially the duality between a LCTVS E and its topological dual E'. A typical and fundamental problem in the theory, first broached and solved by Mackey, is that of determining all the locally convex topologies on a given vector space E relative to which the topological dual coincides (as a vector space) with a preassigned vector subspace of the algebraic dual E^*. Other topics to be discussed include the study of linear maps and their adjoints and biadjoints, bidual spaces and reflexivity, and various topologies on dual spaces. We aim to illustrate these concepts with applications to problems of classical analysis and to integration and distributions theory.

The later portions of the chapter are devoted to some aspects of the theory of integration of vector-valued functions and that of vector-valued Radon measures. These may be regarded in part as extended examples of the use of ideas from duality theory, and at the same time they provide extensions of some of the results obtained in Chapter 4 for integrals of scalar-valued functions and scalar-valued Radon measures.

8.1 Dual Systems and Weak Topologies

By a *dual system* is meant a pair (E, E') of vector spaces (over the same field K) together with a bilinear form $\langle x, x' \rangle$ on $E \times E'$. Such a system defines a linear map f of E into E'^* by

$$f(x)(x') = \langle x, x' \rangle$$

and a linear map f' of E' into E^* by

$$f'(x')(x) = \langle x, x' \rangle.$$

The duality is *separated in* E (resp. in E') if f (resp. f') is one to one, in which case one naturally thinks of E as a vector subspace of E'^* (resp. of E' as a vector subspace of E^*). The duality is *separated* if it is separated both in E and in E'.

By far the most important case is that in which E is a TVS and E' its topological dual, $\langle x, x' \rangle$ being taken as $x'(x)$. This duality is *always* separated in E'; it is separated (in E) whenever E is locally convex and separated (= Hausdorff). This type of duality is, in fact, generally useful only for locally convex spaces.

We have already used the notion of weak topologies for the case of TVSs. In general, if E and E' are in duality (that is, form a dual system), $\sigma(E, E')$ (resp. $\sigma(E', E)$) is the weakest topology on E (resp. on E') for which each linear form $x \to \langle x, x' \rangle$ with x' in E' (resp. $x' \to \langle x, x' \rangle$ with x in E) is continuous. Both these topologies are locally convex; $\sigma(E, E')$ (resp. $\sigma(E', E)$) is separated if the duality is separated in E (resp. in E'). If E is a TVS and E' its topological dual, $\sigma(E, E')$ is spoken of as the *weakened topology* associated with E (or with its given topology); $\sigma(E, E')$ is then separated whenever E is locally convex and separated for its initial topology; and $\sigma(E, E')$ is always weaker than the initial topology.

Our first theorem shows, among other things, that any dual system can be regarded as that obtained by starting from a locally convex space.

8.1.1 **Theorem.** Let E and E' be in duality. The linear forms on E (resp. on E') continuous for $\sigma(E, E')$ (resp. $\sigma(E', E)$) are precisely those of the type $x \to \langle x, x' \rangle$ (resp. $x' \to \langle x, x' \rangle$) with x' (resp. x) fixed but arbitrary in E' (resp. in E).

Proof. The proof is straightforward, using Proposition 1.4.2. ▌

We shall need to know something about the weak topologies defined by the duality between subspaces and quotient spaces of E, or E', or both. Suppose that E and E' are in duality and that M is a vector subspace of E. If $M^0 \subset E'$ is the orthogonal of M—that is, the set of x' in E' satisfying $\langle x, x' \rangle = 0$ for each x in M—the pair $(M, E'/M^0)$ can be put into duality via the bilinear form

$$\langle x, \dot{x}' \rangle = \langle x, x' \rangle,$$

where x is in M and $\dot{x}'$ is the coset modulo M^0 containing x'. Similarly, E/M and M^0 may be put into duality.

8.1.2 **Proposition.** The topology $\sigma(M, E'/M^0)$ is that induced on M by $\sigma(E, E')$. The topology $\sigma(E/M, M^0)$ is the quotient modulo M of the topology $\sigma(E, E')$.

Proof. The first statement follows at once from the remark that if x'_i $(1 = 1, 2, \cdots, n)$ belong to E' and determine the cosets $\dot{x}'_i$ modulo M^0, then the neighbourhood

$$W(\dot{x}'_1, \cdots, \dot{x}'_n) = \left\{x \in M : \operatorname{Sup}_i |\langle x, \dot{x}'_i \rangle| \leq 1\right\}$$

for $\sigma(M, E'/M^0)$ is precisely the intersection with M of the neighbourhood

$$W(x'_1, \cdots, x'_n) = \left\{x \in E : \operatorname{Sup}_i |\langle x, x'_i \rangle| \leq 1\right\}$$

for $\sigma(E, E')$.

The second assertion is less simple. We write σ in place of $\sigma(E/M, M^0)$ and q for the quotient modulo M of $\sigma(E, E')$. Let f be the natural map of E onto E/M. Since $\langle f(x), x'\rangle = \langle x, x'\rangle$ (by definition) for x' in M^0, f is continuous for $\sigma(E, E')$ and σ. It follows that σ is weaker than q. To prove the reverse, let x_i' $(i = 1, \cdots, n)$ belong to E' and $U = W(x_1', \cdots, x_n')$. It will suffice to show that $f(U)$ is a neighbourhood of 0 for σ.

To this end, we consider the vector subspace L of E' generated by M^0 and the x_i', and choose any algebraic complement P of M^0 relative to L. P will be of finite dimension m. We choose a base y_j' $(j = 1, \cdots, m)$ for P. The restrictions $y_j' \mid M$ are linearly independent qua linear forms on M. For if $y' = \sum_j \lambda_j y_j'$ belonged to M^0, then, since $M^0 \cap P = \{0\}$, $y' = 0$ and so $\lambda_j = 0$ for all j. This being so, given any x in E, there exists x_0 in M satisfying $\langle x + x_0, y_j'\rangle = 0$ for all j. If we use the direct sum decomposition $L = M^0 + P$ to write

$$x_i' = t_i' + \sum_j \lambda_{ij} y_j', \qquad t_i' \in M^0,$$

then

$$\langle x, t_i'\rangle = \langle x + x_0, x_i'\rangle \qquad \text{for } i = 1, \cdots, n.$$

If now ξ in E/M satisfies $|\langle \xi, t_i'\rangle| \leq 1$ for all i, we may write $\xi = f(x)$ for some x, and $\langle \xi, t_i'\rangle = \langle x, t_i'\rangle$ by definition. Hence

$$|\langle x + x_0, x_i'\rangle| = |\langle x, t_i'\rangle| = |\langle \xi, t_i'\rangle| \leq 1,$$

showing that $x + x_0$ lies in U. Since x_0 lies in M,

$$f(x + x_0) = f(x) = \xi$$

and so ξ lies in $f(U)$. So $f(U)$ contains all ξ satisfying

$$\operatorname{Sup}_i |\langle \xi, t_i'\rangle| = 1.$$

The latter forming a neighbourhood of 0 for σ, the same is true of $f(U)$. ∎

From this we may derive the following corollary.

8.1.3 Corollary. If N is a vector subspace of E', $\sigma(E/N^0, N)$ is the quotient modulo N^0 of $\sigma(E, E')$ if and only if N is weakly closed in E'.

Proof. Let $\bar{N}$ denote the weak closure of N. Then $N^0 = (\bar{N})^0 = M$, say. Also, $M^0 = \bar{N}$. According to the last proposition, $\sigma(E/M, \bar{N})$ is the quotient of $\sigma(E, E')$ modulo M. On the other hand, if $N \neq \bar{N}$, $\sigma(E/M, \bar{N})$ is strictly stronger than $\sigma(E/M, N)$. Hence the corollary. ∎

8.1.4 Polar Sets. We have already used the notation M^0 to denote the orthogonal of a vector subspace M. It is convenient to extend the definition to arbitrary subsets of either member of a dual system (E, E') in the following way. If A is a subset of E, the *polar set* A^0 (relative to the given dual system) is defined by

$$A^0 = \{x' \in E' : |\langle x, x'\rangle| \leq 1 \text{ for } x \in A\}.$$

Obviously, A^0 is always $\sigma(E, E')$-closed, convex, and balanced. If $B \subset E'$, we define symmetrically its polar set B^0:

$$B^0 = \{x \in E : |\langle x, x' \rangle| \leq 1 \text{ for } x' \in B\}.$$

If E is a TVS and we speak of polars of subsets of E or of E', failing any indication to the contrary, we shall be referring to the canonically associated dual system (E, E') in which E' is the topological dual of E.

From Theorem 2.4.1(c) and Theorem 8.1.1 we at once deduce an important result.

8.1.5 **Theorem (Bipolar Theorem).** Let (E, E') be a dual system, A a subset of E. Then the bipolar set $A^{00} = (A^0)^0$ is the $\sigma(E, E')$-closed, convex and balanced envelope in E of A. (Note also the dual result for subsets of E'.)

Remark. From Theorem 8.2.1 (itself merely a restatement of Corollary 2.2.7) we see that if we are concerned with the dual system (E, E') naturally associated with a LCTVS E, then the bipolar A^{00} of a subset A of E is just the closed, convex, balanced envelope in E of A.

8.2 Properties of the Weakened Topology on a Locally Convex Space

We begin this disussion by recording a result already derived (Corollary 2.2.7) from the Hahn-Banach theorem.

8.2.1 **Theorem.** Let E be a LCTVS, A a convex subset of E. Then A is closed if and only if it is weakly closed (that is, closed for the weakened topology on E).

Yet another property shared by a locally convex topology and its associated weakened topology concerns the concept of bounded set.

8.2.2 **Theorem.** Let E be a LCTVS, A a subset of E. Then A is bounded if and only if it is weakly bounded.

Proof. Since the weakened topology is weaker than the initial topology, boundedness implies weak boundedness.

To prove the converse, we recall that the result has been established (Example 7.2.1) for seminormed TVSs and proceed to reduce the general case to that special one.

Let U be any closed, convex, balanced neighbourhood of 0 in E. It suffices to show that if A is weakly bounded, then $p(A)$ is bounded, p being the gauge function of U. Let $N = p^{-1}(0)$, which is a closed vector subspace of E, and form the quotient space $\dot{E} = E/N$. Define $\dot{p}$ on $\dot{E}$ by $\dot{p}(\dot{x}) = p(x)$ for any x in the coset $\dot{x}$. Then $\dot{p}$ is a norm on $\dot{E}$ and it suffices to show that $\dot{p}(\dot{A})$ is bounded. But any linear form f on $\dot{E}$ continuous for $\dot{p}$ gives rise to a linear form g on E defined by $g(x) = f(\dot{x})$. Then

$$|g(x)| = |f(\dot{x})| \leq \text{const.}\, \dot{p}(\dot{x}) = \text{const.}\, p(x),$$

so that g is continuous on E. By hypothesis, therefore, $g(A)$ is bounded. Hence $f(\dot{A})$ is bounded; that is, $\dot{A}$ is weakly bounded in the seminormed space $\dot{E}$. We now appeal to Example 7.2.1. ▌

8.3 Topologies Compatible with a Given Duality

We turn now to Mackey's problem: Given a dual system (E, E'), how may one characterize those locally convex topologies on E relative to which the topological dual of E is precisely E'? Such topologies do exist, the weakest of them being $\sigma(E, E')$. We shall show that there is also a strongest such topology, and that the others are just those lying between these two.

The solution of this problem will be formulated in terms of the so-called $\mathfrak{S}$-*topologies* on E, $\mathfrak{S}$ being a set of weakly bounded subsets of E'. These topologies are defined by strict analogy with the process used in Subsection 1.10.8 to define topologies on subsets of $L(E, F)$. Given $\mathfrak{S}$, the corresponding topology $T_{\mathfrak{S}}$ is that having a base at 0 formed of the sets $\varepsilon \cdot S^0$ and their finite intersections, where $\varepsilon > 0$, $S \in \mathfrak{S}$, and where

$$S^0 = \{x \in E : |\langle x, x' \rangle| \leq 1 \text{ for } x' \in S\}$$

is the polar in E of S. If $\mathfrak{S}$ forms an increasing directed set (as one may assume without altering $T_{\mathfrak{S}}$), the sets $\varepsilon \cdot S^0$ $(\varepsilon > 0)$ form a base at 0 for $T_{\mathfrak{S}}$.

If we regard elements of E as functions on E', $T_{\mathfrak{S}}$ may be described as the topology of convergence uniform on the sets belonging to $\mathfrak{S}$. Moreover, if E is LCTVS and E' its topological dual, the initial topology of E is a $T_{\mathfrak{S}}$: namely, that obtained by taking for $\mathfrak{S}$ the set of all equicontinuous subsets of E'.

The reader will notice that $T_{\mathfrak{S}}$ is not altered by replacing each S of $\mathfrak{S}$ by its weakly closed, convex, and balanced envelope in E'.

8.3.1 **Theorem (Mackey-Bourbaki).** Let (E, E') be a dual system separated in E' (so that we may regard $E' \subset E^*$). Let $\mathfrak{S}$ be an increasing directed set of weakly bounded, convex, and balanced subsets of E', and suppose that $\mathfrak{S}$ is stable under homothetic maps. The dual of E relative to $T_{\mathfrak{S}}$ is precisely the union of the weak (that is, $\sigma(E^*, E)$-)closures in E^* of sets S in $\mathfrak{S}$.

Proof. Everything depends on showing that if u belongs to the said subspace of E^*, then u is continuous for $T_{\mathfrak{S}}$, and conversely.

If u is weakly adherent to S, however, then it is clear that

$$|u(x)| \leq 1$$

for all x in S^0; S^0 being by definition a $T_{\mathfrak{S}}$-neighbourhood of 0 in E, u is continuous for $T_{\mathfrak{S}}$.

On the other hand, if u is $T_{\mathfrak{S}}$-continuous, there exists a set $S \in \mathfrak{S}$ such that $|u(x)| \leq 1$ for all x in S^0. Since S is convex and balanced, Theorem 8.1.5 shows that u is $\sigma(E^*, E)$-adherent to S. ▌

Remark. The Hahn-Banach theorem shows that E' is $\sigma(E^*, E)$-dense in E^*, so that E^* can be thought of as the completion of E' relative to $\sigma(E', E)$.

8.3.2 Theorem. Let (E, E') be a dual system separated in E'. The locally convex topologies on E compatible with the given duality are exactly those of the type $T_{\mathfrak{S}}$, where $\mathfrak{S}$ is an increasing directed set of weakly compact, convex, and balanced subsets of E' such that $\bigcup_{S \in \mathfrak{S}} S$ generates E'.

Proof. If T is a locally convex topology on E compatible with the duality, then $T = T_{\mathfrak{S}}$, $\mathfrak{S}$ comprising all the weakly compact sets $S = U^0$, U running over a neighbourhood base at 0 for T formed of T-closed, convex, and balanced sets.

Conversely suppose $\mathfrak{S}$ is of the said type. We may add to $\mathfrak{S}$ all homothetic images of its members without altering $T_{\mathfrak{S}}$. Then $\mathfrak{S}$ will cover E', the other conditions remaining as before. Since each S is $\sigma(E', E)$-compact, since $\sigma(E^*, E)$ induces $\sigma(E', E)$ on E', and since $\sigma(E^*, E)$ is separated, it follows that each S is $\sigma(E^*, E)$-closed. Thus Theorem 8.3.1 says that the dual of E relative to $T_{\mathfrak{S}}$ consists of $\bigcup_{S \in \mathfrak{S}} S$, which is E' by assumption. ▌

8.3.3 The Mackey Topology $\tau(E, E')$ and the Arens Topology $\kappa(E', E)$. Theorem 8.3.2 shows that among the locally convex topologies on E compatible with the given duality there is a strongest: this is $T_{\mathfrak{S}}$, where $\mathfrak{S}$ consists of all weakly compact, convex, and balanced (or all weakly compact and convex) subsets of E'. This topology is named after Mackey and is denoted by $\tau(E, E')$. A locally convex topology T on E is compatible with the duality if and only if

$$\sigma(E, E') \leq T \leq \tau(E, E'). \tag{8.3.1}$$

If E is a LCTVS and E' its topological dual, we shall have therefore the relation (8.3.1) holding with T the initial topology of E. Usually the ordering is strict. But the equality $T = \tau(E, E')$ is true for certain important types of LCTVS, which are accordingly said to be *relatively strong* (see Theorem 8.3.5).

Suppose now that E is a TVS, E' its topological dual. Arens [1] introduced the topology on E' having for neighbourhood base at 0 the polars A^0 of compact, convex, and balanced subsets A of E; we call this the *Arens topology* on E' and denote it by $\kappa(E', E)$. It is, of course, locally convex and weaker than the Mackey topology $\tau(E', E)$ (since every compact subset of E is at any rate relatively compact for the weakened topology $\sigma(E, E')$). Since it is obviously stronger than $\sigma(E', E)$, $\kappa(E', E)$ is compatible with the duality between E and E'.

8.3.4 Theorem. Let E be a LCTVS. Then (1) A subset of E is bounded if and only if it is bounded with respect to any one (and hence all) locally convex topologies compatible with the duality between E and E'.

(2) A convex subset of E is closed if and only if it is closed with respect to any one (and hence all) locally convex topologies compatible with the duality between E and E'.

(3) The barrels in E are precisely the polars of weakly bounded subsets of E'.

Proof. (1) follows from Theorem 8.2.2 since all such topologies have the same associated weakened topology $\sigma(E, E')$. (2) follows likewise from Corollary 2.2.7 of the Hahn-Banach theorem. As for (3), suppose first that T is a barrel in E. Since T is absorbent, T^0 is weakly bounded; since T is closed, convex, and balanced, we have $T = T^{00}$ by Theorem 8.1.5. On the other hand, if M is a weakly bounded subset of E', its polar $T = M^0$ is obviously closed, convex, balanced, and absorbent in E—that is, is a barrel in E. ▮

8.3.5 **Theorem.** A LCTVS E is relatively strong if *either* (1) E is barrelled, *or* (2) E is semimetrizable.

Proof. (1) If E is barrelled, any weakly bounded subset of E' is equicontinuous, the same being therefore true a fortiori of any weakly compact subset of E'. The polar of any such set is therefore a neighbourhood of 0 for the initial topology of E, which is consequently stronger than, hence identical with, $\tau(E, E')$.

(2) Let W be any balanced neighbourhood of 0 in E for $\tau(E, E')$. Then (Theorem 8.3.4(1)) W absorbs any subset of E bounded for the initial topology. Since E is bornological (Theorem 7.3.2), W is a neighbourhood of 0 for the initial topology. This last is thus stronger than, hence identical with, $\tau(E, E')$. ▮

8.3.6 **An Example Using Theorem 8.3.2.** Let T be a separated locally compact space, μ a bounded positive Radon measure on T, A a convex subset of $\mathscr{L}^\infty$. We aim to show that

> ▶ If x in $\mathscr{L}^\infty$ is adherent to A for $\sigma(\mathscr{L}^\infty, \mathscr{L}^1)$, then there exists a sequence $(x_n) \in A$ such that $x_n \to x$ in $\mathscr{L}^p$ for every p satisfying $1 \leq p < +\infty$ (Grothendieck, [7], p. 67, Exercice 6).

The duality between $\mathscr{L}^1$ and $\mathscr{L}^\infty$ referred to is the usual one, defined by

$$\langle x, x' \rangle = \int_T x(t)x'(t)\, d\mu(t).$$

Let us denote by $S^{p'}$ the closed unit ball in $\mathscr{L}^{p'}$ (relative to the seminorm $N_{p'}$); since $p < +\infty$, so $p' > 1$, and $S^{p'}$ is compact for $\sigma(\mathscr{L}^{p'}, \mathscr{L}^p)$. Since μ is bounded, $S^{p'} \subset \mathscr{L}^1$; for the same reason $\mathscr{L}^\infty \subset \mathscr{L}^p$ and so $\sigma(\mathscr{L}^1, \mathscr{L}^\infty)$ induces on $\mathscr{L}^{p'}$ a topology weaker than $\sigma(\mathscr{L}^{p'}, \mathscr{L}^p)$. So $S^{p'}$ is $\sigma(\mathscr{L}^1, \mathscr{L}^\infty)$-compact. Let $\mathfrak{S}$ consist of the sets $S^{p'}$ ($1 \leq p < +\infty$). By what we have just established, together with Theorem 8.3.2, the topology $T_\mathfrak{S}$ on $\mathscr{L}^\infty$ is compatible with the given duality between $\mathscr{L}^\infty$ and $\mathscr{L}^1$. On the other hand $T_\mathfrak{S}$ is identical with $T_{\mathfrak{S}'}$, where $\mathfrak{S}' = \{S^1, S^2, \cdots\}$—this is because μ is bounded. Thus $T_\mathfrak{S}$ is metrizable. The assertion now follows from Theorem 8.3.4(2). ▮

See also Exercises 8.5, 8.6, and 8.7.

8.3.7 **The Topology $\tau(E, E^*)$.** Let E be any vector space over K, E^* its algebraic dual. We shall consider the Mackey topology $\tau = \tau(E, E^*)$ and show that it coincides with the strongest locally convex topology on E (see Subsection 1.10.1). Let us temporarily denote the latter topology by t.

Since τ is locally convex, $\tau \leq t$. On the other hand, there exists a neighbourhood base at 0 for t composed of convex, balanced, and absorbent sets A, each

t-closed. Since the topological dual of E relative to t is the same (namely, E^*) as that relative to τ, it follows (Theorem 8.3.4(2)) that each set A is τ-closed. Then the polar $K = A^0$ is convex, balanced, weakly closed, and weakly bounded in E^*; it is therefore weakly compact (Theorem 1.11.4(1)). Consequently, K^0 is a τ-neighbourhood of 0 in E. But, by the bipolar theorem $K^0 = A^{00} = A$. Thus t is weaker than τ. This shows that t and τ are identical, as alleged.

It follows at once that E, taken with τ, is barrelled: in fact, *every* seminorm on E is continuous (for t, hence for τ).

This last remark leads to examples of meagre LCTVSs that are barrelled (see the remark following Corollary 6.2.3). For suppose we take $E = \mathscr{K}(N)$ (the space of all K-valued sequences with finite supports); E is the union of countably many finite-dimensional vector subspaces E_n $(n = 1, 2, \cdots)$; each E_n is closed for τ (since τ is separated); yet each E_n is plainly nowhere-dense for τ. Thus E (endowed with τ) is meagre and barrelled.

8.3.8. The reader might suspect that $\tau = \tau(E, E^*)$ and $\sigma = \sigma(E, E^*)$ are identical topologies. While this is obviously the case if E is finite dimensional, it is false in all other cases. To see this, we suppose that dim $E = \infty$, and take an algebraic basis $(e_i)_{i \in I}$ for E, I being some infinite set. Let (e_i^*) be the dual system in E^*. If we write $x(i) = \langle x, e_i^* \rangle$ for x in E and $x^*(i) = \langle e_i, x^* \rangle$ for x^* in E^*, then each x is a finite sum $\sum_{i \in I} x(i) e_i$, and each x^* is $\sum_{i \in I} x^*(i) e_i^*$, the series being weakly convergent in E^*.

We take any finite, positive-valued function f on I and consider the seminorm

$$p(x) = \sum_{i \in I} f(i)\,|x(i)|$$

on E. p is τ-continuous. If σ and τ were identical, there would exist $x_i^*, \cdots, x_n^*$ in E^* and a number $c \geq 0$ such that

$$\begin{aligned} p(x) &\leq c \cdot \operatorname{Sup}\, \{|\langle x, x_m^* \rangle| : 1 \leq m \leq n\} \\ &= c \cdot \operatorname{Sup} \left\{ \left| \sum_{i \in I} x_m^*(i) x(i) \right| : 1 \leq m \leq n \right\} \end{aligned}$$

for all x in E. In particular, if x is any element of E orthogonal to the x_m^* $(1 \leq m \leq n)$, then $p(x)$ would be 0. Supposing, as we may, that $f(i) > 0$ for all i, $p(x) = 0$ if and only if $x = 0$. Thus we should conclude that the relations $\langle x, x_m^* \rangle = 0$ $(1 \leq m \leq n)$ entail $x = 0$. However, this would entail that dim $E \leq n$ is finite, a contradiction.

We shall return in Subsection 8.10.13 to the topology τ.

8.4 The Strong Topology $\beta(E', E)$ and Reflexivity. Montel Spaces

Throughout this section E will denote a LCTVS, E' its topological dual.

Adopting the notations of Section 8.3, we shall define the so-called *strong topology* $\beta(E', E)$ on E' to be the topology $T_{\mathfrak{S}}$ obtained by taking for $\mathfrak{S}$ the set of all bounded subsets of E. In other words, a neighbourhood base at 0 for $\beta(E', E)$ is formed of the polars B^0, B ranging over the bounded subsets of E. Naturally, the same topology $\beta(E', E)$ is obtained if we restrict B to vary over any base for the bounded subsets of E—for example, the bounded, closed, convex, and balanced subsets of E.

The reader will notice that if E is seminormable, $\beta(E', E)$ is the topology defined by the natural norm on E' (see Subsection 1.10.6).

The topology $\beta(E', E)$ is in general *not* compatible with the duality between E and E'; that is, it is not generally the case that each linear form on E', continuous for $\beta(E', E)$, is generated by an element of E. For example, if $E = c_0(N)$ (the space of scalar-valued sequences converging to 0, endowed with the normable topology derived from the norm $\|x\| = \text{Sup}_n |x(n)|$), then E' is isomorphic with $\ell^1(N)$ (see Theorem 4.2.2) and its strong topology is that defined by the usual norm on $\ell^1(N)$ (see Section 4.2); consequently, the dual of E' is identifiable with $\ell^\infty(N)$ (Theorem 4.2.1) and is thus genuinely "larger" than E itself.

8.4.1 Semireflexive Spaces. Following Bourbaki ([5], p. 87) we shall say that a separated LCTVS E is *semireflexive* if and only if each strongly continuous linear form on E' is generated by an element of E. When this is so, since E is separated and therefore identifiable (as a vector space) with a subset of E'' (see Section 8.7; compare also with Remark 1.9.5(1)), we may write $E'' = E$ as vector spaces, understanding by E'' the topological dual of E' endowed with its strong topology.

From Theorem 8.3.1 we derive a criterion for semireflexivity.

8.4.2 Theorem. A separated LCTVS E is semireflexive if and only if each bounded subset of E is weakly relatively compact.

Examples. From Corollary 1.12.6 it is easily deduced that any Hilbert space is semireflexive. Theorem 4.2.1 shows that $\ell^p(T)$ $(1 < p < \infty)$ is semireflexive; likewise Theorem 4.16.1 shows that $L^p(T, \mu)$ is semireflexive when $1 < p < \infty$. Since in each case these spaces are normable, they are actually reflexive in the sense defined below (see the Remark following Theorem 8.4.5). The examples of Montel spaces given in Subsection 8.4.7 provide instances of nonnormable, and even nonmetrizable, semireflexive spaces.

8.4.3 Theorem. If E is semireflexive, then E' is barrelled for its strong topology.

Proof. Endow E' with $\beta(E', E)$. By hypothesis its dual is then algebraically identifiable with E. Hence (Theorem 8.3.4(3)), the barrels in E' are just the polars of weakly bounded subsets of E. But, by (1) of the same theorem, this says merely that the barrels in E' are neighbourhoods of 0—that is, that E' is barrelled. ∎

8.4.4 Reflexive Spaces. If E is semireflexive, E and E'' may be identified as vector spaces: but what of their respective topologies? E has its initial topology and on E'' we may consider the topology $T_{\mathfrak{S}'}$, where $\mathfrak{S}'$ denotes the set of strongly bounded subsets of E'. Since the initial topology of E is none other than $T_{\mathfrak{S}_1}$, where $\mathfrak{S}_1$ comprises all equicontinuous subsets of E', it is plain that the initial topology on E is generally weaker than $T_{\mathfrak{S}'}$. Again following Bourbaki ([5], p. 88) we shall say that E is *reflexive* if it is semireflexive and if, besides, $T_{\mathfrak{S}'}$ coincides with the initial topology on E; that is, E and E'' may be identified as TVSs (and not merely as vector spaces).

Parallel to Theorem 8.4.2 we have the following theorem.

8.4.5 **Theorem.** The space E is reflexive if and only if it is semireflexive and barrelled.

Proof. If E is semireflexive, the weakly bounded sets in E' are strongly bounded (Theorem 8.3.4(1) applied to E' strong); then (Theorem 8.3.4(3)) the barrels in E are just the polars of strongly bounded subsets of E'—that is, are neighbourhoods of 0 in E''. Thus E will be reflexive if and only if the barrels in E are neighbourhoods of 0—that is, if and only if E is barrelled. ∎

Remark. This result shows that, for barrelled spaces, there is no difference between semireflexivity and reflexivity. Moreover, if E is normable and semireflexive, its closed unit ball (being bounded) is weakly relatively compact (Theorem 8.4.2). (An equivalent assertion is that its closed unit ball is weakly complete.) Then E is necessarily complete and therefore barrelled (Theorem 6.2.2). Thus there is no distinction between semireflexivity and reflexivity for normable spaces. Such a space enjoys these properties if and only if its closed unit ball is weakly compact (or, equivalently, weakly complete).

8.4.6 **Corollary.** If E is reflexive, the same is true of E' for its strong topology.

Proof. By Theorem 8.4.3, E' is barrelled for its strong topology. That E', thus topologized, is semireflexive is an immediate consequence of reflexivity of E—which says that E is identifiable, as a TVS, with the strong dual of E'. Thus E' is reflexive. ∎

Historical note. The concept of reflexivity for normable spaces grew up gradually. Banach, Kakutani, Šmulian, and Eberlein all contributed to its growth. See Dunford and Schwartz [1], Chapter 5, §4 and also p. 463, where the contributions of these and other writers are reviewed.

8.4.7 **Boundedly Compact Spaces and Montel Spaces.** We shall say that a TVS E is *boundedly compact* if each bounded subset of E is relatively compact in E. As we know from F. Riesz' theorem in Subsection 1.9.6 there exist no normable boundedly compact spaces save those that are finite dimensional. However, the space $C^\infty(\Omega)$ (Ω any domain R^n, or more generally any real C^∞ differentiable manifold) is boundedly compact; so too is the space $H(\Omega)$ (Ω a domain in the complex plane, or more generally a complex-analytic manifold), formed of the holomorphic functions on Ω and endowed with the topology of locally uniform convergence. $C^\infty(\Omega)$ is of infinite dimension, and so too in general is $H(\Omega)$.

Bourbaki ([5], p. 89) defines a *Montel space* to be a separated LCTVS that is barrelled and boundedly compact. $C^\infty(\Omega)$ and $H(\Omega)$ are Montel spaces. So too are the spaces $\mathscr{D}(\Omega, S)$ ($S \subset \Omega$ relatively compact), $\mathscr{D}(\Omega)$ and $\mathscr{S}(R^n)$ appearing in the theory of distributions (see Sections 5.1, 6.3, and 5.15).

The preceding results of this section show that any boundedly compact LCTVS is semireflexive, and that any Montel space is reflexive.

We note also the following property of boundedly compact TVSs.

8.4.8 **Theorem.** Let E be a boundedly compact TVS. If also E is separated and locally convex, on each equicontinuous subset Q of E' the weak and the strong topologies coincide. On each bounded subset B of E the weakened and the initial topologies coincide.

Proof. Both statements follow from Ascoli's theorem. For example, if a directed family (x_i') of points of Q converges weakly to $x' \in Q$, then the convergence is uniform on each bounded subset of E; that is, $\lim x_i' = x'$ in the sense of the strong topology $\beta(E', E)$. ▌

8.4.9 **Corollary.** Let E be a Montel space, Q any weakly bounded subset of E'. On Q the weak and strong topologies coincide. In particular, a weakly convergent sequence in E' is strongly convergent.

Proof. Immediate from Theorem 8.4.8 since E is both boundedly compact and barrelled, the latter property ensuring that Q, being weakly bounded, is equicontinuous. ▌

8.4.10 **Application.** In either of the spaces $\mathscr{D}(\Omega)$ or $\mathscr{S}(R^n)$ (see Sections 5.1, 6.3, and 5.15) a weakly convergent sequence is convergent. In either of the spaces $\mathscr{D}'(\Omega)$ or $\mathscr{S}'(R^n)$ (see Sections 5.3 and 5.15) a weakly convergent sequence is strongly convergent.

8.4.11 **Theorem.** If E is a Montel space, then E' is a Montel space for the strong topology $\beta(E', E)$.

Proof. Since E is semireflexive, E' is barrelled for $\beta(E', E)$ (Theorem 8.4.3). Besides this, if Q is a strongly bounded subset of E', then its strongly closed and convex envelope P is also strongly bounded. E being reflexive, P is weakly closed (Theorem 8.3.4). Since E is barrelled, P is equicontinuous and therefore weakly compact (Theorem 1.11.4). Hence (Corollary 8.4.9) P is strongly compact and so Q is strongly relatively compact. Thus E' is boundedly compact for $\beta(E', E)$, hence is a Montel space for the strong topology. ▌

8.4.12 **Application.** The spaces $\mathscr{D}'(\Omega)$ and $\mathscr{S}'(R^n)$ are Montel spaces when endowed with their respective strong topologies.

8.4.13 **Strong Completeness of E'.** From Remark (b) following Theorem 7.3.1 we conclude that E' is strongly complete whenever E is bornological.

While this is perhaps the most important conclusion of this category, it is worth noting that E' is strongly quasi-complete whenever E is infrabarrelled. In fact, the same is true of $L_c(E, F)$, F being any complete LCTVS. For suppose that (u_i) is a strongly Cauchy directed family of elements of $L_c(E, F)$, extracted from a strongly bounded and closed subset H of $L_c(E, F)$. We know already (from Proposition 1.10.9) that (u_i) converges strongly to some u in $L_b(E, F)$, and it will suffice to show that this u is continuous on E. Now, since H is strongly bounded and E is infrabarrelled, H is equicontinuous, hence it follows easily that u is continuous.

8.4.14 **The Strong Dual as an Inductive Limit Space; Distinguished Spaces.** In Remark (3) following Theorem 6.7.2 we have seen the significance of asking under what conditions the strong dual E' is an inductive limit of (a sequence or a more general family of) Fréchet spaces. As we shall proceed to show, this is related to the broader question that asks when E' is bornological or barrelled.

In what follows we denote by β the strong topology $\beta(E', E)$.

We shall begin by constructing an inductive limit topology λ on E', and then consider the relations between β and λ.

Let (U_i) be any family of neighbourhoods of 0 in E, the homothetic images of which form a base at 0. Let P_i be the polar in E' of U_i and $G_i = \bigcup\{\alpha P_i : \alpha > 0\}$, the vector subspace of E' generated by P_i. As in Lemma 6.5.2 we form G_i into a Banach space in which the closed unit ball is P_i; let λ_i be the corresponding topology on G_i. Then (E', λ) is the inductive limit of the spaces (G_i, λ_i) relative to the injection maps of G_i into E'. Thus (E', λ) is an internal inductive limit of Banach spaces. If E is metrizable we may suppose that i runs over the set of natural numbers and we have only a sequence (G_i, λ_i) to handle. In any case (E', λ) is both bornological and barrelled.

Since P_i is strongly bounded, $\beta \mid G_i$ is weaker than λ_i, hence it follows that in all cases β is weaker than λ. The main problem is to determine conditions under which $\beta = \lambda$.

Some worthwhile simplification results if we assume that E is infrabarrelled.

8.4.15 **Proposition.** Suppose that E is infrabarrelled. Then (a) a subset of E' is β-bounded if and only if it is λ-bounded, and (b) (E', β) is bornological if and only if $\beta = \lambda$.

Proof. (a) Since β is weaker than λ, we need only show than any β-bounded set H in E' is λ-bounded. Now, E being infrabarrelled, H is equicontinuous (Theorem 7.1.1(3)). Thus H is contained in αP_i for some i and some $\alpha > 0$. But then it follows that H is absorbed by every λ-neighbourhood of 0; that is, H is λ-bounded.

(b) We must show that λ is weaker than β whenever (E', β) is bornological. But if V is any convex and balanced λ-neighbourhood of 0, it absorbs all the β-bounded sets (by (a)). If (E', β) is bornological, V is therefore necessarily a β-neighbourhood of 0. Thus λ is weaker than β. ▮

A direct application of Theorem 6.2.1(2) leads to the following analogue of part (b) of the last proposition.

8.4.16 **Proposition.** Suppose that (E', β) is barrelled, and that there exists a base of λ-neighbourhoods of 0 each of which is β-closed.

Then $$\lambda = \beta.$$

By striking off in a different direction we can give a necessary and sufficient condition in order that (E', β) be barrelled. For this we need a definition.

Definition. A separated LCTVS E is said to be *distinguished* (Dieudonné and Schwartz [1], p. 78; Bourbaki [5], Chapitre 4, p. 93, Exercice 10)) if and only if each $\sigma(E'', E')$-bounded subset of E'' is contained in the $\sigma(E'', E')$-closure of some bounded subset of E.

8.4.17 **Proposition.** (E', β) is barrelled if and only if E is distinguished.

Proof. If (E', β) is barrelled and Q is $\sigma(E'', E')$-bounded, then Q is equicontinuous; that is, $Q \subset (B^0)^0$, B being a bounded subset of E and the second polar being taken in E''. Assuming (as we may) that B is convex and balanced, B^{00} is none other than the $\sigma(E'', E')$-closure of B. Thus E is distinguished.

Conversely suppose that E is distinguished and that A is a barrel in (E', β). Its polar A^0 in E'' is then $\sigma(E'', E')$-bounded. By hypothesis, therefore, there exists a bounded subset B of E such that A^0 is contained in the $\sigma(E'', E')$-closure of B. But then A^{00} contains B^0, which is a β-neighbourhood of 0. Since A is a β-barrel, $A = A^{00}$, A is a β-neighbourhood of 0, and we have shown that (E', β) is barrelled. ∎

We may summarize the preceding results thus:

▶ (1) (E', β) is barrelled if and only if E is distinguished.

▶ (2) If E is infrabarrelled, (E', β) is bornological if and only if $\lambda = \beta$—in which case (E', β) is also barrelled. It appears therefore that the relation $\lambda = \beta$ entails that E is distinguished. The converse is true for Fréchet spaces, as we shall now show.

8.4.18 **Proposition.** If E is a Fréchet space, then $\lambda = \beta$ if and only if E is distinguished, in which case the strong dual E' is the internal inductive limit of a sequence of Banach spaces.

Proof. It remains only to show that if E is a distinguished Fréchet space, then $\lambda = \beta$. For this, in view of Propositions 8.4.16 and 8.4.17, it suffices to show that the hypotheses of Proposition 8.4.16 are fulfilled. The following argument is due to Bourbaki ([6], Chapitre 6, p. 76, Lemme 1).

Let (U_i) be a decreasing sequence of closed, convex, and balanced sets in E forming a base at 0. Let V be any convex and balanced λ-neighbourhood of 0 in E'. Then to each i corresponds a number $\alpha_i > 0$ such that $\alpha_i U_i^0 \subset \frac{1}{2}V$. Let A_n be the convex envelope of $\bigcup\{\alpha_i U_i^0 : 1 \leq i \leq n\}$ and W the union of the A_n.

Then W is convex, balanced, and absorbent and is contained in $\frac{1}{2}V$. The β-closure $\overline{W}$ of W is a β-barrel and hence a β-neighbourhood of 0. It will suffice to show that $\overline{W}$ is contained in V, the arbitrary nature of V making it then evident that the hypotheses of Proposition 8.4.16 are fulfilled.

To this end, we take any point x' not in V. Each set U_i^0 is $\sigma(E', E)$-compact, and so therefore is A_n. Since x' falls outside $2A_n$, there exists x_n in A_n^0 (polar in E) satisfying $\langle x_n, x' \rangle = 2$. The sequence (x_n) is bounded in E because, for any y' in E', one has $y' \in U_k^0$ for some k, and therefore $|\langle x_n, y' \rangle| \leq \alpha_k^{-1}$ for $n \geq k$, showing that (x_n) is weakly bounded and hence bounded. Let H be the convex balanced envelope in E of the set $\{x_n\}$. H^0 is a β-neighbourhood of 0 in E' and the polar in E'', H^{00}, is $\sigma(E'', E')$-compact. Hence (x_n) has a limiting point x'' in E'' for $\sigma(E'', E')$. Evidently, $\langle x'', x' \rangle = 2$. Yet x'' belongs to A_n^0 (polar in E'') for all n, and so x'' belongs to W^0 (polar in E''). It follows that x' does not belong to W^{00}, hence is not $\sigma(E', E'')$-adherent to W. A fortiori, then, x' is not β-adherent to W; that is, x' is not in $\overline{W}$. Thus $\overline{W} \subset V$, as we had to show. ▮

8.4.19. We see now that both Theorems 6.7.1 and 6.7.2 remain true if we replace throughout E and F by the strong duals E' and F' of distinguished (for example, reflexive) Fréchet spaces. For the details, see Exercises 8.44 and 8.45.

8.5 The Theorem of Banach-Grothendieck

If E is a LCTVS with topological dual E', we know that a linear form on E' is generated by an element of E (that is, is of the type $x' \to \langle x, x' \rangle$ for some x in E) if and only if it is weakly continuous. It is convenient for many applications to have a more manageable formulation of this criterion for restricted types of spaces E.

Let us denote by $j(x)$ the linear form on E' defined by $x' \to \langle x, x' \rangle$.

8.5.1 **Theorem** (Banach-Grothendieck). Let E be a complete LCTVS. A linear form u on E' is of the type $j(x)$ for some $x \in E$ if and only if $u \mid U^0$ is weakly continuous for each U belonging to a neighbourhood base at 0 in E.

Proof. The "only if" assertion is trivial.

For the converse we denote by F the set of all linear forms u on E' for which each $u \mid S$ is weakly continuous for each set of the type U^0. The set of such subsets of E' is denoted by Σ. If $S \in \Sigma$ and $u \in F$, then $u(S)$ is bounded, and we may consider on F the topology T_Σ, making F into a LCTVS. We wish to show that $j(E) = F$.

First of all, it is easy to show that $j(E)$ is closed in F; for if $u_i = j(x_i)$, then the Cauchy nature of (u_i) entails that of (x_i) and one may then appeal to completeness of E. It will thus suffice to show that $j(E)$ is dense in F. Now, if one can show that the dual of F is E', the Hahn-Banach theorem will show that $j(E)$ is dense in F. For any x' in E', for which $x'(j(x)) = 0$ for all $x \in E$, satisfies $x' = 0$ and so $u(x') = 0$ for each $u \in F$.

Thanks to Theorem 8.3.2, it suffices to show that each set $S = U^0$ is compact for $\sigma(E', F)$. Since compactness obtains for $\sigma(E', E)$, it suffices to verify that these two topologies induce the same topology on S. Since the former is stronger than the latter, one has only to verify that $\sigma(E', F) \mid S \leq \sigma(E', E) \mid S$. But the topology $\sigma(E', F) \mid S$ is the weakest on S, which makes each u in F continuous on S. By definition of F, this is weaker than $\sigma(E', E) \mid S$. The proof is thus complete. ▮

Remarks. (1) If E is separated j is one to one and we may as well identify x and $j(x)$ and so regard E as imbedded in F, itself a subset of E'^*.

(2) If E were not complete, the preceding argument would show that F would be identical with $j(\hat{E})$, j being extended to the completion $\hat{E}$ in the obvious way.

8.5.2 **Corollary.** Suppose that E is a separable complete separated LCTVS. A linear form u on E' belongs to $j(E)$ if and only if $\lim_{n\to\infty} u(x'_n) = 0$ for each equicontinuous sequence (x'_n) that converges weakly to 0 in E'.

Proof. The "only if" assertion is trivial. For the rest the reader will observe that if E contains a countable total set, $\sigma(E', E)$ induces on each equicontinuous subset of E' a metrizable topology. It follows thence and from Theorem 8.5.1 that u belongs to $j(E)$ provided $u(x'_n) \to u(x')$ whenever the sequence $(x'_n) \subset S$ converges weakly to x'. It remains only to show that this last follows in turn from the same assumption wherein x' is taken to be 0.

If S is convex and balanced and $u \mid S$ is weakly continuous at 0, however, there exists a weak neighbourhood W of 0 in E' such that $|u(x')| \leq \frac{1}{2}\varepsilon$ for x' in $S \cap W$. If x'_0 is a point of S and x' lies in $(x'_0 + W) \cap S$, then $x' \in S$ and $x' - x'_0 \in W$. Now

$$\tfrac{1}{2}(x' - x'_0) = \tfrac{1}{2}(x' + (-x'_0)) \in S.$$

Assuming as we may that W is convex and balanced, $(x' - x'_0) \in W$. So $\frac{1}{2}(x' - x'_0) \in W \cap S$ and so

$$|u(\tfrac{1}{2}(x' - x'_0))| \leq \tfrac{1}{2}\varepsilon$$

and therefore

$$|u(x') - u(x'_0)| \leq \varepsilon.$$

This establishes continuity of $u \mid S$ at x'_0. So $u \mid S$, if weakly continuous at 0, is weakly continuous on S. ▮

8.6 Transposed and Adjoint Maps

Let E and F be vector spaces (over the same field K), E^* and F^* their algebraic duals. Then (E, E^*) and (F, F^*) are dual systems separated in E^* and F^*, respectively ((E, E^*) is indeed the "largest" dual system (E, E') that is separated in E').

If now u is a linear map of E into F, there is a linear map u^* of F^* into E^* defined by

$$\langle u(x), y^* \rangle = \langle x, u^*(y^*) \rangle.$$

u^* is the *algebraic adjoint* or *transposed* of u.

One may endeavour to generalize this adjoint process to general dual systems (E, E') and (F, F') separated in E' and F', respectively. Given a linear map u of E into F, one asks first for which y' in F' is the linear form $x \to \langle u(x), y' \rangle$ generated by an element of E'—that is, is continuous for $\sigma(E, E')$. The set of such y' is a vector subspace D' of F', and we may define an adjoint map $u' : D' \to E'$ by the requirement that

$$\langle u(x), y' \rangle = \langle x, u'(y') \rangle \tag{8.6.1}$$

for y' in D' and x in E. It follows that $u'(y')$, when it exists, coincides with $u^*(y')$. In other words, D' is the set of y' in $F' \subset F^*$ for which $u^*(y')$ falls into E', and $u' = u^* \mid D'$. The reader will notice that $D' = F'$ if and only if u is weakly continuous (that is, continuous for $\sigma(E, E')$ and $\sigma(F, F')$). Further, if the duality is separated in E and F, and if u is weakly continuous, we may form $u'' = (u')'$: the result is just u. (If u is weakly continuous, u' automatically has the same property.)

Of great importance are two adaptations of the above formulation, both concerned with the case in which E and F are LCTVSs. We discuss this situation in two stages.

(1) If E and F are LCTVSs with duals E' and F', and if u is a continuous linear map of E into F, then u is weakly continuous, so that u' is defined throughout F'. Indeed, for any y' in F', the linear form $x \to \langle u(x), y' \rangle$ is continuous on E for the initial topology and therefore continuous for the weakened topology $\sigma(E, E')$ (by definition of the latter).

(2) Let us turn next to the more general case in which, once again, E and F are LCTVSs, but the map u is not defined everywhere on E. Thus we suppose merely that u has its domain in E and range in F. We do not assume even that u is continuous on its domain.

It is instructive to handle this situation in terms of graphs. To do this we identify the dual of $E \times F$ with $F' \times E'$ in such a way that

$$\langle (x, y), (y', x') \rangle = \langle x, x' \rangle - \langle y, y' \rangle.$$

The appearance of the minus sign is purely a convenience.

Let us consider the polar $(\text{Gr } u)^0$ of the graph $\text{Gr } u$ of u relative to the duality just defined. Thus $(\text{Gr } u)^0$ is the subset of $F' \times E'$ formed of those pairs (y', x') satisfying

$$\langle x, x' \rangle - \langle u(x), y' \rangle = 0$$

for x in $\text{Dom } u$. We ask the question: when is $(\text{Gr } u)^0$ itself a graph? Since $(\text{Gr } u)^0$ is obviously a vector subspace of $F' \times E'$, the answer to this question is: if and only if the relation $(0, x') \in (\text{Gr } u)^0$ implies that $x' = 0$. This means exactly that $x' \in (\text{Dom } u)^0$ (polar relative to the dual system (E, E')) implies $x' = 0$—that is (Hahn-Banach theorem), exactly that $\text{Dom } u$ is weakly dense in E.

Thus $(\text{Gr } u)^0$ is a graph if and only if $\text{Dom } u$ is weakly dense. When this is so, $(\text{Gr } u)^0$ is the graph of a (necessarily linear) map with domain in F' and range in E', and this map has for domain precisely the set of those y' in F' for which an

x' in E' exists so that $\langle x, x'\rangle - \langle u(x), y'\rangle = 0$ for all x in Dom u. Referring to the case already discussed, in which Dom $u = E$, it is clear that the map referred to is none other than the adjoint u'. We continue to use this name and notation even when Dom u is merely weakly dense in E.

To sum up: the adjoint u' exists (that is, (Gr $u)^0$ is a graph) if and only if Dom u is weakly dense in E; if u' exists, its domain is formed of just those y' in F' for which the linear form $x \to \langle u(x), y'\rangle$ is weakly continuous on Dom u; for y' in Dom u', there exists precisely one x' in E' such that

$$\langle x, x'\rangle = \langle u(x), y'\rangle$$

for x in Dom u, and this x' is $u'(y')$. Thus u' is determined completely by the relation

$$\langle x, u'(y')\rangle = \langle u(x), y'\rangle$$

holding for x in Dom u and y' in Dom u'.

We observe that since (Gr $u)^0$ is always weakly closed, so u' is always weakly closed.

It should be noticed that if E and F are TVSs and E' and F' their respective topological duals, it follows that u' is always closed relative to any topologies on F' and E' compatible with the dualities between E and E' and between F and F'. Furthermore, if E is locally convex, u' exists (that is, Dom u is weakly dense) if and only if Dom u is dense (for the given topology on E); in this case one may also say that y' belongs to Dom u' if and only if the linear form $x \to \langle u(x), y'\rangle$ is continuous on Dom u for the topology induced by the given topology on E.

Throughout the remainder of this section we shall be concerned with the case in which (E, E') and (F, F') are dual systems separated in E' and F', respectively, while u is a weakly continuous linear map of E into F. In this situation the following relationships subsist and are applied frequently.

(a) If $A \subset E$ and $B \subset F'$, then

$$u(A)^0 = u'^{-1}(A^0),\ u^{-1}(B^0) = u'(B)^0. \tag{8.6.2}$$

(b) If $A \subset E$ and $B \subset F$, then

$$u(A) \subset B \text{ implies } u'(B^0) \subset A^0, \tag{8.6.3}$$

and the converse is true if B is convex, balanced, and weakly closed.

(c)
$$\overline{u(E)} = (\text{Ker } u')^0 = [u'^{-1}(\{0\})]^0, \tag{8.6.4}$$
$$\overline{u'(F')} = [u^{-1}(\overline{\{0\}})]^0.$$

(d) $u(E)$ is weakly dense in F if and only if u' is one-to-one; if the systems are separated, $u'(F')$ is weakly dense in E' if and only if u is one-to-one. (Failing separatedness in E and F, the requirement becomes

$$u^{-1}(\overline{\{0\}}) = \overline{\{0\}}.)$$

Proof. (a) is a matter of juggling with the formula (8.6.1) and the definition of polar sets (see Subsection 8.1.4). (b) comes from the Bipolar Theorem 8.1.5. (c) is a special case of (a), and (d) is a corollary of (c). ∎

While we consider abstract dual systems and their weak topologies, problems concerning u and u' are fairly straightforward. The situation of greatest interest is that in which E and F are LCTVSs, E' and F' their topological duals, and we wish to connect properties of u relative to the initial topologies with corresponding ones relative to the weak topologies linked with the dual systems (E, E') and (F, F'). This is not a trivial task; it will occupy us for the remainder of this section.

It has already been noted that if a linear map u of E into F is continuous, then it is weakly continuous (so that $u' : F' \to E'$ is defined). The converse is not generally true, but the next two results say something in this direction.

8.6.1 **Theorem.** Let E and F be LCTVSs, u a weakly continuous linear map of E into F. Then

(1) u is continuous for $\tau(E, E')$ and $\tau(F, F')$;

(2) u' is continuous for $\kappa(F', F)$ and $\kappa(E', E)$;

(3) u is bounded.

Proof. (1) If V is a τ-neighbourhood of 0 in F, it contains a set of the type K^0, where K is convex, balanced, and weakly compact in F'. Hence, since u' is weakly continuous, $H = u'(K)$ is convex, balanced, and weakly compact in E', so that $U = H^0$ is a τ-neighbourhood of 0 in E. Now $u^{-1}(V) \supset U$. So $u^{-1}(V)$ is a τ-neighbourhood of 0 in E, and the asserted continuity of u is established.

(2) The proof is almost immediate, using (8.6.3) and the fact that u, being continuous, transforms any compact convex and balanced subset of E into a similar subset of F.

(3) We know that $y' \circ u$ is continuous for each y' in F'. If $A \subset E$ is bounded, $y' \circ u(A)$ is bounded. Hence $u(A)$ is weakly bounded, and therefore bounded (Theorem 8.2.2), in F. Therefore u is bounded. ∎

8.6.2 **Theorem.** Let E and F be LCTVSs, u a weakly continuous linear map of E into F. Then u is continuous in either of the following two cases:

(1) E is relatively strong (for example, barrelled or semimetrizable— Theorem 8.3.5);

(2) E is bornological.

Proof. (1) The proof follows from (1) of the preceding theorem, since $\tau(F, F')$ is stronger than the initial topology of F. (2) The proof here follows from (3) of the preceding theorem combined with Remark (b) following Theorem 7.3.1. ∎

For abstract dual systems the following result is basic.

8.6.3 **Proposition.** Let (E, E') and (F, F') be separated dual systems, and u a weakly continuous linear map of E into F. Then u is weakly open from E onto $u(E)$ if and only if $u'(F')$ is weakly closed in E'.

Proof. (1) Suppose first that u is one to one—that is, that $u'(F')$ is weakly dense in E'. In this case we wish to show that u is a weak isomorphism of E onto $u(E) = M$ if and only if $u'(F') = E'$.

Now if u is a weak isomorphism of E onto M, u^{-1} is continuous for $\sigma(F, F') \mid M$ and $\sigma(E, E')$. Therefore, given x' in E', the linear form $y \to \langle u^{-1}(y), x' \rangle$ on M is continuous for $\sigma(F, F') \mid M$. By the Hahn-Banach theorem, it is therefore the restriction to M of some element y' of F'. Thus

$$\langle u^{-1}(y), x' \rangle = \langle y, y' \rangle \qquad \text{for } y \in M.$$

Taking $y = u(x)$, this reads $\langle x, x' \rangle = \langle u(x), y' \rangle$ for x in E, which shows that $u'(y') = x'$. Thus u' is onto, as we had to show.

Suppose conversely that u' is onto (which implies that u is one-to-one). Then for any finite family (x'_i) in E' there exists a like family (y'_i) in F' such that $u'(y'_i) = x'_i$ for each i. Then the inequalities $|\langle u(x), y'_i \rangle| \leq 1$ are together equivalent to the inequalities $|\langle x, x'_i \rangle| \leq 1$, hence it follows that u is a weak isomorphism of E onto $u(E)$.

(2) Having established the proposition when u is one-to-one, in the general case we introduce the kernel N of u and the one-to-one map v of E/N into F induced by u. To say that u is weakly open into signifies that v is an isomorphism into for the quotient topology $\sigma(E, E')/N$ on E/N and the topology $\sigma(F, F')$. But, by Proposition 8.1.2, this means that v is a weak isomorphism into for the weak topologies on E/N and F. By what we have already proved, this is equivalent to saying that v' maps F' onto N^0 (the latter being the dual of E/N). On the other hand, $u' = i \circ v'$ where i is the injection map of N^0 into E'. Thus we see that if u is weakly open from E into F, then $u'(F') = i(N^0) = N^0$, which is weakly closed.

On the other hand, if $u'(F')$ is weakly closed in E', the relation $u'(F')^0 = N$ shows that $u'(F') = N^0$—that is, that $v'(F') = N^0$, and so that u is weakly open into. ∎

Remark. The reader should note that because of the symmetry displayed by abstract dual systems, Proposition 8.6.3 leads to analogous assertions for weakly continuous linear maps of F' into E'. An instance is provided in (2) of the forthcoming corollary.

8.6.4 **Corollary.** If (E', E) and (E, E') are separated dual systems, and if u is a weakly continuous linear map of E into F, then

(1) u is a weak isomorphism into if and only if $u'(F') = E'$ (that is, u' is onto);

(2) $u(E) = F$ (that is, u is onto) if and only if u' is a weak isomorphism into.

8.6.5 **Proposition.** Let E and F be separated LCTVSs with topological duals E' and F', and let u and v be linear maps of E into F and F' into E', respectively. Consider the assertions

(a) u is continuous;

(b) u is weakly continuous;

(c) v is weakly continuous;

(d) v is strongly continuous.

Then (a) implies (b) and (c) implies (d).

Proof. That (a) implies (b) is already proven. For the remainder, since v is weakly continuous, we may write $v = u'$ for some weakly continuous linear map u of E into F. To show that v is strongly continuous, it suffices to show that if $A \subset E$ is bounded, there exists a bounded $B \subset F$ such that $u'(B^0) \subset A^0$. Now, u being weakly continuous, $u(A)$ is weakly bounded, and therefore bounded (Theorem 8.2.2), in F. It suffices to take $B = u(A)$. ▮

8.6.6 **Corollary.** If E and F are separated LCTVSs with topological duals E' and F', respectively, any continuous linear map u of E into F is weakly continuous, and u' is weakly and strongly continuous.

8.6.7 **Corollary.** Let E and F be separated LCTVSs, and suppose in addition that F is semireflexive. If v is a strongly continuous linear map of F' into E', then v is weakly continuous.

Proof. The implication "(a) implies (b)" of Theorem 8.6.5 shows that v is continuous for $\sigma(F', F'')$ and $\sigma(E', E'')$. Since F is semireflexive, $\sigma(F', F'') = \sigma(F', F)$; and $\sigma(E', E'')$ is stronger than $\sigma(E', E)$ in any case. ▮

8.6.8 **Theorem.** Let E and F be separated LCTVSs with topological duals E' and F', u a continuous linear map of E into F. Then u is an open map of E onto $u(E)$ if and only if

(a) u is weakly open into [that is, $u'(F')$ is weakly closed in E', (Theorem 8.6.3)], and

(b) Each equicontinuous subset of E', contained in $u'(F')$, is the u'-image of some equicontinuous subset of F'.

Proof. Let $N = u^{-1}(0)$, $X = E/N$, $Y = u(E)$, and let v be the one-to-one continuous linear map of X onto Y defined by u. By definition, to say that u is an open map into is equivalent to saying that v is an isomorphism onto. By Proposition 8.1.2, "u is weakly open into" is equivalent to "v is a weak isomorphism onto." Moreover, by the same proposition, X' is identifiable with N^0, whilst the equicontinuous sets in X' are those equicontinuous subsets of E' that are contained in N^0, and the equicontinuous sets in $Y' = F'/Y^0$ are the images (under the natural mapping of F' onto F'/Y^0) of equicontinuous subsets of F'.

Let us suppose now that u is an open map of E onto $u(E)$, so that v is an isomorphism. Proposition 8.6.5 shows that both v and v^{-1} are weakly continuous, and so that v is a weak isomorphism. This signifies that u is weakly open into. But also, since v is a weak isomorphism, $v'(Y')$ is weakly dense and weakly closed in X'; that is, $v'(Y') = X'$, and $u'(F') = N^0$. Since v is an isomorphism, any equicontinuous subset of X' is the v'-image of some equicontinuous subset of Y'; that is, each equicontinuous subset of E' contained in $N^0 = u'(F')$ is the u'-image of some equicontinuous subset of F'. Thus (a) and (b) are both necessary in order that u be open into.

Let us suppose conversely that (a) and (b) are satisfied. By (a), v is a weak isomorphism and so $v'(Y') = X'$; that is, $u'(F') = N^0$. Also (b) signifies that any equicontinuous subset of X' is the v'-image of some equicontinuous subset of Y', which in turn shows that v is an isomorphism (by the dual form of (8.6.3)). Sufficiency of (a) and (b) is thereby established, and the proof is complete. ▮

8.6.9 **Corollary.** Let E and F be separated LCTVSs with topological duals E' and F', u a continuous linear map of E into F. u is an isomorphism into if and only if each equicontinuous subset of E' is the u'-image of some equicontinuous subset of F'.

Proof. If u is an isomorphism into, it is a weak isomorphism into (Theorem 8.6.8) and so (Proposition 8.6.3) $u'(F')$ is weakly closed in E'. Since u is one-to-one, $u'(F')$ is weakly dense in E'. So $u'(F') = E'$. Now we apply Theorem 8.6.8(b). The condition is therefore necessary.

Conversely if u satisfies the said condition, $u'(F') = E'$. So u is one-to-one. But also (Proposition 8.6.3), u is weakly open into. Thus u satisfies (a) and (b) of Theorem 8.6.8 and is therefore an isomorphism into. ▮

8.6.10 **Corollary.** Let E and F be separated LCTVSs, F being relatively strong (for example, metrizable or barrelled). A continuous linear map u of E into F is open if and only if it is weakly open.

Proof. "Open into" implies "weakly open into" in any case (Theorem 8.6.8). For the remainder, if u is weakly open into, Corollary 8.6.4 shows that u' is a weak isomorphism of F' onto $u'(F')$. On the other hand, if $A' \subset u'(F')$ is equicontinuous qua subset of E', it is relatively weakly compact, and so $B' = u'^{-1}(A')$ is relatively weakly compact in F'. Since F is relatively strong, B' is equicontinuous in F'. Since $A' = u'(B')$, condition 8.6.8(b) is satisfied. So u is open into. ▮

8.6.11 **Corollary.** Let E and F be separated LCTVSs, u a continuous linear map of E into F. u is an open map of E onto a *dense* vector subspace of F if and only if u' is one-to-one, $u'(F')$ is weakly closed in E', and the inverse image by u' of any equicontinuous subset of E' is equicontinuous in F'.

Proof. It suffices to use Theorem 8.6.8, together with the fact that $u(E)$ is dense if and only if u' is one-to-one. ▮

Now we shall specialize the preceding results to the important case of Fréchet and Banach spaces. The possible simplifications stem, at least in part, from a lemma.

8.6.12 **Lemma.** Let E be a complete semimetrizable LCTVS, F a semimetrizable LCTVS, and let $A \subset E'$ be equicontinuous, weakly closed, convex, and balanced. Construct the Banach space E'_A (cf. Lemma 6.5.2). If u is a continuous linear map of E into F, and if $u'(F') \cap A$ is closed in E'_A for each A, then $u'(F')$ is weakly closed in E'.

Proof. By Theorem 8.10.5 (whose proof is totally independent of the present arguments), it suffices to show that for each A, $K = u'(F') \cap A$ is weakly closed in E'. Since $\sigma(E', E)$ is separated, this amounts to showing that K is weakly compact. Now, K is convex, balanced, and closed in E'_A (by hypothesis).

On the other hand, F' is the union of a sequence (C_n) of weakly compact, convex, and balanced sets (for example, polars of a neighbourhood base at 0 in F). So $K_n = A \cap u'(C_n)$ is convex and balanced in E'_K and is weakly compact in E' (since u' is weakly continuous). Thus K_n is closed in E'_K. Further, $\bigcup_{n=1}^{\infty} nK_n = E'_K$. So since E'_K is nonmeagre in itself, some K_n is a neighbourhood of 0 in E'_K. Multiplying C_n by a suitable positive scalar, we shall then have

$$K \subset K_n = u'(C_n) \cap A \subset u'(F') \cap A = K.$$

Thus $K = K_n$. Since K_n is weakly compact, so too is K. The proof is complete. ▌

8.6.13 **Theorem.** Let E and F be Fréchet spaces, u a continuous linear map of E into F. The following conditions (1)–(6) are all equivalent:

(1) u is open into;
(2) u is weakly open into;
(3) $u(E)$ is closed in F;
(4) u' is weakly open into;
(5) $u'(F')$ is weakly closed in E';
(6) $u'(F')$ is strongly closed in E'.

Proof. (1) is equivalent to (2) by Corollary 8.6.10, any vector subspace of F being metrizable and therefore relatively strong. (1) is equivalent to (3) by the OMT in the form of Theorem 6.4.4(a), any closed vector subspace of a Fréchet space being itself a Fréchet space. (3) is equivalent to (4) by Proposition 8.6.3 applied to u', since $(u')' = u$ and vector subspaces of E and F are weakly closed if and only if they are closed. (2) is equivalent to (5), again by Proposition 8.6.3. Thus (1)–(5) are all equivalent. (5) implies (6) trivially.

That (6) implies (5) appears from Lemma 8.6.12 since if $u'(F')$ is strongly closed in E', it satisfies the hypothesis of this lemma. Indeed, suppose $u'(F')$ is strongly closed; any A of the type specified in Lemma 8.6.12 is contained in the polar of a neighbourhood U of 0 in E, and if $(x'_n) \subset u'(F') \cap A$ converges to x' in E'_A, then $x' - x'_n \in \lambda_n A$, where $\lambda_n \to 0$. If $B \subset E$ is bounded, B is absorbed by U, and we see that $x'_n \to x'$ uniformly on B. Thus $x'_n \to x'$ strongly in E'. So $x' \in u'(F')$. Since A is weakly closed, and since a fortiori $x'_n \to x'$ weakly, $x' \in A$. Thus $x' \in u'(F') \cap A$, and the latter is weakly closed.

This completes the proof of the theorem. ▌

8.6.14 **Remark.** In any case (6) is implied by

(7) u' is strongly open into;

moreover, if E and F are Banach spaces, so are E' and F' strong, and then (OMT) (6) implies (7). Thus we have the following corollary.

8.6.15 **Corollary.** If E and F are Banach spaces and u a continuous linear map of E into F, the conditions (1)–(7) are all equivalent. In particular, therefore:

(a) u is onto if and only if u' is strong isomorphism into;
(b) u' is onto if and only if u is an isomorphism into.

We proceed to list additional corollaries (cf. Banach, [1], pp. 146–149).

8.6.16 **Corollary.** If E and F are Fréchet spaces and u a continuous linear map of E into F, the following two conditions are equivalent:

(8) u' is one-to-one and u'^{-1} transforms equicontinuous subsets of $u'(F')$ into equicontinuous subsets of E';

(9) $u(E) = F$ and $u'(F') = (u^{-1}(0))^0$.

If also E and F are Banach spaces, (8) is equivalent to

(8′) u' is a strong isomorphism into.

Proof. Suppose (8) holds. Since u' is one to one, $u(E)$ is dense in F. The second part of (8) implies that u' transforms weakly closed vector subspaces into similar subspaces (Theorem 8.10.5); in particular, $u'(F')$ is weakly closed. Thus (5) holds, and therefore (3) holds. So $u(E) = F$. Finally, since $u'(F')$ is weakly closed, it is $(u^{-1}(0))^0$. Thus (8) implies (9).

Reciprocally, if (9) holds, then u' is one-to-one (since u is onto). The second part of (9) shows that $u'(F')$ is weakly closed, so that (5) holds. Therefore (1) holds. So the rest of (8) follows from Theorem 8.6.8(b). ▌

8.6.17 **Corollary.** Let E and F be Fréchet spaces, u a continuous linear map of E into F. The following two conditions are equivalent:

(10) $u'(F') = E'$ and $u(E) = (u'^{-1}(0))^0$;

(11) u is an isomorphism into.

Proof. Let us assume (10). Since $u'(F') = E'$, so u is one to one and (5) holds. Thus (1) holds, hence (11). Conversely, if (11) holds, then $u'(F')$ is weakly dense (u being one-to-one); also, since (11) implies (1), (5) holds and $u'(F')$ is weakly closed. So $u'(F') = E'$. Further, since u is an isomorphism into and E is complete, (3) holds; but then $u(E)$ is weakly closed and coincides with $(u'^{-1}(0))^0$. ▌

8.6.18 **Corollary.** Let E and F be Fréchet spaces, u a continuous linear map of E into F. Then:

(a) u is one-to-one onto if and only if u' is one-to-one onto;

(b) if u is an isomorphism into and u' is one-to-one, then both are onto;

(c) if both u and u' are onto, they are both one-to-one.

Proof. (a) Suppose u is one-to-one onto. Then $u'(F')$ is weakly dense; since (3) holds, so does (5). Hence u' is onto. And since u is onto, u' is one-to-one.

Conversely, if u' is one-to-one onto, then $u(E)$ is dense; also, (5) holds, therefore (3) holds and so u is onto. Since u' is onto, u is one-to-one.

(b) Since u is an isomorphism into, $u(E)$ is closed in F. Since u' is one-to-one, $u(E)$ is dense. Hence u is onto. Again, since u is one-to-one, $u'(F')$ is weakly dense. Further, (1) is true, hence (5) is true, and so u' is onto.

(c) If either of u or u' is onto, the other is one-to-one. ▌

8.7 The Bidual Space and the Biadjoint Mapping

Let E be a separated LCTVS, E' its topological dual. We have defined in Section 8.4 the strong topology $\beta(E', E)$ on E', and we shall henceforth denote by E'' the topological dual of E' relative to this topology; E'' is termed the *strong bidual* of E. There is a natural map j of E into E'' that associates with $x \in E$ the linear form $x' \to \langle x, x' \rangle$ on E'—clearly an element of E''. Obviously j is one-to-one and linear; it is onto if and only if E is semireflexive (Section 8.4). It is customary to identify E with its image $j(E)$ in E''. When E is a normed space, so is E' and $\beta(E', E)$ is just this normed topology; E'' is also normed and j is norm preserving.

On E'' one may consider at least two topologies of the type $T_{\mathfrak{S}}$. The first of these, which shall be denoted by T_e ($e =$ equicontinuous), corresponds to taking for $\mathfrak{S}$ all the equicontinuous subsets of E'; the second, denoted by T_b ($b =$ bounded), corresponds to taking for $\mathfrak{S}$ all the strongly bounded subsets of E'. A neighbourhood base at 0 for T_e is obtained by taking the $\sigma(E'', E')$-closures of a base at 0 for initial topology of E formed of convex and balanced sets; likewise a base at 0 for T_b is obtained by taking the $\sigma(E'', E')$-closures of polars in E of strongly bounded subsets of E'. It follows at once that T_b is always stronger than T_e. If E is bornological, however, T_b and T_e are identical. (For suppose W is a T_b-neighbourhood of 0 in E'', which may be assumed to be the $\sigma(E'', E')$-closure of the polar B'^0 in E of some strongly bounded set B' in E'. If $B \subset E$ is bounded, B' is absorbed by B^0, and so B'^0 absorbs B; that is, $W \cap E$ absorbs B. Since E is bornological, $W \cap E$ must be a neighbourhood of 0 in E, say U. Thus $W \supset U$. Hence also $\bar{W} \supset \bar{U}$ (closures for $\sigma(E'', E')$). Since $W = \bar{W}$ and $\bar{U}$ is a T_e-neighbourhood of 0, W is also such a neighbourhood, showing that T_b is weaker than, hence identical with, T_e.)

In general E is not dense in E'' for T_e (a fortiori not dense for T_b)—for example, $E = c_0$, $E' = \ell^1$, $E'' = \ell^\infty$. As we have seen above, however, E is certainly dense for $\sigma(E'', E')$; in fact, E'' is the union of the $\sigma(E'', E')$-closures of bounded subsets of E.

If E is a normed vector space, this last assertion may be refined:

▶ The unit ball B in E is $\sigma(E'', E')$-dense in the unit ball B'' in E''.

Indeed, according to the bipolar theorem (Theorem 8.1.5) it suffices to show that if $\|x''\| \leq 1$, then

$$|\langle x'', x' \rangle| \leq \text{Sup}\,\{|\langle x, x' \rangle| : \|x\| \leq 1\},$$

that is,

$$|\langle x'', x' \rangle| \leq \|x'\|;$$

but this results from the definition of $\|x''\|$.

Hitherto, when considering abstract separated dual systems (E, E') and (F, F'), we defined the biadjoint u'' of a weakly continuous linear map u of E into F in such a way as to appear as a linear map of E into F coinciding with u. We shall now see that if E and F are separated LCTVSs and u a continuous linear map of E into F, it is altogether natural to extend u'' into a linear map of E'' into F'', its restriction to E being u once again.

8.7.1 **Proposition.** Let E and F be LCTVSs, u a weakly continuous linear map of E into F. Then u' is continuous for $\beta(F', F)$ and $\beta(E', E)$.

Proof. It suffices to show that u transforms bounded sets into bounded sets. This is obvious since, in E or in F, there is an identity between the weakly bounded and the bounded sets, while a weakly continuous map transforms weakly bounded sets into weakly bounded sets. ▌

This being so, we may seek to define u'' as a map of E'' into F'' via the equation

$$\langle u''(x''), y' \rangle = \langle x'', u'(y') \rangle \tag{8.7.1}$$

required to hold for all y' in F' and all x'' in E''. This definition is indeed satisfactory because, as Proposition 8.7.1 shows, $y' \to \langle x'', u'(y') \rangle$ is continuous for $\beta(F', F)$ for each x'' in E'', hence is of the form $y' \to \langle y'', y' \rangle$ for a uniquely determined y'' in F; this y'' is $u''(x'')$. We shall continue to term u'' the biadjoint of u. Obviously u'' is linear and, equally obviously, $u'' \mid E = u$.

8.7.2 **Proposition.** Let E and F be separated LCTVSs, u a continuous linear map of E into F. Its biadjoint u'' is continuous for the T_b-topologies and the T_e-topologies on E'' and F''.

Proof. It suffices to show that u' transforms strongly bounded (resp. equicontinuous) subsets of F' into strongly bounded (resp. equicontinuous) subsets of E'. In respect of strongly bounded sets, this is asserted by Proposition 8.7.1. On the other hand, if V is a neighbourhod of 0 in F, then $U = u^{-1}(V)$ is a neighbourhood of 0 in E; since $u(U) \subset V$, (8.6.3) gives $u'(V^0) \subset U^0$ so that $u'(V^0)$ is equicontinuous in E', q.e.d. ▌

Remark. The equation (8.7.1) shows that u'' is continuous for the weak topologies $\sigma(E'', E')$ and $\sigma(F'', F')$. Combined with the fact that E is $\sigma(E'', E')$-dense in E'', it follows that u'' is the unique result of extending u by continuity with respect to these weak topologies.

Once again there is a refined statement applying when E and F are normed vector spaces:

▶ In this case the map $u \to u''$ of $L_c(E, F)$ into $L_c(E'', F'')$ is norm preserving.

Since E is mapped isometrically into E'' and $u'' \mid E = u$, it suffices to show that $\|u''\| \leq \|u\|$. Now

$$\begin{aligned} \|u''\| &= \operatorname{Sup} \{|\langle u''(x''), y' \rangle| : \|x''\| \leq 1, \|y'\| \leq 1\} \\ &= \operatorname{Sup} \{|\langle x'', u'(y') \rangle| : \|x''\| \leq 1, \|y'\| \leq 1\}. \end{aligned}$$

Now we know that any x'' satisfying $\|x''\| \leq 1$ is the $\sigma(E'', E')$-limit of elements x of E satisfying $\|x\| \leq 1$, and it follows from this that

$$\begin{aligned} \|u''\| &\leq \operatorname{Sup} \{|\langle x, u'(y') \rangle| : \|x\| \leq 1, \|y'\| \leq 1\} \\ &= \operatorname{Sup} |\langle u(x), y' \rangle| = \|u\|, \end{aligned}$$

as we had to show.

We end this section with a result for Banach spaces illustrating the use of the bidual space and the biadjoint map that includes certain interesting applications.

8.7.3 **Theorem.** Let E and F be Banach spaces, u a continuous linear map of E onto a dense vector subspace of F, G, and H strongly closed vector subspaces of E', G_1 the vector subspace $u'^{-1}(G)$ of F', and v the restriction of u' to G_1 regarded as a linear map of G_1 into G. Assume in addition the following

Hypothesis. H contains $u'(F')$ and is contained in the union of the weak closures in E' of bounded subsets of $u'(F')$.

Consider the following conditions:

(a) $u(E) = F$;

(b) $u'(F') = H$;

(c) $u'(F')$ is closed (weakly or strongly) in E';

(d) u' is a strong isomorphism of F' into E';

(e) for each y'' in F'' there exists z' in G' such that $v'(z') = y'' \mid G_1$; or, what is equivalent, there exists x'' in E'' such that $u''(x'') \mid G_1 = y'' \mid G_1$.

The conclusions are:

(1) (a)$\Leftrightarrow$(b)$\Leftrightarrow$(c)$\Leftrightarrow$(d) $\Rightarrow$ (e);

(2) if $G \supset u'(F')$, all five conditions (a)–(e) are equivalent.

Proof. Theorem 8.6.13 informs us that (a), (c), and (d), the second in either form, are equivalent.

If (c) holds, the hypothesis entails (b). Conversely, if (b) holds, $u'(F') = H$ is strongly closed in E' and so (d) is true as a consequence of the OMT. Thus (a) – (d) are equivalent. It remains to show that these imply (e), and conversely if $G \supset u'(F')$—that is, if $G_1 = F'$.

To begin with the reader will note that if z' is an element of G', then (Hahn-Banach theorem) there exists at least one x'' in E'' for which $z' = x'' \mid G$; and that for any such x'' one has $v'(z') = u''(x'') \mid G_1$. In fact the formula

$$\langle x'', u'(y')\rangle = \langle u''(x''), y'\rangle \tag{8.7.2}$$

holds for y' in F'. On the other hand, if y' is in G_1 one has

$$\langle z', v(y')\rangle = \langle v'(z'), y'\rangle,$$

that is,

$$\langle z', u'(y')\rangle = \langle v'(z'), y'\rangle,$$

that is,

$$\langle x'', u'(y')\rangle = \langle v'(z'), y'\rangle. \tag{8.7.3}$$

Comparison of (8.7.2) and (8.7.3) leads to the alleged formula. In this way we see the equivalence of the two formulations of (e).

We equip G and G_1 with the topologies induced by the strong (= normed) topologies of E' and F', respectively, noting that (since G is closed and u' is strongly continuous) both G and G_1 are Banach spaces. On the one hand, it is evident that (d) entails that v is an isomorphism of G_1 into G, and that the converse is true if $G_1 = F'$. On the other hand Corollary 8.6.15(b) informs us that v is an isomorphism into if and only if $v'(G') = G_1'$. Finally (Hahn-Banach theorem), G_1' is identical with the set of restrictions to G_1 of elements of F''. This, together with the preceding remarks, completes the proof. ▌

8.7.4 **Corollary.** If E and F are Banach spaces and u a continuous linear map of E onto a dense vector subspace of F, then $u(E) = F$ if and only if $u''(E'') = F''$.

Proof. We take $G = E'$ and H equal to the strong closure in E' of $u'(F')$ in E' and apply the theorem. ∎

Remark. As will be seen in Section 8.8 there are many practical advantages in taking suitable subspaces G of E' in place of the latter itself, due mainly to the fact that G' is often more manageable than is E''.

8.7.5. By way of illustration of Theorem 8.7.3 we consider the following situation. E is a Banach space and $(e'_t)_{t\in T}$ is a family of elements of E', T being an arbitrary index set. It will be assumed that

$$\lim_{t\to\infty} e'_t = 0 \text{ weakly in } E', \tag{8.7.4}$$

the limit being interpreted by regarding T as a discrete topological space. u will be the mapping of E into $c_0(T) = F$ that transforms an element x of E into the family

$$\hat{x} = (\langle x, e'_t\rangle)_{t\in T}. \tag{8.7.5}$$

It is easily verified that u' transforms the element λ of $F' = \ell^1(T)$ into $\sum \lambda(t)e'_t$, this series being unconditionally weakly convergent in E'. A simple and convenient way of ensuring that u' is one-to-one, and hence that $u(E)$ is dense in F, is to assume that there exists in E a family $(e_t)_{t\in T}$ such that

$$\langle e_t, e'_{t'}\rangle = \delta_{t,t'} \qquad (t, t' \in T). \tag{8.7.6}$$

In this case, to say that an element x' of E' belongs to $u'(F')$ signifies exactly that $\sum_{t\in T} |\langle e_t, x'\rangle| < +\infty$. If G is a strongly closed vector subspace of E' containing all the e'_t, then $G_1 = u'^{-1}(G)$ is none other than $F' = \ell^1(T)$. Moreover, if z' is in G', then $v'(z')$ is the family $\hat{z}' = (\langle e'_t, z'\rangle)_{t\in T}$ belonging to $F'' = \ell^\infty(T)$. Appeal to Theorem 8.7.3 yields the following result.

8.7.6 **Theorem.** Let E be a Banach space, $(e'_t)_{t\in T}$ a family of elements of E' satisfying (8.7.4) and (8.7.6), G a strongly closed vector subspace of E' containing all the e'_t, H a strongly closed vector subspace of E' containing the set

$$\Lambda = \left\{\sum_{t\in T} \lambda(t)e'_t : \lambda \in \ell^1(T)\right\} \tag{8.7.7}$$

and contained in the union of the weak closures in E' of bounded subsets of Λ. Let us assume that the families $\hat{x}$, defined by (8.7.5) with $\hat{x}$ ranging over E, are dense in $c_0(T)$. Then the following assertions are equivalent:

(1) The families $\hat{x}$, with x ranging over E, exhaust $c_0(T)$.
(2) $\sum_{t\in T} |\langle e_t, x'\rangle| < +\infty$ for each x' in H.
(3) Λ is closed (weakly or strongly) in E'.
(4) There exists a number $k \geq 0$ such that for each λ in $\ell^1(T)$ one has

$$\sum_{t\in T} |\lambda(t)| \leq k \cdot \left\|\sum_{t\in T} \lambda(t)e'_t\right\|.$$

(5) Each family β in $\ell^\infty(T)$ can be written

$$\beta(t) = \hat{z}'(t) = (\langle e_t', z' \rangle)_{t \in T}$$

for some z' in G'.

A similar result may be obtained starting from the assumption that for some p satisfying $1 \leq p < +\infty$, the families $\hat{x}$ all lie in $\ell^p(T)$ and form a dense subspace of the latter. In Theorem 8.7.3 we should then take $F = \ell^p(T)$, so that $F' = \ell^{p'}(T)$ $(1/p + 1/p' = 1)$ and $F'' \supset \ell^p(T)$. This would lead us directly to the following theorem.

8.7.7 **Theorem.** Let E be a Banach space, $(e_t')_{t \in T}$ a family of elements of E satisfying (8.7.6) and such that for each x in E the family (8.7.5) lies in $\ell^p(T)$, G a strongly closed vector subspace of E' containing the set

$$\Lambda = \left\{ \sum_{t \in T} \lambda(t) e_t' : \lambda \in \ell^{p'}(T) \right\}^\dagger \tag{8.7.8}$$

and contained in the union of the weak closures in E' of bounded subsets of Λ. Assume that the families $\hat{x}$, defined by (8.7.5) with x ranging over E, are dense in $\ell^p(T)$. Then the following assertions are equivalent:

(1) The families $\hat{x}$, with x ranging over E, exhaust $\ell^p(T)$.

(2) $\sum_{t \in T} |\langle e_t, x' \rangle|^{p'} < +\infty$ for each x' in H.

(3) Λ is closed (weakly or strongly) in E'.

(4) There exists a number $k \geq 0$ such that for each λ in $\ell^{p'}(T)$ one has

$$\left\{ \sum_{t \in T} |\lambda(t)|^{p'} \right\}^{1/p'} \leq k \cdot \left\| \sum_{t \in T} \lambda(t) e_t' \right\|;$$

if $p = 1$, the left-hand side should read $\operatorname{Sup}_{t \in T} |\lambda(t)|$.

(5) The families $\hat{z}' = (\langle e_t', z' \rangle)_{t \in T}$, with z' ranging over G', exhaust $\ell^p(T)$.

In the next section we shall see how the above theorems may be applied to some problems connected with harmonic analysis.

8.7.8. Meanwhile we now wish to consider some extensions of the statement (a) that was made in Corollary 8.6.15. Generalizations may proceed in either or both of two directions, the first of which deals still with Banach spaces while u is not assumed to be continuous, and the second of which seeks to drop the restriction that E and F be Banach spaces. The former type of extension was effected by G.-C. Rota [1], while a much more elaborate treatment has been given by F. E. Browder [1] which incorporates both types of extension and which serves as a foundation for his study of linear partial differential equations (see Section 5.20). Browder's treatment is lengthy and rather complex and we shall restrict attention here to the Banach space case.

---

† Our hypotheses entail that the series $\sum_{t \in T} \lambda(t) e_t'$ is unconditionally weakly convergent in E' for each λ in $\ell^{p'}(T)$.

8.7.9 **Theorem.** Let us suppose that E and F are Banach spaces and u a linear map with a domain D dense in E and with a graph G that is closed in $E \times F$. Then $u(D) = F$ if and only if the adjoint u' has a continuous inverse—that is if and only if there exists a number $k > 0$ such that

$$\|u'(y')\| \geq k \, \|y'\| \qquad (y' \in D'), \tag{8.7.9}$$

where $D' \subset F'$ is the domain of u'.

Remark. The case in which $E = F$ is a Hilbert space is given, and used in connection with the theory of linear partial differential equations, by Hörmander [1]. His proof depends on the special features of this particular case.

Proof. In all cases this involves the consideration of the linear map v defined on the graph G of u by

$$v(x, u(x)) = u(x) \qquad (x \in D).$$

Being closed in $E \times F$, G is itself a Banach space with the induced norm. Also it is obvious that v is continuous on G into F, and that $v(G) = u(D)$. Thus $u(D) = F$ if and only if $v(G) = F$. Moreover, by Corollary 8.6.15 $v(G) = F$ if and only if v' is a strong isomorphism of F' into G'—that is, if and only if there exists a number $c > 0$ such that

$$\|v'(y')\| \geq c \, \|y'\| \qquad (y' \in F'). \tag{8.7.10}$$

Now G' is a quotient of $E' \times F'$, to $(a', b') \in E' \times F'$ corresponding the element $z' = f(a', b')$ of G' defined by

$$(x, u(x)) \to \langle x, a' \rangle + \langle u(x), b' \rangle \qquad (x \in D),$$

and $\|z'\|$ is the infimum of $\|a'\| + \|b'\|$ when (a', b') ranges over $f^{-1}(z')$. On the other hand, if $y' \in F'$ then $z' = v'(y')$ is defined by

$$\begin{aligned} \langle (x, u(x)), z' \rangle &= \langle (x, u(x)), \quad v'(y') \rangle \\ &= \langle v((x, u(x)), y' \rangle \\ &= \langle u(x), y' \rangle \quad (x \in D). \end{aligned}$$

Thus (8.7.10) reads at length:

$$\text{Inf} \, \{\|a'\| + \|b'\| : \langle x, a' \rangle + \langle u(x), b' \rangle = \langle (u(x), y' \rangle \text{ for } x \in D\} \geq c \, \|y'\|,$$

or

$$\text{Inf} \, \{\|a'\| + \|b'\| : \langle x, a' \rangle = \langle u(x), y' - b' \rangle \text{ for } x \in D\} \geq c \, \|y'\|.$$

The condition

$$\langle x, a' \rangle = \langle u(x), y' - b' \rangle \qquad \text{for } x \in D$$

signifies that $w' = y' - b' \in D'$ and $\langle x, a' \rangle = \langle x, u'(w') \rangle$ for $x \in D$. Since D is dense in E, the last relation is equivalent to $a' = u'(w')$. Accordingly (8.7.10) is equivalent to

$$\text{Inf} \, \{\|u'(w')\| + \|y' - w'\| : w' \in D'\} \geq c \, \|y'\| \; (y' \in F'). \tag{8.7.11}$$

It is obvious that (8.7.11) implies (8.7.9), for if $y' \in D'$ then the infimum in (8.7.11) is at most $\|u'(y')\|$ (taking $w' = y'$). Conversely let us suppose that (8.7.9) is true and verify that (8.7.11) follows. This is trivially so if $y' = 0$.

Also and in any case if (8.7.11) holds for a given y', then it continues to hold when y' is replaced by $\lambda y'$ for λ any scalar $\neq 0$. It thus remains only to check that (8.7.11) holds whenever $\|y'\| = 1$. If this were not the case, however, there exist sequences $(y'_n) \in F'$ and $(w'_n) \in D'$ such that

$$\|y'_n\| = 1, \qquad \|u'(w'_n)\| + \|y'_n - w'_n\| \leq \frac{1}{n}.$$

This shows that $\|u'(w'_n)\| \leq 1/n$ and yet

$$1 = \|y'_n\| \leq \|w'_n\| + \|y'_n - w'_n\|,$$

so that

$$\|w'_n\| \geq \frac{n-1}{n}.$$

Thus

$$\|u'(w'_n)\| \leq (n-1)^{-1} \|w'_n\|,$$

which contradicts (8.7.9) and completes the proof. ▮

8.8 Some Applications

We propose to show how Theorems 8.7.3, 8.7.6, and 8.7.7 may be applied to certain problems in harmonic analysis, following Hewitt and Zuckerman and Helson.

(1) Lacunary fourier series. We consider functions of a real variable with period 2π or, what is the same thing, functions defined on the unit circumference S in the complex plane. The notations of Example 7.2.3 are used and extended: if f is a function or a measure on S, $\hat{f}$ denotes its Fourier transform (sequence of Fourier coefficients $\hat{f}(n)$, $n = 0, \pm 1, \pm 2, \cdots$). In what follows T will denote a set of integers—positive, negative, or zero.

A function or measure is said to be T-spectral if $\hat{f}(n) = 0$ for all integers n not in T. The Riemann-Lebesgue lemma shows that the Fourier transformation restricted to $T, u(x) = \hat{x} \mid T$, maps L^1 into $c_0(T)$, and it is well known and simple to prove that $u(L^1)$ is dense in $c_0(T)$. Continuity of u, from L^1 into $c_0(T)$, is trivial.

It was discovered long ago that when T is very sparsely scattered amongst the integers, T-spectral functions and measures have some extraordinary properties. The precise condition on T, termed "lacunarity," was that $T = \{\pm n_k : k = 1, 2, \cdots\}$ where $n_k > 0$ and

$$\operatorname{Inf} \frac{n_{k+1}}{n_k} > 1.$$

Trigonometric series whose nonvanishing harmonics form such a lacunary set are themselvs termed *lacunary*. One typical result about such series is due to Szidon (Zygmund [1], p. 139) and asserts that the Fourier series of a bounded function, if lacunary, is absolutely convergent; another, due to Banach (Zygmund [1], p. 215) asserts that if T is lacunary, then u maps L^1 *onto* $c_0(T)$. Kaczmarz and Steinhaus ([1], pp. 250–255), Hewitt and Zuckerman [2] and, independently, the present writer have shown that without any explicit hypothesis of lacunarity upon T, theorems of the type exemplified by those of Szidon and Banach can be formed into equivalent groups on the basis of functional analytic arguments. The two equivalence theorems of Hewitt and Zuckerman (loc. cit., Theorems 2.1 and 6.1) follow at once from our Theorems 8.7.6 and 8.7.7 in the following way.

In Theorem 8.7.6 we take $E = L^1$, so that E' is L^∞, and e'_t will be the element of L^∞ defined by the function e^{-its} of s. G will be C, the space of continuous functions on S, so that G' is identified with M, the space of Radon measures on S. For e_t we take the element of L^1 defined by the function $(2\pi)^{-1}e^{its}$ of s. It is clear that $u'(F') \subset C = G$. Denoting by $-T$ the set of integers $-t$ as t ranges over T, we may take for H *either* the set of $(-T)$-spectral elements of L^∞ *or* the set of $(-T)$-spectral elements of C: that in either case the Hypothesis of Theorem 8.7.3 is fulfilled, is a consequence of the simplest $(C, 1)$-summability properties of Fourier series. In writing down the condition (2) of Theorem 8.7.6, however, we make use of the obvious fact that the mapping $f \to f'$, where $f'(s) = f(-s)$, is an isomorphism (weak or strong) of L^∞ onto itself that transforms the set of $(-T)$-spectral elements into the set of T-spectral elements. We have finally to notice that v' is again defined by taking the Fourier transform restricted to T. With the above selection and remarks made, an appeal to Theorem 8.7.6 gives the first of the Hewitt-Zuckerman theorems asserting the equivalence of the following five conditions (wherein we use the symbol $\hat{A} \mid T$, A being a set of functions or measures on S, for the set of restrictions to T of Fourier transforms of elements of A):

(a) $\hat{L}^1 \mid T = c_0(T)$;

(b) $\sum_{t \in T} |\hat{f}(t)| < +\infty$ for each T-spectral f in L^∞ (or for each T-spectral f in C);

(c) The set of functions $\sum_{t\in T} \lambda(t)e^{its}$, λ ranging over $\ell^1(T)$, is closed (weakly or strongly) in L^∞;

(d) There exists a number $k \geq 0$ such that for each λ in $\ell^1(T)$

$$\sum_{t \in T} |\lambda(t)| \leq k \cdot \operatorname*{Sup}_s \left| \sum_{t\in T} \lambda(t)e^{its} \right|;$$

(e) $\hat{M} \mid T = \ell^\infty(T)$.

The second of the Hewitt-Zuckerman theorems follows similarly from Theorem 8.7.7 if we make a change in meaning of the symbols. Briefly, one takes $E = C$, so that $E' = M$, $G = L^1$, $p = 2$, and for H one takes the set of $(-T)$-spectral elements of L^1. The e'_t are now the measures with densities e^{-its}. By contrast, u is regarded now as a map of C into $\ell^2(T)$. There results the equivalence of the following four conditions:

(a′) $\hat{C} \mid T = \ell^2(T)$;

(b′) $\sum_{t\in T} |\hat{f}(t)|^2 < +\infty$ for each T-spectral f in M (or for each T-spectral f in L^1);

(c′) The set of functions $\sum_{t\in T} \lambda(t)e^{its}$, λ ranging over $\ell^2(T)$, is closed (weakly or strongly) in M;

(d′) There exists a number $k \geq 0$ such that for each λ in $\ell^2(T)$

$$\left\{\sum_{t\in T} |\lambda(t)|^2\right\}^{1/2} \leq k \cdot \int_{-\pi}^{\pi} \left|\sum_{t\in T} \lambda(t)e^{its}\right| ds.$$

In view of Parseval's formula, (d′) signifies that

$$\|f\|_{L^2} \leq k \cdot \|f\|_{L^1}$$

for each T-spectral f in L^2; by virtue of the CGT this is in turn equivalent to saying that every T-spectral function in L^1 (or even every T-spectral measure) is in L^2 (or has for density a function in L^2). Further developments have been given by Rudin ([3]; [9], Chapter 5).

(2) We next turn to some partial analogues of the preceding results that apply to Fourier integrals, taking a problem studied by Helson [4]. We shall begin with a slightly more general situation than that concerned with Fourier transforms.

Let S and T be locally compact spaces with positive Radon measures ds and dt, respectively, denoting by $L^1(S)$ and $L^1(T)$ the associated Lebesgue spaces. $C_0(S)$ and $C_0(T)$ are the Banach spaces of continuous functions on S and T, respectively, which tend to 0 at infinity, having as duals the corresponding spaces $M(S)$ and $M(T)$ of bounded Radon measures. $L^1(S)$ is imbedded in $M(S)$ by associating with the function class f in $L^1(S)$ the measure with density f with respect to ds; likewise with $L^1(T)$ and $M(T)$.

We take a "kernel" function k on $S \times T$ which we assume to be bounded and continuous and such that

$$\lim_{t\to\infty} k(s, t) = 0 \text{ weakly in } L^\infty(S). \tag{8.8.1}$$

In Theorem 8.7.3 we propose to take $E = L^1(S)$, $F = C_0(T)$ and $u(x) = y$, where

$$y(t) = \int k(s, t)x(s)\, ds. \tag{8.8.2}$$

Accordingly E' will be $L^\infty(S)$ and F' will be $M(T)$. The adjoint u' is given by $x' = u'(y')$, where

$$x'(s) = \int k(s, t)\, dy'(t), \tag{8.8.3}$$

carrying the measure y' in $M(T)$ into the function x' in $L^\infty(S)$. For G we take $C_0(S)$ (or, rather, the set of elements of $L^\infty(S)$ determined by functions in $C_0(S)$) so that G_1 consists of the measures y' in $M(T)$ such that the function x' defined by (8.8.3) tends to 0 as $s \to \infty$; we denote by $M_0(T)$ this latter subset of $M(T)$. Then G' will be $M(S)$ and, for z' in G', $v'(z')$ will be the function $\int k(s, t)\, dz'(s)$ of t. We ignore the role of H in Theorem 8.7.3 since this contributes nothing towards the theorem of Helson at which we aim. The principal deduction we require amounts to the conclusion that, in Theorem 8.7.3, (a) implies (e). In applying this we observe that each bounded, universally measurable function β on T defines an element of $F'' = M(T)'$ that transforms the measure m in $M(T)$ into the number $\int \beta(t)\, dm(t)$. Accordingly we derive the following result:

HELSON'S THEOREM. If u maps $L^1(S)$ onto $C_0(T)$ then, given any bounded universally measurable function β on T, there exists a measure μ in $M(S)$ such that

$$\int \beta(t)\, d\nu(t) = \int d\nu(t) \int k(s, t)\, d\mu(s) \tag{8.8.4}$$

for all ν in $M_0(T)$.

Helson's result is recoverable if we specialize by taking $S =$ real axis, $ds =$ Lebesgue measure, $T =$ a subset of the real axis—which may without loss of generality be assumed to be closed—and dt the restriction to T of Lebesgue measure, and $k(s, t) = e^{-2\pi ist}$. Then $u(x) = \hat{x} \mid T$, where $\hat{x}$ is the usual Fourier transform of x, and u' amounts to the usual inverse Fourier transformation. $M_0(T)$ can be thought of as the set of bounded Radon measures on the real axis that are supported by T and whose inverse Fourier transforms tend to 0 at infinity. It is clear that if ν belongs to $M_0(T)$, so too does $e^{2\pi iat} \cdot \nu$ for any real a. Consequently (8.8.4) entails that $\beta(t) = \int k(s, t)\, d\mu(s)$ a.e. (ν); and Helson's theorem entails that this is true for each ν in $M_0(T)$ whenever $\hat{L}^1 \mid T = C_0(T)$. The remainder of Helson's reasoning is concerned with showing that this cannot be so unless $M_0(T) = \{0\}$. Let us suppose in fact that $M_0(T)$ contains a measure $\nu \neq 0$, and let $K \subset T$ be the support of ν. K cannot be countable: if it were, the inverse Fourier transform of ν would be of the form $\sum_n c_n e^{2\pi i t_n s}$ with $0 < \sum_n |c_n| < +\infty$, while no such function tends to 0 as $s \to \infty$. The open complement of K is the union of a sequence (a_n, b_n) of disjoint open intervals $(a_n < b_n)$. We choose a point t_0 of K distinct from all the a_n and b_n. Then $|\nu|\,(t_0 - e, t_0) > 0$ for each $e > 0$—for otherwise $(t_0 - e, t_0)$ would be contained in (a_n, b_n) for some n, so that $a_n < t_0 \leq b_n$; but t_0 lies in K and so is not less than b_n, and neither can t_0 equal b_n by the way it is chosen. Similarly, $|\nu|\,(t_0, t_0 + e) > 0$ for each $e > 0$. This being so, we take for β a function that is 0 for $t < t_0$ and 1 for $t > t_0$. The continuous function $f(t) = \int e^{-2\pi ist}\, d\mu(s)$ would then have to be such that $f(t) = 0$ for certain values of t arbitrarily close to t_0 on the left and $f(t) = 1$ for certain values of t arbitrarily close to t_0 on the right, which is absurd. We thus conclude with Helson that

▶ If $\hat{L}^1 \mid T = C_0(T)$, then $M_0(T) = \{0\}$; that is, T supports no nonzero bounded Radon measure whose Fourier transform tends to 0 at infinity.

For further developments relating to these ideas, see Kahane and Salem [1] and Rudin [4]. An up-to-date connected account of these problems appears in Chapter 5 of Rudin [9].

8.9 Return to the Open Mapping and Closed Graph Theorems

We now make the promised return to this subject, concentrating on the case of *locally convex* spaces. The theorems in Section 6.4 have been freed of metrizability conditions in respect to *one* of the two spaces involved, the other space remaining subject to this sort of restriction. For locally convex spaces, the situation is more satisfactory, and a new and independent approach is possible and desirable. This is due largely to the work of Pták [1], [2], [3], [4], Collins [1], W. Robertson [1], and A. P. and W. Robertson [1]. W. Robertson's work suggests strongly that some sort of completeness of E is necessary; see Remark (1) following Theorem 6.4.4. This suspicion is confirmed by Pták's analysis, which shows that a form of completeness stronger than the ordinary one is the natural requirement.

8.9.1 Fully Complete and B_r-Complete Spaces. Let E be a LCTVS. A subset M of E' will be termed *almost weakly closed* if $M \cap U^0$ is weakly closed in E' for each neighbourhood U of 0 in E. Since

$$M \cap V^0 = (M \cap U^0) \cap V^0$$

whenever $U \subset V$, it is evidently equivalent to demand that $M \cap U^0$ be weakly closed in E' for each member U of a base at 0 in E.

The concept of almost weakly closed sets will be studied in more detail in the next section. Meanwhile we will reformulate Theorem 8.5.1 in a disguise that will motivate and explain some further definitions.

As we know, a hyperplane H in E' is representable (in many ways) in the form $\{x' \in E'^*: u(x') = 0\}$, where u is a nonzero element of E'^*; and that when u is so chosen, H is weakly closed if and only if u is weakly continuous. There is an analogous statement with "almost weakly closed" replacing "weakly closed" and the demand that u be weakly continuous replaced by the demand that $u \mid U^0$ be weakly continuous for each U.

Indeed, if $u \mid U^0$ is weakly continuous, then obviously $H \cap U^0$ is weakly closed, and so H is almost weakly closed. To prove the converse let us suppose that H is almost weakly closed and define $H(\alpha) = \{x' \in E' : u(x') = \alpha\}$, so that $H = H(0)$. For any U, *either* $U^0 \cap H(\alpha)$ is void, *or* there exists y' belonging to this intersection. In the latter case,

$$H(\alpha) \cap U^0 = (y' + H \cap 2U^0) \cap U^0.$$

So it follows that each $H(\alpha)$ is almost weakly closed. Supposing that $u \mid U^0$ were not weakly continuous, for some $\lambda > 0$ it would be the case that for each weak neighbourhood W of 0 in E' there exists an $x' \in W \cap U^0$ satisfying $|u(x')| \geq \lambda$. On the other hand, $U^0 \cap H(\frac{1}{2}\lambda)$ is weakly closed and does not contain 0. Thus there exists a weak neighbourhood W of 0 in E' such that

$$W \cap U^0 \cap H(\tfrac{1}{2}\lambda) = \emptyset. \tag{8.9.1}$$

We know, however, that there exists an x' in $W \cap U^0$ satisfying $|u(x')| \geq \lambda$. Let $\beta = \lambda/2u(x')$, so that $|\beta| \leq \frac{1}{2}$ and $\beta x' \in W \cap U^0$. At the same time, $u(\beta x') = \frac{1}{2}\lambda$ and so $\beta x' \in H(\frac{1}{2}\lambda)$. Thus $\beta x' \in W \cap U^0 \cap H(\frac{1}{2}\lambda)$, contradicting (8.9.1) and completing the proof.

One may now infer from Theorem 8.5.1 the conclusion that completeness of E is necessary and sufficient in order that any almost weakly closed hyperplane in E' be weakly closed. This in turn motivates and justifies the following definitions.

E will be said to be *fully complete* (Pták's B-completeness) if every almost weakly closed vector subspace M of E' is weakly closed. Pták also introduces *B_r-complete* spaces as those for which the same condition is to be fulfilled for weakly dense vector subspaces M of E' (the conclusion then being, of course, that $M = E'$).

Pták's analysis, which will not be reproduced here in full, shows that (1) E is fully complete if and only if every continuous nearly open linear map u of E onto a separated LCTVS F is open; and that (2) E is B_r-complete if and only if every one-to-one continuous nearly open linear map u of E onto a separated LCTVS is open. We shall content ourselves with giving the proofs of the "only if" portions of (1) and (2), which are naturally the most important fragments from the point of view of applications.

We remark here that Collins [1] introduced the concept of full completeness in a somewhat different fashion. Equivalence with our definition is established by Kelley ([4], Theorem 4). Kelley at the same time introduces the concept of "hypercompleteness" which he shows (loc. cit.) to be equivalent to the demand that each almost weakly closed, convex, balanced subset of E' be weakly closed.

Our starting point will be a general theorem established by Pták [3]. The situation is similar to that in Lemma 6.4.1 and we adopt a similar notation: if S is a subset of $E \times F$, A a subset of E, we write

$$S_F(A) = \pi_F(S \cap \pi_E^{-1}(A)), \qquad S(A) = S \cap \pi_E^{-1}(A).$$

8.9.2 **Theorem.** Let E and F be LCTVSs, S a closed vector subspace of $E \times F$ such that $\pi_F(S) = F$. Suppose that E is fully complete and that for each neighbourhood U of 0 in E, $\overline{S_F(U)}$ is a neighbourhood of 0 in F. Then $S_F(U)$ is itself a neighbourhood of 0 in F. If in addition the set $X = \pi_E(S \cap \pi_F^{-1}(0))$ is $\{0\}$, the conclusion is valid whenever E is B_r-complete.

Proof. Let Z be the set of (x', y') in $E' \times F'$ such that $\langle x, x'\rangle + \langle y, y'\rangle = 0$ for all (x, y) in S; let Q be the set of x' in E' such that (x', y') falls in Z for a suitable y' in F'; and let $X = \pi_E(S \cap \pi_F^{-1}(0)) = \{x \in E : (x, 0) \in S\}$. Q is a vector subspace of E', X a vector subspace of E. The first step is to prove that $Q^{00} = X^0$.

To begin with, $Q \subset X^0$. For if x' is in Q, there exists y' in E' such that (x', y') is in Z. Then $\langle x, x'\rangle + \langle y, y'\rangle = 0$ for all (x, y) in S. If x is in X, $(x, 0)$ is in S, and so $\langle x, x'\rangle = \langle x, x'\rangle + \langle 0, y'\rangle = 0$; that is, x' is in X^0. Thus $Q \subset X^0$.

If on the other hand some x in Q^0 failed to belong to X, then $(x, 0)$ would not belong to S. Since S is closed, there would exist (x', y') in Z such that $\langle x, x'\rangle + \langle 0, y'\rangle \neq 0$; that is, $\langle x, x'\rangle \neq 0$. This would contradict the assumption that x is in Q^0, since (x', y') in Z signifies that x' is in Q. Thus $Q^0 \subset X$.

We have then both $Q \subset X^0$ and $Q^0 \subset X$. Taken together, these give $Q^{00} = X^0$.

The next step is to show, by using full (or B_r-) completeness of E, that Q is weakly closed. When this is done, we shall know that $Q = Q^{00}$ and so $Q = X^0$.

To this end, let U be any neighbourhood of 0 in E. We will show that $Q \cap U^0$ is weakly closed. That Q is weakly closed will then follow from full completeness of E, or from B_r-completeness if we know in advance that $X = \{0\}$. Let us suppose then that (x_i') is a directed family extracted from $Q \cap U^0$ that converges weakly to x': we have to show that x' is in $Q \cap U^0$. Since U^0 is weakly closed, it is certain that x' is in U^0. For each i there exists y_i' in F' such that (x_i', y_i') is in Z and hence

$$\langle x, x_i'\rangle + \langle y, y_i'\rangle = 0 \tag{8.9.2}$$

for all (x, y) in S. We see then that $f(y) = \lim_i \langle y, y_i'\rangle$ exists finitely, for each y in $\pi_F(S) = F$. Plainly f is a linear form on F. It will suffice to show that f is continuous, since it will then be equal to some y' in F'; and from (8.9.2) we shall obtain in the limit $\langle x, x'\rangle + \langle y, y'\rangle = 0$ for all (x, y) in S, hence (x', y') belongs to Z, and so x' belongs to Q.

To show that f is continuous, we use the hypothesis that $V = \overline{S_F(U)}$ is a neighbourhood of 0 in F. For if $y \in V$, it is the limit of a directed family (y_j) extracted from $S_F(U)$. For each j there exists x_j in U such that (x_j, y_j) is in S. Then, by (8.9.2), we see that $\langle y_j, y_i'\rangle = -\langle x_j, x_i'\rangle$ is in modulus at most 1, and this for all i (since x_i' lies in U^0). Keeping i fixed, let j increase: this yields $|\langle y, y_i'\rangle| \leq 1$ for all i. Letting i increase, there follows $|f(y)| \leq 1$. Since this is true for each y in V, f is continuous.

At this stage we have established that $Q = X^0$.

The final step is to show that

$$\overline{S_F(U)} \subset 2 \cdot S_F(U).$$

Suppose that y_0 belonged to the first set but not to the second. Since y_0 is not in $S_F(2U)$, one may choose x_0 in E such that (x_0, y_0) belongs to S, and then the sets $2U$ and $x_0 + X$ would be disjoint. It follows (Corollary 2.2.4) that an x' in U^0 may be found so that

$$m = \operatorname{Inf}\{|\langle x_0 + x, x'\rangle| : x \in X\}$$

satisfies $m > 1$. X being a vector subspace, it follows that $x' \in X^0 = Q$ and one may choose y' so that (x', y') is in Z, and so $\langle x, x'\rangle + \langle y, y'\rangle = 0$ for all (x, y) in S. Now if y is in $S_F(U)$, we may choose x in U so that (x, y) falls into S, and we then obtain from (8.9.2):

$$|\langle y, y'\rangle| = |\langle x, x'\rangle| \leq 1.$$

The same is therefore true by continuity when y is replaced by y_0. On the other hand, since (x_0, y_0) belongs to S, so

$$\langle x_0, x' \rangle + \langle y_0, y' \rangle = 0,$$

which gives $|\langle y_0, y' \rangle| = |\langle x_0, x' \rangle|$, which is at least $m > 1$. This contradiction completes the proof. ▌

It is now an easy matter to read off corresponding formulations of the OMT and the CGT.

8.9.3 Theorem. Let E and F be LCTVSs, u a linear map of a vector subspace E_0 of E onto F that has a graph closed in $E \times F$ and that is nearly open. Then

(1) If E is fully complete, u is open.

(2) If E is B_r-complete and u is one to one, then u is open.

Proof. In Theorem 8.9.2 we take for S the graph of u:

$$S = \{(x, u(x)) : x \in E_0\}.$$

Then

$$S_F(U) = u(U \cap E_0), \qquad \pi_F(S) = u(E_0)$$

and

$$X = u^{-1}(0).$$

The result follows immediately. ▌

8.9.4 Theorem. Let E and F be LCTVSs, E being B_r-complete, v a linear map of F into E having a graph closed in $F \times E$. If v is nearly continuous, it is continuous.

Proof. In Theorem 8.9.2 we take for S the diagonal reflection of the graph of v:

$$S = \{(v(y), y) : y \in F\}.$$

S is closed in $E \times F$. Then

$$S_F(U) = v^{-1}(U), \qquad \pi_F(S) = F,$$

and

$$X = \{v(0)\} = \{0\}.$$

The result follows at once from Theorem 8.9.2. ▌

8.9.5 Remarks. (1) The reader is reminded that if F is barrelled, any linear map u of E_0 onto F is necessarily nearly open; and that any linear map v of F into E is necessarily nearly continuous. See the proof of Theorem 6.4.4.

(2) In so far as LCTVSs are concerned, the present theorems entail our earlier results (labelled 6.4.2 through 6.4.4), once it is known that any complete semimetrizable LCTVS is fully complete. This is established in the next section.

(3) At the expense of imposing further conditions on u (or v) it is possible to reformulate Theorems 8.9.3 and 8.9.4 with the assumption that E has a completion that is fully complete or B_r-complete, respectively. The reader should compare Remark (1) following Theorem 6.4.4.

(4) By combining Theorems 8.9.3 and 8.9.4 with Propositions 8.10.7 and 8.10.11 one can derive versions of the OMT and the CGT that prove to be useful in certain applications. See Exercises 8.41–8.45, and also Grothendieck [9].

8.9.6. In view of Theorem 8.9.4 it is interesting to note that barrelled spaces can be characterized amongst the LCTVSs in terms of the CGT. Mahowald [1] has shown that a LCTVS F is barrelled if and only if it is the case that for any Banach space E, any closed linear map of F into E is continuous.

Proceeding along the same lines, the infrabarrelled spaces F prove to be precisely the LCTVSs with the property that for any Banach space E, any closed linear map of F into E that is bounded (that is, transforms bounded sets into bounded sets) is continuous.

8.9.7. It is not possible in Theorem 8.9.3 to replace the assumption that E is fully complete (or B_r-complete) by ordinary completeness, even when $E_0 = E$ and F is a Banach space. The following example lays any such hopes to rest.

Let E be any infinite dimensional Banach space, the normed topology of which we denote by t. We denote by t_∞ the strongest locally convex topology on E (see Subsection 1.10.1). We have shown (loc. cit.) that (E, t_∞) is complete.

For u let us take the identity map of (E, t_∞) into (E, t). Evidently, u is continuous and onto. Since (E, t) is separated, u has a closed graph. Moreover, u is nearly open. For, if W is any convex, balanced, and absorbent subset of E, the t-closure of $u(W) = W$ is a barrel in (E, t) and therefore a t-neighbourhood of 0.

Despite all this, however, u is not open. If it were, any convex, balanced, and absorbent subset of E would be a t-neighbourhood of 0. In particular, any linear form on E would be t-continuous. This is false: take any infinite sequence (a_n) in E that is linearly independent and assume, as one may, that $\|a_n\| = 1$ for each n. The sequence (a_n) may be extended into an algebraic base for E and a linear form on E may be defined so as to take arbitrarily preassigned values at elements of the base. In particular, there is a linear form f on E such that $f(a_n) = n$ for $n = 1, 2, \cdots$. Evidently, this f is not t-continuous.

This construction yields, of course, an example of a LCTVS E that is separated, complete, and barrelled (see Section 6.2) and yet is not B_r-complete, a fortiori not fully complete. For a more direct construction, see 8.10.3.

8.10 Concerning Fully Complete Spaces

It is possible to consider the topology on E' relative to which the closed sets are precisely those that are almost weakly closed in the sense of Subsection 8.9.1, and it is natural to term this the *almost weak topology* on E'. The name *bounded weak topology* is also current for the case in which E is a Banach space, a situation dealt with by Dieudonné [6]; see also Dunford and Schwartz [1], p. 427. It is by no means clear that this topology is locally convex in general, though we shall verify this for semimetrizable LCTVSs E. Meanwhile we veer slightly in our approach and seek locally convex topologies on E' that induce on each U^0 a topology identical with that induced by $\sigma(E', E)$. Usually there will be many such locally convex topologies. We take stock of three of them:

(1) The topology $T_c = \kappa(E', E)$ (see Subsection 8.3.3) having for a neighbourhood base at 0 the polars of convex, balanced, and compact subsets of E.

(2) The topology T_p having for a neighbourhood base at 0 the polars of precompact subsets of E. (Since the convex balanced envelope of a precompact set is again precompact, it makes no difference if we take polars of convex, balanced, and precompact subsets of E.)

(3) The topology T^* having for a base at 0 the sets W which are convex, balanced, almost weakly closed, and absorbent and such that W^0 is precompact in E.

It is clear that always

$$T_c \leq T_p \leq T^*. \tag{8.10.1}$$

On the other hand, $T_c = T_p$ if E is quasi-complete (since then the closure of any precompact subset of E is compact). Pták has shown ([1], (6.9)) that T^* is the *strongest* locally convex topology on E' inducing on each U^0 the same topology as is induced by $\sigma(E', E)$. We shall not need this fact and omit the proof. It is important, however, to observe that the topology T_c on E' is compatible with the duality between E and E' (Theorem 8.3.2).

Our primary aim is to show that any Fréchet space is fully complete. In doing this the basic result is an adaptation of a lemma given by Dieudonné and Schwartz ([1], p. 84, Lemme).

8.10.1 **Lemma.** Let E be a LCTVS, $(U_n)_{n\geq 0}$ a decreasing sequence of neighbourhoods of 0 in E, $U_0 = E$. Let W be an almost weakly open subset of E' containing 0. For each $n \geq 0$ there exists a finite set $B_n \subset U_n$ such that if $A_n = \bigcup_{p<n} B_p$, then $A_n^0 \cap U_n^0 \subset W$ for $n \geq 1$.

Proof. The argument proceeds by induction on n. We choose a finite set B_0 such that $B_0^0 \cap U_1^0 \subset W$ (possible since W is almost weakly open). Then $A_1^0 \cap U_1^0 \subset W$. We assume now that $n > 1$ and that B_p has been satisfactorily defined for $p \leq n - 1$, so that $A_n^0 \cap U_n^0 \subset W$. Let $H = U_{n+1}^0 \cap (E' \backslash W)$. Then H is weakly closed and therefore weakly compact. If no finite set $B_n \subset U_n$ existed with the desired property, for every finite set $B \subset U_n$, $(A_n \cup B)^0 \cap U_{n+1}^0$ would not be contained in W, so that $A_n^0 \cap B^0 \cap H$ would be nonvoid. These sets are weakly closed and form a decreasing directed family. By weak compactness, therefore, there would exist a point x_0' common to all the sets $A_n^0 \cap B^0 \cap H$ (B finite, $B \subset U_n$). Since U_n is the union of its finite subsets B, x_0' would belong to $A_n^0 \cap U_n^0 \cap H$. But the latter set is empty, since $H \subset E' \backslash W$ and $A_n^0 \cap U_n^0 \subset W$. This contradiction shows that the inductive process can continue and so completes the proof. ▌

8.10.2 **Corollary.** Let E be a semimetrizable LCTVS. If W is an almost weakly open subset of E' containing 0, then W contains a T_p-neighbourhood of 0.

Proof. In the preceding lemma we may take for (U_n) a neighbourhood base at 0 in E. Let $P = \bigcup_{n=1}^{\infty} A_n$. It is plain that P is precompact in E, so that P^0 is a T_p-neighbourhood of 0. For each n we have then $W \supset A_n^0 \cap U_n^0 \supset P^0 \cap U_n^0$. Further, the U_n^0 cover E'. So $W \supset P^0$, which is what we had to show. ▌

8.10.3 **Corollary.** Let E be a complete semimetrizable LCTVS. If W is an almost weakly open subset of E' containing 0, then W contains a T_c-neighbourhood of 0 in E.

Proof. Completeness of E arranges that $T_p = T_c$, so we have only to apply Corollary 8.10.2. ▌

8.10.4 **Corollary.** Let E be a semimetrizable LCTVS. A subset L of E' is almost weakly closed if and only if it is T_p-closed. If E is also complete, this is so if and only if L is T_c-closed.

Proof. Since we know (Proposition 0.4.9) that T_p induces on each U^0 a topology identical with that induced by $\sigma(E', E)$, it follows directly that L, if T_p-closed, is also almost weakly closed.

Suppose on the other hand that L is almost weakly closed and that x_0' is T_p-adherent to L. Since it is easily seen that translation maps almost weakly closed sets into almost weakly closed sets, if x_0' did not belong to L, there would exist an almost weakly open set W containing 0 and not meeting $L - x_0'$. Corollary 8.10.2 shows that then 0 is not T_p-adherent to $L - x_0'$; that is, that x_0' is not T_p-adherent to L—a contradiction. Thus x_0' must belong to L, and the latter must be T_p-closed. ▌

8.10.5 **Theorem.** Let E be a complete semimetrizable LCTVS. A convex subset L of E' is weakly closed if (and only if) it is almost weakly closed. In particular, E is fully complete (and even hypercomplete).

Proof. If L is almost weakly closed, it is T_c -closed by the preceding result. But T_c is compatible with the duality between E and E', so it remains only to apply Theorem 8.3.4(2). ▌

Remark. For Banach spaces Theorem 8.10.5 was established by Krein and Šmulian [1]; for vector subspaces the result was announced earlier by Bourbaki [8]. As stated, the theorem is due to Dieudonné and Schwartz [1], to whom the proof given is due. Extensions have been given by Köthe [1].

We proceed to add a few more properties of fully complete spaces.

8.10.6 **Proposition.** Let E be a LCTVS with topology T, T' a locally convex topology on E such that $T \leq T' \leq \tau(E, E')$. If E is fully (resp. B_r-) complete for T, it is fully (resp. B_r-) complete for T'.

Proof. The dual of E relative to T' is again E'. Let M be a vector subspace of E' that is almost weakly closed relative to T'. Then $M \cap V^0$ is weakly closed for any T'-neighbourhood V of 0 in E. Since T' is stronger than T, it follows directly that $M \cap U^0$ is weakly closed for any T-neighbourhood U of 0 in E. Since E with T is fully complete, M is weakly closed. Thus E is fully complete relative to T'. The proof for B_r-completeness is similar. ▌

8.10.7 **Proposition.** Let E be a Fréchet space. E' is fully complete relative to any locally convex topology T satisfying $T_c \leq T \leq \tau(E', E)$.

Proof. The dual of E' with T_c is E. Thanks to the preceding proposition it therefore suffices to consider the case $T = T_c$. Let M be a vector subspace of E that is almost weakly closed; that is, $M \cap V^0$ is weakly closed for each T_c-neighbourhood V of 0 in E'. This signifies that $M \cap K$ is weakly closed for each convex, balanced, and compact set K in E. We have to show that M is weakly closed—that is (Corollary 2.2.7), that M is closed in E. If x is adherent to M, however, it is adherent to $M \cap K$ for some convex, balanced, and compact subset K of E. (Take a sequence $(x_n) \subset M$ converging to x; the closed, convex, and balanced envelope K of the x_n and x is precompact and hence, E being complete, compact; x is adherent to $M \cap K$.) Now $M \cap K$ is weakly closed, a fortiori closed, and so $x \in K \cap M \subset M$. The proof is thus complete. ▮

8.10.8 **Proposition.** Let E and F be LCTVS, u a continuous linear map of E into F.

(1) If u is nearly open, and if E is fully complete, then F is fully complete.

(2) If u is nearly open and $u^{-1}(\overline{\{0\}}) \subset \overline{\{0\}}$, and if E is B_r-complete, then F is B_r-complete.

Proof. We use the adjoint map u' of F' into E'. Since u is nearly open, $u(E)$ is dense in F, so that u' is one to one. Let M be an almost weakly closed vector subspace of F', and let $L = u'(M)$. Then $M = u'^{-1}(L)$. Since u' is weakly continuous, if we show that L is weakly closed in E', it will follow at once that M is weakly closed in F'. Besides this, the extra condition on u in (2) ensures that if M is weakly dense in F', then L is weakly dense in E'. (Thus M is weakly dense if and only if $M^0 = \overline{\{0\}}$, and analogously for L; also, by (8.6.2) with u replaced by u',

$$L^0 = (u'(M))^0 = u^{-1}(M^0).)$$

It is thus seen that the proof of either (1) or (2) amounts to showing that L is weakly closed whenever M is almost weakly closed. Since E is fully (or B_r-) complete, we need only show that L is almost weakly closed.

To this end, let U be a neighbourhood of 0 in E. By hypothesis, $V = \overline{u(U)}$ is a neighbourhood of 0 in F, so that $M \cap V^0$ is weakly closed. The relation $V = \overline{u(U)}$ shows via (8.6.2) that $U^0 \cap u'(F') = u'(V^0)$. u' being one to one, we have then $L \cap U^0 = u'(M \cap V^0)$. Now u' is weakly continuous and $M \cap V^0$ is weakly compact. Hence $L \cap U^0$ is weakly compact and hence weakly closed. The proof is complete. ▮

8.10.9 **Corollary.** Let E be a LCTVS, M a vector subspace of E. If E is fully complete then so is E/M.

Proof. In the preceding proposition we take $F = E/M$ and u the natural mapping of E onto F. ▮

8.10.10 **Proposition.** In order that a LCTVS E be fully complete it is (a) necessary that E/M be B_r-complete for each vector subspace M of E, and (b) sufficient that E/M be B_r-complete for each closed vector subspace M of E.

Proof. (a) follows from Corollary 8.10.9.

As for (b), let N be an almost weakly closed vector subspace of E'. Consider $F = E/N^0$. We may identify F' with N^{00}, the weak closure of N in E'. It therefore suffices to show that N is almost weakly closed relative to F'—that is, that $N \cap V^0$ is weakly closed in F' for each neighbourhood V of 0 in F. However, V is the projection of some neighbourhood U of 0 in E, and $N \cap V^0 = N \cap U^0$ is weakly closed in E' (by hypothesis), a fortiori weakly closed in F'. ▌

8.10.11 **Proposition.** If F is a B_r-complete LCTVS, E a closed vector subspace of F, then E is B_r-complete.

Proof. Let us consider the injection map u of E into F and its adjoint u'. Here u' consists of restriction of a linear form from F to E. Let L be a weakly dense and almost weakly closed vector subspace of E', and let $M = u'^{-1}(L)$. Since u is an isomorphism into, u' is onto (see, for example, Corollary 8.6.4). We will show first that M is weakly dense in F'—that is, that $M^0 = \overline{\{0\}}$.

Let us suppose that y_0 belongs to M^0. If y_0 did not belong to $E = u(E)$, then, since E is closed, there would exist y' in F' such that $\langle u(E), y' \rangle = 0$ and $\langle y_0, y' \rangle = 1$. The first relation would yield $u'(y') = 0$, which certainly belongs to L, so that y' would belong to M. But then the second relation would contradict our hypothesis that y_0 belongs to M^0. Thus y_0 belongs to E; that is, $y_0 = u(x_0)$ for some x_0 in E. In this case, $0 = \langle y_0, M \rangle = \langle x_0, u'(M) \rangle$. Now if x' belongs to L, $x' = u'(y')$ for some y' (u' being onto); and this y' necessarily belongs to M. But then $\langle x_0, L \rangle = 0$. Since L is weakly dense in E', it follows that x_0 belongs to $\overline{\{0\}}$. Since u is continuous, $y_0 = u(x_0)$ belongs to $\overline{\{0\}}$. Thus M is weakly dense in F'.

Next we show that M is almost weakly closed in F'. Let V be any neighbourhood of 0 in F. It is easily verified that

$$V^0 \cap M = V^0 \cap u'^{-1}(L)$$

is contained in

$$u'^{-1}(u^{-1}(V)^0 \cap L) = u'^{-1}(U^0 \cap L) \subset u'^{-1}(L) = M,$$

where

$$U = u^{-1}(V) = V \cap E$$

is a neighbourhood of 0 in E. Therefore we see that

$$V^0 \cap M = V^0 \cap u'^{-1}(U^0 \cap L).$$

Since L is almost weakly closed in E', $U^0 \cap L$ is weakly compact; u' being weakly continuous, it follows that $V^0 \cap M$ is weakly closed in F'. Thus M is almost weakly closed.

We now know that M is weakly dense and almost weakly closed. Since F is B_r-complete, we conclude that $M = F'$. Since u' is onto, this implies that $L = E'$. This shows that E is B_r-complete, as alleged. ▌

8.10.12 **Proposition.** Let E and F be separated LCTVSs, u a one-to-one open and nearly continuous linear map of E onto a closed vector subspace of F. If F is B_r-complete, so is E.

Proof. Put $F_0 = u(E)$, a subspace of F. By Proposition 8.10.11, F_0 is B_r-complete. Let us consider $v = u^{-1}$, mapping F_0 onto E. v is continuous and nearly open. Also, since F_0 and E are separated and v is continuous, v has a closed graph. Theorem 8.9.2(2) shows that v is open. Thus u is an isomorphism of E onto F_0, and therefore E shares with F_0 the property of B_r-completeness. ▌

8.10.13 **A Complete Space that Is not B_r-Complete.** We shall have recourse to the τ-topology introduced in Subsection 8.3.7—that is, the topology $\tau(E, E^*)$, shown in Subsection 8.3.7 to coincide with the strongest locally convex topology on the given vector space E. Our starting point is a lemma.

Lemma. E is complete for τ.

Proof. The result is due to Kaplan [1], but the proof we give is modelled closely along the lines of that given by Pták ([1], pp. 69–70).

We introduce an algebraic basis $(e_i)_{i \in I}$ for E and the dual system (e_i^*) in E^*. Let y be an element of the τ-completion $\hat{E}$ of E. We shall show that y belongs to E.

Each e_i^* is a τ-continuous linear form on E and may therefore be extended (uniquely) by continuity to $\hat{E}$. Let J be the set of indices i for which $\langle y, e_i^* \rangle \neq 0$. The first step is to show that J is *finite*.

To this end, we define

$$u_i^* = \begin{cases} e_i^* & \text{for } i \in I \backslash J, \\ e_i^* / \langle y, e_i^* \rangle & \text{for } i \in J. \end{cases}$$

For each x in E we have $x = \sum_{i \in I} x(i) e_i$, where $x(i) = \langle x, e_i^* \rangle$ is nonzero only for i in some finite subset $F = F_x$ of I. Thus

$$|\langle x, u_i^* \rangle| = \begin{cases} |x(i)| & \text{if } i \in F \cap (I \backslash J), \\ |x(i)| / |\langle y, e_i^* \rangle| & \text{if } i \in F \cap J. \end{cases}$$

This shows that the family (u_i^*) is weakly bounded in E^*. The same is thus true of the convex envelope H of this family. Thus (Theorem 1.11.4(1)) H is weakly relatively compact in E^*, and H^0 is accordingly a τ-neighbourhood of 0 in E.

Define for x in E the number

$$\begin{aligned} p(x) &= \operatorname{Sup} \{ |\langle x, e_i^* \rangle| / |\langle y, e_i^* \rangle| : i \in J \} \\ &= \operatorname{Sup} \{ |\langle x, u_i^* \rangle| : i \in J \}. \end{aligned}$$

p, in common with every seminorm on E, is τ-continuous; hence p may be extended to $\hat{E}$ by continuity.

Now y is the limit of a directed family (x_α) of points of E, and so $p(y) = \lim_\alpha p(x_\alpha)$. We may therefore find x in E so that $p(y - x) \leq \frac{1}{2}$. It then follows that $\langle x, e_i^* \rangle \neq 0$ for i in J. Finiteness of J is thus established.

Let $x_0 = \sum_{i \in J} \langle y, e_i^* \rangle e_i \in E$. Then $\langle x_0, e_i^* \rangle = \langle y, e_i^* \rangle$ for i in J; for other indices i both sides are 0. Thus agreement obtains for all i.

We put M for the vector subspace of E^* generated by all the e_i^*. We know now that $\langle y, x^*\rangle = \langle x_0, x^*\rangle$ for x^* in M. Our aim is to show that in fact this holds for all x^* in E^*. Now it is easily seen that each x^* in E^* belongs to a set of the form

$$Q = \{x^* \in E^* : |\langle e_i, x^*\rangle| \leq \alpha_i \text{ for } i \in I\},$$

the scalars α_i being suitably selected. Such a set Q is the polar of the set $q(x) \leq 1$, where

$$q(x) = \sum \alpha_i \, |\langle x, e_i^*\rangle|$$

is a seminorm on E. Like any seminorm on E, q is continuous. Hence Q is equicontinuous. Since $y \in \hat{E}$, it follows that $x^* \to \langle y, x^*\rangle$ has a restriction to Q that is weakly (that is, $\sigma(E^*, E)$-) continuous. We have now to show merely that $\langle y, x^*\rangle = \langle x_0, x^*\rangle$ for x^* in any such set Q; and since we know that both sides are continuous, and that agreement obtains on M, it suffices to show that $M \cap Q$ is dense in Q. But this is a very simple consequence of the bipolar theorem, which reduces the problem to showing that for each x^* in Q one has

$$|\langle x, x^*\rangle| \leq \text{Sup}\,\{|\langle x, z^*\rangle| : z^* \in M \cap Q\}$$

for x in E—that is, that

$$q(x) \leq \text{Sup}\,\{|\langle x, z^*\rangle| : z^* \in M \cap Q\}$$

for x in E. And in fact

$$q(x) = \sum_i \alpha_i \, |\langle x, e_i^*\rangle| = \sum \alpha_i \theta_i \langle x, e_i^*\rangle,$$

where the $\theta_i = \theta_i(x)$ satisfy $|\theta_i(x)| \leq 1$, and the right-hand side here is $\langle x, z^*\rangle$ with $z^* = \sum \alpha_i \theta_i e_i^*$ (a finite sum), evidently an element of $M \cap Q$. This completes the proof. ▌

Now we can give the example of a complete LCTVS that is not B_r-complete.

Let E be the vector space $\ell^1(N)$, and let T denote its usual normed topology. Then (E, T) is a Banach space, hence is barrelled.

The dual of (E, T) is $\ell^\infty = \ell^\infty(N)$, which is obviously a proper vector subspace of E^*. The topology τ is thus strictly stronger than T. We will now show that

(a) ℓ^∞ is weakly dense in E^*,

and (b) ℓ^∞ is almost weakly closed in E^*.

Together these assertions amount to saying that (E, τ) is not B_r-complete, even though it is complete (by the preceding lemma).

Now (a) is trivial since ℓ^∞ is total over E.

Let us next consider (b). Suppose that the directed family (x_i^*) belongs to $\ell^\infty \cap U^0$ and converges weakly in E^* to x^*, U being some τ-neighbourhood of 0 in E: we have to show that x^* belongs to ℓ^∞. The family (x_i^*) is clearly weakly bounded in E^*; that is, $\text{Sup}_i \, |\langle x, x_i^*\rangle| < +\infty$ for each $x \in E$. Now, since (E, T) is barrelled, the x_i^* must be norm bounded in ℓ^∞:

$$\|x_i^*\|_\infty \leq k.$$

But then (Theorem 1.11.4(1)), regarding the x_i^* as elements of the dual, ℓ^∞, of (E, T), we know that (x_i^*) has a weak limiting point in ℓ^∞. This limiting point can be none other than x^*. Thus x^* belongs to ℓ^∞, as we had to show.

Remark. Levin [2] studies conditions for B-completeness of ultrabarrelled and barrelled spaces.

8.11 The Hellinger-Toeplitz Theorem

In this section we shall deal with an approach to the closed graph and open mapping theorems (in forms essentially identical with Theorems 8.9.4 and 8.9.3, respectively) that stems from an attempt to analyse and then generalize a theorem due to Hellinger and Toeplitz which, in its original form, concerned itself with self-adjoint endomorphisms of a Hilbert space. Various equivalent formulations of this early result were given by various writers. Some extensions of the result are given in Edwards [8] and Taylor [2], but we shall in the main follow a later treatment by Pták [2].

Throughout this section E and F denote LCTVSs, and u is a linear map of E into F that is not assumed to be continuous; the domain D' of u' is thus generally a proper vector subspace of F' (see Section 8.6). The theorems to be discussed are all of the type that assert the continuity or boundedness of u on the basis of data about D', together perhaps with suitable restrictions upon E and F. The Hellinger-Toeplitz theorem itself is concerned with the case in which D' is known to be total (that is, weakly dense) in F'.

The starting point is an expression of the closed graph property of u in terms of D' which seems to have first been placed on record by Pták ([1], Theorem (3.7)).

8.11.1 Proposition. If u has a closed graph, then D' is total; and conversely, if F is separated.

Proof. If D' is total and F is separated, it is very simple to verify that u has a closed graph. For if a directed family (x_i) in E is such that $x_i \to 0$ in E and $y_i = u(x_i) \to y$ in F, then for each y' in D' one has $\langle y, y' \rangle = \lim \langle y_i, y' \rangle = \lim \langle x_i, u'(y') \rangle = 0$; then, D' being total and F separated, one must have $y = 0$. This shows that the graph of u is closed in $E \times F$.

Let us suppose on the other hand that the graph G of u is closed in $E \times F$; also that y_0, an element of F, is orthogonal to D'. We have to show that $y_0 = 0$, which will be done by showing that $(0, y_0)$ belongs to G. Indeed, if $(0, y_0)$ did not belong to G, then (since G is closed) one could find (x', y') in $E' \times F'$ such that $\langle G, (x', y') \rangle = 0$ and $\langle (0, y_0), (x', y) \rangle \neq 0$. (Here we are using the fact that the dual of $E \times F$ may be identified with $E' \times F'$, the duality being defined by the bilinear form

$$\langle (x, y), (x', y') \rangle = \langle x, x' \rangle + \langle y, y' \rangle.)$$

The first relation shows that y' belongs to D'; the second says that $\langle y_0, y' \rangle \neq 0$. Since y_0 is supposed to be orthogonal to D', one has here a contradiction. The proof is thus complete. ▌

We turn next to examine what can be said when it is given that $D' = F'$. The essential result here is contained in a preliminary proposition.

8.11.2 Proposition. Suppose $D' = F'$. If V is any closed, convex, and balanced neighbourhood of 0 in F, then $U = u^{-1}(V)$ is closed in E.

Proof. Taking any point x_0 of E not in U, we must show that x_0 is not adherent to U. Since $u(x_0)$ is not in V, and since V is convex, balanced, and closed in F, one can choose some y' in F' that belongs to V^0 and for which $|\langle u(x_0), y'\rangle| > 1$. This reads $|\langle x_0, u'(y')\rangle| > 1$. However, since y' belongs to V^0, we have

$$|\langle U, u'(y')\rangle| = |\langle u(U), y'\rangle| \leq |\langle V, y'\rangle| \leq 1.$$

It is thus clear that x_0 is not weakly adherent to U, as we had to show. ▌

8.11.3 Theorem. Suppose that $D' = F'$. Then

(a) u is bounded;

(b) if E is bornological, u is continuous;

(c) if u is nearly continuous, it is continuous.

Proof. The condition $D' = F'$ shows that u transforms bounded subsets of E into subsets of F that are weakly bounded and therefore bounded (Theorem 8.2.2). This establishes (a). (b) follows from (a) and the properties of bornological spaces: see Remark (b) following Theorem 7.3.1. As for (c), we have only to recall the definition of nearly continuous and apply Proposition 8.11.2. ▌

We turn next to some relations between weak closure properties of D' and properties of u and of E.

8.11.4 Proposition. (a) If u is nearly continuous then D' is almost weakly closed in F'.

(b) If E is barrelled, D' is almost weakly closed in F'.

Proof. (a) We have to show that $D' \cap V^0$ is weakly closed in F' for each neighbourhood V of 0 in F. Since u is nearly continuous, $U = \overline{u^{-1}(V)}$ is a neighbourhood of 0 in E. Now

$$\begin{aligned}|\langle u^{-1}(V), u'(D' \cap V^0)\rangle| &\leq |\langle u(u^{-1}(V)), D' \cap V^0\rangle| \\ &\leq |\langle V, D' \cap V^0\rangle| \leq 1,\end{aligned}$$

hence it follows that $U \subset (u'(D' \cap V^0))^0$ and therefore

$$u'(D' \cap V^0) \subset U^0.$$

Suppose that y' is weakly adherent to $D' \cap V^0$: we aim to show that y' belongs to $D' \cap V^0$; since V^0 is weakly closed, this amounts to showing that y' belongs to D'. For arbitrary $x_1, \cdots, x_n$ in E and $\varepsilon > 0$, we denote by $W(x_1, \cdots, x_n; \varepsilon)$ the set of x' in U^0 that satisfy

$$\operatorname*{Sup}_{1 \leq i \leq n} |\langle x_i, x'\rangle - \langle u(x_i), y'\rangle| \leq \varepsilon.$$

This set is plainly weakly closed in E'. It is also nonvoid because, y' being weakly adherent to $D' \cap V^0$, there exists z' in $D' \cap V^0$ such that

$$|\langle u(x_i), z' - y'\rangle| \leq \varepsilon \qquad \text{for } 1 \leq i \leq n;$$

and since z' belongs to D' and to V^0, we have $u'(z') \in U^0$ and further

$$\begin{aligned}\langle u(x_i), y'\rangle &= \langle u(x_i), z'\rangle + \alpha_i\varepsilon \\ &= \langle x_i, u'(z')\rangle + \alpha_i\varepsilon,\end{aligned}$$

where $|\alpha_i| \leq 1$, showing that $u'(z')$ belongs to $W(x_1, \cdots, x_n; \varepsilon)$.

From this it follows that the sets $W(x_1, \cdots, x_n; \varepsilon)$ have the finite intersection property. Since U^0 is weakly compact, we deduce the existence of x' in U^0 belonging to all the $W(x_1, \cdots, x_n; \varepsilon)$. But this signifies that $\langle u(x), y'\rangle = \langle x, x'\rangle$ for all x in E—that is, that y' belongs to D', as we had to establish.

(b) This follows from (a). For, if E is barrelled, any u is nearly continuous (since $\overline{u^{-1}(V)}$ is a barrel in E). ▌

8.11.5 Theorem. (a) Suppose that F is B_r-complete and that u is a nearly continuous linear map of E into F. Then the following three conditions are equivalent:

(1) u is continuous;

(2) $D' = F'$;

(3) D' is total.

(b) If E is barrelled, if F is B_r-complete, and if u is any linear map of E into F, then the conditions (1), (2), and (3) are equivalent.

Proof. If E is barrelled, any linear map u of E into F is nearly continuous. Thus (a) implies (b).

As for (a), the implication (2) $\Rightarrow$ (3) is trivial. If F is B_r-complete, the converse is valid by Proposition 8.11.4(a). Thus (2) and (3) are equivalent. (1) implies (2) trivially, in any case, while the converse comes from Theorem 8.11.3(c). ▌

8.11.6 Remarks. In view of Proposition 8.11.1, part (b) of Theorem 8.11.5 yields an equivalent of the closed graph theorem as the latter is formulated in Theorem 8.9.4 with E and F interchanged and $v : F \to E$ replacing u. We shall return at the end of this section to the problem of deducing the open mapping theorem from this guise of the closed graph theorem, thus providing within the confines of the present section a self-contained treatment of both theorems. Meanwhile we proceed to formulate the generalized Hellinger-Toeplitz theorem stemming directly from Theorem 8.11.5(b).

8.11.7 Theorem. Let S be a total subset of F' and f a mapping of S into E' with the property that for each x in E there exists a y in F such that

$$\langle x, f(s)\rangle = \langle y, s\rangle \qquad (s \in S). \tag{8.11.1}$$

Then if F is separated, y is uniquely determined by x and the mapping $u : x \to y$ so defined is linear. If, further, E is barrelled and F is B_r-complete, then u is continuous from E into F.

Proof. The first assertion is obvious. As for the second, (8.11.1) shows that D' contains S and is therefore total, and that $u'(s) = f(s)$ for s in S. It remains only to apply Theorem 8.11.5(b). ▌

8.11.8 **Remarks.** (1) Taking $S = D'$ and $f = u'$, Theorem 8.11.7 is seen to include Theorem 8.11.5(b)—that is, includes one form of the CGT.

(2) In its original form the Hellinger-Toeplitz theorem dealt with the case in which $E = F$ is a Hilbert space while the condition analogous to (8.11.1) was formulated in terms of the inner product $(\mid)$ on E. We know (see Corollary 1.12.6) that if E is a Hilbert space, there is a conjugate-linear and norm-preserving isomorphism J of E onto E' such that

$$\langle x, Jy \rangle = (x \mid y)$$

for all x and y in E. The Hilbert adjoint u^* is characterized by the relation

$$(u(x) \mid y) = (x \mid u^*(y)),$$

so that u^* and u' are related as follows:

$$u^* = J^{-1} \circ u' \circ J, \; u' = J \circ u^* \circ J^{-1}.$$

The domain D' of u' is total if the domain of u^* is total. The appropriate translation of (8.11.1) is therefore

$$(x \mid f(s)) = (y \mid s) \qquad (s \in S), \tag{8.11.2}$$

f being now a mapping of a total subset S of E into S. We then have f and u^* coinciding on S. The result may therefore be stated thus:

▶ If u is an endomorphism of a Hilbert space, and if the Hilbert adjoint u^* has a domain that is total, then u is continuous. In particular: any self-adjoint endomorphism u is continuous. (It is understood here that the domain of u is the full space.)

Two corollaries of Theorem 8.11.7 are worth recording.

8.11.9 **Corollary.** Suppose E, F, and G are LCTVSs, G being separated, P a set of continuous linear maps of F into F such that

$$y \in F, p(y) = 0 \text{ for all } p \in P \text{ imply } y = 0.$$

Let g be a mapping of P into the set of continuous linear maps of E into G such that to each x in E corresponds y in F satisfying

$$g(p)(x) = p(y) \text{ for all } p \in P.$$

Then y is uniquely determined by x and the mapping $u : x \to y$ so defined is linear. If, in addition, E is barrelled and F is B_r-complete, then u is continuous.

Proof. Apply Theorem 8.11.7, taking for S the set of composite maps $z' \circ p$ obtained when p ranges over P and z' over G', and $f(s) = z' \circ g(p)$ whenever $s = z' \circ p$. ▌

8.11.10 **Corollary.** Let E, F, and G be three LCTVSs, G being separated and F being a vector subspace of G^T, where T is a set. Suppose that for each t in T, the mapping $\varepsilon_t : y \to y(t)$ is continuous from F into G. Suppose, too, that h is a mapping of T into the set of continuous linear maps of E into G with the property that for each x in E, $t \to h(t)(x)$ belongs to F; we denote this element of F by $u(x)$. Then u is a linear map of E into F. If, also, E is barrelled and F is B_r-complete, u is continuous.

Proof. We apply the preceding corollary, taking $P = \{\varepsilon_t : t \,\varepsilon\, T\}$, where ε_t is the continuous linear map of F into G defined by $\varepsilon_t(y) = y(t)$, and defining $g(p)$ for $p = \varepsilon_t$ to be $h(t)$. One then has

$$g(p)(x) = h(t)x = \varepsilon_t(u(x))$$

identically for x in E and p in P. The hypotheses of Corollary 8.11.9 are satisfied, and its conclusion gives the stated result. ▌

The remainder of this section is to be expended on indicating how, starting from Theorem 8.11.5—in fact, from its corollary Theorem 8.9.4—we may rapidly deduce two forms of the OMT as applied to linear maps u of E into F. It is convenient to do this in two stages, in the first of which we assume that u is one to one.

8.11.11 **Theorem.** Suppose that the linear map u of E into F is one-to-one, nearly open onto $F_0 = u(E)$, and has a graph that is closed in $E \times F_0$. Suppose too that E is B_r-complete. Then u is open from E onto $u(E)$.

Proof. We aim to apply Theorem 8.9.4 to the inverse $v = u^{-1}$, which is a linear map of F_0 into E. It is indeed easily verified that v has a closed graph, thanks to the corresponding assumption about u. Further, since u is nearly open from E onto F_0, and since $v^{-1}(U) \supset u(U)$ for any subset U of E, it appears that v is nearly continuous. At this stage we can appeal to Theorem 8.9.4 with F_0 in place of F, to learn that v is continuous. This signifies that u is open from E onto F_0. ▌

8.11.12 **Theorem.** Suppose that a linear map u of E into F is nearly open onto $F_0 = u(E)$, and has a graph closed in $E \times F_0$. We suppose too that E is fully complete. Then u is open from E onto $u(E)$.

Proof. The proof follows from the preceding result on passage to the quotient space $E/\text{Ker}\, u$, taking into account (Proposition 8.10.10(a)) the fact that this quotient space is B_r-complete. It is easily verified that the quotient map fulfills the hypotheses of Theorem 8.11.11. ▌

8.12 Concerning Weak Compactness

Since the weakened topology on a LCTVS in general fails to satisfy the first countability axiom, it is not clear a priori to what extent weak compactness can be formulated in terms of sequences. This question has received attention, and we shall give among other things two results due to Šmulian and to Eberlein, respectively.

8.12.1 **Theorem (Šmulian).** Let E be a metrizable LCTVS, (x_n) a weakly relatively compact sequence in E. Then from (x_n) may be extracted a weakly convergent subsequence.

Proof. We may replace E by the vector subspace generated by the x_n, and so assume that E is separable. This being so, let U_k $(k = 1, 2, \cdots)$ be a neighbourhood base at 0 in E. The sets U_k^0 are then weakly compact in E' and their union is E'. Each U_k^0 is weakly metrizable, hence weakly separable (being weakly compact). Therefore E' too is weakly separable. Let (x_i') be a sequence that is weakly dense in E'. Since (x_n) is weakly relatively compact, for each i the sequence $(\langle x_n, x_i' \rangle)_{n \geq 1}$ contains a convergent subsequence. Using the diagonal process, one may extract a subsequence (y_n) of (x_n) such that $\lim_n \langle y_n, x_i' \rangle$ exists finitely for each i. It follows that the sequence (y_n), which is weakly relatively compact, admits at most one weak limiting point. This sequence must, therefore, be weakly convergent. ▌

8.12.2 **Remark.** The theorem also holds if E is a strict inductive limit of metrizable LCTVSs. For then any weakly relatively compact sequence (x_n), being bounded, must fall within one of the metrizable defining spaces and the above argument applies still. Moreover the proof given works whenever there exists a sequence (U_k) of neighbourhoods of 0 in E having an intersection reduced to $\{0\}$.

Our next result will be based upon a simple lemma concerning continuous functions on a compact space T. We denote (as usual) by $C(T)$ the space of continuous functions on T, regarded as a Banach space with the norm equal to the maximum modulus. It will be necessary also to consider on $C(T)$ the topology of pointwise convergence, and we denote by $C_p(T)$ the corresponding LCTVS.

8.12.3 **Lemma.** Let T be a compact space, H a subset of $C(T)$. Suppose that f_0 is adherent in $C_p(T)$ to H. Then there exists a countable set $H_0 \subset H$ such that f_0 is adherent in $C_p(T)$ to H_0.

Proof. Given two natural numbers m and n, and f in $C(T)$, let $W(f, m, n)$ denote the set of points $(t_i)_{1 \leq i \leq m}$ in T^m satisfying

$$\operatorname*{Sup}_{1 \leq i \leq m} |f(t_i) - f_0(t_i)| < \frac{1}{n} .$$

Each such set is open. As f varies in H, these sets cover T^m. By compactness, there is a finite set $H_{m,n} \subset H$ for which the corresponding $W(f, m, n)$ cover T^m. It suffices to take for H_0 the union of all the $H_{m,n}$. ▌

8.12.4 **Theorem.** Let E be a metrizable LCTVS, H a subset of E, x_0 an element of E. Then:

(a) If x_0 is weakly adherent to H, it is weakly adherent to some countable subset H_0 of H.

(b) If H is weakly relatively compact, and if x_0 is weakly adherent to H, there is a sequence $(x_n) \subset H$ that converges weakly to x_0.

Proof. (a) As in the proof of Theorem 8.12.1, E' is the union of a sequence (T_k) of weakly compact sets. For each k we apply Lemma 8.12.3, each element of E being regarded as an element of $C(T_k)$, and the result follows easily.

(b) According to (a), we may assume that H is countable; and then there is no loss of generality in assuming that E is separable. As in the proof of Theorem 8.12.1, E' will be weakly separable. We take a sequence (x'_i) weakly dense in E'. For each i, we may extract from H a sequence (a_n) such that

$$\lim_n \langle a_n, x'_i \rangle = \langle x_0, x'_i \rangle.$$

By the diagonal process, we may then construct a sequence $(y_n) \subset H$ such that

$$\lim_n \langle y_n, x'_i \rangle = \langle x_0, x'_i \rangle \qquad \text{for all } i.$$

Since H is weakly relatively compact, Šmulian's theorem asserts the existence of a subsequence (x_n) of (y_n) converging weakly to some element of E which, since the x'_i are weakly dense in E', can be none other than x_0. This completes the proof. ∎

8.12.5 **Remark.** Part (b) will again extend to strict inductive limits of metrizable LCTVSs, the required device being as explained in Remark 8.12.2.

Our proof of Eberlein's theorem to follow will depend upon a second lemma concerning the space $C(T)$.

8.12.6 **Lemma.** Let T be a compact space, H a subset of $C(T)$. We suppose that each sequence extracted from H admits a limiting point in $C_p(T)$. Then the pointwise closure $\bar{H}$ of H in K^T is compact for pointwise convergence and is contained in $C(T)$.

Proof. Tychonoff's theorem assures us that $\bar{H}$ is compact for pointwise convergence, so it remains only to show that $\bar{H}$ is contained in $C(T)$.

Suppose if possible that u in $\bar{H}$ were discontinuous at a point a of T. We may then construct inductively a number $d > 0$, a sequence $(t_n) \subset T$, and a sequence $(f_n) \subset H$ with the following properties:

$$|u(t_n) - u(a)| \geq d, \tag{8.12.1}$$

$$|f_m(t_n) - f_m(a)| \leq \frac{d}{8} \qquad (m \leq n), \tag{8.12.2}$$

$$|u(t_n) - f_m(t_n)| \leq \frac{d}{8}, \qquad |u(a) - f_m(a)| \leq \frac{d}{8} \qquad (m > n). \tag{8.12.3}$$

Supposing f to be a limiting point of (f_m) in $C_p(T)$ and b to be a limiting point of (t_n) in T, (8.12.2) would give as $n \to \infty$:

$$|f_m(b) - f_m(a)| \leq \frac{d}{8}.$$

Letting $m \to \infty$, this gives

$$|f(b) - f(a)| \leq \frac{d}{8}. \tag{8.12.4}$$

From (8.12.3) we should derive as $m \to \infty$:

$$|u(t_n) - f(t_n)| \leq \frac{d}{8} \tag{8.12.5}$$

and

$$|u(a) - f(a)| \leq \frac{d}{8}. \tag{8.12.6}$$

Of these, (8.12.5) would imply that

$$|u(t_n) - f(b)| \leq \frac{d}{4}$$

for an infinity of n; with (8.12.4), this would yield

$$|u(t_n) - f(a)| \leq \frac{3d}{8}$$

for an infinity of n. So finally (8.12.6) would yield

$$|u(t_n) - u(a)| \leq \frac{d}{2}$$

for an infinity of n, contradicting (8.12.1). This contradiction completes the proof. ▌

8.12.7 **Theorem** (Eberlein). Let E be a separated LCTVS, H a subset of E. Suppose that (a) each sequence extracted from H admits a limiting point in E, and (b) the closed convex envelope of H is complete for $\tau(E, E')$. Then H is relatively compact in E.

Proof. (a) implies that H is precompact: so $\bar{H}$ will be compact if (and only if) it is complete. It therefore suffices to show that the weak closure of H is weakly complete. This reduces one to the case in which the topology involved is $\sigma(E, E')$.

Let H^* be the weakly closed convex envelope of H. By hypothesis (b), H^* is complete for $\tau(E, E')$. If $\hat{E}$ is the $\tau(E, E')$-completion of E, H^* (being complete) is closed in $\hat{E}$, hence weakly closed therein (being convex). It suffices now to show that the weak closure of H in $\hat{E}$ is weakly compact, this closure being contained in E. This reduces us to the case in which E is complete for $\tau(E, E')$.

Now H is obviously bounded, hence weakly relatively compact in E'^*. Thus we need only show that any linear form u on E' that is weakly adherent in E'^* to H is in E. Now if $K \subset E'$ is weakly compact, $u \mid K$ is pointwise adherent to the set H_K of restrictions $x \mid K$ (x in H). Hypothesis (a) entails that each sequence extracted from H_K admits a limiting point in $C_p(K)$. So Lemma

8.12.6 shows that H_K is relatively compact in $C_p(K)$. Consequently $u \mid K$ is continuous on K. It now remains only to apply Theorem 8.5.1. ∎

8.12.8 **Corollary.** Let T be a compact space. A subset H of $C(T)$ is weakly relatively compact if and only if it is bounded and relatively compact in $C_p(T)$; in that case each sequence extracted from H contains a weakly convergent subsequence.

Proof. The "only if" assertion is trivial.

For the rest, according to Eberlein's theorem we have merely to show that any sequence (f_n) extracted from H admits a weak limiting point in $C(T)$. The sequence (f_n) is relatively compact in $C_p(T)$. Therefore, by Lemma 8.12.9 below, from (f_n) may be extracted a subsequence (g_n) that converges in $C_p(T)$, and Theorem 4.10.2 implies at once that (g_n) is weakly convergent in $C(T)$. Thus (f_n) admits a weak limiting point in $C(T)$. ∎

8.12.9 **Lemma.** Let T be a compact space, $\mathfrak{S}$ a set of subsets of T covering T. Let $C_{\mathfrak{S}}(T)$ be the space of continuous functions on T, equipped with the topology of convergence uniform on the sets belonging to $\mathfrak{S}$. If (f_n) is a relatively compact sequence in $C_{\mathfrak{S}}(T)$, then it contains a subsequence that converges in $C_{\mathfrak{S}}(T)$.

Proof. Let H be the compact closure of (f_n) in $C_{\mathfrak{S}}(T)$. On H the topology induced by $C_{\mathfrak{S}}(T)$ is identical with that induced by the topology of convergence on a dense subset of T. So it suffices to find a subsequence that converges pointwise on a dense subset of T. Moreover, there is no loss of generality in taking for $\mathfrak{S}$ the set of one-point subsets of T, so that $C_{\mathfrak{S}}(T) = C_p(T)$.

Let us suppose first that T is metrizable: it is then necessarily separable, and we suppose (t_i) is a sequence dense in T. Then H is metrizable for the topology of pointwise convergence on the t_i. The conclusion follows in this case by use of the diagonal process.

In general, let T^* be the quotient of T modulo the relation "$f_n(t) = f_n(t')$ for all n." Then T^* is compact and metrizable, its topology being the weakest for which each f_n is continuous on T^*. Moreover, $C_p(T^*)$ may be identified with a closed vector subspace of $C_p(T)$, so that (f_n) may be identified with a relatively compact sequence in $C_p(T^*)$. This reduces the problem to the metrizable case, already dealt with, and so completes the proof. ∎

8.12.10 **Remarks.** Dunford and Schwartz ([1], p. 430) give a sort of combined form of Theorems 8.12.1 and 8.12.7 for Banach spaces. Even in this case the development of the theorems of Šmulian and of Eberlein was gradual; see p. 466 of Dunford and Schwartz. Extensions to more general spaces are due to Collins [1], Dieudonné and Schwartz [1], and Grothendieck [3]. Each of the last two influence the treatment indicated by Bourbaki ([5], p. 82, Exercice 13b); p. 83, Exercice 15b)), which we have followed. See also Grothendieck [7], Chapitre V, pp. 17–19.

Šmulian has given also another "countable" criterion of weak compactness for convex subsets K of a Banach space—namely, that each decreasing sequence of nonvoid closed, convex subsets of K shall have a nonvoid intersection. For a proof see Dunford and Schwartz [1], p. 433. The result appeared first in Šmulian [3]; an alternative proof is due to Klee [2]; and an extension is given by Dieudonné [5].

8.13 The Theorem of Krein

In its general form the theorem runs as follows.

8.13.1 Theorem. Let E be a separated LCTVS, A a compact subset of E, B the closed convex envelope in E of A. Then B is compact if and only if it is complete for $\tau(E, E')$.

Proof. If B is compact, it is complete and a fortiori complete for $\tau(E, E')$.

Let us suppose now that B is complete for $\tau(E, E')$. Certainly B is precompact, so that it is compact if and only if it is complete. It will therefore suffice a fortiori to show that B is weakly compact (or merely weakly relatively compact). This means that we may suppose from the outset that the initial topology of E is $\sigma(E, E')$.

Besides this, since B is τ-complete, it is closed in the τ-completion $\hat{E}$ of E, hence weakly closed in $\hat{E}$ (since it is convex). Thus it suffices to show that B is weakly relatively compact in $\hat{E}$. Accordingly we may assume that E is τ-complete, and our task is to show that B is weakly relatively compact.

According to Eberlein's theorem, it is enough to show that each sequence extracted from B admits a weak limiting point in E. In this way it is seen that there is no harm in assuming that E is separable for τ. (The reader will recall that the closed vector subspaces of E are the same for τ as for σ.)

To sum up, our problem is reduced to the following: E is a complete separated LCTVS that is separable; A is a weakly compact subset of E, B its closed, convex envelope; it is necessary to show that B is weakly relatively compact. This will now be done by representing points of B as integrals over A.

We regard A as a compact space for the weak topology. Let μ be a real Radon measure on A. We shall show that the integral

$$b = \int x \, d\mu(x) \tag{8.13.1}$$

belongs to B. A priori the integral is defined to be that element of E'^*, the algebraic dual of E', given by

$$x' \to \int \langle x, x' \rangle \, d\mu(x),$$

see Subsection 8.14.1; the only properties of this integral we require are immediate consequences of this defining property.

To assert that b belongs to E signifies that this linear form on E' is weakly continuous; once this is established, the obvious inequality

$$\langle b, x' \rangle \leq \operatorname*{Sup}_{x \in A} \langle x, x' \rangle, \tag{8.13.2}$$

holding for $\mu \geq 0$ of total mass unity, will show that b belongs to B.

Weak continuity of b, qua linear form on E', is easily established by appeal to Theorem 8.5.1: since E is separable, the weak topology on U^0 (U any neighbourhood of 0 in E) is metrizable. Continuity of $b \mid U^0$ is thus an immediate consequence of Lebesgue's Theorem 4.8.2.

Thus (8.13.1) gives a mapping t of the set M of normalized positive Radon measures on A into B. We shall next verify that t is *onto*. Indeed, if b belongs to B, then b is the limit of convex combinations

$$\sum_i c_i x_i,$$

where $c_i \geq 0$, $\sum_i c_i = 1$, $x_i \in A$. Each such sum can be expressed as an integral like (8.13.1), μ being the sum of masses c_i at the points x_i. This μ is a normalized positive Radon measure on A; that is, μ is in M. Now M is weakly compact, and it is clear that t is continuous for the weak topology on M and $\sigma(E, E')$. It follows that t maps M onto B.

At the same time, weak compactness of M and continuity of t combine to show that $B = t(M)$ is weakly compact. This completes the proof of Krein's theorem. ▌

8.13.2 **Corollary.** Suppose that E is a separated LCTVS, quasi-complete for $\tau(E, E')$, T a compact space, $t \to x(t)$ a weakly continuous map of T into E, μ a Radon measure on T. Let

$$x_0 = \int_T x(t)\, d\mu(t),$$

a priori an element of E'^*. Then

(a) x_0 belongs to the closed convex balanced envelope in E of $\|\mu\| \cdot x(T)$;

(b) If μ is positive and of unit total mass, then x_0 belongs to the closed convex envelope in E of $x(T)$.

Proof. $A = x(T)$ is weakly compact. Let B and K be, respectively, its closed convex and closed convex balanced envelope in E. Since K is the balanced envelope of B, and since Krein's theorem shows that B is weakly compact, K is weakly compact too.

Now it is clear that x_0 belongs to the weakly closed, convex, balanced envelope *in* E'^* of $\|\mu\| \cdot A$. However, since K is weakly compact, it is weakly closed in E'^* and so the said envelope in E'^* coincides with the corresponding envelope *in* E. Thus $x_0 \in \|\mu\| \cdot K$ and (a) is established.

Part (b) now follows as a special case, since

$$\langle x_0, x' \rangle \leq \operatorname*{Sup}_{t \in T} \langle x(t), x' \rangle$$

for each x' in E', so that x_0 must belong to B in view of the Corollary 2.2.4. ▌

8.13.3 **Corollary.** Let E be separated and quasi-complete for $\tau(E, E')$, $A \subset E$ weakly compact, B its closed convex envelope, K its closed convex balanced envelope. Then

$$B = \left\{\int_A x\, d\mu(x) : \mu \geq 0, \int_A d\mu = 1\right\},$$

$$K = \left\{\int_A x\, d\mu(x) : \mu \text{ real}, \int_A d\,|\mu| \leq 1\right\}.$$

Proof. Use Corollary 8.13.2 and the closing portion of the proof of Krein's theorem, modified in the obvious way in the case of K. ▌

8.13.4 **Remarks.** (1) For Banach spaces, Krein's theorem is given by Dunford and Schwartz ([1], p. 434), who accredit it jointly to Krein and Šmulian, noting that Krein proved it first for a separable Banach space in [1], and that separability was dropped in the joint paper by Krein and Šmulian [1]. The treatment we give for general spaces is due to Grothendieck ([7], Chapitre V, p. 22).

(2) There is a less refined form of Krein's theorem that is a good deal easier to prove, namely, the statement: In a quasi-complete LCTVS E, the closed convex envelope B of a compact set A in E is again compact. Indeed, it is easily seen that in any case B is precompact and bounded; quasi-completeness then finishes the argument.

(3) It is also relatively easy to show that in any TVS E, the closed, convex, and balanced envelope B of a compact, convex set A is itself compact.

To see this one obtains the envelope B by (1) balancing, (2) convexifying, and (3) closing, starting from A; the processes must take place in the order indicated. This shows that for real spaces B is the closed, convex envelope of $A \cup -A$; and that for complex spaces, B is contained in the closed, convex envelope of $2A \cup 2\omega A \cup 2\omega^2 A$, where ω is a cube root of unity, $\omega \neq 1$. The following figure will explain this.

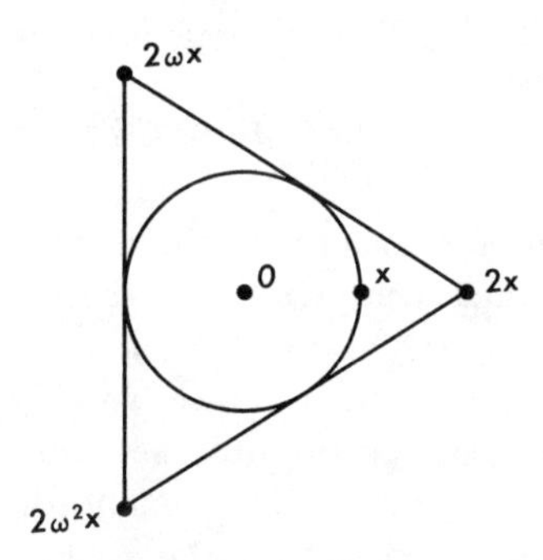

It remains now only to observe that the closed, convex envelope of the union of any finite number of compact, convex sets A_i $(1 \leq i \leq n)$ is itself compact—being in fact the image, by the continuous mapping $(\alpha_1, \cdots, \alpha_n, x_1, \cdots, x_n) \to \sum_{i=1}^n \alpha_i x_i$, of the compact space $[0, 1]^n \times A_1 \times \cdots \times A_n$.

(4) Even in R^2 it is the case that (α) the balanced envelope of a convex set is not necessarily convex; and that (β) the convex envelope of a closed set is not necessarily closed.

For an example to illustrate (α) one has only to take as the convex set a segment not passing through 0. For (β) take the closed set $\{(n, n^{-1}) : n = 1, 2, \cdots\}$, the convex envelope of which is the shaded area in the figure below: The upper horizontal boundary line consists of points adherent to, but not in, the convex envelope.

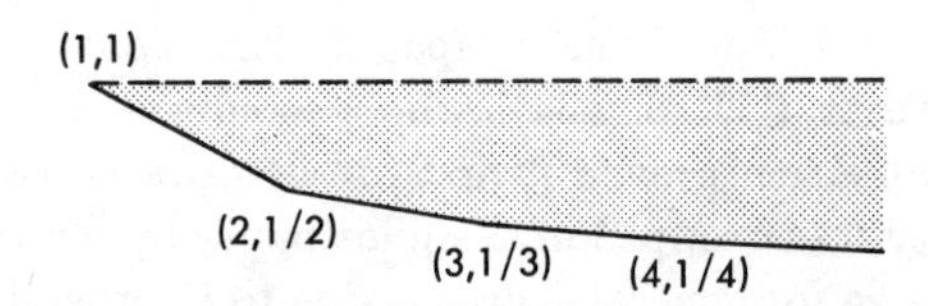

(5) In Corollary 8.13.2(a) we may assume merely that T is locally compact, provided we assume that $\lim_{t\to\infty} x(t) = 0$ weakly. Further, if T is discrete, it suffices to assume merely that $x(t)$ is bounded, that $\lim_{t\to\infty} \langle x(t), x' \rangle = 0$ for each x' in a subset H of E' that is total over E (that is, that separates points of E), and that E is quasi-complete. For then the series $\sum \mu(t)x(t)$ is convergent in E for each $\mu \in \ell^1(T)$. This topic has been dealt with in Section 2.5.

8.13.5 **Example.** Let T denote the real line, let $E = L^\infty(T)$ (constructed relative to the Lebesgue measure), and let $x(t)$ denote the element of $L^\infty(T)$ defined by (that is, the class of) the function $s \to e^{-its}$ of the real variable s. Imagine E to be endowed with the weak topology $\sigma(L^\infty, L^1)$. Then the function $t \to x(t)$ is bounded and weakly continuous, and $\lim_{t\to\infty} x(t) = 0$ weakly (by the Riemann-Lebesgue lemma). If μ is any bounded Radon measure on T, the integral $\int x(t)\, d\mu(t)$ exists as an element of E, being in fact the element of $L^\infty(T)$ that is the class of the bounded continuous function

$$s \to \int e^{-its}\, d\mu(t) = \hat{\mu}(s),$$

which is just the Fourier transform of the measure μ. The preceding general arguments go to show that each member of the closed convex envelope in L^∞ of the exponentials e^{-its} (t real) has the form $\hat{\mu}$ (or, better, the class of $\hat{\mu}$), where μ is positive and $\int d\mu(t) \le 1$. In this case we may do better and claim that necessarily $\int d\mu(t) = 1$: perhaps the quickest route to this conclusion lies in extending the definition of $x(t)$ to the Alexandroff compactification of T by setting $x(\infty) = 0$ (a step suggested by the Riemann-Lebesgue lemma). Then Corollary 8.13.2(b) is directly applicable.

With this example we are brought very close to the topic dealt with Section 10.3, the representation theorem of Bochner for positive definite functions. The essential step missing at present is filled by the Krein-Milman theorem, which shows that the continuous positive definite functions are precisely the multiples of members of the closed, convex envelope of the exponentials e^{-its}.

8.14 Integration of Vector-valued Functions

As was heralded in Subsection 4.10.7, the remainder of this chapter will be devoted to some aspects of the problem of integrating vector-valued functions with respect to scalar-valued Radon measures and the related problem of integrating scalar-valued functions with respect to vector-valued Radon measures. The former topic has been very lightly touched upon as an auxiliary in Section 8.13. Apart from this, the new topics are for the most part unused in this book. Nevertheless they merit some attention inasmuch as they constitute a flourishing field for the application of general duality theory in which there remains much to be done.

Questions arising from the study of the integration of vector-valued functions, usually parallel to results expounded in Chapter 4 for the scalar-valued case, lead rather naturally to the concept of vector-valued Radon measures. The latter objects will be viewed as continuous linear maps of spaces of continuous functions into separated LCTVSs. In this chapter one of the main questions concerning them centres round a type of Lebesgue-Radon-Nikodym theorem. Later, in Subsection 9.4.14, we shall have more to say about their role as continuous linear maps.

Throughout the remainder of this chapter T will, save when it is explicitly specialized, denote a separated locally compact space, μ a positive Radon measure on T, and E a separated LCTVS. Such terms as "measurable," "integrable," "σ-compact," "almost everywhere," "negligible," and so on, are to be understood as referring to μ. It is convenient, and involves no essential loss of generality, to assume that E is a real vector space. (If E is, in fact, a complex vector space, restriction to real scalars introduces no essential modifications.)

The problem before us is that of discussing the possibility of forming the integral

$$\int f\,d\mu = \int f(t)\,d\mu(t),$$

where f is a given (suitably restricted) function on T with values in E. Integrals $\int_M f\,d\mu$ extended over measurable subsets M of T are dealt with in the obvious fashion, setting $\int_M f\,d\mu$ equal by definition to $\int \chi_M f\,d\mu$, where χ_M is the characteristic function of M.

As we hinted in Subsection 4.10.7 there have grown up two different approaches to this problem, one usually qualified by the adjective "weak" (though we shall prefer the term "scalarly" or "scalarwise") and the other a "strong" theory associated, in the case of Banach spaces, with Bochner's name. For one approach to the latter, see Hille [1], Chapter III, and Dunford and Schwartz [1], especially Chapter III. Bourbaki ([6], Chapitre IV) sets forth an approach that strictly parallels his method for scalar-valued functions (given in essence in our Chapter 4) and that leads to the same end. Each of Hille, Dunford and Schwartz, and Bourbaki ([6], Chapitre VI) deal also with various "weak" theories. We shall present a short account of certain aspects only of the subject in a manner which differs in detail from that of any of the references cited above and which serves to present the Bochner-like theory as a natural and workable tool for Fréchet spaces. Outside this category, however, the "weak" theory is definitely simpler and still satisfies most requirements. For general separated LCTVSs E it seems altogether simpler to begin with the weak or scalarwise theory.

Unless the contrary is stated, f will always denote a function on T with values in E.

8.14.1 Various Concepts of Measurability and Integrability. As we have said, these various concepts fall into two categories—the scalarwise concepts (sometimes termed the "weak" concepts, but see Remark (1) below), and the so-called "strong" concepts. We first frame the definitions and then add numerous comments.

(a) SCALARWISE CONCEPTS. In general terms we shall say that a function f mapping T into E possesses scalarly a property P (such as measurability, integrability, or essential integrability) if, for each x' in E' (the topological dual of E), the scalar-valued function $x' \circ f$ possesses the property P.

For brevity we shall often write SM, SI, and SEI in place of the phrases "scalarwise measurable," "scalarwise integrable," and "scalarwise essentially integrable," respectively.

Whenever f is SEI we introduce the integral $\bar{\int} f \, d\mu$ as the unique element z of E'^*, the algebraic dual of E', determined by the relations

$$\langle z, x' \rangle = \bar{\int} (x' \circ f) \, d\mu \tag{8.14.1}$$

holding for all x' in E'. Very often we shall write $\langle f, x' \rangle$ or $f_{x'}$ in place of $x' \circ f$.

Since E is separated, we may identify it with a vector subspace of E'^*. The question as to whether or not $\bar{\int} f \, d\mu$ lies in E, or in other vector subspaces of E'^* (the bidual E'', for instance), is one that often arises and to which we shall devote considerable attention.

If f is SI, it is a fortiori SEI, and in this case we write $\int f \, d\mu$ in place of $\bar{\int} f \, d\mu$.

Some writers have considered the situation that arises if one takes a vector subspace L of E' which separates points of E, and imposes the demand that $f_{x'}$ be essentially integrable for each x' in L. Then $\bar{\int} f \, d\mu$ would be defined a priori as an element of L^*. This situation, however, although apparently more general, is subsumed in that already described if we endow E with the weak topology $\sigma(E, L)$, relative to which the topological dual of E is precisely L. For this reason we leave aside the variation amounting to replacement of E' by L.

(b) MEASURABILITY. The definition here is based upon Lusin's theorem 4.8.5. Thus we shall say that f is measurable if for each integrable set $A \subset T$ (or, what is equivalent, for each compact set $A \subset T$) and each number $\varepsilon > 0$, there exists a compact set $K \subset A$ such that $\mu(A \backslash K) < \varepsilon$ and $f \mid K$ is continuous.

Evidently, measurability in this sense—which is often spoken of as "strong" measurability, although we shall here avoid use of this term—implies scalarwise measurability; see Remark (2) below.

Regarding the history of, and some literature dealing with, the concept of measurability, see the Remark following Theorem 8.15.2.

(c) INTEGRABILITY. The function f, mapping T into E, is said to be integrable (resp. essentially integrable) if it is measurable (see (b)) and if, further,

$$\int^* (p \circ f)\, d\mu < +\infty \qquad \left(\text{resp. } \overline{\int}^* (p \circ f)\, d\mu < +\infty\right) \tag{8.14.2}$$

for each continuous seminorm p on E.

Evidently, an equivalent definition results if we restrict p to belong to any preassigned set of continuous seminorms that defines the topology on E.

As in the case of measurability, the adjective "strong" is often prefixed to this concept of integrability; but as before we eschew this practice.

It is obvious that if f is integrable (resp. essentially integrable), then it is SI (resp. SEI). Later (Theorem 8.15.2) we shall establish partial converses of this implication, valid when E is suitably restricted.

Remarks. (1) The concept of measurability obviously depends to some extent on the given topology on E—to precisely what extent is a nontrivial matter.

An especially important point to notice concerns the concept of measurability associated with the weakened topology $\sigma(E, E')$, which we shall term *weak measurability*. Many writers attach this name to what we here term *scalarwise measurability*, and the likelihood of serious confusion is heightened by the fact that weak continuity (that is, continuity for $\sigma(E, E')$) and scalarwise continuity are indeed the same thing. For measurability, however, the two concepts must be clearly distinguished.

Obviously, weak measurability implies scalarwise measurability. But the converse is false, as is shown by the following example (Bourbaki [6], Ch. VI, p. 81, Exercice 12). Let T be $[0, 1]$ with Lebesgue measure μ and E any Hilbert space having an orthonormal base $(e_t)_{t\in T}$. We define f by $f(t) = e_t$. This f is scalarly negligible, a fortiori scalarly measurable. But f is *not* weakly measurable. For consider which compact sets K in T have the property that $t \to (f(t) \mid x) \mid K$ is continuous for all x in E. It is very simple to see that only the finite sets K have this property. Now if f were weakly measurable, it would necessarily be the case that sets K of the above type exist with $\mu(T\backslash K)$ arbitrarily small, contradicting the finiteness of K.

This example points another useful lesson. Suppose A is a nonmeasurable subset of T, and let g be the function $\chi_A f$. Then $t \to \|g(t)\|$ is nonmeasurable, despite the fact that g (like f) is scalarly negligible and so scalarly measurable. Thus, applying a continuous seminorm to a scalarly measurable function can lead to a nonmeasurable function.

For these reasons we have chosen to follow Bourbaki's terminology in preference to the more usual one.

(2) To balance the negative results spoken of in Remark (1), we shall later (Section 8.15) see that for metrizable spaces E there exist important positive relationships between measurability and scalarwise measurability.

(3) In view of the example given in Remark (1) above, we should perhaps point out the easy result that a continuous (scalar- or vector-valued) function of a measurable (scalar- or vector-valued) function is again measurable.

(4) Scalarwise integrability likewise depends to some extent on the topology carried by E. Consider by way of illustration the following situation (Bourbaki [6], Chapitre VI, p. 80, Exercice 5)). For E we take the space $C(R)$ of complex-valued continuous functions on R, endowed with the topology of locally uniform convergence. For T we take R and for μ the Lebesgue measure. $f(t)$ will be the function (element of E) defined by $\xi \to e^{-2\pi it\xi}$. The function f is evidently bounded and scalarwise measurable, each element of the dual E' being generated by a Radon measure λ on R having a compact support in such a way that

$$\langle f(t), \lambda \rangle = \int e^{-2\pi it\xi} \, d\lambda(\xi),$$

the Fourier transform of λ. This formula shows that f is indeed weakly continuous. However, it is equally evident that f is not scalarwise integrable.

We consider next what happens when we in turn endow E with each of three topologies, namely (a) the weakened topology $\sigma = \sigma(E, E')$, (b) the weak topology $\sigma' = \sigma(E, \mathscr{K})$, and (c) the weak topology $\sigma'' = \sigma(E, \mathscr{D})$. Here $\mathscr{K}$ and $\mathscr{D}$ denote, respectively, the spaces of continuous and indefinitely differentiable complex-valued functions on R that have compact supports, each function g of either of these spaces being identified with the measure $g \cdot \mu$. As is generally the case, scalarwise measurability and integrability signify the same for σ as for the initial topology. The dual of E relative to σ' (resp. σ'') is $\mathscr{K}$ (resp. $\mathscr{D}$), so that f is scalarwise measurable (even weakly continuous) in either case. It is not scalarwise integrable relative to σ or relative to σ', since there exist functions g in $\mathscr{K}$ for which the Fourier transform

$$\hat{g}(t) = \int e^{-2\pi it\xi} g(\xi) \, d\mu(\xi)$$

is not integrable for μ. On the other hand, $\hat{g}$ is μ-integrable for any g in $\mathscr{D}$, so that f is scalarwise integrable relative to σ''.

We give now two simple results leading towards sufficient conditions in order that an integral shall belong to E.

8.14.2 **Lemma.** Suppose that f is SEI, that M is a subset of T carrying μ (that is, $T \backslash M$ is locally negligible), that T_0 is a measurable subset of M such that $f = 0$ l.a.e. on $M \backslash T_0$, and that $m = \mu_*(T_0) < +\infty$. If Γ is the weakly closed convex envelope in E'^* of $f(T_0)$, then $z = \bar{\int} f \, d\mu$ belongs to $m \cdot \Gamma$.

Proof. By Corollary 2.2.4 (1) we have only to show that for each x' in E' one has

$$\langle z, x' \rangle \leq m \cdot \operatorname{Sup} \{\langle f(t), x' \rangle : t \in T_0\}.$$

However, by definition of z it is true that

$$\langle z, x' \rangle = \bar{\int} \langle f(t), x' \rangle \, d\mu(t) = \bar{\int} \chi_M \langle f(t), x' \rangle \, d\mu(t).$$

The essential integrability of $f_{x'}$ shows that there is a σ-finite measurable set $S \subset T_0$ (S possibly depending on x') such that $f_{x'} = 0$ l.a.e. on $M \backslash S$. Thus

$$\langle z, x' \rangle = \int \chi_S(t) \langle f(t), x' \rangle \, d\mu(t) \leq \int_S d\mu(t). \operatorname{Sup} \{\langle f(t), x' \rangle : t \in S\};$$

the first factor on the right here is at most $\mu_*(T_0)$ (S being σ-compact and contained in T_0: see Subsection 4.7.6), and the second factor is at most $\operatorname{Sup} \{\langle f(t), x' \rangle : t \in T_0\}$. The proof is thus complete. ▌

8.14.3 **Proposition.** Suppose that E satisfies the following

Hypothesis. In E the closed convex envelope of a compact (resp. weakly compact) set is $\tau(E, E')$-complete.

If there exists a compact set $K \subset T$ such that $f \mid K$ is continuous (resp. weakly continuous) and $f = 0$ l.a.e. on $T \backslash K$, then $\bar{\int} f \, d\mu$ belongs to E.

Proof. According to the preceding lemma, $z = \bar{\int} f \, d\mu$ belongs to $m\Gamma$, where $m = \mu(K)$ (necessarily finite) and Γ is the weakly closed convex envelope in E'^* of $f(K) \cup \{0\}$. It will therefore suffice to show that $\Gamma \subset E$.

By Krein's Theorem 8.13.1, the closed, convex envelope *in* E of $f(K) \cup \{0\}$ is weakly compact, since our hypotheses on f ensure that $f(K)$, and therefore also $f(K) \cup \{0\}$, is compact (resp. weakly compact). This envelope in E is thus weakly closed in E'^*, and must therefore coincide with Γ, Q.E.D. ▌

Remarks. (1) Our result is suggested by, and is a simple extension of, that given by Bourbaki ([6], Chapitre III, p. 81). The reader will notice that the Hypothesis stated above is certainly satisfied (in either form) if E is quasi-complete, the closed convex envelope of a weakly compact set being bounded and closed, hence complete for the initial topology on E, a fortiori complete for the topology $\tau(E, E')$. (This is because the latter topology is stronger than the initial topology and, in addition, has a neighbourhood base at 0 formed of sets closed for the initial topology; see Proposition 9.1.5.)

(2) It is sometimes convenient to deal directly with a complex LCTVS E (without restricting its scalar field, that is). Here the analogues of Lemma 8.14.2 and Proposition 8.14.3 would involve, in place of the closed convex envelopes, the corresponding closed, convex, and balanced envelopes. However, if f is as in Proposition 8.14.3, there is no need to disturb the hypothesis stated above concerning E. For the closed, convex, and balanced envelope of $f(K) \cup \{0\}$ is none other than the closed, convex envelope of the balanced envelope B of $f(K) \cup \{0\}$; and B is compact whenever $f(K) \cup \{0\}$ is compact (weakly or otherwise in each case). (It is the image of the compact space $S \times (f(K) \cup \{0\})$ under the continuous mapping $(\alpha, x) \to \alpha \cdot x$, S being the set of scalars α satisfying $|\alpha| \leq 1$.) The reader will notice that Proposition 8.14.3, thus modified, will, in fact, handle directly the case in which μ is no longer necessarily positive, but rather any complex Radon measure on T.

8.14.4. We turn next to study the effect of applying a continuous linear map u and a continuous seminorm to an integral. In the former case we suppose that F

is a second separated LCTVS and that u is a continuous linear map of E into F. Let u' be the adjoint of u, mapping F' into E'. Introduce also the species of biadjoint u'^*, mapping E'^* into F'^*, defined by

$$\langle u'^*(z), y' \rangle = \langle z, u'(y') \rangle \tag{8.14.3}$$

for y' in F' and z in E'^*. We notice that $u'^* \mid E'' = u''$ and $u'^* \mid E = u$.

8.14.5 Proposition. Let E and F be separated LCTVSs, u a continuous linear map of E into F. If the function f, mapping T into E, is SEI, then so too is the function $u \circ f$ (mapping T into F), and further

$$\bar{\int} (u \circ f)\, d\mu = u'^*\Big(\bar{\int} f\, d\mu\Big). \tag{8.14.4}$$

Proof. Let $g = u \circ f$. For y' in F' we have

$$y' \circ g(t) = \langle u \circ f(t), y' \rangle = \langle f(t), u'(y') \rangle,$$

which is essentially integrable since f is SEI and $x' = u'(y')$ belongs to E'. Thus g is SEI. Besides this we have, using (8.14.3),

$$\begin{aligned}\Big\langle u'^*\Big(\bar{\int} f\, d\mu\Big), y' \Big\rangle &= \Big\langle \bar{\int} f\, d\mu, u'(y') \Big\rangle \\ &= \bar{\int} \langle f(t), u'(y') \rangle\, d\mu(t) \\ &= \bar{\int} \langle u(f(t)), y' \rangle\, d\mu(t) \\ &= \Big\langle \bar{\int} (u \circ f)\, d\mu, y' \Big\rangle,\end{aligned}$$

thus verifying (8.14.4). ∎

Remarks. Whenever we know on other grounds that $\bar{\int} f\, d\mu$ belongs to E'' (resp. to E), we may in (8.14.4) replace u'^* by u'' (resp. u). A more refined version of Proposition 8.14.5, valid when E and F are complete, and u is merely closed, will appear in Theorem 8.14.15 below.

In dealing with seminorms, we shall find it convenient to extend any seminorm p on E into a function on E'^* (possibly taking the value $+\infty$) by the formula

$$p'^*(z) = \operatorname{Sup} \{|\langle z, x' \rangle| : x' \in W\}, \tag{8.14.5}$$

where W is the polar in E' of the set $\{x \in E : p(x) \leq 1\}$. Apart from the fact that p'^* may take the value $+\infty$ at points of E'^*, it might be termed a seminorm on E'^*. In any case, Corollary 2.2.4(2) ensures that $p'^* \mid E = p$, if p is a continuous seminorm on E.

8.14.6 Proposition. If f is SEI, and if p is any continuous seminorm on E, then

$$p'^*\Big(\bar{\int} f\, d\mu\Big) \leq \int^* (p \circ f)\, d\mu. \tag{8.14.6}$$

Proof. This is almost immediate: If x' belongs to W we have

$$\left|\left\langle \bar{\int} f\,d\mu, x' \right\rangle\right| = \left|\bar{\int} \langle f(t), x' \rangle\, d\mu(t)\right|$$
$$\leq \int |\langle f(t), x' \rangle|\, d\mu(t)$$
$$\leq \int^* p(f(t))\, d\mu(t),$$

whence follows (8.14.6) on taking the supremum as x' ranges over W. ▌

Remark. If we know that $\bar{\int} f\,d\mu$ belongs to E, we may in (8.14.6) replace p'^* by p.

Although the results expressed in Propositions 8.14.3–8.14.6 are elementary, they may be used to illustrate the utility of the idea of integrating vector-valued functions. We do this directly in Subsection 8.14.7. We shall give other examples later, illustrating the use of more refined general theorems.

8.14.7 **Application to the Study of Holomorphic Vector-valued Functions.** The use of holomorphic vector-valued functions is by now a standard device in various branches of functional analysis, as will become apparent to the reader if he will trouble to scan through Hille [1], or Dunford and Schwartz [1], or both (to mention but two possible references). The study of such functions is, as one might predict from knowledge of the theory of scalar-valued holomorphic functions, heavily influenced by integral formulae of the Cauchy type and their consequences. In the case of vector-valued functions it turns out that, thanks to such integral formulae, the weakest concept of holomorphy already implies the strongest that one may reasonably expect. We may reach this conclusion without making any appeal to results about integration beyond the "weak" or "scalarwise" theory.

To be specific, let us suppose that E is a separated LCTVS, that Ω is a domain in the complex plane, and that f is a map of Ω into E. Taking holomorphy in its weakest sense, we shall say that f is holomorphic on Ω if $x' \circ f$ is holomorphic on Ω (in the normal sense for scalar-valued functions) for each x' in E'. For the concept to have any interest, E must be a *complex* vector space: if E were real, $x' \circ f$ would be real valued; being holomorphic, $x' \circ f$ would be constant on Ω, and so f would be constant.

We shall now indicate how one may reach the rather surprising conclusions that follow, provided we impose upon E the Hypothesis of Proposition 8.14.3:

(1) If f is holomorphic on Ω, then it has derivatives of all orders at each point of Ω, these derivatives belonging to E and being themselves holomorphic on Ω.

(2) The aforementioned derivatives are representable by integral formulae exactly analogous to those familiar in the case of complex-valued functions.

(3) There is a Taylor expansion valid and convergent in E throughout a suitable neighbourhood of each point of Ω.

Once these basic facts are established, many other properties of holomorphic functions, familiar in the complex-valued case, may be extracted by standard arguments. The development may be left to the reader; see Exercises 8.15, 8.16 and 8.17.

To reach the above conclusions, we first of all show that any holomorphic f is continuous (and this for any LCTVS E). Indeed, since $x' \circ f$ is scalar-valued holomorphic, there is a number $M = M(x')$ such that

$$|x' \circ f(t) - x' \circ f(t_0)| \leq M \cdot |t - t_0|$$

for t_0 in Ω and t sufficiently close to t_0. Thus the family of elements of E,

$$(t - t_0)^{-1} \{f(t) - f(t_0)\},$$

obtained as t ranges over a suitably small punctured neighbourhood of t_0, is weakly bounded. But then (Theorem 8.2.2) this family is bounded in E. In particular $\lim f(t) = f(t_0)$ in E as $t \to t_0$.

Having established continuity of f, we may introduce the Cauchy integral. Let t_0 be a point of Ω and let c be a piecewise continuously differentiable, closed path in $\Omega \backslash \{t_0\}$ whose winding number relative to t_0 is $+1$. We consider the integral

$$q = (2\pi i)^{-1} \int_c f(t) \frac{dt}{t - t_0}.$$

Thanks to the known continuity of f and the hypothesis on E, Proposition 8.14.3 informs us that q belongs to E.

Combining the definition of the integral (see formula (8.14.1)) with the Cauchy integral formula for scalar-valued functions, it appears at once that q is none other than $f(t_0)$. Thus the Cauchy integral formula holds for f.

This being so, we examine the difference quotient

$$q_1 = (t - t_0)^{-1} \{f(t) - f(t_0)\},$$

representing it as

$$q_1 = (2\pi i)^{-1} \int_c f(s)[(s - t)^{-1} - (s - t_0)^{-1}]\, ds,$$

taking (for example) c to be a positively oriented circumference with centre t_0 and radius $r > 0$ so small that the associated closed disk lies in Ω, and assuming that $t \neq t_0$ lies in the open disk. By using Propositions 8.14.3 and 8.14.6 it follows that

$$(2\pi i)^{-1} \int_c f(s) \frac{ds}{(s - t_0)^2}$$

is an element of E and is, moreover, the limit as $t \to t_0$ of q_1. Thus we conclude that the derivative $f'(t_0)$ exists as an element of E which admits the representation

$$f'(t_0) = (2\pi i)^{-1} \int_c f(t) \frac{dt}{(t - t_0)^2}.$$

Here, as in the Cauchy integral formula itself, c may be chosen to be any piecewise, continuously differentiable, closed path in $\Omega \backslash \{t_0\}$ whose winding number relative to t_0 is $+1$.

In this last formula c may be held fixed and t_0 allowed to vary locally: the conclusion is that f' itself is holomorphic. Indeed we have

$$x' \circ f'(t_0) = (2\pi i)^{-1}\int_c x' \circ f(t) \frac{dt}{(t - t_0)^2},$$

which is (scalar-valued) holomorphic for each x' in E'.

A simple inductive argument on $n = 0, 1, 2, \cdots$ (as for the scalar-valued case) now shows that all the derivatives of f exist as elements of E and are likewise holomorphic on Ω. (1) and (2) are thus established.

From the integral formulae

$$f^{(n)}(t_0) = \left(\frac{n!}{2\pi i}\right)\int_c f(t)\frac{dt}{(t - t_0)^{n+1}},$$

likewise established by induction on n, one may deduce the Cauchy inequalities

$$p(f^{(n)}(t_0)) \leq \frac{n!\, M}{r^n},$$

where M is the supremum of $p(f(t))$ for $|t - t_0| = r$, the closed disk with centre t_0 and radius r being assumed to lie in Ω. (p is any continuous seminorm on E and the inequality flows from use of Proposition 8.14.6 once more.)

From these inequalities follows the Cauchy character of the partial sums of the Taylor series

$$\sum_{n=0}^{\infty} f^{(n)}(t_0)\frac{(t - t_0)^n}{n!}.$$

On the other hand the weak convergence of this series follows from the scalar-valued theory. Combining these facts we are led to conclusion (3), and our limited programme is completed.

Remarks. (1) For further developments and a systematic study of spaces of holomorphic vector-valued functions, see Grothendieck [5] and [6].

(2) The preceding techniques can be applied in limited measure to the case of differentiability of vector-valued functions of a real variable; see Schwartz [17], Edwards [15], and Exercise 8.14. This topic is of importance for the theory of vector-valued distributions (Schwartz [4], [5] and [6]). See also Iino [1].

8.14.8. It is now time to establish some analogues of Proposition 8.14.3 that apply with much lighter requirements on f and heavier ones on E. Even so we shall still avoid becoming involved with measurability and integrability, which are in general relatively difficult to handle.

Let f be SEI and let $z = \bar{\int} f\, d\mu$, an element of E'^*. To say that z belongs to the bidual E'' means simply that the linear form on E' defined by z, namely,

$$x' \to \langle z, x' \rangle = \bar{\int} \langle f, x' \rangle\, d\mu,$$

is strongly continuous on E'. When this is so, we shall refer to $\bar{\int} f\, d\mu$ as a *Gelfand integral*, referring the reader to Gelfand [2] for justification of this name.

If this same linear form is weakly continuous on E', then z will belong to E itself: we then speak of $\overline{\int} f\,d\mu$ as a *Pettis integral* (Pettis [2]).

We give two fairly direct results relating to Gelfand and Pettis integrals.

8.14.9 **Proposition.** Suppose that f is SEI and such that for each compact subset K of T, the closed convex envelope in E of $f(K)$ is weakly compact. Then $\overline{\int} f\,d\mu$ belongs to E''; that is, the integral exists as a Gelfand integral.

Proof. Lemma 8.14.2 shows at once that

$$z_K = \int_K f\,d\mu = \int f\chi_K\,d\mu$$

belongs to Γ, the weakly closed convex envelope in E'^* of $f(K)$. However, the closed convex envelope in E of $f(K)$, say Γ_0, is also the weakly closed convex envelope in E of $f(K)$ and is therefore weakly compact. Γ_0 is thus weakly closed in E'^*. Since in any case Γ_0 is weakly dense in Γ, it follows that $\Gamma_0 = \Gamma$. Thus z_K lies in E.

Now, since f is SEI, $z = \overline{\int} f\,d\mu$ is the weak limit in E'^* of the directed family (z_K), the compact sets K being partially ordered by inclusion. At the same time,

$$|\langle z_K, x'\rangle| \leq \overline{\int_K} |\langle f, x'\rangle|\,d\mu \leq \overline{\int} |\langle f, x'\rangle|\,d\mu < +\infty$$

for each x' in E', showing that the z_K fall into a bounded subset B of E. To complete the proof it therefore suffices to recall that the weak closure in E'^* of any bounded subset B of E is contained in E''. ∎

8.14.10 **Corollary.** If E is semireflexive and f is SEI and locally bounded, then $\overline{\int} f\,d\mu$ belongs to E.

Proof. For now $E'' = E$ (as sets) and, moreover, each bounded subset of E is weakly relatively compact in E. ∎

For normed spaces E there are stronger results.

8.14.11 **Proposition.** (a) If E is normable and f is SEI, then $\overline{\int} f\,d\mu$ lies in E''; that is, the integral exists as a Gelfand integral. (b) If E is a reflexive Banach space and f is SEI, then $\overline{\int} f\,d\mu$ belongs E; that is, the integral exists as a Pettis integral.

Proof. Evidently (b) follows from (a). For (a) it suffices to show that the seminorm

$$q(x') = \overline{\int} |\langle f, x'\rangle|\,d\mu$$

is strongly continuous on E'. Since E' is a Banach space and therefore barrelled, only lower semicontinuity of q need be established. In other words, we are reduced to showing that the subset of E' defined by the inequality $q(x') \leq 1$ is strongly closed. But this is an immediate consequence of Fatou's Lemma 4.5.4. ∎

Remarks. (1) The proof of (a) is seen to be valid whenever the strong dual E' is barrelled and metrizable. If E is a Fréchet space, however, the strong dual E' is metrizable if and only if E is a Banach space, so that there is little room for extension in that direction.

(2) Part (b) applies notably when E is a Hilbert space. The idea of integrating functions with values in a Hilbert space was dealt with in an ad hoc fashion in Subsection 1.12.9.

(3) It is not difficult to give examples of cases in which E is a Banach space, f is SEI, $\bar{\int} f\, d\mu$ belongs to E'', and yet $\bar{\int} f\, d\mu$ does *not* belong to E. Thus (Bourbaki [6], Chapitre 6, p. 80, Exercice 7)) one may take T to be $[0, 1]$, μ the Lebesgue measure restricted to T, E the space c_0 of sequences converging to 0, and f defined by

$$f(t)(n) = n \cdot \chi_{I_n}(t),$$

where $I_n = (0, 1/n)$. Here E' is ℓ^1. For x' in ℓ^1,

$$\langle f(t), x' \rangle = \sum_n n x'(n) \chi_{I_n}(t),$$

which is easily seen to be in $\mathscr{L}^1(T, \mu)$. Thus f is SI. Also $z = \int f\, d\mu$ is that element of $E'' = \ell^\infty$ defined by $z(n) = 1$ for all n, which is visibly not an element of $E = c_0$.

8.14.12 **Remarks and Examples.** Corollary 8.14.10 can be used to deal with simple cases in which f maps T into the dual F' of some LCTVS F, a question to which we shall return later (Theorem 8.16.1). If we suppose that F' is endowed with its weak topology $\sigma(F', F)$, and if F is barrelled, then F' is semireflexive. Indeed the dual of F' will be F and the associated strong topology on F (that of convergence uniform on the weakly bounded subsets of F') will, since F is barrelled and the weakly bounded subsets F' therefore equicontinuous, coincide with the initial topology of F.

From the said corollary we may thus infer that $\bar{\int} f\, d\mu$ lies in F' whenever F is barrelled and $f : T \to F'$ is SEI (in the sense that $\langle y, f \rangle$ is essentially integrable for each y in F) and locally weakly bounded (that is, $f(K)$ is weakly bounded in F' for each compact subset K of T).

Example 1: Product measures. The result just established is strong enough to form the basis of a new definition of the product of two measures, a topic dealt with in Section 4.17.

Let S and T be two separated locally compact spaces, P their product, and let α and β be measures on S and T, respectively. As we know, the space $F = \mathscr{K}(P)$ is barrelled (see Subsection 6.3.3). The dual F' is just the space $M(P)$ of all Radon measures on P. Naturally, the same assertions are valid when P is replaced throughout by S or T.

Given β in $M(T)$ and s in S, denote by ${}_s\beta$ the element of $M(P)$ defined by

$$\langle g, {}_s\beta \rangle \equiv \int_P g \, d_s\beta = \int_T g(s, t)\, d\beta(t),$$

and consider the function f mapping S into $M(P) = F'$ defined by setting $f(s) = {}_s\beta$. It is evident that $\langle g, {}_s\beta \rangle$ belongs to $\mathscr{K}(P)$, so that f is weakly continuous and bounded. Supposing that α (which plays the role of μ in the general theory) is positive, it is also evident that f is SI for α. It follows from our general result that $\bar{\int} f\, d\alpha$ lies in $F' = M(P)$. By definition of the integral one sees directly that $\bar{\int} f\, d\alpha$ is none other than the product measure $\alpha \otimes \beta$.

Naturally, the condition that α be positive is quite pointless and is easily removed.

Example 2: Convolutions of Measures. One may discuss in a similar fashion the convolution of two measures, defined in Section 4.19. Here we take $S = T$, a separated locally compact group. Given β as before, we define the left-translates $L_t\beta$ by the equations

$$\langle g, L_t\beta \rangle = \int g\, d(L_t\beta) = \int g(t^{-1}t')\, d\beta(t')$$

for g in $\mathscr{K}(T)$, and then consider the function $f(t) = L_{t^{-1}}\beta$ mapping T into $F' = M(T)$. Making again the provisional assumption that α is positive, we see that f is scalarwise continuous in any case, and that f is SEI (here equivalent to SI by virtue of scalarwise continuity) if and only if

$$\int^* d\alpha(t) \left| \int g(tt')\, d\beta(t') \right| < +\infty$$

for each g in $\mathscr{K}(T)$.

Now if β also is positive this is precisely the condition used in Section 4.19 to express the existence of the convolution $\alpha * \beta$. When this condition is fulfilled, $\alpha * \beta$ is indeed none other than

$$\bar{\int} f\, d\alpha = \bar{\int} (L_{t^{-1}}\beta)\, d\alpha(t).$$

On the other hand if β is not positive, it seems plausible that $\bar{\int} f\, d\alpha$ may exist (that is, f may be SI) even though $\alpha * \beta$ is not defined (which is so if and only if $\alpha * |\beta|$ is not defined). In any such case we have an extended definition of the convolution.

Returning to the general theory, we consider next some results in which E and f are specialized, the latter by means of conditions approaching those of integrability.

8.14.13 **Proposition.** Suppose that E is quasi-complete and that f fulfils the following two conditions:

(1) f is scalarwise locally integrable and

$$\int_K f\, d\mu = \int f\chi_K\, d\mu$$

belongs to E for each compact set $K \subset T$;

(2) $\bar{\int}^*(p \circ f)\, d\mu < +\infty$ for each p belonging to a family of continuous seminorms defining the topology of E.

Then f is SEI and $\bar{\int} gf\, d\mu$ belongs to E for each g in $\overline{\mathscr{L}}^\infty$. Moreover, $\bar{\int} gf\, d\mu$ is the limit in E of $\int_K gf\, d\mu$ as K expands.

Proof. This falls into two parts.

(a) If f satisfies (1) and (2), then f is SEI and $\bar{\int} f\,d\mu$ is the limit in E of $\int_K f\,d\mu$.

(b) If f satisfies (1) and (2), then $\int_K gf\,d\mu$ belongs to E for each compact K and each g in $\overline{\mathscr{L}}^\infty$.

Once these are established one may apply (a) with f replaced by gf and so complete the proof.

To prove (a), define $z_K = \int_K f\,d\mu$. Then Proposition 8.14.6 and the remark that follows it yield the inequalities

$$p(z_K) \leq \int^* p \circ (f\chi_K)\,d\mu = \int^* (p \circ f) \cdot \chi_K\,d\mu \leq \bar{\int}^* (p \circ f)\,d\mu < +\infty$$

and

$$p(z_{K'} - z_K) = p\Big[\int f\chi_{K'\setminus K}\,d\mu\Big] \leq \int^* (p \circ f) \cdot \chi_{K'\setminus K}\,d\mu$$

for any two compact sets $K \subset K'$. The first shows that the family (z_K) is bounded, and the second that this family is Cauchy. E being quasi-complete, $z = \lim z_K$ exists in E. Since (2) plainly entails that f is SEI, one will have for each x' in E' the relations

$$\langle z, x' \rangle = \lim \langle z_K, x' \rangle = \lim \int_K \langle f, x' \rangle\,d\mu = \bar{\int} \langle f, x' \rangle\,d\mu,$$

showing that $\bar{\int} f\,d\mu$ is none other than z. (a) is thereby established.

As regards (b), gf is plainly SEI for each g in $\overline{\mathscr{L}}^\infty$, so that $\int_K gf\,d\mu$ is certainly defined. In proving that this integral lies in E we may assume that g vanishes on $T\setminus K$. Given K, let G denote the set of g in $\overline{\mathscr{L}}^\infty$ that vanish on $T\setminus K$ and for which $\int gf\,d\mu = \int_K gf\,d\mu$ lies in E. By hypothesis (1), G contains all characteristic functions of compact sets $H \subset K$.

If M is any measurable subset of K, then M is the limit almost everywhere of an increasing sequence of compact sets $H_n \subset M$. As we know, $a_n = \int_{H_n} f\,d\mu$ lies in E. By Proposition 8.14.6 again, if $m < n$,

$$p(a_m - a_n) \leq \int^* (\chi_{H_n} - \chi_{H_m})(p \circ f)\,d\mu \leq \int^* (\chi_M - \chi_{H_m})(p \circ f)\,d\mu$$

which tends to 0 as $m \to \infty$ (by monotone convergence). It follows that $a = \lim a_n$ exists in E and (repeating an argument used in the proof of (a)) can be none other than $\int_M f\,d\mu$. Thus G contains the characteristic functions of all measurable subsets of K, hence contains also the vector space G_0 generated by such functions.

Now if g belongs to $\overline{\mathscr{L}}^\infty$ and vanishes on $T\setminus K$, it is the uniform limit of a sequence (g_n) extracted from G_0. Putting $b_n = \int g_n f\,d\mu$, we see easily that (b_n) is Cauchy, hence convergent, in E. Its limit can be none other than $\int gf\,d\mu$, by the same argument as was used before. Thus g belongs to G and (b) is established. ∎

Remarks. (1) If we are given that E is semireflexive and that f is SEI and satisfies (1) of Proposition 8.14.13, we can conclude again that $\bar{\int} gf\,d\mu$ belongs to E for each g in $\overline{\mathscr{L}}^\infty$. The same conclusion is valid if, in place of assuming that E is semireflexive, we assume that E is weakly sequentially complete and that f vanishes outside a σ-finite set, is SEI, and satisfies (1).

For the sequence (a_n) appearing in part (b) of the preceding proof is weakly Cauchy, and one may conclude that $\int_M f\,d\mu$ lies in E for any relatively compact measurable set M, provided E is weakly sequentially complete—which is certainly the case whenever E is semireflexive.

Considering likewise the sequence (b_n) introduced loc. cit., one sees that $\int gf\,d\mu$ belongs to E for any g in $\overline{\mathscr{L}}^\infty$ vanishing outside a σ-finite set—here again we need only weak sequential completeness. This latter condition therefore suffices (in place of semireflexivity) if f vanishes outside a σ-finite set.

If T is not σ-finite we consider the family

$$c_K = \int_K gf\,d\mu$$

of elements of E. Since f is SEI and g belongs to $\overline{\mathscr{L}}^\infty$, the family (c_K) is weakly bounded and hence bounded. E being semireflexive, E' is barrelled for its strong topology (Theorem 8.4.3). So (c_K) falls into an equicontinuous subset of the dual of E', and the family (c_K) has therefore a weak limiting point z which is strongly continuous on E'. Semireflexivity entails that z must lie in E. And in any case this z must be $\overline{\int} gf\,d\mu$.

(2) Following Bourbaki ([6], Chapitre 6, p. 84, Exercice 19)) one may term *scalarwise well-integrable* (SWI for short) a function f mapping T into E which is SEI and such that $\overline{\int} gf\,d\mu$ belongs to E for each g in $\overline{\mathscr{L}}^\infty$. Proposition 8.14.13 and Remark (1) above thus give sufficient conditions in order that f be SWI. We shall later return to this topic (see the end of Section 9.3) and discover that SWI functions have some remarkable properties.

The next result is one which is the basis for a discussion of integrable functions, especially when one subsequently specializes E to be metrizable (see Section 8.15).

8.14.14 **Theorem.** Suppose that E is complete and that f fulfils the following conditions:

(1) f is locally almost separably valued; that is, each compact set $K \subset T$ contains a negligible set N that $f(K \backslash N)$ is a separable subspace of E;

(2) f is scalarwise measurable;

(3) $\int^*(p \circ f)\,d\mu < +\infty$ for each continuous seminorm p of a family defining the topology of E.

Then $z = \overline{\int} f\,d\mu$ belongs to E.

Proof. According to the last proposition, it suffices to show that $\int_K f\,d\mu$ lies in E for each compact K. Then, since (1) is satisfied, we may assume without loss of generality that T is compact and that $f(T)$ is separable. Moreover, replacing E throughout by the closed vector subspace generated by $f(T)$, we may assume that E itself is separable and complete.

In any case, (2) and (3) together show that f is SEI.

To establish that $z = \overline{\int} f\,d\mu$ belongs to E, it suffices (Theorem 8.5.1) to show that, if Q is any equicontinuous subset of E', then $z \mid Q$ is weakly continuous. Since E is separable, the weak topology of E' induces on Q a metrizable topology.

Thus, all that need be proved is this: if x'_n $(n = 1, 2, \cdots)$ and x' are points of Q, and if $x'_n \to x'$ weakly, then $\langle z, x'_n \rangle \to \langle z, x' \rangle$. Now, by weak convergence, the functions $\langle f, x'_n \rangle$ converge pointwise on T to $\langle f, x' \rangle$. Also, since Q is equicontinuous, there exists a continuous seminorm p on E such that Q lies in the polar of the set defined by $p(x) \leq 1$. Then $|\langle f, x'_n \rangle| \leq p \circ f$ and so condition (3) combines with Lebesgue's Convergence Theorem 4.8.2 to show that

$$\langle z, x'_n \rangle = \int \langle f, x'_n \rangle \, d\mu \to \int \langle f, x' \rangle \, d\mu = \langle z, x' \rangle,$$

which is what we had to verify. ▮

8.14.14bis **Corollary.** Suppose that T is locally separable (for example, locally metrizable), that f is essentially integrable, and that E is complete. Then f is SWI.

Proof. Since gf fulfils our demands on f whenever f itself does so and $g \in \overline{\mathscr{L}}^\infty$, it suffices to show that $\overline{\int} f \, d\mu$ belongs to E. Now essential integrability entails that conditions (2) and (3) of Theorem 8.14.14 are satisfied, so it remains to show only that f is locally almost separably valued.

To see this, we use measurability of f, according to which if K is any compact set in T, there exists a sequence (H_r) of compact sets contained in K and such that $\lim_r \mu(K \backslash H_r) = 0$ and $f \mid H_r$ is continuous for each r. Since T is locally separable, H_r is separable and its continuous image $f(H_r)$ is likewise separable. If $H = \bigcup_r H_r$, then $N = K \backslash H$ is negligible and

$$f(K \backslash N) = f(H) = \bigcup_r f(H_r)$$

is separable. Thus condition (1) of Theorem 8.14.14 is satisfied. ▮

As an application of Theorem 8.14.14 we deduce the promised refinement of Proposition 8.14.5. (Cf. Dunford and Schwartz [1], p. 153, Theorem III. 6.20.)

8.14.15 **Theorem.** Let E and F be complete separated LCTVSs, u a closed linear map from E into F, f a map of T into E with values in Dom u. Suppose that f and $g = u \circ f$ are locally almost separably valued and scalarwise measurable, and that

$$\int^* (p \circ f) \, d\mu < +\infty \qquad \left(\text{resp. } \int^* (q \circ g) \, d\mu < +\infty\right)$$

for each p (resp. q) belonging to a family of seminorms defining the topology of E (resp. F). The conclusion is that $\overline{\int} f \, d\mu$ belongs to Dom u and

$$u\left(\overline{\int} f \, d\mu\right) = \overline{\int} g \, d\mu = \overline{\int} (u \circ f) \, d\mu.$$

Proof. Let G be the graph of u. G is closed in $E \times F$, hence is complete. Let us consider the map h of T into G defined by

$$h(t) = (f(t), g(t)).$$

Since both f and g are locally almost separably valued and scalarwise measurable, the same is true of h. For this to become apparent one need only note that (by the Hahn-Banach theorem) each continuous linear form on G has the form

$$(x, u(x)) \to \langle x, x' \rangle + \langle u(x), y' \rangle \qquad (x \in \text{Dom } u),$$

for a suitably chosen (not necessarily unique) point (x', y') of $E' \times F'$. Besides this, to each continuous seminorm r on G there corresponds a pair (p, q) such that

$$r(x, u(x)) \leq p(x) + q(u(x))$$

for x in Dom u. This is because the seminorms on $E \times F$ of the type $(x, y) \to p(x) + q(y)$ define the product topology. From this it follows that

$$\int^* (r \circ h)\, d\mu < +\infty$$

for each continuous seminorm r on G.

By Theorem 8.14.14 it follows that $\int h\, d\mu$ belongs to G. This signifies that $\int h\, d\mu = (x, u(x))$ for some x in Dom u. By definition of the integral (see Subsection 8.14.1) it follows that

$$\int \langle h(t), (x', y') \rangle\, d\mu(t) = \langle x, x' \rangle + \langle u(x), y' \rangle$$

for each (x', y') in $E' \times F'$. But

$$\langle h(t), (x', y') \rangle = \langle f(t), x' \rangle + \langle g(t), y' \rangle$$

and so we conclude that

$$\left\langle \int f\, d\mu, x' \right\rangle + \left\langle \int g\, d\mu, y' \right\rangle = \langle x, x' \rangle + \langle u(x), y' \rangle$$

for all x' in E' and all y' in F'. Consequently x must coincide with $\int f\, d\mu$ and $u(x)$ with $\int g\, d\mu$. Thus $\int f\, d\mu$ belongs to Dom u and $u(\int f\, d\mu) = \int g\, d\mu$, as asserted. ▮

Remarks. When E and F are Fréchet spaces, our hypotheses on f and $g = u \circ f$ amount simply to their integrability, together with the relationship $f(T) \subset \text{Dom } u$. In this form the theorem is given by Dunford and Schwartz ([1], p. 153).

8.15 The Case in Which E Is Metrizable

It is significant that measurability of f is not assumed in Theorem 8.14.14. In order to fill in the picture we shall, in this section, elucidate to some extent the relations between scalarwise measurability and measurability when E is metrizable. In no other general situation are simple relationships valid.

To begin with we observe that if E is metrizable, then Egoroff's theorem 4.8.3 is true for functions with values in E. To see this we introduce a metric d defining the topology of E. If A is an integrable set and f_n $(n = 1, 2, \cdots)$ and f are measurable functions such that $f_n \to f$ a.e. on A, we consider the numerical functions $g_n = d(f_n, f)$. These are measurable and $g_n \to 0$ a.e. on A. By Theorem 4.8.3, given $\varepsilon > 0$, there exists a measurable set $B \subset A$ (which may be assumed to be compact) such that $\mu(A \backslash B) < \varepsilon$ and such that $g_n \to 0$ uniformly on B. Then $f_n \to f$ uniformly on B. The form of Egoroff's theorem for vector-valued functions is thus established.

We shall need also two preliminary lemmas.

Lemma 1. Let E be a LCTVS, S a separable subspace of E, p a continuous seminorm on E, U the subset of E defined by $p \leq 1$. Then there exists a countable subset W of U^0 such that

$$p(x) = \text{Sup} \{|\langle x, x' \rangle| : x' \in W\}$$

for each x in S.

Proof. Obviously, it suffices to show that

$$p(x) \leq q(x)$$

for x in S, $q(x)$ denoting the specified supremum. We take a sequence (a_m) dense in S. By the Hahn-Banach theorem, for each m one has

$$p(a_m) = \text{Sup} \{|\langle a_m, x' \rangle| : x' \in U^0\}.$$

Hence there exist sequences $(x'_{mn})_{n>0}$ extracted from U^0 such that

$$p(a_m) = \text{Sup}_{n>0} |\langle a_m, x'_{mn} \rangle|.$$

Let $W = \{x'_{mn} : m, n = 1, 2, \cdots\}$. A fortiori one has

$$p(a_m) \leq q(a_m)$$

for all m. Given $\varepsilon > 0$ and x in S we can choose m so that $p(x - a_m) < \varepsilon$ and so

$$p(x) < p(a_m) + \varepsilon \leq q(a_m) + \varepsilon.$$

Since $q \leq p$ is obviously true, $q(a_m) \leq q(x) + q(x - a_m) < q(x) + \varepsilon$. Thus $p(x) < q(x) + 2\varepsilon$. Since ε is arbitrarily small, so $p(x) \leq q(x)$. ▌

Lemma 2. Let E be any LCTVS, f a scalarwise measurable and locally almost separably valued function with values in E. If p is any continuous seminorm on E, then $p \circ f$ is measurable.

Proof. Measurability being a local property, we may assume from the outset that f is separably valued. By Lemma 1, there is a sequence (x'_n) in E' such that

$$p(f(t)) = \text{Sup}_{n} |\langle f(t), x'_n \rangle|$$

for all t. Each function $\langle f, x'_n \rangle$ is measurable by hypothesis, and the result follows. ▌

It is convenient to attach the term *simple function* (with values in E) to any function f with values in E expressible as a finite linear combination of functions $\chi_M a : t \to \chi_M(t)a$, where a is an element of E and M is a measurable subset of T.

Obviously, any simple function is measurable. Notice that if f is simple and vanishes outside a measurable set S, then it is representable as a finite linear combination of functions $\chi_M a$, M being a measurable subset of S.

We can now establish the first of two major results relating measurability and scalarwise measurability.

8.15.1 **Theorem.** In order that a function f, with values in the LCTVS E, be measurable, it is:

(1) Sufficient that for each compact set K there exists a sequence (f_n) of simple functions converging almost everywhere on K to f;

(2) Necessary, when E is metrizable, that to each compact set K shall correspond a sequence (f_n) of simple functions converging almost uniformly on K to f.

Proof. (1) If such a sequence exists, Egoroff's theorem shows that $f_n \to f$ almost uniformly on K. Reference to the definition of measurability (Subsection 8.14.1(b)) shows that f is measurable.

(2) By Egoroff's theorem once more, it is enough to construct a sequence (f_n) of simple functions converging almost everywhere on K to f.

By measurability, there exists a negligible set $N \subset K$ and compact sets. $K_r \subset K$ $(r = 1, 2, \cdots)$ such that $K \backslash N = \bigcup_r K_r$, and $f \mid K_r$ is continuous for each r. Let d be any metric defining the topology on E. We define f_n in the following fashion.

For each $i \leq n$, there is a finite partitition $(A_{ij})_{1 \leq j \leq q_i}$ of K_i into measurable sets on each of which the d-oscillation of f is at most n^{-1}. We choose arbitrarily a point t_{ij} from A_{ij} and put

$$f_n(t) = \begin{cases} f(t_{ij}) & \text{if } t \in A_{ij},\ 1 \leq i \leq n,\ 1 \leq j \leq q_i, \\ 0 & \text{otherwise.} \end{cases}$$

Each f_n is simple, and obviously $f_n \to f$ on $K \backslash N$. ▮

Here now is the main theorem of this section.

8.15.2 **Theorem.** If E is a metrizable LCTVS, a function f with values in E is measurable if and only if it satisfies the following two conditions:

(a) f is locally almost separably valued;

(b) f is scalarwise measurable.

Proof. If f is measurable, then (b) is trivially true. That (a) also holds follows from part (2) of the preceding theorem. The conditions are therefore necessary.

Suppose conversely that (a) and (b) hold. We will show that f is measurable. If K is any compact set, we may assume that $f(K)$ is separable, for alteration of f on a negligible subset of K will not affect its measurability. A sequence (f_n) of simple functions will now be constructed that converges almost everywhere on K to f.

Let (x_n) be a sequence dense in $f(K)$, and let (p_k) be an increasing sequence of seminorms defining the topology on E. We define

$$A_{kn} = \{t \in K : p_k(f(t) - x_n) < k^{-1}\}.$$

By Lemma 2, A_{kn} is measurable. Also, $K = \bigcup_n A_{kn}$ for each k. Let

$$B_{k1} = A_{k1} \quad \text{and} \quad B_{kn} = A_{kn} \backslash \bigcup \{B_{kj} : j < n\}$$

for $n > 1$. For given k, the B_{kn} are measurable and partition K. We define now

$$g_k = \sum_{n=1}^{\infty} x_n \cdot \chi_{B_{kn}}$$

and

$$g_{ki} = \sum_{n=1}^{i} x_n \cdot \chi_{B_{kn}}.$$

Each g_{ki} is a simple function. It is easy to check that

$$p_k(f(t) - g_k(t)) < \frac{1}{k}$$

for all t in K and all k, and therefore

$$p_h(f(t) - g_k(t)) < \frac{1}{k} \tag{8.15.1}$$

for all t in K, all k, and all $h < k$. Moreover,

$$\lim_i g_{ki} = g_k$$

pointwise on K. According to Theorem 8.15.1(1), g_k is measurable. The inequalities (8.15.1) show that $g_k \to f$ uniformly on K, and measurability of f follows—either from Theorem 8.15.1(1) or, more elementarily, by recourse to definition (b) of 8.14.1. ▌

Remarks. For Banach spaces, Theorem 8.15.2 is due to Pettis [2]; see also Hille [1], p. 36 and Dunford and Schwartz [1], p. 149. It is quite easy to give examples showing that hypothesis (a) is essential; that, in fact, there will usually exist scalarwise measurable functions that are not locally almost separably valued.

The property of measurable functions described in Theorem 8.15.1(2) is often used to define measurability, as is done by Hille ([1], p. 36) and by Dunford and Schwartz ([1], p. 106).

Theorem 8.15.2 is essentially the same as Proposition 12 on p. 21 of Bourbaki [6], Chapitre 6. We shall later need a similar result for functions with values in the weak dual of a separable Fréchet space (Bourbaki, loc. cit., p. 21, Proposition 13).

8.15.3 **Proposition.** Let E be a separable Fréchet space, f a function mapping T into the weak dual E'_w. If f is scalarwise measurable, then it is weakly measurable. Moreover, if E is a separable normed space, $\|f\| : t \to \|f(t)\|$ is measurable.

Proof. Let D be a countable set dense in E, and let (V_n) be a decreasing base at 0 in E formed of convex, balanced, and open sets. The sets $T_n = f^{-1}(V_n^0)$ increase and cover T. If

$$S_x = \{t \in T : |\langle x, f(t)\rangle| \le 1\},$$

each S_x is measurable. Moreover, since $D \cap V_n$ is dense in V_n,

$$T_n = \bigcap \{S_x : x \in D \cap V_n\}.$$

This shows that each T_n is measurable.

We take now any compact subset K of T and any number $\varepsilon > 0$. An index n may be chosen so that $\mu[K \backslash (K \cap T_n)] < \varepsilon/4$, and then a compact set $K_1 \subset K \cap T_n$ so that $\mu[(K \cap T_n)\backslash K_1] < \varepsilon/4$. Since each function $\langle x, f\rangle$ is measurable, one may choose a compact set $K_2 \subset K_1$ such that $\mu(K_1 \backslash K_2) < \varepsilon/2$ and such that $\langle x, f\rangle \mid K_2$ is continuous for each x in D.

Now $f(K_2) \subset f(T_n) \subset V_n^0$ is equicontinuous and so $\sigma(E', E)$ and $\sigma(E', D)$ induce on $f(K_2)$ identical topologies. It follows that $f \mid K_2$ is weakly continuous. Since

$$\mu(K \backslash K_2) < \frac{\varepsilon}{4} + \frac{\varepsilon}{4} + \frac{\varepsilon}{2} = \varepsilon,$$

weak measurability of f is established.

Supposing now that E is normed and separable, we may suppose that V_1 is the open unit ball $\|x\| < 1$ in E. Choosing K_2 as above, we shall know that $\langle x, f\rangle \mid K_2$ is continuous for each x in $D \cap V_1$. Now for each t,

$$\|f(t)\| = \text{Sup}\,\{|\langle x, f(t)\rangle| : x \in D \cap V_1\}.$$

It thus appears that $\|f\|$ is lower semicontinuous on K_2. This being the case for compact sets $K_2 \subset K$ with $\mu(K \backslash K_2)$ arbitrarily small, it follows that $\|f\|$ is measurable. ▌

8.16 Functions with Values In E'. The (GDF) Property

We consider briefly the problem of integrating a function f mapping T into the dual E' of a LCTVS E. We denote by E'_w the separated LCTVS obtained by endowing E' with the weak topology $\sigma(E', E)$.

Our limited programme consists first of a result affirming that if E possesses the (GDF) property, then $\int f\, d\mu$ belongs to E' whenever f is SEI from T into E'_w; and second, of arguments exhibiting some important examples of spaces E possessing the (GDF) property. The first result bears close comparison with that in Subsection 8.14.12 above.

The reader will recall (from Subsection 6.7.3) that a separated LCTVS E is said to have the (GDF) property if any linear map of E into a Banach space X, having a graph that is sequentially closed in $E \times X$, is continuous. Also, as was pointed out in Subsection 6.7.3, any Fréchet space and any inductive limit of such spaces possesses the (GDF) property. Other instances and another proof of the second of these assertions are given below.

Our first result is an extension due to Bourbaki ([6], Chapitre 6, p. 18, Théorème 1) of a theorem for Banach spaces noted by Gelfand and Dunford.

8.16.1 Theorem. If E has the (GDF) property, and if f is a SEI function mapping T into E'_w, then $\bar{\int} f\, d\mu$ belongs to E'.

Proof. Our job is to show that the linear form

$$x \to \left\langle x, \bar{\int} f\, d\mu \right\rangle = \bar{\int} \langle x, f \rangle\, d\mu$$

is continuous on E. We shall, in fact, prove more than this—namely, that the linear map u of E into L^1, defined by setting $u(x)$ equal to the class (element of L^1) of the function $\langle x, f\rangle$, is continuous.

To do this, since E has the (GDF) property, it suffices to show that u has a sequentially closed graph. Suppose then that the sequence (x_n) converges to 0 in E and that $(\xi_n) = (u(x_n))$ converges in L^1 to ξ. We must show that $\xi = 0$. But, since $x_n \to 0$, so $\langle x_n, f\rangle \to 0$ pointwise. On the other hand, since $\xi_n \to \xi$ in L^1, we can choose functions g_n and g from the classes ξ_n and ξ, respectively, in such a way that some subsequence (g_{n_k}) converges pointwise to g. Since g_{n_k} and $\langle x_{n_k}, f\rangle$ belong to the same class, they must coincide almost everywhere. It appears therefore that $g = 0$ almost everywhere. Thus $\xi = 0$. ▌

Coming to the second item on the programme, we record the stability of the (GDF) property under the formation of inductive limits.

8.16.2 Proposition. Let E be the separated inductive limit of spaces E_i relative to maps f_i. If each E_i has the (GDF) property, so too has E.

Proof. Let X be a Banach space, u a linear map of E into X having a sequentially closed graph. To show that u is continuous, it is necessary and sufficient to show that $u \circ f_i$ is continuous from E_i into X for each i. But, f_i being continuous from E_i into E, $u \circ f_i$ has a sequentially closed graph. So, each E_i possessing the (GDF) property, $u \circ f_i$ is continuous. ▌

8.16.3 Corollary. Any separated inductive limit of Fréchet spaces has the (GDF) property.

8.16.4 Corollary. The strong dual of a distinguished Fréchet space has the (GDF) property.

Proof. The proof is immediate on combining Propositions 8.4.18 and 8.16.2. ▌

8.16.5 Remarks. As a matter of interest we may note that any space with the (GDF) property is barrelled. For a proof, see Bourbaki [6], Chapitre 6, p. 18, Proposition 11. We shall not need this result and omit any proof.

Regarding Theorem 8.16.1 we add this comment: If E'_w is semireflexive (that is, if each weakly bounded subset of E' is weakly relatively compact, a condition fulfilled whenever E is barrelled), Theorem 8.14.10 shows that $\bar{\int} f\, d\mu$ belongs to E' whenever f is SEI from T into E'_w and furthermore locally weakly bounded.

8.17 The Dunford-Pettis Theorem

The principal result of this section is an integral-representation theorem for certain linear maps of a space $L^1 = L^1(T, \mu)$ into the dual E' of a suitably restricted LCTVS E. When E is a separable Banach space and (T, μ) is an interval in R^n with the induced Lebesgue measure, the result was given by Dunford and Pettis ([1], Theorem 2.1.6), and for arbitrary σ-finite (T, μ) an account is given in Dunford and Schwartz [1], p. 503, Theorem VI.8.6. The approach to the theorem we give here, due to Bourbaki ([6], Chapitre 6, pp. 42–46), makes a preliminary investigation of continuous linear maps of E into $\bar{L}^\infty$ (or, equivalently, into L^∞).

The theorem can be regarded as an extension for vector-valued functions of the result that the dual of L^1 may be identified with $\bar{L}^\infty$ (see Theorem 4.16.1 and Remarks 4.16.2), and one task is to define sets of functions (or function classes) mapping T into E', each of which will, in some measure, play the role of $\bar{L}^\infty$ and which will reduce to the latter when E is one dimensional.

A vector-valued substitute for $\bar{L}^\infty$ suitable in some instances proves to be definable in the following manner. We regard E' as endowed with its weak topology $\sigma(E', E)$ and introduce $\Lambda_{E'}$ as the vector space of maps f of T into E'_w that are scalarwise measurable and equal locally almost everywhere to a map of T into E' taking its values in a equicontinuous subset of E'. We write $\tilde{\Lambda}_{E'}$ for the quotient of $\Lambda_{E'}$ modulo its subspace of functions that are scalarwise locally negligible. If E is a normed space, a scalarwise measurable function f mapping T into E' will belong to $\Lambda_{E'}$ if and only if

$$\bar{N}_\infty(\|f\|) \equiv \operatorname*{loc\,ess\,sup}_{t \in T} \|f(t)\| < +\infty.$$

The expression on the left-hand side is then a seminorm on $\Lambda_{E'}$ which, if E is separable, passes to the quotient space $\tilde{\Lambda}_{E'}$ and induces a norm therein. $\tilde{\Lambda}_{E'}$ is then a Banach space. If E is not separable, $\bar{N}_\infty(\|f\|)$ is in general not dependent solely on the class in $\tilde{\Lambda}_{E'}$ determined by f. If E is a separable Fréchet space, each f in $\Lambda_{E'}$ is weakly measurable; see Proposition 8.15.3.

When we wish to make a notational distinction between a function $f \in \Lambda_{E'}$ and the class (element of $\tilde{\Lambda}_{E'}$) it defines, we shall denote the latter by $\tilde{f}$. For scalar-valued functions we shall use π to denote the natural projection of $\bar{\mathscr{L}}^\infty$ onto $\bar{L}^\infty$. As soon as we seek to establish a relationship between $\tilde{\Lambda}_{E'}$ and continuous linear maps of E into $\bar{L}^\infty$, there arises inevitably the question of partially inverting π, a process we shall term "lifting" (= relèvement in Bourbaki's language).

More specifically, suppose f belongs to $\Lambda_{E'}$ and introduce X_f as an element of $L_c(E, \bar{L}^\infty)$ by

$$X_f(x) = \pi(\langle x, f \rangle). \tag{8.17.1}$$

Clearly, X_f depends only on the class $\tilde{f}$ of f and, by passage to the quotient, we obtain a linear map $X : \tilde{f} \to X_f$ of $\tilde{\Lambda}_{E'}$ into $L_c(E, \bar{L}^\infty)$. The reader will notice that if E is normed, then

$$\bar{N}_\infty(X_f(x)) = \bar{N}_\infty(\langle x, f \rangle) \leq \|x\| \cdot \bar{N}_\infty(\|f\|). \tag{8.17.2}$$

One of our main aims is to show that under certain conditions, X is an onto mapping, and reference to (8.17.1) shows that this will lead to a partial inversion of π. With this aim in mind we turn to a consideration of the problem of lifting.

LIFTING. Let M be a vector subspace of $\bar{L}^\infty$. We say that M can be *lifted* if there exists a linear map λ of M into $\bar{\mathscr{L}}^\infty$, termed a *lift* of M, such that $\pi \circ \lambda$ is the identity on M and

$$\operatorname*{Sup}_{t \in T} |\lambda(f)(t)| \leq \bar{N}_\infty(f)$$

for each f in M. Thus lifting amounts to a suitable selection of a representative function from each function class in M.

8.17.1(a) **Lemma.** Any separable subspace M of $\bar{L}^\infty$ may be lifted.

Proof. Let D be a countable dense subset of M, and let D' be the set of all finite linear combinations with rational coefficients of elements of D. D' is countable. We enumerate as h_n ($n = 1, 2, \cdots$) a base for D' over the rational field Q. For each n we may choose h'_n in $\bar{\mathscr{L}}^\infty$ with $\pi(h'_n) = h_n$. Let λ' be the unique Q-linear map of D' into $\bar{\mathscr{L}}^\infty$ determined by $\lambda'(h_n) = h'_n$. It is clear that $\pi \circ \lambda'$ is the identity on D', and furthermore we shall have

$$|\lambda'(h)(t)| \leq \bar{N}_\infty(h)$$

for each h in D' and all t save those of a set $N(h)$ that is locally negligible. Then $N = \bigcup\{N(h) : h \in D'\}$ is locally negligible. Putting $\lambda(h)$ for the function h'' defined by

$$h''(t) = \begin{cases} \lambda'(h)(t) & \text{if } t \in T \backslash N, \\ 0 & \text{if } t \in N, \end{cases}$$

then λ is defined in D', is Q-linear, $\pi \circ \lambda$ is the identity on D', and

$$|\lambda(h)(t)| \leq \bar{N}_\infty(h)$$

for all h in D' and all t in T. Since D' is dense in M and $\bar{\mathscr{L}}^\infty$ is complete for $\bar{N}_\infty$, it follows that λ can be extended by continuity so as to lift M. ▌

Remark. The problem of lifting is at the heart of many attempts to study the representation in integral form of linear functionals and operators on spaces of vector-valued functions. For a further study see Dinculeanu and Foias [1] and A. and C. Ionescu Tulcea [2], [3].

We shall here merely supplement the preceding lemma by showing (Bourbaki [6], Chapitre 6, p. 92, Exercice 18)) that if T is metrizable (and locally compact), then $\bar{L}^\infty$ itself—and hence every vector subspace thereof—may be lifted, and this relative to any positive Radon measure μ on T. Interesting and useful though it will be, this extension unfortunately does not in itself allow one to remove all traces of separability hypotheses from the main results of this section.

8.17.1(b) **Lemma.** Let T be a locally compact space in which each compact subspace is metrizable, and let μ be any positive Radon measure on T. Then the space $\dot{L}^\infty = \dot{L}^\infty(T, \mu)$ may be lifted.

Proof. By using Proposition 4.14.9 it is easily seen that there is no loss of generality in assuming that T is compact and metrizable. Assuming this to be the case we choose some metric d defining the topology of T.

We choose a sequence π_n $(n = 1, 2, \cdots)$ of finite partitions $\pi_n = (A_{n,k})_{1 \le k \le k_n}$ of T into integrable sets such that

(1) each set $A_{n,k}$ is the union of certain of the sets $A_{n+1,h}$;

(2) $\lim_n \operatorname{Max}_k$ diameter $A_{n,k} = 0$.

For each integrable set A and each integrable function f we define

$$\mu_A(f) = \begin{cases} \mu(A)^{-1} \int_A f \, d\mu & \text{if } \mu(A) > 0. \\ 0 & \text{if } \mu(A) = 0. \end{cases}$$

The reader will notice that $\mu_A(f)$ depends only on the class $\dot{f}$ of f and may thus be written as $\mu_A(\dot{f})$. Define further

$$\lambda_n(f) = \lambda_n(\dot{f}) = \sum_{1 \le k \le k_n} \mu_{A_{n,k}}(f) \cdot \chi_{A_{n,k}}.$$

It is plain that $\lambda_n(f)$ is a measurable function satisfying

$$|\lambda_n(\dot{f})(t)| \le N_\infty(f) = \|\dot{f}\|_\infty$$

whenever f is essentially bounded.

The crux of the proof is to show that

$$\lim_n \lambda_n(f)(t) = f(t) \qquad \text{a.e.} \tag{8.17.3}$$

for each essentially bounded and measurable f. Suppose for the moment that this has been accomplished. One may then proceed to prove the existence of a lift of $\dot{L}^\infty$ in the following fashion.

Let Lim denote a generalized limit, defined for all bounded numerical sequences, of the type described in Section 3.4. If f is essentially bounded and measurable, $\lambda(\dot{f})(t) = \operatorname{Lim} \lambda_n(\dot{f})(t)$ is defined for each t and satisfies

$$|\lambda(\dot{f})(t)| \le \|\dot{f}\|.$$

Moreover, on the basis of (8.17.3), we shall infer that $\lambda(\dot{f})(t) = f(t)$ a.e., which shows that the function $\lambda(\dot{f}) : t \to \lambda(\dot{f})(t)$ is measurable and that $\pi(\lambda(\dot{f})) = \dot{f}$. In other words, $\lambda : \dot{f} \to \lambda(\dot{f})$ is a lift of $\dot{L}^\infty$.

Thus all depends on establishing (8.17.3). We shall in fact do this under the weaker assumption that f is integrable. For this purpose we first establish an inequality.

Let g be any integrable function, α any number > 0. For each n, we denote by B_n the union of those sets $A_{n,k}$ for which

$$\int_{A_{n,k}} g \, d\mu \ge \alpha \cdot \mu(A_{n,k}),$$

and let $B = \bigcup_n B_n$. Thus B is the union of all sets $A_{n,k}$ (both n and k varying) for which the last-written inequality is valid. Because of condition (1), $B_n \subset B_{n+1}$. Moreover, if Σ' denotes summation over those k in $[1, k_n]$ for which $A_{n,k}$ is contained in B_n, one has

$$\mu(B_n) = \sum{}' \mu(A_{n,k}) \leq \sum{}' \alpha^{-1} \int_{A_{n,k}} g \, d\mu \leq \alpha^{-1} \int g \, d\mu.$$

Therefore also

$$\mu(B) \leq \alpha^{-1} \int g \, d\mu.$$

Let us observe next that, neglecting those sets $A_{n,k}$ that are negligible, B_n is the set of t for which $\lambda_n(g)(t) \geq \alpha$. Consequently, if we denote by D the set of points t for which $\lambda_n(g)(t) \geq \alpha$ for some n, then

$$\mu(D) \leq \alpha^{-1} \int g \, d\mu. \tag{8.17.4}$$

Now in view of (2) it is trivial that $\lambda_n(f) \to f$ uniformly on T if f is continuous (and therefore also uniformly continuous). This being so, given any integrable f, we choose a sequence (f_r) of continuous functions such that $\int |f - f_r| \, d\mu \to 0$ and also $\lim f_r = f$ a.e. By dropping terms if necessary we may arrange that, putting $g_r = f - f_r$, the numbers $\beta_r = \int g_r \, d\mu$ are such that $\sum_r \beta_r < +\infty$. Defining D_r in terms of g_r as was D in terms of g—so that D_r is the set of points t for which $\lambda_n(g_r)(t) \geq \alpha$ is true for some n—we have by (8.17.4)

$$\mu(D_r) \leq \alpha^{-1} \beta_r.$$

If $D'_s = \bigcup\{D_r : r \geq s\}$, then $\mu(D'_s) \leq \alpha^{-1} \beta'_s$, where $\beta'_s = \sum_{r \geq s} \beta_r$ tends to 0 as $s \to \infty$. For t in $T \backslash D'_s$ one has

$$\lambda_n(g_r)(t) < \alpha \qquad \text{for all } n \text{ and all } r \geq s.$$

This being so, we may write

$$|\lambda_n(f)(t) - f(t)| \leq |\lambda_n(f)(t) - \lambda_n(f_r)(t)| + |\lambda_n(f_r)(t) - f_r(t)| + |f_r(t) - f(t)|.$$

The first term on the right-hand side is at most $|\lambda_n(g_r)(t)|$ and so if t belongs to $T \backslash D'_s$, one will have

$$|\lambda_n(f)(t) - f(t)| \leq \alpha + |\lambda_n(f_r)(t) - f_r(t)| + |f_r(t) - f(t)|$$

for all n and all $r \geq s$. Using the fact that $\lim f_r = f$ outside a negligible set N, we suppose that t lies in $T \backslash (D'_s \cup N)$ and then choose first $r \geq s$ so that

$$|f_r(t) - f(t)| < \alpha.$$

Fixing this r, we may then invoke the fact that $\lambda_n(f_r) \to f_r$ as $n \to \infty$ in order to deduce that

$$\limsup_{n \to \infty} |\lambda_n(f)(t) - f(t)| \leq 2\alpha,$$

provided t lies in $T \backslash (D'_s \cup N)$. Since $\mu(D'_s)$ tends to 0 as $s \to \infty$, it follows that the term on the left-hand side above is at most 2α, save perhaps on a negligible set N_α which may depend upon α. This being the case for any preassigned number $\alpha > 0$, (8.17.3) is established and with it the lemma. ▮

With these preliminaries over we may state and prove the first of our principal results.

8.17.2 **Theorem.** Let u be a continuous linear map of E into $\bar{L}^\infty$ such that $u(E)$ can be lifted. Then

(a) There exists an f in $\Lambda_{E'}$ such that $u = X_f$; that is, $u(x) = \pi(\langle x, f\rangle)$ for each x in E.

(b) If E is normed, we may choose f as in (a) and such that

$$\operatorname*{Sup}_{t \in T} \|f(t)\| = \|u\|. \tag{8.17.5}$$

(c) If E is separable and f is as in (a), then the class $\tilde{f}$ is uniquely determined.

(d) If E is separable, then (a) and (c) are true, and X is a one-to-one linear map of $\tilde{\Lambda}_{E'}$ onto $L_c(E, \bar{L}^\infty)$, an isometry if in addition E is normed.

Proof. (a) We have to show that there exists an f in $\Lambda_{E'}$ such that $u(x) = \pi(\langle x, f\rangle)$ for all x. Let λ be a lift of $u(E)$. For each t in T, $x \to \lambda(u(x))(t)$ is a continuous linear form on E, so that there exists an element $f(t)$ of E' such that

$$(u(x))(t) = \langle x, f(t)\rangle$$

for all x. In this way one defines a function f mapping T into E'. Since $\lambda(u(x))$ belongs to $\bar{\mathscr{L}}^\infty$, it appears that f is scalarwise measurable. Moreover, if U is the inverse image by u of the unit ball in $\bar{L}^\infty$, the continuity of u arranges that U is a neighbourhood of 0 in E, and for x in U and t in T one has

$$|\langle x, f(t)\rangle| = |\lambda(u(x))(t)| \le \bar{N}_\infty(u(x)) \le 1.$$

This shows that $f(T)$ lies within U^0, so that f belongs to $\Lambda_{E'}$. Finally,

$$\pi(\langle x, f\rangle) = \pi[\lambda(u(x))] = u(x),$$

as we had to show.

(b) Let f be constructed as in the immediately preceding proof of (a). By (8.17.2) we have

$$\|u\| \le \bar{N}_\infty(\|f\|) \le \operatorname*{Sup}_{t \in T} \|f(t)\|.$$

On the other hand we have for each t in T

$$|\langle x, f(t)\rangle| = |\lambda(u(x))(t)| \le \bar{N}_\infty(u(x)) \le \|u\| \cdot \|x\|,$$

by definition of λ as a lift. This entails that $\|f(t)\| \le \|u\|$, so that (8.17.5) holds.

(c) It is necessary merely to show that if

$$\pi(\langle x, f\rangle) = \pi(\langle x, f^*\rangle)$$

for all x, then $f = f^*$ l.a.e. Let S be a countable set dense in E. Applying the above equality for each x in S we conclude that there exists a locally negligible set N such that $\langle x, f(t)\rangle = \langle x, f^*(t)\rangle$ for all x in S and all t in $T \backslash N$. Since S is dense, $f = f^*$ on $T \backslash N$.

(d) If E is separable, so too is $u(E)$. Hence (Lemma 8.17.1(a)) $u(E)$ can be lifted, and thus (a) and (c) are valid. X is therefore one-to-one, linear, and onto. Suppose further than E is normed. Then (b) shows that $\|u\| \ge \bar{N}_\infty(\|f\|)$ for the class $\tilde{f}$ (unique according to (c)) for which $u = X_f = X(f)$. Yet the reverse inequality is affirmed by (8.17.2). Equality must therefore hold. X is thus an isometry. ∎

8.17.3. The remaining principal results will be derived from Theorem 8.17.2 via consideration of a bilinear form associated with X_f according to the formula

$$B_f(x, \dot{g}) = \int g(t) \cdot \langle x, f(t) \rangle \, d\mu(t) \tag{8.17.6}$$

for g in $\mathscr{L}^1$ and x in E, which may be written alternatively as

$$B_f(x, \dot{g}) = \langle \dot{g}, X_f(x) \rangle, \tag{8.17.7}$$

the brackets on the right-hand side referring to the usual duality between L^1 and L^∞. Yet another way of expressing B_f is

$$B_f(x, \dot{g}) = \langle x, Y_f(\dot{g}) \rangle, \tag{8.17.8}$$

where

$$Y_f(\dot{g}) = \int gf \, d\mu \tag{8.17.9}$$

is a linear map of L^1 into E'. Regarding the interpretation of the integral appearing in (8.17.9), we note that gf is SI from T into E' and that $\int gf \, d\mu$ belongs to E' because the continuity on E of the linear form $x \to \int g \cdot \langle x, f \rangle \, d\mu$ is ensured by the fact that there exists a locally negligible set $N \subset T$ such that $f(T \backslash N)$ is equicontinuous in E'.

The rest of the argument may just as well proceed in the general terms set forth below.

8.17.4. Suppose that E and F are LCTVSs with duals E' and F'. E'_w and E'_s denote E' endowed with its weak and strong topologies, respectively, and likewise for F'_w and F'_s. Let B be any separately continuous bilinear form on $E \times F$. We may then write

$$B(x, y) = \langle x, B_R(y) \rangle = \langle y, B_L(x) \rangle,$$

where B_R (resp. B_L) is a continuous linear map of F into E'_w (resp. E into F'_w). Conversely every B_R (or B_L) of the type specified gives rise in this way to a separately continuous bilinear form B. The correspondence is linear and one-to-one.

For B to be continuous (in the pair) it is necessary and sufficient that B_L shall transform a suitable neighbourhood of 0 in E into an equicontinuous subset of F', or that B_R should enjoy the analogous property with E and F interchanged. This implies that B_L is continuous from E into F'_s, and conversely if F is normed.

Thus, if we denote by $B_c(E, F)$ the space of continuous bilinear forms on $E \times F$, and by $L_e(E, F')$ the space of linear maps of E into F' each of which transforms some neighbourhood of 0 in E (depending on the map in question) into an equicontinuous subset of F', then we see that $B \to B_L$ effects a one-to-one linear map of $B_c(E, F)$ onto $L_e(E, F')$, while $B \to B_R$ effects a one-to-one linear map of $B_c(E, F)$ onto $L_e(F, E')$. Moreover, $L_e(E, F')$ is always a subspace of $L_c(E, F'_s)$, with identity of the two if F is normed. If E and F are normed spaces, the above linear maps are isometries.

It is now easy to apply these remarks in the case where $F = L^1$ and B is the bilinear form B_f defined in (8.17.6). When this is done, we may recast Theorem 8.17.2 into either of two equivalent forms, as follows.

8.17.5 **Theorem.** Suppose that E is a separable LCTVS. Each continuous bilinear B form on $E \times L^1$ is expressible as (8.17.6) for a suitably chosen f in $\Lambda_{E'}$, the class $\tilde{f}$ of f being uniquely determined. If E is normed the correspondence $B \leftrightarrow \tilde{f}$ is an isometry.

8.17.6 **Theorem.** (Dunford-Pettis). Suppose that E is a separable LCTVS. Each map v in $L_e(L^1, E')$ has the form $v(\dot{g}) = Y_f(\dot{g}) = \int gf\,d\mu$ for some f in $\Lambda_{E'}$, the class $\tilde{f}$ of f being uniquely determined. If E is normed, this applies to each continuous linear map v of L^1 into E' (with its strong topology), and the correspondence $v \leftrightarrow \tilde{f}$ is an isometry.

Remarks. (1) Amongst the earlier literature one should note, in addition to Dunford and Pettis [1], the works of Gelfand [2], Bochner and Taylor [1], and Kantorovich and Vulikh [1].

(2) It is interesting and important to dispense as far as possible with separability assumptions in Theorems 8.17.5 and 8.17.6. The substance of Subsection 8.17.4 provides the means of passage from one theorem to the other, and so we confine our attention to extensions of Theorem 8.17.6. Two such extensions will be mentioned.

8.17.7. Suppose that T is (locally compact and) locally metrizable. Then, as has been shown in Lemma 8.17.1(b), $\tilde{L}^\infty$ can be lifted. Accordingly, if $v \in L_e(L^1, E')$, and if we define $u : E \to \tilde{L}^\infty$ by

$$\langle \dot{g}, u(x) \rangle = \langle x, v(\dot{g}) \rangle,$$

Theorem 8.17.2(a) shows that $u(x) = \pi(\langle x, f \rangle)$ for some f in $\Lambda_{E'}$. Correspondingly,

$$v(\dot{g}) = \int gf\,d\mu$$

for each g in $\mathscr{L}^1$. Thus the "existence" assertion of Theorem 8.17.6 remains valid. However, the uniqueness of the class $\tilde{f}$ is no longer certain: knowledge of u is a priori adequate only to determine $\langle x, f \rangle$ l.a.e. for each x in E.

8.17.8. The conclusion of Subsection 8.17.7 is again valid whenever E is a Banach space and $v(L^1)$ is strongly separable in E', which is the case whenever L^1 is separable. (Dunford and Schwartz [1], p. 504, Corollary VI.8.7; Bourbaki [6], Chapitre 6, p. 93, Exercice 19)).

The proof of this depends upon reducing this situation to one in which Theorem 8.17.6 itself is applicable. The reduction depends upon two simple lemmas.

Lemma A. Let E be a normed vector space, G a strongly separable and strongly closed vector subspace of E'. Then there exists a closed, separable vector subspace E_0 of E and a linear isometry j of G onto a closed vector subspace of the strong dual E_0' of E_0.

Proof. We choose a sequence (a'_n) that is strongly dense in G, and then a corresponding sequence (a_n) in E such that

$$\|a_n\| \leq 1, \qquad |\langle a_n, a'_n \rangle| = \left(1 - \frac{1}{n}\right) \|a'_n\|.$$

Let E_0 be the closed vector subspace of E generated by the a_n. It is easily verified now that

$$\|x'\| = \operatorname*{Sup}_n |\langle a_n, x' \rangle|$$

for each x' in G.

We define j as follows. If x' lies in G, $j(x')$ is that element of E'_0 defined by

$$\langle y, j(x') \rangle = \langle y, x' \rangle \qquad \text{for } y \in E_0;$$

in other words, $j(x') = x' \mid E_0$. One has

$$\begin{aligned} \|j(x')\| &= \operatorname{Sup} \{|\langle y, j(x') \rangle| : y \in E_0, \|y\| \leq 1\} \\ &= \operatorname{Sup} \{|\langle y, x' \rangle| : y \in E_0, \|y\| \leq 1\}, \end{aligned}$$

which is equal to $\|x'\|$ whenever x' lies in G. Thus j is an isometry from G into E'_0. Since j is also plainly linear, $j(G)$ is a strongly closed vector subspace of E'_0. ▮

Lemma B. Let E be a normed vector space, M a strongly separable vector subspace of E', and ξ a continuous linear form on M. Then there exists a sequence (x_n) extracted from E such that

$$\xi(x') = \lim_n \langle x_n, x' \rangle$$

for x' in M.

Proof. According to the Hahn-Banach theorem, ξ is the restriction to M of some x'' in E''. Moreover, x'' is $\sigma(E'', E')$-adherent to some bounded subset B of E. Now $\sigma(E'', E') \mid B$ is stronger than $\sigma(E, M) \mid B$, while $\sigma(E, M) \mid B$ is semimetrizable since M is strongly separable. It follows that there exists a sequence (x_n) extracted from B such that

$$\langle x_n, x' \rangle \to \langle x'', x' \rangle = \xi(x')$$

for x' in M. ▮

Now we can turn to the proof of the main assertion. Let v be any continuous linear map of L^1 into E' such that $v(L^1)$ is strongly separable. Let G be the strong closure in E' of $v(L^1)$. According to Lemma A there exists a separable closed vector subspace E_0 of E and a linear isometry j of G onto a strongly closed vector subspace of E'_0, this j being such that $j(x') = x' \mid E_0$ for each x' in G. We consider $j \circ v$ as a continuous linear map of L^1 into E'_0. Since E_0 is separable one may apply Theorem 8.17.6 to conclude the existence of a function f_0 in $\Lambda_{E'_0}$ such that $\|f_0(t)\| \leq \|j \circ v\| \leq \|v\|$ and

$$j \circ v(\dot{g}) = \int g f_0 \, d\mu.$$

We put $f = j^{-1} \circ f_0$, so that f maps T into $G \subset E'$ and $\|f(t)\| \leq \|v\|$. It remains only to verify that f belongs to $\Lambda_{E'}$—that is (in view of what we already know), that f is scalarwise measurable.

To check this, let x belong to E. Then $x' \to \langle x, x' \rangle$ is continuous on G and so $y' \to \langle x, j^{-1}(y') \rangle$ is continuous on $j(G) \subset E_0'$. Now $j(G)$ is strongly separable and so Lemma B shows that there exists a sequence (y_n) extracted from E_0 such that

$$\langle x, j^{-1}(y') \rangle = \lim_n \langle y_n, y' \rangle \qquad \text{for } y' \in j(G).$$

In equivalent terms,

$$\langle x, x' \rangle = \lim_n \langle y_n, j(x') \rangle \qquad \text{for } x' \in G.$$

Taking herein $x = f(t)$ one obtains

$$\langle x, f(t) \rangle = \lim_n \langle y_n, f_0(t) \rangle.$$

Since f_0 is scalarwise measurable, each $\langle y_n, f_0 \rangle$ is measurable. Therefore $\langle x, f \rangle$ is measurable, which is what we had to show. ▮

8.18 The Spaces $\mathscr{L}_E^p$ and L_E^p

In Chapter 4 we introduced the spaces $\mathscr{L}^p$ and L^p of numerically valued functions and determined their duals. In doing this we come to the Lebesgue-Radon-Nikodym theorem for numerically valued measures. In this and the next two sections we consider the analogous problems for vector-valued functions and measures. There are many extra complications resulting from admitting vector-valued entities, and few of the known results exhibit the air of finality possessed by their analogues in the numerically valued case.

There is no difficulty in defining the spaces analogous to $\mathscr{L}^p$ and L^p. We work throughout with a separated locally compact space T and a positive Radon measure μ on T; E will denote a separated LCTVS.

If $1 \leq p < \infty$, $\mathscr{L}_E^p = \mathscr{L}_E^p(T, \mu)$ will denote the vector space of functions $\vec{g}$ mapping T into E that are measurable and such that $N \circ \vec{g}$ belongs to $\mathscr{L}^p = \mathscr{L}^p(T, \mu)$ for each continuous seminorm N on E. $L_E^p = L_E^p(T, \mu)$ is the quotient of $\mathscr{L}_E^p$ modulo its subspace of negligible functions. (Since we shall frequently be working simultaneously with vector-valued and numerically valued functions, the arrow in "$\vec{g}$" is intended to remind the reader that it is the symbol denoting a function with values in E; a reversed arrow will be used for functions taking values in the dual space E'.)

For $p = \infty$, $\mathscr{L}_E^\infty = \mathscr{L}_E^\infty(T, \mu)$ denotes the vector space of functions $\vec{g}$ mapping T into E that are measurable and equal almost everywhere to a bounded function (that is, one whose range is a bounded subset of E). The quotient of $\mathscr{L}_E^\infty$ modulo its subspace of negligible functions is denoted by $L_E^\infty = L_E^\infty(T, \mu)$.

There are also the spaces $\bar{\mathscr{L}}_E^p$ and $\bar{L}_E^p$, analogous to $\bar{\mathscr{L}}^p$ and $\bar{L}^p$, in which integrability is replaced by essential integrability and almost everywhere by locally almost everywhere. The modifications necessary for handling these variants will in most cases be left to the reader.

Notice that if E is taken to be the weak dual F'_w of some LCTVS F, one must not confuse $\mathscr{L}_E^\infty$ with the space $\Lambda_{F'}$ introduced in Section 8.17. For one thing, it is not generally true that a member of $\Lambda_{F'}$ is measurable for $\sigma(F', F)$. For another, a member of $\mathscr{L}_E^\infty$ may fail to be equal locally almost everywhere to a function having an equicontinuous range. As one sees from Proposition 8.15.3, however, if E is a separable Fréchet space, then $\Lambda_{F'}$ and $\bar{\mathscr{L}}_E^\infty$ are identical when E is taken to mean F'_w.

8.18.1 The Case of Banach Spaces. If E is a Banach space, we may introduce on $\mathscr{L}_E^p$ the seminorm

$$N_p(\vec{g}) = \begin{cases} \left[\int \|\vec{g}\|^p \, d\mu\right]^{1/p} & \text{if } 1 \le p < \infty, \\ \text{ess sup } \|\vec{g}(t)\| & \text{if } p = \infty. \end{cases}$$

By passage to the quotient one obtains correspondingly a norm on L_E^p.

The study of this species of integration and of the spaces L_E^p is due initially to Bochner [3] and has been pursued subsequently by many writers—for example, Dunford and Pettis [1], Bochner and Taylor [1], Phillips [2], Pettis [2], Birkhoff [4], Dieudonné [8], [9], [10], [11], to mention only a few. Most accounts of the Bochner integral (for example, those in Hille [1] and in Dunford and Schwartz [1]) differ from that which we have adopted, which runs much closer to the approach given by Bourbaki ([6], Chapitre IV, p. 131 and p. 207).

It may be shown (see Exercises 8.28, 8.29) that (if E is a Banach space)

(a) $\mathscr{L}_E^p$ is complete and L_E^p is a Banach space;

(b) The simple functions of the type

$$\sum_{k=1}^{n} a_k \cdot \chi_{A_k},$$

where the $a_k \in E$ and the A_k are integrable sets (which may be assumed to be disjoint), are dense in $\mathscr{L}_E^p$ whenever $1 \le p < \infty$.

The major problem relating to the theory of the spaces L_E^p is the identification of the dual of L_E^p, and with it the subsidiary task of determining when L_E^p is reflexive. Many writers have contributed partial solutions: to the references cited immediately above one should add Mourier [1], Fortet and Mourier [1], and Chatterji (unpublished notes).

By analogy with the case of numerically valued functions, one would conjecture that $(L_E^p)' \approx L_{E'}^{p'}$ if $1 \le p < \infty$, where p' is defined by $p^{-1} + p'^{-1} = 1$, and where E' is the strong dual of E and is itself a Banach space. In this identification one would have in mind the duality obtained by passage to the quotient from the bilinear form

$$\langle \vec{g}, \vec{f} \rangle = \int \langle \vec{g}(t), \vec{f}(t) \rangle \, d\mu(t).$$

One would also conjecture that L_E^p is reflexive provided E is reflexive and $1 < p < \infty$.

Broadly speaking, these conjectures prove to be right, at least whenever suitable separability hypotheses are fulfilled. To what extent such hypotheses may be removed is a matter for further investigation, although it is known that separability may be dropped in various instances when E is reflexive.

As in the numerical case, much depends upon preparing the ground with suitable theorems of the Lebesgue-Radon-Nikodym type.

In the present section we shall examine the case $p = 1$, which is in many respects simpler than the remaining cases. The next section will be devoted to Lebesgue-Radon-Nikodym type theorems as a preliminary for a study of the case $p > 1$.

Before proceeding to the details, let us make two simple negative observations.

If E is not reduced to the zero element, conjecture $(L^p_E)' \approx L^{p'}_{E'}$ is certainly false for $p = \infty$ whenever it is false for numerically valued functions. This follows from the remark that if one chooses $a \in E$ satisfying $\|a\| = 1$, then the mapping that assigns to g in $\mathscr{L}^p$ the function $t \to a \cdot g(t)$ in $\mathscr{L}^p_E$ is an isometric isomorphism of $\mathscr{L}^p$ into $\mathscr{L}^p_E$. From this it follows that $(L^p)' \approx L^{p'}$ whenever $(L^p_E)' \approx L^{p'}_{E'}$.

Again, unless E is trivial, the reflexivity of L^p_E entails that of E. For the above argument shows that L^p_E contains always an isomorphic copy of E as a closed vector subspace (provided $E \neq \{0\}$), and a closed vector subspace of a reflexive Banach space is itself reflexive.

We shall now state and prove a result that specifies the dual of L^1_E in a fairly general situation.

8.18.2 Theorem. Let E be a Banach space.

(1) If $\vec{g}$ belongs to $\mathscr{L}^1_E$ and $\vec{f}$ to $\Lambda_{E'}$, the function $\langle \vec{g}, \vec{f} \rangle$ is essentially integrable and

$$\left| \int \langle \vec{g}, \vec{f} \rangle \, d\mu \right| \leq \bar{N}_1(\|\vec{g}\|) \cdot \bar{N}_\infty(\|\vec{f}\|).$$

Therefore each $\vec{f}$ in $\Lambda_{E'}$ defines, on passage to the quotient from the formula

$$\vec{g} \to \int \langle \vec{g}, \vec{f} \rangle \, d\mu,$$

a continuous linear form on L^1_E of norm at most $\bar{N}_\infty(\|\vec{f}\|)$.

(2) Assume that any one of the following three conditions is fulfilled:

(a) E is separable;

(b) $\bar{L}^\infty$ can be lifted (cf. Subsection 8.17.7);

(c) L^1 is separable.

Then if F is any continuous linear form on L^1_E, there exists an $\vec{f}$ in $\Lambda_{E'}$ such that $\|\vec{f}(t)\| \leq \|F\|$ and such that F is obtained from $\vec{f}$ in the manner described in (1) above.

(3) Suppose E is separable. To each $\tilde{\vec{f}}$ in $\tilde{\Lambda}_{E'}$ we assign the continuous linear form $\beta(\tilde{\vec{f}})$ on L^1_E obtained by passage to the quotient, starting from any member $\vec{f}$ of the class $\tilde{\vec{f}}$. Then β is a linear isometry of $\tilde{\Lambda}_{E'}$ onto the strong dual of L^1_E.

Proof. (1) Let K be any compact subset of T. Since $\vec{g}$ is measurable, it is the limit on K of a sequence of simple functions $\vec{g}_n$ (Theorem 8.15.1). Then $\langle \vec{g}_n, \overleftarrow{f} \rangle \to \langle \vec{g}, \overleftarrow{f} \rangle$ on K. Since $\overleftarrow{f}$ is scalarwise measurable, it is apparent at once that $\langle \vec{g}_n, \overleftarrow{f} \rangle$ is measurable for each n. Measurability of $\langle \vec{g}, \overleftarrow{f} \rangle$ follows. The rest follows from the inequalities

$$|\langle \vec{g}(t), \overleftarrow{f}(t) \rangle| \leq \|\vec{g}(t)\| \cdot \|\overleftarrow{f}(t)\| \leq \|\vec{g}(t)\| \cdot \bar{N}_\infty(\|\overleftarrow{f}\|),$$

the last link in which holds locally almost everywhere.

(2) Given the continuous linear form F on L_E^1, let v be the continuous linear map of L^1 into E' defined by

$$\langle x, v(\dot{g}) \rangle = F(\dot{g}x);$$

here $\dot{g}x$ is the element of L_E^1 obtained as the class of the function $t \to g(t)x$, g being any member of the class $\dot{g} \in L^1$. It is evident that $\|v\| \leq \|F\|$. According as to which of (a), (b), or (c) is satisfied, we appeal to Theorem 8.17.6, to Subsection 8.17.7, or to Subsection 8.17.8. In any case we infer the existence of an $\overleftarrow{f}$ in $\Lambda_{E'}$ such that $\|\overleftarrow{f}(t)\| = \|v\| \leq \|F\|$ and

$$F(\dot{g}x) = \langle x, v(\dot{g}) \rangle = \int g \cdot \langle x, \overleftarrow{f} \rangle \, d\mu$$

for all g in $\mathscr{L}^1$ and all x in E. By linearity, this entails that the restriction of F to the set of classes of simple functions is obtained by passage to the quotient from the mapping

$$\vec{g} \to \int \langle \vec{g}, \overleftarrow{f} \rangle \, d\mu.$$

By result (b) of Subsection 8.18.1, coupled with the continuity of F, it follows that this conclusion extends to arbitrary elements of L_E^1.

(3) According to (2), the map β is certainly onto. Also, since E is separable, the class $\tilde{\overleftarrow{f}}$ is uniquely determined by F, so that β is one to one. Moreover, according to the inequality established in (1),

$$\beta(\tilde{\overleftarrow{f}}) \leq \bar{N}_\infty(\|\overleftarrow{f}\|) = \|\tilde{\overleftarrow{f}}\|.$$

Accordingly we have

$$\|\beta(\tilde{\overleftarrow{f}})\| \leq \bar{N}_\infty(\|\overleftarrow{f}\|) \leq \operatorname{Sup} \|\overleftarrow{f}(t)\| \leq \|F\| \leq \|\beta(\tilde{\overleftarrow{f}})\|,$$

and so in fact $\|\beta(\tilde{\overleftarrow{f}})\| = \|\tilde{\overleftarrow{f}}\|$, that is, β is an isometry. ∎

Remarks. In case (3) we may say that the function $\overleftarrow{f}$ is even weakly measurable ($=$ measurable for $\sigma(E', E)$); see Proposition 8.15.3. Consequently, if E is separable and reflexive (in which case E' is automatically separable, having a separable dual), $\overleftarrow{f}$ will be strongly measurable (Theorem 8.15.2) and therefore a member of $\mathscr{L}_{E'}^\infty$. If E is reflexive, however, one may dispose of separability altogether. This follows from some results established by Phillips [2]. Although the proof will call upon Corollary 8.19.5, it is convenient to deal here with this case in which E is reflexive.

8.18.3 **Theorem.** Suppose that E is a reflexive Banach space. If F is a continuous linear form on L_E^1, there exists a strongly measurable and bounded function $\vec{f}$ from T into E' such that

$$F(\dot{\vec{g}}) = \int \langle \vec{g}, \vec{f} \rangle \, d\mu$$

for all $\vec{g}$ in $\mathscr{L}_E^1$, and $\bar{N}_\infty(\|\vec{f}\|) = \|F\|$. Thus $(L_E^1)' \approx L_{E'}^\infty$.

Proof. If g be fixed in $\mathscr{L}^1$, then $x \to F(\dot{g}x)$ is plainly a continuous linear form E, so that one may write

$$F(\dot{g}x) = \langle x, \vec{m}(\dot{g}) \rangle$$

for some unique element $\vec{m}(\dot{g})$ of E'. Evidently,

$$\|\vec{m}(\dot{g})\| \leq \|F\| \cdot N_1(g).$$

$\vec{m}$ is thus a continuous linear map of L^1 into E'. Since E is reflexive, so too is the strong dual E'. According to Corollary 8.19.5, therefore, there exists a strongly measurable function $\vec{f}$ mapping T into E' such that $\|\vec{f}\| \leq \|F\|$ and

$$F(\dot{g}x) = \langle x, \vec{m}(\dot{g}) \rangle = \int g \cdot \langle x, \vec{f} \rangle \, d\mu$$

for all x and all g in $\mathscr{L}^1$. Since the finite linear combinations of functions of the form $\dot{g}x$ are dense in L_E^1, it follows readily that

$$F(\dot{\vec{g}}) = \int \langle \vec{g}, \vec{f} \rangle \, d\mu$$

for all $\vec{g}$ in $\mathscr{L}_E^1$. This formula itself shows that

$$\|F\| \leq \text{loc ess sup } \|\vec{f}(t)\|,$$

and hence that $\|F\| = \bar{N}_\infty(\|\vec{f}\|)$. ▮

Remarks. Phillips ([2], Theorem 5.1) gives an equivalent form of the crucial step in the above proof, and this without appeal to Corollary 8.19.5, which itself appears as a corollary in Phillips' account. Phillips' Theorem 5.1 is recorded as our Theorem 8.19.4.

8.19 Vector-valued Measures

As an aid in the study of the dual of L_E^1 we shall need to discuss measures taking their values in a given separated LCTVS, and also some analogues of the Lebesgue-Radon-Nikodym theorem for such entities. We shall expound in this section the bare necessities only.

As with numerically valued measures, one may regard a measure either as a linear map defined on $\mathscr{K}(T)$, or as a function of (suitably restricted) subsets of T. The former is the outlook consistently adopted by Bourbaki, and followed thus far in this book. We intend to retain this interpretation in the case of vector-valued measures, but it should be borne in mind that the opposing viewpoint has many adherents and its own advantages in certain connections. It is not until Subsection 9.4.14 that we examine in any detail the connections between the two points of view.

A systematic exposition of the approach we choose to adopt is found in Bourbaki [6], Chapitre 6, pp. 29–57, while the other approach is discussed at length in Dunford and Schwartz [1], pp. 318–328. Dinculeanu [4] discusses a more general species of vector-valued measure.

Our starting point is to define an E-valued Radon measure (E being any separated LCTVS) as a continuous linear map $\vec{m}$ of $\mathscr{K}(T)$ into E. The continuity signifies that $\lim \vec{m}(g_n) = 0$ in the sense of E whenever the continuous functions g_n have supports contained in a compact subset of T (independent of n) and converge uniformly to 0.

Example. Suppose that $\vec{f}$ is a function mapping T into E that is scalarwise locally integrable. One may seek to define a corresponding E-valued Radon measure $\vec{m}$ by writing

$$\vec{m}(g) = \int g\vec{f}\, d\mu \tag{8.19.1}$$

for g in $\mathscr{K}(T)$. Precautions will be necessary to ensure that $\vec{m}(g)$, thus defined and a priori an element of $E'^* \supset E$, does indeed lie in E and depends continuously on g. Various criteria sufficient to ensure this stem from the results in Section 8.14. For instance, from Theorem 8.14.14 one discovers that it is enough to assume that E is complete and that $\vec{f}$ is measurable and $\|\vec{f}\|$ locally integrable.

In view of this example it is entirely natural to pose the problem of discovering conditions—necessary and sufficient, if possible—on the E-valued measure $\vec{m}$ in order that it be of the preceding type. The answer to this will depend in some measure on what one demands of the function $\vec{f}$: for instance, is $\vec{f}$ permitted to be merely scalarwise measurable?; is $\vec{f}$ required to be measurable? In what follows we shall have occasion to deal with both possibilities.

If $\vec{m}$ has the form of (8.19.1), it will generally be the case that there is an obvious extension of $\vec{m}$ to a class of functions wider than $\mathscr{K}(T)$. The same possibility must be considered for measures $\vec{m}$ that are not known to be of this special form. This question is easier to handle in the special case in which E is the weak dual of some LCTVS, a situation that we shall consider in Subsection 8.19.7 onward.

8.19.1 **Extension of a Vector-valued Radon Measure.** Let $\vec{m}$ be an E-valued Radon measure, E being any separated LCTVS. There is a natural and convenient way of extending the definition of $\vec{m}$ from $\mathscr{K}(T)$ to a larger class of functions. The technique is very similar to that employed in defining the integral of an E-valued function with respect to a numerical Radon measure (see Subsection 8.14.1).

A numerically valued function g on T is said to be *essentially integrable for* $\vec{m}$ if it is essentially integrable for each of the numerical Radon measures $x' \circ \vec{m}$, x' ranging over E'. We denote by $\bar{\mathscr{L}}(\vec{m})$ the vector space of such functions.

If g belongs to $\bar{\mathscr{L}}(\vec{m})$, we define $\vec{m}(g)$ to be that element of E'^* defined by

$$\langle \vec{m}(g), x' \rangle = \int g\, d(x' \circ \vec{m})$$

for each x' in E'. It is obvious that this is consistent with the initial definition whenever g belongs to $\mathscr{K}(T)$.

It is evident that $\bar{\mathscr{L}}(\vec{m})$ contains all bounded, universally measurable functions that have compact supports.

If (g_n) is a sequence of functions in $\bar{\mathscr{L}}(\vec{m})$ that is uniformly bounded and vanishes outside a compact set (independent of n), and if $g = \lim g_n$ exists pointwise on T, then

$$\vec{m}(g) = \lim_n \vec{m}(g_n) \tag{8.19.2}$$

weakly in E'^* (that is, for the topology $\sigma(E'^*, E')$). This follows at once from Lebesgue's theorem on dominated convergence.

An important special case is that in which $\vec{m}$ is *scalarwise* LAC(μ) for some positive measure μ on T, by which we mean that each numerical measure $x' \circ \vec{m}$ is LAC(μ). In this case $\bar{\mathscr{L}}(\vec{m})$ contains all functions in $\mathscr{L}^\infty(\mu)$ with compact supports. Moreover, $\vec{m}(g) = 0$ whenever g is locally negligible for μ. The relation (8.19.2) now holds provided there exists a number $c \geq 0$ and a compact set K such that

$$|g_n| \leq c \qquad \text{l.a.e.,}$$
$$g_n = 0 \qquad \text{l.a.e. on } T \backslash K,$$

and

$$g = \lim g_n \qquad \text{l.a.e.}$$

Having extended the definition of $\vec{m}$ so that its values lie in general in E'^*, it is natural to seek conditions under which $\vec{m}(g)$ will lie in $E(\subset E'^*)$. We give two results bearing on this problem.

8.19.2. Suppose that $\vec{m}$ is an E-valued Radon measure having the property that for a certain compact set K in T, the set

$$\{\vec{m}(g) : g \in \mathscr{K}(T, K), |g| \leq 1\} \tag{8.19.3}$$

is weakly relatively compact in E. We shall show that then $\vec{m}(f)$ belongs to E for any f in $\bar{\mathscr{L}}(\vec{m})$ that is everywhere bounded and 0 on $T \backslash K$.

For let A be the set of f in $\bar{\mathscr{L}}(\vec{m})$ that vanish on $T \backslash K$ and satisfy $|f| \leq 1$, and let $B = A \cap \mathscr{K}(T)$. We are given that $\vec{m}(B)$ is weakly relatively compact in E, and we wish to show that $\vec{m}(A) \subset E$. It is thus enough to show that $\vec{m}(A)$ is contained in the weak closure, in E'^*, of $\vec{m}(B)$. Moreover, since $\vec{m}(B)$ is convex and balanced, it comes to the same thing to show that the polar $\vec{m}(B)^0$ is contained in $\vec{m}(A)^0$. However, the relation $x' \in \vec{m}(B)^0$ signifies that

$$\left| \int g \, d(x' \circ \vec{m}) \right| \leq 1$$

for all g in B. This signifies that the restriction to K of the positive measure $|x' \circ \vec{m}|$ has total mass at most unity. But then it is obvious that $|\int f \, d(x' \circ \vec{m})| \leq 1$ for any f in A, which shows that $x' \in \vec{m}(A)^0$. ∎

Remarks. The above argument is due to Bourbaki ([6], Chapitre 6, p. 34). A similar result is obtainable in a different way on the basis of general properties of weakly compact linear maps. For this, see Subsection 9.4.14.

One outcome of any such extension of $\vec{m}$ is the recognition that $\vec{m}$ generates an E-valued set function obtained by defining

$$\vec{m}(A) = \vec{m}(\chi_A).$$

This definition is effective for any subset A of T whose characteristic function belongs to $\bar{\mathscr{L}}(\vec{m})$ and that is contained in some compact set K for which the set (8.19.3) is weakly relatively compact in E. The set function thus obtained is at any rate weakly countably additive. More will be said about this matter in Subsection 9.4.14.

8.19.3. Another important case in which an extension is possible is that in which the set

$$\left\{\vec{m}(g) : g \in \mathscr{K}(T), \quad \int |g| \, d\mu \leq 1\right\} = M$$

is contained in some bounded and sequentially complete subset of E. Here, as we shall verify in a moment, $\vec{m}$ can be extended to $\mathscr{L}^1 = \mathscr{L}^1(\mu)$ in such a way that it yields a continuous mapping of $L^1 = L^1(\mu)$ into E.

Indeed, if this condition is fulfilled, then $\bar{M}$ is bounded, sequentially complete, convex, and balanced. For any g in $\mathscr{K}(T)$,

$$\vec{m}(g) \in N_1(g) \cdot \bar{M}.$$

Now, if h is any member of $\mathscr{L}^1$, one may select a sequence (g_n) in $\mathscr{K}(T)$ such that $\lim N_1(h - g_n) = 0$ and $N_1(g_n) \leq N_1(h)$. It thus appears that each $\vec{m}(g_n)$ lies in $N_1(h) \cdot \bar{M}$, and that $\vec{m}(g_i) - \vec{m}(g_j) \in \beta_{ij}\bar{M}$, where $\beta_{ij} \to 0$ as $i, j \to \infty$. Since $\bar{M}$, and hence any homothetic image of $\bar{M}$, is sequentially complete, $x = \lim \vec{m}(g_n)$ exists in E. The same sort of reasoning shows too that x depends only on h, and not on the particular sequence (g_n) chosen. Thus we may write

$$\vec{m}(h) = x = \lim \vec{m}(g_n).$$

The fact that M is bounded in E entails that for each x' the numerical Radon measure $x' \circ \vec{m}$ is of the form $\varphi_{x'} \cdot \mu$, where $\varphi_{x'}$ is bounded and measurable. From this one concludes that

$$\langle \vec{m}(h), x' \rangle = \langle h, x' \circ \vec{m} \rangle$$

for h in $\mathscr{L}^1$ and x' in E'; this relation itself serves to determine $\vec{m}(h)$ uniquely. It shows moreover that $\vec{m}(h)$ depends only on the class $\dot{h}$ of h, so that one may pass to the quotient space L^1 and view $\vec{m}$ as a linear map of L^1 into E. Considered thus, $\vec{m}$ is continuous since the preceding construction shows that $\vec{m}(h)$ lies in $\bar{M}$ whenever $N_1(h) \leq 1$. ▌

The reader will notice that if the set M is weakly relatively compact in E, the preceding condition is fulfilled and, moreover, $\vec{m}$ is weakly compact from L^1 into E. By starting with this remark we may formulate the first of several results of the Lebesgue-Radon-Nikodym type. It may be regarded as a variant of results given by Phillips ([2], Theorems 5.1 and 5.2), and what is essentially the same conclusion derived by somewhat different methods appears as Theorem 9.4.7. For comparison with Phillips' formulation we remark that if $\vec{m}$ be assumed to satisfy the hypothesis in Subsection 8.19.2, one obtains equivalent conditions on $\vec{m}$ in the present section by replacing M by the set

$$M_1 = \left\{\frac{\vec{m}(A)}{\mu(A)} : \mu(A) > 0\right\},$$

where it is understood that A denotes a relatively compact, universally measurable subset of T.

8.19.4 Theorem. Let $\vec{m}$ be an E-valued Radon measure such that the set

$$M = \left\{\vec{m}(g) : g \in \mathscr{K}(T), \int |g|\, d\mu \leq 1\right\}$$

is weakly relatively compact in E. Then there exists a scalarwise measurable function $\vec{f}$ mapping T into E such that $\vec{f}(T) \subset \bar{M}$ and

$$\vec{m}(g) = \int g\vec{f}\, d\mu$$

for each g in $\mathscr{L}^1$. If also E is metrizable and quasi-complete, one may assume that $\vec{f}$ is measurable.

Proof. According to Subsection 8.19.3 we can extend $\vec{m}$ to $\mathscr{L}^1$, yielding on passage to the quotient a weakly compact linear map of L^1 into E.

We consider now finite families $\pi = (A_k)_{1 \leq k \leq n}$ formed of relatively compact (or μ-integrable), universally measurable sets A_k with $\mu(A_k) > 0$. Such families π may be partially ordered by writing $\pi' > \pi$ if the union of the sets in π' contains the union of sets in π, and if each set in π' is either contained in some set in π or is disjoint from all sets in π. In this way the set of π's is a directed set. To each π we assign a function $\vec{f}_\pi$ mapping T into E, namely,

$$\vec{f}_\pi = \sum_k \mu(A_k)^{-1}\vec{m}(A_k) \cdot \chi_{A_k}.$$

$\vec{f}_\pi$ is measurable and evidently takes its values in $\bar{M}$, since $\bar{M}$ is convex and since $\vec{m}(A_k)\, \mu(A_k)$ belongs to $\bar{M}$ for each k.

As remarked in Subsection 8.19.3, $x' \circ \vec{m} = \varphi_{x'} \cdot \mu$ for some bounded μ-measurable function $\varphi_{x'}$. Consequently

$$\langle \vec{f}_\pi(t), x' \rangle = \sum_k \mu(A_k)^{-1} \int_{A_k} \varphi_{x'}\, d\mu \cdot \chi_{A_k}(t).$$

By an argument like that used in Lemma 8.17.1(b) one may verify that

$$\lim_\pi \langle \vec{f}_\pi(t), x' \rangle = \varphi_{x'}(t) \qquad \text{l.a.e.}$$

On the other hand, since $\bar{M}$ is weakly compact, the directed family $(\vec{f}_\pi(t))$ has for each t a weak limiting point. Let us suppose such a weak limiting point $\vec{f}(t)$ is chosen for each t. The preceding equation shows that $\langle \vec{f}, x' \rangle = \varphi_{x'}$ l.a.e. for each x'. Thus $\vec{f}$ is scalarwise measurable and takes its values in $\bar{M}$. Moreover,

$$\langle \vec{m}(g), x' \rangle = \int g \, d(x' \circ \vec{m}) = \int g \varphi_{x'} \, d\mu = \int g \cdot \langle \vec{f}, x' \rangle \, d\mu$$

for each x' and therefore

$$\vec{m}(g) = \int g \vec{f} \, d\mu,$$

which establishes the first part of the theorem.

To complete the proof we must show that if E is metrizable and quasi-complete, then $\vec{f}$ is measurable. For this it suffices to show that $\vec{f}(K \backslash N)$ is separable for any compact set $K \subset T$ and a suitably chosen negligible subset N of K. This in turn will follow if we show that the set $S = \{\vec{m}(A) : A$ is μ-integrable and $A \subset K\}$ is relatively compact. However, S is the image by $\vec{m}$ of the set $\{\chi_A : A$ is μ-integrable and $A \subset K\} = \Sigma$ in $\mathscr{L}^1$. By Theorem 4.21.2, Σ is weakly relatively compact in $\mathscr{L}^1$. As we shall show in Chapter 9 (specifically Corollary 9.4.5), the weakly compact linear map $\vec{m}$ of L^1 into E necessarily transforms Σ into a relatively compact subset of E, Q.E.D. ∎

For semireflexive spaces E one derives immediately the following corollary, used already in the proof of Theorem 8.18.3.

8.19.5 Corollary. Suppose that E is a semireflexive LCTVS and that $\vec{m}$ is an E-valued Radon measure such that the set

$$M = \left\{ \vec{m}(g) : g \in \mathscr{K}(T), \int |g| \, d\mu \leq 1 \right\}$$

is bounded in E. Then there exists a scalarwise measurable function $\vec{f}$ mapping T into E such that $\vec{f}(T) \subset \bar{M}$ and

$$\vec{m}(g) = \int g \vec{f} \, d\mu$$

for g in $\mathscr{L}^1$. If furthermore E is metrizable, one may suppose that $\vec{f}$ is measurable.

8.19.6 Refinements of Theorem 8.19.4 and Corollary 8.19.5. Reference to the proof of Theorem 8.18.3 shows that in discussing the dual of L^p_E one is led to consider vector-valued Radon measures $\vec{m}$ taking values in the dual space E' that satisfy the condition laid down in Subsection 8.19.2 and possessing a finite p'-variation (a concept we define in a moment). For such measures Phillips ([2], Theorem 5.5 and Corollary 5.6) established refinements of Theorem 8.19.4 and Corollary 8.19.5. These refinements are especially significant when E' is reflexive. This is the case when E itself is reflexive, in which circumstances Phillips' results lead directly to a complete identification of the dual of L^p_E (loc. cit. Theorem 5.7). We shall here formulate these refined results, referring the reader to Phillips' paper for the proofs.

Suppose that E is a Banach space and that $1 \leq p < \infty$. An E-valued Radon measure $\vec{m}$ is said to have a finite p-variation if it satisfies the condition in

Subsection 8.19.2 and if furthermore

$$V_p(\vec{m}) = \operatorname{Sup} \sum \frac{\|\vec{m}(A_k)\|^p}{\mu(A_k)^{p-1}} < +\infty,$$

the supremum being taken over all finite families (A_k) of disjoint, relatively compact, universally measurable subsets A_k of T satisfying $\mu(A_k) > 0$.

Phillips' two results, of which the second is an immediate corollary of the first, may be formulated in the following way.

(1) With the preceding notations, suppose that $\vec{m}$ is an E-valued Radon measure that is scalarwise LAC(μ), of finite p-variation, and such that the set

$$\left\{\frac{\vec{m}(A)}{\mu(A)} : \frac{\|\vec{m}(A)\|}{\mu(A)} \leq n\right\}$$

is weakly relatively compact for each n. Assume finally that (T, μ) is σ-finite. Then there exists a measurable function $\vec{f}$ mapping T into E such that

$$\vec{m}(A) = \int_A \vec{f}\, d\mu$$

for all μ-integrable sets A if $p > 1$, or all μ-measurable sets A if $p = 1$, and such that

$$V_p(\vec{m}) = \int \|\vec{f}\|^p\, d\mu.$$

(2) Suppose that E is a reflexive Banach space and $\vec{m}$ an E-valued Radon measure that is scalarwise LAC(μ) and of finite p-variation. Assume too that (T, μ) is σ-finite. Then the conclusion of (1) is valid once more.

8.19.7 **Measures with Values in a Dual Space.** Since, as we have remarked in Subsection 8.19.6, the discussion of the dual of L_E^p leads to measures with values in E', we intend to look more closely at this special case.

Suppose then that E is an LCTVS, E' its dual. We denote by E'_w the vector space E' endowed with the topology $\sigma(E', E)$, whose dual is E. We shall consider E'_w-valued Radon measures $\bar{m}$. With $\bar{m}$ we now associate the numerical Radon measures $x \circ \bar{m}$ defined by $\int g\, d(x \circ \bar{m}) = \langle x, \bar{m}(g)\rangle$ for g in $\mathscr{K}(T)$. A function g is said to be essentially integrable for $\bar{m}$ if and only if it is essentially integrable for each $x \circ \bar{m}$, in which case $\bar{m}(g)$ is that element of $E^* = ((E'_w)')^*$ for which

$$\langle x, \bar{m}(g)\rangle = \langle g, x \circ \bar{m}\rangle = \int g\, d(x \circ \bar{m})$$

for all x in E.

Once again we are interested in conditions ensuring that $\bar{m}(g)$ lies in E', and not merely in $E^* \supset E'$. Naturally the results of Subsections 8.19.1 and 8.19.3 apply when E is replaced by E'_w. It is usually the case in practice, however, that E is barrelled, and then we have the following stronger and more satisfactory result.

8.19.8 **Proposition.** If E is barrelled, and if $\bar{m}$ is an E'_w-valued Radon measure, then $\bar{m}(g)$ belongs to E' for each g that is essentially integrable for $\bar{m}$.

Proof. This proof will proceed in several steps.

(1) Since $\bar{m}$ is continuous from $\mathscr{K}(T)$ into E'_w, and since E is barrelled, the set $\bar{m}(G)$ is equicontinuous in E' whenever G is a bounded subset of $\mathscr{K}(T)$. Thus there will exist a neighbourhood U of 0 in E such that

$$\left| \int g \, d(x \circ \bar{m}) \right| \leq 1 \qquad \text{for } g \in G \text{ and } x \in U. \tag{8.19.4}$$

Suppose now that g is essentially integrable for $\bar{m}$, bounded, and 0 outside a compact subset K of T. We shall show that $\bar{m}(g)$ lies in E'. To see this we observe that, for any positive measure λ on T relative to which g is measurable, g is the limit in $\mathscr{L}^1(\lambda)$ of functions g_n in $\mathscr{K}(T)$ such that $|g_n| \leq \operatorname{Sup} |g| = c$, say, and $g_n = 0$ outside a compact neighbourhood K_0 of K. If we take G as the set of functions in $\mathscr{K}(T)$ that vanish outside K_0 and are uniformly bounded in modulus by c, and for λ the measure $|x \circ \bar{m}|$, we infer at once from (8.19.4) that

$$\left| \int g \, d(x \circ \bar{m}) \right| \leq 1 \qquad \text{for } x \in U.$$

This shows that the linear form $x \to \langle x, \bar{m}(g) \rangle$ is continuous on E, so that $\bar{m}(g)$ belongs to E'.

(2) Suppose next that g is essentially integrable for $\bar{m}$ and vanishes outside the union of countably many compact sets K_n $(n = 1, 2, \cdots)$. We may assume that $K_n \subset K_{n+1}$. We define

$$g_n(t) = \begin{cases} g(t) & \text{if } t \in K_n \text{ and } |g(t)| \leq n, \\ 0 & \text{elsewhere.} \end{cases}$$

Appeal to (1) shows that $\bar{m}(g_n)$ lies in E' for each n. Besides this one has for each x in E

$$\langle x, \bar{m}(g_n) \rangle = \int g_n \, d(x \circ \bar{m}) \to \int g \, d(x \circ \bar{m}) = \langle x, \bar{m}(g) \rangle,$$

thanks to Lebesgue's theorem on dominated convergence. E being barrelled, it follows that the $\bar{m}(g_n)$ form an equicontinuous subset of E', so that their weak limit, $\bar{m}(g)$, is continuous on E. Thus again $\bar{m}(g)$ lies in E'.

(3) Finally suppose g is merely essentially integrable for $\bar{m}$. For each σ-compact subset S of T, we define $g_S = g \cdot \chi_S$. According to (2), $\bar{m}(g_S)$ lies in E' for each S. Furthermore

$$|\langle x, \bar{m}(g_S) \rangle| \leq \int |g_S| \, d\, |x \circ \bar{m}| \leq \int |g| \, d\, |x \circ \bar{m}| < +\infty.$$

Using once more the fact that E is barrelled, one infers that the $\bar{m}(g_S)$ fall into an equicontinuous, and therefore weakly compact, subset of E'. If we partially order the S's by inclusion, it follows that the directed family $(\bar{m}(g_S))$ admits a weak limiting point x' in E'. Now, for a given x, g vanishes locally almost everywhere with respect to $|x \circ \bar{m}|$ outside some σ-compact set S; and then

$$\langle x, \bar{m}(g_S) \rangle = \langle x, \bar{m}(g) \rangle.$$

It follows at once from this that $\langle x, \bar{m}(g) \rangle = \langle x, x' \rangle$ for each x, and so $\bar{m}(g) = x'$ lies in E'. ∎

8.19.9 **Remark.** Should it be the case that $\overleftarrow{m}$ is scalarwise LAC(μ), then Proposition 8.19.8 shows that $\overleftarrow{m}(g) \in E'$ for any g in $\mathscr{L}^\infty(\mu)$ that vanishes outside a compact subset of T.

8.19.10 **A Lebesgue-Radon-Nikodym Theorem for Measures with Values in E'_w.** If the function $\overleftarrow{f}$ mapping T into E' is scalarwise locally integrable for μ, one may seek to define an E'_w-valued Radon measure $\overleftarrow{m}$ via the formula

$$\overleftarrow{m}(g) = \int g\overleftarrow{f}\, d\mu \qquad (g \in \mathscr{K}(T)); \tag{8.19.5}$$

compare this with (8.19.1). According to Theorem 8.16.1, $\overleftarrow{m}$, as given by (8.19.5), does indeed map $\mathscr{K}(T)$ into E', whenever E possesses the (GDF) property; furthermore, the defining equation

$$\langle x, \overleftarrow{m}(g)\rangle = \int g \cdot \langle x, \overleftarrow{f}\rangle\, d\mu$$

shows at once that the map $g \to \overleftarrow{m}(g)$ is continuous from $\mathscr{K}(T)$ into E'_w. Thus (8.19.5) defines $\overleftarrow{m}$ as an E'_w-valued Radon measure whenever E has the (GDF) property. As we have seen in Section 8.16, any Fréchet space E fulfils this condition.

We wish now to consider the problem of discovering conditions sufficient (and, if possible, also necessary) to ensure that a given $\overleftarrow{m}$ is of the form (8.19.5). To facilitate discussion we denote by $\overleftarrow{f} \cdot \mu$ the measure defined in (8.19.5), and say that such measures are LAC(μ)—thus taking over the terminology used in the case of numerical measures.

An obviously necessary condition is that $\overleftarrow{m}$ be scalarwise LAC(μ)—that is, that each $x \circ \overleftarrow{m}$ be LAC(μ). Supposing this to be the case, $\overleftarrow{m}(g)$ is defined for each g in $\mathscr{L}^\infty(\mu)$ having a compact support (see Remark 8.19.9). It is then tempting to conjecture that if E is a Banach space perhaps the desired necessary and sufficient condition is the following one:

Condition (1). For each compact set $K \subset T$ and each number $\varepsilon > 0$, there exists a number $\delta = \delta(K, \varepsilon) > 0$ such that $\|\overleftarrow{m}(g)\| \leq \varepsilon$ for each g in $\mathscr{L}^\infty(\mu)$ that vanishes outside K and satisfies $\int |g|\, d\mu \leq \varepsilon$.

However, it is known (Pettis [2], p. 303, Example 9.4) that there exist measures $\overleftarrow{m}$ that satisfy Condition (1) above and that are yet not of the form $\overleftarrow{f} \cdot \mu$; and also (Birkhoff [3], p. 377, Example 7) that there exist measures $\overleftarrow{m}$ of the form $\overleftarrow{f} \cdot \mu$ that do not satisfy Condition (1).

To Dieudonné ([11], Théorème 1) one owes what appears to be the appropriate type of condition. The ideas behind Dieudonné's condition are developed by Bourbaki ([6], Chapitre 6, pp. 35–41). For our particular purposes we frame the condition somewhat differently and in the following terms.

Condition (2). Given any $\varepsilon > 0$ and any compact subset K of T, there exists a compact $K' \subset K$ such that $\mu(K \backslash K') < \varepsilon$ and for which the set

$$\left\{\overleftarrow{m}(g) : g \in \mathscr{K}(T, K') \text{ and} \int |g|\, d\mu \leq 1\right\}$$

is weakly relatively compact in E'.

The argument used in Subsection 8.19.2 shows that if $\overleftarrow{m}$ satisfies Condition (2) above, then also the set of $\overleftarrow{m}(g)$, obtained when g ranges over all functions in $\overline{\mathscr{L}}(\overleftarrow{m})$ satisfying $|g(t)| \leq 1$ and vanishing outside K', is also weakly relatively compact in E'.

We aim to show next that Condition (2), together with scalarwise local absolute continuity relative to μ, is sufficient to ensure that $\overleftarrow{m}$ is LAC(μ).

8.19.11 **Proposition.** Suppose that $\overleftarrow{m}$ is an E'_w-valued Radon measure on T that is scalarwise LAC(μ) and satisfies Condition (2). Then $\overleftarrow{m}$ is LAC(μ); that is, there exists a scalarwise locally integrable function $\overleftarrow{f}$ mapping T into E'_w such that

$$\overleftarrow{m}(g) = \int g\overleftarrow{f}\, d\mu$$

for each g in $\mathscr{L}^\infty(\mu)$ having a compact support.

Proof. We divide this into several stages.

(1) Suppose that K and ε are given and that K' is chosen as in Condition (2). One may then apply Theorem 8.19.4, replacing E by E'_w, T by K', and μ by its restriction to K'. The conclusion is that there exists a scalarwise measurable function $\overleftarrow{f}_{K'}$, mapping K' into E'_w, with a range that is weakly relatively compact in E'_w, and such that

$$\overleftarrow{m}(g) = \int g\overleftarrow{f}_{K'}\, d\mu$$

for each g in $\mathscr{L}^1$ that vanishes outside K'.

(2) We next exhaust K by an increasing sequence K_n of compact subsets, together with a negligible set N, each K_n being of the type K' figuring in (1). K is then partitioned by the sets $A_n = K_n \backslash K_{n-1}$ and N, it being agreed that $K_0 = \emptyset$. As follows from (1), for each n there exists a weakly relatively compact-valued and scalarwise measurable function $\overleftarrow{f}_n$ mapping A_n into E'_w such that

$$\overleftarrow{m}(g) = \int g\overleftarrow{f}_n\, d\mu$$

for g in $\mathscr{L}^1(\mu)$ vanishing outside A_n.

Since $\overleftarrow{m}$ is scalarwise LAC(μ), the formula

$$\langle x, \overleftarrow{m}(g)\rangle = \int g\, d(x \circ \overleftarrow{m})$$

shows that $\overleftarrow{m}(g) = 0$ whenever g vanishes outside N.

Suppose now that we define $\overleftarrow{f}_K$ to be equal to $\overleftarrow{f}_n$ on A_n ($n = 1, 2, \cdots$) and to be 0 in N. Then $\overleftarrow{f}_K$ is scalarwise measurable from K into E'_w. Moreover,

$$\overleftarrow{m}(g) = \int g\overleftarrow{f}_K\, d\mu$$

for any g in $\mathscr{L}^1(\mu)$ that vanishes outside A_n for some n, or outside N.

If g belongs to $\mathscr{L}^\infty(\mu)$ and vanishes outside K, we write

$$g = g_0 + \sum (g\chi_{A_n}),$$

where g_0 vanishes outside N. Since $\overleftarrow{m}$ is scalarwise LAC(μ), the substance of Subsection 8.19.1 shows that

$$\overleftarrow{m}(g) = \overleftarrow{m}(g_0) + \sum \overleftarrow{m}(g\chi_{A_n}) = \sum \overleftarrow{m}(g\chi_{A_n})$$

the series converging in E'_w. Now

$$\overleftarrow{m}(g\chi_{A_n}) = \int g\chi_{A_n}\overleftarrow{f}_K \, d\mu$$

for each n, and the weak convergence of the above series entails that of

$$\sum \int g\chi_{A_n} \cdot \langle x, \overleftarrow{f}_K \rangle \, d\mu$$

for each x in E. This being the case for any g in $\mathscr{L}^\infty(\mu)$ that vanishes outside K, one infers that $\langle x, \overleftarrow{f}_K \rangle$ is integrable over K for each x. Knowing this, we may write

$$\begin{aligned}\langle x, \overleftarrow{m}(g) \rangle &= \sum \langle x, \overleftarrow{m}(g\chi_{A_n}) \rangle = \sum \int g\chi_{A_n} \cdot \langle x, \overleftarrow{f}_K \rangle \, d\mu \\ &= \int g\Big(\sum \chi_{A_n}\Big) \cdot \langle x, \overleftarrow{f}_K \rangle \, d\mu = \int g \cdot \langle x, \overleftarrow{f}_K \rangle \, d\mu,\end{aligned}$$

making an appeal to Lebesgue's Theorem 4.8.2 to justify the change in order of summation and integration. Thus

$$\overleftarrow{m}(g) = \int g\overleftarrow{f}_K \, d\mu$$

for any g in $\mathscr{L}^\infty(\mu)$ vanishing outside K.

(3) As a final step we partition T into the union of compact sets K_i and a locally negligible set N', each compact set meeting K_i for at most countably many indices i. Putting $\overleftarrow{f}$ equal to $\overleftarrow{f}_{K_i}$ on K_i for each i and to 0 on N', one obtains a scalarwise measurable function $\overleftarrow{f}$ mapping T into E'_w such that

$$\overleftarrow{m}(g) = \int g\overleftarrow{f} \, d\mu$$

for each g in $\mathscr{L}^\infty(\mu)$ that vanishes outside K_i for some i.

Suppose now that K is any compact set and that g belongs to $\mathscr{L}^\infty(\mu)$ and vanishes outside K. We then have $g = \sum g\chi_{K_i}$ and so, by Subsection 8.19.1,

$$\overleftarrow{m}(g) = \sum \overleftarrow{m}(g\chi_{K_i}) = \sum \int g\chi_{K_i} \cdot \overleftarrow{f} \, d\mu,$$

the series converging in E'_w and all sums extending over the at-most-countable set of i for which K_i meets K. By an argument like that used in (2) it follows that $\langle x, \overleftarrow{f} \rangle$ is integrable over K for each x, so that $\overleftarrow{f}$ is scalarwise locally integrable, and that

$$\overleftarrow{m}(g) = \int g\overleftarrow{f} \, d\mu$$

for each g in $\mathscr{L}^\infty(\mu)$ having a compact support. ▌

A partial converse of the last result is available.

8.19.12 **Proposition.** Suppose that E is a separable Fréchet space and that $\overleftarrow{m}$ is an E'_w-valued Radon measure of the form $\overleftarrow{f} \cdot \mu$, where $\overleftarrow{f}$ is scalarwise locally integrable for μ. Then $\overleftarrow{m}$ is scalarwise LAC(μ) and satisfies Condition (2).

Proof. Since $x \circ \overleftarrow{m} = \langle x, \overleftarrow{f} \rangle \cdot \mu$, it is clear that $\overleftarrow{m}$ is scalarwise LAC(μ). To verify Condition (2) of Subsection 8.19.10 we suppose K is any compact subset of T and that ε is any strictly positive number. By Proposition 8.15.3 and the definition of weak measurability, there exists a compact set $K' \subset K$ such that $\mu(K \backslash K') < \varepsilon$ and $\overleftarrow{f} \mid K'$ is weakly continuous. Thus $\overleftarrow{f}(K')$ is weakly compact and therefore equicontinuous—that is, is contained in the polar U^0 of some neighbourhood U of 0 in E. Then, however, if g belongs to $\mathscr{K}(T, K')$ and $\int |g| \, d\mu \leq 1$, one has for x in U

$$|\langle x, \overleftarrow{m}(g) \rangle| = \left| \int g \cdot \langle x, \overleftarrow{f} \rangle \, d\mu \right| \leq \int |g| \, d\mu \leq 1.$$

Consequently $\overleftarrow{m}(g)$ lies in U^0. Since U^0 is weakly relatively compact, this shows that Condition (2) is fulfilled. ▌

A combination of the last two results leads to the following theorem.

8.19.13 **Theorem.** Let E be a separable Fréchet space, $\overleftarrow{m}$ any E'_w-valued Radon measure on T. In order that $\overleftarrow{m} = \overleftarrow{f} \cdot \mu$ for some scalarwise locally integrable $\overleftarrow{f}$, it is necessary and sufficient that $\overleftarrow{m}$ be scalarwise LAC(μ) and satisfy Condition (2).

8.19.14. The preceding results have some connections with general concepts introduced by Grothendieck ([1], pp. 145 et seq.) and their development in certain directions by Singer [2], [3], and [4].

For definiteness and simplicity, let us assume that T is compact and that E is a Banach space. The general concepts of Grothendieck lead to the consideration of E'_w-valued Radon measures $\overleftarrow{m}$ on T that are "integral" in the sense that there exists a continuous linear form U on $C(T, E)$ (the space of continuous E-valued functions $\vec{\xi}$ on T, the norm being defined by $\|\vec{\xi}\| = \operatorname{Sup} \|\vec{\xi}(t)\|$) such that

$$U(gx) = \langle x, \overleftarrow{m}(g) \rangle \qquad (g \in \mathscr{K}, x \in E) \tag{8.19.6}$$

On the other hand $\overleftarrow{m}$ is majorized (Dieudonné) if there exists a positive Radon measure μ on T such that

$$\|\overleftarrow{m}(g)\| \leq \int |g| \, d\mu \qquad (g \in \mathscr{K}). \tag{8.19.7}$$

Let us place beside these two restrictions a third, amounting to the postulated existence of a positive Radon measure μ and a bounded and scalarwise measurable E'_w-valued function $\overleftarrow{f}$ on T such that

$$\overleftarrow{m}(g) = \int g \overleftarrow{f} \, d\mu \qquad (g \in \mathscr{K}). \tag{8.19.8}$$

It is clear that (8.19.8) implies (8.19.6) on taking

$$U(\vec{\xi}) = \int \langle \vec{\xi}, \vec{f} \rangle \, d\mu.$$

It is also plain that (8.19.8) entails (8.19.7). The proof of Proposition 8.19.11 shows (since T is compact) that (8.19.7) implies (8.19.8).

Thus (8.19.7) and (8.19.8) are equivalent and imply (8.19.6). On the other hand Singer (ibid.) shows that (8.19.6) implies (8.19.7), so that all three conditions are equivalent.

8.20 The Dual of L_E^p when E Is a Banach Space and $1 < p < \infty$

In Section 8.18 we determined the dual of L_E^1 in a number of important cases, and we now turn to the analogous problem for L_E^p, where $1 < p < \infty$. Throughout this section E is assumed to be a Banach space. We write p' for the exponent conjugate to p, so that $1/p + 1/p' = 1$.

8.20.1. Let F be a continuous linear form on L_E^p. One may equally well regard F as being defined on $\mathscr{L}_E^p$, which will avoid complicating the notation. Our aim is to study a vector-valued Radon measure associated with F.

(1) Given g in $\mathscr{K}(T)$, the mapping $x \to F(gx)$ is a continuous linear form on E, so that there exists an element $\vec{m}(g)$ of E' such that

$$\langle x, \vec{m}(g) \rangle = F(gx),$$
$$\|\vec{m}(g)\| \leq \|F\| \cdot N_p(g).$$

Thus $\vec{m}$ is an E'_w-valued Radon measure on T. Actually, it is even an E'-valued measure when E' is endowed with its strong topology.

(2) If we fix x in E, the mapping $g \to F(gx)$ is a continuous linear form on $\mathscr{L}^p$, so that there exists a function h_x in $\mathscr{L}^{p'}$ such that

$$F(gx) = \int g h_x \, d\mu,$$
$$\int |h_x|^{p'} \, d\mu \leq \|F\|^{p'} \cdot \|x\|^{p'}.$$

(3) On combining (1) and (2) one sees that $x \circ \vec{m}$ is the measure $h_x \cdot \mu$, so that $\vec{m}$ is scalarwise LAC(μ). Furthermore, $\overline{\mathscr{L}}(\vec{m})$ contains $\mathscr{L}^p$. If g belongs to $\mathscr{L}^p$, the definition of $\vec{m}(g)$ (see Subsection 8.19.7) gives

$$\langle x, \vec{m}(g) \rangle = \int g \, d(x \circ \vec{m}) = \int g h_x \, d\mu = F(gx).$$

Thus the relations in (1) extend to all g in $\mathscr{L}^p$.

(4) Our next aim is to show that $\vec{m}$ is of p'-bounded variation. Specifically,

$$\sum \frac{\|\vec{m}(A_k)\|^{p'}}{\mu(A_k)^{p'-1}} \leq \|F\|^{p'}$$

for any finite sequence of disjoint integrable sets A_k satisfying $\mu(A_k) > 0$.

To do this we suppose that the x_k belong to E and satisfy $\|x_k\| \leq 1$, and that the ξ_k are scalars. Then

$$\left|\sum \xi_k \langle x_k, \bar{m}(A_k)\rangle\right| = \left|F\left[\sum \xi_k x_k \chi_{A_k}\right]\right| \leq \|F\| \cdot N_p\left(\sum \xi_k x_k \chi_{A_k}\right)$$
$$= \|F\| \cdot \left[\sum |\xi_k|^p \mu(A_k)\right]^{1/p}.$$

Replacing ξ_k by $\xi_k \cdot \mu(A_k)^{-1/p}$, this yields

$$\left|\sum \xi_k \mu(A_k)^{-1/p} \langle x_k, \bar{m}(A_k)\rangle\right| \leq \|F\| \cdot \left[\sum |\xi_k|^p\right]^{1/p}.$$

Varying the x_k and noticing that $p'/p = p' - 1$, one derives the stated inequality.

(5) From (4) one may derive as an application of Hölder's inequality the relation

$$\sum \|\bar{m}(A_k\| \leq \|F\| \cdot \mu(A)^{1/p}$$

whenever A is partitioned by the integrable sets A_k.

(6) We introduce now the positive set function s, defined for all integrable sets by the relation

$$s(A) = \operatorname{Sup} \sum \|\bar{m}(A_k)\|,$$

the supremum being taken over all finite partitions of A into integrable sets A_k. From (5) it appears that

$$s(A) \leq \|F\| \cdot \mu(A)^{1/p}.$$

Let us verify that s is additive. Suppose that

$$A = A' \cup A'',$$

where $A' \cap A'' = \emptyset$. If the A_i' partition A' and the A_j'' partition A'', then taken together they partition A, so that

$$s(A) \geq \sum \|\bar{m}(A_i')\| + \sum \|\bar{m}(A_j'')\|.$$

Since the two sums on the right-hand side can be chosen to approximate $s(A')$ and $s(A'')$ as closely as one chooses, we infer that

$$s(A) \geq s(A') + s(A'').$$

On the other hand, if the A_k partition A, then the $A' \cap A_k$ partition A' and the $A'' \cap A_k$ partition A''. So, since $\bar{m}$ is additive, one has

$$\begin{aligned} s(A') + s(A'') &\geq \sum \|\bar{m}(A' \cap A_k)\| + \sum \|\bar{m}(A'' \cap A_k)\| \\ &\geq \sum \|\bar{m}(A' \cap A_k) + \bar{m}(A'' \cap A_k)\| \\ &= \sum \|\bar{m}(A_k)\|. \end{aligned}$$

The final term on the right-hand side can be made as near to $s(A)$ as one chooses, and therefore $s(A') + s(A'') \geq s(A)$. Additivity is thereby established.

(7) From (6) one may conclude (Exercise 4.39) that there exists a positive locally integrable function L such that

$$s(A) = \int_A L \, d\mu$$

for all integrable sets A. In particular, since $\|\vec{m}(A)\| \leq s(A)$, it appears that

$$\|\vec{m}(A)\| \leq \int_A L \, d\mu$$

for all integrable sets A.

From this it is a simple matter to verify that

$$\|\vec{m}(g)\| \leq \int |g| \, L \, d\mu$$

for each g in $\mathscr{K}(T)$ (and indeed for each g in $\mathscr{L}^\infty(\mu)$ having a compact support).

(8) It is now an easy matter to show that $\vec{m}$, which we know already to be scalarwise LAC(μ), also satisfies Condition (2) of Subsection 8.19.10. Indeed, since L is measurable, given any compact set $K \subset T$ and any $\varepsilon > 0$, there exists a compact set $K' \subset K$ with $\mu(K \backslash K') < \varepsilon$ and such that $L \mid K'$ is continuous. In particular, L is bounded on K', say $L \leq c$ there. Then, if g belongs to $\mathscr{K}(T, K')$ and $\int |g| \, d\mu \leq 1$, the final result of (7) shows that

$$\|\vec{m}(g)\| \leq \int |g| \, L \, d\mu \leq c \int |g| \, d\mu \leq c.$$

It is enough now to remember that norm-bounded subsets of E' are weakly relatively compact.

(9) Appeal to Proposition 8.19.11 shows now that there exists a scalarwise locally integrable function $\vec{f}$ mapping T into E'_w such that

$$\vec{m}(g) = \int g\vec{f} \, d\mu \tag{8.20.1}$$

for each g in $\mathscr{L}^\infty(\mu)$ having a compact support. Comparison with (2) shows that $\langle x, \vec{f} \rangle = h_x$ l.a.e. for each x, so that

$$\bar{\int} |\langle x, \vec{f} \rangle|^{p'} \, d\mu \leq \|F\|^{p'} \cdot \|x\|^{p'}$$

for each x in E.

Since each g in $\mathscr{L}^p$ is the limit in $\mathscr{L}^p$ of a sequence (g_n), where each g_n lies in $\mathscr{L}^\infty(\mu)$ and has a compact support and satisfies $|g_n| \leq |g|$, a simple limiting process starting from (8.20.1) shows that that relation continues to hold for g in $\mathscr{L}^p$.

From (7) we may conclude that

$$\left| \int_A \langle x, \vec{f} \rangle \, d\mu \right| = |\langle x, \vec{m}(A) \rangle| \geq \|x\| \cdot \|\vec{m}(A)\| \geq \|x\| \cdot \int_A L \, d\mu$$

for each x in E and each integrable set A, hence it follows that

$$|\langle x, \vec{f} \rangle| \leq \|x\| \cdot L \qquad \text{l.a.e.}$$

for each x in E.

Now (8.20.1) combines with (3) to show that

$$F(gx) = \int g \cdot \langle x, \vec{f} \rangle \, d\mu$$

and so, by linearity, that

$$F(\vec{g}) = \int \langle \vec{g}, \vec{f} \rangle \, d\mu \tag{8.20.2}$$

is true for any simple $\vec{g}$ in $\mathscr{L}_E^p$. For any such $\vec{g}$ one has therefore

$$\left| \int \langle \vec{g}, \vec{f} \rangle \, d\mu \right| \leq \|F\| \cdot N_p(\vec{g}) \tag{8.20.3}$$

8.20.2. Having established (8.20.2) for simple functions $\vec{g}$ in $\mathscr{L}_E^p$, it is natural to expect and hope that it in fact remains true for *all* $\vec{g}$ in $\mathscr{L}_E^p$. With only the available properties of $\vec{f}$, however, it is not even certain that the right-hand side of (8.20.2) is defined (that is, that $\langle \vec{g}, \vec{f} \rangle$ is integrable). The doubts would evaporate as soon as one knew that

$$\int^* \|\vec{f}\|^{p'} \, d\mu < +\infty,$$

it being easy to see that $\langle \vec{g}, \vec{f} \rangle$ is measurable for each $\vec{g}$ in $\mathscr{L}_E^p$.

These difficulties are surmountable when E is separable, as we shall now show.

(10) If E is separable, then $\vec{f}$ is weakly measurable (that is, measurable from T into E'_w), $\|\vec{f}\|$ is measurable, and

$$\int \|\vec{f}\|^{p'} \, d\mu = \|F\|^{p'}. \tag{8.20.4}$$

In fact Proposition 8.15.3 affirms the first two assertions, and it remains to establish (8.20.4). We will show first that the left-hand side is majorized by the right-hand side.

To do this it suffices to show that

$$\int_{K_0} \|\vec{f}\|^{p'} \, d\mu \leq \|F\|^{p'}$$

for each compact set $K_0 \subset T$. For this, it is enough to show that the same is true when K_0 is replaced by suitable compact sets $K \subset K_0$ for which $\mu(K_0 \backslash K)$ is arbitrarily small. Now we can choose sets K of this type for which it is furthermore the case that $\|\vec{f}\| \mid K$ is continuous and $\vec{f} \mid K$ is weakly continuous. Thanks to these continuity properties, the value of $\int_K \|\vec{f}\|^{p'} \, d\mu$ does not exceed the supremum of sums

$$\sum \|\vec{f}(t_k)\|^{p'} \cdot \mu(A_k),$$

where (A_k) denotes a general partition of K into integrable sets, t_k denoting a point selected freely from A_k. Moreover, only "arbitrarily fine" partitions need be considered. The sum written down is in its turn not bigger than the supremum of

$$\sum |\langle x_k, \vec{f}(t_k) \rangle|^{p'} \mu(A_k),$$

the x_k being arbitrarily chosen from E apart from the condition $\|x_k\| \leq 1$. This sum does not exceed the supremum of

$$\left| \sum \xi_k \mu(A_k)^{1/p} \cdot \langle x_k, \vec{f}(t_k) \rangle \mu(A_k)^{1/p'} \right|^{p'}$$

when the ξ_k vary subject to the restriction $\sum |\xi_k|^p \mu(A_k) \leq 1$. Finally, taking only arbitrarily fine partitions, this is majorized by the supremum of

$$\left| \int \langle \vec{g}, \vec{f} \rangle \, d\mu \right|^{p'}$$

when $\vec{g} = \sum \xi_k x_k \chi_{A_k}$ is simple and satisfies $N_p(\vec{g}) \leq 1$. Reference to (8.20.3) shows then that

$$\int_K \|\vec{f}\|^{p'} \, d\mu \leq \|F\|^{p'}.$$

On the other hand (8.20.2) gives in all cases

$$|F(\vec{g})| \leq \int^* \|\vec{g}\| \cdot \|\vec{f}\| \, d\mu \leq N_p(\vec{g}) \cdot \left[\int^* \|\vec{f}\|^{p'} \, d\mu \right]^{1/p'}$$

and therefore

$$\|F\| \leq \left[\int^* \|\vec{f}\|^{p'} \, d\mu \right]^{1/p'}.$$

Thus (8.20.4) is established for the case in which E is separable.

Since there is no harm done by altering $\vec{f}$ on a locally negligible set, we may summarize the conclusions for the separable case as follows.

8.20.3 **Theorem.** Suppose that E is a separable Banach space, F a continuous linear form on $\mathscr{L}_E^p$, where $1 < p < \infty$. Then there exists a function $\vec{f}$ mapping T into E' that is weakly measurable and such that

$$F(\vec{g}) = \int \langle \vec{g}, \vec{f} \rangle \, d\mu \tag{8.20.5}$$

for each g in $\mathscr{L}_E^p$, and

$$\|F\| = \left[\int \|\vec{f}\|^{p'} \, d\mu \right]^{1/p'} \tag{8.20.6}$$

Conversely, any weakly measurable function $\vec{f}$ for which the right-hand side of (8.20.6) is finite defines, via (8.20.5), a continuous linear form F on $\mathscr{L}_E^p$ for which (8.20.6) holds.

Remark. The continuous linear forms on L_E^p are, in any case, obtainable by passage to the quotient from continuous linear forms on $\mathscr{L}_E^p$. It is therefore superfluous to restate our results in a form referring directly to L_E^p.

8.20.4 **The Case in Which E is Reflexive.** If reflexivity of E is added to the hypotheses of Theorem 8.20.3, then E', possessing a strongly separable dual E, is itself strongly separable. As a result the weakly measurable function $\vec{f}$ is even strongly measurable (Theorem 8.15.2) and therefore belongs to $\mathscr{L}_{E'}^{p'}$. We therefore infer that $(L_E^p)' \approx L_{E'}^{p'}$ if E is a separable and reflexive Banach space. Given reflexivity of E, however, one may drop entirely the separability hypothesis by making use of Phillips' results given in Subsection 8.19.6.

Indeed, if E is reflexive, so also is E' (with its strong topology), so that (3) and (4) of Subsection 8.20.1 combine with (2) of Subsection 8.19.6 (with E

therein replaced by E') to show that, if (T, μ) is σ-finite, there exists an $\overset{*}{f}$ in $\mathscr{L}_{E'}^{p'}$ such that

$$\overset{*}{m}(A) = \int_A \overset{*}{f}\, d\mu$$

for all integrable sets A, and

$$\int \|\overset{*}{f}\|^{p'}\, d\mu \leq \|F\|^{p'}.$$

Granted these two properties of $\overset{*}{f}$, it is easy to see that (8.20.5) and (8.20.6) hold. This yields the following result.

8.20.5 **Theorem.** Suppose that E is a reflexive Banach space, that $1 < p < \infty$, and that (T, μ) is σ-finite. Then the conclusions of Theorem 8.20.3 remain true, even with weak measurability of $\overset{*}{f}$ replaced by strong measurability thereof. In particular, $(L_E^p)' \approx L_{E'}^{p'}$.

Remarks. (1) See Mourier [1] and Fortet and Mourier [1].

(2) Day [3] has shown that the same result obtains for any uniformly convex Banach space E.

(3) The fact that one has a representation (8.20.5) with a strongly measurable $\overset{*}{f}$ implies incidentally that any scalarwise measurable $\overset{*}{f}$ mapping T into E'_w, which satisfies (8.20.6), is equal locally almost everywhere to a strongly measurable function. It is here assumed that E is a reflexive Banach space. For a similar result, see Bourbaki [6], Chapitre 6, p. 95, Exercice 25).

(4) The possibility of integral representations of certain categories of linear operators on spaces L_E^p has attracted a good deal of attention. For this and related matters, see Dinculeanu and Foias [1] and A. and C. Ionescu Tulcea [1], [2], and [3].

EXERCISES

8.1 Let E be a seminormed vector space, E' its dual. Show that if E' is strongly separable, then E is separable. Show also that the converse is generally false.

8.2 Let E be an LCTVS, B a bounded subset of E. Show that if B^0 is strongly separable in E' (hence in particular if E' is strongly separable), then B is separable for the initial topology on E.

[Hint: It may be assumed that B is closed, convex, and balanced. Apply Exercise 8.1 to the seminormed space E_B (compare Lemma 6.5.2).]

8.3 Let E be a separable and reflexive seminormed space. Show that each bounded subset of E is weakly semimetrizable.

[Hint: Show that E' is strongly separable.]

8.4 Let E be a reflexive seminormed space, (x_n) a bounded sequence in E. Show that there is a weakly convergent subsequence of (x_n).

[Hint: Apply Exercise 8.3 to the closed vector subspace of E generated by the x_n. Appeal might alternatively be made to Šmulian's Theorem 8.12.1.]

8.5 The notation and hypothesis are as in Example 8.3.6. Suppose additionally that A is bounded and that p satisfies $1 \leq p < \infty$. Show that f is adherent to A for $\sigma(\mathscr{L}^\infty, \mathscr{L}^1)$ if and only if there exists a sequence (f_n) extracted from A such that $f_n \to f$ in $\mathscr{L}^p$.
[Hint: Necessity is established in Example 8.3.6. To verify sufficiency notice that if $f_n \to f$ in $\mathscr{L}^p$, then $\int gf_n \, d\mu \to \int gf \, d\mu$ for a set of g dense in $\mathscr{L}^1$.]

8.6 Retain the notations and hypotheses of the preceding exercise. Show that if f is $\sigma(\mathscr{L}^\infty, \mathscr{L}^1)$-adherent to A, then there exists a sequence $(f_n) \subset A$ such that $f_n \to f$ a.e. [Hint: Use Exercise 8.5.]

8.7 The result of Exercise 8.6 has been established under the assumption that μ is bounded. Show that it remains true whenever (T, μ) is σ-finite.
[Hint: T is the union of a negligible set and of a sequence K_r of compact sets. Apply Exercise 8.6 to the spaces $(K_r, \mu \mid K_r)$ and use the diagonal process.]

8.8 Let E and F be LCTVSs, u a linear map of E into F, u^* its algebraic adjoint. Consider the following conditions:

(1) u is continuous;
(2) u is bounded;
(3) $u^*(F') \subset E'$.

Prove that (3) implies (2), and that (1) implies both (2) and (3). Deduce that if E is bornological, then all three conditions are equivalent. (For seminormed spaces, see Mackey [2], Theorem II.4.)

8.9 Let E be a vector space, p and q two seminorms on E. Let E'_p and E'_q denote the respective topological duals, each subspaces of E^*. Show that there exists a number c such that $p(x) \leq c \cdot q(x)$ for all $x \in E$ if and only if $E'_p \subset E'_q$.
(The result, for normed spaces, is due to Fichtenholtz [1].)

8.10 Let E and F be LCTVS and u a linear map with domain a vector subspace D of E and range in F. Show that u' is always closed when regarded as a map with domain in F' and range in E', the latter spaces being endowed with their weak topologies.

8.11 Let E be any complete and separated LCTVS that is not semireflexive, and let E'' be its bidual. Show that E is of the first category in E''.

8.12 Let E be an infinite dimensional Banach space. Show that there exists a one-to-one continuous linear map u of ℓ^∞ into E. (Mackey [2], Theorem I.1)
[Hint: Show that one may construct subspaces $M_0 \ (=E) \supset M_1 \supset M_2 \supset \cdots$ and elements $x_n (n \geq 1)$ of E such that $\dim M_n = \infty$, $x_n \in M_{n-1} \backslash M_n$ and $\|x_n\| \leq n^{-2}$. Consider the map

$$u : \xi \to \sum \xi(n) x_n.]$$

8.13 T is a separated locally compact space, μ a positive Radon measure on T, E a normed vector space, f a function on T with values in E such that

$$\int^* |\langle f(t), x' \rangle| \, d\mu(t) < +\infty$$

for each $x' \in E'$. Show that there exists a number $c \geq 0$ such that

$$\int^* |\langle f(t), x' \rangle| \, d\mu(t) \leq c \, \|x'\|$$

for $x' \in E'$.

Discuss extensions of this result to more general separated LCTVSs E; compare Exercise 8.24.
[Hint : Consider the seminorm $N(x') = \int^* |\langle f, x' \rangle| \, d\mu$ and use Fatou's lemma.]

8.14 Consider functions of a real variable, φ. If φ is scalar valued, it is said to be locally Lipschitzian if the ratio

$$\frac{\varphi(t_1) - \varphi(t_2)}{t_1 - t_2}$$

is bounded for $a \leq t_1 < t_2 \leq b$, for any given real numbers a and b satisfying $a < b$ (the bound possibly depending on a and b). A similar definition applies if φ takes its values in a TVS E.

Suppose that E is a quasi-complete separated LCTVS and f a function on R^1 with values in E, n an integer ≥ 0. Assume that for each $x' \in E'$ the function $x' \circ f$ has a locally Lipschitzian nth derivative. Prove that f has an nth derivative in the sense of the topology of E, and that this nth derivative is locally Lipschitzian. [Hint: Proceed by induction on n, using Theorem 8.3.4(1). The inductive step from n to $n + 1$ is made by defining

$$x(h) = h^{-1}[D^n f(t + h) - D^n f(t)] \qquad \left(h \neq 0, \quad D = \frac{d}{dt}\right)$$

and considering the elements $[x(h) - x(k)](h - k)^{-1}$, where 0, h, k are distinct in pairs.]

8.15 Let E be a separated LCTVS satisfying the Hypothesis of Proposition 8.14.3, f an E-valued function that is holomorphic on a domain Ω in the complex plane, p a continuous seminorm on E. Show that if c is a number such that $p \circ f \leq c$ on Ω, then either

(1) $p \circ f = c$ on Ω, or (2) $p \circ f < c$ on Ω.

Remark. If $p \circ f = |x' \circ f|$, where $x' \in E'$, the result may be sharpened to show that $|x' \circ f|$ can admit no *local* maximum in Ω, unless $x' \circ f$ is constant throughout Ω. [Hint: See Subsection 8.14.7. Show first that

$$p \circ f(t_0) \leq (2\pi)^{-1} \int_0^{2\pi} p \circ f(t_0 + re^{i\alpha})\, d\alpha$$

for $t_0 \in \Omega$ and $r \geq 0$ sufficiently small. Then show that the set of points $t \in \Omega$, such that $p \circ f(t) = c$, is both open and closed relative to Ω.]

8.16 The notations and assumptions are as in the preceding exercise. In addition we suppose that Ω is relatively compact.

Prove that if k is a number such that

$$\limsup_{t \in \Omega,\, t \to s} p \circ f(t) \leq k$$

for each point s of the frontier $\dot{\Omega}$ of Ω, then $p \circ f(t) \leq k$ for $t \in \Omega$.

[Hint: Define the function g on $\overline{\Omega}$ by

$$g(t) = \limsup_{t' \in \Omega,\, t' \to t} p \circ f(t').$$

Note that g is USC, that $g \leq k$ on $\dot{\Omega}$, and that $g = p \circ f$ on Ω. g assumes its supremum say m, at some point of $\overline{\Omega}$. Use the preceding exercise to show that necessarily $m \leq k$.]

8.17 Let φ be a (scalar-valued) function holomorphic on the open unit disk centre the origin in the complex plane. Show that if $1 \leq p < \infty$

$$\left[\int_0^{2\pi} |\varphi(re^{i\alpha})|^p\, d\alpha\right]^{1/p}$$

is an increasing function of r for $0 \leq r < 1$.

Note. This is a basic result about the Hardy spaces H^p; for further study of these spaces, see Zygmund [2], Vol. I, Chapter VII, Dunford and Schwartz [1], pp. 364–5, 370–1, and Walters [1] and [2].

[Hint: Apply the preceding exercise.]

8.18 Let E, F, and G be LCTVSs, T a separated locally compact space with a positive Radon measure μ. Suppose that e and f are functions mapping T into E and F, respectively, and that B is a continuous bilinear map of $E \times F$ into G. Define g by

$$g(t) = B(e(t), f(t)),$$

mapping T into G. Prove that

(a) If e and f are measurable, so too is g.

(b) If e is scalarwise measurable, f measurable, and F metrizable, then g is scalarwise measurable.

Show further that in (b) it suffices to assume that B is separately continuous.

8.19 Let T and μ be as in the preceding exercise and let E be a Banach algebra. (This means that E is a complex Banach space with a bilinear map $(x, y) \to xy$ of $E \times E$ into E such that $(xy)z = x(yz)$ and $\|xy\| \leq \|x\| \cdot \|y\|$ for $x, y, z \in E$.) Show that if $f \in \mathscr{L}_E^p$ and $g \in \mathscr{L}_E^{p'}$ (defined as in Section 8.18), then $fg \in \mathscr{L}_E^1$ and

$$\int \|f(t)g(t)\| \, d\mu(t) \leq N_p(f) N_{p'}(g).$$

8.20 Let E be a Banach algebra, $T = R^1$ and μ the Lebesgue measure on T. Show that if f and g belong to $\mathscr{L}_E^1$, then the function $s \to f(t - s)g(s)$ belongs to $\mathscr{L}_E^1$ for almost all t and that if we define

$$h(t) = \int f(t - s)g(s) \, d\mu(s)$$

for such t and put $h(t) = 0$ otherwise, then $h \in \mathscr{L}_E^1$. One may write $h = f * g$, the convolution of f and g. $\mathscr{L}_E^1$ is thus a Banach algebra. (For completeness, see Subsection 8.18.1.)

Investigate what happens when T is separated locally compact group and μ its left Haar measure.

8.21 Let E be a TVS, U a neighbourhood of 0 in E, U^0 its polar in E'. Let E'_U be the vector subspace of E' generated by U^0. Verify that E'_U is a Banach space when one takes as its closed unit ball the set U^0, so that the associated norm is

$$\|x'\| = \operatorname{Inf} \{\lambda : \lambda > 0, x' \in \lambda U^0\}$$

for $x' \in E'_U$.

8.22 Let E be a TVS, E' its topological dual. A vector subspace D of E' is termed a *determining subspace* of E' if the following is true : there exists a family (U_i) of neighbourhoods of 0 in E, whose homothetic images form a base at 0, such that to each i corresponds a number $\alpha_i > 0$ for which $D \cap U_i^0$ is weakly dense in $\alpha_i U_i^0$. Show that this is so if and only if

$$p_i(x) \leq \alpha_i^{-1} \operatorname{Sup} \{|\langle x, x' \rangle| : x' \in D \cap U_i^0\}$$

for all $x \in E$, where p_i is the gauge function of the closed, convex, and balanced envelope in E of U_i.

Verify that if E is a separated LCTVS, then any determining subspace of E' separates points of E.

8.23 Let E be a LCTVS, D a determining subspace of E' that is sequentially strongly closed in E'. Show that if a subset A of E is such that

$$\operatorname{Sup} \{|\langle x, x' \rangle| : x \in A\} < +\infty$$

for each $x' \in D$, then A is bounded in E.

[Hint: Use the preceding two exercises to show that A is absorbed by the closed, convex, balanced envelope of U_i for each i.]

8.24 Let E be a separated LCTVS, D a sequentially strongly closed determining subspace of E'. Let T be a separated locally compact space, μ a positive Radon measure on T, and (f_α) a family of functions mapping T into E. Suppose finally that

$$\operatorname{Sup}_\alpha \overline{\int}^* |\langle f_\alpha, x'\rangle| \, d\mu < +\infty$$

for each $x' \in D$. Show that there exists a base of neighbourhoods U of 0 in E, to each of which corresponds a number $c = c_U$ such that

$$\operatorname{Sup}_\alpha \overline{\int}^* |\langle f_\alpha, x'\rangle| \, d\mu \le c \qquad \text{for } x' \in D \cap U^0.$$

Compare Exercise 8.13.
[Hint: Consider the seminorm

$$N(x') = \operatorname{Sup}_\alpha \overline{\int}^* |\langle f_\alpha, x'\rangle| \, d\mu \qquad \text{on } E'_{U_i} \cap D,$$

the U_i being as in Exercise 8.22.]

8.25 Let E be a sequentially complete separated LCTVS, T a separated locally compact space, μ a positive Radon measure on T, and f a scalarwise locally integrable function on T into E. Assume too that

(1) $\int \varphi f \, d\mu \in E \qquad$ for each $\varphi \in \mathscr{K}(T)$,

(2) $\overline{\int} |\langle f, x'\rangle| \, d\mu < +\infty \qquad$ for $x' \in D$,

where D is a sequentially strongly closed determining subspace of E'. Prove that $\int \varphi f \, d\mu \in E$ for each $\varphi \in C_0(T)$, and that $\varphi \to \int \varphi f \, d\mu$ is continuous from $C(T)$ into E. Deduce in particular that $\overline{\int} |\langle f, x'\rangle| \, d\mu < +\infty$ for each $x' \in E'$.
[Hint: Use Exercise 8.24 to show that if $\varphi \in C_0(T)$ is the uniform limit of a sequence $\varphi_n \in \mathscr{K}(T)$, then the integrals $\int f\varphi_n \, d\mu$ form a Cauchy sequence in E.]

8.26 Let E be a Banach space, D a strongly closed determining subspace of E', (a_n) a sequence of elements of E. Suppose that $\sum |\langle a_n, x'\rangle| < +\infty$ for $x' \in D$. Show that $\sum \xi(n) a_n$ is convergent in E for each $\xi \in c_0$, and that $\xi \to \sum \xi(n) a_n$ is continuous from c_0 into E.
[Hint: Specialize the preceding exercise.]

8.27 Let T be a separated locally compact space, μ a positive Radon measure on T, (u_n) a sequence in $\mathscr{L}^1 = \mathscr{L}^1(T, \mu)$. Define $\hat{f}(n) = \int f u_n \, d\mu$.

Show that if (α_n) is a sequence of positive scalars such that $\sum \alpha_n |\hat{f}(n)| < +\infty$ for each $f \in C_0(T)$, then $\sum \alpha_n \beta_n u_n$ is convergent in $\mathscr{L}^1$ for each sequence (β_n) of scalars such that $\beta_n \to 0$ as $n \to \infty$.

Show similarly that if each $u_n \in C_0(T)$, and if $\sum \alpha_n |\hat{g}(n)| < +\infty$ for each $g \in \mathscr{L}^1$, then $\sum \alpha_n \beta_n u_n$ is convergent in $C_0(T)$ whenever $\beta_n \to 0$ as $n \to \infty$; it is here assumed that the support of μ is T.

Apply these results when $T = [0, 1]$, μ is the Lebesgue measure restricted to T, and $u_n(t) = t^n$. Conclude that $\sum \alpha_n |\int_0^1 f(t) t^n \, dt| < +\infty$ for each $f \in C[0, 1]$ if and only if $\sum (n+1)^{-1} \alpha_n < +\infty$.
[Hint: Apply Exercise 8.26, taking

$$E = \mathscr{L}^1, \qquad E' = \mathscr{L}^\infty, \qquad D = C_0$$
$$E = C_0, \qquad E' = M, \qquad D = \mathscr{L}^1.]$$

8.28 Let E be a Banach space, $1 \le p \le \infty$. Show that $\mathscr{L}^p_E$ is a Banach space.
[Hint: Adapt the proof for scalar-valued functions.]

8.29 Let E be a Banach space, $1 \leq p < \infty$. Show that the simple functions (see Subsection 8.18.1) are dense in $\mathscr{L}_E^p$.

NOTE. The next four exercises are concerned with relations between order boundedness and continuity of linear forms and maps defined on a topological Riesz space. A more systematic study within the framework of ordered vector spaces in general is undertaken by Schaefer [1], [2], [3] and Kist [1].

8.30 Let E be a LCTVS, (λ_n) a sequence of scalars such that $\sum |\lambda_n| < +\infty$, (x_n) a bounded sequence in E. Verify that the series $\sum \lambda_n x_n$ has Cauchy partial sums in E.

SOME DEFINITIONS. Let E be a Riesz space ($=$ a vector lattice; see subsection 4.23.1). A subset A of E is said to be o-bounded (that is, order bounded) if it is contained in some "interval"

$$[x', x''] = \{x \in E : x' \leq x \leq x''\},$$

where x', x'' lie in E; equivalently, if there exists $a \in E$ such that $x \in A$ entails $|x| \leq a$. A is said to be o-convex if $x \in A$, $y \in E$ and $|y| \leq x$ together entail that $y \in A$.

By a topological vector lattice (TVL for short) we shall here mean a vector lattice E that is also a TVS such that (1) the set $E_+ = \{x \in E : x \geq 0\}$ is closed in E, and (2) there exists a base at 0 in E formed of o-convex sets. By a locally convex topological vector lattice (LCTVL for short) we mean a TVL whose topology is locally convex. Compare Birkhoff [2], pp. 238 et seq.

In any TVL any o-bounded set is bounded; the converse is generally false, as is seen by considering $\ell^p (1 \leq p < \infty)$ with its usual topology and order.

If E and F are vector lattices and u a map of E into F, we say that u is positive (written $u \geq 0$) if $u(E_+) \subset F_+$. In addition u is said to be o-bounded if it transforms any o-bounded subset of E into an o-bounded subset of F. These terms are applied also to linear forms on a real vector lattice (that is, a vector lattice whose underlying vector space is real).

8.31 Let E be a TVL. Show that any bounded linear form on E is o-bounded Prove that the converse is true if E is sequentially complete and locally convex. [Hint: For the second part we may assume that the linear form f in question is ≥ 0; see the next exercise. If f were not bounded, there would exist a bounded sequence (x_n) such that $|f(x_n)| > n^2$. Show that one may assume that $x_n \geq 0$ and then consider the series $\sum n^{-2} x_n$, using Exercise 8.30.]

8.32 Let E be a TVL. Show that any o-bounded linear form f on E can be expressed as the difference $f = f^+ - f^-$ of two positive linear forms, and that f^+ and f^- may be assumed to be bounded (resp. continuous) if f is bounded (resp. continuous).

Deduce that $\sigma(E, E')$ has a base at 0 formed of o-convex sets, and that if E is a separated LCTVL, then the positive continuous linear forms on E separate points of E.

[Hint: Define f^+ first for $x \geq 0$ by

$$f^+(x) = \text{Sup}\,\{f(y) : y \in E,\, 0 \leq y \leq x\},$$

showing that f^+ so defined as additive on E_+, hence can be extended so as to be linear on E. Compare the method used in Chapter 4 for Radon measures.]

8.33 Let E be a sequentially complete LCTVL, F a TVL, u an o-bounded linear map of E into F. Prove that u is bounded (and therefore continuous, if E is bornological and F locally convex).

[Hint: Let (x_n) be bounded in E and $y_n = u(x_n)$. It is to be shown that y_n is bounded in F. Use Exercise 8.30 to show that for any sequence (λ_n) with $\sum |\lambda_n| < +\infty$, the sums $\sum_1^N \lambda_n y_n$ are bounded with respect to N.]

Some definitions. Let X and Y be TVSs and $T, T_1, \cdots, T_n$ continuous linear maps of X into Y. Put N for the closure in Y of $\{0\} \subset Y$. In the following three exercises we shall consider a certain condition, which we shall call Condition A. It falls into two parts, the first of which is as follows:

(A1) If $x \in X$ and $T_k x = 0$ for $1 \leq k \leq n$, then $Tx \in N$.

Consider the vector subspace V of Y^n formed of elements $(T_1 x, \cdots, T_n x)$ when x ranges over X. According to (A1) one may define a linear map L of V into Y/N by setting $L(y_1, \cdots, y_n) = Tx \bmod N$ whenever $(y_1, \cdots, y_n) = (T_1 x, \cdots, T_n x) \in V$. The second part of Condition A reads:

(A2) L is continuous from V into Y/N (V being considered as a subspace of Y^n, the latter with its product topology).

8.34 The notations are as immediately above. Suppose that Y is locally convex and that Condition A is fulfilled. Prove that:

$$R(T') \subset R(T_1') + \cdots + R(T_n'), \tag{8.34.1}$$

where R stands for range and where the primes denote the adjoints, so that each of T' and T_k' $(1 \leq k \leq n)$ is a map of Y' into X'.

[Hint: Take any $y' \in Y'$ and consider that linear form on V that carries $(T_1 x, \cdots, T_n x)$ into $\langle Tx, y' \rangle$. Use the Hahn-Banach theorem to extend this to Y^n.]

8.35 The notations being as explained before Exercise 8.34, suppose that $Y'^0 = N$ and that V is bornological. Show that if (8.34.1) holds, then Condition A is satisfied.

[Hint: Use (8.34.1) to show that (A1) holds, so that L is definable, and then that the domain of L' is the whole of Y'. Appeal to Theorem 8.11.3.]

8.36 Elaborate on the preceding two exercises in case Y is a seminormed space. Show in particular that if p is a seminorm on Y defining its topology, in order that there shall exist a number c such that

$$p(Tx) \leq c \cdot \sum_{k=1}^{n} p(T_k x) \text{ for all } x \in X,$$

it is necessary and sufficient that to each $y' \in Y'$ shall correspond $y_1', \cdots, y_n' \in Y'$ such that

$$\operatorname{Sup}_k \|y_k'\| \leq c \cdot \|y'\|$$

and

$$\sum_{k=1}^{n} T_k' y_k' = T' y'.$$

Here $\|y'\|$ denotes the norm of $y' \in Y'$ dual to p, that is,

$$\|y'\| = \operatorname{Sup} \{|\langle y, y' \rangle| : y \in Y, p(y) \leq 1\}.$$

Remark. The case in which $X = Y$, $n = 1$, $T =$ the identity, is closely related to results in Section 8.6; see especially Corollary 8.6.9. Compare also the techniques used in Section 5.14.

8.37 Let E be a Hilbert space, A and B two continuous endomorphisms of E with Hilbert adjoints A^* and B^*, respectively (see Subsection 1.12.8), and suppose that there exists a number $k < 1$ such that

$$|(Ax \mid Bx)| \leq k \, \|Ax\| \, \|Bx\| \qquad (x \in E).$$

Prove that (if R denotes range)

$$R(A^* + B^*) = R(A^*) + R(B^*).$$

[Hint: Verify first that

$$\|Ax\| \leq \text{const.} \, \|Ax + Bx\| \qquad (x \in E),$$

and then modify slightly the methods of the preceding three exercises.]

8.38 Let Ω be an open subset of R^n and $P(D)$ a linear partial differential operator with constant coefficients. Suppose that $1 \leq p < \infty$. Show that the equation

$$P(D)f = \varphi, \ f \in L^{p'}(\Omega)$$

is soluble for each $\varphi \in L^{p'}(\Omega)$ if and only if

$$\|u\|_{L^p(\Omega)} \leq \text{const. } \|P(D)u\|_{L^p(\Omega)}$$

for all $u \in \mathscr{D}(\Omega)$. (The constant may depend on $P(D)$, Ω and p.)
[Hint: Apply Exercises 8.34–8.36, taking $X = \mathscr{D}(\Omega)$ with its own topology, $Y = \mathscr{D}(\Omega)$ with the norm induced by $L^p(\Omega)$, $n = 1$, $Tu = u$, $T_1 u = P(D)u$.]

8.39 The notations being as in Exercise 8.38, assume that Ω is bounded and that $P(D)$ admits on R^n an elementary solution E belonging to $L^q_{\text{loc}}(R^n)$. Verify that the conclusion of Exercise 8.38 is valid for p satisfying $1 \leq p < \infty$, $p \leq q$.
[Hint: Majorize $f = E * \varphi$, having extended φ so as to be 0 outside Ω.]

Remarks. We have seen in Section 5.14 that the conditions of Exercise 8.38 are satisfied for *any* $P(D) \neq 0$ when $p = 2$, and this even for unbounded Ω.

8.40 Let E be a normed vector space, E' its dual, E'' its bidual; regard E as injected into E''. Let Σ and Σ'' be the closed unit balls in E and in E'', respectively. The Bipolar Theorem 8.1.5 implies that Σ is $\sigma(E'', E')$-dense in Σ''.

Show by example that despite this Σ'' will in general contain elements that are not $\sigma(E'', E')$-limits of (countable) sequences of elements of E.
[Hints: Consider $E = L^1(T, \mu)$ for suitably chosen T and μ. Alternatively consider $E = C_0(T)$, T being a suitably chosen separated locally compact space, using the Category Theorem 0.3.15 to show that in general there exist many bounded, universally measurable functions on T that are not limits of sequences of continuous functions on T.]

8.41 Let E be a reflexive Fréchet space. Prove that E' is fully complete for its strong topology, and that any strongly closed vector subspace of E' is B_r-complete when taken with the induced topology.

8.42 Let E be a reflexive Fréchet space, V is a vector subspace of E', and F a barrelled space. Suppose that u is a linear map of V onto F having a graph closed in $E' \times F$ for the product of the strong topology on E' and the initial topology on F. Show that u is open (the topology on V being that induced by the strong topology on E').

8.43 Let E be a reflexive Fréchet space, F a barrelled space, and v a linear map of F into E' with a graph closed in $F \times E'$ for the product of the initial topology on F and the strong topology on E'. Show that v is continuous.

NOTE: Compare the results of the last two exercises with those appearing in Grothendieck [9].

8.44 Let E and F be reflexive (or distinguished) Fréchet spaces and u a linear map of E' onto F'. Suppose that the graph of u is closed in $E' \times F'$ for the product of the strong topologies $\beta(E', E)$ and $\beta(F', F)$.

Prove that u is open for the strong topologies.
[Hint: Use Theorem 6.7.2 and Propositions 8.4.17 and 8.4.18.]

Note. This result, and that of the next exercise, is used by Malgrange [1] in his study of convolution equations.

8.45 Let E and F be reflexive Fréchet spaces, v a linear map of F' into E' having a graph that is sequentially closed in $F' \times E'$ for the product of the strong topologies Show that v is continuous for the strong topologies.

8.46 Let E be a Banach space, M a closed vector subspace of E. Show that if E is reflexive, so too is E/M.

Remark. There exist reflexive Fréchet spaces E possessing closed vector subspaces M such that E/M is not reflexive; see Köthe [5], Kap. V. This entails that there exist bounded subsets of E/M not obtainable as the image, under the canonical mapping u of E onto E/M, of any bounded subset of E—this despite the fact that u is a continuous, open map of E onto E/M. Regarding this last point, see also Bourbaki [5], pp. 123–124, Exercice 21), b).

8.47 Show that any separable Banach space E is isomorphic to a quotient space of ℓ^1.

Deduce that there exists a closed vector subspace M of ℓ^1 with the property that ℓ^1/M contains a sequence converging weakly to 0 that is not the canonical image of any sequence converging weakly to 0 in ℓ^1.

This example is due to Grothendieck [7], p. 129, Exercice 4.

[Hints: For the first part, choose a sequence (a_n) in E such that $\|a_n\| = 1$ and such that the vector subspace of E generated by the a_n is dense in E. Consider the mapping

$$\xi \to \sum_{n=1}^{\infty} \xi(n) a_n \qquad \text{of} \quad \ell^1 \text{ into } E.$$

For the second part apply what precedes to a separable Banach space E in which there exist sequences converging weakly to 0 but not converging to 0, and use Corollary 4.21.7.]

8.48 Let E and E' be vector spaces in duality and let $\mathfrak{S}$ be an increasing directed set of weakly bounded subsets of E whose union is E. Consider the topology $T_{\mathfrak{S}}$ on E' (see Section 8.3). Suppose that M_1 and M_2 are vector subspace of E'. Show that M_1 and M_2 have the same closure (for $T_{\mathfrak{S}}$) if and only if, for each $A \in \mathfrak{S}$, $\sigma(E, M_1)$ and $\sigma(E, M_2)$ induce the same topology on A. (Luxemburg [2].)

8.49 Suppose E, E', and $\mathfrak{S}$ are as in the preceding exercise. Suppose that M is a vector subspace of E'. Show that the closure of M (for $T_{\mathfrak{S}}$) contains precisely those $x' \in E'$ such that $x' \mid A$ is continuous for $\sigma(E, M) \mid A$ for each $A \in \mathfrak{S}$. (Luxemburg [2].)

8.50 Suppose that E is a barrelled space, F a vector space, u a linear map of E into F, F^* the algebraic dual of F, and F' a vector subspace of F^*. Suppose also that $\mathfrak{S}$ is a set of subsets of F^* such that each $A \in \mathfrak{S}$ is contained in the weakly closed, convex, balanced envelope in F^* of some weakly bounded subset of F'. (The adjective "weak" here refers to the topology $\sigma(F^*, F)$.)

Show that if u is continuous for $\sigma(F, F')$, then it is continuous for the topology $T_{\mathfrak{S}}$ on F.

8.51 Let T be a separated locally compact space, $\mathscr{M}(T)$ the space of all Radon measures on T. Let E be any barrelled space and u a linear map of E into $\mathscr{M}(T)$ that is continuous for the vague topology $\sigma(\mathscr{M}(T), \mathscr{K}(T))$ on $\mathscr{M}(T)$. Show that for each compact set $K \subset T$ one has $\lim_{x \to 0} \int_K d\,|u(x)| = 0$.

[Hint: Use the preceding exercise.]

CHAPTER 9

Theory of Compact Operators

9.0 Foreword

This chapter is devoted to a systematic exposition of some general results concerning compact linear maps of one LCTVS into another. The semiclassical Riesz theory will be found in later sections. Originally this applied to Banach spaces and this is still perhaps the most usual field of application. We shall see, however, that some of the results extend naturally and usefully to more general types of LCTVS.

9.1 Compact and Precompact Sets

We collect in this section a few elementary statements about compactness and precompactness for subsets of a LCTVS.

9.1.1 **Proposition.** Let (E, E') be a dual system, A a subset of E. Then A is weakly bounded if and only if it is weakly precompact.

Proof. The proof is immediate. ▮

9.1.2 **Proposition.** Let E be a LCTVS, A a subset of E. A is precompact if and only if (a) A is bounded and (b) the induced uniform structure is the same as the induced weak uniform structure.

Proof. The conditions are sufficient because A, being weakly bounded, is weakly precompact (Proposition 9.1.1).

The necessity of (a) is trivial.

In proving the necessity of (b), one may assume that E is separated and complete. This being done, precompactness of A signifies its relative compactness in E. At the same time, one may view E as a subset of the space $C(E')$ of continuous scalar-valued functions on E', the latter being endowed with its weak topology. Let U be any neighbourhood of 0 in E; U^0 is weakly compact. Since the initial uniformity on E is that of convergence uniform on the sets U^0, $A \mid U^0$ (the set of restrictions to U^0 of the functions on E' defined by elements of A) is relatively compact in the space $C_u(U^0)$. By the converse of Ascoli's theorem (page 34), $A \mid U^0$ is equicontinuous. So, by Proposition 0.4.9, the weak uniformity on $A \mid U^0$ coincides with that of convergence uniform on U^0. From this (b) is easily deducible. ▮

9.1.3 **Proposition.** Let E be a LCTVS, A a precompact subset of E. Then the convex, balanced envelope of A is precompact.

Proof. The proof is immediate from the definition of precompactness. ▌

9.1.4 **Corollary.** Let E be a LCTVS, A a precompact subset of E, B the closed, convex, and balanced envelope of A. In order that B be compact it is sufficient (and necessary, if E is separated) that it be complete. ▌

Remark. This is a weak form of Krein's theorem (Section 8.13) and is much more elementary to prove; see Remark 8.13.4(2).

9.1.5 **Proposition.** Let E be a LCTVS, A a complete subset of E. Then A is complete for any locally convex topology T that is stronger than the initial topology and that has a neighbourhood base at 0 formed of sets closed for the initial topology.

Proof. We denote by T_0 the initial topology on E. Let (x_i) be any T-Cauchy directed family extracted from A. Then (x_i) is a fortiori T_0-Cauchy. Since A is complete, there exists in A a point x to which (x_i) is T_0-convergent. It suffices to show that (x_i) is T-convergent to x. Let V be any T_0-closed T-neighbourhood of 0. By hypothesis, $x_i - x_j$ belongs to V for all sufficiently large i and j. Keeping i fixed and letting j increase, we see that $x_i - x$ belongs to V for all sufficiently large i, hence the result. ▌

9.1.6 **Corollary.** In a LCTVS, any weakly complete set is complete.

9.1.7 **Corollary.** Let E be a LCTVS, A a precompact subset of E, B the closed, convex, and balanced envelope of A. In order that B be compact it is necessary and sufficient that it be weakly compact.

Proof. We apply Corollaries 9.1.4 and 9.1.6. ▌

9.1.8 **Corollary.** In a LCTVS, each weakly compact set is complete.

9.1.9 **Proposition.** Let E be a LCTVS, A and B subsets of E. Then:

(1) If A is compact and B is closed, $A + B$ is closed.

(2) If A and B are compact and convex, the convex envelope C of $A \cup B$ is compact.

Proof. (1) If x does not belong to $A + B$, the compact set $x - A$ is disjoint from B. But then there exists a neighbourhood U of 0 such that $(x - A) + U$ and B are disjoint; that is, $x + U$ is disjoint from $A + B$. Thus $A + B$ is closed.

(2) C is the closure of the set D of sums $\lambda a + (1 - \lambda)b$, where a comes from A, b from B, and λ from the compact interval $I = [0, 1]$. The mapping $(\lambda, a, b) \to \lambda a + (1 - \lambda)b$ is continuous from $I \times A \times B$ into E, so that its image D is compact. In a TVS the closure of a compact set is compact, hence the result. (For suppose the G_i form an open covering of $\bar{D} = C$. For each x in C

there exists then an $i = i(x)$ and a closed neighbourhood U_x of 0 such that $G_{i(x)}$ contains $x + U_x$. The interiors of the $x + U_x$, x ranging over D, cover D; hence a finite number of them, say the $x_k + U_{x_k}$, cover D. Since these latter sets are closed, their union covers $\bar{D} = C$. The same is thus true of the corresponding $G_{i(x_k)}$. This shows that C is compact.) ▌

9.2 Definition and First Properties of Compact and Weakly Compact Linear Maps

Let E and F be TVSs, u a linear map of E into F. We say that u is *compact* (resp. *precompact*) if there exists a neighbourhood U of 0 in E such that $u(U)$ is relatively compact (resp. precompact) in F. Compactness implies precompactness; the two notions coincide if F is quasi-complete.

For LCTVSs E and F we introduce in addition the concept of *weak compactness* of a linear map u of E into F. We say u is weakly compact if there exists a neighbourhood U of 0 in E such that $u(U)$ is weakly relatively compact in F.

It is evident that each of the preceding three properties implies continuity.

NOTES. The study of weakly compact linear operators was initiated by Kakutani [4] and Yosida [1] in connection with ergodic theory. See Section 9.13.

Special instances of compact (or "completely continuous") linear operators arose much earlier. Hilbert [1] expressed the idea in terms of bilinear forms on ℓ^2, while the definition we have given is due to F. Riesz [2] for Banach spaces.

Hilbert's concept is expressible in equivalent form in terms of bounded endomorphisms u of ℓ^2, and is equivalent to the following property:

(H) u transforms weak Cauchy sequences into convergent sequences.

Consider this Hypothesis as applied to arbitrary continuous linear maps u of one LCTVS, E, into a second, F, the latter being assumed to be separated.

In the first place, if u transforms bounded sets into relatively compact sets, then u can easily be seen to satisfy (H) above. For if (x_n) is a weak Cauchy sequence in E, then $(y_n) = (u(x_n))$ is weak Cauchy in F whenever u is continuous. Moreover, $\{x_n\}$ is bounded and so $\{y_n\}$ is relatively compact. It follows that (y_n) has precisely one limiting point, hence is convergent.

On the other hand, if in E each bounded sequence contains a weak Cauchy subsequence, and if F is either metrizable or quasi-complete, then u transforms bounded sets into relatively compact sets whenever it satisfies condition (H). For suppose that A is a bounded subset of E. We wish to show that $u(A)$ is relatively compact in F. If F is metrizable, it suffices to show that each sequence (y_n) extracted from $u(A)$ admits a convergent subsequence. If F is quasi-complete, it suffices to show that $u(A)$ is precompact—that is, that any sequence (y_n) extracted from $u(A)$ admits a Cauchy subsequence. In either case, however, $y_n = u(x_n)$ with x_n in A. By hypothesis on E, (x_n) admits a weak Cauchy subsequence (x_{n_k}). Then (y_{n_k}) is convergent in F by virtue of condition (H) above.

From these remarks we infer that if E is a normed space with the stated properties, and if F is likewise restricted as above, then condition (H) is equivalent to compactness.

Our hypotheses on E are satisfied if *either* (1) the bounded subsets of E are weakly metrizable (for example, if E' is strongly separable), *or* (2) E is a reflexive Banach space (for example, a Hilbert space, or $L^p(T, \mu)$ with $1 < p < \infty$). Regarding case (1), see also Exercises 8.2–8.4.

In this section and the next we shall present two central and very general theorems, numbered 9.2.1 and 9.3.1, both due in their fully general form to Grothendieck [4]; see also Bourbaki [5], Chapitre IV, p. 61, Exercice 12). Each will contain as special cases a number of results proved earlier and by other writers under more restrictive conditions, and each will be used repeatedly in Sections 9.3 and 9.4. Here is the first of these two general results.

9.2.1 **Theorem.** Let us suppose that (E, E') and (F, F') are separated dual systems, $\mathfrak{S}$ a set of weakly bounded subsets of E, $\mathfrak{S}'$ a set of weakly bounded subsets of F', u a weakly continuous linear map of E into F, u' its adjoint. The following six conditions are equivalent:

(1) For each A in $\mathfrak{S}$, $u(A)$ is precompact for $T_{\mathfrak{S}'}$.

(1′) For each A' in $\mathfrak{S}'$, $u'(A')$ is precompact for $T_{\mathfrak{S}}$.

(2) For each A in $\mathfrak{S}$, $u \mid A$ is uniformly continuous for the weak topology on E and $T_{\mathfrak{S}'}$ on F.

(2′) For each A' in $\mathfrak{S}'$, $u' \mid A'$ is uniformly continuous for the weak topology on F' and $T_{\mathfrak{S}}$ on E'.

(3) For each A in $\mathfrak{S}$ and A' in $\mathfrak{S}'$, the restriction to $A \times A'$ of $\langle u(x), y' \rangle = \langle x, u'(y') \rangle$ is uniformly continuous for the product of the weak topologies.

(3_1) As (3), but with the product of the weak topology on A with the strong topology on A' (or vice versa).

If each A in $\mathfrak{S}$ (resp. A' in $\mathfrak{S}'$) is convex and balanced, one can replace in (2) (resp. in (2′)) uniform continuity by continuity at 0.

The preceding conditions imply that

$$\operatorname*{Sup}_{x \in A, y' \in A'} |\langle u(x), y' \rangle| < +\infty$$

for each A in $\mathfrak{S}$ and each A' in $\mathfrak{S}'$.

Finally, if each A in $\mathfrak{S}$ is weakly compact, each of the preceding conditions implies

(4) If A is in $\mathfrak{S}$ and (x_n) is a sequence extracted from A which converges weakly in E, then $(u(x_n))$ is $T_{\mathfrak{S}'}$-convergent in F.

If, further, E is metrizable, then (4) is equivalent to the preceding conditions.

Proof. This proof is lengthy and it is best to describe the plan of action. First we shall prove the implications

$$\text{(a) } (1) \Rightarrow (2'); \qquad \text{(b) } (2') \Rightarrow (1').$$

Then, on account of symmetry, we shall also have $(1') \Rightarrow (2)$ and $(2) \Rightarrow (1)$.

Thus the first four conditions will be equivalent. Next we establish

(c) (2) $\Rightarrow$ (3); (d) (3_1) $\Rightarrow$ (2).

That (3) $\Rightarrow$ (3_1) is trivial. This will yield the equivalence of the first six conditions.

Replacement of uniform continuity by continuity at 0 (when the A, or the A', are convex and balanced) is justified by a simple remark: if G is a vector space, H a convex and balanced subset of G, two locally convex topologies on G induce on H the same uniform structure if and only if they induce the same relative neighbourhoods of 0; compare Exercise 9.2.

It is trivial that (3) implies the boundedness of $\langle u(x), y' \rangle$ for x in A, y' in A', uniform continuity preserving precompactness.

The final steps are

(e) (1) $\Rightarrow$ (4); (f) (4) $\Rightarrow$ (1),

which are established under the stated additional hypotheses.

We begin the proofs of the various implications (a) through (f).

(a) Each set A' is an equicontinuous set of functions on F, the latter being equipped with $T_{\mathfrak{S}'}$. Thus on each A' the weak uniform structure is identical with that of convergence uniform on the subsets of F that are precompact for $T_{\mathfrak{S}'}$. It follows from Proposition 9.1.2 that each A' is precompact for this latter structure. On the other hand, (1) signifies precisely that u' is continuous, hence uniformly continuous for the topology on F' of "precompact convergence" and $T_{\mathfrak{S}}$ on E'. Since the image, by a uniformly continuous map, of a precompact set is precompact, it follows that (1) implies (2').

(b) This follows at once from uniform continuity of the type stated, using Proposition 9.1.2 again.

(c) One has the identity

$$\langle u(x_1), y_1' \rangle - \langle u(x_2), y_2' \rangle = \langle u(x_1 - x_2), y_1' \rangle + \langle u(x_2), y_1' - y_2' \rangle \qquad (9.2.1)$$

whence it is plain that (2) implies (3).

(d) If (3_1) holds, given $\varepsilon > 0$, A, and A', there exists a weak neighbourhood W of 0 in E such that x_1, x_2 in A, $x_1 - x_2$ in W and y' in A' entail

$$|\langle u(x_1), y' \rangle - \langle u(x_2), y' \rangle| \leq \varepsilon;$$

that is, $u(x_1) - u(x_2) \in \varepsilon A'^0$. This means that (2) holds.

(e) Suppose that each A is weakly compact. If (1) holds, $u(A)$ is precompact for $T_{\mathfrak{S}'}$. At the same time, since (2) holds, $(u(x_n))$ is $T_{\mathfrak{S}'}$-Cauchy whenever $(x_n) \subset A$ is weak Cauchy. If therefore (x_n) is weakly convergent, the set $\{x_n\}$ is weakly relatively compact and so, u being weakly continuous, $\{u(x_n)\}$ is weakly relatively compact. The sequence $(u(x_n))$ has therefore a weak limiting point in F. Being $T_{\mathfrak{S}'}$-Cauchy, this sequence is $T_{\mathfrak{S}'}$-convergent in F. Thus (1) implies (4).

(f) Suppose now that each A is weakly compact, that E is metrizable, and that (4) holds: we aim to show that (1) is satisfied. To do this it suffices to show that if $(x_n) \subset A$, one may extract a subsequence (x_{n_k}) such that $(u(x_{n_k}))$ is $T_{\mathfrak{S}'}$-convergent in F. However, by Šmulian's Theorem 8.12.1, from (x_n) we can extract a subsequence (x_{n_k}) that is weakly convergent in E, and it suffices to apply (4) to this subsequence.

This completes the proof of Theorem 9.2.1. ▌

9.2.2 **Corollary.** Let E be a metrizable LCTVS, $\mathfrak{S}$ a set of weakly compact subsets of E, A' a weakly bounded subset of E'. Then:

(1) A' is precompact for $T_{\mathfrak{S}}$ if and only if every weakly convergent sequence extracted from A converges uniformly on A'.

(2) A' is precompact for $\tau(E', E)$ if and only if each weakly convergent sequence in E converges uniformly on A'.

(3) If E is a Fréchet space, A' is relatively compact for $\tau(E', E)$ if and only if each weakly convergent sequence in E converges uniformly on A'.

Proof. (1) In Theorem 9.2.1 we take $E = F$, u = identity, and $\mathfrak{S}' = \{A'\}$, and use the equivalence of (1′) and (4).

(2) In (1) we take $\mathfrak{S}$ to consist precisely of all weakly compact and convex subsets of E, so that $T_{\mathfrak{S}} = \tau(E', E)$.

(3) In view of (2), it suffices merely to show that A', if precompact for τ, is relatively compact for τ. But since E is a Fréchet space, the weakly bounded set A' is equicontinuous (Corollary 7.1.2), hence it follows easily that the τ-closure of A' is τ-complete. Thus A' is relatively compact for τ. ▌

For normed spaces we can deduce two important results from Theorem 9.2.1.

9.2.3 **Corollary.** If E and F are normed spaces, a linear map u of E into F is precompact if and only if its adjoint u' is precompact (for the strong, that is, normed, topologies on F' and E').

Proof. This follows at once from the equivalence of (1) and (1′) of Theorem 9.2.1, if we take for $\mathfrak{S}$ the set of all bounded subsets of E and for $\mathfrak{S}'$ the set of all strongly (that is, norm-) bounded subsets of F'. ▌

Remark. Since E' is always complete, u' is precompact if and only if it is compact. For Banach spaces E and F the result is due to Schauder [4]; see also Kakutani [2]. See also Corollary 9.6.3.

9.2.4 **Corollary.** If E and F are normed spaces, a linear map u of E into F is precompact if and only if u' transforms strongly (that is, norm-) bounded and $\sigma(F', F)$-convergent directed families (y_i') in F' into strongly convergent (that is, norm-convergent) directed families $(u'(y_i'))$ in E'.

Proof. Proof is immediate from the equivalence of (1) and (2′) of Theorem 9.2.1, bearing in mind that the strong topology on E' is definable as $T_{\mathfrak{S}}$, taking for $\mathfrak{S}$ all the *convex* bounded subsets of E. ▌

Remarks. If F is separable, it is enough to demand that u' transform strongly bounded and weakly convergent *sequences* in F' into strongly convergent sequences in E'. (The reader will recall that separability of F ensures that $\sigma(F', F)$ induces a metrizable topology on each strongly bounded subset of F'.) For separable Banach spaces the result is due to Gelfand [2].

We end this section with three results about sets of compact or precompact linear maps from one TVS into another.

9.2.5 **Theorem.** Let E, F, G be separated TVSs.

(a) If $u \in L_c(E, F)$ is compact (resp. precompact), and if $v \in L_c(F, G)$, then $vu \in L_c(E, G)$ is compact (resp. precompact).

(b) If $u \in L_c(E, F)$ and $v \in L_c(F, G)$ is compact (resp. precompact), then $vu \in L_c(E, G)$ is compact (resp. precompact).

Proof. (a) Let U be a neighbourhood of 0 in E such that $u(U)$ is relatively compact (resp. precompact) in F. Then $vu(U) = v(u(U))$ has the same property qua subspace of G, thanks to the fact that v is continuous (resp. uniformly continuous).

(b) Let V be a neighbourhood of 0 in F such that $v(V)$ is relatively compact (resp. precompact) in G. Since u is continuous, there exists a neighbourhood U of 0 in E such that $u(U) \subset V$. Then $vu(U)$, being a subset of $v(V)$, is relatively compact (resp. precompact) in G. ▌

9.2.6 **Theorem.** Let E be a seminormed space, (u_i) a directed family in $L_c(E, F)$ that converges strongly to $u \in L_c(E, F)$. If each u_i is precompact, so is u.

Proof. Let B be the unit ball in E, V any neighbourhood of 0 in F. By strong convergence, we may choose i so that $u(x) - u_i(x) \in V$ for x in B. Since $u_i(B)$ is precompact and $u(B) \subset u_i(B) + V$, it follows that $u(B)$ is precompact. ▌

Remarks. (1) When, as in Theorem 9.2.6, we refer to strong convergence in $L_c(E, F)$, we mean convergence in the sense of the strong topology on $L_c(E, F)$—that is, convergence uniform on the bounded subsets of E (see Subsection 1.10.8); the reader should compare the definition of the strong topology on a dual space in Section 8.4.

Curiously enough, when E and F are normed spaces, there is a prevalent terminology that speaks of this strong topology on $L_c(E, F)$ as the "uniform operator topology"—despite the fact that uniformity refers only to *bounded* subsets of E (not to E itself). Moreover, the so-called "strong operator topology" is that of pointwise convergence. Finally, there is the "weak operator topology," in which convergence means pointwise convergence when F is endowed with the weakened topology $\sigma(F, F')$.

(2) The last two theorems have analogues in which weak compactness replaces compactness (or precompactness). These are delayed until after our second general theorem, appearing as Theorems 9.3.6 and 9.3.7, respectively. Corollary 9.2.3 has its analogue in Corollary 9.3.3.

9.3 A Theorem about Weakly Compact Linear Maps

This section is devoted to the statement and proof of a second general theorem asserting the equivalence of various conditions. It differs from Theorem 9.2.1 in that precompactness is replaced by weak compactness.

9.3.1 Theorem. Let E and F be separated LCTVSs, $\mathfrak{S}$ a set of bounded subsets of E, H the vector subspace of E'' generated by the weak closures in E'' of sets A in $\mathfrak{S}$, u a continuous linear map of E into F. Assume that $E \subset H$. Let us consider the following conditions:

(1) For each A in $\mathfrak{S}$, $u(A)$ is weakly relatively compact in F.

(2) $u''(H) \subset F$.

(2a) u'' transforms equicontinuous subsets of H (qua subset of the dual of $(E', T_{\mathfrak{S}})$) into weakly relatively compact subsets of F.

(3a) u' is continuous for $\tau(F', F)$ and $T_{\mathfrak{S}}$.

(3b) u' is continuous for $\sigma(F', F)$ and $\sigma(E', H)$.

(4) u' transforms weakly compact subsets of F' into $\sigma(E', H)$-compact subsets of E'.

(4a) u' transforms equicontinuous and weakly closed subsets of F' into $\sigma(E', H)$-compact subsets of E'.

The conclusions are:

(i) Conditions (1), (2), (2a), (3a), (3b) are equivalent and imply (4); and (4) implies (4a).

(ii) If F is quasi-complete, all the preceding conditions are equivalent.

Proof. Since the proof is lengthy we again sketch the plan of attack. We begin by establishing the implications:

(a) (1) $\Rightarrow$ (2a); (b) (2a) $\Rightarrow$ (2); (c) (2) $\Rightarrow$ (1).

This will show that (1), (2), and (2a) are equivalent. Next we establish:

(d) (3a) $\Rightarrow$ (2a) and (3b) $\Rightarrow$ (2);

(e) (2a) $\Rightarrow$ (3a); (f) (2) $\Rightarrow$ (3b).

So (1), (2), (2a), (3a), (3b) are equivalent. The final steps will be:

(g) (3b) $\Rightarrow$ (4) $\Rightarrow$ (4a);

(h) (4a) + F quasi-complete $\Rightarrow$ (2).

(a) Suppose (1) is true. Let A belong to $\mathfrak{S}$; $u(A)$ is weakly relatively compact in F. Let $\bar{A}$ be the $\sigma(H, E')$-closure of A. Since u'' is continuous for $\sigma(E'', E')$ and $\sigma(F'', F')$, $u''(\bar{A})$ is contained in the $\sigma(F'', F')$-closure of $u(A)$. Hence, $u(A)$ being weakly relatively compact in F, it follows that $u''(\bar{A})$ is contained in the $\sigma(F, F')$-closure of $u(A)$, and so is weakly relatively compact in F. Now the $\bar{A}$ form a base for the equicontinuous subsets of H, so that (2a) follows.

(b) The arguments are very similar to those just given.

(c) Let us suppose that (2) holds. The equicontinuous subsets of H are relatively compact for $\sigma(H, E')$. If A is in $\mathfrak{S}$, $A \subset A^{00} \cap H$; and A^{00} is equicontinuous and compact for $\sigma(E'', E')$. As we have noted already, u'' is continuous for $\sigma(E'', E')$ and $\sigma(F'', F')$. Thus (2) entails that $u(A) = u''(A)$ is weakly relatively compact in F, which is (1).

(d) Suppose (3a) is valid. If A is in $\mathfrak{S}$, there exists a $\tau(F', F)$-neighbourhood V of 0 such that $u'(V) \subset A^0$. The relation

$$\langle x'', u'(y') \rangle = \langle u''(x''), y' \rangle \tag{9.3.1}$$

shows then that $|\langle u''(x''), y' \rangle| \leq 1$ for x'' in A^{00} and y' in V. So we see that $u''(x'')$ belongs to F and that $u''(A^{00})$ is weakly relatively compact in F. (The reader will recall the definition of $\tau(F', F)$ and the fact that it is compatible with the duality between F and F'.) Since the A^{00} form a base for the equicontinuous sets in E'' (relative to $T_{\mathfrak{S}}$ on E'), (2a) is established.

That (3b) $\Rightarrow$ (2) follows by a similar argument.

(e) Let us suppose (2a) holds. If A is in $\mathfrak{S}$, A^0 is a $T_{\mathfrak{S}}$-neighbourhood of 0 and so $A \subset A^{00}$ is equicontinuous. Thus $u''(A)$ is weakly relatively compact in F. The relation $\langle x, u'(y') \rangle = \langle u''(x), y' \rangle$ then shows that $u'(y') \to 0$ for $T_{\mathfrak{S}}$ whenever $y' \to 0$ for $\tau(F', F)$, so that (3a) holds.

(f) Suppose (2) holds. Equation (9.3.1) shows at once that u' is continuous for $\sigma(F', F)$ and $\sigma(E', H)$.

(g) It is clear that (3b) $\Rightarrow$ (4) and that (4) $\Rightarrow$ (4a).

(h) Let us suppose (4a) holds. Let $B \subset F'$ be equicontinuous and weakly closed in F'. Then $u'(B)$ is $\sigma(E', H)$-compact. So, on $u'(B)$, $\sigma(E', H)$ and $\sigma(E', E)$ induce the same topology. Hence $u' \mid B$ is continuous for $\sigma(F', F)$ and $\sigma(E', H)$. Therefore, if x'' belongs to H, $u''(x'') = x'' \circ u'$ is a linear form on F' whose restrictions to the equicontinuous sets are weakly continuous. Thus there exists a weak neighbourhood W of 0 in F' such that, for $y' \in W \cap B$,

$$|\langle u''(x''), y' \rangle| = |\langle x'', u'(y') \rangle| \leq 1.$$

Remark (2) to Theorem 8.5.1 now shows that $u''(x'')$ belongs to the completion $\hat{F}$ of F. Besides this, since u'' is continuous for $\sigma(E'', E')$ and $\sigma(F'', F')$, $u''(x'')$ belongs to the weak closure in $\hat{F}$ of some set $u(A)$ with A in $\mathfrak{S}$. Now $u(A)$ is bounded in F, and we may assume that A, and therefore $u(A)$ as well, is convex and balanced. Then the weak closure in $\hat{F}$ of $u(A)$ is the same as its closure for the natural topology of $\hat{F}$ (qua completion of F). If F is quasi-complete, this closure of $u(A)$ in $\hat{F}$ is identical with its closure in F, and so $u''(x'')$ belongs to F. Thus (2) holds.

The proof of Theorem 9.3.1 is thus complete. ∎

9.3.2 **Corollary.** Let E and F be two separated LCTVSs, u a continuous linear map of E into F. The following two conditions are equivalent:

(1) u transforms bounded subsets of E into weakly relatively compact subsets of F.

(2) $u''(E'') \subset F$.

These imply

(3) u' transforms equicontinuous subsets of F' into subsets of E' that are relatively compact for $\sigma(E', E'')$.

If F is quasi-complete, all three conditions are equivalent.

Proof. The proof is immediate from Theorem 9.3.1, taking for $\mathfrak{S}$ the set of all bounded subsets of E. ▌

9.3.3 **Corollary.** Let E be a normed vector space, F a Banach space, u a continuous linear map of E into F. Then u is weakly compact if and only if u' is weakly compact (qua map of the Banach space F' into the Banach space E').

Proof. u is weakly compact if and only if (1) of the last corollary holds. On the other hand, condition (3) of the same corollary signifies in this case precisely that u' is weakly compact from the Banach space F' into the Banach space E'. ▌

Remarks. The equivalence of conditions (1) and (2) of Corollary 9.3.2 is due, for separable Banach spaces, to Gantmacher [1]; the nonseparable case is due to Nakamura [1]. Similar remarks apply to Corollary 9.3.3. When E is a normed space, condition (1) says exactly that u is weakly compact; so in this case we conclude that u is weakly compact if and only if $u''(E'') \subset F$.

9.3.4 **Theorem.** Let E be a separated LCTVS, $\mathfrak{S}$ a set of bounded subsets of E. The following conditions are equivalent:

(1) If F is separated LCTVS and u a continuous linear map of E into F that transforms bounded sets into weakly relatively compact sets, then u transforms sets belonging to $\mathfrak{S}$ into precompact subsets of F.

(1′) As (1), but F is assumed to be a Banach space.

(2) Each A in $\mathfrak{S}$ is precompact for the topology T of convergence uniform on the equicontinuous, convex, balanced, and $\sigma(E', E'')$-compact subsets of E'.

(2a) Each equicontinuous, convex, balanced, and $\sigma(E', E'')$-compact subset of E' is precompact for $T_{\mathfrak{S}}$.

Proof. (2) and (2a) are equivalent by virtue of Theorem 9.2.1, taking therein $E = F$ and u the identity map. Evidently, (1) $\Rightarrow$ (1′).

We next show that (1′) $\Rightarrow$ (2). Let A belong to $\mathfrak{S}$. Let E_T be the space E equipped with the topology T. Note that T is compatible with the duality between E and E' (Mackey's Theorem 8.3.2). By Exercise 6.17, E_T is isomorphic with a vector subspace of a product of complete seminormed spaces. It is therefore enough to show that any continuous linear map u of E_T into a complete seminormed space, F, transforms A into a precompact set; and in doing this we may assume that F is separated—that is, is a Banach space. Now, since it is evident that T is weaker than the initial topology of E, u will be continuous from E into F; and in addition u' transforms weakly closed and equicontinuous subsets of F' into weakly closed and equicontinuous subsets of $(E_T)'$—that is, into subsets of E' that are compact for $\sigma(E', E'')$. But then (Theorem 9.3.1) u transforms bounded subsets of E into weakly relatively compact subsets of F. Consequently, assuming (1′), $u(A)$ is precompact in F.

It remains to show that (2) implies (1). In (1) we may assume that F is complete. Let u be as in (1), and let A belong to $\mathfrak{S}$. We have to show that $u(A)$ is precompact. By hypothesis, A is precompact for T; so it suffices to show that u is continuous from E_T into F—that is, that u' transforms equicontinuous, convex, balanced, and weakly compact subsets of F' into equicontinuous and $\sigma(E', E'')$-compact subsets of E'. But since u is continuous from E into F, u' transforms equicontinuous sets into equicontinuous sets; and, by Theorem 9.3.1, u' furthermore transforms weakly compact subsets of F' into $\sigma(E', E'')$-compact subsets of E'. This completes the proof. ∎

Remark. If each set $A \in \mathfrak{S}$ is weakly compact, we may in (1), (1'), and (2) replace the term "precompact" by "compact." For a set that is precompact and weakly compact is precompact and weakly complete, hence precompact and complete, hence compact.

9.3.5 **Corollary.** Let E be a separated LCTVS. The following conditions are equivalent:

(1) If F is a separated LCTVS and u a continuous linear map of E into F that transforms bounded sets into weakly relatively compact sets, then u transforms weak Cauchy sequences into Cauchy sequences.

(1') As (1), but F is assumed to be a Banach space.

(2) Each weak Cauchy sequence in E is T-Cauchy.

Proof. We apply Theorem 9.3.4, taking $\mathfrak{S}$ to consist of all sets $A = \{x_n\}$, where (x_n) is a weak Cauchy sequence in E. It is necessary merely to note this: if G is a separated LCTVS, (z_n) a weak Cauchy sequence in G, P the set of points z_n, then P is precompact if and only if (z_n) is Cauchy. Indeed, if (z_n) is Cauchy, P is obviously precompact. On the other hand, if P is precompact, its closure $\bar{P}$ in the completion $\hat{G}$ of G is compact. On $\bar{P}$ the weak uniform structure coincides with that induced by the natural structure of $\hat{G}$. Hence (z_n) is Cauchy for the initial structure of G. ∎

9.3.6 **Theorem.** Let E, F, and G be LCTVSs, E being normable, and let $u \in L_c(E, F)$ and $v \in L_c(F, G)$. Then $vu \in L_c(E, G)$ is weakly compact provided *either* (a) u is weakly compact *or* (b) v is weakly compact.

Proof. We have

$$(vu)''(E'') = v''(u''(E'')).$$

If (a) is true, Corollary 9.3.2 shows that $u''(E'') \subset F$. Since in any case $v'' \mid F = v$, it appears that $(vu)''(E'') \subset G$. Then the Remark following Corollary 9.3.3 shows that vu is weakly compact. In case (b) the above equality shows that in any case $(vu)''(E'') \subset v''(F'')$, which is contained in G by virtue of Corollary 9.3.2. So, by the same token as before, vu is weakly compact. ∎

9.3.7 **Theorem.** Let E be a normed space, F a separated and quasi-complete LCTVS. Suppose that a directed family $(u_i) \subset L_c(E, F)$ converges in the strong topology of $L_c(E, F)$ to u. If each u_i is weakly compact, so too is u.

Proof. We know (Corollary 9.3.2) that $u_i''(E'') \subset F$ for each i, and we have to show that $u''(E'') \subset F$ also. Given any convex and balanced neighbourhood V of 0 in F, we can choose i so that $u(x) - u_i(x) \in V$ whenever $\|x\| \leq 1$.

Suppose that V is weakly closed. Since u'' is obtained from u by extension by weak continuity, it follows that $u''(x'') - u_i''(x'') \in \bar{V}$ for $\|x''\| \leq 1$, $\bar{V}$ denoting the closure of V in F''. Now the strong convergence of (u_i) shows that, for some i_0 and some bounded closed subset H of F, one has $u_i(B) \subset H$ for all $i \geq i_0$, B denoting the unit ball in E. Moreover, if i and j are sufficiently large, we have for $\|x''\| \leq 1$

$$u_i''(x'') - u_j''(x'') \in 2\bar{V} \cap F;$$

this shows that the family $(u_i''(x''))$ $(i \geq i_0)$ is Cauchy in H. F being quasi-complete, $\lim u_i''(x'')$ exists in H, a fortiori in F. Since F, and therefore F'', is separated, this limit can be none other than $u''(x'')$. Hence $u''(E'') \subset F$, as we had to show. ▌

Some illustrations and supplements. The Banach spaces c_0 and ℓ^1 are separable, and it follows that:

(a) The unit ball in c_0 is weakly metrizable

(b) The unit ball in ℓ^1 is metrizable for $\sigma(\ell^1, c_0)$.

In addition one derives easily from Theorem 4.21.2 the conclusion:

(c) For any index set I, any weakly compact set in $\ell^1(I)$ is compact.

Direct Proof. Since it suffices (Theorem 8.12.7) to show that from any weakly relatively compact sequence (x_n) in $\ell^1(I)$ one may extract a convergent subsequence, one may assume that I is countable and so deal directly with ℓ^1. We then apply (b), extracting from (x_n) a subsequence (y_n) that converges for $\sigma(\ell^1, c_0)$. It is then easily shown that (y_n) converges for $\sigma(\ell^1, \ell^\infty)$ also, using the weak relative compactness of (x_n). Using these observations we proceed to establish the next proposition.

9.3.8 **Proposition.** Let I be any set of indices, E any separated LCTVS, u a continuous linear map of $c_0(I)$ into E. If u is weakly compact, it is compact.

Remark. See also Proposition 9.4.13.

Proof. By Corollary 9.3.2, to say that u is weakly compact signifies that u' transforms equicontinuous sets in E' into weakly relatively compact subsets of $\ell^1(I)$. But then by (c), u' transforms equicontinuous subsets of E' into relatively compact subsets of $\ell^1(I)$. Let V be any neighbourhood of 0 in E, (x_n) any sequence in $c_0(I)$ with $\|x_n\| \leq 1$. We know that $(u(x_n))$ has a weak limiting point y in E. So, if W is any weak neighbourhood of 0 in E and n any integer, there exists an integer $N > n$ such that $u(x_N) - y$ belongs to W. We consider the set of ordered pairs $i = (n, W)$, partially ordered in the obvious way, and set $a_i = x_N$. The directed family $(u(a_i))$ is then weakly convergent to y, hence certainly weakly Cauchy in E. Now

$$\langle a_i - a_j, u'(y') \rangle = \langle u(a_i - a_j), y' \rangle,$$

so that (a_i) is Cauchy for the structure of pointwise convergence on $u'(V^0)$. Since $u'(V^0)$ is relatively compact and $\|a_i\| \leq 1$, Proposition 0.4.9 shows that (a_i) is Cauchy for the structure of convergence *uniform* on u (V^0). This means that $(u(a_i))$ is Cauchy in E. Since $u(a_i) \to y$ weakly in E, it follows that $u(a_i) \to y$ in E, which implies that y is a limit point in E of $(u(x_n))$. Thus u is compact. ▌

9.3.9 **Proposition.** Let I be any set of indices, E any separated LCTVS, u a linear map of $\ell^\infty(I)$ into E that is continuous for $\sigma(\ell^\infty, \ell^1)$ and $\sigma(E, E')$. Then u transforms the closed unit ball of $\ell^\infty(I)$ into a compact subset of E.

Proof. Let S be the closed unit ball in ℓ^∞. S is compact for $\sigma(\ell^\infty, \ell^1)$. Therefore $A = u(S)$ is weakly compact in E. On the other hand, if S_0 is the unit ball in c_0, then S_0 is weakly dense in S, so that $u(S_0)$ is weakly dense in A. Since $u(S_0)$ is convex, it follows that $u(S_0)$ is dense in A. Applying Proposition 9.3.8 to $u \mid c_0$, which is weakly compact, we see that $u(S_0)$ is relatively compact in E. Hence A is compact in E, as we had to show. ▌

Remarks. (1) As the above argument shows, $u_0 = u \mid c_0(I)$ transforms the unit ball $S_0 \subset c_0(I)$ into a relatively compact subset of E. Applying Theorem 9.2.1, we conclude that u_0' transforms equicontinuous subsets of E' into precompact (that is, in this case, relatively compact) subsets of $\ell^1(I)$.

(2) Examples of such maps u arise by taking a family $(a_i)_{i\in I}$ of elements of E having the assumed property that $\sum_i s(i)a_i$ is weakly convergent in E for each s in $\ell^\infty(I)$. (It may be added parenthetically that if we take a *sequence* (a_n), and if we assume that E is weakly sequentially complete, it suffices already to assume that $\sum_n s(n)a_n$ is weakly convergent in E for each s in c_0. For semireflexive spaces E, this same milder condition will suffice for arbitrary families (a_i).) This assumption implies that $\sum |\langle a_i, x'\rangle| < +\infty$ for each x' in E'; it then follows that the map u defined by

$$u(s) = \sum s(i)a_i$$

is continuous for $\sigma(\ell^\infty, \ell^1)$ and $\sigma(E, E')$. Proposition 9.3.9 and Remark (1) apply to u. Since u_0' maps x' in E' onto the family $(\langle a_i, x'\rangle)$, the conclusion of Remark (1) entails (via Theorem 4.21.2, for example) that for any neighbourhood U of 0 in E and any $\varepsilon > 0$, there exists a finite set $F \subset I$ such that

$$\sum_{i\in I\setminus F} |\langle a_i, x'\rangle| \le \varepsilon$$

uniformly for x' ranging over U^0. Assuming that U is convex, balanced, and closed, this signifies that

$$\sum_{i\in F'\setminus F} s(i)a_i \in \varepsilon \cdot U$$

for any finite F', $F \subset F' \subset I$, and any s in S_0. The finite partial sums of $\sum s(i)a_i$ therefore form a bounded, Cauchy directed family in E. If E is quasi-complete, the series $\sum s(i)a_i$ is therefore unconditionally convergent in E for each s in $\ell^\infty(I)$ (see also Proposition 9.3.10).

(3) As an application of Proposition 9.3.9 to summability theory we give a proof of a theorem due to Schur. This concerns matrices A that transform

convergent-to-zero sequences into sequences of the same nature; see Examples 6.1.4, 6.4.10, and 7.2.4. Schur's result is this:

▶ In order that a matrix A shall transform each convergent-to-zero sequence into a sequence of the same type, it is necessary and sufficient that

$$\lim_m \sum_n |A(m, n)| = 0. \tag{9.3.1}$$

Indeed, the sufficiency of this condition is evident since it obviously implies (7.2.10) and (7.2.11) of Example 7.2.4, (1) and (2). Assume therefore that A has the said property and let $u = u_A$ be the corresponding linear map. As we know from Example 7.2.4, u is extendible into a continuous linear map of ℓ^∞ into itself and (by part (3) of the cited example) u is continuous for $\sigma(\ell^\infty, \ell^1)$. By Proposition 9.3.9 and Remark (1) immediately above, u transforms the unit ball of c_0 into a relatively compact subset P of c_0 (since we know that u maps c_0 into the subspace c_0 of ℓ^∞). Now we can regard P as a subset of the space of continuous functions on the compact space N^* formed of the set N of natural numbers with ∞ adjoined as Alexandroff point at infinity. As such, the relative compactness of P implies equicontinuity at ∞. Thus, given $\varepsilon > 0$, there must exist a natural number m_0 such that $m > m_0$ entails $|y(m)| \leq \varepsilon$ for all y in P—that is,

$$\left| \sum_n A(m, n)x(n) \right| \leq \varepsilon$$

for $m > m_0$ and x in c_0 satisfying $\|x\|_\infty \leq 1$. Keeping $m > m_0$ fixed and varying x, this is seen to be equivalent to

$$\sum_n |A(m, n)| \leq \varepsilon \quad (m > m_0).$$

Thus equation (9.3.1) is established. ▮

Applications to scalarwise well-integrable functions. We shall now examine a sort of continuous analogue of Proposition 9.3.9, following Grothendieck [4] (especially his Corollaire to Proposition 3 and his Théorème 5) and also Bourbaki ([6], Chapitre 6, pp. 84–85, Exercices 19) and 21)).

9.3.10 **Proposition.** Let F be a separated LCTVS, u a linear map of $L^\infty = L^\infty(T, \mu)$ into F that is continuous for $\sigma(L^\infty, L^1)$ and $\sigma(F, F')$. Then:

(a) u is weakly compact from the Banach space L^∞ into F.

(b) u' transforms equicontinuous subsets of F' into weakly relatively compact subsets of L^1 (that is, relatively compact for $\sigma(L^1, L^\infty)$).

(c) If a bounded sequence (f_n) in $\mathscr{L}^\infty$ converges locally in measure to f (necessarily in $\mathscr{L}^\infty$), then $u(f_n)$ converges in F to $u(f)$.

Proof. The unit ball in L^∞ is compact for $\sigma(L^\infty, L^1)$, hence (a). Then (b) follows from Corollary 9.3.2. As for (c), we examine the formula

$$\langle u(f_n), x' \rangle = \langle f_n, u'(x') \rangle$$

and apply (b) in conjunction with the Dunford-Pettis criterion (Theorem 4.21.2) to deduce that $u(\dot{f}_n) \to u(\dot{f})$ in the sense of F (that is, in the sense of convergence uniform on equicontinuous subsets of F'). The details of this step may be left to the reader. ▮

An interesting category of maps u of the type figuring in the last proposition arises from functions Φ mapping T into F that are scalarwise well integrable in Bourbaki's terminology (loc. cit.); see also Remark (2) following Proposition 8.14.13. We recall that Φ is said to be scalarwise well integrable (SWI for short) if it is SEI and if further $\bar{\int} f\Phi \, d\mu$ belongs to F for each f in $\mathscr{L}^\infty$ (or, what is equivalent, for each bounded and μ-measurable function f). We shall then denote by u_Φ the map of L^∞ into F defined by

$$u_\Phi(\dot{f}) = \bar{\int} f\Phi \, d\mu,$$

f being any function belonging to the class $\dot{f}$. At the same time we introduce v_Φ whenever Φ is SEI, v_Φ being the linear map of F' into L^1 carrying $x' \in F'$ into the class of the function $\langle \Phi, x' \rangle$.

It is very easy to check that Φ is SWI if and only if it is SEI and v_Φ is continuous for $\sigma(F', F)$ and $\sigma(L^1, L^\infty)$. Also, whenever Φ is SWI we have the identity

$$\langle u_\Phi(\dot{f}), x' \rangle = \langle \dot{f}, v_\Phi(x') \rangle$$

for $\dot{f}$ in L^∞ and x' in F'. In particular, u_Φ is continuous for $\sigma(L^\infty, L^1)$ and $\sigma(F, F')$ and its adjoint is v_Φ.

Proposition 9.3.10 therefore applies to u_Φ and constitutes an extension of a theorem of Orlicz. However, we can prove more.

9.3.11 **Proposition.** Let us suppose that Φ is SWI from T into F. Let π denote the set of subsets H of F' with the property that from each sequence extracted from H one may extract a weak Cauchy sequence, and let F_π be the space F endowed with the topology T_π (of convergence uniform on the sets belonging to π). Then u_Φ is precompact from L^∞ into F_π.

Proof. By Theorem 9.2.1 we have to show that $v_\Phi(H)$ is precompact in L^1 for each H in π. For this we need only show that, if (x'_n) is a sequence extracted from H, then there exists a subsequence (x'_{n_k}) such that $v_\Phi(x'_{n_k})$ is convergent in L^1. By hypothesis, we may choose the subsequence so as to be weak Cauchy. Since v_Φ is continuous for $\sigma(F', F)$ and $\sigma(L^1, L^\infty)$, then, $(v(x'_{n_k})) = (h_k)$ is weak Cauchy in L^1 and therefore weakly convergent in L^1 (Theorem 4.21.4). Besides this, h_k is the class of the function $\langle \Phi, x'_{n_k} \rangle$, and these functions converge pointwise. Hence (Corollary 4.21.6) the sequence (h_k) is convergent in L^1. ▮

It is an open problem whether or not u_Φ is always precompact from L^∞ into F (with its initial topology). But this conclusion follows from Proposition 9.3.11 in a number of important special cases.

9.3.12 Corollary. If Φ is SWI from T into F, then u_Φ is precompact from L^∞ into F in each of the following cases:

(a) F is separable;

(b) F is a reflexive Banach space;

(c) F is metrizable and Φ is measurable.

Proof. In each of cases (a) and (b) we have only to note that each equicontinuous subset H of F' belongs to π, showing that π is stronger than the initial topology of F. In case (a) this is so since each such set H is weakly relatively compact and weakly metrizable. In case (b) we appeal to Šmulian's Theorem 8.12.1 for the Banach space F'.

As for (c), let us consider first the case in which T is σ-compact. By Theorem 8.15.2 we may assume without loss of generality that Φ is separably valued and we may then argue with F replaced by the separable vector subspace generated by $\Phi(T)$ and appeal to (a). Otherwise we adopt a different device, leading to a reduction to the σ-finite case. Thus, the closed unit ball B_0 in $C_0(T)$ is dense for $\sigma(L^\infty, L^1)$ in the unit ball B of L^∞ and so, since u is continuous for $\sigma(L^\infty, L^1)$ and $\sigma(F', F)$, it suffices to show that $u_\Phi(B_0)$ is precompact in F. [The reader will notice that $u_\Phi(B)$ is the weak closure of $u_\Phi(B_0)$ and hence, being convex, is also the closure of $u_\Phi(B_0)$.] To establish this point it suffices to show that for each sequence (f_n) in B_0, the set of $u_\Phi(f_n)$ is precompact. Now the f_n collectively vanish outside a σ-compact subset of T, hence the desired reduction. ∎

Proposition 9.3.11 and its corollary assume a rather striking form if we impose further hypotheses on F, as is shown by a result announced by Bourbaki (loc. cit., p. 86, Exercice 25)).

9.3.13 Proposition. If F is a separable and weakly sequentially complete Fréchet space, any SEI function Φ mapping T into F is SWI. In particular, u_Φ transforms the unit ball of L^∞ into a relatively compact subset of F, and v_Φ transforms equicontinuous subsets of F' into relatively compact subsets of L^1.

Proof. The proof proceeds in three stages, of which the first two are relatively simple and themselves finish the proof if (T, μ) is σ-finite. In step (1) we show that $\int_K \Phi \, d\mu \in F$ for any compact subset K of T. Since $f\Phi$ is SEI whenever Φ has that property and f belongs to $\mathscr{L}^\infty$, it follows from this that $u_\Phi(f)$ belongs to F whenever f belongs to $\mathscr{L}^\infty_c$, the set of functions in $\mathscr{L}^\infty$ that vanish outside compact sets. Next, in step (2), we show that $u_\Phi(f)$ belongs to F whenever f belongs to $\mathscr{L}^\infty_0$, the set of functions in $\mathscr{L}^\infty$ that vanish outside σ-finite sets. Finally, in step (3), we deal with an arbitrary f in $\mathscr{L}^\infty$.

(1) According to Theorem 8.15.2, Φ is measurable. So the given compact set K can be written as $(\bigcup_{n=1}^\infty K_n) \cup N$, where each K_n is compact and where N is negligible, the restriction $\Phi \mid K_n$ being continuous. Then

$$\int_K \Phi \, d\mu = \lim_n \int_{K_n} \Phi \, d\mu$$

for $\sigma(F'^*, F')$. On the other hand, Proposition 8.14.3 shows that $x_n = \int_{K_n} \Phi \, d\mu$ belongs to F for each n. The sequence (x_n) is moreover weakly Cauchy in F, hence weakly convergent to some x in F. This weak limit x can be none other than $\int_K \Phi \, d\mu$ (as one sees by applying an arbitrary x' in F' to the integrals), so that indeed $\int_K \Phi \, d\mu$ belongs to F.

(2) Let us suppose now that f belongs to $\overline{\mathscr{L}}_0^\infty$. Altering f on a negligible set, we may assume that f vanishes outside the union of an increasing sequence (K_n) of compact sets. Putting $f_n = f\chi_{K_n}$ we learn from (1) that $x_n = \int f_n \Phi \, d\mu$ belongs to F. Moreover it is evident that (x_n) is $\sigma(F'^*, F')$-convergent to $\int f\Phi \, d\mu$. It now suffices to repeat the argument used in (1) and based on the weak sequential completeness of F.

(3) Here we utilize the map v_Φ and show that it transforms any given equicontinuous subset H of F' into a weakly relatively compact subset of L^1. Supposing this accomplished, given any f in $\overline{\mathscr{L}}^\infty$ and any σ-finite set S, we denote by f_S the function $f\chi_S$. f_S belongs to $\overline{\mathscr{L}}_0^\infty$ and so, by (2), $u_\Phi(\dot{f}_S) = \int_S f\Phi \, d\mu$ belongs to F. Moreover we have

$$\langle u_\Phi(\dot{f}_S), x' \rangle = \langle \dot{f}_S, v_\Phi(x') \rangle = \int_S f \cdot \langle \Phi, x' \rangle \cdot d\mu$$

for each S. Now, using Theorem 4.21.2, it is apparent that $\int_S fh \, d\mu \to \overline{\int} fh \, d\mu$ as S expands, and this uniformly when h ranges over any weakly relatively compact set in $\mathscr{L}^1$. Consequently, by what we are assuming to be true, we see that the directed family (x_S), where $x_S = u_\Phi(\dot{f}_S)$, is Cauchy in F. Completeness of F ensures the existence of the limit x of (x_S). For each x' in F' we then have

$$\begin{aligned} \langle x, x' \rangle = \lim \langle x_S, x' \rangle &= \lim \int_S f \cdot \langle \Phi, x' \rangle \cdot d\mu \\ &= \overline{\int} f \cdot \langle \Phi, x' \rangle \cdot d\mu, \end{aligned}$$

showing that x is none other than $\overline{\int} f\Phi \, d\mu$. The latter integral thus belongs to F and our job will be done.

It remains for us to show that $G = v_\Phi(H)$ is weakly relatively compact in L^1 whenever H is equicontinuous in F'. According to Eberlein's Theorem 8.12.7 it suffices to show that, for any sequence (x'_n) extracted from H, the sequence $(v_\Phi(x'_n))$ has a weak limiting point in L^1. Now, since F is separable, H is weakly metrizable. Since H is in any case weakly relatively compact in F', from (x'_n) we may extract a subsequence (y'_n) weakly convergent in F',—to x', say. We will show that $(\dot{h}_n) = (v_\Phi(y'_n))$ is then weakly convergent in L^1 to $\dot{h} = v_\Phi(x')$. This signifies that

$$\overline{\int} f \cdot \langle \Phi, y'_n \rangle \cdot d\mu \to \overline{\int} f \cdot \langle \Phi, x' \rangle \cdot d\mu$$

for each f in $\overline{\mathscr{L}}^\infty$. Since the functions $\langle \Phi, y'_n \rangle$ and $\langle \Phi, x' \rangle$ collectively vanish outside the union of some σ-finite set and a locally negligible set, we may here assume that f lies in $\overline{\mathscr{L}}_0^\infty$. Then, however, the relation we must establish reads simply

$$\langle u_\Phi(\dot{f}), y'_n \rangle \to \langle u_\Phi(\dot{f}), x' \rangle,$$

where we know from (2) that $u_\Phi(\dot{f})$ belongs to F. This relation is a trivial consequence of the weak convergence of (y'_n) to x'. ▌

At this point the reader should see Exercise 9.14.

9.4 The Dunford-Pettis and Dieudonné Properties: Weakly Compact and Compact Linear Maps of Spaces L^1 and C

In their joint paper [1] Dunford and Pettis made (amongst other things) a systematic study of representation theorems for compact and weakly compact linear maps of L^1 into a Banach space F. More recently Grothendieck [4] examined the problems from a more general point of view. His analysis shows that many of the earlier results stem from the fact that the space $E = L^1$ satisfies the equivalent conditions stated in Theorem 9.3.4. Besides this, he showed that a similar situation obtains when E is a space L^∞ or a space $C(T)$ (continuous functions on a separated compact space T) and exploited this in connection with the study of vector-valued measures.

In this section we aim to give an account of some of these ideas, in doing which we shall lean heavily on Grothendieck's work (loc. cit.) and also take advantage of ideas proposed by Bourbaki ([6], Chapitre 6).

Until further specialization E will denote a separated LCTVS, Σ the set of convex, balanced, and $\sigma(E, E')$-compact subsets of E, Σ' the set of convex, balanced, equicontinuous, and $\sigma(E', E'')$-compact subsets of E', $T_{\Sigma'}$ the topology on E having as a base at 0 the sets S^0 when S ranges over Σ'. The analogous topology T_Σ on E' is none other than the Mackey topology $\tau(E', E)$ (see Subsection 8.3.3).

From Theorem 9.3.4 may be read off the equivalence of the following three conditions on E:

(DP_1) Any continuous linear map of E into a separated and quasi-complete LCTVS F, which transforms bounded sets into weakly relatively compact sets, transforms each set in Σ into a relatively compact subset of F. (Or the same assertion, F being restricted to being a Banach space.)

(DP_2) Each set A in Σ is precompact for $T_{\Sigma'}$.

(DP_3) Each set in Σ' is precompact (or, equivalently, relatively compact) for $\tau(E', E)$.

Likewise Corollary 9.3.5 yields the equivalence of the following three conditions on E:

(SDP_1) Any continuous linear map of E into a separated and quasi-complete LCTVS F, which transforms bounded sets into weakly relatively compact sets, transforms weak Cauchy sequences into convergent (or, equivalently, Cauchy) sequences in F. (Or the same assertion, F being restricted to be a Banach space.)

(SDP_2) Each weak Cauchy sequence in E is Cauchy for $T_{\Sigma'}$.

(SDP_3) Each set in Σ' is precompact (or, equivalently, relatively compact) for the topology of convergence uniform on the weak Cauchy sequences in E.

The reader should bear in mind various other equivalent conditions derivable, for example, from Theorem 9.2.1.

As has been remarked, Grothendieck's analysis in [4] showed that the work of Dunford and Pettis on linear maps with domain L^1 rests heavily on the fact that the space $E = L^1$ satisfies the equivalent conditions (DP_1)–(DP_3); and that the space $E = C(T)$ has the equivalent properties (SDP_1)–(SDP_3), which fact has a bearing on the study of vector-valued measures. For these reasons Grothendieck introduced the following definition.

Definition. A separated LCTVS E is said to have the (DP) (= Dunford-Pettis) [resp. the (SDP) (= strict Dunford-Pettis)] property if any one of the equivalent conditions (DP_1)–(DP_3) [resp. (SDP_1)–(SDP_3)] is satisfied.

9.4.1 **Remarks.** (1) If E and F are LCTVSs, and if u is a continuous linear map of E into F, then the following two assertions are equivalent (cf. condition (SDP_1) above):

(a) u transforms weak Cauchy sequences into Cauchy sequences;

(b) u transforms bounded and weakly metrizable sets into precompact sets.

Thus, if (a) holds and A is bounded and weakly metrizable, then A is weakly precompact (being bounded). Being weakly metrizable too, A has the property that any sequence extracted from it possesses a weak Cauchy subsequence, which is transformed by u into a Cauchy sequence. Thus any sequence extracted from $u(A)$ admits a Cauchy subsequence, which implies that $u(A)$ is precompact. Thus (a) implies (b).

On the other hand, assuming that (b) is true, suppose that (x_n) is a weak Cauchy sequence in E. The set $S = \{x_n\}$ is then bounded and also weakly metrizable (Exercise 9.11). By hypothesis, therefore, $u(S)$ is precompact. But $(u(x_n))$ is a weak Cauchy sequence in F, thanks to continuity of u. Being precompact, this sequence must be Cauchy. (b) therefore implies (a), and the two conditions are indeed equivalent.

(2) In Remark (1) we have called upon the fact (Exercise 9.11) that in a uniform space E, any Cauchy sequence forms a semimetrizable subspace (a metrizable subspace if E is separated). If E is a separated LCTVS, a stronger result is valid and will be used from time to time—namely, that the closed, convex, balanced envelope of the set of points of a Cauchy sequence is metrizable. As a consequence, a linear map of E into a LCTVS F, which transforms sequences converging to 0 into convergent sequences (necessarily converging to 0), transforms Cauchy sequences into Cauchy sequences. (See Exercises 9.12 and 9.13.) Naturally, this may be applied when either or both of E and F is endowed with its weakened topology.

(3) If E is complete (or merely quasi-complete for $\tau(E, E')$), no effective change is made if in (DP_1)–(DP_3) we interpret Σ as the set of all weakly compact subsets of E. Moreover (see the Remark following Theorem 9.3.4) we may in (DP_2) replace the word "precompact" by "compact."

9.4.2 **Examples.** Those semireflexive spaces E possessing the (DP) property are precisely the boundedly compact spaces (see Subsection 8.4.7).

For let us suppose that E is semireflexive and has the (DP) property, and consider the identity map u of E into $F = E$. u transforms bounded sets into weakly relatively compact sets (Theorem 8.4.2) and therefore transforms bounded sets into relatively compact sets. Thus E is boundedly compact.

Conversely, if E is boundedly compact, we may verify that (DP_1) holds. In fact, each set A in Σ is bounded, hence relatively compact, and so $u(A)$ is relatively compact whenever u is continuous. (It is, of course, equally simple to verify that (DP_2) holds.)

We see therefore that the only semireflexive Banach spaces possessing the (DP) property are finite dimensional. On the other hand, the Montel spaces $C^\infty(\Omega)$, $H(\Omega)$, $\mathscr{D}(\Omega, S)$, $\mathscr{D}(\Omega)$, and $\mathscr{D}'(\Omega)$ all possess the (DP) property (see Subsection 8.4.7 again).

We note further that any boundedly compact space possesses the (SDP) property.

These examples are all rather "crude" and Theorem 9.4.4 below is a good deal more interesting, both in itself and by virtue of its consequences.

9.4.3 **Some General Properties.** We collect here a number of general statements relating to the (DP) and (SDP) properties.

(a) If (E_i) is an arbitrary family of LCTVSs, each having the (DP) [resp. (SDP)] property, then the product space $E = \prod E_i$ possesses the (DP) [resp. (SDP)] property.

Proof. Take the case of the (DP) property, for example. We verify that (DP_1) is true in its apparently weaker form. If u is a continuous linear map of E into F, and if j_i is the natural map of E_i into E, then $u_i = u \circ j_i$ is continuous from E_i into F. Since F is a Banach space, the continuity of u entails that $u_i = 0$ for all but a finite set of indices i. It therefore suffices to handle the case of a finite product. Now in any case, if u transforms bounded sets into weakly relatively compact sets, the same is true of each u_i. Then, by hypothesis, u_i transforms sets in Σ_i (the set of convex, balanced, and $\sigma(E_i, E_i')$-compact subsets of E_i) into relatively compact sets in F. Since $\sigma(E, E')$ is the product of the topologies $\sigma(E_i, E_i')$, it follows easily that u transforms sets in Σ into relatively compact subsets of F. ▌

(b) If E possesses the (DP) [resp. (SDP)] property, and if H is a vector subspace of E admitting a topological complement relative to E, then H possesses the (DP) [resp. (SDP)] property.

Proof. The hypothesis on H signifies that there exists a continuous projection π of E onto H. If u is linear and continuous from H into F and transforms bounded sets into weakly relatively compact sets, then $v = u \circ \pi$ is linear and continuous from E into F and transforms bounded sets into weakly relatively compact sets. If $A \subset H$ is convex, balanced, and $\sigma(H, H')$-compact, then it is $\sigma(E, E')$-compact. Assuming that E has the (DP) property, it follows that $u(A) = v(A)$ is relatively compact in F, and hence that H has the (DP) property. Similarly, H will have the (SDP) property whenever E does. ▌

Remark. The assertion is false for *arbitrary* vector subspaces H of E.

(c) Suppose that E and F are complete separated LCTVSs, and that E possesses the (DP) property. If f is a continuous bilinear map of $E \times F$ into a TVS G, then $f(x_n, t_n) \to f(x, y)$ weakly in G whenever the sequences (x_n) and (y_n) converge weakly in E and F, respectively, to limits x and y.

Proof. It is enough to handle the case in which G is the scalar field and f a continuous bilinear form on $E \times F$. In this case we may write $f(x, y) = \langle x, u(y) \rangle$, where u is a continuous linear map of F into the strong dual E' of E. If $x'_n = u(y_n)$, the sequence (x'_n) is $\sigma(E', E'')$-convergent to $x' = u(y)$, and we must show that $\langle x_n, x'_n \rangle \to \langle x, x' \rangle$ as $n \to \infty$. Since E possesses the (DP) property reference to Theorem 9.2.1 indicates that it is sufficient to verify that (1) the closed, convex, balanced envelope A of $\{x_n\}$ is weakly compact in E, and that (2) the $\sigma(E', E'')$-closed, convex, balanced envelope A' of $\{x'_n\}$ is $\sigma(E', E'')$-compact.

The truth of (1) follows directly from Krein's Theorem 8.13.1. As for (2), we apply the same theorem to conclude that the closed, convex, balanced envelope in F of $\{y_n\}$ is weakly compact, and then use the continuity of u. ▮

(c′) In order that a Banach space E possess the (DP) property, it is necessary and sufficient that $\langle x_n, x'_n \rangle \to 0$ whenever the sequence (x_n) converges weakly to 0 in E and (x'_n) is $\sigma(E', E'')$-convergent. The stated condition is plainly unaltered if we add to it the assumption that $x'_n \to 0$ for $\sigma(E', E'')$.

Proof. The necessity is apparent from (c). To establish the sufficiency, we argue that if E failed to possess the (DP) property, then (see Theorem 9.2.1) there would exist a sequence (x_n) converging weakly to 0 in E, and a $\sigma(E', E'')$-compact set A' in E', such that (x_n) does not converge to 0 uniformly on A'. Thus, there would exist a number $\varepsilon > 0$ and a sequence (x'_n) extracted from A', such that $|\langle x_n, x'_n \rangle| > \varepsilon$ for all n. By Šmulian's Theorem 8.12.1 applied to the Banach space E', we may assume that the sequence (x'_n) is $\sigma(E', E'')$-convergent. This would amount to an infringement of the hypothesis, hence the sufficiency. ▮

Remark. Brace [1] has made a study of Banach spaces E for which it is true that $x_n \to x$ weakly in E and $x'_n \to x'$ for $\sigma(E', E'')$ together imply that $\langle x_n, x'_n \rangle \to \langle x, x' \rangle$.

(d) If E is metrizable, then (SDP_2) implies (DP_2). If, furthermore, E is weakly sequentially complete, then (DP_2) and (SDP_2) are equivalent.

Proof. Suppose that A belongs to Σ. To show that A is precompact for $T_{\Sigma'}$, it suffices to show that from each sequence (x_n) extracted from A one may extract a subsequence, the set of terms of which is precompact for $T_{\Sigma'}$. Now the set of x_n is weakly relatively compact and so Šmulian's Theorem 8.12.1 entails that there is a subsequence of (x_n) that converges weakly. But then (SDP_2) affirms that this subsequence is Cauchy, and hence precompact, for $T_{\Sigma'}$. Thus (SDP_2) implies (DP_2)

Supposing now that E is also weakly sequentially complete, and that (DP_2) holds. Any weak Cauchy sequence (x_n) in E is weakly convergent and the set $\{x_n\}$ is weakly relatively compact. By (DP_2), the set $\{u(x_n)\}$ is relatively compact in F. But continuity of u shows that $(u(x_n))$ is weakly convergent in F, and it follows that this sequence is convergent for the initial topology of F. This verifies (SDP_2). ▮

(d′) In order that a metrizable LCTVS E possess the (DP) property, it is necessary and sufficient that whenever (x_n) is a sequence converging weakly to

0 in E, then (x_n) converges to 0 uniformly on each set in Σ' (that is, converges to 0 for $T_{\Sigma'}$).

Proof. This stems from Theorem 9.2.1 and the definition of the (DP) property. The reader should compare the proof of (c'). ▮

We shall have need also of the following general result (Grothendieck [4], Proposition 1, Corollaire; Bourbaki [6], Chapitre 6, p. 94, Exercice 22c)).

(e) If E is infrabarrelled, and if the strong dual E' possesses the (DP) property, then so too does E.

Proof. If the strong dual E' has the (DP) property, then each set in Σ' is precompact for $T_{\Sigma''}$, where Σ'' is the set of convex, balanced, equicontinuous and $\sigma(E'', E''')$-compact subsets of E''. According to Theorem 9.2.1, therefore, each set in Σ'' is precompact for $T_{\Sigma'}$ (or, more precisely, for the topology on E'' having for base at 0 the polars S^0 *in* E'' of sets S in Σ', which topology induces $T_{\Sigma'}$ on E). It suffices now to verify that $\Sigma \subset \Sigma''$, that is, that each A in Σ is $\sigma(E'', E''')$-compact. (The reader will note that A, being bounded in E, is equicontinuous in E'', the dual of E'.) Now, since E is infrabarrelled, the strong topology of E'' induces on E its initial topology, so that each element of E''' is the extension of some element of E'. But then $\sigma(E'', E''')$ induces on A a topology weaker than that induced by $\sigma(E, E')$. For the latter topology A is compact by definition, so that A is compact for $\sigma(E'', E''')$. ▮

Prior to stating one of Grothendieck's key results ([4], Théorème 1) we remind the reader that we use $C_0(T)$ to denote the Banach space of continuous functions vanishing at infinity on the separated locally compact space T, the norm being defined as the maximum modulus of the function in question.

9.4.4 **Theorem.** Let T be a separated locally compact space, μ a positive Radon measure on T. Each of the spaces $C_0(T)$ and $L^1 = L^1(T, \mu)$ possesses the (SDP) property and hence (by Subsection 9.4.3(d)) the (DP) property as well.

Proof. We can complete all the details in the case of $C_0(T)$, but for L^1 we have space only to indicate how this may be reduced to the former case.

(a) THE CASE OF $C_0(T)$. We have to show that if the sequence (f_n) is weak Cauchy in $C_0(T)$, then $\int f_n \, d\lambda$ converges uniformly on each subset H of M, the Banach space of bounded Radon measures on T, which is norm bounded and compact for $\sigma(M, M')$. Now the Cauchy nature of (f_n) signifies that the functions f_n are uniformly bounded, say $|f_n(t)| \leq c$ for all n and all t, and pointwise convergent to some function f. And we must show that $\int f_n \, d\lambda \to \int f \, d\lambda$ as $n \to \infty$, uniformly as λ ranges over H. To do this, it suffices to establish that uniformity obtains relative to any countable subset of H. So we may assume that H is countable and weakly relatively compact. According to the Remark preceding the proof of Theorem 4.22.1, the problem is reducible to the case in which $H = \{h \cdot \alpha : h \in P\}$, where α is a certain bounded positive measure and P is a weakly relatively compact subset of $\mathscr{L}^1(\alpha)$. All that has to be shown is that $\int f_n h \, d\alpha \to \int fh \, d\alpha$ uniformly for h in P. But, by virtue of Egoroff's Theorem 4.8.3, this follows from Theorem 4.21.10.

(b) THE CASE OF L^1. Using Subsection 9.4.3(d), we have merely to show that L^1 has the (DP) property, the fact that L^1 is weakly sequentially complete being established in Theorem 4.21.4. By Subsection 9.4.3(e), it suffices to show that the Banach space L^∞ has the (DP) property.

At this stage we must fall back on known results lying outside the scope of this book. L^∞ is an instance of a commutative Banach algebra with involution, and the theory of such structures (Loomis [1], p. 88; Rickart [1], p. 190, Theorem (4.2.2)) affirms that L^∞ is isometrically isomorphic to a space $C(S)$, where S is a suitable separated compact space, the isomorphism mapping real-valued functions in L^∞ into real-valued functions in $C(S)$. This, together with (a) above, shows that L^∞ has the (DP) property. ▌

Remark. The concluding paragraph of the above proof shows that if T is any completely regular topological space, the Banach space $C_{bd}(T)$ of all bounded, continuous functions on T (with norm equal to the supremum of the modulus of the function in question) likewise possess the (DP) and the (SDP) properties. Indeed,

$$C_{bd}(T) \approx C(\beta T)$$

where βT is the Stone-Čech compactification of T (compare Exercise 1.28).

Let us state explicitly some consequences of the preceding theorem.

9.4.5 **Corollary.** Suppose that u is a weakly compact linear map of $C_0(T)$ [or of L^1] into a quasi-complete separated LCTVS F. Then

(1) u transforms weakly relatively compact sets into relatively compact sets, and

(2) u transforms weak Cauchy sequences into convergent sequences.

Proof. As for (1) we need only note that either space $C_0(T)$ or L^1 has the (DP) property, and that in any Banach space the convex and balanced envelope of any weakly relatively compact set is again weakly relatively compact. This last property follows from Krein's Theorem 8.13.1; in the case of L^1 it follows also from Theorem 4.21.2.

On the other hand (2) follows from the fact that each of $C_0(T)$ and L^1 possesses the (SDP) property. ▌

Remarks. (1) For the case of L^1, see Dunford and Schwartz [1], p. 508. Theorem VI. 8.12.

(2) THE (DP) PROPERTY AND THE PRODUCT OF WEAKLY COMPACT LINEAR MAPS. Let us suppose that X, E, and F are separated LCTVSs, E possessing the (DP) property and F being quasi-complete. We suppose too that $v : X \to E$ and $u : E \to F$ are linear and continuous and transform bounded sets into weakly relatively compact sets. It then appears that $uv : X \to F$ transforms bounded sets into relatively compact sets.

In particular, if X and E are normed (the other hypotheses being undisturbed), then uv is compact if u and v are each weakly compact. For the case in which $X = E = F = L^1$, see Dunford and Schwartz [1], p. 510, Theorem VI.8.13, and also Exercise 9.18.

Similarly, if we suppose that E has the (SDP) property (X, F, u, and v being as before), then uv transforms weak Cauchy sequences into convergent sequences. Moreover, we could here assume that it is X (rather than E) that has the (SDP) property.

In particular, if X and E are normed, and if at least one of them has the (SDP) property, then uv transforms weak Cauchy sequences into convergent sequences, provided each of u and v is weakly compact.

9.4.6 **Other Spaces Having the (DP) and (SDP) Properties.** Starting from Theorem 9.4.4 one may establish that certain other important specific spaces enjoy the (DP) and the (SDP) properties.

(a) Let T be any normal topological space, π a set of subsets of T, $C_\pi(T)$ the space of functions continuous on T and bounded on each set belonging to π. If $C_\pi(T)$ is endowed with the topology of convergence uniform on each set in π, the resulting LCTVS has both the (DP) and the (SDP) properties.

(b) Let T be a separated locally compact space, μ a positive Radon measure on T, π a set of relatively compact and μ-measurable subsets of T. Let $L_\pi = L_\pi(T, \mu)$ denote the space of (equivalence classes of) functions f that are μ-measurable and such that $\int_P^* |f| \, d\mu < +\infty$ for each P in π, endowed with the topology defined by the seminorms $f \to \int_P |f| \, d\mu$. Then L_π possesses both the (DP) and the (SDP) properties.

Proof. We give the details for (a) only. The proof of (b) is closely analogous, using that part of Theorem 9.4.4 which refers to L^1.

We may assume without loss of generality that each P in π is closed, and that π forms an increasing directed set. It suffices to show that if (f_i) is a weak Cauchy sequence (or a weakly convergent directed family of points of some weakly compact subset) in $E = C_\pi$, then (f_i) converges uniformly on each set $A' \subset E'$ that is equicontinuous and $\sigma(E', E'')$-compact.

Since A' is equicontinuous, there exists a set P in π such that, putting $U = \{f \in C_\pi : |f(t)| \leq 1 \text{ for } t \in P\}$, one has $A' \subset k \cdot U^0$ for some number $k > 0$.

Let u be the continuous linear map of C_π into $C_{bd}(P) = F$ that assigns to each f in C_π its restriction $f \mid P$. Then u maps U onto the closed unit ball of F, thanks to normality of T and Urysohn's Theorem 0.2.13. Accordingly, u' is a one-to-one linear map of F' into E' that maps the unit ball of F' onto the polar set U^0. Hence each x' in A' is the u'-image of precisely one y' in F'.

Let us verify next that $u'^{-1}(A')$ is $\sigma(F', F'')$-compact. To do this, it suffices to show that u' is an isomorphism of F' into E' (the latter with its strong topology). And for this it is enough to show that there exists a bounded set B in E such that $u(B)$ is dense in the unit ball in F. But, by Urysohn's theorem once again, one may indeed take for B the set of f in C_π having an absolute value of at most 1 everywhere.

This being so, to show that $\lim_i \langle f_i, x' \rangle$ exists uniformly for x' in A', we write

$$\langle f_i, x' \rangle = \langle u(f_i), y' \rangle,$$

where $y' = u'^{-1}(x')$. Since u is continuous, $(u(f_i))$ is a weak Cauchy sequence [or a weakly convergent directed family extracted from a weakly compact subset of F], and so the desired uniformity stems from the fact that $F = C_{bd}(P)$ possesses both the (DP) and the (SDP) properties (see the Remark following Theorem 9.4.4). ▌

(c) Let Ω be an open subset of R^n, $C^m(\Omega)$ the space of m times continuously differentiable functions on Ω (the topology being that of locally uniform convergence on Ω of the partial derivatives $\partial^p f$ for each fixed p with $|p| \leq m$—see Chapter 5). The space $C^m(\Omega)$ possesses both the (DP) and the (SDP) properties.

Proof. There is no harm in assuming that Ω is a domain, since in any case $C^m(\Omega)$ is isomorphic with the product $\prod C^m(\Omega_i)$, where the Ω_i are the connected components of Ω, and we may appeal to Subsection 9.4.3(a).

On the other hand, our proof will invoke convexity of the domain Ω. Failing this, a fairly complicated argument employing partitions of unity is necessary to reduce the problem of one in which convexity obtains. We shall omit the details. Suppose, then, that Ω is a convex domain.

Consider $C^m(\Omega)$ as a vector subspace of the product $E = \prod_{|p| \leq m} E_p$, each E_p being $C(\Omega)$, via the isomorphism that carries f into the family $(\partial^p f)_{p \leq m}$. In view of Subsection 9.4.3(a) and (b), it is enough to show that $C^m(\Omega)$ admits a topological complement in E—that is, that there exists a continuous projection of E onto $C^m(\Omega)$. If we choose t_0 in Ω and denote by $C_0^m(\Omega)$ the vector subspace of $C^m(\Omega)$ formed of those f in $C^m(\Omega)$ for which $\partial^p f(t_0) = 0$ for $|p| < m$, then $C_0^m(\Omega)$ is of finite codimension in $C^m(\Omega)$. Because of this, it suffices to exhibit a continuous projection π of E onto $C_0^m(\Omega)$.

A suitable projection π is obtained by setting $\pi((f_p)) = (g_p)$, the family (g_p) being defined as follows. If $|p| = m$, $g_p = f_p$. Supposing that g_p is defined for each index p satisfying $p = k$ (where $0 < k \leq m$), and that q is an index with $|q| = k - 1$, we define

$$g_q(t) = g_{(q_1, \cdots, q_n)}(t) = \int_{t_0 t} [g_{q'}\, dt_1 + g_{q''}\, dt_2 + \cdots + g_{q^{(n)}}\, dt_n],$$

where $q' = (q_1 + 1, q_2, \cdots, q_n)$, $q'' = (q_1, q_2 + 1, \cdots, q_n)$, and so on, are indices p satisfying $|p| = k$. The path of integration is the segment joining t_0 to t, a choice that is admissible since Ω is convex. (It is here, in the choice of the path of integration, that difficulties are encountered when Ω is not convex.) One may verify step by step that $g_p(t_0) = 0$ for $|p| < m$, and that $\partial_k g_{(p_1, \cdots, p_n)} = g_{(p_1, \cdots, p_k + 1, \cdots, p_n)}$, whence it appears that π is a continuous projection of E onto $C_0^m(\Omega)$. ▌

(d) In view of Theorem 9.4.4 and Theorem 4.23.2 (unproved in this book), the space $M(T)$ (bounded Radon measures on a separated locally compact space T) possesses the (DP) and the (SDP) properties.

A THEOREM OF DUNFORD, PETTIS, AND PHILLIPS. As an application of Theorem 9.4.4 we intend giving a proof of an important representation theorem due to Dunford and Pettis ([1], Theorem 3.1.7) and later modified by Phillips ([2], Theorem 5.2). The proof we give follows in large measure that sketched out by Bourbaki ([6], Chapitre 6, p. 95, Exercice 24)).

9.4.7 **Theorem.** Let F be a Banach space, T a separated locally compact space, μ a positive Radon measure on T, and u a weakly compact linear map of $L^1 = L^1(T, \mu)$ into F. Then there exists a measurable function f mapping T into F such that

(1) $\|f(t)\| \leq \|u\|$ for $t \in T$;

(2) f is weakly compact valued (that is, $f(T)$ is weakly relatively compact in F); and

(3) $u(\dot{g}) = \int gf \, d\mu \qquad$ for $g \in \mathscr{L}^1$.

Conversely, if f is a measurable function from T into F that satisfies (2) (or is equal locally almost everywhere to such a function), then (3) defines a weakly compact linear map u of L^1 into F.

Proof. This is rather lengthy and will be split into stages. In the first and most important stage we show that if u is weakly compact, it admits a representation (3) in which f is measurable and satisfies (1). In the second stage we conclude that this f must, after modification on a locally negligible set (which can always be done leaving (1) undisturbed), take its values in a weakly relatively compact subset of F. Thus (2) will be attained. In the final stage we establish the converse assertion.

Stage I. Suppose that the representation were established when T is compact. In the general case we know (Proposition 4.14.9) that T is the union of a locally negligible set N and of a locally countable family of disjoint compact sets K_i. If the special case were available, we should know that for each i there is a measurable $f_i : K_i \to F$ such that $\|f_i(t)\| \leq \|u\|$ for t in K_i and

$$u(\dot{g}) = \int gf_i \, d\mu$$

for each g in $\mathscr{L}^1$ vanishing on $T \backslash K_i$. We could then define f to be f_i on each K_i and to be 0 on N, thus obtaining a measurable $f : T \to F$ for which $\|f(t)\| \leq \|u\|$ everywhere, and for which

$$u(\dot{g}) = \int gf \, d\mu$$

for all g in $\mathscr{L}^1$. In other words, we may reduce the discussion to the case in which T is compact, which assumption we make henceforth.

If T is compact, $L^\infty \subset L^1$, and the closed unit ball B in L^∞ is bounded in L^1. Hence $u(B)$ is weakly relatively compact in F. Besides this, B is compact for $\sigma(L^1, L^\infty)$ and so, since L^1 has the (DP) property, $u(B)$ is relatively compact in F. In particular, $u(B)$ is separable. Hence $u(L^1)$ also is separable.

This being so, we may everywhere replace F by the closed vector subspace generated by $u(L^1)$, and so assume from the outset that F is separable and that $u(L^1)$ is dense in F.

We put A for the closure in F of the u-image of the unit ball in L^1. A is weakly compact. Since F is separable and metrizable, F' is weakly separable (being the union of the weakly compact and weakly metrizable, hence weakly separable, sets V_n^0, (V_n) being a countable base at 0 in F). Thus there exists a sequence (x_n') that is total in F'. From this we infer that A is weakly metrizable in terms of the metric

$$d(x, y) = \sum_n 2^{-n} |\langle x - y, x'_n \rangle| / (1 + |\langle x - y, x'_n \rangle|).$$

In fact this metric defines on A a topology that is separated and weaker than that induced by $\sigma(F, F')$. Since A is compact for the latter, the two topologies are identical. Knowing that A is both compact and metrizable for $\sigma(F, F')$, we conclude that A is separable for that topology.

Now we introduce the normed space Z consisting of the points of F' and whose closed unit ball is A^0. We may then identify A with the closed unit ball of Z' and (by the Hahn-Banach theorem) Z is isometric and isomorphic with $C(A)$, A being endowed with the topology induced by $\sigma(F, F')$. A being compact and metrizable, it follows that Z is separable. To see this we appeal to the Weierstrass-Stone theorem of Subsection 4.10.5 according to which if (a_n) is a sequence dense in A, then the subalgebra of $C(A)$ generated by the functions $\varphi_n(x) = d(x, a_n)$ is dense in $C(A)$.

In this manner u appears as a linear map of L^1 into Z', where Z is a separable normed space. It is obvious that u is continuous from L^1 into Z'. By the Dunford-Pettis Theorem 8.17.6, there exists a scalarwise measurable function $f : T \to Z'$ ($\subset F$ as a vector space) such that

$$u(\dot{g}) = \int gf \, d\mu$$

for g in $\mathscr{L}^1$. To say that f is scalarwise measurable signifies that $z \circ f$ is measurable for each z in Z—that is, for each z in F'. So f is scalarwise measurable from T into F. Since F is separable, f is measurable from T into F (Theorem 8.15.2).

It remains to show that f may be chosen so that $\|f(t)\| \leq \|u\|$ holds everywhere. Now the representation formula itself shows at once that $|\langle f(t), x' \rangle| \leq \|u\|$ for each x' satisfying $\|x'\| \leq 1$ and each t in $T \backslash N_{x'}$, where $N_{x'}$ is locally negligible. On the other hand we have seen already that the unit ball in F' is weakly separable. It follows that there exists a locally negligible set N such that $|\langle f(t), x'| \leq \|u\|$ for all t in $T \backslash N$ and all x' satisfying $\|x'\| \leq 1$. Hence $\|f(t)\| \leq \|u\|$ for t in $T \backslash N$. If we redefine $f(t)$ to be 0 for t in N, the representation formula is left undisturbed and this new f will fulfill all our requirements.

Stage II. In order to show that f may be chosen to satisfy (2), it will be more than enough to establish the following statement.

If the function f, mapping T into F, is measurable and scalarwise locally integrable, and if the set

$$Q = \{\mu(A)^{-1} \int_A f \, d\mu : A \text{ integrable}, \mu(A) > 0\}$$

is contained in F, then there exists a locally negligible set $N \subset T$ such that $f(T \backslash N)$ is contained in the closure in F of Q. (Here it suffices to assume that F is any separated LCTVS.)

To see this, we decompose T into the union of a locally negligible set N_0 and of a locally countable family (K_i) of disjoint compact sets (see Proposition 4.14.9). It will then evidently suffice to show that for each i there exists a negligible set

$N_i \subset K_i$ such that $f(K_i \backslash N_i) \subset \bar{Q}$. (The reader will recall that each compact subset of T meets at most countably many sets K_i.)

Since f is measurable, for each i one may express K_i as the union of a sequence (H_n) of compact sets and a negligible set N' such that $f \mid H_n$ is continuous. Consequently, it would be enough to show that $f(H_n \backslash M_n) \subset \bar{Q}$ for a suitably chosen negligible set $M_n \subset H_n$.

Suppose then that H is compact and $f \mid H$ is continuous. We wish to show that $f(H \backslash M) \subset \bar{Q}$ for some negligible set $M \subset H$. To this end, we define M to be the set of points t in H for which $\mu(H \cap U) = 0$ for some neighbourhood U of t. M is open relative to H, hence is measurable. It is claimed that M is negligible. Indeed, if L is any compact subset of M, to each t in L corresponds an open neighbourhood U_t of t such that $\mu(H \cap U_t) = 0$. A finite number of the sets U_t cover L, so that L is contained in the union of finitely many negligible sets $H \cap U_t$ and is therefore itself negligible. This being true of any compact set $L \subset M$, it follows that M is negligible.

If now t lies in $H \backslash M$, then $\mu(H \cap U) > 0$ for each neighbourhood U of t. Since $f \mid H$ is continuous, the integrals

$$\mu(H \cap U)^{-1} \int_{H \cap U} f \, d\mu,$$

which belong to Q, converge to $f(t)$ as the neighbourhood U shrinks indefinitely. Thus $f(H \backslash M) \subset \bar{Q}$, as we wished to show.

Stage III. Suppose that f is measurable from T into F and that f is equal locally almost everywhere to a function satisfying (2). We wish to show that (3) defines u as a weakly compact linear map of L^1 into F. In doing this we may clearly assume that f itself satisfies (2). But then $f(T)$ is weakly bounded and hence bounded in F. Theorems 8.14.14 and 8.15.2 show that (3) defines u as a linear map of L^1 into F. Our task is to show that u is weakly compact.

Thanks to (2) and Krein's Theorem 8.13.1, there exists a convex, balanced, weakly compact set P containing $f(T)$. If g belongs to $\mathscr{L}^1$, $\int |g| \, d\mu \leq 1$, and x' belongs to P^0, then

$$|\langle u(\dot{g}), x' \rangle| = \left| \int g \cdot \langle f, x' \rangle \cdot d\mu \right| \leq \int |g| \cdot 1 \cdot d\mu \leq 1.$$

The bipolar theorem implies that $u(\dot{g})$ lies in $P^{00} = P$. Thus u transforms the unit ball of L^1 into a subset of P, showing that u is weakly compact. ∎

Remarks. The theorem is given by Dunford and Schwartz ([1], p. 507, Theorem VI.8.10) under the additional hypothesis that $u(L^1)$ is separable.

The last portion of Stage I of the proof can be used in the more general situation in which F is any separated LCTVS and u transforms the unit ball of L^1 into a subset of some compact, convex, and balanced set $A \subset F$ that is metrizable for the induced topology. One would then conclude the existence of a representation (3) in which f is scalarwise measurable and such that $x' \circ f$ is locally essentially bounded for each x' in F'. For more details see Dunford and Schwartz [1], p. 499, Theorem VI.8.2.

Granted Theorem 9.4.7, it is a relatively easy matter to discover conditions on f in order that the associated map u be compact from L^1 into F. This also is discussed by Dunford and Pettis ([1], Theorem 3.1.2); see also Dunford and Schwartz [1], p. 507, Corollary VI.8.11.

9.4.8 **Theorem.** Suppose F, T, and μ are as in Theorem 9.4.7. If u is a compact linear map of L^1 into F, there exists a measurable function f mapping T into F such that $\|f(t)\| \leq \|u\|$ everywhere, $f(T)$ is relatively compact in F, and

$$u(\dot{g}) = \int gf\, d\mu \tag{9.4.1}$$

for g in $\mathscr{L}^1$. Conversely, if f is measurable from T into F and is equal locally almost everywhere to a function whose range is relatively compact in F, then the formula (9.4.1) defines u as a compact linear map of L^1 into F.

Proof. Dealing with the direct assertion, one may apply Theorem 9.4.7 to obtain a measurable f such that $\|f(t)\| \leq \|u\|$ everywhere and such that (9.4.1) holds. The result established in Stage II of the preceding proof then shows that there exists a locally negligible set N such that $f(T\backslash N) \subset H$, where H is the compact closure of the u-image of the unit ball in L^1. It suffices to redefine f so that $f(N) = 0$.

As for the converse, one may suppose that $f(T)$ is relatively compact. One may then follow closely the lines of Stage III of the proof of Theorem 9.4.7, taking for P the closed, convex, balanced envelope in F of $f(T)$. One concludes that $u(\dot{g})$ lies in P whenever $g \in \mathscr{L}^1$ and $\int |g|\, d\mu \leq 1$, showing that u is compact. ▮

THE DIEUDONNÉ PROPERTY. The Dunford-Pettis properties afford a sort of "axiomatization" of relatively deep characteristics of weakly relatively compact linear maps acting on spaces of continuous or of integrable functions (see Corollary 9.4.5). For spaces of continuous functions, Grothendieck ([4], pp. 157–161) isolated another similar property and subjected it to a similar process of "axiomatization," terming it the Dieudonné property (property (D) for short). Before using the (DP) property of $C_0(T)$ as an aid in discussing vector-valued Radon measures, we shall examine this new property in general terms, for it too has some bearing on the topic of vector-valued measures.

9.4.9 **Proposition.** Let E be a separated LCTVS, $\mathscr{F}$ a set of directed families of points of E, each of which is weakly (that is, $\sigma(E'', E')$-) convergent in E'', and H the vector subspace of E'' generated by E and the limits of members of $\mathscr{F}$. The following conditions on E are equivalent:

(1) Any continuous linear map u of E into a complete separated LCTVS F which transforms members of $\mathscr{F}$ into weakly convergent directed families in F, transforms bounded sets into weakly relatively compact sets.

(1′) As (1), F being restricted to be a Banach space.

(2) Any continuous linear map u of E into F (F as in (1)), for which $u''(H) \subset F$, satisfies $u''(E'') \subset F$.

(2′) As (2), F being restricted to be a Banach space.

(3) Any equicontinuous, convex, balanced, and $\sigma(E', H)$-compact set in E' is also $\sigma(E', E'')$-compact.

Proof. This proof will make repeated use of Corollary 9.3.2. It is evident that (1) implies (1′) and that (2) implies (2′).

(a) (1) $\Leftrightarrow$ (2). Suppose that E satisfies (1) and that the hypotheses of (2) are fulfilled. Let (x_i) be a member of $\mathscr{F}$. By hypothesis, (x_i) is weakly convergent in E'' to some $x'' \in H$. Then $u(x_i) = u''(x_i) \to u''(x'')$ weakly in F, u'' being continuous for $\sigma(E'', E')$ and $\sigma(F'', F')$ in all cases while $u''(H) \subset F$ by special hypothesis. Thus u fulfills the hypotheses of (1) and so transforms bounded sets into weakly relatively compact sets. Hence (Corollary 9.3.2) $u''(E'') \subset F$, which is the conclusion of (2). Thus (1) implies (2). The argument is reversible to show that (1) and (2) are equivalent.

(a′) A similar argument shows that (1′) and (2′) are equivalent.

(b) (2′) implies (3). We suppose that $A' \subset E'$ is equicontinuous, convex, balanced, and $\sigma(E', H)$-compact, and that (2′) holds. We must show that A' is $\sigma(E', E'')$-compact, or, what is here equivalent, that it is relatively compact for $\sigma(E', E'')$. Let p be the seminorm on E obtained as the gauge of A'^0, form the quotient space $E/\text{Ker}\, p$, on which p yields a norm, and then complete this quotient space to obtain a Banach space F. Let u be the natural map of E into F. If B denotes the closed unit ball in F', it is easy to verify that $u'(B)$ contains A'. Using Corollary 9.3.2 once more, to show that A' is relatively compact for $\sigma(E', E'')$, it is therefore enough to show that u'' maps E'' into F. Since E is assumed to satisfy (2), this in turn will follow from the inclusion $u''(H) \subset F$. Now E is dense in H for the Mackey topology $\tau(H, E')$, as one sees by appeal to the Hahn-Banach theorem. Moreover, by hypothesis on A', $u'' \mid H$ is continuous for $\tau(H, E')$ and the topology on F'' having for base at 0 the polars in F'' of equicontinuous subsets of F'. The latter topology on F'' induces on F its initial topology. Since $u''(E) \subset F$ and F is complete, the desired inclusion $u''(H) \subset F$ follows.

(c) (3) implies (1). Let u be a continuous linear map of E into F for which $u''(H) \subset F$. We must show that if E satisfies (3), then u transforms bounded sets into weakly relatively compact sets. According to Corollary 9.3.2, this last signifies that u' transforms equicontinuous subsets B' of F' into sets which are relatively compact for $\sigma(E', E'')$. Now the relation $u''(H) \subset F$ entails that u' is continuous for $\sigma(F', F)$ and $\sigma(E', H)$, whence it follows that $u'(B')$ is relatively compact for $\sigma(E', H)$. But then, on the basis of (3), we conclude that $u'(B')$ is relatively compact for $\sigma(E', E'')$.

We now have the scheme

$$(1) \Rightarrow (1') \Leftrightarrow (2') \Rightarrow (3) \Rightarrow (1) \Leftrightarrow (2),$$

which exhibits the equivalence of all the conditions. ∎

Remark. In one form of Proposition 9.4.9 $\mathscr{F}$ is replaced by a set Σ of bounded subsets of E, in which case the assertion in condition (1) referring to $\mathscr{F}$ is replaced by the assertion "$\cdots$, which transforms sets A in Σ into weakly relatively compact subsets of F, $\cdots$." H is then to be replaced by the vector subspace of E'' generated by E and the weak closures in E'' of sets belonging to Σ. No new proof is required, however, for the new proposition follows from the old if we take for $\mathscr{F}$ the set of all directed families (x_i) that are (1) weakly convergent in E'', and (2) extracted from some set belonging to Σ.

On the basis of Proposition 9.4.9 we frame a definition.

Definition. We say that E has the *Dieudonné property* (= (D) property) if the equivalent conditions of Proposition 9.4.9 are fulfilled when for $\mathscr{F}$ we take the set of all weak Cauchy sequences in E. In this case, H is termed the Baire subspace of E'' of the first class.

Remark. One might of course introduce a "Dieudonné property" relative to any set $\mathscr{F}$ of directed families satisfying the conditions imposed in Proposition 9.4.9. It is then plain that the smaller the set $\mathscr{F}$, the stronger is the associated Dieudonné property.

Examples. (1) Every semireflexive space E enjoys the Dieudonné property, since now $E \subset H \subset E'' = E$ and so $H = E''$, and thus condition (2) of Proposition 9.4.9 must hold.

If E is complete and weakly sequentially complete, it can possess the Dieudonné property only if it is semireflexive. For the identity map u of E onto itself maps weak Cauchy sequences into weakly convergent sequences and, hence, if E has the Dieudonné property, bounded sets into weakly relatively compact sets. This means that E is semireflexive (Theorem 8.4.2).

In particular, therefore, a space L^1 has the Dieudonné property only when it is finite dimensional. For, as one sees most easily on combining Theorem 8.4.2 with Theorem 4.21.2, L^1 is semireflexive only in the stipulated trivial case.

(2) If in E the bounded sets are weakly metrizable (for example, if E is a normed space for which E' is strongly separable), then E possesses the Dieudonné property. For in this case $H = E''$ (E'' being in any case the union of the $\sigma(E'', E')$-closures of bounded subsets of E) and so condition (2) of Proposition 9.4.9 is again satisfied.

(3) As with the (DP) and (SDP) properties (see Subsection 9.4.3(a) and (b)) the product of spaces with the Dieudonné property possesses that property; and if E possesses the Dieudonné property, so too does any vector subspace H of E that admits a topological complement relative to E. The proofs of these statements are much as before.

Here now is the main result (Grothendieck [4], p. 160, Théorème 6).

9.4.10 **Theorem.** (a) If T is any separated locally compact space, $C_0(T)$ possesses the Dieudonné property. In fact (what is more), if u is any continuous linear map of $C_0(T)$ into a complete separated LCTVS F, then the following conditions are equivalent:

(1) u is weakly compact;

(2) $u''(\chi_A)$ belongs to F for each closed subset A of T;

(2 bis) as (2), but A restricted to have a countable neighbourhood base;

(3) u transforms any bounded, increasing sequence in $C_0(T)$ into a sequence converging weakly in F.

(b) If Ω is an open subset of R^n, then $C^m(\Omega)$ has the Dieudonné property.

Proof. (a) It is clear that (1) implies (3), any sequence extracted from $C_0(T)$ of the type specified in (3) being a weak Cauchy sequence and being transformed by u into a weak Cauchy sequence in F as a consequence of the mere continuity of u.

(1) implies (2) by Corollary 9.3.2; and (2) implies (2 bis) trivially.

On the other hand (2 bis) is equivalent to the assertion obtained from it by supposing that A is an open set that is the union of a sequence of closed sets. If A is of this type, χ_A is the weak limit in $C_0(T)''$ of a sequence extracted from $C_0(T)$ and of the type specified in (3). It follows that (3) implies (2 bis).

The cycle will be completed by establishing that (2 bis) implies (1), to do which it suffices to verify that $E = C_0(T)$ satisfies any one of the equivalent conditions (1)–(3) of Proposition 9.4.9 when we take for $\mathscr{F}$ the set of bounded, monotone increasing sequences extracted from $C_0(T)$. But then the space H is identical with that denoted by Q in Theorem 4.22.3, which theorem affirms that condition (3) of Proposition 9.4.9 is fulfilled, thus completing the proof of (a).

(b) By virtue of statement (3) preceding the present theorem, combined with the proof of statement 9.4.6(c), it is enough to show that $C(\Omega)$ possesses the Dieudonné property. We shall in fact verify that the space $E = C(\Omega)$ satisfies condition (3) of Proposition 9.4.9.

In this case E' is the space of Radon measures on Ω having compact supports in Ω, and H is the set of functions on Ω obtainable as limits of pointwise convergent and locally uniformly bounded sequences extracted from $C(\Omega)$. Since E is here a Fréchet space (and therefore barrelled), if A' is an equicontinuous subset of E', then A' is identifiable with a bounded subset of $M(K)$ for some compact set $K \subset \Omega$. If also A' is $\sigma(E', H)$-compact, Theorem 4.22.3 shows that it is weakly compact in $M(K)$. Now any strongly continuous linear form on E' has a restriction to $M(K)$ that is continuous, because it is easily verified that a sequence that converges to 0 in $M(K)$ converges to 0 strongly in E'. It follows that $\sigma(E', E'')$ induces on $M(K)$ a topology weaker than $\sigma(M, M')$, hence it follows that A' is relatively compact for $\sigma(E', E'')$. The said condition (3) is thereby verified. ▌

At this point the reader should see Exercise 9.14.

It is worth recording separately a property of continuous linear maps defined on a space with the Dieudonné property.

9.4.11 Theorem. Suppose that E has the Dieudonné property, that F is complete and weakly sequentially complete, and that u is any continuous linear map of E into F. Then u transforms:

(a) weak Cauchy sequences into weakly convergent sequences;

(b) bounded sets into weakly relatively compact sets.

Proof. Continuity of u implies (for any E and F) that u transforms weak Cauchy sequences into sequences of the same type. If F is weakly sequentially complete, weak Cauchy sequences in F are weakly convergent. Whence (a). From this (b) follows whenever E has the Dieudonné property. ▌

9.4.12 Remarks. If it is known that E has both the Dieudonné and the (DP) [or (SDP)] properties—for example, if $E = C_0(T)$—then even more remarkable statements may be made about any continuous linear map u of E into F (the latter being assumed complete and weakly sequentially complete).

Thus if E possesses both the Dieudonné and (DP) properties, then u transforms

(c) weakly relatively compact sets into relatively compact sets;

and if E possesses both the Dieudonné and (SDP) properties, then u transforms

(d) weak Cauchy sequences into convergent sequences, and

(e) bounded and weakly metrizable sets into relatively compact sets.

In view of (e) it is worth considering spaces E in which each bounded set is weakly metrizable, among which are to be found all the spaces E with strongly separable duals. The following result is given by Grothendieck ([4], p. 170, Proposition 15).

9.4.13 Proposition. Let E and F be complete separated LCTVSs, and suppose that in E each bounded set is weakly metrizable. Then the following statements are true:

(1) E possesses the Dieudonné property;

(2) Any continuous linear map u of E into F, which is such that $u(x_n) \to 0$ in F whenever the sequence (x_n) converges weakly to 0 in E, transforms bounded sets into relatively compact sets;

(3) If E possesses the (DP) property, any continuous linear map u of E into F, which transforms bounded sets into weakly relatively compact sets, transforms bounded sets into relatively compact sets; and each equicontinuous subset of E', which is relatively compact for $\sigma(E', E'')$, is strongly relatively compact.

Proof. (1) has already been noted in Example (2) preceding Theorem 9.4.10.

As for (2), let A be any convex, balanced, and bounded subset of E. Since A is weakly metrizable, the hypothesis of (2) implies that $u \mid A$ is continuous at 0. Then, however, since A is convex and balanced, $u \mid A$ is uniformly continuous. In each case here we picture A endowed with the induced weak topology. Thus, since A is weakly precompact, $u(A)$ is precompact and therefore relatively compact.

Finally, if u transforms bounded sets into weakly relatively compact sets, and if E has the (DP) property, then u transforms weakly relatively compact sets into relatively compact sets. In particular, u will transform weakly convergent sequences into convergent sequences. Then (2) shows that u transforms bounded sets into relatively compact sets. The remainder of (3) ensues from Theorem 9.3.4, taking therein $\mathfrak{S}$ to consist of all bounded subsets of E. The reader will notice that if $A' \subset E'$ is $\sigma(E', E'')$-compact, it is complete for this structure, hence strongly complete; so if A' is strongly precompact it is strongly relatively compact. ∎

Remark. Grothendieck ([4], p. 171, Théorème 10) has shown that the conclusions of Proposition 9.4.13 apply whenever E is a quotient space, or a closed vector subspace, of c_0. See Exercise 9.16.

9.4.14 **Applications to Vector-valued Radon Measures.** In Section 8.19 (see especially the remarks following Subsection 8.19.2) we have broached the problem of extending the domain of definition of a vector-valued Radon measure. As was remarked, a different approach from the one adopted there is possible if results from the general theory of weakly compact linear maps are utilized. These results are those given in the preceding portions of the present section. It is now convenient for us to give some account of this alternative approach.

Throughout the discussion we assume that the separated LCTVS E is complete. To simplify matters we assume too that the separated base space T is compact, so that $\mathscr{K}(T) = C_0(T) = C(T)$ is a Banach space. The term "Radon measure" will mean "scalar-valued Radon measure on T." For the rest we adopt the notation of Section 8.19. In particular, $\vec{m}$ will denote an E-valued Radon measure on T; according to our adopted definition, then, $\vec{m}$ is a continuous linear map of $C(T)$ into E. Ideas similar to those to be discussed are set forth by Dunford and Schwartz ([1], pp. 492–498) for the case in which E is a Banach space; broadly speaking, we cover much the same ground in a rather more general setting.

(A) Let us begin by assuming merely that $\vec{m}$ is continuous. It is then clear that for each x' in E', the linear form $\vec{m}_{x'} = x' \circ \vec{m}$,

$$f \to \langle \vec{m}(f), x' \rangle,$$

on $C(T)$ is a Radon measure on T. Moreover, identifying the dual $C(T)'$ with the space $M(T)$ of Radon measures on T (see Theorem 4.10.1), the adjoint $\vec{m}'$ carries x' into the measure $\vec{m}_{x'}$.

To proceed further we must introduce the bidual $C(T)'' = M(T)'$. There is no known simple "concrete" representation of this bidual space, but an important closed vector subspace thereof may be represented as the space $U(T)$ of bounded, universally measurable functions g on T. Such a function g is identified with the element of $M(T)'$ defined by $\lambda \to \int g\, d\lambda$, so that we shall write

$$\langle g, \lambda \rangle = \int g\, d\lambda.$$

Clearly, $C(T)$ is a closed vector subspace of $U(T)$.

Although $\vec{m}(g)$ is defined only when $g \in C(T) \subset U(T)$, yet $\vec{m}''(g)$ is defined for each g in $U(T)$ and is indeed characterized by the relation

$$\langle \vec{m}''(g), x' \rangle = \langle g, \vec{m}'(x') \rangle = \int g\, d\vec{m}_{x'},$$

which must hold for each g in $U(T)$ and each x' in E'. $\vec{m}''(g)$ is an element of the bidual E''. According to the general theory (Section 8.7) we may determine $\vec{m}''(g)$ as the $\sigma(E'', E')$-limit of the directed family $(\vec{m}(f_i))$ whenever the directed family (f_i) in $C(T)$ is bounded and $\sigma(C'', M)$-convergent to g. This last requirement demands that $\lim f_i = g$ pointwise on T. Moreover by Lebesgue's Theorem 4.8.2, a *sequence* (g_i) is $\sigma(C'', M)$-convergent to g if and only if it is bounded and converges pointwise to g on T.

Let us temporarily write $\mathfrak{m}$ for the set of all universally measurable subsets of T; this is a σ-algebra (that is, it is stable under countable unions and complementation); $\mathfrak{m}$ contains all Borel subsets of T. For M in $\mathfrak{m}$ we may define $\mu(M) = \vec{m}''(\chi_M) \in E''$, so deriving an E''-valued set function defined on $\mathfrak{m}$. For each x' in E' one has

$$\langle \mu(M), x' \rangle = \vec{m}_{x'}(M),$$

hence it appears that μ is weakly countably additive. From this it follows via a theorem due to Pettis (see Dunford and Schwartz [1], p. 318), a proof of which we give in (B) below (with E in place of E''), that μ is actually countably additive.

Once this stage is reached, it is easy to show that the set function μ serves to represent the functional $\vec{m}$, in the sense that

$$\vec{m}(f) = \int f(t)\, d\mu(t),$$

where we may define the integral $\int g(t)\, d\mu(t)$ as an element of E'' for any g in $U(T)$ as the $\sigma(E'', E')$-limit of Lebesgue sums of the type

$$\sum_{i=1}^{n} c_i \cdot \mu(M_i).$$

If g is complex valued, it is convenient to deal separately with the real and imaginary parts of g. If g is real valued, a Lebesgue sum is associated with each finite sequence $c_0 < c_1 < \cdots < c_n$ of real numbers such that the interval $[c_0, c_n)$ contains the range of g and where

$$M_i = \{t \in T : c_{i-1} \leq g(t) < c_i\}.$$

The sums approximate the integral if we take a sequence of such "ladders" (c_i) for which $\text{Sup}_i\,(c_i - c_{i-1})$ tends to 0. The argument is well known and elementary when scalar-valued measures are involved (see Exercise 4.5).

These last remarks suggest a partial converse, in which we suppose given a countably additive E-valued set function μ defined for all Borel subsets of T. Then, thanks to the assumed completeness of E, the formula $\vec{m}(f) = \int f\, d\mu$ will define a continuous linear map of $C(T)$ into E (for more details, see (B) below).

The reader will notice, however, this unsatisfactory feature: Starting from $\vec{m}$, we have been led to a countably additive vector-valued set function μ for which $\vec{m}(f) = \int f\, d\mu$ for f in $C(T)$; yet we know only that μ takes its values in E'', not necessarily in E. As we shall proceed to show, this disappointing misfit proves to be symptomatic of those $\vec{m}$ that are not weakly compact.

(B) THE CASE IN WHICH $\vec{m}$ IS WEAKLY COMPACT. Let us consider the following conditions on the E-valued Radon measure $\vec{m}$:

(1) $\vec{m}$ is weakly compact;

(2) $\vec{m}''(\chi_A) \in E''$ for each closed set $A \subset T$ (or for each closed set $A \subset T$ having a countable neighbourhood base);

(2′) $\vec{m}''(C'') \subset E$;

(3) $\vec{m}'$ transforms equicontinuous subsets of E' into sets relatively compact for $\sigma(M', M)$;

(4) $\vec{m}$ transforms bounded increasing sequences in $C(T)$ into weakly convergent sequences in E;

(4′) $\vec{m}$ transforms weak Cauchy sequences in $C(T)$ into weakly convergent sequences in E;

(5) $\vec{m}$ transforms weakly relatively compact subsets of $C(T)$ into relatively compact subsets of E.

We allege that

(a) conditions (1)–(4′) are equivalent;

and that

(b) any one of (1)–(4′) implies (5).

Indeed, Theorem 9.4.10 says that (1), (2), and (4) are equivalent. Theorem 9.3.2 says that (1), (2′), and (3) are equivalent. Thus (1), (2), (2′), (3), and (4) are equivalent. (4′) trivially implies (4). On the other hand, (4) is equivalent to (2′), which implies (4′)—because any weak Cauchy sequence in $C(T)$ is weakly convergent in $C(T)''$, hence is transformed by u'' into a sequence that is weakly convergent in E'', hence (4′) when (2′) is granted. Thus (1)–(4′) are equivalent. Finally, since $C(T)$ possesses the (DP) property, (1) implies (5).

We have still to show that if $\vec{m}$ is weakly compact, then μ, which we know to be weakly countably additive, is indeed countably additive. Suppose then that (M_n) is a sequence of disjoint, universally measurable subsets of T, and let $M = \bigcup_{n=1}^{\infty} M_n$. We wish to show that the series $\sum_{n=1}^{\infty} \mu(M_n)$ is convergent in E to the sum $\mu(M)$. (So far we know only that this series converges weakly in E to the sum $\mu(M)$.) We put $a_n = \mu(M_n)$, and let $\phi \in \ell^\infty$. We define

$$g_r(t) = \sum_{n \le r} \phi(n)\chi_{M_n}(t);$$

the function g_r belongs to $U(T)$ for $r = 1, 2, \cdots$, the g_r are bounded, and they converge pointwise on T to $\sum_{n=1}^{\infty} \phi(n)\chi_{M_n}$. It follows by the remarks made in (A) that $\vec{m}''(g_r) \to \vec{m}''(g)$ for $\sigma(E, E')$. In particular, therefore, $\sum_n \phi(n)a_n$ converges weakly in E for each ϕ in ℓ^∞. From Proposition 9.3.10 (or from Remark (2) following Proposition 9.3.9, since E is quasi-complete) we infer that $\sum_n a_n$ is convergent in E; that is, that $\sum_{n=1}^{\infty} \mu(M_n)$ is convergent. Since it converges weakly to $\mu(M)$, its sum must be $\mu(M)$ (E being separated by hypothesis). This establishes Pettis' theorem.

Conversely, assume we are given a countably additive E-valued set function μ, defined on some σ-algebra $\mathfrak{m}_0 \subset \mathfrak{m}$ of subsets of T that contains all Borel sets. Let $U_0(T) \subset U(T)$ be the space of all bounded, complex-valued functions

g on T that are $\mathfrak{m}_0$-measurable (by which we mean that if g' and g'' are the real and imaginary parts of g, then the inverse image, by each of g' and g'', of any real interval belongs to $\mathfrak{m}_0$). Since E is complete by hypothesis, it is easy to show that the integral

$$v(g) = \int g \, d\mu$$

exists as an element of E for each g in $U_0(T)$. In fact, thanks to completeness, the integral exists as the limit in E of the Lebesgue sums described in (A) above. In particular, the restriction $\vec{m}$ of v to $C(T)$ is a linear map of $C(T)$ into E, and $v = \vec{m}'' \mid U_0(T)$. Since $\vec{m}''$ is continuous for $\sigma(C'', C)$ and $\sigma(E'', E')$, and since the closed unit ball of C'' is $\sigma(C'', C')$-compact, it follows at once that $\vec{m}$ is weakly compact from $C(T)$ into E. A moment's study will show that the vector-valued set function derived from this map $\vec{m}$ in the manner described in (A) above is none other than the measure μ from which we have started.

Thus we may say that the weakly compact linear maps of $C(T)$ into E—any separated and complete LCTVS—are in one-to-one correspondence with the countably additive E-valued measures defined on σ-algebras of universally measurable subsets of T that contain all Borel sets (or, equivalently, contain all open—or all closed—subsets of T). This conclusion is due to Grothendieck ([4], p. 167, Proposition 14).

We note finally that condition (2a) of Theorem 9.3.1 shows that whenever $\vec{m}$ is weakly compact, the corresponding set function μ is weakly compact valued; that is, $\mu(\mathfrak{m})$ is weakly relatively compact in E.

For another discussion of the weakly compact case, see Bartle, Dunford and Schwartz [1].

(C) THE CASE IN WHICH $\vec{m}$ IS COMPACT. In this case we may appeal to Theorem 9.2.1, taking for $\mathfrak{S}$ the bounded subsets of E and for $\mathfrak{S}'$ the equicontinuous subsets of E'. We then learn that if Q is any equicontinuous subset of E', the set $P = \{\vec{m}_{x'} : x' \in Q\}$ is relatively compact in $M(T)$. It is left as an exercise for the reader to show that this implies that $\mu(\mathfrak{m})$ is precompact in E, hence relatively compact since E is quasi-complete. We describe this state of affairs by saying that μ is compact valued (compare the situation in (B)).

As in (B), there is a converse: If μ is a compact valued, countably additive E-valued set function, and if E is quasi-complete, then $\vec{m}(f) = \int f \, d\mu$ is a compact linear map of $C(T)$ into E (as follows easily from Krein's Theorem 8.13.1), and the set function derived from this $\vec{m}$ by the procedure described in (A) is none other than μ itself.

9.4.15 **Remarks.** (1) Dinculeanu [3] has examined the case of set functions, defined on Borel subsets of a compact space T, and taking values in $L_c(E, F)$ (E and F Banach spaces) with a view to their use in integral representations of linear maps of the space of continuous E-valued functions on T into F. In his earlier papers [1] and [2], the same author discusses other such representation theorems.

(2) A good deal of attention has been directed to the study of continuous linear maps with domain the space $C(S)$, in the very special case in which S is separated, compact, and Stonian. (A topological space S is Stonian if, in S, the closure of any open set is again open.)

Grothendieck ([4], p. 168, Théorème 9 and Corollaire 1) shows that in $M(S)$ (the space of bounded Radon measures on S) every sequence which is $\sigma(M, C)$-convergent is weakly (that is, $\sigma(M, M')$-) convergent; and, as a consequence, that every continuous linear map of $C(S)$ into a complete, separated, and separable LCTVS is weakly compact.

Grothendieck's arguments call upon certain other interesting properties of $C(S)$, in particular the property that if $C(S)$ is a closed vector subspace of any Banach space X, then $C(S)$ admits a topological complement relative to X. This property characterizes the spaces $C(S)$ among all Banach spaces. For these aspects, see Nachbin [3], Goodner [1], and Kelley [3].

It is, of course, important to be able to recognize spaces that are isomorphic to $C(S)$. An aid to this is a theorem of Kakutani running parallel Theorem 4.23.2, also due to him and characterizing abstractly the L^1 spaces. A Riesz space E with a norm satisfying the conditions laid out in Subsection 4.23.1, save that (L) is replaced by

$$(M) \text{ if } x \geq 0 \text{ and } y \geq 0, \text{ then } \|x \vee y\| = \text{Sup}\,(\|x\|, \|y\|),$$

is termed an *abstract* (M)*-space*. A *unit* e in an abstract (M)-space is an element possessing the following two properties:

$$(U)\ e \geq 0 \text{ and } \|e\| = 1,$$

$$(U')\ \text{if } \|x\| \leq 1, \text{ then } x \leq e.$$

Kakutani ([6], Theorem 2) showed that if E is an abstract (M)-space with a unit e, then E is isometric and isomorphic (linearly and with regard to order) with a space $C_R(T)$ for some separated compact space T, the space T being unique up to homeomorphism. Here, $C_R(T)$ is the real Banach space of real-valued continuous functions on T, with norm $\|f\| = \text{Sup}\,\{|f(t)| : t \in T\}$ and with its natural partial order ($f \leq g$ signifying that $f(t) \leq g(t)$ for all t in T).

On the other hand, it may be shown (Bourbaki [6], Chapitres I–IV, p. 32, Exercice 13f)) that $C_R(T)$ is order-complete (that is, has the property that any nonempty majorized subset has a supremum) if and only if T is Stonian.

One infers, then, that any order-complete abstract (M)-space with unit is isometric and isomorphic with $C_R(S)$ for some separated, compact, Stonian S that is uniquely determined up to homeomorphism amongst all separated, compact spaces.

There are two notable instances. (a) $E = \ell^\infty(I)$, I being any set (regarded as a discrete space); the appropriate S is in this case the Stone-Čech compactification of I. (b) $E = \underline{L}^\infty = \underline{L}^\infty(T, \mu)$, T being any separated locally compact space and μ any positive Radon measure on T. The reader may try his hand (Exercise 9.17) at verifying that the real elements of $\underline{L}^\infty$ constitute an order-complete abstract (M)-space with unit. An element $\dot{f}$ of $\dot{L}^\infty$ is real if it has a representative function that is real valued, and the relation $\dot{f} \leq \dot{g}$ subsists between two real elements if and only if they possess representative functions f and g such that $f(t) \leq g(t)$ l.a.e.

9.5 Integral Operators and Kernel Representations

Many of the earliest examples of compact linear maps are to be found in connection with integral equations and problems which can be thrown into that form. The associated linear maps carry one function space into another and are of the type

$$\begin{aligned} y &= u(x), \\ y(t) &= \int k(t, s)x(s)\, d\sigma(s). \end{aligned} \tag{9.5.1}$$

It is here supposed that s and t denote variable points of two separated locally compact spaces S and T with positive Radon measures σ and τ, respectively, and that k is a given function or "kernel" defined on $T \times S$ which we shall assume outright to be measurable for $\tau \otimes \sigma$. Under suitable conditions, examples of which will comprise the first half of this section, (9.5.1) serves to define u as a continuous linear map of $L^p(S, \sigma)$ into $L^q(T, \tau)$, or from $C(S)$ into $C(T)$, or from and into various other familiar function spaces. More refined conditions on k will in turn arrange that u is compact. In discussing these matters we shall make no notational distinction between a function and the class (element of $L^p(S, \sigma)$ or $L^q(T, \tau)$, as the case may be) that it defines, leaving the reader to make the necessary mental adjustment.

In the second half of this section we shall comment briefly on the converse problem—namely, that of representing a given map u in the form of (9.5.1). Generally speaking, this is a more delicate problem. Some pioneering work is due to Dunford and Pettis [1] (especially Part 3), see also Dunford and Schwartz [1], p. 505, Corollary VI.8.9. In both cases, maps from L^p into L^q are considered. Kernel representations for weakly compact or compact linear maps from $C(S)$ into a Banach space have been studied by Bartle [1], Bartle, Dunford, and Schwartz [1], and by Wada [1] in a very general setting. We shall give some account of these developments.

Integral operators of the type in (9.5.1) and the associated integral equations arise in connection with differential equations, both ordinary and partial, and in potential theory. For the most part we must leave the reader to explore these connecting links; see, for example, Smithies [2], Courant-Hilbert [1], [2], Coddington and Levinson [1]. However, Section 9.12 will sketch an approach to differential equations via the introduction of compact linear maps and the associated theory in such a way that the kernel representation of the maps is of secondary importance.

We shall begin the first and more concrete part of our programme by dealing with some inequalities of Young bearing upon the continuity of maps of (9.5.1) considered as acting from $L^p(S, \sigma)$ into $L^q(T, \tau)$.

9.5.1 Theorem. (a) Suppose that $0 < a < +\infty$, $0 < b < +\infty$, and that

$$\tau\text{-ess sup}\left\{\int |k(t, s)|^a \, d\sigma(s)\right\}^{1/a} = M_1 < +\infty, \tag{9.5.2}$$

$$\sigma\text{-ess sup}\left\{\int |k(t, s)|^b \, d\tau(t)\right\}^{1/b} = M_2 < +\infty. \tag{9.5.3}$$

Then u, defined by (9.5.1), is a continuous linear map of $L^p(S, \sigma)$ into $L^q(T, \tau)$ whenever

$$\frac{1}{p} - \left(\frac{b}{a}\right)\frac{1}{q} = 1 - \frac{1}{a}, \qquad 1 \le p \le q < +\infty; \tag{9.5.4}$$

and then

$$\|u\| \le M_1^{1-b/a} M_2^{b/a}. \tag{9.5.5}$$

(b) Suppose that $1 \le a \le +\infty$ and that (9.5.2) holds. Then u is a continuous linear map from $L^a(S, \sigma)$ into $L^\infty(T, \tau)$, and $\|u\| \le M_1$.

(c) Suppose that $1 \le b \le +\infty$ and that (9.5.3) holds. Then u is a continuous linear map of $L^1(S, \sigma)$ into $L^b(T, \tau)$, and $\|u\| \le M_2$.

(a′) Suppose that $\sigma(S)$ and $\tau(T)$ are finite, and that (9.5.2) and (9.5.3) hold with $0 < a < +\infty$, $0 < b < +\infty$. Then u is a continuous linear map of $L^p(S, \sigma)$ into $L^q(T, \tau)$ whenever there exists $q_1 \ge q$ satisfying

$$\frac{1}{p} - \left(\frac{b}{a}\right)\frac{1}{q_1} \le 1 - \frac{1}{a}, \qquad 1 \le p \le q_1 \le +\infty. \tag{9.5.4′}$$

(b′) Suppose that $\sigma(S)$ and $\tau(T)$ are finite and that (9.5.2) holds with $1 \le a \le +\infty$. Then u is a continuous linear map of $L^p(S, \sigma)$ into $L^q(T, \tau)$ whenever

$$\frac{1}{p} \le 1 - \frac{1}{a}, \qquad 0 < p \le +\infty, \qquad 0 < q \le +\infty;$$

and then $\|u\| \le M_1 \tau(T)^{1/q} \sigma(S)^{1-1/a-1/p}$.

(c′) Suppose that $\sigma(S)$ and $\tau(T)$ are finite and that (9.5.3) holds with $1 \le b \le +\infty$. Then u is a continuous linear map of $L^1(S, \sigma)$ into $L^q(T, \tau)$ whenever $0 < q \le b$; and then $\|u\| \le M_2 \cdot \tau(T)^{1/q-1/b}$.

Proof. (a) Hölder's inequality shows that (for any positive measure μ)

$$\int |f \cdot g \cdots h| \, d\mu \le \left(\int |f|^\alpha \, d\mu\right)^{1/\alpha} \left(\int |g|^\beta \, d\mu\right)^{1/\beta} \cdots \left(\int |h|^\gamma \, d\mu\right)^{1/\gamma}$$

whenever $0 < \alpha, \beta, \cdots, \gamma \le +\infty$ and $1/\alpha + 1/\beta + \cdots + 1/\gamma = 1$. (One or more of the $\alpha, \beta, \cdots, \gamma$ may be $+\infty$, the corresponding integral on the right being replaced by the appropriate ess sup.) We take here $1/\alpha = 1/q$, $1/\beta = 1/a - b/aq$, $1/\gamma = 1/p - 1/q$. Then (9.5.4) shows that Hölder's inequality is applicable and gives

$$\begin{aligned}
|y(t)| &\le \int \{|k(t, s)|^{b/q} |x(s)|^{p/q}\}\{|k(t, s)|^{a(1/a-b/aq)}\}\{|x(s)|^{p(1/p-1/q)}\} \, d\sigma(s) \\
&\le \left\{\int |k(t, s)|^b |x(s)|^p \, d\sigma(s)\right\}^{1/q} \left\{\int |k(t, s)|^a \, d\sigma(s)\right\}^{1/\beta} \left\{\int |x(s)|^p \, d\sigma(s)\right\}^{1/\gamma} \\
&\le \left\{\int |k(t, s)|^b |x(s)|^p \, d\sigma(s)\right\}^{1/q} \cdot M_1^{a/\beta} \cdot (\|x\|_p)^{p/\gamma},
\end{aligned}$$

the last step by use of (9.5.2), and the inequality holding for almost all t. Hence

$$\begin{aligned}\|y\|_q^q &\le M_1^{aq/\beta}\,\|x\|_p^{pq/\gamma}\int d\tau(t)\int|k(t,s)|^b\,|x(s)|^p\,d\sigma(s)\\ &= M_1^{aq/\beta}\,\|x\|_p^{pq/\gamma}\int|x(s)|^p\,d\sigma(s)\int|k(t,s)|^b\,d\tau(t)\\ &\le M_1^{aq/\beta}\,\|x\|_p^{pq/\gamma}\,\|x\|_p^p\,M_2^b.\end{aligned}$$

Since $a/\beta = 1 - b/q$ and $p/\gamma + p/q = 1 - p/q + p/q = 1$, the stated result is established.

(b) If $a \ge 1$ and (9.5.2) holds, the ordinary form of Hölder's inequality gives

$$|y(t)| \le \int|k(t,s)|\,|x(s)|\,d\sigma(s) \le \|x\|_{a'}\cdot\left\{\int|k(t,s)|^a\,d\sigma(s)\right\}^{1/a}$$

for almost all t, hence the result.

(c) If $b \ge 1$, (9.5.3) holds, and x is in $L^1(S,\sigma)$, we can choose z in $L^{b'}(T,\tau)$ so that $\|z\|_{b'} = 1$ and

$$\|y\|_b = \int y(t)z(t)\,d\tau(t).$$

Hölder's inequality then gives

$$\begin{aligned}\|y\|_b &= \int z(t)\,d\tau(t)\int k(t,s)x(s)\,d\sigma(s)\\ &= \int x(s)\,d\sigma(s)\int k(t,s)z(t)\,d\tau(t)\\ &\le \int |x(s)|\,d\sigma(s)\cdot\|z\|_{b'}\cdot\left\{\int|k(t,s)|^b\,d\tau(t)\right\}^{1/b},\end{aligned}$$

and the assertion follows.

(a′) For finite measure spaces, continuity from L^p into L^{q_1} implies that from L^p into L^q, whenever $q_1 \ge q$. Let p, q_1 satisfy (9.5.4′), so that

$$\frac{1}{p} - \left(\frac{b}{a}\right)\frac{1}{q_1} = 1 - \frac{1}{a} - d, \qquad d \ge 0.$$

We define q_2 by the relation $1/p - (b/a)1/q_2 = 1 - 1/a$. There are two cases to consider.

Suppose first that $1/p > 1 - 1/a$. Then $0 < q_2 < +\infty$ and $q_1 \le q_2$. By (a), u is continuous from L^p into L^{q_2}, hence a fortiori from L^p into L^{q_1} and from L^p into L^q.

In the contrary case, $1/p \le 1 - 1/a$ and so (b′) applies.

(b′) Suppose that $1 \le a \le +\infty$, $1/p \le 1 - 1/a$, and that k satisfies (9.5.2). According to (b), u is continuous from $L^{a'}$ into L^∞. Now $1/p \le 1 - 1/a = 1/a'$; so $a' \le p$. Hence, since the measure spaces involved are finite,

$$\begin{aligned}\|y\|_q &\le \tau(T)^{1/q}\cdot\|y\|_\infty \le \tau(T)^{1/q}\cdot M_1\cdot\|x\|_{a'}\\ &\le \tau(T)^{1/q}M_1\sigma(S)^{1/a'-1/p}\,\|x\|_p,\end{aligned}$$

as we had to show.

(c′) If $1 \le b \le +\infty$ and k satisfies (9.5.3), then (c) shows that u is continuous from L^1 into L^b. Since $q \le b$ and the measure spaces are finite, u is continuous from L^1 into L^q; and

$$\|y\|_q \le \tau(T)^{1/q-1/b} \|y\|_b \le \tau(T)^{1/q-1/b} M_2 \|x\|_1.$$

The proof is complete. ∎

Regarding compactness we have two relatively simple results.

9.5.2 Proposition. Suppose S and T are compact and k continuous on $T \times S$. Then u, defined by (9.5.1), is compact from $C(S)$ into $C(T)$, from $L^p(S)$ into $C(T)$, from $C(S)$ into $L^q(T)$, and from $L^p(S)$ into $L^q(T)$ whenever $1 \le p \le +\infty$ and $0 < q \le +\infty$.

Proof. Since the injection maps of C into L^r and of L^p into L^1 are continuous for any $r > 0$ and any $p \ge 1$, it suffices to show that u is compact from $L^1(S)$ into $C(T)$. Now S and T may be endowed with uniform structures defining their respective topologies, and continuity of k implies its uniform continuity. It is then clear that u transforms the unit ball of $L^1(S)$ into a subset of K^T that is bounded and equicontinuous, hence (Ascoli's theorem) relatively compact in $C(T)$. ∎

Remark. For a deeper result, see Proposition 9.5.17 to follow.

9.5.3 Proposition (Banach). Suppose that S and T are compact, that $1 \le r < +\infty$, and that

$$M = \left\{\iint |k(t, s)|^r \, d\sigma(s) \, d\tau(t)\right\}^{1/r} < +\infty. \tag{9.5.6}$$

Then u is compact from $L^p(S, \sigma)$ into $L^q(T, \tau)$ whenever

$$r' \le p \le +\infty, \qquad 0 < q \le r. \tag{9.5.7}$$

Proof. Since $p \ge r'$, so $p' \le r$. Hence, $\sigma(S)$ being finite,

$$\begin{aligned} |y(t)| &\le \|x\|_p \cdot \left\{\int |k(t, s)|^{p'} \, d\sigma(s)\right\}^{1/p'} \\ &\le \|x\|_p \cdot \left\{\int |k(t, s)|^r \, d\sigma(s)\right\}^{1/r} \cdot \text{const.} \end{aligned}$$

Since $q \le r$ and $\tau(T)$ is finite, so

$$\begin{aligned} \|y\|_q &\le \text{const} \, \|y\|_r \le \text{const} \, \|x\|_p \cdot \left\{\int d\tau(t) \int |k(t, s)|^r \, d\sigma(s)\right\}^{1/r} \\ &\le \text{const} \, M \cdot \|x\|_p. \end{aligned} \tag{9.5.8}$$

This shows that u is continuous from L^p into L^q.

The compactness of u follows from Proposition 9.5.2 whenever k is continuous. Otherwise, we approximate k in $L^r(T \times S, \tau \otimes \sigma)$ by continuous kernels k_i, noting that (9.5.8) shows that the corresponding maps u_i converge strongly to u, and appeal to the following simple lemma. ∎

9.5.4 Lemma. Let E be a normed vector space, F a TVS, (u_i) a directed family of precompact linear maps of E into F that converge for the strong topology on $L_c(E, F)$ (that is, uniformly on bounded subsets of E) to u. Then u is precompact.

Proof. Let B be the unit ball in E, V any neighbourhood of 0 in F. We choose i so that $u(x) - u_i(x) \in V$ for x in B. Since $u_i(B)$ is precompact, it appears that $u(B) \subset u_i(B) + V$ has the same property. Thus u is precompact. ▌

9.5.5 **Remark.** The reader will verify easily that if S and T are not compact, the assertions of Propositions 9.5.2 and 9.5.3 remain valid, provided we assume additionally that k has a compact support.

The next result is less obvious.

9.5.6 **Theorem.** Let us suppose that $\sigma(S)$ and $\tau(T)$ are finite.

(a) Assume that k satisfies (9.5.2) and (9.5.3) with $0 < a \le +\infty$ and $0 < b < +\infty$. Then u, defined by (9.5.1), is compact from $L^p(S, \sigma)$ into $L^q(T, \tau)$ whenever there exists $q_1 \ge q$ such that

$$\frac{1}{p} - \left(\frac{b}{a}\right)\frac{1}{q_1} < 1 - \frac{1}{a}, \qquad 1 < p \le q_1 < +\infty, \qquad b < q_1. \tag{9.5.9}$$

(b) Suppose that $1 < a \le +\infty$ and that k satisfies (9.5.2). Then u is compact from $L^p(S, \sigma)$ into $L^q(T, \tau)$ whenever $0 < p \le +\infty$, $0 < q \le +\infty$ and $1/p < 1 - 1/a$.

Proof. (a) Since $\tau(T)$ is finite, it is enough to show that u is compact from L^p into L^{q_1} whenever p and q_1 satisfy (9.5.9). We define a_1 by

$$\frac{1}{p} - \left(\frac{b}{a_1}\right)\frac{1}{q_1} = 1 - \frac{1}{a_1}.$$

Then $0 < a_1 < +\infty$. Further, (9.5.9) can be written

$$\frac{1}{p} - \left(\frac{b}{a}\right)\frac{1}{q_1} = 1 - \frac{1}{a} - d$$

for some $d > 0$. Since $q_1 - b > 0$, so $a_1 < a$.

We define the kernel k_n ($n = 1, 2, \cdots$) to be equal to k whenever this has absolute value at most n and to be 0 elsewhere, and let u_n be the map defined by k_n. Since k_n is bounded and $\sigma(S)$ and $\tau(T)$ are finite, u_n is compact from L^p into L^{q_1}: this follows from Proposition 9.5.3 on taking therein any finite $r \ge \operatorname{Max}(p', q_1)$. In view of Lemma 9.5.4, it will now suffice to show that

$$\|u(x) - u_n(x)\|_{q_1} \le M_n \|x\|_p$$

where $M_n \to 0$ as $n \to \infty$.

Now Theorem 9.5.1(a) shows that we may assume that

$$M_n \le (M_n^1)^{1-b/q_1}(M_n^2)^{b/q_1},$$

where

$$M_n^1 = \tau\text{-ess sup}\left\{\int |k(t, s) - k_n(t, s)|^{a_1}\, d\sigma(s)\right\}^{1/a_1},$$

$$M_n^2 = \sigma\text{-ess sup}\left\{\int |k(t, s) - k_n(t, s)|^{b}\, d\tau(t)\right\}^{1/b}.$$

Since $|k - k_n| \leq |k|$, so M_n^2 is bounded. On the other hand, since $a_1 < a$, we may write $a - a_1 = c > 0$. Then, in view of the fact that $|k - k_n|$ is either 0 or larger than n, we have

$$|k - k_n|^{a_1} \leq n^{-c}\, |k - k_n|^a \leq n^{-c}\, |k|^a$$

and therefore

$$M_n^1 \leq \tau\text{-ess sup}\, \left\{n^{-c} \int |k(t, s)|^a \, d\sigma(s)\right\}^{1/a_1}$$
$$\leq (n^{-c} M_1^a)^{1/a_1} \to 0$$

as $n \to \infty$. This completes the proof of (a).

(b) The proof here is similar to that of (a), using Theorem 9.5.1(b′). ▌

9.5.7. We turn next to a useful refinement of Proposition 9.5.2, weakening the demand that k be continuous. We shall say that k is quasi-equicontinuous at a point t_0 of T if for each $\varepsilon > 0$ there exists a set $S_0 \subset S$ with $\sigma(S_0) < \varepsilon$, and a neighbourhood U of t_0, such that

$$|k(t, s) - k(t_0, s)| < \varepsilon$$

whenever $(t, s) \in U \times (S \backslash S_0)$.

9.5.8 **Theorem** (Kantorovich). Let us suppose that $\sigma(S)$ and $\tau(T)$ are finite, that k satisfies (9.5.2) for some $a > 0$, and that k is quasi-equicontinuous at each point of T. Then u is compact from $L^p(S, \sigma)$ into $C_{bd}(T)$† whenever $1/p < 1 - 1/a$, $0 < p \leq +\infty$.

Proof. In view of Theorem 9.5.6(b), with $q = \infty$, it suffices to show that u maps L^p into $C_{bd}(T)$. Now

$$y(t) - y(t_0) = \int \{k(t, s) - k(t_0, s)\} x(s)\, d\sigma(s).$$

We choose c satisfying $1 < c < a$, $1/p < 1 - 1/c$. In Theorem 9.5.1(b′) we take $T = \{t\}$, τ the measure placing unit mass at t, $q = +\infty$, and kernel $k(t, s) - k(t_0, s)$, obtaining

$$|y(t) - y(t_0)| \leq \left\{\int |k(t, s) - k(t_0, s)|^c \, d\sigma(s)\right\}^{1/c} \cdot \|x\|_p.$$

We aim to show that the first factor on the right-hand side tends to 0 as $t \to t_0$. This factor is majorized by

$$\left\{\int_{S \backslash S_0} |\cdots|^c \, d\sigma\right\}^{1/c} + \left\{\int_{S_0} |\cdots|^c \, d\sigma\right\}^{1/c}.$$

The first term here is at most $\varepsilon \cdot \sigma(S)^{1/c}$ when t belongs to U; and by Hölder's inequality the second term is at most

$$\left\{\int_{S_0} |\cdots|^{c \cdot a/c} \, d\sigma\right\}^{1/a} \cdot \left\{\int_{S_0} d\sigma(s)\right\}^{(1-c/a)/c} \leq 2M_1 \cdot \varepsilon^{1/c - 1/a},$$

which tends to 0 with ε because $1/c - 1/a > 0$. The proof is thus complete. ▌

---

† $C_{bd}(T)$ is the space of bounded, continuous functions on T with the norm $\|y\| = \text{Sup}\, \{|y(t)| : t \in T\}$.

9.5.9. With the aid of the last theorem we can establish, by way of example, some properties of the important kernels of "potential type." For this, $S = T$ is a bounded subset D of R^n, $\sigma = \tau$ being the restriction to D of n-dimensional Lebesgue measure. Extending the idea of the kernel r^{2-n} defining the Newtonian potential in R^n, M. Riesz introduced the potentials of order α, $0 < \alpha \leq n$, generated by the kernel

$$k(t, s) = |t - s|^{\alpha - n} = r^{\alpha - n}.$$

Formal partial differentiation of the corresponding potentials leads one to consider the more general kernel

$$k(t, s) = B(t, s) \cdot |t - s|^{-\lambda} \tag{9.5.10}$$

in which B is bounded and also continuous for $t \neq s$, and $0 < \lambda < n$. We consider briefly what some of the preceding results have to say about integral operators of this type.

Since the function B and the set D are bounded, (9.5.2) and (9.5.3) hold whenever $\lambda a < n$ and $\lambda b < n$, respectively. (Without further assumptions about the behaviour of $B(t, s)$ near the diagonal $t = s$, neither of these conditions would hold for $D = R^n$ itself.)

We proceed to interpret Theorems 9.5.1(a′), 9.5.1(c′), and 9.5.8 in the case under discussion.

9.5.10 Theorem. Let D be a bounded subset of R^n, k the kernel (9.5.10), the function $B(t, s)$ being bounded everywhere and continuous for $t \neq s$, and let u be defined by (9.5.1) with $T = S = D$ and σ and τ the Lebesgue measure.

(a) If $1 < p \leq +\infty$, $0 < q < +\infty$ and

$$\frac{1}{p} - \frac{1}{q} < 1 - \frac{\lambda}{n}, \tag{9.5.11}$$

then u is continuous from $L^p(D)$ into $L^q(D)$.

(b) If $0 < q < n/\lambda$, then u is continuous from $L^1(D)$ into $L^q(D)$.

(c) If $p > n/(n - \lambda)$, then u is compact from $L^p(D)$ into $C_b(D)$.

Proof. (a) We apply Theorem 9.5.1(a′), taking therein $a = (1 - \alpha)n/\lambda$ and $b = (1 - \beta)n/\lambda$, where α and β are to be chosen later subject to $0 < \alpha < 1$, $0 < \beta < 1$. We write temporarily $q_0 = \text{Max}\,(q, p)$. The possible choice of q_1 is then described by $q_1 = q_0(1 + \gamma)$ where $0 \leq \gamma < +\infty$. It is then found that (cf. (9.5.4′))

$$\frac{1}{p} - \left(\frac{b}{a}\right)\frac{1}{q_1} - \left(1 - \frac{1}{a}\right) = \frac{1}{p} - \frac{1}{q_0} - \left(1 - \frac{\lambda}{n}\right) + R,$$

where

$$R = \frac{q_0^{-1}\gamma}{(1 + \gamma)} + \frac{q_0^{-1}(\beta - \alpha)}{(1 - \alpha)(1 + \gamma)} + \left(\frac{\lambda}{n}\right)\left(\frac{\alpha}{1 - \alpha}\right).$$

Now the relation

$$\frac{1}{p} - \frac{1}{q_0} < 1 - \frac{\lambda}{n}$$

is equivalent to

$$\frac{1}{p} < \frac{1}{q_0} + 1 - \frac{\lambda}{n} = \operatorname{Min}\left(\frac{1}{p}, \frac{1}{q}\right) + 1 - \frac{\lambda}{n}$$
$$= \operatorname{Min}\left(1 - \frac{\lambda}{n} + \frac{1}{p}, 1 - \frac{\lambda}{n} + \frac{1}{q}\right).$$

Since $1 - \lambda/n > 0$, this signifies that

$$\frac{1}{p} < 1 - \frac{\lambda}{n} + \frac{1}{q},$$

which is (9.5.11). Thus

$$\frac{1}{p} - \frac{1}{q_0} - \left(1 - \frac{\lambda}{n}\right) = -r$$

where $r > 0$ is independent of α, β, and γ. So

$$\frac{1}{p} - \left(\frac{b}{a}\right)\frac{1}{q_1} - \left(1 - \frac{1}{a}\right) = -r + R.$$

It remains merely to verify that α, β, and γ can be chosen so as to make $R \leq r$. The verification is trivial, and our assertion (a) follows from Theorem 9.5.1(a′).

(b) In Theorem 9.5.1(c′) we take b satisfying $0 < b < n/\lambda$, $1 \leq b < +\infty$, taking b as close as we wish to n/λ (which is >1) and exceeding q.

(c) To apply Theorem 9.5.8 we must arrange that $a < n/\lambda$ and $0 < p \leq +\infty$, $1/p < 1 - 1/a$. This is clearly possible whenever $p > 0$ and $1/p < 1 - \lambda/n$, that is, $p > n/(n - \lambda)$.

The proof is complete. ▮

9.5.11 Notes. A few more results of the type dealt with so far in the present section are included as Exercises 9.18–9.22.

We add some notes about, and references to, further work connected with integral transforms that has no direct bearing upon the main theme of this chapter.

The study of integral operators has been pursued vigorously and there is a large and specialized literature. Apart from what has been written specifically about integral equations (see the references mentioned at the beginning of this section), the reader may consult the relevant portions of Zaanen [1] and Dunford and Schwartz [1]. In the examples we have studied so far the integrals involved exist (at least almost everywhere) as Lebesgue-Stieltjes integrals; that is, the integrals are absolutely convergent. For several integral transforms important in analysis this is no longer the case: the integrals are "singular" and exist (perhaps almost everywhere) as Cauchy principal values. For work in this direction see Cotlar [1], [2], Calderón and Zygmund [1], [3], [4] and [5], and Koizumi [1], and the references there cited.

One of the most important instances of singular integral transforms is the Fourier transform when one wishes to regard it as defined for certain non-integrable functions: see, for example, Titchmarsh [1] and Bochner [1]. In

Chapter 10 we shall by way of an example consider what is in one sense the simplest (L^2) case of this problem for locally compact Abelian groups.

Another most important instance of more recent origin is the Hilbert transform, which has been studied in considerable detail. The study of Fourier integrals itself leads to that of the Hilbert transform in one dimension, the kernel of which is

$$k(t, s) = (t - s)^{-1}.$$

For this case see Titchmarsh [1] and the references there cited. Results were in this case first obtained by the use of complex variable methods. In $n > 1$ dimensions the Hilbert transform arises out of potential theory and, in fact, is a generic term for various integral transforms with kernels

$$k(t, s) = |t - s|^{-n} \cdot \omega\left(\frac{t - s}{|t - s|}\right),$$

where ω is one of a certain restricted class of functions defined on the unit sphere in R^n. Real variable methods (which cover also the case $n = 1$) are discussed in Cotlar [1]; see also Koizumi [2] and the references there listed.

Important tools in all these investigations are the concepts of strong and weak types for transforms from (subsets of) L^p into L^q, the Riesz-Thorin convexity theorem for strong types (Dunford and Schwartz [1], p. 520 et seq., and Cotlar [1]), and a similar interpolation theorem of Marcinkiewicz and Zygmund for weak types (Zygmund [3], Cotlar [1], Koizumi [1]). See also Section 4.24.

9.5.12. At this point we embark on the second half of the programme to be covered in the present section. Before dealing with the kernel representation of linear maps from $C(S)$ into $C(T)$, we shall give some results about maps from a fairly general separated LCTVS E into a space $C(T)$. The main source is Wada [3]. Whereas Wada deals with the case in which S and T are not necessarily separated locally compact, we shall retain this hypothesis. As usual, $C(S)$ [resp. $C(T)$] denotes the space of continuous complex-valued functions on S [resp. T] equipped with the topology of locally uniform convergence. For brevity we shall often write F in place of $C(T)$. The dual F' is then identifiable with the space of complex Radon measures on T having compact supports.

Throughout the rest of Section 9.5, u will denote a continuous linear map of E into $F = C(T)$. The adjoint u' maps F' into E' and is continuous for the topologies $\sigma(F', F)$ and $\sigma(E', E)$. We let ε_t denote the Dirac measure at t, regarded as an element of F', and put $\xi(t) = u'(\varepsilon_t)$. Then ξ is a mapping of T into E' that is continuous for $\sigma(E', E)$. As will be shown directly, u and certain of its properties are expressible in terms of ξ.

To begin with one has

$$u(x)(t) = \langle u(x), \varepsilon_t \rangle = \langle x, u'(\varepsilon_t) \rangle$$

and so u finds its expression through the formula

$$u(x)(t) = \langle x, \xi(t) \rangle. \tag{9.5.12}$$

It is evident that if a given linear map u of E into $C(T)$ is at all expressible via (9.5.12), for some map ξ of T into E', then ξ is uniquely determined. A little less obvious is the next assertion.

9.5.13 **Proposition.** Suppose that E is barrelled. A linear map u of E into $C(T)$ is continuous if and only if there exists a map ξ of T into E' for which (9.5.12) holds and that is continuous for $\sigma(E', E)$.

Proof. In view of the preceding remarks, it remains only to show that if E and ξ satisfy the stated conditions, then u, defined by (9.5.12) is continuous. However, continuity of ξ entails that $\xi(K)$ is $\sigma(E', E)$-compact for each compact set K in T. If E is barrelled, $\xi(K)$ is therefore equicontinuous. Hence there exists a neighbourhood U of 0 in E such that $\xi(K) \subset U^0$ and so $|u(x)(t)| \leq 1$ for x in U and t in K, which signifies that u is continuous from E into $C(T)$. ∎

In the next result we see how to express weak compactness of u in terms of ξ.

9.5.14 **Proposition.** Let E be any separated LCTVS, u a continuous linear map of E into $F = C(T)$, and ξ the map of T into E' for which (9.5.12) is true. Then u is weakly compact if and only if ξ has the following property: there exists a convex, balanced, equicontinuous, and weakly closed subset Q of E' such that ξ maps T into E'_Q (the Banach space $\bigcup \{\alpha Q : \alpha > 0\}$ in which the closed unit ball is Q) and is continuous for $\sigma(E'_Q, (E'_Q)')$.

Proof. We first reduce ourselves to the case in which E is a normed space by the following device: If u is weakly compact, there exists a closed, convex, balanced neighbourhood U of 0 in E such that $u(U)$ is weakly relatively compact in F. We put $N = \bigcap \{n^{-1}U : n = 1, 2, \cdots\}$ and let $\dot{E}$ be the quotient space E/N endowed with the norm obtained by passage to the quotient from the gauge of U. If j is the natural map of E onto $\dot{E}$, then $u = v \circ j$, where v is weakly compact from $\dot{E}$ into F. Granted the proposition for normed spaces, it would follow that there exists a map ξ_1 of T into $\dot{E}'$ that is continuous for $\sigma(\dot{E}', \dot{E}'')$ and such that

$$u(x)(t) = v(\dot{x})(t) = \langle \dot{x}, \xi_1(t) \rangle$$

for $\dot{x} = j(x)$. One could then take $Q = U^0$ (polar in E') and $\xi = \xi_1$, having identified $\dot{E}'$ with E'_Q.

Conversely, if ξ has the stated property, we may repeat the preceding construction with $U = Q^0$ (which will be a neighbourhood of 0 in E) to yield an expression $u = v \circ j$. If the proposition is established for the normed case, v will be weakly compact. Hence so also will be u.

In this way it appears sufficient to handle the case in which E is a normed space, which restriction we impose hereafter.

In this case the condition involving ξ amounts merely to its continuity for $\sigma(E', E'')$.

Now if u is weakly compact, then (Theorem 9.3.1) u' is continuous for $\sigma(F', F)$ and $\sigma(E', E'')$, hence it appears at once that ξ is continuous for $\sigma(E', E'')$.

Conversely, assume that ξ is continuous for $\sigma(E', E'')$. We must prove (Theorem 9.3.1 again) that u' is continuous for $\sigma(F', F)$ and $\sigma(E', E'')$.

The proof of this depends upon showing that

$$u'(\mu) = \int \xi \, d\mu \tag{9.5.13}$$

for each μ in F'—that is, each Radon measure μ on T having a compact support. The existence of the integral and its membership to E' may be seen by endowing E' with the topology $\sigma(E', E'')$, using the continuity of ξ for that topology, and appealing to Proposition 8.14.3. By definition of the integral one has for any x'' in E''

$$\left\langle x'', \int \xi \, d\mu \right\rangle = \int \langle x'', \xi(t) \rangle \, d\mu(t).$$

In particular, replacing x'' by an arbitrary element x of E, one has by (9.5.12)

$$\left\langle x, \int \xi \, d\mu \right\rangle = \int \langle x, \xi(t) \rangle \, d\mu(t) = \int u(x)(t) \, d\mu(t)$$
$$= \langle u(x), \mu \rangle = \langle x, u'(\mu) \rangle,$$

whence (9.5.13).

Now, if a directed family (μ_i) converges to μ for the topology $\sigma(F', F)$, then (9.5.13) and the continuity of ξ for $\sigma(E', E'')$ combine to give

$$\langle x'', u'(\mu_i) \rangle = \left\langle x'', \int \xi \, d\mu_i \right\rangle = \int \langle x'', \xi \rangle \, d\mu_i \to \int \langle x'', \xi \rangle \, d\mu$$
$$= \left\langle x'', \int \xi \, d\mu \right\rangle = \langle x'', u'(\mu) \rangle,$$

which establishes the required continuity of u'. ▮

As a companion to the last proposition is one which expresses compactness of u in terms of ξ.

9.5.15 **Proposition.** Let E, u, and ξ be as in Proposition 9.5.14. Then u is compact if and only if ξ has the following property: there exists a convex, balanced, equicontinuous, and weakly closed subset Q of E' such that ξ maps T into E'_Q and is continuous for the normed topology on E'_Q.

Proof. If ξ fulfils the stated condition, an application of Ascoli's theorem shows that $u(U)$ is relatively compact in $F = C(T)$ for any neighbourhood U of 0 in E for which $Q \subset U^0$.

Conversely we suppose that u is compact and let U be any neighbourhood of 0 in E such that $u(U)$ is relatively compact in $F = C(T)$. According to the reverse of Ascali's theorem (page 34), for each t in T one has from (9.5.12)

$$\operatorname*{Sup}_{x \in U} |\langle x, \xi(t) \rangle| = \operatorname*{Sup}_{x \in U} |u(x)(t)| < +\infty,$$

showing that ξ maps T into E'_Q, Q being U^0. Again by the converse of Ascoli's theorem (page 34), $u(U)$ is equicontinuous at each point t_0 of T. This signifies that to each $\varepsilon > 0$ corresponds a neighbourhood W of t_0 such that

$$|u(x)(t) - u(x)(t_0)| < \varepsilon$$

for t in W and x in U. By virtue of (9.5.12) and the definition of $Q = U^0$, this means that $\xi(t) - \xi(t_0) \in \varepsilon \cdot Q$ for t in W—that is, that ξ is continuous for the normed topology on E'_Q. ▌

At this stage we specialize by assuming that $E = C(S)$ and employ the last two propositions as aids in establishing kernel representations of weakly compact and compact linear maps u of $C(S)$ into $C(T)$, valid whenever T is σ-compact.

9.5.16 **Proposition.** Suppose that T is σ-compact, and that u is a continuous linear map of $E = C(S)$ into $F = C(T)$. In order that u be weakly compact it is necessary and sufficient that there exist a compact subset K of S, a function k on $T \times K$, and a positive Radon measure σ on K, such that

$$u(x)(t) = \int_K k(t, s)x(s)\, d\sigma(s), \tag{9.5.14}$$

k satisfying the following three conditions:

(1) for each t in T, $s \to k(t, s)$ is σ-integrable;

(2) for each Borel subset B of K, the function

$$t \to \int_B k(t, s)\, d\sigma(s)$$

is continuous;

(3) for each compact subset K' of T,

$$\operatorname*{Sup}_{t \in K'} \int_K |k(t, s)|\, d\sigma(s) < +\infty.$$

9.5.17 **Proposition.** Suppose that S, T, and u are as in the preceding proposition. Then u is compact if and only if it admits a representation (9.5.14), the kernel k satisfying (1) above and also

(4) $$\lim_{t \to t_0} \int_K |k(t, s) - k(t_0, s)|\, d\sigma(s) = 0$$

for each t_0 in T.

Proof of Proposition 9.5.16. The sufficiency may be seen by writing $u = v \circ j$, where j is the map of $C(S)$ into $C(K)$ obtained by restricting functions from S to K, and v is the map of $C(K)$ into $C(T)$ defined by the right-hand side of (9.5.14), wherein we may take for x any member of $C(K)$. The weak compactness of u will follow from that of v, since j transforms suitably chosen neighbourhoods of 0 in $C(S)$ into bounded subsets of $C(K)$. On the other hand, to show that v is weakly compact it suffices (Theorem 9.4.10) to verify that $v''(g)$ lies in F (and not merely in F'') for each function g on K that is the pointwise limit of a bounded sequence in $C(K)$. Now it is easily verified that $v''(g)$ is that element of F'' defined by the function $t \to \int_K k(t, s)g(s)\, d\sigma(s)$, so that all depends upon showing that this function is continuous. However, (2) and (3) combine to show that this function has a restriction to K' that is continuous, g being the limit, uniformly on K, of finite linear combinations of characteristic functions of Borel sets $B \subset K$. Since T is locally compact, the function in question is therefore continuous on T. Sufficiency is thereby established.

To establish the necessity part, we suppose that u is weakly compact. Then there exists a compact set K in S such that if U is the set of x in $C(S)$ satisfying $|x(s)| \leq 1$ for s in K, then $u(U)$ is contained in some weakly compact subset P of $C(T)$. Let

$$A = \bigcap \{n^{-1}U : n = 1, 2, \cdots\} = \{x \in C(S) : x \mid K = 0\}.$$

Then $u(A) \subset \bigcap \{n^{-1}P : n = 1, 2, \cdots\}$, which intersection is $\{0\}$ since P is bounded. Thus $u = v \circ j$, j being the restriction map of $C(S)$ into $C(K)$ and v being weakly compact from $C(K)$ into $C(T)$. By Proposition 9.5.14, there exists a map ξ of T into $C(K)'$, continuous for the topology $\Sigma = \sigma(C(K)',C(K)'')$, such that $v(x)(t) = \langle x, \xi(t)\rangle$ for x in $C(K)$ and t in T. So, for x in $C(S)$ one has

$$u(x)(t) = \int_K x \, d[\xi(t)],$$

$\xi(t)$ being for each t a Radon measure on K.

T being σ-compact, it is the union of a sequence (T_n) of compact sets. Then by continuity of ξ, $\xi(T_n)$ is weakly compact in $C(K)' = M(K)$. By Corollary 4.23.4 there exists a positive Radon measure σ_n on K relative to which each member of $\xi(T_n)$ is absolutely continuous. Putting

$$\sigma = \sum_{n=1}^{\infty} 2^{-n} \, \|\sigma_n\|^{-1} \cdot \sigma_n,$$

it is easily seen that every member of $\xi(T)$ is absolutely continuous relative to σ. By the Lebesgue-Radon-Nikodym theorem, therefore, we may write symbolically $d[\xi(t)](s) = k(t, s) \, d\sigma(s)$, where $k(t, s)$ is σ-integrable qua function of s for each t. Thus

$$u(x)(t) = \int_K k(t, s)x(s) \, d\sigma(s),$$

and it remains to show that k has the properties (2) and (3).

To this end we consider the subset L of $C(K)' = M(K)$ formed of measures $f \cdot \sigma$, where f is σ-integrable. Each bounded and σ-measurable function g on K defines a continuous linear form on L, and is therefore the restriction to L of some element x'' of $C(K)''$. For t in T one has

$$v''(x'')(t) = \langle v''(x''), \varepsilon_t\rangle = \langle x'', v'(\varepsilon_t)\rangle = \langle x'', \xi(t)\rangle = \langle x'', k_t \cdot \sigma\rangle,$$

where k_t is the function $s \to k(t, s)$. Since $k_t \cdot \sigma$ belongs to L, we infer that

$$v''(x'')(t) = \int_K g(s)k(t, s) \, d\sigma(s).$$

On the other hand, since v is weakly compact, $v''(x'')$ lies in $F = C(T)$. Thus $\int_K g(s)k(t, s) \, d\sigma(s)$ is a continuous function of t for each bounded and σ-measurable g. From this (2) follows at once as a special case, taking for g the characteristic function of B.

Besides this, v'' transforms bounded subsets of $C(K)''$ into weakly relatively compact, hence bounded, subsets of $F = C(T)$. It follows that if K' is any compact subset of T, then $|\int_K g(s)k(t, s)\, d\sigma(s)|$ is bounded when t ranges over K' and g over all σ-measurable functions satisfying $|g| \leq 1$ on K. This leads at once to (3), and the proof is compete. ▌

Proof of Proposition 9.5.17. This is a simple development of Proposition 9.5.16 and its proof, the details of which are left to the reader. The sufficiency is evident. As for necessity, we invoke Proposition 9.5.16 to ensure a representation (9.5.14). Condition (4) is then derivable either on the basis of Proposition 9.5.15, or by a direct appeal to Ascoli's theorem. The reader will recall that $\xi(t)$ is the measure $k_t \cdot \sigma$, so that (3) expresses the continuity of ξ for the normed topology on the space of bounded Radon measures on K. ▌

9.6 Return to the Theory of Compact Linear Maps

Almost all books on functional analysis devote space to a presentation of the Riesz-Schauder theory of linear maps of the type $1 - u$, where u is compact, leading to precise information about the spectral theory of u. The early sources are F. Riesz [2] and Schauder [4]. Among the presentations in book form, incorporating interim modifications and extensions, we single out those of Banach ([1], Chapitre VI), Taylor ([1], Chapters 5 and 6), Zaanen ([1], Chapter 11), and Dieudonné ([13], Chapter 11). This list is not exhaustive.

Naturally enough, the original formulations were in terms of Banach spaces. This setting is sufficiently general to cover many important applications Extensions have been made, however, and are proving themselves useful in applications (see Subsection 9.10.6). Such extensions were begun by Schwartz [13]. The presentation we follow is borrowed in large measure from that of Grothendieck [7]. See also Kelley and Namioka [1], Problems 21A–21D.

The key theorems fall into two groups comprising Subsections 9.6.5–9.6.8 and 9.10.1–9.10.4, respectively. Those in the first group show that in many respects, and neglecting subspaces of finite dimension, the addition to an isomorphism of a compact linear map leaves unaltered several important features (the open nature of the mapping, and the closed-ness of the image, for example). In the second group these results are applied to the spectral theory of a compact endomorphism u; that is, the study of the family of endomorphisms $\lambda 1 - u$ obtained when λ ranges over the scalars. The resulting theory embraces that associated with the name of Fredholm [1] and dealing with the case in which u is a kernel operator (see Section 9.5). Spectral theory for more general linear maps is left untouched in this book; it has its own large stock of literature, for some of which see, for example, Dunford and Schwartz [1], Chapter VII, and Lorch [1].

9.6.1 **Proposition.** Let E and F be separated LCTVSs, u a continuous linear map of E into F. If u is compact, then u' is compact for $\kappa(F', F)$ and $\kappa(E', E)$. (These topologies are defined in Section 8.10.)

Proof. Let U be a neighbourhood of 0 in E such that $K = u(U)$ is relatively compact in F. Then $u'(K^0) \subset U^0$. Now K^0 is a neighbourhood of 0 for $\kappa(F', F)$; and U^0 is equicontinuous and weakly compact, hence compact for $\kappa(E', E)$. This completes the proof. ▌

It is natural to ask under what conditions u' is *strongly* compact whenever u is compact.

9.6.2 **Proposition.** Let E be a separated LCTVS. The following three conditions are equivalent:

(1) If $A \subset E$ is compact, convex, and balanced, there exists in E a bounded, closed, convex, and balanced set B containing A such that A is compact in the normed space E_B.

(2) For any separated LCTVS F and any compact linear map u of F into E, u' is strongly compact.

(3) As (2), with F a Banach space.

Proof. Since it is obvious that (2) implies (3), it remains only to show that (a) (1) implies (2); (b) (3) implies (1).

(a) Let us assume (1). Let $u : F \to E$ be linear and compact. Thus there exists a neighbourhood V of 0 in F such that $u(V)$ is relatively compact in E. Let $A = \overline{u(V)}$. Since we may assume that V is convex and balanced, (1) assures us that there exists a bounded, closed, convex, and balanced set B containing A such that A is compact in E_B. We aim to show that $u'(B^0)$ is strongly relatively compact in F'.

Each y' in $u'(B^0)$ can be written $y' = u'(x')$ for some x' in B^0. The restriction $x' \mid E_B$ belongs to the unit ball in $(E_B)'$, which is weakly compact and hence (Proposition 0.4.9) compact for the convergence uniform on compact subsets of E_B. On the other hand, if a directed family (x_i') is such that $x_i' \mid E_B \to x' \mid E_B$ uniformly on compact subsets of E_B, then

$$\langle y, u'(x_i')\rangle = \langle u(y), x_i'\rangle \to \langle u(y), x'\rangle = \langle y, u'(x')\rangle$$

uniformly for y in V, since $A = \overline{u(V)}$ is compact in E_B. Hence $u'(x_i')$ converges uniformly on bounded subsets of F—that is, $u'(x_i') \to u'(x')$ strongly in F'. It follows that $u'(B^0)$ is strongly relatively compact, and (a) is established.

(b) Let us assume that (3) holds. Let $A \subset E$ be as in (1). Let $F = E_A$, a Banach space, and let u be the injection map of F into E. u is compact, and linear. By (3), u' is strongly compact from E' into $(E_A)'$. Hence there exists a bounded set $B \subset E$ such that $u'(B^0)$ is strongly relatively compact in $(E_A)'$. Clearly, we may assume that B is closed, convex, balanced, and contains A. (Enlarging B only diminishes B^0.) We will show that A is compact in E_B, in doing which it suffices to show that E and E_B induce the same topology on A; and, since A is convex and balanced, it remains only to show that E and E_B induce the same neighbourhoods of 0 in A.

Since the topology of E_B is stronger than that induced by the topology of E, we are reduced to showing that: If a directed family $(x_i) \subset A$ is such that $u(x_i) \to 0$ (that is, $x_i \to 0$ in E), then $x_i \to 0$ in E_B. We have $\langle u(x_i), x'\rangle \to 0$ for each x' in E'; that is, $\langle x_i, u'(x')\rangle \to 0$ for each x'. Hence $x_i \to 0$ pointwise on $u'(B^0)$. Since $\|x_i\|_A \le 1$ and $u'(B^0)$ is strongly relatively compact in $(E_A)'$, Proposition 0.4.9 implies that $\langle x_i, u'(x')\rangle \to 0$ uniformly for x' in B^0. Putting

$$c_i = \text{Sup}\,\{|\langle x_i, u'(x')\rangle| : x' \in B^0\},$$

we have $c_i \to 0$ and $|\langle u(x_i), x'\rangle| \le c_i$ for x' in B^0. Since u is an injection map, this reads $|\langle x_i, x'\rangle| \le c_i$ for x' in B^0. Now B is convex, balanced, and closed in E, and so these inequalities show that $x_i \in c_i \cdot B$; that is, $\|x_i\|_B \le c_i$. Thus $x_i \to 0$ in E_B, as we had to show.

The proof of the proposition is complete. ▌

9.6.3 **Corollary.** Let E be a separated LCTVS, F a metrizable LCTVS, u a compact linear map of E into F. Then u' is strongly compact.

Proof. It suffices to show that condition (1) of the preceding proposition is satisfied when F replaces E, assumed here to be metrizable. This follows from a lemma. ▌

9.6.4 **Lemma.** Let F be a metrizable LCTVS, A a bounded subset of F. There exists in F a bounded, closed, convex, and balanced set B containing A such that the normed space F_B induces on A the same uniform structure as that induced by F.

Proof. We may assume that A is convex and balanced. It is then sufficient to show that the induced neighbourhoods of 0 are the same; that is, that if a neighbourhood base $V_1, V_2, \cdots$ at 0 in F is chosen, for each $\lambda > 0$ one has $A \cap V_n \subset \lambda B$ for some n. Now, since A is bounded in F, there exist positive numbers $\lambda_i \to \infty$ such that $A \subset \bigcap_i \lambda_i V_i$. We choose numbers $\mu_i \ge \lambda_i$ such that $\lambda_i/\mu_i \to 0$, and let $B = \bigcap_i \mu_i V_i$. If $\lambda > 0$, we shall have $\lambda_i \le \lambda\mu_i$ for $i \ge i_0$. Then $A \subset \lambda\mu_i V_i$ $(i \ge i_0)$. There exists n such that $V_n \subset \bigcap_{i<i_0} \lambda\mu_i V_i$. So $A \cap V_n \subset \lambda\mu_i V_i$ for all i; that is, $A \cap V_n \subset \lambda B$. This completes the proof. ▌

We turn now to the central theorems concerning compact operators, generalizing those of Riesz motivated by certain important types of integral equations.

9.6.5 **Theorem.** Let E and F be two separated LCTVSs; u and v linear maps of E into F such that

(1) u is an isomorphism of E onto a closed vector subspace of F;

(2) v is compact.

Then $w = u + v$ is an open map of E onto $w(E)$, $w(E)$ is closed in F, and $w^{-1}(\{0\})$ is of finite dimension.

Proof. Let U be a closed, convex, and balanced neighbourhood of 0 in E such that $v(U)$ is relatively compact in F, and let $N = w^{-1}(\{0\})$ and $U_1 = U \cap N$. Then $u(U_1) = -v(U_1) \subset v(U)$, so that $u(U_1)$ is relatively compact in F and therefore precompact. Since u is an isomorphism, U_1 is precompact. Now U_1 is a relative neighbourhood of 0 in N. Therefore N is of finite dimension (see Subsection 1.9.6).

We take any topological complement M of N in E; this choice is possible since N is finite dimensional (see Subsection 1.9.7). The restrictions $u_0 = u \mid M$, $v_0 = v \mid M$ and $w_0 = w \mid M$ then satisfy conditions (1) and (2) with M in place of E; moreover

$$w(E) = w(M + N) = w(M) = w_0(M),$$
$$w_0^{-1}(\{0\}) = M \cap w^{-1}(\{0\}) = M \cap N = \{0\},$$

so that w_0 is one to one. It thus suffices to deal with the case in which w is one to one, and to show that then w is an isomorphism of E onto a closed vector subspace of F. Now this signifies that, if (x_i) is a directed family in E such that $(w(x_i))$ is convergent in F, then (x_i) has a limiting point in E. For if this condition is fulfilled, since w is one to one, (x_i) has just one limiting point, hence is convergent. We proceed to verify this condition.

Letting p denote the gauge function of U, there are two cases to examine.

(a) Let us suppose that $(p(x_i))$ has a finite limiting point. By dropping terms (that is, by passage to a cofinal subfamily) one may then assume that $(p(x_i))$ is bounded, say $p(x_i) \le m$. So $x_i \in mU$ and so the set of $v(x_i)$ is relatively compact. So $(v(x_i))$ has a limiting point, and therefore $u(x_i) = w(x_i) - v(x_i)$ has a limiting point y. Since $u(E)$ is closed, $y = u(x)$ for some x; and since u is an isomorphism, x is a limiting point of (x_i).

(b) Otherwise $p(x_i) \to \infty$. We show that this possibility cannot in fact arise. For we should have as a consequence $w(x_i)/p(x_i) \to 0$. Then, by (a), $x_i/p(x_i)$ would have a limiting point x; $p(x)$ would be a limiting point of $p(x_i/p(x_i)) = 1$, hence $p(x) = 1$. At the same time, however, $w(x)$ would be a limiting point of $w(x_i/p(x_i)) = w(x_i)/p(x_i)$ and must therefore be 0. The relations $p(x) = 1$, $w(x) = 0$, together with the one-to-one character of w, constitute a contradiction.

The proof is thus complete. ▌

9.6.6 Theorem. Let E and F be separated LCTVSs, u and v linear maps of E into F. Let us suppose that

(1a) u is a weakly open map of E onto F;

(1b) each closed, convex, balanced, and compact subset of F is contained in the u-image of a similar subset of E;

(2) v is compact.

Then $w = u + v$ is a weakly open map of E onto a closed vector subspace of F, and $w(E)$ is of finite codimension in F.

Proof. The central idea is to equip E' and F' with their κ-topologies and to apply the preceding theorem to u' and v' making essential use of the results of Section 8.6. Note that these κ-topologies are compatible with the dualities between E and E' and F and F', respectively.

By Proposition 9.6.1, v' is compact. By Proposition 8.6.3, $u'(F')$ is weakly closed and therefore closed for $\kappa(E', E)$. u' is continuous for the κ-topologies since, if H is a compact, convex, and balanced subset of E, then $K = u(H)$ is likewise in F, and (8.6.3) gives $u'(K^0) \subset H^0$. Moreover, u' is an isomorphism for the κ-topologies, thanks to (1b) and Corollary 8.6.9 (wherein E and F are replaced by E' and F' with their κ-topologies). Thus u' and v' satisfy the conditions of the preceding theorem.

It follows that $w' = u' + v'$ is an open map of F' onto a closed subspace of E', and that w' has a kernel of finite dimension. $w'(F')$ is therefore weakly closed in E' and so (Proposition 8.6.3) w is weakly open. Since w' is open, it is weakly open (Theorem 8.6.8); and $(w')' = w$, so that another application of Proposition 8.6.3 shows that $w(E)$ is closed in F. Finally, the fact that $w(E)$ is closed in F arranges that it is the orthogonal of the kernel of w', hence is of finite codimension. This completes the proof. ▌

9.6.7 **Theorem.** Let E and F be Fréchet spaces, u and v continuous linear maps of E into F. If $u(E) = F$ and v is compact, then $w = u + v$ is an open map of E onto a closed vector subspace of finite codimension in F.

Proof. The conditions ensure that u is open, a fortiori weakly open (Theorems 6.4.4(a) and 8.6.8). Assuming for a moment that condition (1b) of Theorem 9.6.6 is fulfilled, we conclude that w is weakly open. But then it is open (Theorem 8.6.13).

Condition (1b) is obviously satisfied if u is one-to-one, which is perhaps the most important case anyhow. Otherwise we need Lemma 9.6.9 below. ▌

Meanwhile we note another derivative of Theorems 9.6.5 and 9.6.6.

9.6.8 **Theorem.** Let E and F be separated LCTVSs, u an isomorphism of E onto F, v a compact linear map of E into F. Then $w = u + v$ is an open map of E onto a closed vector subspace of finite codimension in F, and $w^{-1}(\{0\})$ is of finite dimension.

Proof. Both Theorems 9.6.5 and 9.6.6 apply, whence the stated conclusions. ▌

Remark. It will be shown later (Corollary 9.7.2) that in fact

$$\dim w^{-1}(\{0\}) = \operatorname{codim} w(E) \ (<\infty).$$

This corollary applies to the case in which $E = F$ and u is the identity map, but it is clear that the general case is immediately reducible to this one.

It remains to deal with the lemma required in the proof of Theorem 9.6.7.

9.6.9 **Lemma.** Let E and F be metrizable TVSs, E complete, u an open map of E onto F. Each compact set K in F is the u-image of some compact set H in E.

Proof. We introduce an invariant metric d on E and write $\|x\| = d(x, 0)$. For y in F we define

$$\|y\| = \operatorname{Inf}\{\|x\| : x \in E,\ u(x) = y\}.$$

The metric $\|y_1 - y_2\|$ defines on F a topology that, since u is open, coincides with the initial topology. We write $B(y_0, r) = \{y \in F : \|y - y_0\| < r\}$.

The compact set K can be covered by finitely many sets $B_i = B(y_i, \frac{1}{2})$, the y_i being chosen from K. Since u is onto, we may choose x_i in E so that $y_i = u(x_i)$. Likewise one may choose finitely many points y_{ij} in $B_i \cap K$ such that the sets $B_{ij} = B(y_{ij}, 1/2^2)$ cover $B_i \cap K$; and then x_{ij} in E so that $u(x_{ij}) = y_{ij}$ and $\|x_{ij} - x_i\| < \frac{1}{2}$. Proceeding to the next stage, one chooses finitely many y_{ijk} in $B_{ij} \cap K$ so that the $B_{ijk} = B(y_{ijk}, 1/2^3)$ cover $B_{ij} \cap K$; and then x_{ijk} in E so that $u(x_{ijk}) = y_{ijk}$ and $\|x_{ijk} - x_{ij}\| < 1/2^2$. This process is continued indefinitely.

Let $H_0 = \{x_i, x_{ij}, x_{ijk}, \cdots\}$. For any x in H_0 we can find a pair (i, j) so that

$$\|x - x_{ij}\| < 1/2^2 + 1/2^3 + \cdots = \tfrac{1}{2},$$

and a triplet (i, j, k) so that

$$\|x - x_{ijk}\| < 1/2^3 + 1/2^4 + \cdots = 1/2^2,$$

and so on. Thus H_0 is precompact and so, E being complete, the closure H of H_0 is compact. Further, $u(H_0) = \{y_i, y_{ij}, y_{ijk}, \cdots\}$ is dense in K. Since u is open and continuous, $u(H)$ is closed and so

$$u(H) = \overline{u(H_0)} = K,$$

and the proof is complete. ▌

Remark. The above proof is due to J.-P. Serre.

9.7 Endomorphism of Vector Spaces

By an endomorphism of a vector space we mean, as usual, a linear map of that vector space into itself. This section is devoted to some purely algebraic results about the kernels and images of the iterates of an endomorphism.

9.7.1 **Proposition.** Let E be a vector space, u an endomorphism of E. Let E_n be the kernel of u^n, I_n the image of u^n; here n is an integer ≥ 0 and $E_0 = \{0\}$, $I_0 = E$. The following assertions are true:

(1) The sequence (E_n) is either strictly increasing, or is strictly increasing up to some index p and thereafter stationary.

(2) The sequence (I_n) is either strictly decreasing, or is strictly decreasing up to some index q and thereafter stationary.

(3) If both (E_n) and (I_n) are ultimately stationary, then $p = q$. In this case, putting $E_\infty = E_p = \bigcup_n E_n$, $I_\infty = I_p = \bigcap_n I_n$, E is the direct sum $E_\infty + I_\infty$, $u \mid E_\infty$ is nilpotent, and $u \mid I_\infty$ is an automorphism of I_∞.

Proof. We have $E_{n+m} = (u^m)^{-1}(E_n)$ and $I_{m+n} = u^m(I_n)$. Since u^n commutes with u, both E_n and I_n are stable under u. Hence $E_{n+1} = u^{-1}(E_n)$ contains E_n, and likewise $I_{n+1} \subset I_n$, and so monotonicity is clear.

If for some n one has $E_n = E_{n+1}$, then $(u^m)^{-1}(E_n) = (u^m)^{-1}(E_{n+1})$; that is, $E_{n+m} = E_{n+m+1}$ and so (E_i) is stationary for $i \geq n$. Similarly for the sequence (I_i). This proves (1) and (2).

For (3) we show first that $q \leq p$. To do this it suffices to show that $E_n = E_{n+1}$ and $I_n \neq I_{n+1}$ together entail $I_{n+1} \neq I_{n+2}$. Now if in fact $I_{n+1} = I_{n+2}$, let $u^n(x)$ belong to I_n. Then $u^{n+1}(x)$ belongs to I_{n+1}, hence to I_{n+2}, hence may be written $u^{n+2}(y)$ for some y. Then $u^{n+1}(x - u(y)) = 0$ and so, since $E_n = E_{n+1}$, $u^n(x - u(y)) = 0$; that is, $u^n(x) = u^{n+1}(y)$ belongs to I_{n+1}. Thus $I_n = I_{n+1}$, contrary to hypothesis. The relation $q \leq p$ is thereby established. That $p \leq q$ is proved similarly. Thus $p = q$ whenever they are both defined.

Assuming still that p and q are both defined, let us show that $E_\infty \cap I_\infty = \{0\}$. Any element x of this intersection is of the form $x = u^p(y)$, and satisfies further $u^p(x) = 0$. So $u^{2p}(y) = 0$ which, since $E_p = E_{2p}$, entails $u^p(y) = 0$; that is, $x = 0$.

Let $\tilde{u}$ be the endomorphism of E/E_∞ obtained from u by passage to the quotient. Since $E_\infty = E_p = E_{p+1}$, $\tilde{u}$ is one to one. We claim that $\tilde{u}$ is onto. By the first part of (3) applied to $\tilde{u}$ (for which $p = 0$), if $\tilde{u}$ were *not* onto, the sequence of images $\tilde{u}^n(E/E_\infty)$ would be strictly decreasing. This would entail that the $u^n(E) + E_\infty = I_n + E_\infty$ would likewise decrease strictly. But this would conflict with the assumption that (I_n) is ultimately stationary. Thus $\tilde{u}$ is an automorphism. The same is therefore true of $\tilde{u}^p$, and in particular $\tilde{u}^p$ maps, E/E_∞ onto itself. This implies that $u^p(E) + E_\infty = E$; that is, $I_\infty + E_\infty = E$, so that E is the direct sum of E_∞ and I_∞.

Finally, to say that $\tilde{u}$ is an automorphism, signifies that $u \mid I_\infty$ is an automorphism; and, since $u^p(E_\infty) = \{0\}$, $u \mid E_\infty$ is plainly nilpotent. The proof is complete. ▮

9.7.2 **Corollary.** Let us suppose that E and u are as in Proposition 9.7.1, that (3) is the case, and that E_∞ has finite dimension. Then

$$\dim u^{-1}(\{0\}) = \operatorname{codim} u(E).$$

Proof. In view of (3) it suffices to verify the equality separately for $u \mid E^\infty$ and for $u \mid I_\infty$. Since E_∞ is assumed to be of finite dimension, the former case is elementary. On the other hand, since $u \mid I_\infty$ is an automorphism, both terms are 0 in the second case. ▮

9.7.3 **Remark.** Let us suppose that E and u are as in Proposition 9.7.1. Suppose too that the sequence (E_n) is ultimately stationary and that p the least natural number for which $E_{n+1} = E_n$ whenever $n \geq p$. Then, as we know, $u \mid E_\infty$ is nilpotent. We define k to be the least natural number such that $u^k(E_\infty) = \{0\}$. We allege that $k = p$.

Indeed, $u^p(E_p) = \{0\}$ by definition of E_p; and $E_\infty = E_p$ by the stationary character of (E_n) from $n = p$ onward. Hence $u^p(E_\infty) = \{0\}$ and therefore $k \leq p$. Moreover, if k were less than p, the relation $u^k(E_\infty) = \{0\}$ yields $E_\infty \subset E_k$, hence $E_\infty = E_k$ (the sequence of E_n being increasing). But also $E_\infty = E_p$. Thus $E_k = E_p$. The increasing character of the sequence then entails that

$$E_k = E_{k+1} = \cdots = E_p = E_{p+1} = \cdots,$$

which contradicts the definition of p. Accordingly, $k = p$, as alleged.

9.8 Eigenvalues and Spectra

Let us suppose again that E is a vector space and u an endomorphism of E. An *eigenvalue* of u is a scalar λ such that $u(x) = \lambda x$ holds for some $x \neq 0$ in E. Thus, if 1 denotes the identity endomorphism of E, λ is an eigenvalue if and only if $\lambda 1 - u$ is *not* one-to-one. For any scalar λ, the kernel of $\lambda 1 - u$ is termed the *eigenmanifold* of u associated with λ. This eigenmanifold is $\{0\}$ unless λ is an eigenvalue.

By an *algebraic spectral value* of u is meant a scalar λ such that $\lambda 1 - u$ has no inverse in the algebra of all endomorphisms of E: this means that *either* $\lambda 1 - u$ is not one-to-one (that is, that λ is an eigenvalue of u), *or* $\lambda 1 - u$ is not onto (or perhaps both alternatives obtain). Accordingly we see that any eigenvalue is an algebraic spectral value; the converse is generally false, but is valid if E is of finite dimension.

A third classification results when E is a TVS and u is a *continuous* endomorphism of E: in this case a scalar λ is said to be a (topological) *spectral value* of u if $\lambda 1 - u$ is not inversible in the algebra $L_c(E)$ of all continuous endomorphisms of E.

Let $\sigma_0(u)$, $\sigma_a(u)$, and $\sigma(u)$ denote, respectively, the set of all eigenvalues, of all algebraic spectral values, and of all spectral values of u. We will call $\sigma_a(u)$ and $\sigma(u)$ the *algebraic spectrum* and *spectrum*, respectively, of u. One always has the inclusions

$$\sigma_0(u) \subset \sigma_a(u) \subset \sigma(u).$$

As we have remarked, $\sigma_0(u) = \sigma_a(u)$ if E has finite dimension, in which case $\sigma_a(u) = \sigma(u)$ too (every endomorphism of E being continuous for any separated TVS structure on E). Besides this, if E is a complete metrizable TVS, then $\sigma_a(u) = \sigma(u)$ for any continuous endomorphism u of E : this follows from the inversion theorem in the form of Theorem 6.4.5. (It may nevertheless still be true that $\sigma_0(u)$ is distinct from $\sigma_a(u)$, if E is of infinite dimension.)

For each scalar λ and each integer $n \geq 0$ we write $E_{\lambda,n}$ and $I_{\lambda,n}$ for the kernel and image, respectively, of $(\lambda 1 - u)^n$, and put $E_\lambda = \bigcup_{n\geq 0} E_{\lambda,n}$ and $I_\lambda = \bigcap_{n\geq 0} I_{\lambda,n}$. Then $E_{\lambda,1}$ is the eigenmanifold of u associated with λ, as already defined. E_λ is termed the *spectral manifold of u associated with λ*. For $\lambda = 0$, the subspaces just introduced are those appearing in Proposition 9.7.1. The dimension of E_λ (which may, of course, be infinite) is termed the *spectral multiplicity* of λ.

From elementary linear algebra we know that if E is finite dimensional and the scalar field is algebraically closed, then E is the direct sum of the E_λ with λ ranging over the eigenvalues of u; these eigenvalues are precisely the zeros of the so-called "characteristic polynomial" $P(\lambda) = \det(\lambda 1 - u)$ of degree equal to the dimension of E; and the spectral multiplicity of an eigenvalue λ (that is, dimension of E_λ) is equal to the multiplicity of λ qua zero of the polynomial P.

9.8.1 **Proposition.** Let E be a vector space, u an endomorphism of E. If $\lambda \neq \lambda'$, then $E_\lambda \subset I_{\lambda'}$.

Proof. Let x belong to E_λ, so that $(\lambda 1 - u)^n(x) = 0$ for some n. We have to show that $x = (\lambda' 1 - u)^m(y)$ for all m and some y (possibly depending on m).

Let G be the vector subspace of E generated by the elements x, $u(x)$, $u^2(x)$, $\cdots$; G is stable under u and $\dim G \le n$. The restriction $u \mid G$ is an endomorphism of G having λ as its only eigenvalue, so that $(\lambda' 1 - u) \mid G$ is an inversible endomorphism of G. The same is therefore true of $(\lambda' 1 - u)^m \mid G$ for any m. But then, since $\dim G$ is finite, $(\lambda' 1 - u)^m$ maps G onto itself and so $x = (\lambda' 1 - u)^m(y)$ for some y. This completes the proof. ▌

9.9 Results Concerning Spectra of Endomorphisms

We are concerned in this section with a few generalities concerning the spectrum of a continuous endomorphism u of a TVS E. The simplest case is that in which E is a Banach space. As we shall see in Proposition 9.9.3, the general case can sometimes be reduced to this one.

9.9.1 **Proposition.** Let E be a Banach space. Then $L_c(E)$ is a Banach space when furnished with the norm

$$\|u\| = \operatorname*{Sup}_{\|x\| \le 1} \|u(x)\| = \operatorname*{Sup}_{\|x\| = 1} \|u(x)\|.$$

The set of inversible elements of $L_c(E)$ is open and contains all elements u of $L_c(E)$ satisfying $\|u - 1\| < 1$.

Proof. The first statement is trivial. Let us suppose first that u is an element of $L_c(E)$ satisfying $\|u - 1\| < 1$. Then the series $\sum_{n \ge 0} (1 - u)^n$ converges in $L_c(E)$ and multiplication on the right or left by $u = 1 - (1 - u)$ leads to 1, showing that u is inversible in $L_c(E)$.

Finally, let us suppose that u_0 is inversible in $L_c(E)$, and that u is an element of $L_c(E)$ satisfying

$$\|u - u_0\| < \|u_0^{-1}\|^{-1} \tag{9.9.1}$$

we shall show that u is inversible. In fact

$$u = u_0 - (u_0 - u) = u_0[1 - u_0^{-1}(u_0 - u)].$$

Since the norm of the product of two endomorphisms does not exceed the product of their norms, the second factor is inversible as a result of (9.9.1) and what we have already proved. The first factor, u_0, is inversible by hypothesis. Hence u, the product of two inversible endomorphisms, is inversible, as we had to show. ▌

9.9.2 **Corollary.** If E is a Banach space and u a continuous endomorphism of E, then $\sigma(u)$ is a compact set of scalars; and any λ in $\sigma(u)$ satisfies $|\lambda| < \|u\|$.

Proof. The spectrum $\sigma(u)$ is the inverse image, under the map $\lambda \to \lambda 1 - u$, of the set of noninversible elements of $L_c(E)$. By Proposition 9.9.1, this latter set is closed. Since the map $\lambda \to \lambda 1 - u$ is clearly continuous, $\sigma(u)$ is closed. If λ is a scalar satisfying $|\lambda| > \|u\|$, then $\lambda 1 - u = \lambda(1 - u/\lambda)$. Since $\|u/\lambda\| < 1$, Proposition 9.9.1 shows that $1 - u/\lambda$ is inversible—that is, that λ does *not* belong to $\sigma(u)$. The rest of the corollary follows directly. ▌

Remark. If E is not normable, $\sigma(u)$ may well be neither bounded nor closed. However, some information may be obtained in suitable cases by use of the next proposition.

9.9.3 **Proposition.** Let E be a TVS, F a vector subspace of E equipped with a TVS structure stronger than that of E, and let j be the injection map of F into E. Let u be a continuous linear map of E into F. Then:

(1) With the possible exception of 0, the spectrum of ju in $L_c(E)$ is identical with that of uj in $L_c(F)$.

(2) If $\lambda \neq 0$ belongs to this common spectrum, then the eigenmanifold (resp. spectral manifold) of uj is the same as that of ju.

Proof. Let λ be nonzero. We begin by showing that if $\lambda 1 - ju$ is inversible in $L_c(E)$, the same is true of $\lambda 1 - uj$ in $L_c(F)$. A similar argument suffices to prove the converse. Dividing by λ and replacing u by $-\lambda^{-1}u$, it suffices to consider the inversibility of $1 + ju$ and $1 + uj$.

Suppose then that $1 + ju$ is inversible in $L_c(E)$. Its inverse may be written $1 + v$ with v in $L_c(E)$. Then

$$ju + v + juv = ju + v + vju = 0. \tag{9.9.2}$$

From the first equation we get $v = jw$, where $w = -u(1 + v)$ is a continuous linear map of E into F. Replace v by this expression in (9.9.2), the result being

$$j(u + w + ujw) = j(u + w + wju) = 0.$$

Since j is one-to-one, this entails

$$u + w + ujw = u + w + wju = 0.$$

Multiplying on the right by j we obtain

$$uj + wj + (uj)(wj) = uj + wj + (wj)(uj) = 0,$$

which signifies that $1 + wj$ is an inverse of $1 + uj$ in $L_c(F)$. This establishes (1).

Passing to (2) we shall suppose that a nonzero scalar λ lies in the spectrum of ju. The associated eigenmanifold of uj is the intersection of F with the associated eigenmanifold of ju; and this last is already contained in F because $ju(x) = \lambda x$ gives $x = \lambda^{-1}ju(x)$, which belongs to F. Thus the eigenmanifolds are identical. This same argument applies to show that the kernels of $(\lambda 1 - ju)^n$ in E and of $(\lambda 1 - uj)^n$ in F are identical—for, apart from a factor λ^n, these maps can be written $1 + ju_n$ and $1 + u_n j$ for a suitable continuous linear map u_n of E into F. The identity of the spectral manifolds is then clear from their definitions. ▌

9.9.4 **Corollary.** Let E be a separated TVS, u an endomorphism of E having the following property: there exists a neighbourhood U of 0 in E and a bounded, convex, balanced subset B of E such that $u(U) \subset B$ and the normed vector space E_B is complete. Then $\sigma(u)$ is compact. The conclusion holds in particular if u is compact.

Proof. Two cases are to be distinguished.

Suppose first that 0 does *not* belong to $\sigma(u)$. Then u is an isomorphism of E onto itself; thus B is a neighbourhood of 0 in E and so E and E_B are isomorphic. Corollary 9.9.2 may therefore be applied.

Suppose on the other hand that 0 belongs to $\sigma(u)$. By Proposition 9.9.3, applied with $F = E_B$, apart from 0 the spectrum $\sigma(u)$ of u in $L_c(E)$ is identical with that of the continuous endomorphism uj of the Banach space E_B. If σ' denotes the spectrum of the latter, then σ' is compact by Corollary 9.9.2; and we have

$$\sigma(u)\backslash\{0\} = \sigma'\backslash\{0\}.$$

Hence

$$\sigma(u) = (\sigma'\backslash\{0\}) \cup \{0\}$$

is bounded and closed, hence compact. This completes the proof. ▌

Further information about the spectrum of a *compact* endomorphism is obtained in the next section.

9.10 Spectral Theory of Compact Endomorphisms

We come now to a generalized formulation of F. Riesz's discoveries about compact endomorphisms. For Banach spaces the results were presented by F. Riesz [2] and subsequently generalized by Leray [2] to LCTVSs.

9.10.1 Theorem. Let E be a separated LCTVS, u a compact endomorphism of E, $v = 1 + u$. We put

$$E_\infty = \bigcup_{n\geq 0} (v^n)^{-1}(\{0\}), \qquad E'_\infty = \bigcup_{n\geq 0} (v'^n)^{-1}(\{0\}),$$

$$I_\infty = \bigcap_{n\geq 0} v^n(E), \qquad I'_\infty = \bigcap_{n\geq 0} v'^n(E').$$

Then the following statements are true.

(1) v is an open map of E onto a closed vector subspace $v(E)$ whose codimension is finite and equal to the dimension of the kernel $v^{-1}(\{0\})$ of v.

(2) E is the topological direct sum of E_∞ and I_∞; E_∞ is of finite dimension; E_∞ and I_∞ are each stable under u and v, and $v \mid E_\infty$ is nilpotent and $v \mid I_\infty$ a topological automorphism of I_∞;

$$v(E) = (v'^{-1}(\{0\}))^0, \qquad v'(E') = (v^{-1}(\{0\}))^0; \tag{9.10.1}$$

$$\text{codim } v(E) = \text{codim } v'(E') = \dim v^{-1}(\{0\}) = \dim v'^{-1}(\{0\}) < \infty; \tag{9.10.2}$$

$$E'_\infty = I^0_\infty, \qquad I'_\infty = E^0_\infty; \tag{9.10.3}$$

$$\dim E_\infty = \dim E'_\infty < \infty. \tag{9.10.4}$$

(3) The following conditions are equivalent:

(a) v is one to one; (a′) v' is one-to-one;

(b) v is onto; (b′) v' is onto;

(c) v is a topological automorphism, (c′) v' is a topological automorphism.

(4) The sequences $E_n = (v^n)^{-1}(\{0\})$ and $I_n = v^n(E)$ $(n = 1, 2, \cdots)$ are each ultimately stationary, and the least natural number p for which $E_{n+1} = E_n$ for $n \geq p$ is also the least natural number p for which $I_{n+1} = I_n$ for $n \geq p$. This same p is also the least natural number satisfying $v^p(E_\infty) = \{0\}$.

Proof. This proof is lengthy and will be broken into several steps.

Step (1). We begin by showing that the sequence of kernels $E_n = (v^n)^{-1}(\{0\})$ is ultimately stationary. To do this, we choose a neighbourhood U of 0 in E and a compact, convex, balanced set A such that $u(U) \subset A \subset E$. Then E_A is a Banach space and the injection map j of E_A into E is compact. u defines a continuous linear map u_0 of E into E_A and $u_0 j$ is compact. In view of Proposition 9.9.3, it suffices to establish our assertion for the case in which u is replaced by $u_0 j$. In other words, we are reduced to the case in which E is a Banach space. In this case, we proceed by reduction ad absurdum. Suppose if possible that the E_n increase strictly. Then one could find a sequence (y_n) such that y_n belongs to E_{n+1}, $\|y_n\| = 1$, and y_n has distance from E_n not less than $\frac{1}{2}$. (Indeed, E_n is closed and a proper subset of E_{n+1}; we choose a in E_{n+1} not in E_n; the distance d of a from E_n is nonzero, and we can choose z in E_n so that $d \leq \|a - z\| \leq d(1 + \varepsilon)$ for any preassigned $\varepsilon > 0$. We put $y_n = (a - z)/\|a - z\|$. It is then easily verified that y_n has norm unity and has distance from E_n not less than $(1 + \varepsilon)^{-1}$.) Choosing the sequence (y_n) thus we shall have for $m < n$

$$u(y_n) - u(y_m) = (1 - v)(y_m) - (1 - v)(y_n) = y_m - x$$

where $x = v(y_m) + (1 - v)(y_n)$ belongs to E_m. So $\|u(y_n) - u(y_m)\| \geq \frac{1}{2}$ for $m \neq n$. Since the y_n are normalized and u is compact, this leads to a contradiction. The E_n are therefore ultimately stationary.

Step (2). The adjoint $v' = 1' + u'$. Here u' is a compact endomorphism of E' equipped with $\kappa(E', E)$ (Proposition 9.6.1). So step (1) shows that the sequence of kernels $E'_n = (v'^n)^{-1}(\{0\})$ is ultimately stationary.

Step (3). We have $v'^n = (v^n)'$ and $v^n = (1 + u)^n = 1 + u_n$, where u_n is compact. So (Theorem 9.6.8) v^n is an open map of E onto a closed subspace I_n of finite codimension that is the orthogonal of $(v^n)'^{-1}(\{0\}) = (v'^n)^{-1}(\{0\}) = E'_n$. By (2), this latter sequence is ultimately stationary. So we see that the $I_n = v^n(E)$ are ultimately stationary and $I_\infty = \bigcap_{n \geq 0} I_n$ is a closed subspace of finite codimension. By (1) and Proposition 9.7.1, E_n and I_n are ultimately stationary from some point $n = p$ onwards, $E_\infty = E_p$, $I_\infty = I_p$, and E is the algebraic direct sum of E_∞ and I_∞. Since I_∞ is closed and of finite codimension, the said direct sum decomposition is necessarily topological. Further (Proposition 9.7.1 again) $v \mid E_\infty$ is nilpotent and $v \mid I_\infty$ is an algebraic automorphism. By Theorem 9.6.8, $v \mid I_\infty$ is open from I_∞ onto itself; hence $v \mid I_\infty$ is a topological automorphism.

Statement (4) of Theorem 9.10.1 now follows from Proposition 9.7.1 and Remark 9.7.3.

Step (4). The truth of (9.10.1) follows because $v(E)$ is closed in E, and $v'(E')$ is closed for $\kappa(E', E)$ and therefore weakly closed. Then (9.10.2) follows from this and Corollary 9.7.2. To derive (9.10.3) and (9.10.4) it suffices to apply (9.10.1) and (9.10.2) to the powers v^n of v, and to recall that $E_\infty = E_p$, $I_\infty = I_p$, et cetera.

At this point we have completed the proofs of statements (1), (2), and (4) of the theorem.

Turning to statement (3), we note that the equivalence of (a) and (b) and of (a′) and (b′) appears from (9.10.2). On the other hand, since v is open, (a) and (b) together imply (c); and the converse is obvious. Likewise (a′) and (b′) together are equivalent to (c′). Thus (a), (b), and (c) are equivalent; and likewise for (a′), (b′), and (c′). It suffices now to show that (b) and (a′) are equivalent: but this too appears from (9.10.2).

The proof is therefore complete. ▌

We can now deduce some important properties of the spectrum of a compact endomorphism.

9.10.2 **Theorem.** Let E be a separated LCTVS and u a compact endomorphism of E.

(1) Any nonzero λ in $\sigma(u)$ is an eigenvalue of u, and the associated spectral manifold E_λ is of finite dimension equal to that of the associated spectral manifold E'_λ of u'.

(2) u and u' have the same nonzero eigenvalues and these have the same spectral multiplicities.

(3) The spectral manifolds E_λ $(\lambda \neq 0)$ are topologically independent; that is, for each $\lambda \neq 0$ there exists a closed vector subspace of E containing all the $E_{\lambda'}$ with $\lambda' \neq 0$ distinct from λ and meeting E_λ only in $\{0\}$.

(4) $\sigma(u)$ is compact and has no limit point different from 0 (that is, $\sigma(u)$ is either finite, or is a denumerable sequence converging to 0).

Proof. We may apply Theorem 9.10.1 with u there replaced by $U = -\lambda^{-1}u$, and so v by $V = \lambda^{-1}(\lambda 1 - u)$. If λ is in $\sigma(u)$, then V is *not* an isomorphism of E onto itself. Since V is in any case an open map (Theorem 9.10.1(1)), this means that *either* (a) V is not one-to-one, or (b) V is not onto. If (b) holds, then codim $V(E) > 0$ and so (Theorem 9.10.1(1)) dim $V^{-1}(\{0\}) > 0$, and therefore (a) holds. So (a) holds in either case, and shows that λ is an eigenvalue of u. The rest of (1) follows from (9.10.2), as also does (2).

As for (3) it is necessary merely to note that Proposition 9.8.1 and Theorem 9.10.1 combine to show that I_λ is a vector subspace of E enjoying the stated properties.

Turning to (4), reference to Corollary 9.9.4 shows that $\sigma(u)$ is compact. For the rest, we suppose that $\lambda \neq 0$ is a point of $\sigma(u)$. We know (Theorem 9.10.1(2)) that E is the topological direct sum of E_λ and I_λ, each of which is stable under u. Let u_1 and u_2 be, respectively, the restrictions of u to E_λ and to I_λ. It is trivial to verify that $\sigma(u) = \sigma(u_1) \cup \sigma(u_2)$. Now $\sigma(u_1) = \{\lambda\}$, as we see by using (1) and the fact that $(\lambda 1 - u) \mid E_\lambda$ is nilpotent (Theorem 9.10.1(2)). On the other hand, by (1) again, $\sigma(u_2)$ is compact. Moreover, λ does *not* belong to $\sigma(u_2)$: for otherwise I_λ would contain a nonzero eigenvector x associated with the eigenvalue λ, and this x would belong to E_λ, which cannot be because $E_\lambda \cap I_\lambda = \{0\}$. It follows at once that λ is an isolated point of $\sigma(u)$.

The proof is complete. ▌

9.10.3 The Fredholm Alternative. It appears from Theorem 9.10.2(1) that for any $\lambda \neq 0$, just one of the following alternatives can obtain:

(a) There exists $x \neq 0$ satisfying $u(x) = \lambda x$.

(b) For each y in E the equation $\lambda x - u(x) = y$ has a unique solution x in E, and this x depends continuously on y.

Moreover, by Theorem 9.10.1(3), for the same $\lambda \neq 0$ and the corresponding alternatives (a′) and (b′) involving E' and u', the corresponding choice of (a′) and (b′) is appropriate.

This state of affairs was first pointed out by Fredholm for the case in which u is a compact integral operator of Hilbert-Schmidt type (see Section 9.5; Smithies [2], pp. 51–52; Banach [1], Chapter X).

At the risk of labouring the obvious we shall reformulate parts of Theorems 9.10.1 and 9.10.2 in a slightly different form, which is sometimes preferred in connection with integral equations while at the same time making things somewhat more explicit.

In this presentation one takes again a compact endomorphism u of E and considers the endomorphism $1 - \mu u = v$ for various values of the parameter μ; μ corresponds to λ^{-1} in the earlier notation. The principal results may then be expressed in the following way.

9.10.4 Theorem. Let E be a separated LCTVS, u a compact endomorphism of E. To each scalar μ corresponds an integer $n_\mu \geq 0$ such that the following assertions are true:

(1) $n_0 = 0$; the scalars μ for which $n_\mu > 0$ are either finite in number or, if infinite in number, may be arranged as a sequence (μ_r) such that $\lim_{r\to\infty} |\mu_r| = +\infty$.

(2) The kernel of $1 - \mu u$ is of dimension n_μ, as also is the kernel of $1' - \mu u'$. Let us suppose we choose a basis $(a_i)_{1 \leq i \leq n_\mu}$ for the kernel of $1 - \mu u$ and a basis $(a'_i)_{1 \leq i \leq n_\mu}$ for the kernel of $1' - \mu u'$.

(3) Given y in E (resp. y' in E'), the equation

$$x - \mu u(x) = y \quad (\text{resp. } x' - \mu u'(x') = y') \tag{9.10.5}$$

(resp. (9.10.5′))

is soluble for x in E (resp. x' in E') if and only if

$$\langle y, a'_i \rangle = 0 \quad (\text{resp. } \langle a_i, y' \rangle = 0) \quad \text{for} \quad 1 \leq i \leq n_\mu \tag{9.10.6}$$

(resp. (9.10.6′))

When these conditions are fulfilled, the set of solutions x (resp. x') of (9.10.5) (resp. (9.10.5′)) is a linear variety in E (resp. E') of dimension n_μ. More specifically, if x_0 (resp. x'_0) is a solution of (9.10.5) (resp. (9.10.5′)), any other solution can be expressed in the form $x_0 + \sum_{1 \leq i \leq n_\mu} \alpha_i a_i$ (resp. $x'_0 + \sum_{1 \leq i \leq n_\mu} \alpha_i a'_i$) for suitably chosen scalars α_i; and, conversely, whatever the scalars α_i, this latter element is always a solution of the said equation.

(4) For any μ such that $n_\mu = 0$, $1 - \mu u$ is a topological automorphism of E and $1' - \mu u'$ is a topological automorphism of E'.

9.10.5 Potentially Compact Endomorphisms. We use the term "potentially compact" to describe a continuous endomorphism u, some one of whose iterates is compact. It is worth noting that properties (1) and (4) in Theorem 9.10.2 remain true of any such endomorphism u.

To see this, assume that u^p is compact, p being some natural number. On the one hand, it is almost evident that if λ belongs to $\sigma(u)$, then λ^p belongs to $\sigma(u^p)$. Applying Theorem 9.10.2(4) to u^p, we infer that $\sigma(u)$ is compact and has no limiting point other than 0. Again, since $(\lambda 1 - u)^n(x) = 0$ entails

$$(\lambda^p 1 - u^p)^n(x) = 0, \quad \text{so} \quad E_\lambda(u) \subset E_{\lambda^p}(u^p).$$

Thus Theorem 9.10.2(1), applied to u^p, shows that $E_\lambda(u)$ is finite dimensional. For another similar property, see Exercise 9.24.

The reader will notice that if E has the strict Dunford-Pettis property, any weakly compact endomorphism u is potentially compact, u^2 being compact (Remark (2) following Corollary 9.4.5).

9.10.6 Notes. (1) Before leaving generalities we should mention the exposition of A. Deprit [1], where a full study is made of linear maps that are "close to the identity" in various senses (for example, finite dimensional kernels, or ranges of finite codimension, and that are also open onto their ranges). Attention is also paid to those linear maps for which the sequence of iterated kernels (or iterated ranges) are ultimately stationary.

The properties of a linear map described in the conclusion (2) of Theorem 9.10.1, together with the ultimately stationary character of the sequences (E_n) and (I_n), are sometimes briefly described as the "Riesz properties." If we relax the stationary properties, the linear map is sometimes termed "finite." For the case of Banach and Hilbert spaces, maps of this type and their spectral properties have been studied in considerable detail. See, for example, Heuser [1], [2], [3].

(2) An elegant account of the Riesz theory for normed spaces appears in Chapter XI of Dieudonné [13]. Indications are given there of possible extensions to maps of the form $v = f - u$ from one space, E, into a second, F, u being compact and f an isomorphism; see especially the problems at the end of section 11.3 of Dieudonné's book just cited.

(3) As has been said, and as will be illustrated in Sections 9.11 and 9.12, the general theory developed in Section 9.6 onward to this point finds its more immediate applications to cases in which E and F are Banach spaces. Applications in which the spaces E and F are of a more general type are both more recent and more difficult to cover within a reasonable space. One striking application—unfortunately too lengthy to include when accompanied by the necessary preliminaries—lies in the field of cohomology theory. A typical "finiteness theorem" in that theory asserts that the cohomology groups $H^n(X, F)$ are of finite dimension (considered as vector spaces over the complex field) whenever X is a compact complex-analytic manifold, F a coherent analytic

sheaf (= stack, faisceau) over X, and n any integer ≥ 0. A proof of this, based upon Theorem 9.6.7, is presented in notes of the Séminaire Cartan (Paris, 1953–54), Exposés XVI and XVII.

9.11 Spectral Theory of Compact Endomorphisms of a Hilbert Space

Failing explicit indications to the contrary, E will denote throughout this section a Hilbert space with inner product $(x \mid y)$ and norm $\|x\|$. As we know (see Remarks to Corollary 1.12.6 and Subsection 8.11.8), there is a conjugate-linear and norm-preserving map J of E onto E' such that

$$(x \mid y) = \langle x, Jy \rangle \tag{9.11.1}$$

identically for x and y in E.

Let u be a continuous endomorphism of E. As in Subsection 8.11.8 we denote by u^* its Hilbert adjoint, defined by

$$(u(x) \mid y) = (x \mid u^*(y)) \tag{9.11.2}$$

for x and y in E. u^* is related to the adjoint u' by the relation $u^* = J^{-1}u'J'$ and u is said to be self-adjoint if $u^* = u$—that is, if

$$(u(x) \mid y) = (x \mid u(y)) \tag{9.11.3}$$

identically for x and y in E.

A number of elementary spectral properties of self-adjoint endomorphisms stem at once from (9.11.3). Thus, since (9.11.3) yields the conclusion that $(u(x) \mid x)$ is real valued, any eigenvalue of a self-adjoint endomorphism u is real. If, furthermore, u is positive self-adjoint, its eigenvalues are nonnegative real numbers. Again, (9.11.3) shows that eigenvectors of a self-adjoint u belonging to distinct eigenvalues are necessarily orthogonal. Finally, if u is self-adjoint, (9.11.3) combines with an easy inductive argument to show that $u^n(x) = 0$ entails $u(x) = 0$. Consequently the spectral manifold $\bigcup_{n \geq 0}$ Ker $(u - \lambda 1)^n$ is identical with the eigenmanifold Ker $(u - \lambda 1)$, λ being any real number.

We plan to apply Theorem 9.10.2 to a compact self-adjoint endomorphism u of E, incorporating as we go some facts peculiar to the present specialization. These last are consequences of the following preliminary result, which possesses already some intrinsic interest inasmuch as the method of proof indicates a variational method of locating the numerically largest eigenvalue of u.

9.11.1 Lemma. If E is a Hilbert space and u a compact self-adjoint endormorphism of E, $u \neq 0$, then u admits at least one eigenvalue $\lambda \neq 0$.

Proof. u is necessarily continuous. Since $u \neq 0$ and self-adjoint,

$$\lambda = \operatorname*{Sup}_{\|x\|=1} |(u(x), x)|$$

is strictly positive. Since $(u(x), x)$ is real, we may assume (by changing u into $-u$ if necessary) that λ is in fact the supremum of $(u(x) \mid x)$ for $\|x\| = 1$. We take a sequence (x_n) satisfying $\|x_n\| = 1$, $(u(x_n) \mid x_n) \to \lambda$. Since u is compact, we may assume that $(u(x_n))$ is convergent to z, say. Then:

$$\|u(x_n) - \lambda x_n\|^2 = \|u(x_n)\|^2 - 2\lambda(u(x_n) \mid x_n) + \lambda^2 \|x_n\|^2$$

tends to $\|z\|^2 - \lambda^2$. Hence $\|z\| \geq \lambda$. However, $|(u(x), x)| \leq \lambda \|x\|^2$ for all x and so also $|(u(x) \mid y)| \leq \lambda \|x\| \cdot \|y\|$ for all x and all y. Hence $\|u(x)\| \leq \lambda \|x\|$ for all x. So $\|z\| = \lim \|u(x_n)\| \leq \lambda$. Thus $\|z\| = \lambda$ and so $\|u(x_n) - \lambda x_n\| \to 0$. It appears then that $x_n \to \lambda^{-1}z$ and so $u(x_n) \to \lambda^{-1}u(z)$, showing that $u(z) = \lambda z$—that is, that z is an eigenvector of u associated with λ. This establishes the lemma. ▌

Here now is the main theorem.

9.11.2 **Theorem.** Let E be a Hilbert space and u a compact self-adjoint endomorphism of E. Then:

(1) The spectrum $\sigma(u)$ of u is a set of real numbers, nonvoid if $u \neq 0$, positive if u is positive self-adjoint, and in any case identical with the set of eigenvalues of u. The eigenmanifold of u associated any eigenvalue λ is identical with the spectral manifold of u associated with λ. Eigenmanifolds of u associated with distinct eigenvalues are orthogonal.

(2) The nonzero eigenvalues of u, if existent, can be enumerated as a finite or infinite sequence $\lambda_1, \lambda_2, \cdots$ which, if infinite, satisfies $\lim_{k\to\infty} \lambda_k = 0$. Nonzero eigenvalues exist whenever $u \neq 0$. If 0 is an eigenvalue of u (which situation obtains whenever E is of infinite dimension), we include it in the enumeration as λ_0. Each eigenmanifold associated with a nonzero eigenvalue is of finite dimension.

(3) Putting $E_k = \operatorname{Ker}(u - \lambda_k 1)$ for $k = 0, 1, 2, \cdots$, E is the Hilbertian direct sum of the E_k $(k = 0, 1, 2, \cdots)$. If P_k is the orthogonal projector with range E_k, each P_k commutes with u,

$$u = \sum_{k\geq 0} \lambda_k P_k = \sum_{k\geq 1} \lambda_k P_k, \tag{9.11.4}$$

$$1 = \sum_{k\geq 0} P_k, \tag{9.11.5}$$

and

$$P_k P_{k'} = 0 \qquad \text{if } k \neq k'. \tag{9.11.6}$$

Remarks. Taken together, (9.11.5) and (9.11.6) may be expressed by saying that the sequence $(P_k)_{k\geq 0}$ of orthogonal projectors is a "resolution of the identity (endomorphism)." The fact that the P_k commute with u and that (9.11.4) is true is expressed by saying that (P_k) is a resolution of the identity "belonging to" u, or that (9.11.4) is the "spectral resolution" of u.

Proof of Theorem 9.11.2. Statements (1) and (2) follow from the preliminary remarks opening this section, together with Theorem 9.10.2.

Granted that E is the Hilbertian direct sum of the E_k, the remaining portions of (3) are immediate.

Let us suppose then that M is the vector subspace of E generated by the union of the E_k with $k > 0$. We must show that its orthogonal complement $M^{\perp}$ is identical with $E_0 = \text{Ker } u$. By (1), E_0 is contained in $M^{\perp}$. So we have only to show that $u(M^{\perp}) = \{0\}$. However, were this not the case, Lemma 9.11.1 asserts the existence of a scalar $\lambda \neq 0$ and an element $x \neq 0$ in $M^{\perp}$ such that $u(x) = \lambda x$. But then λ is necessarily equal to λ_k for some $k > 0$, and correspondingly x must belong to E_k. Since E_k and $M^{\perp}$ are orthogonal, this contradicts the relation $x \neq 0$. The proof is thus complete. ▮

9.11.3 **Supplements to Theorem 9.11.2.** Concerning the equation

$$u(x) - \lambda x = y \tag{9.11.7}$$

for the unknown x, y being given in E, we may make the following statements. (We assume the hypothesis of Theorem 9.11.2 to be fulfilled.)

(a) If λ is distinct from all the λ_k ($k = 0, 1, 2, \cdots$), then (9.11.7) has the unique solution

$$x = \sum_{k \geq 0} (\lambda_k - \lambda)^{-1} P_k y. \tag{9.11.8}$$

(b) If $\lambda = \lambda_h$ for some $h > 0$, then (9.11.7) has a solution if and only if $P_h y = 0$, in which case the solutions are given by

$$x = a + \sum_{k \geq 0, k \neq h} (\lambda_k - \lambda)^{-1} P_k y,$$

a denoting an arbitrary element of $E_h = \text{Ker } (u - \lambda_h 1)$.

(c) If $\lambda = 0$, then (9.11.7) has a solution if and only if $P_0 y = 0$ and $\sum_{k>0} \lambda_k^{-2} \|P_k y\|^2 < +\infty$, in which case the solutions are given by

$$x = a + \sum_{k>0} \lambda_k^{-1} P_k y,$$

a denoting an arbitrary element of $E_0 = \text{Ker } u$.

Verification of these statements is left as an exercise for the reader. No convergence criterion need appear in cases (a) and (b), the appropriate series being visibly convergent in E.

We add an alternative formula for the solution in case (a). From (9.11.7) we obtain for the solution x the expression $x = -\lambda^{-1} y + \lambda^{-1} u(x)$. If in the term $u(x)$ we substitute the series appearing in (a), there appears the expression

$$x = -\lambda^{-1} y + \sum_{k>0} \frac{\lambda_k P_k y}{\lambda(\lambda_k - \lambda)}. \tag{9.11.9}$$

In obtaining this version it should be noted that $uP_0 = 0$.

An alternative version of (9.11.4) also calls for mention. This depends on a choice of an orthonormal base in each E_k with $k > 0$. Thus, we define $n_0 = 0$ and natural numbers n_k ($k > 0$) and vectors e_i ($i = 1, 2, \cdots$) in E in such a way that (e_i), with $n_{k-1} < i \leq n_k$, is an orthonormal base in E_k $(k > 0)$. The dimension of E_k is thus $n_k - n_{k-1}$. We define (λ'_i) to be the sequence $(\lambda_k)_{k>0}$ with "repetitions" according to the formula

$$\lambda'_i = \lambda_k \qquad \text{for} \qquad n_{k-1} < i \leq n_k.$$

It is still the case that (λ'_i), if an infinite sequence, converges to 0. Furthermore in place of (9.11.4) and (9.11.5) we may write

$$u(x) = \sum_{i>0} \lambda'_i(x \mid e_i)e_i \tag{9.11.4'}$$

and

$$x = P_0 x + \sum_{i>0} (x \mid e_i)e_i. \tag{9.11.5'}$$

There is a converse: If the sequence (λ'_i) is given, tending to 0 if it is an infinite sequence, and if (e_i) is any given orthonormal family in E, then (9.11.4') yields an endomorphism u of E that is compact and that admits the λ'_i and the e_i as eigenvalues and eigenvectors. This u is self-adjoint if and only if all the λ'_i are real, positive self-adjoint if and only if the λ'_i are positive. As we shall see in Subsection 9.11.6 the case of nonreal λ'_i corresponds exactly to the compact normal endomorphisms of E.

9.11.4 The Situation When E is not Complete. Some applications place a premium on a formulation of the preceding results—or at any rate of as many of them as is possible—when E is a (not necessarily complete) pre-Hilbert space. We therefore examine this situation at some length.

In this subsection, E denotes a pre-Hilbert space. Let $\hat{E}$ be the completion of E, which is a Hilbert space. u is again a compact (hence continuous) endomorphism of E that is self-adjoint in the sense that (9.11.3) is to be valid for x and y in E.

Now Theorem 9.10.2 does not assume the completeness of E, so that statements (1) and (2) of Theorem 9.11.2 are in no way disturbed by dropping the hypothesis of completeness of E. Again, Lemma 9.11.1 nowhere uses completeness of E, the possible absence of this property being compensated by the compactness of u. Accordingly it is still true that E_0 is the orthocomplement in E of $\bigcup_{k>0} E_k$.

If, however, we wish to proceed further by taking orthogonal projections P_k onto the E_k, some forethought is necessary. Such projection onto a vector subspace of a pre-Hilbert space is in general possible only if the vector subspace is complete. This is true of the E_k with $k > 0$, each of these being of finite dimension. Thus $P_k x$ is defined as an element of E for each $k > 0$. If, as is generally the case, E_0 is of infinite dimension, we cannot be sure of the existence of $P_0 x$ as an element of E (but merely as an element of $\hat{E}$ which belongs to the closure in $\hat{E}$ of E_0). Looked at in another way, given x in E we may consider the series $\sum_{k>0} P_k x$. Since the P_k are orthogonal, the partial sums $z_n = \sum_{0<k<n} P_k x$ of this

series form a Cauchy sequence in E. The sequence of partial sums (z_n) will converge in $\hat{E}$ (but perhaps not in E) to a sum z, say. The element $x - z = x_0$ of $\hat{E}$ must then replace P_0x. Concerning x_0 we may say that it is orthogonal to $\bigcup_{k>0}E_k$, but we do not know, in general, that x_0 belongs to $E_0 = \text{Ker } u$.

Thus we shall have

$$x = x_0 + \sum_{k>0} P_k x, \tag{9.11.4''}$$

where x_0 is orthogonal to $\bigcup_{k>0}E_k$ and P_kx belongs to E_k for $k > 0$. Any such decomposition is necessarily unique for given x. Neither x_0 nor the sum $\sum_{k>0}P_kx$ will in general belong to E, though we can be sure that they do if E_0 is of finite dimension (a fortiori if E itself is of finite dimension). On the other hand, if we know that x_0 belongs to E, then it must belong to E_0 (the latter being the orthocomplement in E of $\bigcup_{k>0}E_k$). Unfortunately, both (9.11.4) and (9.11.5) break down in general.

In view of the preceding remarks it is rather surprising to find that (9.11.9), which involves no reference to the terms x_0 and y_0, may be salvaged. That this is so depends upon a remark that is a special case of the following lemma.

Lemma. E being a pre-Hilbert space and u a compact self-adjoint endomorphism of E, let us suppose the E_k and P_k are defined for $k > 0$ as above. Then, whatever the bounded sequence (ξ_k) of scalars, the series $\sum_{k>0}\xi_kP_kx$ is convergent in E whenever $x \in u(E)$. Moreover for any such x one has

$$x = \sum_{k>0} P_k x.$$

Proof. Suppose that $x = u(z)$ for some $z \in E$. If t_n and s_n are, respectively, the finite partial sums of the series $\sum_{k>0} \xi_kP_kx$ and $\sum_{k>0} \xi_kP_kz$, the fact that each P_k commutes with u shows that $t_n = u(s_n)$. Now, since the P_k are orthogonal, the sequence (s_n) is Cauchy, and therefore bounded, in E. Hence (t_n) is Cauchy. But, u being compact, (t_n) possesses a convergent subsequence. Since (t_n) is Cauchy, it must be convergent in E. This establishes the first assertion.

By (9.11.4″) we have $x = x_0 + \sum_{k>0}P_kx$ where what we have just established about the convergence of the series $\sum_{k>0}P_kx$ shows that x_0 belongs to E and therefore to Ker $u = E_0$. Since E_0 is the orthocomplement of $\bigcup_{k>0} E_k$,

$$\|x_0\|^2 = (x_0 \mid x_0) = (x \mid x_0) = (u(z) \mid x_0) = (z \mid u(x_0) = (z \mid 0) = 0,$$

showing that $x_0 = 0$ and thereby establishing the second statement in the lemma. ▮

Returning to the consideration of (9.11.9), from the relation (9.11.7) (the unique solubility of which is not in doubt) we find as before that

$$P_k x = (\lambda_k - \lambda)^{-1} P_k y \qquad \text{for } k > 0.$$

Moreover, since $y + \lambda x = u(x)$ belongs to $u(E)$, the lemma shows that

$$y + \lambda x = \sum_{k>0} (P_k y + \lambda P_k x).$$

Substituting in here the expression for $P_k x$ in terms of $P_k y$ and then solving for x, one obtains (9.11.9). Notice that the convergence of the series in (9.11.9) is ensured by the preceding argument.

9.11.5 **Application to Integral Equations.** We shall here consider briefly an integral operator $y = u(x)$ of the form

$$y(t) = \int_0^1 K(t, s)x(s)\, ds,$$

writing ds in place of $d\mu(s)$ when μ is Lebesgue measure on the real axis. The kernel function $K(t, s)$ is thus defined on the square $[0, 1] \times [0, 1]$ in R^2.

Depending upon the nature of $K(t, s)$, which is assumed in any case to be measurable for Lebesgue measure on R^2, there are several natural settings for a study of u. For example, one might seek to study it as an endomorphism of the Hilbert space $L^2 = L^2([0, 1], \mu)$, in which case criteria for the compactness of u follow from Propositions 9.5.2, 9.5.3, and 9.5.8. Or again one might seek to treat u as an endomorphism of $C = C[0,1]$, the Banach space of continuous scalar-valued functions on $[0, 1]$ with the "uniform norm" $\|x\|_\infty = \text{Sup}\,\{|x(t)| : t \in [0, 1]\}$. In this case, to make certain that u is indeed an endomorphism of C, the following two conditions are obviously sufficient:

$$\int_0^1 |K(t, s)|\, ds < +\infty \qquad \text{for } t \in [0, 1], \tag{9.11.10}$$

$$\lim_{t \to t_0} \int_0^1 |K(t, s) - K(t_0, s)|\, ds = 0 \qquad \text{for } t_0 \in [0, 1]. \tag{9.11.11}$$

It then follows that

$$\operatorname*{Sup}_{t \in [0,1]} \int_0^1 |K(t, s)|\, ds < +\infty.$$

As a consequence, Ascoli's theorem shows that u is even a compact endomorphism of C whenever (9.11.10) and (9.11.11) are fulfilled.

Despite these two attractive possibilities we shall, in fact, examine a sort of "hybrid" formulation in which we wish to picture u as an endomorphism of a pre-Hilbert space and call upon the results (Subsection 9.11.4) peculiar to this case. The pre-Hilbert space we shall choose, and which we shall denote by E, is the vector space C equipped with the inner product induced on it as a subset of L^2, viz.,

$$(x \mid y) = \int_0^1 x(t)\overline{y(t)}\, dt.$$

For the preceding theory to apply we must now add the demand that u be self-adjoint. In terms of K this is easily seen to require that

$$K(t, s) = \overline{K(s, t)} \tag{9.11.12}$$

for almost all points (t, s) of the square $[0, 1] \times [0, 1]$. In this case we say that the kernel K is "Hermitian symmetric."

With the replacement of the Banach space C by the pre-Hilbert space E, it is no longer certain that (9.11.10) and (9.11.11) suffice to ensure that u is a compact endomorphism of E. (The reader will note that the topology on E is weaker than that on C.) For reasons that will appear later we shall, in fact, impose upon K hypotheses strong enough to ensure that u is compact from E into C (and so a fortiori compact from E into E). Sufficient conditions are (as one sees easily by using the Cauchy-Schwarz inequality and Ascoli's theorem) that

$$\int_0^1 |K(t, s)|^2 \, ds < +\infty \qquad \text{for } t \in [0, 1] \tag{9.11.13}$$

and

$$\lim_{t \to t_0} \int_0^1 |K(t, s) - K(t_0, s)|^2 \, ds = 0 \qquad \text{for } t_0 \in [0, 1]. \tag{9.11.14}$$

Granted (9.11.12) through (9.11.14), one may apply all the preceding theory. There is in the present instance, however, and due partly to the "hybrid" nature of the context, a valuable opportunity to refine slightly the lemma in Subsection 9.11.4. As we have said, (9.11.13) and (9.11.14) ensure not only that u is a compact endomorphism of E, but even that u is compact from E into C. The proof of the said lemma is thus seen to show that for any x in $u(E)$ and any bounded sequence (ξ_k) of scalars, the series $\sum_{k>0} \xi_k P_k x$ is convergent in C (and a fortiori convergent in E). What is more, the same conclusion applies to the series obtained by replacing each P_k (the orthogonal projector onto the finite dimensional eigenmanifold E_k associated with λ_k) by the orthogonal projector Q_k onto any vector subspace of E_k.

To see what these refinements entail, let $n_k - n_{k-1}$ be the dimension of E_k for $k > 0$, where $n_0 = 0$. In E_k we choose any orthonormal base (φ_i), the index i ranging over the natural numbers satisfying $n_{k-1} < i \leq n_k$. It is now convenient to replace the sequence (λ_k) of distinct nonzero eigenvalues of u by the sequence obtained from it by repeating λ_k precisely n_k times. In other words, we shall replace (λ_k) by a sequence (λ_i), where λ_i is the old λ_k for $n_{k-1} < i \leq n_k$ and $k = 1, 2, \cdots$. In the same way we shall pass from (ξ_k) to the sequence (ξ_i) "with repetitions." With these notational changes, the series $\sum_{k>0} \xi_k P_k x$ is

$$\sum_{k>0} \left[\sum_{n_{k-1} < i \leq n_k} \xi_i (x \mid \varphi_i) \varphi_i \right].$$

This is the series $\sum_{i>0} \xi_i (x \mid \varphi_i) \varphi_i$ with a particular grouping of terms, indicated by the bracketing $[\cdots]$. Ordinarily, of course, there would be no justification for removing these brackets. But justification is at hand in this instance in the shape of our freedom to pass from the P_k to the Q_k. For this means that, for any choice of the finite subsets I_k of the intervals $(n_{k-1}, n_k]$, the bracketed series $\sum_{k>0}[\sum_{i\in I_k} \xi_i(x \mid \varphi_i)\varphi_i]$ is convergent in C. By separating each term into its real and imaginary parts, and then choosing the I_k suitably, it appears that the bracketed series of the absolute values $|\xi_i(x \mid \varphi_i)\varphi_i(t)|$ is convergent for each t in $[0, 1]$. We thus reach the conclusion that the series $\sum_{i>0} \xi_i(x \mid \varphi_i)\varphi_i$ is absolutely and uniformly convergent on $[0, 1]$.

Reference to the relevant portion of Subsection 9.4.11 now yields the following conclusion:

(a) Suppose that the kernel K satisfies (9.11.12) through (9.11.14). Then if λ is nonzero and different from all the λ_i, the equation

$$y(t) = \int_0^1 K(t, s)x(s)\,ds - \lambda x(t)$$

has for each given continuous y a unique continuous solution x that is expressed by the formula

$$x(t) = -\lambda^{-1}y(t) + \lambda^{-1}\sum_{i>0} \lambda_i(\lambda_i - \lambda)^{-1}(y \mid \varphi_i)\varphi_i(t),$$

the series on the right-hand side being absolutely and uniformly convergent for t in $[0, 1]$. (The λ_i and the φ_i are the nonzero eigenvalues and eigenfunctions, chosen and labelled as described above.)

Having reached this stage, we proceed to derive some further properties of the integral equation.

(b) Assuming that the kernel K satisfies (9.11.12) through (9.11.14), one has

$$\sum_{i>0} \lambda_i^2 \le \int_0^1 dt \int_0^1 |K(t, s)|^2\,ds < +\infty.$$

Proof. According to Bessel's inequality, applied to the function $t \to K(t, s)$ and the orthonormal system (φ_i), one has for each n the inequality

$$\sum_{0<i<n} \left| \int_0^1 K(t, s)\overline{\varphi_i(t)}\,dt \right|^2 \le \int_0^1 |K(t, s)|^2\,dt.$$

Using the Hermitian symmetry of K and the fact that φ_i is an eigenfunction associated with the eigenvalue λ_i, the series on the left-hand side is seen to be $\sum_{0<i<n} \lambda_i^2\,|\varphi_i(s)|^2$. The stated inequality now follows by integrating with respect to s, using the orthonormality of the φ_i, the Fubini-Tonelli theorem, and finally (9.11.13) and (9.11.14) in combination (showing that $\int_0^1 |K(t, s)|^2\,ds$ is bounded with respect to t). The proof is complete. ▌

(b′) Assume the hypotheses in (a) and also that $t \to K(t, s)$ is continuous for each fixed s. Then the series $\sum_{i>0} \lambda_i^2\,|\varphi_i(t)|^2$ is uniformly convergent to the sum $\int_0^1 |K(t, s)|^2\,ds$.

Proof. We introduce the "iterated kernel"

$$H(t, s) = \int_0^1 K(t, \xi)K(\xi, s)\,d\xi.$$

The reader will note that

$$H(t, t) = \int_0^1 |K(t, \xi)|^2\,d\xi,$$

which is continuous by (9.11.14). Since $K(t, s)$ is a continuous function of t, so $t \to H(t, s)$ is for a fixed s a member of $u(E)$. By the lemma of Subsection 9.11.4 (refined as above), it follows that

$$H(t, s) = \sum_{i>0} \left[\int_0^1 H(t', s)\overline{\varphi_i(t')}\, dt' \right] \varphi_i(t),$$

the series being absolutely and uniformly convergent. Now the integral $[\cdots]$ appearing in the ith term of the series on the right-hand side is easily seen to be $\lambda_i^2\overline{\varphi_i(s)}$. Thus we conclude that

$$H(t, s) = \sum_{i>0} \lambda_i^2 \varphi_i(t)\overline{\varphi_i(s)},$$

the series converging for each pair (t, s). Taking $t = s$, we see that $\sum_{i>0} \lambda_i^2 \, |\varphi_i(t)|^2$ is convergent to the continuous sum $H(t, t)$. By Dini's theorem, the convergence is necessarily uniform, which is what we wished to establish. ▌

(c) The hypotheses are as in (b′). The conclusion is that for each nonzero λ distinct from all the eigenvalues λ_i, the series

$$R_\lambda(t, s) = -\lambda^{-2}K(t, s) + \lambda^{-2} \sum_{i>0} \lambda_i^2(\lambda_i - \lambda)^{-1}\varphi_i(t)\overline{\varphi_i(s)}$$

is absolutely and uniformly convergent in the pair (t, s). Furthermore the solution x of the equation dealt with in (a) can be written

$$x(t) = -\lambda^{-1}y(t) + \int_0^1 R_\lambda(t, s)y(s)\, ds.$$

Proof. We use the solution formula established in (a) as a starting point. Besides this we know (from the refined form of the lemma once again) that

$$u(y)(t) = \int_0^1 K(t, s)y(s)\, ds = \sum_{i>0} \lambda_i \int_0^1 y(s)\overline{\varphi_i(s)}\, ds \cdot \varphi_i(t),$$

the series converging absolutely and uniformly. Using the identity

$$\lambda^{-1}(\lambda_i - \lambda)^{-1} + \lambda^{-2} = \lambda^{-2}\lambda_i(\lambda_i - \lambda)^{-1},$$

one derives for x the formula

$$x(t) = -\lambda^{-1}y(t) - \lambda^{-2} \int K(t, s)y(s)\, ds + \lambda^{-2} \sum_{i>0} \lambda_i^2(\lambda_i - \lambda)^{-1} \times \int y(s)\overline{\varphi_i(s)}\, ds \cdot \varphi_i(t),$$

all integrals being extended over the interval $[0, 1]$.

It therefore suffices to show that the series defining $R_\lambda(t, s)$ is absolutely and uniformly convergent in the pair (t, s), for then we may make the interchange $\sum \int = \int \sum$.

Now we know from (b′) that $\sum \lambda_i^2 \, |\varphi_i(s)|^2$ is uniformly convergent. On the other hand,

$$\operatorname*{Inf}_{i>0} |\lambda_i - \lambda| = c > 0.$$

By the Cauchy-Schwarz inequality we shall therefore have, for sums over any finite range, the inequality

$$\sum \lambda_i^2 |\lambda_i - \lambda|^{-1} |\varphi_i(t)\overline{\varphi_i(s)}| \le c^{-1}\Big[\sum \lambda_i^2 |\overline{\varphi_i(t)}|^2\Big]^{1/2}\Big[\sum \lambda_i^2 |\varphi_i(s)|^2\Big]^{1/2},$$

which makes apparent the desired conclusion. ▌

(d) The hypotheses are as in (b′). The conclusion is that

$$\lim_{n\to\infty} \int_0^1 |K(t,s) - \sum_{0<i<n} \lambda_i \varphi_i(t)\overline{\varphi_i(s)}|^2 \, ds = 0$$

uniformly with respect to t.

Proof. A direct computation shows that the term on the left-hand side, of which the limit is to be taken, is none other than

$$\int_0^1 |K(t,s)|^2 \, ds - \sum_{0<i<n} \lambda_i^2 |\varphi_i(t)|^2.$$

The assertion now follows from (b′). ▌

Remark. In particular, therefore $K(t, s)$ is the limit in L^2 (over the square $[0,1] \times [0, 1]$) of the series $\sum_{i>0} \lambda_i \varphi_i(t)\overline{\varphi_i(s)}$. It is not true in general, however, that the series is pointwise convergent throughout the square; see Smithies [2], p. 126. According to a well-known theorem of Mercer, pointwise convergence obtains if the endomorphism u defined by K is *positive* self-adjoint. We shall not give the proof of this, referring the reader to Dieudonné [13], p. 338, Smithies [2], p. 128, or Zaanen [1], p. 534.

9.11.6 **Compact Normal Endomorphisms.** We now consider briefly the matter mentioned at the end of Subsection 9.11.3—namely, the possibility of removing the hypothesis of self-adjointness so far imposed upon u. For the sake of simplicity we shall confine our attention to the case in which u is a compact endomorphism of a Hilbert (rather than a pre-Hilbert) space E.

The principal conclusion will be that the spectral resolution, expressed via (9.11.4) through (9.11.6), is still possible for those compact u which are *normal* in the sense that u and u^* commute. That this restriction is essential is quite plain, for any u that is expressible as a sum (9.11.4), wherein the P_k are commuting orthogonal projectors is visibly normal.

Before we can embark upon the extension envisaged, it is necessary to make an additional remark concerning the known (self-adjoint) case. If we use the spectral resolution of u, it is plain that

$$F(u) = \sum_{k>0} F(\lambda_k) P_k \tag{9.11.15}$$

for any polynomial F on the real axis, where, if

$$F(\lambda) = \sum c_n \lambda^n,$$

$F(u)$ denotes the endomorphism $\sum c_n u^n$. On the other hand, if F is any bounded, complex-valued function defined on $S = \sigma(u)\backslash\{0\} = \{\lambda_1, \lambda_2, \cdots\}$, the series $\sum_{k>0} F(\lambda_k)P_k x$ is convergent for each x of E and thus serves to define an endomorphism of E that we continue to denote by $F(u)$. It is also easily verified that

$$\|F(u)\| \leq \operatorname{Sup}_{k>0} |F(\lambda_k)|. \tag{9.11.16}$$

This shows in particular that $F(u)$ is compact whenever (u is compact and) $\lim_{k\to\infty} F(\lambda_k) = 0$. Moreover, when this last condition is fulfilled, the function F can be continuously extended to the compact set $\sigma(u)$ and hence to the entire real axis. But then we can approximate F uniformly on $\sigma(u)$ by polynomials (Weierstrass' theorem). Accordingly $F(u)$ is the limit in $L_c(E)$ of polynomials in u. This applies in particular if F is taken to be equal to 1 at λ_k and to be 0 at $\lambda_{k'}$ for $k' \neq k$, in which case $F(u)$ is simply P_k. Thus, each P_k with $k > 0$ is the limit of polynomials in u. (The preceding argument shows even that these polynomials may be chosen to have a zero constant term, but this is irrelevant for our purposes.)

Now let us return to the case in which u is compact and normal. We define

$$u_1 = \tfrac{1}{2}(u + u^*), \qquad u_2 = -\tfrac{1}{2}i(u - u^*),$$

corresponding to the resolution of a complex number into its real and imaginary parts. Since u and u^* commute, u_1 and u_2 are self-adjoint. And since u and u^* are compact, so too are u_1 and u_2. We write the resolution of the identity belonging to u_n ($n = 1, 2$) in the shape of a family $P_{n,\lambda}$ (λ real) of orthogonal projectors, 0 except for λ in $\sigma(u_n)$, such that $P_{n,\lambda}P_{n,\lambda'} = 0$ whenever $\lambda \neq \lambda'$,

$$\sum_\lambda P_{n,\lambda} = 1, \qquad \sum_\lambda \lambda P_{n,\lambda} = u_n.$$

It is crucial to notice that in addition each $P_{1,\lambda}$ commutes with each $P_{2,\mu}$. This follows because, as we have proved above, $P_{n,\lambda}$ is the limit of polynomials in u_n, while u_1 and u_2 commute owing to the normal character of u.

Thanks to this, if we define

$$P_\xi = P_{1,\lambda}P_{2,\mu}$$

for each complex $\xi = \lambda + i\mu$ (λ and μ real), then the P_ξ form a family of orthogonal projectors, $P_\xi P_{\xi'} = 0$ if $\xi \neq \xi'$, $P_\xi \neq 0$ save when ξ belongs to the set $\sigma(u_1) + i\sigma(u_2)$ (which is countable). Moreover,

$$\sum_\xi P_\xi = 1.$$

and

$$u = u_1 + iu_2 = \sum_\lambda \lambda P_{1,\lambda} + i\sum_\mu \mu P_{2,\mu},$$

which can be written in the form

$$u = \sum_\xi \xi P_\xi.$$

So the family (P_ξ) constitutes a resolution of the identity belonging to u. The reader will notice that each P_ξ with $\xi \neq 0$ is the limit of polynomials in u and u^*, hence commutes with each of u and u^*; the same is true of P_0 (which is equal to $1 - \sum_{\xi \neq 0} P_\xi$). We have thus obtained the desired spectral resolution of u.

At this point we may add that the same arguments can be employed to handle the case of any finite number of commuting normal endomorphisms, leading to a simultaneous spectral resolution for the endomorphisms involved. One can even deal with an infinite commuting set of continuous normal endomorphisms, though here sums must be replaced by integrals. For this problem more sophisticated methods are needed. The case of discontinuous self-adjoint endomorphisms is still more complicated and itself forms a very large portion of Hilbert space lore. All of these developments lie beyond the scope of this account, and the reader must be referred to the appropriate references, for example: Stone [1], Riesz-Nagy [1], Cooke [2], Halmos [2] (representative of the earlier approaches), Loomis [1], Godement [5], Rickart [1] (representative of the modern approach founded on the study of Banach algebras).

9.12 Partial Differential Equations and Compact Linear Maps

In Subsection 5.13.6 we made reference to the implications in the theory of linear partial differential equations of the study of compact linear maps of one TVS into another. In this section we propose to set forth some of the relevant details. Regarding linear partial differential equations we shall retain the notations introduced in Section 5.13, summarizing them here for the reader's convenience.

Throughout Ω will denote a domain in R^n; μ denotes Lebesgue measure on R^n. $p_0, \cdots, p_n$ are positive measurable functions on Ω subject to the conditions

$$p_0, \cdots, p_n \in \mathscr{L}^1_{\text{loc}}(\Omega). \tag{9.12.1}$$

V denotes the space of (equivalence classes of) functions f on Ω such that

$$f,\ \partial_k f \in L^1_{\text{loc}}(\Omega),\ p_0^{1/2} f,\ p_k^{1/2}\,\partial_k f \in L^2(\Omega) \qquad (1 \leq k \leq n). \tag{9.12.2}$$

V is made into an inner-product space by defining

$$(f \mid g)_V = (p_0^{1/2} f \mid p_0^{1/2} g)_{L^2(\Omega)} + \sum_{k=1}^{n} (p_k^{1/2}\,\partial_k f \mid p_k^{1/2}\,\partial_k g)_{L^2(\Omega)}. \tag{9.12.3}$$

It is evident that $\mathscr{D}(\Omega) \subset V$.

Our linear partial differential operator $P = P(x, \partial_k)$ has the property that

$$(v \mid \varphi)_V = \langle \bar{\varphi}, Pv \rangle \tag{9.12.4}$$

for v in V and φ in $\mathscr{D}(\Omega)$, the right-hand side referring to the duality between $\mathscr{D}(\Omega)$ and $\mathscr{D}'(\Omega)$. This involves (see Subsection 5.13.1) that P is the map of V into $\mathscr{D}'(\Omega)$ defined by

$$Pf = p_0 f + \sum_{k=1}^{n} \partial_k (p_k\,\partial_k f). \tag{9.12.5}$$

For definiteness we shall consider the Dirichlet problem and, in the notations of Subsection 5.13.2, take V_0 to be the closure in V of $\mathscr{D}(\Omega)$. We assume further (cf. (5.13.8)) that V_0 is complete and

$$\|v\|_{L^2(\Omega)} \leq c \cdot \|v\|_V \qquad (v \in V_0). \tag{9.12.6}$$

As before, N will denote the set of u in V_0 such that $Pu \in L^2(\Omega)$. We remark that, on the basis of (9.12.6), N is complete for the structure defined by the inner product

$$(u \mid v)_N = (u \mid v)_V + (Pu \mid Pv)_{L^2(\Omega)}. \tag{9.12.7}$$

Conditions sufficient to ensure that (9.12.6) is satisfied are given in Subsection 5.13.4. Let us recall that if (9.12.1) holds, and if furthermore

$$p_1^{-1}, \cdots, p_n^{-1} \in \mathscr{L}^1_{\text{loc}}(\Omega), \tag{9.12.8}$$

then V and hence also V_0 are complete. By Subsection 5.13.9(b), if

$$m_r = \text{ess inf}_\Omega\, p_r > 0 \qquad (0 \leq r \leq n), \tag{9.12.9}$$

then V is a subspace of $\mathscr{E}^1_{L^2}(\Omega)$ and has a structure stronger than that induced by the latter space; if moreover

$$\text{ess sup}_\Omega\, p_r < +\infty \qquad (0 \leq r \leq n),$$

then V and $\mathscr{E}^1_{L^2}(\Omega)$ are identical as sets and their structures are equivalent.

The final requirement, permitting an appeal to the theory of compact operators, will be that the injection map of V_0 into $L^2(\Omega)$ be compact. Plainly, if (9.12.9) holds, it is sufficient for this that the injection map of $\mathscr{E}^1_{L^2}(\Omega)$ (or merely of $\mathscr{D}^1_{L^2}(\Omega)$) into $L^2(\Omega)$ be compact. We propose to give three simple sets of conditions ensuring that this is the case. For $n = 1$, a few special remarks will be in order.

9.12.1 Compactness of the Injection of V_0 into $L^2(\Omega)$.

The three results alluded to above will be stated in succession, the proofs following later.

(a) Suppose that Ω is bounded and that (9.12.9) holds. Then the injection of V_0 into $L^2(\Omega)$ is compact.

This result and its proof are due to Gårding [2].

(b) Suppose that (9.12.9) holds and that

$$\operatorname*{Sup}_{K}\,(\operatorname*{ess\,inf}_{\Omega\backslash K} p_0) = +\infty,$$

K ranging over the compact subsets of Ω. Then the injection of V into $L^2(\Omega)$ is compact.

For this result, see Lions [1], Proposition 4.3.

(c) Suppose that $\mu(\Omega) < +\infty$, that (9.12.9) holds, and that Ω has the following property:

(S) There exists a number $q > 2$ such that $\mathscr{E}^1_{L^2}(\Omega) \subset L^q(\Omega)$.

Then the injection of V into $L^2(\Omega)$ is compact.

This result and its proof are due to Deny and Lions; see Deny-Lions [1], Chapter I, Théorème 9.1. We shall comment further on the property (S) in Subsection 9.12.2 below.

(d) As we saw when we discussed, in Subsection 5.13.4, the completeness of V, the case $n = 1$ can be treated separately and more elementarily. We do not propose to set forth the details here, leaving this as an exercise for the reader. Let us merely note a couple of simple results available in this case.

(d′) If $\mu(\Omega) < +\infty$, p_0 is nonnegligible, and p_1^{-1} belongs to $\mathscr{L}^1(\Omega)$, then the injection map of V into $C_{bd}(\Omega)$ is compact. (Here $C_{bd}(\Omega)$ is the Banach space of bounded, continuous functions on Ω with the uniform norm.) If further Ω is bounded, this injection is compact from V into $C(\bar{\Omega})$.

(d″) If $\mu(\Omega) < +\infty$ and $p_1^{-1} \in \mathscr{L}^1(\Omega)$, and if V_1 consists of those f in V that have limit 0 at the left (or the right) extremity of Ω, then the injection is compact from V_1 into $C_{bd}(\Omega)$.

In either case, since $\mu(\Omega) < +\infty$, compactness in the sense specified in (d′) or (d″) implies compactness of the said injection regarded as a mapping into $L^2(\Omega)$.

Without any assumptions on Ω, it is easily established too that the injection of V into $L^2(\Omega)$ is compact whenever $m_0 > 0$ and p_1^{-1} is in $\mathscr{L}^1_{\text{loc}}(\Omega)$; this sharpens (b).

The reader's attention must also be directed to the results of Hörmander ([1], Section 2.9) about the compactness of the inverse of a linear partial differential operator with constant coefficients. These results are framed in a context somewhat different from that in which we have chosen to operate.

The proofs of (b) and (c) rest upon the use of Weil's criterion of compactness in Lebesgue spaces; see Section 4.20.

Throughout the ensuing proofs, if f is a function on Ω we denote by f^* the function on R^n that is equal to f on Ω and to 0 elsewhere. B denotes the closed unit ball in $\mathscr{E}^1_{L^2}(\Omega)$. For brevity we shall write simply $|||f|||_1$ for the norm of f in $\mathscr{E}^1_{L^2}(\Omega)$, $\mathscr{E}$ for $\mathscr{E}^1_{L^2}(\Omega)$, and L^p for $L^p(R^n)$.

Proof of (a). Under the stated conditions on m_r $(0 \le r \le n)$, V is a subspace of $\mathscr{E}$ with a structure stronger than that induced by $\mathscr{E}$. So it follows that here V_0 is a subspace of $\mathscr{D}^1_{L^2}(\Omega)$ with a topology stronger than that induced by the latter space. Accordingly it will suffice to show that the injection of $\mathscr{D}^1_{L^2}(\Omega)$ into $L^2(\Omega)$ is compact.

To do this we have to show that, given a sequence (f_k) extracted from $B \cap \mathscr{D}^1_{L^2}(\Omega)$, there exists a subsequence that converges in $L^2(\Omega)$. Since $\mathscr{D} = \mathscr{D}(\Omega)$ is dense in $\mathscr{D}^1_{L^2}(\Omega)$, it is clearly enough to deal with the case in which (f_k) is extracted from $B \cap \mathscr{D}$.

We introduce the Fourier transforms

$$F_k(\xi) = \int_\Omega e^{-2\pi i\xi\cdot x} f_k(x)\, d\mu(x) = \int e^{-2\pi i\xi\cdot x} f_k^*(x)\, d\mu(x).$$

We know that

$$\|f_k\|_{L^2(\Omega)} \le |||f_k|||_1 \le 1$$

and so there exists a subsequence (f_{k_p}) that converges weakly in $L^2(\Omega)$. Since Ω is bounded, Ascoli's theorem shows that the sequence (F_{k_p}) converges locally uniformly on R^n. By Parseval's formula

$$\begin{aligned}\|f_{k_p} - f_{k_q}\|^2_{L^2(\Omega)} &= \|f^*_{k_p} - f^*_{k_q}\|^2_{L^2} = \|F_{k_p} - F_{k_q}\|^2_{L^2} \\ &= \int_{|\xi|\le R} |F_{k_p}(\xi) - F_{k_q}(\xi)|^2 \, d\mu(\xi) + \int_{|\xi|>R} |\cdots|^2 \, d\mu(\xi).\end{aligned}$$

Applying the Parseval formula once again, we have (remembering that $f_k \in \mathscr{D}(\Omega)$)

$$\begin{aligned}\int_{|\xi|>R} |\cdots|^2 \, d\mu(\xi) &\le (2\pi R)^{-2} \int 4\pi^2 \, |\xi|^2 \, |\cdots|^2 \, d\mu(\xi) \\ &= (2\pi R)^{-2} \sum_{i=1}^{n} \|\partial_i f_k\|^2_{L^2(\Omega)} \\ &\le (2\pi R)^{-2} \, |||f_k|||_1^2 \le (2\pi R)^{-2}.\end{aligned}$$

On the other hand, for any given R,

$$\int_{|\xi|\le R} |\cdots|^2 \, d\mu(\xi) \to 0 \qquad \text{as } p, q \to \infty$$

thanks to local uniform convergence. It thus appears that (f_{k_p}) is Cauchy, and therefore convergent, in $L^2(\Omega)$. ▌

Proof of (c). We deal with (c) first since the proof of (b) follows similar lines.

Our hypotheses arrange that $V \subset \mathscr{E}$ with a topology finer than that induced by $\mathscr{E}$, so that it suffices to prove the compactness of the injection of $\mathscr{E}$ into $L^2(\Omega)$. It is equivalent to show that B^*, the image of B under the mapping $f \to f^*$, is compact in L^2.

To begin with, the condition (S) combines with the closed graph theorem to show that there exists a number C such that

$$\|f\|_{L^q(\Omega)} \le C \, |||f|||_1 \qquad (f \in \mathscr{E}).$$

Now if $f \in B$ an application of Hölder's inequality shows that, on writing $r = q/(q-2)$, one has

$$\begin{aligned}\int_{R^n \setminus K} |f^*|^2 \, d\mu = \int_{\Omega\setminus K} |f|^2 \, d\mu &\le \| \, |f|^2 \, \|_{L^{q/2}(\Omega)} \cdot \|\chi_{\Omega\setminus K}\|_{L^r(\Omega)} \\ &= \|f\|_{L^q(\Omega)} \cdot \mu(\Omega\setminus K)^{1/r} \\ &\le C \cdot \mu(\Omega\setminus K)^{1/r}.\end{aligned}$$

Since $q > 2$ and $\mu(\Omega)$ is finite, we conclude that

$$\int_{R^n\setminus K} |f^*|^2 \, d\mu \le \varepsilon \qquad (f \in B),$$

provided that the compact set K is sufficiently large, $\varepsilon > 0$ being arbitrarily preassigned.

Thus, given $\varepsilon > 0$, we can choose a compact set $K' \subset \Omega$ such that

$$\|f^*\|_{L^2(R^n\setminus K')} \le \frac{\varepsilon}{3} \qquad (f \in B).$$

Having selected ε and K', we choose and fix a compact set $K'' \subset \Omega$ such that $\text{Int } K'' \supset K'$. Then there exists a number $\delta' = \delta'(\varepsilon) > 0$ such that

$$x \in R^n \backslash K'', \quad |h| \leq \delta' \Rightarrow x + h \in R^n \backslash K'.$$

If T_h denotes translation by amount h, we shall then have

$$\|T_h f^*\|_{L^2(R^n \backslash K')} \leq \frac{\varepsilon}{3} \qquad (f \in B, \quad |h| \leq \delta').$$

Fixing ε (and hence K', K'', and δ'), we choose ϕ in $\mathscr{D}(\Omega)$ such that $0 \leq \phi \leq 1$, $\phi = 1$ on K'; let

$$C(\varepsilon) = \sup_{1 \leq i \leq n} \|\partial_i \phi\|_{L^\infty}.$$

It then appears after a simple calculation that $g = \phi f^* \in \mathscr{E}^1_{L^2}(R^n)$ and

$$\|g\|_{\mathscr{E}^1_{L^2(R^n)}} \leq C'(\varepsilon) \qquad (f \in B).$$

Moreover

$$\begin{aligned} \|T_h f^* - f^*\|_{L^2} &= \|T_h g - g + T_h((1 - \phi)f^*) - (1 - \phi)f^*\|_{L^2} \\ &\leq \|T_h g - g\|_{L^2} + \frac{2\varepsilon}{3}. \end{aligned}$$

According to Parseval's formula,

$$\|T_h g - g\|^2_{L^2} = \int |e^{2\pi i h \cdot \xi} - 1|^2 \, |G(\xi)|^2 \, d\mu(\xi),$$

G being the Fourier transform of g. Now $|h \cdot \xi| \leq |h| \cdot |\xi|$ and $|e^{it} - 1| \leq C' \, |t|$ for t real. Hence, using Parseval's formula once again in reverse,

$$\begin{aligned} \|T_h g - g\|^2_{L^2} &\leq C''^2 \, |h|^2 \int |\xi|^2 \cdot |G(\xi)|^2 \, d\mu(\xi) \\ &\leq C''^2 \, |h|^2 \sum_{i=1}^{n} \|\partial_i g\|_{L^2} \leq C''^2 \, |h|^2 \, \|g\|^2_{\mathscr{E}^1_{L^2(R^n)}} \end{aligned}$$

and so

$$\|T_h g - g\|_{L^2} \leq C'' \, |h| \cdot C'(\varepsilon).$$

Hence

$$\|T_h f^* - f^*\|_{L^2} \leq C'' \, |h| \, C'(\varepsilon) + 2 \, \frac{\varepsilon}{3}$$

for all f in B and all $|h| \leq \delta'$. This is in turn at most ε if we suppose that f is in B and $|h| \leq \delta$, where

$$\delta = \delta(\varepsilon) = \text{Min} \left[\delta'(\varepsilon), \frac{\varepsilon}{3C''C'(\varepsilon)} \right].$$

In this way we have verified that Weil's criterion is satisfied by B^* and the proof is complete. ▌

Proof of (b). Once again we seek to verify that B^* satisfies Weil's criterion. As before we may replace V by $\mathscr{E}$.

To begin with, given $\varepsilon > 0$, there exists a compact set $K \subset \Omega$ such that $p_0 > \varepsilon^{-1}$ almost everywhere on $\Omega \backslash K$. Then, for f in B,

$$1 \geq |||f|||_1^2 \geq \int_\Omega p_0 \, |f|^2 \, d\mu \geq \varepsilon^{-1} \int_{\Omega \backslash K} |f|^2 \, d\mu$$

and therefore

$$\int_{R^n \backslash K} |f^*|^2 \, d\mu \leq \varepsilon,$$

for f in B.

Having reached this stage, we can verify the second half of Weil's criterion exactly as in the preceding proof of (c). ▮

9.12.2 **Concerning Condition** (S). If $n > 2$ and q is restricted to the specific value $q = 2n/(n - 2)$ (always > 2), condition (S) applied to domains in R^n of finite measure, characterizes the so-called *Soboleff domains*; see the remarks in Subsection 5.11.1. The role played by this condition in the study of Beppo Levi functions has been known for some time (see the bibliography attached to Deny-Lions [1]). We now see how it bears upon the study of partial differential equations.

9.12.3 **Concerning the Green's Operator** $\mathscr{G}$. We return to the set-up described at the outset of Section 9.12. Assume that Ω, the p_r $(0 \leq r \leq n)$, and V are such that the injection of V_0 into $L^2(\Omega)$ is compact. We assume further that (9.12.6) is fulfilled, so that V_0 is a Hilbert space. The same is true of N, furnished with the inner product (9.12.7). If G has the significance attached to it in Subsection 5.13.5, and if $\mathscr{G} = G \mid V_0$, then $\mathscr{G}$ is a compact, positive, self-adjoint endomorphism of V_0. To $\mathscr{G}$ we may thus apply the theory developed in Section 9.11.

The outcome of this application is that $\mathscr{G}$ has eigenvalues that may be arranged as a sequence $1/\lambda_k$ $(k = 1, 2, \cdots)$, where $\lambda_k > 0$ and $\lambda_k \uparrow +\infty$; the associated eigenfunctions of $\mathscr{G}$ form a total set in V_0. We may enumerate the latter as u_k $(k = 1, 2, \cdots)$, repeating the λ_k's if necessary so as to arrange that u_k is associated with $1/\lambda_k$:

$$\mathscr{G}u_k = \frac{u_k}{\lambda_k}.$$

It is then possible to orthonormalize the u_k in $L^2(\Omega)$, after which we shall have

$$(u_i \mid u_j)_V = (\lambda_i \mathscr{G} u_i \mid u_j)_V = \lambda_i (\mathscr{G} u_i \mid u_j)_V = \lambda_i (u_i \mid u_j)_{L^2(\Omega)},$$

the last step by (5.13.19), and so

$$(u_i \mid u_j)_V = \lambda_i \delta_{ij}. \tag{9.12.10}$$

Now $\mathscr{D}(\Omega)$ is dense in $L^2(\Omega)$ and so V_0, which contains $\mathscr{D}(\Omega)$, is dense in $L^2(\Omega)$. The u_k forming an orthonormal base in $L^2(\Omega)$, therefore, the $\lambda_k^{-1/2} u_k$ form an orthonormal base in V_0. Besides this (see Subsection 5.13.5 once more) the relations

$$u \in V_0, \qquad u \neq 0, \qquad \mathscr{G}u = \lambda_k^{-1} u \tag{9.12.11}$$

are equivalent to the relations

$$u \in N, \qquad u \neq 0, \qquad Pu = \lambda_k u \tag{9.12.12}$$

and it then appears that the λ_k and the u_k are eigenvalues and eigenfunctions (in the sense customary in discussing differential equations) of P, considered as a map of N into $L^2(\Omega)$. (The restriction of P to N amounts, as we have seen earlier, to imposing boundary conditions on the solutions regarded as acceptable.)

Finally, the $(\lambda_k + \lambda_k^2)^{-1/2}u_k$ form an orthonormal basis in N for the structure defined by (9.12.7).

If we impose on the p_r and f suitable smoothness restrictions, the u_k (and indeed all the solutions of the Dirichlet-Neumann problem) will likewise be "smooth": for example, if all the p_r and f belong to $C^\infty(\Omega)$, the same is true of the u_k, as appears from the relation $u_k = \lambda_k^{-1}Pu_k$.

It is easily verified formally that $\mathscr{G}$, if representable as an integral operator

$$\mathscr{G}u(x) = \int_\Omega \mathscr{G}(x, y)u(y)\, d\mu(y), \tag{9.12.13}$$

must have for kernel the function

$$\mathscr{G}(x, y) = \sum_k \lambda_k^{-1} u_k(x)\overline{u_k(y)}. \tag{9.12.14}$$

Closer examination shows that such a representation is possible when the p_r are sufficiently smooth. To check the formal correctness of (9.12.14) on the basis of (9.12.13), one has only to use the relation

$$(\mathscr{G}u \mid v)_V = (u \mid v)_{L^2(\Omega)},$$

which is (5.13.19), perform the indicated integrations term by term, and recall that the u_k are orthonormal in $L^2(\Omega)$ and the relation (9.12.10). Naturally, it is easier to establish (9.12.13) whenever the sum in (9.12.14) is known to be convergent in a suitable sense (for example, convergent in $L^2(\Omega \times \Omega; \mu \otimes \mu)$); this will be verifiable a priori as soon as some knowledge is available concerning the rate of growth of the λ_k. But this is a distinct problem into which we do not propose to enter.

Dieudonné ([13], Section 11.7) gives a detailed account of the one-dimensional case, specifically the so-called Sturm-Liouville problem. The method employed there amounts to reformulating the given differential equation and boundary conditions as an equivalent integral equation, to which the arguments of Subsection 9.11.5 may be applied.

9.13 Return to Ergodic Theorems

In Subsection 1.12.9 we discussed a mean ergodic theorem for certain endomorphisms of a Hilbert space and commented on its applications in statistical mechanics. It was pointed out there that ergodic theorems of various types are known for spaces more general than a Hilbert space, and an undertaking was given that this matter would receive some attention. It is at this point that we may conveniently carry out this programme.

We shall formulate and prove one general theorem, from which several special cases of importance and interest may be derived.

9.13.1 Theorem. Let E be a separated LCTVS, u a continuous endomorphism of E, and

$$A_n = \frac{1 + u + u^2 + \cdots + u^{n-1}}{n}.$$

Suppose that

(a) $\{A_n(x) : n = 1, 2, \cdots\}$ is weakly relatively compact in E for each $x \in E$;

(a′) the A_n are equicontinuous;

(b) $\lim_n u^n(x)/n = 0$ weakly for each x in E.

We define $N = \text{Ker}\,(1 - u)$ and $R = Im(1 - u)$. Then

(1) E is the topological direct sum of N and $\bar{R}$;

(2) if π is the projection of E onto N and parallel to $\bar{R}$, then $\lim_n A_n(x) = \pi(x)$ weakly for each x in E.

Finally, if (b) holds with the initial topology replacing the weakened one, the same is true of (2).

Proof. Let x be chosen in E and denote by x_0 any weak limiting point of the sequence $(A_n(x))$, such x_0 existing because of (a). The identity

$$(1 - u)A_n(x) = \frac{u^n(x) - x}{n}$$

combines with (b) to show that x_0 belongs to N.

We show next that $x - x_0$ lies in $\bar{R}$. For let us suppose that $x_0' \in E'$ is orthogonal to $\bar{R}$. If $x - x_0$ were not in $\bar{R}$, x_0' could be chosen so that $\langle x - x_0, x_0' \rangle = 1$. However, $\langle (1 - u)u^n(x), x_0' \rangle = 0$ and so $\langle u^n(x), x_0' \rangle = \langle u^{n+1}(x), x_0' \rangle$, hence it follows that $\langle A_n(x), x_0' \rangle = \langle x, x_0' \rangle$ for all n. But then on passage to the limit one finds that $\langle x_0, x_0' \rangle = \langle x, x_0' \rangle$; that is, $\langle x - x_0, x_0' \rangle = 0$. This is a contradiction, and it follows that $x - x_0$ lies in $\bar{R}$.

The relation $N \cap \bar{R} = \{0\}$ will now be verified. To begin with, if z lies in R, the preceding identity combines with (b) to show that $A_n(z) \to 0$ weakly. Thanks to (a′), this remains true for each z in $\bar{R}$. On the other hand, if z lies in N, one has $u(z) = z$ and therefore $A_n(z) = z$ for all n. It follows that the only z common to N and $\bar{R}$ is 0, as alleged.

Since $x = (x - x_0) + x_0$, it appears now that E is the algebraic direct sum of $\bar{R}$ and N. Moreover x_0 is uniquely determined when x is given, so that $(A_n(x))$ has precisely one weak limiting point and is therefore weakly convergent. Its weak limit is $\pi(x)$, where π is the projection of E onto N parallel to $\bar{R}$. Condition (a′) now implies that π is continuous. So, finally, the direct sum $E = N + \bar{R}$ is topological as well as algebraic.

If (b) holds in the sense of the initial topology, then one sees that $A_n(z) \to 0$ in the same sense, when z lies in $\bar{R}$. It follows that (2) holds in the sense of the initial topology. ▮

The result we have proved calls for several comments.

9.13.2. If E is barrelled, (a) already implies (a′) and the latter condition may be dropped altogether.

9.13.3. Let us suppose that E is a normed vector space, that u is weakly compact and that its iterates u^n are equicontinuous. If one writes

$$A_n(x) = \frac{x}{n} + u(B_n(x)),$$

where

$$B_n = \frac{1 + u + \cdots + u^{n-2}}{n},$$

one sees that the B_n and the A_n are equicontinuous, and that weak compactness of u ensures that (a) holds. Moreover, (b) holds in the sense of the initial topology. Thus one infers that (2) holds in the sense of the initial topology.

This form of the theorem was given, for the case in which E is a Banach space, by Yosida [1]. See also Dunford [2].

9.13.4. We notice that if u is compact, then R is closed in E (Theorem 9.10.1). The same is true if E is a normed vector space and u is potentially compact (that is, some positive power of u is compact); see Exercise 9.24.

9.13.5. Let us suppose that E is a normed vector space and that (b) is strengthened to $\|u^n\|/n \to 0$. For z in R one has then $z = x - u(x)$ for some x and so $A_n(z) = [x - u^n(x)]/n$, showing that

$$\|A_n(z)\| \leq \left(n^{-1} + \frac{\|u^n\|}{n}\right) \|x\|.$$

If furthermore $1 - u$ is an isomorphism of E onto R (which, if E is a Banach space, is tantamount to assuming that R is closed), then

$$\|A_n(z)\| \leq \text{const}\left(\frac{1}{n} + \frac{\|u^n\|}{n}\right) \|z\|$$

for each z in R. Since $A_n(y) = y$ for all n and all y in N, the decomposition

$$x = \pi(x) + \pi'(x),$$

π' being the projection of E onto R parallel to N, shows that

$$A_n(x) = \pi(x) + A_n\pi'(x)$$

and so

$$\begin{aligned}\|A_n(x) - \pi(x)\| &\leq \text{const}\left(\frac{1}{n} + \frac{\|u^n\|}{n}\right) \cdot \|\pi'(x)\| \\ &\leq \alpha_n \|x\|,\end{aligned}$$

where $\lim \alpha_n = 0$. Thus in this case

$$\lim_n \|A_n - \pi\| = 0.$$

Naturally, the hypothesis that $\|u^n\|/n \to 0$ entails that (a′) holds.

EXERCISES

9.1 Let F be a separated LCTVS, A a subset of F such that *either* (1) A is relatively compact in F, *or* (2) F is quasi-complete and A is precompact in F. Prove that on A the weak and the initial topologies are identical.

Deduce that if E is a TVS and u a continuous linear map of E into F for which it is known that *either* (a) u is compact, *or* (b) F is quasi-complete and u is precompact, then u transforms any bounded directed family (x_i) in E converging weakly to x into a directed family that converges in F to $u(x)$.

9.2 Suppose that E and F are LCTVSs, that the sets $A \subset E$ and $B \subset F$ are convex and balanced, and that f is a bilinear form on $E \times F$. Prove that:

(a) If $f \mid A \times B$ is continuous at $(0, 0)$, then $f \mid A \times B$ is continuous.

(b) If A is precompact, B compact, and $f \mid A \times B$ is continuous at $(0, 0)$, then $f \mid A \times B$ is uniformly continuous.

(Grothendieck [7], p. 95, Exercice 4.)

9.3 Let E and F be LCTVSs, u a weakly continuous linear map of E into F, $\mathfrak{S}$ a set of bounded, convex, and balanced subsets of E. Prove that the following two assertions are equivalent:

(1) $u(A)$ is precompact for each $A \in \mathfrak{S}$.

(2) $\langle u(x), y' \rangle \mid A \times B$ is continuous at $(0, 0)$ for the product of the weak topologies for each $A \in \mathfrak{S}$ and each set $B \subset F'$ that is convex, balanced, weakly closed, and equicontinuous.

[Hint: Use Theorem 9.2.1 and the preceding exercise.]

9.4 Let E be a separable Banach space. Prove that the following two assertions are equivalent:

(1) In E each weakly compact set is compact.

(2) If $\lim x_n = 0$ weakly in E and $\lim x'_n = 0$ weakly in E', then $\langle x_n, x'_n \rangle \to 0$.

[Hint: Apply Exercise 9.3, taking for $\mathfrak{S}$ the set of all weakly compact subsets of E, $F = E$, and u the identity mapping of E into itself.]

9.5 Let E and F be LCTVSs and suppose that in E the bounded sets are weakly metrizable (as is the case if E' is strongly separable). Let u be a weakly continuous linear map of E into F. Prove that:

(a) u transforms bounded subsets of E into precompact subsets of F, if and only if $\lim u(x_n) = 0$ in F whenever $\lim x_n = 0$ weakly in E.

(Cf. Proposition 9.4.13(2).)

(b) If further each weakly compact subset of F is compact, then u transforms bounded sets into precompact sets.

9.6 Let E be a normed vector space in which each bounded set is weakly metrizable. Show that if in E each weakly compact set is compact, then $\dim E < \infty$.

[Hint: Apply (b) of Exercise 9.5, taking $F = E$ and u the identity map of E into itself.]

9.7 Suppose that E is a Banach space of infinite dimension whose dual E' is strongly separable. Prove that there exist sequences (x_n) in E and (x'_n) in E' such that $\lim x_n = 0$ weakly, $\lim x'_n = 0$ weakly, and yet $\langle x_n, x'_n \rangle = 1$ for all n.

[Hint: Use Exercises 9.4 and 9.6.]

9.8 Let E be a Banach space. Prove the equivalence of the following two assertions:

(1) Each sequence (x'_n) of elements of E' that converges weakly to 0 converges to 0 for the topology $\sigma(E', E'')$.

(2) Each continuous linear map u of E into a separable, complete, and separated LCTVS F is weakly compact.

[Hint: Use Theorem 9.2.1 and Eberlein's theorem.]

9.9 Let E and F be separated LCTVSs, F being semireflexive, and let u be a continuous linear map of E into F. Show that $u''(E'') \subset F$.

9.10 Consider the problem of Fourier factor sequences introduced in Example 6.4.9. Show that if (f_n) is a factor sequence of class (C, L^p) $(1 < p < \infty$, C the space of continuous periodic functions), then (f_n) is also of type (L^∞, L^p).

[Hint: Use Exercise 9.9 with $E = C$, $F = L^p$ and u the map defined by (f_n). Observe that each function in L^∞ is equal almost everywhere to a bounded, universally measurable function, and that the latter functions form a subspace of E''.]

9.11 Let E be a uniform space and (x_n) a Cauchy sequence in E. Show that the subspace $S = \{x_n\}$ is semimetrizable as a uniform space.

[Hint: Show that the induced structure on S has a countable base of vicinities of the diagonal.]

9.12 Let E be a separated LCTVS, (x_n) a Cauchy sequence in E, A the closed, convex, and balanced envelope in E of the set $\{x_n\}$. Show that the induced uniform structure on A is metrizable. (Grothendieck [7], Ch. V, p. 24, Exercice 1.)

[Hint: Reduce the problem to the case in which $\lim x_n = 0$. Then use Corollary 8.13.3 and Krein's theorem.]

9.13 Let E and F be separated LCTVSs, u a linear map of E into F. Show that if u is sequentially continuous at 0, then u transforms Cauchy sequences into Cauchy sequences. (Grothendieck [7], Ch. V, p. 24, Exercice 2.)

[Hint: Let (x_n) be a Cauchy sequence in E, A the closed, convex, and balanced envelope in E of $\{x_n\}$. Using Exercise 9.12 show that $u \mid A$ is continuous at 0, hence that $u \mid A$ is uniformly continuous.]

9.14 Let T be a separated locally compact space, μ a positive Radon measure on T, and F a Banach space. Suppose that Φ is a mapping of T into F that is scalarwise essentially integrable (see Subsection 8.14.1). Show that in order that Φ be scalarwise well integrable (see Remark (2) following Proposition 8.14.13), it is necessary and sufficient that $\bar{\int}_A \Phi \, d\mu \in F$ for each closed subset A of T (or for each closed subset A of T that has a countable neighbourhood base). (Grothendieck [4], p. 162, Proposition 13.)

[Hints: The necessity is trivial. As for the sufficiency show first, by using Proposition 8.14.11 and the CGT, that $u(f) = \bar{\int} f\Phi \, d\mu$ is continuous from L^∞ into F''. Notice that the unit ball B_0 in $C_0(T)$ is $\sigma(L^\infty, L^1)$-dense in the unit ball of L^∞. The continuity of u shows that u maps $C_0(T)$ into F, each function in $C_0(T)$ being the uniform limit of finite linear combinations of functions χ_A. It is enough to show that the restriction of u to $C_0(T)$ is weakly compact from $C_0(T)$ into F. Finally use Theorem 9.4.10.]

9.15 Let T, μ, F, and Φ satisfy the hypotheses of Exercise 9.14. Show by example that the further assumption that $\bar{\int} f\Phi \, d\mu$ $(= \int f\Phi \, d\mu) \in F$ for each $f \in C_0(T)$ is not adequate to ensure that Φ is scalarwise well integrable.

[Hint: Take $F = c_0$, T any separated locally compact and noncompact space, and μ such that there exists a sequence (I_n) of disjoint integrable sets such that $\mu(I_n) > 0$, and such that each compact subset of T meets only a finite number of the I_n. Consider Φ defined by

$$\Phi(t)(n) = \frac{\chi_{I_n}(t)}{\mu(I_n)}.]$$

9.16 Let E be a Banach space isomorphic to a quotient space of c_0. Prove the following statements:

(1) E has the Dieudonné property.
(2) Any continuous linear map of E into a separated and complete LCTVS F, which transforms weakly convergent sequences into convergent, is compact.
(3) E possesses the (DP) property.
(4) Any weakly compact linear map of E into a separated LCTVS is compact.
(5) Each subset of E' that is relatively compact for $\sigma(E', E'')$ is strongly relatively compact.
(6) Any sequence in E' that is $\sigma(E', E'')$-Cauchy is strongly convergent.

[Hint: Use Proposition 9.4.13 combined with the Dunford-Pettis criterion in ℓ^1.]

9.17 Let T be a separated locally compact space, μ a positive Radon measure on T. Let $\bar{L}_R^\infty = \bar{L}_R^\infty(T, \mu)$ be the space of equivalence classes of bounded measurable real-valued functions on T, partially ordered by the relation $\dot{f} \leq \dot{g}$ defined to signify that $f \leq g$ l.a.e. Show that $\bar{L}_R^\infty$ is a complete lattice (see Remark (2) at the end of Subsection 9.4.14).

[Hint: It is plain that each finite subset of $\bar{L}_R^\infty$ admits a supremum, hence the sufficiency of verifying that any majorized increasing directed family $(\dot{f}_i)$ admits a supremum. Observe that $\dot{f} \leq \dot{g}$ holds if and only if $\int fh \, d\mu \leq \int gh \, d\mu$ for each positive h in $\mathscr{L}^1(\mu)$, consider the linear forms on $\mathscr{L}^1(\mu)$ defined by the $\dot{f}_i$, and use Theorem 4.16.1.]

9.18 Using the notation of Section 9.5, suppose that

$$\int^* \|K_t\|_{L^\infty(S,\sigma)} \, d\tau(t) < +\infty$$

K_t denoting the function $s \to k(t, s)$ on S. Show that u is weakly compact from $L^1(S, \sigma)$ into $L^1(T, \tau)$, so that u^2 is compact from $L^1(S,\sigma)$ into (L^1T, τ). (Counterexamples are known which show that u itself need not be compact.)

9.19 Using the notation of Theorem 9.5.10, show that if D is bounded and $B(t, s)$ is bounded on $D \times D$, then u is continuous from $L^p(D)$ into $L^p(D)$ for $1 \leq p \leq \infty$.

9.20 Suppose that $1 < p < \infty$ and that

$$\int d\tau(t) \Big[\int |k(t, s)|^{p'} \, d\sigma(s) \Big]^{p/p'} < +\infty$$

Show that u is compact from $L^p(S, \sigma)$ into $L^p(T, \tau)$.

9.21 Using the notation of Theorem 9.5.10, suppose that D is bounded subset of R^n, and that

$$k(t, s) = B(t, s) \cdot |t - s|^{-\lambda},$$

where $0 \leq \lambda < n$ and where B is bounded on $D \times D$. We know (Exercise 9.19) that u is a continuous endomorphism of $L^p(D)$ for $1 \leq p \leq \infty$. Assume now that $1 \leq p < \infty$, and show that given $\varepsilon > 0$ there exists $\delta > 0$ such that the relations

$$y = u(x), \qquad \|x\|_{L^p} \leq 1, \qquad \mu(M) < \delta$$

together entail that

$$\int_M |y|^p \, d\mu < \varepsilon.$$

9.22 Let D, k, B, and λ be as in Exercise 9.21. Show that u is a compact endomorphism of $L^p(D)$ whenever $1 < p < \infty$.

[Hint: Consider first the case $p > n/(n - \lambda)$, using Exercise 9.20. Then deal with the case $1 < p < n/\lambda$ by considering the adjoint mapping u' of $L^{p'}(D)$ into itself, using Corollary 9.2.3. If $\lambda < \frac{1}{2}n$, this covers all cases. Finally deal with the case $\frac{1}{2}n \leq \lambda < n$ by using Exercise 9.21.]

9.23 Let E be a Hilbert space and u an endomorphism of E such that

$$|||u|||^2 = \sum_{i,j} |(u(e_i) \mid e_j)|^2 = \sum_i \|u(e_i)\|^2 < +\infty$$

for some one orthonormal base (e_i) in E. Prove that the same is true for any orthonormal base in E. Prove that u is compact. Deduce from Theorem 9.11.2 that if u is normal and if the λ_i are the nonzero eigenvalues of u (repeated according to their multiplicity), then

$$|||u|||^2 = \sum_i |\lambda_i|^2.$$

Remarks. The inequality $\sum_i |\lambda_i|^2 \leq |||u|||^2$ is true for any u with $|||u||| < +\infty$, normal or not; this result is due to Schur [1]; see also Zaanen [1], p. 353. A similar theorem of Lalesco (Lalesco [1], Gheorghiu [1], Hille and Tamarkin [1]) asserts that if $u = u'u''$, where $|||u'||| < +\infty$ and $|||u''||| < +\infty$, then $\sum |\lambda_i| \leq |||u'||| \cdot |||u''|||$; see also Zaanen, loc. cit. For the relationship between those integral operators u which are compact and those for which $|||u^n||| < +\infty$ for some n, see Zaanen [1], p. 360.

9.24 Let E be a normed vector space, E an endomorphism of u such that u^p is compact for some natural number p (see Subsection 9.10.5). Show that $v = 1 - u$ transforms bounded, closed sets into closed sets. Deduce that v transforms closed vector subspaces into closed vector subspaces.

9.25 Let E be a reflexive Banach space, u a continuous linear map of L^1 into the strong dual E'. Show that there exists a strongly measurable function f mapping T into E' such that $\|f(t)\| \leq \|u\|$ and

$$u(\dot{g}) = \int gf \, d\mu$$

for each g in $\mathscr{L}^1$.
[Hint: Apply Theorem 9.4.7, noting that, since u is continuous and E is reflexive, u is weakly compact from L^1 into E'.]

CHAPTER 10

The Krein-Milman Theorem and Its Applications

10.0 Foreword

In this chapter we discuss an important theorem due to Krein and Milman ([1]) concerning compact convex sets which proves to be the abstract setting of several representation theorems first discovered in classical analysis by methods that tended to disguise their common source. Perhaps the most important among these representation theorems are those of Bochner about positive definite functions and of Bernstein about completely monotone functions. The approach to Bochner's theorem afforded by the Krein-Milman theorem has provided the means (though not the only one) of extending the former theorem to locally compact Abelian groups, as was shown first by Cartan and Godement ([1]). The first to show how one might similarly approach Bernstein's theorem was Choquet ([1], Chapter VII). We shall turn to these applications in the later sections of this chapter. None of these applications is immediate and the necessarily rather lengthy expositions have led to the formation of a separate chapter rather than a continuation of Chapter 8, despite the obvious close affinity of the Krein-Milman theorem with the circle of ideas dealt with in Section 8.13.

10.1 Extreme Points and the Krein-Milman Theorem

Throughout this section E will denote a real TVS. If A is a convex subset of E, by an open segment of A we mean a subset of the type

$$\{(1-\lambda)a + \mu b : 0 < \lambda < 1\},$$

where a and b are *distinct* points of A.

The Krein-Milman theorem concerns the notion of extreme (or extremal) points of a convex set A. A point x_0 of A is said to be an *extreme point* of A if it belongs to no open segment of A; that is, equivalently, if the relations

$$a \in A, \qquad b \in A, \qquad x_0 = (1-\lambda)a + \lambda b, \qquad 0 \leq \lambda \leq 1$$

together entail that x_0 is either a or b. The theorem indicates how in certain cases A may be recovered from its extreme points.

For the purposes of our proof we shall need to extend the concept a little. A closed subset B of a convex set A in E is termed an *extreme subset* of A if the relations

$$a \in A, \qquad b \in A, \qquad \lambda a + (1 - \lambda)b \in B, \qquad 0 < \lambda < 1$$

together entail that both a and b belong to B. It is a simple matter to verify that a singleton $\{x_0\}$ is an extreme subset of A if and only if x_0 is an extreme point of A.

For our proof we shall need a preliminary lemma.

10.1.1 **Lemma.** Let E be a real TVS, A a nonvoid closed compact and convex subset of E, f a continuous linear form on E, and $\beta = \operatorname{Inf} f(A)$. The set $B = A \cap f^{-1}(\{\beta\})$ is a nonvoid, compact, extreme subset of A.

Proof. Continuity and linearity of f ensure that B is closed and convex. Since A is compact, its closed subset B is compact. Compactness of A also ensures that f, being continuous, assumes its infimum on A. So B is nonvoid. It remains to show that B is an extreme subset of A. But suppose that $(1 - \lambda)a + \lambda b \in B$, where $a \in A$, $b \in A$, and $0 < \lambda < 1$. If, for example, a did not belong to B, then $f(a) > \beta$ and so

$$f[(1 - \lambda)a + \lambda b] = (1 - \lambda)f(a) + \lambda f(b) > (1 - \lambda)\beta + \lambda\beta = \beta,$$

which contradicts the hypothesis that $(1 - \lambda)a + \lambda b \in B$. Thus a must belong to B. Similarly $b \in B$. Thus B is an extreme subset of A. ▮

10.1.2 **Theorem** (Krein-Milman). Let E be a real separated LCTVS, A a nonvoid compact convex subset of E. Then A is the closed, convex envelope in E of the set of extreme points of A.

Proof. The proof proceeds in two steps.

(a) Each nonvoid extreme subset X of A contains an extreme point of A.

To see this, we write $\mathscr{X}$ for the set of all extreme subsets of A. By Lemma 10.1.1, $\mathscr{X}$ is nonvoid. We partially order $\mathscr{X}$ by reversed set inclusion. Then $\mathscr{X}$ is inductively ordered. Indeed if $\mathscr{L}$ is a totally ordered subset of $\mathscr{X}$, then $Y = \bigcap \{X : X \in \mathscr{L}\}$ is nonvoid since A is compact, and it is easy to see that Y is a lower bound of $\mathscr{L}$ in $\mathscr{X}$. By Zorn's Lemma 0.1.5, $\mathscr{X}$ possesses a minimal member, say Y. It remains to show that Y is a singleton. Otherwise, however, Y would contain at least two distinct points x and y. Then, since E is separated and locally convex, there is a continuous linear form f on E such that $f(x) < f(y)$. By the preceding lemma, $Z = Y \cap f^{-1}[\operatorname{Inf} f(Y)]$ is a nonvoid extreme subset of A that does not contain y. Thus Z is an extreme subset of A satisfying $Z \subset Y$, $Z \neq Y$, which would contradict the minimality of Y. Therefore, Y is a singleton, which is what we had to show.

(b) Let B be the convex envelope in E of the set of all extreme points of A; then $\bar{B}$ is compact, convex, and contained in A. We must show that $A \backslash \bar{B}$ is void. Suppose, however, that $x_0 \in A \backslash \bar{B}$. By the Hahn-Banach theorem, there would exist a continuous linear form f on E such that $\operatorname{Inf} f(\bar{B}) > f(x_0)$. Then

$W = A \cap f^{-1}[\operatorname{Inf} f(A)]$ would be a nonvoid extreme subset of A disjoint from $\bar{B}$ (Lemma 10.1.1). Then, by (a), W would contain an extreme point of A, which is absurd since $W \cap B \subset W \cap \bar{B} = \emptyset$. ▌

NOTES. For an alternative discussion of the Krein-Milman theorem in a slightly more general form, see Dunford and Schwartz [1], pp. 424, 463. The earliest proof is due to Krein and Milman [1], and an improvement followed from Milman and Rutman [1]. Another proof is due to Godement [4], which doubtless influenced that given by Bourbaki ([4], p. 84). The proof given is due to Kelley [5].

In respect of special spaces, Arens and Kelley [1] determined the extreme points of the closed unit ball in a closed vector subspace of $C(T)$ (T a compact space) and made applications to a theorem of Banach ([1], p. 173; Stone [2], p. 469) concerning isometric isomorphisms between spaces $C(T)$ and $C(T')$.

The Krein-Milman theorem has a partial converse which is set out below. We may note here that apart from the applications dealt with in this chapter, the theorem and the said partial converse have interesting applications in the study of doubly stochastic and substochastic matrices (Kendall [1]).

Here now is the supplement and partial converse to the Krein-Milman theorem referred to above.

10.1.3 **Proposition.** Suppose that E is a separated real LCTVS, K a compact subset of E whose closed convex envelope A is compact. Then each extreme point of A belongs to K.

Proof. Let x be an extreme point of A. If U is any closed, convex, and circled neighbourhood of 0 in E, there exist finitely many points a_i of K ($1 \leq i \leq n$) such that the sets $a_i + U$ cover K. Let A_i be the closed, convex envelope of $K \cap (a_i + U)$. Each A_i is compact. The convex envelope of the union of the A_i, being compact, contained in A, and containing K, must be A itself. Hence $x = \sum_{i=1}^{n} \lambda_i x_i$ with x_i in A_i, $\lambda_i \geq 0$, and $\sum_{i=1}^{n} \lambda_i = 1$. Since x is an extreme point of A, x must coincide with x_i for some i. Thus x belongs to $A_i \subset a_i + U$, and so x belongs to $K + U$. Since K is closed and U is arbitrary, it follows that x belongs to K, as alleged. ▌

10.1.4 **Remarks and Examples**

(1) Even when $\dim E$ is finite, the set of extreme points of a compact, convex set A need not be closed. As an example: Let $E = R^3$ and let A be the convex envelope of the union of circle $\{(x, y, z) : z = 0, x^2 + y^2 - 2x = 0\}$ and the two points $(0, 0, 1)$ and $(0, 0, -1)$. It is easily seen that the origin is not an extreme point of A but is the limit of many sequences of such points.

(2) A compact, convex set A need not be the convex envelope of its extreme points. An example is obtained by taking $E = \ell^\infty$, putting e_n for the element of E defined by $e_n(k) = 1$ if $k = n$, 0 if $k \neq n$, and taking for A the closed, convex envelope in E of the points e_n/n ($n = 1, 2, \cdots$). It may be deduced from Proposition 10.1.3 that the extreme points of A are 0 and the points e_n/n ($n = 1, 2, \cdots$). Moreover, A is compact and contains all points

$x = \sum_{n \geq 1} \lambda_n e_n/n$, (λ_n) being any sequence satisfying $\lambda_n \geq 0$, $\sum_{n \geq 1} \lambda_n = 1$. If we suppose that $\lambda_n > 0$ for all n, it is clear that x cannot belong to the convex envelope of the extreme points of A (any point of the convex envelope being a sequence with a finite support).

(3) A closed, noncompact, convex set A may possess no extreme points at all. For example, let $E = c_0$ and let A be the closed unit ball in E. If x is any element of A, we consider two cases. Suppose first that x has a finite support, say $x(n) = 0$ for $n > r$. We define a and b by

$$a(n) = x(n) \quad (n \leq r), \qquad \frac{1}{n} \quad (n > r)$$

$$b(n) = x(n) \quad (n \leq r), \qquad -\frac{1}{n} \quad (n > r).$$

Then a and b belong to A, $a \neq b$, $x = \frac{1}{2}(a + b)$, and so x is not an extreme point of A. Suppose next that x has not a finite support. Choose r so that $|x(n)| \leq \frac{1}{2}$ for $n > r$, and define a and b by

$$a(n) = x(n) \quad (n \leq r), \qquad 2x(n) \quad (n > r,\, n \text{ even}), \qquad 0 \quad (n > r,\, n \text{ odd})$$

$$b(n) = x(n) \quad (n \leq r), \qquad 0 \quad (n > r,\, n \text{ even}), \qquad 2x(n) \quad (n > r,\, n \text{ odd}).$$

Then again a and b belong to A, $a \neq b$, $x = \frac{1}{2}(a + b)$, and so x is not an extreme point of A.

(4) Let T be a separated locally compact space, $M(T)$ the space of real bounded Radon measures on T. We may view $M(T)$ as the dual of $C_0(T)$ and equip it with the associated weak topology. The set $M_+^1(T)$ of positive measures in $M(T)$ having total mass at most unity is compact and convex.

We will show that the extreme points of $M_+^1(T)$ are 0 and the Dirac measures $\varepsilon_t (t \in T)$. It is, in fact, clear that 0 is an extreme point of $M_+^1(T)$. Let $\mu \neq 0$ be any other. It will suffice to show that the support K of μ is reduced to a single point. Now, if K contained t_1 and t_2, $t_1 \neq t_2$, we choose disjoint neighbourhoods U_1 and U_2 of t_1 and t_2. Then $m = \mu(U_1)$ satisfies $0 < m < 1$. We put $\alpha = (\mu \mid U_1)/m$, $\beta = (\mu - m\alpha)/(1 - m)$. Then it is evident that α and β belong to $M_+^1(T)$, $\alpha \neq \beta$, $\mu = m\alpha + (1 - m)\beta$, contradicting the fact that μ is extremal. Hence K is reduced to a single point.

Similar arguments show that if T is compact, the set of positive measures of unit total mass is compact and convex, and that its extreme points are the Dirac measures ε_t. This result should be reviewed in the light of Theorem 10.1.7 to follow, being a sort of complement to the latter.

We turn next to a result about the extreme points of a convex subset of an ordered vector space which is sometimes a useful auxiliary in the application of the Krein-Milman theorem.

10.1.5 **Theorem.** Let E be an ordered vector space and E_+ the set of positive elements of E, P a nonvoid subset of E_+. A point x of P is said to be P-minimal (simply minimal, if $P = E_+$) if the relations

$$y \in P, \qquad y \leq x$$

imply that y is a scalar multiple of x. Let us suppose we are given a nonvoid convex set $A \subset P$ and a real-valued function f defined on P and assume that the following conditions are fulfilled:

(1) If x is in A, y in P, and $y \leq x$, then y and $x - y$ are in A;

(2) $f(\lambda x) = \lambda \cdot f(x)$ and $f(x + y) = f(x) + f(y)$ whenever $\lambda \geq 0$ and x, y, λx and $x + y$ all belong to P;

(3) if x is in A and $f(x) = 0$, then $x = 0$;

(4) $f(x) \leq 1$ for each x in A;

(5) if x is in A and $f(x) \neq 0$, then $x/f(x)$ is in A.

The conclusions are:

(a) 0 is an extreme point of A;

(b) the extreme points of A other than 0 are precisely the P-minimal points x belonging to A that satisfy $f(x) = 1$.

Proof. Since A is nonvoid, (1) shows that 0 is in A.

(a) 0 is an extreme point of A. For suppose $0 = \lambda x + (1 - \lambda)y$ for some x, y in A and $0 \leq \lambda \leq 1$. Then (2) gives

$$0 = f(0) = \lambda f(x) + (1 - \lambda)f(y).$$

On the other hand, (5) entails that $f(A) \geq 0$. It follows that

$$\lambda f(x) = (1 - \lambda)f(y) = 0.$$

If $\lambda \neq 0$ this gives $f(x) = 0$ and so, by (3), $x = 0$; likewise, if $1 - \lambda \neq 0$, we get $y = 0$; and if $0 < \lambda < 1$, we get $x = y = 0$. In any case, at least one of $x = 0$ or $y = 0$ is true, showing that 0 is extremal.

(b) To begin with we observe that if $x \neq 0$ is an extreme point of A, then, since x belongs to the segment joining the points 0 and $x/f(x)$ (by (3) and (4)), so must x coincide with $x/f(x)$, and therefore $f(x)$ must be unity.

Supposing still that $x \neq 0$ is an extreme point of A, we prove now that x is P-minimal. Suppose that y belongs to P and $y \leq x$. In the proof we may exclude the trivial cases in which y is 0 or x. By (1), y and $x - y$ belong to A and differ from 0. So by (3) and (5) the elements

$$u = \frac{y}{f(y)}, \qquad v = \frac{x - y}{f(x - y)} = \frac{x - y}{(1 - f(y))}$$

belong to A. But $0 < f(y) < 1$ and

$$x = f(y)u + (1 - f(y))v.$$

Since x is extremal, we must have *either* $x = u = y/f(y)$ *or* $x = v = (x - y)/(1 - f(y))$. In either case, $y = f(y)x$, showing thus that x is P-minimal.

Conversely assume that x is P-minimal, that x belongs to A, and that $f(x) = 1$: we show that x is an extreme point of A. For suppose that $x = \alpha a + (1 - \alpha)b$ where $0 \leq \alpha \leq 1$ and a and b belong to A. Then $\alpha a \leq x$ and, if $\alpha \neq 0$, minimality of x entails that $a = \lambda x$ for some λ necessarily ≥ 0. (If λ were negative, a would be both ≥ 0 and ≤ 0, hence 0; then x would be

$(1 - \alpha)b$; if $\alpha = 1$, this contradicts $x \neq 0$; and if $\alpha \neq 1$, it would give $f(b) = (1 - \alpha)^{-1}f(x) = (1 - \alpha)^{-1} > 1$, conflicting with (4) since b belongs to A.) Furthermore, $1 \geq f(a) = \lambda f(x) = \lambda$, by (2) and (4), so that $0 \leq \lambda \leq 1$.

One sees likewise that if $\alpha \neq 1$, then $b = \mu x$ with $0 \leq \mu \leq 1$.

Thus, if $0 < \alpha < 1$, then $x = \alpha\lambda x + (1 - \alpha)\mu x$, hence $1 = \alpha\lambda + (1 - \alpha)\mu$. This entails $\lambda = \mu = 1$ and so $x = a = b$. If on the other hand $\alpha = 0$ (or 1), we have $x = a$ (or $x = b$). In all cases therefore x coincides with a or with b (or both), showing that x is extremal. The proof is complete. ▌

10.1.6 **Integral Representations in Terms of Extreme Points.** Assume once more that E is a separated real LCTVS and A a nonvoid compact convex set in E. Let $e(A)$ denote the set of extreme points of A. The Krein-Milman theorem asserts that each x_0 in A is the limit in E of convex combinations $\sum_{i=1}^n \lambda_i x_i$, with the x_i in $e(A)$, $\lambda_i \geq 0$, and $\sum_{i=1}^n \lambda_i = 1$. On the other hand we have seen (Remark (1) in Subsection 10.1.4) that $e(A)$ need not be compact. If $e(A)$ is compact, we can deduce that each point of A is the "mass centre" of some positive measure of unit total mass on $e(A)$. More generally we have the following theorem.

10.1.7 **Theorem.** Let E be a separated real LCTVS that is quasi-complete for $\tau(E, E')$, A a nonvoid compact convex set in E, $e(A)$ the set of extreme points of A. Then A consists precisely of the points x_0 of E of the type

$$x_0 = \int x \, d\mu(x), \tag{10.1.1}$$

where μ is a positive Radon measure of unit total mass on $\overline{e(A)}$ (closure in E).

Proof. $B = \overline{e(A)}$ is compact. Let M denote the set of positive Radon measures of unit total mass on B. Regarded as a subset of $C(B)'$, M is weakly compact. Reference to Corollary 8.13.2 and Corollary 8.13.3 shows that the mapping $\mu \to m(\mu) = \int x \, d\mu(x)$ maps M into the closed, convex envelope C in E of B, and that this mapping is continuous for the weak topology of M (qua subset of the dual of $C(B)$) and $\sigma(E, E')$. Thus on the one hand (since $B \subset A$) we have $m(M) \subset A$; and on the other hand $m(M)$ is weakly closed in E. But also $m(M)$ is dense in A as a consequence of the Krein-Milman theorem. Thus $m(M) = A$, as we had to show. ▌

Remark. A variant of the last theorem is as follows: If T is a separated locally compact space, f a weakly continuous mapping of T into E such that $e(A) \subset f(T)$, then each element x_0 of A admits a representation of the type $x_0 = \int_T f(t) \, d\mu(t)$, where μ is a positive Radon measure on T of total mass at most unity. This integral representation theorem can be usefully supplemented in the following way.

10.1.8 **Proposition.** With the hypotheses and notations of Theorem 10.1.7, if x_0 is a nonextreme point of A, there exists a representation (10.1.1) in which $\mu(\{x_0\}) = 0$.

Proof. We begin with any one representation (10.1.1). If it be assumed that $\mu \neq \varepsilon_{x_0}$ (Dirac measure at x_0), then we shall have $m = \mu(\{x_0\}) < 1$. The measure

$$\nu = \frac{\mu - m\varepsilon_{x_0}}{1 - m}$$

is then positive and has unit total mass, and $\nu(\{x_0\}) = 0$. Moreover

$$\begin{aligned}\int x\, d\nu(x) &= (1-m)^{-1}\int x\, d\mu(x) - m(1-m)^{-1}x_0 \\ &= (1-m)^{-1}x_0 - m(1-m)^{-1}x_0 = x_0,\end{aligned}$$

so that ν effects a representation of the desired type. Thus we are reduced to showing merely that at least one representation (10.1.1) exists in which $\mu \neq \varepsilon_{x_0}$.

To verify this we use the fact that x_0 is nonextremal: There exist distinct points a and b of A and a number λ satisfying $0 < \lambda < 1$ for which $x_0 = (1-\lambda)a + \lambda b$. Now the representation theorem itself permits us to write

$$a = \int x\, d\alpha(x), \qquad b = \int x\, d\beta(x)$$

with α and β positive measures on B of unit total mass. This leads to (10.1.1) with $\mu = (1-\lambda)\alpha + \lambda\beta$. Since $a \neq b$, it is impossible that both $\alpha(\{x_0\})$ and $\beta(\{x_0\})$ should be unity (otherwise both α and β would be ε_{x_0} and a and b would not be distinct). Hence at least one of $\alpha(\{x_0\})$ and $\beta(\{x_0\})$ is less than unity, showing that then the same is true of

$$\mu(\{x_0\}) = (1-\lambda)\alpha(\{x_0\}) + \lambda\beta(\{x_0\}).$$

This completes the construction. ▌

Remarks. Choquet [2]–[8] has considered the existence and uniqueness of integral representations of points of a convex cone in terms of its extreme points. See also Choquet and Meyer [1].

10.2 Application to Bernstein's Theorem

Throughout this subsection we use T to denote an additively written Abelian semigroup with neutral element 0, and E will be the real vector space R^T. E is a separated LCTVS when endowed with the topology of pointwise convergence on T. The iterated difference operators are endomorphisms of E defined by $d^0x = x$, $d_a^1x(t) = x(t+a) - x(t)$ and

$$d^{n+1}_{a_1,a_2,\cdots,a_n,a_{n+1}} = d^1_{a_1} \circ d^n_{a_1,\cdots,a_n},$$

where $a, a_1, \cdots, a_{n+1}$ are elements of T. For brevity we shall sometimes write simply d^n in place of $d^n_{a_1,\cdots,a_n}$ when the a_i are assumed to be unspecified.

A function x belonging to $E = R^T$ is said to be *completely monotone* if $(-1)^n\, d^nx \geq 0$ for $n = 0, 1, 2, \cdots$. These functions form a convex subset M of E.

10.2.1 **Example.** Suppose that $T = R_+$, with addition having its usual meaning. It is then easily seen that if x is continuous for $t \geq 0$ and indefinitely differentiable for $t > 0$, it is completely monotone if and only if $(-1)^n x^{(n)}(t) \geq 0$ for $n = 0, 1, 2, \cdots$ and for $t > 0$. It will follow from Bernstein's theorem, yet to be discussed, that any completely monotone function on R_+ is necessarily indefinitely differentiable for $t > 0$, though it may have a simple jump discontinuity at $t = 0$.

By condensing some arguments given by Choquet ([1], Chapter VII) we aim to apply the Krein-Milman theorem in order to obtain an integral representation for completely monotone functions that reduces to Bernstein's theorem when $T = R_+$.

If x is completely monotone, then $x(t + a) - x(t) \leq 0$ and so $x(a) \leq x(0)$. It follows readily that the set M_0, consisting of the completely monotone functions x satisfying $x(0) \leq 1$, is a compact convex subset of E. Our method is to apply Theorem 10.1.5 in order to determine the extreme points of M_0, and then to appeal to Theorem 10.1.7. For this purpose, we partially order E by the convention that x precedes y if and only if $y - x$ belongs to M. We shall then take, in Theorem 10.1.5, $P = M = E_+$, $A = M_0$, and $f(x) = x(0)$. The problem is now to determine the minimal elements of M_0 satisfying $x(0) = 1$. These will prove to be the so-called "exponentials," which we now introduce.

By an *exponential* on T we mean a function u in $E = R^T$ that is not identically 0 on T and that satisfies the functional equation

$$u(t + t') = u(t)u(t') \tag{10.2.1}$$

and the inequality

$$0 \leq u(t) \leq 1, \tag{10.2.2}$$

t and t' being arbitrary points of T. It then follows that $u(0) = 1$. Moreover one has

$$(-1)^n d^n_{a_1, \cdots, a_n} u(t) = u(t) \cdot (1 - u(a_1)) \cdots (1 - u(a_n)),$$

showing that u is completely monotone, and hence that u belongs to M_0.

10.2.2 **Example.** Returning to the situation of Example 10.2.1, where $T = R_+$, it is easy to identify the exponentials. Indeed, if u is an exponential that is nonvanishing, one can extend it by the definition $u(-t) = u(t)^{-1}$ $(t > 0)$ into a character of R; being a decreasing function of t, it must be of the form $u_s(t) = e^{-st}$ for some number $s \geq 0$. If, on the other hand u vanishes for some $t > 0$, we let t_0 be the infimum of such zeros of u; then $u(t) = 0$ for $t > t_0$. If t_0 were strictly positive, the relation $u(2t_0) = u(t_0)^2$ shows that $u(t_0) = 0$; so $u(t) = 0$ for $t \geq t_0$, $u(t) > 0$ for $0 \leq t < t_0$. Then, however, the relation $u(t_0) = u(\frac{1}{2}t_0)^2$ is self-contradictory. Thus t_0 must be 0 and u is the function $u_\infty(t) = 1$ for $t = 0$, $= 0$ for $t > 0$. The exponentials on R_+ are therefore the functions u_s $(0 \leq s \leq \infty)$.

10.2.3 **Proposition.** Let x be an M-minimal element of M_0 satisfying $x(0) = 1$. Then x is an exponential on T.

Proof. We define x_1 and x_2 by

$$x_1(t) = -d_a^1 x(t), \qquad x_2(t) = x(t + a),$$

a being any preassigned element of T. Then x_1 and x_2 belong to M and $x = x_1 + x_2$. Minimality of x entails the existence of a number λ_a satisfying $0 \leq \lambda_a \leq 1$ and $x_2 = \lambda_a x$; that is, $x(t + a) = \lambda_a x(t)$ for all t. Taking $t = 0$ we see that $\lambda_a = x(a)$. Thus x satisfies (10.2.1). On the other hand, (10.2.2) follows from the fact that $x(0) = 1$ and x belongs to M. The proof is complete. ∎

We are now able to state and prove the main result.

10.2.4 Theorem. The extreme points of M_0 are 0 and the exponentials on T. The set S of exponentials is compact for the topology of pointwise convergence on T, and each completely monotone function x on T admits a representation

$$x(t) = \int_S u(t)\, dm(u), \tag{10.2.3}$$

m being a positive Radon measure on S.

Proof. It is clear that S is compact. By Proposition 10.2.3, the set of extreme points of M_0 is contained in $S \cup \{0\}$. That a representation (10.2.3) is possible follows at once from Theorem 10.1.7. To complete the proof, it has to be shown that each exponential u is indeed an extreme point of M_0. For this we shall use Proposition 10.1.8.

Let us suppose if possible that the exponential u_0 were not an extreme point of M_0. Then, according to the proposition just cited, we should have

$$u_0(t) = \int_S u(t)\, dm(u), \tag{10.2.4}$$

where $m(\{u_0\}) = 0$. Given t and $\varepsilon > 0$, the closed set of u in S such that $u(t) \geq u_0(t) + \varepsilon$ is m-negligible: for if the measure of this set is p, then

$$(u_0(t))^n = u_0(nt) = \int_S u(nt)\, dm(u) = \int u(t)^n\, dm(u)$$
$$\geq p \cdot (u_0(t) + \varepsilon)^n,$$

which shows (on letting $n \to \infty$) that $p = 0$. Thus $u(t) \leq u_0(t)$ for almost all u. Since (10.2.4) gives $1 = u_0(0) = \int_S dm(u)$, one has, again by (10.2.4),

$$\int_S (u_0(t) - u(t))\, dm(u) = u_0(t) - \int_S u(t)\, dm(u) = 0.$$

The integrand is positive almost everywhere and so $u_0(t) = u(t)$ a.e. By continuity, $u_0(t) = u(t)$ for all u in the support of m. It follows that this support is reduced to a single point, which can only be u_0. This contradicts the assumption that $m(\{u_0\}) = 0$ and completes the proof. ∎

10.2.5 Corollary (Bernstein). Let x be a completely monotone function on R_+. There exists a bounded positive Radon measure μ on R_+ and a number $c \geq 0$ such that

$$x(t) = c \cdot u_\infty(t) + \int e^{-ts}\, d\mu(s) \tag{10.2.5}$$

for $t \geq 0$, where $u_\infty(t) = 1$ if $t = 0$, $= 0$ if $t > 0$.

Proof. We have seen in Example 10.2.2 that S consists of the functions e^{-st} ($s \in R_+$) and u_∞. It suffices to apply Theorem 10.2.4. ▌

10.3 Application to Bochner's Theorem

Bochner ([1]) was the first to investigate systematically the continuous positive definite functions of a real variable, having been led to them through the study of Fourier integrals. According to Bochner a continuous (complex-valued) function ϕ on R is positive definite if the inequality

$$\sum_{p=1}^{n}\sum_{q=1}^{n} \phi(t_p - t_q) z_p \bar{z}_q \geq 0 \tag{10.3.1}$$

holds for all finite families $(t_p)_{1 \leq p \leq n}$ in R and all finite families $(z_p)_{1 \leq p \leq n}$ of complex numbers. It follows easily that ϕ is "Hermitian symmetric"—that is, that

$$\phi(t) = \tilde{\phi}(t) \equiv \overline{\phi(-t)}, \tag{10.3.2}$$

and that ϕ is bounded:

$$|\phi(t)| \leq \phi(0). \tag{10.3.3}$$

The continuity of ϕ being assumed, (10.3.1) is equivalent to the requirement that

$$\phi * f * \tilde{f}(0) = \iint \phi(t_1 - t_2) f(t_1) \overline{f(t_2)}\, dt_1\, dt_2 \geq 0 \tag{10.3.4}$$

for each f in $\mathscr{K} = \mathscr{K}(R)$; and if boundedness of ϕ is assumed, it amounts to the same thing to demand that (10.3.4) holds for all f in $\mathscr{L}^1(R)$. Now (10.3.4) is applicable for any locally integrable ϕ and will be adopted here as the basic characterization of (not necessarily continuous) positive definite functions.

It is a very simple matter to verify that if μ is any positive bounded Radon measure on R, the function

$$\phi(t) = \int e^{2\pi i t s}\, d\mu(s), \tag{10.3.5}$$

the inverse Fourier transform of μ, is continuous and positive definite. Bochner's principal result was to the effect that the converse is true: every continuous positive definite function on R is uniquely obtainable in this way. His proof involves the use of basic but nontrivial properties of the Fourier transformation. Bochner's theorem has since been used (Cooper [1], [2]) to study one-parameter groups of unitary endomorphisms of a Hilbert space; and various writers (Cooper [3]; Crum [1]) have considered in detail the form of the representation necessary to cover discontinuous positive definite functions. None of these extensions and developments constitutes our main concern, which is the possibility of applying the Krein-Milman theorem and its corollaries to the proof of Bochner's theorem itself. More accurately one should say that only the "existence" part of Bochner's theorem is directly deducible from the Krein-Milman theorem; the "uniqueness" assertion follows from other considerations.

A proof of this nature has been given by Cartan and Godement in collaboration (Cartan and Godement [1]). An outcome of this method of proof is an extension of the theorem to functions on a group. It is obvious that the concept of positive definite function is meaningful if R is replaced by any group T whatsoever; for continuity and the criterion (10.3.4) to retain sense, one will assume that T is separated and locally compact. Godement [4] and, independently, Gelfand and Raikov [1] in collaboration have examined the case of non-Abelian groups. It appears that here there appear complexities which justify the view that the separated locally compact Abelian (LCA) group provides the "natural domain" of Bochner's theorem. At any rate, we shall confine our remarks to this case. The analogue of the representation (10.3.5) is then clear: One expects to replace the exponentials $e^{2\pi ist}$ by the bounded, continuous characters of T, hoping that these will be formable into a separated, locally compact space carrying the measure μ. As we shall show, these expectations are fully justified.

It should be observed that the Krein-Milman theorem is not the only functional analytic tool leading to a proof of this general form of Bochner's theorem. Another is based upon the theory of commutative normed algebras and was discovered by several people independently and almost simultaneously early in World War II, at a time when suspended communications hindered the exchange of news concerning research. A connected account of this approach is now readily available (see, for example, Loomis [1], Naĭmark [1], Rudin [9]).

In view of the extension of Bochner's theorem to LCA groups, a much earlier result of Herglotz [1] about positive definite sequences appears as the special case of Bochner's theorem when the group T is the discrete additive group of integers. There is, however, one sensible barrier separating the case in which T is either discrete or compact from the general one, the latter being essentially more difficult. For compact groups, in fact, even the non-Abelian case is relatively easy to deal with.

The central difficulty in this (and most other) applications of the Krein-Milman theorem is the identification of the extreme points of the appropriate convex set. In the present instance, E may be taken to be the space of all Hermitian-symmetric functions in $L^\infty(T)$, A being the set of continuous positive definite functions ϕ (identified with their classes in $L^\infty(T)$) satisfying $\phi(0) \leq 1$. It is then necessary to show that the extreme points of A are 0 and the bounded continuous characters of T. Once this is done it is a relatively easy matter to bring Theorem 10.1.7 into action and so derive Bochner's theorem.

The identification is effected by relating the continuous positive definite functions in a quasi-algebraic fashion with the unitary representations of the group in question, an idea that apparently occurred almost simultaneously and independently to Cartan and Godement and to Gelfand and Raikov. This relationship seems quite fundamental and is the source of almost all that is known about positive definite functions on groups.

After these lengthy preliminaries, we now begin the details.

10.3.1 Characters. In what follows $T = \{t\}$ is a separated LCA group, additively written, and with neutral element 0. We choose and fix any Haar measure on T (see Section 4.18) and denote the corresponding integral by $\int \cdots dt$.

By a *character* of T is meant a representation of T in the multiplicative group of nonzero complex numbers. In other words, a character of T is a complex-valued function χ on T such that $\chi(0) = 1$ and $\chi(t_1 + t_2) = \chi(t_1)\chi(t_2)$ for t_1, t_2 in T. The set of characters itself forms a group under pointwise multiplication. By $\hat{T}$ we denote the subgroup formed of bounded continuous characters of T. It is usual to write $\hat{T}$ additively (despite the fact that its composition law is pointwise multiplication); $\hat{T}$ is termed the *group dual* to T. If $\hat{t}$ is an element of $\hat{T}$, we shall write $[t, \hat{t}]$ in place of $\hat{t}(t)$, t being in T. Thus

$$[0, \hat{t}] = 1, \ [t_1 + t_2, \hat{t}] = [t_1, \hat{t}] \cdot [t_2, \hat{t}],$$
$$[t, \hat{t}_1 + \hat{t}_2] = [t, \hat{t}_1] \cdot [t, \hat{t}_2]$$

and boundedness implies that $|[t, \hat{t}]| = 1$ identically, so that $[-t, \hat{t}] = \overline{[t, \hat{t}]} = [t, -\hat{t}]$. From this it follows that each $\hat{t}$ in $\hat{T}$ is a continuous positive definite function on T.

We shall see later that $\hat{T}$ can be made in a natural way into a separated LCA group, and this in such a way that T is (isomorphic as a topological group with) the dual of $\hat{T}$. The verification of this assertion is trivial when T is the real line, the real line modulo 2π (or any other nonzero number), or a finite product of such groups. These cases cover all applications in classical analysis.

10.3.2 Unitary Representations. By a *unitary representation* of a group T we shall mean a pair (H, U) composed of a Hilbert space H and a homomorphism $U : t \to U(t)$ of T into the multiplicative group of unitary endomorphisms of H. H is spoken of as the *representation space* of the representation, the dimension of H being also spoken of as the dimension of the representation. Unless the contrary is explicitly stated, the term "representation" will henceforth mean "unitary representation"; see the notes at the end of this section.

A representation (H, U) of separated locally compact group T is said to be continuous (resp. measurable) if for each $x \in H$ the scalar-valued function $t \to (U(t)x \mid x)$ is continuous (resp. measurable for Haar measure on T). It is equivalent to demand that for each pair x, y of elements of H the function $t \to (U(t)x \mid y)$ is continuous (resp. measurable).

It is important to observe that the continuous one-dimensional representations of T are of the form $U(t) = \chi(t) \cdot 1$, where 1 denotes the identity endomorphism and χ is an element of $\hat{T}$.

As has been said, part of our task lies in associating the continuous representations of T with the continuous positive definite functions on T. In this association it is crucial to pick out those representations corresponding to the extreme points of a certain convex set of positive definite functions. It turns out that the appropriate property of a representation is that of irreducibility, a concept we examine forthwith.

IRREDUCIBILITY. Let (H, U) be a representation of T. We consider the following three assertions:

(a) There exists in H no closed vector subspace other than $\{0\}$ and H that is stable under each $U(t)$ $(t \in T)$.

(b) There exists no orthogonal projector P on H other than 0 and 1 that commutes with each $U(t)$ $(t \in T)$.

(c) There exists no continuous normal endomorphism A of H other than scalar multiples of 1 that commutes with each $U(t)$ $(t \in T)$.

The reader will recall that a continuous endomorphism A of H is said to be normal if it commutes with its adjoint; thus each self-adjoint endomorphism, and each unitary endomorphism, is normal.

It is quite simple to show that (a) and (b) are equivalent (see Exercise 10.1). It is trivial to assert that (c) implies (b). That (b) implies (c), and hence that all three conditions are equivalent, is true but far from trivial. The proof depends on the spectral theory of continuous normal endomorphisms. In the sequel we shall assume their equivalence and use any one of them as the definition of *irreducibility* of the representation (H, U). For a proof of equivalence, any one of a number of standard texts may be consulted—for example, Riesz and Nagy [1], Chapitre VII, or Lorch [1]. There exists also a very elegant approach via the general theory of commutative Banach algebras with involution; see Loomis [1], Chapter V, especially pp. 90–95.

Examples. It is evident that any one-dimensional representation is irreducible. As we shall see in a moment, any irreducible representation of an Abelian group T is one dimensional. If T is not Abelian, an irreducible representation of T is in general not one dimensional (see Exercises 10.3, 10.4 and 10.5).

MAIN CONSEQUENCE OF IRREDUCIBILITY. Any irreducible representation of an Abelian group T is one dimensional.

In fact, since U is a homomorphism of T, $U(s)U(t) = U(st) = U(ts) = U(t)U(s)$ follows from the Abelian character of T. Thus each $U(s)$ is a normal endomorphism of H that commutes with each $U(t)$. Irreducibility therefore entails that $U(s) = \xi(s)1$, where $\xi(s)$ is some scalar. Moreover irreducibility implies (a), which in turn entails that for any $x \neq 0$ in H, the closed vector subspace of H generated by the transforms $U(t)x$ $(t \in T)$ is H itself. In the present case, this entails that H is one dimensional.

We turn now to examine the connection between measurable representations and positive definite functions.

If (H, U) is a measurable unitary representation of T, and if h is any element of H, then $(U(t)h \mid h)$ is a bounded positive definite function on T, continuous if the representation is continuous. A fundamental technique in the theory is based upon showing that the converse is true—that is, that with each positive definite function in $L^{\infty}(T)$ may be associated a measurable unitary representation of T from which the function may be derived in the above way. We proceed to describe this association in detail.

Let ϕ be a given positive definite function on T. For f and g in $\mathscr{K} = \mathscr{K}(T)$, we define

$$(f \mid g)_\phi = f * \phi * \tilde{g}(0) = \iint \phi(t_1 - t_2) f(t_1)\overline{g(t_2)}\, dt_1\, dt_2. \tag{10.3.6}$$

By (10.3.4), $(f, g)_\phi$ is a positive Hermitian sesquilinear form on $\mathscr{K}$. Let N_ϕ be the set of f in $\mathscr{K}$ satisfying $(f \mid f)_\phi = 0$; N_ϕ is a vector subspace of $\mathscr{K}$ and we may form the quotient space $\mathscr{K}_\phi = \mathscr{K}/N_\phi$. Denoting by f^ϕ the element of $\mathscr{K}_\phi$ determined by f, $(f \mid g)_\phi$ depends only on f^ϕ and g^ϕ and can thus be used to define $(f^\phi \mid g^\phi)_\phi$. In this way $\mathscr{K}_\phi$ appears as a pre-Hilbert space. Let H_ϕ be the completion of $\mathscr{K}_\phi$. The unitary endomorphisms of H_ϕ forming the second component of the representation associated with ϕ are obtained from the translation operators in $\mathscr{K}$ itself: for t in T and f in $\mathscr{K}$, f_t is the t-translate of f, defined by $f_t(t') = f(t' - t)$. The reader will verify without difficulty that f_t belongs to N_ϕ whenever f belongs to N_ϕ, so that the class $f_t^\phi = (f_t)^\phi$ depends only on the class f^ϕ. We may thus define an endomorphism $U_\phi(t)$ of $\mathscr{K}_\phi$ by the equation $U_\phi(t)f^\phi = f_t^\phi$ for each f in $\mathscr{K}$, verify that $U_\phi(t)$ is unitary on $\mathscr{K}_\phi$, and then extend $U_\phi(t)$ by continuity into a unitary endomorphism of H_ϕ. There is also no difficulty in verifying that $t \to U_\phi(t)$ is a representation of T, so that the pair (H_ϕ, U_ϕ) constitutes a unitary representation of T. This representation is continuous: to see this it suffices to check that $(f_t \mid f)_\phi$ is a continuous function of t for each f in $\mathscr{K}$, which is clear from (10.3.6).

When ϕ is understood we shall often write H, U, $(\ \mid\ )$, $\cdots$ in place of H_ϕ, U_ϕ, $(\ \mid\ )_\phi$, $\cdots$.

On the basis of this construction we may establish an important result due to Gelfand and Raikov.

10.3.3 **Theorem.** Suppose that ϕ is a positive definite function on T with the property that for some number $k \geq 0$,

$$f * \phi * \tilde{f}(0) \leq k\Big(\int |f|\, dt\Big)^2 \tag{10.3.7}$$

for all f in $\mathscr{K}$ vanishing outside some preassigned neighbourhood of 0 in T. Then ϕ is equal locally almost everywhere on T to a continuous positive definite function.

Proof. Assume that (10.3.7) holds for all f that vanish outside a neighbourhood V_0 of 0 in T. For each neighbourhood V of 0 in T satisfying $V \subset V_0$, we take a function g_V in $\mathscr{K}$ that is positive, vanishes outside V, and is such that $\int g_V\, dt = 1$. We partially order the neighbourhoods V by set-inclusion reversed, getting a directed set. The directed family (g_V^ϕ) is then weakly convergent in H. Indeed, (10.3.7) shows that this family is bounded in H, so that it is enough to show that $\lim_V (f^\phi \mid g_V^\phi)$ exists finitely for each f in $\mathscr{K}$ (the f^ϕ being dense in H). But

$$\begin{aligned}(f^\phi \mid g_V^\phi) &= \iint \phi(t_1 - t_2) f(t_1) g_V(t_2)\, dt_1\, dt_2 \\ &= \int g_V(t_2)\, dt_2 \int \phi(t_1 - t_2) f(t_1)\, dt_1;\end{aligned}$$

the inner integral is a continuous function of t_2 and so, by choice of the g_V, the said limit exists and is equal to $\int \phi f\,dt$. Suppose then that h is the weak limit in H of (g_V^ϕ). We have just shown that

$$(f^\phi \mid h) = \int \phi f\,dt \quad (f \in \mathscr{K}).$$

More generally we shall have that $U(t)g_V^\phi \to U(t)h$ weakly in H, hence by direct calculation it is found that

$$\begin{aligned}(f^\phi \mid U(t)h) &= \lim_V\,(f^\phi \mid U(t)g_V^\phi)\\ &= \lim_V \int\!\!\int \phi(t_1 - t_2) f(t_1) g_V(t_2 - t)\,dt_1\,dt_2\\ &= \lim_V \int g_V(t_2 - t)\,dt_2 \int \phi(t_1 - t_2) f(t_1)\,dt_1\\ &= \int \phi(t_1 - t) f(t_1)\,dt_1.\end{aligned}$$

Thus

$$(f^\phi \mid g^\phi) = \int (f^\phi \mid U(t)h) \cdot \overline{g(t)}\,dt \tag{10.3.8}$$

holds for f and g in $\mathscr{K}$. This extends by continuity when one replaces f^ϕ by any element of H and then shows (via the Hahn-Banach theorem) that the elements $U(t)h$ $(t \in T)$ are total in H. Further, if we replace f^ϕ in (10.3.8) by h, we derive

$$(h \mid g^\phi) = \int (h \mid U(t)h) \cdot \overline{g(t)}\,dt \quad (g \in \mathscr{K}),$$

or equivalently

$$(g^\phi \mid h) = \int (U(t)h \mid h) \cdot g(t)\,dt \quad (g \in \mathscr{K}).$$

The left-hand side here is, as we have shown, $\int \phi g\,dt$. Comparison thus leads to the conclusion that

$$\phi(t) = (U(t)h \mid h) \qquad \text{l.a.e.,} \tag{10.3.9}$$

and the right-hand side is a continuous, positive definite function on T. The proof is complete. ▮

10.3.4 **Corollary.** A positive definite function that is essentially bounded on some neighbourhood of 0 in T is equal locally almost everywhere to a continuous positive definite function.

Each element of L^∞ contains at most one continuous function, so that one may identify a bounded continuous function with its class in L^∞. Moreover, if ϕ is a continuous positive definite function, then $\|\phi\|_\infty = \phi(0)$ (see (10.3.3)). Corollary 10.3.4 shows now that the set P_0 of continuous positive definite functions ϕ satisfying $\phi(0) \le 1$ is weakly compact in L^∞. Obviously P_0 is convex. It is to P_0 that we shall apply the Krein-Milman theorem. The main problem remaining is to identify the extreme points of P_0, in which task we are assisted by Theorem 10.1.5. In this theorem we shall take E to be the set of all Hermitian symmetric functions in L^∞, E_+ the set of all positive definitive functions in L^∞, P the set of all continuous positive definite functions in L^∞, $A = P_0$ and $f(\phi) = \phi(0)$. It is clear that the conditions (1)–(5) of Theorem 10.1.5 are fulfilled, and we conclude that the extreme points of P_0 are (a) 0, and (b) the P-minimal points of P_0 satisfying $\phi(0) = 1$. The reader will note that an element ϕ of P is P-minimal if and only if each continuous positive definite function ψ, for which $\phi - \psi$ is positive definite, is a scalar multiple of ϕ. The required identification is expressed in the following proposition.

10.3.5 **Proposition.** A continuous positive definite function ϕ satisfying $\phi(0) = 1$ is P-minimal if and only if the associated unitary representation (H_ϕ, U_ϕ) is irreducible, which† in turn signifies that ϕ is an element of $\hat{T}$.

Proof. (a) Let us suppose that ϕ is P-minimal and $\phi(0) = 1$. Let Q be any projector in H commuting with all the $U(t)$. Then

$$\begin{aligned}\phi(t) &= (U(t)h \mid h) = (U(t)h \mid QU(t)h) + (U(t)h \mid h - QU(t)h) \\ &= (U(t)Qh \mid Qh) + (U(t)Q'h \mid Q'h),\end{aligned}$$

where $Q' = 1 - Q$, using the fact that Q and Q' commute with the $U(t)$ while $(Sh \mid h) = (Sh \mid Sh)$ obtains in any Hilbert space in which S is a projector. Now both terms last written are continuous positive definite functions in P_0, and so the P-minimality of ϕ entails the existence of a scalar c such that

$$(U(t)Qh \mid Qh) = c \cdot (U(t)h \mid h).$$

Using again the fact that Q commutes with each $U(t)$, this may be written as $(U(t)h \mid Qh) = c \cdot (U(t)h \mid h)$. The proof of Theorem 10.3.3 disclosed that the $U(t)h$ are total in H, hence it follows that $Qh = c \cdot h$. Q being a projector, c must be 0 or 1. Thus (H, U) is irreducible.

(b) Suppose conversely that (H, U) is irreducible, that ψ belongs to P, and that $\phi - \psi$ has the same property. We have to show that ψ is a scalar multiple of ϕ. If f is in $\mathscr{K}$, the fact that $\phi - \psi$ belongs to P entails that $(f^\psi \mid f^\psi)_\psi \leq (f^\phi \mid f^\phi)$. Thus $(f^\psi \mid g^\psi)_\psi$ defines a bounded Hermitian sesquilinear form on H, and so there exists a bounded self-adjoint endomorphism A of H such that

$$(f^\psi \mid g^\psi)_\psi = (Af^\phi \mid g^\phi) \qquad (f, g \in \mathscr{K}). \tag{10.3.10}$$

Also

$$\begin{aligned}(U(t)Af^\phi \mid g^\phi) &= (Af^\phi \mid U(-t)g^\phi) = (f^\psi \mid U_\psi(-t)g^\psi)_\psi \\ &= (U_\psi(t)f^\psi \mid g^\psi)_\psi\end{aligned}$$

and, by similar juggling,

$$\begin{aligned}(AU(t)f^\phi \mid g^\phi) &= (U(t)f^\phi \mid Ag^\phi) = \overline{(Ag^\phi \mid U(t)f^\phi)} \\ &= \overline{(g^\psi \mid U_\psi(t)f^\psi)_\psi} = (U_\psi(t)f^\psi \mid g^\psi)_\psi.\end{aligned}$$

Thus

$$(U(t)Af^\phi \mid g^\phi) = (AU(t)f^\phi \mid g^\phi).$$

---

† This is the first point at which essential use is made of the fact that T is Abelian.

This entails that $U(t)A = AU(t)$—that is, that A commutes with the $U(t)$. Irreducibility of (H, U) forces the conclusion that $A = c \cdot 1$ and so

$$\psi(t) = (U_\psi(t)h_\psi \mid h_\psi)_\psi = (AU(t)h \mid h)$$

(by (10.3.10) continuously extended)

$$= c \cdot (U(t)h \mid h) = c \cdot \phi(t),$$

as had to be shown.

(c) Finally we use the Abelian nature of T: thanks to this, the $U(t)$ commute. If, therefore, (H, U) is irreducible, each $U(t)$ must be a scalar multiple of $1 : U(t) = \chi(t) \cdot 1$. Then χ is necessarily a bounded continuous character of T, reflecting the fact that U is a continuous unitary representation thereof. Hence $\phi(t) = (U(t)h \mid h) = \chi(t) \cdot (h \mid h) = \phi(0) \cdot \chi(t)$. If $\phi(0) = 1$, ϕ is thus an element of $\hat{T}$. Conversely, if ϕ is a bounded continuous character, the construction of H shows that it is one-dimensional and (H, U) is irreducible. This completes the proof. ▌

We now know that the set of extreme points of P_0 can be identified with $T' = \hat{T} \cup \{0\} \subset P_0 \subset L^\infty$. The Krein-Milman theorem is now applicable and shows that:

▶ Any weakly closed convex subset of L^∞ containing 0 and all the bounded continuous characters of T must contain P_0 itself.

However, Bochner's theorem will come from applying Theorem 10.1.7. For this purpose, we wish to show that $\hat{T}$ may be made into a separated locally compact space.

10.3.6 **Proposition.** (a) T' is weakly closed in P_0, hence is compact for the weak topology of L^∞.

(b) $\hat{T}$ is locally compact for the weak topology of L^∞.

(c) On $\hat{T}$ the weak topology of L^∞ coincides with that of convergence uniform on compact subsets of T.

Proof. It suffices for all three allegations to prove that if a directed family (t_i) in $\hat{T}$ converges weakly in L^∞ to a limit $\phi \neq 0$ in P_0, then $(\hat{t}_i) \to \phi$ uniformly on compact subsets of T. Now since $\phi \neq 0$, f may be chosen from $\mathscr{K}$ so that $\int f\phi \, dt \neq 0$. We put $\phi_i(t) = [t, \hat{t}_i]$. Then

$$\phi_i * f(t) = \int \phi_i(t - t')f(t') \, dt' = \phi_i(t) \int \overline{\phi_i(t')} f(t') \, dt' = c_i \cdot \phi_i(t),$$

the scalars c_i converging to $\int f\phi \, dt \neq 0$. The first expression for $\phi_i * f(t)$ shows that $\phi_i * f \to \phi * f$ pointwise on T. In addition, however, since the ϕ_i are uniformly bounded, the $\phi_i * f$ are equicontinuous:

$$|\phi_i * f(t_1) - \phi_i * f(t_2)| \leq \|\phi_i\|_\infty \cdot \int |f_1(t_1 - t) - f(t_2 - t)| \, dt$$
$$\leq \int |f(t') - f(t' + t_2 - t_1)| \, dt'$$

by invariance of the Haar measure. So (Proposition 0.4.9), $\phi_i * f \to \phi * f$ uniformly on compact subsets of T. The result follows immediately. ▮

Remarks. The topology on $\hat{T}$ defined by locally uniform convergence on T was introduced by Pontryagin ([1], Chapter V) and is often known after him. It is easy to verify that $\hat{T}$, so topologized, is a separated topological group.

An important consequence of Proposition 10.3.6 is the following generalized version of the famous Riemann-Lebesgue lemma.

10.3.7 **Proposition.** If f is an integrable function on T, its Fourier transform

$$\hat{f}(\hat{t}) = \int_T f(t) \cdot \overline{[t, \hat{t}]}\, dt$$

belongs to $C_0(\hat{T})$ (the space of continuous, complex-valued functions on $\hat{T}$ which tend to 0 at infinity). Moreover, as f ranges over $\mathscr{L}^1(T)$, the $\hat{f}$ form a dense vector subspace of $C_0(\hat{T})$, relative to the topology of convergence uniform on $\hat{T}$.

Proof. Since $\hat{f}$ is the uniform limit of functions

$$\int_K f(t) \cdot \overline{[t, \hat{t}]}\, dt,$$

K being a compact subset of T, Proposition 10.3.6 (c) shows that $\hat{f}$ is continuous on $\hat{T}$. Besides this, given $\varepsilon > 0$, there exists an open neighbourhood W of 0 in T' such that $|\hat{f}(\hat{t})| < \varepsilon$ if $\hat{t} \in T' \backslash W$: this is so by very definition of the weak topology on L^∞. By Proposition 10.3.6 once more, $K = T' \backslash W$ is a compact subset of $\hat{T}$. Since $|\hat{f}(\hat{t})| < \varepsilon$ for $\hat{t} \in \hat{T} \backslash K$, it follows that $\hat{f} \in C_0(\hat{T})$.

To prove the remainder of the proposition, we imagine each $\hat{f}$ to be defined on the compact space $T' = \hat{T} \cup \{0\}$ (see Proposition 10.3.6) and satisfying $\hat{f}(0) = 0$. Each $\hat{f}$ belongs to $C(T')$. We observe also that the Fourier transform of $\tilde{f}$ is the complex conjugate of $\hat{f}$, and that the Fourier transform of $f * g$ (see Subsection 4.19.12) is the product $\hat{f} \cdot \hat{g}$. This being so, one may appeal to the Weierstrass-Stone theorem (Theorem B of Subsection 4.10.5) to conclude that the functions of the type $\lambda + \hat{f}$, where λ is a complex constant and $f \in \mathscr{L}^1(T)$, form a dense subalgebra of $C(T')$. It follows directly that the functions $\hat{f}$ form a dense subalgebra of $C_0(T)$, the latter being identifiable with the set of functions in $C(T')$ that vanish at 0. ▮

Remark. Although it is true that $\hat{f}$ tends to 0 at infinity on $\hat{T}$ for each $f \in \mathscr{L}^1(T)$, it can be seen that whenever $\hat{T}$ is not compact (which means that T is not discrete), there exist integrable functions f on T whose Fourier transforms tend to 0 arbitrarily slowly as $\hat{t}$ tends to infinity. See Exercise 10.8.

Our goal is now within reach.

10.3.8 **Theorem.** If ϕ is a locally essentially bounded and measurable positive definite function on T, there exists a unique bounded positive Radon measure μ on $\hat{T}$ such that

$$\phi(t) = \int [t, \hat{t}]\, d\mu(\hat{t}) \qquad \text{l.a.e. on } T;$$

if ϕ is continuous, equality holds at all points $t \in T$.

Proof. In proving the existence of μ we may, by virtue of Corollary 10.3.4, assume that ϕ is continuous. Then if $\phi(0) = 0$, ϕ is identically 0 and $\mu = 0$ serves our purpose. Otherwise one may arrange that $\phi \in P_0$, in which case the existence of μ follows from Theorems 10.1.5, 10.1.7, and Proposition 10.3.5.

To prove the uniqueness of μ amounts to showing that if

$$\int [t, \hat{t}] \, d\mu(\hat{t}) = 0$$

for all $t \in T$, then $\mu = 0$. This hypothesis, however, implies that

$$\int \hat{f}(\hat{t}) \, d\mu(\hat{t}) = 0$$

for all integrable functions f on T, as one sees by use of the Lebesgue-Fubini Theorem 4.17.4. That $\mu = 0$ now follows from the density assertion in the preceding proposition. ▌

10.3.9 **Notes.** Classical representation theory applied originally to finite groups and it was customary to speak of matricial representations or, what is equivalent, to representations in terms of invertible endomorphisms of a finite-dimensional vector space H. Naturally, H may be made into a Hilbert space in many ways. It can be shown (Maak [1], §§4, 30) that any bounded representation of this sort is equivalent to a unitary representation, if one makes H into a Hilbert space in a suitable way.

On the discovery of the existence of Haar measure for separated compact groups, much of the theory was extended to representations of these; see Weil [1], Chapitres IV and V. Here it was still adequate to consider finite-dimensional representations. The same technique is employed for arbitrary groups by Maak [1] by exploiting the existence of a mean value for almost periodic functions and relating the latter to bounded representations; a small facet of this is dealt with in Exercise 10.7. As is shown by Weil (loc. cit. Chapitre IV) any measurable representation of finite dimension of a separated compact group is necessarily continuous; and any finite-dimensional representation of a separated locally compact group that belongs to L^p is in fact bounded. So, for separated compact groups, any measurable finite-dimensional representation is equivalent to a continuous unitary representation, and there is no loss entailed in considering only the latter.

For separated locally compact groups, experience shows that it is still adequate to consider unitary representations, but one must then admit infinite-dimensional ones.

For finite-dimensional representations the main consequence of irreducibility was obtained via Schur's lemma (see Exercises 10.5 and 10.6). This lemma cannot be extended directly to infinite-dimensional representations, a major obstacle being the unfortunate fact that an endomorphism of an infinite-dimensional vector space may possess no eigenvalues at all.

The explicit computation of the irreducible unitary representations of a given non-Abelian group is usually a task of the first magnitude. Those for the rotation group in R^3 are closely related to spherical harmonics, but the details are quite complex; see Maak [1], § 48. The same problem for the Lorentz group is of interest to physicists as well as to mathematicians. The interested reader may consult Chapter XV of Lyubarskiĭ [1].

10.4 The Plancherel Theorem and the Duality Law

The direct application of the Krein-Milman theorem is completed with Theorem 10.3.8, but we intend to exploit the Bochner theorem in a manner similar to that devised by Cartan and Godement in collaboration (Cartan and Godement [1]). This leads to an extended concept of the Fourier transformation, so far defined only for functions integrable over T, applying to functions that are square integrable over T. As a derivative of this, we shall establish the Pontryagin duality law, which says that $\hat{\hat{T}} = (\hat{T})^\wedge$ is (isomorphic as a topological group with) T.

If T is specialized to the additive group R^n, the results we shall obtain cover those used in various places in Chapter 5, especially Subsection 5.15.1. The reader will recall that at that point we called upon the Parseval formula for restricted functions in order to motivate and develop some of the properties of the Schwartz-Fourier transform of temperate distributions.

The starting point is a refinement of Bochner's theorem, deducible from and actually equivalent to, the latter.

10.4.1 Theorem. By suitable normalization of the Haar measure on $\hat{T}$ one can arrange that for each continuous and positive definite function ϕ in $\mathscr{L}^1(T)$, the Fourier transform $\hat{\phi}$ belongs to $\mathscr{L}^1(\hat{T})$ and the inversion formula

$$\phi(t) = \int \hat{\phi}(\hat{t}) \cdot [t, \hat{t}]\, d\hat{t} \tag{10.4.1}$$

is valid.

Proof. Let us denote by P^1 the set of continuous positive definite functions in $\mathscr{L}^1(T)$. We know (Theorem 10.3.8) that to each ϕ in P^1 corresponds a unique bounded Radon measure μ_ϕ on T such that

$$\phi(t) = \int [t, \hat{t}]\, d\mu_\phi(\hat{t}); \tag{10.4.2}$$

this measure μ_ϕ is positive. Thus we have to show that, in fact, one can normalize the Haar measure m on $\hat{T}$ in such a way that $\mu_\phi = \hat{\phi} \cdot m$.

The procedure is to show that a measure m exists on $\hat{T}$ such that $\mu_\phi = \hat{\phi} \cdot m$ for each ϕ in P^1, and that this m is nontrivial and translation invariant. It will then be the appropriately normalized Haar measure.

Now from (10.4.2) one infers easily that

$$\phi * f(0) = \int \phi(t) f(-t)\, dt = \mu_\phi(\hat{f}) \tag{10.4.3}$$

for all f in $\mathscr{L}^1(T)$. Moreover, if (10.4.3) holds for all f in a dense subset of $\mathscr{L}^1(T)$, then (10.4.2) follows. Thus μ_ϕ is the unique bounded Radon measure on T for which (10.4.3) holds for all f in $\mathscr{L}^1(T)$, or for all f in a dense subset of $\mathscr{L}^1(T)$. Equation (10.4.2) shows also that $\mu_\phi(\hat{T}) = \phi(0)$.

The characterization of μ_ϕ afforded by (10.4.3) enables us to conclude that

$$\mu_{\phi*\phi'} = \hat{\phi}' \cdot \mu_\phi = \hat{\phi} \cdot \mu_{\phi'} \tag{10.4.4}$$

whenever ϕ and ϕ' belong to P^1. We shall now study this relation when ϕ' is allowed to vary in a suitable way.

We take a directed family (ϕ_i') in P^1 such that $\phi_i' \geq 0$, $\int \phi_i'\, dt = 1$, and ϕ_i' vanishes outside a neighbourhood of 0 that becomes arbitrarily small for all sufficiently large i. (For example: $\phi_i' = f_i * \tilde{f}_i$, where f_i belongs to $\mathscr{K}_+(T)$, $\int f_i\, dt = 1$, and f_i vanishes outside arbitrarily small neighbourhoods of 0 as i grows.) In particular, the measures $\phi_i'\, dt$ converge vaguely to ε, the Dirac measure at 0. If one could take $\phi' = \varepsilon$ in (10.4.4), no problem would remain. Clearly, unless T is discrete, this cannot be done. But the limiting process of allowing ϕ' to vary along the directed family (ϕ_i') will suffice. In fact, we have in the first place

$$\phi * \phi_i'(f) = \phi * \phi_i' * f(0) \to \phi * f(0) = \mu_\phi(\hat{f}),$$

since $\phi * f$ is a bounded continuous function on T. Further, $\mu_{\phi*\phi_i'}(\hat{T}) = \phi * \phi_i'(0) \to \phi(0)$ and so remains bounded. Proposition 10.3.7 shows therefore that $\mu_{\phi*\phi_i'} \to \mu_\phi$ weakly in the dual of $C_0(\hat{T})$. On the other hand, $\mu_{\phi*\phi_i'} = \hat{\phi} \cdot \mu_{\phi_i'}$ for each i. Using Proposition 10.3.7 again, one sees the possibility of choosing ϕ in P^1 so that $\hat{\phi}$ is bounded away from 0 on any preassigned compact subset of $\hat{T}$. So, since we know that the $\hat{\phi} \cdot \mu_{\phi_i'}$ converge weakly, it follows that the $\mu_{\phi_i'}$ must themselves converge vaguely to some measure m on $\hat{T}$, m being necessarily positive. Then necessarily $\mu_\phi = \lim \mu_{\phi*\phi_i'} = \lim \hat{\phi} \cdot \mu_{\phi_i'}$, weakly in the dual of $C_0(\hat{T})$, must be none other than the vague limit of the $\hat{\phi} \cdot \mu_{\phi_i'}$, that is, $\hat{\phi} \cdot m$.

Thus $\mu_\phi = \hat{\phi} \cdot m$. Since μ_ϕ is bounded and $\hat{\phi}$ is continuous, so $\hat{\phi}$ must be integrable for m. We now have in place of (10.4.3)

$$\phi * f(0) = \int \hat{\phi} \hat{f}\, dm, \tag{10.4.5}$$

which relation holds for all ϕ in P^1 and all f in $\mathscr{L}^1(T)$, and serves to determine m uniquely. Evidently, m is not 0.

In order to prove that m is translation invariant, it is convenient to denote by S the vector subspace of $\mathscr{L}^1(T)$ generated by P^1. Both sides of (10.4.5) being bilinear in the pair (ϕ, f), this same (10.4.5) holds for $\phi \in S$ and $f \in S$. Since S contains all functions of the form $u * \tilde{v}$ with $u, v \in \mathscr{K}(T)$, it is dense in $\mathscr{L}^1(T)$. Hence, by Proposition 10.3.7, the set of Fourier transforms of members of S is uniformly dense in $C_0(\hat{T})$. In addition, if in (10.4.5) we multiply each of ϕ and f by a character $\hat{a} \in \hat{T}$, the left-hand side of this equation remains unaltered, hence we may conclude that

$$\int \hat{\phi}(\hat{t} - \hat{a}) \hat{f}(\hat{t} - \hat{a})\, dm(\hat{t}) = \int \hat{\phi}(\hat{t}) \hat{f}(\hat{t})\, dm(\hat{t}) \tag{10.4.5'}$$

for $\phi, f \in S$.

Now let $g \in \mathscr{K}(\hat{T})$. We can choose $\hat{\phi} \in S$ such that $\hat{\phi}$ is nonvanishing on some neighbourhood of the support of g. Consequently, $g = \hat{\phi}h$ for some $h \in \mathscr{K}(\hat{T})$. This function h is the uniform limit of a sequence $(\hat{f}_n)$, where $f_n \in S$. Correspondingly, $g = \lim \hat{\phi}\hat{f}_n$ uniformly. At the same time,

$$|\hat{\phi}(\hat{t} - \hat{a})\hat{f}_n(\hat{t} - \hat{a})| \leq \text{const } |\hat{\phi}(\hat{t} - \hat{a})|,$$

and the function on the right is known to be integrable for m. So, by Theorem 4.8.2,

$$\int g(\hat{t} \quad \hat{a})\, dm(\hat{t}) = \lim_{n \to \infty} \int \hat{\phi}(\hat{t} - \hat{a})\hat{f}_n(\hat{t} - \hat{a})\, dm(\hat{t}).$$

Reference to (10.4.5′) shows that the right-hand side of this expression is independent of $\hat{a}$. Thus m is indeed translation invariant, and the proof is complete. ∎

10.4.2 **Remark.** If ϕ belongs to P^1, the fact that $\mu_\phi = \hat{\phi} \cdot m$ is positive signifies that the continuous function $\hat{\phi}$ is positive. Thus, if ϕ is continuous, positive definite, and integrable, then $\hat{\phi}$ is continuous, positive, and integrable.

10.4.3 **Corollary.** If $t' \neq t''$ belong to T, there exists $\hat{t}$ in $\hat{T}$ such that $[t', \hat{t}] \neq [t'', \hat{t}]$.

Proof. It suffices to show that if $t' \neq 0$, there exists $\hat{t}$ such that $[t', \hat{t}] \neq 1$. Now we can, as in the proof of Theorem 10.4.1, construct a ϕ in P^1 satisfying $\phi(0) = 1$ and $\phi(t') = 0$. The inversion formula (10.4.1) then shows that one cannot have $[t', \hat{t}] = 1 = [0, \hat{t}]$ for all $\hat{t}$, Q.E.D. ∎

10.4.4 We shall now derive from Theorem 10.4.1 a restricted form of the Parseval formula, which in turn forms the basis for the Plancherel theorem.

Suppose f belongs to $\mathscr{L}^1(T) \cap \mathscr{L}^2(T)$ and consider $\phi = f * \tilde{f}$. This function ϕ is continuous, a property it shares with the convolution of any two functions in $\mathscr{L}^2(T)$ by virtue of the Cauchy-Schwarz inequality; it is integrable, being the convolution of two integrable functions; and it is positive definite. Thus ϕ belongs to P^1. Its Fourier transform $\hat{\phi} = \hat{f} \cdot \bar{\hat{f}} = |\hat{f}|^2$. From Theorem 10.4.1 we conclude therefore that $\hat{f}$ belongs to $\mathscr{L}^2(\hat{T})$ and that

$$\int |f|^2\, dt = \phi(0) = \int \hat{\phi}(\hat{t})\, d\hat{t} = \int |\hat{f}(\hat{t})|^2\, d\hat{t}. \tag{10.4.6}$$

More generally, if f and g belong to $\mathscr{L}^1(T) \cap \mathscr{L}^2(T)$, (10.4.6) applies when f is replaced by $f + \alpha g$, α being any complex number. It follows that

$$\int f\bar{g}\, dt = \int \hat{f}\bar{\hat{g}}\, d\hat{t}, \tag{10.4.7}$$

which is Parseval's formula for the restricted class of functions f and g considered.

It is now a simple matter to derive the Plancherel theory. Let us suppose f belongs to $\mathscr{L}^2(T)$ and choose any sequence $(f_n) \subset \mathscr{L}^1(T) \cap \mathscr{L}^2(T)$ converging in $\mathscr{L}^2(T)$ to f. Then (10.4.6) shows that $(\hat{f}_n)$ is a Cauchy sequence in $\mathscr{L}^2(\hat{T})$. So there exists an F in $\mathscr{L}^2(\hat{T})$ to which $(\hat{f}_n)$ is convergent; F is uniquely determined modulo negligible sets—that is, the class $\dot{F}$ is uniquely determined. Not only is

$\dot{F}$ determined uniquely by the sequence $(\hat{f}_n)$, but it is uniquely determined when f is given: this follows by applying (10.4.6) to the differences $f'_n - f''_n$ of any two approximating sequences (f'_n) and (f''_n). Moreover (10.4.6) gives in the limit

$$\int |f|^2 \, dt = \int |F|^2 \, d\hat{t}.$$

It is also evident that $\dot{F}$ depends only upon the class $\dot{f}$. The passage from f (or $\dot{f}$) to $\dot{F}$ thus defines a mapping u of $\mathscr{L}^2(T)$ (or $L^2(T)$) into $L^2(\hat{T})$. Plainly, u is linear and isometric. Furthermore, u constitutes an extension of the Fourier transformation: if f belongs to $\mathscr{L}^1(T) \cap \mathscr{L}^2(T)$, then $\dot{F}$ is none other than the class determined by $\hat{f}$. We shall therefore denote by $\hat{f}$ both the class $\dot{F}$ and also (by abuse of language) a representative F of that class. In this way (10.4.6) and (10.4.7) retain their validity for functions f and g in $\mathscr{L}^2(R)$. For purposes of distinction, one might refer to $u(f)$ as the Plancherel-Fourier transform of f.

To complete the picture we wish to show that u maps $L^2(T)$ *onto* $L^2(\hat{T})$. Consider the u-image of $L^2(T)$, say Q. Since u is an isometry and $L^2(\hat{T})$ is complete, Q is closed in $L^2(\hat{T})$. To show that $Q = L^2(\hat{T})$ it therefore suffices to show that it is dense in $L^2(\hat{T})$. This appears from the following lemma.

10.4.5 **Lemma.** Let F belong to $\mathscr{L}^1(\hat{T}) \cap \mathscr{L}^2(\hat{T})$. Then

$$f(t) = \int F(\hat{t}) \cdot [t, \hat{t}] \, d\hat{t}$$

belongs to $\mathscr{L}^2(T)$ and $\hat{f} = F$ a.e.

Proof. If g belongs to $\mathscr{K}(T)$ we have

$$\int F\bar{\hat{g}} \, d\hat{t} = \int F(\hat{t}) \, d\hat{t} \int \overline{g(t)} \cdot [t, \hat{t}] \, dt$$
$$= \int f(t)\overline{g(t)} \, dt$$

and so

$$\left| \int f\bar{g} \, dt \right| \le \left[\int |F|^2 \, d\hat{t}\right]^{1/2} \cdot \left[\int |\hat{g}|^2 \, d\hat{t}\right]^{1/2}$$
$$= \left[\int |F|^2 \, d\hat{t}\right]^{1/2} \cdot \left[\int |g|^2 \, dt\right]^{1/2}$$

the last step by (10.4.6) applied with g in place of f. This shows that f belongs to $\mathscr{L}^2(T)$. Applying (10.4.7) we have $\int \hat{f}\bar{\hat{g}} \, d\hat{t} = \int f\bar{g} \, dt = \int F\bar{\hat{g}} \, d\hat{t}$, this for all g in $\mathscr{K}(T)$. The set of $\hat{g}$ is dense in $C_0(\hat{T})$, and it follows that $\hat{f} = F$ a.e. ▌

The results can be summarized as follows.

10.4.6 **Theorem.** The Plancherel-Fourier transformation defines a unitary map of $L^2(T)$ onto $L^2(\hat{T})$, the unitary character being expressed more concretely by (10.4.7).

10.4.7 **Remark.** From Lemma 10.4.5 it appears that any function F in $\mathscr{L}^1(\hat{T}) \cap \mathscr{L}^2(\hat{T})$ is equal almost everywhere to the transform $\hat{f}$ of some f that is bounded, continuous, and in $\mathscr{L}^2(T)$. It follows at once that any function

$H = F * G$, where F and G belong to $\mathscr{L}^1(\hat{T}) \cap \mathscr{L}^2(\hat{T})$, is equal (everywhere) to the transform of a function h that is bounded, continuous, and in $\mathscr{L}^1(T)$. This observation will prove to be useful in connection with the duality law.

A very rapid application of the Riesz convexity theorem (see Section 4.24) shows that the Fourier transformation may be extended into one of $L^p(T)$ into $L^{p'}(\hat{T})$ whenever $1 < p < 2$ and $1/p + 1/p' = 1$ (see Exercise 10.26).

10.4.8 The Duality Law. If t is a point of T, then $\hat{t} \to [t, \hat{t}]$ is a bounded continuous character of $\hat{T}$; and to different t's correspond thus different characters of $\hat{T}$. In this way we may regard T as a subgroup of $\hat{\hat{T}}$. Moreover, from Proposition 10.3.6 with $\hat{T}$ replacing T, it results easily that the topology of $\hat{\hat{T}}$ (qua dual of $\hat{T}$) induces the initial topology on T. Since T is locally compact and therefore complete, it must be closed in $\hat{\hat{T}}$. The duality law, asserting that $T = \hat{\hat{T}}$, will therefore be established as soon as it is shown that T is dense in $\hat{\hat{T}}$.

Let us suppose if possible that T were not dense in $\hat{\hat{T}}$. Then there would exist a point $\hat{\hat{t}}_0$ of $\hat{\hat{T}}$ and a neighbourhood N of 0 in $\hat{\hat{T}}$ such that $\hat{\hat{t}}_0 + 2N$ does not meet T. We could then take F in $\mathscr{K}_+(\hat{\hat{T}})$, $F \neq 0$, $F = 0$ outside N. Then $F * \tilde{F}$ is positive, does not vanish identically, and is 0 outside $N + N$. So $G(\hat{\hat{t}}) = F * \tilde{F}(\hat{\hat{t}} - \hat{\hat{t}}_0)$ is not identically vanishing and is 0 on T. By Remark 10.4.7, there would exist a function g in $\mathscr{L}^1(\hat{T})$ such that

$$G(\hat{\hat{t}}) = \int g(\hat{t}) \cdot \overline{[\hat{\hat{t}}, \hat{t}]}\, d\hat{t}.$$

This would give $\int g(\hat{t}) \cdot [t, \hat{t}]\, d\hat{t} = 0$ for all t in T, hence (Exercise 10.10) $g = 0$ a.e. But then G would be identically 0 on $\hat{\hat{T}}$, a contradiction. ▌

10.4.9 The Inverse Fourier Transform. Having identified $\hat{\hat{T}}$ with T it is now permissible to interchange T and $\hat{T}$ in all the results in Section 10.3 and the preceding portions of the present section. In doing this one's first inclination is perhaps to define the Fourier transformation from $\hat{T}$ to T as that carrying a function F on $\hat{T}$ into the function

$$\int F(\hat{t})\overline{[t, \hat{t}]}\, d\hat{t}$$

on T. Yet a rather trivial change is suggested by Theorem 10.4.1, which says that if we replace $\overline{[t, \hat{t}]}$ in the above integral by $[t, \hat{t}]$, one obtains a transformation that, at least for suitably restricted functions, is inverse to the Fourier transformation $f \to \hat{f}$.

Taking this hint, we write $\mathscr{F}$ for the Fourier transformation $f \to \hat{f}$ studied thus far and define a transformation $\overline{\mathscr{F}}$ of functions on $\hat{T}$ into functions on T by the formula

$$\overline{\mathscr{F}} F(t) = \int F(\hat{t})[t, \hat{t}]\, d\hat{t}. \tag{10.4.8}$$

This is initially defined only for $F \in \mathscr{L}^1(\hat{T})$, and then its properties may be read off from the corresponding ones of $\mathscr{F}$. In particular, Theorem 10.4.1 says that if f is continuous, positive definite, and belongs to $\mathscr{L}^1(T)$, then $F = \mathscr{F}f \in \mathscr{L}^1(\hat{T})$ and $f = \overline{\mathscr{F}}F$; that is, that $f = \overline{\mathscr{F}}\mathscr{F}f$. Actually (see Exercise 10.14) a little more argument shows that the formula

$$f = \overline{\mathscr{F}}\mathscr{F}f \tag{10.4.9}$$

holds in the almost everywhere sense for any $f \in \mathscr{L}^1(T)$ for which $\mathscr{F}f \in \mathscr{L}^1(\hat{T})$.

In view of the Plancherel Theorem 10.4.6, it is natural to hope and expect that (10.4.9) will hold for all $f \in \mathscr{L}^2(T)$. Let us examine this question.

To begin with, an interchange of T and $\hat{T}$ leads, via the arguments used to establish Theorem 10.4.6, to the conclusion that $\overline{\mathscr{F}}$ may be extended from $\mathscr{L}^1(\hat{T}) \cap \mathscr{L}^2(\hat{T})$ into a unitary map of $\mathscr{L}^2(\hat{T})$ onto $\mathscr{L}^2(T)$. It follows that $\overline{\mathscr{F}}\mathscr{F}$ is a continuous endomorphism of $\mathscr{L}^2(T)$ so that, in order to establish (10.4.9) for all $f \in \mathscr{L}^2(T)$, it suffices to verify it for those f belonging to a subset of $\mathscr{L}^2(T)$ whose finite linear combinations are dense in $\mathscr{L}^2(T)$. Let M consist of all continuous, positive definite and integrable functions on T, together with their translates. Then, as we have remarked, (10.4.9) is true for all $f \in M$. It remains to show that M generates a dense vector subspace of $\mathscr{L}^2(T)$.

According to the Hahn-Banach theorem it suffices to show that if $g \in \mathscr{L}^2(T)$ satisfies $f * g = 0$ for all f that are continuous, positive definite and integrable, then $g = 0$ a.e. For each compact neighbourhood V of 0 in T, we choose a symmetric neighbourhood of 0 satisfying $V' + V' \subset V$. Next we choose $k \in \mathscr{K}_+(T)$ vanishing outside V' and such that $\int k\, dt = 1$. The function $f_V = k * \tilde{k}$ is continuous, positive definite, integrable, positive, and $\int f_V\, dt = 1$. In these circumstances it is easy to show that $f_V * g \to g$ in $\mathscr{L}^2(T)$ as V shrinks. But $f_V * g = 0$ for all V, and it follows that $g = 0$ a.e.

Thus we have established that (10.4.9) holds almost everywhere for all $f \in \mathscr{L}^2(T)$. An exactly similar argument shows that

$$F = \mathscr{F}\overline{\mathscr{F}}F \tag{10.4.9'}$$

almost everywhere for all $F \in \mathscr{L}^2(\hat{T})$. One is thus fully justified in referring to $\overline{\mathscr{F}}$ as the inverse Fourier transform.

10.4.10 **Remarks on the Inversion Formula.** As we have already remarked, the inversion formula

$$f(t) = \int \hat{f}(\hat{t}) \cdot [t, \hat{t}]\, d\hat{t} \tag{10.4.10}$$

may be shown without much difficulty to hold in the almost everywhere pointwise sense whenever $f \in \mathscr{L}^1(T)$ is such that $\hat{f} \in \mathscr{L}^1(\hat{T})$. When $f \in \mathscr{L}^2(T)$, the formula holds in a different sense: in this case (10.4.9) means that if one takes a sequence $F_n \in \mathscr{L}^1(\hat{T}) \cap \mathscr{L}^2(\hat{T})$ converging in $\mathscr{L}^2(\hat{T})$ to $\hat{f}$, then the functions

$$f_n(t) = \int F_n(\hat{t}) \cdot [t, \hat{t}]\, d\hat{t},$$

which are members of $C_0(T)$, converge in $\mathscr{L}^2(T)$ to f.

In respect of a genuine pointwise (everywhere or almost everywhere) inversion formula, the author knows of no completely satisfactory result. Troubles arise even for $T = R^n$. It is not in the least surprising that difficulties are encountered in the general case, if only because the classical results for $T = R^1$ involve conditions on f of the type of locally bounded variation, which depend on order properties of R^1. See Edwards and Hewitt [1].

Despite this it is relatively easy to mimic various classical summability procedures for Fourier series and integrals. To manufacture suitable summability factors, we construct functions s_V on T (like the f_V appearing in Subsection 10.4.9), V running over a base of compact neighbourhoods of 0 in T, each s_V being positive, continuous, integrable, positive definite, and vanishing outside V. (For many purposes it would suffice that the last property be weakened to the demand that

$$\lim_V \int_{T \setminus U} s_V \, dt = 0$$

for each neighbourhood U of 0 in T.) The s_V are to be normalized by the condition

$$\int s_V \, dt = 1.$$

The transform $S_V = \hat{s}_V$ will then be positive, continuous, integrable, and positive definite. It is a relatively easy matter to show that

$$\lim_V s_V * f = f \tag{10.4.11}$$

in the sense of $\mathscr{L}^p(T)$ when $f \in \mathscr{L}^p(T)$ and $1 \leq p < \infty$; if $p = \infty$, the relation still holds in the sense of the weak topology of $\mathscr{L}^\infty(T)$ defined by its duality with $\mathscr{L}^1(T)$; if f is bounded and continuous, the relation holds in the sense of locally uniform convergence, or in the sense of uniform convergence if f is bounded and uniformly continuous.

On the other hand, if $f \in \mathscr{L}^1(T)$, then the transform of $s_V * f$, namely, $S_V \hat{f}$, is integrable. So, by the inversion formula (10.4.10),

$$s_V * f(t) = \int S_V(\hat{t}) \hat{f}(\hat{t}) \cdot [t, \hat{t}] \, d\hat{t} \tag{10.4.12}$$

holds for all $t \in T$. Consequently,

$$f = \lim_V \int S_V(\hat{t}) \hat{f}(\hat{t}) \cdot [t, \hat{t}] \, d\hat{t} \tag{10.4.13}$$

holds in the sense of $\mathscr{L}^1(T)$. It also holds in the pointwise sense at each point of continuity of f. This is the desired summation formula generalizing (10.4.10).

The reader will notice too that if $f \in \mathscr{L}^2(T)$, then $S_V \hat{f}$ is again integrable (as the product of two functions in $\mathscr{L}^2(\hat{T})$), so that again (10.4.12) holds, while (10.4.13) holds with the limit taken in the $\mathscr{L}^2(T)$ sense, and in the pointwise sense at each point of continuity of f. The same is true for $1 < p < 2$ if one has defined the Fourier transform for functions $f \in \mathscr{L}^p(T)$ in such a way that $\hat{f} \in \mathscr{L}^{p'}(\hat{T})$.

In brief, then, the summability theory is broadly satisfactory.

10.4.11 **Miscellaneous Notes.** So far in this chapter we have defined the Fourier transform for $\mathscr{L}^1(T)$ and for $\mathscr{L}^2(T)$. By using the properties so far established, together with the M. Riesz convexity theorem (see, for example, Section 4.24), it is possible to extend the Fourier transform yet further into a continuous linear map of $L^p(T)$ into $L^{p'}(\hat{T})$ for $1 < p < 2$ and $p' = p/(p-1)$. The classical case of this extension applies to ordinary Fourier series and is fully discussed in Zygmund [2], Vol. II, Chapter XII. The case of separated locally compact Abelian groups in general is treated by Weil [1], Chapitre VI. For $1 < p < 2$, the Fourier image of $L^p(T)$ is generally not the whole of $L^{p'}(\hat{T})$.

For the classical groups we have, in Section 5.15, discussed the extension of the Fourier transformation to arbitrary temperate distributions. It is natural to seek an extension of Bochner's theorem applying to positive definite distributions: such an extension is given by Schwartz ([2], Théorème XVIII).

There are many challenging unsolved problems connected with the finer aspects of harmonic analysis, and this even if one restricts oneself to R^1 or to the circle group $R^1/2\pi$. For a survey, see Hewitt [11] and Rudin [9]. A few of the problems are hinted at in the exercises at the end of this chapter. Besides this, one might mention that the natural question of extending the Riemann theory of general trigonometric series and integrals in one real variable (Zygmund [2], Vol. I, Chapter IX and Vol. II, Chapter XVI) has scarcely been touched.

EXERCISES

10.1 Let E be a Hilbert space, A a continuous endomorphism of E, M a closed vector subspace of E, and P the orthogonal projector onto M. Show that $A(M) \subset M$ and $A^*(M) \subset M$ if and only if $PA = AP$.

10.2 Give an example of a vector space E, a direct sum decomposition $E = M + N$, and an endomorphism u of E such that if P is the projection onto M parallel to N, one has $u(M) \subset M$ and yet $uP \neq Pu$. (Compare with the preceding exercise.)

10.3 Give an example of a continuous unitary representation of a non-Abelian compact group T that is irreducible and of dimension greater than 1.

10.4 Let T be the additive group of real numbers modulo 2π. For $s \in T$ define $U(s)$ to be that endomorphism of $L^2(T)$ that carries the function class $\dot{f}$ into the class of the function $t \to f(t - s)$.

Let H be a closed vector subspace of $L^2(T)$ that is invariant under translations (that is, $U(s)H \subset H$ for each $s \in T$).

Show that (H, U) is a continuous unitary representation of T and that if $H \neq \{0\}$, this representation is irreducible if and only if H consists solely of scalar multiples of some function e^{int} for some integer n.

Repeat the procedure on replacing T by the additive group of real numbers, and show that if $H \neq \{0\}$, then (H, U) is in this case never irreducible.

10.5 Let E and F be vector spaces and $\mathscr{E}$ and $\mathscr{F}$ irreducible sets of endomorphisms of E and of F, respectively. (Irreducibility of $\mathscr{E}$ means here that there exists no vector subspace of E, other than $\{0\}$ and E, that is left stable by every member of $\mathscr{E}$.) Suppose that $A \neq 0$ is a linear map of E into F such that

$$\mathscr{F} \cdot A = A \cdot \mathscr{E},$$

by which it is meant that for each $T \in \mathscr{F}$ there exists $S \in \mathscr{E}$ such that $TA = AS$, and that for each $S \in \mathscr{E}$ there exists $T \in \mathscr{F}$ such that the same relation holds.

Show that A is an isomorphism of E onto F. (This is one form of Schur's lemma.)

10.6 Let E be a finite dimensional vector space over the complex field, $\mathscr{E}$ an irreducible set of endomorphisms of E. Show that if T is an endomorphism of E that commutes with $\mathscr{E}$—that is, is such that $\mathscr{E} \cdot T = T \cdot \mathscr{E}$ (see the preceding exercise), then there exists a complex number λ such that $T = \lambda 1$.

Consequently, if we assume further that $\mathscr{E}$ is commutative (any two members of $\mathscr{E}$ commute), then $\dim E = 1$.

10.7 Let T be a group that possesses a right-invariant mean (see Section 3.5)—that is a positive linear form μ on the space of all bounded complex-valued functions on T such that $\mu(1) = 1$ and $\mu(f_s) = \mu(f)$ when $f_s(t) = f(ts)$. Let $t \to A(t)$ be any bounded representation of T by continuous endomorphisms of a Hilbert space H; thus $A(st^{-1}) = A(s)A(t)^{-1}$ and $\operatorname{Sup}_{t \in T} \|A(t)\| < +\infty$. Show that this representation is equivalent to a unitary representation; that is, that there exists an automorphism S of H such that $SA(t)S^{-1}$ is unitary for each $t \in T$.
[Hint: Consider the new inner product on H defined by

$$(x \mid y)_1 = \mu[(A(t)x \mid A(t)y)];$$

show that this satisfies an inequality of the type

$$c^{-1} \|x\| \leq \|x\|_1 \leq c \|x\|$$

for some $c > 0$. Some research additional to material covered in this book is necessary to show that there exists an automorphism S of H such that $(x \mid y)_1 = (Sx \mid Sy)$ for $x, y \in H$.]

10.8 Let T be a separated locally compact Abelian group, $\hat{T}$ its dual group. Let p be a strictly positive function on $\hat{T}$ whose infimum is 0, and let Ω be any nonvoid open subset of T. Show that there exists on T a continuous integrable function f with support contained in Ω and such that

$$\operatorname{Sup}\left\{\frac{|\hat{f}(\hat{t})|}{p(\hat{t})} : \hat{t} \in \hat{T}\right\} = +\infty$$

(Notice that if p is bounded away from 0 on each compact subset of $\hat{T}$, then the conclusion may be written

$$\limsup_{\hat{t} \to \infty} \frac{|\hat{f}(\hat{t})|}{p(\hat{t})} = +\infty.)$$

[Hint: Attempt a proof by contradiction, considering the seminorm

$$N(f) = \operatorname{Sup} \frac{|\hat{f}(\hat{t})|}{p(\hat{t})}$$

on the space of continuous functions vanishing outside a selected relatively compact nonvoid open subset of Ω.]

10.9 Let T be a separated locally compact Abelian group, $\hat{T}$ its dual. Show that
(a) If T is discrete, then $\hat{T}$ is compact;
(b) If T is compact, then $\hat{T}$ is discrete.

Note. In view of the duality law of Subsection 10.4.8, both assertions are in fact "if and only if" statements.

10.10 T and $\hat{T}$ are as in Exercise 10.9. Use Proposition 10.3.7 to prove the following uniqueness theorem: If μ is a bounded Radon measure on $\hat{T}$, and if the inverse Fourier transform

$$\int [t, \hat{t}]\, d\mu(\hat{t}) = 0$$

for all $t \in T$, then $\mu = 0$.

Prove that conversely this uniqueness theorem implies that the transforms $\hat{f}$ ($f \in \mathscr{L}^1(T)$) are uniformly dense in $C_0(\hat{T})$.

10.11 T is a separated locally compact Abelian group. Show that the product of two continuous positive definite functions on T is again positive definite.

10.12 T is a separated locally compact group and $f \in \mathscr{L}^p(T) \cap \mathscr{L}^{p'}(T)$, where $1 \leq p \leq \infty$ and $1/p + 1/p' = 1$. Show that the function

$$g(t) = f * \tilde{f}(t) = \int f(t')\overline{f(tt')}\, dt'$$

is continuous and positive definite on T.

10.13 T is a separated locally compact Abelian group and $f \in L^2(T)$. Show that the finite linear combinations of translates of f are dense in $L^2(T)$ if and only if the set of zeros of $\hat{f}$ is locally negligible.
[Hint: Use the Hahn-Banach theorem and the Parseval formula.]

Notes. For $T = R^1$ the result is due to Wiener ([1], p. 100), though his proof is quite different. The problem for $L^p(T)$, $p \neq 2$ and T noncompact, is much more difficult. For $p = 1$ the solution is known (Wiener [1], pp. 97–99, for $T = R^1$; for more general groups T one can apply an approach based upon Banach algebras, as, for example, in Loomis [1]). For $p = \infty$ the solution is known for the weak topology (Helson [1]). For other values of p the available results are rather fragmentary; see Segal [2], [3], Agnew [1], [2] and Edwards [4]. Similar problems have been studied in other spaces; see, for example, Schwartz [9] and [11].

10.14 Let T be a separated locally compact Abelian group, and suppose that $f \in \mathscr{L}^1(T)$ has an integrable Fourier transform $\hat{f}$. Prove that

$$f(t) = \int \hat{f}(\hat{t})[t, \hat{t}]\, d\hat{t}$$

holds almost everywhere on T.

10.15 T is as in the preceding exercise and $f \in \mathscr{L}^1(T)$. Prove that if $\hat{f} \geq 0$, then f is positive definite, and conversely.

10.16 Establish the analogue for $\mathscr{L}^2$ of the results of the last exercise.

10.17 Suppose that T is a separated locally compact Abelian group that is σ-compact and satisfies the first countability axiom. Suppose further that each function in $C_0(T)$ is of the form $\hat{f}$ for some $f \in \mathscr{L}^1(T)$. Show that T is finite.

Notes. Compare some similar results in Edwards [5], where countability restrictions are dropped. See also Edwards [10] and Helson [4].

10.18 Take $T = R^1$. Identify the summation factors $S_n(\xi)$ ($\xi \in \hat{T}$) obtained by the process described in Subsection 10.4.10 starting from the functions s_n on T

defined as follows:

(a) $s_n(t) = \left(\frac{\sin \pi nt}{\pi nt}\right)^2$,

(b) $s_n(t) = \pi^{-1}n(1 + n^2t^2)^{-1}$,

(c) $s_n(t) = 2\pi^{-1/2}n \exp(-n^2t^2)$.

(These lead to the summability methods associated with the names of Césaro, Abel-Poisson, and Gauss, respectively.)

10.19 Let T be a separated locally compact Abelian group, (μ_i) a norm-bounded directed family of bounded Radon measures on T, μ a bounded Radon measure on T. Prove that $\lim \mu_i = \mu$ weakly in the dual of $C_0(T)$ if and only if $\lim \hat{\mu}_i = \hat{\mu}$ weakly in the dual of $\mathscr{L}^1(\hat{T})$.

10.20 T is as in the preceding exercise. Let (P_i) be a directed family of functions $\hat{T}$, each continuous, positive definite, and integrable, and suppose that $\lim P_i = 1$ boundedly on $\hat{T}$. Let F be a bounded and continuous function on $\hat{T}$. Define

$$f_i(t) = \int P_i(\hat{t})F(\hat{t})\,[t, \hat{t}]\,d\hat{t}.$$

Show that F is the Fourier transform of a bounded Radon measure on T if and only if the functions f_i belong to, and are bounded in, $\mathscr{L}^1(T)$.

10.21 Let T be the circle group $(= R^1/2\pi)$, so that $\hat{T}$ is the additive group of integers. Let F be a function on $\hat{T}$ of the type

$$F(k) = A(k) + \hat{\alpha}(k),$$

where α is a Radon measure on T and

$$A(k) = \begin{cases} a & \text{if } k < 0, \\ b & \text{if } k \geq 0. \end{cases}$$

Show that if $a \neq b$, then F is not the Fourier transform of any Radon measure on T.

10.22 Adopt the notations and terminology used in Exercises 1.19–1.28. In $m(T)$, let m_1 denote the set of positive additive set functions μ satisfying $\mu(T) = 1$. Verify that

(a) m_1 is convex and compact for the topology induced by $\sigma(m(T), \ell^\infty(T))$;

(b) the set of extreme points of m_1 is precisely M (as defined in Exercise 1.26).

Deduce that to each $\mu \in m_1$ there corresponds a positive Radon measure μ^* on T^* of total mass 1 such that

$$\int_T f(t)\,d\mu(t) = \int_{T^*} f^*(t^*)\,d\mu^*(t^*)$$

for $f \in \ell^\infty(T)$ (T^* and f^* being defined as in Exercises 1.27 and 1.28).
[Hints: For last part use Theorem 10.1.7.]

Remark. The existence of such a representation formula follows also from the isomorphism between $\ell^\infty(T)$ and $C(T^*)$ established in Exercise 1.28. One may say in brief that the additive set functions of finite total variation on T correspond exactly to the Radon measures on T^*.

10.23 Let T be a separated compact Abelian group. For each n let c_n be a complex-valued function on the dual group $\hat{T}$ such that $\sum_{\hat{t} \in \hat{T}} |c_n(\hat{t})| < +\infty$.

Suppose furthermore that $\lim_n c_n = c$ pointwise on $\hat{T}$. Define the functions s_n on T by

$$s_n(t) = \sum_{\hat{t} \in \hat{T}} c_n(\hat{t}) \cdot [t, \hat{t}].$$

Suppose finally that $1 \leq p < \infty$.
Prove that the following three conditions are equivalent:
(1) $\lim_n \sum_{\hat{t} \in \hat{T}} c_n(\hat{t}) f(\hat{t})$ exists finitely for each $f \in L^p = L^p(T)$.
(2) $\text{Sup}_n \|s_n\|_{L^{p'}} < +\infty$
(3) $\lim_n s_n$ exists weakly in $L^{p'} = L^{p'}(T)$.
Show also that each of these conditions implies each of the following two:
(4) $\lim_n \sum_{\hat{t} \in \hat{T}} c_n(\hat{t}) \hat{f}(\hat{t}) [t, \hat{t}]$ exists uniformly with respect to $t \in T$, for each $f \in L^p$.
(5) There exists $s \in L^{p'}(T)$ such that $\hat{s} = c$.

Remark. If T is the circle group and $1 < p < \infty$, and if $c_n(\hat{t}) = 1$ or 0 according as $|\hat{t}| \leq n$ or $|\hat{t}| > n$, then (5) implies all the others: this is because now s_n is the nth symmetric partial sum of the Fourier series of s and is known to converge in $L^{p'}$ to s. This is no longer true if $p = 1$. (See Zygmund [2], Vol. II, p. 266.)

10.24 State and prove an analogue of the results in the preceding exercise when $L^p(T)$ is replaced by $C(T)$ and $L^{p'}$ by $M(T)$ (the space of Radon measures on T).

10.25 Investigate possible extensions of the results of Exercises 10.23 and 10.24 to the case in which T is noncompact (but merely separated locally compact and Abelian).

10.26 Assuming the Riesz convexity theorem (described in Section 4.24), show that the Fourier transformation may be extended into a continuous linear mapping of $L^p(T)$ into $L^{p'}(\hat{T})$ whenever $1 < p < 2$ and $1/p + 1/p' = 1$.
[Hint: Observe that the said transformation, regarded as being initially defined on $\mathscr{K}(T)$, is simultaneously of strong types $(1, \infty)$ and $(2,2)$.]

Bibliography

ACHIESER, N. I. AND GLASSMAN, I. M.
[1] *Theorie der Linearen Operatoren im Hilbert-Räum.* Akademie Verlag G.m.b.H., Berlin, 1954.

AGNEW, R. P.
[1] Spans in Lebesgue and uniform spaces of translations of step functions. *Bull. Am. Math. Soc.* (4) **51** (1945), 229–233.
[2] Spans in Lebesgue and uniform spaces of translations of peak functions. *Am. J. Math.* **67** (1945), 431–436.

AGNEW, R. P. AND MORSE, A. P.
[1] Extensions of linear functionals, with applications to limits, integrals, measures and densities. *Ann. Math.* **39** (1938), 20–30.

AGRANOVIČ, M. S.
[1] On partial differential equations with constant coefficients. (In Russian) *Usp. Mat. Nauk.* **16** (1961), no. 2, (98), 27–93.

ALEXANDROFF, A. D.
[1] Additive set functions in abstract space, I. *Mat. Sb. N.S.* **8** (50) (1940), 307–348.
[2] Additive set functions in abstract space, II. *Mat. Sb. N.S.* **9** (51) (1941), 563–628.
[3] Additive set functions in abstract space, III. *Mat. Sb. N.S.* **13** (55) (1943), 169–238.

ALTMAN, M.
[1] An extension to locally convex spaces of Borsuk's theorem on antipodes. *Bull. Acad. Polon. Sci.* **6** (1958), 293–295.
[2] Continuous transformations of open sets in locally convex spaces. *Bull. Acad. Polon. Sci.* **6** (1958), 297–301.

AMBROSE, W.
[1] *Lectures on Topological Groups.* Unpublished lecture notes, Univ. of Michigan, Ann Arbor, 1943.
[2] Measures on locally compact topological groups. *Trans. Am. Math. Soc.* **61** (1947), 106–121.

ARENS, R. F.
[1] Duality in linear spaces. *Duke Math. J.* (3) **14** (1947), 787–794.
[2] Approximation in, and representation of, certain Banach algebras. *Am. J. Math.* **71** (1947), 763–790.
[3] Representation of functionals by integrals. *Duke Math. J.* **17** (1950), 499–506.

ARENS, R. F. AND KELLEY, J. L.
[1] Characterization of the space of continuous functions over a compact Hausdorff space. *Trans. Am. Math. Soc.* **62** (1947), 499–508.

ARSOVE, M. G.
[1] Similar bases and isomorphisms in Fréchet spaces. *Math. Ann.* **135** (1958), 283–293.
[2] The Paley-Wiener theorem in metric linear spaces. *Pacific J. Math.* **10** (1960), 365–379.

ARSOVE, M. G. AND EDWARDS, R. E.
[1] Generalized bases in topological linear spaces. *Studia Math.* **XIX** (1960), 95–113.

ARTIN, E.
[1] *Geometric Algebra.* Interscience Publishers, Inc., New York-London, 1957.

BANACH, S.
[1] *Opérations Linéaires.* Monografje Matematyczne, **I,** Warszawa, 1932.
[2] Sur les opérations dans les ensembles abstraits et leurs applications aux équations intégrales. *Fund. Math.* **3** (1922), 133–181.

BARTLE, R. G.
[1] On compactness in functional analysis. *Trans. Am. Math. Soc.* **79** (1955), 35–57.

BARTLE, R. G., DUNFORD, N. AND SCHWARTZ, J.
[1] Weak compactness and vector measures. *Can. J. Math.* **7** (1955), 289–305.

BASS, R.
[1] *Contributions to the theory of nonlinear oscillations,* Vol. IV (ed. S. Lefschetz), p. 208.

BEHNKE, H. AND STEIN, K.
[1] Entwicklung analytischer Funktionen auf Riemannschen Flächen. *Math. Ann.* **120** (1948), 430–461.

BELLMAN, R.
[1] On an application of the Banach-Steinhaus theorem to the study of boundedness of solutions of nonlinear differential and difference equations. *Ann. Math.* **49** (1948), 515–522.

BERGMANN, S.
[1] *Sur les fonctions orthogonales de plusieurs variables complexes avec les applications à la théorie des fonctions analytiques.* Mém. Sci. Math. Fasc. **CVI,** Paris, 1947.

BESSAGA, C. AND PEŁCZYŃSKI, A.
[1] Spaces of continuous functions (IV). *Studia Math.* **XIX** (1960), 53–62.

BEURLING, A.
[1] On the spectral synthesis of bounded functions. *Acta Math.* **81** (1948), 225–238.
[2] Sur les spectres de fonctions. Analyse Harmonique: *Colloq. Intern. Centre Natl. Rech. Sci.* (15) (1949), 9–29.

BEURLING, A. AND DENY, J.
[1] Espaces de Dirichlet. I. Le cas élémentaire. *Acta Math.* **99** (1958), 203–224.
[2] Dirichlet spaces. *Proc. Natl. Acad. Sci. U.S.* (2) **45** (1959), 208–215.

BIELICKI, A.
[1] Une remarque sur la méthode de Banach-Cacciopoli-Tikhonov. *Bull. Acad Polon. Sci.* **IV** (1956), 261–268.

BIRKHOFF, G. D.
[1] Proof of the ergodic theorem. *Proc. Natl. Acad. Sci. U.S.* **17** (1931), 656–660.
[2] *Lattice Theory.* Am. Math. Soc. Colloquium Publications, **25,** 1948.
[3] Integration of functions with values in a Banach space. *Trans. Am. Math. Soc.* **38** (1935), 357–378.

BIRKHOFF, G. D. AND KELLOGG, O. D.
[1] Invariant points in function space. *Trans. Am. Math. Soc.* **23** (1922), 96–115.

Bledsoe, W. W. and Morse, A. P.
[1] Product measures. *Trans. Am. Math. Soc.* (1) **79** (1955), 173–215.

Bochner, S.
[1] *Vorlesungen über Fouriersche Integralen.* Akademie Verlag, Leipzig, 1932; reprinted by Chelsea Publishing Co., New York, 1948.
[2] Dirichlet problem for domains bounded by spheres. *Ann. Math.* Study No. 25, Princeton 1950, 24–45.
[3] Integration von Funktionen, deren Werte die Elemente eines Vectorraumes sind. *Fund. Math.* **20** (1933), 262–276.

Bochner, S. and Chandrasekharan, K.
[1] *Fourier Transforms.* Ann. Math. Study No. 19, Princeton, 1949.

Bochner, S. and Martin, W. T.
[1] *Several Complex Variables.* Princeton Univ. Press, Princeton, N.J., 1948.

Bochner, S. and Taylor, A. E.
[1] Linear functionals on certain spaces of abstractly valued functions. *Ann. Math.* (2) **39** (1938), 913–944.

Bogolyubov, N. N. and Shirkov, D. V.
[1] *Introduction to the Theory of Quantized Fields.* Interscience Publishers Inc., New York, 1959.

Bogolyubov, N. N. and Vladimirov, V. S.
[1] On some mathematical problems of quantum field theory. *Proc. Intern. Congr. Math.*, Edinburgh, 1958; Cambridge Univ. Press, London, 1960.

Bohnenblust, F. and Sobczyk, A.
[1] Extensions of functionals on complex linear spaces. *Bull. Am. Math. Soc.* **44** (1938), 91–93.

Bonsall, F. F.
[1] *Lectures on Some Fixed Point Theorems of Functional Analysis.* Tata Institute, Bombay, 1962.

Bosanquet, L. S. and Kestelman, H.
[1] The absolute convergence of a series of integrals. *Proc. London Math. Soc.* (2) **45** (1939), 88–97.

Bourbaki, N.
References [1]–[6] and [8] of the following are in the series, *Éléménts de Mathématique,* Actualités Sci. et Ind., Hermann et Cie, Paris; each title is identified by a serial number.
[1] *Théorie des Ensembles.* Chapitres I and II, No. 1212 (1954).
[1a] *Théorie des Ensembles.* Chapitre III, No. 1243 (1956).
[2] *Topologie générale.* Chapitres I and II, No. 858 (1940).
[2a] *Topologie générale.* Chapitres III and IV, No. 1143 (1951).
[3] *Algèbre.* Chapitre II, No. 1032 (1947).
[3a] *Algèbre.* Chapitres IV and V, No. 1102. (1950)
[4] *Espaces Vectoriels Topologiques.* Chapitres I and II, No. 1189 (1953).
[5] *Espaces Vectoriels Topologiques.* Chapitres III and IV, No. 1229 (1955).
[6] *Intégration.* Chapitres I–IV, 5, 6, 7, 8, Nos. 1175 (1952), 1244 (1956), 1281 (1959), 1306 (1963).
[7] Sur certains espaces vectoriels topologiques. *Ann. Inst. Fourier.* **II** (1950), 5–16.
[8] Sur les espaces de Banach. *Compt. Rend. Acad. Sci., Paris.* **206** (1938), 1071–1074.

BRACE, J. W.
[1] Transformations in Banach spaces. Dissertation. Cornell University Press, Ithaca, N.Y., 1953.

BRELOT, M.
[1] La théorie moderne du potentiel. *Ann. Inst. Fourier.* **IV** (1954), 113–140.
[2] *Eléments de la Théorie du Potentiel.* Cours de Sorbonne, Paris, 1959.
[3] *Lectures on Potential Theory.* Tata Institute, Bombay, 1960.

BROWDER, F. E.
[1] Functional analysis and partial differential equations, I. *Math. Ann.* **138** (1959), 55–79.
[1a] Functional analysis and partial differential equations, II. *Math. Ann.* **145** (1961/62), 81–226.
[2] On the spectral theory of elliptic differential equations, I. *Math. Ann.* **142** (1961), 22–130.
[3] Approximation by solutions of partial differential equations. *Am. J. Math.* (1) **LXXXIV** (1962), 134–160.
[4] Analyticity and partial differential equations, I. *Am. J. Math.* (4) **LXXXIV** (1962), 666–710.
[5] Variational boundary value problems for quasi-linear elliptic equations of arbitrary order. *Proc. Natl. Acad. Sci. U.S.* (1) **50** (1963), 31–37.

BUCK, R. C. (editor)
[1] *Studies in Modern Analysis.* Studies in Mathematics, **1.** The Math. Assoc. of America. Prentice-Hall, Inc., Englewood Cliffs, N.J., 1962.

BURKILL, J. C.
[1] *The Lebesgue Integral.* Cambridge Tracts in Math. and Math. Phys. no. 40, Cambridge Univ. Press, New York, 1951.

CACCIOPPOLI, R.
[1] Un teorema generale sull'esistenza di elementi uniti in una transformazione funzionale. *Rend. Accad. Naz. Lincei.* **11** (1930), 794–799.

CALDERÓN, A. P.
[1] Lebesgue spaces of differentiable functions and distributions. Proc. of Symposia in Pure Mathematics. **IV,** *Am. Math. Soc.* (1961), 33–49.

CALDERÓN, A. P. AND ZYGMUND, A.
[1] Singular integral operators and differential equations. *Am. J. Math.* **79** (1957), 901–921.
[2] On the theorem of Haussdorff-Young and its extensions. *Ann. Math.* Study 25, Princeton (1950), 166–188.
[3] On singular integrals. *Acta Math.* **88** (1952), 85–139.
[4] On singular integrals. *Am. J. Math.* **78** (1956), 289–309.
[5] Algebras of certain singular operators. *Am. J. Math.* **78** (1956), 310–320.
[6] A note on the interpolation of sublinear operations. *Am. J. Math.* **78** (1956), 282–288.

CARLEMAN, T.
[1] *L'intégrale de Fourier et questions qui s'y rattachent.* Publications Scientifiques de l'Institut Mittag-Leffler. Uppsala, 1944.

CARTAN, H.
[1] *Séminaire.* École Norm. Sup. 1953/4.
[2] Sur la mesure de Haar. *Compt. Rend. Acad. Sci., Paris.* **211** (1940), 759–762.
[3] *Séminaire.* École Norm. Sup. 1951/2.

[4] Idéaux et modules de fonctions analytiques de variables complexes. *Bull. Soc. Math. France.* **78** (1950), 28–64.
[5] Sur les fondements de la théorie du potentiel. *Bull. Soc. Math. France.* **69** (1941), 71–96.
[6] Théorie du potentiel newtonien: énergie, capacité, suites de potentiels. *Bull. Soc. Math. France.* **73** (1945), 74–106.
[7] Théorie générale du balayage en potentiel newtonien. *Ann. Univ. Grenoble.* **XXII** (1946), 221–280.

CARTAN, H. AND GODEMENT, R.
[1] Théorie de la dualité et analyse harmonique dans les groupes Abéliens localement compacts. *Ann. Sci. École Norm. Sup.* (3) **64** (1947), 79–99.

CHANDRASEKHARAN, K: *see* Bochner, S.

CHEVALLEY, C.
[1] *Theory of Distributions.* Lectures at Columbia Univ. Columbia University Press, New York, 1950–51.

CHOQUET, G.
[1] *Theory of Capacities.* University of Kansas Press, Lawrence, 1954.
[2] Unicité des représéntations intégrales au moyen des points extrémaux dans les cônes convexes réticulés. *Compt. Rend. Acad. Sci., Paris.* **243** (1956), 555–557.
[3] Existence des représentations intégrales au moyen des points extrémaux dans les cônes convexes. *Compt. Rend. Acad. Sci., Paris.* **243** (1956), 699–702.
[4] Limites projectives d'ensembles convexes et éléments extrémaux. *Compt. Rend. Acad. Sci., Paris.* **250** (1960), 2495–2497.
[5] Le théorème de représentation intégrale dans les ensembles convexes compacts. *Ann. Inst. Fourier* (Grenoble). **X** (1960), 333–344.
[6] Représentations intégrales dans les cônes convexes sans base compacte. *Compt. Rend. Acad. Sci., Paris.* **253** (1961), 1901–1903.
[7] Axiomatiques des mesures maximales. Applications aux cônes convexes faiblement complets. *Compt. Rend. Acad. Sci., Paris.* **255** (1962), 37–39.
[8] Étude des mesures coniques. Cônes convexes saillants faiblement complets sans génératrices extrémales. *Compt. Rend. Acad. Sci., Paris.* **255** (1962), 445–447.

CHOQUET, G. AND DENY, J.
[1] Aspects linéaires de la théorie du potentiel, I. *Compt. Rend. Acad. Sci., Paris.* **243** (1956), 222–225.
[2] Aspects linéaires de la théorie du potentiel, II. *Compt. Rend. Acad. Sci., Paris.* **243** (1956), 764–767.

CHOQUET, G. AND MEYER, P.-A.
[1] Existence et unicité des représentations intégrales dans convexes compacts quelconques. *Ann. Inst. Fourier* (*Grenoble*). **XIII** (1963), 139–154.

CHRISTIAN, R. R.
[1] On integration with respect to a finitely additive measure whose values lie in a Dedekind complete partially ordered vector space. Dissertation, Yale University Press, New Haven, Conn. 1954.

CLARKSON, J. A.
[1] Uniformly convex spaces. *Trans. Am. Math. Soc.* **40** (1936) 396–414.

CODDINGTON, E. A. AND LEVINSON, N.
[1] *Theory of Ordinary Differential Equations.* McGraw-Hill Book Co., Inc., New York-Toronto-London, 1955.

COLLINS, H. S.
[1] Completeness and compactness in linear topological spaces. *Trans. Am. Math. Soc.* **79** (1955), 256–280.

COOKE, R. G.
[1] *Infinite Matrices and Sequence Spaces.* The Macmillan Co., New York, 1950.
[2] *Linear Operators.* The Macmillan Co., New York, 1953.

COOPER, J. L. B.
[1] One-parameter semigroups of isometric operators in Hilbert space. *Ann. Math.* **48** (1947), 827–842.
[2] Symmetric Operators in Hilbert space. *Proc. London Math. Soc.* **50** (1948), 11–55.
[3] Positive definite functions of a real variable. *Proc. London Math. Soc.* (37) **10** (1960), 53–66.

CORDUNEANU, C.
[1] Sisteme diferenţiale care admit solutii mărginite. *Analele Ştiint. Univ. "Al. I. Cuza," Iasi.* (2) **8** (1957), 1–20.
[2] Une application du théorème du point fixe à la théorie des équations différentielles. *Analele Ştiint. Univ. "Al. I. Cuza," Iaşi* (2) **4** (1958), 43–47.
[3] Sur la stabilité conditionelle par rapport aux perturbations permanentes. *Acta Sci. Math. Szeged.* **19** (1958), 229–236.
[4] Asupra existenţei soluţiilor mărginite pentru uncele sisteme differentiale neliniare. *Acad. Rep. Populare Romine Fil. Iaşi Stud. Cerc. Şti. Mat.* **XI** (1960), 271–282.
[5] Sur certains systèmes différentiels non-linéaires. *Analele Ştiint. Univ. "Al. I. Cuza," Iaşi.* **6** (1960), 257–260.

COTLAR, M.
[1] *Lecture notes on integral operators.* Universidad de Buenos Aires. Facultad de Ciencias Exactas y Naturales. Departamento de Matemáticas, Buenos Aires, 1959.
[2] A unified theory of Hilbert transforms and ergodic theorems. *Rev. Mat. Cuyana.* **1** (1955), 105–167.

COURANT, R. AND HILBERT, D.
[1] *Methoden der mathematischen Physik.* **II.** Springer, Berlin, 1937.
[2] *Methods of Mathematical Physics. Vol. II. Partial Differential Equations.* Interscience Publishers, Inc., New York, 1962.

CRUM, M. M.
[1] On positive definite functions. *Proc. London Math. Soc.* (24) **6** (1956), 548–560.

DANIELL, P. J.
[1] A general form of integral. *Ann. Math.* (2) **19** (1917–1918), 279–294.

DAY, M. M.
[1] *Normed Linear Spaces.* Ergebnisse der Mathematik und ihrer Grenzgebiete, Neue Folge, Heft 21, Berlin-Göttingen-Heidelberg, 1962. 2d. printing corrected.
[2] The spaces L^p with $0 < p < 1$. *Bull. Am. Math. Soc.* **46** (1940), 816–823.
[3] Some uniformly convex spaces. *Bull. Am. Math. Soc.* **47** (1941), 504–507.

DE BRANGES, L.
[1] The Stone-Weierstrass theorem. *Proc. Am. Math. Soc.* **10** (1959), 822–824.

DE LEEUW, K. AND MIRKIL, H.
[1] Majorations dans L_∞ des opérateurs différentiels à coéfficients constants. *Compt. Rend. Acad. Sci., Paris.* **254** (1962), 2286–2288.

DENY, J.: *see also* Beurling, A. and Choquet, G.
[1] Potential theory. Lecture notes. *London Math. Soc. Intern. Conference.* (Easter, 1961).

DENY, J. AND LIONS, J. L.
[1] Les espaces du type de Beppo Levi. *Ann. Inst. Fourier.* **V** (1953–4), 305–370.

DEPRIT, A.
[1] Contributions à l'étude de l'algèbre des applications linéaires continues d'un espace localement convexe séparé; Théorie de Riesz—théorie spectrale. *Acad. Roy. Belg. Classe Sci. Mém.* Collection in 8°. (31) (1959), no. 2.

DE RHAM, G.
[1] *Variétés Différentiables.* Act. Sci. et Ind., No. 1222, Hermann et Cie, Paris, 1955.

DE RHAM, G. AND KODAIRA, K.
[1] *Harmonic Integrals.* Institute for Advanced Study, Princeton, N.J., 1950.

DIEUDONNÉ, J.
[1] La dualité dans les espaces vectorielles topologiques. *Ann. Sci. École Norm. Sup.* (3) **59** (1942), 107–139.
[2] Recent developments in the theory of locally convex vector spaces. *Bull. Am. Math. Soc.* **59** (1953), 495–512.
[3] Sur les espaces de Köthe. *J. Anal. Math.* **I** (1951), 81–115.
[4] Sur les espaces de Montel métrisables. *C.R. Acad. Sci., Paris.* **238** (1954), 194–5.
[5] Sur un théorème de Šmulian. *Arch. Math.* **III** (1952), 436–440.
[6] Natural homomorphisms in Banach spaces. *Proc. Am. Math. Soc.* (1) **1** (1950), 54–59.
[7] Sur le théorème de Lebesgue-Nikodym. *Ann. Math.* **42** (1941), 547–555.
[8] Sur le théorème de Lebesgue-Nikodym, II. *Bull. Soc. Math. France.* **72** (1944), 193–239.
[9] Sur le théorème de Lebesgue-Nikodym, III. *Ann. Univ. Grenoble* **XXIII** (1947–8) 25–53.
[10] Sur le théorème de Lebesgue-Nikodym, IV. *J. Indian Math. Soc.* **XV** (1951), 77–86.
[11] Sur le théorème de Lebesgue-Nikodym, V. *Can. J. Math.* (2) **3** (1951), 129–139.
[12] Sur le produit de composition. *Comp. Math.* (1) **12** (1954), 17–34.
[12a] Sur le produit de composition, II. *J. Math. Pures Appl.* (3) **29** (1960), 275–292.
[13] *Foundations of Modern Analysis.* Academic Press, Inc., New York, 1960.
[14] Sur la convergence des suites de mesures de Radon. *Anais Acad. Brasil. Cienc.* **23** (1951), 21–38.
[15] On biorthogonal systems. *Mich. Math. J.* **2** (1954), 7–20.
[16] *La Géométrie des Groupes Classiques.* Springer-Verlag, Berlin, Vienna, 1955.

DIEUDONNÉ, J. AND SCHWARTZ, L.
[1] La dualité dans les espaces (F) et (LF). *Ann. Inst. Fourier.* **I** (1949), 61–101.

DINCULEANU, N.
[1] Sur la représentation intégrale des certains opérateurs linéaires, I. *Compt. Rend. Acad. Sci., Paris.* **245** (1957), 1203–1205.
[2] Sur la représentation intégrale des certains opérateurs linéaires, II. *Composito Math.* **14** (1959), 1–22.
[3] Sur la représentation intégrale des certains opérateurs linéaires, III. *Proc. Am. Math. Soc.* **10** (1959), 59–68.
[4] Mesures vectorielles sur les espaces localement compacts. *Bull. Math. Soc. Sci. Math. Phys. Rep. Populare Roumaine (N.S.).* **2** (50) (1958), 137–164.

DINCULEANU, N. AND FOIAS, C.
[1] Sur la représentation intégrale des certaines opérateurs linéaires, IV, Opérateurs linéaires sur l'espace $L_{\mathscr{A}}$. *Can. J. Math.* **13** (1961), 529–556.

DIXMIER, J.
[1] Sur un théorème de Banach. *Duke Math. J.* **15** (1948), 1057–1071.
[2] Sur les bases orthonormales dans les espaces prehilbertiens. *Acta Sci. Math. Szeged.* **XV** (1953), 29–30.
[3] Les moyennes invariantes dans les semi-groupes et leurs applications. *Acta Sci. Math. Szeged.* **XII** (1950), 213–227.

DUNFORD, N.: *see also* Bartle, R. G.
[1] Integration and linear operations. *Trans. Am. Math. Soc.* **40** (1936), 474–494.
[2] Spectral theory I; Convergence to projections. *Trans. Am. Math. Soc.* **54** (1943), 185–217.
[3] Direct decompositions of Banach spaces. *Bol. Soc. Mat. Mex.* **3** (1946), 1–12.

DUNFORD, N. AND PETTIS, B. J.
[1] Linear operators on summable functions. *Trans. Am. Math. Soc.* **47** (1940), 323–392.

DUNFORD, N. AND SCHWARTZ, J. T.
[1] *Linear Operators, Part* 1*: General Theory.* Interscience Publishers, Inc., New York & London, 1958.
[2] *Linear Operators, Part 2: Spectral Theory, Self-Adjoint Operators in Hilbert Space.* Interscience Publishers Inc., New York & London, 1964.

DU PLESSIS, N.
[1] Concerning the validity of finite difference operations. *J. London Math. Soc.* **34** (1959), 208–214.

DVORETZKY, A. AND ROGERS, C. A.
[1] Absolute and unconditional convergence in normed linear spaces. *Proc. Natl. Acad. Sci. U.S.* (3) **36** (1950), 192–197.

EBERLEIN, W. F.
[1] Weak compactness in Banach spaces, I. *Proc. Natl. Acad. Sci. U.S.* **33** (1947), 51–53.
[2] Banach-Haussdorff limits. *Proc. Am. Math. Soc.* (5) **1** (1950), 662–665.
[3] Closure, convexity, and linearity in Banach spaces. *Ann. Math.* **47** (1946), 688–703.

EDELSTEIN, M.
[1] An extension of Banach's contraction principle. *Proc. Am. Math. Soc.* **12** (1) (1961), 7–10.
[2] On fixed and periodic points under contractive mappings. *J. London Math. Soc.* **37** (1) (1962), 74–79.

EDWARDS, R. E.: *see also* Arsove, M. G.
[1] On functions whose translates are independent. *Ann. Inst. Fourier.* **III** (1951), 31–72.
[2] On convex spans of translates of functions on a group. *Proc. London Math. Soc.* (3) **3** (1953), 222–242.
[3] A theory of Radon measures on locally compact spaces. *Acta Math.* **89** (1953), 133–164.
[4] The exchange formula for distributions and spans of translates. *Proc. Am. Math. Soc.* **4** no. 6 (1953), 888–894.
[5] On the functions which are Fourier transforms. *Proc. Am. Math. Soc.* **5** no. 1 (1954), 71–78.

[6] On factor functions. *Pacific J. Math.* **5** (1955), 367–378.
[7] Note on two theorems about function algebras. *Mathematika.* **4,** (1957), 138–139.
[8] The Hellinger-Toeplitz theorem. *J. London Math. Soc.* **32,** (1957), 499–501.
[9] Representation theorems for certain functional operators. *Pacific J. Math.* **7** (1957), 1333–1339.
[10] Bounded functions and Fourier transforms. *Proc. Am. Math. Soc.* **9** (1958), 440–446.
[11] Simultaneous analytic continuation. *J. London Math. Soc.* **34** (1959), 264–272.
[12] Approximation theorems for translates. *Proc. London Math. Soc.* (3) **9** (1959), 321–342.
[13] Integral bases in inductive limit spaces. *Pacific J. Math.* **10** (1960), 797–812.
[14] Algebras of holomorphic functions. *Proc. London Math. Soc.* (3) **7** (1957), 510–517.
[15] Derivatives of vector-valued functions. *Mathematika.* **5** (1958), 58–61.
[16] The stability of weighted Lebesgue spaces. *Trans. Am. Math. Soc.* **93** (1959), 369–394.

Edwards, R. E. and Hewitt, E.
[1] Pointwise limits for sequences of convolution operators. *Acta Math.* 113 (1965), 181–218.

Ehrenpreis, L.
[1] *The Theory of Distributions for Locally Compact Spaces.* Mem. Am. Math. Soc. No. 21, 1956.
[2] Solution of some problems of division I, II. *Am. J. Math.* **76** (1954), 883–903; **77** (1955), 286–292.
[3] Solution of some problems of division III. *Am. J. Math.* **78** (1956), 685–715.
[4] (with Mautner, F. J.) Some properties of distributions on Lie groups. *Pacific J. Math.* **6** (1956), 591–606.
[5] (with Mautner, F. J.) Some properties of the Fourier transform on semisimple Lie groups, II. *Trans. Am. Math. Soc.* **84** (1957), 1–55.
[6] (with Mautner, F. J.) Some properties of the Fourier transform on semisimple Lie groups, III. *Trans. Am. Math. Soc.* **90** (1959), 431–484.
[7] Analytic functions and the Fourier transform of distributions, I. *Ann. Math.* **63** (1956), 129–159, II, *Trans. Am. Math. Soc.* **89** (1958), 450–483.
[8] Mean periodic functions. *Am. J. Math.* **78** (1955), 292–328.
[9] General theory of elliptic equations. *Proc. Natl. Acad. Sci. U.S.* (1) **42** (1956), 39–41.
[10] The division problem for distributions. *Proc. Natl. Acad. Sci. U.S.* **42** (1956), 756–758.
[11] Completely inversible operators. *Proc. Natl. Acad. Sci. U.S.* **42** (1956), 945–946.
[12] Theory of infinite derivatives. *Am. J. Math.* **81** (1959), 799–845.
[13] A new proof and extension of Hartog's theorem. *Bull. Am. Math. Soc.* **67** (1961), 507–509.
[14] Analytically uniform spaces and some applications. *Trans. Am. Math. Soc.* **101** (1961), 52–74.
[15] A fundamental principle for systems of linear differential equations with constant coefficients and some of its applications. *Proc. Intern. Symp. Linear Spaces* (Jerusalem, 1960), pp. 161–174. Pergamon Press, Inc., New York, 1961.
[16] On the theory of kernels of Schwartz. *Proc. Am. Math. Soc.* **7** (1956), 713–718.

Eilenberg, S. and Steenrod, N.
[1] *Foundations of Algebraic Topology.* Princeton University Press, Princeton, N.J., 1952.

Epstein, B.
[1] *Partial Differential Equations.* McGraw-Hill Book Co., Inc., New York, 1962.

Fan, Ky.
[1] Fixed point and minimax theorems in locally convex topological linear spaces. *Proc. Natl. Acad. Sci. U.S.* **38** (1952), 121–126.

Fichtenholtz, G.
[1] Sur les fonctionelles linéaires continues au sens generalisée. *Mat. Sb.* **4** (1938), 193–213.

Fichtenholtz, G. and Kantorovich, L.
[1] Sur les opérations dans l'espace des fonctions bornées. *Studia Math.* **V** (1934), 69–98.

Foias, C.: *see* Dinculeanu, N.

Fomin, S. V.: *see* Kolmogorov, A.

Fortet, R. and Mourier, E.
[1] Loi des grandes nombres et théorie ergodique. *Compt. Rend. Acad. Sci., Paris.* **232** (1954), 923.

Fréchet, M.
[1] Sur les opérations linéaires, I. *Trans. Am. Math. Soc.* **5** (1904), 493–499.
[2] Sur les opérations linéaires, II. *Trans. Am. Math. Soc.* **6** (1905), 134–140.
[3] Sur les opérations linéaires, III. *Trans. Am. Math. Soc.* **8** (1907), 433–446.

Fredholm, I.
[1] Sur une classe d'équations fonctionelles. *Acta Math.* **27** (1903), 365–390.

Friedman, A.
[1] *Generalized Functions and Partial Differential Equations.* Prentice-Hall, Inc., Englewood Cliffs, N.J., 1963.

Friedrichs, K. O.
[1] *Mathematical Aspects of the Quantum Theory of Fields.* Interscience Publishers, Inc., New York-London, 1953.
[2] On differential operators in Hilbert spaces. *Am. J. Math.* **61** (1939), 523–544.
[3] The identity of strong and weak extensions of differential operators. *Trans. Am. Math. Soc.* **55** (1944), 132–151.

Frostman, O.
[1] *Potentiel d'équilibre et capacité des ensembles avec quelques applications à la théorie de fonctions.* Thèse (Lund, 1935).
[2] Sur le balayage des masses. *Acta Szeged.* **9** (1938), 43–51.

Fuglede, B.
[1] *Extremal Length and Closed Extensions of Partial Differential Operators.* Jul. Gjellerups Boghandel, København, 1960.

Gantmacher, V.
[1] Über swache totalstetige Operatoren. *Mat. Sb.* **7** (49) (1940), 301–308.

Gårding, L.
[1] Linear hyperbolic partial differential equations with constant coefficients. *Acta Math.* **35** (1950), 1–62.
[2] Dirichlet's problem for linear elliptic partial differential equations. *Math. Scand.* **1** (1953), 55–72.
[3] *Cauchy's Problem for Hyperbolic Equations.* Lecture notes, Univ. of Chicago Press, Chicago, 1957.

[4] *Some Trends and Problems in Linear Partial Differential Equations.* Proc. Intern. Congr. Math. Edinburgh, 1958. Cambridge Univ. Press, London, 1960.
[5] Distributions and their applications to partial differential equations. Lecture notes, *Conf. on Functional Analysis*, London, 1961.
[6] Dirichlet's problem and the vibration problem for linear elliptic partial differential equations with constant coefficients. *Proc. Symp. on Spectral Theory and Differential Problems.* Stillwater, 1951.

GÅRDING, L. AND LIONS, J. -L.
[1] Functional analysis. *Nuovo Cimento.* **XIV**, Ser. X (1959), Supplemento, 19–66.

GÅRDING, L. AND MALGRANGE, B.
[1] Opérateurs différentiels partiellement hypoelliptiques. *Compt. Rend. Acad. Sci., Paris.* **247** (1958), 2083–2085.

GARNIR, H. G.
[1] *Les Problèmes aux Limites de la Physique Mathématique.* Birkhäuser Verlag, Basel-Stuttgart, 1958.

GELFAND, I. M.
[1] Sur un lemme de la théorie des espaces linéaires. *Comm. Soc. Math., Kharkov.* **13** (1936), 35–40.
[2] Abstrakte Funktionen und lineare Operatoren. *Mat. Sb.* **4** (1938), 235–284.

GELFAND, I. M. AND RAIKOV, D.
[1] Irreducible unitary representation of locally bicompact groups. *Mat. Sb.* **13** (1943), 301–316.

GELFAND, I. M. AND ŠILOV, G.
[1] Uber verschiedene methoden der einfuhrung der topologie in die menge der maximalen ideale eines normierten ringes. *Mat. Sb.* **9** (51) (1941), 25–40.
[2] *Verallgemeinerte Funktionen (Distributionen), I.* VEB Deutscher Verlag der Wissenschaften, Berlin, 1960.
[3] *Verallgemeinerte Funktionen (Distributionen), II.* VEB Deutscher Verlag der Wissenschaften, Berlin, 1962.

GHEORGHIU, R.
[1] *Sur l'équation de Fredholm.* Thèse. Paris, 1928.

GIL DE LAMADRID, J.
[1] Completion of seminormed spaces and the Daniell process of extending an integral. *Math. Mag.* **33** (1959/60), 199–210.

GILLMAN, L. AND JERISON, M.
[1] *Rings of Continuous Functions.* D. Van Nostrand Co., Inc., Princeton, N.J. 1960.

GLASSMAN, I. M: *see* Achiezer, N. I.

GLICKSBERG, I.
[1] The representation of functionals by integrals. *Duke Math. J.* **19** (1952), 253–261.

GODEMENT, R.: *see also* Cartan, H.
[1] *L'Analyse Harmonique dans les Groupes non Abéliens.* Suppl. au Colloq. Anal. Harmonique. Nancy, 1952.
[2] Extension à un groupe quelconque des théorèmes taubériens de N. Wiener et d'un théorème de A. Beurling. *Compt. Rend. Acad. Sci., Paris.* **223** (1946), 16–18.
[3] Théorèmes taubériens et théorie spectrale. *Ann. Sci. École Norm. Sup.* (3) **64** (1947/8), 119–138.

[4] Les fonctions de type positif et la théorie des groupes. *Trans. Am. Math. Soc.* **63** (1948), 1–84.
[5] Sur la théorie des représentations unitaires. *Ann. Math.* (2) **53** (1951), 68–124.

GOFFMAN, C.
[1] *Real Functions.* Holt, Rinehart and Winston, Inc., New York, 1953.

GOODNER, D. B.
[1] Projections in normed linear spaces. *Trans. Am. Math. Soc.* **69** (1950), 89–108.

GOULD, G. G.
[1] On a class of integration spaces. *J. London Math. Soc.* **34** (1959), 161–172.

GOULD, G. G. AND MAHOWALD, M.
[1] Quasi-barrelled locally convex spaces. *Proc. Am. Math. Soc.* **11** (1960), 811–816.
[2] Measures on completely regular spaces. *J. London Math. Soc.* **37** (1962), 103–111.

GRANAS, A.
[1] Extension homotopy theorem in Banach spaces and some of its applications to the theory of nonlinear equations. *Bull. Acad. Polon. Sci.* **7** (1959), 387–394. (Russian with English summary.)

GRAVES, L. M.
[1] *The Theory of Functions of Real Variables,* 2d. ed., McGraw-Hill Book Co., Inc., New York, 1956.
[2] Topics in the functional calculus. *Bull. Am. Math. Soc.* **41** (1935), 641–662. Errata, ibid. **42** (1936), 381–382.

GROTHENDIECK, A.
[1] *Produits Tensoriels Topologiques et Éspaces Nucléaires.* Mem. Am. Math. Soc. No. 16, 1955.
[2] Sur la complétion du dual d'un espace vectoriel localement convexe. *Compt. Rend. Acad. Sci., Paris.* **230** (1950), 605–606.
[3] Critères de compacité dans les espaces généraux. *Am. J. Math.* **LXXIV** (1952), 168–186.
[4] Sur les applications linéaires faiblement compactes des espaces du type C(K). *Can. J. Math.* **5** (1953), 129–173.
[5] Sur certains espaces de fonctions holomorphes, I. J. f. die reine u. ang. Math. (1) 192 (1953), 35–64.
[6] Sur certains espaces de fonctions holomorphes, II. J. f. die reine u. ang. Math. (2) 192 (1953), 77–95.
[7] *Espaces Vectoriels Topologiques.* Instituto de Matemática Pura e Aplicada, Universidade de São Paulo, São Paulo, 1954.
[8] Sur les espaces de solutions d'une classe générale d'équations aux dérivées partielles. *J. Ann. Math.* **II** (1952/53), 243–280.
[9] Sur les espaces (F) et (DF). *Sum. Brasil. Mathematicae.* **3** (1954), 57–123.

HADAMARD, J.
[1] Sur les opérations fonctionelles. *Compt. Rend. Acad. Sci., Paris.* **136** (1903), 351–354.
[2] *Le Problème de Cauchy et les Équations aux Dérivées Partielles.* Hermann et Cie, Paris, 1932.

HAHN, H.
[1] Über Folgen linearen Operationen. *Monatsch. Math. Phys.* **32** (1922), 3–88.
[2] Über lineare Gleichungsysteme in linearen Räumen. *J. F. Math.* **157** (1927). 214–229.

HALMOS, P.
[1] *Measure Theory*. D. Van Nostrand Co., Inc., Princeton, N.J., 1950.
[2] *Introduction to Hilbert Space and the Theory of Spectral Multiplicity*. Chelsea Publishing Co., New York, 1951.
[3] *Finite Dimensional Vector Spaces* 2d. ed., D. Van Nostrand Co., Inc., Princeton, N.J., 1958.
[4] *Naive Set Theory*. D. Van Nostrand Co., Inc., Princeton, N.J., 1960.
[5] *Lectures on Ergodic Theory*. The Math. Soc. of Japan, Tokyo, 1956.

HALPERIN, I.
[1] Closures and adjoints of linear differential operators. *Ann. Math.* (4) **38** (1937), 880–919.
[2] Function spaces, *Can. J. Math.* **5** (1953), 273–288.
[3] *Introduction to the Theory of Distributions*. Univ. of Toronto Press, Toronto, 1952.

HARDY, G. H. AND LITTLEWOOD, J.
[1] A maximal theorem with function-theoretic applications. *Acta Math.* **54** (1930), 81–116.

HARDY, G. H., LITTLEWOOD, J. E. AND PÓLYA, G.
[1] *Inequalities*. Cambridge Univ. Press, New York, 1934.

HELLINGER, E. AND TOEPLITZ, O.
[1] Grundlagen für eine Theorie der unendlichen Matrizen. *Math. Ann.* **69** (1910), 289–330.

HELLY, E.
[1] Über lineare Funktionaloperationen. *S.-B. K. Akad. Wiss. Wien Math. Naterwiss.* Kl. 121 **IIa** (1912), 265–297.

HELSON, H.
[1] Spectral synthesis of bounded functions. *Ark. Mat.* **1** (1951), 497–502.
[2] On the ideal structure of group algebras. *Ark. Mat.* **2** (1952), 83–86.
[3] Isomorphisms of group algebras. *Ark. Mat.* **2** (1953), 475–487.
[4] Fourier transforms on perfect sets. *Studia Math.* **XIV** (1954/5), 209–213.

HENSTOCK, R.
[1] Difference sets and the Banach-Steinhaus theorem. *Proc. London Math. Soc.* **13** (1963), 305–321.

HERGLOTZ, G.
[1] Über Potenzreihen mit positivem, reellem Teil in Einheitskreis. *S.-B. Sachs. Akad. Wiss.* **63** (1911), 501–511.

HEUSER, H.
[1] Über die iteration Rieszcher operatoren. *Arch. Math.* **9** (1958), 202–210.
[2] Über eigenwerte und eigenlösungen symmetrisierbar finiter operatoren. *Arch. Math.* **10** (1959), 12–20.
[3] On the spectral theory of symmetric finite operators. *Trans. Am. Math. Soc.* **94** (1960), 327–336.

HEWITT, E.: *see also* Edwards, R. E.
[1] Linear functionals on spaces of continuous functions. *Fund. Math.* **37** (1950), 161–189.
[2] Integration on locally compact spaces I. *Univ. of Wash. Publ. Math.* (2) **3** (1952), 71–75.
[3] Certain generalizations of the Weierstrass approximation theorem. *Duke Math. J.* (2) **14** (1947), 419–427.

[4] Rings of real-valued continuous functions. *Trans. Am. Math. Soc.* (1) **64** (1948), 45–99.
[5] A problem concerning finitely additive measures. *Mat. Tidssk.* B (1951), 81–95.
[6] Integral representation of certain linear functionals. *Ark. Mat.* (11) **2** (1952), 269–282.
[7] The role of compactness in analysis. *Am. Math. Monthly.* (6) **67** (1960), 499–516.
[8] The assymetry of certain algebras of Fourier-Stieltjes transforms. *Mich. Math. J.* **5** (1958), 149–158.
[9] Linear functionals on almost periodic functions. *Trans. Am. Math. Soc.* **74** (1953), 303–322.
[10] Representation of functions as absolutely convergent Fourier-Stieltjes transforms. *Proc. Am. Math. Soc.* (4) **4** (1953), 663–670.
[11] *Abstract Harmonic Analysis. Surveys in Applied Mathematics IV.* 107–168. Wiley and Sons, Inc., New York, 1958.
[12] Two notes on measure theory. *Bull. Am. Math. Soc.* **49** (1943), 719–721.

HEWITT, E. AND ROSS, K. A.
[1] *Abstract Harmonic Analysis, I.* Grundlehren der Math. Wiss., Band 115. Springer-Verlag, Heidelberg, 1963.

HEWITT, E. AND YOSIDA, K.
[1] Finitely additive measures. *Trans. Am. Math. Soc.* (1) **72** (1952), 46–66.

HEWITT, E. AND ZUCKERMAN, H. S.
[1] Integration in locally compact spaces II. *Nagoya Math. J.* **3** (1951), 7–22.
[2] Some theorems on lacunary Fourier series, with extensions to compact groups. *Trans. Am. Math. Soc.* (1) **93** (1959), 1–19.
[3] Finite-dimensional convolution algebras. *Acta Math.* **93** (1955), 67–119.
[4] The l^1-algebra of a commutative semigroup. *Trans. Am. Math. Soc.* (1) **83** (1956), 70–97.
[5] Structure theory for a class of convolution algebras. *Pacific J. Math.* (1) **7** (1957), 913–941.

HILBERT, D.
[1] Grundzuge einer allgemeinen Theorie der linearen Integralgleichungen, I–VI:
I, *Nachr. Akad. Wiss. Göttingen Math.-Phys. Kl.* (1904), 49–91.
II, ibid. (1905), 213–259.
III, ibid. (1905), 307–338.
IV, ibid. (1906), 157–227.
V, ibid. (1906), 439–480.
VI, ibid. (1910), 355–417.
Published as a book by Teubner Verlagsgesellschaft, m.b.H, Leipzig, 1912; reprinted by Chelsea Publ. Co., New York, 1952.

HILDEBRANDT, T. H.
[1] Integration in abstract spaces. *Bull. Am. Math. Soc.* **59** (1953), 111–139.

HILLE, E.
[1] *Functional Analysis and Semigroups.* Am. Math. Soc. Colloquium Publications **31,** New York, 1948.

HILLE, E. AND TAMARKIN, J. D.
[1] On the characteristic values of linear integral equations. *Acta Math.* **57** (1931), 1–76.

HIRATA, Y.
[1] On convolutions in the theory of distributions. *J. Sci. Hiroshima Univ.* Ser. A **22** (1958), 89–98.

HIRATA, Y. AND OGATA, H.
[1] On the exchange formula for distributions. *J. Sci. of Hiroshima Univ.* Ser. A **22** (1958), 147–152.

HOFFMAN, K.
[1] *Banach Spaces of Analytic Functions.* Prentice-Hall, Inc., Englewood Cliffs, N.J., 1962.

HÖRMANDER, L.
[1] On the theory of general partial differential operators. *Acta Math.* **94** (1955), 160–248.
[2] On the interior regularity of the solutions of partial differential equations. *Comm. Pure Appl. Math.* **11** (1958), 197–218.
[3] On the interior regularity of the solutions of boundary problems. *Acta Math.* **99** (1958), 225–264.
[4] On the division of distributions by polynomials. *Ark. Mat.* **3** (1958), 555–568.
[5] Local and global properties of the fundamental solutions. *Math. Scand.* **5** (1957), 27–39.
[6] Definitions of maximal differential operators. *Ark. Mat.* **3** (1958), 501–504.
[7] *Linear Partial Differential Operators.* Grundlehren der Math. Wiss., Band 116. Springer-Verlag, Heidelberg, 1963.
[8] Hypoelliptic convolution equations. *Math. Scand.* **9** (1961), 178–184.
[9] On the range of convolution operators. *Ann. Math.* (2) **76** (1962), 148–170.

HUNT, G. A.
[1] Markoff processes and potentials, I. *Illinois J. Math.* **1** (1957), 44–93.
[2] Markoff processes and potentials, II. *Illinois J. Math.* **1** (1957), 316–369.
[3] Markoff processes and potentials, III. *Illinois J. Math.* **2** (1958), 151–213.

HUSAIN, T.
[1] S-Spaces and the open mapping theorem. *Pacific J. Math.* (1) **12** (1962), 253–271.

HUSAIN, T. AND MAHOWALD, M.
[1] Barrelled spaces and the open mapping theorem. *Proc. Am. Math. Soc.* (3) **13** (1962), 423–424.

IINO, R.
[1] Sur les dérivations dans les espaces vectoriels topologiques sur le corps des nombres complexes, I, II, and III. *Proc. Japan. Acad.* **35** (1959), 343–348, 530–535; **36** (1960), 27–32.

IONESCU TULCEA, A. AND IONESCU TULCEA, C.
[1] On the decomposition and integral representation of continuous linear operators. *Ann. Math. Pura Appl.* (4) **53** (1961), 63–87.
[2] On the lifting property, I. *J. Math. Anal. Appl.* **3** (1961), 537–546.
[3] On the lifting property, II. Representations of linear operators on spaces L^r_E, $1 \leq r < \infty$. *J. Math. Mech.* **11** (1962), 773–795.

ISEKI, K.
[1] On a theorem on function spaces of A. Grothendieck. *Proc. Japan. Acad.* **33** (1957), 605–607.
[2] On complete orthonormal sets in Hilbert space. *Proc. Japan. Acad.* **33** (1957), 450–452.

ISEKI, K. AND KASAHARA, S.
[1] Some properties of convex sets in linear spaces. *Rev. Fac. Cien. Lisboa Univ.* (2) **3** (1954/5), 238–242.

JERISON, M.: *see* Gillman, L.

JERISON, M. AND RUDIN, W.
[1] Translation-invariant functionals. *Proc. Am. Math. Soc.* (3) **13** (1962), 417–423.

JOHN, F.
[1] General properties of solutions of elliptic partial differential equations. *Symposium on Spectral Theory and Differential Problems.* Mathematics Dept., Oklahoma Agricultural and Mechanical College, Stillwater, Oklahoma (1951), 113–175.

KACZMARZ, S. AND STEINHAUS, H.
[1] *Theorie der Orthogonalreihen.* Chelsea Publ. Co., New York, 1951.

KAHANE, J. P. AND SALEM, R.
[1] Sur les ensembles linéaires ne portant pas de pseudomesures. *Compt. Rend. Acad. Sci., Paris.* **243** (1956), 1185–1187.

KAKUTANI, S
[1] Two fixed-point theorems concerning bicompact convex sets. *Proc. Imp. Acad., Tokyo.* **14** (1938), 242–245.
[2] A proof of Schauder's theorem. *J. Math Soc. Japan.* (1) **3** (1951), 228–231.
[3] Rings of analytical functions. *Lectures on Functions of a Complex Variable.* Univ. of Michigan Press, Ann Arbor, 1955, 71–83.
[4] Iteration of linear operators in complex Banach spaces. *Proc. Imp. Acad., Tokyo.* **14** (1938), 295–300.
[5] Concrete representations of abstract (L)-spaces and the mean ergodic theorem. *Ann. Math.* (2) **42** (1941), 523–537.
[6] Concrete representations of abstract (M)-spaces. (A characterization of the space of continuous functions.) *Ann. Math.* (2) **42** (1941), 994–1024.
[7] A generalization of Brouwer's fixed point theorem. *Duke Math. J.* **8** (1941), 457–459.

KAKUTANI, S. AND KLEE, V. L.
[1] The finite topology of a linear space. *Arch. Math.*, Fasc. 1, **XIV** (1963), 55–58.

KAKUTANI, S. AND KODAIRA, K.
[1] Über das Haarsche Mass in der lokal bikompakten Gruppe. *Proc. Imp. Acad., Tokyo.* **20** (1944), 444–450.

KANTOROVICH, L. V: *see also* Fichtenholtz, G.
[1] The method of successive approximations for functional equations. *Acta Math.* **71** (1939), 63–97.

KANTOROVICH, L. V., PINSKER, A. G. AND VULIKH, B. Z.
[1] *Functional Analysis in Partially Ordered Spaces.* Gosudarstr. Izdat. Tehn.-Teor. Lit., Moscow-Leningrad, 1950.

KANTOROVICH, L. V. AND VULIKH, B. Z.
[1] Sur la représentation des opérations linéaires. *Compositio Math.* **5** (1938), 119–165.

KAPLAN, S.
[1] Cartesian products of reals. *Am. J. Math.* **74** (1952), 936–954.

KAPLANSKY, I.
[1] *Functional Analysis.* Surveys in Applied Math. IV, 3–30. John Wiley and Sons, Inc., New York, 1958.
[2] The structure of certain operator algebras. *Trans. Am. Math. Soc.* **70** (1951), 219–255.
[3] The Weierstrass theorem in fields with valuations. *Proc. Am. Math. Soc.* **1** (1950), 356–357.

KASAHARA, S: *see* Iseki, K.

KELLOGG, O. D.: *see also* Birkhoff, G. D.
[1] *Foundations of Potential Theory.* Frederick Ungar Publishing Co., New York, 1929.

KELLEY, J. L: *see also* Arens, R. F.
[1] *General Topology.* D. Van Nostrand Co., Inc., Princeton, N.J., 1955.
[2] Convergence in topology. *Duke Math. J.* **17** (1950), 277–283.
[3] Banach spaces with the extension property. *Trans. Am. Math. Soc.* **72** (1952), 323–326.
[4] Hypercomplete linear topological spaces. *Mich. Math. J.* **5** (1958), 235–246.
[5] Note on a theorem of Krein and Milman. *J. Osaka Inst. of Sci. Tech.* **3** (1951), 1–2.

KELLEY, J. L. AND NAMIOKA, I. (and co-authors)
[1] *Linear Topological Spaces.* D. Van Nostrand Co. Inc., Princeton, N.J., 1963.

KENDALL, D. G.
[1] On infinite doubly stochastic matrices and Birkhoff's problem III. *J. London Math. Soc.* **35** (1960), 81–84.

KESTELMAN, H: *see* Bosanquet, L. S.

KHINCHIN, A. I.
[1] *Mathematical Foundations of Statistical Mechanics.* Dover Publications, New York, 1959.

KIST, J.
[1] Locally *o*-convex spaces. *Duke Math. J.* **25** (1958), 569–582.

KLEE, V. L: *see also* Kakutani, S.
[1] Convex sets in linear spaces. *Duke Math. J.* **18** (1951), 443–466.
[2] Convex sets in linear spaces, II. *Duke Math. J.* **18** (1951), 875–893.
[3] Invariant extensions of linear functionals. *Pacific J. Math.* **4** (1954), 37–46.
[4] Leray-Schauder theory without local convexity. *Math. Ann.* **141** (1960), 286–296.
Corrections to "Leray-Schauder theory without local convexity." *Math. Ann.* **145** (1962), 464–465.

KODAIRA, K: *see* de Rham, G. and Kakutani, S.

KOIZUMI, S.
[1] On the singular integrals, I–VI. *Proc. Japan Acad.* **34** (1958), 193–198, 235–240, 594–598, 653–656; **35** (1959), 1–6, 323–328.
[2] On the Hilbert Transform, I and II. *J. Fac. Sci. Hokkaido Univ.* Ser. I, **XIV** (1959), 153–224; **XV** (1960), 93–130.

KOLMOGOROV, A.
[1] Zur Normierbarkheit eines allgemeinen topologischen linearen Räumes. *Studia Math.* **V** (1934), 29–33.

KOLMOGOROV, A. AND FOMIN, S. V.
[1] *Introduction to Functional Analysis.* Vols. 1 and 2. Graylock Press, New York, 1957 and 1961.

KÖNIG, H.
[1] Neue begründung der theorie der distributionen von L. Schwartz. *Math. Nach.* (3) **9** (1953), 129–148.
[2] Multiplikation von distributionen. *Math. Ann.* **128** (1955), 420–452.

KOREVAAR, J.

[1] Distributions defined by fundamental sequences, I–V. *Ned. Akad. Wetenschap. Proc.* A **58** (1955).

KÖTHE, G.

[1] Uber zwei Sätze von Banach. *Math. Z.* **53** (1950), 203–209.

[2] Zur theorie der kompakten operatoren in lokalconvexen Räumen. *Port. Math.* **13** (1954), 97–104.

[3] A teoria dos espaços localemente convexos e as suas appliçacões à Análise. *Acad. das Cien. de Lisboa.* 1954.

[4] Die Quotientenräume eines linearen vollkommenen Räumes. *Math. Z.* **51** (1947), 17–55.

[5] *Topologische lineare Räume* I. Die Grund. der Math. Wiss. Band 107, Springer-Verlag, Berlin, Vienna, 1960.

[6] Une caractérisation des espaces bornologiques. *Colloque sur l'Analyse Functionelle,* Louvain (1960), 39–45. *Librairie Universitaire,* Louvain (1961).

KREIN, M. G.

[1] Sur quelques questions de la géometrie des ensembles convexes situés dans un espace linéaire normé et complet. *Dokl. Akad. Nauk SSSR* (*N.S.*). **14** (1937), 5–7.

KREIN, M. G. AND MILMAN, D. P.

[1] On extreme points of regularly convex sets. *Studia Math.* **IX** (1940), 123–138.

KREIN, M. G. AND ŠMULIAN, V.

[1] On regularly convex sets in the space conjugate to a Banach space. *Ann. Math.* (2) **41** (1940), 556–583.

KURATOWSKI, C.

[1] *Topologie, I.* Monografje Matematyczne, **III,** Warszawa-Lwów, 1933.

LALESCO, T.

[1] Un théorème sur les noyeaux composés. *Bull. Acad. Roumaine.* (1915), 271–272.

LEFSCHETZ, S.

[1] *Introduction to Topology.* Princeton Univ. Press, Princeton, N.J., 1949.

[2] *Algebraic Topology.* Am. Math. Soc. Colloquium Publications, **27,** New York, 1942.

LERAY, J.

[1] *Hyperbolic Differential Equations.* Inst. for Advance Study, Princeton, N.J., 1953.

[2] Valeurs propres et vecteurs propres d'un endomorphisme complètement continu d'un espace vectoriels à voisinage convex. *Acta Sci. Math. Szeged.* **12** (1950), 177–186.

[3] La théorie des points fixes et ses applications en analyse. *Proc. Intern. Congress Math.,* Cambridge, Mass. (1950), 202–208.

LERAY, J. AND SCHAUDER, J.

[1] Topologie et équations fonctionelles. *Ann. Sci. École Norm. Sup.* (3) **51** (1934), 45–78.

LEVIN, V. L.

[1] On a class of locally convex spaces. *Soviet Mat. Dokl.* (4) **3** (July, 1962), 929–931.

[2] Conditions for B-completeness of ultrabarrelled and barrelled spaces. *Soviet Mat. Dokl.* (4) **3** (July, 1962), 984–985.

LEVINSON, N: *see under* Coddington, E. A.

LIGHTHILL, M. J.
[1] *An Introduction to Fourier Analysis and Generalized Functions.* Cambridge Univ. Press, New York, 1958.

LIONS, J. L: *see also* Deny, J. and Gårding, L.
[1] Problèmes aux limites en théorie des distributions. *Acta Math.* **94** (1955), 13–153.
[2] Supports de produits de composition. *Compt. Rend. Acad. Sci., Paris.* **232** (1951), 1530–1532.
[3] Supports de produits de composition. *Compt. Rend. Acad. Sci., Paris.* **232** (1951), 1622–1624.
[4] Supports dans la transformation de Laplace. *J. Anal. Math.* **II** (1952/3), 369–380.
[5] Problèmes aux limites de type mixte. *Second colloque sur les équations aux dérivées partielles, C.B.R.M.* (*Centre Belge de la Recherche Mathématique*), *Bruxelles, 1954.*
[6] Sur quelques problèmes aux limites relatifs à des opérateurs différentiels elliptiques. *Bull. Soc. Math. France.* **83** (1955), 225–250.
[7] Sur les problèmes aux limites du type dérivée oblique. *Ann. Math.* (2) **64** (1956), 207–239.
[8] Opérateurs de Delsarte et Problèmes Mixtes. *Bull. Soc. Math. France.* **84** (1956), 9–95.
[9] Problèmes mixtes abstracts. *Proc. Intern. Congr. Math.* Edinburgh, 1958; Cambridge Univ. Press, New York, 1960.

LITTLEWOOD, J. E.: *see* Hardy, G. H.

LIVINGSTONE, A. E.
[1] The space H^p, $0 < p < 1$, is not normable. *Pacific J. Math.* **3** (1953), 613–616.

LJUSTERNIK, L. A. and SOBOLEFF, W. I.
[1] *Elemente der Funktionalanalysis.* Akademie-Verlag G.m.b.H., Berlin, 1955.

LOJASIEWICZ, S.
[1] Division d'une distribution par une fonction analytique de variables réelles. *Compt. Rend. Acad. Sci., Paris.* **246** (1958), 683–686.
[2] Sur le problème de la division. *Studia Math.* **XVIII** (1959), 87–136.

LOOMIS, L. H.
[1] *Abstract Harmonic Analysis.* D. Van Nostrand Co., Inc., Princeton, N.J., 1953.
[2] Abstract congruence and the uniqueness of Haar measure. *Ann. Math.* (2) **46** (1945), 348–355.
[3] Haar measure in uniform structures. *Duke Math. J.* **16** (1949), 193–208.
[4] Linear functionals and content. *Am. J. Math.* **76** (1954), 168–182.

LORCH, E. R.
[1] *Spectral Theory.* University Texts in the Math. Sciences. Oxford Univ. Press, New York, 1962.

LORENTZ, G. G. AND WERTHEIM, D. G.
[1] Representation of linear functionals on Köthe spaces. *Can. J. Math.* **5** (1953), 568–575.

LOVE, E. R.
[1] A Banach space of distributions (I). *J. London Math. Soc.* **32** (1957), 483–498.

LUXEMBURG, W. A. J.
[1] *Banach Function Spaces.* Proefschrift, Delft, 1955.
[2] On closed linear subspaces and dense linear subspaces of locally convex topological linear spaces. *Proc. Intern. Symp. on Linear Spaces, Jerusalem.* (July 1960), 307–318.

LUXEMBURG, W. A. J. AND ZAANEN, A. C.
[1] Conjugate spaces of Orlicz spaces. *Koninkl. Ned. Akad. Wetenschap.* A(2) **59** (1956), 217–228.

LYUBARSKIĬ, G. YA.
[1] *The Application of Group Theory in Physics.* Pergamon Press, Inc., New York, 1960.

MAAK, W.
[1] *Fastperiodische Funktionen.* Die Grundlehren der Math. Wiss., Band LXI. Springer-Verlag, Berlin, 1950.

MACKEY, G. W.
[1] Note on a theorem of Murray. *Bull. Am. Math. Soc.* (4) **52** (1946), 322–325.
[2] On infinite dimensional linear spaces. *Trans. Am. Math. Soc.* **57** (1945), 155–207.
[3] On convex topological linear spaces. *Trans. Am. Math. Soc.* **60** (1946), 519–537.
[4] Functions on locally compact groups. *Bull. Am. Math. Soc.* **56** (1950), 385–412.
[5] Equivalence of a problem in measure theory to a problem in the theory of vector lattices. *Bull. Am. Math. Soc.* **50** (1944), 719–722.

MAHOWALD, M: *see also* Gould, G. and Husain, T.
[1] Barrelled spaces and the closed graph theorem. *J. London Math. Soc.* **36** (1961), 108–111.

MALGRANGE, B: *see also* Gårding, L.
[1] Existence et approximation des solutions des équations aux dérivées partielles et des équations de convolution. Thèse, Paris, 1956.
[2] Sur une classe d'opérateurs différentiels hypoelliptiques. *Bull. Soc. Math. France.* **85** (1957), 283–306.
[3] Sur les équations de convolution. *Rend. Seminario Mat. Univ. Torino.* **19** (1959/60), 19–27.

MALLIAVIN, P.
[1] Sur l'impossibilité de la synthèse spectrale dans une algèbre de fonctions presque périodiques. *Compt. Rend. Acad. Sci., Paris.* **248** (1959), 1756–1759.
[2] Sur l'impossibilité de la synthèse spectrale sur la droite. *Compt. Rend. Acad. Sci., Paris.* **248** (1959), 2155–2157.
[3] Impossibilité de la synthèse spectrale dans les groupes abéliens non compacts. *Publ. Math. Inst. Hautes Études Sci. Paris.* (1959), 61–68.

MARCINKIEWICZ, J.
[1] Sur l'interpolation d'opérations. *Compt. Rend. Acad. Sci., Paris.* **208** (1939), 1272–3.

MARKOV, A.
[1] On mean values and exterior densities. *Mat. Sb.* (46) **4** (1938), 165–191.
[2] Quelques théorèmes sur les ensembles abéliens. *Dokl. Akad. Nauk. SSSR* (*N.S.*). **10** (1936), 311–314.

MARKUSHEVICH, A. I.
[1] Sur les bases (au sens large) dans les espaces linéaires. *Dokl. Akad. Nauk* (*N.S.*). **41** (1943), 227–229.
[2] On bases in the space of analytic functions. *Mat. Sb.* (59) **17** (1945), 221–252.

MARTIN, W. T: *see* Bochner, S.

MASSERA, J. L. AND SCHÄFFER, J. J.
[1] Linear differential equations and functional analysis, I. *Ann. Math.* **67** (1958), 517–573.
[2] Linear differential equations and functional analysis, II. Equations with periodic coefficients. *Ann. Math.* **69** (1959), 88–104.

[3] Linear differential equations and functional analysis, III. Lyapunov's second method in the case of conditional stability. *Ann. Math.* **69** (1959), 535–574.
[4] Linear differential equations and functional analysis, IV. *Math. Ann.* **139** (1960), 287–342.

MAUTNER, F. J.: *see* Ehrenpreis, L.

MAZUR, S. AND ORLICZ, W.
[1] On linear methods of summability. *Studia Math.* **XIV** (1955), 129–160.

McKINSEY, J. C. C.
[1] *Introduction to the Theory of Games.* McGraw-Hill Book Company, Inc., New York, 1952.

McSHANE, E. J.
[1] *Integration.* Princeton Univ. Press, Princeton, N.J., 1947.
[2] *Order Preserving Maps and Integration Processes.* Ann. Math. Study No. 31. Princeton Univ. Press, Princeton, N.J., 1953.

MEHDI, M. R.
[1] Continuity of seminorms on topological vector spaces. *Studia Math.* **XVIII** (1959), 81–86.

MEYER, P.-A.: *see* Choquet, G.

MIKUSIŃSKI, J.
[1] Une définition de distributions. *Bull. Acad. Polon. Sci.* Cl. III, **3** (1955), 589–591.
[2] Sur la méthode de généralisation de M. Laurent Schwartz et sur la convergence faible. *Fund. Math.* **35** (1948), 235–239.
[3] Sur les fondements du calcul opératoire. *Studia Math.* **XI** (1949), 41–70.
[4] L'anneau algébrique et ses applications dans l'analyse fonctionelle. *Ann. Univ. Mariae Curie-Sklodowska.* **II** (1), section A (1947), 1–48; ibid. **III** (1), section A (1949), 1–82.
[5] *Operational Calculus.* Pergamon Press, Inc., New York, 1959.

MILMAN, D. P.: *see* Krein, M. G.

MILMAN, D. P. AND RUTMAN, M. A.
[1] On a more precise theorem about the completeness of the system of extremal points of a regularly convex set. *Dokl. Akad. Nauk SSSR* (*N.S.*). **60** (1948), 25–86.

MIRANDA, C.
[1] *Equazioni alle derivate parziali di tipo ellittico.* Ergeb. der Math. und ihrer Grenzgebeite (N.F.) Heft 2. Springer-Verlag, Berlin, 1955.

MIRKIL, H: *see* de Leeuw, K.

MIZOHATA, S.
[1] Hypoellipticité des équations paraboliques. *Bull. Soc. Math. France.* **85** (1957), 15–50.

MOORE, E. H.
[1] Definition of limit in general integral analysis. *Proc. Natl. Acad. Sci. U.S.* **1** (1915), 628 et seq.

MOORE, E. H. AND SMITH, H. L.
[1] A general theory of limits. *Am. J. Math.* **44** (1922), 102–121.

MORGENSTERN, O. *see* Neumann, J. von

MORSE, A. P: *see also* Bledsoe, W. W.
[1] Squares are normal. *Fund. Math.* **36** (1949), 35–39.

MOURIER, EDITH.: *see also* Fortet, R.
[1] Eléments aléatoires dans un espace de Banach. *Ann. Inst. Henri Poincaré.* **XIII** (1952), 161–244.

MYERS, S. B.
[1] Normed linear spaces of continuous functions. *Bull. Am. Math. Soc.* **56** (1950), 233–241.

NACHBIN, L.
[1] *Topological Vector Spaces.* Rio de Janeiro, 1948.
[2] Topological vector spaces of continuous functions. *Proc. Natl. Acad. Sci. U.S.* **40** (1954), 471–474.
[3] A theorem of the Hahn-Banach type for linear transformations. *Trans. Am. Math. Soc.* **68** (1950), 28–46.

NAGUMO, M.
[1] Degree of mapping in convex linear topological spaces. *Am. J. Math.* **73** (1951), 497–511.

NAGY, B. Sz: *see* Riesz, F.

NAǏMARK, M. A.
[1] *Normed Rings.* P. Noordhoff Ltd., Groningen, The Netherlands, 1959.

NAKAMURA, M.
[1] Complete continuities of linear operators. *Proc. Japan Acad.* **27** (1951), 544–547.

NAKANO, H.
[1] *Topology and Topological Linear Spaces.* Maruzen Co., Ltd., Tokyo, 1951.
[2] *Modern Spectral Theory.* Maruzen Co., Ltd., Tokyo, 1950.

NAMIOKA, I: *see* Kelley, J. L.

NATANSON, I. P.
[1] *Theory of Functions of a Real Variable.* Frederick Ungar Publishing Co., New York, 1955.

NEUMANN, J. VON
[1] *Mathematical Foundations of Quantum Mechanics.* Princeton Univ. Press, Princeton, N.J., 1955.
[2] On complete topological spaces. *Trans. Am. Math. Soc.* **37** (1935), 1–20.
[3] *Functional Operators.* Princeton Univ. Press, Princeton, 1933–35.
[4] The uniqueness of Haar's measure. *Mat. Sb.* **1** (1936), 721–734.
[5] *Lectures on Invariant Measures.* Princeton Univ., 1940 (unpublished).
[6] Proof of the quasi-ergodic hypothesis. *Proc. Natl. Acad. Sci. U.S.* **18** (1932), 70–82.

NEUMANN, J. VON, AND MORGENSTERN, O.
[1] *Theory of Games and Economic Behaviour.* 2d. ed. Princeton Univ. Press, Princeton, N.J., 1947.

NEWMAN, D. J.
[1] The non-existence of projections from L^1 to H^1. *Proc. Am. Math. Soc.* **12** (1961), 98–99.

NEWNS, W. F.
[1] On the representation of analytical functions by infinite series. *Phil. Trans. Roy. Soc., London.* (A) **245** (1953), 429–468.

NIKODYM, O.
[1] Sur une généralisation des intégrales de M. J. Radon. *Fund. Math.* **15** (1930), 131–179.

NIRENBERG, L.
[1] Remarks on strongly elliptic partial differential equations. *Comm. Pure Appl. Math.* **8** (1955), 649–675.

OGATA, H.: *see* Hirata, Y.

OHTSUKA, M.
[1] On potentials in locally compact spaces. *J. Sci. Hiroshima Univ.* (2) **25** (1961) 135–353.

ORLICZ, W: *see also* Mazur, S.
[1] Beiträge zur Theorie der Orthogonalentwicklungen. *Studia Math.* **I** (1928), 1–39.

ORLICZ, W. AND PTÁK, V.
[1] Some remarks on Saks spaces. *Studia Math.* **XVI** (1957), 56–68.

PALEY, R. E. A. C. AND WIENER, N.
[1] *Fourier Transforms in the Complex Domain.* Am. Math. Soc. Colloquium Publications, **19,** New York, 1934.

PEANO, G.
[1] Intégration par séries des équations différentielles linéaires. *Math. Ann.* **32** (1888), 450–456.
[2] Démonstration de l'intégrabilité des équations différentielles ordinaires. *Math. Ann.* **37** (1890), 182–228.

PEETRE, J.
[1] Une charactérisation abstraite des opérateurs différentiels. *Math. Scand.* **7** (1959), 211–218. Rectification à l'article "Une charactérisation abstraite des opérateurs différentiels." *Math. Scand.* **8** (1960), 116–120.
[2] Théorèmes de régularité pours quelques classes d'opérateurs différentiels. *Medd. Lunds Univ. Mat. Sem.* **16** (1959).

PELCZYŃSKI, A.: *see* Bessaga, C.

PETROWSKY, I. G.
[1] Sur l'analyticité des solutions des systèmes d'équations différentiels. *Mat. Sb.* (47) **5** (1939), 3–68.
[2] On some problems in the theory of partial differential equations. *Usp. Mat. Nauk.* **1** (3–4), 44–70; Am. Math. Soc. Transl. (12) (1946).

PETTIS, B. J: *see also* Dunford, N.
[1] On continuity and openness of homomorphisms in topological groups. *Ann. Math.* **52** (1950), 293–308.
[2] On integration in vector spaces. *Trans. Am. Math. Soc.* **44** (1938), 277–304.
[3] Differentiation in Banach spaces. *Duke Math. J.* **5** (1939), 254–269.

PHILLIPS, R. S.
[1] On linear transformations. *Trans. Am. Math. Soc.* **48** (1940), 516–541.
[2] On weakly compact subsets of a Banach space. *Am. J. Math.* **65** (1943), 108–136.

PICARD, E.
[1] Mémoire sur la théorie des équations aux dérivées partielles et la méthode des approximations successives. *J. Math.* (4) **6** (1890), 145–210.

PINSKER, A. G: *see* Kantorovich, L. V.

PLANCHEREL, M.
[1] Contributions à l'étude de la représentation d'une fonction arbitraire par les intégrales définies. *Rend. Circ. Mat., Palermo.* **30** (1910), 289–335.

PÓLYA, G.: *see* Hardy, G. H.

PONTRYAGIN, L. S.
[1] *Topological Groups.* Princeton Univ. Press, Princeton, N.J., 1946.

PRICE, G. B.
[1] The theory of integration. *Trans. Am. Math. Soc.* **47** (1940), 1–50.

PTÁK, V: *see also* Orlicz, W.
[1] Completeness and the open mapping theorem. *Bull. Soc. Math. France.* **86** (1958), 41–74.
[2] The Principle of Uniform Boundedness and the Closed Graph Theorem. *Czech. Math. J.* (87) **12** (1962), 523–528.
[3] On the closed graph theorem. *Czech. Math. J.* (84) **9** (1959), 523–527.
[4] On complete topological linear spaces. *Čehoslovack. Mat. Ž.* (78) **3** (1953), 285–290, 301–364.
[5] On a theorem of W. F. Eberlein. *Studia Math.* **XIV** (1954), 276–284.
[6] Weak compactness in convex topological linear spaces. *Čehoslovack. Mat. Z.* **4** (79) (1954), 175–186.

RADON, J.
[1] Theorie und Anwendungen der absolut additiven Mengenfunktionen. *S.-B. Akad. Wiss. Wien.* **128** (1913), 1295–1438.

RAIKOV, D. A: *see also* Gelfand, I. M.
[1] Harmonic analysis on commutative groups with Haar measure and the theory of characters. *Tr. Mat. Inst. Steklov.* **14** (1945); German translation in *Sowjet Arbeiten zur Funktionanalysis.* Verlag Kultur und Fortschr., Berlin 1954, 11–87.

RAKOTCH, E.
[1] A note on contractive mappings. *Proc. Am. Math. Soc.* (3) **13** (1962), 459–465.

RAUCH, H. E.
[1] Harmonic and analytic functions of several variables and the maximal theorem of Hardy and Littlewood. *Can. J. Math.* **8** (1956), 171–183.

REITER, H.
[1] Über L^1-Räume auf Gruppen, I. *Monatsh. Math.* (2) **58** (1954), 73–76.
[2] Uber L^1-Räume auf Gruppen, II. *Monatsh. Math.* (3) **58** (1954), 172–180.
[3] Investigations in harmonic analysis. *Trans. Am. Math. Soc.* **73** (1952), 401–427.
[4] On a certain class of ideals in the L^1-algebra of a locally compact Abelian group. *Trans. Am. Math. Soc.* **75** (1953), 505–509.
[5] Contributions to harmonic analysis. *Acta Math.* **96** (1956), 254–263.
[6] Contributions to harmonic analysis, III. *J. London. Math. Soc.* **32** (1957), 477–483.

RICKART, C. E.
[1] *General Theory of Banach Algebras.* D. Van Nostrand Co., Inc., Princeton, N.J., 1960.

RIESZ, F.
[1] *Les systèmes d'équations à une infinité d'inconnues.* Gauthier-Villars, Paris, 1913.
[2] Über lineare Funktionalgeichungen. *Acta Math.* **41** (1918), 71–98.
[3] Sur les opérations fonctionelles linéaires. *Compt. Rend. Acad. Sci., Paris.* **149** (1909), 974–977.

[4] Sur certains systèmes singuliers d'équations intégrales. *Ann. Sci. École Norm. Sup.* (3) **28** (1911), 33–62.
[5] Démonstration nouvelle d'un théorème concernant les opérations fonctionelles linéaires. *Ann. Sci. École Norm. Sup.* (3) **31** (1914), 9–14.
[6] Sur la représentation des opérations fonctionelles linéaires par des intégrales de Stieltjes. *Proc. Roy. Physiol. Soc. Lund* **21** (16) (1962), 145–151.
[7] Sur la théorie ergodique. *Comm. Math. Helv.* **17** (1945), 221–239.
[8] Sur quelques notions fondamentales dans la théorie générale des opérations linéaires. *Ann. Math.* (2) **41** (1940), 174–206.

RIESZ, F. AND SZ.-NAGY, B.
[1] *Leçons d'analyse fonctionelle.* Académie des Sciences de Hongrie, Budapest, 1952.

RIESZ, M.
[1] Sur les ensembles compacts de fonctions sommables. *Acta Sci. Math. Szeged.* **6** (1933), 136–142.
[2] Sur le maxima des formes bilinéaires et sur les fonctionelles linéaires. *Acta Math.* **49** (1926), 465–497. *Trans. Am. Math. Soc.* **83** (1956), 482–492.

RINGROSE, J. R.
[1] A note on uniformly convex spaces. *J. London Math. Soc.* **34** (1959), 92.

RISS, J.
[1] Éléments de calcul différentiel et théorie de distributions sur les groupes abéliens localement compacts. *Acta Math.* **89** (1953), 46–105.

ROBERTSON, A. P. AND ROBERTSON, W.
[1] On the closed graph theorem. *Proc. Glasgow Math. Assoc.* **3** (1956), 9–12.

ROBERTSON, W.: *see also* Robertson, A. P.
[1] Completions of topological vector spaces. *Proc. London Math. Soc.* (3) **8** (1958), 242–257.

ROGERS, C. A.: *see* Dvoretzky, A.

ROGOSINSKI, W. W.
[1] Continuous linear functionals on subspaces of L^p and C. *Proc. London Math. Soc.* (22) **VI** (1956), 175–189.
[2] Linear extremum problems for real polynomials and trigonometric polynomials. *Arch. Math.* **V** (1954), 182–190.
[3] Linear extremum problems for real polynomials and trigonometric polynomials. Corrigenda, *Arch. Math.* **VI** (1955), 87.
[4] On finite systems of linear equations with an infinity of unknowns. *Math. Z.* **63** (1955), 97–108.
[5] Extremum problems for polynomials and trigonometric polynomials. *J. London Math. Soc.* **29** (1954), 259–275.

ROSENBLOOM, P.
[1] *The Elements of Mathematical Logic.* Dover Publications, New York, 1960.

ROSS, K. A: *see* Hewitt, E.

ROTA, G.-C.
[1] Extension theory of ordinary differential operators. Thesis. Yale Univ. Press, 1956.

ROYDEN, H. L.
[1] On a paper of Rogosinski. *J. London Math. Soc.* **35** (1960), 225–228.

RUDIN, W.: *see also* Jerison, M.
[1] The automorphisms and endomorphisms of the group algebra of the unit circle. *Acta Math.* **95** (1956), 39–55.
[2] Representation of functions by convolutions. *J. Math. Mech.* **7** (1958), 103–116.
[3] Fourier-Stieltjes transforms of measures on independent sets. *Bull. Am. Math. Soc.* **66** (1960), 199–202.
[4] Trigonometric series with gaps. *J. Math. Mech.* **9** (1960), 203–228.
[5] Independent measures on Abelian groups. *Pacific J. Math.* **9** (1959), 195–209.
[6] Measure algebras on Abelian groups. *Bull. Am. Math. Soc.* **65** (1959), 227–247.
[7] Closed ideals in group algebras. *Bull. Am. Math. Soc.* **66** (1960), 81–83.
[8] Projections on invariant subspaces. *Proc. Am. Math. Soc.* (3) **13** (1962), 429–432.
[9] *Fourier Analysis on Groups.* Interscience Publishers Inc., New York, 1962.

RUTMAN, M. A: *see* Milman, D. P.

SAKS, S.
[1] *Theory of the Integral*, 2d. edition. Monografje Matematyczne, Warszawa-Lwów, 1937.
[2] Integration in abstract metric spaces. *Duke Math. J.* **4** (1938), 408–411.

SALEM, R.: *see* Kahane J.-P.

SARGENT, W. L. C.
[1] Some sequence spaces related to l^p spaces. *J. London Math. Soc.* **35** (1960), 161–171.
[2] On some theorems of Hahn, Banach and Steinhaus. *J. London Math. Soc.* **28** (1953), 438–451.

SCHAEFER, H.
[1] Halbgeordnete lokalkonvexe Vektorräume. *Math. Ann.* **135** (1958), 115–141.
[2] Halbgeordnete lokalkonvexe Vektorräume II. *Math. Ann.* **138** (1959), 259–286.
[3] Halbgeordnete lokalkonvexe Vektorräume III. *Math. Ann.* **141** (1960), 113–142.

SCHÄFFER, J. J: *see* Massera, J. L.

SCHAUDER, J: *see also* Leray, J.
[1] Das Anfangswertproblem einer quasilinearen hyperbolischen Differentialgleichung zweiter Ordnung in beliebiger Anzahl von unabhängigen Veränderlichen. *Fund. Math.* **24** (1935), 213–246.
[2] Équations du type elliptique, problèmes linéaires. *Enseign. Math.* **35** (1936), 126–139.
[3] Zur Theorie stetiger Abbildungen in Funktionalräumen. *Math. Z.* **26** (1927), 47–65.
[4] Über lineare, vollstetige Funktionaloperatoren. *Studia Math.* **II** (1930), 183–196.
[5] Eine Eigenschaft des Haarschen Orthogonalsystemes. *Math. Z.* **28** (1928), 317–320.
[6] Die Fixpunktsatz in Funktionalräumen. *Studia Math.* **II** (1930), 171–180.

SCHECHTER, M.
[1] Some L^p estimates for partial differential equations. *Bull. Am. Math. Soc.* **68** (1962), 470–474.

SCHMEIDLER, W.
[1] *Lineare Operatoren im Hilbertschen Raum.* Teubner Verlagsgesellschaft, m.b.H., Stuttgart, 1954.

SCHUR, I.

[1] Über die charakterische Wurzeln einer linearen Substitution mit ein Anwendung auf der Theorie der Integralgleichungen. *Math. Ann.* **66** (1909), 488–510.

SCHWARTZ, J. T.: *see also* Bartle, R. G. and Dunford, N.

[1] A remark on inequalities of the Calderón-Zygmund type for vector-valued functions. *Comm. Pure Appl. Math.* **14** (1961), 785–799.

SCHWARTZ, L: *see also* Dieudonné, J.

[1] *Théorie des Distributions, I.* Act. Sci. et Ind. No. 1091; Hermann et Cie, Paris, 1950.

[1a] *Théorie des Distributions, I.* 2d. ed. Act. Sci. et Ind. No. 1245; Hermann et Cie, Paris, 1957.

[2] *Théorie des Distributions, II.* Act. Sci. et Ind. No. 1122; Hermann et Cie, Paris, 1951.

[3] Sur certains familles non-fondamentales de fonctions continues. *Bull. Soc. Math. France.* **72** (1944), 141–145.

[4] Distributions à valeurs vectorielles, I. *Ann. Inst. Fourier.* **VII** (1957), 1–141.

[5] Distributions à valeurs vectorielles II. *Ann. Inst. Fourier.* **VIII** (1959), 1–207.

[6] Espaces de fonctions différentiables à valeurs vectorielles. *J. Anal. Math.* **IV** (1954–55), 88–148.

[7] Sur une propriété de synthèse spectrale dans les groupes non-compacts. *Compt. Rend. Acad. Sci., Paris.* **227** (1948), 421–426.

[8] *Équations aux dérivées partielles.* Séminaire Inst. Henri Poincaré, 1954–55, Paris, 1955.

[9] Théorie générale des fonctions moyenne-périodiques. *Ann. Math.* **48** (1947), 857–929.

[10] Problèmes aux limites dans les équations aux dérivées partielles elliptiques. *Second Coll. sur les équations aux dérivées partielles. C.B.R.M.* (*Centre Belge de la Recherche Mathématique*), Bruxelles, 1954.

[11] Analyse et synthèse harmoniques dans les espaces de distributions. *Can. J. Math.* (4) **III** (1951), 503–512.

[12] Les équations d'évolution liées au produit de composition. *Ann. Inst. Fourier.* **II** (1950), 19–49.

[13] Homomorphismes et applications complètement continues. *Compt. Rend. Acad. Sci., Paris.* **236** (1953), 2472–2473.

[14] *Mathematica y Fisica Cuantica.* Lecture notes, Universidad de Buenos Aires, Faculdad de Ciençias Exactas y Naturales, Buenos Aires, 1958.

[15] *Étude des sommes d'exponentielles réelles. Departamento de Matematicas.* Hermann et Cie., Paris, 1943.

[16] Approximation d'une fonction quelconque par des sommes d'exponentielles imaginaires. *Ann. Fac. Sci. Univ. Toulouse.* **6** (1943), 111–174.

[16a] Étude des sommes d'exponentielles. *Act. Sci. et Ind.* (959). Paris, 1959. (This work combines the contents of [15] and [16].)

[17] Un lemme sur la dérivation des fonctions vectorielles d'une variable réelle. *Ann. Inst. Fourier.* **II** (1950), 17–18.

[18] Théorie des noyaux. *Proc. Intern. Congr. Mathematicians.* Cambridge, Mass., **I** (1950), 220–230.

[19] Sur l'impossibilité de la multiplication des distributions. *Compt. Rend. Acad. Sci., Paris.* **239** (1954), 847–848.

[20] *Some Applications of the Theory of Distributions.* (*Lectures on Modern Mathematics, Vol. I.* Edited by T. L. Saaty. John Wiley & Sons, Inc., New York-London (1963), pp. 23–58.)

SCHWARZ, S.

[1] On weak and strong extensions of partial differential operators with constant coefficients. *Ark. Mat.* **3** (1958), 515–526.

SEBASTIÃO E SILVA, J.

[1] Su certi spazi localmente convessi importanti per le applicazioni. Rend. Math. Univ. Roma (5) **14** (1955), 388–410.

[2] Sur une construction axiomatique de la théorie des distributions. Univ. Lisboa. Revista Fac. Ci. (2) **4** (1955), 79–186; **5** (1956), 169–170.

SEGAL, I. E.

[1] The group ring of a locally compact group, I. *Proc. Natl. Acad. Sci. U.S.* (7) **27** (1941), 348–352.

[2] The span of translations of a function in a Lebesgue space. *Proc. Natl. Acad. Sci. U.S.* (7) **30** (1944), 165–169.

[3] The group algebra of a locally compact group. *Trans. Am. Math. Soc.* **61** (1947), 69–105.

[4] The class of functions which are absolutely convergent Fourier transforms. *Acta Sci. Math. Szeged.* **XII** (1950), 157–161.

SHIRAISHI, R.

[1] On the definition of convolutions for distributions. *J. Sci. Hiroshima Univ.* Ser. A **23** (1959), 19–32.

SHIRKOV, D. V: *see* Bogolyubov, N. N.

SHIROTA, T.

[1] On locally convex vector spaces of continuous functions. *Proc. Japan. Acad.* **30** (1954), 294–299.

ŠILOV, G: *see also* Gelfand, I. M.

[1] Homogeneous rings of functions. *Usp. Mat. Nauk.* **6** (1951). Am. Math. Soc. Transl. (92).

SILVERMAN, R. J.

[1] Invariant linear functions. *Trans. Am. Math. Soc.* **81** (1956), 411–424.

[2] Means on semigroups and the Hahn-Banach extension property. *Trans. Am. Math. Soc.* **83** (1956), 222–237.

SIMONS, S.

[1] The bornological space associated with R^I. *J. London Math. Soc.* **36** (1961), 461–473.

SINGER, I.

[1] On a theorem of J. D. Weston. *J. London Math. Soc.* **34** (1959), 320–324.

[2] Sur les applications linéaires intégrales des espaces de fonctions continues, I. *Rev. Math. Pures Appl.* **4** (1959), 391–401.

[3] Sur les applications linéaires intégrales des espaces de fonctions continues à valeurs vectorielles. *Acta Math. Acad. Sci. Hung.* **11** (1960), 3–13.

[4] Sur les applications linéaires intégrales majorées des espaces de fonctions continues. *Acta Acad. Naz. Lincei. Rend. Cl. Sci. Fis. Mat. Nat.* (27) **8** (1959), 35–41.

[5] Sur une classe d'applications linéaires continues des espaces L_F^p $(1 \leq p < \infty)$. *Ann. Sci. École Norm. Sup.* (3) **77** (1960), 235–256.

SMITH, H. L: *see* Moore, E. H.

SMITH, K. T.

[1] A generalization of an inequality of Hardy and Littlewood. *Can. J. Math.* **8** (1956), 157–170.

SMITHIES, F.
[1] A note on completely continuous transformations. *Ann. Math.* (3) **38** (1937), 626–630.
[2] *Integral Equations.* Cambridge Tracts in Math. and Math. Phys., (49). Cambridge Univ. Press, New York, 1958.

ŠMULIAN, V: *see also* Krein, M. G.
[1] Sur les ensembles régulièrement fermés et faiblement compacts à l'espace du type (B). *Dokl. Akad. Nauk. SSSR.* **18** (1938), 405–407.
[2] Uber lineare topologischen Räume. *Mat. Sb.* **7** (1940), 425–448.
[3] On the principle of inclusion in the space of type (B). *Mat. Sb.* (47) **5** (1939), 317–328.

SOBCZYK, A: *see* Bohnenblust, F.

SOBOLEFF S. L.
[1] *Applications d'analyse fonctionelle à la physique mathématique.* Leningrad, 1950.
[1a] *Applications of Functional Analysis in Mathematical Physics.* Amer. Math. Soc. Transl. of Math. Monographs, **7**, Providence, Rhode Island, 1963.
[2] Méthode nouvelle à résoudre le problème de Cauchy pour les équations linéaires hyperboliques. *Mat. Sb.* (1) **43** (1936), 39–71.
[3] Sur un théorème d'analyse fonctionelle. *Mat. Sb.* **4** (1938), 471–496.

SOBOLEFF, W. I: *see* Ljusternik, L. A.

ŠREĬDER, Y. A.
[1] On an example of a generalized character. *Mat. Sb.* **29** (1951), 419–426.

STEENROD, N: *see* Eilenberg, S.

STEINHAUS, H: *see* Kaczmarz, S.

STEIN, E. M.
[1] Interpolation of linear operators. *Trans. Am. Math. Soc.* **83** (1956), 482–492.
[2] On limits of sequences of operators. *Ann. Math.* (1) **74** (1961), 140–170.

STEIN, K: *see* Behnke, H.

STONE, M. H.
[1] *Linear Transformations in Hilbert Space.* Am. Math. Soc. Colloquium Publications, **15,** New York, 1932.
[2] Applications of the theory of Boolean rings to general topology. *Trans. Am. Math. Soc.* **41** (1937), 375–481.
[3] The generalized Weierstrass approximation theorem. *Math. Mag.* **21** (1947/8), 167–183, 237–254.
[4] Notes on integration, I–IV. *Proc. Natl. Acad. Sci. U.S.* **34** (1948), 336–342, 447–455, 483–490; **35** (1949), 50–58.

SUPPES, P.
[1] *Axiomatic Set Theory.* D. Van Nostrand Co., Inc., Princeton, N.J., 1960.

TAKENOUCHI, O.
[1] Sur les espaces linéaires localement convexes. *Math. J. Okayama Univ.* (1) **2** (1952), 57–84.
[2] Une démonstration directe d'un théorème de M. G. W. Mackey. *Kōdai Math. Sem. Rep.* (3/4) (1951), 49–50.

TAMARKIN, J. D.: *see* Hille, E.

TATCHELL, J. B.
[1] A note on matrix summability of unbounded sequences. *J. London Math. Soc.* **34** (1959), 27–36.

Taylor, A. E.: *see also* Bochner, S.
[1] *Introduction to Functional Analysis*. John Wiley and Sons, Inc., New York, 1958.
[2] Extensions of a theorem of Hellinger and Toeplitz. *Math. Z.* **66** (1956), 53–57.
[3] Banach spaces of functions analytic in the unit circle. *Studia Math.* **XI** (1950), 145–170; **12** (1951), 25–50.

Temple, G.
[1] Generalized functions. *Proc. Roy. Soc.* A **228** (1955), 175–190.
[2] Theories and applications of generalized functions. *J. London Math. Soc.* **28** (1953), 134–148.

Ter Haar, D.
[1] Foundations of statistical mechanics. *Rev. Mod. Phys.* (3) **27** (1955), 289–338.

Thorin, G. O.
[1] An extension of a convexity theorem due to M. Riesz. *Comm. Sém. Math. Univ. Lund.* (4), (1939).
[2] Convexity theorems. ibid. (9), (1948).

Titchmarsh, E. C.
[1] *Theory of Fourier Integrals*. Oxford Univ. Press, Oxford, 1937.

Ti Yen
[1] The Hahn-Banach theorem and the least upper bound property. *Trans. Am. Math. Soc.* **90** (1959), 523–526.

Toeplitz, O: *see* Hellinger, E.

Tolstov, G. P.
[1] On an abstract integral considered by S. Banach. (Russian), *Mat. Sb.* **57** (1962), 319–322.

Trèves, F.
[1] Domination et problèmes aux limites du type mixte. *Compt. Rend. Acad. Sci., Paris.* **245** (1947), 2454–2457.
[2] Domination et opérateurs hyperboliques. *Compt. Rend. Acad. Sci., Paris.* **246** (1958), 680–683.
[3] Domination et opérateurs paraboliques. *Compt. Rend. Acad. Sci., Paris.* **246** (1958), 867–870.
[4] Relations de domination entre opérateurs différentiels. Thèse. Paris 1958; *Acta Math.* **101** (1959), 1–139.
[5] Opérateurs différentiels hypoelliptiques. *Ann. Inst. Fourier.* **9** (1959), 1–73.
[6] Differential equations in Hilbert space. *Proc. Symp. Pure Math., Am. Math. Soc.* **IV,** 83–89 (1961).

Tsuji, M.
[1] On the compactness of space L^p $(p > 0)$ and its application to integral equations. *Kōdai Math. Sem. Rep.* (1951), 33–36.

Tukey, J. W.
[1] *Convergence and Uniformity in Topology* Ann. Math. Study No. 2, Princeton Univ. Press, Princeton, N.J., 1940.

Tychonoff, A.
[1] Ein Fixpunktzatz. *Math. Ann.* **111** (1935), 767–776.

Ulam, S.
[1] Zur Masstheorie in der allgemeinen Mengenlehre. *Fund. Math.* **16** (1930), 140–150.

VALLÉE POUSSIN, C. DE LA,
[1] *Les Nouvelles Méthodes de la Théorie du Potentiel et le Problème Généralisé de Dirichlet*. Act. Sci. Ind. (578); Hermann et Cie, Paris, 1937.

VARADARAJAN, V. S.
[1] On a theorem of F. Riesz concerning the form of linear functionals. *Fund. Math.* **46** (1958), 209–220.

VISCHIK, M. J.
[1] The method of orthogonal and direct decomposition in the theory of elliptic differential equations. (Russian). *Mat. Sb.* **25** (1949), 189–234.

VLADIMIROV, V. S.: *see* Bogolyubov, N. N.

VULIKH, B. Z: *see* Kantorovich, L. V.

WADA, J.
[1] Positive linear functionals on ideals of continuous functions. *Osaka Math. J.* **11** (1959), 173–185.
[2] Strict convexity and smoothness of normed spaces. *Osaka Math. J.* **10** (1958), 221–230.
[3] Weakly compact linear operators on function spaces. *Osaka Math. J.* **13** (1961), 169–183.

WAELBROECK, L.
[1] Les espaces à bornes complets. *Colloque sur l'Analyse Functionelle*, Louvain, (1960), 51–55. *Librairie Universitaire*, Louvain (1961).

WALTERS, S. S.
[1] The space H^p with $0 < p < 1$. *Proc. Am. Math. Soc.* **1** (1950), 800–805.
[2] Remarks on the spaces H^p. *Pacific J. Math.* **1** (1951), 455–471.

WARNER, S.
[1] The topology of compact convergences on continuous function spaces. *Duke Math. J.* **25** (1958), 265–282.

WEHAUSEN, J. V.
[1] Transformations in topological linear spaces. *Duke Math. J.* **4** (1938), 157–169.

WEIL, A.
[1] *L'intégration dans les groupes topologiques et ses applications*. Act. Sci. et Ind., Nos. 869, 1145; Hermann et Cie, Paris, 1940, 1951.
[2] *Sur les espaces à structure uniforme*. Act. Sci. et Ind. No. 551; Hermann et Cie, Paris, 1937.

WEISS, G.
[1] Análisis armónica en varias variables. Teoría de los espacios H^p. *Cursos y Seminar. de Mat.* Fasc. 9., Universidad de Buenos Aires (1960).

WENDEL, J.
[1] Left centralizers and isomorphisms of group algebras. *Pacific J. Math.* **2** (1952), 251–261.

WERTHEIM, D. G: *see* Lorentz, G. G.

WESTON, J. D.
[1] On the representation of operators by convolution integrals. *Pacific J. Math.* [4] **10** (1960), 1453–1468.
[2] The representation of linear functionals by sequences of functions. *J. London Math. Soc.* **33** (1958), 123–125.

WEYL, H.
[1] The method of orthogonal projection in potential theory. *Duke Math. J.* **7** (1940), 411–444.

WHITNEY, H.
[1] On ideals of differentiable functions. *Am. J. Math.* **70** (1948), 635–658.

WIENER, N: *see also* Paley, R. E. A. C.
[1] *The Fourier integral and certain of its applications.* Cambridge Univ. Press, Cambridge, 1933.

WIGHTMAN, A. S.
[1] Les problèmes mathématiques de la théorie quantique des champs. *Colloq. Intern. Centre Nat. Rech. Sci., Lille* (1957).

WILLIAMSON, J. H.
[1] Two conditions equivalent to normability. *J. London Math. Soc.* **31** (1956), 111–113.
[2] A third condition equivalent to normability. *J. London Math. Soc.* **32** (1957), 231–232.
[3] Linear transformations in arbitrary linear spaces. *J. London Math. Soc.* **28** (1953), 203–210.
[4] Compact linear operators in linear topological spaces. *J. London Math. Soc.* **29** (1954), 149–156.

WOODBURY, M. A.
[1] Invariant functionals and measures. *Bull. Am. Math. Soc.* **56** (1950), 172, Abstract 168t.

YOOD, B.
[1] On fixed points for semigroups of linear operations. *Proc. Am. Math. Soc.* **2** (1951), 225–233.

YOSIDA, K: *see also* Hewitt, E.
[1] Mean ergodic theorem in Banach spaces. *Proc. Imp. Acad. Tokyo.* **14** (1938), 292–294.

ZAANEN, A. C: *see also* Luxemburg, W. A. J.
[1] *Linear Analysis.* North Holland Publishing Co., Amsterdam, 1953.
[2] A note on the Daniell-Stone integral. *Colloque sur l'Analyse Fonctionelle,* Louvain (1960), 63–69. *Librairie Universitaire,* Louvain (1961).

ZELLER, K.
[1] Allgemeine Eigenschaften von Limitierungsverfahren. *Math. Z.* **53** (1951), 463–487.
[2] FK-Räume und Matrixtransformationen. *Math. Z.* **58** (1953), 46–48.
[3] Matrixtransformationen von Folgenräumen. *Univ. Roma Ist. Naz. Alta Mat. Rend. Mat. e App.* (5) **12** (1953), 340–346.
[4] *Theorie der Limitierungsverfahren.* Ergebnisse der Mathematik und Ihrer Grenzgebeite, N.F., Heft 15. Springer-Verlag, Berlin-Vienna, 1958.

ZUCKERMAN, H. S.: *see* Hewitt, E.

ZYGMUND, A: *see also* Calderón, A. P.
[1] *Trigonometrical Series.* Monografje Matematyczme, **V,** Warszawa-Lwów, 1935.
[2] *Trigonometric series, I and II.* Cambridge Univ. Press, New York, 1959.
[3] On a theorem of Marcinkiewicz concerning interpolation of operations. *J. Math.* **35** (1956), 223–248.

Symbols and Abbreviations

Apart from occasional temporary variations, the following symbols retain the meanings assigned to them at the indicated points of reference. A few symbols (Δ, for example) are, according to custom, required to play several roles; in such cases the role is usually abundantly clear from the context.

Abbreviations

AC, absolutely continuous
a.e., almost everywhere
BV, bounded variation
CGT, closed graph theorem
FPT, fixed-point theorem
(GDF), graphe dénombrablement fermée
LAC, locally absolutely continuous
l.a.e., locally almost everywhere
LBV, locally bounded variation
LCA, locally compact Abelian
LCTVL, locally convex topological vector lattice
LCTVS, locally convex topological vector space
LPDE, linear partial differential equation
LPDO, linear partial differential operator
L-S, Lebesgue-Stieltjes
LSC, lower semicontinuous
OGBK, Orlicz-Gelfand-Bosanquet-Kestelman
OMT, open mapping theorem
OVS, ordered vector space
R-S, Riemann-Stieltjes
RRT, Riesz representation theorem
SEI, scalarwise essentially integrable
SI, scalarwise integrable
SM, scalarwise measurable
SWI, scalarwise well integrable
TVL, topological vector lattice
TVS, topological vector space
USC, upper semicontinuous

Appendix

Additions to the Dover Edition

ADDITION TO PAGE 41: If H is the hyperplane $f^{-1}(\{\alpha\})$ in a real vector space, the sets $f^{-1}([\alpha, \infty))$, $f^{-1}((\alpha, \infty))$, $f^{-1}((-\infty, \alpha])$, $f^{-1}((-\infty, \alpha))$ are termed *half-spaces* defined by H. A set A is said to *lie (strictly) on one side of* H if it is contained in one of $f^{-1}([\alpha, \infty))$ or $f^{-1}((-\infty, \alpha])$ (or in one of $f^{-1}((\alpha, \infty))$ or $f^{-1}((-\infty, \alpha))$).

ADDITION TO PAGE 65: If E is a real TVS and $H = f^{-1}(\{\alpha\})$ is a closed hyperplane in E, the sets $f^{-1}([\alpha, \infty))$ and $f^{-1}((-\infty, \alpha])$ are the *closed half-spaces* defined by H; the sets $f^{-1}((\alpha, \infty))$ and $f^{-1}((-\infty, \alpha))$ are the *open half-spaces* defined by H.

ADDITION TO PAGE 146: (In a real vector space E, a *hyperplane of support* of a subset A of E is a hyperplane H in E meeting A and such that A lies on one side of H.)

ADDITION TO PAGE 281: This proof requires a change from $|\mu_{n+1}|(S)$ to $|\mu_{n+1}|'(S)$ as defined in the corrected version of Exercise 4.42 below.

Index

Notes References in bold typeface refer to the exercises. All named theorems are collected under "Theorem."